AS YOU EXAMINE

HOLT ALGEBRA 2
WITH TRIGONOMETRY

PLEASE NOTICE This Algebra 2 with Trigonometry text provides a balanced course designed to be mastered by all second-year algebra students. All the basic concepts that are usually covered in a second-year algebra course, along with the fundamental concepts of trigonometry, are found in HOLT ALGEBRA 2 WITH TRIGONOMETRY.

FORMAT

A significant innovation in HOLT ALGEBRA 2 WITH TRIGONOMETRY is the format of each lesson involving a five-part presentation (see any lesson):

- Lesson Objectives
- Review Capsule
- Worked-out Examples in black type (right-hand column)
- Hints, Suggestions, and Comments in green type (left-hand column)
- Graded Exercises

CONTENT

1. Reading is kept to a minimum through the use of a two-column format. (See any lesson.)
2. Worked-out examples focus on the mathematics, thus eliminating long, wordy explanations. (See pp. 88, 493.)
3. Suggestions, hints, and comments in color on the left-hand side of the page help clarify the development of the mathematics in the right-hand column. (See any lesson.)
4. Oral exercises appear before many written exercise sets. These provide readiness for the written exercises. (See pp. 8, 144, 373, 495.)
5. Exercises are plentiful and are grouped in three levels of difficulty: A, B, C. Examples in the lesson match every type of exercise in Part A and Part B. (See pp. 21, 183, 344, 345, 446.)
6. Each chapter ends with a Chapter Review and Test. The Review is diagnostic and keyed to lesson pages. The Test is an alternate form of the Review.
7. Special Topics pages vary the pace and enrich the course. They include applications of algebra (and trigonometry) in the everyday world, as well as careers that use mathematics. (See pp. viii, 12, 217, 358, 359.)
8. Flow-chart approach to calculator and computer activities. (See pp. 96, 167.)
9. Chapters 16-19 are devoted to the basic concepts of trigonometry. (See pp. 404-523.)
10. Answers to odd-numbered exercises are provided in the back of the pupil edition. (See pp. 546-565.)
11. Teacher's Edition with Commentary for Teachers and Lesson Commentaries with Cumulative Review exercises (see pp. T-15 to T-53), Pacing Charts with assignments for three levels of difficulty (see pp. T-55 to T-72), and a complete set of answers (see pp. T-73 to T-112).

SUPPLEMENTARY MATERIALS ON DUPLICATING MASTERS

- A set of alternate chapter tests.
- A set of 60 Skillmasters provides two daily quizzes or additional exercise sets on each master, as well as two Cumulative Reviews.
- For individualizing HOLT ALGEBRA 2 WITH TRIGONOMETRY, there is a set of duplicating masters containing assignment sheets, placement tests, and record-keeping forms, as well as a Teacher's Guide.

HOLT ALGEBRA WITH TRIGONOMETRY 2

TEACHER'S EDITION

Eugene D. Nichols
Mervine L. Edwards
E. Henry Garland
Sylvia A. Hoffman
Albert Mamary
William F. Palmer

HOLT ALGEBRA 2 WITH TRIGONOMETRY

TEACHER'S EDITION

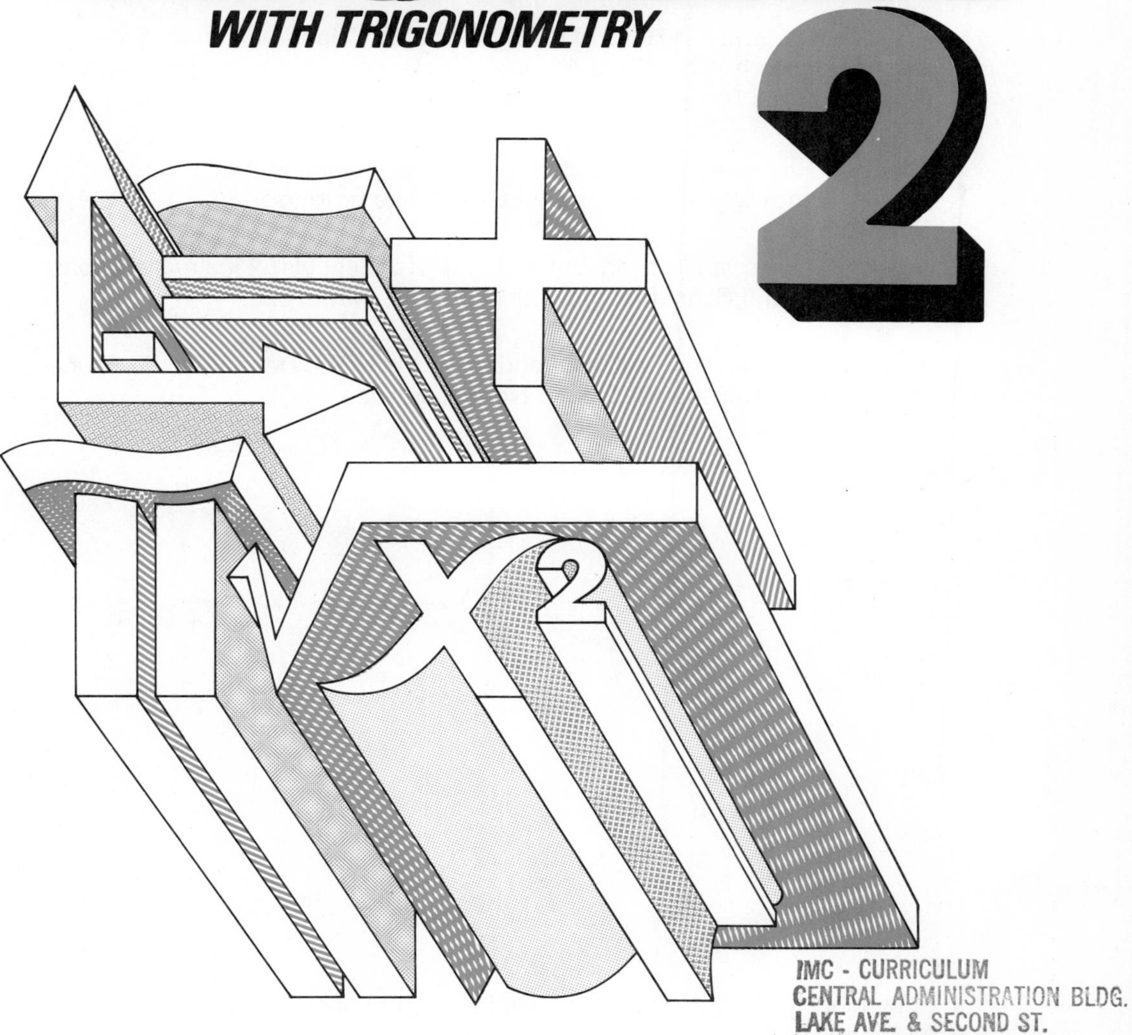

HOLT, RINEHART AND WINSTON, PUBLISHERS
New York · Toronto · London · Sydney

ABOUT THE AUTHORS

Eugene D. Nichols is Professor of Mathematics Education and Lecturer in the Department of Mathematics at Florida State University, Tallahassee, Florida.

Mervine L. Edwards is Chairman of the Mathematics Department, Shore Regional High School, West Long Branch, New Jersey.

E. Henry Garland is Head of the Mathematics Department at the Developmental Research School, and Associate Professor of Mathematics Education at Florida State University, Talahassee, Florida.

Sylvia A. Hoffman is Curriculum Coordinator for the Metropolitan Chicago Region of the Illinois Office of Education, State of Illinois.

Albert Mamary is Assistant Superintendent of Schools for Instruction, Johnson City Central School District, Johnson City, New York.

William F. Palmer is Professor and Chairman of the Department of Education, Catawba College, Salisbury, North Carolina.

Printed in the United States of America

ISBN: 0-03-018916-0

7890123456 071 987654321

CONTENTS

HOLT ALGEBRA 2

WITH TRIGONOMETRY

BENEFITS

1. CONTAINS ALL BASIC SKILLS AND CONCEPTS
2. TEACHES BY WORKED-OUT EXAMPLES
3. SINGLE CONCEPT LESSONS
4. FIVE-PART PRESENTATION OF LESSONS
5. OPEN FORMAT
6. REDUCED READING
7. SPECIAL TOPICS WITH PROJECTS
8. GRADED EXERCISES
9. CHAPTER REVIEWS AND TESTS
10. EXTENSIVELY FIELD TESTED
11. SUPPLEMENTARY TESTS AND SKILLMASTERS
12. HELPFUL TEACHER'S EDITION
13. MANAGEMENT SYSTEM FOR INDIVIDUALIZING

HOW THE TEXT

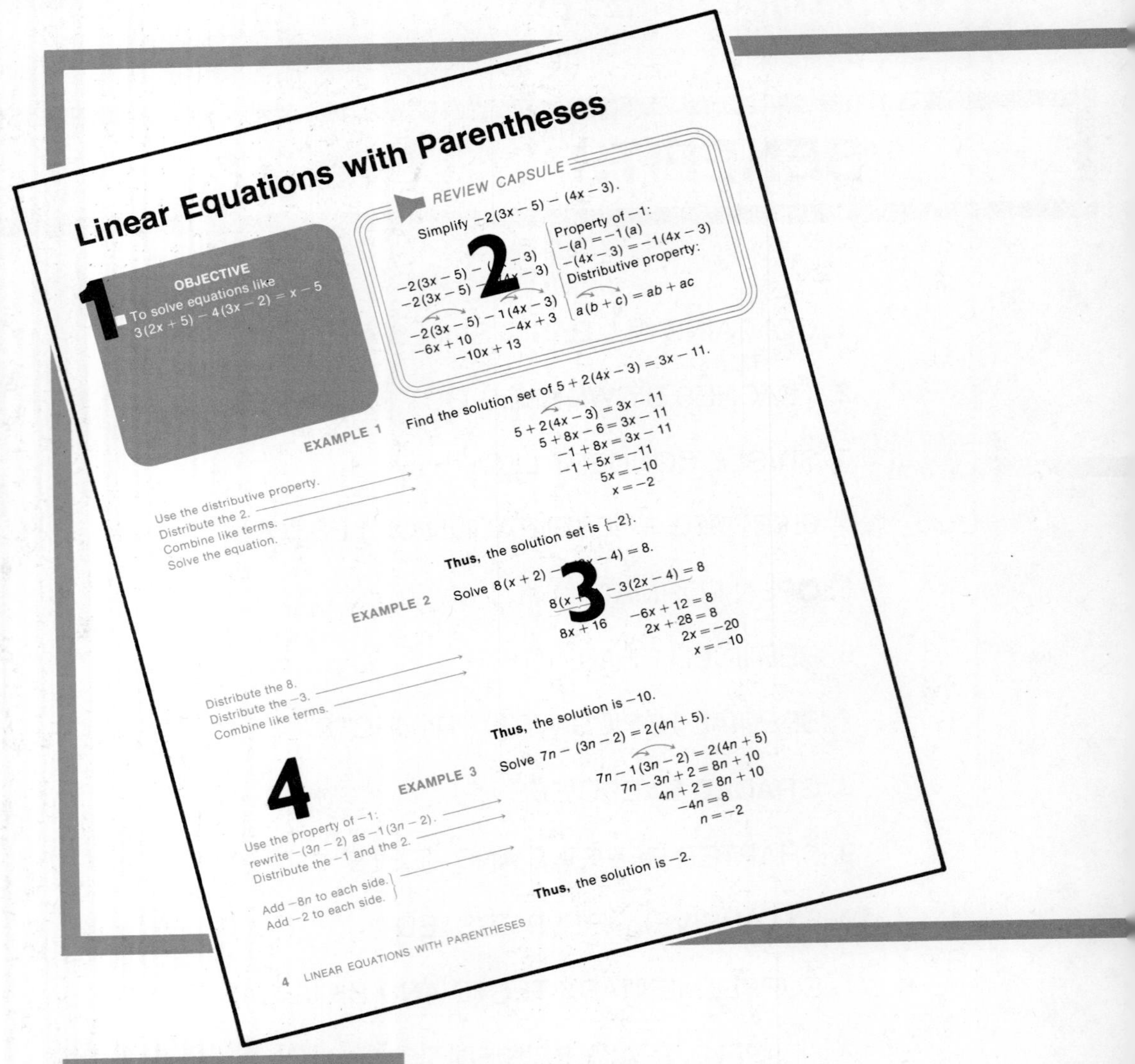

Linear Equations with Parentheses

OBJECTIVE

To solve equations like $3(2x + 5) - 4(3x - 2) = x - 5$

REVIEW CAPSULE

Simplify $-2(3x - 5) - (4x - 3)$.

$$-2(3x - 5) - (4x - 3)$$
$$-2(3x - 5) - 1(4x - 3)$$
$$-2(3x - 5) - 1(4x - 3)$$
$$-6x + 10 \quad -4x + 3$$
$$-10x + 13$$

Property of −1: $-(a) = -1(a)$
$-(4x - 3) = -1(4x - 3)$
Distributive property:
$a(b + c) = ab + ac$

EXAMPLE 1 Find the solution set of $5 + 2(4x - 3) = 3x - 11$.

$$5 + 2(4x - 3) = 3x - 11$$
$$5 + 8x - 6 = 3x - 11$$
$$-1 + 8x = 3x - 11$$
$$-1 + 5x = -11$$
$$5x = -10$$
$$x = -2$$

Use the distributive property.
Distribute the 2.
Combine like terms.
Solve the equation.

Thus, the solution set is $\{-2\}$.

EXAMPLE 2 Solve $8(x + 2) - 3(2x - 4) = 8$.

$$8(x + 2) - 3(2x - 4) = 8$$
$$8x + 16 \quad -6x + 12 = 8$$
$$2x + 28 = 8$$
$$2x = -20$$
$$x = -10$$

Distribute the 8.
Distribute the −3.
Combine like terms.

Thus, the solution is −10.

EXAMPLE 3 Solve $7n - (3n - 2) = 2(4n + 5)$.

$$7n - 1(3n - 2) = 2(4n + 5)$$
$$7n - 3n + 2 = 8n + 10$$
$$4n + 2 = 8n + 10$$
$$-4n = 8$$
$$n = -2$$

Use the property of −1: rewrite $-(3n - 2)$ as $-1(3n - 2)$.
Distribute the −1 and the 2.
Add $-8n$ to each side.
Add −2 to each side.

Thus, the solution is −2.

4 LINEAR EQUATIONS WITH PARENTHESES

FIVE-PART PRESENTATION OF EACH LESSON

1. OBJECTIVE
2. REVIEW CAPSULE
3. WORKED-OUT EXAMPLES
4. HINTS, COMMENTS, AND SUGGESTIONS
5. GRADED EXERCISES

____ IS ORGANIZED

1. Expected performance is stated simply in the language of the student.
2. Summarizes a skill or concept the student needs to recall for this lesson.
3. The use of words is cut to a minimum so the mathematical ideas stand out.
4. Hints on the left-hand side of the page direct attention to ke parts of the development.
5. Oral exercises provide readiness for the written exercises.
   Exercises are graded according to level of difficulty, A, B, and C.

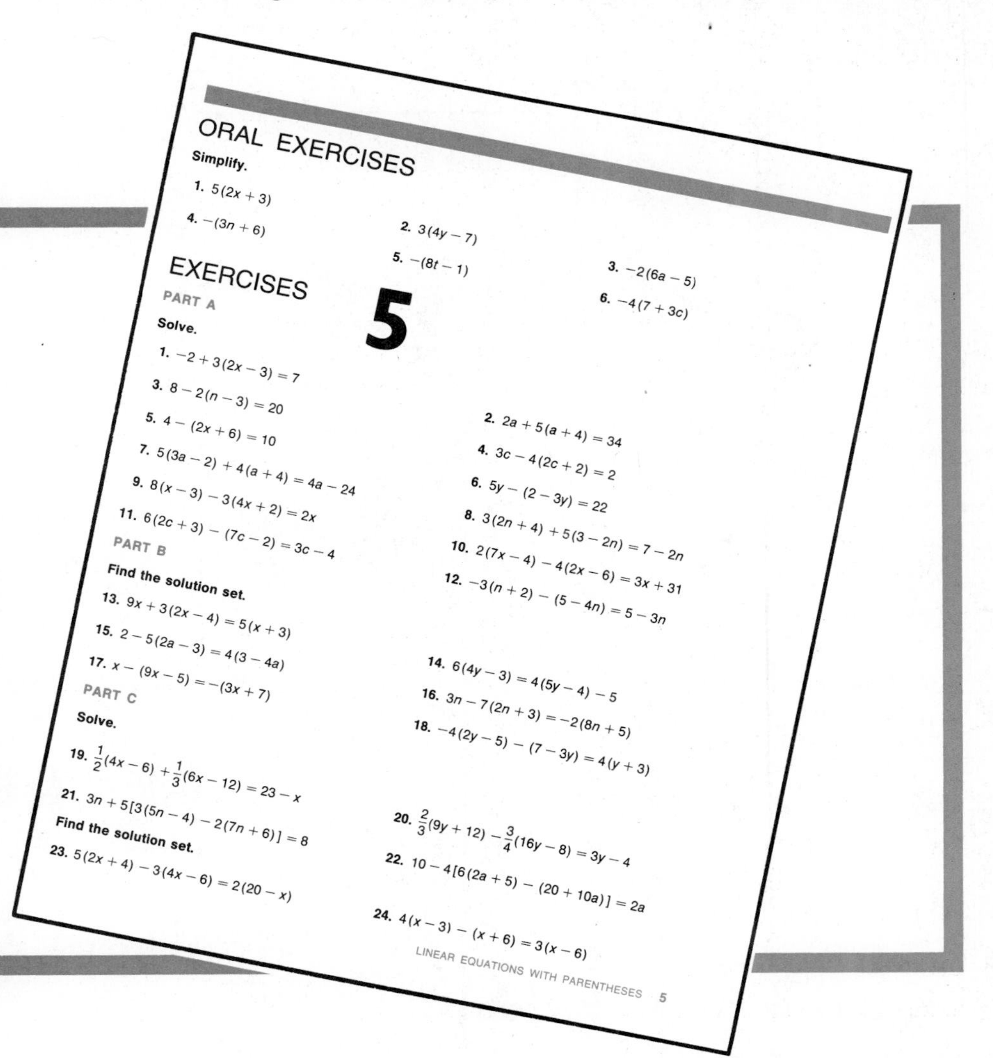

ORAL EXERCISES

Simplify.

1. $5(2x+3)$ **2.** $3(4y-7)$ **3.** $-2(6a-5)$

4. $-(3n+6)$ **5.** $-(8t-1)$ **6.** $-4(7+3c)$

EXERCISES

5

PART A

Solve.

1. $-2+3(2x-3)=7$ **2.** $2a+5(a+4)=34$

3. $8-2(n-3)=20$ **4.** $3c-4(2c+2)=2$

5. $4-(2x+6)=10$ **6.** $5y-(2-3y)=22$

7. $5(3a-2)+4(a+4)=4a-24$ **8.** $3(2n+4)+5(3-2n)=7-2n$

9. $8(x-3)-3(4x+2)=2x$ **10.** $2(7x-4)-4(2x-6)=3x+31$

11. $6(2c+3)-(7c-2)=3c-4$ **12.** $-3(n+2)-(5-4n)=5-3n$

PART B

Find the solution set.

13. $9x+3(2x-4)=5(x+3)$ **14.** $6(4y-3)=4(5y-4)-5$

15. $2-5(2a-3)=4(3-4a)$ **16.** $3n-7(2n+3)=-2(8n+5)$

17. $x-(9x-5)=-(3x+7)$ **18.** $-4(2y-5)-(7-3y)=4(y+3)$

PART C

Solve.

19. $\frac{1}{2}(4x-6)+\frac{1}{3}(6x-12)=23-x$ **20.** $\frac{2}{3}(9y+12)-\frac{3}{4}(16y-8)=3y-4$

21. $3n+5[3(5n-4)-2(7n+6)]=8$ **22.** $10-4[6(2a+5)-(20+10a)]=2a$

Find the solution set.

23. $5(2x+4)-3(4x-6)=2(20-x)$ **24.** $4(x-3)-(x+6)=3(x-6)$

TRIGONOMETRY

- Covers fundamental concepts of trigonometry

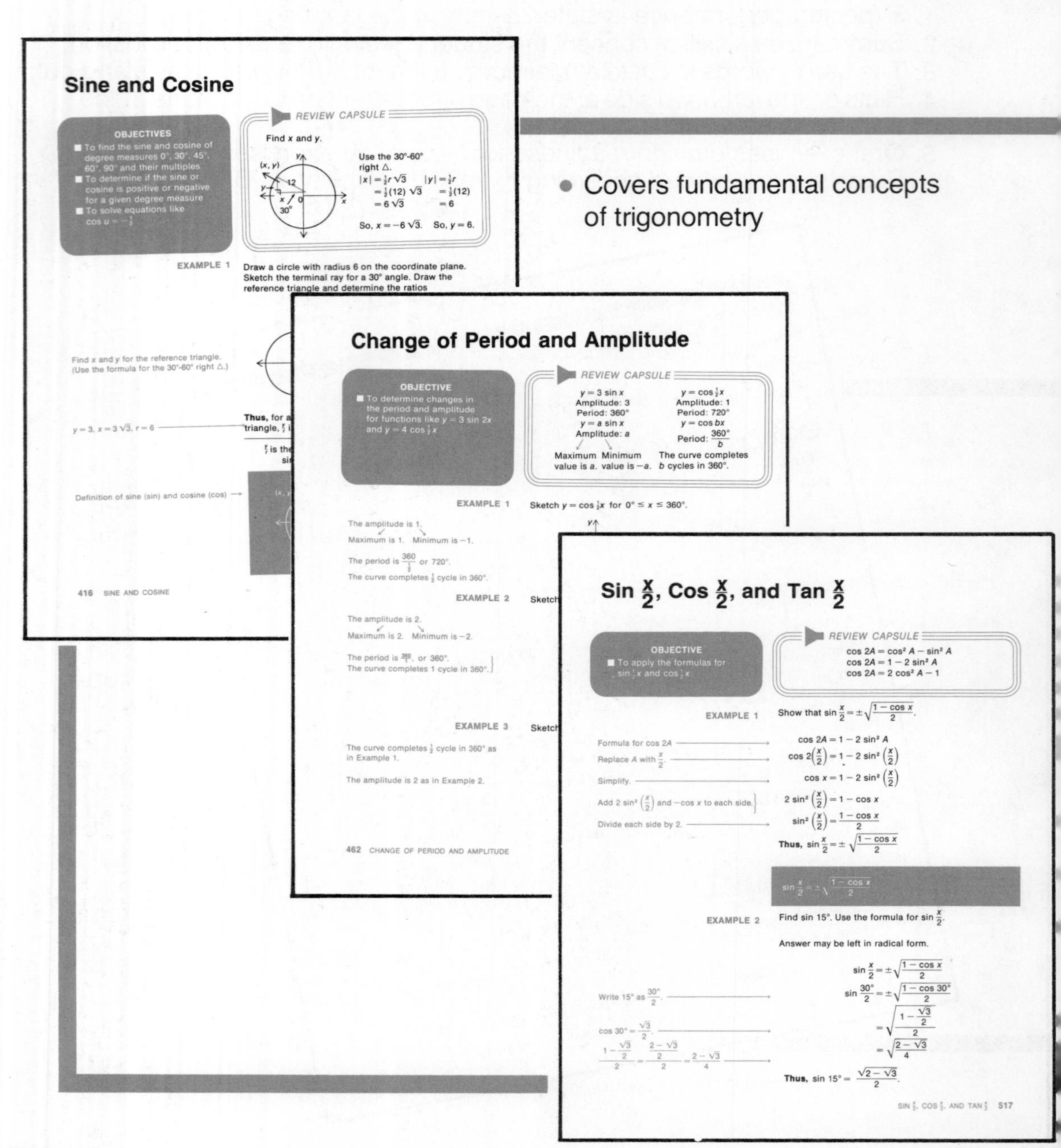

Sine and Cosine

OBJECTIVES

- To find the sine and cosine of degree measures 0°, 30°, 45°, 60°, 90° and their multiples
- To determine if the sine or cosine is positive or negative for a given degree measure
- To solve equations like $\cos u = -\frac{1}{2}$

REVIEW CAPSULE

Find x and y.

Use the 30°-60° right △.

$|x| = \frac{1}{2}r\sqrt{3}$ $\quad |y| = \frac{1}{2}r$

$= \frac{1}{2}(12)\sqrt{3}$ $\quad = \frac{1}{2}(12)$

$= 6\sqrt{3}$ $\quad = 6$

So, $x = -6\sqrt{3}$. So, $y = 6$.

EXAMPLE 1 Draw a circle with radius 6 on the coordinate plane. Sketch the terminal ray for a 30° angle. Draw the reference triangle and determine the ratios

Find x and y for the reference triangle. (Use the formula for the 30°-60° right △.)

$y = 3$, $x = 3\sqrt{3}$, $r = 6$ ——— **Thus,** for a ... triangle, $\frac{y}{r}$ i...

$\frac{y}{r}$ is the ... si...

Definition of sine (sin) and cosine (cos) →

416 SINE AND COSINE

Change of Period and Amplitude

OBJECTIVE

- To determine changes in the period and amplitude for functions like $y = 3 \sin 2x$ and $y = 4 \cos \frac{1}{2}x$

REVIEW CAPSULE

$y = 3 \sin x$	$y = \cos \frac{1}{2}x$
Amplitude: 3	Amplitude: 1
Period: 360°	Period: 720°
$y = a \sin x$	$y = \cos bx$
Amplitude: a	Period: $\frac{360°}{b}$
Maximum value is a. Minimum value is $-a$.	The curve completes b cycles in 360°.

EXAMPLE 1 Sketch $y = \cos \frac{1}{2}x$ for $0° \le x \le 360°$.

The amplitude is 1.
Maximum is 1. Minimum is −1.

The period is $\frac{360}{\frac{1}{2}}$ or 720°.
The curve completes $\frac{1}{2}$ cycle in 360°.

EXAMPLE 2 Sketch...

The amplitude is 2.
Maximum is 2. Minimum is −2.

The period is $\frac{360}{1}$, or 360°.
The curve completes 1 cycle in 360°.

EXAMPLE 3 Sketch...

The curve completes $\frac{1}{2}$ cycle in 360° as in Example 1.

The amplitude is 2 as in Example 2.

462 CHANGE OF PERIOD AND AMPLITUDE

Sin $\frac{x}{2}$, Cos $\frac{x}{2}$, and Tan $\frac{x}{2}$

OBJECTIVE

- To apply the formulas for $\sin \frac{1}{2}x$ and $\cos \frac{1}{2}x$

REVIEW CAPSULE

$\cos 2A = \cos^2 A - \sin^2 A$
$\cos 2A = 1 - 2\sin^2 A$
$\cos 2A = 2\cos^2 A - 1$

EXAMPLE 1 Show that $\sin \frac{x}{2} = \pm\sqrt{\frac{1-\cos x}{2}}$.

Formula for cos 2A → $\cos 2A = 1 - 2\sin^2 A$

Replace A with $\frac{x}{2}$. → $\cos 2\left(\frac{x}{2}\right) = 1 - 2\sin^2\left(\frac{x}{2}\right)$

Simplify. → $\cos x = 1 - 2\sin^2\left(\frac{x}{2}\right)$

Add $2\sin^2\left(\frac{x}{2}\right)$ and $-\cos x$ to each side. → $2\sin^2\left(\frac{x}{2}\right) = 1 - \cos x$

Divide each side by 2. → $\sin^2\left(\frac{x}{2}\right) = \frac{1-\cos x}{2}$

Thus, $\sin \frac{x}{2} = \pm\sqrt{\frac{1-\cos x}{2}}$

$$\sin \frac{x}{2} = \pm\sqrt{\frac{1-\cos x}{2}}$$

EXAMPLE 2 Find sin 15°. Use the formula for $\sin \frac{x}{2}$.

Answer may be left in radical form.

$\sin \frac{x}{2} = \pm\sqrt{\frac{1-\cos x}{2}}$

Write 15° as $\frac{30°}{2}$. → $\sin \frac{30°}{2} = \pm\sqrt{\frac{1-\cos 30°}{2}}$

$\cos 30° = \frac{\sqrt{3}}{2}$. → $= \sqrt{\frac{1-\frac{\sqrt{3}}{2}}{2}}$

$\frac{1-\frac{\sqrt{3}}{2}}{2} = \frac{\frac{2-\sqrt{3}}{2}}{2} = \frac{2-\sqrt{3}}{4}$ → $= \sqrt{\frac{2-\sqrt{3}}{4}}$

Thus, $\sin 15° = \frac{\sqrt{2-\sqrt{3}}}{2}$.

SIN $\frac{x}{2}$, COS $\frac{x}{2}$, AND TAN $\frac{x}{2}$ 517

ALGEBRAIC SKILLS APPLIED TO PROBLEM SOLVING

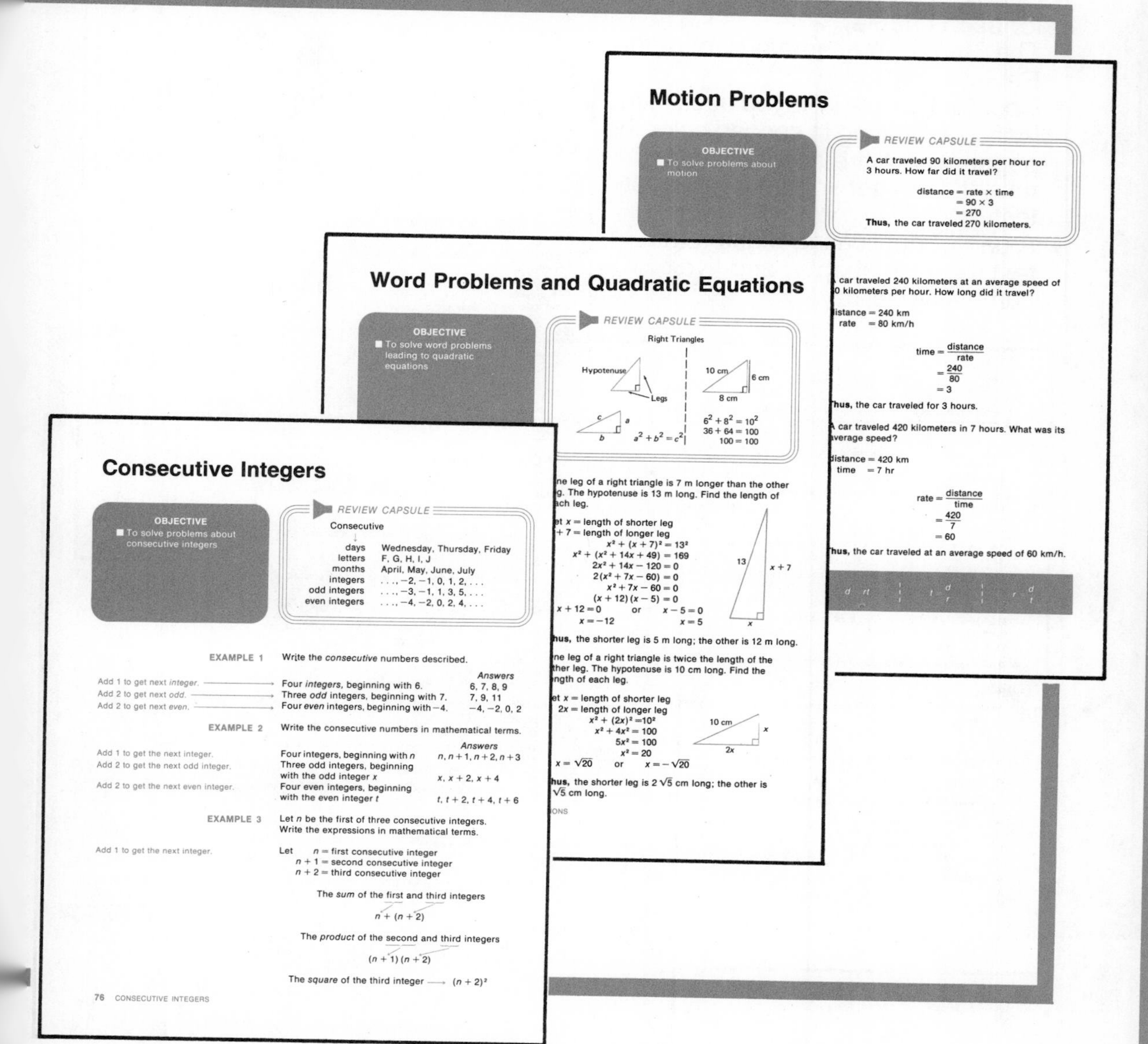

Motion Problems

OBJECTIVE
- To solve problems about motion

REVIEW CAPSULE

A car traveled 90 kilometers per hour for 3 hours. How far did it travel?

distance = rate × time
= 90 × 3
= 270

Thus, the car traveled 270 kilometers.

A car traveled 240 kilometers at an average speed of 0 kilometers per hour. How long did it travel?

istance = 240 km
rate = 80 km/h

$\text{time} = \frac{\text{distance}}{\text{rate}} = \frac{240}{80} = 3$

hus, the car traveled for 3 hours.

A car traveled 420 kilometers in 7 hours. What was its verage speed?

istance = 420 km
time = 7 hr

$\text{rate} = \frac{\text{distance}}{\text{time}} = \frac{420}{7} = 60$

hus, the car traveled at an average speed of 60 km/h.

Word Problems and Quadratic Equations

OBJECTIVE
- To solve word problems leading to quadratic equations

REVIEW CAPSULE

Right Triangles

Hypotenuse, Legs, 10 cm, 6 cm, 8 cm, c, a, b

$a^2 + b^2 = c^2$

$6^2 + 8^2 = 10^2$
$36 + 64 = 100$
$100 = 100$

ne leg of a right triangle is 7 m longer than the other g. The hypotenuse is 13 m long. Find the length of ch leg.

t x = length of shorter leg
$+ 7$ = length of longer leg

$x^2 + (x + 7)^2 = 13^2$
$x^2 + (x^2 + 14x + 49) = 169$
$2x^2 + 14x - 120 = 0$
$2(x^2 + 7x - 60) = 0$
$x^2 + 7x - 60 = 0$
$(x + 12)(x - 5) = 0$
$x + 12 = 0$ or $x - 5 = 0$
$x = -12$ $\quad$ $x = 5$

13, $x + 7$, x

hus, the shorter leg is 5 m long; the other is 12 m long.

ne leg of a right triangle is twice the length of the ther leg. The hypotenuse is 10 cm long. Find the ngth of each leg.

et x = length of shorter leg
$2x$ = length of longer leg

$x^2 + (2x)^2 = 10^2$
$x^2 + 4x^2 = 100$
$5x^2 = 100$
$x^2 = 20$
$x = \sqrt{20}$ or $x = -\sqrt{20}$

10 cm, x, $2x$

hus, the shorter leg is $2\sqrt{5}$ cm long; the other is $\sqrt{5}$ cm long.

ONS

Consecutive Integers

OBJECTIVE
- To solve problems about consecutive integers

REVIEW CAPSULE

Consecutive

days	Wednesday, Thursday, Friday
letters	F, G, H, I, J
months	April, May, June, July
integers	. . ., −2, −1, 0, 1, 2, . . .
odd integers	. . ., −3, −1, 1, 3, 5, . . .
even integers	. . ., −4, −2, 0, 2, 4, . . .

EXAMPLE 1 Write the *consecutive* numbers described.

		Answers
Add 1 to get next *integer.*	Four *integers,* beginning with 6.	6, 7, 8, 9
Add 2 to get next *odd.*	Three *odd* integers, beginning with 7.	7, 9, 11
Add 2 to get next *even.*	Four *even* integers, beginning with −4.	−4, −2, 0, 2

EXAMPLE 2 Write the consecutive numbers in mathematical terms.

		Answers
Add 1 to get the next integer.	Four integers, beginning with n	$n, n+1, n+2, n+3$
Add 2 to get the next odd integer.	Three odd integers, beginning with the odd integer x	$x, x+2, x+4$
Add 2 to get the next even integer.	Four even integers, beginning with the even integer t	$t, t+2, t+4, t+6$

EXAMPLE 3 Let n be the first of three consecutive integers. Write the expressions in mathematical terms.

Add 1 to get the next integer.

Let n = first consecutive integer
$n + 1$ = second consecutive integer
$n + 2$ = third consecutive integer

The *sum* of the first and third integers

$n + (n + 2)$

The *product* of the second and third integers

$(n + 1)(n + 2)$

The *square* of the third integer ⟶ $(n + 2)^2$

76 CONSECUTIVE INTEGERS

A COMPLETE EVALUATION SYSTEM

- Chapter Review and Chapter Test for each chapter.
- The Test is an alternate form of the Review.
- Chapter Review is keyed back to the lesson indicating in second color the pages where the concepts were taught, thus serving as a diagnostic test.
- A pad of duplicating masters has a test for each chapter and five cumulative tests.

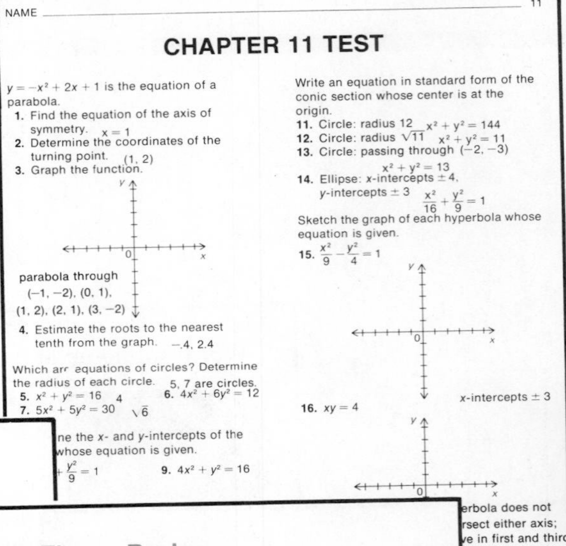
NAME ______________________________ 11

CHAPTER 11 TEST

$y = -x^2 + 2x + 1$ is the equation of a parabola.

1. Find the equation of the axis of symmetry. $x = 1$
2. Determine the coordinates of the turning point. $(1, 2)$
3. Graph the function.

parabola through $(-1, -2)$, $(0, 1)$, $(1, 2)$, $(2, 1)$, $(3, -2)$

4. Estimate the roots to the nearest tenth from the graph. $-.4$, 2.4

Which are equations of circles? Determine the radius of each circle. 5, 7 are circles.

5. $x^2 + y^2 = 16$ 4
6. $4x^2 + 6y^2 = 12$
7. $5x^2 + 5y^2 = 30$ $\sqrt{6}$

...ne the x- and y-intercepts of the ...whose equation is given.

... $\frac{y^2}{9} = 1$ 9. $4x^2 + y^2 = 16$

Write an equation in standard form of the conic section whose center is at the origin.

11. Circle: radius 12 $x^2 + y^2 = 144$
12. Circle: radius $\sqrt{11}$ $x^2 + y^2 = 11$
13. Circle: passing through $(-2, -3)$ $x^2 + y^2 = 13$
14. Ellipse: x-intercepts ± 4, y-intercepts ± 3 $\frac{x^2}{16} + \frac{y^2}{9} = 1$

Sketch the graph of each hyperbola whose equation is given.

15. $\frac{x^2}{9} - \frac{y^2}{4} = 1$ x-intercepts ± 3
16. $xy = 4$

...erbola does not ...rsect either axis; ...e in first and third ...drants.

...hose equation

18. $\frac{x^2}{20} + \frac{y^2}{20} = 1$ circle

20. $y - x^2 = 10$

Chapter Eleven Test

Find the equation of the axis of symmetry for the parabola with the given equation. Then determine the coordinates of the turning point.

1. $y = x^2 + 8x - 2$ 2. $y = -3x^2 + 1$

Graph each function. Draw the axis of symmetry and label the coordina... turning point.

3. $y = x^2 + 3x - 4$ 4. $y = x^2 - 6x + 5$

Graph each function. Estimate the roots to the nearest tenth from the g...

5. $y = 5x^2 + 9x - 1$ 6. $y = -3x^2 + 5x + 1$

7. Graph the family of parabolas described by $y = -x^2$, $y = -2x^2$, and $y =$ ...

Write an equation of the circle whose center is at the origin and with ra...

8. 10 9. $\sqrt{13}$ 10. 30

Write an equation of the circle whose center is at the origin and which ... given point.

11. $(-3, -4)$ 12. $(6, -1)$

Determine which of the following are equations of circles. Determine th... circle. Then sketch the graph.

13. $x^2 + y^2 = 25$ 14. $6x^2 + 6y^2 = 24$ 15. $4x^2 +$ ...

Write an equation in standard form of the ellipse whose center is at the ... intercepts are given.

16. $x = \pm 3, y = \pm 2$ 17. $x = \pm 6, y = \pm 2$

Determine the x- and y-intercepts of the ellipse whose equation is given... the graph.

18. $\frac{x^2}{64} + \frac{y^2}{36} = 1$ 19. $\frac{x^2}{4} + \frac{y^2}{81} = 1$ 20. $x^2 +$ ...

Sketch the graph of the hyperbola whose equation is given.

21. $\frac{x^2}{64} - \frac{y^2}{16} = 1$ 22. $xy = 9$ 23. $9x^2 -$ ...

Identify the conic section whose equation is given.

24. $\frac{x^2}{4} + \frac{y^2}{16} = 1$ 25. $\frac{x^2}{2} - \frac{y^2}{6} = 1$ 26. $\frac{x^2}{18} + \frac{y^2}{18} = 1$ 2...

Graph each of the following.

28. $y = |x - 6|$ 29. $y = (x - 3)^2 + 1$ 30. $y = 3$...

Complete the square and graph.

31. $y = x^2 - 8x + 11$

CHA...

Chapter Eleven Review

Find the equation of the axis of symmetry for the parabola with the given equation. Then determine the coordinates of the turning point. [p. 273]

1. $y = 3x^2 - 6x + 8$ 2. $y = -4x^2 + 8x - 7$ 3. $y = x^2 + 4x - 3$

Graph each function. Draw the axis of symmetry and label the coordinates of the turning point. [p. 273]

4. $y = x^2 - 6x + 12$ 5. $y = -2x^2 + 8x + 2$ 6. $y = -x^2 - 12x - 30$

Graph each function. Estimate the roots to the nearest tenth from the graph. [p. 273]

7. $y = 3x^2 + 8x - 2$ 8. $y = -2x^2 + 7x + 1$ 9. $y = x^2 + 4x - 3$

10. Graph the family of parabolas described by $y = x^2$, $y = 2x^2$, and $y = 4x^2$. [p. 273]

Write an equation of the circle whose center is at the origin and with radius as given. [p. 277]

11. 13 12. $\sqrt{15}$ 13. 17 14. $\sqrt{8}$ 15. $\sqrt{10}$ 16. 25

Write an equation of the circle whose center is at the origin and which passes through the given point. [p. 277]

17. $(-2, -1)$ 18. $(3, -2)$ 19. $(7, -4)$ 20. $(0, -4)$ 21. $(-5, -3)$

Determine which of the following are equations of circles. Determine the radius of each circle. Then sketch the graph. [p. 277]

22. $x^2 + y^2 = 49$ 23. $4x^2 + 4y^2 = 16$ 24. $8x^2 + 8y^2 = 16$ 25. $4x^2 + 6y^2 = 36$

Write an equation in standard form of the ellipse whose center is at the origin and whose intercepts are given. [p. 280]

26. $x = \pm 4, y = \pm 2$ 27. $x = \pm 7, y = \pm 3$ 28. $x = \pm 5, y = \pm 5$ 29. $x = \pm 2, y = \pm 8$

Determine the x- and y-intercepts of the ellipse whose equation is given. Then sketch the graph. [p. 280]

30. $\frac{x^2}{36} + \frac{y^2}{4} = 1$ 31. $\frac{x^2}{49} + \frac{y^2}{16} = 1$ 32. $x^2 + 4y^2 = 16$ 33. $5x^2 + 25y^2 = 125$

Sketch the graph of the hyperbola whose equation is given. [p. 283]

34. $\frac{x^2}{36} - \frac{y^2}{4} = 1$ 35. $xy = -6$ 36. $xy = 12$ 37. $16x^2 - 4y^2 = 64$

Identify the conic section whose equation is given. [p. 286]

38. $\frac{x^2}{16} - \frac{y^2}{16} = 1$ 39. $\frac{x^2}{16} + \frac{y^2}{36} = 1$ 40. $4x^2 + 4y^2 = 12$ 41. $x^2 - y = 8$

Graph each of the following. [p. 288]

42. $y = |x| - 4$ 43. $y = (x - 2)^2 + 3$ 44. $y = 2 + |x + 6|$ 45. $y = -3 + x^2$

Complete the square and graph each of the following. [p. 288]

46. $y = x^2 + 6x + 4$ 47. $y = x^2 - 10x + 20$ 48. $y = x^2 + 12x + 30$

292 CHAPTER ELEVEN REVIEW

ADDITIONAL EVALUATION IN TEACHER'S EDITION AND ON SKILLMASTERS

- Cumulative Review exercises in the Teacher's Edition provide lesson-by-lesson practice or quiz.
- A pad of duplicating masters, SKILLMASTERS, provides two short quizzes for every lesson in the student book.
- Cumulative Reviews for every two lessons are also provided on the SKILLMASTERS.

MIDPOINT FORMULA **[241]**

The formula for finding the midpoint of a line segment is developed by determining the midpoints of horizontal and vertical line segments. In Part B of the exercises, students find the coordinates of an endpoint when the coordinates of the other endpoint and the midpoint are given. In Part C, students may observe the generalization that the segments joining the midpoints of any quadrilateral in order form a parallelogram. Dividing a line segment into any given ratio may be an optional approach to finding a midpoint. For example, if the endpoints of $\overline{AB}$ are (x_1, y_1) and (x_2, y_2), then the coordinates of a point which divides the segment into the ratio 1 to 2 are $x_1 + \frac{1}{3}(x_2 - x_1)$ and $y_1 + \frac{1}{3}(y_2 - y_1)$. In general, if the endpoints of $\overline{AB}$ are (x_1, y_1) and (x_2, y_2), then the coordinates of a point which divides the segment into the ratio 1 to 1 (midpoint) are $x_1 + \frac{1}{2}(x_2 - x_1)$ and $y_1 + \frac{1}{2}(y_2 - y_1)$, or $x_1 + \frac{1}{2}x_2 - \frac{1}{2}x_1$ and $y_1 + \frac{1}{2}y_2 - \frac{1}{2}y_1$, or $\left(\frac{x_1 + x_2}{2}, \frac{y_1 + y_2}{2}\right)$.

Cumulative Review

1. Factor $9x^2 - 1$. **2.** Find and graph the solution set of $x^2 - 16 > 0$. **3.** Factor $3x^2 + 5x + 2$.

1. *$(3x + 1)(3x - 1)$* **2.** *$\{x \mid x < -4 \text{ or } x > 4\}$*
3. *$(3x + 2)(x + 1)$*

PARALLEL AND PERPENDICULAR LINES **[244]**

Students are asked to find the relationship between slopes of parallel lines and between slopes of perpendicular lines. The converses of these relationships are also true. These are given and applied to the problem of writing an equation of a line passing through a given point and parallel or perpendicular to a given line.

Two lines may be perpendicular to each other with one of the lines having undefined slope. For example, a vertical line and a horizontal line are always perpendicular, yet the slope of the vertical line is undefined.

If the slope of a line is $-\frac{3}{2}$ then the slope of a line that is perpendicular to it is $\frac{2}{3}$ and $\left(-\frac{3}{2}\right)\left(\frac{2}{3}\right) = -1$. Hence, another approach is to state that two perpendicular lines have slopes whose product is -1 (no vertical lines). Also, if the product of the slopes of two lines is -1, the lines are perpendicular.

Cumulative Review

1. Solve $x^2 - 5x + 4 = 0$. **2.** Find and graph the solution set of $x^2 - 7x + 12 \leq 0$.
3. Factor $16 - 4x^2$ completely.

HAT [After p. 241]
Form A

Find the midpoint of $\overline{PQ}$.

1. $P(2, 4)$, $Q(8, 12)$ *$(5, 8)$*
2. $P(-4, 2)$, $Q(8, -10)$ *$(2, -4)$*
3. $P(-2, -6)$, $Q(8, -4)$ *$(3, -5)$*
4. $P(3, 2)$, $Q(7, 5)$ *$\left(5, \frac{7}{2}\right)$*
5. Find the coordinates of the endpoint Q of $\overline{PQ}$, given endpoint P and midpoint M. $P(6, 2)$, $M(7, -2)$
$(8, -6)$

HAT [After p. 241]
Form B

Find the midpoint of $\overline{PQ}$.

1. $P(6, 8)$, $Q(10, 14)$ *$(8, 11)$*
2. $P(-8, 4)$, $Q(6, -12)$ *$(-1, -4)$*
3. $P(-4, -2)$, $Q(6, -10)$ *$(1, -6)$*
4. $P(5, 4)$, $Q(6, 7)$ *$\left(\frac{11}{2}, \frac{11}{2}\right)$*
5. Find the coordinates of the endpoint Q of $\overline{PQ}$, given endpoint P and midpoint M. $P(-8, 4)$, $M(1, 5)$
$(10, 6)$

HAT [After p. 244]
Form A

For the line joining the points $A(2, -3)$ and $B(-4, 5)$ find the slope of the indicated line.
1. parallel to $\overline{AB}$ *$-\frac{4}{3}$* **2.** perpendicular to $\overline{AB}$ *$\frac{3}{4}$*

Write an equation of the line parallel to the given line and passing through the given point.
3. $3x + 5y = 15$; $(-2, 1)$ *$3x + 5y = -1$*

Write an equation of the line perpendicular to the given line and passing through the given point.
4. $2x - 3y = 9$; $(1, -2)$ *$3x + 2y = -1$*

Determine whether $\overleftrightarrow{PQ} \parallel \overleftrightarrow{RS}$, $\overleftrightarrow{PQ} \perp \overleftrightarrow{RS}$, or neither.
5. $P(1, -2)$, $Q(-2, 3)$, $R(2, 3)$, $S(7, 6)$
perpendicular

HAT [After p. 244]
Form B

For the line joining the points $A(1, -4)$ and $B(-3, 2)$ find the slope of the indicated line.
1. parallel to $\overline{AB}$ *$-\frac{3}{2}$* **2.** perpendicular to $\overline{AB}$ *$\frac{2}{3}$*

Write an equation of the line parallel to the given line and passing through the given point.
3. $2x + 5y = 25$; $(-4, 2)$ *$2x + 5y = 2$*

Write an equation of the line perpendicular to the given line and passing through the given point.
4. $3x - 2y = 4$; $(1, -3)$ *$2x + 3y = -7$*

Determine whether $\overleftrightarrow{PQ} \parallel \overleftrightarrow{RS}$, $\overleftrightarrow{PQ} \perp \overleftrightarrow{RS}$, or neither.
5. $P(3, -1)$, $Q(-6, 5)$, $R(1, -3)$, $S(4, -5)$
parallel

HAT CUMULATIVE REVIEW [After p. 244]
Form A

1. Find the slope of $3y - 7x = 2$. *$\frac{7}{3}$*

2. Solve $12 - (5 - x) \geq 2x + 7$. *$x \leq 0$*

3. Find the value of $\frac{4}{8^{-\frac{2}{3}}}$. *16*

HAT CUMULATIVE REVIEW [After p. 244]
Form B

1. Find the slope of $4x - 3y = 5$. *$\frac{4}{3}$*

2. Solve $8 - (3 - x) \geq 3x + 5$. *$x \leq 0$*

3. Find the value of $\frac{3}{27^{-\frac{2}{3}}}$. *27*

Holt, Rinehart and Winston—Algebra 2 with Trigonometry—Skillmaster—32

SPECIAL TOPICS PAGES

- Careers using Mathematics
- Practical Applications of Algebra
- Extensions of Concepts

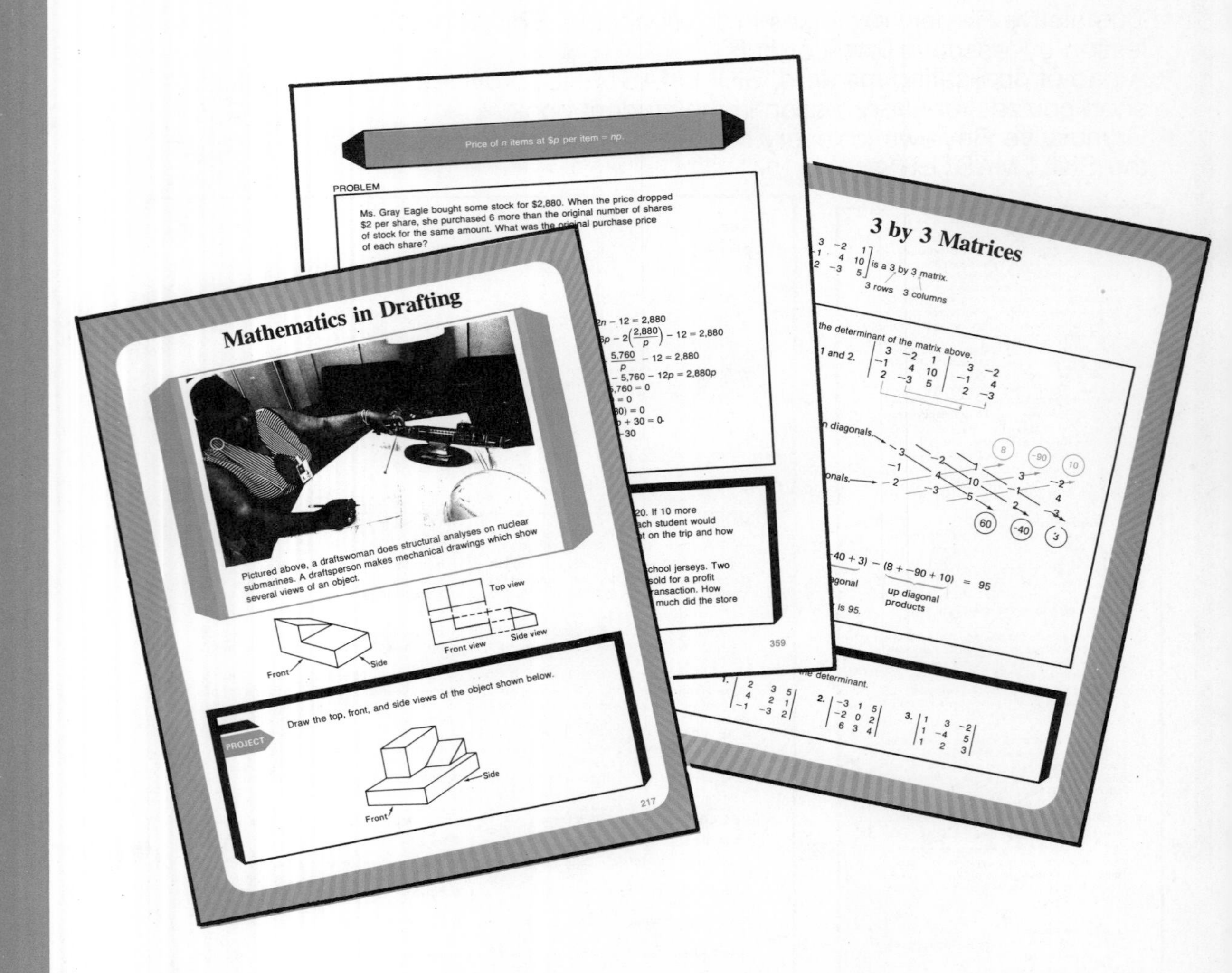

COMMENTARIES

The Commentary for Teachers describes the features of the text and offers suggestions for individualizing the course.

The Lesson Commentaries present a brief overview of each chapter followed by a short paragraph for each lesson.

With each lesson a Cumulative Review, with answers, is included. You may use the Cumulative Review before or after the lesson is studied. The purpose of the Cumulative Reviews is to sharpen students' algebraic skills.

COMMENTARY FOR TEACHERS

Holt Algebra 2 with Trigonometry is a balanced course presented in such a manner that it can be mastered by all high school students enrolled in second-year algebra. It can be studied immediately after a first-year algebra course, or immediately after a geometry course, completed after the first course in algebra.

It is a well established fact that students entering the second year of study of algebra do so with varied degrees of understanding of concepts that were covered in the first-year algebra course. They also differ a great deal in their ability to perform algebraic manipulations. For this reason, some effort is devoted to the review of concepts and skills covered in the first-year algebra course.

The course covers not only all of the fundamental concepts of algebra ordinarily studied in the second year, but the fundamental concepts of trigonometry as well. These constitute the contents of the last four chapters, making it convenient to include or exclude the study of trigonometry as desired.

One of the most significant innovations in *Holt Algebra 2* is the format, which results in a course that is easy to teach and lends itself easily to an individualized style of presentation.

NEW FORMAT

Following are the features of the book that elicited much enthusiasm from the pilot teachers and the thousands of users of the first edition. Because of the format's unusual effectiveness with both the teachers and students, this format is being retained in this revised edition.

1. The *objectives* of each lesson are given on the pupil page. These state very briefly and concisely the student's expected performance at the completion of the lesson.
2. A *Review Capsule* is supplied for each lesson. The capsule reintroduces a skill or concept that the student needs to recall for that lesson.
3. Each lesson is developed through a sequence of *worked-out examples.* The use of words is cut to the minimum so that mathematical ideas stand out.
4. *Hints and suggestions* along the left-hand side of the page are used to direct the students' attention to the key parts of the development. The hints and suggestions not only help the students who need help, but allow the students who grasp ideas quickly to move on without being slowed down by explanations that they do not need.
5. *Definitions* and other *key statements* are clearly set out in color for easy identification. This facilitates study and review.
6. There is an abundance of *worked-out examples* to introduce concepts and establish patterns. These serve as reference for problems of a similar nature that are to be solved by students on their own. Over the years, teachers have ascertained that students understand ideas better and feel more confident with mathematics if they see worked-out examples.
7. *Exercises* are plentiful in each lesson. They are presented in Parts A, B, and C according to level of difficulty. In this way, all students may work on material suited to their abilities. Part C exercises consist of more challenging problems and thus provide an enriched course. Columns of exercises are similar to each other. Successful performance in a given column implies that the student is ready to advance to the next level. Unsuccessful performance indicates the necessity for the

student to restudy the illustrative examples in the lesson and then to proceed to a corresponding column in the same level.

8. *Oral Exercises* precede the exercise set in many lessons. The purpose of the Oral Exercises is to help the student recall basic facts that will be used in the exercise set. The Oral Exercises serve for practice on simple concepts and skills. Many of these simple concepts and skills can be practiced in a short time.
9. *Chapter Reviews* are keyed into lesson pages for quick reference. *Chapter Tests* exactly parallel the Chapter Reviews. This enables students to quickly locate the pages on which the items they need to study further were taught.

It is a well known fact that a significant number of high school students do not succeed in mathematics. The authors of *Holt Algebra 2* believe that part of the fault lies with the way in which mathematics is presented. For this reason, the radically different approach described above was adopted.

The use of the first edition of the book has proved the effectiveness of the new format. Students can read and understand what they are reading. They are able to solve problems on their own. Thus, very carefully sequenced experiences followed by immediate reinforcement do result in greater success.

IMPROVEMENTS IN THE PRESENT EDITION

While the thousands of teachers who used the first edition of *Holt Algebra 2* found their students succeeding remarkably well, they made suggestions for making the book even more teachable. These suggestions for improvements were incorporated in the 1978 edition of the book. Some of these improvements are the following:

1. The number and frequency of Part C exercises have been increased. In some instances they have been made more challenging. Students are asked to apply the concepts to some novel situations.
2. Additional material concerning careers was incorporated.
3. Applications of algebra to real-life problems were added. Many of these are directly tied in with Career Awareness pages.
4. A lesson on dry-mixture problems was added to Chapter 2. An optional lesson on wet-mixture problems was added to Chapter 5.
5. The lesson on $\sin(x \pm y)$ was revised to include $\sin 2x$. Similarly, the lesson on $\cos(x \pm y)$ was revised to include $\cos 2x$.
6. A lesson on $\sin\frac{x}{2}$, $\cos\frac{x}{2}$, and $\tan\frac{x}{2}$ was added.
7. Suggestions for easier or alternate methods of handling some concepts have been provided in several lessons.
8. Answers to all odd-numbered exercises are provided in the student text.
9. The metric system of measurement is now used exclusively.

INDIVIDUALIZING THE COURSE

Holt Algebra 2 lends itself easily to an individualized approach. Each lesson is designed so that it can be read and understood by all students.

Holt Algebra 2 provides for three different ways to individualize the course.

1. **By Lesson** Each lesson has exercises arranged in three levels of difficulty: Part A, Part B, and Part C. Those students who wish to pursue the basic course would do some Part A exercises and some Part B exercises —usually the least difficult exercises in each section. Students who wish to pursue the average course would do some Part A exercises and some Part B exercises—usually those of medium difficulty. Students who are capable of handling the enriched course would do some exercises from each group: Part A, Part B, and Part C.

2. **By Course** At the end of the Lesson Commentaries in the front of the Teacher's Edition, the Pacing Chart divides the year's work into three levels:

   Level A provides the minimal course and allows time for reinforcement and recycling.

Level B	provides the regular course.
Level C	provides the enriched course.

You may choose to suggest that the basic student follow the Level A assignments; the average student, the Level B assignments; and the capable student, the Level C assignments. Each of the three groups of students would complete a different course throughout the year.

3. **Using the Individualized Guide** The Teacher's Guide for Individualizing the Holt Algebras is a rich source of ideas and aids for providing for individual differences. Other components, in addition to the text, that can be used with the Individualized Guide are Test Masters, Skillmasters, and Individualization Masters.

PILOTING THE PROGRAM

While the first edition of this program was under development, it was tested in order to ascertain its effectiveness. Schools scattered throughout the United States used the materials under ordinary classroom conditions. Both average and below-average classes were included in the pilot testing. The testing of materials extended over a two-year period. A systematic evaluation of results for each lesson, by the teachers, established that the material is highly teachable to all students. One of the recurring comments from teachers was that students' achievement in the *Holt Algebra 2* program was higher than in other programs used previously or concurrently with other classes.

ORGANIZATION OF THE COURSE

The book begins with a review of some basic concepts necessary for further study of algebra: problem solving using equations and inequalities, and using the concept of absolute value. The contents of Chapters 1 and 2 are devoted to this review. Chapter 3 proceeds with exponents, polynomials, multiplication of polynomials, factoring, solving quadratic equations and inequalities, and some verbal problems. Rational expressions is the subject of Chapter 4, including addition of fractions. Rational expressions are studied in their applications to such topics as complex fractions, literal equations, equations with decimals, repeating decimals, and division of polynomials, in Chapter 5. A thorough study of the set of real numbers takes place in Chapter 6. The study of quadratic equations in Chapter 7 includes radical equations, operations with complex numbers, and verbal problems leading to quadratic equations. In Chapters 8 and 9, the plane is studied by means of algebra. The slope of a line, Pythagorean theorem, distance formula, and midpoint formulas are covered. The very important concept of function and the related concepts of direct variation, inverse variation, and combined and joint variation are studied in Chapter 10. Next, conic sections are explored: parabola, circle, ellipse, and hyperbola. Systems of equations and inequalities are solved in Chapter 12. This includes both linear and quadratic systems. Problems leading to these systems are also considered. Arithmetic progressions, geometric progressions, and finite and infinite geometric series are presented in Chapter 13. Chapters 14 and 15 are devoted to the presentation of logarithms and their applications, including scientific notation. Chapters 16 through 19 are devoted to the study of the basic concepts of trigonometry: trigonometric functions and their graphs, trigonometric equations, logarithms of trigonometric functions, solving triangles by the use of trigonometric laws, and special identity formulas.

The contents, as described above, give a full course in second-year algebra and a substantial portion of a course in trigonometry. In addition to the course described above, 51 pages of material are provided to vary the pace and enrich the course. These pages contain biographies, recreational mathematics, historical data, mathematical curiosities, career awareness, applications, flowcharting, and measurement. For a listing of the Special Topics, see page viii in the front of the pupil edition.

HOW TO USE THE BOOK

The format of *Holt Algebra 2* was developed on the basis of the conviction that no student should be penalized in the study of mathematics because of the lack of reading skills. Inability to read should not stand in the way of learning mathematics.

It is also recognized that teachers succeed with their students in many different ways. Each teacher's style of teaching should be respected. *Holt Algebra 2* provides the freedom to style your teaching in accordance with your individual preferences. The authors would not presume to impose any specific best way of teaching from this textbook. Some general guidelines, however, are offered.

The textbook can be used in an individualized instruction situation. Individualization is made possible by guiding the student through the lesson developments by means of many examples. Exercises are arranged from simple to more complex. A multitude of practice material, much of a very basic nature, facilitates individual study.

Where group instruction takes place, the examples can be discussed in small groups or with an entire class. The exercises can be used in the usual manner, some for practicing new work and some for review and quizzes.

There should be no doubt that students will be attracted to the study of algebra if they are encouraged frequently. A definite and positive attitude as put forth in *Holt Algebra 2* will help keep students interested and active.

SUPPLEMENTARY MATERIALS

Holt Algebra 2 is a system that supplies the teacher with a variety of materials, making it easy to adjust the course to each individual student.

Holt Algebra 2 Tests The Test Masters are a means of evaluating a student's abilities to perform the objectives of the *Holt Algebra 2* program. The items on the Chapter Test sample the objectives covered in that chapter only. Each of the Chapter Test Masters is an alternate form of the Chapter Test found in the textbook.

The Chapter Test Masters can be used with students who have not satisfactorily succeeded with the Chapter Tests found in the textbook. Thus, the tests found in the Duplicating Masters can serve as a second chance for the students to succeed. They also provide an opportunity to use one of the tests (e.g., Chapter Test in the textbook) as a pre-test, and the other test (Chapter Test Master) as a post-test.

The Test Masters also contain five cumulative tests that can be used for semester or trimester schedules.

Holt Algebra 2 Skillmasters Skillmasters are usable in a class or in an individualized situation. Each sheet has two parallel forms (A and B) for every lesson. They may be used as pre- or post-lesson self-tests, pre-tests for diagnosis, extra practice to set a skill, quizzes to test mastery, or assignments. At the bottom of each sheet are two parallel forms of a Cumulative Review that tests skills and concepts previously presented.

REFERENCE CHART FOR ALGEBRA 2 SKILLMASTERS

LESSON PAGES	SKILLMASTER NUMBER	LESSON PAGES	SKILLMASTER NUMBER
1, 4	1	235, 238	31
6, 9	2	241, 244	32
14, 19	3	247, 253	33
22, 28	4	256, 258	34
32, 37	5	261, 264	35
40, 45	6	267, 273	36
49, 53	7	277, 280	37
56, 59	8	283, 286	38
62, 64	9	288, 295	39
69, 73	10	298, 304	40
76, 80	11	308, 311	41
85, 87	12	314, 318	42
90, 94	13	323, 327	43
98, 101	14	333, 337	44
109, 113	15	339, 342	45
116, 119	16	346, 350	46
124, 127	17	360, 363	47
134, 137	18	365, 368	48
141, 145	19	370, 374	49
148, 153	20	376, 378	50
156, 161	21	383, 386	51
164, 171	22	389, 392	52
174, 176	23	395, 398	53
179, 182	24	399, 405	54
185, 188	25	407, 409	55
193, 196	26	413, 416	56
201, 205	27	420, 424	57
211, 214	28	429, 431	58
218, 221	29	435, 437	59
226, 228	30	441, 443	60

Teacher's Guide for Individualizing Holt Algebra 1 and Holt Algebra 2 with Trigonometry The teacher's guide presents the different approaches for individualizing and the ways for implementing each approach. There is also an Activity Reservoir that contains suggestions for research projects, activities, and games.

The activities can supply students with interesting diversions of topics that can satisfy the unique desires of various students.

Duplicating Masters for Individualizing Holt Algebra 1 and Holt Algebra 2 These duplicating masters assist the teacher in classroom management when individualizing the Holt Algebra programs. They contain placement tests which provide the teacher with the ready means to assess the student's standing and thus place her or him at the level where she or he will succeed.

Also included are an assignment sheet for each chapter, a teacher record sheet for the individual student, a class record sheet, and a parent report form to be sent to parents at marking periods. The assignment sheet for each chapter contains specific references to several sources of additional development and practice aids. Thus, they provide the work that is commensurate with the individual student's capabilities, interests, and needs.

ACKNOWLEDGEMENTS

We wish to thank and acknowledge the help of the pilot teachers and students for testing the first edition of *Holt Algebra 2.* Their comments and suggestions have been of invaluable help in making this a better and more teachable text. The following teachers and schools took part in the pilot program.

Mary Dozier, Central High School, Macon, Georgia; Shirley Freeman, Waller High School, Chicago, Illinois; Henry Baldassari, Salem Classical High School, Salem, Massachusetts; E. Henry Garland, The University School, Tallahassee, Florida; John McConnaughhay, The University School, Tallahassee, Florida; Andrew Arner, San Pasqual High School, Escondido, California; Jean Henrichs, East High School, Lincoln, Nebraska.

We also want to thank the hundreds of teachers, supervisors, and administrators who participated in our surveys used for this new series.

LESSON COMMENTARIES

CHAPTER 1 *Linear Equations and Inequalities*

Solving first-degree equations and inequalities with parentheses and with a variable on both sides provides a review of several skills and concepts needed for an Algebra 2 course. Applying the distributive property to remove parentheses and to combine like terms are two examples of such skills. A first review of word problems is kept simple by representing and finding *only one* number. Formulas are introduced as a means of solving *practical* problems.

PUZZLERS [x]

These puzzlers may be used as an optional assignment limited to several weeks. Extra credit can be given if an answer *and the work* are submitted. You may wish to assign this topic to all students.

LINEAR EQUATIONS [1]

The two different directions *Solve* and *Find the solution set* give answers that are *numbers* and a *set of numbers* respectively. Simple fractional equations, like $\frac{2}{3}x - 5 = 2$, are solved. Part B exercises require *combining like terms* as a first step.

Cumulative Review

1. Compute $7 - 9 - (-5) - 10$. **2.** Compute $6(-3) + (-4)(-2)$. **3.** Simplify $7x - 5y - x + y$.

1. *−7* **2.** *−10* **3.** *6x − 4y*

LINEAR EQUATIONS WITH PARENTHESES [4]

Parentheses are removed from equations, like $8x \pm 3(2x \pm 5) = 43$ and $5 - (4x \pm 3) = 18$. In the first instance, ± 3 is *distributed* over $(2x \pm 5)$. In the second instance, $-(4x \pm 3)$ becomes $-1(4x \pm 3)$ by the property of *−1 for multiplication* $[-(a) = -1(a)]$; and -1 is then *distributed* over $(4x \pm 3)$.

Cumulative Review

1. Compute $3 + 2 \cdot 5 - 2$. **2.** Solve $\frac{2}{3}x - 1 = 5$. **3.** Simplify $-4(2x - 3) - (5 - 10x)$.

1. *11* **2.** *9* **3.** *2x + 7*

LINEAR INEQUALITIES [6]

Students should become aware that solving equations is similar to solving inequalities *with one exception.* That exception is that *dividing each side by a negative number* reverses the order: $<$ to $>$, or $>$ to $<$. Beginning with Example 1, many students need practice in reading set notation, like $\{x | x > -4\}$. In Part C exercises 23 and 24 note that *multiplying* each side by a *negative* number reverses the order just as *dividing* each side by a *negative* number reverses the order.

Cumulative Review

1. Simplify $7x - (5x + 3)$. **2.** Solve $3(5x - 4) - (7x + 2) = 2$. **3.** Compute $3(-4) + (-7)(-2)$.

1. *2x − 3* **2.** *2* **3.** *2*

FINDING A NUMBER [9]

The word problems are restricted to those with only one unknown so that students may practice representing the unknown and writing an equation in a simple setting. Students should give special attention to the phrase *less than: a less than b* is $b - a$, not $a - b$.

Cumulative Review
1. Compute $12 - 5(2) - 4 + 3(-2)$. **2.** Simplify $3x + 2(4x - 3)$. **3.** Solve $3 - 4x = 5x - 4$.
1. *−8* **2.** *$11x - 6$* **3.** $\frac{7}{9}$

THE METRIC SYSTEM **[12]**
You should decide how much emphasis you wish to give to these pages. Word problems in this text use the following metric units but do not involve changing from one unit to another:

meter (m)	gram (g)
kilometer (km)	kilogram (kg)
centimeter (cm)	liter (L)
decimeter (dm)	milliliter (mL)

Another way of thinking of the chart is that each unit of length has a place value of one tenth of the unit of length immediately to its left. Thus, 1 mm = .1 cm and 1 hm = .1 km.

USING FORMULAS **[14]**
Two formulas involving simple interest and one formula involving resistances in an electric circuit are used to show applications of algebra to problem solving. Part C Exercises 19 and 20 involve Ohm's law. A hand-held mini-calculator can be used to advantage in all of the exercises.

Cumulative Review
1. Solve $8x - 3(2x + 5) = 7$. **2.** Find the solution set of $9 - 3x < 5x + 1$. **3.** Two less than a number is the same as 3 times the sum of the number and 4. Find the number.
1. *11* **2.** $\{x|x > 1\}$ **3.** *−7*

CHAPTER 2 *Absolute Value and Word Problems*

Equations and inequalities involving absolute value are solved. Then five types of word problems are analyzed and solved.

ALGEBRAIC PROOFS **[18]**
This is the first in a series of Special Topics on number systems and their properties. The series consists of *Algebraic Proofs,* page 18; *Modulo 5,* page 26; *Groups,* page 35; and *Fields,* page 132. You may wish to assign these topics separately as they occur or as a unit.

COMPOUND INEQUALITIES **[19]**
Inequalities of the forms $a < bx \pm c < d$ and $[bx \pm c < a$ or $bx \pm c > d]$ are solved and graphed as preparation for the next lesson, solving inequalities involving absolute value.

Cumulative Review
1. Solve $4n - (2n + 3) = 12 - 3n$. **2.** Find the solution set of $5(x + 2) < 7x - 4$. **3.** Compute $-5 - 2(-3) - 8(-2)$.
1. *3* **2.** $\{x|x > 7\}$ **3.** *17*

ABSOLUTE VALUE **[22]**
Thinking of absolute value as a distance and visualizing relationships on a number line will help in solving absolute value equations and inequalities.

Cumulative Review
1. Compute $-8 - 4(-3)$. **2.** Simplify $8x + y - x + 3y$. **3.** Solve $7x - 2 = 10 + x$.
1. *4* **2.** *$7x + 4y$* **3.** *2*

MATHEMATICS IN THE RESTAURANT **[25]**
This is the first in a series of Special Topics on career awareness. An opportunity for students to better understand the mathematics involved in planning a menu for a restaurant or school cafeteria is presented. This project can be extended to determine the cost of other menus. You may wish to have students interview restaurant or school cafeteria managers to gain an understanding of how managers estimate the number of dinners to be served, how they determine the costs of a menu, and how they allow for overhead expenses such as energy and employee expenses. An oral report can be made by those students working on this project. This project is appropriate for all levels.

MODULO 5 **[26]**
This is the second in a series of Special Topics on number systems and their properties. For suggestions on using these pages, see the commentary for page 18 on page T-24.

NUMBER AND PERIMETER PROBLEMS [28]

Encourage students to *write* the two or three unknowns in mathematical terms. In a geometry type (perimeter) problem, they should sketch and label a figure.

Cumulative Review

1. Compute $\frac{-18}{3} - 6(-4) + 5(-5)$. **2.** Solve $|3n - 4| = 2$. **3.** Solve $2n - 3(2n - 4) = 8$.

1. -7 **2.** $2, \frac{2}{3}$ **3.** 1

COIN PROBLEMS [32]

After representing the *numbers* of coins in mathematical terms, the terms must be multiplied by a factor like 5, 10, 25, and so on, to express the *value.* The Review Capsule can be used to emphasize this.

Cumulative Review

1. Compute $-3 + 5(7 - 3 \cdot 4)$. **2.** Simplify $2x - (7x - 3)$. **3.** Solve $5 + \frac{1}{4}x = 7$.

1. -28 **2.** $-5x + 3$ **3.** 8

GROUPS [35]

This is the third in a series of Special Topics on number systems and their properties. See the commentary for page 18 on page T-24 for a listing of the other topics in this series and for suggestions on using these pages.

DRY MIXTURE PROBLEMS [37]

In Example 1 students should understand why the number of kg of almonds is $30 - x$, rather than $x - 30$. This can be done by using numerical examples such as the following. If there are 3 kg of peanuts, then there are 27, or $(30 - 3)$, kg of almonds. The last column in each table (value in cents) is found by multiplying the *number of items* by *the cost of one item in cents,* as suggested in the Review Capsule. The equations are written by using the concept that a *number of cents* plus another *number of cents* equals the *total number of cents.*

Cumulative Review

1. Find the solution set of $-5 < 2x + 3 < 7$.
2. Find the solution set of $|x - 5| > 3$.

1. $\{x | -4 < x < 2\}$ **2.** $\{x | x < 2 \text{ or } x > 8\}$

AGE PROBLEMS [40]

Students frequently have trouble with algebraic representation. Encourage them to first express the ages now in mathematical terms. Then, ages in the past (ago) are represented by *subtracting* the given number of years, and ages in the future (from now) are represented by *adding* the given number of years.

Cumulative Review

1. Find the solution set of $|4x - 2| < 6$.
2. Simplify $5x - 8y - 6x + 9y$. **3.** Solve $-6 - 3y = 4y + 1$.

1. $\{x | -1 < x < 2\}$ **2.** $-x + y$ **3.** -1

CHAPTER 3 *Polynomials and Factoring*

Properties of positive exponents are developed for product of powers, power of a power, and power of a product. Addition, subtraction, and multiplication of polynomials are practiced. The quotient of powers and division of polynomials occur in Chapters 4 and 5. Major emphasis is on factoring. Applications of factoring are treated in lessons on solving quadratic equations and inequalities.

METRIC SYSTEM: AREA AND TEMPERATURE [44]

You should decide how much emphasis to give to this page. The following units of area are used in problem-solving situations.

square meter (m^2)
square centimeter (cm^2)
square decimeter (dm^2)

You may wish to review metric units of length at this time. This topic is discussed on page 12.

EXPONENTS [45]

The summary may help students distinguish the three different situations.

Cumulative Review

1. Simplify $5x - 3y + 2x - 4$. **2.** Solve $|x - 4| = 5$. **3.** Write $\frac{10}{15}$ in the simplest form.

1. $7x - 3y - 4$ **2.** 9 and -1 **3.** $\frac{2}{3}$

POLYNOMIALS [49]

Subtracting polynomials in horizontal form $p - q$ is done by using the property of -1, $[p - 1(q)]$, then distributing the -1 and combining like terms.

Cumulative Review

1. Simplify $3a + 5(2a - 4)$. **2.** Find the solution set of $5x - 4 < 2x + 8$. **3.** Multiply $\frac{3}{5} \times \frac{5}{3}$.

1. $13a - 20$ **2.** $\{x|x < 4\}$ **3.** 1

MULTIPLYING POLYNOMIALS [53]

Vertical form is used except for multiplying by a monomial. Rules for the special products $(a + b)(a - b)$, $(a + b)^2$, and $(a - b)^2$ are given later in the chapter.

Cumulative Review

1. Simplify $7x - 2(3x + 5)$. **2.** Find the solution set of $|x| < 5$. **3.** Divide $\frac{2}{3} \div \frac{3}{4}$.

1. $x - 10$ **2.** $\{x|-5 < x < 5\}$ **3.** $\frac{8}{9}$

COMMON FACTORS [56]

To find the GCF of a polynomial, first find the GCF of the coefficients and then find the GCF of the variables.

Cumulative Review

1. Simplify $(-7a)(4c)$. **2.** Solve $|n + 3| = 8$. **3.** Add $\frac{2}{7} + \frac{1}{7} + \frac{3}{7}$.

1. $-28ac$ **2.** 5 and -11 **3.** $\frac{6}{7}$

FACTORING INTO BINOMIALS [59]

Trinomials are restricted to $ax^2 \pm bx \pm c$, where a and c are prime numbers, to give the minimum number of possible binomial factors. The next lesson covers $ax^2 \pm bx \pm c$, where a or c, or both, are not prime numbers.

Cumulative Review

1. Simplify $3x + 2 - (6 + 4x)$. **2.** Find the solution set of $-3x < 6$. **3.** Add $\frac{1}{2} + \frac{1}{4} + \frac{1}{8}$.

1. $-x - 4$ **2.** $\{x|x > -2\}$ **3.** $\frac{7}{8}$

FACTORING TRINOMIALS [62]

The special cases of $x^2 - y^2$ and $x^2 \pm 2xy + y^2$ occur in the next lesson.

Cumulative Review

1. Factor $8x^2 - 4x$. **2.** Find the solution set of $|x| > 3$. **3.** Rewrite $\frac{12}{15}$ in simplest form.

1. $4x(2x - 1)$ **2.** $\{x|x < -3 \text{ or } x > 3\}$ **3.** $\frac{4}{5}$

SPECIAL PRODUCTS AND FACTORING [64]

Learning the special patterns illustrated here will save much time and effort. However, for students who do not wish to memorize the forms, there are these alternatives:

1. Multiply binomials of the form $(a + b)(a - b)$ as they would multiply any two binomials.
2. Factor a difference of squares like $4x^2 - 9$ as a trinomial, $4x^2 + 0 \cdot x - 9$.
3. Square binomials of the form $(a \pm b)^2$ as they would multiply any two binomials.
4. Factor a perfect square trinomial like $9x^2 + 24x + 16$ as they would factor any trinomial.

Cumulative Review

1. Simplify $(3x^2y)(-5xy^3)$. **2.** Solve $|2n - 3| = 7$. **3.** Multiply $\frac{2}{3} \times \frac{5}{7}$.

1. $-15x^3y^4$ **2.** 5 and -2 **3.** $\frac{10}{21}$

FACTOR THEOREM [68]

The factor theorem depends on the following. A polynomial $P(x)$ may be divided by a binomial $(x - a)$ to give a polynomial quotient $Q(x)$ and a real number remainder r. Quotient times divisor plus remainder equals dividend.

Thus, $P(x) = (x - a) \cdot Q(x) + r$.
Substitute a for x. $P(a) = (a - a) \cdot Q(a) + r = 0 + r$.
So, $P(a) = r$.

And if $P(a) = r = 0$,
then, $P(x) = (x - a) \cdot Q(x) + 0$ and $x - a$ is a factor of $P(x)$.

QUADRATIC EQUATIONS [69]

In an exercise like Example 6, students sometimes forget to set the monomial factor equal to 0. Emphasize that students should generally expect two solutions except when both factors are the same, as in Example 7.

Cumulative Review

1. Simplify $(-5)(-2x)(3y)$.
2. Find the solution set of $5x - 4 \geq 3x$.
3. Divide $3 \div \frac{1}{2}$.

1. $30xy$ 2. $\{x|x \geq 2\}$ 3. 6

QUADRATIC INEQUALITIES [73]

To solve the *inequality* $x^2 + bx + c < 0$ (or $c > 0$), solve the *equation* $x^2 + bx + c = 0$ and then decide if solutions of the inequality are between or outside the solutions of the equation by testing a number in each region.

Cumulative Review

1. Factor $3x^2 + 16x - 12$.
2. Find the solution set of $|2x| < 8$.
3. Add $\frac{1}{4} + \frac{2}{3}$.

1. $(3x - 2)(x + 6)$ 2. $\{x|-4 < x < 4\}$
3. $\frac{11}{12}$

CONSECUTIVE INTEGERS [76]

Examples 1–5 concentrate on representing consecutive integers, consecutive odd integers, and consecutive even integers in mathematical terms. Problems leading to both linear and quadratic equations are presented.

Cumulative Review

1. Simplify $x - y - 3x + 4y$.
2. Solve $|3n + 1| = 10$.
3. Rewrite $\frac{32}{48}$ in simplest form.

1. $-2x + 3y$ 2. 3 and $-3\frac{2}{3}$ 3. $\frac{2}{3}$

FACTORING COMPLETELY [80]

Students should *look for a GCF first.* Then they should try to factor the remaining polynomial using any of the methods or special patterns studied earlier.

Cumulative Review

1. Simplify $10x - (8 + 2x)$.
2. Find the solution set of $3x - 4 \leq 6 - 2x$.
3. Divide $\frac{1}{5} \div \frac{2}{3}$.

1. $8x - 8$ 2. $\{x|x \leq 2\}$ 3. $\frac{3}{10}$

CHAPTER 4 *Rational Expressions*

Methods of simplifying algebraic fractions are developed. Then the four basic operations are practiced with algebraic fractions. Applications of these topics are found in Chapter 5.

MARIA GAETANA AGNESI [84]

This is the first in a series of Special Topics on famous mathematicians. Some students may be interested in reading more detailed biographies to learn more about the person's successes as well as failures in life aside from mathematics. You may suggest these references:

Bell, E. T. *Men of Mathematics.* Simon and Schuster, 1937.

Osen, Lynn M. *Women in Mathematics.* Massachusetts Institute of Technology, 1974.

IDENTIFYING RATIONAL EXPRESSIONS [85]

The denominator of a fraction cannot be zero. For example, $\frac{8}{2} = 4$ because $4 \cdot 2 = 8$; but $\frac{5}{0} = n$ has no solution since $n \cdot 0 = 5$ has no solution.

Cumulative Review

1. Simplify $(a^3b)(a^2b^2)$.
2. Find the solution set of $8x + 5 > 5x - 1$.
3. Factor $x^2 - 22x - 48$.

1. a^5b^3 2. $\{x|x > -2\}$ 3. $(x - 24)(x + 2)$

SIMPLIFYING RATIONAL EXPRESSIONS [87]

Finding the quotient of powers is developed along with writing algebraic fractions in simplest form.

Cumulative Review

1. Simplify $(2x^4)^3$.
2. Find the solution set of $|2n + 7| = 5$.
3. Factor $2x^2 - 18$ completely.

1. $8x^{12}$ 2. $\{-1, -6\}$ 3. $2(x + 3)(x - 3)$

MULTIPLYING AND DIVIDING [90]

For multiplying algebraic fractions, students should follow a procedure similar to the one they use for arithmetic fractions. Example 1 emphasizes this.

Cumulative Review

1. Evaluate $5x^2y - xy$ if $x = 2$ and $y = -3$.
2. Find the solution set of $-4x < -12$.
3. Factor $4x^3 + 6x$ completely.

1. -54 2. $\{x|x > 3\}$ 3. $2x(2x^2 + 3)$

ADDING: SAME DENOMINATORS [94]

The direction *add* automatically implies that students should write the result in simplest form.

Cumulative Review

1. Subtract $(4x^3 + 2x^2) - (5x^2 - 3x)$.
2. Find the solution set of $|2x + 3| < 9$.
3. Simplify $5x - 3(3x - 4)$.

1. $4x^3 - 3x^2 + 3x$ 2. $\{x|-6 < x < 3\}$
3. $-4x + 12$

SEQUENCE SUSPENSE [96]

Page 96 may be assigned independently of page 97. The tenth fraction and those beyond are always .618 to 3 decimal places. You may wish to have students use a mini-calculator to change the last fraction to a decimal. Students who have access to a mini-calculator may find it interesting to change *each* fraction, beginning with the first one, to a decimal. The sequence of decimals will get ever closer to .61803 to five decimal places.

ADDING: DIFFERENT DENOMINATORS [98]

Students should observe that the LCD has all of the factors contained in each denominator. Emphasis is on rewriting each fraction using the LCD before the fractions are combined.

Cumulative Review

1. Multiply $(3x + 4y)(2x - 5y)$.
2. Solve $x^2 - 7x + 12 = 0$.
3. Add $(5x^3 - 3x^2 + 4) + (x^2 + x - 6)$.

1. $6x^2 - 7xy - 20y^2$ 2. 3 and 4
3. $5x^3 - 2x^2 + x - 2$

RATIONAL EXPRESSIONS: $\frac{a}{b} - \frac{c}{d}$ [101]

The property $-\frac{x}{y} = \frac{-1x}{y}$ is used to rewrite expressions of the form $\frac{a}{b} - \frac{c}{d}$ as $\frac{a}{b} + \frac{-1c}{d}$. Then, the example becomes an addition exercise like those in the previous lesson.

Cumulative Review

1. Square $(2x + 3)^2$.
2. Solve $x^2 - 4x - 12 = 0$.
3. Find the solution set of $(x - 2)(x - 5) < 0$.

1. $4x^2 + 12x + 9$ 2. 6 and -2
3. $\{x|2 < x < 5\}$

FRACTURED FRACTIONS [105]

Any fraction of the form $\frac{ax + b}{(x + c)(x + d)}$ can be written as an indicated sum of the form $\frac{A}{x + c} + \frac{B}{x + d}$. In Step 4 of the example, the numerator $(A + B)x + (2A + B)$ comes from Step 3; that is, $Ax + 2A + Bx + B = (Ax + Bx) + (2A + B) = (A + B)x + (2A + B)$.

CHAPTER 5 *Using Rational Expressions*

The operations on rational expressions are applied to solving fractional equations, literal equations, and work problems. Rational number is defined when repeating decimals are introduced. Complex fractions and division of polynomials are also treated. Wet Mixture Problems are treated as an optional lesson.

IS 2 EVER EQUAL TO 1? [108]

The only error occurs in Step 6 where dividing by $a - b$ is *dividing by zero,* since $a = b$. This dramatically illustrates why dividing by zero should be left undefined. Most students find this puzzle interesting but challenging to solve.

FRACTIONAL EQUATIONS [109]

To avoid excessive arithmetic in checking, it is only necessary to substitute apparent solutions to see if any denominator is zero. A proportion is a fractional equation, but it is solved more readily by "cross-multiplying" than by the general method. Decimal equations are solved.

Cumulative Review

1. Square $(x - 6)^2$.
2. Find the solution set of $-2x + 3 < 11$.
3. Simplify $\frac{6m^4n^2}{8mn^5}$.

1. $x^2 - 12x + 36$ 2. $\{x|x > -4\}$ 3. $\frac{3m^3}{4n^3}$

WORK PROBLEMS [113]

The key is that the *sum* of the fractional parts of the job done is 1, signifying the whole job.

Cumulative Review

1. Evaluate $2x^2y + xy^2$ if $x = -1$ and $y = 3$.
2. Solve $|x - 3| = 7$.
3. Multiply $\frac{x^2 - 16}{2y - 6} \cdot \frac{3y - 9}{2x + 8}$.

1. -3 2. 10 and -4 3. $\frac{3(x - 4)}{4}$

COMPLEX RATIONAL EXPRESSIONS [116]

Multiplying the numerator and the denominator by the LCD is more efficient than combining in the numerator and denominator and then dividing.

Cumulative Review

1. Solve $x^2 - 5x = 14$.
2. Simplify $10x - (x + 5)$.
3. Add $\frac{4}{3x^2} + \frac{5}{6x}$.

1. 7 and -2 2. $9x - 5$ 3. $\frac{8 + 5x}{6x^2}$

LITERAL EQUATIONS AND FORMULAS [119]

Modeling an example by the solution of a similar equation with numerical coefficients may help students who have difficulty.

Cumulative Review

1. Find the solution set of $(x + 3)(x - 2) < 0$.
2. Simplify $5x^2 - 4y^2 - x^2 + y^2$.
3. Divide $\frac{3m}{10x^2} \div \frac{6m^4}{5x^3}$.

1. $\{x|-3 < x < 2\}$ 2. $4x^2 - 3y^2$ 3. $\frac{x}{4m^3}$

WET MIXTURE PROBLEMS [122]

You may wish to assign all or part of this optional lesson to all students.

RATIONAL NUMBERS AND DECIMALS [124]

Students should become accustomed to both symbols, $.35\overline{35}$ and .353535 . . . , for repeating decimals. Both are used again in the study of irrational numbers.

Cumulative Review

1. Simplify $(x^2y^3)^4$. 2. Find the solution set of $|x| > 2$. 3. Add $\frac{4}{x + 3} + \frac{5x}{2x + 6}$.

1. x^8y^{12} 2. $\{x|x < -2 \text{ or } x > 2\}$
3. $\frac{8 + 5x}{2(x + 3)}$

DIVIDING POLYNOMIALS [127]

Dividing $(600 + 70 + 9)$ by $(30 + 2)$ provides an analogy for dividing $(3x^2 - 5x - 30)$ by $(x - 4)$. You may wish to assign the flow chart on page 167 at this time. For enrichment, the Special Topic *Synthetic Division* on page 184 may be included here for some classes or some students.

Cumulative Review

1. Simplify $(6a^3b^2)(2ab^3)$. 2. Solve $11 - 2x = 2x - 9$. 3. Multiply $(x^2 - 4) \cdot \frac{1}{3x + 6}$.

1. $12a^4b^5$ 2. 5 3. $\frac{x - 2}{3}$

CHAPTER 6 *Real Numbers*

Radicals are developed as solutions of equations like $x^2 = 7$, $x^3 = 10$, and $x^4 = 5$. Exponents are extended to include negative, fractional, and zero exponents. Irrational numbers are introduced. Operations with square roots, as well as their approximations, are treated.

FIELDS [132]

This is the fourth and last in a series of Special Topics on number systems and their properties. See the commentary for page 18 on page T-24 for suggestions on using these pages.

SOLVING $x^2 = k$ [134]

Square roots are introduced as solutions of $x^2 = k$. Emphasize that the symbol $\sqrt{\ }$ indicates only the positive, or principal, square root.

Cumulative Review

1. Simplify $\frac{2x - 8}{x^2 - 7x + 12}$. 2. Evaluate $x^3 + x^2 - 4x$ if $x = -3$. 3. Solve $\frac{n}{2} - \frac{n - 1}{3} = 2$.

1. $\frac{2}{x-3}$ 2. -6 3. 10

IRRATIONAL NUMBERS [137]

An *irrational number* is defined in terms of its decimal numeral, which is nonrepeating and nonterminating. A divide-and-average method is used to approximate square roots. The phrase *real number* is introduced and means a number that is either *rational* or *irrational.*

Cumulative Review

1. Multiply $\frac{x^2-25}{3x-15} \cdot \frac{6}{x+5}$. **2.** What is the degree of $5x^2y + 2xy^3$? **3.** Solve $\frac{5}{x-3} = \frac{2}{x-3}$.

1. 2 2. 4 3. No solution

MULTIPLYING AND SIMPLIFYING SQUARE ROOTS [141]

Students should note that the property $\sqrt{x} \cdot \sqrt{y} = \sqrt{x \cdot y}$ holds only for nonnegative values of x and y. Similarly, all variables in the remainder of the chapter represent positive numbers so that we may work freely on expressions with even-indexed roots.

Cumulative Review

1. Divide $\frac{x^2y^3}{ab^2} \div \frac{x^3y}{a^3b^2}$. **2.** Add $(2x + 3x^2 - 4) + (4x^2 - 2 - 5x)$. **3.** Solve $.04n = 2.4$.

1. $\frac{a^2y^2}{x}$ 2. $7x^2 - 3x - 6$ 3. 60

ADDING AND MULTIPLYING SQUARE ROOTS [145]

Multiplication is extended to include binomial factors. Students should recognize that combining like roots is again an application of the distributive property.

Cumulative Review

1. Simplify $(a^2b^2)(ac^3)(b^3c)$. **2.** Add $\frac{3a}{2a+6} + \frac{9}{2a+6}$. **3.** Solve $\frac{2}{3x} + \frac{5}{6x^2} = \frac{1}{2x}$.

1. $a^3b^5c^4$ 2. $\frac{3}{2}$ 3. -5

DIVIDING SQUARE ROOTS [148]

In Example 4, $\frac{2}{\sqrt{3}}$ is approximated by two methods to motivate the process of rationalizing the denominator. In Examples 10 and 11, introduce the word *conjugate,* given in the side column. *Conjugate* will be used again with complex numbers.

Cumulative Review

1. Simplify $(10x^2y^4)^3$. **2.** Subtract $(4n^3 + 2n^2 - 3n) - (5n^2 + 2n)$. **3.** Solve $\frac{2}{x+3} = \frac{3}{x-2}$.

1. $1{,}000\, x^6y^{12}$ 2. $4n^3 - 3n^2 - 5n$ 3. -13

METHOD FOR FINDING SQUARE ROOTS [152]

A centuries-old algorithm for finding square roots is shown.

CUBE ROOTS AND FOURTH ROOTS [153]

Cube roots and fourth roots are introduced as solutions of $x^3 = k$ and $x^4 = k$, respectively. Students should observe that the radicand can be negative for cube roots. A table of frequently used third and fourth powers of numbers should be listed on the chalkboard during the lesson as shown in the Review Capsule.

Cumulative Review

1. Simplify $\frac{a^5bc^2}{a^2b^3c^4}$. **2.** Find the solution set of $x^2 + 20 = 9x$. **3.** Simplify $\frac{2 + \frac{1}{2x}}{\frac{1}{3x} + 3}$.

1. $\frac{a^3}{b^2c^2}$ 2. $\{4, 5\}$ 3. $\frac{12x+3}{2+18x}$

ZERO AND NEGATIVE EXPONENTS [156]

Zero and negative integer exponents are introduced by extending a pattern. The power of a quotient is developed. The four previously developed properties of exponents are given in the Review Capsule and are applied to these new exponents.

Cumulative Review

1. Simplify $\frac{5n+6}{3a-2} - \frac{2n+4}{3a-2}$. **2.** Square $(x+3y)^2$. **3.** Solve $ax = bx + c$ for x.

1. $\frac{3n+2}{3a-2}$ **2.** $x^2 + 6xy + 9y^2$

3. $x = \frac{c}{a-b}$

FRACTIONAL EXPONENTS **[161]**

Fractional exponents are defined in terms of radicals.

Cumulative Review

1. Divide $(2x^3 - 5x^2 - 4x + 3) \div (x - 3)$.

2. Factor $9n^2 - 6n + 1$. **3.** Find the solution set of $|6x - 3| < 15$.

1. $2x^2 + x - 1$ **2.** $(3n - 1)^2$

3. $\{x|-2 < x < 3\}$

USING FRACTIONAL EXPONENTS **[164]**

The properties of exponents, which have been previously developed, are applied to an enlarged set, including fractional exponents.

Cumulative Review

1. Show that .777 . . . is a rational number.

2. Give the reciprocal of $\frac{5x}{3a}$.

3. Solve $|2n + 5| = 17$.

1. $\frac{7}{9}$ **2.** $\frac{3a}{5x}$ **3.** 6 and −11

FLOW CHART: DIVIDING POLYNOMIALS **[167]**

This flow chart serves as a review of division of polynomials.

CHAPTER 7 *Quadratic Equations*

The quadratic formula is developed. The solving of quadratic equations is applied to solving word problems and radical equations. The relationship between the coefficients of a quadratic equation and the sum and product of its solutions is treated. The nature of the solutions is examined through the discriminant. The desire to solve $x^2 = -1$ motivates the introduction of complex numbers.

FUN FOR PHILATELISTS **[170]**

A detailed article, "Mathematics in Use, as Seen on Postage Stamps" by William L. Schaaf, appears in *The Mathematics Teacher,* January 1974, on pages 16–24. Pictures of thirty-three such stamps are shown, and each stamp is discussed. You may wish to read the article, examine the stamp pictures, and recommend the article to some of your students. The centerfold of *The Mathematics Teacher,* April 1975, is a stamp poster suitable for your bulletin board. The stamps illustrated there are enlargements of some of those shown in the January issue cited above.

AREA PROBLEMS **[171]**

These word problems lead to quadratic equations which are solved by factoring.

Cumulative Review

1. Solve $x^2 = 12$. **2.** Multiply $(3\sqrt{5} - 5)(3\sqrt{5} + 5)$. **3.** Simplify $\frac{4x+12}{x^2 - 2x - 15}$.

1. $2\sqrt{3}$ and $-2\sqrt{3}$ **2.** 20 **3.** $\frac{4}{x-5}$

SOLVING $(x - a)^2 = k$ **[174]**

Equations are restricted to those where the square has already been completed. In the next lesson, students will learn to complete the square.

Cumulative Review

1. Rationalize the denominator of $\frac{2}{\sqrt{3}}$.

2. Find the solution set of $-4x > 24$.

3. Square $(n + 6)^2$.

1. $\frac{2\sqrt{3}}{3}$ **2.** $\{x|x < -6\}$ **3.** $n^2 + 12n + 36$

SOLVING BY COMPLETING THE SQUARE **[176]**

The process of completing the square will be used to derive the quadratic formula. This process will also be used again in the study of conic sections.

Cumulative Review

1. Rationalize the denominator of $\frac{5}{3-\sqrt{2}}$.
2. Simplify $(-2xy^4)^3$. 3. Solve $\frac{a}{x}=b$ for x.

1. $\frac{15+5\sqrt{2}}{7}$ 2. $-8x^3y^{12}$ 3. $x=\frac{a}{b}$

QUADRATIC FORMULA [179]

The steps shown in the Review Capsule may be used as a numerical model for the steps in deriving the quadratic formula shown in Example 1.

Cumulative Review

1. Simplify $\sqrt{8a^{10}}$. 2. Multiply $(5\sqrt{2}+\sqrt{3})(\sqrt{2}-\sqrt{3})$. 3. Solve $.6x=.42$.

1. $2a^5\sqrt{2}$ 2. $7-4\sqrt{6}$ 3. $.7$

THE DISCRIMINANT [182]

If $b^2-4ac<0$, there are no real number solutions. This case will be reexamined later in the chapter after complex numbers are introduced. In the set of complex numbers, if $b^2-4ac<0$, there are two complex number solutions which are not real numbers.

Cumulative Review

1. Rationalize the denominator of $\frac{-3}{2\sqrt{5}}$.
2. Multiply $(7x-y)(2x+y)$. 3. Show that $.\overline{444}$ is a rational number.

1. $\frac{-3\sqrt{5}}{10}$ 2. $14x^2+5xy-y^2$ 3. $\frac{4}{9}$

SYNTHETIC DIVISION [184]

Synthetic division is accomplished by not writing the variable and not repeating any coefficients. Also, subtractions are changed to additions by using the additive inverse of the divisor.

SUM AND PRODUCT OF SOLUTIONS [185]

An inductive approach is used to suggest the relationships between solutions and coefficients. Interested students might like to see a direct proof. Let r and s be the solutions of $ax^2+bx+c=0$. Then $(x-r)(x-s)=0$ and the equation, in standard form, is $x^2-(r+s)x+rs=0$. Rewriting the original equation with leading coefficient 1 and equating coefficients shows the desired result.

Cumulative Review

1. Rationalize the denominator of $\frac{4}{4+\sqrt{3}}$.
2. Factor $n^2+4n-21$. 3. Solve $\frac{x-4}{x+2}=\frac{4}{5}$.

1. $\frac{16-4\sqrt{3}}{13}$ 2. $(n+7)(n-3)$ 3. 28

WORD PROBLEMS AND QUADRATIC EQUATIONS [188]

The student has to make two decisions with each problem: (1) whether to solve the equation by factoring or formula and (2) whether to use or reject a negative solution. Part A exercises lead to factoring. Part B exercises require the quadratic formula.

Cumulative Review

1. Simplify $3\sqrt{7}+8\sqrt{5}+2\sqrt{7}-\sqrt{5}$.
2. Square $(5x+3)^2$. 3. Locate $\sqrt{51}$ between consecutive integers.

1. $5\sqrt{7}+7\sqrt{5}$ 2. $25x^2+30x+9$
3. $7<\sqrt{51}<8$

RADICAL EQUATIONS [193]

Emphasize the need to check apparent solutions. Example 1 shows a situation where an extra solution is introduced by squaring the original equation. Students should see that the key to solving radical equations lies in rewriting them so that one radical appears alone on one side. It may be necessary to square twice, as in Example 4.

Cumulative Review

1. Rationalize the denominator of $\frac{-\sqrt{7}}{3\sqrt{2}}$.
2. Find the solution set of $(x+4)(x-3)<0$.
3. Simplify $\frac{7x+14}{5x+10}$.

1. $\frac{-\sqrt{14}}{6}$ 2. $\{x|-4<x<3\}$ 3. $\frac{7}{5}$

COMPLEX NUMBERS [196]

Solving $x^2=-1$ is used to motivate the need for a new number i.

Cumulative Review

1. Rationalize the denominator of $\frac{10}{2\sqrt{5} - 3}$.
2. Evaluate $xy^2 - 5xy$ if $x = -2$ and $y = -1$.
3. Simplify $\frac{8x + 16}{4x + 12}$.

1. $\frac{20\sqrt{5} + 30}{11}$ 2. -12 3. $\frac{2(x + 2)}{x + 3}$

POWERS OF *i* **[200]**

Students may be interested in this graph which shows that the cycle i, -1, $-i$, 1 occurs for consecutive powers of i.

MULTIPLYING AND DIVIDING COMPLEX NUMBERS **[201]**

In Example 3, students can see why the property $\sqrt{x} \cdot \sqrt{y} = \sqrt{x \cdot y}$ was restricted to nonnegative values of x and y. Multiplying complex numbers like $(a + bi)(c + di)$ is similar to multiplying binomials like $(3x + 4)(2x + 5)$. Relate division of complex numbers $\frac{a + bi}{c + di}$ to rationalizing the denominator of a fraction like $\frac{3}{4 + 2\sqrt{3}}$.

Cumulative Review

1. Simplify $(m^3n^{-2})^{-2}$. 2. Simplify $\frac{5}{3a} + \frac{5}{4a}$.
3. Write $\sqrt[3]{x}$ in exponent form.

1. $\frac{n^4}{m^6}$ 2. $\frac{35}{12a}$ 3. $x^{\frac{1}{3}}$

COMPLEX NUMBER SOLUTIONS OF EQUATIONS **[205]**

The case of the negative discriminant is reexamined in the light of an expanded number system.

Cumulative Review

1. Find the value of $8^{\frac{2}{3}}$. 2. Add $\frac{7}{x - 4} + \frac{2}{4 - x}$. 3. Show that $2\frac{1}{3}$ is a rational number.

1. 4 2. $\frac{5}{x - 4}$ 3. $\frac{7}{3}$

CHAPTER 8 *Linear Sentences*

The concept of slope is developed and used to write and graph linear equations. Other applications of slope are treated in the next chapter. The method for finding the slope of a line is motivated through directed distances.

GRAPHS OF COMPLEX NUMBERS **[210]**

Distinguish between vector and real number addition.

The graph of $a + bi$ is the point (a, b). The sum of $a + bi$ and $c + di$ is $(a + c) + (b + d)i$; thus, the points (a, b), (c, d), $(a + c, b + d)$, and $(0, 0)$ form a parallelogram for constructing the sum graphically. The ray (directed segment) from the origin to the graph of $a + bi$ is the *vector* $a + bi$. To subtract $c + di$ graphically, draw the additive inverse of $c + di$, which is $-c - di$, and proceed as in addition:

$$(a + bi) - (c + di) = (a + bi) + [-(c + di)] = (a + bi) + (-c - di)$$

This topic is suitable for above average students. A good project for some able students is to study vectors independently or in small groups.

VERTICAL AND HORIZONTAL LINES **[211]**

To be able to find directed distances and slopes of lines, students must be able to find the distance between two points on a horizontal or vertical line.

Cumulative Review

1. Simplify $(x^5y^3)(x^3y^4)$. 2. Evaluate x^3y^5 if $x = 2$, $y = 3$. 3. Simplify $(x^4)^2$.

1. x^8y^7 2. 1,944 3. x^8

DIRECTED DISTANCES **[214]**

The key idea here is to determine the direction of distances. This is important in determining whether slopes are positive or negative. For certain students it might be meaningful to discuss the arbitrariness in which the sign of the direction was determined. To facilitate the development of the slope formula, students should become accustomed to using the formulas given here, even though an analysis from the diagram might seem easier.

Cumulative Review

1. Simplify $\sqrt{8}$. **2.** Simplify $-\sqrt{27}$.
3. Simplify $\sqrt{300}$.
1. $2\sqrt{2}$ **2.** $-3\sqrt{3}$ **3.** $10\sqrt{3}$

MATHEMATICS IN DRAFTING **[217]**

As an extension of the project on this page, you may wish to have students draw the top, front, and side views of objects of their own choosing.

SLOPE OF A LINE **[218]**

Students can be guided to discover that lines which slant up to the right have a positive slope and those slanting down to the right have a negative slope. A puzzling point to students is that a vertical line has an undefined slope, *not* a zero slope. Emphasize that vertical lines do not have a slope. It is incorrect to say that vertical lines have a slope of infinity. Avoid the use of the word *infinity* in this context.

The concept of the slope of a line can also be taught through a discussion of the ratio *vertical change to horizontal change* $\left(\frac{rise}{run}\right)$. Either approach leads to the same definition of slope $m = \dfrac{y_2 - y_1}{x_2 - x_1}$.

Cumulative Review

1. Simplify $2a - 2(7 - 3a)$. **2.** Graph $x + 3 > 10$. **3.** Simplify $\dfrac{\frac{4}{3x^2} + \frac{3}{2x}}{\frac{4}{x} + \frac{2}{x^2}}$.

1. $8a - 14$ **2.** *All points to the right of 7*
3. $\dfrac{9x + 8}{24x + 12}$

EQUATION OF A LINE: SLOPE-INTERCEPT **[221]**

Since the slope of a line is the same no matter what two points of the line are used, use this idea to write an equation of the line. By representing the slope of a line twice using two different pairs of points and then setting the slopes equal, we are able to determine an equation of the line. The slope-intercept form, $y = mx + b$, is used to write an equation of a line given its slope m and y-intercept b. It is also used to find the slope and y-intercept by rewriting the equation in this form.

For the equation $y = mx + b$, have students discuss its graph for different values of x, y, m, and b. For example if $x = 0$, $y = b$. If $b < 0$, the line crosses the y-axis below the origin. Consider positions of the line when $b = 0$ and m assumes different values.

Cumulative Review

1. Graph $2x \leq 8$. **2.** Simplify $4b - (3 - 2b)$.
3. Find the solution set of $|x| \geq 3$.
1. *-4 and all points to the left* **2.** $6b - 3$
3. $\{x | x \leq -3 \text{ or } x \geq 3\}$

PERMUTATIONS **[224]**

With some students, you may wish to show that another expression for ${}_nP_r$ is $\dfrac{n!}{(n-r)!}$.

$n! = n(n-1)(n-2)\cdots(n-r+1)(n-r)!$ Then

$$\frac{n!}{(n-r)!} = \frac{n(n-1)(n-2)\cdots(n-r+1)(n-r)!}{(n-r)!}$$
$$= n(n-1)(n-2)\cdots(n-r+1).$$

The more able students may be directed to study permutations further.

EQUATION OF A LINE: POINT SLOPE **[226]**

The slope of a line is again represented using any two points on the line. The point-slope form $y - y_1 = m(x - x_1)$ is used to write an equation given any point on the line and the slope. Choosing a general point $G(x, y)$ may cause difficulty. Discuss the meaning of the term *any point* $G(x, y)$. Although it is not developed here, you may wish to have students use the two-point form, $\dfrac{y - y_1}{x - x_1} = \dfrac{y_2 - y_1}{x_2 - x_1}$.

Cumulative Review

1. Find the solution set of $|x| \leq 4$. **2.** Simplify $\dfrac{\frac{3}{2y} + 2}{1 - \frac{3}{8y}}$. **3.** Graph $6 - n \leq 9$.

1. $\{x | -4 \leq x \leq 4\}$ **2.** $\dfrac{12 + 16y}{8y - 3}$
3. *-3 and all points to the right*

GRAPHING LINES [228]

The approach used to graph lines is to find the slope and y-intercept of the line. Students should first plot the y-intercept and then use the numerator and denominator of the slope to locate another point on the line. Other approaches to graphing lines such as plotting points and using intercepts should be encouraged.

Cumulative Review

1. Factor $5x^2 - 14x - 3$. **2.** Find the solution set of $x^2 + 2x - 15 = 0$. **3.** Factor $4a^2 - 25$.

1. $(5x + 1)(x - 3)$ **2.** $\{-5, 3\}$
3. $(2a - 5)(2a + 5)$

COMBINATIONS [231]

It is not likely that students will experience difficulty when dealing with combinations or permutations separately. They may experience difficulty when a choice must be made as to whether the problem involves a permutation or a combination. You may wish to provide problems where students must make the choice. Emphasize that a permutation is used when order is important and a combination is used when order does not make any difference. Some students may be directed to continue their study of combinations. This topic is for the average and above average students.

CHAPTER 9 *Analytic Geometry*

Analytic geometry involving the Pythagorean theorem, the distance formula, the midpoint formula, and slopes of parallel and perpendicular lines is treated. In a later chapter, the conic sections will be presented.

MATHEMATICS AND THE ELECTRIC COMPANY [234]

Have a group of students interview a manager of an electric company to determine how consumers can save energy and money. Consumers of electricity are not only people who own their homes or rent apartments, but can also be corporations and other businesses. In some localities, companies are given a lower energy rate if they use most of their energy at other than peak-demand times. Some students may wish to explore this topic further. This topic is appropriate for all levels of ability.

PYTHAGOREAN THEOREM AND ITS CONVERSE [235]

The Pythagorean theorem is used to find the length of a side of a right triangle when the lengths of two sides are given, while the converse is used to determine whether a triangle is a right triangle. Students can make models to demonstrate the Pythagorean relationship. For example, they can cut out small, congruent squares covering the square on side a and the square on side b and then show that the sum of those squares covers the square on side c.

Cumulative Review

1. Simplify $\dfrac{\frac{3}{4} + \frac{1}{5}}{\frac{4}{5} + \frac{1}{2}}$. **2.** Find the solution set of $|3x - 7| \leq 2$. **3.** Simplify $3(4a - 2) + 2(3 - 2a)$.

1. $\frac{19}{26}$ **2.** $\left\{x \middle| \frac{5}{3} \leq x \leq 3\right\}$ **3.** $8a$

THE DISTANCE BETWEEN TWO POINTS [238]

The difficult concept in this lesson is to find the coordinates of point C, the intersection of the perpendicular lines from A and B. See Example 2 for the diagram. This important idea is used to develop a formula for finding the distance between any two points.

You may wish to discuss with students that the distance formula works for these special cases: when both points fall on a vertical line and when both points fall on a horizontal line.

Cumulative Review

Factor out the GCF.

1. $3x^2y^3 + 9xy^2$ **2.** $10xy^2 - 20x^2y$
3. $3a^2 - 6a$

1. $3xy^2(xy + 3)$ **2.** $10xy(y - 2x)$
3. $3a(a - 2)$

COST OF ELECTRICITY [240]

It would not be surprising to find out that most consumers do not know how to determine the cost of electricity. The problem presented is a realistic one. Have students determine how their parents' electric bill is determined by calling the electric company. Have them determine how they can reduce the amount of electricity used in their house or apartment. This can lead to an interesting group discussion. This topic is appropriate for all levels of ability.

MIDPOINT FORMULA [241]

The formula for finding the midpoint of a line segment is developed by determining the midpoints of horizontal and vertical line segments. In Part B of the exercises, students find the coordinates of an endpoint when the coordinates of the other endpoint and the midpoint are given. In Part C, students may observe the generalization that the segments joining the midpoints of any quadrilateral in order form a parallelogram. Dividing a line segment into any given ratio may be an optional approach to finding a midpoint. For example, if the endpoints of $\overline{AB}$ are (x_1, y_1) and (x_2, y_2), then the coordinates of a point which divides the segment into the ratio 1 to 2 are $x_1 + \frac{1}{3}(x_2 - x_1)$ and $y_1 + \frac{1}{3}(y_2 - y_1)$. In general, if the endpoints of $\overline{AB}$ are (x_1, y_1) and (x_2, y_2), then the coordinates of a point which divides the segment into the ratio 1 to 1 (midpoint) are $x_1 + \frac{1}{2}(x_2 - x_1)$ and $y_1 + \frac{1}{2}(y_2 - y_1)$, or $x_1 + \frac{1}{2}x_2 - \frac{1}{2}x_1$ and $y_1 + \frac{1}{2}y_2 - \frac{1}{2}y_1$, or $\left(\frac{x_1 + x_2}{2}, \frac{y_1 + y_2}{2}\right)$.

Cumulative Review

1. Factor $9x^2 - 1$. **2.** Find and graph the solution set of $x^2 - 16 > 0$. **3.** Factor $3x^2 + 5x + 2$.

1. $(3x + 1)(3x - 1)$ **2.** $\{x|x < -4 \text{ or } x > 4\}$
3. $(3x + 2)(x + 1)$

PARALLEL AND PERPENDICULAR LINES [244]

Students are asked to find the relationship between slopes of parallel lines and between slopes of perpendicular lines. The converses of these relationships are also true. These are given and applied to the problem of writing an equation of a line passing through a given point and parallel or perpendicular to a given line.

Two lines may be perpendicular to each other with one of the lines having undefined slope. For example, a vertical line and a horizontal line are always perpendicular, yet the slope of the vertical line is undefined.

If the slope of a line is $-\frac{3}{2}$ then the slope of a line that is perpendicular to it is $\frac{2}{3}$ and $\left(-\frac{3}{2}\right)\left(\frac{2}{3}\right) = -1$. Hence, another approach is to state that two perpendicular lines have slopes whose product is -1 (no vertical lines). Also, if the product of the slopes of two lines is -1, the lines are perpendicular.

Cumulative Review

1. Solve $x^2 - 5x + 4 = 0$. **2.** Find and graph the solution set of $x^2 - 7x + 12 \leq 0$.
3. Factor $16 - 4x^2$ completely.

1. 1; 4 **2.** $\{x|3 \leq x \leq 4\}$
3. $4(2 - x)(2 + x)$

APPLICATIONS OF FORMULAS [247]

This lesson applies all of the concepts which were developed in the chapter. A review of some geometric definitions and properties can appropriately be given in conjunction with the lesson.

Cumulative Review

1. Find and graph the solution set of $(x - 2)(x + 3) > 0$. **2.** Factor $x^2 - 3x - 28$.
3. Find the solution set of $x^2 - x - 2 = 0$.

1. $\{x|x < -3 \text{ or } x > 2\}$ **2.** $(x - 7)(x + 4)$
3. $\{-1, 2\}$

CHAPTER 10 *Functions*

The chapter proceeds from an ordered pair concept, to the definition of a function, to finding function values, and finally to identifying and graphing functions. Different types of variations—direct, inverse, combined, and joint—are also treated.

SONYA KOVALEVSKY [252]

Some students may think that it is not unusual to study calculus while still in high school as Sonya Kovalevsky did, but for the 1870's this was an exceptional achievement. For enrichment, some students may wish to read more about the lives of Sonya Kovalevsky and Karl Weierstrass. They may be directed to Chapter 22 of *Men of Mathematics*. Some students might be directed to study biographies of other famous scientists and mathematicians.

RELATIONS AND FUNCTIONS [253]

A function is defined as a special kind of relation. Both a visual and a graphic approach are used to develop the key ideas of relation, domain, range, and function.

Cumulative Review

Find the slope of a line parallel to $\overleftrightarrow{AB}$.
1. $A(-2, 3)$, $B(4, 5)$ **2.** $A(-1, -3)$, $B(8, -2)$
3. $A(4, -6)$, $B(-6, 3)$
1. $\frac{1}{3}$ **2.** $\frac{1}{9}$ **3.** $-\frac{9}{10}$

FUNCTION VALUES [256]

An understanding of function values is essential for sketching and interpreting graphs of functions. This is a difficult concept for many students. Correct terminology should be stressed. Exercises dealing with the composition of functions are presented in Part C.

Cumulative Review

Find the slope of a line perpendicular to $\overleftrightarrow{PQ}$.
1. $P(-3, -1)$, $Q(2, 5)$ **2.** $P(2, -5)$, $Q(3, -4)$ **3.** $P(-3, 4)$, $Q(-6, -5)$
1. $-\frac{5}{6}$ **2.** -1 **3.** $-\frac{1}{3}$

IDENTIFYING AND GRAPHING FUNCTIONS [258]

The vertical line test is used to determine whether a graph is that of a function. Students should be able to determine whether a relation is a function by looking at the ordered pairs or by using a vertical line test. In Part C, the greatest integer function is discussed.

With some students, you may wish to discuss other special classes of functions such as the following.

1. A function f is an *increasing function* in some interval if and only if for all $x_1 < x_2$ in the interval, $f(x_1) \leq f(x_2)$.
2. A function g is a *decreasing function* in some interval if and only if for all $x_1 < x_2$ in the interval, $f(x_1) \geq f(x_2)$.
3. A function h is called an *even function* if and only if $h(-x) = h(x)$ for all x in the domain.
4. A function f is called an *odd function* if and only if $f(x) = -f(-x)$ for all x in the domain.
5. A function g is called a *periodic function* if and only if there is some real number p such that $g(x + p) = g(x)$ for all x in the domain.

Cumulative Review

Find the midpoint of $\overline{AB}$.
1. $A(6, -2)$, $B(-4, 4)$ **2.** $A(-4, -3)$, $B(6, 9)$
3. $A(6, -7)$, $B(-2, 9)$
1. *(1, 1)* **2.** *(1, 3)* **3.** *(2, 1)*

DIRECT VARIATION [261]

The graph of a direct variation is a line that passes through the origin. Another way to determine whether a relation is a direct variation is to plot the ordered pairs and see if the graph is a line that passes through the origin. Students may observe that the slope of the line is the constant of proportionality. The constant of proportionality depends on the order in which the variables are compared. Thus, the answers to the Exercises can be the reciprocals of those given.

A direct variation is always linear, but a linear function does not necessarily yield a direct variation. If a straight line does not pass through the origin, the graph is linear but the function does not yield a direct variation.

An alternate approach to teaching direct variation is to do so without using the constant of proportionality. See Examples 6 and 7.

Cumulative Review

Is the triangle a right triangle?
1. 3, 4, 4 **2.** 5, 12, 13 **3.** 6, 8, 10
1. *no* **2.** *yes* **3.** *yes*

INVERSE VARIATION [264]

Inverse variation is characterized by the concept that when one variable doubles, the other is halved; when one variable triples, the other is divided by three; and so forth. Contrast this type of variation with direct variation. Some students might like to examine the graph of an inverse variation, although they have not yet studied this curve (an equilateral hyperbola).

Cumulative Review

Find the length of the missing side of each right triangle. (c is the hypotenuse.)

1. $a = 12, b = 5$ **2.** $c = \sqrt{5}, a = \sqrt{2}$
3. $c = 5, b = 4$

1. *13* 2. $\sqrt{3}$ 3. *3*

COMBINED VARIATIONS [267]

Note the similarities between joint variation and direct variation. A joint variation is a function in which one variable varies directly as the product of the two others. Combined variations are variations which are combinations of direct and inverse variations.

Cumulative Review

Write an equation of a line parallel to the line described by the equation and passing through the given point.

1. $2x - 3y = 12$; (2, 1) **2.** $3y + 7x = 9$; (−2, −3) **3.** $x - y = 10$; (−7, 2)

1. $2x - 3y = 1$ 2. $7x + 3y = -23$
3. $x - y = -9$

CHAPTER 11 *Conic Sections*

Graphs of conics are sketched by plotting points from a table of values and drawing a smooth curve through the points. A discussion of this procedure, as well as its application as an alternate method for graphing a line, would be appropriate. The equation of a parabola $y = ax^2 + bx + c$ is treated. Then students examine ellipses and hyperbolas centered at the origin. After the lesson on translations, you may wish to extend the material on these to figures not centered at the origin.

For advanced students with some knowledge of trigonometry, you may wish to discuss the topic of rotations. Also, you may wish to demonstrate how conics are formed by passing planes through a doublenapped cone.

KARL FRIEDRICH GAUSS [272]

The biography of Gauss is interesting, amazing, and fascinating. Groups of students can cooperate on a project concerning the life of Gauss and his contributions to mathematics and science. Some students may wish to derive the sum of an arithmetic progression in the same manner that Gauss found the sum of the first hundred natural numbers. This can be done by letting a be the first term, $a + d$ the second term, l the last term, and $l - d$ the next to the last term.

THE PARABOLA [273]

The basic ideas of turning point, axis of symmetry, roots, and graph of a parabola are covered. An interesting project might be to have students determine why the headlamps on a car are paraboloid in shape.

You may wish to have some students write an equation of a parabola by making use of the definition in terms of directrix and focus. By completing the square, represent $y = ax^2 + bx + c$ in the form $y - k = a(x - h)^2$. The coordinates of the turning point (maximum or minimum) are (h,k). The equation of the axis of symmetry is $x = h$. If $h > 0$, the graph is moved from the origin to the right h units. If $h < 0$, the graph is moved from the origin to the left h units. If $k > 0$, the graph is moved up k units. If $k < 0$, the graph is moved down k units.

Have students discover what happens to the graph if $h = 0$ and $k \neq 0$, and if $h \neq 0$ and $k \neq 0$. If $a > 0$, the graph of the parabola will open upward and have a minimum point. If $a < 0$, the graph of the parabola will open downward and have a maximum point.

Cumulative Review

1. Factor out the GCF from $3x - 9$. **2.** Factor $2x^2 + 2x - 12$ completely. **3.** Simplify $\frac{x^2 + x - 6}{x^2 + 2x - 8}$.

1. $3(x - 3)$ 2. $2(x - 2)(x + 3)$ 3. $\frac{x + 3}{x + 4}$

THE CIRCLE

[277]

Students are expected to write an equation of a circle whose center is at the origin given a radius or a point through which the circle passes. They are also expected to find the length of a radius of a circle and to sketch its graph given the equation.

Part C exercises involve writing the equation of a circle whose center is not at the origin.

Cumulative Review

1. Factor $5x^3 - 20x$ completely. **2.** Simplify $\frac{10x^0y^5z}{5xy^0z}$. **3.** Factor out the GCF from $3x^6 + 4x^4 + x^2$.

1. $5x(x + 2)(x - 2)$ **2.** $\frac{2y^5}{x}$ **3.** $x^2(3x^4 + 4x^2 + 1)$

THE ELLIPSE

[280]

Example 1 motivates the definition of an ellipse and the standard form of its equation. Interested students can derive the equation of an ellipse through the definition. (See Part C.) Encourage students to graph the ellipse by finding both the x and y intercepts rather than by plotting points.

Cumulative Review

1. Simplify $\frac{a^2 + a - 6}{2a + 6}$. **2.** Factor out the GCF from $4a^4 - 12a^2 + 16a^5$. **3.** Simplify $\frac{3}{x - 5} - \frac{2}{5 - x}$.

1. $\frac{a - 2}{2}$ **2.** $4a^2(a^2 - 3 + 4a^3)$ **3.** $\frac{5}{x - 5}$

THE HYPERBOLA

[283]

Again, more-able students should have the experience of using the definition of a hyperbola to write its equation. Encourage students to use asymptotes and the rectangle method of sketching a hyperbola.

With some students, you may wish to show that $\sqrt{(x + 5)^2 + y^2} - \sqrt{(x - 5)^2 + y^2} = 8$ in Example 1 simplifies to $\frac{x^2}{16} - \frac{y^2}{9} = 1$.

$$\sqrt{(x + 5)^2 + y^2} = 8 + \sqrt{(x - 5)^2 + y^2}$$
$$(x + 5)^2 + y^2 = 64 + 16\sqrt{(x - 5)^2 + y^2} + (x - 5)^2 + y^2$$
$$(x + 5)^2 - (x - 5)^2 - 64 = 16\sqrt{(x - 5)^2 + y^2}$$
$$20x - 64 = 16\sqrt{(x - 5)^2 + y^2}$$
$$5x - 16 = 4\sqrt{(x - 5)^2 + y^2}$$
$$25x^2 - 160x + 256 = 16[(x - 5)^2 + y^2]$$
$$25x^2 - 160x + 256 = 16x^2 - 160x + 400 + 16y^2$$
$$9x^2 - 16y^2 = 144$$
$$\frac{x^2}{16} - \frac{y^2}{9} = 1$$

In Part C, students are asked to derive the equations of the asymptotes. They can first solve the equation of the hyperbola for y, $y = \pm \frac{b}{a}\sqrt{x^2 - a^2}$. As x gets larger and larger (that is, as the hyperbola gets closer and closer to the asymptotes), the constant a^2 becomes negligible in $\sqrt{x^2 - a^2}$. The equation becomes approximately $y = \pm \frac{b}{a}\sqrt{x^2}$; thus the equations of the asymptotes are $y = \pm \frac{b}{a}x$.

Cumulative Review

Evaluate if $x = -2$, and $y = 3$.

1. x^2y^3 **2.** $-2xy^2$ **3.** $3x^2y$

1. 108 **2.** 36 **3.** 36

IDENTIFYING CONICS

[286]

Students should be able to identify the type of conic by examing the equation. The coefficients of the variables are the key indicators.

Cumulative Review

1. Simplify $\frac{4a^4bc}{8abc}$. **2.** Factor $-3x^2 - 9x + 12$ completely. **3.** Simplify $\frac{3x^2 + 3x - 6}{x^2 - x - 6}$.

1. $\frac{a^3}{2}$ **2.** $-3(x + 4)(x - 1)$ **3.** $\frac{3(x - 1)}{x - 3}$

TRANSLATIONS [288]

The lesson lends itself to a discovery approach. Students should be encouraged to make generalizations. The Part B Example is of particular interest. By completing the square, students can easily obtain the turning point, the axis of symmetry, the roots, and make a reasonably good sketch.

Cumulative Review

For what values of the variable is the expression undefined?

1. $\frac{2}{5y}$ **2.** $\frac{x}{3x-12}$ **3.** $\frac{2a}{a^2+3a-18}$

1. *0* **2.** *4* **3.** *−6 and 3*

CONIC SECTIONS [291]

This topic can be extended nicely into the topic of curve sketching. To make a parabola in two dimensions, draw an angle on a sheet of paper. Starting at the vertex, mark off equal segments on both rays. Next, draw straight lines connecting the marks on the rays. Connect the first mark on one ray to the last on the second, the second mark on the first ray to the next to the last on the second, and so on. Some students might enjoy making the conics and other straight-line curves using nails, colored string, and wood.

CHAPTER 12 *Systems of Equations and Inequalities*

Solutions, both graphic and algebraic, of systems of equations in both two and three variables are discussed. Both linear and quadratic systems, as well as systems of inequalities, are covered. Digit and motion problems are solved as applications of systems of equations.

HOW THE U.S. USES ITS ENERGY [294]

A discussion of this topic might include questions such as these: What suggestions can students make to reduce the United States' dependence on other countries for its energy? Can students formulate an energy policy? How do students feel about the rationing of fuel? How would they implement a rationing system? Obviously, with a topic like uses of energy, many dilemmas are present, none of which have easy solutions. You may wish to have some students debate or discuss their supporting positions in reducing energy or conserving energy. This topic is appropriate for all levels.

SOLVING LINEAR SYSTEMS BY GRAPHING [295]

Since the intersection of the lines represented by the equations is the solution of a linear system, students should fully understand and have sufficient experience with graphing lines in order to understand algebraic solutions. It is sometimes difficult for students to read the solution from a graph. Encourage students to estimate to the nearest tenth. Consistent and inconsistent systems are discussed in Part B.

Cumulative Review

1. Write the value of 9^0. **2.** Simplify $\sqrt{2x} \cdot \sqrt{18x^3}$. **3.** Write $3\sqrt[5]{a^2}$ in exponent form.

1. *1* **2.** $6x^2$ **3.** $3a^{\frac{2}{5}}$

SOLVING LINEAR SYSTEMS ALGEBRAICALLY [298]

The method of substitution is treated first, followed by the method of addition. You may wish to reverse the order of presentation. A third method which is not discussed is to solve both equations for the same variable and then use the transitive property of equality to write an equation in one unknown. Systems of fractional equations are given in Part C.

Cumulative Review

1. Write $\sqrt{5}$ in exponent form. **2.** Simplify $\frac{\sqrt{18a^7}}{\sqrt{2a^3}}$. **3.** Simplify $3\sqrt{x} + \sqrt{4x} - \sqrt{x}$.

1. $5^{\frac{1}{2}}$ **2.** $3a^2$ **3.** $4\sqrt{x}$

2 BY 2 MATRICES [302]

Elementary matrix algebra has long been a tool for mathematicians but recently has assumed an increasing significance in finding solutions of applied problems. Modern computers have extended the uses of matrix algebra into a host of areas, among which are statistics, business, psychology, atomic physics, and most forms of engineering. Students with good mathematical ability should be encouraged to study matrix theory further.

SYSTEMS OF THREE LINEAR EQUATIONS [304]

Students may choose any two equations and eliminate one of the variables and then choose two other equations and eliminate the same variable. Students frequently forget with which equations they are working. Numbering equations may help. Discuss what the three numbers of the solution represent. A graphic description involving planes may be enlightening.

Cumulative Review

1. Simplify $3\sqrt{5} \cdot 2\sqrt{5}$. **2.** Rationalize the denominator of $\frac{3}{\sqrt{x}}$. **3.** Solve $-\frac{3}{4}x = 6$.

1. 30. **2.** $\frac{3\sqrt{x}}{x}$ **3.** -8

3 BY 3 MATRICES [306]

Students can pursue the study of matrices and determinants in depth according to the amount of time and interest they have. Have some of your better students launch such a study. You may wish to assign other systems of equations to be solved by using determinants. This topic is appropriate for the above average students.

LINEAR-QUADRATIC SYSTEMS [308]

A brief review of the methods of graphing conics will be useful. By examining different situations, like those illustrated in Example 2, students should see why there are linear-quadratic systems with one solution, two solutions, and no solutions. Again, it may be difficult for some students to gain confidence in estimating solutions to the nearest tenth from the graph. Both graphic and algebraic solutions are discussed.

Cumulative Review

1. Simplify $\frac{\sqrt{35}}{\sqrt{7}}$. **2.** Simplify $8\sqrt{12} + 2\sqrt{3} + 2\sqrt{27}$. **3.** Find the value of 4^{-2}.

1. $\sqrt{5}$ **2.** $24\sqrt{3}$ **3.** $\frac{1}{16}$

QUADRATIC SYSTEMS [311]

Again, the graphic solutions illustrated in Example 1 should help students see why quadratic systems may have no, one, two, three, or four solutions. Examples 3 and 4 show a system, solved algebraically and graphically, where there are four solutions. Students should be careful to include all of them. In Part B, fourth degree equations are encountered. However, these are in quadratic form and can be solved by factoring.

Cumulative Review

1. Rationalize the denominator of $\frac{1}{\sqrt{2}}$.

2. Simplify $\frac{\sqrt{24x^4}}{2x^2}$. **3.** Find the value of $3^2 \cdot 2^{-3}$.

1. $\frac{\sqrt{2}}{2}$ **2.** $\sqrt{6}$ **3.** $\frac{9}{8}$

DIGIT PROBLEMS [314]

Phrases such as *digits reversed*, *the sum of*, *more than*, and *less than* may give difficulty. Practice in interpreting these is encouraged.

Cumulative Review

1. Simplify $3\sqrt{2} + 2\sqrt{3} + 4\sqrt{2}$.

2. Rationalize the denominator of $\frac{4}{2\sqrt{3} - \sqrt{5}}$.

3. Solve $\frac{a+1}{2} - \frac{a-3}{3} = 7$.

1. $7\sqrt{2} + 2\sqrt{3}$ **2.** $\frac{8\sqrt{3} + 4\sqrt{5}}{7}$ **3.** 33

MOTION PROBLEMS [318]

The basic formulas are motivated through problems similar to those encountered by students. Many different types of motion problems are treated, and it may be difficult for students to handle all of these at once. You may wish to cover some now and return to the topic later.

Cumulative Review

1. Solve $\frac{3}{x+2} = \frac{2}{x-1}$. **2.** Write $6\sqrt{b}$ in exponent form. **3.** Simplify $(-3\sqrt{b})^2$.

1. *7* **2.** $6b^{\frac{1}{2}}$ **3.** *9b*

SYSTEMS OF LINEAR INEQUALITIES [323]

Whenever the equation of a line is graphed, the plane is separated into three regions: the region above the line, the region below the line, and the line itself. Discuss the use of dashed lines and solid lines when graphing inequalities. Students should be encouraged to use a check point.

Cumulative Review

Find the slope and the *y*-intercept.

1. $\{(x,y)|y = 3x - 1\}$ **2.** $\left\{(x,y)|y = \frac{1}{2}x + 3\right\}$

3. $\{(x,y)|y = -5\}$

1. *3; −1* **2.** $\frac{1}{2}$*; 3* **3.** *0; −5*

SYSTEMS OF QUADRATIC INEQUALITIES [327]

Single inequalities and systems of inequalities are discussed. Students should graph the inequalities as though they were equations and then determine the appropriate region by using a test point.

Cumulative Review

Graph the line described by each equation.

1. $3x + 4y = 8$ **2.** $x - 5 = 0$

3. $2y - 3x = -6$

1. *line through (0, 2) and (4, −1)* **2.** *vertical line through (5, 0)* **3.** *line through (0, −3) and (2, 0)*

CROSSNUMBER PUZZLE [329]

This topic provides practice and review of previously learned concepts. The use of a calculator might be appropriate. This topic is appropriate for all students.

CHAPTER 13 *Progressions and Series*

Arithmetic and geometric progressions are introduced with emphasis on finding a given *n*th term and finding means between two given terms. The corresponding arithmetic and geometric *series* are developed with emphasis on finding the sum of the first *n* terms of a series. The sum of an *infinite* geometric series is included. The use of subscripts is avoided.

In this chapter there is much arithmetic computation. You may wish to let students use an inexpensive pocket or mini- calculator. It will save time and increase accuracy. Warn such students to guard against loss or theft of their calculators.

An interesting paperback booklet to have on hand is *Games Calculators Play* by Wallace Judd, published by Warner Books, Inc., New York, N.Y. 10019. The book contains many games, patterns, puzzles, and tricks to be done with a mini-calculator. One section shows pictures of the parts inside a calculator and describes their functions.

FERMAT'S LAST THEOREM [332]

Students who are interested in reading more about Fermat may be directed to Chapter 4 of *Men of Mathematics*.

ARITHMETIC PROGRESSIONS [333]

Students should be able to recognize an arithmetic progression and to determine the common difference. A formula for the *n*th term is suggested through a pattern approach. You may wish to introduce the term *sequence* as a synonym for *progression*, since some students will meet it in their continuing mathematical studies.

Cumulative Review

1. Find the slope of $\overleftrightarrow{AB}$ for $A(-2, 4)$, $B(3, 3)$. **2.** Solve $x^2 + 3x - 1 = 0$. **3.** Simplify $x^{-3} \cdot x^5$.

1. $-\frac{1}{5}$ **2.** $\frac{-3 \pm \sqrt{13}}{2}$ **3.** x^2

ARITHMETIC MEANS [337]

The process of inserting arithmetic means involves applying the formula for the *n*th term in order to find the common difference *d*.

Cumulative Review

1. Find the slope and the y-intercept of the line with equation $2y = -6x + 8$. **2.** Multiply $(3 + 2i)(2 + i)$. **3.** Simplify $(2x^3)^4$.

1. $-3; 4$ **2.** $4 + 7i$ **3.** $16x^{12}$

ARITHMETIC SERIES [339]

In Example 4, a numerical exercise is used to develop the sum formula that follows the example.

Cumulative Review

1. Find the distance between $P(1, -1)$ and $Q(-2, 3)$. **2.** Simplify $6 - \sqrt{-12}$.
3. Find the solution set of $|x| < 7$.

1. 5 **2.** $6 - 2i\sqrt{3}$ **3.** $\{x | -7 < x < 7\}$

GEOMETRIC PROGRESSIONS [342]

The development parallels the one used in the initial lesson on arithmetic progressions.

Cumulative Review

1. For $f = \{(x, y) | y = -3x - 2\}$, find $f(-2)$.
2. Solve $.12x = 6$. **3.** Simplify $\frac{a^5b^3c^{-2}}{a^2b^{-3}c^4}$.

1. 4 **2.** 50 **3.** $\frac{a^3b^6}{c^6}$

GEOMETRIC MEANS [346]

We cannot speak of *the* geometric mean between two numbers since there is a positive mean and a negative mean. The *positive* geometric mean between two numbers is asked for in the exercises, except in Part C where both solutions are called for.

Cumulative Review

1. Find the radius of the circle with equation $2x^2 + 2y^2 = 18$. **2.** Solve $3ax - b = cx$ for x.
3. Write $\sqrt[3]{7}$ in exponent form.

1. 3 **2.** $x = \frac{b}{3a - c}$ **3.** $7^{\frac{1}{3}}$

SUMS OF EVEN AND ODD NUMBERS [348]

Patterns leading to the formulas for the sum of the first k even numbers and the sum of the first k odd numbers are presented. This topic is appropriate for all students.

GEOMETRIC SERIES [350]

In Example 3, a numerical exercise is used to develop the sum formula that follows the example. After completing Example 6, compare the answers of Examples 5 and 6. Example 5 calls for the sum of the first 10 terms of the same infinite geometric series shown in Example 6. For more able students, you may wish to provide a discussion of convergence of infinite series. Notice that Examples 9 and 10 provide a new approach to the problem of writing a repeating decimal in the form $\frac{a}{b}$.

Cumulative Review

1. Solve $\begin{matrix} 3x + 2y = 8 \\ 5x - 2y = 8 \end{matrix}$. **2.** Simplify $\frac{x}{x^2 - 5x + 6} - \frac{2}{x^2 - 5x + 6}$. **3.** Find the value of $16^{-\frac{1}{2}}$

1. $(2, 1)$ **2.** $\frac{1}{x - 3}$ **3.** $\frac{1}{4}$

CHAPTER 14 *Logarithms*

The rules for finding the log of a product, a quotient, a power, and a root are developed and applied to numbers greater than 1. Numbers between 0 and 1 are used in the next chapter. Numbers are considered as powers of 10 before the words *log* and *antilog* are introduced.

MATHEMATICS IN BUSINESS [358]

You may wish to give students experience in graphing as it is used in business and in predicting trends. Have students visit local businesses to gather information for this purpose. Encourage them to graph sales, profit, and supply and demand. Encourage students to forecast trends. Students can also get statistical information about national business trends from local stockbrokers. Students can also chart revenues, expenditures, consumer price index trends, and cost of running local government.

NUMBERS AS POWERS OF 10 [360]

To emphasize that a logarithm is an exponent, the table of mantissas is introduced *without* using the words *characteristic*, *mantissa*, and *logarithm*. In the next lesson, students will multiply and divide numbers as powers of 10.

Cumulative Review

1. Find the solution set of $5(3x + 2) \leq 11x - 4$.
2. Show that $.53\overline{53}$ is a rational number.
3. Find the slope of $\overleftrightarrow{AB}$ for $A(2,4)$, $B(-3,-3)$.

1. $\{x | x \leq -3\frac{1}{2}\}$ **2.** $\frac{53}{99}$ **3.** $\frac{7}{5}$

MULTIPLYING AND DIVIDING POWERS OF 10 [363]

Students rewrite numbers as powers of 10. Then they use the now familiar properties of exponents to do the calculations. Final answers are converted back to ordinary notation.

Cumulative Review

1. Solve $a - 2x = bx - cx$ for x. **2.** Simplify $(2ab^2)^3$. **3.** Give the x-intercepts of the ellipse with equation $\frac{x^2}{25} + \frac{y^2}{16} = 1$.

1. $x = \frac{a}{b - c + 2}$ **2.** $8a^3b^6$ **3.** $5, -5$

LOGARITHMS [365]

Writing numbers as powers of 10 is translated into the more customary language of logarithms. Students should be aware that only the terminology is new in this lesson.

Cumulative Review

1. Multiply $(3x - 5y)(2x + 3y)$. **2.** Write the equation of line $\overleftrightarrow{PQ}$ for $P(2,5)$, $Q(1,3)$.
3. Write $\sqrt[3]{y}$ in exponent form.

1. $6x^2 - xy - 15y^2$ **2.** $y = 2x + 1$ **3.** $y^{\frac{1}{3}}$

LOGARITHM OF A PRODUCT [368]

Since logarithms are exponents and since when multiplying numbers with the same base the exponents are added, the logarithm of a product is the sum of the logarithms of the factors.

Cumulative Review

1. Factor $5x^2 + 2x - 3$. **2.** Find $f(-2)$ if $f(x) = 4 - 3x$. **3.** Simplify $(a^{-3}b)^{-2}$.

1. $(5x - 3)(x + 1)$ **2.** 10 **3.** $\frac{a^6}{b^2}$

SCIENTIFIC NOTATION [370]

Scientific notation identifies the characteristic readily and facilitates placing the decimal point in the antilog.

Cumulative Review

1. Solve $x^2 - 6x + 9 = 0$. **2.** Write $\sqrt{7}$ in exponent form. **3.** Divide $(2x^3 - 4x^2 - 11x + 15) \div (x - 3)$.

1. 3 **2.** $7^{\frac{1}{2}}$ **3.** $2x^2 + 2x - 5$

LOGARITHM OF A QUOTIENT [374]

Division like $275 \div 684$ ($a \div b$, where $a < b$) is not included since the quotient is less than 1. This type is developed in the next chapter.

Cumulative Review

1. Simplify $\frac{6a^5b}{8a^2b^4}$. **2.** In which quadrant is $P(-3, 5)$? **3.** Solve $x^2 + 3x - 2 = 0$.

1. $\frac{3a^3}{4b^3}$ **2.** *second* **3.** $\frac{-3 \pm \sqrt{17}}{2}$

LOGARITHM OF A POWER [376]

Again, an analogy to the corresponding property of exponents, $(10^m)^n = 10^{m \cdot n}$, may be helpful.

Cumulative Review

1. Add $\frac{x + 4}{3x + 9} + \frac{x + 2}{3x + 9}$. **2.** Find the slope of the line with equation $3x + 5y = 7$. **3.** Write $\sqrt[4]{x}$ in exponent form.

1. $\frac{2}{3}$ **2.** $-\frac{3}{5}$ **3.** $x^{\frac{1}{4}}$

LOGARITHM OF A ROOT [378]

The log of a root is developed as the log of a power, a fractional power, rather than giving a *fourth* rule $\left(\log \sqrt[n]{x} = \frac{1}{n} \cdot \log x \right)$ to remember.

Cumulative Review

1. Multiply $(3-2i)(4+i)$. **2.** Find the 31st term of 3, 7, 11, 15, ... **3.** Divide $\frac{3x^2}{2n} \div \frac{12x}{5n}$.

1. *14 − 5i* **2.** *123* **3.** $\frac{5x}{8}$

CHAPTER 15 *Using Logarithms*

Computing with logs is extended to numbers between 0 and 1. Interpolation, logs with other bases, and powers of a binomial are developed. Exponential equations are solved.

FIBONACCI SEQUENCE [382]

This topic may be assigned in conjunction with the Fibonacci Lab, which is found on page 404.

LOGARITHMS OF DECIMALS [383]

The log of a number between 0 and 1 is developed as negative characteristics are introduced. Scientific notation is again a useful aid for determining the characteristic and the mantissa. Review writing numbers between 0 and 1 in scientific notation.

Cumulative Review

1. Solve $\frac{x}{3} = \frac{x+4}{5}$. **2.** Find the one arithmetic mean between 7 and 11.

3. Simplify $\frac{5+\frac{2}{x}}{\frac{3}{y}+7}$.

1. *6* **2.** *9* **3.** $\frac{5xy+2y}{3x+7xy}$

USING LOGARITHMS OF DECIMALS [386]

Division involves manipulations to keep the mantissa positive. These are shown in Example 4 and the Part B Example. Finding a root requires manipulations to keep the characteristic an integer, as is shown in Example 5. Each exercise involves a single operation. In the next lesson, each exercise involves at least two operations.

Cumulative Review

1. Solve $\frac{7}{10} = \frac{x}{9}$. **2.** Simplify $(-3a^2b^3)(7ab^2)$. **3.** Find the distance between $A(2, 3)$ and $B(4, 2)$.

1. *6.3* **2.** $-21a^3b^5$ **3.** $\sqrt{5}$

COMPUTATIONS WITH LOGARITHMS [389]

Carefully writing the log equation and outlining the various computations neatly will help students avoid errors.

Cumulative Review

1. Square $(3x+5)^2$. **2.** Solve $\frac{x}{10} = \frac{3}{4}$.

3. Write an equation of the circle with radius 4 and center at the origin.

1. $9x^2+30x+25$ **2.** $7\frac{1}{2}$ **3.** $x^2+y^2=16$

INTERPOLATION [392]

The process will be used again later to find trigonometric values and their logarithms.

Cumulative Review

1. Find the midpoint of segment $\overline{PQ}$ for $P(-3, 4)$, $Q(7, 6)$. **2.** Factor $25x^2-4$.

3. Find the sum of the first 40 terms of $1+2+3+4+\cdots$.

1. $M(2, 5)$ **2.** $(5x+2)(5x-2)$ **3.** *820*

LOGARITHMS: BASES OTHER THAN 10 [395]

The emphasis placed on logarithms as exponents should make the transition to other bases fairly easy. The key is to be able to write log equations in exponential form. For enrichment, students can do research on natural logarithms.

Cumulative Review

1. Find the solution set of $x^2+2x=8$.

2. Give the name of the conic section whose equation is $\frac{x^2}{16}+\frac{y^2}{36}=1$. **3.** Rationalize the denominator of $\frac{3}{2+i}$.

1. $\{-4, 2\}$ **2.** *ellipse* **3.** $\frac{6-3i}{5}$

EXPONENTIAL EQUATIONS [398]

Before studying Example 1, challenge students to tell what they can about x in the equations $3^{2x} = 9$; $(x = 1)$ and $3^x = 17$; $(2 < x < 3)$. You may wish to have students use a pocket calculator for this lesson. Then the complicated division of two logarithms to four digits can be done easily.

Cumulative Review

1. Find the solution set of $|2x| > 10$. **2.** Give the next 2 terms of the geometric progression 5, 10, 20, ... **3.** Simplify $\frac{x^2 - 25}{4x + 20}$.

1. $\{x|x < -5 \text{ or } x > 5\}$ **2.** *40, 80* **3.** $\frac{x-5}{4}$

BINOMIAL EXPANSIONS [399]

For enrichment, see the Special Topics on pages 382 and 404.

Cumulative Review

1. Add $\frac{3}{4x^2} + \frac{5}{6x}$. **2.** Find the positive geometric mean between 3 and 12. **3.** Solve $\sqrt{x+2} = 2\sqrt{x-1}$.

1. $\frac{9 + 10x}{12x^2}$ **2.** *6* **3.** *2*

CHAPTER 16 *Trigonometric Functions*

The earlier stages of the chapter build a base for defining the six trigonometric functions. The flow proceeds from determining the quadrant of the terminal side of an angle of rotation, to determining reference angles in triangles, to examining relationships in special right triangles, to discussing symmetric points, and finally to defining the six trigonometric functions. Trigonometry of right triangles is treated in the next chapter.

FIBONACCI LAB [404]

Much has been written by and about Fibonacci and his famous rabbit problem. In the rabbit problem, it is assumed that a pair of rabbits will produce a pair every month and that a pair of rabbits begin to bear the young two months after their own birth. Ask students to find out how many pairs will be born in a year. In the first month, there are two pairs; in the second month, three pairs; in the third month, five pairs; in the fourth month, eight pairs. By continuing the pattern, there should be 377 pairs of rabbits.

Students who wish to read more about Fibonacci and the sequences will find a discussion in *The World Book Encyclopedia* under "Mathematics (The Middle Ages)." Much of Fibonacci's work can be understood by good high school students.

ANGLES OF ROTATION AND THEIR MEASURES [405]

Students should distinguish between an angle and an angle of rotation. An *angle* is a set of points, while an *angle of rotation* is the angle associated with a rotating ray.

The important idea in this lesson is that students are able to determine the quadrant of the terminal side of any angle of rotation.

Cumulative Review

1. Find the sum of the first 16 terms of the arithmetic series: $3 + 7 + 11 + 15 + \cdots$.
2. Find the sum of the first 6 terms of the geometric series: $243 + 81 + 27 + \cdots$.
3. Find log 3.87.

1. *528* **2.** *364* **3.** *.5877*

REFERENCE ANGLES AND TRIANGLES [407]

To determine trigonometric values, students must be able to locate reference angles and triangles. They should get used to sketching the rotation in determining the reference angle of the triangle rather than having to memorize all the reduction formulas later.

Cumulative Review

1. Find the sum of the first 5 terms of the geometric series: $3 + 9 + 27 + 81 + \cdots$
2. Solve $\begin{matrix} 2x^2 + y^2 = 36 \\ 3x - y = 10 \end{matrix}$. **3.** Find x if $\log x = 2.9809$.

1. *363* **2.** $(4, 2)$ *and* $\left(\frac{16}{11}, -\frac{62}{11}\right)$ **3.** *957*

SPECIAL RIGHT TRIANGLES [409]

Students should memorize the properties of the 30°–60° right triangle and the isosceles right triangle since they will be used throughout the study of trigonometry.

Cumulative Review

1. Find the sum of the infinite geometric series: $8 + 2 + \frac{1}{2} + \frac{1}{8} + \cdots$. **2.** Find the sum of the first 30 terms of the arithmetic series: $8 + 2 - 4 - 10 - \cdots$. **3.** Solve $\begin{matrix} 3x + 4y \geq 8 \\ 2x - y < 6 \end{matrix}$ graphically.

1. $\frac{32}{3}$ *2.* *−2,370* *3.* *above and including the line containing (0, 2), (4, −1) and above the line containing (0, −6), (3, 0)*

SYMMETRIC POINTS IN A PLANE [413]

A thorough understanding of the properties of symmetry will help students visualize reduction formulas such as $\sin(180 + u) = -\sin u$, which will be developed later. Although it is not discussed here, you may wish to have students determine whether functions have x-axis, y-axis, origin, or no symmetry. For example, the graph of $y = x^2$, a parabola, has y-axis symmetry and the graph of $y = \frac{1}{x}$, a hyperbola, has origin symmetry.

Symmetry is a valuable tool for graphing. If a graph is symmetric with respect to the x-axis, it is sufficient to sketch it in the 1st and 2nd quadrants and then reproduce it in the 3rd and 4th quadrants. Ask students to determine how symmetry applies to graphs with y-axis and origin symmetry.

Cumulative Review

1. Solve $\begin{matrix} 2x + 3y = -4 \\ x - 2y = 5 \end{matrix}$. **2.** Find the sum of the first 12 terms of the arithmetic series: $3 + 3.4 + 3.8 + \cdots$. **3.** Find the sum of the infinite geometric series: $64 + 48 + 36 + 27 + \cdots$.

1. *(1, −2)* *2.* *62.4* *3.* *256*

SINE AND COSINE [416]

This lesson provides a foundation for the rest of the study of trigonometry. A good formative assessment may be in order before going on. Tangent, cotangent, secant, and cosecant functions are defined in terms of sine and cosine.

Cumulative Review

1. Solve $\begin{matrix} y \leq -3x + 4 \\ y > 2x - 1 \end{matrix}$ graphically. **2.** Find the antilog of .9053. **3.** Find the sum of the first 8 terms of the geometric series: $64 - 32 + 16 - 8 + \ldots$.

1. *Below and including the line containing (0, 4), (1, 1) and above the line containing (0, −1), (1, 1)* *2.* *8.04* *3.* $42\frac{1}{2}$

TANGENT AND COTANGENT [420]

Many of the properties of the tangent and cotangent functions can be derived quite easily by using the relations $\tan u = \frac{\sin u}{\cos u}$ and $\cot u = \frac{\cos u}{\sin u}$.

Cumulative Review

1. Find log 438. **2.** Solve $\begin{matrix} -2x - 3y \leq 9 \\ 3x + 4y > 8 \end{matrix}$ graphically. **3.** Find the sum of the infinite series: $4 + .4 + .04 + \cdots$.

1. *2.6415* *2.* *above and including the line containing (0, −3), (−3, −1) and above the line containing (0, 2), (4, −1)* *3.* $4\frac{4}{9}$

SECANT AND COSECANT [424]

Students should be able to work with the definitions in terms of sine and cosine and in terms of the reference triangle.

Since trig ratios are true for all nonzero denominators, it might be interesting to approach the formulas by using a unit circle (a circle with a radius 1 unit long). Then $\sin u = y$ and $\cos u = x$. By locating any point on the unit circle, the first coordinate is the cos and the second coordinate the sin. The ratio y to x is the tan; the ratio x to y is the cot; the reciprocal of x is the sec; and the reciprocal of y is the csc. While this approach has many applications, especially in deriving trig formulas, one immediate help is in remembering the quadrantal angles. For example, sketch a unit circle with terminal ray for a 90° angle. The intersection of the ray and circle has coordinates (0, 1). Hence, $\sin 90° = 1$, $\cos 90° = 0$, $\tan 90°$ is undefined, $\cot 90° = 0$, $\sec 90°$ is undefined and $\csc 90° = 1$.

Cumulative Review

1. Find x if $\log x = 1.6454$.

2. Solve $\begin{aligned} 2x - 3y + z &= -7 \\ x + 2y + 2z &= 5. \\ x + 5y - 3z &= 6 \end{aligned}$

3. Find log .0354.

1. *44.2* 2. *(−1, 2, 1)* 3. *8.5490 − 10*

CHAPTER 17 *Using Trigonometric Functions*

Both the special right triangles and the trigonometric tables are used to find the trigonometric value given an angle measure or to find the angle measure given a trigonometric value. Interpolation is used for values which cannot be found in the table. Linear trigonometric equations are solved, and traditional right triangle trigonometry is treated.

Students should memorize lengths of sides of special right triangles such as 3–4–5 and 5–12–13. It will save them much time. They should also understand that multiples of these sides also yield right triangles. For example, if a triangle has sides with lengths 6 $(2 \cdot 3)$, 8 $(2 \cdot 4)$, and 10 $(2 \cdot 5)$, the triangle is a right triangle. Also, if a right triangle has two sides with lengths 15 $(3 \cdot 5)$ and 36 $(3 \cdot 12)$, then the third side has length 39 $(3 \cdot 13)$.

MAKING A SINE TABLE [428]

Discuss with students that in making a sine table, it is assumed that the circle is a unit circle (a circle with a radius one unit long). If the length of a side of an angle is equal to the length of the opposite side divided by the length of the hypotenuse, which is one unit, the sine of the angle in the unit circle is the second coordinate, or the function value. This leads to the drawing of the vertical line next to the circle. Explain that to construct a cosine table, it would be necessary to place the measure line above the circle and to draw vertical segments to the measure. The reason for this is that on a unit circle, the first value of the ordered pair is the value of the cosine. You may wish to have students construct a cosine table. This topic is appropriate for the average and above average students.

FINDING TRIGONOMETRIC VALUES [429]

Discuss with students why two trigonometric conditions must be given in order to determine a unique solution.

Cumulative Review

Compute. Use logarithms. Interpolate to four digits.

1. 47.3×4.14 2. $\dfrac{134}{25.1}$

1. *195.8* 2. *5.339*

USING TRIGONOMETRIC TABLES [431]

Difficulty occurs when students read tables from bottom to top, and they must use the right-hand column. You may wish to indicate that the tables were constructed in this form to conserve space and to show how the cofunction relationships permit this format. Note that tables provided are essentially to four significant digits.

Cumulative Review

Compute. Use logarithms. Interpolate to four digits.

1. $\dfrac{\sqrt{87.3}}{1.05}$ 2. $\dfrac{(41.8 \times 21.3)^2}{45}$

1. *8.898* 2. *17,620*

LINEAR TRIGONOMETRIC EQUATIONS [435]

Solving linear trigonometric equations is like solving algebraic equations with the additional steps of determining the reference measure and then the angle. Students should again be encouraged to make diagrams showing the reference measure in the indicated quadrants.

Cumulative Review

1. Find log 63.65. 2. Solve $7^{x+1} = 12$ for x to 3 significant digits.

1. *1.8038* 2. *.277*

TRIGONOMETRIC INTERPOLATION [437]

The procedure is the same as for logarithms and will be used again in our study of logs and trigonometric functions.

An alternate approach to interpolation that can be used with some of the better students is to explain the theory of linear interpolation so that it applies to interpolation which is not linear. Trig interpolation is not linear. Assume that the tabular entries are close together so that the trig values do not introduce severe errors.

Consider the idea of a slope in the study of linear interpolation. For example, to find the sin 33°37′, assume that the graph of the sin function is linear between the points (33°30′, sin 33°30′) and (33°40′, sin 33°40′). Find the value of sin 33°37′ or (33°37′, y). Since the slope of a line is the same between any points on the line, we get

$$\frac{y - \sin 33°30'}{33°37' - 33°30'} = \frac{\sin 33°40' - \sin 33°30'}{33°40' - 33°30'}.$$

Solve for y where $y = \sin 33°37'$.

Cumulative Review

1. Find $\log_3 27$. **2.** Find log .5037.

1. *3* **2.** *9.7022 − 10*

RIGHT TRIANGLE TRIGONOMETRY [440]

Although some courses in trigonometry begin with the study of right triangles, right triangle trigonometry is just a special case of the trigonometry of circular functions. Some of the work encountered here will be a review for many students.

Cumulative Review

1. Solve $4^x = 27$ for x to 3 significant digits.

2. Find $(x^2 - y)^4$. Simplify each term.

1. *2.38* **2.** $x^8 - 4x^6y + 6x^4y^2 - 4x^2y^3 + y^4$

USING RIGHT TRIANGLE TRIGONOMETRY [443]

The most difficult task for students seems to be deciding which function to use. Drawing a diagram and listing which measures are given and which are to be found should help.

Cumulative Review

1. Find $(x - 2y)^5$. Then simplify each term.

2. Find log .03564 **3.** Find $\log_2 \frac{1}{8}$.

1. $x^5 - 10x^4y + 40x^3y^2 - 80x^2y^3 + 80xy^4 - 32y^5$

2. *8.5519 − 10* **3.** *−3*

POLAR FORM OF COMPLEX NUMBERS [447]

Multiplication using polar representation and De Moivre's theorem are logical extensions of this topic. Most advanced algebra books detail these extensions. Some able students should be directed into studying this topic further.

CHAPTER 18 *Graphs and Identities*

A foundation for advanced mathematics is presented. A thorough analysis is made of the graphs of trigonometric functions. The use of log-trig tables is introduced to aid students in doing complex calculations. Radian measure is treated, and students learn to solve trigonometric equations in quadratic form. Basic trigonometric identities are used to prove more complicated ones. Finally, inverses of the trigonometric functions are considered.

MAKING A TANGENT TABLE [450]

An alternate approach to making a tangent table is to make a sine table and a cosine table using a vertical and horizontal measure bar. To find the tangent, draw a horizontal segment to the vertical measure bar and a vertical segment to the horizontal measure bar and then divide the sine value by the cosine value. You may wish to have some of your students construct a physical model to determine trigonometric function values. This topic is appropriate for average and above average students.

TRIGONOMETRIC FUNCTIONS OF NEGATIVE ANGLES [451]

Although the reduction formulas for negative angles are developed through examples, you may wish to have interested students prove them, using the properties of symmetric points.

Many formulas for trig can be developed by considering the concepts of symmetry and any point (x, y) on a unit circle. For example, if u is any point on a unit circle and if u is in the 1st quadrant with coordinates (x, y), its symmetric point in the 4th quadrant is $-u$ with coordinates $(x, -y)$. Then $\sin u = y$ and $\sin(-u) = -y$. Hence, $\sin(-u) = -\sin u$. Similarly, $\cos u = x$ and $\cos(-u) = x$. Hence, $\cos(-u) = \cos u$.

Cumulative Review
Find the indicated term of the product. Then simplify the term.
1. 4th term of $(x + y)^6$ **2.** 3rd term of $(x - 2y)^7$
1. $20x^3y^3$ **2.** $84x^5y^2$

GRAPHS OF $y = \sin x$ and $y = \cos x$ **[453]**
Understanding the graphs of $y = \sin x$ and $y = \cos x$ is important since many of the properties of the sine and cosine functions can be observed from their graphs. After graphing these functions, students can discuss their properties.

Cumulative Review
Determine the reference angle.
1. 125° **2.** 240° **3.** −300°
1. 55° **2.** 60° **3.** 60°

CHANGE OF AMPLITUDE **[456]**
In Part B, negative coefficients are considered. A complete discussion of the effects on the graphs of $y = a \sin x$ and $y = a \cos x$ as a assumes different values, both positive and negative and both fractional and integral, might be in order.

Cumulative Review
Find the points symmetric with respect to the x-axis, the y-axis, and the origin.
1. (7, −2) **2.** (−4, 3)
1. (7, 2); (−7, −2); (−7, 2) **2.** (−4, −3); (4, 3); (4, −3)

CHANGE OF PERIOD **[459]**
Changing the coefficient b causes a change in the period of $y = \sin bx$ and $y = \cos bx$. Students should be able to discuss the effects on the sine and cosine functions caused by assuming different values for b.

Cumulative Review
1. Find sin 60°. **2.** Simplify $-3n - 2n(4 - n)$.
3. Evaluate $-2xy^2$ if $x = 2$ and $y = -3$.
1. $\frac{\sqrt{3}}{2}$ **2.** $2n^2 - 11n$ **3.** −36

CHANGE OF PERIOD AND AMPLITUDE [462]
Students should be able to discuss how the graphs of $y = a \sin bx$ and $y = a \cos bx$ change for different values of a and b. You may wish to compare the graphs of the trig functions with algebraic graphs. For example, the graph $y = 2x^2$ might be compared to the graph of $y = 2 \sin x$. Students can observe the effect that the coefficient 2 has on the basic graphs.

An alternate approach to plotting points when sketching the sin and cos functions with variations in amplitude and period is to construct a trig tracer. Sketch the sin curve on an index card and cut along the graph. Label the axes. To avoid drawing a separate cos curve, draw a vertical line perpendicular to the x-axis at $\frac{\pi}{2}$.

To sketch the graph of $y = \sin\left(x - \frac{\pi}{2}\right)$, move the tracer so the y-axis is at $-\frac{\pi}{2}$. Trace the sin curve.

Cumulative Review
1. Simplify $8t - 3(5t - 2)$. **2.** Find the solution set of $3n + 2 \geq 14$. **3.** Write $a^{\frac{1}{3}}$ in radical form.
1. $-7t + 6$ **2.** $\{n|n \geq 4\}$ **3.** $\sqrt[3]{a}$

GRAPHS OF $y = \tan x$ and $y = \cot x$ **[465]**
A similar study of the graphs of these two functions is made. The material provides good reinforcement of previously treated concepts.

Cumulative Review
1. Find the solution set of $-4x \leq 24$. **2.** Find the solution set of $|2t - 3| = 9$. **3.** Find $16^{\frac{3}{2}}$.
1. $\{x|x \geq -6\}$ **2.** {6, −3} **3.** 64

LOGARITHMS OF TRIGONOMETRIC FUNCTIONS **[468]**
Example 1 shows how the new table combines the trigonometry table with the logarithm table. Interpolation is considered in Examples 4 and 5.

Cumulative Review

1. Find the solution set of $|x - 2| < 7$.
2. Find cos 60°. **3.** Write $\sqrt[3]{19}$ in exponent form.
1. $\{x|-5 < x < 9\}$ *2.* $\frac{1}{2}$ *3.* $19^{\frac{1}{3}}$

QUADRATIC FORM OF TRIGONOMETRIC EQUATIONS [471]

Solving quadratic trigonometric equations is essentially the same as solving quadratic algebraic equations. However, for trigonometric equations in quadratic form, there may be as many as four solutions.

You may wish to reintroduce algebraic quadratic equations and give students practice.

Cumulative Review

Graph the solution set.
1. $3x - 6 \geq 21$ **2.** $|2x + 5| \leq 7$
3. $|x + 4| > 9$
1. 9 and all points to the right *2. the segment with endpoints −6 and 1* *3. all points left of −13 or right of 5*

APPLICATION OF THE SINE FUNCTION [474]

As a project you may have students find applications of the concept presented. This can be done by looking in physics and other science books. Look for Snell's Law. This topic is designed for the above average students.

BASIC TRIGONOMETRIC IDENTITIES [475]

Identities are developed for the general case. Then, for practice, students verify them for a particular degree measure. All students should memorize these identities since they are basic to the study of trigonometry.

Cumulative Review

1. Simplify $(x^4)^3$. **2.** Simplify $6 - (2a - 3b)$.
3. Simplify $(2a^2b^3)^2$.
1. x^{12} *2.* $-2a + 3b + 6$ *3.* $4a^4b^6$

PROVING TRIGONOMETRIC IDENTITIES [478]

Whenever students have trouble proving an identity, they should try to express each side in terms of sine and cosine. Caution students that since they have not yet proved that the equality holds, they must work on each side separately.

Cumulative Review

1. Evaluate x^2y if $x = -1$ and $y = 4$. **2.** Solve $\frac{x}{3} - 3 = 10$. **3.** Write $8b^{\frac{1}{2}}$ in radical form.
1. 4 *2.* 39 *3.* $8\sqrt{b}$

RADIAN MEASURE [480]

An alternate standard for measuring angles of rotation is introduced. Since radian measure is a real number, this method of measuring angles is used almost exclusively in higher mathematics. A proportion is used for converting from degrees to radians and from radians to degrees.

Cumulative Review

1. Find $27^{\frac{1}{3}}$. **2.** Write $5\sqrt[4]{7}$ in exponent form.
3. Find tan 45°.
1. 3 *2.* $5 \cdot 7^{\frac{1}{4}}$ *3.* 1

APPLYING RADIAN MEASURE [483]

Previously studied ideas are reinforced using radian measure instead of degree measure.

Cumulative Review

1. Write $\sqrt{7}$ in exponent form. **2.** Simplify $-3x \cdot 2x^4$. **3.** Find $8^{\frac{5}{3}}$.
1. $7^{\frac{1}{2}}$ *2.* $-6x^5$ *3.* 32

INVERSES OF TRIGONOMETRIC FUNCTIONS [486]

The inverses of trigonometric functions are not functions. It is possible, however, to make the inverses into functions by restricting their range. The restricted range is called the *principal value*. Two notations are shown. For example, the inverse sine function can be written as Arc sin x or $\text{Sin}^{-1}x$.

Students should be careful not to confuse the latter with a negative exponent.

Cumulative Review

1. Write $15^{\frac{1}{2}}$ in radical form. **2.** Evaluate $-x^2y^2$ if $x = -2$ and $y = -3$. **3.** Find the solution set of $|3x + 6| > 15$.
1. $\sqrt{15}$ *2.* -36 *3.* $\{x|x < -7 \text{ or } x > 3\}$

CHAPTER 19 *Solving Triangles*

Students will be concerned with finding measures of sides and angles of any triangle using the law of cosines and the law of sines. Trigonometric formulas for the sum and difference of two angles, for double angles, and for half angles are developed. Applications are included.

SIN (T) USING INFINITE SERIES [492]

You may wish to have some of your able students express the cosine function and other trigonometric functions using infinite series. Some students may find it interesting to explore the topic of infinite series in general. Students should be allowed to use calculators.

LAW OF COSINES [493]

The law of cosines is used to find the measure of a side when the measures of two sides and the included angle are known. The derivation is motivated through an example. Careful examination of Example 1 will make the derivation in Example 2 easier.

Cumulative Review

Evaluate and simplify.

1. $\dfrac{\cos 60° + \sin 30°}{\tan 30°}$ **2.** $\dfrac{\tan 45° + \cot 45°}{\sec 45°}$

3. $\dfrac{\sin 150° + \cos 120°}{\tan 225°}$

1. *$\sqrt{3}$* 2. *$\sqrt{2}$* 3. *0*

FINDING ANGLES BY THE LAW OF COSINES [496]

To find the measure of an angle using the law of cosines, first solve for the cosine of the angle, then use the trigonometric tables.

Cumulative Review

Evaluate and simplify.

1. $\dfrac{\sin \frac{\pi}{3} + \cos \frac{\pi}{3}}{\tan \frac{\pi}{4}}$ **2.** $\dfrac{\tan \frac{\pi}{6} + \cos \frac{\pi}{3}}{\sin \frac{\pi}{6}}$

3. $\dfrac{\cot \frac{\pi}{4} + \tan \frac{\pi}{3}}{\cos \frac{\pi}{4}}$

1. *$\dfrac{\sqrt{3}+1}{2}$* 2. *$\dfrac{2\sqrt{3}+3}{3}$* 3. *$\sqrt{2} + \sqrt{6}$*

AREA OF A TRIANGLE [498]

A formula for finding the area of a triangle given the measures of two sides and the included angle is developed. This formula for area of a triangle will be used to derive the law of sines as well as to derive the formulas for sin $(x \pm y)$.

Cumulative Review

1. Find cos 42°18′. **2.** Find cot u if cos $u = \frac{-4}{5}$ and sin $u < 0$. **3.** Find PQ if $P(-3, -2)$, $Q(-7, -1)$.

1. *.7396* 2. *$\frac{4}{3}$* 3. *$\sqrt{17}$*

LAWS OF SINES [500]

The law of sines is a useful proportion for solving triangles when the measures of two angles and a side opposite one of them are given or when two sides and an angle opposite one of them are given.

Cumulative Review

1. Find u to the nearest minute ($u < 90°$) if sin $u = .7365$. **2.** Find $\log \dfrac{m^2n}{t^3}$. **3.** Find tan 134°14′.

1. *47°26′* 2. *(2 log m + log n) −3 log t*
3. *−1.0271*

THE AMBIGUOUS CASE [503]

Students may wish to create a physical model which demonstrates the ambiguous case. Although students can memorize all the rules of the ambiguous case, it seems more appropriate to have them understand the concepts and determine the number of constructable triangles directly from the given conditions.

Cumulative Review

Find all values of u to the nearest minute ($0° < u < 360°$).

1. $4 \cos u + 3 = -2 \cos u + 5$ **2.** $-3 \sin u + 1 = 2 \sin u - 2$ **3.** $4 \tan u + 4 = 0$

1. *70°32′. 289°28′* 2. *36°52′, 143°8′*
3. *135°, 315°*

SIN ($x \pm y$) and SIN 2x [506]

The derivation shown for $\sin(x + y)$ is valid only if $x + y < 180°$. However, a rigorous proof of the more general case can be developed using the appropriate reduction formulas. This formula is basic to the development of the other sum and difference formulas. The formula for $\sin 2x$ is derived from $\sin (x + y)$, where $x = y$.

Cumulative Review

1. Find $\log \frac{ab^3}{c}$. **2.** Find the midpoint of $\overline{PQ}$ if $P(2, 5)$, $Q(-1, 3)$. **3.** Find u to the nearest minute ($u < 90°$) if $\cos u = .5649$.

1. $(\log a + 3 \log b) - \log c$ **2.** $\left(\frac{1}{2}, 4\right)$
3. $55°36'$

COS ($x \pm y$) and COS 2x [510]

The formulas are easily derived from previous results, and students should try to derive them on their own. They should become familiar with the three ways of expressing $\cos 2x$ since all three ways will be used.

Cumulative Review

Write an equation of the line that is parallel to the line whose equation is given and that passes through the given point.
1. $2x + 4y = 7$, $(2, 1)$ **2.** $-3x + 5y = 10$, $(4, 2)$ **3.** $8x - y = 6$, $(-1, -5)$
1. $x + 2y = 4$ **2.** $-3x + 5y = -2$
3. $8x - y = -3$

TAN ($x \pm y$) AND TAN 2x [514]

After deriving the formulas, the examples are similar to those used for the sine and cosine functions. Better students will be able to develop the formulas for themselves.

Cumulative Review

1. Find PQ if $P(-1, -2)$, $Q(6, -1)$. **2.** Find $\sec u$ if $\tan u = 1$ and $\sin u < 0$. **3.** Find the midpoint $\overline{PQ}$ if $P(2, -3)$, $Q(-3, 1)$.
1. $5\sqrt{2}$ **2.** $-\sqrt{2}$ **3.** $\left(-\frac{1}{2}, -1\right)$

SIN $\frac{x}{2}$, COS $\frac{x}{2}$, and TAN $\frac{x}{2}$ [517]

The half-angle formulas are derived from $\cos 2A$.

Cumulative Review

1. Find $\tan u$ $\sin u = \frac{-3}{5}$ and $\cos u > 0$.
2. Find PQ if $P(-1, -2)$, $Q(3, 1)$. **3.** Find $\sin 228°12'$.
1. $-\frac{3}{4}$ **2.** 5 **3.** $-.7455$

DOUBLE AND HALF ANGLE IDENTITIES [520]

Double angle formulas are used to prove identities. Students should express functions of double angles as functions of single angles when difficulty arises. Expressing functions in terms of sine and cosine may help also in proving difficult identities.

Cumulative Review

1. Find the midpoint of $\overline{PQ}$ if $P(6, 1)$, $Q(-2, 3)$.
2. Find $\log \sqrt{5x^3}$. **3.** Find u to the nearest minute ($u < 90°$) if $\tan u = 1.7297$.
1. $(2, 2)$ **2.** $\frac{1}{2}(\log 5 + 3 \log x)$ **3.** $59°58'$

PACING CHART

The Pacing Chart is designed as an aid to help you plan differentiated assignments for your class over the course of 170 days. The entries under the DAY heading indicate the day of the regular school year. The entries under each LEVEL heading give the daily assignments: first the page and then the exercises of the assignment. The column in color gives a minimum course, the middle column an average course, and the gray column an above-average course.

The Pacing Chart is meant only as a guide and may be adapted by you to better meet the needs of your students.

PACING CHART

DAY	LEVEL A	LEVEL B	LEVEL C
1	1–3 Orals All Col 1, 2 Ex 1–23 Ex 25–28	1–3 Orals All Col 2, 3 Ex 2–24 Ex 25–32	1–3 Ex 22–24; 29–40 4–5 Ex Even 6–18 Ex 19–24
2	4–5 Orals All Ex 1–12	4–5 Orals All Ex 5–18	6–8 Orals 5–8 Col 3 Ex 3–15 Ex 19–24
3	6–8 Orals All Col 1, 2 Ex 1–14 Ex 16–18	6–8 Orals All Col 1, 3 Ex 1–15 Ex 16–21	9–11 Ex 17–24 14–15 Ex Even Ex 19, 20
4	9–11 Orals All Ex 1–16	9–11 Orals All Ex Even 2–14 Ex 15–20	16 Chapter Review
5	14–15 Ex 1–12	14–15 Ex Even 2–12 Ex 13–18	17 Chapter Test
6	16 Chapter Review	16 Chapter Review	19–21 Ex 2–5; 14–17 Ex Even 20–26 Ex 27–36
7	17 Chapter Test	17 Chapter Test	22–24 Ex 8–16; 25–36 28–31 Ex Even 2–18 Ex 19, 20
8	19–21 Col 1, 2 Ex 1–17 Ex Odd 19–25	19–21 Col 2, 3 Ex 2–18 Ex 22–26	32–34 Ex 9–16
9	22–24 Orals All Ex 1–24	22–24 Orals All Ex 5–28	37–39 Ex Even 2–10 Ex 11, 12
10	28–31 Orals All Ex 1–10	28–31 Orals All Ex Even 2–10 Ex 11–17	40–41 Ex 1; 4–10

DAY	LEVEL A	LEVEL B	LEVEL C
11	32–34 Orals All Ex 1–12	32–34 Orals All Ex 1–14	42 Chapter Review
12	37–39 Ex 1–6	37–39 Ex 2–5; 7–9	43 Chapter Test
13	40–41 Orals All Ex 1–5; 7	40–41 Orals All Ex 1–8	45–48 Col 3 Ex 3–39 Ex Odd 41–47 Ex 48–69
14	42 Chapter Review	42 Chapter Review	49–52 Orals All Ex 1–3 Ex Even 4–18 Ex 20–29
15	43 Chapter Test	43 Chapter Test	53–55 Ex Even 2–40 Ex 41–49
16	45–48 Orals All Col 1, 2 Ex 1–38 Ex Even 40–46	45–48 Orals All Col 2, 3 Ex 2–39 Ex Odd 41–47	56–58 Orals All Ex Even 2–34 Ex 35–38
17	49–52 Orals All Ex 1–19	49–52 Orals All Ex 1–3 Ex Even 4–18 Ex 20–29	59–61 Ex Odd 13–39 Ex 40–45
18	53–55 Orals All Ex 1–16 Ex Odd 17–27	53–55 Orals All Ex Even 2–28 Ex 29–40	62–63 Col 1 Ex 16–55 Ex 58–68
19	56–68 Orals All Ex 1–26	56–58 Orals All Ex Even 2–20 Ex 24–34	64–67 Orals All Col 1, 2 Ex 1–38 Ex 40–45
20	59–61 Orals All Col 1, 3 Ex 1–27 Col 3 Ex 30–39	59–61 Col 1, 2 Ex 1–26 Ex 28–34	69–72 Orals 4–13 Ex Even 30–56 Ex 58–68
21	62–63 Orals All Col 1, 2 Ex 1–35 Col 1 Ex 43–55	62–63 Col 3 Ex 3–36 Col 2, 3 Ex 38–57	73–75 Orals All Col 2, 3 Ex 2–15 Ex 18, 21, 24–34
22	64–67 Orals All Col 1, 2 Ex 1–26	64–67 Orals All Col 2, 3 Ex 2–39	76–79 Orals 13–20 Ex 2, 3; 9–12; 19–28

DAY	LEVEL A	LEVEL B	LEVEL C
23	69–72 Orals All Ex 1–24	69–72 Orals All Ex Odd 7–57	80–81 Orals All Ex 10–18; 21–30 82 Chapter Review
24	73–75 Orals All Ex 1–15	73–75 Orals All Col 1, 2 Ex 1–23	83 Chapter Test
25	76–79 Orals All Ex 1–6	76–79 Orals All Ex 2, 3, 5, 8, 10 11, 13; 16–23	85–86 Ex 1–10; 16–25 87–89 Col 3 Ex 3–33 Ex 34–39
26	79 Ex 7, 8; 12–16	80–81 Orals All Col 2, 3 Ex 2–18 Ex 19–28	90–93 Orals 7–15 Col 2, 3 Ex 2–15 Ex Odd 17–29 Ex 30–33
27	80–81 Orals All Ex 1–18	82 Chapter Review	94–95 Ex 6, 9, 12, 15; 18–21 98–100 Col 3 Ex 6–24 Ex 28, 30, 32, 36
28	82 Chapter Review	83 Chapter Test	101–104 Orals All Ex Even 12–24 Ex 25–32
29	83 Chapter Test	85–86 Ex 1–10; 16–25 87–89 Col 2 Ex 2–23 Col 1, 2 Ex 25–32	106 Chapter Review
30	85–86 Ex 1–25	90–93 Orals All Col 2 Ex 2–14 Ex Even 16–22 Ex 24–29	107 Chapter Test
31	87–89 Col 1, 2 Ex 1–23	94–95 Col 1, 3 Ex 1–15 Ex 16–23	109–112 Ex 3–6; 10–12; 17–29; 32–37
32	90–93 Orals All Ex 1–9	98–100 Orals All Col 2, 3 Ex 2–24 Ex Odd 25–35	113–115 Ex 1–5; 7–12
33	93 Ex 10–17	101–104 Orals All Ex Even 2–14 Ex 15–24	116–118 Col 2 Ex 2–26 119–121 Col 3 Ex 9–24 Ex 29–45

DAY	LEVEL A	LEVEL B	LEVEL C
34	94–95 Col 1, 2 Ex 1–14 Ex 16–19	106 Chapter Review	124–126 Orals All Ex 4–6; 17–22; 27–30 127–129 Orals All Ex 7–20
35	98–100 Orals All Col 1, 3 Ex 1–24	107 Chapter Test	130 Chapter Review
36	101–104 Orals All Ex 1–18	109–112 Col 2, 3 Ex 2–27 Ex 30–33	131 Chapter Test
37	106 Chapter Review	113–115 Ex 1–4; 6–12	134–136 Ex 9–12; 17–28 137–140 Ex 13–25
38	107 Chapter Test	116–118 Col 1 Ex 1–13 Col 1, 3 Ex 16–27	141–144 Orals All Col 3, 4 Ex 3–52 Ex 53–60
39	109–112 Col 1, 2 Ex 1–26	119–121 Col 3 Ex 3–24 Ex 29–35	145–147 Ex Even 2–24 Col 2, 3 Ex 26–39 Ex 40–45
40	113–115 Ex 1–8	124–126 Orals All Ex 3–10; 14–26	148–151 Col 4, 5 Ex 4–40 Col 3–5 Ex 43–55
41	116–118 Col 1, 3 Ex 1–15	127–129 Orals All Ex 5–18	153–155 Col 3, 4 Ex 3–36 Ex 43–48 156–160 Ex 5–14; 21–26 Col 3 Ex 29–49
42	119–121 Col 1, 2 Ex 1–23 Ex 25–28	130 Chapter Review	156–160 Ex 51–54 Col 3 Ex 57–69 161–163 Col 5, 6 Ex 5–36 Ex Odd 37–59
43	124–126 Orals All Ex 1–14	131 Chapter Test	164–166 Orals All Col 3, 4 Ex 31–48 Ex 49–60
44	127–129 Orals All Ex 1–12	134–136 Ex 9–12; 17–28 137–140 Orals All Ex 11, 12; 15–20	168 Chapter Review

DAY	LEVEL A	LEVEL B	LEVEL C
45	130 Chapter Review	141–144 Orals All Ex Odd 1–31 Col 1–3 Ex 33–51	169 Chapter Test
46	131 Chapter Test	145–147 Ex 1–24 Col 1, 2 Ex 25–38	171–173 Orals All Ex 7–10; 12–16 174–175 Orals All Col 3, 4 Ex 3–16
47	134–136 Ex 1–28	148–151 Col 3, 4 Ex 3–34 Ex 36–39 Col 1–3 Ex 41–53	176–178 Orals All Col 2 Ex 2–23 Ex 25–30
48	137–140 Orals All Ex 1–10; 15–20	153–155 Col 1–3 Ex 1–31 Ex 33–48	179–181 Orals 5–8 Ex 1–5 Col 1 Ex 6–24 Ex 34–36; 39–45
49	141–144 Orals All Ex 1–16	156–160 Ex 2–11; 15–17 Col 3, 4 Ex 21–50 Col 1, 2 Ex 51–68	182–183 Ex 14–23; 25–33
50	144 Ex 17–34	161–163 Col 3–6 Ex 3–36 Ex Even 38–60	185–187 Orals All Ex 3–5; 11–13 Col 3, 4 Ex 23–44 Ex 45, 46
51	145–147 Ex 1–27	164–166 Orals All Col 3, 4 Ex 3–28 Col 1–3 Ex 29–47 Ex Even 50–54	188–192 8, 10, 12 Ex Even 18–38 Ex 39–41
52	148–151 Col 1–3 Ex 1–28	168 Chapter Review	193–195 Orals All Col 4 Ex 4–24 Col 3 Ex 27–39 Ex 40–47

DAY	LEVEL A	LEVEL B	LEVEL C
53	151 Ex 31–38	169 Chapter Test	196–199 Orals All Col 2, 3 Ex 2–9; 17–42 Ex 45–50
54	153–155 Ex 1–32	171–173 Orals All Ex Odd 1–7 Ex 9–15	201–204 Orals All Col 1 Ex 1–28 Col 1, 2 Ex 31–56 Ex 58–61
55	156–160 Orals All Ex 1–26	174–175 Orals All Ex 5–16	205–207 Ex 1–3 Col 3 Ex 12–30 Ex 31–36; 40–54
56	156–160 Col 1–3 Ex 27–49	176–178 Orals All Col 1, 3 Ex 1–24	208 Chapter Review
57	161–163 Col 1–4 Ex 1–34	179–181 Orals All Ex 1–5 Col 2, 3 Ex 7–32 Ex 33–35; 37–39	209 Chapter Test
58	164–166 Orals All Ex 1–28	182–183 Ex 13–24 185–187 Orals 1–6	211–213 Orals All Ex 1–12; 16–21
59	168 Chapter Review	185–187 Orals 7, 8 Col 2, 3 Ex 2–19 Col 1, 2 Ex 21–42	214–216 Orals All Col 1 Ex 1–14 Ex 15–24
60	169 Chapter Test	188–191 Ex 2–10 Ex Odd 11–19	218–220 Col 1, 4 Ex 1–17 Ex 22–25
61	171–173 Orals All Ex 1–10	192 Ex Odd 21–37	221–223 Orals All Col 1, 3 Ex 1–18 Ex 19, 22, 23, 26, 27, 29; 32–34
62	174–175 Orals All Ex 1–8	193–195 Orals All Col 3 Ex 3–23 Col 1, 2 Ex 25–38	226–227 Col 1, 4 Ex 1–16 Ex 14, 19, 20

DAY	LEVEL A	LEVEL B	LEVEL C
63	176–178 Orals All Col 1, 2 Ex 1–14	196–199 Orals All Col 1, 2 Ex 1–41 Ex 43–48	228–230 Orals All Ex 1, 2, 6, 8, 10, 11, 14, 16, 17, 18, 20, 21, 23; 26–32
64	179–181 Orals All Ex 1–5 Col 1, 2 Ex 6–31	201–204 Orals All Col 2, 3 Ex 2–57	232 Chapter Review
65	182–182 Ex 1–15	205–207 Orals All Ex 1–3 Col 3 Ex 6–24 Ex 31–39	233 Chapter Test
66	185–187 Orals All Ex 1–4; 7–10; 15–18	208 Chapter Review	235–237 Col 1, 4, 5 Ex 1–20 Ex 21, 22
67	188–191 Ex Odd 1–11	209 Chapter Test	238–239 Col 1, 4 Ex 1–20 Ex 24–29
68	191 Ex Even 10–18	211–213 Orals All Ex 1–19	241–243 Ex 1, 19 Col 3 Ex 3–42 Ex 31, 34, 37, 40, 43, 44
69	193–195 Orals All Col 1, 2 Ex 1–22	214–216 Orals All Ex 1–14	244–246 Ex 1–8; 12–14; 18–25
70	196–199 Orals All Ex 1–30	218–220 Col 1, 3 Ex 1–15 Ex 18–21	247–249 Ex 1, 7, 10 11, 14; 15–20
71	201–204 Orals All Col 1, 2 Ex 1–29	221–223 Orals All Col 1, 2 Ex 1–17 Ex 19, 21, 23, 25; 27–31	250 Chapter Review
72	204 Ex 31–36; 40–42	226–227 Col 1–3 Ex 1–15 Ex 17, 18	251 Chapter Test

DAY	LEVEL A	LEVEL B	LEVEL C
73	205–207 Orals All Col 1, 2 Ex 1–29	228–230 Orals All Ex 1–7; 10–28; 30, 31	253–255 Ex 1, 3; 8–19
74	208 Chapter Review	232 Chapter Review	256–257 Ex 1–12; 13, 15; 17–28
75	209 Chapter Test	233 Chapter Test	258–260 Ex 1–4; 23–28 Col 1, 3 Ex 5–22
76	211–213 Orals All Ex 1–12	235–237 Orals All Col 1, 2, 4 Ex 1–19 Ex 21, 22	261–263 Ex 1, 2, 5, 6, 7, 10, 11, 13, 14, 15; 18–25
77	214–216 Orals All Ex 1–11	238–239 Col 1–3 Ex 1–19 Ex 21–26	264–266 Ex 1–3; 6, 10, 11, 12, 15; 16–23
78	218–220 Col 1, 2 Ex 1–14 Ex 17–20	241–243 Ex 1, 7, 16, 22, 25, 28 Col 2 Ex 2–41	267–269 Ex 1–3; 6, 8 10, 13, 14, 19, 20
79	221–223 Orals All Col 1, 2 Ex 1–17 Ex 19, 20, 23, 24	244–246 Ex 1–11; 15–17; 21, 22	270 Chapter Review
80	226–227 Col 1–3 Ex 1–15	247–249 Ex 1, 7, 11, 13 Col 2 Ex 3–14 Ex 15–19	271 Chapter Test
81	228–230 Orals All Ex 1–6; 9, 12, 15, 16; 18–22	250 Chapter Review	273–276 Col 1, 3 Ex 1–18 Ex 19–21
82	232 Chapter Review	251 Chapter Test	277–279 Col 1, 6 Ex 1–18 Col 1, 5 Ex 19–38 Col 3 Ex 41–53 Ex 54–57
83	233 Chapter Test	253–255 Orals All Col 2 Ex 2–10 Ex 1, 5, 9; 12–16	280–282 Ex 1, 3, 5 Col 1, 3 Ex 7–18 Ex 19–24

DAY	LEVEL A	LEVEL B	LEVEL C
84	235–237 Orals All Col 1–3 Ex 1–18	256–257 Ex 1–20	283–285 Col 1, 4 Ex 1–16 Ex 17, 19; 21–33
85	238–239 Col 1–3 Ex 1–19	258–260 Ex 1–4 Col 1, 2 Ex 5–21	286–187 Col 1, 3 Ex 1–33
86	241–243 Col 1 Ex 1–19 Ex 2, 5, 20	261–263 Orals All Ex 1–6; 8, 10, 12; 14–17	288–290 Col 1, 4 Ex 1–12 Col 1, 3 Ex 13–18 Ex 19–24
87	244–246 Ex 1–10; 12, 15, 16, 18	264–266 Orals All Ex 1–8; 12–15	292 Chapter Review
88	247–249 Col 1 Ex 1–13 Ex 3	267–269 Orals All Ex 1–8; 13–18	293 Chapter Test
89	250 Chapter Review	270 Chapter Review	295–297 Col 1, 4 Ex 1–12 Ex 13–19
90	251 Chapter Test	271 Chapter Test	298–301 Orals All Col 1, 4 Ex 1–24 Ex 25–30
91	253–255 Orals All Col 1 Ex 1–11 Ex 4, 6	273–276 Col 1, 2 Ex 1–17 Ex 19, 20	304–305 Col 1, 3 Ex 1–9
92	256–257 Ex 1–16	277–279 Orals All Col 1, 5 Ex 1–17 Col 1, 4 Ex 19–37 Col 1, 2 Ex 39–52	308–310 Ex 1, 9, 17, 18 Col 4 Ex 4–20
93	258–260 Ex 1–13	280–282 Orals All Col 1, 2 Ex 1–17 Ex 19–21	311–313 Col 1, 4 Ex 1–12 Ex 10
94	261–263 Orals All Ex 1–13	283–285 Orals All Col 1–3 Ex 1–15 Ex 17–20	314–317 Col 1 Ex 1–15 Ex 4, 8, 12, 14, 16

DAY	LEVEL A	LEVEL B	LEVEL C
95	264–266 Orals All Ex 1–11	286–287 Col 1, 2 Ex 1–32	318–322 Col 1 Ex 1–11 Ex 4, 6, 12
96	267–269 Orals All Ex 1–13; 15	288–290 Orals All Col 1, 3 Ex 1–11 Ex 13–18	323–326 Col 1, 3 Ex 1–35 Ex 13, 17; 36–38
97	270 Chapter Review	292 Chapter Review	327–328 Col 1 Ex 1–16 Ex 17–21
98	271 Chapter Test	293 Chapter Test	330 Chapter Review
99	273–276 Col 1, 2 Ex 1–17	295–297 Orals All Col 1, 3 Ex 1–11 Ex 13–16	331 Chapter Test
100	277–279 Orals All Col 1–3 Ex 1–36	298–301 Orals All Col 1, 3 Ex 1–23 Ex 18, 22	333–336 Col 3 Ex 3–30 Ex Even 32–44 Col 2 Ex 51–65 Ex Even 66–70
101	280–282 Orals All Ex 1–6 Col 1 Ex 7–16	304–305 Col 1, 2 Ex 1–8	337–338 Orals All Ex Odd 11–31 Ex 24
102	283–285 Orals All Col 1–3 Ex 1–15	308–310 Col 1, 3 Ex 1–19	339–341 Ex 1–9 Ex Odd 11–31 Ex 33, 34
103	286–287 Col 1, 2 Ex 1–20	311–313 Orals All Col 1, 3 Ex 1–11	342–345 Orals All Ex 4–6; 11–13 Col 2 Ex 17–46 Ex 47–57
104	288–290 Orals All Ex 1–12	314–317 Orals All Col 1 Ex 1–15 Ex 4, 8, 12, 14, 16	346–347 Col 3, 4 Ex 3–40 Ex 41–51
105	292 Chapter Review	318–322 Orals All Col 1 Ex 1–11 Ex 2, 4, 6, 12	350–354 Orals All Ex Even 10–48
106	293 Chapter Test	323–326 Orals All Col 1, 3 Ex 1–35 Ex 25, 28, 31, 34	355 Ex Even 50–82 Ex 83–86

DAY	LEVEL A	LEVEL B	LEVEL C
107	295–297 Orals All Col 1, 2 Ex 1–10	327–328 Orals All Col 1, 3 Ex 1–18 Ex 14, 17	356 Chapter Review
108	298–301 Ex 1–8	330 Chapter Review	357 Chapter Test
109	301 Orals All Ex 9–16	331 Chapter Test	360–362 Col 3, 4 Ex 3–64
110	304–305 Col 1, 2 Ex 1–5	333–336 Orals All Col 2 Ex 2–17 Col 2, 3 Ex 20–45 Ex Even 46–52; 58–64	363–364 Orals All Ex Even 18–24 Ex 27–36
111	308–310 Col 1, 2 Ex 1–14	337–338 Orals All Ex 1–16; 25 Ex Even 18–30	365–367 Orals All Col 3 Ex 3–48 368–369 Orals All Ex Even 8–22 Ex 23, 24
112	311–313 Orals All Col 1, 2 Ex 1–6	339–341 Ex 1–9 Ex Even 10–32	370–373 Orals All Col 3, 4 Ex 27–48 Ex 49–52 374–375 Orals All Ex 4, 8, 11; 14–17
113	314–317 Orals All Col 1 Ex 1–9 Ex 2, 6, 10	342–345 Orals All Ex Odd 1–15 Col 2, 3 Ex 17–36 Ex Odd 37–45	376–377 Ex 6, 12, 18 Ex Even 20–26 Ex 27, 28
114	318–322 Orals All Ex 1–4	346–347 Col 2, 3 Ex 2–23 Col 1, 2 Ex 25–38	378–379 Col 3, 4 Ex 7–20 Ex 21–26
115	322 Ex 5–10	350–354 Ex Even 2–40	380 Chapter Review
116	323–326 Orals All Col 1, 2 Ex 1–22	354–355 Ex Odd 41–81	381 Chapter Test

DAY	LEVEL A	LEVEL B	LEVEL C
117	327–328 Orals All Col 1, 2 Ex 1–11	356 Chapter Review	383–385 Orals All Col 2, 3 Ex 2–36
118	330 Chapter Review	357 Chapter Test	386–388 Col 3 Ex 6–21 Col 4 Ex 25–37 Ex 38–43
119	331 Chapter Test	360–362 Col 2, 3 Ex 2–63	389–391 Orals All Ex 8, 15, 21 Ex Even 32–46 Ex Odd 47–53
120	333–336 Orals All Col 1, 2 Ex 1–5; 10–14; 19–23; 31–44	363–364 Orals All Col 3 Ex 3–12 Ex 15–22	392–394 Col 2, 3 Ex 5–9; 17–21 Col 2 Ex 35–47 Ex 48, 49
121	337–338 Orals All Col 1–3 Ex 1–23 Ex 25	365–367 Orals All Col 2 Ex 2–26 Col 1, 2 Ex 28–47	395–397 Orals All Col 2, 3 Ex 2–30 Ex 31–42
122	339–341 Ex 1–9 Col 1, 2 Ex 10–24	368–369 Orals All Ex Even 2–8 Ex Odd 13–21	398 Ex 4–12
123	342–345 Orals All Ex 1–12 Col 1, 2 Ex 16–35	370–373 Col 3, 4 Ex 3–24 Col 1, 2 Ex 25–46	399–401 Col 3 Ex 3–30
124	346–347 Ex 1–24	374–375 Orals All Col 1, 2 Ex 1–13	401 Ex 33, 36, 39, 41, 43; 45–49
125	350–351 353 Orals All 354 Ex Odd 1–17	376–377 Col 2 Ex 2–17 Ex Odd 19–25	402 Chapter Review
126	352 354 Ex Odd 19–47	378–379 Col 2–4 Ex 2–12 Col 1, 2 Ex 13–18	403 Chapter Test
127	356 Chapter Review	380 Chapter Review	405–406 Col 1, 6, 7, 8 Ex 1–24 Ex 25–28

DAY	LEVEL A	LEVEL B	LEVEL C
128	357 Chapter Test	381 Chapter Test	407–408 Col 1, 5, 6 Ex 1–24 Ex 25–36
129	360–362 Col 1–3 Ex 1–43	383–385 Orals All Col 1, 2 Ex 1–35	409–412 Ex 1–19
130	363–364 Orals All Col 1, 2 Ex 1–17	386–388 Col 2 Ex 2–35 Ex 18, 21, 24, 28	413–415 Col 1, 3 Ex 1–9 Ex 17–29
131	365–367 Orals All Ex 1–26	389–391 Orals All Col 2 Ex 4–23 Col 2, 4 Ex 26–40 Ex 41, 43, 45	416–419 Col 1, 3, 6 Ex 1–36 Ex 37–50
132	368–369 Orals All Ex Odd 1–11	392–394 Ex 2, 3, 14, 15 Col 3 Ex 24–33 Ex 38, 40, 46	420–423 Col 1, 4, 6 Ex 1–42 Ex 43; 46–55; 60–63
133	370–373 Orals All Col 1, 2 Ex 1–22	395–397 Orals All Col 1, 3 Ex 1–30 Ex 31–42	424–425 Col 1, 2, 5, 6 Ex 1–18 Ex 19–26
134	373 Col 1–3 Ex 25–39	398 Ex 4–12	426 Chapter Review
135	374–375 Orals All Ex 1–9	399–401 Col 2 Ex 2–23 Ex 25	427 Chapter Test
136	376–377 Col 1, 3 Ex 1–18	401 Col 1, 2 Ex 28–38 Ex 40, 42, 44	429–430 Ex 1–14; 19–24
137	378–379 Col 1–3 Ex 1–11	402 Chapter Review	431–434 Ex 4–9; 10–12; 16–20; 22–25; 29–31; 36–47
138	380 Chapter Review	403 Chapter Test	435–436 Col 1, 4 Ex 1–12 Ex 18–24; 28–30
139	381 Chapter Test	405–406 Col 1, 2, 6, 7, 8 Ex 1–24 Ex 25–28	437–439 Ex 9–12; 18–21; 25–28; 31, 32, 34; 35–37

DAY	LEVEL A	LEVEL B	LEVEL C
140	383–385 Orals All Ex 1–21	407–408 Col 1, 2, 4, 5 Ex 25–36	440–442 Ex 1–11
141	386–388 Col 1 Ex 1–34 Ex 18, 21	409–412 Ex 1–17	443–446 Ex 4–6; 10–12; 15–17; 20, 21
142	389–391 Orals All Col 1 Ex 1–22 Col 1, 3 Ex 25–39	413–415 Orals All Col 1, 3, Ex 1–9 Ex 10–16; 20–22	448 Chapter Review
143	392–394 Col 1 Ex 1–31	416–419 Orals All Col 1, 3, 5, 6 Ex 1–36 Ex 37–48	449 Chapter Test
144	395–397 Orals All Col 1, 2 Ex 1–29	420–423 Orals All Col 1, 2, 4, 5 Ex 1–41 Ex 43; 46–59	451–452 Ex 1–10; 16–25
145	398 Ex 1–6	424–425 Col 1, 2, 3, 5 Ex 1–18 Ex 19–26	453–455 Ex 1–16; 19–22
146	399–401 Col 1 Ex 1–22	426 Chapter Review	456–458 Ex 5–10; 15–20
147	401 Col 2 Ex 2–26	427 Chapter Test	459–461 Col 1, 4 Ex 1–24 Ex 26; 28–30
148	402 Chapter Review	429–430 Ex 1–20	462–464 Col 1, 4 Ex 1–24 Ex 26, 28; 34–38
149	403 Chapter Test	431–434 Orals All Ex 1–6; 10–15; 18–21; 26–28; 32–39	465–467 Ex 1–10; 15–18; 23, 24, 29, 30, 32; 35–38
150	405–406 Col 1–5 Ex 1–21	435–436 Col 1–4 Ex 1–12 Ex 16–27	468–470 Col 1, 4 Ex 1–12 Col 1, 3 Ex 13–27 Ex 28, 29
151	407–408 Col 1–4 Ex 1–22	437–439 Ex 5–12; 16–22; 26–34	471–473 Ex 4–9; 13–17

DAY	LEVEL A	LEVEL B	LEVEL C
152	409–412 Ex 1–8	440–442 Ex 1–11	475–577 Ex 1–5 Col 1, 2, 4 Ex 6–25 Ex 26–29
153	412 Ex 9–12	443–446 Ex 4–6; 10–12; 15–20	478–479 Ex 1–16
154	413–415 Orals All Col 1, 2 Ex 1–8 Ex 10–13	448 Chapter Review	480–482 Ex 1–6; 10–14; 17–20; 25–30; 33–35
155	416–419 Orals All Col 1–4 Ex 1–22	449 Chapter Test	483–485 Ex 1–6; 11–14; 17, 18; 24–27; 31–33; 40–44
156	419 Ex 25–40	451–452 Orals All Ex 1–5; 11–25	486–489 Ex 1–18
157	420–423 Orals All Col 1–4 Ex 1–28	453–455 Ex 1–18	490 Chapter Review
158	423 Col 1–4 Ex 31–40 Ex 43–45; 48–51	456–458 Orals All Ex 1–4; 9–14	491 Chapter Test
159	424–425 Col 1–4 Ex 1–16 Ex 19–21	459–461 Orals All Col 1, 3 Ex 1–23 Ex 25–29	493–495 Col 1 Ex 1–11 Ex 8, 10; 12–14
160	426 Chapter Review	462–464 Orals All Col 1, 3 Ex 1–23 Ex 25, 27, 29; 31–34	496–497 Ex 1–11
161	427 Chapter Test	465–467 Ex 1–14; 16, 18 Col 2, 4 Ex 21–30 Ex 31–34	498–499 Ex 1–12
162	429–430 Ex 1–16	468–470 Col 1, 3 Ex 1–30	500–502 Ex 1–6; 9–12
163	431–434 Orals All Ex 1–9	471–473 Ex 1–6; 10–12	503–505 Ex 6–12
164	434 Ex 10–13; 18–21; 26–28	475–477 Ex 1–5 Col 1–3 Ex 6–24	506–509 Ex 1–7; 12–14 Col 2 Ex 16–24 Ex 25–33

DAY	LEVEL A	LEVEL B	LEVEL C
165	435–436 Col 1–3 Ex 1–11 Ex 13–18	478–479 Ex 1–15	510–513 Ex 4–8; 10–15; 19, 20 Col 1 Ex 22–30 Ex 31–39
166	437–439 Ex 1–8; 13–16; 22–25	480–482 Orals All Ex 1–12; 15–25; 29–35	514–516 Ex 1–3; 5–7; 9–11; 13–19
167	440–442 Ex 1–8	483–485 Ex 1–8; 15, 16; 19–23; 28–30; 34–39	517–519 Ex 1–10; 13–18
168	443–446 Ex 1–3; 7–9; 13, 14	486–489 Ex 1–18	520–521 Ex 1–12
169	448 Chapter Review	490 Chapter Review	522 Chapter Review
170	449 Chapter Test	491 Chapter Test	523 Chapter Test

ANSWERS

This section contains a complete set of answers for the Exercise sets (including Oral Exercises), the Special Topics, the Chapter Reviews, and the Chapter Tests.

ANSWERS

Page x

1. 1:05 $\frac{5}{11}$ p.m. **2.** 4,624; 6,084; 6,400; 8,464
3. 120 km
4. Sue: Betty's plate, Donna's cup
Betty: Donna's plate, Jill's cup
Donna: Jill's plate, Sue's cup
Jill: Sue's plate, Betty's cup

Page 3

ORALS **1.** -6 **2.** 3 **3.** 12 **4.** -7 **5.** -4 **6.** $\frac{8}{3}$ or $2\frac{2}{3}$ **7.** $\frac{2}{9}$ **8.** 18

EXERCISES **1.** $\{-4\}$ **2.** $\{6\}$ **3.** $\{-6\}$ **4.** $\{-2\}$ **5.** $\{2\}$ **6.** $\{-3\}$ **7.** $\{-3\}$ **8.** $\{2\}$ **9.** $\{3\}$ **10.** $\{4\}$ **11.** $\{-11\}$ **12.** $\{-3\}$ **13.** $-3\frac{1}{3}$ **14.** $-\frac{3}{4}$ **15.** $1\frac{3}{5}$ **16.** $-3\frac{1}{4}$ **17.** $1\frac{1}{3}$ **18.** $\frac{7}{8}$ **19.** 10 **20.** -6 **21.** 12 **22.** $10\frac{1}{2}$ **23.** $9\frac{1}{3}$ **24.** $12\frac{1}{2}$ **25.** $\{5\}$ **26.** $\{-3\}$ **27.** $\left\{\frac{3}{4}\right\}$ **28.** $\left\{2\frac{2}{5}\right\}$ **29.** $\{3\}$ **30.** $\{-1\}$ **31.** $\left\{-1\frac{5}{6}\right\}$ **32.** $\left\{\frac{2}{3}\right\}$ **33.** $\frac{a+b}{5}$ **34.** $-a$ **35.** $\frac{c-b}{5a}$ **36.** $\frac{3a}{2}$ **37.** $\frac{5b-5a}{4}$ **38.** $\frac{bc}{a}$ **39.** ϕ **40.** {all signed numbers}

Page 5

ORALS **1.** $10x+15$ **2.** $12y-21$ **3.** $-12a+10$ **4.** $-3n-6$ **5.** $-8t+1$ **6.** $-28-12c$

EXERCISES **1.** 3 **2.** 2 **3.** -3 **4.** -2 **5.** -6 **6.** 3 **7.** -2 **8.** 10 **9.** -5 **10.** 5 **11.** -12 **12.** 4 **13.** $\left\{2\frac{7}{10}\right\}$ **14.** $\left\{-\frac{3}{4}\right\}$ **15.** $\left\{-\frac{5}{6}\right\}$ **16.** $\left\{2\frac{1}{5}\right\}$ **17.** $\left\{2\frac{2}{5}\right\}$ **18.** $\left\{\frac{1}{9}\right\}$ **19.** 6 **20.** 2 **21.** 16 **22.** -3 **23.** ϕ **24.** {all signed numbers}

Page 8

ORALS **1.** T **2.** T **3.** T **4.** T **5.** F **6.** F **7.** T **8.** T

EXERCISES Exercises **1–6** require *graphs* and solution sets. **1.** $\{y|y \geq -3\}$ **2.** $\{x|x < -4\}$ **3.** $\{x|x \leq 4\}$ **4.** $\{x|x < 2\}$ **5.** $\{y|y \geq -2\}$ **6.** $\{y|y < 5\}$ **7.** $\{x|x > -2\}$ **8.** $\{x|x > 5\}$ **9.** $\{x|x \leq 1\}$ **10.** $\{x|x < 3\}$ **11.** $\{x|x \geq -2\}$ **12.** $\{x|x \geq -2\}$ **13.** $\{x|x \leq -2\}$ **14.** $\{x|x > 2\}$ **15.** $\{x|x < -3\}$ Exercises **16–21** require *graphs* and solution sets. **16.** $\{x|x < 5\}$ **17.** $\{x|x > -2\}$ **18.** $\{x|x \leq 1\}$ **19.** $\{x|x < 1\}$ **20.** $\{x|x < 1\}$ **21.** $\{x|x \leq 3\}$ **22.** $\{x|x < 8\}$ **23.** $\{x|x > -7\}$ **24.** $\{x|x > 6\}$

Page 10

ORALS **1.** $n-2$ **2.** $n+3$ **3.** $10+n$ **4.** $n-6$ **5.** $2n+5$ **6.** $3n-8$ **7.** $6(3+n)$ **8.** $5(n-2)$

EXERCISES **1.** $6(n+4)+7$ **2.** $12-5(n+7)$ **3.** $5+(2n-3)$ **4.** $8(2n-4)-10$ **5.** -2 **6.** 2 **7.** 2 **8.** 4 **9.** -7 **10.** -5 **11.** -8 **12.** -2 **13.** 7 **14.** 7 **15.** 2 **16.** -3 **17.** 2 **18.** -1 **19.** -7 **20.** -9 **21.** $\{x|x > -2\}$ **22.** $\{x|x > -1\}$ **23.** $\{x|x \geq 5\}$ **24.** $\{x|x \geq 4\}$

Page 12

1. 639,000 m **2.** .639 m **3.** 6.39 m **4.** 783.9 cm **5.** .007839 km **6.** 78.39 dm **7.** 780 m **8.** 2,370 mm **9.** 83.6 m **10.** 5,000 L **11.** 13 L **12.** .0029 L **13.** .086 L **14.** 10.5 L **15.** .003 L **16.** 4,600 L **17.** 300 L **18.** 15,000 g **19.** 6,800,000 mg **20.** .25 g

Page 15

1. \$1,680 **2.** 4.5 yr **3.** \$8,000 **4.** 7% **5.** 12 yr **6.** \$1,400 **7.** 18.75 ohms **8.** 48 ohms **9.** 40 ohms **10.** 30 ohms **11.** \$9,780 **12.** \$2,500 **13.** 3.84 ohms **14.** 8.12 ohms **15.** \$4,000 **16.** \$5,000 **17.** 6.5 yr **18.** 6% **19.** 157.5 volts **20.** 50 ohms

Page 16

1. $\{-4\}$ **2.** $\left\{-1\frac{1}{3}\right\}$ **3.** $\{-2\}$ **4.** $\left\{1\frac{1}{6}\right\}$ **5.** $\{9\}$
6. $\left\{-12\frac{1}{2}\right\}$ **7.** $\{2\}$ **8.** $\left\{-3\frac{1}{4}\right\}$ **9.** 2 **10.** -2 **11.** 1
12. $\frac{1}{2}$ **13.** 2 **14.** $2\frac{1}{10}$ Exercises **15** and **16** require *graphs* and solution sets.
15. $\{n|n \geq -4\}$ **16.** $\{x|x < -3\}$ **17.** $\{x|x < 4\}$
18. $\{y|y \leq 2\}$ **19.** $\{a|a > -3\}$ **20.** $\{x|x \geq -3\}$
21. $6 + 2n$ **22.** $4n - 5$ **23.** $5(7 + n) + 3$
24. $9 - 6(n + 3)$ **25.** -3 **26.** 2 **27.** 40 ohms
28. 4.8 ohms **29.** \$4,000 **30.** \$2,680

Page 17

1. $\{-5)$ **2.** $\left\{\frac{2}{3}\right\}$ **3.** $\{-2\}$ **4.** $\left\{1\frac{3}{5}\right\}$ **5.** $\{9\}$ **6.** $\{-2\}$
7. 2 **8.** -7 **9.** 3 **10.** 2 Exercises **11** and **12** require *graphs* and solution sets.
11. $\{n|n \geq -2\}$ **12.** $\{x|x < 4\}$ **13.** $\{x|x < 4\}$
14. $\{a|a \geq -5\}$ **15.** $8n - 4$ **16.** $8 + 2(4 + n)$
17. -4 **18.** 5 **19.** 60 ohms **20.** \$3,000

Page 18

1. $ac + cb = ca + cb$ Comm. Prop. Mult.
$= c(a + b)$ Dist. Prop.
$= (a + b)c$ Comm. Prop. Mult.
$ac + cb = (a + b)c$ Trans. Prop. Equality

2. $a(b + c) = ab + ac$ Dist. Prop.
$= ac + ab$ Comm. Prop. Add.
$= ca + ba$ Comm. Prop. Mult.
$a(b + c) = ca + ba$ Trans. Prop. Equality

3. $ac + b = b + ac$ Comm. Prop. Add.
$= b + ca$ Comm. Prop. Mult.
$ac + b = b + ca$ Trans. Prop. Equality

4. $a + (b + c) = (b + c) + a$ Comm. Prop. Add.
$= b + (c + a)$ Assoc. Prop. Add.
$a + (b + c) = b + (c + a)$ Trans. Prop. Equality

5. $ab + ac = a(b + c)$ Dist. Prop.
$= (b + c)a$ Comm. Prop. Mult.
$= (c + b)a$ Comm. Prop. Add.
$ab + ac = (c + b)a$ Trans. Prop. Equality

6. $a + [b(c + d)] = a + (bc + bd)$ Dist. Prop.
$= a + (bd + bc)$ Comm. Prop. Add.
$= (a + bd) + bc$ Assoc. Prop. Add.
$= (bd + a) + bc$ Comm. Prop. Add.
$a + [b(c + d)] = (bd + a) + bc$ Trans. Prop. Equality

Page 21

(Graphs for Exercises **1-9** are described.)
1. between -4 and 2 **2.** -1, between -1 and 3
3. left of -2, right of 3 **4.** -4, left of -4, right of 1 **5.** 6, between 2 and 6 **6.** -5, -2, between -5 and -2 **7.** left of 3, 5, right of 5
8. -5 and left, -2 and right **9.** all points except zero **10.** $\{x|-2 < x < 3\}$ **11.** $\{y|-5 \leq y < 5\}$
12. $\{n|-2 < n \leq 4\}$ **13.** $\{a|-3 \leq a \leq 5\}$
14. $\{x|-1 < x < 4\}$ **15.** $\{y|-2 < y \leq 3\}$
16. $\{n|-1 \leq n < 3\}$ **17.** $\{a|1 \leq a \leq 2\}$
18. $\{x|-1 < x < 3\}$ Exercises **19–26** require *graphs* and solution sets. **19.** $\{x|x < 2 \text{ or } x > 5\}$
20. $\{y|y \leq -4 \text{ or } y \geq 2\}$ **21.** $\{a|a < -2 \text{ or } a \geq 3\}$
22. $\{n|n < -3 \text{ or } n > 2\}$ **23.** $\{x|x < -1 \text{ or } x > 3\}$
24. $\{y|y \leq -3 \text{ or } y \geq 3\}$ **25.** $\{a|a \leq -2 \text{ or } a \geq 3\}$
26. $\{n|n < -2 \text{ or } n > 3\}$ **27.** $\{x|-3 < x < 2\}$
28. $\{x|-2 \leq x \leq 1\}$ **29.** $\{x|-3 < x < 5\}$
30. $\{x|-6 \leq x \leq -3\}$ **31.** $\{x|-6 < x < 2\}$
32. $\{x|-6.5 \leq x \leq 1.5\}$ **33.** $\{x|x \leq -3 \text{ or } x \geq 2\}$
34. $\{x|x < -4 \text{ or } x > 2\}$
35. $\{x|-1 < x < 2 \text{ or } x > 4\}$
36. $\{x|-5 < x < -2 \text{ or } -1 < x < 2\}$

Page 24

ORALS **1.** 12 **2.** 3 **3.** 6 **4.** 12 **5.** 8 **6.** 10 **7.** 14 **8.** 5

EXERCISES **1.** $\{10, -4\}$ **2.** $\{-4, -8\}$
3. $\{5, -5\}$ **4.** ϕ **5.** $\{4, -4\}$ **6.** $\{5, -2\}$
7. $\{5, -5\}$ **8.** $\{-1, -2\}$ **9.** $\left\{5, -1\frac{2}{3}\right\}$ **10.** $\left\{\frac{1}{2}, -2\right\}$
11. $\left\{2, -2\frac{2}{5}\right\}$ **12.** $\left\{1, \frac{3}{5}\right\}$ **13.** $\{x|-3 < x < 3\}$
14. $\{n|n \leq -6 \text{ or } n \geq 6\}$ **15.** $\{y|-8 \leq y \leq 8\}$
16. $\{a|a < -5 \text{ or } a > 5\}$ Exercises **17–36** require *graphs* and solution sets. **17.** $\{x|2 < x < 4\}$
18. $\{y|y \leq -11 \text{ or } y \geq 3\}$ **19.** $\{a|-2 \leq a \leq 6\}$
20. $\{n|n < -8 \text{ or } n > -2\}$ **21.** $\{a|-6 < a < 6\}$
22. $\{y|y \leq -4 \text{ or } y \geq 4\}$ **23.** $\{n|-3 \leq n \leq 3\}$
24. $\{x|x < -2 \text{ or } x > 2\}$ **25.** $\{x|-1 < x < 6\}$
26. $\{y|y \leq -5 \text{ or } y \geq 3\}$ **27.** $\{a|-1 \leq a \leq 5\}$
28. $\{n|n < -4 \text{ or } n > 1\}$ **29.** $\{y|-2 < y < 1\}$
30. $\{a|a \leq 1 \text{ or } a \geq 5\}$ **31.** $\{x|-3 \leq x \leq -1\}$
32. $\{n|n < 1 \text{ or } n > 2\}$ **33.** $\{x|-1 < x < 5\}$
34. $\{x|x < 1 \text{ or } x > 4\}$
35. $\{x|x < -4 \text{ or } -2 < x < 4 \text{ or } x > 6\}$
36. $\{x|-5 \leq x \leq -1 \text{ or } 1 \leq x \leq 5\}$

Page 25
Answers may vary.

Page 26
1. T **2.** T **3.** F **4.** 4 **5.** 3 **6.** 1

7.

$\oplus$	0	1	2	3	4
0	0	1	2	3	4
1	1	2	3	4	0
2	2	3	4	0	1
3	3	4	0	1	2
4	4	0	1	2	3

8.

$\otimes$	0	1	2	3	4
0	0	0	0	0	0
1	0	1	2	3	4
2	0	2	4	1	3
3	0	3	1	4	2
4	0	4	3	2	1

9. $3 \oplus 4 = 2$; $4 \oplus 3 = 2$ **10.** $2 \otimes 3 = 1$; $3 \otimes 2 = 1$ **11.** $4 \otimes 2 = 3$; $2 \otimes 4 = 3$ **12.** $(3 \oplus 2) \oplus 4 = (0) \oplus 4 = 4$; $3 \oplus (2 \oplus 4) = 3 \oplus (1) = 4$ **13.** $(2 \otimes 3) \otimes 1 = (1) \otimes 1 = 1$; $2 \otimes (3 \otimes 1) = 2 \otimes (3) = 1$ **14.** 3 **15.** 3 **16.** 3 **17.** 3

18.

$\oplus$	0	1	2	3
0	0	1	2	3
1	1	2	3	0
2	2	3	0	1
3	3	0	1	2

$\otimes$	0	1	2	3
0	0	0	0	0
1	0	1	2	3
2	0	2	0	2
3	0	3	2	1

Page 30

ORALS **1.** f = 1st; $3f + 5$ = 2nd **2.** s = smaller; $2s - 4$ = greater **3.** g = greater; $4g - 17$ = smaller **4.** f = 1st; $6 - 2f$ = 2nd

EXERCISES **1.** 6; 11 **2.** 10; 70 **3.** 7; 11 **4.** 2; 14 **5.** 2; 7 **6.** 3; 10 **7.** 5 m; 11 m **8.** 8 m; 12 m **9.** 6 cm; 10 cm; 12 cm **10.** 15 m; 10 m; 20 m **11.** 2; 14 **12.** 2; −1 **13.** 3; 18 **14.** 5; 7 **15.** 4 cm; 17 cm **16.** 8 m; 10 m; 10 m **17.** 3; 11; 21 **18.** 7; 14; 10 **19.** 3 cm; 9 cm **20.** 32 m

Page 34

ORALS **1.** $5n$ ¢ **2.** $10(n + 2)$ or $10n + 20$ ¢ **3.** $25(3n)$ or $75n$ ¢ **4.** $10(5n - 4)$ or $50n - 40$ ¢

EXERCISES **1.** $15n + 40$ **2.** $35d - 50$ **3.** $130n$ **4.** $85q$ **5.** $20d + 15$ **6.** $40q - 30$ **7.** 7 nickels; 15 dimes **8.** 9 nickels; 3 quarters **9.** 4 dimes; 12 quarters **10.** 3 dimes; 21 nickels **11.** 3 quarters; 68 dimes **12.** 6 quarters; 18 nickels **13.** 5 at 15¢; 10 at 8¢; 14 at 20¢ **14.** 8 quarters; 20 dimes; 40 nickels **15.** 22 nickels; 18 dimes **16.** 12 dimes; 8 quarters

Page 35
1. no **2.** yes **3.** no **4.** no **5.** no; yes; no; no

Page 39
1. 12 at 50¢; 8 at 30¢ **2.** 15 cheese; 25 meat **3.** 6 kg **4.** 50 kg **5.** 4 kg soft; 8 kg hard **6.** 10 kg brand A; 6 kg brand B **7.** 20 kg milk; 4 kg cocoa **8.** 20 kg wheat; 10 kg bran **9.** 16 at 60¢; 8 at 80¢ **10.** 7 at 30¢; 21 at 20¢ **11.** 3 g chemical A; 6 g B; 9 g C **12.** 16 kg brand A; 6 kg B; 8 kg C

Page 41

ORALS **1.** $(j + 7)$ yrs **2.** $(f - 5)$ yrs **3.** $5b$ yrs **4.** $(s - 6)$ yrs

EXERCISES **1.** Walter 8; Brenda 12 **2.** Cindy 16; Byron 14 **3.** Conrad 6; Denise 18 **4.** Edna 3; Samson 15 **5.** Ted 2; Wanda 22 **6.** Susan 20; Ray 8 **7.** Paula 8; Mel 24 **8.** Amos 6; Gloria 24 **9.** Keith 14; Phyllis 20 **10.** Ruby 14; Eric 10

Page 42

1. -3 and all points between -3 and 5
2. 6 and all points between -4 and 6 **3.** all points to the left of -4; 2 and all points to the right of 2 **4.** all points to the left of 2; all points to the right of 5 **5.** $\{x|-1 < x < 4\}$
6. $\{y|-3 \le y \le -1\}$ **7.** $\{a|a < -4 \text{ or } a > 1\}$
8. $\{n|n \le -1 \text{ or } n \ge 2\}$ **9.** $\{5, 2\}$ **10.** $\left\{3, -5\frac{2}{3}\right\}$
11. $\{a|-5 < a < 5\}$ **12.** $\{n|n \le -3 \text{ or } n \ge 3\}$
Exercises **13–16** require *graphs* and solution sets. **13.** $\{x|-8 \le x \le 2\}$
14. $\{y|y < -3 \text{ or } y > 3\}$ **15.** $\{a|a < -1 \text{ or } a > 4\}$
16. $\{n|-5 < n < 1\}$ **17.** 4; 19 **18.** 7 m; 8 m
19. 2; 11 **20.** 4 cm; 12 cm; 10 cm
21. 12 nickels; 5 dimes **22.** 7 quarters; 30 dimes **23.** 3 kg caramels; 5 kg peppermints
24. 4 kg **25.** Conrad 9; Edna 14 **26.** Ann 12; Ted 3

Page 43

1. 2 and all points between -5 and 2 **2.** -3 and all points to the left of -3; all points to the right of 4 **3.** $\{x|-3 \le x \le 2\}$
4. $\{y|y < -2 \text{ or } y > 3\}$ **5.** $\{5, -11\}$ **6.** $\left\{2, \frac{2}{3}\right\}$
Exercises **7–12** require *graphs* and solution sets. **7.** $\{x|-2 < x < 2\}$ **8.** $\{y|y \le -4 \text{ or } y \ge 4\}$
9. $\{a|a < 1 \text{ or } a > 7\}$ **10.** $\{n|-3 \le n \le 3\}$
11. $\{x|-4 < x < 3\}$ **12.** $\{y|y \le 2 \text{ or } y \ge 4\}$
13. 2; 7 **14.** 4 m; 17 m **15.** 7 nickels; 11 dimes
16. 6 kg brand A; 9 kg B **17.** Melvin 18; Phyllis 3

Page 44

1. .36 m^2 **2.** .07 m^2 **3.** 420,000 m^2 **4.** 2,400 m^2
5. **–9.** (Answers may vary.) **5.** 30° **6.** $-5°$
7. 39° **8.** 35° **9.** 180°

Page 47

ORALS **1.** 5^9 **2.** 5^{18} **3.** 6^{15} **4.** 6^8 **5.** x^9
6. x^{18} **7.** c^{15} **8.** c^8 **9.** x^6 **10.** $2x^3$ **11.** $7x^2$
12. $12x^4$ **13.** x^5 **14.** $2x^5$ **15.** x^2 **16.** $2x$ **17.** 1
18. -1 **19.** 1 **20.** -1 **21.** 1 **22.** 4 **23.** -8
24. 16 **25.** -32 **26.** 64

EXERCISES **1.** $12x^5$ **2.** $-10a^9$ **3.** $20c^6$
4. $64x^3$ **5.** $16a^4$ **6.** $49c^2$ **7.** $5x^4$ **8.** $-3a^2$
9. $40n^5$ **10.** $3x^8$ **11.** $-5a^{12}$ **12.** $-c^6$ **13.** $81x^8$
14. $7{,}000x^6$ **15.** $100x^7$ **16.** $2x^3 + 6x^2$
17. $-5x^3 + 7x$ **18.** $3x^2 - 2x$ **19.** -27 **20.** 75
21. -16 **22.** 64 **23.** 1,600 **24.** 144 **25.** x^5y^6
26. $3a^5b^5$ **27.** $-10x^5y^7$ **28.** a^3b^3 **29.** $25x^2y^2$
30. $16c^4d^4$ **31.** a^6b^4 **32.** x^6y^{12} **33.** x^8y^{12}
34. $25x^4y^6$ **35.** $-8a^3b^{12}$ **36.** $81c^{12}d^4$
37. $8a^2 + 5b^2$ **38.** $6a^2 + 3b^3$ **39.** $6x^2 - 4y^2$
40. 80 **41.** 32 **42.** 1,000 **43.** 900 **44.** 81
45. $-64{,}000$ **46.** -4 **47.** 52 **48.** x^{2a} **49.** x^{5a}
50. x^{6a+3} **51.** x^4 **52.** $x^{3a}y^{7a}$ **53.** $x^{2a+1}y^{2a+2}$
54. $x^{7a}y^{5b}$ **55.** $x^{3a+4}y^{4b}$ **56.** x^{2a} **57.** x^{3a}
58. x^{12a} **59.** x^{a^2} **60.** x^{6a^2} **61.** 5^ax^a **62.** $2^{3a}x^{3a}$
63. $3^{a+1}x^{a+1}$ **64.** $4^{2a}x^{4a}$ **65.** $2^{a+2}x^{3a+6}$
66. $x^{3a}y^{3b}$ **67.** $x^{4ac}y^{4bc}$ **68.** $x^{4a+8}y^{4b+12}$
69. $x^{3a+9}y^{4a+12}$

Page 52

ORALS **1.** yes **2.** yes **3.** no **4.** yes **5.** no
6. yes **7.** yes **8.** yes **9.** binom. **10.** trinom.
11. monom. **12.** polynom. with 5 terms
13. 5 **14.** 2 **15.** 4 **16.** 5 **17.** $-4x^3 + 2x - 2$
18. $7y^5 - y^2 + 6$ **19.** $-2x^5 + x^3 + 2x - 3$
20. $4y^5 - 7y^4 + 2y^3$

EXERCISES **1.** -192 **2.** -72 **3.** 11
4. $4x^2 - 2x + 2$ **5.** $2a^2 + 9$ **6.** $2x^2 + 3x - 3$
7. $5y^2 - y + 2$ **8.** $7x^2 + 4x + 1$ **9.** $-9b^2 - 5b - 6$ **10.** $-5x + 11$ **11.** $6y^2 + y - 3$
12. $3x^2 + 7x + 1$ **13.** $-12a^2 + a - 1$
14. $-4x^2 + 10x + 4$ **15.** $5y^2 - 2y + 6$
16. $6x^2 + 3x + 9$ **17.** $12a - 9$ **18.** $-4y^2 - 6y + 7$ **19.** $-3n^2 + 3n - 6$ **20.** $9xy + 5xy^2 + 3x^2$ **21.** $-3ab + 4a^3b - 3b^2$ **22.** $4x^2y^2 - 5xy^2 + 2x^2y$ **23.** $2a^2 + 7ab + b^2 - 3a^2b$
24. $4x^3 - 5x^2 + 6x + 14$ **25.** $4y^6 + 4y^4 - 2y^2 - 5$ **26.** $7x^2y^2 - 7xy + 12$ **27.** $8a^2b^2 - 2ab + 14$ **28.** $2rs + 2r^2 - 3s^2 - 9 + r^2s^2$
29. $7xy + 8x^2$

Page 54

ORALS **1.** $20y^2 - 12y + 8$ **2.** $-6x^2 - 8x + 10$ **3.** $-8x^2 + 6x - 7$ **4.** $10x^3 - 5x^2 + 15x$
5. $-6xy - 15y^2 + 3y$ **6.** $6x^2 - 4xy + 8x$

EXERCISES **1.** $8x^2 - 14x - 15$ **2.** $15x^2 + 4x - 4$ **3.** $8x^2 - 2x - 15$ **4.** $10x^2 - 11x + 3$
5. $2x^2 + 3x - 35$ **6.** $6x^2 + 19x + 3$ **7.** $4x^2 - 9$
8. $9x^2 - 16$ **9.** $5x^3 - 11x^2 - 2x + 8$ **10.** $8x^3 + 18x^2 - 22x + 3$ **11.** $9x^3 - 9x^2 + 14x - 8$

12. $8x^3 + 20x^2 + 2x - 12$ **13.** $x^2 + 10x + 25$ **14.** $x^2 - 4x + 4$ **15.** $4x^2 - 12x + 9$ **16.** $9x^2 + 24x + 16$ **17.** $6x^2 + 8xy - 8y^2$ **18.** $15x^2 + 22xy + 8y^2$ **19.** $12x^2 - xy - y^2$ **20.** $5x^2 - 7xy + 2y^2$ **21.** $4x^2 - 9y^2$ **22.** $9x^2 - 25y^2$ **23.** $12x^2 + 7xy + 9x - 10y^2 - 6y$ **24.** $2x^2 + 3xy + 6x - 5y^2 + 15y$ **25.** $16x^2 + 8xy + y^2$ **26.** $x^2 - 4xy + 4y^2$ **27.** $9x^2 - 24xy + 16y^2$ **28.** $4x^2 + 12xy + 9y^2$ **29.** $8x^2 + 4x - 60$ **30.** $-24x^2 + 6x + 45$ **31.** $24x^3 + 76x^2 + 40x$ **32.** $-40x^3 + 14x^2 + 12x$ **33.** $16x^2 - 4$ **34.** $75x^3 - 27x$ **35.** $18x^2 - 12x + 2$ **36.** $50x^3 + 40x^2 + 8x$ **37.** $20x^2 - 2xy - 6y^2$ **38.** $27x^2 - 12y^2$ **39.** $45x^2 - 80y^2$ **40.** $10x^2 + 80xy + 160y^2$ **41.** $x^3 + 15x^2 + 75x + 125$ **42.** $x^3 + 3x^2y + 3xy^2 + y^3$ **43.** $x^3 - 3x^2y + 3xy^2 - y^3$ **44.** $4x^2 + 12xy - 20x + 9y^2 - 30y + 25$ **45.** $25a^2 - 20ab + 40a + 4b^2 - 16b + 16$ **46.** $64c^2 - 16cd + 64ce + d^2 - 8de + 16e^2$ **47.** $x^{2n} - 2x^n - 15$ **48.** $3x^{4a} + 7x^{2a} + 4$ **49.** $x^{2c} - y^{4c}$

Page 57

ORALS **1.** 4 **2.** 6 **3.** 3 **4.** 2 **5.** x **6.** x **7.** x^2 **8.** x^2 **9.** x **10.** $3x$ **11.** $2x^2$ **12.** a **13.** 5 **14.** x **15.** 3 **16.** x^2

EXERCISES **1.** $5(t - 2)$ **2.** $8(a + 1)$ **3.** $5(x^2 - 3x + 2)$ **4.** $3(2x^2 + x - 3)$ **5.** $2(3a^2 + 4a - 6)$ **6.** $3(4x - 5y - 1)$ **7.** $x(4x^2 + x - 7)$ **8.** $x^2(2x^3 - 5x + 3)$ **9.** $a(2a + 1)$ **10.** $x(7x - 1)$ **11.** $c(c + 1)$ **12.** $n(n - 1)$ **13.** $x(7x^2 + 2x + 1)$ **14.** $x^2(3x^3 - 5x + 1)$ **15.** $3n(n + 2)$ **16.** $5x(2x^2 + 4x - 3)$ **17.** $5x^2(2x - 3)$ **18.** $3x^2(x^2 - 2x + 4)$ **19.** $4a^3(a - 3)$ **20.** $4x^3(2x^2 - 3x + 4)$ **21.** $-1(5a - 3b)$ **22.** $-1(2x^2 + 7m)$ **23.** $-1(3x^2 - 4x + 2)$ **24.** $-1(x^2 - 9)$ **25.** $-1(x + 4)$ **26.** $-1(x^2 - x + 12)$ **27.** $3xy(x + 2)$ **28.** $7xy(2x - 3y)$ **29.** $x^2y^2(y^2 - xy + x^2)$ **30.** $ax(x^2 + ax + a^2)$ **31.** $2ay(2y + 3a + 4)$ **32.** $5x^2y(3 - 7xy + 4x^2y^2)$ **33.** $4xy(11x^2 - 25y^2)$ **34.** $8x^2y^2(1 - 2xy)$ **35.** $x^4(x^{a+2} + 1)$ **36.** $x^{3c}(x^{2c} + 1)$ **37.** $x^{n+2}(x^3 + 1)$ **38.** $x^a(x^a + x^3)$

Page 61

ORALS **1.** $-16x$ **2.** $+8x$ **3.** $-5x$ **4.** $-9x$

EXERCISES **1.** $(3x + 2)(x + 1)$ **2.** $(2x + 1)(x + 5)$ **3.** $(5x + 7)(x + 1)$ **4.** $(2x - 1)(x - 3)$ **5.** $(7x - 2)(x - 1)$ **6.** $(3x - 1)(x - 7)$ **7.** $(2x - 1)(x + 2)$ **8.** $(5x - 3)(x + 1)$ **9.** $(3x - 2)(x + 1)$ **10.** $(3x + 1)(x - 3)$ **11.** $(7x + 2)(x - 1)$ **12.** $(5x + 3)(x - 1)$ **13.** $(x + 1)(x - 3)$ **14.** $(x + 5)(x - 1)$ **15.** $(x - 7)(x + 1)$ **16.** $(5x + 1)(x + 3)$ **17.** $(7x + 1)(x + 3)$ **18.** $(x + 1)(x + 13)$ **19.** $(7x - 5)(x - 1)$ **20.** $(5x - 1)(x - 7)$ **21.** $(x - 17)(x - 1)$ **22.** $(5x - 1)(x + 3)$ **23.** $(7x + 11)(x - 1)$ **24.** $(x - 1)(x + 11)$ **25.** $(5x + 1)(x - 11)$ **26.** $(3x - 13)(x + 1)$ **27.** $(x - 19)(x + 1)$ **28.** $(3x + y)(x + 2y)$ **29.** $(5x - 3y)(x - y)$ **30.** $(x - 5y)(x - y)$ **31.** $(2x - y)(x + 3y)$ **32.** $(5x + 7y)(x - y)$ **33.** $(x + 3y)(x - y)$ **34.** $(5x + 2y)(x - y)$ **35.** $(3x + 2y)(x - y)$ **36.** $(x - 5y)(x + y)$ **37.** $(3x - y)(x + 3y)$ **38.** $(7x + 3y)(x - y)$ **39.** $(2x + 3y)(x - y)$ **40.** $(x^m + 3)(x^m + 1)$ **41.** $(2x^m - 1)(x^m + 5)$ **42.** $(2x^{2m} + 7)(x^{2m} - 1)$ **43.** $(x^a + y^b)(x^a + 2y^b)$ **44.** $(2x^{2a} - y^b)(x^{2a} - 3y^b)$ **45.** $(3x^a + y^{3b})(x^a - 3y^{3b})$

Page 63

ORALS **1.** $-29x$ **2.** $+29x$ **3.** $-x$ **4.** $+7x$ **5.** $-7x$

EXERCISES **1.** $(2a + 5)(a - 2)$ **2.** $(2x - 7)(x - 2)$ **3.** $(2x + 5)(x + 2)$ **4.** $(6x - 1)(x - 2)$ **5.** $(4n + 3)(2n + 1)$ **6.** $(4t - 5)(t - 1)$ **7.** $(3a - 8)(a + 1)$ **8.** $(5c + 9)(c - 1)$ **9.** $(5x - 6)(x - 1)$ **10.** $(6c + 5)(c + 1)$ **11.** $(2x + 5)(2x - 1)$ **12.** $(10n + 1)(n - 1)$ **13.** $(7x - 10)(x + 1)$ **14.** $(3x + 2)(x + 3)$ **15.** $(3x - 10)(x - 1)$ **16.** $(5c - 3)(2c - 1)$ **17.** $(5x + 3)(3x - 1)$ **18.** $(2x + 5)(2x + 1)$ **19.** $(3x - 5)(x - 3)$ **20.** $(5x + 3)(x + 5)$ **21.** $(7n + 6)(n + 1)$ **22.** $(8x + 7)(x - 1)$ **23.** $(5a - 3)(3a - 1)$ **24.** $(3x + 2)(3x + 1)$ **25.** $(n - 8)(n + 3)$ **26.** $(x - 8)(x - 1)$ **27.** $(n - 4)(n + 3)$ **28.** $(a + 6)(a - 2)$ **29.** $(x + 5)(x + 3)$ **30.** $(a - 6)(a - 3)$ **31.** $(x - 9)(x + 2)$

32. $(x-12)(x-1)$ **33.** $(n+6)(n-4)$
34. $(c-12)(c+3)$ **35.** $(x+10)(x+1)$
36. $(a+9)(a+4)$ **37.** $(5x+4)(2x+3)$
38. $(5n+4)(3n-1)$ **39.** $(4c-3)(c-3)$
40. $(3a+2)(3a+4)$ **41.** $(3x+2)(2x-5)$
42. $(4n-3)(3n-2)$ **43.** $(2a-3b)(a+2b)$
44. $(3x-y)(x+4y)$ **45.** $(3m-n)(2m+3n)$
46. $(5x-6y)(x+2y)$ **47.** $(7a+4b)(a+2b)$
48. $(3y-4z)(2y+3z)$ **49.** $(5r-4s)(3r+s)$
50. $(2r+s)(r-4s)$ **51.** $(2x+5y)(x-2y)$
52. $(2a+5b)(2a+b)$ **53.** $(2x+3y)(x-2y)$
54. $(5b-3c)(3b-c)$ **55.** $(5x-3y)(2x-y)$
56. $(3r+5s)(3r-2s)$ **57.** $(4a+3b)(2a-3b)$
58. $(3x^m+5)(2x^m+3)$
59. $(3x^{2m}+10)(3x^{2m}-2)$
60. $(4x^{3m}-7)(2x^{3m}+3)$
61. $(5x^a+3y^b)(x^a+4y^b)$
62. $(4x^{2a}-3y^{2b})(2x^{2a}-5y^{2b})$
63. $(6x^{3a}+5y^{2b})(2x^{3a}-3y^{2b})$
64. $(x^{n+2}+3)(x^{n+2}+4)$
65. $(x^{n-1}+5)(x^{n-1}-2)$
66. $(x^{4n+3}-3)(x^{4n+3}+1)$
67. $(x^{a+3}+y^{a-2})(x^{a+3}+y^{a-2})$
68. $(2x^{a+1}+3y^a)(x^{a+1}+y^a)$

Page 66

ORALS **1.** yes **2.** no **3.** yes **4.** yes
5. $(10n)^2-(3)^2$ **6.** $(5)^2-(2y)^2$ **7.** $(2x)^2-(3y)^2$
8. $(x^2)^2-(y^3)^2$

EXERCISES **1.** $16x^2-25$ **2.** $100y^2-49$
3. n^2-36 **4.** $25x^2-4y^2$ **5.** a^2-9b^2
6. $16n^4-1$ **7.** $(2x+3)(2x-3)$
8. $(3a+10)(3a-10)$ **9.** $(c+1)(c-1)$
10. $(5+4y)(5-4y)$ **11.** $(7+b)(7-b)$
12. $(1+6n)(1-6n)$ **13.** $4x^2+20x+25$
14. $36n^2-12n+1$ **15.** $9+24x+16x^2$
16. $9x^2-24xy+16y^2$ **17.** $a^2+14ab+49b^2$
18. $4x^2-4xy+y^2$ **19.** no **20.** no **21.** yes
22. $(5x+2)(5x+2)$ **23.** $(3n-4)(3n-4)$
24. $(2x+1)(2x+1)$ **25.** $(x+4)(x+4)$
26. $(n-3)(n-3)$ **27.** $(1+10a)(1+10a)$
28. $(3a+5b)(3a-5b)$ **29.** $(9x+y)(9x-y)$
30. $(x+4y)(x-4y)$ **31.** $(10x+7y^2)(10x-7y^2)$
32. $(m^2+n)(m^2-n)$ **33.** $(7cd+1)(7cd-1)$
34. $(x+8y)(x+8y)$ **35.** $(5c-2d)(5c-2d)$
36. $(2a+b)(2a+b)$ **37.** $(6x-5y)(6x-5y)$
38. $(3a+2b)(3a+2b)$ **39.** $(xy-1)(xy-1)$
40. $(x-2)(x^2+2x+4)$
41. $(x+5)(x^2-5x+25)$
42. $(3x-1)(9x^2+3x+1)$
43. $(4y+1)(16y^2-4y+1)$
44. $(5x-4y)(25x^2+20xy+16y^2)$
45. $(2x^2+y)(4x^4-2x^2y+y^2)$

Page 68

1. yes **2.** yes **3.** no **4.** yes **5.** yes **6.** no

Page 71

ORALS **1.** 7 **2.** 3; -3 **3.** -6; 2
4. $x^2-7x+12=0$ **5.** $x^2-x-6=0$
6. $x^2-6x+9=0$ **7.** $3x^2+10x-8=0$
8. $2x^2-7x+5=0$ **9.** $x^2-9x+14=0$
10. T **11.** F **12.** T **13.** F

EXERCISES **1.** $\{2, 5\}$ **2.** $\{-3, 6\}$
3. $\{-4, -1\}$ **4.** $\left\{\frac{1}{2}, 3\right\}$ **5.** $\left\{1\frac{2}{3}, -2\right\}$
6. $\left\{-1\frac{1}{2}, -\frac{2}{5}\right\}$ **7.** $\{0, 4\}$ **8.** $\{0, -5\}$ **9.** $\{6\}$
10. 1; 2 **11.** -5; 3 **12.** -4; -2 **13.** $\frac{1}{2}$; 4
14. $-\frac{1}{3}$; 3 **15.** $\frac{2}{3}$; -1 **16.** 2; -2 **17.** 6; -6
18. 4; -4 **19.** 0; 5 **20.** 0; -4 **21.** 0; 3 **22.** -3
23. 4 **24.** -6 **25.** 5; 8 **26.** 7; -6 **27.** -5; -10
28. $\frac{1}{4}$; $-\frac{1}{2}$ **29.** $-\frac{1}{3}$; $-\frac{1}{4}$ **30.** $\frac{1}{5}$; $\frac{1}{2}$ **31.** $\{-4, 3\}$
32. $\{-2, 3\}$ **33.** $\{3, -3\}$ **34.** $\{2, 4\}$ **35.** $\{-3, 5\}$
36. $\{0, 6\}$ **37.** $\frac{1}{2}$; 3 **38.** $\frac{1}{3}$; -2 **39.** $-\frac{1}{2}$; -1
40. $\left\{1\frac{1}{2}, 2\right\}$ **41.** $\left\{1\frac{2}{3}, -3\right\}$ **42.** $\left\{-3\frac{1}{2}, -1\right\}$
43. $\left\{-1\frac{2}{3}, 1\frac{2}{3}\right\}$ **44.** $\left\{-\frac{1}{4}, \frac{1}{4}\right\}$ **45.** $\left\{-1\frac{1}{5}, 1\frac{1}{5}\right\}$
46. $\left\{0, \frac{2}{3}\right\}$ **47.** $\left\{-1\frac{3}{4}, 0\right\}$ **48.** $\left\{2\frac{1}{2}, 0\right\}$ **49.** $\{9\}$
50. $\{-11\}$ **51.** $\{7\}$ **52.** $\left\{2\frac{2}{3}, \frac{1}{2}\right\}$ **53.** $\left\{1\frac{1}{4}, -2\frac{1}{2}\right\}$
54. $\left\{-1\frac{2}{3}, \frac{3}{4}\right\}$ **55.** 2; 5 **56.** -3; 4 **57.** -6; 1
58. $\frac{1}{2}$; -4 **59.** $-\frac{3}{4}$; 2 **60.** $\frac{2}{3}$; 3
61. $\{-3, -1, -2, 4\}$ **62.** $\{3, 4, 1, 2\}$
63. $\{-4, 4, -10, 10\}$ **64.** $\{2, -5\}$ **65.** $\{0, 4, -5\}$
66. $\{-7, 2, 7, -1\}$ **67.** $\{-3, 3, -2, 2\}$
68. $\{-5, 5\}$

Page 75

ORALS **1.** T **2.** F **3.** F **4.** T **5.** T **6.** F

EXERCISES **1.** $\{x|3<x<6\}$
2. $\{x|x<-4 \text{ or } x>-1\}$ **3.** $\{x|-5\le x\le 2\}$
4. $\{x|-2<x<4\}$ **5.** $\{x|x\le 1 \text{ or } x\ge 3\}$

6. $\{x|-6 < x < -2\}$ **7.** $\{x|x < -3 \text{ or } x > 6\}$
8. $\{x|2 \le x \le 5\}$ **9.** $\{x|x < -5 \text{ or } x > -3\}$
10. $\{x|-3 \le x \le 3\}$ **11.** $\{x|x < -5 \text{ or } x > 5\}$
12. $\{x|-1 < x < 1\}$ **13.** $\{x|0 < x < 5\}$
14. $\{x|x \le -3 \text{ or } x \ge 0\}$ **15.** $\{x|x < 0 \text{ or } x > 6\}$
16. $\{x|x < -2 \text{ or } x > 2\}$ **17.** $\{x|-6 \le x \le 6\}$
18. $\{x|-7 < x < 7\}$ **19.** $\{x|-4 < x < 3\}$
20. $\{x|x < 2 \text{ or } x > 4\}$ **21.** $\{x|x \le -3 \text{ or } x \ge -1\}$
22. $\{x|x \le 0 \text{ or } x \ge 2\}$ **23.** $\{x|-5 < x < 0\}$
24. $\{x|x < 0 \text{ or } x > 7\}$ **25.** $\{x|x < 1 \text{ or } 4 < x < 7\}$
26. $\{x|2 < x < 4 \text{ or } x > 8\}$
27. $\{x|x \le -3 \text{ or } 2 \le x \le 5\}$
28. $\{x|-5 \le x \le -2 \text{ or } x \ge 3\}$
29. $\{x|x < 2 \text{ or } x > 2\}$ **30.** $\{x|x = 3\}$
31. $\{x|x < -2 \text{ or } x > 4\}$ **32.** $\{x|-3 < x < 6\}$
33. $\{x|x \le -5 \text{ or } x \ge 5\}$ **34.** $\{x|0 < x < 8\}$

Page 78

ORALS **1.** 14, 15, 16, 17 **2.** −1, 0, 1, 2 **3.** −7, −6, −5, −4 **4.** $x, x + 1, x + 2, x + 3$ **5.** 15, 17, 19, 21 **6.** −1, 1, 3, 5 **7.** −7, −5, −3, −1 **8.** $n, n + 2, n + 4, n + 6$ **9.** 14, 16, 18, 20 **10.** −2, 0, 2, 4 **11.** −8, −6, −4, −2 **12.** $n, n + 2, n + 4, n + 6$ **13.** $x + (x + 2)$ **14.** $x(x + 4)$ **15.** $(x + 2)^2 + 4$ **16.** $2x - 5(x + 4)$ **17.** $n^2 + (n + 1)^2 + (n + 2)^2$ **18.** $4(n + 1) + 6(n + 2)$ **19.** $n(n + 1) - 2$ **20.** $(n + 2)^2 - n^2$

EXERCISES **1.** 57, 58, 59 **2.** 24, 26, 28 **3.** 23, 24, 25 **4.** 45, 47, 49 **5.** 41, 42, 43 **6.** 34, 36, 38 **7.** 32, 33, 34 **8.** 51, 53, 55 **9.** 64, 65, 66 **10.** 60, 62, 64 **11.** 70, 71, 72 **12.** 71, 73, 75 **13.** 5, 6, 7; −8, −7, −6 **14.** 4, 6, 8; −6, −4, −2 **15.** 6, 7, 8; −8, −7, −6 **16.** 7, 9, 11; −9, −7, −5 **17.** 4, 5, 6 **18.** −8, −6, −4 **19.** 7, 8, 9; −7, −6, −5 **20.** 5, 7, 9; −1, 1, 3 **21.** 18, 19, 20 **22.** 16, 18, 20 **23.** 8, 9, 10; −10, −9, −8 **24.** 3, 5, 7; −7, −5, −3 **25.** 3, 4, 5; 9, 10, 11 **26.** 8, 10, 12 **27.** any 3 consec. integers **28.** no solution

Page 81

ORALS **1.** no **2.** yes **3.** no **4.** yes **5.** no **6.** no

EXERCISES **1.** $3(x + 3)(x + 4)$ **2.** $3(2y + 1)(2y - 1)$ **3.** $2(x + 4)^2$ **4.** $2(n + 5)(n - 3)$ **5.** $2(3x + 5)(3x - 5)$ **6.** $3(n + 5)^2$ **7.** $x(x - 4)(x - 2)$ **8.** $y(2x + 3)(2x - 3)$ **9.** $a(2a + b)^2$ **10.** $x(x - 6)$ **11.** $x^2(x + y)(x - y)$ **12.** $a(5x - 2)^2$ **13.** $2(3x - 1)(x - 2)$ **14.** $2(5x + 2y)(3x - y)$ **15.** $25(2x + y)^2$ **16.** $4a(2a - 5)(a + 2)$ **17.** $3a(a + 5b)(a - 5b)$ **18.** $2x(x - 1)^2$ **19.** $-1(2x - 3)(2x + 1)$ **20.** $-1(y + 7)(y + 2)$ **21.** $-1(n - 1)^2$ **22.** $-1(x - 4)^2$ **23.** $-3(x + 3)(x - 3)$ **24.** $-2(5x + 3)(4x - 1)$ **25.** $-1(3n + 5)^2$ **26.** $-c(2c + 1)(2c - 1)$ **27.** $-a(2c + 1)^2$ **28.** $-x(x - 3y)^2$ **29.** $a(x - 5a)(x + 2a)$ **30.** $r(\pi r - 1)$

Page 82

1. $15x^5$ **2.** $-8x^5$ **3.** a^4b^5 **4.** $-10x^3y^4$ **5.** x^{12} **6.** $-x^{12}$ **7.** $25c^2$ **8.** $27a^3b^{12}$ **9.** −144 **10.** 576 **11.** 28 **12.** 30 **13.** no **14.** no **15.** yes **16.** 5th **17.** trinom. **18.** monom. **19.** binom. **20.** polynom. with 4 terms **21.** $4x^3 + 2x^2 - 5x + 4$ **22.** $2x^2y + 2xy + 5xy^2$ **23.** $6x^2 - 9x$ **24.** $x^3 - 4x - 2$ **25.** $2x^3 + 2xy^2 + 6x^2y$ **26.** $6x^2 - 31x + 35$ **27.** $4x^2 - 9y^2$ **28.** $3x^3 + 13x^2 - 3x - 20$ **29.** $4x^2 - 7xy - 6x + 3y^2 + 6y$ **30.** $3(2x^2 + 3x - 1)$ **31.** $n(3n^2 - 2n + 1)$ **32.** $4x^2(2x - 3)$ **33.** $-1(x^2 - 7x + 3)$ **34.** $16x^2 + 24xy + 9y^2$ **35.** $(x - 7)(x + 2)$ **36.** $(6n + 5)(n + 1)$ **37.** $(3y - 2)(3y - 4)$ **38.** $(3x + 4y)(3x - 4y)$ **39.** $(3n + 2)^2$ **40.** $(2x - 3y)(x - 4y)$ **41.** $4(x + 1)^2$ **42.** $3(2y + 3)(2y - 3)$ **43.** $x(x - 5)(x + 3)$ **44.** $3x(3x - 1)(2x - 1)$ **45.** $-1(x - 5)^2$ **46.** $-1(2x - 7)(x + 2)$ **47.** $\left\{\frac{2}{3}, -4\right\}$ **48.** $\left\{2\frac{1}{2}, 3\right\}$ **49.** $\{0, -7\}$ **50.** $\{x|-4 < x < 2\}$ **51.** $\{x|x \le -3 \text{ or } x \ge 5\}$ **52.** $\{x|-5 < x < 5\}$ **53.** 65; 67; 69 **54.** 4, 5, 6; −7, −6, −5

Page 83

1. $6a^4b^4$ **2.** $16n^4$ **3.** $25x^4y^6$ **4.** −32 **5.** −512 **6.** −36 **7.** yes **8.** no **9.** no **10.** 3 **11.** binom. **12.** trinom. **13.** monom. **14.** $2x^3 + 4x^2 + 7x - 10$ **15.** $-3xy^2 + 9xy + 2x^2y$ **16.** $-3x^2 - 2x + 8$ **17.** $-9y^2 - 14y + 3$ **18.** $6x^3 - 3x^2 + 3x$ **19.** $10x^2 + 11xy - 6y^2$ **20.** $2x^3 - 5x^2 - 22x - 15$ **21.** $5(x^2 - 4x + 2)$ **22.** $n(2n^2 - 6n + 3)$ **23.** $3y(2y^2 - 2y + 3)$

24. $-1(x^2 - 9x + 4)$ **25.** $4x^2 - 20x + 25$
26. $(5x + 2)(3x - 1)$ **27.** $(x + 6)^2$
28. $(3x - 2y)(2x - y)$ **29.** $(4x + y)(4x - y)$
30. $5(n + 3)(n - 3)$ **31.** $2x(y - 3)(y + 2)$
32. $\{3, 7\}$ **33.** $\left\{1\frac{1}{2}, -4\right\}$ **34.** $\{-6, 4\}$
35. $\{x|-2 < x < 5\}$ **36.** $\{x|x \leq -1 \text{ or } x \geq 3\}$
37. 24; 25; 26 **38.** 4, 6, 8; -10, -8, -6

Page 86

1. yes **2.** yes **3.** no **4.** no **5.** -4 **6.** 0 **7.** 0 **8.** does not exist **9.** does not exist **10.** does not exist **11.** 3 **12.** 0 **13.** $\frac{7}{8}$ **14.** does not exist **15.** does not exist **16.** 5 **17.** 0 **18.** $-\frac{1}{6}$ **19.** does not exist **20.** does not exist **21.** 0 **22.** -2 **23.** 3 **24.** -5; 5 **25.** 3; 4

Page 89

1. $\frac{x^2}{y^4}$ **2.** $\frac{2b^4}{3a^4}$ **3.** $\frac{c^3}{a^2}$ **4.** $\frac{-2x^3}{3y^3}$ **5.** $\frac{3}{2b^6c^3}$ **6.** $\frac{2xy^2}{3z}$
7. $\frac{3x+4}{2}$ **8.** $\frac{2}{2a-1}$ **9.** $\frac{2x-3}{a}$ **10.** $\frac{3}{5}$ **11.** $\frac{a+2}{a-3}$
12. $\frac{2}{3}$ **13.** -2 **14.** $\frac{3}{-2}$ **15.** $\frac{1}{-2}$ **16.** $\frac{n+5}{3}$
17. $\frac{x+2}{5}$ **18.** $\frac{a+4}{2}$ **19.** $\frac{3}{3n+1}$ **20.** $\frac{x+4}{2}$
21. $\frac{5}{4a-5}$ **22.** $\frac{x-5}{x-1}$ **23.** $\frac{a+3}{a-6}$ **24.** $\frac{c-1}{c-6}$
25. $\frac{5}{-1(x-4)}$ **26.** $\frac{a-3}{-1(a-2)}$ **27.** $\frac{c+6}{-1(c-5)}$
28. $\frac{-2(x-4)}{5}$ **29.** $\frac{3(a-2)}{a+3}$ **30.** $\frac{n-2}{-2(n-5)}$
31. $\frac{a+8}{4(a-1)}$ **32.** $\frac{-5(x+3)}{x-7}$ **33.** $\frac{3(c-2)}{c+6}$
34. $x^{4a}y^{2b}$ **35.** $x^{2c+3}y^{3d-4}$ **36.** $x^{3a+2b}y^{a-b}$
37. $\frac{x^2+2x+4}{3}$ **38.** $x + 4$ **39.** $\frac{x^2+3x+9}{4}$

Page 92

ORALS **1.** $\frac{6}{55}$ **2.** $\frac{5}{14}$ **3.** $\frac{y}{6}$ **4.** $\frac{6}{7}$ **5.** $\frac{3x}{y}$ **6.** 1
7. $\frac{4}{3}$ **8.** $\frac{1}{7}$ **9.** $\frac{5y}{-2x}$ **10.** $\frac{x^2-9}{x+2}$ **11.** $x - y$
12. $\frac{1}{3x+6}$ **13.** no reciprocal **14.** -1 **15.** 1

EXERCISES **1.** $\frac{3y^2}{4x^2}$ **2.** $\frac{2}{3a^2yz}$ **3.** $\frac{3b}{2c}$ **4.** $-\frac{1}{6}$
5. $-\frac{6}{c}$ **6.** $-\frac{15}{8}$ **7.** $\frac{3(x-3)}{2}$ **8.** $\frac{3(c+5)}{-2}$
9. $\frac{3c}{x-4}$ **10.** $\frac{6y^2}{5x^3}$ **11.** $\frac{-4ac}{3b^4}$ **12.** $\frac{3}{4n}$ **13.** $\frac{5x^3}{-6}$
14. $\frac{2(x-5)}{-3}$ **15.** $\frac{3}{5x(x-4)}$ **16.** $\frac{2(x-3)}{3(x+5)}$
17. $\frac{n+6}{2(n+4)}$ **18.** $\frac{2(3c+1)}{3c^2}$ **19.** $\frac{2(x+1)}{-1(2x+3)}$
20. $\frac{(a-2)(a-1)}{3}$ **21.** $\frac{4x}{-1(x-5)}$ **22.** $2(x - 3y)$
23. $\frac{2x(x+2y)}{3}$ **24.** $\frac{3(2n+3)}{4a^3(n+3)}$ **25.** $\frac{2(3x-1)}{a(2x+1)}$
26. $\frac{x-2y}{8(x+2y)}$ **27.** $\frac{-1(c+d)}{2(a+b)}$ **28.** $2n^2(n-1)$
29. $\frac{1}{2(5x-4)}$ **30.** $\frac{x^{2a}}{y^{5a}}$ **31.** $\frac{(x^2+2x+4)(x+4)}{3}$
32. $\frac{(x^a-3)(x^{2a}-3)}{6}$ **33.** $\frac{4(x^a-y^b)}{3(2x^c-5)}$

Page 95

1. $\frac{a}{2}$ **2.** $2x$ **3.** $\frac{5a-3}{4}$ **4.** $\frac{3}{y}$ **5.** $\frac{2}{b}$ **6.** $\frac{3c}{x^2y}$ **7.** $\frac{3x}{4}$
8. $\frac{17}{5a}$ **9.** $\frac{3}{y}$ **10.** $\frac{3}{n-4}$ **11.** $\frac{1}{2}$ **12.** $\frac{1}{2}$ **13.** $\frac{3}{5}$ **14.** $\frac{4}{3}$
15. $\frac{3(n+1)}{n+4}$ **16.** $\frac{2}{x-4}$ **17.** $\frac{3}{x-5}$ **18.** $\frac{4}{a+5}$
19. $\frac{n+1}{n+3}$ **20.** $\frac{1}{n(n-4)}$ **21.** $\frac{2}{x+2}$ **22.** $\frac{2}{c-5}$
23. $\frac{1}{3(a+6)}$

Page 96

Answers may vary. The last fraction is always equal to .618 to three decimal places.

Page 97

$\frac{123}{199} \doteq .618$

Page 100

ORALS **1.** 24 **2.** x^2y^3 **3.** $12(x + 2)(x + 3)$ **4.** $2(x + 3)(x + 3)$ **5.** $a - 2$ **6.** $3(x + 2)(x - 2)$

EXERCISES **1.** $\frac{11c}{6}$ **2.** $\frac{9x+10y}{24}$ **3.** $\frac{12n-1}{35}$
4. $\frac{3b^2-2}{ab}$ **5.** $\frac{7y+4x}{x^2y}$ **6.** $\frac{9x+2y}{12x^2y^2}$ **7.** $\frac{19x-6}{30}$
8. $\frac{4ab^2-2b^3+3a}{a^2b^3}$ **9.** $\frac{3y^3+24+x^2y}{6xy^2}$

10. $\frac{26x}{15(x+4)}$ 11. $\frac{2a^2-9a+8}{2(a-3)(a-4)}$
12. $\frac{4n^2-8n+6}{(3n-2)(4n+1)}$ 13. $\frac{4x^2-12x+1}{4x}$
14. $\frac{2a^2+4a+3}{a+2}$ 15. $\frac{2n^2-3}{n-3}$ 16. $\frac{6x-8}{(x+2)(x-2)}$
17. $\frac{11a+13}{2(a+3)(a-3)}$ 18. $\frac{4n^2-10n+9}{3(n+4)(n-4)}$
19. $\frac{4}{3}$ 20. $\frac{5}{4}$ 21. $\frac{3x}{2}$ 22. $\frac{x+5}{x}$ 23. $\frac{a-2}{a}$
24. $\frac{n-1}{n}$ 25. $\frac{5x+1}{3(x-2)(x-4)}$
26. $\frac{11x+13}{2(x+3)(x+5)}$ 27. $\frac{3x^2-6x+1}{(x+3)(x-1)(x-3)}$
28. $\frac{3x^2+14x-2}{(x+6)(x-4)(x+4)}$
29. $\frac{x^2-7x}{(2x-5)(x+2)(x-1)}$
30. $\frac{2x^2+2}{(3x+1)(x-1)(x+1)}$
31. $\frac{8n+1}{(n+1)(n-1)}$ 32. $\frac{4a+17}{(a-2)(a+2)}$
33. $\frac{12a+3b}{5(a+b)(a-b)}$ 34. $\frac{6a^2-7ab+2b^2}{(a+b)(2a-3b)}$
35. $\frac{7x+2y}{3(x-y)(x-y)}$ 36. $\frac{13x-8y}{2(3x+2y)(3x-2y)}$

Page 103

ORALS 1. T 2. T 3. F 4. T 5. T 6. T 7. F 8. T 9. T 10. F 11. T 12. T

EXERCISES 1. $\frac{2x}{3}$ 2. $\frac{2}{c}$ 3. $\frac{2}{x+3}$ 4. $\frac{3}{5}$
5. $\frac{2(a-3)}{a-4}$ 6. 2 7. $\frac{5n}{24}$ 8. $\frac{9+10cd}{12c^2d^2}$
9. $\frac{2x^2-3x}{2(x-4)(x-3)}$ 10. $\frac{x^2-5x-6}{(x+3)(x-3)}$ 11. $\frac{4}{x-5}$
12. 2 13. $\frac{13}{x-4}$ 14. 4 15. $\frac{3}{a-4}$ 16. $\frac{5}{n+6}$
17. $\frac{4n^2-2n-3}{3(n+2)(n-2)}$ 18. $\frac{a+39}{2(a+5)(a-5)}$
19. $\frac{3}{x-3}$ 20. $\frac{5}{n-5}$ 21. $\frac{a+9}{2(a+3)(a-1)}$
22. $\frac{5c-5}{3(c+2)(c-4)}$ 23. $\frac{4}{x-5}$ 24. $\frac{3x+1}{(x+3)(x-3)}$
25. 2 26. $\frac{4x}{x-5}$ 27. $\frac{7yz-3xz-5xy}{xyz}$
28. $\frac{3x^2-10x-24}{x(x+4)(x+2)}$ 29. $\frac{9x-51}{4(x-3)(x+3)}$

30. $\frac{-82x-26}{15(x-2)(x+2)}$ 31. $\frac{a+7}{(a+2)(a+2)(a+3)}$
32. $\frac{c^2-6c}{(3c-1)(c+2)(c-2)}$

Page 105

1. $\frac{1}{x+5}+\frac{1}{x+3}$ 2. $\frac{2}{x+1}+\frac{1}{x+7}$
3. $\frac{5}{4(x+1)}+\frac{11}{4(x+5)}$, or $\frac{1.25}{x+1}+\frac{2.75}{x+5}$

Page 106

1. yes 2. no 3. yes 4. $\frac{1}{6}$ 5. $\frac{3}{10}$ 6. does not exist
7. 0 8. 3 9. 2, 3 10. $-\frac{2a^3d}{3b^2}$ 11. $\frac{4x-5}{3}$ 12. $\frac{3}{5}$
13. $\frac{x-5}{2x-1}$ 14. $\frac{a-5}{3}$ 15. $\frac{1}{-5}$ 16. $\frac{a+4}{2a-1}$
17. $\frac{1}{x^2-9}$ 18. $\frac{y}{-3x}$ 19. $\frac{2b^3c^2}{-9a}$ 20. $\frac{4(x+1)}{3}$
21. $\frac{3(3c-2)}{4}$ 22. $\frac{3y^2z}{-8x^2}$ 23. $\frac{5(x+y)}{6(c+2)}$
24. $\frac{1}{2(n+4)}$ 25. $\frac{2c}{xy}$ 26. $\frac{3}{5}$ 27. $\frac{5}{x-3}$
28. $\frac{15x+4y^2}{18x^2y}$ 29. $\frac{6a^2-16a}{3(a-2)(a-4)}$
30. $\frac{n^2+9n+6}{n+3}$ 31. $\frac{4}{x-4}$ 32. $\frac{7}{2n+5}$
33. $\frac{11x+7}{3(x+5)(x-3)}$ 34. $\frac{5n^2+16n}{(n+2)(n+2)(n+5)}$
35. $\frac{5}{3}$ 36. $\frac{11}{x-4}$ 37. $\frac{8}{2x-3}$ 38. $\frac{2x-26}{(x+2)(x-4)}$
39. $\frac{4x+18}{(x-5)(x+5)}$ 40. $\frac{2x-1}{3(x-4)(x+4)}$

Page 107

1. yes 2. yes 3. no 4. 4 5. $\frac{7}{3}$ 6. does not exist
7. 0 8. 2 9. 6, −6 10. $\frac{a^2}{2b^3}$ 11. $\frac{x+5}{2x+1}$ 12. −4
13. $\frac{3y+2}{x-5}$ 14. $\frac{1}{x^2-25}$ 15. $\frac{3b^3}{2a^2}$ 16. $\frac{5}{6}$
17. $\frac{3(x-3)}{2(c-5)}$ 18. $\frac{2(n-5)}{3}$ 19. $\frac{2(n-2)}{5}$
20. $\frac{a-5}{4}$ 21. $\frac{5}{3}$ 22. $\frac{3}{x-5}$ 23. $\frac{9b+10a}{12a^2b^2}$
24. $\frac{6a+3}{2(a-3)(a+4)}$ 25. $\frac{x^2-x}{x+2}$
26. $\frac{3c+2}{2(c-2)(c-3)}$ 27. 3 28. $\frac{2}{x-7}$
29. $\frac{3x+7}{(x-1)(x+1)}$

Page 108

Since $a = b$, $a - b = 0$; thus step 6 is invalid since you can not divide by zero.

Page 112

1. 1 **2.** -2 **3.** 2 **4.** 4 **5.** $\frac{1}{2}$ **6.** $-\frac{5}{8}$ **7.** $\{1, 6\}$ **8.** $\left\{-1\frac{1}{2}, 2\right\}$ **9.** $\{1, 4\}$ **10.** $\{-6\}$ **11.** $\{-8\}$ **12.** $\{3\}$ **13.** 11 **14.** 4 **15.** 2 **16.** $1\frac{3}{5}$ **17.** $\frac{3}{10}$ **18.** $-\frac{1}{7}$ **19.** 2; 5 **20.** -5; 3 **21.** 2 **22.** 300 **23.** 12 **24.** 240 **25.** 40 **26.** 95.5 **27.** 7 **28.** $\{5\}$ **29.** $\left\{\frac{7}{3}\right\}$ **30.** $\{7\}$ **31.** $\{4\}$ **32.** $\{-6, 5\}$ **33.** $\{-3, 4\}$ **34.** -7 **35.** $\frac{3}{2}$ **36.** no solution **37.** all numbers except -4 and 3

Page 115

1. $\frac{1}{15}$; $\frac{4}{15}$; $\frac{x}{15}$ **2.** $\frac{1}{x}$; $\frac{4}{x}$ **3.** 6 **4.** $4\frac{4}{5}$ **5.** 15 **6.** 12 **7.** $3\frac{3}{7}$ **8.** 30 **9.** 42 **10.** 4 **11.** $6\frac{6}{11}$ **12.** 6

Page 118

1. $\frac{13}{44}$ **2.** $\frac{19}{16}$ **3.** $\frac{51}{32}$ **4.** $\frac{5y + 2x}{3y + 4x}$ **5.** $\frac{b - a}{b + a}$ **6.** $\frac{c^2d - d}{c + cd^2}$ **7.** $\frac{4xy - 6x^2}{7y^2 + 2xy}$ **8.** $\frac{20b + 18a}{2b - 9a}$ **9.** $\frac{4y + 20xy}{10xy - 3x}$ **10.** $\frac{8 + 9x}{5x + 6}$ **11.** $\frac{2b^2 - 20a^2b}{7ab^2 + 15a^2}$ **12.** $\frac{12ac^2 - 3}{2c - 36ac^2}$ **13.** $\frac{7x - 6}{8x - 10}$ **14.** $\frac{2a - 36}{7a - 14}$ **15.** $x + 3$ **16.** $\frac{5x + 4}{9x + 2}$ **17.** $\frac{6a - 2}{5a + 6}$ **18.** $\frac{-4c - 4}{8c - 5}$ **19.** 1 **20.** $\frac{7a}{8a + 26}$ **21.** $\frac{3c - 55}{7c}$ **22.** $\frac{x + 5}{x + 4}$ **23.** $\frac{a + 2}{a - 3}$ **24.** $\frac{2c + 3}{3c + 2}$ **25.** $\frac{x + 4y}{x + y}$ **26.** $\frac{a - 3b}{a + 2b}$ **27.** $\frac{3}{4}$

Page 121

1. $x = \frac{b + c}{a}$ **2.** $x = \frac{c - a}{b}$ **3.** $x = \frac{c + ab}{a}$ **4.** $x = ab$ **5.** $x = a$ **6.** $x = \frac{bc}{a}$ **7.** $x = \frac{b}{a + c}$ **8.** $x = \frac{b + 2}{a - 3}$ **9.** $x = \frac{2b}{3a}$ **10.** $x = \frac{b}{a + 1}$ **11.** $x = b$ **12.** $x = \frac{c + 3}{a - b}$ **13.** $x = \frac{ac + d}{ab}$ **14.** $x = \frac{ab + cd}{a - c}$ **15.** $x = \frac{-ab - c}{2a - cn}$ **16.** $x = \frac{ac - cd}{b}$ **17.** $x = a - bc$ **18.** $x = \frac{-c}{ad - b}$ **19.** $x = \frac{ab}{3}$ **20.** $x = \frac{ab}{a + b}$ **21.** $x = \frac{3ab}{4a + 2b}$ **22.** $x = \frac{bc - a}{1 + c}$ **23.** $x = \frac{3a}{2 - a}$ **24.** $x = \frac{b}{ad - c}$ **25.** $p = \frac{i}{rt}$; 600 **26.** $w = \frac{p - 2l}{2}$; 17 **27.** $h = \frac{2A}{b}$; 4.1 **28.** $C = \frac{5F - 160}{9}$; 20 **29.** $x = \frac{ab}{c - a}$ **30.** $x = \frac{ad - bc}{d - c}$ **31.** $x = \frac{a^2 - 2c}{a - 2b}$ **32.** $h = \frac{A}{a + b + c}$; 9 **33.** $h = \frac{3V}{B}$; 11 **34.** $c = \frac{ab}{b - a}$; 24 **35.** $m = \frac{T}{g - f}$; 7 **36.** $x = a$ **37.** $x = 8a - 14$ **38.** $x = \frac{bc + de}{a + b - d}$ **39.** c; $-d$ **40.** $2a$; $3a$ **41.** $5a$; $-5a$ **42.** $-3a$ **43.** $-a$; $-b$ **44.** $c + d$; $-c - d$ **45.** $a - b$; $b - a$

Page 123

1. 80 cm³ **2.** 30 cm³ **3.** 450 L **4.** 20 dL **5.** 40 cm³ **6.** 40 mL

Page 126

ORALS **1.** $\frac{5}{1}$ **2.** $\frac{3}{100}$ **3.** $\frac{12}{5}$ **4.** $\frac{-3}{1}$ **5.** $\frac{-43}{10}$ **6.** $\frac{0}{1}$ **7.** $\frac{-7}{2}$

EXERCISES **1.** .25; term. **2.** $.\overline{6}$; rep. **3.** .625; term. **4.** $.\overline{4}$; rep. **5.** 1.6; term. **6.** $.6\overline{1}$; rep. **7.** $\frac{7}{9}$ **8.** $\frac{32}{99}$ **9.** $\frac{8}{9}$ **10.** $\frac{47}{99}$ **11.** $\frac{8}{99}$ **12.** $\frac{31}{90}$ **13.** $\frac{65}{90}$ **14.** $\frac{26}{90}$ **15.** $\frac{48}{90}$ **16.** $\frac{7}{90}$ **17.** $\frac{74}{9}$ **18.** $\frac{-43}{99}$ **19.** $\frac{-4}{9}$ **20.** $\frac{267}{99}$ **21.** $\frac{393}{90}$ **22.** 1 **23.** 3 **24.** -5 **25.** $\frac{1}{2}$ **26.** $\frac{-225}{90}$ **27.** $\frac{361}{999}$ **28.** $\frac{-438}{999}$ **29.** $\frac{346}{990}$ **30.** $\frac{4{,}517}{990}$

Page 129

ORALS **1.** yes **2.** no **3.** yes **4.** yes **5.** yes **6.** no

EXERCISES **1.** $5x + 6$ **2.** $3y - 2$, rem. 2 **3.** $n + 7$ **4.** $2a - 3$, rem. -4 **5.** $3n^3 + n^2 - 2$ **6.** $2x^2 + 5x - 2$, rem. 3 **7.** $4n - 5$ **8.** $3x^2 - x + 2$, rem. -3 **9.** $a^2 - 3a - 4$ **10.** $3y^3 + 2y^2 + 3y + 2$, rem. -2 **11.** yes; $3x^2 - x - 2$, rem. 0 **12.** no; $2n^2 - 5n + 3$, rem. -3 **13.** no; $2n^2 + n - 1$, rem. 2 **14.** yes; $a^2 - a + 1$, rem. 0 **15.** $2n^2 - 4n + 1$, rem. 6 **16.** $3x^2 + 2x + 12$ **17.** $3x + 2$, rem. -3 **18.** $3a^3 - a^2 + 1$, rem. -3 **19.** 6; 6; yes **20.** 100; 100; yes

Page 130

1. 1 **2.** 4 **3.** 27 **4.** 4 **5.** -5 **6.** $-1\frac{1}{4}$ **7.** 300 **8.** 6 **9.** 80 **10.** 2 **11.** -2, 5 **12.** 1 **13.** $\{-1, 4\}$ **14.** $\{-4\}$ **15.** $\{6\}$ **16.** $\{4\}$ **17.** $1\frac{7}{8}$ **18.** 18 **19.** $\frac{2y^2 + 9x^2y}{5xy^2 - 12x^2}$ **20.** $\frac{12a^3b - 2b}{a^2 + 18a^2b^2}$ **21.** $\frac{2n + 4}{8n - 19}$ **22.** $x = \frac{b + d}{a - c}$ **23.** $x = \frac{ab + 3c}{a - 2c}$ **24.** $b = \frac{ac}{c - a}$; 6 **25.** $l = \frac{p - 2w}{2}$; 7 **26.** $\frac{4}{9}$ **27.** $\frac{72}{99}$ **28.** $\frac{38}{90}$ **29.** $\frac{-10}{3}$ **30.** $\frac{7}{100}$ **31.** $3n^2 + 6n + 1$, rem. 5 **32.** $2x - 3$ **33.** yes; $3x^2 + 2x - 2$, rem. 0 **34.** no; $n^2 + n + 1$, rem. -6

Page 131

1. 3 **2.** 5 **3.** -7 **4.** 7 **5.** 9 **6.** -4, 3 **7.** $\{1, 6\}$ **8.** $\{-4\}$ **9.** $4\frac{4}{9}$ **10.** 12 **11.** $\frac{36a^2 + 3}{2a - 24a^2}$ **12.** $\frac{7x - 15}{6x - 6}$ **13.** $x = \frac{b + 3d}{2a - c}$ **14.** $x = \frac{ac}{ab - d}$ **15.** $c = \frac{2A - hb}{h}$; 8 **16.** $c = \frac{T - 2ab}{2a + 2b}$; 3 **17.** $\frac{59}{99}$ **18.** $\frac{34}{90}$ **19.** $\frac{-15}{2}$ **20.** $\frac{451}{100}$ **21.** $3n - 9$, rem. 5 **22.** yes; $x^2 + 3x - 2$, rem. 0

Page 132

1. yes **2.** no **3.** yes **4.** no

Page 136

1. $\{-6, 6\}$ **2.** $\{-\sqrt{7}, \sqrt{7}\}$ **3.** $\{-10, 10\}$ **4.** $\{-\sqrt{2}, \sqrt{2}\}$ **5.** yes; $4 = 2^2$ **6.** no **7.** yes; $81 = 9^2$ **8.** no **9.** 5 **10.** not possible **11.** -2 **12.** not possible **13.** $2 < \sqrt{8} < 3$ **14.** $1 < \sqrt{3} < 2$ **15.** $4 < \sqrt{20} < 5$ **16.** $3 < \sqrt{15} < 4$ **17.** $\{-20, 20\}$ **18.** $\{-\sqrt{35}, \sqrt{35}\}$ **19.** $\{-12, 12\}$ **20.** $\{-\sqrt{70}, \sqrt{70}\}$ **21.** -11 **22.** undefined **23.** -30 **24.** undefined **25.** $5 < \sqrt{35} < 6$ **26.** $7 < \sqrt{61} < 8$ **27.** $6 < \sqrt{39} < 7$ **28.** $9 < \sqrt{97} < 10$

Page 140

ORALS **1.** 5.4 **2.** 7.4 **3.** 2.8 **4.** 4.0

EXERCISES **1.** 4.1 **2.** 3.2 **3.** 6.7 **4.** 5.5 **5.** irrat. **6.** irrat. **7.** rat. **8.** irrat. **9.** rat. **10.** irrat. **11.** 6.3 **12.** 7.7 **13.** 8.7 **14.** 9.4 **15.** rat. **16.** irrat. **17.** rat., $.242\overline{242}$ **18.** irrat. **19.** rat **20.** irrat. **21.** 6.32 **22.** 7.68 **23.** 8.72 **24.** 9.38 **25.** 6.633 **26.** 7.348

Page 144

ORALS **1.** 12 **2.** 10 **3.** $3\sqrt{7}$ **4.** 11 **5.** c **6.** x **7.** x^5 **8.** $x^4\sqrt{x}$

EXERCISES **1.** $3\sqrt{3}$ **2.** $-3\sqrt{5}$ **3.** $-2\sqrt{6}$ **4.** $3\sqrt{10}$ **5.** $4\sqrt{2}$ **6.** $-4\sqrt{5}$ **7.** $-6\sqrt{10}$ **8.** $4\sqrt{6}$ **9.** $8\sqrt{5}$ **10.** $-10\sqrt{6}$ **11.** $-15\sqrt{7}$ **12.** $30\sqrt{5}$ **13.** $3\sqrt{2}$; 4.2 **14.** $2\sqrt{5}$; 4.5 **15.** $4\sqrt{3}$; 6.9 **16.** $3\sqrt{5}$; 6.7 **17.** $\sqrt{77}$ **18.** $6\sqrt{21}$ **19.** $\sqrt{10ab}$ **20.** $-12\sqrt{5mn}$ **21.** 8 **22.** 6 **23.** $2\sqrt{3}$ **24.** $5\sqrt{2}$ **25.** $3y^3$ **26.** $a^2\sqrt{7}$ **27.** $2b\sqrt{3}$ **28.** $2x^5\sqrt{6}$ **29.** $x^4\sqrt{x}$ **30.** $4y^3\sqrt{y}$ **31.** $a^2\sqrt{5a}$ **32.** $2b\sqrt{2b}$ **33.** x^2y^3 **34.** $6ab^4$ **35.** $2c^5d^2\sqrt{6}$ **36.** $x^3y^2\sqrt{y}$ **37.** $ab\sqrt{a}$ **38.** $3m^4n^4\sqrt{2n}$ **39.** $xy^3\sqrt{xy}$ **40.** $2a^2b\sqrt{3ab}$ **41.** $6x^2$ **42.** $3y^4\sqrt{2}$ **43.** $10x^2\sqrt{6}$ **44.** $2x^2y^2\sqrt{6}$ **45.** 28 **46.** a^2b **47.** $9c^2d$ **48.** $12x^2y$ **49.** $x + 5$ **50.** $a - 3$ **51.** $4x + 4$ **52.** $18a + 45$ **53.** x^m **54.** $x^m\sqrt{x}$ **55.** x^{2m} **56.** $x^{3m}\sqrt{x}$ **57.** x^{2m} **58.** x^{3m+1} **59.** $x^{2m}y^n\sqrt{y}$ **60.** $x^{3m+2}y^{2n+1}\sqrt{y}$

Page 147

1. $-4\sqrt{10} + 5\sqrt{5}$ **2.** $2\sqrt{5} - 6\sqrt{2}$ **3.** $5\sqrt{5}$ **4.** $15\sqrt{2}$ **5.** $3\sqrt{n}$ **6.** $-2\sqrt{c}$ **7.** $6\sqrt{3} - 3\sqrt{5}$ **8.** $-8 - 6\sqrt{2}$ **9.** $4\sqrt{3} + 3$ **10.** $-10 + \sqrt{10}$ **11.** $6 - 6\sqrt{10}$ **12.** $-20 - 6\sqrt{5}$ **13.** $4 - 2\sqrt{3}$ **14.** $-10\sqrt{5} + 20$ **15.** $2 - 3\sqrt{3}$ **16.** $29 - 5\sqrt{35}$ **17.** $34 - 10\sqrt{14}$ **18.** $46 - 9\sqrt{30}$ **19.** 2 **20.** 53 **21.** $8 + 2\sqrt{15}$ **22.** $25 - 6\sqrt{14}$ **23.** $22 - 4\sqrt{30}$ **24.** $68 + 16\sqrt{15}$ **25.** $5\sqrt{2} + 5\sqrt{3}$ **26.** $8\sqrt{2} - 5\sqrt{5}$ **27.** $3\sqrt{3} + 5\sqrt{5}$ **28.** $5c\sqrt{d}$ **29.** $6b\sqrt{ab}$ **30.** $7mn\sqrt{n}$ **31.** $4c + \sqrt{cd} - 5d$ **32.** $15a - 8\sqrt{ab} + b$ **33.** $2c - 5\sqrt{cd} - 3d$

34. $a - b$ **35.** $9c - 4d$ **36.** $25x - y^2$
37. $a + 2\sqrt{ab} + b$ **38.** $9x - 12y\sqrt{x} + 4y^2$
39. $4c + 12\sqrt{cd} + 9d$ **40.** $x + 13 + 7\sqrt{x + 3}$
41. $2x - 14$ **42.** y **43.** $13 + x - 8\sqrt{x - 3}$
44. $2x + 2 - 2\sqrt{x^2 + 2x}$ **45.** $13x - 81 + 12\sqrt{x^2 - 9x}$

Page 151

1. $\sqrt{3}$ **2.** 2 **3.** $2\sqrt{2}$ **4.** $\sqrt{5}$ **5.** $2\sqrt{5}$ **6.** a^3 **7.** $2b$
8. $c^2\sqrt{2c}$ **9.** $3n^2\sqrt{n}$ **10.** $2x^2\sqrt{3}$ **11.** $\frac{\sqrt{6}}{6}$
12. $\frac{2\sqrt{5}}{5}$ **13.** $\frac{\sqrt{6}}{2}$ **14.** $\frac{-5\sqrt{7}}{14}$ **15.** $\frac{-\sqrt{15}}{4}$
16. $\frac{\sqrt{2}}{4}$ **17.** $\frac{5\sqrt{3}}{6}$ **18.** $\frac{\sqrt{14}}{6}$ **19.** $\frac{-3\sqrt{2}}{8}$
20. $\frac{-7\sqrt{15}}{10}$ **21.** $\frac{5\sqrt{a}}{a}$ **22.** $\frac{3\sqrt{b}}{b^2}$ **23.** $\frac{3a\sqrt{c}}{5c}$
24. $\frac{-4\sqrt{d}}{7d^2}$ **25.** $\frac{\sqrt{xy}}{xy^2}$ **26.** $\frac{9 + 3\sqrt{5}}{4}$
27. $\frac{-6 + 2\sqrt{2}}{7}$ **28.** $\frac{\sqrt{30} + 2\sqrt{3}}{6}$
29. $\frac{-\sqrt{14} + 2\sqrt{2}}{3}$ **30.** $\frac{3\sqrt{10} + 3\sqrt{6}}{2}$
31. $\frac{3\sqrt{2}}{2}$; 2.1 **32.** $\frac{10\sqrt{3}}{3}$; 5.8 **33.** $\frac{11\sqrt{2}}{10}$; 1.6
34. $\frac{5\sqrt{3}}{6}$; 1.4 **35.** $\frac{5\sqrt{2}}{4}$; 1.8 **36.** $\frac{\sqrt{6}}{3}$ **37.** $\frac{\sqrt{35}}{7}$
38. $\frac{\sqrt{cd}}{d}$ **39.** $\frac{\sqrt{10ab}}{5b}$ **40.** $\frac{\sqrt{6x - 12}}{6}$
41. $\frac{15\sqrt{2} + 5\sqrt{7}}{11}$ **42.** $\frac{\sqrt{30} - 2\sqrt{6}}{3}$
43. $\frac{6\sqrt{15} + 9\sqrt{6}}{2}$ **44.** $\frac{6\sqrt{x} - 2\sqrt{y}}{9x - y}$
45. $\frac{2\sqrt{a} + b}{4a - b^2}$ **46.** $\frac{a\sqrt{10b}}{2b}$ **47.** $\frac{2\sqrt{3xy}}{3y^2}$
48. $\frac{\sqrt{10mn}}{4n}$ **49.** $\frac{a\sqrt{7ab}}{7b^3}$ **50.** $\frac{\sqrt{9a - 6}}{3}$
51. $\frac{\sqrt{3ab}}{ab^2}$ **52.** $\frac{2\sqrt{2d}}{cd^2}$ **53.** $\frac{2\sqrt{cy}}{x^2y^3}$ **54.** $\frac{\sqrt{5amn}}{m^2n^3}$
55. $\frac{\sqrt{10c + 8}}{2}$

Page 152

1. 113 **2.** 675 **3.** 98 **4.** 376 **5.** .51 **6.** .91
7. 14.23 **8.** 18.03

Page 155

1. 2 **2.** −3 **3.** 4 **4.** −5 **5.** 5 **6.** −10 **7.** 1 **8.** −5
9. −4 **10.** 10 **11.** −1 **12.** 5 **13.** x^4 **14.** x
15. x^3 **16.** x^5 **17.** $\sqrt[3]{20}$ **18.** $\sqrt[3]{6ab}$ **19.** $\sqrt[3]{14c^2}$
20. $\sqrt[3]{12x^2y^2}$ **21.** $\sqrt[4]{15}$ **22.** $\sqrt[4]{8ab}$ **23.** $\sqrt[4]{30c^3}$
24. $\sqrt[4]{21x^2y^3}$ **25.** $7\sqrt[3]{4}$ **26.** $9\sqrt[4]{7}$ **27.** $6\sqrt[3]{c}$
28. $5\sqrt[4]{x}$ **29.** $2\sqrt[3]{2}$ **30.** $2\sqrt[4]{2}$ **31.** $-2\sqrt[3]{3}$
32. $-2\sqrt[4]{3}$ **33.** $a\sqrt[3]{a^2}$ **34.** $b\sqrt[4]{b^3}$ **35.** $x^3\sqrt[3]{x}$
36. $y^2\sqrt[4]{y}$ **37.** $6\sqrt[3]{2}$ **38.** $5\sqrt[4]{2}$ **39.** $4\sqrt[3]{2}$
40. $x^2\sqrt[3]{x}$ **41.** $y^2\sqrt[4]{y}$ **42.** $a^3\sqrt[3]{a^2}$ **43.** $3a\sqrt[3]{a}$
44. $5c^2\sqrt[4]{c}$ **45.** $5x^2\sqrt[3]{x^2}$ **46.** $5a\sqrt[3]{a^2}$ **47.** $5c\sqrt[4]{c^3}$
48. $3x\sqrt[3]{2x}$

Page 160

ORALS **1.** 1 **2.** 5 **3.** 1 **4.** $\frac{1}{2}$ **5.** $\frac{3}{2}$ **6.** $\frac{1}{6}$

EXERCISES **1.** 1 **2.** 4 **3.** 1 **4.** $\frac{1}{125}$ **5.** $\frac{1}{16}$
6. $\frac{3}{16}$ **7.** $\frac{1}{144}$ **8.** 125 **9.** 160 **10.** $\frac{16}{3}$ **11.** 144
12. 72 **13.** $\frac{1}{72}$ **14.** $\frac{9}{8}$ **15.** $\frac{64}{25}$ **16.** $\frac{1}{49}$ **17.** 49
18. 25 **19.** $\frac{8}{x^3}$ **20.** $\frac{1}{16x^4}$ **21.** $\frac{-10}{c^3}$ **22.** $\frac{1}{-8c^3}$
23. $\frac{x^3}{4}$ **24.** $64x^3$ **25.** $\frac{3a^2}{5}$ **26.** $100a^2$ **27.** $\frac{a^4d^5}{c^2b^3}$
28. $\frac{x^2z^3}{y^4}$ **29.** $\frac{2b^5d^4}{3c^3a^2}$ **30.** $\frac{3y^2z}{4x^5}$ **31.** x^3 **32.** $\frac{1}{a^3}$
33. $\frac{1}{n^6}$ **34.** $\frac{1}{y^3}$ **35.** x^6 **36.** $\frac{1}{a^5}$ **37.** $\frac{3n^3}{4}$ **38.** $\frac{-1}{2c^2}$
39. $\frac{1}{x^6}$ **40.** $\frac{1}{a^{12}}$ **41.** n^{12} **42.** $\frac{5}{y^8}$ **43.** $\frac{16}{x^{12}}$ **44.** $\frac{1}{9a^6}$
45. $\frac{n^8}{16}$ **46.** $\frac{-125}{y^6}$ **47.** $\frac{x^{12}}{125}$ **48.** a^6b^4 **49.** $\frac{1}{c^8d^{12}}$
50. $\frac{y^{10}}{x^{15}}$ **51.** $\frac{120y^4}{x^2}$ **52.** $\frac{60a^2}{b^2}$ **53.** $\frac{40y^7}{x^5}$
54. $\frac{-60a^4}{b^3}$ **55.** $\frac{x^5}{y^6}$ **56.** $\frac{b^2}{a^3}$ **57.** $\frac{2d^6}{3c^4}$ **58.** $\frac{x^9}{y^6}$
59. $\frac{1}{a^6b^8}$ **60.** $\frac{16d^{12}}{c^8}$ **61.** $\frac{y^6}{x^8}$ **62.** $\frac{a^6}{b^{12}}$ **63.** $\frac{c^9d^6}{-27}$
64. $\frac{27a^6}{-125b^9}$ **65.** $\frac{16z^4}{x^6y^6}$ **66.** $\frac{b^{12}c^4}{16a^8}$ **67.** $\frac{a^6c^4}{b^8d^{10}}$
68. $\frac{-8y^3z^9}{27x^6w^3}$ **69.** $\frac{a^4c^8f^{12}}{b^8d^4e^{12}}$

Page 163

1. 3 **2.** 2 **3.** 2 **4.** 125 **5.** 4 **6.** 8 **7.** 4 **8.** 3 **9.** 4
10. 32 **11.** 16 **12.** 32 **13.** $6^{\frac{1}{2}}$ **14.** $20^{\frac{1}{3}}$ **15.** $10^{\frac{1}{4}}$
16. $4a^{\frac{1}{3}}$ **17.** $5b^{\frac{1}{2}}$ **18.** $3b^{\frac{1}{4}}$ **19.** $\sqrt{10}$ **20.** $\sqrt{a}$
21. $5\sqrt{a}$ **22.** $\sqrt{5a}$ **23.** $3\sqrt{b}$ **24.** $\sqrt{3b}$ **25.** $\sqrt[3]{7}$
26. $\sqrt[4]{a}$ **27.** $6\sqrt[3]{c}$ **28.** $\sqrt[3]{6c}$ **29.** $2\sqrt[4]{n}$ **30.** $\sqrt[4]{2n}$
31. $c^{\frac{2}{3}}$ **32.** $9^{\frac{4}{3}}$ **33.** $x^{\frac{3}{4}}$ **34.** $7^{\frac{3}{4}}$ **35.** $a^{\frac{5}{2}}$ **36.** $10^{\frac{3}{2}}$
37. $7^{\frac{2}{3}}$ **38.** $2^{\frac{4}{3}}$ **39.** $2^{\frac{3}{2}}$ **40.** $3^{\frac{3}{2}}$ **41.** $3^{\frac{3}{4}}$ **42.** $2^{\frac{5}{4}}$

43. $\sqrt{27}$ **44.** $\sqrt[4]{125}$ **45.** $4\sqrt[3]{x^2}$ **46.** $\sqrt[3]{16x^2}$ **47.** $5\sqrt[4]{x^3}$ **48.** $\sqrt[4]{125x^3}$ **49.** $\frac{1}{5}$ **50.** $\frac{1}{2}$ **51.** $\frac{1}{2}$ **52.** $\frac{1}{3}$ **53.** $\frac{1}{3}$ **54.** $\frac{1}{5}$ **55.** $\frac{1}{125}$ **56.** $\frac{1}{9}$ **57.** $\frac{1}{125}$ **58.** $\frac{1}{32}$ **59.** $\frac{1}{16}$ **60.** $\frac{1}{32}$

Page 165

ORALS **1.** $a^{\frac{2}{3}}$ **2.** b^3 **3.** c^3 **4.** $x^{\frac{2}{5}}$ **5.** $c^{\frac{2}{3}}$ **6.** $\frac{1}{c^{\frac{2}{3}}}$

EXERCISES **1.** $7^{\frac{4}{5}}$ **2.** $5^{\frac{2}{3}}$ **3.** $3^{\frac{3}{7}}$ **4.** $2^{\frac{5}{2}}$ **5.** $x^{\frac{7}{3}}$ **6.** $a^{\frac{2}{7}}$ **7.** $n^{\frac{3}{5}}$ **8.** $b^{\frac{5}{4}}$ **9.** x^3 **10.** a^6 **11.** n^4 **12.** b^4 **13.** $x^{\frac{3}{5}}$ **14.** $a^{\frac{6}{7}}$ **15.** $n^{\frac{2}{3}}$ **16.** $b^{\frac{3}{2}}$ **17.** xy^2 **18.** a^2b^3 **19.** $9mn^5$ **20.** $8c^3d^6$ **21.** $\frac{x^2}{y^3}$ **22.** $\frac{a^9}{b^8}$ **23.** $\frac{5m^2}{3n^3}$ **24.** $\frac{4c^2}{9d^6}$ **25.** 3 **26.** 4 **27.** 10 **28.** -1250 **29.** $\frac{1}{x^{\frac{2}{5}}}$ **30.** $\frac{1}{a^{\frac{1}{6}}}$ **31.** $\frac{1}{n^{\frac{5}{4}}}$ **32.** $\frac{1}{b^{\frac{3}{2}}}$ **33.** $\frac{1}{x^4}$ **34.** $\frac{1}{a^6}$ **35.** $\frac{1}{n^3}$ **36.** $\frac{1}{b^2}$ **37.** $\frac{1}{x^{\frac{1}{3}}}$ **38.** $\frac{1}{a^{\frac{1}{6}}}$ **39.** $\frac{1}{n^{\frac{1}{3}}}$ **40.** $\frac{1}{b^{\frac{5}{4}}}$ **41.** $\frac{y^{\frac{1}{3}}}{x^{\frac{1}{2}}}$ **42.** $\frac{b^{\frac{3}{4}}}{a^{\frac{2}{3}}}$ **43.** $m^{\frac{1}{3}}m^{\frac{1}{4}}$ **44.** $\frac{1}{c^{\frac{3}{5}}d^{\frac{2}{5}}}$ **45.** $\frac{y}{x^2}$ **46.** $\frac{b}{a^2}$ **47.** $\frac{3n}{2m^2}$ **48.** $\frac{125d^9}{8c^6}$ **49.** $\frac{3}{4}$ **50.** $\frac{1}{2}$ **51.** $\frac{1}{1250}$ **52.** $\frac{2}{3}$ **53.** $\frac{27}{8}$ **54.** $\frac{8}{27}$ **55.** $\{2\}$ **56.** $\{-2\}$ **57.** $\{-2, 2\}$ **58.** ϕ **59.** $\{\sqrt[3]{7}\}$ **60.** $\{\sqrt[5]{7}\}$

Page 167

$2x^2 - x$, rem. 2

Page 168

1. $\{-10, 10\}$ **2.** $\{-\sqrt{14}, \sqrt{14}\}$ **3.** $\{-8, 8\}$ **4.** $\{-\sqrt{17}, \sqrt{17}\}$ **5.** $4 < \sqrt{21} < 5$ **6.** 7.4 **7.** rat. **8.** irrat. **9.** irrat. **10.** rat. **11.** $-2\sqrt{7}$ **12.** $-10\sqrt{2}$ **13.** $2a^3b^2\sqrt{b}$ **14.** $cd^2\sqrt{cd}$ **15.** $5\sqrt{7} - 2\sqrt{5}$ **16.** $4\sqrt{2}$ **17.** $6\sqrt{a}$ **18.** $5a\sqrt{b}$ **19.** $30\sqrt{2}$ **20.** $4x^3$ **21.** 18 **22.** $8 + \sqrt{15}$ **23.** 4 **24.** $32 - 6\sqrt{15}$ **25.** $x + 10\sqrt{xy} + 25y$ **26.** $\frac{3\sqrt{7}}{7}$ **27.** $\frac{-\sqrt{5a}}{2a}$ **28.** $\frac{5\sqrt{2x}}{6x^2}$ **29.** $\frac{6\sqrt{3} + 3\sqrt{10}}{2}$ **30.** $2\sqrt{2}$ **31.** $c^2\sqrt{2}$ **32.** $3a\sqrt{a}$ **33.** $2x^3\sqrt{2}$ **34.** $\frac{\sqrt{15}}{5}$ **35.** $\frac{\sqrt{6cd}}{3d}$ **36.** $\frac{\sqrt{21bc}}{6bc^2}$ **37.** $\frac{\sqrt{6n-3}}{3}$ **38.** -2 **39.** $3x^3$ **40.** $3b^2\sqrt[4]{b^3}$ **41.** $3\sqrt[3]{2}$ **42.** 3 **43.** $\frac{8}{9}$ **44.** $\frac{1}{6}$ **45.** 25 **46.** $\frac{3}{2}$ **47.** $\frac{2}{3}$ **48.** $\frac{5}{x^3}$ **49.** $\frac{6}{x^2}$ **50.** $\frac{x^8}{16y^{12}}$ **51.** $\frac{1}{a^{\frac{3}{7}}}$ **52.** a^3 **53.** $\frac{a^3c^4}{b^2d^5}$ **54.** $\frac{3y^2}{4x^3}$ **55.** $\frac{9y^6}{x^4}$ **56.** $c^{\frac{3}{4}}$ **57.** $4x^2y^4$ **58.** $3^{\frac{1}{2}}$ **59.** $a^{\frac{2}{3}}$ **60.** $5^{\frac{3}{4}}$ **61.** $\sqrt{6}$ **62.** $5\sqrt[3]{x}$ **63.** $\sqrt[4]{8a^3}$

Page 169

1. $\{-4, 4\}$ **2.** $\{-\sqrt{15}, \sqrt{15}\}$ **3.** $6 < \sqrt{38} < 7$ **4.** 3.3 **5.** rat. **6.** irrat. **7.** irrat. **8.** rat. **9.** $6\sqrt{2}$ **10.** $2a^3\sqrt{2}$ **11.** $x^2y\sqrt{y}$ **12.** $6\sqrt{7} - 5\sqrt{3}$ **13.** $4\sqrt{3}$ **14.** $6x\sqrt{y}$ **15.** $20\sqrt{3}$ **16.** $6x^2$ **17.** $12a$ **18.** $-5\sqrt{6}$ **19.** 11 **20.** $9x + 6\sqrt{xy} + y$ **21.** $\frac{\sqrt{3}}{3}$ **22.** $\frac{-5\sqrt{c}}{2c^2}$ **23.** $\frac{6\sqrt{2} - 2\sqrt{7}}{11}$ **24.** $2\sqrt{2}$ **25.** $2a\sqrt{3a}$ **26.** $\frac{\sqrt{30}}{6}$ **27.** $\frac{\sqrt{6x}}{4x^2}$ **28.** -3 **29.** $2x^2$ **30.** $3\sqrt[3]{3}$ **31.** $2x\sqrt[4]{x}$ **32.** 6 **33.** $\frac{27}{16}$ **34.** 27 **35.** $\frac{5}{3}$ **36.** $\frac{6}{a^4}$ **37.** $\frac{y^8}{x^6}$ **38.** $\frac{1}{a^{\frac{3}{5}}}$ **39.** $\frac{1}{x^2}$ **40.** $\frac{x^6}{y^5}$ **41.** $\frac{8d^9}{c^6}$ **42.** $x^{\frac{4}{5}}$ **43.** $5a^2b^3$ **44.** $x^{\frac{3}{4}}$ **45.** $4\sqrt[3]{x^2}$

Page 173

ORALS **1.** 40 m² **2.** 25 *cm*² **3.** 15 dm²

EXERCISES **1.** $w = 5$ mm, $l = 15$ mm **2.** $w = 5$ cm, $l = 9$ cm **3.** $b = 4$ m, h = 8 m **4.** $b = 2$ mm, $h = 12$ mm **5.** 6 dm **6.** 12 cm **7.** $w = 4$ cm, $l = 5$ cm **8.** $w = 3$ m, $l = 14$ m **9.** $b = 6$ dm, $h = 2$ dm **10.** $b = 8$ cm, $h = 11$ cm **11.** 5 mm; 10 mm **12.** 2 m; 6 m **13.** square, 9 mm²; rect., 36 mm² **14.** square, 16 cm²; rect., 52 cm² **15.** tri., 36 m²; rect., 72 m² **16.** tri., 70 dm²; rect., 100 dm²

Page 175

ORALS **1.** $\frac{5}{4}, -\frac{1}{4}$ **2.** $-1, -4$ **3.** $\frac{2}{3}, 0$ **4.** $\frac{4}{5}, -2$ **5.** $\frac{1 + 3\sqrt{2}}{2}, \frac{1 - 3\sqrt{2}}{2}$

EXERCISES **1.** $\left\{\pm \frac{\sqrt{5}}{3}\right\}$ **2.** $\{\pm 2\sqrt{2}\}$ **3.** $\{7, -1\}$ **4.** $\{-5 \pm 2\sqrt{3}\}$ **5.** $\{4, 1\}$ **6.** $\left\{\frac{1}{3}, -3\right\}$ **7.** $\left\{\frac{2 \pm \sqrt{10}}{3}\right\}$ **8.** $\left\{\frac{-5 \pm \sqrt{3}}{4}\right\}$ **9.** $\left\{\frac{-3 \pm 2\sqrt{2}}{5}\right\}$ **10.** $\left\{\frac{7 \pm 2\sqrt{3}}{2}\right\}$ **11.** $\{3\sqrt{5}, \sqrt{5}\}$ **12.** $\{-2\sqrt{2}, -4\sqrt{2}\}$ **13.** $\{1 \pm \sqrt{5}\}$ **14.** $\{4 \pm 2\sqrt{2}\}$ **15.** $\{2, -4\}$ **16.** $\left\{\pm \frac{\sqrt{7}}{3}\right\}$

Page 178

ORALS **1.** $a^2 + 12a + 36$ **2.** $n^2 - 6n + 9$ **3.** $x^2 + 3x + \frac{9}{4}$ **4.** $b^2 - b + \frac{1}{4}$ **5.** $c^2 + \frac{3}{2}c + \frac{9}{16}$ **6.** $x^2 - \frac{2}{3}x + \frac{1}{9}$

EXERCISES **1.** $\{2, -8\}$ **2.** $\{2, -5\}$ **3.** $\{5, 3\}$ **4.** $\{3 \pm \sqrt{14}\}$ **5.** $\{-4 \pm \sqrt{7}\}$ **6.** $\{2 \pm \sqrt{11}\}$ **7.** $\left\{\frac{-3 \pm \sqrt{21}}{2}\right\}$ **8.** $\left\{\frac{5 \pm \sqrt{17}}{2}\right\}$ **9.** $\left\{\frac{-1 \pm \sqrt{21}}{2}\right\}$ **10.** $\{-4 \pm \sqrt{19}\}$ **11.** $\{5 \pm \sqrt{10}\}$ **12.** $\{-6 \pm \sqrt{38}\}$ **13.** $\left\{\frac{5 \pm 3\sqrt{5}}{2}\right\}$ **14.** $\left\{\frac{-7 \pm 3\sqrt{5}}{2}\right\}$ **15.** $\left\{\frac{1 \pm 3\sqrt{5}}{2}\right\}$ **16.** $\frac{2}{5}, -1$ **17.** $4, -\frac{1}{2}$ **18.** $-\frac{1}{3}, -3$ **19.** $\frac{-5 \pm \sqrt{65}}{4}$ **20.** $\frac{7 \pm \sqrt{29}}{10}$ **21.** $\frac{-2 \pm \sqrt{13}}{3}$ **22.** $3 \pm 2\sqrt{3}$ **23.** $-5 \pm 3\sqrt{2}$ **24.** $4 \pm 2\sqrt{5}$ **25.** $3 \pm \sqrt{15}$ **26.** $-3 \pm \sqrt{23}$ **27.** $\frac{5 \pm 2\sqrt{5}}{6}$ **28.** $\frac{-5 \pm \sqrt{73}}{6}$ **29.** $\frac{3 \pm \sqrt{17}}{4}$ **30.** $\frac{-2 \pm 3\sqrt{2}}{4}$

Page 181

ORALS **1.** $a = 5, b = 2, c = -3$ **2.** $a = 1, b = -5, c = 7$ **3.** $a = 2, b = 1, c = 0$ **4.** $a = 1, b = 0, c = -7$ **5.** $x^2 - 8x - 7 = 0$ **6.** $5x^2 + x + 4 = 0$ **7.** $4x^2 - x - 2 = 0$ **8.** $2x^2 + 3x + 0 = 0$

EXERCISES **1.** $\frac{-3 \pm 2\sqrt{2}}{2}$ **2.** $\frac{2 \pm \sqrt{5}}{3}$ **3.** $-3 \pm 2\sqrt{2}$ **4.** $-1 \pm \sqrt{3}$ **5.** $\frac{3}{2}$ **6.** $\frac{-5 \pm \sqrt{41}}{4}$ **7.** $\frac{-1 \pm \sqrt{33}}{8}$ **8.** $\frac{3 \pm \sqrt{5}}{2}$ **9.** $\frac{-7 \pm \sqrt{5}}{2}$ **10.** $\frac{-7 \pm \sqrt{13}}{6}$ **11.** $\frac{1 \pm \sqrt{17}}{4}$ **12.** $\frac{2 \pm \sqrt{2}}{2}$ **13.** $-2 \pm \sqrt{5}$ **14.** $\frac{-3 \pm \sqrt{3}}{2}$ **15.** $\frac{3 \pm \sqrt{2}}{2}$ **16.** $\frac{-2 \pm \sqrt{3}}{2}$ **17.** $\frac{1 \pm 2\sqrt{2}}{3}$ **18.** $\frac{-5 \pm \sqrt{37}}{2}$ **19.** $\frac{-1 \pm \sqrt{41}}{10}$ **20.** $\frac{1 \pm \sqrt{41}}{4}$ **21.** $\frac{-3 \pm \sqrt{15}}{3}$ **22.** $\frac{3 \pm \sqrt{7}}{2}$ **23.** $\frac{-1 \pm \sqrt{6}}{5}$ **24.** 3 **25.** $-\frac{2}{3}$ **26.** $\frac{5}{2}$ **27.** 5, 0 **28.** $0, -\frac{2}{3}$ **29.** $\frac{7}{2}, 0$ **30.** $\pm\sqrt{7}$ **31.** $\pm 2\sqrt{3}$ **32.** $\pm\frac{\sqrt{22}}{2}$ **33.** $\left\{\frac{-1 \pm \sqrt{17}}{2}\right\}$ **34.** $\left\{\frac{9 \pm \sqrt{21}}{6}\right\}$ **35.** $\left\{\frac{2 \pm \sqrt{10}}{2}\right\}$ **36.** $\{5\}$ **37.** $\left\{\frac{7 \pm \sqrt{37}}{2}\right\}$ **38.** $\left\{\frac{-5 \pm \sqrt{33}}{4}\right\}$ **39.** $\left\{\frac{-1 \pm \sqrt{7}}{2}\right\}$ **40.** $\left\{\frac{3 \pm \sqrt{3}}{3}\right\}$ **41.** $\left\{\frac{7 \pm \sqrt{29}}{2}\right\}$ **42.** $\left\{\frac{-4 \pm \sqrt{7}}{3}\right\}$ **43.** $\{3\sqrt{2}, -\sqrt{2}\}$ **44.** $\left\{\sqrt{6}, -\frac{\sqrt{6}}{3}\right\}$ **45.** $\left\{\frac{-\sqrt{10} \pm \sqrt{15}}{5}\right\}$

Page 183

1. 36; 2 **2.** 0; 1 **3.** −8; none **4.** 16; 2 **5.** 0; 1 **6.** 29; 2 **7.** −7; none **8.** 8; 2 **9.** 13; 2 **10.** 0; 1 **11.** 108; 2 **12.** −11; none **13.** 1; 2 **14.** 0; 1 **15.** 33; 2 **16.** 33; 2 **17.** −7; none **18.** 49; 2 **19.** 60; 2 **20.** 0; 1 **21.** 69; 2 **22.** 121; 2 **23.** −19; none **24.** 32; 2 **25.** 81; 2 **26.** 0; 1 **27.** −4; none **28.** 32; 2 **29.** 0; 1 **30.** −2; none **31.** 121; 2 **32.** 0; 1 **33.** −8; none

Page 184

1. $5x^3 + 7x^2 + 16x + 30$, rem. 20 **2.** $2x^3 - 2x^2 - x + 3$, rem. −1 **3.** $3x^2 - x - 2$, rem 2

Page 187

ORALS **1.** $\frac{7}{2}; 2$ **2.** $-\frac{2}{3}; -\frac{2}{3}$ **3.** $2; -\frac{3}{4}$ **4.** $-8; 12$ **5.** $-\frac{3}{5}; 0$ **6.** $0; -2$ **7.** $\frac{2}{5}; -\frac{3}{5}$ **8.** $-\frac{7}{4}; -\frac{1}{4}$

EXERCISES **1.** $x^2 - 5x + 6 = 0$ **2.** $x^2 + 3x - 10 = 0$ **3.** $x^2 - 3x - 10 = 0$ **4.** $x^2 + 7x + 12 = 0$ **5.** $x^2 - 6x + 9 = 0$ **6.** $x^2 + 8x + 16 = 0$ **7.** $x^2 - 10x + 25 = 0$ **8.** $x^2 + 4x + 4 = 0$ **9.** $8x^2 - 6x + 1 = 0$ **10.** $9x^2 + 3x - 2 = 0$ **11.** $16x^2 - 8x - 3 = 0$ **12.** $25x^2 + 25x + 6 = 0$ **13.** $4x^2 - 4x + 1 = 0$ **14.** $9x^2 + 6x + 1 = 0$ **15.** $25x^2 - 20x + 4 = 0$ **16.** $16x^2 + 24x + 9 = 0$ **17.** $2x^2 - 5x + 2 = 0$ **18.** $4x^2 + 3x - 1 = 0$ **19.** $3x^2 - 5x - 2 = 0$ **20.** $3x^2 + 11x + 6 = 0$ **21.** $x^2 - 6x + 4 = 0$ **22.** $x^2 + 4x + 1 = 0$ **23.** $x^2 - 4x - 3 = 0$ **24.** $x^2 + 6x - 1 = 0$ **25.** $4x^2 - 8x + 1 = 0$

26. $9x^2 + 18x + 7 = 0$ **27.** $4x^2 - 2x - 1 = 0$
28. $3x^2 + 4x - 1 = 0$ **29.** $x^2 - 10x + 13 = 0$
30. $x^2 + 10x + 7 = 0$ **31.** $x^2 - 6x - 11 = 0$
32. $x^2 + 12x - 4 = 0$ **33.** $4x^2 - 4x - 7 = 0$
34. $3x^2 + 6x - 1 = 0$ **35.** $4x^2 - 12x + 1 = 0$
36. $9x^2 + 24x - 4 = 0$ **37.** $x^2 - 15x + 50 = 0$
38. $x^2 + 5x - 300 = 0$ **39.** $x^2 - 5x - 500 = 0$
40. $x^2 + 30x + 200 = 0$ **41.** $10x^2 - 29x + 10 = 0$
42. $10x^2 - 3x - 1 = 0$ **43.** $10x^2 + 7x - 6 = 0$
44. $100x^2 + 100x + 21 = 0$
45. $\frac{(-b + \sqrt{b^2 - 4ac}) + (-b - \sqrt{b^2 - 4ac})}{2a} =$
$\frac{-2b + 0}{2a} = \frac{-2b}{2a} = \frac{-b}{a} = -\frac{b}{a}$
46. $\frac{(-b + \sqrt{b^2 - 4ac})(-b - \sqrt{b^2 - 4ac})}{(2a)(2a)} =$
$\frac{b^2 - (b^2 - 4ac)}{4a^2} = \frac{4ac}{4a^2} = \frac{c}{a}$

Page 191

1. $x = 4$; $x - 1 = 3$ **2.** $x = 6$; $x + 2 = 8$
3. $x = 12$; $x - 7 = 5$ **4.** $x = 3$; $2x - 1 = 5$
5. $x = 8$; $2x - 6 = 10$ **6.** $x = 5$; $2x + 3 = 13$
7. $x = 12$; $x - 3 = 9$; $x + 3 = 15$ **8.** $x = 12$;
$x + 4 = 16$; $x + 8 = 20$ **9.** $x = 5$; $2x + 2 = 12$;
$2x + 3 = 13$ **10.** 5 **11.** 11 **12.** 4 **13.** 8 **14.** 5 m
15. 41 cm **16.** 7, 3; −3, −7 **17.** 3, 11;
$-3\frac{2}{3}$, −9 **18.** 3, 8; −6, −1 **19.** 4, 5; $\frac{1}{2}$, −2
20. $x = 3\sqrt{2}$ **21.** $x = 3\sqrt{5}$; $2x = 6\sqrt{5}$
22. $x = \sqrt{10}$; $3x = 3\sqrt{10}$ **23.** $x = 2 + \sqrt{5}$;
$2x = 4 + 2\sqrt{5}$; $2x + 1 = 5 + 2\sqrt{5}$
24. $x = 2 + 2\sqrt{3}$; $x + 1 = 3 + 2\sqrt{3}$;
$x + 3 = 5 + 2\sqrt{3}$ **25.** $x = 2 + \sqrt{7}$;
$2x + 1 = 5 + 2\sqrt{7}$; $2x + 2 = 6 + 2\sqrt{7}$
26. $2\sqrt{2}$ cm **27.** $4\sqrt{2}$ mm **28.** $2 + 2\sqrt{2}$ m;
$4 + 2\sqrt{2}$ m **29.** $3 + 3\sqrt{2}$ dm; $6 + 3\sqrt{2}$ dm
30. $l = 1 + \sqrt{5}$ cm; $w = -1 + \sqrt{5}$ cm
31. $l = 2 + \sqrt{10}$ m; $w = -2 + \sqrt{10}$ m
32. $1 + \sqrt{5}, -1 + \sqrt{5}$; $1 - \sqrt{5}, -1 - \sqrt{5}$
33. $-2 + \sqrt{5}, 2 + \sqrt{5}$; $-2 - \sqrt{5}, 2 - \sqrt{5}$
34. $\frac{1 + \sqrt{7}}{2}, -1 + \sqrt{7}$; $\frac{1 - \sqrt{7}}{2}, -1 - \sqrt{7}$
35. $\frac{-2 + \sqrt{10}}{2}, 2 + \sqrt{10}$; $\frac{-2 - \sqrt{10}}{2}, 2 - \sqrt{10}$
36. $w = \frac{-1 + \sqrt{7}}{2}, l = 1 + \sqrt{7}$
37. $w = \frac{-2 + 3\sqrt{2}}{2}, l = 2 + 3\sqrt{2}$
38. $w = 4 + 2\sqrt{5}, l = 8 + 4\sqrt{5}, d = 10 + 4\sqrt{5}$
39. $w = 6 + 2\sqrt{10}, l = 18 + 6\sqrt{10}$,
$d = 20 + 6\sqrt{10}$ **40.** $w = 1 + \sqrt{6}, l = 3 + \sqrt{6}$,
$d = 4 + \sqrt{6}$ **41.** $w = 3 + 2\sqrt{3}, l = 10 + 6\sqrt{3}$,
$d = 11 + 6\sqrt{3}$

Page 195

ORALS **1.** {5} **2.** ϕ **3.** {5} **4.** {3} **5.** {3, 5}

EXERCISES **1.** 25 **2.** 11 **3.** no sol. **4.** 33
5. 8 **6.** 7 **7.** −1, −2 **8.** 6 **9.** 3 **10.** 2 **11.** $\frac{1}{2}$
12. −2 **13.** {9} **14.** ϕ **15.** $\left\{\frac{1}{4}\right\}$ **16.** ϕ **17.** {12}
18. {7} **19.** {−2, −3} **20.** {2} **21.** {5} **22.** $\left\{\frac{1}{2}\right\}$
23. {3} **24.** {−1} **25.** {10} **26.** {2, 6} **27.** $\left\{-1, \frac{1}{2}\right\}$
28. 3, 7 **29.** 2 **30.** 3 **31.** 4 **32.** −1, 3 **33.** 2, 3
34. 5, 8 **35.** 2, 6 **36.** 2 **37.** 9 **38.** 0, 4 **39.** 1
40. 4 **41.** 4 **42.** 81 **43.** 3 **44.** 7 **45.** 5 **46.** 6
47. no sol.

Page 198

ORALS **1.** $0 + 2i$ **2.** $6 + 0i$ **3.** $-4 + 0i$
4. $0 + 1i$ **5.** $0 + (-4)i$ **6.** $-2 + 0i$ **7.** $0 + 0i$
8. $-3.5 + 0i$ **9.** $0 + 3i$ **10.** $-3 + 0i$ **11.** $0 + i\sqrt{2}$
12. $\sqrt{5} + 0i$

EXERCISES **1.** $5i$ **2.** $-4i$ **3.** $8i$ **4.** $i\sqrt{5}$
5. $-i\sqrt{2}$ **6.** $i\sqrt{21}$ **7.** $2i\sqrt{2}$ **8.** $-3i\sqrt{3}$
9. $3i\sqrt{5}$ **10.** $-3 - 2i$ **11.** $4 + i\sqrt{2}$
12. $7 - i\sqrt{15}$ **13.** $3 + 3i\sqrt{2}$ **14.** $-2 - 2i\sqrt{3}$
15. $6 - 5i\sqrt{2}$ **16.** $-i, i$ **17.** $-i\sqrt{2}, i\sqrt{2}$
18. $-2i\sqrt{3}, 2i\sqrt{3}$ **19.** $-2i\sqrt{5}, 2i\sqrt{5}$
20. $-i\sqrt{30}, i\sqrt{30}$ **21.** $-6i, 6i$ **22.** $5 + 7i$
23. $2 + 4i$ **24.** $4 + 3i$ **25.** $-1 - 2i$ **26.** $6 + 2i$
27. $-2 + 4i$ **28.** -2 **29.** $-7 + 4i$ **30.** $5 + 5i$
31. $6i$ **32.** $-10i\sqrt{5}$ **33.** $8i\sqrt{3}$ **34.** $5 + 6i$
35. $-2 - 15i$ **36.** $2 + 10i\sqrt{2}$ **37.** $7 - 6i\sqrt{3}$
38. $-1 + 12i\sqrt{2}$ **39.** $-4 - 6i\sqrt{5}$ **40.** $-5i, 5i$
41. $-3i\sqrt{5}, 3i\sqrt{5}$ **42.** $-i\sqrt{35}, i\sqrt{35}$
43. $8 + 7i\sqrt{2}$ **44.** $-3 + 3i$ **45.** $5 + 7i\sqrt{3}$
46. $3 + 4i\sqrt{5}$ **47.** $2 + 4i\sqrt{2}$ **48.** $4 + 3i\sqrt{2}$
49. $0 + 0i$; $0 + 0i$ **50.** $-a - bi$; $-a - bi$

Page 200

1. $-i$ **2.** i **3.** −1 **4.** 1 **5.** −1 **6.** $-i$

Page 203

ORALS **1.** −1 **2.** $-i$ **3.** 1 **4.** i **5.** −1 **6.** i
7. $4 + 3i$ **8.** $1 - 5i$ **9.** $-7 + 4i$ **10.** $-5 - 2i\sqrt{3}$

EXERCISES **1.** -6 **2.** 20 **3.** $-\sqrt{6}$ **4.** $-\sqrt{15}$ **5.** $\sqrt{21}$ **6.** $-\sqrt{14}$ **7.** $-6\sqrt{10}$ **8.** $12\sqrt{22}$ **9.** $-20\sqrt{30}$ **10.** $2i\sqrt{6}$ **11.** $3i\sqrt{10}$ **12.** $-6i$ **13.** -9 **14.** -16 **15.** -5 **16.** $-4 + 7i$ **17.** $14 - 2i$ **18.** $-8 - 9i$ **19.** $14 - 2i$ **20.** $9 + 7i$ **21.** $-5 + 5i$ **22.** $-7 + 24i$ **23.** $21 - 20i$ **24.** $-2i$ **25.** 13 **26.** 41 **27.** 25 **28.** 5 **29.** 10 **30.** 61 **31.** $\frac{7i}{-2}$ **32.** $\frac{3i}{4}$ **33.** $\frac{5i}{3}$ **34.** $\frac{2i}{7}$ **35.** $\frac{i}{6}$ **36.** $-8i$ **37.** $\frac{5+i}{13}$ **38.** $\frac{7-i}{2}$ **39.** i **40.** $\frac{8-12i}{13}$ **41.** $\frac{-6-3i}{10}$ **42.** $\frac{1+3i}{2}$ **43.** $4 + i$ **44.** $\frac{6+17i}{25}$ **45.** $\frac{7-i}{10}$ **46.** $-2\sqrt{6}$ **47.** $6\sqrt{10}$ **48.** $-4\sqrt{7}$ **49.** $-8i$ **50.** $32i$ **51.** $-i$ **52.** $4 + 14i\sqrt{2}$ **53.** 21 **54.** $-14 + 12i\sqrt{2}$ **55.** $4 - 6i\sqrt{5}$ **56.** $-28 - 7i\sqrt{5}$ **57.** 11 **58.** $a + bi$; $1 + 0i$ **59.** 1 or $1 + 0i$; $\frac{a-bi}{a^2+b^2}$ **60.** $(a + bi)(a - bi) = a^2 - b^2i^2 = a^2 + b^2$; $a^2 + b^2$ is a real number since a and b are real numbers.
61. $(a + bi) + (a - bi) = 2a + 0 = 2a$; $2a$ is a real number since a is a real number.

Page 207

ORALS **1.** yes **2.** no **3.** yes **4.** no **5.** no

EXERCISES **1.** $\frac{1 \pm 2i}{2}$ **2.** $\frac{-2 \pm i\sqrt{5}}{2}$ **3.** $3 \pm i\sqrt{2}$ **4.** $\left\{\frac{-3 \pm i\sqrt{3}}{2}\right\}$ **5.** $\left\{\frac{3 \pm i\sqrt{7}}{2}\right\}$ **6.** $\left\{\frac{-1 \pm i\sqrt{7}}{2}\right\}$ **7.** $\{-2 \pm i\}$ **8.** $\{1 \pm i\}$ **9.** $\{4 \pm 2i\}$ **10.** $\left\{\frac{-3 \pm i\sqrt{15}}{4}\right\}$ **11.** $\left\{\frac{1 \pm i\sqrt{7}}{4}\right\}$ **12.** $\left\{\frac{-5 \pm i\sqrt{7}}{4}\right\}$ **13.** $\left\{\frac{-1 \pm i}{2}\right\}$ **14.** $\left\{\frac{1 \pm 3i}{2}\right\}$ **15.** $\left\{\frac{1 \pm 5i}{2}\right\}$ **16.** $\left\{\frac{-3 \pm i\sqrt{3}}{6}\right\}$ **17.** $\left\{\frac{1 \pm i\sqrt{11}}{6}\right\}$ **18.** $\left\{\frac{-1 \pm i\sqrt{23}}{6}\right\}$ **19.** $\{-3 \pm i\sqrt{2}\}$ **20.** $\left\{\frac{2 \pm i\sqrt{2}}{2}\right\}$ **21.** $\left\{\frac{3 \pm i\sqrt{5}}{2}\right\}$ **22.** $\left\{\frac{-2 \pm i\sqrt{2}}{3}\right\}$ **23.** $\left\{\frac{1 \pm i\sqrt{2}}{3}\right\}$ **24.** $\left\{\frac{-1 \pm i\sqrt{3}}{4}\right\}$ **25.** $\left\{\frac{\pm i\sqrt{10}}{2}\right\}$ **26.** $\left\{\frac{\pm i\sqrt{3}}{2}\right\}$ **27.** $\left\{\frac{\pm i\sqrt{15}}{5}\right\}$ **28.** $\left\{\frac{\pm 3i\sqrt{2}}{2}\right\}$ **29.** $\left\{\frac{\pm 2i\sqrt{3}}{3}\right\}$ **30.** $\left\{\frac{\pm i}{2}\right\}$ **31.** -7; yes **32.** 9; no **33.** -23; yes **34.** -16; yes **35.** 32; no **36.** 0; no **37.** $\frac{7 \pm i\sqrt{3}}{2}$ **38.** $\frac{3 \pm i\sqrt{15}}{6}$ **39.** $\frac{1 \pm i\sqrt{15}}{8}$ **40.** $\frac{3 \pm i}{2}$ **41.** $3 \pm i$ **42.** $\frac{1 \pm i\sqrt{2}}{2}$ **43.** $\{-3i, i\}$ **44.** $\{3i \pm i\sqrt{17}\}$ **45.** $\left\{\frac{i}{2}, 2i\right\}$ **46.** $\{-9\}$ **47.** $\{-3\}$ **48.** $\{-6i\}$ **49.** $\{2, -1 \pm i\sqrt{3}\}$ **50.** $\left\{3, \frac{-3 \pm 3i\sqrt{3}}{2}\right\}$ **51.** $\{-4, 2 \pm 2i\sqrt{3}\}$ **52.** $\{2, -2, 2i, -2i\}$

53.
$$\begin{aligned}&(-1 + i\sqrt{3})(-1 + i\sqrt{3})(-1 + i\sqrt{3})\\ &= (1 - 2i\sqrt{3} + 3i^2)(-1 + i\sqrt{3})\\ &= (-2 - 2i\sqrt{3})(-1 + i\sqrt{3})\\ &= 2 - 6i^2 = 2 + 6 = 8\end{aligned}$$

54.
$$\begin{aligned}&\frac{-3 - 3i\sqrt{3}}{2} \cdot \frac{-3 - 3i\sqrt{3}}{2} \cdot \frac{-3 - 3i\sqrt{3}}{2}\\ &= \frac{9 + 18i\sqrt{3} + 27i^2}{4} \cdot \frac{-3 - 3i\sqrt{3}}{2}\\ &= \frac{-18 + 18i\sqrt{3}}{4} \cdot \frac{-3 - 3i\sqrt{3}}{2}\\ &= \frac{-9 + 9i\sqrt{3}}{2} \cdot \frac{-3 - 3i\sqrt{3}}{2}\\ &= \frac{27 - 81i^2}{4} = \frac{27 + 81}{4}\\ &= \frac{108}{4} = 27\end{aligned}$$

Page 208

1. $\{-2\sqrt{3}, 2\sqrt{3}\}$ **2.** $\{4, -8\}$ **3.** $\left\{\frac{-1 \pm \sqrt{7}}{3}\right\}$ **4.** $\{2 \pm \sqrt{6}\}$ **5.** $\{3\}$ **6.** $\{4\}$ **7.** $\frac{3 \pm \sqrt{5}}{2}$ **8.** 41; two **9.** 0; one **10.** $-\frac{3}{4}$; $-\frac{3}{4}$ **11.** $x^2 + 6x + 2 = 0$ **12.** $x = 3\sqrt{10}$, $3x = 9\sqrt{10}$ **13.** $x = 3 + 2\sqrt{6}$, $x + 1 = 4 + 2\sqrt{6}$, $x + 4 = 7 + 2\sqrt{6}$ **14.** $4 + 4\sqrt{2}$ cm; $8 + 4\sqrt{2}$ cm **15.** 6 **16.** $1\frac{1}{3}$, 3; -3, -10 **17.** 11 m; 3 m **18.** $-4i\sqrt{7}$ **19.** $6 + 10i\sqrt{3}$ **20.** $-4 - 6i\sqrt{2}$ **21.** $4 + 2i$ **22.** $-2 + 5i$ **23.** $512i$ **24.** -20 **25.** $8\sqrt{15}$ **26.** -3 **27.** $21 - i$ **28.** 25 **29.** $-16 - 30i$ **30.** $\frac{2i}{3}$ **31.** $\frac{-8-i}{5}$ **32.** $-4i$, $4i$; yes **33.** $-6i$, $6i$; yes **34.** $\frac{-3 \pm i\sqrt{7}}{2}$; yes **35.** $-2 \pm i\sqrt{2}$; yes **36.** $\frac{3 \pm i\sqrt{7}}{4}$; yes

Page 209
1. $4 \pm \sqrt{21}$ **2.** $\{-\sqrt{3}, \sqrt{3}\}$ **3.** $\{-2, 8\}$
4. $\left\{\frac{-3 \pm \sqrt{33}}{4}\right\}$ **5.** $\{2 \pm \sqrt{3}\}$ **6.** $\{5\}$ **7.** $\{5\}$
8. 0; one **9.** 41; two **10.** -3; $-\frac{1}{2}$ **11.** $x^2 - 7x + 12 = 0$ **12.** $x = 4\sqrt{5}$; $2x = 8\sqrt{5}$ **13.** $x = 9$; $x + 3 = 12$; $x + 6 = 15$ **14.** 3 dm; 7 dm **15.** 4 **16.** $1 + \sqrt{2}$ m; $2 + \sqrt{2}$ m **17.** 7, 1; -4, -10 **18.** $i\sqrt{6}$ **19.** $-12i$ **20.** $5 - 6i\sqrt{5}$ **21.** $6 - 3i$ **22.** $4 + 6i$ **23.** -18 **24.** $\sqrt{10}$ **25.** -12 **26.** -6 **27.** $-8 + 9i$ **28.** 11 **29.** $21 - 20i$
30. $32i$ **31.** $\frac{5i}{-2}$ **32.** $\frac{-4 + 7i}{13}$ **33.** $-i\sqrt{6}$, $i\sqrt{6}$; yes **34.** $-3i$, $3i$; yes **35.** $\frac{-1 \pm i\sqrt{7}}{4}$; yes
36. $3 \pm i$; yes

Page 210
1. $5 - i$ **2.** $9 - 3i$ **3.** $1 + 6i$ **4.** $2i$

Page 213

ORALS **1.** 2 **2.** 1 **3.** 3 **4.** 2 **5.** 4

EXERCISES **1.** quad. 1 **2.** quad. 2 **3.** quad. 3 **4.** quad. 4 **5.** quad. 1 **6.** $A(3, 2)$ $B(3, -1)$ $C(5, -1)$ $D(5, 4)$ $E(-5, -1)$ $F(-5, 2)$ $G(-3, 4)$ $H(-3, -4)$ $I(1, -4)$ $J(1, 3)$ $K(-2, 3)$ **7.** *A* 3 units from vertical axis, 2 units from horizontal axis; *B* 3 units from vertical axis, 1 unit from horizontal axis; *C* 5 units from vertical axis, 1 unit from horizontal axis; *D* 5 units from vertical axis, 4 units from horizontal axis; *E* 5 units from vertical axis, 1 unit from horizontal axis; *F* 5 units from vertical axis, 2 units from horizontal axis; *G* 3 units from vertical axis, 4 units from horizontal axis; *H* 3 units from vertical axis, 4 units from horizontal axis; *I* 1 unit from vertical axis, 4 units from horizontal axis; *J* 1 unit from vertical axis, 3 units from horizontal axis; *K* 2 units from vertical axis, 3 units from horizontal axis. **8.** $AF = 8$, $JK = 3$, $DG = 8$, $BC = 2$, $BE = 8$ **9.** $FE = 3$, $GH = 8$, $JI = 7$, $AB = 3$, $DC = 5$ **10.** 5 **11.** 3 **12.** 5 **13.** 4 **14.** 4 **15.** 1 **16.** quad. 1 **17.** quad. 2 **18.** quad. 4 **19.** quad. 3 **20.** $|a_2 - a_1|$ **21.** $|b_2 - b_1|$

Page 216

ORALS **1.** negative **2.** negative **3.** positive **4.** positive

EXERCISES **1.** -5 **2.** -4 **3.** -4 **4.** -10 **5.** -9 **6.** 3 **7.** 7 **8.** -10 **9.** -4 **10.** 8 **11.** -2 **12.** -4 **13.** $7\frac{3}{4}$ **14.** $7\frac{3}{4}$ **15.** $7\frac{3}{4}$ **16.** $-7\frac{7}{10}$ **17.** $3a$ **18.** $-11x$ **19.** $-12y$ **20.** $5b$ **21.** $-7b$ **22.** $-5a$ **23.** $14y$ **24.** $13x$

Page 217
Check students' drawings.

Page 220
1. negative **2.** zero **3.** undefined **4.** positive **5.** -1 **6.** 3 **7.** $\frac{6}{7}$ **8.** 0 **9.** 0 **10.** undefined **11.** $-\frac{1}{2}$ **12.** $-\frac{1}{2}$ **13.** undefined **14.** $\frac{5}{2}$ **15.** 1 **16.** 1 **17.** slopes equal, $\frac{3}{2}$ **18.** slopes equal, $\frac{1}{2}$ **19.** slopes equal, $\frac{2}{3}$ **20.** $-\frac{16}{9}$ **21.** 3 **22.** $\frac{5}{4}$
23. $-\frac{b}{a}$ **24.** $\frac{2y}{x}$ **25.** $\frac{y_2 - y_1}{x_2 - x_1}$

Page 223

ORALS **1.** $m = 2, b = 3$ **2.** $m = -\frac{1}{2}$, $b = -3$ **3.** $m = -4, b = 5$ **4.** $m = 1, b = -3$ **5.** $m = 4, b = -\frac{1}{2}$

EXERCISES **1.** $y = 4x - 2$ **2.** $y = -7x - 4$ **3.** $y = -\frac{1}{2}x - 2$ **4.** $y = 1$ **5.** $y = -x - 1$ **6.** $y = -4x - 4$ **7.** $y = 2$ **8.** $y = -x - 3$ **9.** $y = 4$ **10.** $y = 2x + 3$ **11.** $y = -\frac{1}{2}x - 2$ **12.** $y = -2x - \frac{1}{2}$ **13.** $y = \frac{1}{2}x - \frac{1}{3}$ **14.** $y = -\frac{1}{3}x + 2$ **15.** $y = -4x + 4$ **16.** $y = 2x - 1$ **17.** $y = -2x + 3$ **18.** $y = -\frac{1}{2}x - \frac{1}{2}$ **19.** $m = \frac{1}{2}, b = 2$ **20.** $m = 3, b = -\frac{1}{2}$ **21.** $m = -4, b = 7$ **22.** $m = -7, b = -3$ **23.** $m = -\frac{2}{3}, b = 3$ **24.** $m = \frac{3}{2}, b = -5$ **25.** $m = -\frac{2}{3}, b = \frac{3}{2}$ **26.** $m = -\frac{3}{2}, b = \frac{7}{2}$ **27.** $m = 2, b = \frac{4}{3}$ **28.** $m = -\frac{7}{3}, b = 4$ **29.** $m = -\frac{4}{5}, b = 2$ **30.** $m = 1, b = -2$ **31.** $m = -2, b = -\frac{4}{3}$ **32.** $m = -\frac{2}{3}, b = -\frac{7}{9}$ **33.** and **34.** Solve for y. $y = \frac{-A}{B}x - \frac{C}{B}$; $m = \frac{-A}{B}$; $b = -\frac{C}{B}$

Page 224
1. 30 **2.** 2,520 **3.** 6 **4.** 3,024 **5.** 6 **6.** 6,720 **7.** 3,024

Page 227
1. $y - 3 = 5(x - 2)$; $y = 5x - 7$
2. $y + 3 = \frac{8}{5}(x + 2)$; $y = \frac{8}{5}x + \frac{1}{5}$
3. $y - 1 = -\frac{1}{3}(x - 4)$; $y = -\frac{1}{3}x + \frac{7}{3}$
4. $y - 2 = \frac{1}{6}(x - 4)$; $y = \frac{1}{6}x + \frac{4}{3}$
5. $y - 4 = \frac{1}{2}(x - 8)$; $y = \frac{1}{2}x$ **6.** $y - 3 = \frac{1}{4}(x - 6)$; $y = \frac{1}{4}x + \frac{3}{2}$ **7.** $y + 6 = \frac{9}{7}(x + 5)$; $y = \frac{9}{7}x + \frac{3}{7}$
8. $y + 5 = \frac{5}{3}(x + 4)$; $y = \frac{5}{3}x + \frac{5}{3}$
9. $y - 5 = 3(x - 1)$; $y = 3x + 2$
10. $y + 3 = -2(x - 4)$; $y = -2x + 5$
11. $y + 3 = -\frac{1}{2}(x + 2)$; $y = -\frac{1}{2}x - 4$
12. $y - 4 = -4(x - 2)$; $y = -4x + 12$
13. $y - 4 = -4(x + 2)$; $y = -4x - 4$
14. $y - 1 = 0$; $y = 1$ **15.** $y - 2 = -3(x + 4)$; $y = -3x - 10$ **16.** $y - 3 = -\frac{1}{2}(x + 1)$; $y = -\frac{1}{2}x + \frac{5}{2}$ **17.** $y + \frac{1}{2} = -\frac{1}{3}(x + 3)$; $y = -\frac{1}{3}x - \frac{3}{2}$ **18.** $y - 2 = -2\left(x + \frac{1}{3}\right)$; $y = -2x + \frac{4}{3}$ **19.** $y - 6 = 3\left(x - \frac{5}{3}\right)$; $y = 3x + 1$
20. Find slope. Use $y - y_1 = m(x - x_1)$.

Page 230

ORALS **1.** horizontal **2.** vertical **3.** vertical **4.** vertical **5.** horizontal **6.** horizontal

EXERCISES Three points are given for each line. **1.** (0, 3), (2, 4), (4, 5) **2.** (0, −2), (3,−4), (6, −6) **3.** (0, 1), (3, 2), (6, 3) **4.** (0, 4), (1, 6), (2, 8) **5.** (0, −1), (2, −2), (4, −3) **6.** (0, 1), (1, −1), (2, −3) **7.** (0, −2), (1, −5), (2, −8) **8.** (0, 3), (3, 5), (6, 7) **9.** (0, 4), (1, 2), (2, 0) **10.** (0, −1), (1, −4), (2, −7) **11.** (0, 3), (1, 7), (2, 11) **12.** (0, −5), (2, −4), (4, −3) **13.** (0, 7), (3, 8), (6, 9) **14.** (3, 0), (3, 1), (3, 2) **15.** (−1, 0), (−1, 1), (−1, 2) **16.** (0, −3), (1, −3), (2, −3) **17.** (0, 2), (1, 2), (2, 2) **18.** yes **19.** yes **20.** no **21.** no **22.** no **23.** yes **24.** (0, 2), (−2, 5), (−4, 8) **25.** (0, −2), (3, −1), (6, 0) **26.** (0, 2), (−2, 3), (−4, 4) **27.** (0, 5), (1, 8), (2, 11) **28.** (0, 3), (1, 3), (2, 3) **29.** (−7, 0), (−7, 1), (−7, 2) **30.** (0, −4), (1, −3), (2, −2) **31.** (0, 2), (7, 0), (14, −2) **32.** (2, 4)

Page 231
1. 42 **2.** 720 **3.** 35 **4.** 36 **5.** 3,060 **6.** 2,380

Page 232
1. quad. 2 **2.** quad. 1 **3.** quad. 3 **4.** quad. 4 **5.** 11 **6.** 5 **7.** 6 **8.** 2 **9.** 1 **10.** 3 **11.** 4 **12.** +9 **13.** +1 **14.** −12 **15.** $\frac{2}{9}$ **16.** $-\frac{1}{2}$ **17.** undefined
18. $y = -\frac{7}{3}x - 2$ **19.** $y = 8x - 18$
20. $y = -\frac{1}{3}x + 5$ **21.** $m = \frac{2}{3}$, $b = -1$ **22.** $m = 4$, $b = -7$ **23.** $m = -12$, $b = 14$ **24.** (0, −1), (1, 4), (2, 9) **25.** (0, 4), (2, 7), (4, 10) **26.** (0, −1), (1, 6), (2, 13) **27.** (0, −3), (1, −3), (2, −3) **28.** (7, 0), (7, 1), (7, 2) **29.** yes **30.** yes

Page 233
1. quad. 4 **2.** *y*-axis **3.** quad. 2 **4.** quad. 3 **5.** 4 **6.** 12 **7.** 4 **8.** 3rd quadrant **9.** 1st quadrant **10.** 2nd quadrant **11.** −4 **12.** 4 **13.** 10 **14.** $-\frac{1}{4}$ **15.** $-\frac{1}{7}$ **16.** 0 **17.** undefined
18. $y = \frac{1}{3}x - 3$ **19.** $y = 4x - 26$ **20.** $y = 4x - 26$
21. $y = -8x + 3$ **22.** $m = -\frac{3}{5}$, $b = 7$
23. $m = -\frac{3}{2}$, $b = 4$ **24.** $m = 2$, $b = -6$
25. (0, −3), (1, −5), (2, −7) **26.** (0, 4), (2, 1), (4, −2) **27.** (0, −1), (3, 1), (6, 3) **28.** (0, −1), (1, −1), (2, −1) **29.** (4, 0), (4, 1), (4, 2) **30.** yes **31.** no

Page 237

ORALS **1.** yes **2.** no **3.** yes **4.** yes **5.** yes

EXERCISES **1.** $4\sqrt{6}$ **2.** $\sqrt{161}$ **3.** $\sqrt{65}$ **4.** $7\sqrt{3}$ **5.** $\sqrt{61}$ **6.** $2\sqrt{3}$ **7.** $\sqrt{89}$ **8.** 8 **9.** $\sqrt{5}$ **10.** 4 **11.** no **12.** yes **13.** yes **14.** yes **15.** no **16.** yes **17.** yes **18.** yes **19.** yes **20.** yes **21.** 12 m **22.** 1,200 cm^2

Page 239
1. $2\sqrt{5}$ **2.** $\sqrt{89}$ **3.** $\sqrt{97}$ **4.** 15 **5.** $3\sqrt{13}$ **6.** $\sqrt{58}$ **7.** $\sqrt{149}$ **8.** 5 **9.** $\sqrt{41}$ **10.** $5\sqrt{5}$ **11.** $2\sqrt{5}$ **12.** $\sqrt{146}$ **13.** $\sqrt{10}$ **14.** $\sqrt{85}$ **15.** $2\sqrt{2}$ **16.** $\sqrt{433}$ **17.** $6\sqrt{2}$ **18.** $3\sqrt{10}$ **19.** $\sqrt{178}$ **20.** $3\sqrt{29}$ **21.** $\frac{\sqrt{229}}{6}$ **22.** $\frac{7\sqrt{5}}{4}$ **23.** $\frac{\sqrt{221}}{6}$ **24.** $\frac{\sqrt{5}}{2}$ **25.** $\frac{17}{20}$ **26.** $\frac{\sqrt{2}}{3}$ **27.** $\sqrt{4a^2 + b^2}$ **28.** $\sqrt{25x^2 + 4y^2}$ **29.** $\sqrt{16s^2 + 9p^2}$

Page 240
\$37.23

Page 243
1. (4, 2) **2.** (3, 4) **3.** (2, 0) **4.** $\left(5, \frac{3}{2}\right)$ **5.** (4, 5) **6.** (−1, 3) **7.** (2, −3) **8.** (5, 2) **9.** (2, 1) **10.** (4, 3) **11.** (6, 4) **12.** (1, 3) **13.** (3, 2) **14.** (4, −5) **15.** (−4, −3) **16.** (−5, −6) **17.** (−5, −5) **18.** (−5, −3) **19.** (−5, −3) **20.** (−5, −5) **21.** $\left(-\frac{13}{2}, -\frac{13}{2}\right)$ **22.** (8, 7) **23.** (9, 8) **24.** (0, 13) **25.** (−1, −2) **26.** (5, 6) **27.** (4, −2) **28.** (−12, 0) **29.** (8, 1) **30.** (5, 12) **31.** (10, 3) **32.** (0, −2) **33.** (9, 9) **34.** (4, −20) **35.** (8, −12) **36.** (0, 4) **37.** (−9, −14) **38.** (8, 9) **39.** (4, 7) **40.** (−14, −12) **41.** (9, −3) **42.** (11, 13) **43.** $M = \left(0, \frac{1}{2}\right)$, $N = (3, 0)$, $T = (3, 2)$, $V = \left(0, 2\frac{1}{2}\right)$; $MN = \frac{\sqrt{37}}{2}$ and $VT = \frac{\sqrt{37}}{2}$ therefore $MN = VT$; $MV = 2$ and $NT = 2$ therefore $MV = NT$ **44.** $T = \left(\frac{d+f}{2}, \frac{e+g}{2}\right)$, $V = \left(\frac{f+a}{2}, \frac{g}{2}\right)$, $M = \left(\frac{a+b}{2}, \frac{c}{2}\right)$, $N = \left(\frac{d+b}{2}, \frac{e+c}{2}\right)$; $MN = \frac{\sqrt{a^2 - 2ad + d^2 + e^2}}{2}$ and $VT = \frac{\sqrt{a^2 - 2ad + d^2 + e^2}}{2}$ therefore $MN = VT$; $MV = \frac{\sqrt{b^2 - 2bf + f^2 + c^2 - 2cg + g^2}}{2}$ and $NT = \frac{\sqrt{b^2 - 2bf + f^2 + c^2 - 2cg + g^2}}{2}$ therefore $MV = NT$

Page 246
1. $\frac{2}{3}$ **2.** $\frac{6}{5}$ **3.** $\frac{1}{2}$ **4.** $-\frac{3}{4}$ **5.** $-\frac{5}{3}$ **6.** $\frac{2}{7}$ **7.** $-\frac{6}{5}$ **8.** −2 **9.** $y = \frac{3}{2}x + 4$ **10.** $y = \frac{2}{3}x$ **11.** $y = \frac{1}{2}x + 2$ **12.** $y = -\frac{1}{2}x + 5$ **13.** $y = 2$ **14.** $y = 6$ **15.** $y = 2x - 5$ **16.** $y = -\frac{2}{3}x - \frac{7}{3}$ **17.** $y = -\frac{4}{3}x + \frac{8}{3}$ **18.** $x = -3$ **19.** $y = -\frac{3}{2}x + \frac{5}{2}$ **20.** $y = -\frac{4}{3}x + \frac{5}{3}$ **21.** neither **22.** perpendicular **23.** $y = -\frac{3}{2}x + \frac{13}{2}$ **24.** $y = \frac{3}{4}x - 2$ **25.** $y = \frac{1}{5}x + 1$

Page 249
1. $m(\overline{BC}) = -\frac{1}{3}$, $m(\overline{AB}) = 3$ **2.** $m(\overline{AB}) = \frac{1}{2}$, $m(\overline{BC}) = -2$ **3.** $AB = \sqrt{68}$, $BC = \sqrt{68}$, $AC = \sqrt{136}$; $(\sqrt{136})^2 = (\sqrt{68})^2 + (\sqrt{68})^2$ **4.** $AB = \sqrt{20}$, $BC = \sqrt{20}$, $AC = \sqrt{40}$; $(\sqrt{40})^2 = (\sqrt{20})^2 + (\sqrt{20})^2$ **5.** yes **6.** yes **7.** yes **8.** yes **9.** no **10.** no **11.** $m(\overline{AB}) = m(\overline{DC}) = 2$; $m(\overline{AD}) = m(\overline{BC}) = \frac{3}{5}$ **12.** $AB = BC = \sqrt{34}$ **13.** midpoint $\overline{AB}$(5, 2), midpoint $\overline{BC}$(6, 4) **14.** midpoint $\overline{BC}$(5, 7), midpoint $\overline{AC}$(1, 3) **15.** $m(\overline{RV}) = m(\overline{ST}) = \frac{3}{2}$; $m(\overline{RS}) = m(\overline{VT}) = 0$ **16.** no **17.** $RV = ST = \sqrt{13}$, $VT = RS = 6$ **18.** midpoint $\overline{RT}$ = midpoint $\overline{VS} = \left(4, \frac{3}{2}\right)$ **19.** parallelogram **20.** parallelogram

Page 250
1. $b = 12$ **2.** $c = 3$ **3.** $a = 4\sqrt{10}$ **4.** $c = 2\sqrt{13}$ **5.** $a = 4$ **6.** no **7.** yes **8.** yes **9.** yes **10.** no **11.** $15\sqrt{2}$ m **12.** 50 km **13.** $\sqrt{53}$ **14.** $\sqrt{10}$ **15.** $\sqrt{113}$ **16.** $\sqrt{449}$ **17.** (5.5, .5) **18.** (2.5, −4) **19.** (5.5, 6) **20.** (8, 3.5) **21.** (5, −2) **22.** (2, 1) **23.** (12, 7) **24.** (−9, −3) **25.** $\frac{3}{4}$ **26.** $\frac{8}{5}$ **27.** $\frac{1}{3}$ **28.** $\frac{1}{2}$ **29.** $-\frac{6}{5}$ **30.** $\frac{1}{2}$ **31.** $\frac{3}{2}$ **32.** $\frac{5}{4}$ **33.** $y = \frac{3}{5}x + \frac{7}{5}$ **34.** $y = -\frac{2}{3}x + \frac{5}{3}$ **35.** $y = \frac{7}{3}x + \frac{11}{3}$ **36.** $y = \frac{5}{2}x - \frac{1}{2}$ **37.** $y = \frac{1}{2}x - 8$ **38.** $y = -\frac{2}{3}x + 2\frac{1}{3}$ **39.** $\overleftrightarrow{PQ} \parallel \overleftrightarrow{RS}$ **40.** $\overleftrightarrow{PQ} \perp \overleftrightarrow{RS}$ **41.** $m(\overline{BC}) = \frac{1}{2}$; $m(\overline{AC}) = -2$ **42.** $BC = \sqrt{97}$, $AC = \sqrt{97}$, $AB = \sqrt{194}$ **43.** $m(\overline{AB}) = m(\overline{BC}) = m(\overline{AC}) = -2$ **44.** $m(\overline{AD}) = m(\overline{BC}) = 3$, $m(\overline{AB}) = m(\overline{DC}) = -\frac{1}{3}$ **45.** $M\left(\frac{1}{2}, \frac{3}{2}\right)$, $N\left(\frac{9}{2}, 4\right)$, $m(\overline{MN}) = m(\overline{AB}) = \frac{5}{8}$, $MN = \frac{1}{2}\sqrt{89}$, $AB = \sqrt{89}$

Page 251
1. $\sqrt{15}$ **2.** $\sqrt{66}$ **3.** yes **4.** yes **5.** $50\sqrt{5}$ **6.** $\sqrt{13}$ **7.** $\sqrt{5}$ **8.** (.5, −1) **9.** (14, 10) **10.** $\frac{2}{5}$ **11.** $-\frac{1}{3}$ **12.** $-\frac{2}{3}$ **13.** −4 **14.** $y = -\frac{1}{2}x$ **15.** $y = -\frac{5}{3}x - 6$ **16.** $y = \frac{3}{2}x + \frac{1}{2}$ **17.** $y = \frac{2}{3}x - \frac{13}{3}$ **18.** $\overleftrightarrow{PQ} \parallel \overleftrightarrow{RS}$ **19.** $\overleftrightarrow{PQ} \perp \overleftrightarrow{RS}$ **20.** $m(\overline{BC}) = -\frac{4}{3}$, $m(\overline{AC}) = \frac{3}{4}$ **21.** $(\sqrt{25})^2 + (\sqrt{25})^2 = (\sqrt{50})^2$ **22.** yes **23.** $m(\overline{AB}) = \frac{2}{3}$ and $m(\overline{CD}) = \frac{2}{3}$ therefore $\overleftrightarrow{AB} \parallel \overleftrightarrow{CD}$; $m(\overline{BC}) = -\frac{5}{3}$ and $m(\overline{AD}) = -\frac{5}{3}$ therefore $\overleftrightarrow{BC} \parallel \overleftrightarrow{AD}$; quadrilateral $ABCD$ is a parallelogram since its opposite sides are parallel. **24.** $M\left(\frac{1}{2}, 2\right)$, $N\left(3, \frac{11}{2}\right)$

Page 255

ORALS **1.** $D = \{3, -1\}$, $R = \{5, 3, -4\}$ **2.** $D = \{-2, 6, 7\}$, $R = \{-3, 1\}$ **3.** $D = \{8, -7\}$, $R = \{-1, 6, -6\}$

EXERCISES **1.** $D = \{0, 2, -1, 3, 4\}$, $R = \{0, 3, 2\}$; yes **2.** $D = \{8, 7, -1, -7, 0\}$, $R = \{0, 6, -1, 7\}$; yes **3.** $D = \{-1, -2, -3\}$, $R = \{-2, -1, -3\}$; no **4.** $D = \{3, 4, -4, 2\}$, $R = \{-2, 2, -3\}$; yes **5.** $D = \{4, 6, -2, 5, 0\}$, $R = \{-1, 3, 0\}$; yes **6.** $D = \{4, -4, 0\}$, $R = \{0, -4, 4\}$; no **7.** $D = \{-1, -2, -3, 4\}$, $R = \{4\}$; yes **8.** $D = \{-1\}$, $R = \{2, -2, -3, -4\}$; no **9.** $D = \{2, 3, -2, -3\}$, $R = \{-3\}$; yes **10.** $D = \{5, 4, -1, -4, -5\}$, $R = \{1, 2, -2, -5\}$; yes **11.** $\{(-2, 1), (-1, 2), (1, 1), (2, 2), (3, -1)\}$, $D = \{-2, -1, 1, 2, 3\}$, $R = \{1, 2, -1\}$; yes **12.** $\{(-1, 1), (-1, -1), (2, 2), (3, 3)\}$, $D = \{-1, 2, 3\}$, $R = \{1, -1, 2, 3\}$; no **13.** $\{(-2, -1), (-1, 1), (1, 1), (2, -1), (4, 3)\}$, $D = \{-2, -1, 1, 2, 4\}$, $R = \{-1, 1, 3\}$; yes **14.** $D = \{x | -2 \leq x \leq 2\}$, $R = \{y | -2 \leq y \leq 2\}$; yes **15.** $D = \{x | 0 \leq x < 3\}$, $R = \{y | -2 < y < 2\}$; no **16.** $D = \{x | -3 \leq x \leq 3\}$, $R = \{y | -3 \leq y \leq 3\}$; no **17.** $D = \{x | -2 < x \leq 2 \text{ and } x \neq -2, -1, 0, 1\}$, $R = \{y | -2 \leq y \leq 2\}$; yes **18.** $D = \{x | 0 \leq x\}$, $R = \{y\}$; no **19.** $D = \{x | x \neq -2, 2\}$, $R = \{y | -3 < y \text{ and } y \neq -2\}$; yes

Page 257

1. 6 **2.** 3 **3.** -7 **4.** 8 **5.** 1 **6.** 4 **7.** 8 **8.** 4 **9.** 5 **10.** 13 **11.** 5 **12.** 20 **13.** $\{5, 11, 20\}$ **14.** $\{8, 3, -7\}$ **15.** $\{26, 8, 98\}$ **16.** $\{0, 12\}$ **17.** $\{9, 4, 1\}$ **18.** $\left\{-1, \frac{3}{5}, -\frac{29}{5}, \frac{23}{40}\right\}$ **19.** $\{4, 1, 0, 2\}$ **20.** $\{0, \sqrt{2}, 2, 3\}$ **21.** 10 **22.** 8 **23.** -9 **24.** $a + t$ **25.** $2a + b^2 - 3$ **26.** $2a + 2h - 3$ **27.** $2ah + h^2$ **28.** 15

Page 260

1. linear function **2.** constant function **3.** not a function **4.** not a function **5.** linear function **6.** linear function **7.** linear function **8.** linear function **9.** linear function **10.** linear function **11.** not a function **12.** constant function **13.** constant function **14.** linear function **15.** linear function **16.** linear function **17.** function **18.** function **19.** function **20.** function **21.** function **22.** not a function **23.** function **24.** function **25.** function **26.** function **27.** function **28.** function

Page 263

ORALS **1.** yes; $k = 15$ **2.** no **3.** yes; $k = \frac{1}{4}$ **4.** yes; $k = 6$ **5.** yes; $k = -4$ **6.** yes; $k = 3.14$ **7.** no **8.** yes; $k = 6$ **9.** no **10.** yes; $k = 7$

EXERCISES **1.** yes; $k = -\frac{1}{2}$ **2.** no **3.** no **4.** yes; $k = 5$ **5.** yes; $k = -1$ **6.** -32 **7.** -32 **8.** -63 **9.** -6 **10.** 10 **11.** 6 **12.** 36 **13.** 40 **14.** 100 **15.** 120 **16.** 18 **17.** 24 **18.** line **19.** decreases, increases **20.** line **21.** increases, decreases **22.** 21.98 **23.** 360 **24.** 18 **25.** 360 km

Page 266

ORALS **1.** yes; $k = 12$ **2.** yes; $k = 16$ **3.** no **4.** yes; $k = 12$ **5.** no **6.** yes; $k = 12$ **7.** yes; $k = 2$ **8.** no **9.** no **10.** no

EXERCISES **1.** yes; $k = 45$ **2.** no **3.** no **4.** yes; $k = 100$ **5.** yes; $k = 1$ **6.** 6 **7.** 3 **8.** -2 **9.** 3 **10.** 3 **11.** 1 **12.** 12 **13.** $1\frac{7}{9}$ **14.** $4\frac{1}{2}$ **15.** $\frac{35}{6}$ **16.** 1st and 3rd quadrants **17.** decreases, increases **18.** 1st and 3rd quadrants **19.** increases **20.** 100 **21.** 4 **22.** 400 **23.** 12

Page 269

ORALS **1.** inverse **2.** direct **3.** inverse **4.** joint **5.** direct **6.** joint

EXERCISES **1.** joint **2.** joint **3.** no **4.** joint **5.** no **6.** 112 **7.** $\frac{20}{3}$ **8.** 18 **9.** 128 **10.** 60 **11.** 144 **12.** 120 **13.** 18 **14.** $\frac{7}{5}$ **15.** $\frac{400}{9}$ **16.** 12 **17.** 128 **18.** 240 **19.** 2.22×10^{-6} dynes **20.** 28,160 kg

Page 270

1. $D = \{4, -1, 3, -5\}$, $R = \{3, 2\}$; function **2.** $D = \{0, 5, -3, 7\}$, $R = \{3, 0, 6\}$; function **3.** $D = \{-1, -2, -3\}$, $R = \{2, 3\}$; function **4.** $D = \{-2, 1, 2, 3\}$, $R = \left\{-1, 1, 1\frac{1}{2}, 2\right\}$; function **5.** $D = \{-1, 1, 2, 3\}$, $R = \{1, -1, 2\}$; not a function **6.** $D = \{x | -1 \leq x \leq 3\}$, $R = \{y | -1 \leq y \leq 3\}$; function

7. $D = \{x|0 \le x < 2\}$, $R = \{y|-2 < y < 2\}$; not a function **8.** 3 **9.** 1 **10.** 0 **11.** 7 **12.** 2 **13.** 1 **14.** 5 **15.** 10 **16.** $R = \{-4, -1, 8\}$ **17.** $R = \{-1, 0, 3\}$ **18.** $R = \left\{4, 1, \frac{1}{4}\right\}$ **19.** not a function **20.** linear function **21.** function **22.** function **23.** function **24.** linear function **25.** constant function **26.** direct; $k = 6$ or $\frac{1}{6}$ **27.** inverse; $k = 12$ **28.** inverse; $k = 24$ **29.** direct; $k = -3$ or $-\frac{1}{3}$ **30.** inverse; $k = -14$ **31.** 30 **32.** 16 **33.** 2 **34.** 5 **35.** 180 **36.** 9 **37.** 120 **38.** 2

Page 271

1. $D = \{-2, -3, 4, 6\}$, $R = \{1, -3\}$; function **2.** $D = \{0, 3, 5, 1\}$, $R = \{3, 0\}$; function **3.** $D = \{-2, -1, 0, 1, 2, 3\}$, $R = \{-1, 0, 1, 2\}$; function **4.** $D = \{-2 < x < 2\}$, $R = \{0 \le y < 2\}$; function **5.** $D = \{-2 \le x \le 2\}$, $R = \{-2 \le y \le 2\}$; function **6.** 4 **7.** -3 **8.** 0 **9.** 1 **10.** 3 **11.** 9 **12.** $R = \{4, 3, 12\}$ **13.** $R = \{3, 19, 23\}$ **14.** linear function **15.** not a function **16.** constant function **17.** linear function **18.** constant function **19.** function **20.** inverse; $k = 24$ **21.** direct; $k = 5$ **22.** inverse; $k = 18$ **23.** 30 **24.** 2 **25.** 18 **26.** $\frac{25}{12}$

Page 276

1. $x = -\frac{3}{2}$; $\left(-\frac{3}{2}, \frac{9}{2}\right)$ **2.** $x = -1$; $(-1, 8)$ **3.** $x = \frac{3}{4}$; $\left(\frac{3}{4}, -\frac{49}{4}\right)$ **4.** $x = -3$; $(-3, 3)$ **5.** $x = -2$; $(-2, -7)$ **6.** $x = 3$; $(3, 13)$ **7.** $x = 2$; $(2, -1)$ **8.** $x = 0$; $(0, 0)$ **9.** $x = 0$; $(0, 0)$ **10.** $x = 0$; $(0, 5)$ **11.** $x = 1$; $(1, 3)$ **12.** $x = -4$; $(-4, -13)$ **13.** .4; -3.9 **14.** 2.7; $-.7$ **15.** 4.2; $-.2$ **16.** 3.5; $-.5$ **17.** 4.4; $-.4$ **18.** -2; .3 **19.** $(0, 0)$; $x = 0$ **20.** $(0, -2)$; $x = 0$ **21.** Turning point is on axis of symmetry $x = -\dfrac{b}{2a}$. Substitute $x = -\dfrac{b}{2a}$ in $y = ax^2 + bx + c$. Solve for y.

Page 278

ORALS **1.** 5 **2.** 8 **3.** 4 **4.** 7 **5.** $\sqrt{15}$

EXERCISES **1.** $x^2 + y^2 = 64$ **2.** $x^2 + y^2 = 25$ **3.** $x^2 + y^2 = 121$ **4.** $x^2 + y^2 = 1$ **5.** $x^2 + y^2 = 169$ **6.** $x^2 + y^2 = 400$ **7.** $x^2 + y^2 = 16$ **8.** $x^2 + y^2 = 9$ **9.** $x^2 + y^2 = 4$ **10.** $x^2 + y^2 = 36$ **11.** $x^2 + y^2 = 81$ **12.** $x^2 + y^2 = 49$ **13.** $x^2 + y^2 = 100$ **14.** $x^2 + y^2 = 144$ **15.** $x^2 + y^2 = 225$ **16.** $x^2 + y^2 = 6$ **17.** $x^2 + y^2 = 7$ **18.** $x^2 + y^2 = 3$ **19.** $x^2 + y^2 = 169$ **20.** $x^2 + y^2 = 34$ **21.** $x^2 + y^2 = 5$ **22.** $x^2 + y^2 = 25$ **23.** $x^2 + y^2 = 25$ **24.** $x^2 + y^2 = 16$ **25.** $x^2 + y^2 = 13$ **26.** $x^2 + y^2 = 100$ **27.** $x^2 + y^2 = 9$ **28.** $x^2 + y^2 = 41$ **29.** 4 **30.** 10 **31.** 2 **32.** 3 **33.** 6 **34.** 9 **35.** $\sqrt{17}$ **36.** $2\sqrt{3}$ **37.** $\sqrt{6}$ **38.** $\sqrt{3}$ **39.** circle, $r = \sqrt{6}$ **40.** not a circle **41.** circle, $r = \sqrt{7}$ **42.** not a circle **43.** circle, $r = \frac{1}{2}\sqrt{35}$ **44.** not a circle **45.** circle, $r = \frac{1}{2}\sqrt{17}$ **46.** not a circle **47.** circle, $r = 2$ **48.** circle, $r = \dfrac{4\sqrt{3}}{3}$ **49.** circle, $r = \sqrt{2}$ **50.** circle, $r = 2\sqrt{2}$ **51.** circle, $r = \frac{1}{4}$ **52.** circle, $r = \dfrac{\sqrt{6}}{6}$ **53.** not a circle **54.** $(x - 3)^2 + (y - 2)^2 = 36$ **55.** $(x + 1)^2 + (y - 2)^2 = 16$ **56.** $(x - 4)^2 + (y - 1)^2 = 25$ **57.** $(x + 1)^2 + (y + 2)^2 = 50$

Page 282

ORALS **1.** $x = \pm 4$, $y = \pm 2$ **2.** $x = \pm 9$, $y = \pm 4$ **3.** $x = \pm 3$, $y = \pm 5$ **4.** $x = \pm 5$, $y = \pm 5\sqrt{3}$ **5.** $x = \pm 10$, $y = \pm 5$ **6.** $x = \pm 3$, $y = \pm\sqrt{6}$

EXERCISES **1.** $\dfrac{x^2}{25} + \dfrac{y^2}{9} = 1$ **2.** $\dfrac{x^2}{49} + \dfrac{y^2}{64} = 1$ **3.** $\dfrac{x^2}{4} + \dfrac{y^2}{3} = 1$ **4.** $\dfrac{x^2}{64} + \dfrac{y^2}{3} = 1$ **5.** $\dfrac{x^2}{8} + \dfrac{y^2}{36} = 1$ **6.** $\dfrac{x^2}{5} + \dfrac{y^2}{7} = 1$ **7.** $x = \pm 5$, $y = \pm 2$ **8.** $x = \pm 4$, $y = \pm 9$ **9.** $x = \pm 5$, $y = \pm 3$ **10.** $x = \pm 10$, $y = \pm 5$ **11.** $x = \pm 4$, $y = \pm 2$ **12.** $x = \pm 3$, $y = \pm 10$ **13.** $x = \pm 5$, $y = \pm 10$ **14.** $x = \pm\sqrt{5}$, $y = \pm 1$ **15.** $x = \pm 5$, $y = \pm 5\sqrt{3}$ **16.** $x = \pm 6$, $y = \pm 2$ **17.** $x = \pm 4$, $y = \pm 4\sqrt{2}$ **18.** $x = \pm 4$, $y = \pm 2\sqrt{6}$ **19.** $a = \pm\sqrt{10}$, $b = \pm\sqrt{15}$ **20.** $a = \pm 2$, $b = \pm\sqrt{\dfrac{16}{5}}$ **21.** $a = \pm\sqrt{\dfrac{7}{8}}$,

$b = \pm\sqrt{\frac{7}{2}}$ **22.** After squaring equation twice, equation simplifies to $400 = 16x^2 + 25y^2$, which is then simplified to $1 = \frac{x^2}{25} + \frac{y^2}{16}$.

23. $\frac{(x+1)^2}{16} + \frac{y^2}{12} = 1$ **24.** $\frac{x^2}{36} + \frac{y^2}{32} = 1$

Page 285

ORALS **1.** ± 2 **2.** ± 6 **3.** ± 4 **4.** ± 3

EXERCISES **1.** $a = \pm 2, b = \pm 4$ **2.** $a = \pm 6, b = \pm 3$ **3.** $a = \pm 3, b = \pm 4$ **4.** $a = \pm 5, b = \pm 6$ **5.** $a = \pm 10, b = \pm 5$ **6.** $a = \pm 9, b = \pm 4$ **7.** $a = \pm 5, b = \pm 2$ **8.** $a = \pm 7, b = \pm 2$ **9.** $a = \pm 3, b = \pm 6$ **10.** $a = \pm 3, b = \pm 2$ **11.** $a = \pm 3, b = \pm 4$ **12.** $a = \pm 2, b = \pm 4$ **13.** 1st and 3rd quadrants **14.** 1st and 3rd quadrants **15.** 1st and 3rd quadrants **16.** 1st and 3rd quadrants **17.** $a = \pm 2\sqrt{2}, b = \pm\sqrt{2}$ **18.** $a = \pm 2\sqrt{5}, b = \pm 4$ **19.** 2nd and 4th quadrants **20.** 2nd and 4th quadrants

21. $y = \pm\frac{b}{a}x$ **22.** $y = \pm 2x$ **23.** $y = \pm\frac{1}{2}x$

24. $y = \pm\frac{4}{3}x$ **25.** $y = \pm\frac{6}{5}x$ **26.** $y = \pm\frac{1}{2}x$

27. $y = \pm\frac{4}{9}x$ **28.** $y = \pm\frac{2}{5}x$ **29.** $y = \pm\frac{2}{7}x$

30. $y = \pm 2x$ **31.** $y = \pm\frac{2}{3}x$ **32.** $y = \pm\frac{4}{3}x$

33. $y = \pm 2x$

Page 287

1. ellipse **2.** circle **3.** hyperbola **4.** hyperbola **5.** ellipse **6.** hyperbola **7.** ellipse **8.** circle **9.** hyperbola **10.** circle **11.** parabola **12.** hyperbola **13.** hyperbola **14.** circle **15.** ellipse **16.** hyperbola **17.** hyperbola **18.** circle **19.** parabola **20.** parabola **21.** hyperbola **22.** ellipse **23.** hyperbola **24.** circle **25.** ellipse **26.** ellipse **27.** circle **28.** hyperbola **29.** ellipse **30.** circle **31.** circle **32.** hyperbola **33.** ellipse

Page 290

ORALS **1.** same **2.** 5 units to right **3.** 1 unit up **4.** 3 units to left, 2 units down **5.** same **6.** 3 units to left **7.** 2 units down **8.** 1 unit to right, 3 units down

EXERCISES **1.** 2 units to left **2.** 2 units to right **3.** 3 units to left, 1 unit up **4.** 3 units to right, 2 units down **5.** 2 units to right **6.** 2 units to left **7.** 3 units to left, 3 units down **8.** 3 units to right, 1 unit up **9.** 2 units up **10.** 1 unit to left, 2 units down **11.** 4 units to right, 3 units down **12.** 2 units to left, 4 units up

13. $y = (x+4)^2 - 12$ **14.** $y = (x+2)^2 - 5$ **15.** $y = (x+3)^2 - 13$ **16.** $y = (x+1)^2 - 6$ **17.** $y = (x+5)^2 - 3$ **18.** $y = (x+2)^2 - 2$

19. circle, center $(1, -1)$, $r = 4$ **20.** circle, center $(-2, 1)$, $r = 2\sqrt{3}$ **21.** circle, center $(-1, 2)$, $r = 1$ **22.** ellipse, center $(-2, 1)$, $a = \pm 4$, $b = \pm 2$ **23.** hyperbola, center $(-1, -2)$, $a = \pm 5$, $b = \pm 2$ **24.** parabola, $x = 3$

Page 291

1. hyperbola **2.** circle

Page 292

1. $(1, 5)$, $x = 1$ **2.** $(1, -3)$, $x = 1$ **3.** $(-2, -7)$, $x = -2$ **4.** $x = 3$, $(3, 3)$ **5.** $x = 2$, $(2, 10)$ **6.** $x = -6$, $(-6, 6)$ **7.** -2.9, $.2$ **8.** 3.6, $-.2$ **9.** -4.7, $.7$ **10.** $x = 0$, $(0, 0)$ **11.** $x^2 + y^2 = 169$ **12.** $x^2 + y^2 = 15$ **13.** $x^2 + y^2 = 289$ **14.** $x^2 + y^2 = 8$ **15.** $x^2 + y^2 = 10$ **16.** $x^2 + y^2 = 625$ **17.** $x^2 + y^2 = 5$ **18.** $x^2 + y^2 = 13$ **19.** $x^2 + y^2 = 65$ **20.** $x^2 + y^2 = 16$ **21.** $x^2 + y^2 = 34$ **22.** circle, $r = 7$ **23.** circle, $r = 2$ **24.** circle, $r = \sqrt{2}$

25. not a circle **26.** $\frac{x^2}{16} + \frac{y^2}{4} = 1$

27. $\frac{x^2}{49} + \frac{y^2}{9} = 1$ **28.** $\frac{x^2}{25} + \frac{y^2}{25} = 1$

29. $\frac{x^2}{4} + \frac{y^2}{64} = 1$ **30.** $x = \pm 6, y = \pm 2$

31. $x = \pm 7, y = \pm 4$ **32.** $x = \pm 4, y = \pm 2$ **33.** $x = \pm 5, y = \pm\sqrt{5}$ **34.** $a = \pm 6, b = \pm 2$ **35.** 2nd and 4th quadrants **36.** 1st and 3rd quadrants **37.** $a = \pm 2, b = \pm 4$. **38.** hyperbola **39.** ellipse **40.** circle **41.** parabola **42.** 4 units down **43.** 2 units to right, 3 units up **44.** 6 units to left, 2 units up **45.** 3 units down

46. $y = (x+3)^2 - 5$ **47.** $y = (x-5)^2 - 5$ **48.** $y = (x+6)^2 - 6$

Page 293

1. $(-4, -18)$, $x = -4$ **2.** $(0, 1)$, $x = 0$ **3.** $x = -\frac{3}{2}$, $\left(-\frac{3}{2}, -\frac{25}{4}\right)$ **4.** $x = 3$, $(3, -4)$ **5.** $.1$, -1.9 **6.** 1.9, $-.2$ **7.** $x = 0$, $(0, 0)$ **8.** $x^2 + y^2 = 100$ **9.** $x^2 + y^2 = 13$ **10.** $x^2 + y^2 = 900$ **11.** $x^2 + y^2 = 25$ **12.** $x^2 + y^2 = 37$ **13.** circle, $r = 5$ **14.** circle,

$r = 2$ **15.** not a circle **16.** $\frac{x^2}{9} + \frac{y^2}{4} = 1$

17. $\frac{x^2}{36} + \frac{y^2}{4} = 1$ **18.** $x = \pm 8$, $y = \pm 6$
19. $x = \pm 2$, $y = \pm 9$ **20.** $x = \pm 6$, $y = \pm 3$
21. $a = \pm 8$, $b = \pm 4$ **22.** 1st and 3rd quadrants
23. $a = \pm 3$, $b = \pm 9$ **24.** ellipse **25.** hyperbola
26. circle **27.** hyperbola **28.** 6 units to right
29. 3 units to right, 1 unit up **30.** 1 unit to right, 3 units up **31.** $y = (x - 4)^2 - 5$

Page 294
1. 6.58% **2.** 11.0% **3.** .29% **4.** 1.74% **5.** 1.04%
6. 5.64%

Page 297

ORALS **1.** (1, −1) **2.** (−1, −2) **3.** (2, 1)
4. (−1, 1)

EXERCISES **1.** (3, 1) **2.** (3, −4) **3.** (1, −5)
4. (2.5, 5) **5.** (0, 1) **6.** (2, 3) **7.** (.5, 0)
8. (3, 1.5) **9.** (2, 1) **10.** (−3, −2) **11.** (3, −2)
12. (7, 2) **13.** inconsistent **14.** consistent
15. inconsistent **16.** consistent **17.** same slopes
18. different slopes **19.** same line

Page 301

ORALS **1.** (2 and 3) or (5 and −2)
2. (7 and 3) or (−2 and 5) **3.** (2 and 5) or (7 and −4) **4.** (−2 and 3) or (−5 and 6)
5. (−4 and 9) or (−3 and 5) **6.** (−2 and 1) or (−7 and 6)

EXERCISES **1.** (1, 2) **2.** (−1, −2)
3. (2, −1) **4.** (4, −1) **5.** (4, 1) **6.** (−3, −1)
7. (7, 1) **8.** (−2, 1) **9.** (3, 0) **10.** (10, 2)
11. (−1, −3) **12.** (−3, −1) **13.** (2, −3)
14. (−1, 2) **15.** (2, 3) **16.** (1, 3) **17.** (1, 2)
18. (−1, 2) **19.** (3, −2) **20.** (1, −2) **21.** (2, 0)
22. (0, 3) **23.** $\left(\frac{b - c}{2}, \frac{b - 3c}{2}\right)$
24. $\left(\frac{a - 3b}{5}, \frac{a + 2b}{5}\right)$ **25.** (4, 6) **26.** (10, 12)
27. (−21, 21) **28.** (40, 21) **29.** $\left(-\frac{8}{5}, \frac{18}{5}\right)$
30. (−5, 49)

Page 302
1. 3 **2.** −2 **3.** 2 **4.** −4 **5.** $x = 5$, $y = -1$
6. $x = -4$, $y = -2$ **7.** $x = 3$, $y = \frac{1}{2}$

Page 305
1. (3, −2, 1) **2.** (2, 3, −1) **3.** (2, 1, −1)
4. (−3, 0, 4) **5.** (7, −3, 6) **6.** (7, 3, 0)
7. (−1, −1, −2) **8.** (2, −3, 4) **9.** (4, −5, 8)

Page 306
1. −63 **2.** 8 **3.** −28 **4.** $y = -2$, $z = 1$

Page 310
1. (4, 1), (3, 0) **2.** (1.6, 1.3), (.7, −2)
3. (−7.8, 4.9), (6.4, −2.2) **4.** (4, 2) **5.** (.7, −1.8), (2.5, .9) **6.** (4.0, .3), (16.5, −8) **7.** no points in common **8.** (0, −2), (1.2, −1.6) **9.** (1, 2), (−1, −2) **10.** (4, 4), (−4, −4) **11.** (5, 4), (4, 5)
12. (4, −1), (14, 19) **13.** (5, 3), (2, 0)
14. (1, −3), $\left(-\frac{11}{19}, \frac{63}{19}\right)$ **15.** (0, −3), (3, 0)
16. (3, −2), (2, −3) **17.** (0, −3), (2, 1)
18. (3, −3), $\left(-\frac{3}{5}, \frac{21}{5}\right)$ **19.** (1, −1), $\left(-\frac{3}{11}, \frac{17}{11}\right)$
20. (2, −3), $\left(-\frac{2}{3}, -\frac{1}{3}\right)$

Page 313

ORALS **1.** 4, circle and ellipse; (3.2, 2.5), (−3.2, 2.5), (−3.2, −2.5), (3.2, −2.5) **2.** 4, parabola and hyperbola; approx. (4, 3), (−4, 3), (−1.5, −1), (1.5, −1) **3.** 2, parabola; (0, 0) and (3, 0)

EXERCISES **1.** (4, 0), (−4, 0) **2.** $(\sqrt{3}, 2)$, $(-\sqrt{3}, 2)$, $(\sqrt{3}, -2)$, $(-\sqrt{3}, -2)$ **3.** (0, 2), (0, −2) **4.** $(\sqrt{5}, 2)$, $(-\sqrt{5}, 2)$, $(\sqrt{5}, -2)$, $(-\sqrt{5}, -2)$ **5.** (3, 2), (−3, 2), (3, −2), (−3, −2)
6. $\left(\frac{\sqrt{3}}{3}, 4\right)$, $\left(-\frac{\sqrt{3}}{3}, 4\right)$, $\left(\frac{\sqrt{3}}{3}, -4\right)$, $\left(-\frac{\sqrt{3}}{3}, -4\right)$ **7.** (5, 0), (−5, 0) **8.** (5, 1), (−5, 1), (5, −1), (−5, −1) **9.** (4, 3), (3, 4), (−4, −3), (−3, −4) **10.** $(3, \frac{1}{2})$, $(\sqrt{6}, -1)$, $(-3, \frac{1}{2})$, $(-\sqrt{6}, -1)$ **11.** (5, 1), (1, 5), (−5, −1), (−1, −5) **12.** (2, 1), $(\sqrt{3}, 0)$, (−2, 1), $(-\sqrt{3}, 0)$

Page 316

ORALS **1.** $t = 7$, $u = 2$ **2.** $t = 3$, $u = 6$
3. $t = 9$, $u = 1$ **4.** $t = 8$, $u = 6$ **5.** $t = 4$, $u = 3$
6. $t = 1$, $u = 2$ **7.** 63, $t = 6$, $u = 3$ **8.** 12, $t = 1$, $u = 2$ **9.** 76, $t = 7$, $u = 6$ **10.** 49, $t = 4$, $u = 9$
11. 31, $t = 3$, $u = 1$ **12.** 7, $t = 0$, $u = 7$ **13.** 7
14. 3 **15.** 10 **16.** 10 **17.** 11 **18.** 17

EXERCISES **1.** 52 **2.** 86 **3.** 65 **4.** 82 **5.** 72 **6.** 24 **7.** 35 **8.** 42 **9.** 72 **10.** 45 **11.** 39 **12.** 23 **13.** 27 **14.** 48 **15.** 64 **16.** 52

Page 322

ORALS **1.** 270 **2.** 30 **3.** 45 **4.** 70 **5.** 80 **6.** 80 **7.** 5 **8.** 6 **9.** 8

EXERCISES **1.** 4 **2.** 4 **3.** 6 **4.** 50, 150 **5.** 1 **6.** 650 **7.** 180 **8.** 20 **9.** 4 **10.** 6, 3 **11.** 46 **12.** 60

Page 326

ORALS **1.** yes, left **2.** yes, below **3.** no, above **4.** no, above **5.** yes, below **6.** no, right **7.** yes, below **8.** no, below **9.** no, above **10.** yes, below **11.** yes, left **12.** yes, right

EXERCISES
For each of the following, the two points given lie on the line. They are not necessarily part of the graph.
1. below and including; (0, −1), (1, 2)
2. above and including; (0, 4), (−1, 1)
3. below; (0, 1), (1, 3) **4.** above; (0, 3), (1, 1)
5. below; (0, −4), (−4, 0) **6.** above and including; (0, −3), (3, 0) **7.** below and including; (0, 4), (−1, 1) **8.** below and including; (0, 7), (1, 2) **9.** below and including; (0, 6), (2, 0)
10. below and including; (0, −1), (3, −1)
11. above; (0, 3), (3, 3) **12.** below and including; (0, 4), (3, 4) **13.** to the right and including; (2, 0), (2, 3) **14.** to the left; (−3, 0), (−3, 3) **15.** above and including; (0, −2), (1, 0) **16.** below and including; (0, −3), (−2, −1) **17.** below; (0, 3), (1, 1) **18.** below the line containing (0, 3), (1, 2) *and* above the line containing (0, −2), (1, 0)
19. above the line containing (0, 4), (1, 1) *and* below the line containing (0, −3), (1, −1)
20. above the line containing (0, −3), (1, −2) *and* below the line containing (0, −1), (−1, 3)
21. above and including the line containing (0, 6), (−3, 0) *and* below and including the line containing (0, −1), (−1, 2) **22.** below and including the line containing (−4, 0), (−3, 2) *and* below and including the line containing (0, 3), (1, 5) **23.** above and including the line containing (0, 4), (−1, 1) *and* above and including the line containing (0, −1), (1, 2)
24. below and including; (1, −1), $\left(2, -\frac{1}{3}\right)$
25. below and including; (0, 4), (−4, 2)
26. above and including; (−3, −4), (−6, −2)
27. below; (0, 2), (−2, 1) **28.** below; (−2, −2), (0, −4) **29.** above; (0, 5), (2, 0) **30.** below the line containing (0, 3), (1, 1) *and* above and including the line containing (0, 0), (2, 2)
31. below and including the line containing (0, −6), (−2, 0) *and* above the line containing (0, 1), (2, 5) **32.** below and including the line containing (0, −3), $\left(4, -\frac{1}{3}\right)$ *and* above the line containing (0, 3), (6, 0) **33.** below and including the line containing (0, −2), (−3, 0) *and* above the line containing (0, −2), (2, −1) **34.** below and including the line containing (0, 2), (2, 2) *and* to the left and including the line containing (4, 4), (4, 0) **35.** below the line containing (1, −4), (3, −3) *and* below and including the line containing (0, 2), (4, 0) **36.** below and including the line containing (0, 3), (1, 2) *and* below and including the line containing $\left(0, -\frac{1}{5}\right)$, $\left(\frac{1}{4}, 0\right)$ *and* above the line containing (0, 2), (1, 3) **37.** below and including the line containing (4, 0), (0, −8) *and* below the line containing $\left(0, \frac{7}{3}\right)$, $\left(3, \frac{4}{3}\right)$ *and* above and including the line containing (0, 2), (4, 0)
38. below the line containing (−5, 0), (−2, 1) *and* below and including the line containing (2, 0), (0, −3) *and* below the line containing (0, 3), (−1, −1)

Page 328

ORALS **1.** $y \geq -2x + 1$ **2.** $xy \leq 4$ **3.** $x^2 + y^2 \leq 9$

EXERCISES For each of the following, the two points given lie on the curve. They are not necessarily part of the graph. **1.** Inside the circle; (0, 2), (−2, 0) **2.** inside and including the branches of the hyperbola; (2, 2), (−1, −4) **3.** outside and including the circle; (0, 4), (−4, 0) **4.** outside and including the ellipse; (−3, 0), (0, 2.4) **5.** outside the branches of the hyperbola; (4, 2.8), (−3.5, −2) **6.** inside the ellipse; (0, 2), (−4, 0) **7.** inside the branches of the hyperbola; (−1, −5), (5, 1) **8.** inside and including the circle; (−2, 0), (0, 2) **9.** inside and including the branches of the hyperbola; (4, 5.2), (−4.5, 6) **10.** outside and including the branches of the hyperbola; (−2.4, 0), (2.4, 0) **11.** inside the ellipse; (3, 0), (0, −1.7) **12.** inside the branches of the hyperbola; (1, 1), $\left(-2, -\frac{1}{2}\right)$ **13.** inside and including the circle on which lie the points (3, 0), (0, −3) *and* below and including the line on which lie the points (0, 3), (2, 1) **14.** inside and including the ellipse on which lie the points (−3, 0), (0, 2) *and* above and including the line on which lie the points (0, 3), (3, 1) **15.** inside and including the branches of the hyperbola on which lie the points (2, 2), (−1, −4), *and* below the line on which lie the points (0, 4), (2, 3) **16.** inside and including the branches of the hyperbola on which lie the points (4, 2), (−4, 2) *and* below the line on which lie the points (0, 4), (3, 3) **17.** inside and including the circle on which lie the points (2, 0), (0, −2) *and* above the line on which lie the points (0, −2), (2, −1) **18.** inside and including the ellipse on which lie the points (4, 0), (0, −2) *and* below the line on which lie the points (0, −3), (−3, −1) **19.** inside and including the circle on which lie the points (−4, 0), (0, 4) *and* inside and including the branches of the hyperbola on which lie the points (−1, −4), (2, 2) **20.** inside and including the branches of the hyperbola on which lie the points (2, 0), (−2, 0) *and* inside and including the circle on which lie the points (3, 0), (0, 3) **21.** inside the ellipse on which lie the points (−3, 0), (0, −2) *and* inside and including the circle on which lie the points (−6, 0), and (0, 6)

Page 329

Across **1.** 41; 43; 47 **7.** 4 **8.** 1,492 **9.** 100 **11.** 14,007 **13.** 101 **16.** 510 **19.** 3 **20.** 2 **21.** 79 **22.** 25 **24.** 12 **25.** 666 **27.** 1,024 **29.** 102 **31.** 1 Down **1.** 411 **2.** 144 **3.** 490 **4.** 320 **5.** 4 **6.** 71 **7.** 40 **10.** 0 **12.** 7 **13.** 13 **14.** 0 **15.** 1,224 **16.** 576 **17.** 196 **18.** 0 **22.** 2 **23.** 512 **24.** 121 **26.** 61 **28.** 0 **30.** 0

Page 330

1. (5, 2) **2.** (−3, 7) **3.** (4, −1) **4.** $\left(\frac{1}{2}, 0\right)$ **5.** (1, 1) **6.** (2, 1) **7.** (2, 1) **8.** (−1, 2, 3) **9.** (−2, −1, 2) **10.** (−1.6, 5.8), (5.6, 2.2) **11.** (−3, 0), (5, 4) **12.** (4, 0), (0, −4) **13.** (3, 0), (−1, 4) **14.** $(2\sqrt{2}, 2)$, $(2\sqrt{2}, -2)$, $(-2\sqrt{2}, 2)$, $(-2\sqrt{2}, -2)$ **15.** (3, 2), (2, 3), (−3, −2), (−2, −3) **16.** $(2, \sqrt{2})$, $(-2, \sqrt{2})$, $(2, -\sqrt{2})$, $(-2, -\sqrt{2})$ **17.** 93 **18.** 37 **19.** 4 **20.** 1 **21.** below and including; (0, 1), (1, −1) **22.** above; (0, 4), (−1, −1) **23.** below; (0, 2), (2, 1) **24.** below and including; $\left(-\frac{5}{3}, -3\right)$, $\left(-\frac{2}{3}, -4\right)$ **25.** below the line containing (0, 1), (1, 3) *and* above the line containing (0, −5), (−3, −2) **26.** below and including the line containing (0, 6), (2, 0) *and* above and including the line containing (0, −3), (1, −1) **27.** above the line containing (0, −1), (3, 1) *and* above and including the line containing (0, 0), (3, 3) **28.** above and including the line containing (0, 4), (−2, 1) *and* below the line containing (0, 6), (−1, 4) **29.** inside and including the circle on which lie the points (0, 4), (−4, 0) **30.** inside and including the branches of the hyperbola on which lie the points (−1, −3), (3, 1) **31.** outside and including the ellipse on which lie the points (−3, 0), (0, 2) **32.** inside and including the circle on which lie the points (2, 0), (0, −2) *and* above the line containing the points (0, 4), (4, 2) **33.** outside and including the branches of the hyperbola on which lie the points (2, 3), (−6, −1) *and* above the line containing the points (0, −4), (3, −2)

Page 331

1. (5, 5) **2.** (−3, 1) **3.** (6, 7) **4.** (11, 0) **5.** (−3, −1, 2) **6.** (5, 4), (−3, 0) **7.** (4, 3), (3, 4), (−4, −3), (−3, −4) **8.** (0, −5), (3, 4) **9.** (4, 3), (4, −3), (−4, 3), (−4, −3) **10.** 24 **11.** 4 hours **12.** above; (0, 6), (2, 0) **13.** above and including; (0, −2), (2, −1) **14.** above and including the line containing (0, 3), (1, 1) *and* below the line containing (0, −2), (1, 1) **15.** below and including the line containing (0, −2), (3, 0) *and* below the line containing (1, 2), (−6, −2) **16.** inside the circle on which lie the points (6, 0), (0, −6) **17.** outside and including the branches of the hyperbola on which lie the points (2, 0), (−2, 0) **18.** outside and including the ellipse on which lie the points (2, 2), (0, −2) *and* below the line containing the points (−2, 4), (−3, 2)

Page 335

ORALS **1.** 19, 23, 27 **2.** $-3, -5, -7$ **3.** 5, 8, 11 **4.** 5, $5\frac{1}{2}$, 6 **5.** 2, 6, 10, 14 **6.** $-4, -6, -8, -10$ **7.** $-10, -5, 0, 5$ **8.** 20, 10, 0, -10 **9.** $d = 2$ **10.** $d = -3$ **11.** not arith. **12.** $d = 2$

EXERCISES **1.** 5, 5.5, 6 **2.** 6.8, 8, 9.2 **3.** 5, 3.5, 2 **4.** $2\frac{1}{3}$, $2\frac{2}{3}$, 3 **5.** $-2, -1\frac{1}{2}, -1$ **6.** 5, $4\frac{3}{4}$, $4\frac{1}{2}$ **7.** $2 + 3\sqrt{5}$, $2 + 4\sqrt{5}$, $2 + 5\sqrt{5}$ **8.** $4 + 3\sqrt{3}$, $5 + 4\sqrt{3}$, $6 + 5\sqrt{3}$ **9.** $9 - 3\sqrt{2}$, $11 - 4\sqrt{2}$, $13 - 5\sqrt{2}$ **10.** 8, 15, 22, 29 **11.** 9, 3, -3, -9 **12.** $-7, -3, 1, 5$ **13.** 5, 7.4, 9.8, 12.2 **14.** $-3, -\frac{2}{3}, 1\frac{2}{3}, 4$ **15.** $-6, -6.3, -6.6, -6.9$ **16.** $-2, \sqrt{5}, 2 + 2\sqrt{5}, 4 + 3\sqrt{5}$ **17.** 3, $4 - \sqrt{3}$, $5 - 2\sqrt{3}$, $6 - 3\sqrt{3}$ **18.** i, $2 + 2i$, $4 + 3i$, $6 + 4i$ **19.** $d = 4$ **20.** $d = -9$ **21.** not arith. **22.** $d = 2.6$ **23.** not arith. **24.** $d = -\frac{2}{3}$ **25.** $d = \sqrt{2}$ **26.** $d = -2 + \sqrt{3}$ **27.** $d = 2 - i$ **28.** $d = 4$ **29.** not arith. **30.** $d = \sqrt{2}$ **31.** 87 **32.** -110 **33.** 52 **34.** 10 **35.** -27 **36.** 17 **37.** -70 **38.** 203 **39.** -85 **40.** 10 **41.** -13 **42.** 14 **43.** 169 **44.** -179 **45.** 98 **46.** 23 **47.** 32 **48.** 14 **49.** 21 **50.** 45 **51.** 33 **52.** 31 **53.** 26 **54.** 31 **55.** 11 **56.** 61 **57.** 46 **58.** $4 + 40\sqrt{3}$ **59.** $22 + 20\sqrt{5}$ **60.** $23 - 20\sqrt{2}$ **61.** $23 - 18i$ **62.** $21 - 25\sqrt{2}$ **63.** $48 + 25\sqrt{3}$ **64.** $-69 + 46\sqrt{5}$ **65.** $22 + 45i$ **66.** \$895 **67.** \$9,000 **68.** 2 **69.** 3; $-\frac{1}{2}$ **70.** 5 **71.** -4

Page 338

ORALS **1.** 6 **2.** not arith. **3.** 1, -2 **4.** not arith. **5.** $-3, -1, 1, 3$ **6.** 2.5, 3, 3.5, 4

EXERCISES **1.** 8 **2.** 16 **3.** 5.5 **4.** -1 **5.** 18, 25 **6.** 24, 18 **7.** -6, 2 **8.** 3.5, 5 **9.** 9, 15, 21, 27 **10.** 33, 26, 19, 12 **11.** 2, 6, 10, 14 **12.** 12, 9, 6, 3 **13.** 10.4, 10.8, 11.2, 11.6 **14.** 30.6, 31.2, 31.8, 32.4 **15.** 6.2, 8.4, 10.6, 12.8 **16.** 7.2, 5.4, 3.6, 1.8 **17.** 2.8 **18.** 7.5 **19.** 7.9 **20.** 9.9 **21.** $3\sqrt{3}$ **22.** $2 + \sqrt{5}$ **23.** $4 + 2\sqrt{2}$ **24.** $\frac{x + y}{2}$ **25.** yes **26.** 5.7, 8 **27.** 2, 3.9 **28.** 7.7, 11.4 **29.** $3\sqrt{5}$, $4\sqrt{5}$ **30.** $5 + \sqrt{2}$, $7 + 2\sqrt{2}$ **31.** $3 - \sqrt{3}$, $5 - 3\sqrt{3}$

Page 341

1. $5 + 8 + 11 + 14 + \cdots$ **2.** $4 + 1 - 2 - 5 - \cdots$ **3.** $-1 - 2 - 3 - 4 - \cdots$ **4.** 3 **5.** -2 **6.** not arith. **7.** .4 **8.** .5 **9.** $2\sqrt{2}$ **10.** 1,350 **11.** $-2,260$ **12.** 2,500 **13.** $-4,815$ **14.** 1,050 **15.** $-1,475$ **16.** 418 **17.** $-1,809$ **18.** 3,503.5 **19.** 2,000 **20.** $-2,880$ **21.** 546 **22.** $-9,660$ **23.** 774 **24.** 999 **25.** 170.3 **26.** 930.7 **27.** 385 **28.** 185 **29.** -60 **30.** -26 **31.** $2,501\sqrt{3}$ **32.** $-192\sqrt{5}$ **33.** 210 **34.** \$46.50

Page 344

ORALS **1.** 3 **2.** -2 **3.** $\frac{1}{2}$ **4.** not geom. **5.** $-\frac{1}{3}$ **6.** .1 **7.** 2 **8.** -1

EXERCISES **1.** 24, 48, 96 **2.** 1, $\frac{1}{4}$, $\frac{1}{16}$ **3.** .008, .0016, .00032 **4.** 9, -27, 81 **5.** -4, 2, -1 **6.** $-8, -16, -32$ **7.** 10, 20, 40, 80 **8.** $-\frac{1}{4}, -1, -4, -16$ **9.** .2, 1, 5, 25 **10.** 16, 8, 4, 2 **11.** $-9, -3, -1, -\frac{1}{3}$ **12.** 5, .5, .05, .005 **13.** $\frac{1}{2}, -1, 2, -4$ **14.** $4, -2, 1, -\frac{1}{2}$ **15.** $-2, 6, -18, 54$ **16.** 1,536 **17.** $-5,120$ **18.** -256 **19.** $\frac{1}{128}$ **20.** $-\frac{1}{32}$ **21.** $-\frac{1}{256}$ **22.** $-4,096$ **23.** $-5,120$ **24.** 256 **25.** $-\frac{1}{128}$ **26.** $-\frac{1}{256}$ **27.** $\frac{1}{32}$ **28.** -810 **29.** -375 **30.** 4,000,000 **31.** .00009 **32.** $-.0016$ **33.** $-.007$ **34.** $-\frac{5}{81}$ **35.** $-\frac{1}{8}$ **36.** $\frac{16}{25}$ **37.** 54 **38.** $-100\sqrt{5}$ **39.** 80 **40.** $54\sqrt{3}$ **41.** $324\sqrt{2}$ **42.** -200 **43.** $-3i$ **44.** -4 **45.** $160i$ **46.** $81i$ **47.** $\frac{1}{4}$ dm **48.** $15\frac{5}{8}$L **49.** $10\frac{2}{3}$ m **50.** $21\frac{1}{3}$L **51.** -2; 2 **52.** $-\frac{1}{6}$; $\frac{1}{6}$ **53.** $\frac{1}{3}$; 3 **54.** $-4i$; $4i$ **55.** 11th **56.** 8th **57.** 11th

Page 347

1. 6 **2.** 12 **3.** 2 **4.** 2 **5.** 4 **6.** 3 **7.** 1 **8.** 3 **9.** .4 **10.** .6 **11.** .3 **12.** .04 **13.** 6, 18 **14.** 6, -12 **15.** $-8, -32$ **16.** -1, 5 **17.** 3, 1 **18.** -4, 2 **19.** 10, 2 **20.** 2, $-\frac{1}{2}$ **21.** .6, .06 **22.** .8, .16 **23.** .3, .09 **24.** .08, .032 **25.** $3\sqrt{2}$ **26.** 10 **27.** $4\sqrt{3}$ **28.** 15 **29.** $2\sqrt{2}$ **30.** 6 **31.** $6\sqrt{2}$ **32.** 10 **33.** $2\sqrt{3}$, 6 **34.** $5\sqrt{2}$, 10 **35.** $3\sqrt{7}$, 21 **36.** $4\sqrt{5}$, 20 **37.** 2, $2\sqrt{2}$ **38.** 6, $6\sqrt{3}$ **39.** 7, $7\sqrt{7}$ **40.** 15, $15\sqrt{5}$ **41.** $2i$; $-2i$ **42.** 3; -3 **43.** $2i$; $-2i$ **44.** $6i$; $-6i$ **45.** $i\sqrt{2}$; $-i\sqrt{2}$ **46.** $\sqrt{5}$; $-\sqrt{5}$ **47.** $2i\sqrt{5}$; $-2i\sqrt{5}$ **48.** $3i\sqrt{2}$; $-3i\sqrt{2}$ **49.** $2 + 2i$; $-2 - 2i$ **50.** $3 - 3i$; $-3 + 3i$ **51.** $\sqrt{xy}$; $-\sqrt{xy}$

Page 348

1. 110 **2.** 272 **3.** 870 **4.** 10,100 **5.** 100 **6.** 144 **7.** 81 **8.** 400 **9.** 10,000 **10.** 1,600 **11.** 5 **12.** 5 **13.** 55 **14.** 5,050

Page 353

ORALS **1.** $1 + 3 + 9 + 27 + \cdots$ **2.** $3 - 1 + \frac{1}{3} - \frac{1}{9} + \cdots$ **3.** $4 + 2 + 1 + \frac{1}{2} + \cdots$ **4.** $-\frac{1}{5} + 1 - 5 + 25 - \cdots$ **5.** 3 **6.** 2 **7.** not geom. **8.** $-\frac{1}{2}$ **9.** -4 **10.** $\frac{1}{5}$ **11.** .1 **12.** .01 **13.** .1

EXERCISES **1.** 5,115 **2.** $-1{,}533$ **3.** 171 **4.** 1,364 **5.** 5,465 **6.** $-511\frac{1}{2}$ **7.** $31\frac{7}{8}$ **8.** $21\frac{1}{4}$ **9.** 333,333.3 **10.** 90,909 **11.** 6,666,666 **12.** -728 **13.** 77.777 **14.** 31 **15.** 24.8 **16.** 22.222 **17.** -3.3333 **18.** 3.72 **19.** 16 **20.** -4 **21.** 27 **22.** $13\frac{1}{2}$ **23.** 1 **24.** $\frac{1}{6}$ **25.** $2\frac{2}{9}$ **26.** $7\frac{7}{9}$ **27.** $\frac{3}{4}$ **28.** $\frac{1}{3}$ **29.** $.5 + .05 + .005 + \cdots$; $a = .5$; $r = .1$ **30.** $.3 + .03 + .003 + \cdots$; $a = .3$; $r = .1$ **31.** $6 + .6 + .06 + \cdots$; $a = 6$; $r = .1$ **32.** $4 + .4 + .04 + \cdots$; $a = 4$; $r = .1$ **33.** $.02 + .002 + .0002 + \cdots$; $a = .02$; $r = .1$ **34.** $20 + 2 + .2 + \cdots$; $a = 20$; $r = .1$ **35.** $\frac{2}{9}$ **36.** $\frac{7}{9}$ **37.** $\frac{5}{90}$ **38.** $\frac{4}{90}$ **39.** $\frac{70}{9}$ **40.** $\frac{30}{9}$ **41.** $-21\frac{3}{8}$ **42.** $781\frac{6}{25}$ **43.** $682\frac{1}{2}$ **44.** $-2{,}730$ **45.** $728\frac{2}{3}$ **46.** $7{,}812\frac{2}{5}$ **47.** 13,021 **48.** 781.2 **49.** -364 **50.** $40\frac{4}{9}$ **51.** $393\frac{3}{4}$ **52.** $-102\frac{3}{8}$ **53.** 51 **54.** 416 **55.** 6,825 **56.** $-11{,}718$ **57.** -547 **58.** -205 **59.** $r = 5$; no sum **60.** $2\frac{2}{3}$ **61.** $-85\frac{1}{3}$ **62.** $-31\frac{1}{4}$ **63.** $-20\frac{1}{4}$ **64.** $4\frac{1}{6}$ **65.** $-9\frac{1}{7}$ **66.** $17\frac{6}{7}$ **67.** $16\sqrt{5}$ **68.** $4\sqrt{3}$ **69.** $81\sqrt{2}$ **70.** $256\sqrt{7}$ **71.** $.38 + .0038 + .000038 + \cdots$; $a = .38$; $r = .01$ **72.** $.65 + .0065 + .000065 + \cdots$; $a = .65$; $r = .01$ **73.** $53 + .53 + .0053 + \cdots$; $a = 53$; $r = .01$ **74.** $74 + .74 + .0074 + \cdots$; $a = 74$; $r = .01$ **75.** $.026 + .00026 + .0000026 + \cdots$; $a = .026$; $r = .01$ **76.** $.029 + .00029 + .0000029 + \cdots$; $a = .029$; $r = .01$ **77.** $\frac{35}{99}$ **78.** $\frac{28}{99}$ **79.** $\frac{1700}{99}$ **80.** $\frac{6200}{99}$ **81.** $\frac{47}{990}$ **82.** $\frac{32}{990}$ **83.** 382 m **84.** 384 m **85.** $362\frac{1}{3}$ m **86.** 405 m

Page 356

1. $-3, 2, 7, 12$ **2.** $6, 2, -2, -6$ **3.** 4, 5.6, 7.2, 8.8 **4.** $-3, -1 - \sqrt{3}, 1 - 2\sqrt{3}, 3 - 3\sqrt{3}$ **5.** -99 **6.** -90 **7.** 23 **8.** 15 **9.** 5.4, 5.8, 6.2, 6.6 **10.** $5 + 2 - 1 - 4 - \cdots$ **11.** 105 **12.** 500 **13.** -546 **14.** 273 **15.** $5, -10, 20, -40$ **16.** $\frac{1}{2}, 2, 8, 32$ **17.** $-16, -8, -4, -2$ **18.** $2, 2\sqrt{3}, 6, 6\sqrt{3}$ **19.** 3,072 **20.** $\frac{1}{128}$ **21.** 512 **22.** 324 **23.** -54 **24.** 6 **25.** $-4, 8$ **26.** 222, $222\frac{1}{5}$ **27.** $10\frac{5}{8}$ **28.** -4.4444 **29.** -84 **30.** -8 **31.** $25\frac{3}{5}$ **32.** $5\frac{5}{9}$ **33.** $.35 + .0035 + .000035 + \cdots$; $a = .35$; $r = .01$ **34.** $\frac{2}{9}$

Page 357

1. $3, 0, -3, -6$ **2.** 2, 3.5, 5, 6.5 **3.** -91 **4.** 12 **5.** 7.2, 7.4, 7.6, 7.8 **6.** $-3 - 1 + 1 + 3 + \cdots$ **7.** 380 **8.** 210 **9.** 168 **10.** $-8{,}000$ **11.** $3, -6, 12, -24$ **12.** $9, 3, 1, \frac{1}{3}$ **13.** 256 **14.** $\frac{1}{256}$ **15.** -486 **16.** -12 **17.** 6, 12 **18.** $511\frac{1}{2}$ **19.** $65\frac{5}{8}$ **20.** 3.333 **21.** $-4{,}545$ **22.** -4 **23.** $12\frac{4}{5}$ **24.** $.27 + .0027 + .000027 + \cdots$; $a = .27$; $r = .01$ **25.** $\frac{4}{9}$

Page 358

1. 30; \$4 **2.** 20; \$4

Page 362

1. 10^2 **2.** 10^4 **3.** 10^1 **4.** 10^0 **5.** $10^{1.7372}$ **6.** $10^{3.7372}$ **7.** $10^{0.7372}$ **8.** $10^{2.7372}$ **9.** $10^{2.3284}$ **10.** $10^{0.5416}$ **11.** $10^{1.6637}$ **12.** $10^{3.2430}$ **13.** $10^{1.4871}$ **14.** $10^{2.8463}$ **15.** $10^{0.7042}$ **16.** $10^{3.6107}$ **17.** $10^{2.3096}$ **18.** $10^{0.7803}$ **19.** $10^{1.9036}$ **20.** $10^{4.0374}$ **21.** $10^{0.8142}$ **22.** $10^{2.8686}$ **23.** $10^{4.9159}$ **24.** $10^{1.9943}$ **25.** 8.32 **26.** 83.2 **27.** 832 **28.** 8,320 **29.** 27.7 **30.** 7.46 **31.** 372 **32.** 6,810 **33.** 439 **34.** 92,300 **35.** 1.68 **36.** 54.4 **37.** 8,150 **38.** 32.5 **39.** 773 **40.** 4.61 **41.** 9.76 **42.** 23,400 **43.** 12.2 **44.** 577 **45.** $10^{1.6335}$ **46.** $10^{0.8808}$ **47.** $10^{2.7634}$ **48.** $10^{3.3222}$ **49.** $10^{0.7924}$ **50.** $10^{1.2788}$ **51.** $10^{3.9243}$ **52.** $10^{2.5441}$ **53.** $10^{0.6990}$ **54.** $10^{1.4771}$ **55.** $10^{3.3010}$ **56.** $10^{2.9542}$ **57.** 240 **58.** 7,500 **59.** 560 **60.** 400 **61.** 1,700 **62.** 380 **63.** 83,000 **64.** 6,000

Page 364

ORALS **1.** 100 **2.** 1,000 **3.** 100 **4.** 1,000

EXERCISES **1.** 74.9 **2.** 826 **3.** 959 **4.** 36,300 **5.** 23.6 **6.** 76,800 **7.** 384,000 **8.** 47,600 **9.** 1,780,000 **10.** 3,930 **11.** 300,000

12. 4,420,000 **13.** 17.3 **14.** 155 **15.** 14.6
16. 138 **17.** 5.93 **18.** 6.99 **19.** 68.6 **20.** 22.7
21. 58,100 **22.** 1,200 **23.** 10,400
24. 1,890,000 **25.** 13,900 **26.** 5,550 **27.** 181
28. 22.2 **29.** 30.7 **30.** 3.79 **31.** 9.60 **32.** 5.26
33. 824 **34.** 179 **35.** 895,000,000 **36.** 14,500

Page 367

ORALS **1.** 2 **2.** 1 **3.** 4 **4.** 0 **5.** 2.7661
6. 100 **7.** 1,000 **8.** 10 **9.** 1 **10.** 10,000 **11.** 2
12. 3 **13.** 1 **14.** 0 **15.** 1 **16.** 2 **17.** 0 **18.** 4
19. 1; .6920 **20.** 3; .4533 **21.** 0; .8739

EXERCISES **1.** 0.5465 **2.** 1.5465 **3.** 3.5465
4. 2.6656 **5.** 0.4456 **6.** 1.2504 **7.** 3.9047
8. 1.7042 **9.** 0.7810 **10.** 23.5 **11.** 2.35 **12.** 235
13. 5.78 **14.** 326 **15.** 9,770 **16.** 123 **17.** 83.3
18. 16,700 **19.** 77.4 **20.** 6.29 **21.** 4,620
22. 2; .6739 **23.** 0; .3345 **24.** 1; .8476
25. 0; .9671 **26.** 3; .7803 **27.** 4; .5132
28. 1.7243 **29.** 0.2788 **30.** 2.8573 **31.** 0.9731
32. 1.5798 **33.** 3.8062 **34.** 4.4216 **35.** 4.9069
36. 5.6365 **37.** 34 **38.** 1.8 **39.** 730 **40.** 5.7
41. 86 **42.** 300 **43.** 430 **44.** 6,500 **45.** 7,000
46. 51,200 **47.** 22,300 **48.** 920,000

Page 369

ORALS **1.** T **2.** T **3.** F **4.** T **5.** T **6.** F

EXERCISES **1.** 864 **2.** 9.76 **3.** 26,000
4. 4,330 **5.** 210,000 **6.** 203,000 **7.** 141,000
8. 8,720 **9.** 35,500 **10.** 2,750,000 **11.** 5,700
12. 250,000 **13.** $\log l + \log w$ **14.** $\log B + \log h$ **15.** $\log p + \log r + \log t$ **16.** $\log \pi + \log r + \log s$ **17.** $\log 2 + \log \pi + \log r$
18. $\log 2 + \log \pi + \log r + \log h$ **19.** 80,100
20. 852,000 **21.** 4,520,000 **22.** 224
23. $\log (x^2 - y^2) = \log (x + y)(x - y) = \log (x + y) + \log (x - y)$
24. $\log (x^2 + 2xy + y^2) = \log (x + y)(x + y) = \log (x + y) + \log (x + y) = 2 \cdot \log (x + y)$

Page 373

ORALS **1.** 2.64×10^3 **2.** 2.64×10^0
3. 2.64×10^{-1} **4.** 2.64×10^{-3} **5.** 47,300
6. 4.73 **7.** .0473 **8.** .473

EXERCISES **1.** 3.45×10^2 **2.** 7.26×10^1
3. 2.79×10^4 **4.** 4.62×10^0 **5.** 2.16×10^{-2}
6. 5.13×10^{-1} **7.** 3.78×10^{-2} **8.** 7.35×10^{-3}
9. 2.3×10^1 **10.** 5.7×10^{-3} **11.** 7.6×10^3
12. 4.8×10^{-1} **13.** 817 **14.** 32.2 **15.** 5,630
16. 67,500 **17.** .624 **18.** .00789 **19.** .0352
20. 1.53 **21.** 29,000 **22.** .36 **23.** 630 **24.** .078
25. 2.9258 **26.** 1.8669 **27.** 3.3747 **28.** 0.4942
29. 0.7993 **30.** 4.8351 **31.** 1.7218 **32.** 3.6335
33. 582 **34.** 1,880 **35.** 64.7 **36.** 2.72 **37.** 8.49
38. 72.3 **39.** 5,200 **40.** 44,600 **41.** 6.7193
42. 5.5705 **43.** 6.9345 **44.** 7.4472 **45.** 487,000
46. 6,300,000 **47.** 625,000 **48.** 45,000,000
49. 1.5×10^8 km **50.** 2×10^{-10} cm
51. 2.998×10^{10} cm per sec **52.** 2×10^{-20} erg

Page 375

ORALS **1.** T **2.** F **3.** T **4.** F **5.** T **6.** F

EXERCISES **1.** 14.3 **2.** 159 **3.** 15.1 **4.** 175
5. 6.91 **6.** 6.98 **7.** 49.0 **8.** 40.2 **9.** 173
10. 29.4 **11.** 227 **12.** 26.0 **13.** 4.54 **14.** 13.1
15. 1.54 **16.** 247 **17.** 26.5

Page 377

1. 3,170 **2.** 25,400 **3.** 329 **4.** 108,000 **5.** 350
6. 64.3 **7.** 183,000 **8.** 142,000 **9.** 29,600
10. 314,000 **11.** 575 **12.** 11,500 **13.** 228
14. 15,200 **15.** 2,760 **16.** 326 **17.** 694 **18.** 115
19. 434,000 **20.** 18,000,000 **21.** 605,000
22. 36,900 **23.** 10.6 **24.** 17,400 **25.** 16.2
26. 39.7 **27.** $\log \left(\frac{ab}{c}\right)^r = r \cdot \log \frac{ab}{c} = r(\log ab - \log c) = r(\log a + \log b - \log c)$
28. $\log \left(\frac{a}{bc}\right)^r = r \cdot \log \frac{a}{bc} = r(\log a - \log bc) = r[\log a - (\log b + \log c)]$

Page 379

1. 19.2 **2.** 4.35 **3.** 71.6 **4.** 2.10 **5.** 19.1 **6.** 4.07
7. 45.5 or 45.6 **8.** 19.2 **9.** 12.0 **10.** 12.9
11. 3.68 **12.** 11.1 **13.** 13.8 **14.** 8.44 **15.** 5.18
16. 2.73 **17.** 238 **18.** 63.2 **19.** 5.72 **20.** 8.96
21. 2.13 **22.** 3.16 **23.** 2.28 **24.** 2.40 **25.** 1.18
26. 75.4

Page 380
1. 10^3 **2.** $10^{2.9138}$ **3.** $10^{0.7007}$ **4.** 16.8 **5.** 7,730 **6.** 3.72 **7.** 56.7 **8.** 20.1 **9.** 0.7505 **10.** 4.4886 **11.** 1.4314 **12.** 578 **13.** 1.67 **14.** 70,000 **15.** 3; .6920 **16.** 1; .8739 **17.** 0; .4533 **18.** 8.43×10^3 **19.** 4.68×10^1 **20.** 2.07×10^{-1} **21.** 6.2×10^{-3} **22.** 31,500 **23.** 503 **24.** .078 **25.** .914 **26.** 17,800 **27.** 19,600 **28.** 367 **29.** 60.6 **30.** 30,300 **31.** 18,200 **32.** 735,000,000 **33.** 135,000 **34.** 62.9 **35.** 1,480 **36.** 79.9 **37.** 1.73 **38.** 13.0 **39.** 12.2 **40.** 15.3 **41.** 10.4 **42.** $\log i = \log 500 + \log r + \log t$ **43.** $V = \log 2 + \log w + \log h$

Page 381
1. $10^{1.7193}$ **2.** 277 **3.** 70.3 **4.** 78.1 **5.** 3.9243 **6.** 60.5 **7.** 3; .6405 **8.** 0; .5611 **9.** 6.25×10^3 **10.** 3.72×10^{-2} **11.** 462 **12.** .208 **13.** 46,800 **14.** 18,300 **15.** 236 **16.** 59.3 **17.** 753 **18.** 15.4 **19.** 19.7 **20.** 2.67 **21.** $\log i = \log 7 + \log p + \log r$

Page 382
1. 21, 34, 55, 89 **2.** Add 2 consecutive terms to get the next term in the sequence. **3.** $y_n = y_{n-1} + y_{n-2}$

Page 385

ORALS **1.** -2 **2.** -3 **3.** -1 **4.** -4 **5.** 0 **6.** .01 **7.** .1 **8.** 1 **9.** .001 **10.** .0001 **11.** $8 - 10$; -2 **12.** $9 - 10$; -1 **13.** $7 - 10$; -3 **14.** $6 - 10$; -4 **15.** 3.72×10^{-2} **16.** 2.14×10^{-1} **17.** 7.6×10^{-3} **18.** 5.3×10^{-4} **19.** -3 or $7 - 10$ **20.** -1 or $9 - 10$ **21.** -2 or $8 - 10$ **22.** -2 or $8 - 10$ **23.** -4 or $6 - 10$ **24.** 0 **25.** .563 **26.** .0087 **27.** .0206 **28.** .00042

EXERCISES **1.** $9.5490 - 10$ **2.** $8.5490 - 10$ **3.** $6.5490 - 10$ **4.** $7.7875 - 10$ **5.** $8.2227 - 10$ **6.** $7.9149 - 10$ **7.** $9.3160 - 10$ **8.** $8.6075 - 10$ **9.** $9.7042 - 10$ **10.** .0326 **11.** .326 **12.** .00326 **13.** .235 **14.** .0578 **15.** .000167 **16.** .0462 **17.** .123 **18.** .00833 **19.** .629 **20.** .0977 **21.** .000774 **22.** $9.8633 - 10$ **23.** $7.7782 - 10$ **24.** $8.4914 - 10$ **25.** $6.7324 - 10$ **26.** $6.9031 - 10$ **27.** $5.3802 - 10$ **28.** .43 **29.** .034 **30.** .0057 **31.** .00018 **32.** .0073 **33.** .00086 **34.** .00003 **35.** .0007 **36.** .00002

Page 388
1. .262 **2.** 15.5 **3.** .0738 **4.** .0421 **5.** 2.46 **6.** .180 **7.** .0233 **8.** .0492 **9.** .00677 **10.** .151 **11.** .00757 **12.** .443 **13.** 237 **14.** 2,700 **15.** 45.4 **16.** .419 **17.** .0768 **18.** .956 **19.** .798 **20.** .518 **21.** .189 **22.** .242 **23.** .0310 **24.** .0310 **25.** .00616 **26.** .176 **27.** .299 **28.** .0295 **29.** .0709 **30.** 2.29 **31.** .442 **32.** 28.3 **33.** .566 **34.** .263 **35.** .0225 **36.** .00897 **37.** .000296 **38.** $\log \frac{7m}{n}$ **39.** $\log \frac{t^2}{vw}$ **40.** $\log 3f^2h$ **41.** $\log \left(\frac{mn}{p}\right)^3$ **42.** $\log \frac{7\sqrt{t}}{y}$ **43.** $\log \sqrt[3]{\frac{xy}{z}}$

Page 390

ORALS **1.** $\log 2 + \log a$ **2.** $2 \cdot \log a$ **3.** $\log 7 - \log c$ **4.** $\log a - \log 3$ **5.** $\frac{1}{2} \cdot \log a$ **6.** $\frac{1}{3} \cdot \log a$

EXERCISES **1.** .966 **2.** 13.8 **3.** .0677 **4.** 1.69 **5.** 143 **6.** 14.9 **7.** .0645 **8.** .0281 **9.** .0663 **10.** .00209 **11.** .00395 **12.** .0393 **13.** 3.83 **14.** .266 **15.** .102 **16.** .00232 **17.** .00121 **18.** .0000705 **19.** 1.60 **20.** .909 **21.** .708 **22.** .0255 **23.** .664 **24.** .138 **25.** $\log 3 + \log f + \log g$ **26.** $2 \cdot \log m + \log n$ **27.** $\log 2 + \log t + 3 \cdot \log v$ **28.** $\log x + \left(\frac{1}{2}\right) \cdot \log y$ **29.** $\log 5 + \log f - \log g$ **30.** $2 \cdot \log v - \log 2 - \log h$ **31.** $\log 3 + \log d - 2 \cdot \log t$ **32.** $\log w + 3 \cdot \log v - \log 2 - \log g$ **33.** $2(\log 7 + \log f + \log w)$ **34.** $3(2 \cdot \log t + \log v)$, or $6 \cdot \log t + 3 \cdot \log v$ **35.** $\frac{1}{2}(\log 2 + \log g + \log h)$ **36.** $\frac{1}{3}(\log 5 + 2 \cdot \log d)$ **37.** $2(\log 3 + \log d - \log g)$ **38.** $3(2 \cdot \log v + \log w - \log m - \log n)$ **39.** $\frac{1}{2}(\log 2 + \log d - \log g)$ **40.** $\frac{1}{3}(\log t + 2 \cdot \log v - \log g - \log h)$ **41.** 105 **42.** 24.9 **43.** 30.8 **44.** 2.13 **45.** 378 **46.** .436 **47.** .0415 **48.** .960 **49.** 2.59 **50.** .507 **51.** .370 **52.** .264 **53.** no **54.** mult., div., powers, roots, and combinations of these

Page 394
1. 1.2838 **2.** 3.2842 **3.** $8.2849 - 10$ **4.** 0.4016 **5.** 2.5496 **6.** 1.7831 **7.** $9.8661 - 10$

8. 8.5998 − 10 **9.** 7.3166 − 10 **10.** 16.32 **11.** 2.624 **12.** 412.7 **13.** 4,787 **14.** 18.15 **15.** 896.6 **16.** 60.06 **17.** 4.843 **18.** 2,185 **19.** .1854 **20.** .03275 **21.** .007268 **22.** 169.7 **23.** 33,880 **24.** 9.495 **25.** 24.67 **26.** 5.103 **27.** .03693 **28.** 3,435 **29.** .1062 **30.** 33.33 **31.** 96.68 **32.** 3.741 **33.** .08382 **34.** 92.62 **35.** 174.4 **36.** 599.5 **37.** .04506 **38.** 356.3 **39.** .3039 **40.** 23.65 **41.** .06813 **42.** 9.267 **43.** 1.526 **44.** 8.004 **45.** .6805 **46.** 61.60 **47.** 18.81 **48.** $5,409 **49.** $2,565

Page 397

ORALS **1.** 2 **2.** 1 **3.** 0 **4.** $\frac{1}{2}$ **5.** −1 **6.** $32 = 2^y$ **7.** $x = 3^5$ **8.** $125 = n^3$ **9.** $\frac{1}{9} = 3^y$ **10.** $x = 5^{-2}$ **11.** $\sqrt{7} = n^{\frac{1}{2}}$

EXERCISES **1.** 2 **2.** 6 **3.** 1 **4.** 2 **5.** 3 **6.** 0 **7.** 4 **8.** 7 **9.** 2 **10.** $\frac{1}{2}$ **11.** $\frac{1}{3}$ **12.** $\frac{1}{4}$ **13.** −1 **14.** −2 **15.** −2 **16.** −2 **17.** −2 **18.** −4 **19.** 16 **20.** 9 **21.** 125 **22.** $\frac{1}{3}$ **23.** $\frac{1}{25}$ **24.** $\frac{1}{8}$ **25.** 2 **26.** 3 **27.** 5 **28.** 3 **29.** 2 **30.** 5 **31.** $\frac{2}{3}$ **32.** $\frac{4}{5}$ **33.** $\frac{5}{3}$ **34.** $\frac{3}{2}$ **35.** $\frac{2}{5}$ **36.** $\frac{3}{4}$ **37.** $\frac{2}{5}$ **38.** $\frac{2}{3}$ **39.** 2 **40.** 3 **41.** 1 **42.** 2

Page 398

1. 2.20 **2.** 3.20 **3.** 1.20 **4.** 2.47 **5.** 1.23 **6.** .465 **7.** .816 **8.** 2.82 **9.** .908 **10.** .827 **11.** .414 **12.** 1.41

Page 401

1. $a^7 + 7a^6b + 21a^5b^2 + 35a^4b^3 + 35a^3b^4 + 21a^2b^5 + 7ab^6 + b^7$ **2.** $a^8 + 8a^7b + 28a^6b^2 + 56a^5b^3 + 70a^4b^4 + 56a^3b^5 + 28a^2b^6 + 8ab^7 + b^8$ **3.** $a^9 + 9a^8b + 36a^7b^2 + 84a^6b^3 + 126a^5b^4 + 126a^4b^5 + 84a^3b^6 + 36a^2b^7 + 9ab^8 + b^9$ **4.** $x^6 + 12x^5 + 60x^4 + 160x^3 + 240x^2 + 192x + 64$ **5.** $x^5 - 15x^4 + 90x^3 - 270x^2 + 405x - 243$ **6.** $x^7 + 7x^6 + 21x^5 + 35x^4 + 35x^3 + 21x^2 + 7x + 1$ **7.** $x^5 - 5x^4y + 10x^3y^2 - 10x^2y^3 + 5xy^4 - y^5$ **8.** $m^5 - 10m^4n + 40m^3n^2 - 80m^2n^3 + 80mn^4 - 32n^5$ **9.** $x^4 + 12x^3y + 54x^2y^2 + 108xy^3 + 81y^4$ **10.** $16x^4 + 32x^3y + 24x^2y^2 + 8xy^3 + y^4$ **11.** $243x^5 - 405x^4y + 270x^3y^2 - 90x^2y^3 + 15xy^4 - y^5$ **12.** $64r^6 + 192r^5t + 240r^4t^2 + 160r^3t^3 + 60r^2t^4 + 12rt^5 + t^6$ **13.** $x^8 + 4x^6y + 6x^4y^2 + 4x^2y^3 + y^4$ **14.** $x^{10} - 15x^8 + 90x^6 - 270x^4 + 405x^2 - 243$ **15.** $x^5 + 5x^4y^2 + 10x^3y^4 + 10x^2y^6 + 5xy^8 + y^{10}$ **16.** $x^4 + .4x^3 + .06x^2 + .004x + .0001$ **17.** $x^4 + .8x^3 + .24x^2 + .032x + .0016$ **18.** $32x^5 + 8x^4 + .8x^3 + .04x^2 + .001x + .00001$ **19.** $35a^3b^4$ **20.** $80x^3y^2$ **21.** $-540x^3$ **22.** $-270c^2d^3$ **23.** $15x^4y^4$ **24.** $560x^9$ **25.** $32x^5 + 240x^4 + 720x^3 + 1{,}080x^2 + 810x + 243$ **26.** $81x^4 - 216x^3 + 216x^2 - 96x + 16$ **27.** $32x^5 - 400x^4 + 2{,}000x^3 - 5{,}000x^2 + 6{,}250x - 3{,}125$ **28.** $x^{10} + 15x^8y + 90x^6y^2 + 270x^4y^3 + 405x^2y^4 + 243y^5$ **29.** $x^{12} - 8x^9y + 24x^6y^2 - 32x^3y^3 + 16y^4$ **30.** $x^{10} + 5x^8y^2 + 10x^6y^4 + 10x^4y^6 + 5x^2y^8 + y^{10}$ **31.** $243x^5 + 810x^4y + 1{,}080x^3y^2 + 720x^2y^3 + 240xy^4 + 32y^5$ **32.** $16x^4 - 160x^3y + 600x^2y^2 - 1{,}000xy^3 + 625y^4$ **33.** $32x^5 - 160x^4y + 320x^3y^2 - 320x^2y^3 + 160xy^4 - 32y^5$ **34.** $x^6 + 6x^5\sqrt{2} + 30x^4 + 40x^3\sqrt{2} + 60x^2 + 24x\sqrt{2} + 8$ **35.** $x^5 - 5x^4\sqrt{3} + 30x^3 - 30x^2\sqrt{3} + 45x - 9\sqrt{3}$ **36.** $x^4 - 4x^3\sqrt{y} + 6x^2y - 4xy\sqrt{y} + y^2$ **37.** $x^4 + 2x^3y + \frac{3x^2y^2}{2} + \frac{xy^3}{2} + \frac{y^4}{16}$ **38.** $\frac{1}{x^5} + \frac{5}{x^4y} + \frac{10}{x^3y^2} + \frac{10}{x^2y^3} + \frac{5}{xy^4} + \frac{1}{y^5}$ **39.** $\frac{1}{m^6} - \frac{6}{m^5n} + \frac{15}{m^4n^2} - \frac{20}{m^3n^3} + \frac{15}{m^2n^4} - \frac{6}{mn^5} + \frac{1}{n^6}$ **40.** $720x^3y^2$ **41.** $-20{,}000x^3y^3$ **42.** $80x^6y^3$ **43.** $135x^4y^2$ **44.** $15x^4$ **45.** $-10x^3$ **46.** $x^{11} + 11x^{10}y + 55x^9y^2 + 165x^8y^3 + \cdots$ **47.** $x^{10} - 20x^9y + 180x^8y^2 - 960x^7y^3 + \cdots$ **48.** $x^{20} + 10x^{18}y^3 + 45x^{16}y^6 + 120x^{14}y^9 + \cdots$ **49.** $x^{12} + 6x^{11}y + \frac{33x^{10}y^2}{2} + \frac{55x^9y^3}{2} + \cdots$

Page 402

1. −2 **2.** 7.5391 − 10 **3.** 9.9058 − 10 **4.** .0167 **5.** .00123 **6.** .977 **7.** 217 **8.** .000443 **9.** .0220 **10.** .0546 **11.** .0768 **12.** .352 **13.** 6.96 **14.** .574 **15.** .0883 **16.** 404 **17.** .891 **18.** 14.3 **19.** $\log 5 + 2 \cdot \log t + \log w$ **20.** $\log 2 + \log d - \log m - \log n$ **21.** $3(\log 7 + 4 \cdot \log t)$ **22.** $\frac{1}{2}(\log 3 + \log v + \log w - \log x)$ **23.** 1.7732 **24.** 8.4414 − 10 **25.** 484.3 **26.** .02185 **27.** 149.3 **28.** 451.7 **29.** 11.23 **30.** 4 **31.** −2 **32.** $\frac{1}{2}$ **33.** $\frac{4}{3}$ **34.** 2.97 **35.** .315

36. 2.76 **37.** $x^6 + 12x^5y + 60x^4y^2 + 160x^3y^3 + 240x^2y^4 + 192xy^5 + 64y^6$ **38.** $x^8 - 12x^6 + 54x^4 - 108x^2 + 81$ **39.** $32x^5 - 40x^4y + 20x^3y^2 - 5x^2y^3 + \frac{5xy^4}{8} - \frac{y^5}{32}$ **40.** $540m^3n^3$ **41.** $60x^4$

Page 403

1. 8.5729 − 10 **2.** .463 **3.** 2.23 **4.** 36.5 **5.** .218 **6.** .000547 **7.** .0406 **8.** 5.22 **9.** $2 \cdot \log m + \log n - \log p$ **10.** $2(\log 5 + 3 \cdot \log t)$ **11.** 2.4833 **12.** .4787 **13.** 308.3 **14.** 42.23 **15.** 2 **16.** −2 **17.** $\frac{4}{5}$ **18.** 3.40 **19.** $x^5 + 10x^4y + 40x^3y^2 + 80x^2y^3 + 80xy^4 + 32y^5$ **20.** $x^8 - 4x^6y + 6x^4y^2 - 4x^2y^3 + y^4$ **21.** $-160x^3y^3$

Page 404

1. $\frac{8}{32}, \frac{13}{64}, \frac{21}{128}, \frac{34}{256}$ **2.** Numerators are numbers of the Fibonacci sequence; denominators are powers of 2. **3.** $\frac{y_{n-1} + y_{n-2}}{2^{n-2}}$ where y_{n-1} and y_{n-2} are terms of the Fibonacci sequence.

Page 406

1. 1 **2.** 3 **3.** 4 **4.** 4 **5.** 3 **6.** 2 **7.** 2 **8.** 4 **9.** 3 **10.** 1 **11.** 2 **12.** negative x-axis **13.** positive y-axis **14.** 2 **15.** 4 **16.** negative y-axis **17.** 1 **18.** positive x-axis **19.** 3 **20.** 1 **21.** 1 **22.** 2 **23.** 1 **24.** 2 **25.** 405°, −315° **26.** 300°, −420° **27.** 600°, −120° **28.** 135°, −585°

Page 408

1. 20° **2.** 40° **3.** 45° **4.** 15° **5.** 30° **6.** 70° **7.** 60° **8.** 10° **9.** 65° **10.** 30° **11.** 45° **12.** 60° **13.** 75° − 15° rt. △ **14.** 40° − 50° rt. △ **15.** 30° − 60° rt. △ **16.** 75° − 15° rt. △ **17.** 30° − 60° rt. △ **18.** 60° − 30° rt. △ **19.** 30° − 60° rt. △ **20.** 80° − 10° rt. △ **21.** 70° − 20° rt. △ **22.** 65° − 25° rt. △ **23.** 30° − 60° rt. △ **24.** 60° − 30° rt. △ **25.** 30° **26.** 45° **27.** 45° **28.** 30° **29.** 70° **30.** 45° **31.** 45° **32.** 45° **33.** 75° **34.** 80° **35.** 15° **36.** 30°

Page 412

1. $2\sqrt{2}$ **2.** 3 **3.** $5\sqrt{3}$ **4.** $3\sqrt{2}$ **5.** $b = 5$, $c = 5\sqrt{2}$ **6.** $b = 8\sqrt{3}$, $c = 16$ **7.** $b = 7$, $c = 14$ **8.** $a = 6\sqrt{2}$, $b = 6\sqrt{2}$ **9.** $x = -3$, $y = 3\sqrt{3}$ **10.** $r = 3\sqrt{2}$ **11.** $x = 2\sqrt{3}$, $y = 2$ **12.** 3, $3\sqrt{3}$ **13.** 4, $2\sqrt{3}$ **14.** $\frac{8\sqrt{3}}{3}, \frac{4\sqrt{3}}{3}$ **15.** $3\sqrt{2}$, 3 **16.** $2\sqrt{2}$, 2 **17.** $\frac{5\sqrt{2}}{2}, \frac{5\sqrt{2}}{2}$ **18.** In 30° − 60° rt. △ABC, the hypotenuse $AB = x$. Build an equilateral △ so that $BD = AB$. $BC = \frac{1}{2}BD = \frac{1}{2}x$, $AC^2 + BC^2 = AB^2$. $AC^2 + \left(\frac{1}{2}x\right)^2 = x^2$. $AC^2 = \frac{3}{4}x^2$. $AC = \frac{1}{2}x\sqrt{3}$. **19.** In a 45° − 45° rt. △ABC, the hypotenuse $AB = x$. $BC^2 + AC^2 = AB^2 = x^2$. $2BC^2 = x^2$. $BC^2 = \frac{x^2}{2}$. $BC = \sqrt{\frac{x^2}{2}} = \frac{x}{\sqrt{2}}$, or $\frac{x\sqrt{2}}{2}$.

Page 415

ORALS **1.** symmetric **2.** not symmetric **3.** symmetric **4.** symmetric **5.** not symmetric **6.** symmetric **7.** symmetric **8.** not symmetric **9.** symmetric

EXERCISES **1.** x-axis **2.** y-axis **3.** origin **4.** origin **5.** y-axis **6.** not symmetric **7.** origin **8.** origin **9.** x-axis and origin **10.** (2, −3), (−2, 3), (−2, −3) **11.** (−3, −5), (3, 5), (3, −5) **12.** (−1, 2), (1, −2), (1, 2) **13.** (4, 3), (−4, −3), (−4, 3) **14.** (−2, 3), (2, −3), (2, 3) **15.** (−2, −1), (2, 1), (2, −1) **16.** (2, 4), (−2, −4), (−2, 4) **17.** (4, −3), (−4, 3), (−4, −3) **18.** (−5, −2), (5, 2), (5, −2) **19.** (4, 0), (−4, 0), (−4, 0) **20.** $(2\sqrt{2}, 2\sqrt{2})$, $(2\sqrt{2}, -2\sqrt{2})$, $(-2\sqrt{2}, 2\sqrt{2})$, $(-2\sqrt{2}, -2\sqrt{2})$ **21.** (1, −1), (1, 1), (−1, −1), (−1, 1) **22.** $(-\sqrt{3}, -1)$, $(-\sqrt{3}, 1)$, $(\sqrt{3}, -1)$, $(\sqrt{3}, 1)$ **23.** A graph is symmetric with respect to the x-axis if the portion of the graph above the x-axis is a reflection of the portion of the graph below the x-axis. A graph is symmetric with respect to the y-axis if the portion of the graph to the right of the y-axis is a reflection of the portion of the graph to the left of the y-axis. A graph is symmetric with respect to the origin if the portion of the graph in the 1st quadrant is a reflection of the portion of the graph in the 3rd quadrant, and the portion of the graph in the 2nd quadrant is a reflection of the portion of the graph in the 4th quadrant. **25.** even **26.** odd **27.** even **28.** odd **29.** odd

Page 419

ORALS **1.** positive **2.** negative **3.** negative **4.** positive **5.** positive **6.** negative **7.** positive **8.** negative **9.** negative **10.** positive **11.** positive **12.** negative

EXERCISES **1.** $\frac{1}{2}$ **2.** $\frac{\sqrt{3}}{2}$ **3.** $\frac{\sqrt{2}}{2}$ **4.** $\frac{\sqrt{2}}{2}$ **5.** 0 **6.** 0 **7.** 0 **8.** 0 **9.** 0 **10.** $\frac{1}{2}$ **11.** $\frac{\sqrt{3}}{2}$ **12.** $\frac{1}{2}$ **13.** $-\frac{\sqrt{3}}{2}$ **14.** $\frac{\sqrt{2}}{2}$ **15.** $-\frac{\sqrt{3}}{2}$ **16.** $-\frac{1}{2}$ **17.** $-\frac{\sqrt{2}}{2}$ **18.** $-\frac{1}{2}$ **19.** $\frac{\sqrt{3}}{2}$ **20.** $-\frac{\sqrt{2}}{2}$ **21.** $\frac{\sqrt{3}}{2}$ **22.** $-\frac{1}{2}$ **23.** $\frac{1}{2}$ **24.** $\frac{\sqrt{3}}{2}$ **25.** sin 30° **26.** −cos 30° **27.** −cos 45° **28.** sin 45° **29.** −sin 30° **30.** −cos 30° **31.** −cos 45° **32.** −sin 45° **33.** −sin 30° **34.** cos 30° **35.** cos 45° **36.** −sin 45° **37.** 1, 2 **38.** 3, 4 **39.** 1, 4 **40.** 2, 3 **41.** 30°, 150° **42.** 45°, 315° **43.** 45°, 135° **44.** 120°, 240° **45.** 150°, 210° **46.** 240°, 300° **47.** 135°, 225° **48.** 210°, 330° **49.** 330°, 690° **50.** 210°, 570°

Page 423

ORALS **1.** positive **2.** negative **3.** negative **4.** positive **5.** negative **6.** negative **7.** negative **8.** positive **9.** negative **10.** positive **11.** positive **12.** negative

EXERCISES **1.** $\frac{\sqrt{3}}{3}$ **2.** $\sqrt{3}$ **3.** $\frac{\sqrt{3}}{3}$ **4.** $\sqrt{3}$ **5.** 0 **6.** undefined **7.** 0 **8.** undefined **9.** $-\sqrt{3}$ **10.** $\frac{-\sqrt{3}}{3}$ **11.** −1 **12.** −1 **13.** $\frac{-\sqrt{3}}{3}$ **14.** $-\sqrt{3}$ **15.** undefined **16.** 0 **17.** $\frac{\sqrt{3}}{3}$ **18.** $\sqrt{3}$ **19.** 1 **20.** 1 **21.** $\sqrt{3}$ **22.** $\frac{\sqrt{3}}{3}$ **23.** 0 **24.** undefined **25.** $-\sqrt{3}$ **26.** −1 **27.** $-\sqrt{3}$ **28.** 0 **29.** −1 **30.** −1 **31.** −tan 30° **32.** −cot 30° **33.** −cot 45° **34.** −tan 45° **35.** tan 30° **36.** cot 30° **37.** cot 45° **38.** tan 45° **39.** −tan 30° **40.** −cot 45° **41.** −cot 60° **42.** −tan 60° **43.** −3 **44.** $\frac{5}{3}$ **45.** $\tan x = \frac{\sqrt{3}}{3}$, $\cot x = \sqrt{3}$ **46.** $\tan x = \frac{3}{4}$, $\cot x = \frac{4}{3}$ **47.** $\tan x = -\frac{5}{12}$, $\cot x = -\frac{12}{5}$ **48.** 1, 3 **49.** 2, 4 **50.** 1, 3 **51.** 2, 4 **52.** 45°, 225° **53.** 135°, 315° **54.** 30°, 210° **55.** 120°, 300° **56.** 60°, 240° **57.** 150°, 330° **58.** 30°, 210° **59.** 135°, 315° **60.** 150°, 330° **61.** 240°, 600° **62.** 330°, 690° **63.** 45°, 405°

Page 425

1. $\frac{2\sqrt{3}}{3}$ **2.** 2 **3.** $\sqrt{2}$ **4.** $\sqrt{2}$ **5.** 2 **6.** $\frac{2\sqrt{3}}{3}$ **7.** $\sqrt{2}$ **8.** $-\sqrt{2}$ **9.** $\frac{-2\sqrt{3}}{3}$ **10.** 2 **11.** $-\sqrt{2}$ **12.** $-\sqrt{2}$ **13.** −2 **14.** $\frac{-2\sqrt{3}}{3}$ **15.** $\frac{-2\sqrt{3}}{3}$ **16.** 2 **17.** $\sqrt{2}$ **18.** $-\sqrt{2}$ **19.** −sec 30° **20.** −csc 10° **21.** −sec 40° **22.** −csc 40° **23.** −sec 50° **24.** 0°, 360° **25.** 210°, 330° **26.** 30°, 330°

Page 426

1. 1 **2.** 3 **3.** 4 **4.** 3 **5.** 1 **6.** 2 **7.** 2 **8.** 30° **9.** 60° **10.** 40° **11.** 50° **12.** 60° − 30° rt. △ **13.** 45° − 45° rt △ **14.** 60° − 30° rt. △ **15.** 20° − 70° rt. △ **16.** 60° **17.** 60° **18.** 30° **19.** 50° **20.** 20° **21.** 75° **22.** 20° **23.** $4\sqrt{2}$ **24.** $5\sqrt{3}$ **25.** $(6, 6\sqrt{3})$, $r = 12$; $(6, -6\sqrt{3})$, $(-6, 6\sqrt{3})$, $(-6, -6\sqrt{3})$ **26.** $(3\sqrt{2}, 3\sqrt{2})$; $(3\sqrt{2}, -3\sqrt{2})$, $(-3\sqrt{2}, 3\sqrt{2})$, $(-3\sqrt{2}, -3\sqrt{2})$ **27.** y-axis **28.** origin **29.** (6, 1), (−6, −1), (−6, 1) **30.** (−3, −1), (3, 1), (3, −1) **31.** (−2, 5), (2, −5), (2, 5) **32.** $\frac{1}{2}$ **33.** $\sqrt{3}$ **34.** −2 **35.** $\frac{-\sqrt{3}}{2}$ **36.** $\frac{-2\sqrt{3}}{3}$ **37.** $\frac{\sqrt{3}}{3}$ **38.** sin 60° **39.** −tan 40° **40.** −sec 30° **41.** −cos 30° **42.** −cot 45° **43.** csc 60° **44.** 3 and 4 **45.** 1 and 4 **46.** 2 and 4 **47.** 1 and 3 **48.** 60°, 120° **49.** 60°, 300° **50.** 135°, 315° **51.** 225°, 315° **52.** 30°, 150° **53.** 135°, 225° **54.** 120°, 300° **55.** 60°, 240° **56.** 0 **57.** undefined **58.** −1 **59.** 0 **60.** 0 **61.** undefined

Page 427

1. 1 **2.** 3 **3.** 3 **4.** 4 **5.** 3 **6.** 60° **7.** 30° **8.** 50° **9.** 60° **10.** 45° − 45° rt. △ **11.** 60° − 30° rt. △ **12.** 60° − 30° rt. △ **13.** 70° − 20° rt. △ **14.** 30° **15.** 30° **16.** 30° **17.** 20° **18.** $a = 6\sqrt{2}$, $b = 6\sqrt{2}$ **19.** $a = 4$, $b = 4\sqrt{3}$ **20.** $(-4, 4)$, $r = 4\sqrt{2}$; $(-4, -4)$, $(4, 4)$, $(4, -4)$ **21.** $(8\sqrt{3}, -8)$; $(8\sqrt{3}, 8)$, $(-8\sqrt{3}, -8)$, $(-8\sqrt{3}, 8)$ **22.** x-axis **23.** origin **24.** (3, −4), (−3, 4), (−3, −4) **25.** (−4, 5), (4, −5), (4, 5)

26. $\frac{\sqrt{2}}{2}$ **27.** $\frac{\sqrt{3}}{2}$ **28.** -1 **29.** $\frac{\sqrt{3}}{3}$ **30.** 2
31. $-\sqrt{2}$ **32.** $-\cos 30°$ **33.** $-\sin 60°$
34. $-\tan 30°$ **35.** $\cot 45°$ **36.** $-\sec 60°$
37. $-\csc 30°$ **38.** 1, 2 **39.** 2, 3 **40.** 1, 3
41. 2, 4 **42.** 210°, 330° **43.** 30°, 330° **44.** 45°, 225° **45.** 120°, 300° **46.** 150°, 210° **47.** 1
48. 1 **49.** undefined **50.** undefined
51. undefined **52.** undefined

Page 428
1. .77 **2.** $-.77$ **3.** .87 **4.** 0 **5.** $-.77$ **6.** -1
7. horizontally above the circle

Page 430
1. 3, 4 **2.** 2, 3 **3.** 1, 4 **4.** 1, 2 **5.** 1, 3 **6.** 2, 4
7. 2, 3 **8.** 1, 2 **9.** 2, 4 **10.** 1, 3 **11.** 4
12. no quad. **13.** $\cot u = -\frac{5}{12}$, $\sin u = \frac{12}{13}$, $\csc u = \frac{13}{12}$, $\cos u = -\frac{5}{13}$, $\sec u = -\frac{13}{5}$
14. $\csc u = \frac{5}{3}$, $\tan u = -\frac{3}{4}$, $\cot u = -\frac{4}{3}$, $\cos u = -\frac{4}{5}$, $\sec u = -\frac{5}{4}$ **15.** $\tan u = -1$, $\sin u = -\frac{\sqrt{2}}{2}$, $\csc u = -\sqrt{2}$, $\cos u = \frac{\sqrt{2}}{2}$, $\sec u = \sqrt{2}$ **16.** $\sin u = -\frac{5}{13}$, $\cos u = -\frac{12}{13}$, $\tan u = \frac{5}{12}$, $\sec u = -\frac{13}{12}$, $\cot u = \frac{12}{5}$
17. $\cot u = -\frac{\sqrt{3}}{3}$, $\sin u = \frac{\sqrt{3}}{2}$, $\cos u = -\frac{1}{2}$, $\csc u = \frac{2\sqrt{3}}{3}$, $\sec u = -2$ **18.** $\cot u = \sqrt{3}$, $\sin u = -\frac{1}{2}$, $\cos u = -\frac{\sqrt{3}}{2}$, $\sec u = -\frac{2\sqrt{3}}{3}$, $\csc u = -2$ **19.** $\cot u = \sqrt{3}$, $\sin u = -\frac{1}{2}$, $\cos u = -\frac{\sqrt{3}}{2}$, $\tan u = \frac{\sqrt{3}}{3}$, $\csc u = -2$
20. $\sec u = -\sqrt{3}$, $\sin u = \frac{\sqrt{6}}{3}$, $\csc u = \frac{\sqrt{6}}{2}$, $\tan u = -\sqrt{2}$, $\cot u = -\frac{\sqrt{2}}{2}$ **21.** $\cos u = -\frac{\sqrt{3}}{2}$, $\cot u = \sqrt{3}$, $\sec u = \frac{-2\sqrt{3}}{3}$, $\csc u = -2$
22. $\sin u = \frac{\sqrt{2}}{2}$, $\tan u = 1$, $\cot u = 1$, $\csc u = \sqrt{2}$ or $\sin u = \frac{-\sqrt{2}}{2}$, $\tan u = -1$, $\cot u = -1$, $\csc u = -\sqrt{2}$ **23.** $\sin u = \frac{1}{2}$, $\tan u = \frac{\sqrt{3}}{3}$, $\cot u = \sqrt{3}$, $\sec u = \frac{2\sqrt{3}}{3}$
24. $\sin u = -\frac{\sqrt{2}}{2}$, $\cos u = \frac{\sqrt{2}}{2}$, $\sec u = \sqrt{2}$, $\csc u = -\sqrt{2}$ or $\sin u = \frac{\sqrt{2}}{2}$, $\cos u = -\frac{\sqrt{2}}{2}$, $\sec u = -\sqrt{2}$, $\csc u = \sqrt{2}$

Page 433

ORALS **1.** .5446 **2.** .3249 **3.** .2924
4. .9325 **5.** 1.0786 **6.** .1822 **7.** 2.3183
8. .4462 **9.** .9426 **10.** .0058 **11.** .7030
12. 1.9210 **13.** 12° **14.** 56° **15.** 59° **16.** 19°
17. 67°10′ **18.** 12°50′ **19.** 80°50′
20. 83°20′ **21.** 80°30′

EXERCISES **1.** 27° **2.** 64° **3.** 37° **4.** 9°
5. 72°10′ **6.** 11°10′ **7.** 50°10′ **8.** 73°30′
9. 2°10′ **10.** .7314 **11.** 2.7475 **12.** .9063
13. 5.5764 **14.** $-.1703$ **15.** .7254 **16.** -5.9758
17. $-.8646$ **18.** cos 63° **19.** sin 44°
20. tan 72° **21.** cot 17° **22.** cot 26°30′
23. tan 10°50′ **24.** cos 49°10′ **25.** sin 1°20′
26. 9°, 171° **27.** 32°, 212° **28.** 162°, 198°
29. 130°10′, 310°10′ **30.** 19°20′, 199°20′
31. 19°20′, 160°40′ **32.** $-\sin 30°$
33. $-\cos 40°$ **34.** $-\cot 35°$ **35.** cot 10°
36. $-\cos 35°10'$ **37.** cos 10°30′
38. cos 39°50′ **39.** $-\cot 30°30'$
40. $\frac{1}{\cos 35°}$ **41.** $\frac{1}{\tan 40°}$ **42.** $\frac{1}{\sin 7°10'}$
43. $-\frac{1}{\tan 10°10'}$ **44.** 1.0230 **45.** 3.7420
46. 23°50′ **47.** 53°10′

Page 436
1. 30°, 150° **2.** 150°, 210° **3.** 45°, 225° **4.** 30°, 210° **5.** 135°, 315° **6.** 225°, 315° **7.** 90°, 270°
8. 150°, 330° **9.** 210°, 330° **10.** 30°, 330°
11. 135°, 225° **12.** 60°, 120° **13.** 270° **14.** 180°
15. 172°50′, 352°50′ **16.** 60°, 300° **17.** 56°30′, 123°30′ **18.** 45°, 225° **19.** not possible
20. 71°30′, 251°30′ **21.** no answer between 0° and 360° **22.** 41°20′, 401°20′, 318°40′, 678°40′ **23.** 53°10′, 413°10′, 126°50′, 486°50′ **24.** 143°10′, 323°10′, 503°10′, 683°10′ **25.** 120°, 300°, 480°, 660° **26.** 240°, 300°, 600°, 660° **27.** 150°, 330°, 510°, 690°
28. 45°, 225° **29.** 135°, 315° **30.** 45°, 135°, 225°, 315°

Page 439

1. .9791 **2.** 4.1023 **3.** .5993 **4.** .5106 **5.** .9068 **6.** .7615 **7.** .1239 **8.** .1444 **9.** .1771 **10.** 1.4650 **11.** 1.9883 **12.** .8546 **13.** $27°58'$ **14.** $30°57'$ **15.** $59°20'$ **16.** $56°19'$ **17.** $57°28'$ **18.** $86°17'$ **19.** $18°38'$ **20.** $63°30'$ **21.** $9°03'$ **22.** 120°, 240° **23.** $203°35'$, $336°25'$ **24.** $39°48'$, $219°48'$ **25.** $116°34'$, $296°34'$ **26.** $33°34'$, $326°26'$ **27.** $14°29'$, $165°31'$ **28.** .6996 **29.** −.6961 **30.** −7.8069 **31.** 3.5340 **32.** $156°35'$, $203°25'$ **33.** $235°18'$, $304°42'$ **34.** $23°08'$, $203°08'$ **35.** 45°, 225° **36.** $22°50'$, $202°50'$ **37.** all except 90° and 270°

Page 442

1. $\sin G = \frac{g}{i}$, $\cos G = \frac{h}{i}$, $\tan G = \frac{g}{h}$
2. $\sin B = \frac{b}{c}$, $\cos B = \frac{a}{c}$, $\tan B = \frac{b}{a}$
3. $\sin O = \frac{o}{m}$, $\cos O = \frac{n}{m}$, $\tan O = \frac{o}{n}$
4. $\sin F = \frac{f}{e}$, $\cos F = \frac{d}{e}$, $\tan F = \frac{f}{d}$
5. 71° **6.** 57° **7.** 70° **8.** 30° **9.** 1 **10.** 0
11. $\frac{3\sqrt{2}}{2}$

Page 446

1. $b = 12$, $c = 14$ **2.** $b = 23$, $c = 26$ **3.** $a = 20$, $c = 21$ **4.** $a = 7$, $c = 11$ **5.** $b = 4$, $c = 7$ **6.** $a = 22$, $c = 24$ **7.** 37°, 53° **8.** 67°, 23° **9.** 53°, 37° **10.** 29°, 61° **11.** 42°, 48° **12.** 34°, 56° **13.** 96 m **14.** 11 m **15.** 66° **16.** 1,082 **17.** 10.9 **18.** 35.5 **19.** 17.3 **20.** 110 m **21.** 1,208 m

Page 447

1. $2(\cos 30° + i \sin 30°)$
2. $\sqrt{2}(\cos 315° + i \sin 315°)$
3. $1(\cos 45° + i \sin 45°)$ **4.** $\sqrt{2} + \sqrt{2}i$
5. $-\frac{5}{2} + \frac{5\sqrt{3}}{2}i$ **6.** $\frac{3\sqrt{3}}{2} + \frac{3}{2}i$

Page 448

1. 2, 3 **2.** 1, 2 **3.** 2, 4 **4.** 1, 3 **5.** 3, 4 **6.** $\sin u = -\frac{5}{13}$, $\cos u = -\frac{12}{13}$, $\cot u = \frac{12}{5}$, $\sec u = -\frac{13}{12}$, $\csc u = -\frac{13}{5}$ **7.** $\sin u = \frac{3}{5}$, $\tan u = -\frac{3}{4}$, $\cot u = -\frac{4}{3}$, $\sec u = -\frac{5}{4}$, $\csc u = \frac{5}{3}$ **8.** $\sin u = \frac{2\sqrt{5}}{5}$, $\cos u = \frac{\sqrt{5}}{5}$, $\tan u = 2$, $\sec u = \sqrt{5}$, $\csc u = \frac{\sqrt{5}}{2}$ **9.** .4695 **10.** 1.2876 **11.** −.7986 **12.** −.8541 **13.** $10°40'$ **14.** $82°10'$ **15.** 106°, 286° **16.** $29°40'$, $209°40'$ **17.** cot 17° **18.** $\cos 27°50'$ **19.** sin 30° **20.** −tan 10° **21.** 120°, 240° **22.** 135°, 315° **23.** $111°50'$, $291°50'$ **24.** $48°40'$, $131°20'$ **25.** .6060 **26.** −.5820 **27.** $37°43'$, $322°17'$ **28.** $132°36'$, $312°36'$ **29.** 90° **30.** $26°34'$, $206°34'$ **31.** $1 + \sqrt{3}$ **32.** $\frac{2(\sqrt{2} - 1)}{\sqrt{3}}$ **33.** $\sqrt{6} + 1$ **34.** $a = 3$, $c = 4$ **35.** $b = 9$, $c = 10$ **36.** $b = 46$, $c = 47$ **37.** 36°, 54° **38.** 23°, 67° **39.** 69°, 21° **40.** 97 **41.** 18

Page 449

1. 3, 4 **2.** 1, 3 **3.** 2, 3 **4.** 2, 4 **5.** $\sin u = -\frac{4}{5}$, $\tan u = \frac{4}{3}$, $\cot u = \frac{3}{4}$, $\sec u = -\frac{5}{3}$, $\csc u = -\frac{5}{4}$ **6.** $\sin u = \frac{4}{5}$, $\cos u = \frac{3}{5}$, $\cot u = \frac{3}{4}$, $\sec u = \frac{5}{3}$, $\csc u = \frac{5}{4}$ **7.** .6561 **8.** .6202 **9.** −1.1778 **10.** $38°20'$ **11.** $55°20'$ **12.** $198°10'$, $341°50'$ **13.** 32°, 212° **14.** cos 72° **15.** $\tan 16°50'$ **16.** −cot 40° **17.** $\sin 19°40'$ **18.** 135°, 315° **19.** 150°, 330° **20.** $56°30'$, $123°30'$ **21.** .7849 **22.** −.6397 **23.** $160°22'$, $340°22'$ **24.** $154°9'$, $205°51'$ **25.** $108°26'$, $288°26'$ **26.** 1 **27.** $\sqrt{3}(\sqrt{2} - 2)$ **28.** $a = 3$, $c = 5$ **29.** $b = 10$, $c = 10$ **30.** 37°, 53° **31.** 39°, 51° **32.** 120 **33.** 27

Page 450

1. .18 **2.** 1 **3.** .70 **4.** 1.19 **5.** 1.43 **6.** undefined

Page 452

ORALS **1.** −sin 20° **2.** −cos 55° **3.** −tan 15° **4.** sin 50° **5.** cos 20° **6.** tan 10° **7.** sin 60° **8.** −cos 10°

EXERCISES **1.** −tan 10° **2.** −sin 35° **3.** cos 20° **4.** tan 70° **5.** −sin 50° **6.** −cos 57° **7.** −cot 18° **8.** sin 6° **9.** −csc 38° **10.** sec 50° **11.** sin 35° **12.** cos 10° **13.** tan 80° **14.** $\sin 16°18'$ **15.** $\cos 44°27'$ **16.** −cos 10° **17.** −sin 10° **18.** −tan 10° **19.** cos 40° **20.** sin 10° **21.** sec 35° **22.** tan 20° **23.** sec 20° **24.** $-\sin 9°30'$ **25.** $-\cot 35°37'$

Page 455
1. 1, 1 **2.** 360°, 360° **3.** 1, 1 **4.** −1, −1 **5.** 1 and 4, 3 and 4 **6.** 2 and 3, 1 and 2 **7.** .9 **8.** .5 **9.** .3 **10.** .7 **11.** .9 **12.** .3 **13.** 45° and 225° **14.** −315°, −135°, 45°

Page 458

ORALS **1.** 3 **2.** $\frac{1}{3}$ **3.** $\frac{1}{5}$ **4.** 4 **5.** $\frac{1}{2}$ **6.** 10 **7.** $\frac{1}{8}$ **8.** 24

EXERCISES **9.** 360°, 2 and 360°, $\frac{1}{2}$ **10.** 360°, 3 and 360°, $\frac{1}{2}$ **11.** 360°, $\frac{1}{2}$ **12.** 360°, 2 **13.** 360°, $\frac{1}{2}$ **14.** 360°, 3 **15.** 2 **16.** 2

Page 461

ORALS **1.** 2, 180° **2.** $\frac{1}{3}$, 1,080° **3.** $\frac{1}{2}$, 720° **4.** 3,120° **5.** $\frac{1}{6}$, 2,160° **6.** 9, 40° **7.** 6, 60° **8.** $\frac{1}{9}$, 3,240°

EXERCISES **1.** 360° **2.** 360° **3.** 180° **4.** 180° **5.** 720° **6.** 720° **7.** 120° **8.** 120° **9.** 90° **10.** 90° **11.** 1,080° **12.** 1,080° **25.** 5 **26.** 3 **27.** 180° **28.** 720° **29.** 180° **30.** 4

Page 463

ORALS **1.** 180°, 3 **2.** 180°, $\frac{1}{2}$ **3.** 720°, 2 **4.** 120°, $\frac{1}{2}$ **5.** 120°, $\frac{1}{3}$ **6.** 1,080°, 3 **7.** 60°, 2 **8.** 2,160°, 6

EXERCISES **1.** 720°, 3 **2.** 720°, 2 **3.** 720°, $\frac{1}{2}$ **4.** 1,080°, 2 **5.** 180°, $\frac{1}{2}$ **6.** 1,080°, 3 **7.** 720°, 2 **8.** 90°, $\frac{1}{6}$ **9.** 60°, $\frac{1}{4}$ **10.** 2,160°, 6 **11.** 40°, 6 **12.** 30°, 4 **13.** 720°, 2 **14.** 720°, 2 **15.** 180°, $\frac{1}{2}$ **16.** 180°, $\frac{1}{2}$ **17.** 120°, 3 **18.** 120°, 3 **19.** 120°, $\frac{1}{3}$ **20.** 120°, $\frac{1}{3}$ **21.** 1,080°, 3 **22.** 1,080°, 3 **23.** 60°, 2 **24.** 40°, $\frac{1}{2}$ **25.** 2 **26.** 2 **27.** 2 **28.** 3 **29.** 720°, 2 **30.** 180°, $\frac{1}{2}$ **31.** 720°, 2 **32.** 180°, $\frac{1}{2}$ **33.** 720°, 2 **34.** 180°, $\frac{1}{2}$ **35.** 2 **36.** 4 **37.** 4 **38.** 4

Page 467
1. −180° to 360°, none **2.** none, none **3.** 180° **4.** none **6.** none, −180° to 360° **7.** none, none **8.** 180° **9.** none **11.** 360° **12.** 90° **13.** 30° **14.** 60° **15.** 15° **16.** 90° **17.** 30° **18.** 60° **31.** 4 **32.** 8 **33.** 360° **34.** 90° **35.** 90° **36.** 360° **37.** 2 **38.** 8

Page 470
1. .5719 **2.** 9.9052 − 10 **3.** 9.8699 − 10 **4.** 8.6101 − 10 **5.** 9.8699 − 10 **6.** 9.2536 − 10 **7.** 9.9225 − 10 **8.** 9.4218 − 10 **9.** 9.9519 − 10 **10.** .5111 **11.** 9.8495 − 10 **12.** 9.7338 − 10 **13.** 3°0′ **14.** 63°20′ **15.** 87°0′ **16.** 19°10′ **17.** 82°10′ **18.** 64°17′ **19.** 9°49′ **20.** 18°12′ **21.** 56°53′ **22.** 2.1 **23.** 2.4 **24.** 17.7 **25.** 1.4 **26.** .8 **27.** 3233.1 **28.** 19.8 **29.** 1,790,000.0 **30.** .0006

Page 473
1. 45°, 135°, 225°, 315° **2.** 45°, 135°, 225°, 315° **3.** 45°, 135°, 225°, 315° **4.** 90°, 270°, 120°, 240° **5.** 90°, 270°, 135°, 315° **6.** 210°, 330°, 90° **7.** 45°, 225° **8.** 16°, 164° **9.** no values between 0° and 360° **10.** 259°16′, 100°44′ **11.** 190°34′, 349°26′ **12.** 8°29′, 188°29′, 73°23′, 253°23′ **13.** 18°35′, 161°25′ **14.** 73°05′, 286°55′ **15.** 210°, 330°
15. 210°, 330° **16.** $\frac{1}{3-k}$ **17.** 0, b

Page 474
1. 17°0′ **2.** 19°0′ **3.** 42°50′

Page 477
1. $\frac{1}{\sin 30°}$ **2.** $\frac{\cos 135°}{\sin 135°}$ or $\frac{-\cos 45°}{\sin 45°}$ **3.** $\frac{1}{\sin 320°}$ or $\frac{-1}{\sin 40°}$ **4.** $\frac{1}{\cos 75°}$ **5.** $\frac{-\sin 310°}{\cos 310°}$ or $\frac{\sin 50°}{\cos 50°}$ **22.** $\pm\sqrt{1-\sin^2 x}$ **23.** $\frac{\pm\sqrt{1-\sin^2 x}}{\sin x}$ **24.** $\frac{\pm 1}{\sqrt{1-\sin^2 x}}$ **25.** $1+\frac{\sin^2 x}{1-\sin^2 x}$ **26.–29.** Answers may vary.

Page 479
1.–16. Answers may vary.

Page 482

ORALS **1.** 2π **2.** $\frac{\pi}{2}$ **3.** $\frac{\pi}{4}$ **4.** π **5.** $\frac{\pi}{6}$ **6.** $\frac{\pi}{3}$ **7.** $\frac{5\pi}{3}$ **8.** 90° **9.** 45° **10.** 270° **11.** 180° **12.** 150° **13.** 30° **14.** 210°

EXERCISES **1.** $\frac{\pi}{12}$ **2.** $\frac{\pi}{9}$ **3.** $\frac{4\pi}{9}$ **4.** $\frac{7\pi}{6}$ **5.** $\frac{3\pi}{4}$ **6.** $\frac{11\pi}{6}$ **7.** $\frac{7\pi}{4}$ **8.** $\frac{-\pi}{4}$ **9.** $\frac{-2\pi}{3}$ **10.** $\frac{-3\pi}{4}$ **11.** $\frac{-7\pi}{6}$ **12.** $\frac{-3\pi}{2}$ **13.** $\frac{-5\pi}{3}$ **14.** -2π **15.** 60° **16.** 20° **17.** 45° **18.** 135° **19.** 240° **20.** 300° **21.** 162° **22.** 36° **23.** −120° **24.** −330° **25.** −270° **26.** −90° **27.** −360° **28.** −40° **29.** 1 **30.** $\frac{\sqrt{2}}{2}$ **31.** −1 **32.** $-\frac{1}{2}$ **33.** −1 **34.** $\frac{1+\sqrt{2}}{2}$ **35.** $\frac{2+\sqrt{3}}{\sqrt{3}}$

Page 485

1. 1, −1 **2.** π **3.** 1 **4.** 2, −2 **5.** 2π **6.** 2 **7.** 2π, 3 **8.** 4π, 1 **9.** 2π, $\frac{1}{3}$ **10.** 2π, 4 **11.** 4π, 2 **12.** π, $\frac{1}{2}$ **13.** π, 2 **14.** 4π, $\frac{1}{2}$ **15.** 2 **16.** 4 **17.** $\frac{\pi}{2}$, $\frac{-\pi}{2}$ **18.** 0, 2π, APP $\frac{5\pi}{6}$, $\frac{-5\pi}{6}$ **19.** $\frac{\pi}{6}$, $\frac{5\pi}{6}$ **20.** $\frac{5\pi}{6}$, $\frac{7\pi}{6}$ **21.** $\frac{3\pi}{4}$, $\frac{7\pi}{4}$ **22.** $\frac{2\pi}{3}$, $\frac{4\pi}{3}$ **23.** $\frac{\pi}{6}$, $\frac{5\pi}{6}$, $\frac{11\pi}{6}$, $\frac{7\pi}{6}$ **24.** $\frac{5\pi}{4}$, $\frac{\pi}{4}$, $\frac{7\pi}{4}$, $\frac{3\pi}{4}$ **25.** 0, π, 2π, $\frac{\pi}{6}$, $\frac{5\pi}{6}$ **26.** 0, 2π, $\frac{\pi}{3}$, $\frac{5\pi}{3}$ **27.** $\frac{\pi}{2}$, $\frac{7\pi}{6}$, $\frac{11\pi}{6}$ **34.** 2π, 2 **35.** 2π, $\frac{1}{2}$ **36.** π, 1 **37.** 4π, 1 **38.** π, none **39.** π, none **40.** π, $\frac{1}{2}$ **41.** π, 2 **42.** π, $\frac{1}{2}$ **43.** 2 **44.** 5

Page 489

1. {(6, −1), (3, −2), (6, 5)}, no **2** {(7, 0), (−1, 7), (2, −8)}, yes **3.** $2y - 3x = 8$, yes **4.** $x = y^2$, no **5.** $x = \sin 2y$, no **6.** $\frac{\pi}{3}$ or 60° **7.** $\frac{-\pi}{4}$ or −45° **8.** $\frac{\pi}{4}$ **9.** $\frac{\pi}{6}$ **10.** $\frac{1}{2}$ **11.** $\frac{-\pi}{2}$ **12.** $\frac{5\pi}{6}$ or 150° **13.** $\frac{5}{13}$ **14.** $\frac{4}{5}$ **15.** −1

Page 490

1. −cos 10° **2.** tan 30° **3.** sin 40° **4.** −sin 10° **5.** cos 30° **6.** 360°, 2 **7.** 360°, none **8.** 120°, 1 **9.** 720°, 3 **10.** 4 **11.** 4 **12.** 9.8175 − 10 **13.** 72°08′ **14.** .00000276 **15.** 14,450 **16.** .2 **17.** 28°, 332° **18.** 90°, 270° **19.** 0, π, $\frac{3\pi}{4}$, $\frac{7\pi}{4}$, 2π **20.** $\frac{\pi}{2}$, $\frac{7\pi}{6}$, $\frac{11\pi}{6}$ **21.** $\frac{1}{\cos 45°}$ **22.** $\frac{\sin 10°}{\cos 10°}$ **23.** $\frac{\cos 80°}{\sin 80°}$ **24.** $\frac{-1}{\sin 320°}$ or $\frac{1}{\sin 40°}$ **26.** $(\sin x)\left(\frac{1}{\sin x}\right) = 1$ **27.** $\pm\frac{\sin x}{\sqrt{1-\sin^2 x}}$ **28.** $\pm\frac{\sqrt{1-\sin^2 x}}{\sin x}$ **29.–31.** Answers may vary. **32.** $\frac{\pi}{4}$ **33.** $\frac{-\pi}{6}$ **34.** $\frac{5\pi}{3}$ **35.** $\frac{-7\pi}{6}$ **36.** 20° **37.** −150° **38.** 270° **39.** −720° **40.** $\frac{-\sqrt{3}}{3}$ **41.** $\frac{-\sqrt{3}}{2}$ **42.** 1 **43.** $\frac{2(3-\sqrt{3})}{3}$ **44.** {(7, −3), (−3, 7), (4, 6)}, yes **45.** {(2, −1), (3, −2), (2, 4)}, no **46.** $-2y + x = 6$, yes **47.** $x = \sin y$, no **48.** 30° or $\frac{\pi}{6}$ **49.** −30° or $-\frac{\pi}{6}$ **50.** $\frac{3}{5}$ **51.** $\frac{4}{5}$

Page 491

1. −sin 30° **2.** −cot 35° **3.** sin 40° **4.** 360°, $\frac{1}{2}$ **5.** 120°, 2 **6.** 90°, none **7.** 2 **8.** 4 **9.** 9.9432 − 10 **10.** .0516 **11.** 10°50′ **12.** 20°06′ **13.** 368.8 **14.** .5 **15.** 45°, 135°, 225°, 315° **16.** 19°, 161° **17.** 0, $\frac{\pi}{2}$, π, 2π **18.** $\frac{1}{\sin 30°}$ **19.** $\frac{-\sin 340°}{\cos 340°}$ or $\frac{\sin 20°}{\cos 20°}$ **20.** $\frac{\cos 230°}{\sin 230°}$ or $\frac{\cos 50°}{\sin 50°}$ **22.** $\pm\frac{\sqrt{1-\sin^2 x}}{\sin x}$ **23.** $\frac{1-\sin^2 x}{\sin x}$ **24.–25.** Answers may vary. **26.** $\frac{\pi}{3}$ **27.** $\frac{-5\pi}{3}$ **28.** 150° **29.** −120° **30.** $-\frac{1}{2}$ **31.** $\frac{3\sqrt{2}+2\sqrt{6}}{6}$ **32.** {(2, −3), (6, 4), (2, 3)}, no **33.** $y - x = 3$, yes **34.** $x = \cos y$, no **35.** 45° or $\frac{\pi}{4}$ **36.** $\frac{4}{3}$ **37.** 30° or $\frac{\pi}{6}$

Page 492

1. .2955 **2.** .9553

Page 495

ORALS **1.** $x^2 = a^2 = b^2 + c^2 - 2bc \cos A = (50)^2 + (40)^2 - 2(50)(40) \cos 30°$ **2.** $x^2 = c^2 = a^2 + b^2 - 2ab \cos C = (20)^2 + (15)^2 - 2(20)(15) \cos 120°$ **3.** $x^2 = b^2 = a^2 + c^2 - 2ac \cos B = (12)^2 + (15)^2 - 2(12)(15) \cos 150°$

EXERCISES **1.** $\sqrt{19}$ or 4 **2.** $\sqrt{85 + 42\sqrt{3}}$ **3.** $\sqrt{85 + 42\sqrt{2}}$ **4.** $\sqrt{65 - 28\sqrt{3}}$ **5.** $\sqrt{13 - 6\sqrt{2}}$ **6.** $\sqrt{129}$ or 11 **7.** $\sqrt{29}$ or 5 **8.** $\sqrt{284}$ or 17 **9.** $\sqrt{132}$ or 11 **10.** 7 **11.** 495 km **12.** 626 km **13.–14.** See Example 2, page 493.

Page 497

1. .759 **2.** .983 **3.** .806 **4.** −.702 **5.** 22° **6.** 19° **7.** 67° **8.** 54° **9.** 29° **10.** 43° **11.** See Example 2, page 496.

Page 499

1. 12 **2.** 19 **3.** 22 **4.** 7 **5.** 28° **6.** 29° **7.** 25° **8.** 15° **9.** 4 **10.** 35 **11.–12.** See Example 2, page 498.

Page 502

ORALS **1.** $\frac{\sin A}{a} = \frac{\sin C}{c}$ **2.** $\frac{\sin A}{a} = \frac{\sin C}{c}$ **3.** $\frac{\sin C}{c} = \frac{\sin B}{b}$

EXERCISES **1.** 18 **2.** 29 **3.** 26 **4.** 24 **5.** .8999 **6.** .7 **7.** 22 **8.** 42 **9.** 19 **10.** 50.93 **11.–12.** See Example 1, page 500.

Page 505

1. 2 **2.** 1 **3.** none **4.** 1 **5.** 2 **6.** none **7.** none **8.** none **9.** 1 **10.** none **11.** none **12.** none

Page 509

1. $\frac{\sqrt{2}}{4}(\sqrt{3} + 1)$ **2.** $\frac{\sqrt{3}}{2}$ **3.** $\frac{\sqrt{2}}{4}(\sqrt{3} + 1)$ **4.** $\frac{\sqrt{2}}{4}(\sqrt{3} - 1)$ **5.** $\frac{\sqrt{3}}{2}$ **6.** $\frac{\sqrt{2}}{2}$ **7.** $\frac{1}{2}$ **8.** 1 **9.** $\frac{1}{2}$ **10.** 0 **11.** 1 **12.** $\frac{\sqrt{3}}{2}$ **13.** 0 **14.** $\frac{\sqrt{3}}{2}$ **15.** $\frac{33}{65}, \frac{63}{65}$ **16.** $\frac{63}{65}, \frac{33}{65}$ **17.** $\frac{84}{85}, -\frac{36}{85}$ **18.** $\frac{120}{169}$, 0 **19.** $-\frac{21}{221}, \frac{171}{221}$ **20.** $-\frac{33}{65}, \frac{63}{65}$ **21.** $\frac{24}{25}$ **22.** $-\frac{4\sqrt{5}}{9}$ **23.** $\frac{20\sqrt{6}}{49}$ **24.** $-\frac{12}{13}$ **25.** $\sec A$ **26.** $\frac{1}{2} \csc A$ **27.** $2 \tan A$ **28.** $\sin (90° + u) = \sin 90° \cos u + \cos 90° \sin u = 1(\cos u) + 0(\sin u) = \cos u$ **29.** $\sin (90° - u) = \sin 90° \cos u - \cos 90° \sin u = 1(\cos u) - 0(\sin u) = \cos u$ **30.** $\sin (180° + u) = \sin 180° \cos u + \cos 180° \sin u = 0(\cos u) + (-1)(\sin u) = -\sin u$ **31.** $\sin (180° - u) = \sin 180° \cos u - \cos 180° \sin u = 0(\cos u) - (-1)(\sin u) = \sin u$ **32.** $\sin (270° - u) = \sin 270° \cos u - \cos 270° \sin u = -(1)(\cos u) - 0(\sin u) = -\cos u$ **33.** $\sin (360° - u) = \sin 360° \cos u - \cos 360° \sin u = 0(\cos u) - 1(\sin u) = -\sin u$

Page 513

1. 0 **2.** $\frac{1}{2}$ **3.** $-\frac{1}{2}$ **4.** $\frac{\sqrt{2}}{4}(1 - \sqrt{3})$ **5.** $\frac{\sqrt{2}}{4}(\sqrt{3} - 1)$ **6.** $-\frac{\sqrt{2}}{2}$ **7.** 0 **8.** −1 **9.** $\frac{1}{2}$ **10.** $\frac{\sqrt{3}}{2}$ **11.** $-\frac{1}{2}$ **12.** 1 **13.** $\frac{\sqrt{2}(1 - \sqrt{3})}{4}$ **14.** $\frac{\sqrt{2}(\sqrt{3} - 1)}{4}$ **15.** $\frac{\sqrt{2}}{4}(\sqrt{3} + 1)$ **16.** 0 **17.** $\frac{1}{2}$ **18.** 0 **19.** $-\frac{1}{2}$ **20.** −1 **21.** $-\frac{63}{65}, -\frac{33}{65}$ **22.** $-\frac{56}{65}, \frac{16}{65}$ **23.** 0, $\frac{120}{169}$ **24.** $\frac{44 - \sqrt{759}}{84}, \frac{44 + \sqrt{759}}{84}$ **25.** $-\frac{120}{169}$, 0 **26.** $-\frac{56}{65}, -\frac{16}{65}$ **27.** $\frac{119}{169}$ **28.** $-\frac{119}{169}$ **29.** $\frac{7}{25}$ **30.** $-\frac{7}{25}$ **31.** 1 **32.** 1 **33.** −1 **34.** $\cos (90° + u) = \cos 90° \cos u - \sin 90° \sin u = 0(\cos u) - 1(\sin u) = -\sin u$ **35.** $\cos (180° - u) = \cos 180° \cos u + \sin 180° \sin u = -1(\cos u) + 0(\sin u) = -\cos u$ **36.** $\cos (90° - u) = \cos 90° \cos u + \sin 90° \sin u = 0 (\cos u) + 1(\sin u) = \sin u$ **37.** $\cos (270° + u) = \cos 270° \cos u - \sin 270° \sin u = 0(\cos u) - (-1)(\sin u) = \sin u$ **38.** $\cos (180° + u) = \cos 180° \cos u - \sin 180° \sin u = -1(\cos u) - 0(\sin u) = -\cos u$ **39.** $\cos (360° - u) = \cos 360° \cos u + \sin 360° \sin u = 1(\cos u) + 0(\sin u) = \cos u$

Page 516

1. $\frac{3+\sqrt{3}}{3-\sqrt{3}}$ **2.** doesn't exist **3.** $\frac{3-\sqrt{3}}{3+\sqrt{3}}$ **4.** 0
5. $\frac{3\sqrt{3}+4}{3-4\sqrt{3}}$ **6.** $-\frac{1}{8}$ **7.** $\frac{-3+\sqrt{3}}{3+\sqrt{3}}$ **8.** $-\frac{63}{16}$ **9.** $-\frac{95}{59}$
10. $\frac{-(\sqrt{3}+1)}{(1-\sqrt{3})}$ **11.** $\frac{5\sqrt{3}-36}{15+12\sqrt{3}}$ **12.** $-\frac{24}{7}$
13. $\frac{120}{119}$ **14.** $\sqrt{3}$ **15.** doesn't exist **16.** $-\frac{56}{33}$
17. $-\frac{24}{7}$ **18.** $\frac{63}{16}, \frac{33}{56}, -\frac{24}{7}, \frac{120}{119}$ **19.** $\frac{3\sqrt{3}-4}{3+4\sqrt{3}}$,
$\frac{3\sqrt{3}+4}{3-4\sqrt{3}}$, $-\sqrt{3}$, $\frac{24}{7}$

Page 518

ORALS **1.** ± 1 **2.** $\pm\frac{\sqrt{2}}{2}$ **3.** ± 1 **4.** 0
5. $\pm\frac{\sqrt{3}}{2}$ **6.** $\pm\frac{\sqrt{3}}{2}$

EXERCISES **1.** $\frac{\sqrt{2+\sqrt{3}}}{2}$ **2.** $\frac{\sqrt{2-\sqrt{2}}}{2}$
3. $-\frac{\sqrt{2}}{2}$ **4.** $\frac{\sqrt{2}}{2}$ **5.** $\frac{\sqrt{2+\sqrt{2}}}{2}$ **6.** $\frac{\sqrt{2-\sqrt{2}}}{2}$
7. $\frac{\sqrt{2+\sqrt{2}}}{2}$ **8.** $\frac{\sqrt{2+\sqrt{3}}}{2}$ **9.** $\frac{\sqrt{2}}{4}, \frac{\sqrt{14}}{4}$
10. $\frac{\sqrt{2+\sqrt{3}}}{2}, \frac{\sqrt{2-\sqrt{3}}}{2}$ **11.** $\frac{\sqrt{3}}{3}, \frac{\sqrt{6}}{3}$
12. $\frac{\sqrt{6}}{4}, \frac{\sqrt{10}}{4}$ **13.** $\frac{\sqrt{10}}{4}, -\frac{\sqrt{6}}{4}$ **14.** $\frac{\sqrt{2+2a}}{2}$,
$-\frac{\sqrt{2-2a}}{2}$ **15.** $-\sqrt{7-4\sqrt{3}}$ **16.** $-\frac{2\sqrt{3}}{3}$
17. $\sqrt{7-4\sqrt{3}}$ **18.** $\frac{3}{4}$

Page 521

1.–12. Answers may vary.

Page 522

1. $\sqrt{22}$ **2.** $\sqrt{193}$ **3.** $\frac{53}{80}$ **4.** 44° **5.** 16 **6.** 10
7. 34° **8.** 56° **9.** 37 **10.** 77° **11.** 2 **12.** 1
13. $\frac{\sqrt{2}}{4}(\sqrt{3}+1)$ **14.** $\frac{\sqrt{3}}{2}$ **15.** $\frac{\sqrt{2}}{4}(\sqrt{3}-1)$
16. $\frac{\sqrt{2}}{4}(\sqrt{3}+1)$ **17.** $\frac{\sqrt{2}}{4}(\sqrt{3}-1)$
18. $\frac{\sqrt{2}}{4}(1-\sqrt{3})$ **19.** $-\frac{33}{65}, \frac{63}{65}$ **20.** $-\frac{24}{25}$, 0
21. $\frac{120}{169}$ **22.** $\frac{\sqrt{7}}{4}$ **23.** $-\frac{16}{63}$ **24.** $-\frac{73}{129}$ **25.** $-\frac{120}{119}$
26. $\frac{3+\sqrt{3}}{3-\sqrt{3}}$ **27.** csc *A* **28.** Answers may vary.
29. 0, $\frac{24}{7}$, $\frac{24}{7}$

Page 523

1. $\sqrt{58.864}$ or 8 **2.** $\sqrt{136+60\sqrt{3}}$ **3.** $-\frac{5}{28}$
4. 131° **5.** 59 **6.** 13 **7.** 42° **8.** 34 **9.** 19° **10.** 2
11. 1 **12.** $\frac{\sqrt{2}}{4}(\sqrt{3}+1)$ **13.** $\frac{1}{2}$ **14.** $\frac{\sqrt{2-\sqrt{2}}}{2}$
15. $-\frac{\sqrt{2}}{2}$ **16.** $\frac{1}{2}$ **17.** $\frac{20\sqrt{2}+14\sqrt{6}}{63}$ **18.** $\frac{33}{65}$
19. $-\frac{120}{169}$ **20.** $-\frac{119}{169}$ **21.** $-\frac{13}{47}$ **22.** $-\frac{50}{49}$ **23.** $-\frac{24}{7}$
24. $\frac{\sqrt{3}+1}{1-\sqrt{3}}$ **25.** 2 **26.** Answers may vary.
27. $\frac{16}{63}$

HOLT ALGEBRA
WITH TRIGONOMETRY
2

Eugene D. Nichols
Mervine L. Edwards
E. Henry Garland
Sylvia A. Hoffman
Albert Mamary
William F. Palmer

HOLT ALGEBRA 2

WITH TRIGONOMETRY

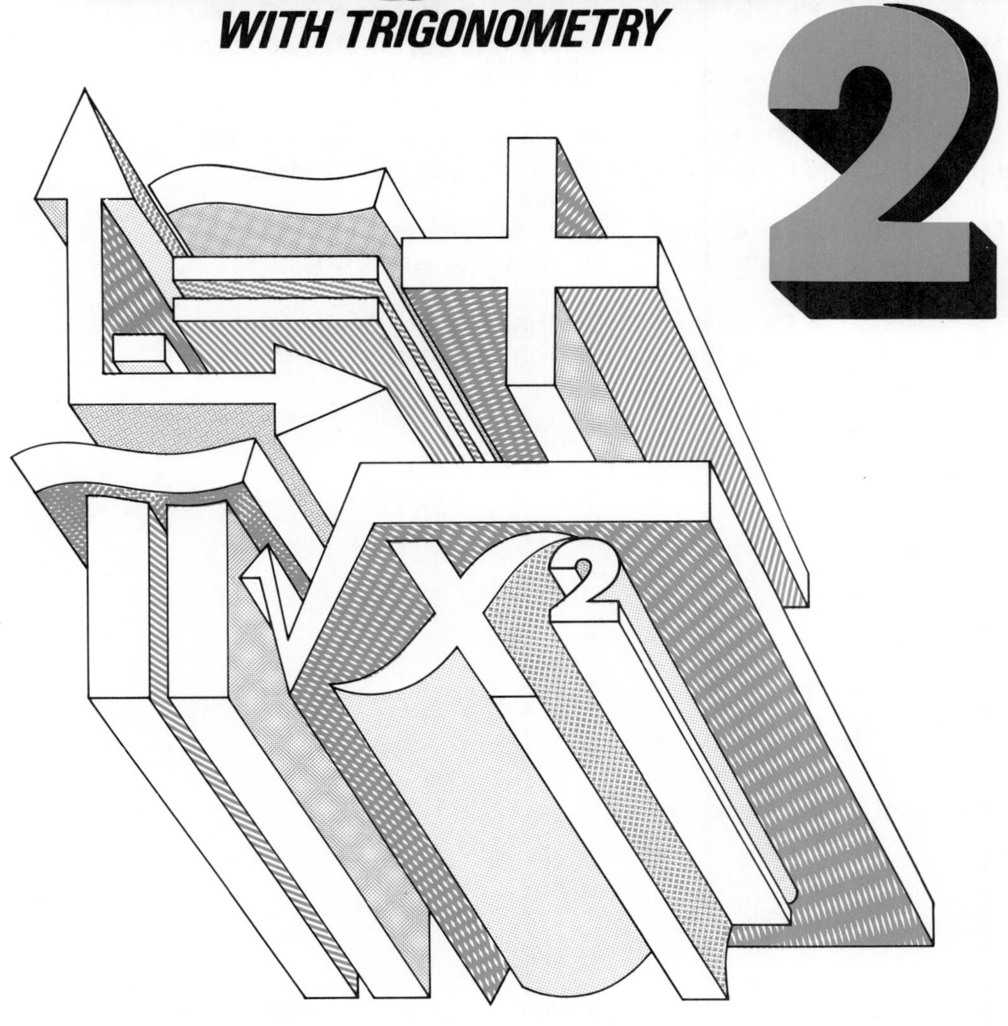

HOLT, RINEHART AND WINSTON, PUBLISHERS
New York · Toronto · London · Sydney

About the Authors

Eugene D. Nichols is Professor of Mathematics Education and Lecturer in the Department of Mathematics at Florida State University, Tallahassee, Florida.

Mervine L. Edwards is Chairman of the Mathematics Department, Shore Regional High School, West Long Branch, New Jersey.

E. Henry Garland is Head of the Mathematics Department at the Developmental Research School, and Associate Professor of Mathematics Education at Florida State University, Tallahassee, Florida.

Sylvia A. Hoffman is Curriculum Coordinator for the Metropolitan Chicago Region of the Illinois Office of Education, State of Illinois.

Albert Mamary is Assistant Superintendent of Schools for Instruction, Johnson City Central School District, Johnson City, New York.

William F. Palmer is Professor and Chairman of the Department of Education, Catawba College, Salisbury, North Carolina.

Photo credits are on page viii.

Printed in the United States of America

ISBN: 0-03-018911-x

7890123456 071 987654321

CONTENTS

1 LINEAR EQUATIONS AND INEQUALITIES

2 ABSOLUTE VALUE AND PROBLEM SOLVING

3 POLYNOMIALS AND FACTORING

4 RATIONAL EXPRESSIONS

5 USING RATIONAL EXPRESSIONS

6 REAL NUMBERS

SPECIAL TOPICS

ACKNOWLEDGEMENTS FOR PHOTOGRAPHS

Page 25 Cooking For Profit, Peoples Gas Co.
Page 217 © 1977 Betty Medsger
Page 234 New York Power Pool
Page 358 HRW Photo by Russell Dian

SYMBOL LIST

Symbol	Meaning	Page
$=$	is equal to	1
$<$	is less than	6
$>$	is greater than	6
$\{x \mid x > 5\}$	the set of all numbers x such that $x > 5$	6
$\leq$	is less than or equal to	7
$\geq$	is greater than or equal to	7
$a < x < b$	x is between a and b	19
$\lvert x \rvert$	the absolute value of x	22
x^n	x to the nth power	46
GCF	greatest common factor	56
$\cdots$	the pattern continues on and on	76
LCD	least common denominator	98
$.624\overline{24}$	.6242424 . . .	125
$\sqrt{x}$	the principal square root of x	134
ϕ	the empty set	135
$\doteq$	is approximately equal to	137
$\sqrt[n]{x}$	the nth root of x	153
$a \pm \sqrt{b}$	$a + \sqrt{b}$, $a - \sqrt{b}$	174
i	$\sqrt{-1}$	196
(x, y)	the ordered pair x, y	211
$P(x, y)$	the point P with coordinates x, y	211
x_2	x sub two	212
$\overline{AB}$	line segment AB	212
$\vec{d}(AB)$	the directed distance from A to B	214
x_P	x sub P (the x-coordinate of point P)	215
$\overleftrightarrow{AB}$	line AB	218
$\{(x, y) \mid y = 2x - 3\}$	the set of all (x, y) such that $y = 2x - 3$	228
AB	the distance between A and B	238
$m(\overline{AB})$	slope of line segment AB	244
$\parallel$	is parallel to	244
$\perp$	is perpendicular to	245
$f(x)$	f at x (the value of function f at x)	256
$[x]$	the greatest integer $\leq x$	260
$\log x$	the logarithm of x	365
$\log_n x$	the log, base n, of x	395
$m\angle A$	degree measure of $\angle A$	444
$\sin^{-1}$	arc sine	487

1 LINEAR EQUATIONS AND INEQUALITIES

Puzzlers

PUZZLE 1.

At noon the hands of a clock are together. At exactly what time will they be together again for the first time?

PUZZLE 2.

Do you know that there are only 4 four-digit perfect squares with all even digits? Find as many of these 4 numbers as you can.

PUZZLE 3.

A bird flew south alternating the distance it traveled each day. One day the bird would fly $\frac{1}{10}$ of the remaining distance. The next day it would fly $\frac{2}{3}$ the total distance it had already flown. The bird followed this pattern for 7 days at which time it had 22.5 kilometers remaining to fly. What was the total distance of the bird's journey south? [Hint: Let x = the total distance of the journey south. The first day the bird flew $\frac{1}{10}x$ km and then $\frac{9}{10}x$ km remained.]

PUZZLE 4.

Four friends attended a school picnic where each had brought his or her own plate and cup. After some confusion no one left with the cup and plate that he or she had brought, but each with the cup of one friend and the plate of another. Jill took the cup belonging to the friend whose plate was taken by Sue. Betty took Donna's plate. Sue's cup was taken by the girl who took Jill's plate. Whose cup and plate did each girl take home?

Linear Equations

OBJECTIVE

- To solve equations like $4x + 3 - 7x = 8 + 2x - 14$

REVIEW CAPSULE

Sign Rules for Computing

$-5(-3) = +15$	$-4(+3) = -12$
$\frac{-8}{-2} = +4$	$\frac{+6}{-3} = -2$
Like signs give *positive* products and quotients.	Unlike signs give *negative* products and quotients.

$$8 - (-3) = 8 + (+3), \text{ or } 11$$
$$-2 - 7 = -2 + (-7), \text{ or } -9$$
$$a - b = a + (-b)$$

To subtract a number, add its additive inverse (opposite).

EXAMPLE 1 Solve $8x - 3 = 2x - 21$.

Both sides have an x-term. → $8x - 3 = 2x - 21$

Add $-2x$ to each side. The opposite of $2x$ is $-2x$. → $\underline{-2x \qquad -2x}$; $6x - 3 = 0 - 21$

Add $+3$ to each side. The opposite of -3 is $+3$. → $\underline{+3 \qquad +3}$; $6x + 0 = -18$

Divide each side by 6. → $\frac{6x}{6} = \frac{-18}{6}$

$\frac{-18}{+6} = -3$ → $x = -3$

Check.

$8x - 3$	$2x - 21$
$8(-3) - 3$	$2(-3) - 21$
$-24 - 3$	$-6 - 21$
-27	-27

Substitute -3 for x.

Results are the same; $-27 = -27$

Thus, -3 is the solution.

EXAMPLE 2 Find the solution set of $23 = 7 - 8x$.

It is easier here to get the x-term alone on the *right* side.

$$23 = 7 - 8x$$

Add -7 to each side. → $16 = -8x$

Divide each side by -8. → $\frac{16}{-8} = \frac{-8x}{-8}$

$$-2 = x$$

The *solution set* is the *set* of all solutions. → **Thus,** the solution set is $\{-2\}$.

EXAMPLE 3 Find the solution set of $6 + 3y = 20 - 2y$.

$$6 + 3y = 20 - 2y$$

Add $2y$ to each side. → $6 + 5y = 20$

Add -6 to each side. → $5y = 14$

Divide each side by 5. → $\frac{5y}{5} = \frac{14}{5}$

$\frac{14}{5} = 2\frac{4}{5}$ → $y = \frac{14}{5}$, or $2\frac{4}{5}$

Thus, the solution set is $\left\{2\frac{4}{5}\right\}$.

EXAMPLE 4 Solve $\frac{3}{4}n = 6$.

$$\frac{3}{4}n = 6$$

Multiply each side by the denominator 4. → $\frac{4}{1} \cdot \frac{3}{4} \cdot n = 6 \cdot 4$

$\frac{\not{4}^{1}}{1} \cdot \frac{3}{\not{4}_{1}} \cdot n = 3n$ → $3n = 24$ ← No fractions

Divide each side by 3. → $n = 8$

Thus, the solution is 8.

EXAMPLE 5 Find the solution set of $\frac{3}{5}a - 3 = 2$.

$$\frac{3}{5}a - 3 = 2$$

Add 3 to each side. → $\frac{3}{5}a = 5$

Multiply each side by 5. → $\frac{5}{1} \cdot \frac{3}{5} \cdot a = 5 \cdot 5$

$\frac{\not{5}^{1}}{1} \cdot \frac{3}{\not{5}_{1}} \cdot a = 3a$ → $3a = 25$

Divide each side by 3. → $a = \frac{25}{3}$, or $8\frac{1}{3}$

Thus, the solution set is $\left\{8\frac{1}{3}\right\}$.

EXAMPLE 6 Solve $3x + 4 - 8x = 5 + 2x - 15$.

$$3x + 4 - 8x = 5 + 2x - 15$$

Combine like terms on each side: $3x - 8x = -5x$; $5 - 15 = -10$. → $-5x + 4 = -10 + 2x$

$$-7x + 4 = -10$$

$$-7x = -14$$

$$\frac{-7x}{-7} = \frac{-14}{-7}$$

$(-) \div (-) = +$ → $x = 2$

Thus, the solution is 2.

ORAL EXERCISES

Answers for all exercises are given in the front of the Teacher's Edition.

Solve.

1. $x + 9 = 3$ *−6*
2. $8c = 24$ *3*
3. $7 = a - 5$ *12*
4. $35 = -5x$ *−7*
5. $-n = 4$ *−4*
6. $3x = 8$ *$\frac{8}{3}$ or $2\frac{2}{3}$*
7. $9a = 2$ *$\frac{2}{9}$*
8. $\frac{1}{3}n = 6$ *18*

EXERCISES

PART A

Find the solution set.

1. $29 = 5 - 6x$ *{−4}*
2. $26 = 5a - 4$ *{6}*
3. $3 = 4w + 27$ *{−6}*
4. $9x - 6 = 12x$ *{−2}*
5. $10 - 2a = 3a$ *{2}*
6. $8y + 30 = -2y$ *{−3}*
7. $7x + 5 = 5x - 1$ *{−3}*
8. $2y - 3 = 9 - 4y$ *{2}*
9. $23 - 5a = 2 + 2a$ *{3}*
10. $n + 7 = 4n - 5$ *{4}*
11. $-3a - 7 = 15 - a$ *{−11}*
12. $16 + 9c = -20 - 3c$ *{−3}*

Solve.

13. $-8 = 3x + 2$ *$-3\frac{1}{3}$*
14. $7 = 4 - 4y$ *$-\frac{3}{4}$*
15. $-10 = -2 - 5c$ *$1\frac{3}{5}$*
16. $-3a + 5 = a + 18$ *$-3\frac{1}{4}$*
17. $7 - 5x = -2x + 3$ *$1\frac{1}{3}$*
18. $6t + 3 = 10 - 2t$ *$\frac{7}{8}$*
19. $\frac{3}{5}y = 6$ *10*
20. $10 + \frac{2}{3}x = 6$ *−6*
21. $7 = \frac{3}{4}n - 2$ *12*
22. $\frac{2}{3}x = 7$ *$10\frac{1}{2}$*
23. $\frac{3}{4}n + 2 = 9$ *$9\frac{1}{3}$*
24. $\frac{2}{5}n - 4 = 1$ *$12\frac{1}{2}$*

PART B

Find the solution set.

25. $2x + 7 + x = 5 + 4x - 3$ *{5}*
26. $5y - 3 - 2y = 4y - 18 - 6y$ *{−3}*
27. $3n - 8 - 5n = 2 - 6n - 7$ *$\{\frac{3}{4}\}$*
28. $2m + 7 - 4m = -3m - 5 + 6m$ *$\{2\frac{2}{5}\}$*
29. $a + 3a + a + 3a = 12 + 2a + 6$ *{3}*
30. $3 - 5x - 7 - 2x = 3x + 5 - x$ *{−1}*
31. $3t - 7 - 8t + 4 = 4t + 8 - 3t$ *$\{-1\frac{5}{6}\}$*
32. $8 - 6c + 2 + 4c = 14 - 5c - 2$ *$\{\frac{2}{3}\}$*

PART C

Solve for x.

33. $5x - a = b$
34. $-3x = 3a$
35. $7ax + b = 2ax + c$
36. $\frac{2}{3}x = a$
37. $a + \frac{4}{5}x = b$
38. $\frac{a}{b} \cdot x = c$

Find the solution set.

39. $2x + 3 + 5x = 10x + 6 - 3x$
40. $7x - 2 - 2x = 8 + 5x - 10$

Linear Equations with Parentheses

OBJECTIVE

- To solve equations like $3(2x + 5) - 4(3x - 2) = x - 5$

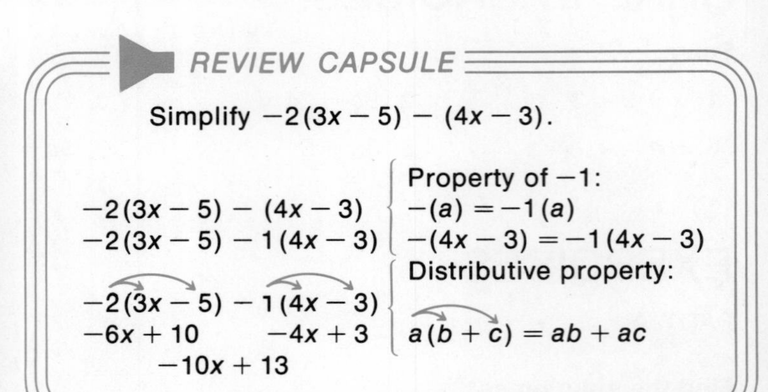

REVIEW CAPSULE

Simplify $-2(3x - 5) - (4x - 3)$.

$-2(3x - 5) - (4x - 3)$

$-2(3x - 5) - 1(4x - 3)$

Property of -1:
$-(a) = -1(a)$
$-(4x - 3) = -1(4x - 3)$

$-2(3x - 5) - 1(4x - 3)$

$-6x + 10 \qquad -4x + 3$

$-10x + 13$

Distributive property:
$a(b + c) = ab + ac$

EXAMPLE 1 Find the solution set of $5 + 2(4x - 3) = 3x - 11$.

Use the distributive property.

$$5 + 2(4x - 3) = 3x - 11$$

Distribute the 2. →

$$5 + 8x - 6 = 3x - 11$$

Combine like terms. →

$$-1 + 8x = 3x - 11$$

Solve the equation.

$$-1 + 5x = -11$$
$$5x = -10$$
$$x = -2$$

Thus, the solution set is $\{-2\}$.

EXAMPLE 2 Solve $8(x + 2) - 3(2x - 4) = 8$.

$$8(x + 2) - 3(2x - 4) = 8$$

Distribute the 8.
Distribute the -3. →

$$8x + 16 \quad -6x + 12 = 8$$

Combine like terms. →

$$2x + 28 = 8$$
$$2x = -20$$
$$x = -10$$

Thus, the solution is -10.

EXAMPLE 3 Solve $7n - (3n - 2) = 2(4n + 5)$.

Use the property of -1:
rewrite $-(3n - 2)$ as $-1(3n - 2)$. →

$$7n - 1(3n - 2) = 2(4n + 5)$$

Distribute the -1 and the 2. →

$$7n - 3n + 2 = 8n + 10$$
$$4n + 2 = 8n + 10$$

Add $-8n$ to each side.
Add -2 to each side. →

$$-4n = 8$$
$$n = -2$$

Thus, the solution is -2.

ORAL EXERCISES

Simplify.

1. $5(2x+3)$ *$10x+15$*
2. $3(4y-7)$ *$12y-21$*
3. $-2(6a-5)$ *$-12a+10$*
4. $-(3n+6)$ *$-3n-6$*
5. $-(8t-1)$ *$-8t+1$*
6. $-4(7+3c)$ *$-28-12c$*

EXERCISES

PART A

Solve.

1. $-2+3(2x-3)=7$ *3*
2. $2a+5(a+4)=34$ *2*
3. $8-2(n-3)=20$ *-3*
4. $3c-4(2c+2)=2$ *-2*
5. $4-(2x+6)=10$ *-6*
6. $5y-(2-3y)=22$ *3*
7. $5(3a-2)+4(a+4)=4a-24$ *-2*
8. $3(2n+4)+5(3-2n)=7-2n$ *10*
9. $8(x-3)-3(4x+2)=2x$ *-5*
10. $2(7x-4)-4(2x-6)=3x+31$ *5*
11. $6(2c+3)-(7c-2)=3c-4$ *-12*
12. $-3(n+2)-(5-4n)=5-3n$ *4*

PART B

Find the solution set.

13. $9x+3(2x-4)=5(x+3)$ *$\{2\frac{7}{10}\}$*
14. $6(4y-3)=4(5y-4)-5$ *$\{-\frac{3}{4}\}$*
15. $2-5(2a-3)=4(3-4a)$ *$\{-\frac{5}{6}\}$*
16. $3n-7(2n+3)=-2(8n+5)$ *$\{2\frac{1}{5}\}$*
17. $x-(9x-5)=-(3x+7)$ *$\{2\frac{2}{5}\}$*
18. $-4(2y-5)-(7-3y)=4(y+3)$ *$\{\frac{1}{9}\}$*

PART C

Solve.

19. $\frac{1}{2}(4x-6)+\frac{1}{3}(6x-12)=23-x$
20. $\frac{2}{3}(9y+12)-\frac{3}{4}(16y-8)=3y-4$
21. $3n+5[3(5n-4)-2(7n+6)]=8$
22. $10-4[6(2a+5)-(20+10a)]=2a$

Find the solution set.

23. $5(2x+4)-3(4x-6)=2(20-x)$
24. $4(x-3)-(x+6)=3(x-6)$

Linear Inequalities

OBJECTIVE

- To find and graph the solution sets of inequalities like $3(2x + 1) < 2x - 5$

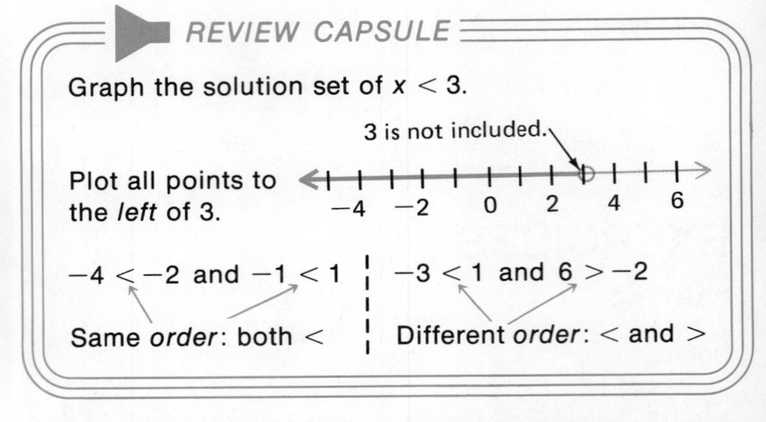

Linear *inequalities* are solved like linear equations with only one difference. Begin with $-8 < 6$.

Divide each side by −2.	Divide each side by 2.
Divide each side by the same *negative* number.	Divide each side by the same *positive* number.
$-8 < 6$	$-8 < 6$
$\frac{-8}{-2} \downarrow \frac{6}{-2}$	$\frac{-8}{2} \downarrow \frac{6}{2}$
$4 > -3$	$-4 < 3$
The order is *reversed* from < to >.	The order is *not* reversed.

$c < 0$ means c is negative.
$c > 0$ means c is positive.

Division Property for Inequalities

If $a < b$ and $c < 0$, then $\frac{a}{c} > \frac{b}{c}$.	If $a < b$ and $c > 0$, then $\frac{a}{c} < \frac{b}{c}$.
Dividing each side by the same *negative* number *reverses* the order.	Dividing each side by the same *positive* number does *not* change the order.

EXAMPLE 1 Find and graph the solution set of $-3x < 12$.

Divide each side by −3.
Reverse the order from < to >.

$$-3x < 12$$
$$\frac{-3x}{-3} \downarrow \frac{12}{-3} \quad \text{reverse order}$$
$$x > -4$$

$\{x \mid x > -4\}$ means "the set of all x such that $x > -4$." ⟶ The solution set is $\{x \mid x > -4\}$.

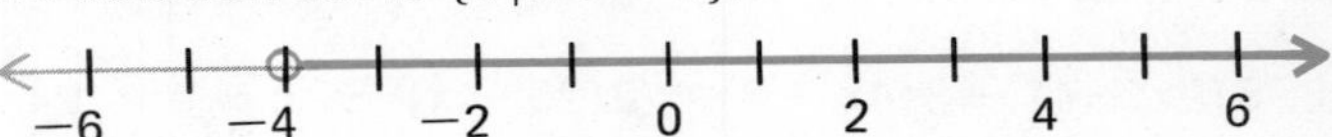

EXAMPLE 2 Find the solution set of $5x \geq 10$.

Divide each side by $+5$. →
Order is *not* reversed. →

$$\begin{array}{c} 5x \geq 10 \\ \dfrac{5x}{5} \downarrow \dfrac{10}{5} \\ x \geq 2 \end{array} \quad \text{same order}$$

Thus, $\{x \mid x \geq 2\}$ is the solution set.

Begin with $-8 < 6$.
Add -3 to each side:
$-8 \quad < 6$
$-8 + (-3) \downarrow 6 + (-3)$
$-11 < 3$
Order is *not* reversed.

Addition Property for Inequalities

If $a < b$,
then $a + c < b + c$.
Adding the same number to each side does *not* change the order.

EXAMPLE 3 Find and graph the solution set of $x + 5 \geq 3$.

Add -5 to each side. →
Order is not reversed. →

$$\begin{array}{rcr} x + 5 & \geq & 3 \\ -5 & & -5 \\ \hline x & \geq & -2 \end{array}$$

The solution set is $\{x \mid x \geq -2\}$.

Plot -2 and all points to the *right* of -2. →

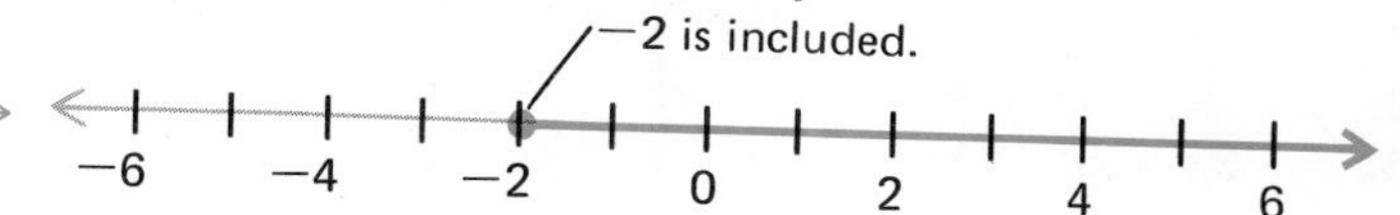

EXAMPLE 4 Find the solution set of $3 - 5x \leq 10 + 2x$.

Add $-2x$ to each side. →

Add -3 to each side. →

Divide each side by -7. →

Reverse the order from $\leq$ to $\geq$. →

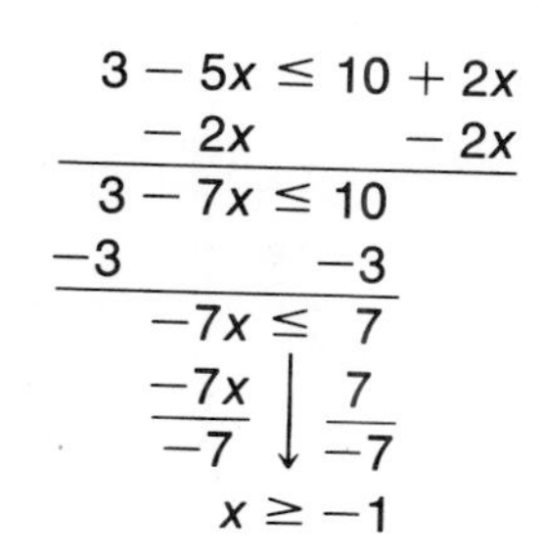

$$\begin{array}{rcl} 3 - 5x & \leq & 10 + 2x \\ -2x & & -2x \\ \hline 3 - 7x & \leq & 10 \\ -3 & & -3 \\ \hline -7x & \leq & 7 \\ \dfrac{-7x}{-7} & \downarrow & \dfrac{7}{-7} \\ x & \geq & -1 \end{array}$$

Thus, $\{x \mid x \geq -1\}$ is the solution set.

EXAMPLE 5 Find the solution set of $2 > x - 1$.

Add 1 to each side. →
Rewrite $3 > x$ with x on the left side.
Use the property: If $a > b$, then $b < a$. →

$$\begin{array}{c} 3 > x \\ \downarrow \\ x < 3 \end{array} \quad \text{reverse order}$$

Thus, $\{x \mid x < 3\}$ is the solution set.

ORAL EXERCISES

True or false?

1. $-6 < 2$ *T*
2. $-5 \leq -3$ *T*
3. $-8 \leq 4$ *T*
4. $5 \leq 5$ *T*
5. If $a < b$, then $a + 5 > b + 5$. *F*
6. If $a < b$, then $\frac{a}{-3} < \frac{b}{-3}$. *F*
7. If $5 < x$, then $x > 5$. *T*
8. If $-2x < 8$, then $x > -4$. *T*

EXERCISES

PART A

Find and graph the solution set.

1. $-2y \leq 6$ $\{y \mid y \geq -3\}$
2. $3x < -12$ $\{x \mid x < -4\}$
3. $8 \geq 2x$ $\{x \mid x \leq 4\}$
4. $x - 4 < -2$ $\{x \mid x < 2\}$
5. $y + 3 \geq 1$ $\{y \mid y \geq -2\}$
6. $2 > y - 3$ $\{y \mid y < 5\}$

Find the solution set.

7. $7 - 4x < 15$ $\{x \mid x > -2\}$
8. $2x - 3 > 7$ $\{x \mid x > 5\}$
9. $2 \leq 5 - 3x$ $\{x \mid x \leq 1\}$
10. $7x - 2 < 5x + 4$ $\{x \mid x < 3\}$
11. $4 - 2x \leq 3x + 14$
12. $4x - 3 \geq 2x - 7$
13. $1 + 3x \geq 8x + 11$ $\{x \mid x \leq -2\}$
14. $7 - 5x < 2x - 7$ $\{x \mid x > 2\}$
15. $10 - x > 8x + 37$ $\{x \mid x < -3\}$

PART B

EXAMPLE Find the solution set of $2(3 - 2x) < 10$.

$$2(3 - 2x) < 10$$

Distribute the 2. →

$$6 - 4x < 10$$
$$-4x < 4$$

Divide each side by -4. Reverse the order. →

$$x > -1$$

Thus, $\{x \mid x > -1\}$ is the solution set.

Find and graph the solution set.

16. $3(x + 4) < 27$ $\{x \mid x < 5\}$
17. $4(2x + 1) > 6x$ $\{x \mid x > -2\}$
18. $2(5 - 3x) \geq 5 - x$ $\{x \mid x \leq 1\}$
19. $8 - 2(2x + 1) > x + 1$ $\{x \mid x < 1\}$
20. $2x + 3(x - 2) < 3 - 4x$ $\{x \mid x < 1\}$
21. $5x - (2 - 3x) \leq 22$ $\{x \mid x \leq 3\}$

PART C

Find the solution set.

22. $\frac{3}{4}x < 6$
23. $\frac{2x - 1}{-3} < 5$
24. $\frac{2(2x + 3)}{-5} > 6 - 2x$

Finding a Number

OBJECTIVES

- To write word phrases in mathematical terms
- To solve word problems about one number

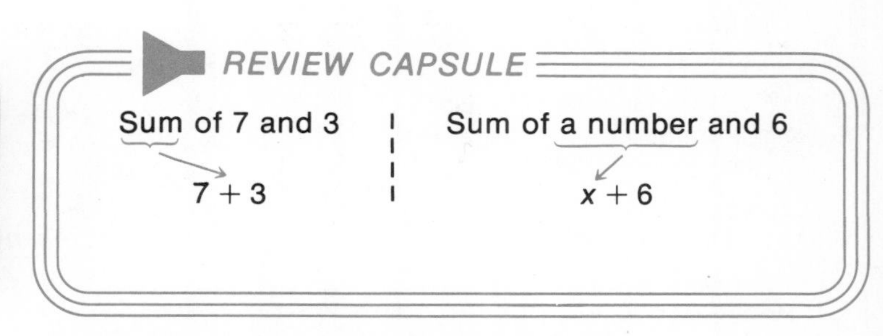

EXAMPLE 1 Write the phrase in mathematical terms.

Phrase in words → 9 increased by a number | A number decreased by 5

Phrase in mathematical terms → $9 + n$ | $x - 5$

EXAMPLE 2 Write in mathematical terms.

4 less than x → $x - 4$; *not* $4 - x$

4 less than a number | 8 more than a number

$x - 4$ | $n + 8$

EXAMPLE 3 Write each in mathematical terms.

Use parentheses, $(n + 7)$, to show times *the sum*. →

3 times the sum of n and 7

$3(n + 7)$

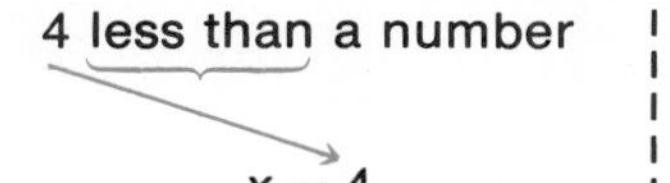

2 less than means subtract 2. →

2 less than 6 times a number

$6n - 2$

EXAMPLE 4 Three more than 5 times a number is 23. Find the number.

Let n be the number. → 3 more than 5 times n is 23.

Write an equation. Write = for is. → Solve the equation.

$$5n + 3 = 23$$
$$5n = 20$$
$$n = 4$$

Check 4 in the problem.

3 more than 5 times 4 is 23.

$5(4) + 3$	23
23	

$5(4) + 3 = 20 + 3$, or 23 →

Thus, the number is 4.

EXAMPLE 5 Four increased by twice a number is 5 less than the number. Find the number.

Let x be the number. → 4 increased by twice x is 5 less than x.

Twice x means 2 times x. → $4 + 2x = x - 5$

Solve the equation.

$$4 + 2x = x - 5$$
$$4 + x = -5$$
$$x = -9$$

Check −9 in the problem. → 4 increased by twice −9 is 5 less than −9.

$4 + 2(-9) = 4 + (-18)$ or -14 →

$4 + 2(-9)$	$-9 - 5$
-14	-14

Thus, the number is −9.

EXAMPLE 6 Four increased by 5 times the sum of a number and 2 is the same as 6 decreased by 3 times the number. Find the number.

Let n be the number. → 4 increased by 5 times the sum of n and 2

First part of the equation → $4 + 5(n + 2)$

6 decreased by 3 times n

Second part of the equation → $6 - 3n$

$4 + 5(n + 2)$ is the same as $6 - 3n$.

Write an equation. Write = for is the same as. →

$$4 + 5(n + 2) = 6 - 3n$$
$$4 + 5n + 10 = 6 - 3n$$
$$14 + 5n = 6 - 3n$$
$$14 + 8n = 6$$
$$8n = -8$$
$$n = -1$$

Check −1 in the problem. → **Thus,** the number is −1.

ORAL EXERCISES

State in mathematical terms.

1. Two less than a number $n - 2$

2. The sum of a number and 3 $n + 3$

3. Ten increased by a number $10 + n$

4. A number decreased by 6 $n - 6$

5. Five more than twice a number $2n + 5$

6. Eight less than 3 times a number $3n - 8$

7. Six times the sum of 3 and n $6(3 + n)$

8. Five times the sum of n and −2 $5(n - 2)$

EXERCISES

PART A

Write in mathematical terms.

1. Seven more than 6 times the sum of a number and 4 $6(n+4)+7$
2. Twelve decreased by 5 times the sum of a number and 7 $12-5(n+7)$
3. Five increased by 3 less than twice a number $5+(2n-3)$
4. Ten less than 8 times the sum of twice a number and -4 $8(2n-4)-10$

Find the number.

5. Eight more than 3 times a number is 2. Find the number. -2
6. Five less than 6 times a number is 7. Find the number. 2
7. Twelve increased by 4 times a number is 20. Find the number. 2
8. Nine decreased by twice a number is 1. Find the number. 4
9. Seven less than a number is twice the number. Find the number. -7
10. A number increased by 5 times the number is -30. Find the number. -5
11. Four times the sum of a number and 10 is 8. Find the number. -8
12. Eight decreased by 5 times a number is 18. Find the number. -2
13. Fourteen more than a number is 3 times the number. Find the number. 7
14. Three times the sum of a number and -2 is 15. Find the number. 7

PART B

Find the number.

15. Seven times a number is the same as 10 more than twice the number. Find the number. 2
16. A number increased by 16 is the same as 4 decreased by 3 times the number. Find the number. -3
17. Nine less than 5 times a number is the same as the number decreased by 1. Find the number. 2
18. Four more than a number is the same as 3 times the sum of the number and 2. Find the number. -1
19. Five times the sum of 6 and a number is the same as 2 increased by the number. Find the number. -7
20. Nine more than 8 times a number is the same as 7 times the number. Find the number. -9

PART C

Write an inequality for each sentence. Then find the solution set.

21. Six decreased by 3 times a number is less than 12.
22. Seven more than 4 times a number is greater than 3.
23. Twenty increased by twice a number is less than or equal to 6 times the number.
24. Eight less than 5 times a number is greater than or equal to 8 more than the number.

The Metric System of Measurement

LENGTH

The simplicity of the metric system is based on two things: (1) It is a decimal system, and (2) each unit of measure is named by using a base unit and a prefix. These are illustrated below. Units and prefixes that are used most commonly are screened.

The meter (m) is the basic unit of length. A baseball bat is about 1 meter long; a slice of bread, about 1 centimeter thick; a dime, about 1 millimeter thick.

Metric Units of Length

kilometer km	hectometer hm	dekameter dam	meter m	decimeter dm	centimeter cm	millimeter mm
1,000 m	100 m	10 m	1 m	.1 m	.01 m	.001 m

Prefix and Meaning

kilo	hecto	deka		deci	centi	milli
1,000	100	10		.1	.01	.001

PROBLEMS

Change 7.4 km to m.

1 km = 1,000 m

7.4 × 1 km = 7.4 × 1,000 m

7.4 km = 7.400. m, or 7,400 m

Change 53.4 mm to m.

1 mm = .001 m

53.4 × 1 mm = 53.4 × .001 m

53.4 mm = .053.4 m, or .0534 m

Change 27.2 m to cm.

.01 m = 1 cm

So, 1 m = 100 cm

27.2 × 1 m = 27.2 × 100 cm

27.2 m = 2,720 cm

Change 308.5 m to km.

1,000 m = 1 km

So, 1 m = .001 km

308.5 × 1 m = 308.5 × .001 km

308.5 m = .3085 km

PROJECT

Change as indicated.

1. 639 km to m
2. 639 mm to m
3. 639 cm to m
4. 7.839 m to cm
5. 7.839 m to km
6. 7.839 m to dm
7. .78 km to m
8. 2.37 m to mm
9. 836 dm to m

CAPACITY (VOLUME)

Capacity is the amount of liquid that a container will hold. The liter is used to measure capacity. As illustrated below, all units of capacity are defined in terms of the liter. The capacity of 3 soft-drink cans is about 1 liter; the capacity of an eyedropper is about 1 milliliter.

kiloliter kL	hectoliter hL	dekaliter daL	liter L	deciliter dL	centiliter cL	milliliter mL
1,000 L	100 L	10 L	1 L	.1 L	.01 L	.001 L

A cube of sugar has a volume of about 1 cubic centimeter (cm^3). A cubic decimeter (dm^3) is 1,000 cm^3, or 1 liter.

$1\ m^3 = 1{,}000$ L, or 1 kL
$1\ dm^3 = 1$ L
$1\ cm^3 = .001$ L, or 1 mL

PROJECT

Change to liters.

10. 5 kL
11. 1.3 daL
12. 2.9 mL
13. 8.6 cL
14. 10.5 dm^3
15. 3 cm^3
16. 4.6 m^3
17. .3 m^3

WEIGHT (MASS)

The gram is the basic unit of weight. All other units of weight are defined in terms of the gram.

kilogram kg	hectogram hg	dekagram dag	gram g	decigram dg	centigram cg	milligram mg
1,000 g	100 g	10 g	1 g	.1 g	.01 g	.001 g

A raisin weighs about 1 gram; your algebra book weighs about 1 kilogram.

PROJECT

Change as indicated.

18. 15 kg to g
19. 6.8 kg to mg
20. 250 mg to g

Using Formulas

OBJECTIVE

- To find the value of a variable in a formula for given values of the other variables

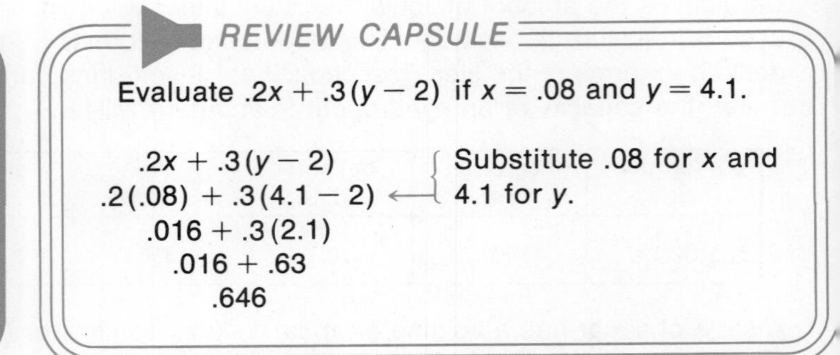

REVIEW CAPSULE

Evaluate $.2x + .3(y - 2)$ if $x = .08$ and $y = 4.1$.

$$.2x + .3(y - 2)$$
$$.2(.08) + .3(4.1 - 2)$$
$$.016 + .3(2.1)$$
$$.016 + .63$$
$$.646$$

Substitute .08 for x and 4.1 for y.

$3,000 invested at 8% for 6 years earns $1,440 in interest.

A principal p invested at r% per year for t years earns a simple interest i.

Formula →

$$i = prt$$

$3,000 + $1,440 = $4,440

The simple interest prt added to the principal p gives the total amount A in the account.

Formula →

$$A = p + prt$$

EXAMPLE 1

Find the time t needed to earn a simple interest of $962 if $2,000 is invested at $6\frac{1}{2}$%.

Formula →

Substitute: $i = \$962$, $p = \$2,000$, $r = 6\frac{1}{2}\% = .065$.

Solve the equation.

$$i = prt$$
$$962 = 2,000(.065)t$$
$$962 = 130t$$
$$7.4 = t$$

Thus, t is 7.4 years.

Find p if $A = \$2,340$, $r = 8\%$, and $t = 7$ years.

$$A = p + prt$$
$$2,340 = 1p + p(.08)(7)$$
$$2,340 = 1p + .56p$$
$$2,340 = 1.56p$$
$$1,500 = p$$

Thus, p is $1,500.

Parallel circuit with 3 resistances (lamps)

Source

In a parallel electric circuit having two resistances, the total resistance R, measured in ohms, is related to the two resistances r_1 and r_2, measured in ohms. The formula is $R(r_1 + r_2) = r_1 \cdot r_2$.

EXAMPLE 2

Find r_2 if $R = 4$ ohms and $r_1 = 6$ ohms.

Formula →

$$R(r_1 + r_2) = r_1 \cdot r_2$$

Substitute for R and r_1.

$$4(6 + r_2) = 6 \cdot r_2$$

Distribute the 4. →

$$24 + 4r_2 = 6r_2$$

Add $-4r_2$ to each side. →

$$24 = 2r_2$$
$$12 = r_2$$

In a parallel circuit the total resistance R is less than any resistance $r_1, r_2, r_3, \ldots$.

Thus, r_2 is 12 ohms.

EXERCISES

PART A

Use the formula $i = prt$.

1. Find the simple interest earned on a \$4,000 principal invested at 7% for 6 years. *\$1,680*
2. How many years will it take to earn a simple interest of \$1,152 on an investment of \$3,200 at 8%? *4.5 yr*
3. What principal must be invested at 6% for 10 years to earn a simple interest of \$4,800? *\$8,000*
4. At what interest rate will \$6,500 earn a simple interest of \$3,640 in 8 years? *7%*
5. How many years will it take to earn a simple interest of \$6,528 on an investment of \$6,400 at $8\frac{1}{2}$%? *12 yr*
6. What principal must be invested at $5\frac{1}{4}$% for 10 years to earn a simple interest of \$735? *\$1,400*

Use the formula $R(r_1 + r_2) = r_1 \cdot r_2$.

7. Find R in ohms if $r_1 = 30$ ohms and $r_2 = 50$ ohms. *18.75 ohms*
8. Find r_2 in ohms if $R = 19.2$ ohms and $r_1 = 32$ ohms. *48 ohms*
9. Find r_2 in ohms if $R = 24$ ohms and $r_1 = 60$ ohms. *40 ohms*
10. Find r_1 in ohms if $R = 21$ ohms and $r_2 = 70$ ohms. *30 ohms*

Use the formula $A = p + prt$.

11. Find A in dollars if $p = \$6{,}000$; $r = 9\%$; and $t = 7$ years. *\$9,780*
12. Find p in dollars if $A = \$4{,}300$; $r = 6\%$; and $t = 12$ years. *\$2,500*

PART B

Use the formula $R(r_1 + r_2) = r_1 \cdot r_2$.

13. Find r_2 in ohms if $R = 2.4$ ohms and $r_1 = 6.4$ ohms. *3.84 ohms*
14. Find r_1 in ohms if $R = 4.2$ ohms and $r_2 = 8.7$ ohms. *8.12 ohms*

Use the formula $A = p + prt$.

15. Find p in dollars if $A = \$6{,}720$; $r = 8\frac{1}{2}\%$; and $t = 8$ years. *\$4,000*
16. Find p in dollars if $A = \$8{,}625$; $r = 7\frac{1}{4}\%$; and $t = 10$ years. *\$5,000*
17. Find t in years if $A = \$12{,}160$; $p = \$8{,}000$; and $r = 8\%$. *6.5 yr*
18. Find the interest rate r if $A = \$616$; $p = \$400$; and $t = 9$ years. *6%*

PART C

Ohm's Law states that in a *series* circuit the intensity I of the electric current, measured in amperes, is equal to the electromotive force E, in volts, divided by the total resistance R, in ohms.

19. Use Ohm's law to find E in volts if $I = 3.5$ amperes and $R = 45$ ohms.
20. Use Ohm's law to find R in ohms if $I = 2.4$ amperes and $E = 120$ volts.

Chapter One Review

Find the solution set. [p. 1]

1. $16 = 4 - 3x$
2. $7a - 4 = 10a$
3. $5x + 9 = 2x + 3$
4. $4y - 3 = 4 - 2y$
5. $\frac{2}{3}n = 6$
6. $7 + \frac{2}{5}x = 2$
7. $3y - 4 + 2y = 5 - 3y + 7$
8. $5a + 8 - 3a = 4a - 5 - 6a$

Solve. [p. 4]

9. $-7 + 4(2x + 1) = 13$
10. $4n - 3(3n - 2) = 16$
11. $10 - (5a + 3) = 2a$
12. $-2x - (4 - 5x) = x - 3$
13. $4(2y - 5) + 3(6 - y) = y + 6$
14. $4c - 2(4c - 3) = 3(2c - 5)$

Find and graph the solution set. [p. 6]

15. $-5n \leq 20$
16. $5 > 17 + 4x$

Find the solution set. [p. 6]

17. $5x - 8 < 2x + 4$
18. $8 - 3y \geq 2y - 2$
19. $2(3a + 1) > 2a - 10$
20. $3(x - 2) \leq 5x$

Write in mathematical terms. [p. 9]

21. Six increased by twice a number
22. Five less than 4 times a number
23. Three more than 5 times the sum of 7 and a number
24. Nine decreased by 6 times the sum of a number and 3

Find the number. [p. 9]

25. Seven more than 4 times a number is -5. Find the number.
26. Twice the sum of a number and 3 is the same as 5 times the number. Find the number.

Use the formula $R(r_1 + r_2) = r_1 \cdot r_2$. [p. 14]

27. Find r_2 in ohms if $R = 8$ ohms and $r_1 = 10$ ohms.
28. Find R in ohms if $r_1 = 8$ ohms and $r_2 = 12$ ohms.

Use the formula $A = p + prt$. [p. 14]

29. Find p in dollars if $A = \$5,200$, $r = 6\%$, and $t = 5$ years.
30. Find A in dollars if $p = \$2,000$, $r = 8\frac{1}{2}\%$, and $t = 4$ years.

Chapter One Test

Find the solution set.

1. $1 = 3a + 16$
2. $3x + 2 = 6x$
3. $3y + 14 = 4 - 2y$
4. $4c - 6 = 9c - 14$
5. $6 + \frac{2}{3}x = 12$
6. $n - 5 - 4n = 5 + 3n + 2$

Solve.

7. $3 + 5(2x - 1) = 7x + 4$
8. $2n - (5n + 4) = 17$
9. $4y - 3(3y - 4) = 3 - 2y$
10. $-2(x + 5) + 4(2x + 3) = 14$

Find and graph the solution set.

11. $-4n \leq 8$
12. $15 > 7 + 2x$

Find the solution set.

13. $7x - 3 < 2x + 17$
14. $3(2a + 1) \geq 4a - 7$

Write in mathematical terms.

15. Four less than 8 times a number
16. Eight increased by twice the sum of 4 and a number

Find the number.

17. Nine increased by 5 times a number is -11. Find the number.
18. Eight less than twice a number is the same as 7 decreased by the number. Find the number.

Use the formula $R(r_1 + r_2) = r_1 \cdot r_2$.

19. Find r_2 in ohms if $R = 15$ ohms and $r_1 = 20$ ohms.

Use the formula $A = p + prt$.

20. Find p in dollars if $A = \$4,440$, $r = 8\%$, and $t = 6$ years.

2 ABSOLUTE VALUE AND PROBLEM SOLVING

Algebraic Proofs

ALGEBRAIC AXIOMS

- COMMUTATIVE PROPERTY OF ADDITION
- COMMUTATIVE PROPERTY OF MULTIPLICATION
- ASSOCIATIVE PROPERTY OF ADDITION
- ASSOCIATIVE PROPERTY OF MULTIPLICATION
- INVERSE PROPERTY FOR ADDITION
- INVERSE PROPERTY FOR MULTIPLICATION
- IDENTITY PROPERTY FOR ADDITION
- IDENTITY PROPERTY FOR MULTIPLICATION
- DISTRIBUTIVE PROPERTY
- TRANSITIVE PROPERTY OF EQUALITY IF $a=b$, and $b=c$, then $a=c$

Axioms are assumed to be true without proof. Axioms are used in the proofs of theorems.

PROBLEMS

Prove: $a + bc = cb + a$

Proof

Statement	Reason
$a + bc = a + cb$	Comm. Prop. Mult.
$= cb + a$	Comm. Prop. Add.
Thus, $a + bc = cb + a$	Trans. Prop. Equality

Prove: $a(b \cdot c) = c(a \cdot b)$

Proof

Statement	Reason
$a(b \cdot c) = a(c \cdot b)$	Comm. Prop. Mult.
$= (a \cdot c)b$	Assoc. Prop. Mult.
$= (c \cdot a)b$	Comm. Prop. Mult.
$= c(a \cdot b)$	Assoc. Prop. Mult.
Thus, $a(b \cdot c) = c(a \cdot b)$	Trans. Prop. Equality

PROJECT

Use the known axioms to prove each statement.

1. $ac + cb = (a + b)c$

2. $a(b + c) = ca + ba$

3. $ac + b = b + ca$

4. $a + (b + c) = b + (c + a)$

5. $ab + ac = (c + b)a$

6. $a + [b(c + d)] = (bd + a) + bc$

Compound Inequalities

OBJECTIVES

- To find and graph the solution sets of inequalities like $-3 < 2x - 5 < 7$
- To graph the solution sets of sentences like $[x \leq -1 \text{ or } x > 5]$

REVIEW CAPSULE

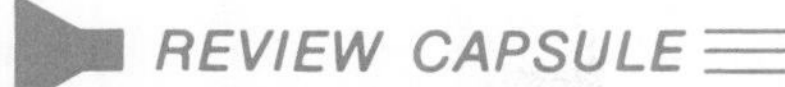

Find the solution set of $5x - 3 < 7$.

$$\begin{array}{rcl} 5x - 3 & < & 7 \\ +3 & & +3 \\ \hline 5x & < & 10 \\ \frac{5x}{5} & < & \frac{10}{5} \\ x & < & 2 \end{array}$$

Thus, the solution set is $\{x \mid x < 2\}$.

EXAMPLE 1 Graph the solution set of $-2 < x < 4$.

$-2 < x < 4$ means $[-2 < x \text{ and } x < 4]$.

Plot -2 and 4 on a number line.

-2 and 4 separate the number line into 3 parts. →

Numbers less than −2	**Numbers between −2 and 4**	**Numbers greater than 4**

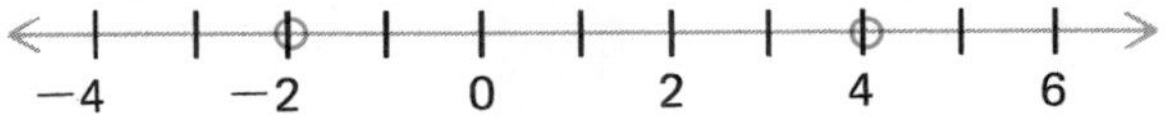

Try a number from each part. →
Substitute for x. →
And means *both* parts of the sentence must be true. →

Try −5 for x:	Try 1 for x:	Try 6 for x:
$-2 < x$ and $x < 4$	$-2 < x$ and $x < 4$	$-2 < x$ and $x < 4$
$-2 < -5$ and $-5 < 4$	$-2 < 1$ and $1 < 4$	$-2 < 6$ and $6 < 4$
False	True	False

All other numbers make $[-2 < x \text{ and } x < 4]$ false. → Graph all the numbers *between* -2 and 4.

Graph of $\{x \mid -2 < x < 4\}$ →

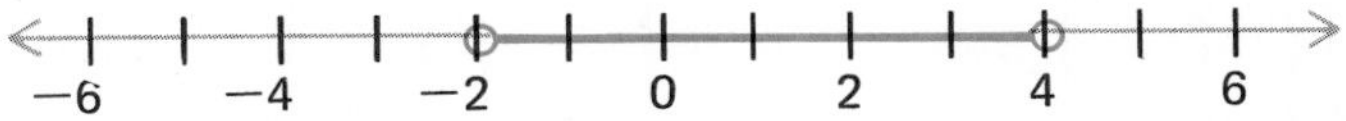

> **$a < x < b$ means $[a < x \text{ and } x < b]$.**
> The solutions of $a < x < b$ are the numbers between a and b, if $a < b$.

EXAMPLE 2 Graph the solution set of $-3 \leq x < 2$.

-3 is also a solution: $-3 \leq -3$ and $-3 < 2$.

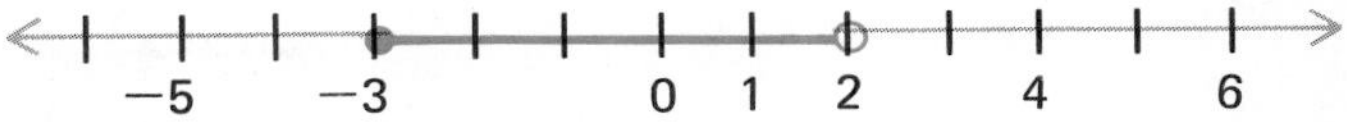

EXAMPLE 3 Find the solution set of $-3 < 2y - 7 \leq 1$.

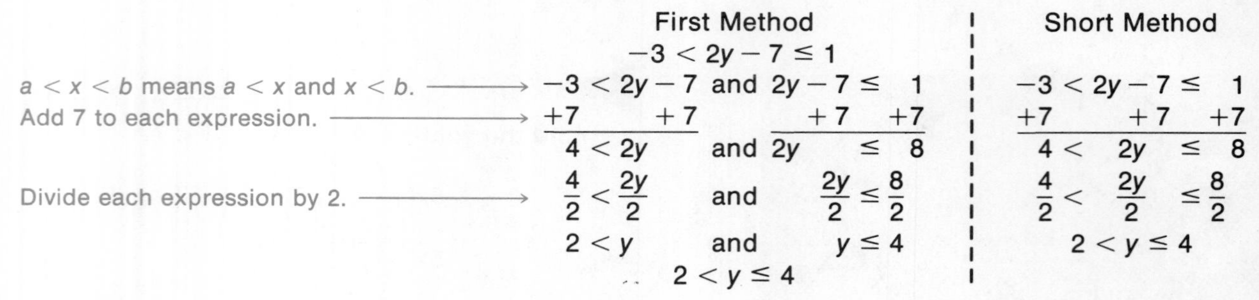

Thus, the solution set is $\{y \mid 2 < y \leq 4\}$.

EXAMPLE 4 Find and graph the solution set of $-5 \leq 4x + 3 \leq 11$.

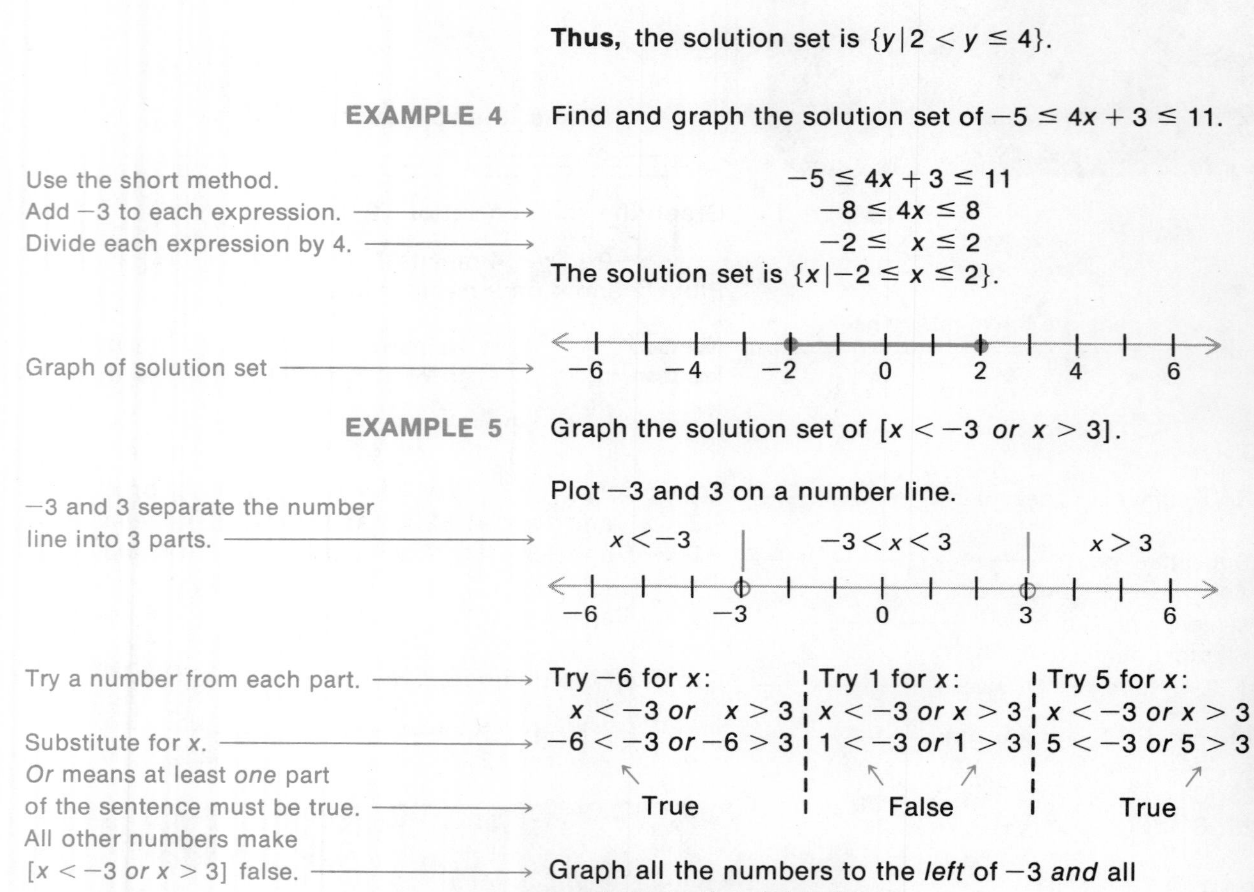

EXAMPLE 5 Graph the solution set of $[x < -3$ *or* $x > 3]$.

EXAMPLE 6 Graph the solution set of $[x \leq -2$ *or* $x \geq 4]$.

Graph all $x \leq -2$.
Then graph all $x \geq 4$.

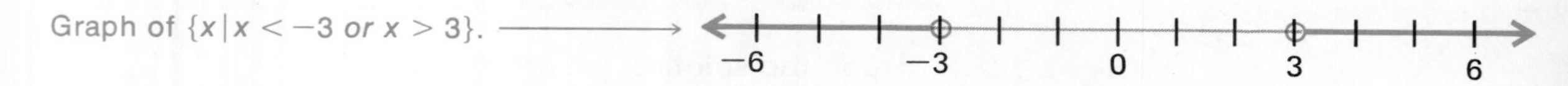

EXERCISES

PART A

Graph the solution set.

1. $-4 < x < 2$ **2.** $-1 \leq y < 3$ **3.** $a < -2$ or $a > 3$
4. $n \leq -4$ or $n > 1$ **5.** $2 < x \leq 6$ **6.** $-5 \leq y \leq -2$
7. $a < 3$ or $a \geq 5$ **8.** $n \leq -5$ or $n \geq -2$ **9.** $x < 0$ or $x > 0$

Find the solution set.

10. $-4 < x - 2 < 1$
11. $-2 \leq y + 3 < 8$ $\{y|-5 \leq y < 5\}$
12. $-6 < 3n \leq 12$ $\{n|-2 < n \leq 4\}$
13. $-6 \leq 2a \leq 10$
14. $-3 < 2x - 1 < 7$
15. $-4 < 3y + 2 \leq 11$
16. $1 \leq 5 + 4n < 17$ $\{n|-1 \leq n < 3\}$
17. $-1 \leq 6a - 7 \leq 5$ $\{a|1 \leq a \leq 2\}$
18. $3 < 10 + 7x < 31$ $\{x|-1 < x < 3\}$

PART B

EXAMPLE Find and graph the solution set of $[2x - 3 < -7 \text{ or } 2x - 3 > 5]$.

Add 3 to each expression.
Divide each expression by 2.

$$2x - 3 < -7 \text{ or } 2x - 3 > 5$$
$$\quad +3 \quad +3 \qquad\quad +3 \quad +3$$
$$2x < -4 \text{ or } 2x > 8$$
$$\frac{2x}{2} < \frac{-4}{2} \text{ or } \frac{2x}{2} > \frac{8}{2}$$
$$x < -2 \qquad x > 4$$

Thus, the solution set is $\{x|x < -2 \text{ or } x > 4\}$.

Find and graph the solution set.

19. $x - 4 < -2$ or $x - 4 > 1$
20. $y + 3 \leq -1$ or $y + 3 \geq 5$ $\{y|y \leq -4 \text{ or } y \geq 2\}$
21. $5a < -10$ or $5a \geq 15$
22. $4n < -12$ or $4n > 8$
23. $3x - 2 < -5$ or $3x - 2 > 7$
24. $2y + 5 \leq -1$ or $2y + 5 \geq 11$
25. $3 + 4a \leq -5$ or $3 + 4a \geq 15$ $\{a|a \leq -2 \text{ or } a \geq 3\}$
26. $5n - 6 < -16$ or $5n - 6 > 9$ $\{n|n < -2 \text{ or } n > 3\}$

PART C

Find the solution set.

27. $-6 < -3x < 9$
28. $-4 \leq 1 - 5x \leq 11$
29. $-3 < 2 - x < 5$
30. $3 \leq -2x - 3 \leq 9$
31. $2x - 1 < 3x + 5 < 2x + 7$
32. $3x - 7 \leq 5x + 6 \leq 3x + 9$
33. $-4x \leq -8$ or $-4x \geq 12$
34. $2 - 3x < -4$ or $2 - 3x > 14$
35. $-4 < 3x - 1 < 5$ or $3x - 1 > 11$
36. $-7 < 2x + 3 < -1$ or $1 < 2x + 3 < 7$

Absolute Value

OBJECTIVES

- To solve equations like $|3x-2|=7$
- To find and graph the solution sets of inequalities like $|2x-3|<5$

REVIEW CAPSULE

What numbers are 3 units from zero, the origin?

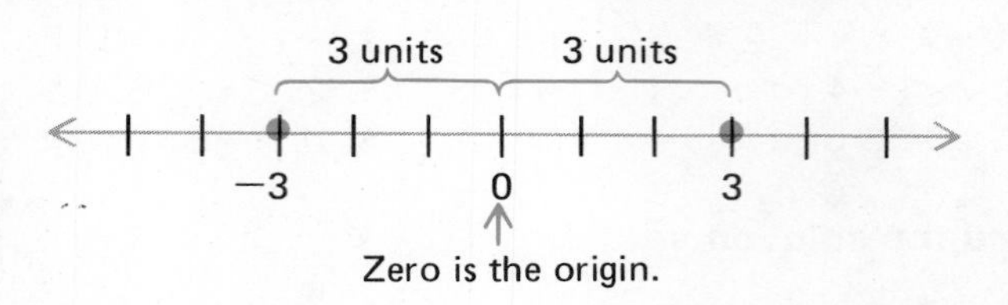

Both −3 and 3 are 3 units from zero, the origin.

The *absolute value* of x, $|x|$, is the distance between x and the origin, 0.

EXAMPLE 1

−2 is 2 units from 0, the origin.
+2 is 2 units from 0, the origin.

Find $|-2|$.
The distance between −2 and the origin, 0, is 2.

Thus, $|-2|=2$.

Find $|2|$.
The distance between 2 and the origin, 0, is 2.

Thus, $|2|=2$.

$|x|=-x$ if x is *negative*.
$|-2|=-(-2)$, or 2. ⟶

Notice that $|-2|=-(-2)$, or 2.

Definition of $|x|$, the *absolute value* of x ⟶

$|x|=$ ↗ $-x$, if x is *negative*.
↘ x, if x is *positive*, or *zero*.

EXAMPLE 2

Find the solution set of $|x|=4$.

$|x|$ is the distance between x and the origin, 0.
Both −4 and 4 are 4 units from the origin, 0.

$|x|=4$

$x=-4$ or $x=4$.

If $|x|=4$

$|-4|=4$ $\quad$ $|4|=4$

then $x=-4$ or $x=4$.

Thus, $\{-4,4\}$ is the solution set of $|x|=4$.

Equation property for absolute value ⟶

If $|x|=k$, then $x=-k$ or $x=k$, $[k\geq 0]$.

EXAMPLE 3 Find the solution set of $|3n - 5| = 7$.

$|x| = k$ → $x = -k$ or $x = k$

Solve each equation.

$$|3n - 5| = 7$$

$$3n - 5 = -7 \quad \text{or} \quad 3n - 5 = 7$$

$$3n = -2 \qquad\qquad 3n = 12$$

$$n = -\frac{2}{3} \qquad\qquad n = 4$$

Thus, $\{-\frac{2}{3}, 4\}$ is the solution set.

EXAMPLE 4 Find the solution set of $|y| < 4$.

$|y|$ is distance between y and the origin, 0.

Graph all y *less than* 4 units from 0.

Graph of solution set →

−6 −4 −2 0 2 4 6

The graph shows the solutions.

Solutions are all numbers *between* −4 and 4.

Thus, $\{y \mid -4 < y < 4\}$ is the solution set of $|y| < 4$.

EXAMPLE 5 Find the solution set of $|n| > 4$.

$|n|$ is distance between n and the origin, 0.

Graph all n *more than* 4 units from 0.

Graph of solution set →

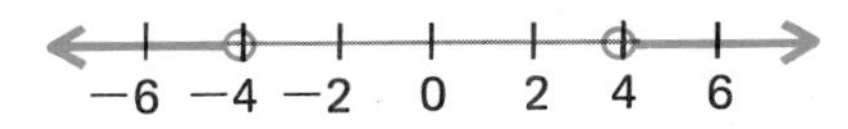

The graph shows the solutions.

Solutions are to the *left of* −4 or to the *right of* 4.

Thus, $\{n \mid n < -4 \text{ or } n > 4\}$ is the solution set of $|n| > 4$.

Inequality properties for absolute value →

If $|x| < k$, then $-k < x < k$, $[k > 0]$.	If $|x| > k$, then $x < -k$ or $x > k$, $[k > 0]$.

EXAMPLE 6 Find the solution set of $|4y + 2| \leq 10$.

If $|x| < k$, →

then $-k < x < k$. →

Add −2 to each expression.

Divide each expression by 4.

$$|4y + 2| \leq 10$$

$$-10 \leq 4y + 2 \leq 10$$

$$-12 \leq 4y \leq 8$$

$$-3 \leq y \leq 2$$

Thus, $\{y \mid -3 \leq y \leq 2\}$ is the solution set.

EXAMPLE 7 Find and graph the solution set of $|2n - 1| > 5$.

If $|x| > k$, then $x < -k$ or $x > k$. Solve both inequalities.

$$|2n - 1| > 5$$
$$2n - 1 < -5 \text{ or } 2n - 1 > 5$$
$$2n < -4 \text{ or } 2n > 6$$
$$n < -2 \text{ or } n > 3$$

The solution set is $\{n \mid n < -2 \text{ or } n > 3\}$.

Graph of the solution set

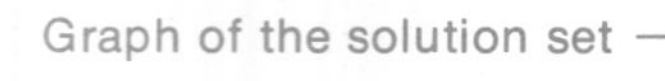

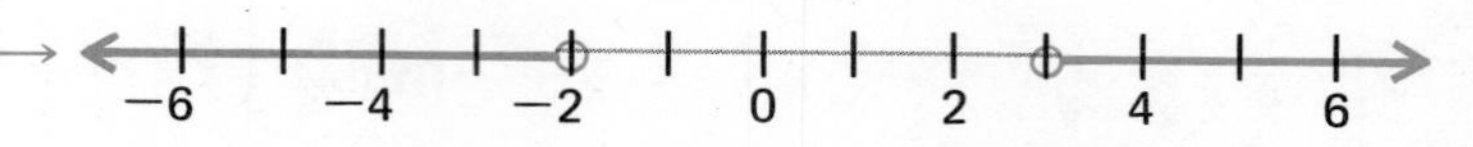

ORAL EXERCISES

Find the absolute value.

1. $|-12|$ *12*
2. $|4 - 7|$ *3*
3. $|-2 + 8|$ *6*
4. $|-3(4)|$ *12*
5. $|-2(-4)|$ *8*
6. $|3(-4) + 2|$ *10*
7. $|2(-5) - 4|$ *14*
8. $|-5(-3) - 10|$ *5*

EXERCISES

PART A

Find the solution set.

1. $|x - 3| = 7$
2. $|a + 6| = 2$ *$\{-4, -8\}$*
3. $|n| = 5$ *$\{5, -5\}$*
4. $|x - 2| = -3$
5. $|3a| = 12$
6. $|2x - 3| = 7$
7. $|4t| = 20$
8. $|4n + 6| = 2$
9. $|3x - 5| = 10$
10. $|4y + 3| = 5$
11. $|5n + 1| = 11$
12. $|4 - 5n| = 1$
13. $|x| < 3$
14. $|n| \geq 6$ *$\{n \mid n \leq -6 \text{ or } n \geq 6\}$*
15. $|y| \leq 8$ *$\{y \mid -8 \leq y \leq 8\}$*
16. $|a| > 5$ *$\{a \mid a < -5 \text{ or } a > 5\}$*

PART B

Find and graph the solution set.

17. $|x - 3| < 1$
18. $|y + 4| \geq 7$
19. $|a - 2| \leq 4$ *$\{a \mid -2 \leq a \leq 6\}$*
20. $|n + 5| > 3$ *$\{n \mid n < -8 \text{ or } n > -2\}$*
21. $|2a| < 12$
22. $|3y| \geq 12$
23. $|5n| \leq 15$
24. $|4x| > 8$
25. $|2x - 5| < 7$
26. $|3y + 3| \geq 12$
27. $|3a - 6| \leq 9$
28. $|2n + 3| > 5$
29. $|4y + 2| < 6$
30. $|5a - 15| \geq 10$ *$\{a \mid a \leq 1 \text{ or } a \geq 5\}$*
31. $|5x + 10| \leq 5$ *$\{x \mid -3 \leq x \leq -1\}$*
32. $|4n - 6| > 2$ *$\{n \mid n < 1 \text{ or } n > 2\}$*

PART C

Find and graph the solution set.

33. $|6 - 3x| < 9$
34. $|5 - 2x| > 3$
35. $|x - 1| < 3$ or $|x - 1| > 5$
36. $|x + 3| \leq 2$ or $|x - 3| \leq 2$

Mathematics in the Restaurant

The restaurant manager above is calculating the prices for the menu. The price includes cost of food, cost of preparing food, and profit.

PROJECT

Find the cost for each serving of salad, entree, dessert, and beverage; the cost of each meal; and the total cost of serving 700 students. Consult your cafeteria manager for prices. Disregard profit, plant operation, and employee costs. Use the following school menu.

Salad: Lettuce, celery, spinach, and dressing
Entree: 1 Hamburger on a roll
1 Serving of a vegetable
1 Serving of French fries
Dessert: Butterscotch pudding
Beverage: Milk

Modulo 5

$8 \div 5 = 1$ r 3	$13 \div 5 = 2$ r 3	$18 \div 5 = 3$ r 3	$23 \div 5 = 4$ r 3
$2 \div 5 = 0$ r 2	$12 \div 5 = 2$ r 2	$22 \div 5 = 4$ r 2	$32 \div 5 = 6$ r 2
$4 \div 5 = 0$ r 4	$19 \div 5 = 3$ r 4	$34 \div 5 = 6$ r 4	$49 \div 5 = 9$ r 4

Notice that when each number was divided by 5, the remainder in each row was the same. Any pair of numbers, like 8 and 13, from one of the rows is *congruent modulo 5*.

Two integers a and b are congruent modulo 5 if dividing a and b by 5 gives the same remainder.

$a \equiv b \pmod 5$ means $a \div 5$ and $b \div 5$ have the same remainder.

↑ Read: a is congruent to b, mod 5.

PROBLEM 1.

True or false: $12 \equiv 7 \pmod 5$?

$12 \div 5 = 2$ r 2
$7 \div 5 = 1$ r 2

Thus, $12 \equiv 7 \pmod 5$ is true.

True or false: $28 \equiv 2 \pmod 5$?

$28 \div 5 = 5$ r 3
$2 \div 5 = 0$ r 2

Thus, $28 \equiv 2 \pmod 5$ is false.

PROBLEM 2.

Find a value for b so that $34 \equiv b \pmod 5$.

$34 \div 5$ gives remainder 4.
Try $b = 4$. $4 \div 5$ gives remainder 4.

Thus, $b = 4$ and $34 \equiv 4 \pmod 5$.

PROJECT

True or false?

1. $13 \equiv 3 \pmod 5$ **2.** $18 \equiv 8 \pmod 5$ **3.** $29 \equiv 13 \pmod 5$

Find a value for b from {0, 1, 2, 3, 4} to make the equation true.

4. $29 \equiv b \pmod 5$ **5.** $43 \equiv b \pmod 5$ **6.** $26 \equiv b \pmod 5$

Addition Mod 5

$c \oplus d = n \pmod 5$, where n is from $\{0, 1, 2, 3, 4\}$.

Examples: $2 \oplus 1 = 3 \pmod 5$ $\quad$ $2 \oplus 4 = 1 \pmod 5$

↗

$6 \equiv 1 \pmod 5$

Multiplication Mod 5

$c \otimes d = m \pmod 5$, where m is from $\{0, 1, 2, 3, 4\}$.

Examples: $2 \otimes 2 = 4 \pmod 5$ $\quad$ $2 \otimes 4 = 3 \pmod 5$

↗

$8 \equiv 3 \pmod 5$

PROJECT

Complete the table.

7.

⊕	0	1	2	3	4
0	0				
1					
2		3			
3				1	
4		0			

8.

⊗	0	1	2	3	4
0			0		
1					
2			4		
3		3			
4				2	

Check these samples of the commutative properties in mod 5.

9. $3 \oplus 4 = 4 \oplus 3$ **10.** $2 \otimes 3 = 3 \otimes 2$ **11.** $4 \otimes 2 = 2 \otimes 4$

Check these samples of the associative properties in mod 5.

12. $(3 \oplus 2) \oplus 4 = 3 \oplus (2 \oplus 4)$ **13.** $(2 \otimes 3) \otimes 1 = 2 \otimes (3 \otimes 1)$

Write the answer in mod 5.

14. $2 \otimes (3 \oplus 1)$ **15.** $(2 \otimes 3) \oplus (2 \otimes 1)$

16. $3 \otimes (4 \oplus 2)$ **17.** $(3 \otimes 4) \oplus (3 \otimes 2)$

18. Two integers a and b are congruent mod n, $a \equiv b \pmod n$ if dividing both a and b by n gives the same remainder. If $n = 4$, construct an addition table and a multiplication table (mod 4). Use $\{0, 1, 2, 3\}$.

Number and Perimeter Problems

OBJECTIVES

- To solve word problems about numbers
- To solve perimeter problems

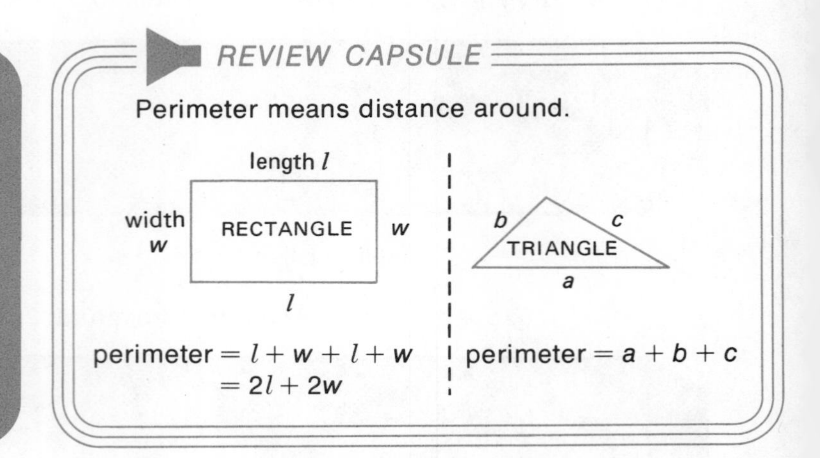

EXAMPLE 1 Mary is thinking of two numbers. The second number is 7 more than twice the first number. Represent the two numbers in mathematical terms.

Let $f =$ first number

7 more than twice the first → $2f + 7 =$ second number

Thus, f and $2f + 7$ represent the two numbers.

EXAMPLE 2 The greater of two numbers is 10 less than 3 times the smaller. Represent the two numbers in mathematical terms.

The smaller is mentioned *after* the greater. Represent the smaller with a variable. → Let $s =$ smaller number

10 less than 3 times the smaller → $3s - 10 =$ greater number

Thus, s and $3s - 10$ represent the two numbers.

EXAMPLE 3 John is thinking of three numbers. The second number is 5 times the first. The third number is 8 more than twice the second. Represent the three numbers in mathematical terms.

Let $f =$ first number

5 times the first → $5f =$ second number

8 more than twice the second → $2(5f) + 8 =$ third number

$2(5f) + 8 = 10f + 8$ → **Thus,** the three numbers are f, $5f$, and $10f + 8$.

EXAMPLE 4 The second of two numbers is 6 less than 4 times the first. The sum of the numbers is 29. Find the two numbers.

Represent the numbers in mathematical terms.

Let f = first number

6 less than 4 times the first → $4f - 6$ = second number

Write an equation.

Their sum is 29. → $f + (4f - 6) = 29$

Solve the equation.

$$5f - 6 = 29$$
$$5f = 35$$
$$f = 7$$

Find both numbers.

First number, f is 7.
Second number, $4f - 6$ is $4(7) - 6$, or 22.

Check in the problem. →

Second is 6 less than 4 times first.	
22	$4(7) - 6$ 22

Their sum is 29.	
$7 + 22$ 29	29

Thus, 7 and 22 are the two numbers.

EXAMPLE 5 The smaller of two numbers is 12 less than 3 times the greater. If 6 times the greater is decreased by the smaller, the result is 27. Find the two numbers.

The greater is mentioned *after* the smaller. Represent the greater with a variable. → Let g = greater number

12 less than 3 times greater → $3g - 12$ = smaller number

6 times greater, decreased by smaller is 27.

Use parentheses around $3g - 12$ to show subtracting the smaller. → $6g - (3g - 12) = 27$

$-(3g - 12) = -1(3g - 12)$ → $6g - 1(3g - 12) = 27$

$$6g - 3g + 12 = 27$$
$$3g + 12 = 27$$
$$3g = 15$$
$$g = 5$$

Find both numbers.

Greater number, g is 5.
Smaller number, $3g - 12$ is $3(5) - 12$, or 3.

$3(5) - 12 = 15 - 12$, or 3

Smaller is 12 less than 3 times greater.	
3	$3(5) - 12$ 3

$6(5) - 3 = 30 - 3$, or 27

6 times greater, decreased by smaller is 27.	
$6(5) - 3$ 27	27

Thus, 3 and 5 are the two numbers.

EXAMPLE 6 The length of a rectangle is 4 meters more than twice the width. The perimeter is 26 meters. Find the length and the width.

$2w + 4$

w w

$2w + 4$

Sketch a rectangle. Represent width and length in mathematical terms.

Let $w = \text{width}$
$2w + 4 = \text{length}$

Formula ⟶ $l + w + l + w = \text{perimeter}$

Write an equation. ⟶

$$(2w + 4) + w + (2w + 4) + w = 26$$
$$6w + 8 = 26$$
$$6w = 18$$
$$w = 3$$

Check.
Perimeter $= 3 + 10 + 3 + 10$
$= 26$

Width, w is 3 m.
Length, $2w + 4$ is $2(3) + 4$, or 10 m.
Thus, the rectangle is 3 m wide and 10 m long.

EXAMPLE 7 Side b of a triangle is twice as long as side a. Side c is 1 cm more than twice the length of side a. The perimeter is 16 cm. Find the lengths of the three sides.

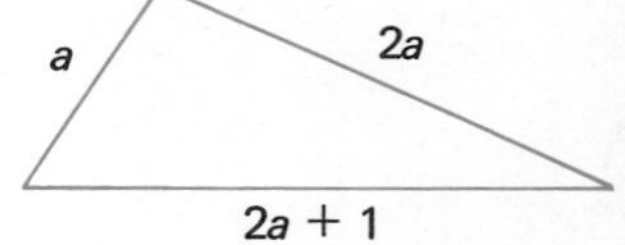

Represent the 3 lengths in mathematical terms.

Let $a = \text{side } a$
$2a = \text{side } b$
$2a + 1 = \text{side } c$

Formula ⟶ $a + b + c = \text{perimeter}$

Write an equation. ⟶

$$a + 2a + (2a + 1) = 16$$
$$5a + 1 = 16$$
$$5a = 15$$
$$a = 3$$

Find the 3 lengths.

Length of side a, a is 3 cm.
Length of side b, $2a$ is $2(3)$, or 6 cm.
Length of side c, $2a + 1$ is $2(3) + 1$, or 7 cm.

Check.
Perimeter $= 3 + 6 + 7$
$= 16$

Thus, the sides are 3 cm, 6 cm, and 7 cm long.

ORAL EXERCISES

Represent both numbers in mathematical terms.

1. The second of two numbers is 5 more than 3 times the first number.
2. The greater of two numbers is 4 less than twice the smaller number. *s = smaller, $2s - 4$ = greater*
3. The smaller of two numbers is 17 less than 4 times the greater number. *g = greater, $4g - 17$ = smaller*
4. The second of two numbers is 6 decreased by twice the first. *f = 1st, $6 - 2f$ = 2nd*

EXERCISES

PART A

1. The second of two numbers is 5 more than the first. Their sum is 17. Find the two numbers. *6; 11*
2. The second of two numbers is 7 times the first number. Their sum is 80. Find the two numbers. *10; 70*
3. The second of two numbers is 3 less than twice the first. Their sum is 18. Find the two numbers. *7; 11*
4. The second of two numbers is 6 more than 4 times the first. Their sum is 16. Find the two numbers. *2; 14*
5. The smaller of two numbers is 5 less than the greater. If the smaller is increased by 4 times the greater, the result is 30. Find the two numbers. *2; 7*
6. The greater of two numbers is 7 more than the smaller. If the greater is decreased by 5 times the smaller, the result is −5. Find the two numbers. *3; 10*
7. The length of a rectangle is 6 cm more than the width. The perimeter is 32 cm. Find the length and width.
8. The width of a rectangle is 4 m less than the length. The perimeter is 40 m. Find the length and width.
9. Side b of a triangle is 4 cm more than side a. Side c is twice the length of side a. The perimeter is 28 cm. Find the lengths of the three sides. *6 cm; 10 cm; 12 cm*
10. Side b of a triangle is 5 m less than side a. Side c is 5 m more than side a. The perimeter is 45 m. Find the lengths of the three sides. *15 m; 10 m; 20 m*

PART B

11. The second of two numbers is 4 more than 5 times the first. Seven times the first is equal to the second. Find the two numbers. *2; 14*
12. The second of two numbers is 5 less than twice the first. Three times the first, decreased by 4 times the second, is 10. Find the two numbers.
13. The greater of two numbers is 3 less than 7 times the smaller. Twelve more than the greater is the same as 10 times the smaller. Find the two numbers. *3; 18*
14. The smaller of two numbers is 2 less than the greater. If 4 times the greater is decreased by the smaller, the result is 23. Find the two numbers.
15. The length of a rectangle is 5 cm more than 3 times the width. The perimeter is 42 cm. Find the length and width. *4 cm; 17 cm*
16. Side b of a triangle is 6 m less than twice side a. Sides b and c have the same length. The perimeter is 28 m. Find the lengths of the three sides.
17. The second of three numbers is 8 more than the first. The third is 7 times the first. The sum of the three numbers is 35. Find the three numbers. *3; 11; 21*
18. The second of three numbers is twice the first. The third is 4 less than the second. The sum of the first two is 1 more than twice the third. Find the numbers. *7; 14; 10*

PART C

19. A rectangle and a square have the same width. The rectangle is 6 cm longer than the square. One perimeter is twice the other. Find the rectangle's dimensions.
20. One side of an equilateral triangle has the same length as a square. The sum of the two perimeters is 56 m. Find the perimeter of the square.

Coin Problems

OBJECTIVE

- To solve problems about coins

REVIEW CAPSULE

Number of Dimes	Value in Cents
1	10
4	40
d	$10d$
$3x$	$30x$

Number of Nickels	Value in Cents
1	5
7	35
n	$5n$
$(x+3)$	$5(x+3)$

EXAMPLE 1

Isabel has some dimes. José has some nickels. The number of dimes is 6 more than the number of nickels. Represent the total value in cents in mathematical terms.

The number of nickels is mentioned *after* the number of dimes. Represent the number of nickels with a variable. ⟶ Let $n =$ number of nickels

6 more than number of nickels ⟶ $n + 6 =$ number of dimes

Number of coins ⟶
Value in cents ⟶

	Nickels	Dimes
Number	n	$n+6$
Value	$5n$	$10(n+6)$

Total value in cents ⟶ $5n + 10(n+6)$

$15n + 60$

Thus, $15n + 60$ cents is the total value in cents, where n is the number of nickels.

EXAMPLE 2

A box has some dimes and quarters. The number of quarters is 4 times the number of dimes. Write the total value in cents in mathematical terms.

Represent *how many coins* in mathematical terms.

Let $d =$ number of dimes

4 times number of dimes ⟶ $4d =$ number of quarters

Number of coins ⟶
Value in cents ⟶

	Dimes	Quarters
Number	d	$4d$
Value	$10d$	$25(4d)$

Total value in cents ⟶ $10d + 25(4d)$, or $110d$

$$10d + 25(4d) = 10d + 100d = 110d$$

Thus, $110d$ cents is the total value in cents, where d is the number of dimes.

EXAMPLE 3 Martha has only nickels and dimes in her coin collection. The number of dimes is 4 less than the number of nickels. The value of her collection is \$1.40. Find the number of nickels and dimes.

4 less than number of nickels ⟶

Let n = number of nickels
$n - 4$ = number of dimes

	Nickels	Dimes
Number	n	$n-4$
Value	$5n$	$10(n-4)$

Write an equation. ⟶
\$1.40 = 140 cents

Solve the equation.

$$5n + 10(n-4) = 140$$
$$5n + 10n - 40 = 140$$
$$15n - 40 = 140$$
$$15n = 180$$
$$n = 12$$

Find both numbers.

Number of nickels, n is 12.
Number of dimes, $n - 4$ is $12 - 4$, or 8.

Check in the problem.

Number of dimes is 4 less than number of nickels.

8	$12 - 4$
	8

Value of coins is \$1.40.

$5(12) + 10(8)$	140
$60 + 80$	
140	

Thus, there are 12 nickels and 8 dimes.

EXAMPLE 4 Maria has some quarters. Alberto has some nickels. The number of quarters is 4 more than 3 times the number of nickels. The total value of their money is \$5.80. Find the number of nickels and quarters.

4 more than 3 times number of nickels ⟶

Let n = number of nickels
$3n + 4$ = number of quarters

	Nickels	Quarters
Number	n	$3n+4$
Value	$5n$	$25(3n+4)$

Write an equation. ⟶
\$5.80 = 580 cents

$$5n + 25(3n+4) = 580$$
$$5n + 75n + 100 = 580$$
$$80n + 100 = 580$$
$$80n = 480$$
$$n = 6$$

Find both numbers.

Number of nickels, n is 6.
Number of quarters, $3n + 4$ is $3(6) + 4$, or 22.

Check in the problem. ⟶ **Thus,** there are 6 nickels and 22 quarters.

ORAL EXERCISES

State the value in cents in mathematical terms.

1. n nickels *$5n$¢*
2. $(n + 2)$ dimes *$10(n + 2)$ or $(10n + 20)$¢*
3. $3n$ quarters *$25(3n)$ or $75n$¢*
4. $(5n - 4)$ dimes *$10(5n - 4)$ or $(50n - 40)$¢*

EXERCISES

PART A

Represent the total value in cents in mathematical terms.

1. n nickels and $(n + 4)$ dimes *$15n + 40$*
2. d dimes and $(d - 2)$ quarters *$35d - 50$*
3. n nickels and $5n$ quarters *$130n$*
4. q quarters and $6q$ dimes *$85q$*
5. d dimes and $(2d + 3)$ nickels *$20d + 15$*
6. q quarters and $(3q - 6)$ nickels *$40q - 30$*
7. Margaret has some dimes and Paul has some nickels. The number of dimes is 8 more than the number of nickels. The total value of their money is $1.85. Find the number of nickels and dimes. *7 nickels; 15 dimes*
8. Tom has some quarters and Ann has some nickels. The number of quarters is 6 less than the number of nickels. The total value of their money is $1.20. Find the number of nickels and quarters. *9 nickels; 3 quarters*
9. Adriana has only dimes and quarters in her coin collection. The number of quarters is 3 times the number of dimes. The value of her collection is $3.40. Find the number of dimes and quarters. *4 dimes; 12 quarters*
10. Conrad has only nickels and dimes in his coin collection. The number of nickels is 7 times the number of dimes. The value of his collection is $1.35. Find the number of nickels and dimes. *3 dimes; 21 nickels*

PART B

11. A parking meter takes only dimes and quarters. At the end of the day the number of dimes is 8 more than 20 times the number of quarters. The value of the coins is $7.55. Find the number of dimes and quarters. *3 quarters; 68 dimes*
12. A coin box has only nickels and quarters. The number of nickels is 6 less than 4 times the number of quarters. The value of the coins is $2.40. Find the number of nickels and quarters. *6 quarters; 18 nickels*
13. One page of a stamp collection has only 8¢, 15¢, and 20¢ stamps. The number of 8¢ stamps is twice the number of 15¢ stamps. There are 4 more 20¢ stamps than 8¢ stamps. The total value is $4.35. Find the number of each kind of stamp.
14. A soft drink machine takes dimes, nickels, and quarters. One day its coin box had 12 more dimes than quarters and the number of nickels was twice the number of dimes. The total value of the money was $6.00. Find the number of each kind of coin.
15. Regina has only nickels and dimes in her coin collection. There are 40 coins worth $2.90. Find the number of nickels and dimes. *22 nickels; 18 dimes*
16. There are 20 coins in a parking meter. It takes only dimes and quarters. The value of the coins is $3.20. Find the number of dimes and quarters. *12 dimes; 8 quarters*

Groups

A set of elements is a group if for a defined operation:

- the set is closed.
- the associative property holds.
- there is an identity.
- each element in the set has an inverse.

If the commutative property also holds, then it is a *commutative group*, or an *Abelian group*, after the Norwegian mathematician Niels Henrik Abel.

PROBLEM 1.

Does the set of integers, $I = \{\ldots, -3, -2, -1, 0, 1, 2, 3, 4, \ldots\}$, form a group for addition?

Closed: Is the sum of two integers another integer?

$2 + -7 = -5$ $-8 + -3 = -11$

Integers Integers

Yes

Associative: Does $(a + b) + c = a + (b + c)$?

$(2 + -7) + -6 = 2 + (-7 + -6)$

$-11 = -11$

Same

Yes

Identity: Does $a + 0 = a$?

$2 + 0 = 2$ $-7 + 0 = -7$

Identity

Yes

Inverse: Does $a + (-a) = 0$?

$-2 + 2 = 0$ $3 + -3 = 0$

Additive Inverses Additive Inverses

Yes

We are sure there are no exceptions.
Thus, the set of integers forms a group for addition.

PROBLEM 2.

Does the set $I = \{\ldots, -3, -2, -1, 0, 1, 2, 3, \ldots\}$ form an Abelian group for addition?

Commutative: Does $a + b = b + a$?

$-3 + -4 = -4 + -3$
$-7 = -7$
Same

Yes

We are sure there are no exceptions.
Thus, the set of integers forms a commutative (Abelian) group for addition.

PROBLEM 3.

Does the set $I = \{\ldots, -3, -2, -1, 0, 1, 2, 3, \ldots\}$ form a group for multiplication?

Closed: Is the product of two integers another integer?

$(-3)(5) = -15$
Integers

Yes

Associative: Does, $(ab)c = a(bc)$?

$(2 \times -4) \times -3 = 2 \times (-4 \times -3)$
$24 = 24$
Same

Yes

Identity: Does $a \cdot 1 = a$?

$-3 \cdot 1 = -3 \quad 5 \cdot 1 = 5$
Identity

Yes

Inverse: Is there an integer b such that $a \cdot b = 1$?

No integer b such that $3 \cdot b = 1$.

No

Thus, the integers do not form a group for multiplication.

PROJECT

Does each set below form a group for the given operation?

1. Odd integers for addition
2. Even integers for addition
3. Integers for division
4. Integers for subtraction
5. Does each set above form an Abelian group for the given operation?

Dry Mixture Problems

OBJECTIVE

- To solve problems about mixtures of dry items

REVIEW CAPSULE

7 pencils at 6¢ per pencil are worth 42¢.

$$\begin{pmatrix}\text{Number}\\ \text{of items}\end{pmatrix} \times \begin{pmatrix}\text{Cost per}\\ \text{item}\end{pmatrix} = \begin{pmatrix}\text{Value}\\ \text{of items}\end{pmatrix}$$

$x + 5$ books at \$3 per book are worth $3(x + 5)$ dollars.

EXAMPLE 1 Peanuts costing \$1.20 per kilogram are mixed with almonds costing \$1.80 per kg to make 30 kg of a mixture costing \$1.40 per kg. How many kg of peanuts and how many kg of almonds should be used?

$x + (30 - x) = 30$ →

Let $x =$ number of kg of peanuts
$30 - x =$ number of kg of almonds

(No. of kg) × (cost per kg) = value of nuts

	Number of kg	Cost per kg in ¢	Value in ¢
Peanuts	x	120	$120x$
Almonds	$30 - x$	180	$180(30 - x)$
Mixture	30	140	$30(140)$

Write an equation. →
Solve the equation.

$$\begin{pmatrix}\text{Value of}\\ \text{peanuts}\end{pmatrix} + \begin{pmatrix}\text{Value of}\\ \text{almonds}\end{pmatrix} = \begin{pmatrix}\text{Value of}\\ \text{mixture}\end{pmatrix}$$

$$\begin{aligned} 120x + 180(30 - x) &= 30(140) \\ 120x + 5{,}400 - 180x &= 4{,}200 \\ -60x + 5{,}400 &= 4{,}200 \\ -60x &= -1{,}200 \\ x &= 20 \end{aligned}$$

Number of kg of peanuts, x is 20.
Number of kg of almonds, $30 - x$ is $30 - 20$, or 10.

Check the values:

(20 kg at \$1.20) + (10 kg at \$1.80)	30 kg at \$1.40
20(1.20) + 10(1.80)	30(1.40)
24.00 + 18.00	42.00
42.00	

Thus, 20 kg of peanuts and 10 kg of almonds are used in the mixture.

EXAMPLE 2 Some brand A coffee beans costing 80¢ per kg are added to 20 kg of brand B costing 60¢ per kg to make a mixture costing 75¢ per kg. How many kg of brand A should be added?

20 kg added to x kg →

Let x = number of kg of brand A
$x + 20$ = number of kg in mixture

(No. of kg) × (cost per kg) = value of beans

	Number of kg	Cost per kg	Value in ¢
Brand A	x	80	$80x$
Brand B	20	60	$60(20)$
Mixture	$x + 20$	75	$75(x + 20)$

Write an equation. →
Solve the equation.

$$\begin{pmatrix}\text{Value of}\\\text{brand A}\end{pmatrix} + \begin{pmatrix}\text{Value of}\\\text{brand B}\end{pmatrix} = \begin{pmatrix}\text{Value of}\\\text{mixture}\end{pmatrix}$$

$$\begin{aligned} 80x + 60(20) &= 75(x + 20) \\ 80x + 1{,}200 &= 75x + 1{,}500 \\ 5x &= 300 \\ x &= 60 \end{aligned}$$

Check in the problem.

60 kg at 80¢ + 20 kg at 60¢	80 kg at 75¢
4,800 + 1,200	6,000
6,000	

Thus, 60 kg of brand A should be added.

EXAMPLE 3 Some 25¢ comic books are added to some 40¢ comic books to make a package worth \$4.70. The number of 40¢ books is 2 more than the number of 25¢ books. How many books of each kind should be included?

2 more than x →

Let x = number of 25¢ books
$x + 2$ = number of 40¢ books

\$4.70 = 470¢

	No. of books	Cost per book	Value in ¢
25¢ books	x	25	$25x$
40¢ books	$x + 2$	40	$40(x + 2)$
Package	–	–	470

Write an equation. →
Solve the equation.

$$\begin{pmatrix}\text{Value of}\\\text{25¢ books}\end{pmatrix} + \begin{pmatrix}\text{Value of}\\\text{40¢ books}\end{pmatrix} = \begin{pmatrix}\text{Value of}\\\text{package}\end{pmatrix}$$

$$\begin{aligned} 25x + 40(x + 2) &= 470 \\ 25x + 40x + 80 &= 470 \\ 65x &= 390 \\ x &= 6 \end{aligned}$$

Number of 25¢ books, x is 6.
Number of 40¢ books, $x + 2$ is $6 + 2$, or 8.

Check in the problem.

(6 at 25¢) + (8 at 40¢)	470¢
6(25) + 8(40)	470
150 + 320	
470	

Thus, 6 of the 25¢ books and 8 of the 40¢ books should be included in the package.

EXERCISES

PART A

1. Some 50¢ red pens and some 30¢ blue pens are mixed to make a package of 20 pens. The package is worth $8.40. How many 50¢ and how many 30¢ pens are in the package? *50¢: 12; 30¢: 8*

2. Some 8¢ slices of cheese and some 16¢ slices of meat are mixed to make a package of 40 slices. The package is worth $5.20. How many slices of each kind are in the package? *cheese: 15; meat: 25*

3. Some brand A dog food costing 70¢ per kg is added to 9 kg of brand B costing 50¢ per kg to make a mixture costing 58¢ per kg. How many kg of brand A should be added? *6 kg*

4. Some corn costing 60¢ per kg is added to 50 kg of oats costing 90¢ per kg to make an animal feed costing 75¢ per kg. How many kg of corn should be added? *50 kg*

5. Soft candy costing $3.50 per kg is mixed with hard candy costing $2.00 per kg to make 12 kg of a mixture costing $2.50 per kg. How many kg of each kind should be used? *soft: 4 kg; hard: 8 kg*

6. Brand A of bulk tea costing $1.40 per kg is mixed with brand B costing $3.00 per kg to make 16 kg of a mixture costing $2.00 per kg. How many kg of each brand should be used? *A: 10 kg; B: 6 kg*

PART B

7. Powdered milk costing $2.00 per kg is mixed with dry cocoa costing $2.60 per kg. The number of kg of milk is 16 more than the number of kg of cocoa. The mixture is worth $50.40. How many kg of milk and how many kg of cocoa should be mixed? *milk: 20 kg; cocoa: 4 kg*

8. Wheat flour costing 40¢ per kg is mixed with bran flour costing 60¢ per kg. The number of kg of bran is 10 less than the number of kg of wheat. The mixture is worth $14.00. How many kg of each kind of flour should be used? *wheat: 20 kg; bran: 10 kg*

9. Some 60¢ paper tablets are added to some 80¢ tablets to make a box of tablets worth $16.00. The number of 60¢ tablets is twice the number of 80¢ tablets. How many tablets of each kind should be included? *60¢: 16; 80¢: 8*

10. Some 30¢ birthday cards are mixed with some 20¢ birthday cards to make a box of cards worth $6.30. The number of 20¢ cards is 3 times the number of 30¢ cards. How many cards of each kind should be included? *30¢: 7; 20¢: 21*

PART C

11. Chemicals A, B, and C cost 60¢, 40¢, and 80¢ per gram, respectively. They are mixed so that the number of grams of B is twice the number of grams of A and is 3 less than the number of grams of C. The mixture is worth $11.40. How many grams of each chemical should be used?

12. Brands A, B, and C of charcoal cost 30¢, 60¢, and 40¢, respectively. They are mixed so that the number of kg of A is 2 less than 3 times the number of kg of B and is 8 more than the number of kg of C. The mixture is worth $11.60. How many kg of each brand should be used?

Age Problems

OBJECTIVE

- To solve problems about ages

REVIEW CAPSULE

Rita's age now 14 years
5 years *ago* $14 - 5$, or 9 years
8 years *from now* $14 + 8$, or 22 years

EXAMPLE 1 Diane is 4 years older than Fred. Represent their ages 7 years ago in mathematical terms.

Fred is mentioned *after* Diane. Represent Fred's age now with a variable. → Let f = Fred's age now

4 years older than Fred → $f + 4$ = Diane's age now

Ages now →
Subtract 7 from ages now. →

	Fred	Diane
Now	f	$f + 4$
Then	$f - 7$	$(f + 4) - 7$

$(f + 4) - 7 = f + 4 - 7 = f - 3$

Thus, Fred was $(f - 7)$ years old and Diane was $(f + 4) - 7$, or $(f - 3)$ years old, where f is Fred's age now.

EXAMPLE 2 Manuel's age is 3 times Teresa's age. Represent their ages 5 years from now in mathematical terms.

Write their ages *now*. Let t = Teresa's age now

3 times Teresa's age → $3t$ = Manuel's age now

Ages now →
Add 5 to ages now. →

	Teresa	Manuel
Now	t	$3t$
Then	$t + 5$	$3t + 5$

Thus, Teresa will be $(t + 5)$ years old and Manuel will be $(3t + 5)$ years old, where t is Teresa's age now.

EXAMPLE 3 Mike is 3 years younger than Hazel. Four years ago the sum of their ages was 21 years. How old is each now?

Let Hazel's age now = h.
Ages now →
Ages 4 years ago →

	Hazel	Mike
Now	h	$h - 3$
Then	$h - 4$	$(h - 3) - 4$

Write an equation.
Sum of their ages was 21. →

$$(h - 4) + (h - 7) = 21$$

Solve the equation.

$$2h - 11 = 21$$
$$h = 16$$

Find both ages now.

Thus, Hazel's age now, h is 16 and Mike's age now, $h - 3$ is $16 - 3$, or 13.

ORAL EXERCISES

1. Karen is 7 years older than Jane. State Karen's age if Jane is j years old. *$(j + 7)$ yr*
2. Ed is 5 years younger than Frank. State Ed's age if Frank is f years old. *$(f - 5)$ yr*
3. Eloise is 5 times as old as Bobby. State Eloise's age if Bobby is b years old. *$5b$ yr*
4. Sarah is s years old now. State her age 6 years ago. *$(s - 6)$ yr*

EXERCISES

PART A

1. Brenda is 4 years older than Walter. Two years ago the sum of their ages was 16. How old is each now?
2. Byron is 2 years younger than Cindy. Five years from now the sum of their ages will be 40. How old is each now?
3. Denise is 3 times as old as Conrad. Five years ago the sum of their ages was 14. How old is each now? *Conrad 6; Denise 18*
4. Samson is 5 times as old as Edna. Seven years from now the sum of their ages will be 32. How old is each now? *Edna 3; Samson 15*

PART B

EXAMPLE Howard is 4 times as old as his niece, Selma. Ten years from now he will be twice as old as she will be then. How old is each now?

Let Selma's age now $= s$.

Ages 10 years from now →

	Selma	Howard
Now	s	$4s$
Then	$s + 10$	$4s + 10$

Write an equation.

His age will be 2 times her age. → $4s + 10 = 2(s + 10)$

Solve the equation.

$4s + 10 = 2s + 20$

Add $-2s$ and -10 to each side. → $2s = 10$

$s = 5$

Check in the problem. → **Thus,** Selma's age now, s is 5 and Howard's age now, $4s$ is 20.

5. Wanda is 20 years older than Ted. Eight years from now she will be 3 times as old as he will be then. How old is each now? *Ted 2; Wanda 22*
6. Raymond is 12 years younger than Susan. Four years ago she was 4 times as old as he was then. How old is each now? *Susan 20; Ray 8*
7. Melvin is 3 times as old as Paula. In 8 years he will be twice as old as she will be then. How old is each now?
8. Gloria is 4 times as old as Amos. Three years ago she was 7 times as old as he was then. How old is each now?
9. Phyllis is 6 years older than Keith. If his age is added to 3 times her age, the result is 74. How old is each now? *Keith 14; Phyllis 20*
10. Eric is 4 years younger than Ruby. If twice her age is added to 4 times his age, the result is 68. How old is each now? *Ruby 14; Eric 10*

Chapter Two Review

Graph the solution set. [p. 19]

1. $-3 \leq x < 5$
2. $-4 < y \leq 6$
3. $a < -4$ or $a \geq 2$
4. $n < 2$ or $n > 5$

Find the solution set. [p. 19]

5. $-5 < 3x - 2 < 10$
6. $-6 \leq 4y + 6 \leq 2$
7. $2a + 5 < -3$ or $2a + 5 > 7$
8. $5n - 2 \leq -7$ or $5n - 2 \geq 8$

Find the solution set. [p. 22]

9. $|2x - 7| = 3$
10. $|3y + 4| = 13$
11. $|a| < 5$
12. $|n| \geq 3$

Find and graph the solution set. [p. 22]

13. $|x + 3| \leq 5$
14. $|4y| > 12$
15. $|2a - 3| > 5$
16. $|3n + 6| < 9$

Solve.

17. The second of two numbers is 3 more than 4 times the first. Their sum is 23. Find both numbers. [p. 28]
18. The length of a rectangle is 6 meters less than twice the width. The perimeter is 30 meters. Find the length and the width. [p. 28]
19. The greater of two numbers is 5 less than 8 times the smaller. One more than the greater is the same as 6 times the smaller. Find both numbers. [p. 28]
20. Side *b* of a triangle is 3 times as long as side *a*. Side *c* is 2 cm shorter than side *b*. The perimeter is 26 cm. Find the lengths of the sides. [p. 28]
21. Adriana has only nickels and dimes in her coin collection. The number of dimes is 7 less than the number of nickels. The value of her collection is $1.10. Find the number of nickels and dimes. [p. 32]
22. A coin box has only dimes and quarters. The number of dimes is 9 more than 3 times the number of quarters. The value of the coins is $4.75. Find the number dimes and quarters. [p. 32]
23. Caramels costing $2.60 per kg are mixed with peppermints costing $3.00 per kg to make 8 kg of a mixture costing $2.85 per kg. How many kg of each kind should be used? [p. 37]
24. Some brand A cat food costing 85¢ per kg is added to 10 kg of brand B costing 50¢ per kg to make a mixture costing 60¢ per kg. How many kg of brand A should be added? [p. 37]
25. Edna is 5 years older than Conrad. Three years ago the sum of their ages was 17. How old is each now? [p. 40]
26. Ted is 9 years younger than Ann. Six years from now her age will be twice his age. How old is each now? [p. 40]

Chapter Two Test

Graph the solution set.

1. $-5 < x \leq 2$

2. $y \leq -3$ or $y > 4$

Find the solution set.

3. $-3 \leq 2x + 3 \leq 7$

4. $3y - 5 < -11$ or $3y - 5 > 4$

5. $|a + 3| = 8$

6. $|3n - 4| = 2$

Find and graph the solution set.

7. $|x| < 2$

8. $|y| \geq 4$

9. $|a - 4| > 3$

10. $|2n| \leq 6$

11. $|2x + 1| < 7$

12. $|3y - 9| \geq 3$

Solve.

13. The greater of two numbers is 3 less than 5 times the smaller. The sum of the two numbers is 9. Find both numbers.

14. The length of a rectangle is 5 meters more than 3 times the width. The perimeter is 42 meters. Find the length and the width.

15. A coin collection contains only nickels and dimes. The number of dimes is 4 more than the number of nickels. The value of the collection is $1.45. Find the number of nickels and dimes.

16. Brand A cleaning powder costing 80¢ per kg is mixed with brand B costing 60¢ per kg to make 15 kg of a mixture costing 68¢ per kg. How many kg of each brand should be used?

17. Melvin is 15 years older than Phyllis. Two years from now he will be 4 times as old as she will be then. How old is each now?

3 POLYNOMIALS AND FACTORING

Metric System: Area and Temperature

AREA

The basic unit of area is the square meter (m^2). A door of a sports car has an area of about 1 m^2.

1 square centimeter (cm^2) = .0001 m^2
1 square decimeter (dm^2) = .01 m^2
1 square dekameter (dam^2) = 100 m^2
1 are (a) = 100 m^2, or 1 dam^2
1 hectare (ha) = 100 ares = 10,000 m^2

TEMPERATURE

Temperature is measured in degrees Celsius (°C).

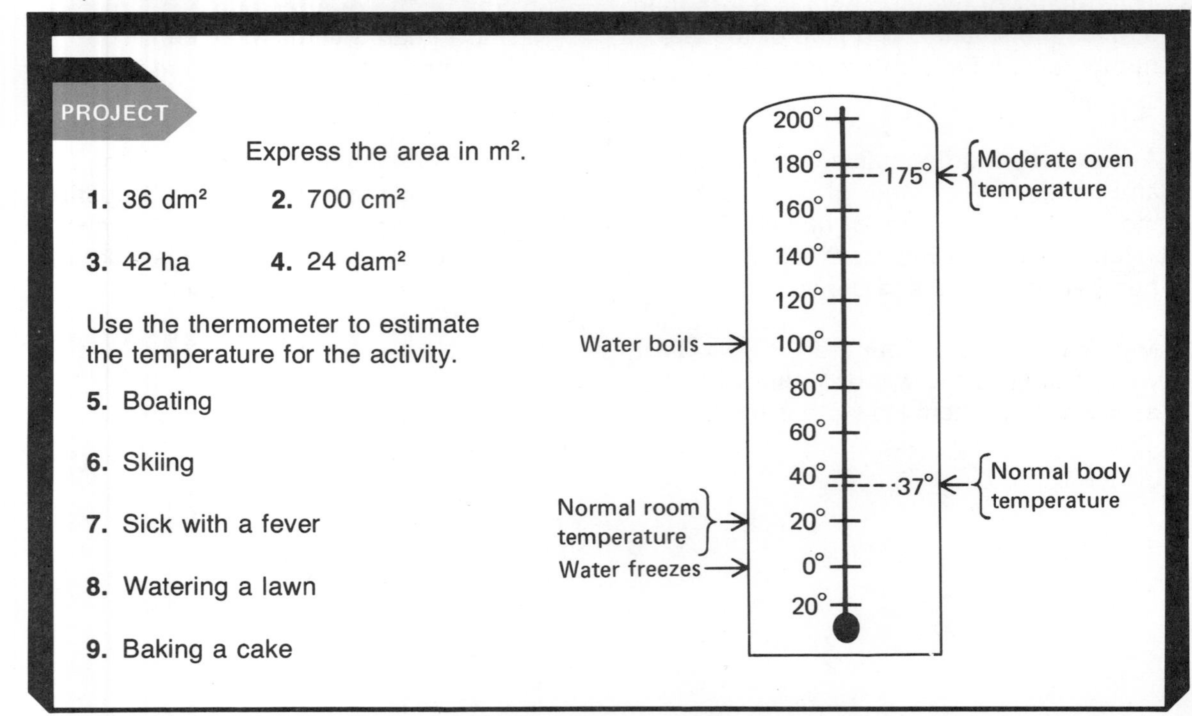

PROJECT

Express the area in m^2.

1. 36 dm^2 **2.** 700 cm^2

3. 42 ha **4.** 24 dam^2

Use the thermometer to estimate the temperature for the activity.

5. Boating

6. Skiing

7. Sick with a fever

8. Watering a lawn

9. Baking a cake

Exponents

OBJECTIVE

- To simplify and evaluate expressions like $(3x^2y)(-2x^3y^2)$, $(5a^2b^3)^4$, and $4x^3 + 5x^2 - 2x^3 + 3x^2$

REVIEW CAPSULE

Base → 5^3 ← Exponent

Third *power* of 5 = $\underbrace{5 \cdot 5 \cdot 5}_{\text{3 factors}}$

Simplify $(-2x)(3y)$.

$(-2x)(3y)$

$= -2 \cdot 3 \cdot x \cdot y$ ← Rearrange factors.

$= -6xy$

EXAMPLE 1 Simplify $x^2 \cdot x^3$.

$x^2 \cdot x^3$

$(x \cdot x)(x \cdot x \cdot x)$

x^5

$\overbrace{(x \cdot x)(x \cdot x \cdot x)}^{\text{5 factors}} = x^5$ ⟶

Add exponents of *like* bases.

Product of Powers

$x^m \cdot x^n = x^{m+n}$

EXAMPLE 2 Simplify $(a^2b)(a^3b^2)$.

	Simplify $(a^2b)(a^3b^2)$.	Simplify $(-2x^2)(5x^4)$.
	$(a^2b)(a^3b^2)$	$(-2x^2)(5x^4)$
$b = b^1$ ⟶	$(a^2b^1)(a^3b^2)$	
Group like factors together. ⟶	$(a^2 \cdot a^3)(b^1 \cdot b^2)$	$(-2 \cdot 5)(x^2 \cdot x^4)$
Add exponents of *like* bases. ⟶	a^5b^3	$-10x^6$

EXAMPLE 3 Simplify $x^4 \cdot y^3$.

x and *y* are *different bases.* ⟶

$x^4 \cdot y^3 = (x \cdot x \cdot x \cdot x)(y \cdot y \cdot y)$

$= x^4 \cdot y^3$

Thus, $x^4 \cdot y^3$ cannot be simplified.

EXAMPLE 4 Show that $(2^3)^4 = 2^{3 \cdot 4}$.

Base → $\boxed{2^3}$ ④ ← Exponent

$(2^3)^4$	$2^{3 \cdot 4}$
$2^3 \cdot 2^3 \cdot 2^3 \cdot 2^3$	2^{12}
2^{12}	

Exponents are multiplied. ⟶ **Thus,** $(2^3)^4 = 2^{3 \cdot 4}$, or 2^{12}.

Multiply exponents.

Power of a Power

$(x^m)^n = x^{m \cdot n}$

EXAMPLE 5 Simplify $(x^5)^4$.

$(x^m)^n = x^{m \cdot n}$ → Multiply exponents.

$$(x^5)^4 = x^{5 \cdot 4} = x^{20}$$

EXAMPLE 6 Simplify $(xy)^3$.

$$(xy)^3$$
$$xy \cdot xy \cdot xy$$

Regroup. →

$$(x \cdot x \cdot x)(y \cdot y \cdot y)$$
$$x^3 \cdot y^3$$

Each factor of the base is raised to the third power. → **Thus,** $(xy)^3 = x^3 \cdot y^3$.

Raise each factor to the power m.

Power of a Product

$(xy)^m = x^m \cdot y^m$

EXAMPLE 7 Simplify $(-5b^2)^4$.

$(xy)^m = x^m \cdot y^m$ →
$(b^2)^4 = b^{2 \cdot 4}$, or b^8

$$(-5b^2)^4$$
$$(-5)^4 \cdot (b^2)^4$$
$$625b^8$$

EXAMPLE 8 Simplify $(-4cd^5)^3$.

Each factor is raised to third power. →

$$(-4cd^5)^3$$
$$(-4)^3(c)^3(d^5)^3$$
$$-64c^3d^{15}$$

EXAMPLE 9 Simplify $6x^3 + 5x^2 + 2x^3 - 2x^2$.

Regroup like terms. →
$6x^3 + 2x^3 = (6+2)x^3$
Combine like terms. →

$$6x^3 + 5x^2 + 2x^3 - 2x^2 = 6x^3 + 2x^3 + 5x^2 - 2x^2$$
$$= 8x^3 + 3x^2$$

EXAMPLE 10 Simplify $8x^3 + 3x^2$.

$8x^3$ and $3x^2$ are not like terms.

Thus, $8x^3 + 3x^2$ cannot be simplified.

EXAMPLE 11 Evaluate $4x^3$ if $x = -2$.

Write a *negative base,* like -2, inside parentheses and exponent outside $(-2)^3$.

$$4x^3 = 4 \cdot (-2)^3 = 4(-2)(-2)(-2) = 4(-8) = -32$$

Evaluate $(-3n^3)^2$ if $n = 2$.

$$(-3n^3)^2 = [-3(2^3)]^2 = [-3(8)]^2 = (-24)^2 = (-24)(-24) = 576$$

EXAMPLE 12 Evaluate $-x^4y^2$ if $x = -3$ and $y = 5$.

$-a = -1a$ ⟶ $-x^4y^2 = -1 \cdot x^4 \cdot y^2$

Substitute -3 for x, 5 for y. ⟶ $= -1(-3)^4(5^2)$

$= -1(81)(25)$

$= -2{,}025$

EXAMPLE 13 Evaluate $5a^2 - 4b^3$ if $a = 3$ and $b = -2$.

Substitute for a and b. ⟶ $5a^2 - 4b^3 = 5(3^2) - 4(-2)^3$

$= 5(9) - 4(-8)$

$= 45 + 32$

$= 77$

SUMMARY

Product of Powers	$x^m \cdot x^n = x^{m+n}$
Power of a Power	$(x^m)^n = x^{m \cdot n}$
Power of a Product	$(xy)^m = x^m \cdot y^m$

ORAL EXERCISES

Simplify.

1. $5^3 \cdot 5^6$ 5^9 **2.** $(5^3)^6$ 5^{18} **3.** $(6^5)^3$ 6^{15} **4.** $6^5 \cdot 6^3$ 6^8

5. $x^3 \cdot x^6$ x^9 **6.** $(x^3)^6$ x^{18} **7.** $(c^5)^3$ c^{15} **8.** $c^5 \cdot c^3$ c^8

9. $x^3 \cdot x^3$ x^6 **10.** $x^3 + x^3$ $2x^3$ **11.** $3x^2 + 4x^2$ $7x^2$ **12.** $3x^2 \cdot 4x^2$ $12x^4$

13. $x^4 \cdot x$ x^5 **14.** $2x \cdot x^4$ $2x^5$ **15.** $x \cdot x$ x^2 **16.** $x + x$ $2x$

Compute.

17. $(-1)^2$ 1 **18.** $(-1)^3$ -1 **19.** $(-1)^4$ 1 **20.** $(-1)^5$ -1 **21.** $(-1)^6$ 1

22. $(-2)^2$ 4 **23.** $(-2)^3$ -8 **24.** $(-2)^4$ 16 **25.** $(-2)^5$ -32 **26.** $(-2)^6$ 64

EXERCISES

PART A

Simplify.

1. $3x^2 \cdot 4x^3$ $12x^5$
2. $-2a^5 \cdot 5a^4$ $-10a^9$
3. $(-10c^3)(-2c^3)$ $20c^6$
4. $(4x)^3$ $64x^3$
5. $(-2a)^4$ $16a^4$
6. $(7c)^2$ $49c^2$
7. $5x^3 \cdot x$ $5x^4$
8. $a(-3a)$ $-3a^2$
9. $(4n^3)(2n)(5n)$ $40n^5$
10. $3(x^2)^4$ $3x^8$
11. $-5(a^4)^3$ $-5a^{12}$
12. $-(c^3)^2$ $-c^6$
13. $(-3x^2)^4$ $81x^8$
14. $7(10x^2)^3$ $7{,}000x^6$
15. $4x(5x^3)^2$ $100x^7$
16. $7x^3 + 4x^2 - 5x^3 + 2x^2$ $2x^3 + 6x^2$
17. $3x^3 + 2x - 8x^3 + 5x$ $-5x^3 + 7x$
18. $4x^2 - 3x - x^2 + x$ $3x^2 - 2x$

Evaluate.

19. x^3 if $x = -3$ -27
20. $3n^2$ if $n = 5$ 75
21. $-c^4$ if $c = -2$ -16
22. $(x^2)^3$ if $x = -2$ 64
23. $(5a^3)^2$ if $a = 2$ $1{,}600$
24. $5x^3 + x^2$ if $x = 3$ 144

PART B

Simplify.

25. $(x^3y^2)(x^2y^4)$ x^5y^6
26. $3a^2b(a^3b^4)$ $3a^5b^5$
27. $(-2x^4y^2)(5xy^5)$ $-10x^5y^7$
28. $(ab)^3$ a^3b^3
29. $(5xy)^2$ $25x^2y^2$
30. $(-2cd)^4$ $16c^4d^4$
31. $(a^3b^2)^2$ a^6b^4
32. $(x^2y^4)^3$ x^6y^{12}
33. $(x^2y^3)^4$ x^8y^{12}
34. $(5x^2y^3)^2$ $25x^4y^6$
35. $(-2ab^4)^3$ $-8a^3b^{12}$
36. $(3c^3d)^4$ $81c^{12}d^4$
37. $3a^2 + 2b^2 + 5a^2 + 3b^2$ $8a^2 + 5b^2$
38. $4a^2 + 6b^3 + 2a^2 - 3b^3$ $6a^2 + 3b^3$
39. $x^2 - 3y^2 + 5x^2 - y^2$ $6x^2 - 4y^2$

Evaluate.

40. $-x^4y$ if $x = 2$ and $y = -5$ 80
41. $2a^2b^4$ if $a = 4$ and $b = -1$ 32
42. $(xy)^3$ if $x = 5$ and $y = 2$ $1{,}000$
43. $(5ab)^2$ if $a = -2$ and $b = 3$ 900
44. $(x^2y^3)^2$ if $x = 3$ and $y = -1$ 81
45. $(-2ab^2)^3$ if $a = 5$ and $b = 2$ $-64{,}000$
46. $5x^3 + 4y^2$ if $x = -2$ and $y = 3$ -4
47. $3a^2 - 5b^3$ if $a = 2$ and $b = -2$ 52

PART C

Simplify.

48. $(x^a)(x^a)$
49. $(x^{3a})(x^{2a})$
50. $(x^{4a})(x^{2a+3})$
51. $(x^{2a+1})(x^{3-2a})$
52. $(x^ay^{3a})(x^{2a}y^{4a})$
53. $(x^{a+2}y^{a-3})(x^{a-1}y^{a+5})$
54. $(x^{3a}y^{2b})(x^{4a}y^{3b})$
55. $(x^{2a+3}y^{3b-1})(x^{a+1}y^{b+1})$
56. $(x^a)^2$
57. $(x^3)^a$
58. $(x^{4a})^3$
59. $(x^a)^a$
60. $(x^{2a})^{3a}$
61. $(5x)^a$
62. $(2x)^{3a}$
63. $(3x)^{a+1}$
64. $(4x^2)^{2a}$
65. $(2x^3)^{a+2}$
66. $(x^ay^b)^3$
67. $(x^ay^b)^{4c}$
68. $(x^{a+2}y^{b+3})^4$
69. $(x^3y^4)^{a+3}$

Polynomials

OBJECTIVES

- To identify polynomials and their degrees
- To evaluate polynomials
- To add and subtract polynomials

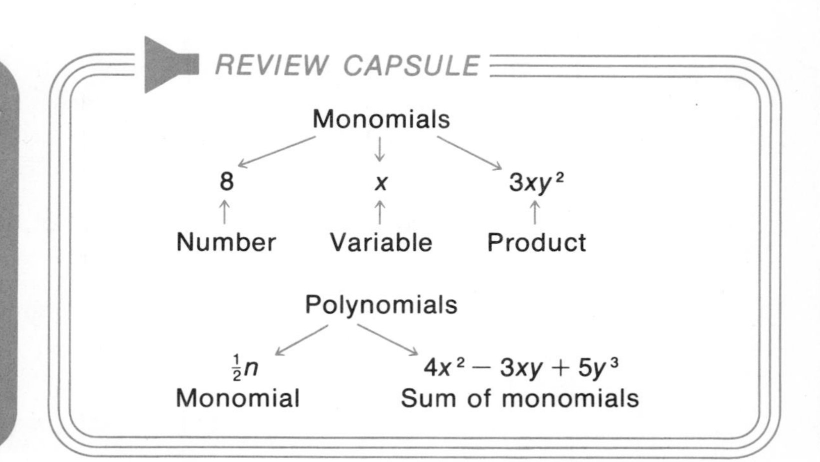

EXAMPLE 1 Which expressions are monomials?

	Expression	*Answers*
	$-5cd$	Yes
Dividing by a variable →	$\frac{3}{x}$	No
	$-t$	Yes
Square root of a variable →	$\sqrt{y}$	No

Definition of monomial →

> A *monomial* is a number, or a variable, or the product of several such factors.

EXAMPLE 2 Which expressions are polynomials?

	Expression	*Answers*
Square root of a variable →	$x + \sqrt{y}$	No
Sum of monomials →	$3x^2y - 5xy + 4$	Yes
Dividing by a variable →	$7y + \frac{2}{x}$	No
Monomial →	-2	Yes

Definition of polynomial →
Bi means *two.*
Tri means *three.*

> A *polynomial* is a monomial or the sum of monomials.
> Binomial: sum of two monomials
> Trinomial: sum of three monomials

EXAMPLE 3 Classify each polynomial.

	Polynomial	*Answers*
Sum of 3 terms →	$3x^2y - 2xy + y^2$	Trinomial
Sum of 2 terms →	$4n^2 - 1$	Binomial
Single term →	$-7c$	Monomial
More than 3 terms, just give number	$2a - b + 3c - 4d + 6$	Polynomial with 5 terms

EXAMPLE 4 Find the sum of the exponents of each monomial.

	Monomial	Sum of Exponents
Add the exponents of the variables.	$3x^2y^3$	$2 + 3$ or 5
$b = b^1$ →	$-5a^2b$	$2 + 1$ or 3
	7	0

Definition of degree of a monomial →

> The *degree* of a monomial is the sum of the exponents of its variables.

EXAMPLE 5 Which term of the polynomial has the greatest degree?

Add the exponents of the variables.

$$4xy^2 + 7x^2y^3 - 3x + 6$$

Degree: 3, 5, 1, 0

Thus, the term with the greatest degree is $7x^2y^3$.

Definition of degree of a polynomial →

> The *degree of a polynomial* is the degree of the term with the greatest degree.

EXAMPLE 6 Write the degree of each polynomial.

	Polynomial	Degree
$-7ab^3 = -7a^1b^3$ →	$a^3 - 7ab^3$	4
	$9f + 2g - h$	1
	$3x^2 + 4y^2 + x^3y^2 - 6$	5
	$3x^4 + 5x^7 + 2x - 3$	7

EXAMPLE 7 Arrange the terms of $-2x^3 + 5x^4 + x^5 + 4 + x$ so that the degrees are in descending order.

Descending means decreasing.

$$-2x^3 + 5x^4 + x^5 + 4 + x$$

Degrees are 5, 4, 3, 1, 0. →

$$x^5 + 5x^4 - 2x^3 + x + 4$$

Thus, the polynomial arranged in descending order is $x^5 + 5x^4 - 2x^3 + x + 4$.

EXAMPLE 8 Evaluate $2x^3y - 5x^2y^2 + 4$ if $x = -2$ and $y = 3$.

$$2x^3y - 5x^2y^2 + 4$$

Substitute -2 for x, 3 for y. →

$$2(-2)^3(3) - 5(-2)^2(3)^2 + 4$$
$$2(-8)(3) - 5(4)(9) + 4$$

$-48 - 180 + 4 = -224$

$$-224$$

Thus, the value is -224 if $x = -2$ and $y = 3$.

EXAMPLE 9 Add $\begin{array}{r} 5y^2 + 3y - 8 \\ \underline{2y^2 - 7y + 9} \end{array}$.

$(5y^2 + 3y - 8) + (2y^2 - 7y + 9)$ is written in vertical form. Like terms are in the same column.

Combine like terms. →

$$\begin{array}{r} 5y^2 + 3y - 8 \\ \underline{2y^2 - 7y + 9} \\ 7y^2 - 4y + 1 \end{array}$$

Thus, the sum is $7y^2 - 4y + 1$.

EXAMPLE 10 Add $(5xy - 2xy^2 + x^2) + (3xy^2 - 4x^2 - 2xy)$.

Group like terms. → $(5xy - 2xy) + (-2xy^2 + 3xy^2) + (1x^2 - 4x^2)$

Combine like terms. → $3xy + xy^2 - 3x^2$

EXAMPLE 11 Subtract $(5x^2 - 2x + 4) - (2x^2 - 4x - 3)$.

$$(5x^2 - 2x + 4) - (2x^2 - 4x - 3)$$

$-(a) = -1(a)$ → $(5x^2 - 2x + 4) - 1(2x^2 - 4x - 3)$

Distribute the -1. → $5x^2 - 2x + 4 - 2x^2 + 4x + 3$

Group like terms. $(5x^2 - 2x^2) + (-2x + 4x) + (4 + 3)$

Combine like terms. $3x^2 + 2x + 7$

EXAMPLE 12 Subtract $\begin{array}{r} 3x^2 - 2x + 6. \\ \underline{5x^2 + 5x - 8} \end{array}$

$(3x^2 - 2x + 6) - (5x^2 + 5x - 8)$ is written in vertical form.

$a - (b) = a + (-1 \cdot b)$ →

$$\begin{array}{r} 3x^2 - 2x + 6 \\ \underline{5x^2 + 5x - 8} \end{array} \quad \text{Multiply by } -1 \text{ and add.} \rightarrow \quad \begin{array}{r} 3x^2 - 2x + 6 \\ \underline{-5x^2 - 5x + 8} \\ -2x^2 - 7x + 14 \end{array}$$

Thus, the difference is $-2x^2 - 7x + 14$.

EXAMPLE 13 Subtract $(-6x + 6x^2y + 2y^2) - (-4x^2y + 3)$.

$$(-6x + 6x^2y + 2y^2) - (-4x^2y + 3)$$

$-(a) = -1(a)$ → $(-6x + 6x^2y + 2y^2) - 1(-4x^2y + 3)$

Distribute the -1. → $-6x + 6x^2y + 2y^2 + 4x^2y - 3$

Group like terms. → $-6x + (6x^2y + 4x^2y) + 2y^2 - 3$

Combine like terms. → $-6x + 10x^2y + 2y^2 - 3$

ORAL EXERCISES

Which expressions are polynomials?

1. $4x^3y$ *yes*
2. -5 *yes*
3. $a + b + \sqrt{c}$ *no*
4. $3a - 4b + 2$ *yes*
5. $\frac{2x^2}{y}$ *no*
6. $\frac{1}{2}x$ *yes*
7. $\frac{1}{4}x^2 - 9$ *yes*
8. $xy + 2x^2 - y^2$ *yes*

Classify the polynomial. *(Monomial, M; Binomial, B; Trinomial, T; Polynomial, P)*

9. $n^2 - 4$ *B*
10. $x^2 + x + 1$ *T*
11. $-4a$ *M*
12. $x^3 + y^3 - x - y + 3$ *P with 5 terms*

Give the degree of the polynomial.

13. $4x^2y^3$ *5*
14. $2xy$ *2*
15. $x^2 - 9x^2y^2$ *4*
16. $x^2y + xy - x^2y^3$ *5*

Arrange the terms in descending order.

17. $2x - 4x^3 - 2$ *$-4x^3 + 2x - 2$*
18. $-y^2 + 6 + 7y^5$ *$7y^5 - y^2 + 6$*
19. $2x - 3 + x^3 - 2x^5$ *$-2x^5 + x^3 + 2x - 3$*
20. $2y^3 + 4y^5 - 7y^4$ *$4y^5 - 7y^4 + 2y^3$*

EXERCISES

PART A

Evaluate the polynomial if $x = 2$ and $y = -3$.

1. $3x^2y - 4x^2y^2 + 2xy$ *-192*
2. $4x^3 + 4y^3 + 4$ *-72*
3. $x - xy - y$ *11*

Add.

4. $3x^2 - 2x - 7$; $x^2 + 9$
5. $a^2 + 2a + 6$; $a^2 - 2a + 3$
6. $2x^2 - 3x$; $6x - 3$
7. $3y^2 + 2$; $2y^2 - y$
8. $(2x^2 - 3x + 9) + (5x^2 + 7x - 8)$
9. $(-6b^2 - b + 2) + (-3b^2 - 4b - 8)$
10. $(3x - 2x^2 + 4) + (2x^2 - 8x + 7)$
11. $(6 + y^2 - 2y) + (3y + 5y^2 - 9)$ *$6y^2 + y - 3$*

Subtract.

12. $(5x^2 + 3x - 4) - (2x^2 - 4x - 5)$
13. $(-9a^2 - 5a + 6) - (3a^2 - 6a + 7)$
14. $(2x - 3x^2 + 9) - (5 + x^2 - 8x)$
15. $(2y - 3y^2 + 11) - (5 - 8y^2 + 4y)$
16. $8x^2 - 4x + 3$; $2x^2 - 7x - 6$
17. $5a^2 + 9a - 2$; $5a^2 - 3a + 7$ *$12a - 9$*
18. $-3y^2 + 4$; $y^2 + 6y - 3$ *$-4y^2 - 6y + 7$*
19. $n^2 + 3n$; $4n^2 + 6$ *$-3n^2 + 3n - 6$*

PART B

Add.

20. $(5xy + 2xy^2 + x^2) + (3xy^2 + 2x^2 + 4xy)$
21. $(2ab - 3a^3b + b^2) + (7a^3b - 4b^2 - 5ab)$
22. $(3x^2y^2 - 5xy^2) + (2x^2y + x^2y^2)$
23. $(2a^2 + ab - 3b^2) + (4b^2 + 6ab - 3a^2b)$
24. $(5x + 3x^2 + 7 + 6x^3) + (7 - 2x^3 - 8x^2 + x)$ *$4x^3 - 5x^2 + 6x + 14$*
25. $(-3 + y^2 - 2y^4 + 5y^6) + (6y^4 - y^6 - 2 - 3y^2)$ *$4y^6 + 4y^4 - 2y^2 - 5$*

Subtract.

26. $(4x^2y^2 - 2xy + 4) - (-3x^2y^2 + 5xy - 8)$
27. $(9a^2b^2 - 2ab + 6) - (a^2b^2 - 8)$
28. $(3rs + 2r^2 - 3s^2) - (9 + rs - r^2s^2)$
29. $(8xy + 9x^2 - y^2) - (xy + x^2 - y^2)$

Multiplying Polynomials

OBJECTIVE

- To multiply polynomials like $(5x + 3y)(2x - y)$ and $(3x - 2y)(4x + 3y - 5)$

REVIEW CAPSULE

Monomial times Monomial	Monomial times Binomial
$(-2x)(4x^2y)$	$3x(2x + 5y)$
$(-2 \cdot 4)(x \cdot x^2)(y)$	$3x(2x) + 3x(5y)$
$-8x^3y$	$6x^2 + 15xy$

EXAMPLE 1 Multiply $2x(3x^2 - 4bx + 5)$.

Monomial × Polynomial → $2x(3x^2 - 4bx + 5)$

Distribute the $2x$. → $2x(3x^2) + 2x(-4bx) + 2x(5)$

Simplify. → $6x^3 \quad -8bx^2 \quad +10x$

EXAMPLE 2 Multiply $(2x + 5)(3x - 4)$.

Binomial × Binomial → $(2x + 5)(3x - 4)$

Distribute $(2x + 5)$.

$\square(3x - 4) = \square(3x) + \square(-4)$ → $(2x + 5)(3x) + (2x + 5)(-4)$

Distribute $3x$; distribute -4. → $(2x)(3x) + 5(3x) + 2x(-4) + 5(-4)$

Simplify. → $6x^2 \quad +15x \quad -8x \quad -20$

Combine like terms. → $6x^2 + 7x - 20$

The same product may be found in the following manner.

Step 1	Step 2	Step 3	Step 4
$2x + 5$	$2x \; + 5$	$2x + 5$	$2x \quad + 5$
$3x - 4$	$3x$	-4	$3x \quad - 4$
	$6x^2 + 15x$	$-8x - 20$	$6x^2 + 15x$
			$- 8x - 20$
			$6x^2 + 7x - 20$
Write one binomial below the other.	Multiply $3x(2x + 5)$.	Multiply $-4(2x + 5)$.	Add.

A convenient arrangement for multiplying binomials →

$$\begin{array}{r} 2x \quad + 5 \\ 3x \quad - 4 \\ \hline 6x^2 + 15x \\ - \;\; 8x - 20 \\ \hline 6x^2 + \;\; 7x - 20 \end{array}$$

Product: $3x(2x + 5)$ → $6x^2 + 15x$

Product: $-4(2x + 5)$ → $-8x - 20$

Sum of Products → $6x^2 + 7x - 20$

EXAMPLE 3 Multiply $(5x - 2y)(2x - y)$.

$2x - y = 2x - 1y$ ⟶ $\begin{array}{r} 5x \quad -2y \\ 2x \quad -1y \end{array}$

$2x(5x - 2y) = 10x^2 - 4xy$ ⟶ $10x^2 - 4xy$

$-1y(5x - 2y) = -5xy + 2y^2$ ⟶ $-5xy + 2y^2$

Combine like terms. ⟶ $10x^2 - 9xy + 2y^2$

EXAMPLE 4 Multiply $(2x + 3)(5x^2 - x - 2)$.

Write the trinomial above the binomial. ⟶ $\begin{array}{r} 5x^2 - 1x - 2 \\ 2x + 3 \end{array}$

Multiply.

$2x(5x^2 - 1x - 2)$ ⟶ $10x^3 - 2x^2 - 4x$

$+3(5x^2 - 1x - 2)$ ⟶ $+ 15x^2 - 3x - 6$

Combine like terms. ⟶ $10x^3 + 13x^2 - 7x - 6$

EXAMPLE 5 Multiply $(3x + 2)(3x - 2)$.

$\begin{array}{r} 3x + 2 \\ 3x - 2 \end{array}$

Multiply. $3x(3x + 2)$ ⟶ $9x^2 + 6x$

$-2(3x + 2)$ ⟶ $- 6x - 4$

$+6x - 6x = 0x = 0$ ⟶ $9x^2 + 0x - 4$

$9x^2 + 0x - 4 = 9x^2 - 4$ ⟶ **Thus,** $(3x + 2)(3x - 2) = 9x^2 - 4$.

EXAMPLE 6 Multiply $(3x + 5)(3x + 5)$.

Write in vertical form. ⟶ $\begin{array}{r} 3x + 5 \\ 3x + 5 \end{array}$

Multiply. $3x(3x + 5)$ ⟶ $9x^2 + 15x$

$+5(3x + 5)$ ⟶ $+ 15x + 25$

Combine like terms. ⟶ $9x^2 + 30x + 25$

Thus, $(3x + 5)(3x + 5) = 9x^2 + 30x + 25$.

ORAL EXERCISES

Multiply.

1. $4(5y^2 - 3y + 2)$ *$20y^2 - 12y + 8$*
2. $-2(3x^2 + 4x - 5)$ *$-6x^2 - 8x + 10$*
3. $-1(8x^2 - 6x + 7)$ *$-8x^2 + 6x - 7$*
4. $5x(2x^2 - x + 3)$
5. $-3y(2x + 5y - 1)$ *$-6xy - 15y^2 + 3y$*
6. $2x(3x - 2y + 4)$ *$6x^2 - 4xy + 8x$*

EXERCISES

PART A

Multiply.

1. $(2x - 5)(4x + 3)$ $8x^2 - 14x - 15$
2. $(3x + 2)(5x - 2)$ $15x^2 + 4x - 4$
3. $(4x + 5)(2x - 3)$ $8x^2 - 2x - 15$
4. $(5x - 3)(2x - 1)$ $10x^2 - 11x + 3$
5. $(2x - 7)(x + 5)$ $2x^2 + 3x - 35$
6. $(6x + 1)(x + 3)$ $6x^2 + 19x + 3$
7. $(2x + 3)(2x - 3)$ $4x^2 - 9$
8. $(3x - 4)(3x + 4)$ $9x^2 - 16$
9. $(5x + 4)(x^2 - 3x + 2)$ $5x^3 - 11x^2 - 2x + 8$
10. $(4x - 3)(2x^2 + 6x - 1)$ $8x^3 + 18x^2 - 22x + 3$
11. $(3x - 2)(3x^2 - x + 4)$ $9x^3 - 9x^2 + 14x - 8$
12. $(2x + 4)(4x^2 + 2x - 3)$ $8x^3 + 20x^2 + 2x - 12$
13. $(x + 5)(x + 5)$ $x^2 + 10x + 25$
14. $(x - 2)(x - 2)$ $x^2 - 4x + 4$
15. $(2x - 3)(2x - 3)$ $4x^2 - 12x + 9$
16. $(3x + 4)(3x + 4)$ $9x^2 + 24x + 16$

PART B

Multiply.

17. $(3x - 2y)(2x + 4y)$ $6x^2 + 8xy - 8y^2$
18. $(5x + 4y)(3x + 2y)$ $15x^2 + 22xy + 8y^2$
19. $(4x + y)(3x - y)$ $12x^2 - xy - y^2$
20. $(x - y)(5x - 2y)$ $5x^2 - 7xy + 2y^2$
21. $(2x + 3y)(2x - 3y)$ $4x^2 - 9y^2$
22. $(3x - 5y)(3x + 5y)$ $9x^2 - 25y^2$
23. $(3x - 2y)(4x + 5y + 3)$
24. $(2x + 5y)(x - y + 3)$
25. $(4x + y)(4x + y)$ $16x^2 + 8xy + y^2$
26. $(x - 2y)(x - 2y)$ $x^2 - 4xy + 4y^2$
27. $(3x - 4y)(3x - 4y)$ $9x^2 - 24xy + 16y^2$
28. $(2x + 3y)(2x + 3y)$ $4x^2 + 12xy + 9y^2$

EXAMPLE Multiply $3x\underbrace{(2x + 1)(5x - 4)}$.

Multiply the binomials first. → $3x(10x^2 - 3x - 4)$

Multiply by 3x. → $30x^3 - 9x^2 - 12x$

Multiply.

29. $4(2x - 5)(x + 3)$ $8x^2 + 4x - 60$
30. $-3(4x + 5)(2x - 3)$ $-24x^2 + 6x + 45$
31. $4x(3x + 2)(2x + 5)$ $24x^3 + 76x^2 + 40x$
32. $-2x(4x - 3)(5x + 2)$ $-40x^3 + 14x^2 + 12x$
33. $4(2x + 1)(2x - 1)$ $16x^2 - 4$
34. $3x(5x - 3)(5x + 3)$ $75x^3 - 27x$
35. $2(3x - 1)(3x - 1)$ $18x^2 - 12x + 2$
36. $2x(5x + 2)(5x + 2)$ $50x^3 + 40x^2 + 8x$
37. $2(5x - 3y)(2x + y)$ $20x^2 - 2xy - 6y^2$
38. $3(3x - 2y)(3x + 2y)$ $27x^2 - 12y^2$
39. $5(3x + 4y)(3x - 4y)$ $45x^2 - 80y^2$
40. $10(x + 4y)(x + 4y)$ $10x^2 + 80xy + 160y^2$

PART C

Multiply.

41. $(x + 5)(x + 5)(x + 5)$
42. $(x + y)^3$
43. $(x - y)^3$
44. $(2x + 3y - 5)^2$
45. $(5a - 2b + 4)^2$
46. $(8c - d + 4e)^2$
47. $(x^n + 3)(x^n - 5)$
48. $(3x^{2a} + 4)(x^{2a} + 1)$
49. $(x^c + y^{2c})(x^c - y^{2c})$

Common Factors

OBJECTIVE

- To factor out the greatest common factor (GCF) from a polynomial like $8x^3 - 10x^2 + 6x$

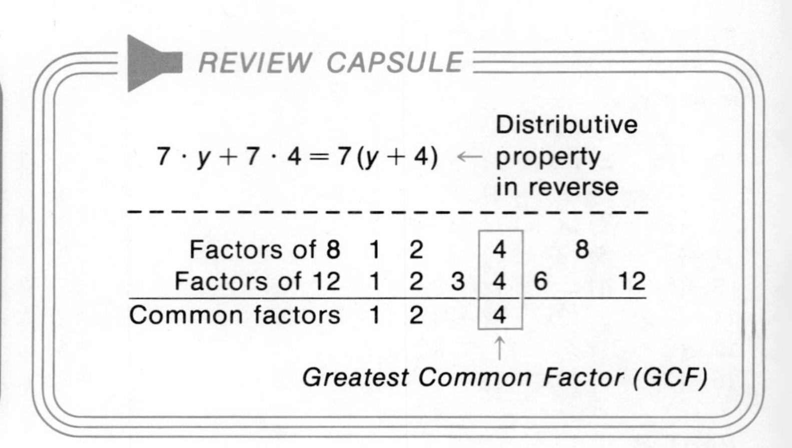

REVIEW CAPSULE

$7 \cdot y + 7 \cdot 4 = 7(y + 4)$ ← Distributive property in reverse

Factors of 8	1	2		4		8	
Factors of 12	1	2	3	4	6		12
Common factors	1	2		4			

↑ *Greatest Common Factor (GCF)*

EXAMPLE 1 Find the GCF of x^3 and x^2.

$x^3 = x \cdot x^2$ ⟶
$x^2 = x \cdot x$ ⟶
Factors of *both* x^3 and x^2 ⟶

Factors of x^3	x	x^2	x^3
Factors of x^2	x	x^2	
Common factors	x	x^2	

↑ GCF of x^3 and x^2

Thus, x^2 is the GCF of x^3 and x^2.

EXAMPLE 2 Determine the missing factor.

$$8x + 12y = (?)(2x + 3y)$$

$$8x + 12y$$

4 is the GCF of $8x$ and $12y$. ⟶ $4(2x) + 4(3y)$

Use the distributive property. ⟶ $4(2x + 3y)$

Check by multiplying. ⟶ $4(2x + 3y) = 4(2x) + 4(3y)$
$= 8x + 12y$

Thus, the missing factor is 4.

EXAMPLE 3 Factor the GCF from $5x^3 + x^2$.

$$5x^3 + x^2$$

x^2 is the GCF of $5x^3$ and x^2. ⟶ $(x^2)(5x) + (x^2)(1)$

Check by multiplying. ⟶ $x^2(5x + 1)$

The greatest common factor of a polynomial is the greatest common factor of all of its terms.

EXAMPLE 4 Factor the GCF from $8x^3 - 4x^2 + 12x$.

Find the GCF of the numbers. → The GCF of 8 and −4 and 12 is 4.
Find the GCF of the variables. → The GCF of x^3 and x^2 and x is x.

Find the GCF of the terms. → GCF of $8x^3$ and $-4x^2$ and $12x$ is $4x$

$$8x^3 \quad -4x^2 \quad +12x$$
$$4x(2x^2) + (4x)(-1x) + (4x)(3)$$

4x is the GCF of the terms. → $4x(2x^2 - 1x + 3)$

Thus, $8x^3 - 4x^2 + 12x = 4x(2x^2 - x + 3)$.

EXAMPLE 5 Factor the GCF from $4x^4 + 6x^3 - 8x^2$.

$$4x^4 \quad +6x^3 \quad -8x^2$$

The GCF of $4x^4$ and $6x^3$ and $-8x^2$ is $2x^2$. → $2x^2(2x^2) + (2x^2)(3x) + (2x^2)(-4)$
Use the distributive property. $2x^2(2x^2 + 3x - 4)$

Thus, $4x^4 + 6x^3 - 8x^2 = 2x^2(2x^2 + 3x - 4)$.

EXAMPLE 6 Factor −1 from $-5x^3 + 3x^2 - x + 2$.

$$-5x^3 \quad +3x^2 \quad -1x \quad +2$$

Factor −1 from each term. → $-1(5x^3) - 1(-3x^2) - 1(+1x) - 1(-2)$
$-x = -1(x) = -1(1x)$ → $-1(5x^3 - 3x^2 + 1x - 2)$

Thus, $-5x^3 + 3x^2 - x + 2 = -1(5x^3 - 3x^2 + x - 2)$.

ORAL EXERCISES

Find the GCF of the polynomial.

1. $4 + 8a$ 4
2. $6y + 6$ 6
3. $12a - 9b$ 3
4. $8x^2 - 6x + 10$ 2
5. $x^3 + x^2 + x$ x
6. $5x^4 - 2x^2 + 7x$ x
7. $x^4 + x^3 + x^2$ x^2
8. $5x^3 - 3x^2$ x^2
9. $3x^4 + 5x^3 + x$ x
10. $9x^3 - 6x^2 + 3x$ $3x$
11. $4x^5 - 6x^3 + 2x^2$ $2x^2$
12. $2ax - 5ay - 4a$ a

Determine the missing factor.

13. $10x + 5 = (?)(2x + 1)$ 5
14. $x^3 + x = (?)(x^2 + 1)$ x
15. $6x + 9y = (?)(2x + 3y)$ 3
16. $x^4 + x^2 = (?)(x^2 + 1)$ x^2

EXERCISES

PART A

Factor out the GCF.

1. $5t - 10$ $5(t-2)$
2. $8a + 8$ $8(a+1)$
3. $5x^2 - 15x + 10$ $5(x^2 - 3x + 2)$
4. $6x^2 + 3x - 9$ $3(2x^2 + x - 3)$
5. $6a^2 + 8a - 12$ $2(3a^2 + 4a - 6)$
6. $12x - 15y - 3$ $3(4x - 5y - 1)$
7. $4x^3 + x^2 - 7x$ $x(4x^2 + x - 7)$
8. $2x^5 - 5x^3 + 3x^2$ $x^2(2x^3 - 5x + 3)$
9. $2a^2 + a$ $a(2a+1)$
10. $7x^2 - x$ $x(7x-1)$
11. $c^2 + c$ $c(c+1)$
12. $n^2 - n$ $n(n-1)$
13. $7x^3 + 2x^2 + x$ $x(7x^2 + 2x + 1)$
14. $3x^5 - 5x^3 + x^2$ $x^2(3x^3 - 5x + 1)$
15. $3n^2 + 6n$ $3n(n+2)$
16. $10x^3 + 20x^2 - 15x$ $5x(2x^2 + 4x - 3)$
17. $10x^3 - 15x^2$ $5x^2(2x-3)$
18. $3x^4 - 6x^3 + 12x^2$ $3x^2(x^2 - 2x + 4)$
19. $4a^4 - 12a^3$ $4a^3(a-3)$
20. $8x^5 - 12x^4 + 16x^3$ $4x^3(2x^2 - 3x + 4)$

Factor out −1.

21. $-5a + 3b$ $-1(5a - 3b)$
22. $-2x^2 - 7m$ $-1(2x^2 + 7m)$
23. $-3x^2 + 4x - 2$ $-1(3x^2 - 4x + 2)$
24. $-x^2 + 9$ $-1(x^2 - 9)$
25. $-x - 4$ $-1(x+4)$
26. $-x^2 + x - 12$ $-1(x^2 - x + 12)$

PART B

EXAMPLE Factor the GCF from $6x^5y^2 + 9x^2y^3$.

Find the GCF of the numbers. →	GCF of	6	and	9	is	3.
Find the GCF of x^5 and x^2. →	GCF of	x^5	and	x^2	is	x^2.
Find the GCF of y^2 and y^3. →	GCF of	y^2	and	y^3	is	y^2.
Find the GCF of the terms. →	GCF of	$6x^5y^2$	and	$9x^2y^3$	is	$3x^2y^2$.

$3x^2y^2$ is the GCF.
Factor $3x^2y^2$ from each term. →

$$6x^5y^2 + 9x^2y^3$$
$$3x^2y^2(2x^3) + 3x^2y^2(3y)$$
$$3x^2y^2(2x^3 + 3y)$$

Factor out the GCF.

27. $3x^2y + 6xy$ $3xy(x+2)$
28. $14x^2y - 21xy^2$ $7xy(2x - 3y)$
29. $x^2y^4 - x^3y^3 + x^4y^2$ $x^2y^2(y^2 - xy + x^2)$
30. $ax^3 + a^2x^2 + a^3x$ $ax(x^2 + ax + a^2)$
31. $4ay^2 + 6a^2y + 8ay$
32. $15x^2y - 35x^3y^2 + 20x^4y^3$
33. $44x^3y - 100xy^3$ $4xy(11x^2 - 25y^2)$
34. $8x^2y^2 - 16x^3y^3$ $8x^2y^2(1 - 2xy)$

PART C

Factor out the GCF.

35. $x^{a+6} + x^4$
36. $x^{5c} + x^{3c}$
37. $x^{n+5} + x^{n+2}$
38. $x^{2a} + x^{a+3}$

Factoring into Binomials

OBJECTIVE

- To factor trinomials like $2x^2 - 5x - 3$ and $3x^2 + 5xy - 2y^2$ into two binomials

$$\begin{array}{lrl}
\text{Binomials} \rightarrow & 3x + 2 & \leftarrow \text{a factor} \\
\searrow & 2x - 5 & \leftarrow \text{a factor} \\
\hline
 & 6x^2 + 4x \qquad & \\
 & -15x - 10 & \\
\hline
\text{Trinomial} \rightarrow & 6x^2 - 11x - 10 & \leftarrow \text{the product}
\end{array}$$

EXAMPLE 1 Find two binomials whose product is $5x^2 + 11x + 2$.

Write the problem in vertical form.

$$\begin{array}{rl}
\underline{?}\,x + \underline{?} & \leftarrow \text{binomial factor} \\
\underline{?}\,x + \underline{?} & \leftarrow \text{binomial factor} \\
\hline
5x^2 + 11x + 2 & \leftarrow \text{product}
\end{array}$$

There are *two* possibilities.

$$\begin{array}{c}
5x + 2 \\
\underline{1x + 1}
\end{array}
\qquad
\begin{array}{c}
5x + 1 \\
\underline{1x + 2}
\end{array}$$

$$\begin{array}{ll}
5x & +2 \\
1x & +1 \\
\hline
5x^2 + __ & +2
\end{array}
\qquad
\begin{array}{ll}
5x & +1 \\
1x & +2 \\
\hline
5x^2 + __ & +2
\end{array}$$

Both products have $5x^2$ and $+2$.

Add the *cross products* to find the middle term, $+11x$.

Multiply to find which gives the correct middle term.

First cross product →
Second cross product →
Sum of cross products →

$5x \quad +2$ $1x \quad +1$	$5x \quad +1$ $1x \quad +2$
$+2x$	$+1x$
$+5x$	$+10x$
$+7x$	$+11x$
Not correct middle term	Correct middle term

Check by multiplying $(5x + 1)(x + 2)$. → **Thus,** the two binomials whose product is $5x^2 + 11x + 2$ are $5x + 1$ and $x + 2$.

EXAMPLE 2 Factor $2x^2 + 5x + 3$ into two binomials.

There are two possibilities.

$$\begin{array}{ll}
2x & +1 \\
1x & +3 \\
\hline
2x^2 + __ & +3
\end{array}
\qquad
\begin{array}{ll}
2x & +3 \\
1x & +1 \\
\hline
2x^2 + __ & +3
\end{array}$$

$2x \quad +1$ $1x \quad +3$	$2x \quad +3$ $1x \quad +1$
$+1x$	$+3x$
$+6x$	$+2x$
$+7x$	$+5x$
Not correct middle term	Correct middle term

Sum of cross products →

$+5x$ is middle term of $2x^2 + 5x + 3$. →

Check by multiplying. → **Thus,** $2x^2 + 5x + 3 = (2x + 3)(x + 1)$.

EXAMPLE 3 Factor $3x^2 - 8x + 5$ into two binomials.

Use $3x$ and x to give $3x^2$, -5 and -1 to give $+5$.

Use *negatives* so that middle term will be negative.

There are *two* possibilities.

$$\begin{array}{c} 3x - 1 \\ \underline{1x - 5} \end{array} \qquad \begin{array}{c} 3x - 5 \\ \underline{1x - 1} \end{array}$$

Add the cross products to find the middle term, $-8x$.

$$\begin{array}{cc} 3x & -1 \\ 1x & -5 \\ \hline 3x^2 & +5 \end{array} \qquad \begin{array}{cc} 3x & -5 \\ 1x & -1 \\ \hline 3x^2 & +5 \end{array}$$

Sum of cross products →

$$\begin{array}{c} 3x \times -1 \\ \underline{1x \times -5} \\ -1x \\ \underline{-15x} \\ -16x \end{array} \qquad \begin{array}{c} 3x \times -5 \\ \underline{1x \times -1} \\ -5x \\ \underline{-3x} \\ -8x \end{array}$$

Not correct middle term | Correct middle term

Thus, $3x^2 - 8x + 5 = (3x - 5)(x - 1)$.

EXAMPLE 4 Factor $5x^2 + 9x - 2$ into two binomials.

Factors of -2, third term, must be a *positive* and a *negative*. Use $(+2)(-1)$, or $(-2)(+1)$.

All four products have $5x^2$ and -2 for first and third terms.

$$\begin{array}{c} 5x + 2 \\ \underline{1x - 1} \\ +2x \\ \underline{-5x} \\ -3x \end{array} \quad \begin{array}{c} 5x - 2 \\ \underline{1x + 1} \\ -2x \\ \underline{+5x} \\ +3x \end{array} \quad \begin{array}{c} 5x + 1 \\ \underline{1x - 2} \\ +1x \\ \underline{-10x} \\ -9x \end{array} \qquad \begin{array}{c} 5x - 1 \\ \underline{1x + 2} \\ -1x \\ \underline{+10x} \\ +9x \end{array}$$

$+9x$ is middle term of $5x^2 + 9x - 2$. → Not correct middle term | Correct middle term

Thus, $5x^2 + 9x - 2 = (5x - 1)(x + 2)$.

EXAMPLE 5 Factor $3x^2 - 2x - 5$ into two binomials.

All four products have $3x^2$ and -5 for first and third terms. Use $(-5)(+1)$, or $5(-1)$ for -5.

There are *four* possibilities.

$$\begin{array}{c} 3x - 1 \\ \underline{1x + 5} \\ -1x \\ \underline{+15x} \\ +14x \end{array} \quad \begin{array}{c} 3x + 1 \\ \underline{1x - 5} \\ +1x \\ \underline{-15x} \\ -14x \end{array} \quad \begin{array}{c} 3x - 5 \\ \underline{1x + 1} \\ -5x \\ \underline{+3x} \\ -2x \end{array} \quad \begin{array}{c} 3x + 5 \\ \underline{1x - 1} \end{array}$$

Sum of cross products →

$-2x$ is middle term of $3x^2 - 2x - 5$. → Not correct middle term | Correct middle term

Check by multiplying $(3x - 5)(x + 1)$.

Thus, $3x^2 - 2x - 5 = (3x - 5)(x + 1)$.

ORAL EXERCISES

Add the cross products and give the middle term of the product.

1. $\begin{array}{c} 5x \quad -1 \\ x \quad -3 \end{array}$ $-16x$

2. $\begin{array}{c} 3x \quad +5 \\ x \quad +1 \end{array}$ $+8x$

3. $\begin{array}{c} 2x \quad -7 \\ x \quad +1 \end{array}$ $-5x$

4. $\begin{array}{c} 2x \quad +1 \\ x \quad -5 \end{array}$ $-9x$

EXERCISES

PART A

Factor into two binomials.

1. $3x^2 + 5x + 2$
2. $2x^2 + 11x + 5$ *$(2x + 1)(x + 5)$*
3. $5x^2 + 12x + 7$ *$(5x + 7)(x + 1)$*
4. $2x^2 - 7x + 3$
5. $7x^2 - 9x + 2$
6. $3x^2 - 22x + 7$
7. $2x^2 + 3x - 2$
8. $5x^2 + 2x - 3$
9. $3x^2 + x - 2$
10. $3x^2 - 8x - 3$
11. $7x^2 - 5x - 2$
12. $5x^2 - 2x - 3$
13. $x^2 - 2x - 3$
14. $x^2 + 4x - 5$
15. $x^2 - 6x - 7$
16. $5x^2 + 16x + 3$
17. $7x^2 + 22x + 3$
18. $x^2 + 14x + 13$
19. $7x^2 - 12x + 5$
20. $5x^2 - 36x + 7$
21. $x^2 - 18x + 17$
22. $5x^2 + 14x - 3$
23. $7x^2 + 4x - 11$
24. $x^2 + 10x - 11$
25. $5x^2 - 54x - 11$ *$(5x + 1)(x - 11)$*
26. $3x^2 - 10x - 13$ *$(3x - 13)(x + 1)$*
27. $x^2 - 18x - 19$ *$(x - 19)(x + 1)$*

PART B

EXAMPLE Factor $3x^2 - 2xy - 5y^2$ into two binomials.

There are four possibilities. All four products have $3x^2$ and $-5y^2$ for first and third terms.

$3x + 1y$	$3x - 1y$	$3x + 5y$	$3x - 5y$
$1x - 5y$	$1x + 5y$	$1x - 1y$	$1x + 1y$
$+1xy$	$-1xy$	$+5xy$	$-5xy$
$-15xy$	$+15xy$	$-3xy$	$+3xy$
$-14xy$	$+14xy$	$+2xy$	$-2xy$
Not correct middle term			Correct middle term

$-2xy$ is middle term of $3x^2 - 2xy - 5y^2$.

Check by multiplying. **Thus,** $3x^2 - 2xy - 5y^2 = (3x - 5y)(x + y)$.

Factor into two binomials.

28. $3x^2 + 7xy + 2y^2$
29. $5x^2 - 8xy + 3y^2$ *$(5x - 3y)(x - y)$*
30. $x^2 - 6xy + 5y^2$ *$(x - 5y)(x - y)$*
31. $2x^2 + 5xy - 3y^2$
32. $5x^2 + 2xy - 7y^2$
33. $x^2 + 2xy - 3y^2$
34. $5x^2 - 3xy - 2y^2$
35. $3x^2 - xy - 2y^2$
36. $x^2 - 4xy - 5y^2$
37. $3x^2 + 8xy - 3y^2$ *$(3x - y)(x + 3y)$*
38. $7x^2 - 4xy - 3y^2$ *$(7x + 3y)(x - y)$*
39. $2x^2 + xy - 3y^2$ *$(2x + 3y)(x - y)$*

PART C

Factor into two binomials.

40. $x^{2m} + 4x^m + 3$
41. $2x^{2m} + 9x^m - 5$
42. $2x^{4m} + 5x^{2m} - 7$
43. $x^{2a} + 3x^a y^b + 2y^{2b}$
44. $2x^{4a} - 7x^{2a}y^b + 3y^{2b}$
45. $3x^{2a} - 8x^a y^{3b} - 3y^{6b}$

Factoring Trinomials

OBJECTIVE

- To factor trinomials like $6x^2 + x - 15$ into two binomials

REVIEW CAPSULE

Factor $10x^2$ *two* different ways. → $10x^2 = (10x)(1x)$ → $10x^2 = (5x)(2x)$

Factor -6 *four* different ways.

$(-6)(+1)$ $(+6)(-1)$ $(-3)(+2)$ $(+3)(-2)$

EXAMPLE 1 Factor $15x^2 - 7x - 2$ into two binomials.

$15x^2 - 7x - 2$: $15x^2$ → $(15x)(1x)$ or $(5x)(3x)$; -2 → $(-2)(1)$ or $(2)(-1)$

Use a factor from each to form binomials.

There are *eight* possibilities.

$15x + 2$ $1x - 1$	$15x - 2$ $1x + 1$	$15x + 1$ $1x - 2$	$15x - 1$ $1x + 2$
$5x + 2$ $3x - 1$	$5x - 2$ $3x + 1$	$5x + 1$ $3x - 2$	$5x - 1$ $3x + 2$

All eight products have $15x^2$ and -2 for first and third terms.

Check each for the correct middle term.

Add the cross products of each possibility *until* you get the correct middle term.

$-7x$ is the middle term of $15x^2 - 7x - 2$. →

$$\begin{array}{r} 5x + 1 \\ \underline{3x - 2} \\ +3x \\ \underline{-10x} \\ -7x \end{array}$$

↑ Correct middle term

Thus, $15x^2 - 7x - 2 = (5x + 1)(3x - 2)$.

EXAMPLE 2 Factor $3x^2 + x - 10$ into two binomials.

There are eight possibilities.
Use $(+10)(-1)$ and $(-10)(+1)$.
Also, use $(+5)(-2)$ and $(-5)(+2)$.

$3x + 10$ $1x - 1$	$3x - 1$ $1x + 10$	$3x + 1$ $1x - 10$	$3x - 10$ $1x + 1$
$3x + 5$ $1x - 2$	$3x - 2$ $1x + 5$	$3x + 2$ $1x - 5$	$3x - 5$ $1x + 2$ $-5x$ $+6x$ $+1x$

Add the cross products of each possibility until you get the correct middle term, $+1x$.

Thus, $3x^2 + x - 10 = (3x - 5)(x + 2)$.

ORAL EXERCISES

Add the cross products and give the middle term of the product.

1. $\begin{array}{r} 6x + 1 \\ \underline{x - 5} \\ -29x \end{array}$
2. $\begin{array}{r} 6x - 1 \\ \underline{x + 5} \\ +29x \end{array}$
3. $\begin{array}{r} 6x + 5 \\ \underline{x - 1} \\ -x \end{array}$
4. $\begin{array}{r} 3x + 5 \\ \underline{2x - 1} \\ +7x \end{array}$
5. $\begin{array}{r} 3x - 5 \\ \underline{2x + 1} \\ -7x \end{array}$

EXERCISES

PART A

Factor into two binomials.

1. $2a^2 + a - 10$
2. $2x^2 - 11x + 14$ $(2x - 7)(x - 2)$
3. $2x^2 + 9x + 10$ $(2x + 5)(x + 2)$
4. $6x^2 - 13x + 2$
5. $8n^2 + 10n + 3$
6. $4t^2 - 9t + 5$
7. $3a^2 - 5a - 8$
8. $5c^2 + 4c - 9$
9. $5x^2 - 11x + 6$
10. $6c^2 + 11c + 5$
11. $4x^2 + 8x - 5$
12. $10n^2 - 9n - 1$
13. $7x^2 - 3x - 10$
14. $3x^2 + 11x + 6$
15. $3x^2 - 13x + 10$
16. $10c^2 - 11c + 3$
17. $15x^2 + 4x - 3$
18. $4x^2 + 12x + 5$
19. $3x^2 - 14x + 15$
20. $5x^2 + 28x + 15$
21. $7n^2 + 13n + 6$
22. $8x^2 - x - 7$ $(8x + 7)(x - 1)$
23. $15a^2 - 14a + 3$ $(5a - 3)(3a - 1)$
24. $9x^2 + 9x + 2$ $(3x + 2)(3x + 1)$

Factor into two binomials.

25. $n^2 - 5n - 24$
26. $x^2 - 9x + 8$ $(x - 8)(x - 1)$
27. $n^2 - n - 12$ $(n - 4)(n + 3)$
28. $a^2 + 4a - 12$
29. $x^2 + 8x + 15$
30. $a^2 - 9a + 18$
31. $x^2 - 7x - 18$
32. $x^2 - 13x + 12$
33. $n^2 + 2n - 24$
34. $c^2 - 9c - 36$ $(c - 12)(c + 3)$
35. $x^2 + 11x + 10$ $(x + 10)(x + 1)$
36. $a^2 + 13a + 36$ $(a + 9)(a + 4)$

PART B

Factor into two binomials.

37. $10x^2 + 23x + 12$
38. $15n^2 + 7n - 4$ $(5n + 4)(3n - 1)$
39. $4c^2 - 15c + 9$ $(4c - 3)(c - 3)$
40. $9a^2 + 18a + 8$ $(3a + 2)(3a + 4)$
41. $6x^2 - 11x - 10$ $(3x + 2)(2x - 5)$
42. $12n^2 - 17n + 6$ $(4n - 3)(3n - 2)$

Factor into two binomials.

43. $2a^2 + ab - 6b^2$
44. $3x^2 + 11xy - 4y^2$ $(3x - y)(x + 4y)$
45. $6m^2 + 7mn - 3n^2$ $(3m - n)(2m + 3n)$
46. $5x^2 + 4xy - 12y^2$
47. $7a^2 + 18ab + 8b^2$
48. $6y^2 + yz - 12z^2$
49. $15r^2 - 7rs - 4s^2$
50. $2r^2 - 7rs - 4s^2$
51. $2x^2 + xy - 10y^2$
52. $4a^2 + 12ab + 5b^2$
53. $2x^2 - xy - 6y^2$
54. $15b^2 - 14bc + 3c^2$
55. $10x^2 - 11xy + 3y^2$ $(5x - 3y)(2x - y)$
56. $9r^2 + 9rs - 10s^2$ $(3r + 5s)(3r - 2s)$
57. $8a^2 - 6ab - 9b^2$ $(4a + 3b)(2a - 3b)$

PART C

Factor into two binomials.

58. $6x^{2m} + 19x^m + 15$
59. $9x^{4m} + 24x^{2m} - 20$
60. $8x^{6m} - 2x^{3m} - 21$
61. $5x^{2a} + 23x^ay^b + 12y^{2b}$
62. $8x^{4a} - 26x^{2a}y^{2b} + 15y^{4b}$
63. $12x^{6a} - 8x^{3a}y^{2b} - 15y^{4b}$
64. $x^{2n+4} + 7x^{n+2} + 12$
65. $x^{2n-2} + 3x^{n-1} - 10$
66. $x^{8n+6} - 2x^{4n+3} - 3$
67. $x^{2a+6} + 2x^{a+3}y^{a-2} + y^{2a-4}$
68. $2x^{2a+2} + 5x^{a+1}y^a + 3y^{2a}$

Special Products and Factoring

OBJECTIVES

- To multiply binomials like $(a + b)(a - b)$ and $(x + y)^2$
- To factor polynomials like $a^2 - b^2$ and $x^2 + 2xy + y^2$

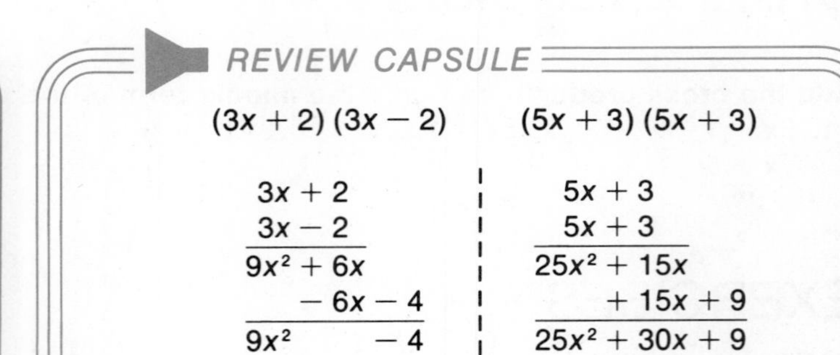

EXAMPLE 1 Multiply $(3x + 5)(3x - 5)$.

$$\begin{array}{r} 3x + 5 \\ 3x - 5 \\ \hline 9x^2 + 15x \\ - 15x - 25 \\ \hline 9x^2 + 0x - 25 \end{array}$$

Middle term is $0x$, or 0. →

$9x^2 - 25$ is a *difference of squares.* → $(3x)^2 - (5)^2$

Thus, $(3x + 5)(3x - 5) = (3x)^2 - (5)^2$, or $9x^2 - 25$.

A special product →

$$(a + b)(a - b) = a^2 - b^2$$

Product of a sum and difference — Difference of squares

EXAMPLE 2 Multiply $(5x + 3)(5x - 3)$. Multiply $(3x - 4y)(3x + 4y)$.

Binomials in form $(a + b)(a - b)$ → $(5x + 3)(5x - 3)$ | $(3x - 4y)(3x + 4y)$

$(a + b)(a - b) = (a)^2 - (b)^2$ → $(5x)^2 - (3)^2$ | $(3x)^2 - (4y)^2$

$25x^2 - 9$ | $9x^2 - 16y^2$

EXAMPLE 3 Factor $n^2 - 25$ into two binomials.

$$n^2 - 25$$

Difference of squares → $(n)^2 - (5)^2$

Product of sum and difference → $(n + 5)(n - 5)$

Thus, $n^2 - 25 = (n + 5)(n - 5)$.

Factoring a difference of squares →

$$a^2 - b^2 = (a + b)(a - b)$$

Difference of Squares   Product of Sum and Difference

EXAMPLE 4 Factor $16c^2 - 49$ into two binomials.

Write $16c^2 - 49$ as the difference of squares. →
Product of sum and difference →

$$16c^2 - 49$$
$$(4c)^2 - (7)^2$$
$$(4c + 7)(4c - 7)$$

Thus, $16c^2 - 49 = (4c + 7)(4c - 7)$.

EXAMPLE 5 Factor $25x^4 - 9y^2$. Factor $1 - n^2$.

$(5x^2)^2 = 25x^4$

$$25x^4 - 9y^2 \qquad\qquad 1 - n^2$$
$$(5x^2)^2 - (3y)^2 \qquad\qquad (1)^2 - (n)^2$$
$$(5x^2 + 3y)(5x^2 - 3y) \qquad\qquad (1 + n)(1 - n)$$

EXAMPLE 6 Square the binomial $(4x + 5)^2$.

$(y)^2$ means $(y)(y)$. →
Distributive property →
Distributive property again →

$$\begin{aligned}(4x + 5)^2 &= (4x + 5)(4x + 5)\\ &= (4x + 5)(4x) + (4x + 5)(5)\\ &= (4x)(4x) + \underbrace{(5)(4x) + (4x)(5)} + (5)(5)\\ &= (4x)^2 + 2(4x)(5) + (5)^2\end{aligned}$$

$(4x)^2 + 2(4x)(5) + (5)^2$
$= 16x^2 + 40x + 25$ →

Thus, $(4x + 5)^2 = 16x^2 + 40x + 25$.

Squaring a binomial →

Square of a Binomial		Square of First term		Twice the Product		Square of Last Term
$(a + b)^2$	$=$	a^2	$+$	$2ab$	$+$	b^2

$a^2 + 2ab + b^2$ is a perfect square trinomial.

EXAMPLE 7 Square the binomial $(3x - 10)^2$.

Squaring a binomial →
Let $a = 3x$, $b = -10$. →
Simplify. →

$$(a + b)^2 = a^2 + \underbrace{2ab} + b^2$$
$$\begin{aligned}(3x - 10)^2 &= (3x)^2 + 2(3x)(-10) + (-10)^2\\ &= 9x^2 - 60x + 100\end{aligned}$$

Thus, $(3x - 10)^2 = 9x^2 - 60x + 100$.

EXAMPLE 8 Square the binomial $(5x + 3y)^2$.

$(a + b)^2 = a^2 + 2ab + b^2$
Let $a = 5x$, $b = 3y$.

$$\begin{aligned}(5x + 3y)^2 &= (5x)^2 + 2(5x)(3y) + (3y)^2\\ &= 25x^2 + 30xy + 9y^2\end{aligned}$$

Thus, $(5x + 3y)^2 = 25x^2 + 30xy + 9y^2$.

EXAMPLE 9 Determine whether $16x^2 + 24x + 9$ is a perfect square trinomial.

A perfect square trinomial is the square of a binomial.

$$16x^2 + 24x + 9$$
$$(4x)^2 + \_\_\_ + (3)^2$$
$$(4x + 3)^2$$

Check: $(4x + 3)^2 = (4x)^2 + 2(4x)(3) + (3)^2$
$= 16x^2 + 24x + 9$

Product gives correct middle term, $+24x$. →

Thus, $16x^2 + 24x + 9$ is a perfect square trinomial.

EXAMPLE 10 Factor $16x^2 + 24x + 9$ into two binomials.

$$16x^2 + 24x + 9$$

Perfect square trinomial → $(4x)^2 + 2(4x)(3) + (3)^2$

Square of binomial → $(4x + 3)(4x + 3)$

Thus, $16x^2 + 24x + 9 = (4x + 3)(4x + 3)$, or $(4x + 3)^2$.

Factoring a perfect square trinomial →

$a^2 + 2ab + b^2 =$	$(a + b)^2$
Perfect Square Trinomial	Square of a Binomial

EXAMPLE 11 Factor into two binomials.

	$49n^2 - 14n + 1$	$x^2 + 12xy^2 + 36y^4$
Perfect Square trinomial →	$(7n)^2 + 2(7n)(-1) + (-1)^2$	$(x)^2 + 2(x)(6y^2) + (6y^2)^2$
Square of binomial →	$(7n - 1)(7n - 1)$	$(x + 6y^2)(x + 6y^2)$

SUMMARY

Multiplying	**Factoring**
$(a + b)(a - b) = a^2 - b^2$	$a^2 - b^2 = (a + b)(a - b)$
$(a + b)^2 = a^2 + 2ab + b^2$	$a^2 + 2ab + b^2 = (a + b)^2$

ORAL EXERCISES

Determine if the expression is of the form $(a + b)(a - b)$.

1. $(5x + 6)(5x - 6)$ *yes*
2. $(7x + 4)(7x - 3)$ *no*
3. $(x + y^2)(x - y^2)$ *yes*
4. $(4x - 2)(4x + 2)$ *yes*

Give the binomial as a difference of squares, like $9x^2 - 16 = (3x)^2 - (4)^2$.

5. $100n^2 - 9$
6. $25 - 4y^2$ $(5)^2 - (2y)^2$
7. $4x^2 - 9y^2$ $(2x)^2 - (3y)^2$
8. $x^4 - y^6$ $(x^2)^2 - (y^3)^2$

EXERCISES

PART A

Multiply.

1. $(4x+5)(4x-5)$ *$16x^2-25$*
2. $(10y-7)(10y+7)$ *$100y^2-49$*
3. $(n+6)(n-6)$ *n^2-36*
4. $(5x+2y)(5x-2y)$ *$25x^2-4y^2$*
5. $(a-3b)(a+3b)$ *a^2-9b^2*
6. $(4n^2+1)(4n^2-1)$ *$16n^4-1$*

Factor into two binomials.

7. $4x^2-9$
8. $9a^2-100$ *$(3a+10)(3a-10)$*
9. c^2-1 *$(c+1)(c-1)$*
10. $25-16y^2$ *$(5+4y)(5-4y)$*
11. $49-b^2$ *$(7+b)(7-b)$*
12. $1-36n^2$ *$(1+6n)(1-6n)$*

Square the binomial.

13. $(2x+5)^2$
14. $(6n-1)^2$ *$36n^2-12n+1$*
15. $(3+4x)^2$ *$9+24x+16x^2$*
16. $(3x-4y)^2$ *$9x^2-24xy+16y^2$*
17. $(a+7b)^2$ *$a^2+14ab+49b^2$*
18. $(2x-y)^2$ *$4x^2-4xy+y^2$*

Determine whether the trinomial is a perfect square.

19. $25x^2+50x+16$ *no*
20. $9x^2-13x+4$ *no*
21. $4x^2-20xy+25y^2$ *yes*

Factor into two binomials.

22. $25x^2+20x+4$ *$(5x+2)^2$*
23. $9n^2-24n+16$ *$(3n-4)^2$*
24. $4x^2+4x+1$ *$(2x+1)^2$*
25. $x^2+8x+16$ *$(x+4)^2$*
26. n^2-6n+9 *$(n-3)^2$*
27. $1+20a+100a^2$ *$(1+10a)^2$*

PART B

Factor into two binomials.

28. $9a^2-25b^2$
29. $81x^2-y^2$ *$(9x+y)(9x-y)$*
30. x^2-16y^2 *$(x+4y)(x-4y)$*
31. $100x^2-49y^4$ *$(10x+7y^2)(10x-7y^2)$*
32. m^4-n^2 *$(m^2+n)(m^2-n)$*
33. $49c^2d^2-1$ *$(7cd+1)(7cd-1)$*

Factor into two binomials.

34. $x^2+16xy+64y^2$
35. $25c^2-20cd+4d^2$ *$(5c-2d)^2$*
36. $4a^2+4ab+b^2$ *$(2a+b)^2$*
37. $36x^2-60xy+25y^2$ *$(6x-5y)^2$*
38. $9a^2+12ab^2+4b^4$ *$(3a+2b^2)^2$*
39. $x^2y^2-2xy+1$ *$(xy-1)^2$*

PART C

Factoring differences and sums of cubes

$$a^3-b^3=(a-b)(a^2+ab+b^2)$$
$$a^3+b^3=(a+b)(a^2-ab+b^2)$$

EXAMPLE Factor $8x^3-27$.

Write as a difference of cubes. ⟶ $8x^3-27=(2x)^3-(3)^3$

$a^3-b^3=(a-b)(a^2+ab+b^2)$ ⟶ $=(2x-3)[(2x)^2+(2x)(3)+(3)^2]$

Simplify. ⟶ $=(2x-3)(4x^2+6x+9)$

Factor.

40. x^3-8
41. x^3+125
42. $27x^3-1$
43. $64y^3+1$
44. $125x^3-64y^3$
45. $8x^6+y^3$

Factor Theorem

$$\begin{array}{r} x^2 + 4x \ + 2 \\ x \ - 2 \\ \hline x^3 + 4x^2 + 2x \\ - 2x^2 - 8x - 4 \\ \hline x^3 + 2x^2 - 6x - 4 \end{array}$$

($x^2 + 4x + 2$ and $x - 2$: Factors; $x^3 + 2x^2 - 6x - 4$: Product)

Thus, the factors of $x^3 + 2x^2 - 6x - 4$ are $x - 2$ and $x^2 + 4x + 2$

Evaluate $x^3 + 2x^2 - 6x - 4$ for $x = 2$.

$$2^3 + 2(2)^2 - 6(2) - 4$$
$$8 + 8 - 12 - 4$$
$$0$$

Evaluate $(x - 2)(x^2 + 4x + 2)$ for $x = 2$.

$$(2 - 2)(x^2 + 4x + 2)$$
$$0(x^2 + 4x + 2)$$
$$0$$

Factor Theorem
$x - a$ is a factor of a polynomial if the value of the polynomial is zero for $x = a$.

PROBLEM

Is $x + 3$ a factor of $2x^2 + 5x - 9$?

Let $x + 3 = x - (-3)$,
then $x - a = x - (-3)$.
Use the factor theorem:

For $x = -3$

$$\begin{aligned} & 2x^2 + 5x - 9 \\ &= 2(-3)^2 + 5(-3) - 9 \\ &= 18 - 15 - 9 \\ &= -6 \end{aligned}$$

Thus, $x + 3$ is *not* a factor.

Is $x - 3$ a factor of $2x^2 - 3x - 9$?

Let $x - a = x - 3$.
Use the factor theorem:

For $x = 3$

$$\begin{aligned} & 2x^2 - 3x - 9 \\ &= 2(3)^2 - 3(3) - 9 \\ &= 18 - 9 - 9 \\ &= 0 \end{aligned}$$

Thus, $x - 3$ *is* a factor.

PROJECT

Use the factor theorem to determine whether the first expression is a factor of the second.

1. $x - 3;\ x^2 + 6x - 27$

2. $x + 4;\ x^2 - 3x - 28$

3. $x - 2;\ 3x^2 - 5x - 4$

4. $x + 1;\ 4x^2 + 7x + 3$

5. $x - 2;\ x^3 + 2x^2 - 9x + 2$

6. $x + 3;\ x^3 + 2x^2 - x + 9$

Quadratic Equations

OBJECTIVE

- To solve equations like $3x^2 + 5x - 2 = 0$ by factoring

REVIEW CAPSULE

Solve $2x - 5 = 0$.

$2x - 5 = 0$

Add 5 to each side. → $2x = 5$

Divide each side by 2. → $x = \frac{5}{2}$, or $2\frac{1}{2}$

EXAMPLE 1 Find the missing factor. What conclusion can you draw?

Multiplying by 0 gives 0.

$(3)(?) = 0 \rightarrow (3)(0) = 0$

$(?)(-5) = 0 \rightarrow (0)(-5) = 0$

$(0)(?) = 0 \rightarrow (0)(0) = 0$

If a product is 0, at least one of the factors must be 0.

If $(a)(b) = 0$, then $a = 0$ or $b = 0$.
If a product is 0, then at least one factor is 0.

EXAMPLE 2 Find the solution set of $(4x - 3)(x + 2) = 0$.

$(a)(b) = 0$, → $(4x - 3)(x + 2) = 0$

$a = 0$ or $b = 0$. → $4x - 3 = 0$ or $x + 2 = 0$

Solve each equation.

$4x = 3 \qquad x = -2$

$x = \frac{3}{4}$

Check both solutions. → $x = \frac{3}{4}$ | $x = -2$

$(4x - 3)(x + 2)$	0
$(4 \cdot \frac{3}{4} - 3)(\frac{3}{4} + 2)$	0
$(3 - 3)(2\frac{3}{4})$	
$(0)(2\frac{3}{4})$	
0	

$(4x - 3)(x + 2)$	0
$[4(-2) - 3](-2 + 2)$	0
$(-8 - 3)(-2 + 2)$	
$(-11)(0)$	
0	

The equation has *two* solutions. → **Thus,** $\{\frac{3}{4}, -2\}$ is the solution set.

EXAMPLE 3 Solve $3x^2 - 5x - 2 = 0$.

The right side is 0.
Factor the left side. →

$$3x^2 - 5x - 2 = 0$$
$$(3x + 1)(x - 2) = 0$$

Set each factor equal to 0. →
Solve each equation.

$$3x + 1 = 0 \quad \text{or} \quad x - 2 = 0$$
$$3x = -1 \qquad\qquad x = 2$$
$$x = -\frac{1}{3}$$

Check *both* solutions. →

$x = -\frac{1}{3}$

$3x^2 - 5x - 2$	0
$3(-\frac{1}{3})^2 - 5(-\frac{1}{3}) - 2$	0
$3(\frac{1}{9}) + \frac{5}{3} - 2$	
$\frac{1}{3} + \frac{5}{3} - \frac{6}{3}$	
0	

$x = 2$

$3x^2 - 5x - 2$	0
$3(2)^2 - 5(2) - 2$	0
$3(4) - 10 - 2$	
$12 - 10 - 2$	
0	

Thus, the solutions are $-\frac{1}{3}$ and 2.

Quadratic equation →

a, b, and c are numbers.

> $ax^2 + bx + c = 0$, $a \neq 0$, is a *quadratic equation* in standard form.
>
> To solve: First, factor.
> Second, set each factor equal to 0.
> Third, solve each equation.

EXAMPLE 4 Solve $x^2 + 5x + 6 = 0$.

Quadratic equation →
Factor. →
Set each factor equal to 0. →
Solve each equation. →

$$x^2 + 5x + 6 = 0$$
$$(x + 3)(x + 2) = 0$$
$$x + 3 = 0 \quad \text{or} \quad x + 2 = 0$$
$$x = -3 \qquad\qquad x = -2$$

Thus, the solutions are -3 and -2.

EXAMPLE 5 Find the solution set of $4x^2 - 25 = 0$.

Factor $4x^2 - 25$ as a difference of squares: $(2x)^2 - (5)^2$.
Set each factor equal to 0.
Then solve each equation.

$$(2x + 5)(2x - 5) = 0$$
$$2x + 5 = 0 \quad \text{or} \quad 2x - 5 = 0$$
$$2x = -5 \qquad\qquad 2x = 5$$
$$x = -2\tfrac{1}{2} \qquad\qquad x = 2\tfrac{1}{2}$$

Thus, the solution set is $\{-2\frac{1}{2}, 2\frac{1}{2}\}$.

EXAMPLE 6 Solve $5x^2 + 15x = 0$.

$5x$ is the GCF of $5x^2$ and $15x$. →
Set each factor equal to 0. →
Solve each equation. →

$$5x(x + 3) = 0$$
$$5x = 0 \quad \text{or} \quad x + 3 = 0$$
$$x = 0 \qquad\qquad x = -3$$

Thus, the solutions are 0 and -3.

EXAMPLE 7 Find the solution set of $x^2 - 10x + 25 = 0$.

$x^2 - 10x + 25$ is a perfect square trinomial.

$$x^2 - 10x + 25 = 0$$
$$(x - 5)(x - 5) = 0$$
$$x - 5 = 0 \quad \text{or} \quad x - 5 = 0$$
$$x = 5 \qquad\qquad x = 5$$

There is only *one* solution.

Thus, the solution set is $\{5\}$.

EXAMPLE 8 Solve $x^2 + 4x = 21$.

Write in standard form.
Add -21 to each side.

$$x^2 + 4x = 21$$
$$x^2 + 4x - 21 = 0$$
$$(x - 3)(x + 7) = 0$$
$$x - 3 = 0 \quad \text{or} \quad x + 7 = 0$$
$$x = 3 \qquad\qquad x = -7$$

Thus, the solutions are 3 and -7.

EXAMPLE 9 Find the solution set of $7x + 6x^2 = 3$.

Write in standard form. Arrange terms in descending order. ⟶

$$7x + 6x^2 = 3$$
$$7x + 6x^2 - 3 = 0$$
$$6x^2 + 7x - 3 = 0$$
$$(2x + 3)(3x - 1) = 0$$
$$2x + 3 = 0 \quad \text{or} \quad 3x - 1 = 0$$
$$2x = -3 \qquad\qquad 3x = 1$$
$$x = -\frac{3}{2} \qquad\qquad x = \frac{1}{3}$$

Thus, the solution set is $\{-\frac{3}{2}, \frac{1}{3}\}$.

ORAL EXERCISES

Solve.

1. $(x - 7)(x - 7) = 0$ *7*
2. $(x - 3)(x + 3) = 0$ *3; −3*
3. $(x + 6)(x - 2) = 0$ *−6; 2*

Give in standard form.

4. $12 + x^2 - 7x = 0$ *$x^2 - 7x + 12 = 0$*
5. $x^2 - x = 6$ *$x^2 - x - 6 = 0$*
6. $x^2 + 9 = 6x$ *$x^2 - 6x + 9 = 0$*
7. $10x + 3x^2 = 8$ *$3x^2 + 10x - 8 = 0$*
8. $5 + 2x^2 = 7x$ *$2x^2 - 7x + 5 = 0$*
9. $x^2 = 9x - 14$ *$x^2 - 9x + 14 = 0$*

True or false?

10. If $x(x - 2) = 0$, then $x = 0$ or $x - 2 = 0$. *T*
11. If $x(x - 3) = 5$, then $x = 5$ or $x - 3 = 5$. *F*
12. $4x^2 - 8x = 0$ is a quadratic equation. *T*
13. $5x - 10 = 0$ is a quadratic equation. *F*

EXERCISES

PART A

Find the solution set.

1. $(x-2)(x-5)=0$ $\{2,5\}$
2. $(x+3)(x-6)=0$ $\{-3,6\}$
3. $(x+4)(x+1)=0$ $\{-4,-1\}$
4. $(2x-1)(x-3)=0$ $\{\frac{1}{2},3\}$
5. $(3x-5)(x+2)=0$ $\{1\frac{2}{3},-2\}$
6. $(2x+3)(5x+2)=0$
7. $3x(x-4)=0$ $\{0,4\}$
8. $x(x+5)=0$ $\{0,-5\}$
9. $(x-6)(x-6)=0$ $\{6\}$

Solve and check.

10. $x^2-3x+2=0$ $1;\ 2$
11. $x^2+2x-15=0$ $-5;\ 3$
12. $x^2+6x+8=0$ $-4;\ -2$

Solve.

13. $2x^2-9x+4=0$ $\frac{1}{2};\ 4$
14. $3x^2-8x-3=0$ $-\frac{1}{3};\ 3$
15. $3x^2+x-2=0$ $\frac{2}{3};\ -1$
16. $x^2-4=0$ $2;\ -2$
17. $x^2-36=0$ $6;\ -6$
18. $x^2-16=0$ $4;\ -4$
19. $2x^2-10x=0$ $0;\ 5$
20. $3x^2+12x=0$ $0;\ -4$
21. $x^2-3x=0$ $0;\ 3$
22. $x^2+6x+9=0$ -3
23. $x^2-8x+16=0$ 4
24. $x^2+12x+36=0$ -6
25. $x^2-13x+40=0$ $5;\ 8$
26. $x^2-x-42=0$ $7;\ -6$
27. $x^2+15x+50=0$ $-5;\ -10$
28. $8x^2+2x-1=0$ $\frac{1}{4};\ -\frac{1}{2}$
29. $12x^2+7x+1=0$ $-\frac{1}{3};\ -\frac{1}{4}$
30. $10x^2-7x+1=0$ $\frac{1}{5};\ \frac{1}{2}$

Find the solution set.

31. $x^2+x=12$ $\{-4,3\}$
32. $x^2-x=6$ $\{-2,3\}$
33. $x^2=9$ $\{-3,3\}$
34. $x^2+8=6x$ $\{2,4\}$
35. $x^2-15=2x$ $\{-3,5\}$
36. $x^2=6x$ $\{0,6\}$

PART B

Solve.

37. $2x^2-7x+3=0$ $\frac{1}{2};\ 3$
38. $3x^2+5x-2=0$ $\frac{1}{3};\ -2$
39. $2x^2+3x+1=0$ $-\frac{1}{2};\ -1$

Find the solution set.

40. $2x^2-7x+6=0$ $\{1\frac{1}{2},2\}$
41. $3x^2+4x-15=0$ $\{1\frac{2}{3},-3\}$
42. $2x^2+9x+7=0$ $\{-3\frac{1}{2},-1\}$
43. $9x^2-25=0$ $\{-1\frac{2}{3},1\frac{2}{3}\}$
44. $16x^2-1=0$ $\{-\frac{1}{4},\frac{1}{4}\}$
45. $25x^2-36=0$ $\{-1\frac{1}{5},1\frac{1}{5}\}$
46. $3x^2-2x=0$ $\{0,\frac{2}{3}\}$
47. $4x^2+7x=0$ $\{-1\frac{3}{4},0\}$
48. $2x^2-5x=0$ $\{2\frac{1}{2},0\}$
49. $x^2-18x+81=0$ $\{9\}$
50. $x^2+22x+121=0$ $\{-11\}$
51. $x^2-14x+49=0$ $\{7\}$
52. $6x^2-19x+8=0$ $\{2\frac{2}{3},\frac{1}{2}\}$
53. $8x^2+10x-25=0$ $\{1\frac{1}{4},-2\frac{1}{2}\}$
54. $12x^2+11x-15=0$ $\{-1\frac{2}{3},\frac{3}{4}\}$

Solve.

55. $x^2=7x-10$ $2;\ 5$
56. $x^2=12+x$ $-3;\ 4$
57. $x^2=6-5x$ $-6;\ 1$
58. $2x^2=4-7x$ $\frac{1}{2};\ -4$
59. $4x^2=5x+6$ $-\frac{3}{4};\ 2$
60. $3x^2=11x-6$ $\frac{2}{3};\ 3$

PART C

Find the solution set.

61. $(x+3)(x+1)(x-2)(x-4)=0$
62. $(x^2-7x+12)(x^2-3x+2)=0$
63. $(x^2-16)(x^2-100)=0$
64. $(x^2-4x+4)(x^2+10x+25)=0$
65. $(x^2-4x)(x^2+5x)=0$
66. $(x^2+5x-14)(x^2-6x-7)=0$
67. $x^4-13x^2+36=0$
68. $x^4-24x^2-25=0$

Quadratic Inequalities

OBJECTIVE

- To find and graph the solution sets of inequalities like $x^2 + 2x - 8 < 0$ and $-x^2 + 5x \geq 0$

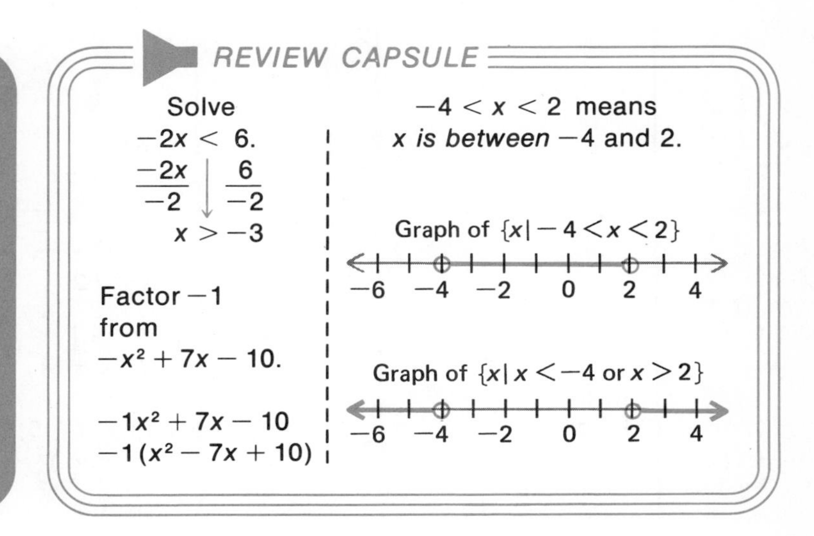

EXAMPLE 1 Find and graph the solution set of $(x + 3)(x - 4) < 0$.

First, solve $(x + 3)(x - 4) = 0$.

$x + 3 = 0$ or $x - 4 = 0$

$x = -3 \qquad x = 4$

The solutions are -3 and 4.

Graph the solutions of $(x + 3)(x - 4) = 0$ on a number line.

-3 and 4 divide the number line into 3 parts.

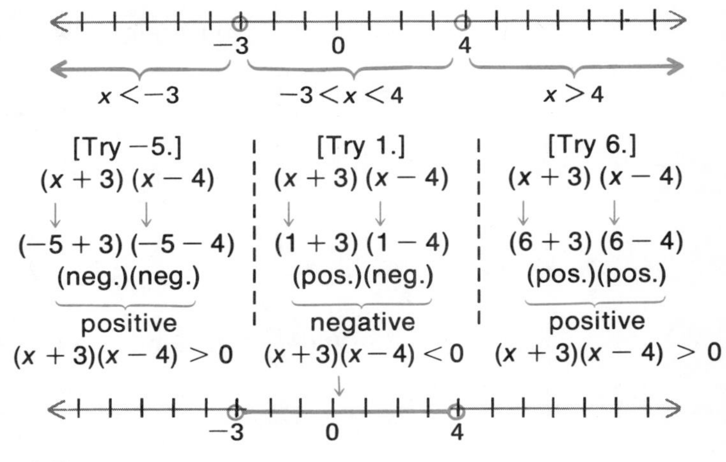

Try a number from each part in $(x + 3)(x - 4) < 0$.

$(-5 + 3)(-5 - 4)$
$(-2)(-9)$
(neg.)(neg.)
positive

Solutions are those numbers which yield a *negative* product.

Thus, The solution set of $(x + 3)(x - 4) < 0$ is $\{x \mid -3 < x < 4\}$.

EXAMPLE 2 Find and graph the solution set of $x^2 - 3x - 10 > 0$.

Solve and graph the solutions of $x^2 - 3x - 10 = 0$.

Factor. → $(x + 2)(x - 5) = 0$

$x + 2 = 0$ or $x - 5 = 0$

The solutions are -2 and 5. → $x = -2$ $x = 5$

-2 and 5 divide the number line into 3 parts. →

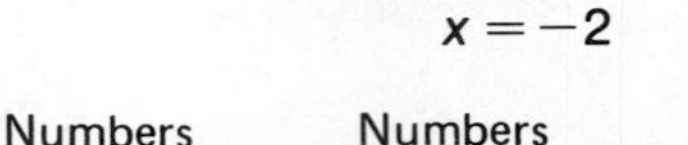

Numbers less than -2 | Numbers between -2 and 5 | Numbers greater than 5

-2 0   5

$x < -2$ $-2 < x < 5$ $x > 5$

Try a number from each part in $(x + 2)(x - 5) > 0$. →

[Try -4.]	[Try 3.]	[Try 8.]
$(x + 2)(x - 5)$	$(x + 2)(x - 5)$	$(x + 2)(x - 5)$
$(-4 + 2)(-4 - 5)$	$(3 + 2)(3 - 5)$	$(8 + 2)(8 - 5)$
(neg.)(neg.)	(pos.)(neg.)	(pos.)(pos.)
positive	negative	positive
$(x + 2)(x - 5) > 0$	$(x + 2)(x - 5) < 0$	$(x + 2)(x - 5) > 0$

$(-4 + 2)(-4 - 5)$
$(-2)(-9)$
(neg.)(neg.)
positive

Solutions are those numbers which yield a positive product. →

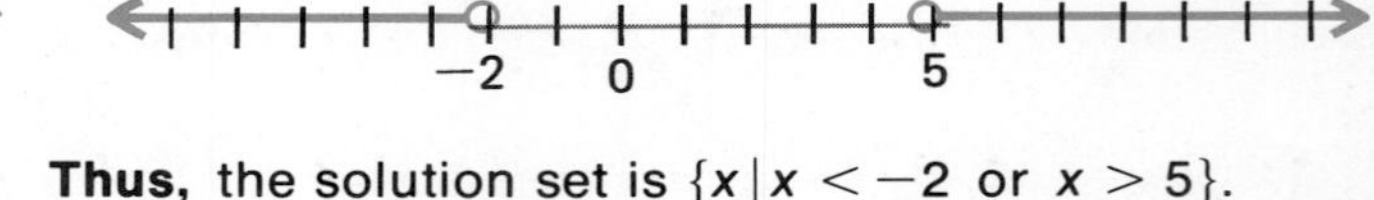

Thus, the solution set is $\{x \mid x < -2 \text{ or } x > 5\}$.

Finding solution sets of quadratic inequalities →

> If r and s are the solutions of $ax^2 + bx + c = 0$, where a is positive and $r < s$, then the solution set of
> $ax^2 + bx + c < 0$ is $\{x \mid r < x < s\}$.
> $ax^2 + bx + c > 0$ is $\{x \mid x < r \text{ or } x > s\}$.

EXAMPLE 3 Find and graph the solution set of $x^2 - 4 \le 0$.

Solve $x^2 - 4 = 0$. →

$x^2 - 4 = 0$
$(x + 2)(x - 2) = 0$
$x = -2$, or $x = 2$

−2   0   2   4   6

$-2 < 2$ and $x^2 - 4 \le 0$ → **Thus,** the solution set is $\{x \mid -2 \le x \le 2\}$.

EXAMPLE 4 Find and graph the solution set of $x^2 - 5x - 6 \ge 0$.

$x^2 - 5x - 6 = 0$
$(x + 1)(x - 6) = 0$
$x = -1$, or $x = 6$

−3   −1   1   3   6

$-1 < 6$ and $x^2 - 5x - 6 > 0$
So, $x < -1$ or $x > 6$. → **Thus,** the solution set is $\{x \mid x \le -1 \text{ or } x \ge 6\}$.

ORAL EXERCISES

True or false?

1. If $x = 3$, then $(x-1)(x-5) < 0$. *T*
2. If $x = -2$, then $(x+3)(x-2) > 0$. *F*
3. If $x = 6$, then $(x-1)(x-5) < 0$. *F*
4. If $x = 4$, then $(x+3)(x-2) > 0$. *T*
5. If $(x-1)(x-5) < 0$, then $1 < x < 5$. *T*
6. If $(x+3)(x-2) > 0$, then $-3 < x < 2$. *F*

EXERCISES

PART A

Find and graph the solution set.

1. $(x-3)(x-6) < 0$
2. $(x+4)(x+1) > 0$ $\{x \mid x < -4 \text{ or } x > -1\}$
3. $(x+5)(x-2) \leq 0$ $\{x \mid -5 \leq x \leq 2\}$
4. $x^2 - 2x - 8 < 0$
5. $x^2 - 4x + 3 \geq 0$
6. $x^2 + 8x + 12 < 0$
7. $x^2 - 3x - 18 > 0$
8. $x^2 - 7x + 10 \leq 0$
9. $x^2 + 8x + 15 > 0$
10. $x^2 - 9 \leq 0$
11. $x^2 - 25 > 0$
12. $x^2 - 1 < 0$
13. $x^2 - 5x < 0$ $\{x \mid 0 < x < 5\}$
14. $x^2 + 3x \geq 0$ $\{x \mid x \leq -3 \text{ or } x \geq 0\}$
15. $x^2 - 6x > 0$ $\{x \mid x < 0 \text{ or } x > 6\}$

PART B

EXAMPLE Find and graph the solution set of $-x^2 + 16 < 0$.

$-x^2 = -1x^2$ →	$-1x^2 + 16 < 0$
Factor -1 from $-1x^2 + 16$. →	$-1(x^2 - 16) < 0$
Divide each side by -1. →	$\frac{-1(x^2-16)}{-1} \downarrow \frac{0}{-1}$
Dividing by a negative number *reverses* the order. →	$x^2 - 16 > 0$
Factor $x^2 - 16$. →	$(x+4)(x-4) > 0$

Thus, the solution set is $\{x \mid x < -4 \text{ or } x > 4\}$.

Graph of solution set →

−4 0 4

Find and graph the solution set.

16. $-x^2 + 4 < 0$
17. $-x^2 + 36 \geq 0$ $\{x \mid -6 \leq x \leq 6\}$
18. $-x^2 + 49 > 0$ $\{x \mid -7 < x < 7\}$
19. $-x^2 - x + 12 > 0$
20. $-x^2 + 6x - 8 < 0$
21. $-x^2 - 4x - 3 \leq 0$
22. $-x^2 + 2x \leq 0$ $\{x \mid x \leq 0 \text{ or } x \geq 2\}$
23. $-x^2 - 5x > 0$ $\{x \mid -5 < x < 0\}$
24. $-x^2 + 7x < 0$ $\{x \mid x < 0 \text{ or } x > 7\}$

PART C

Graph the solution set.

25. $(x-1)(x-4)(x-7) < 0$
26. $(x-2)(x-4)(x-8) > 0$
27. $(x+3)(x-2)(x-5) \leq 0$
28. $(x+5)(x+2)(x-3) \geq 0$
29. $(x-2)(x-2) > 0$
30. $(x-3)(x-3) \leq 0$
31. $x^2 - 2x > 8$
32. $x^2 - 18 < 3x$
33. $x^2 \geq 25$
34. $x^2 < 8x$

Consecutive Integers

OBJECTIVE

- To solve problems about consecutive integers

REVIEW CAPSULE

Consecutive
↓

days	Wednesday, Thursday, Friday
letters	F, G, H, I, J
months	April, May, June, July
integers	. . ., $-2, -1, 0, 1, 2,$ . . .
odd integers	. . ., $-3, -1, 1, 3, 5,$ . . .
even integers	. . ., $-4, -2, 0, 2, 4,$ . . .

EXAMPLE 1 Write the *consecutive* numbers described.

		Answers
Add 1 to get next *integer*. →	Four *integers,* beginning with 6.	6, 7, 8, 9
Add 2 to get next *odd*. →	Three *odd* integers, beginning with 7.	7, 9, 11
Add 2 to get next *even*. →	Four *even* integers, beginning with -4.	$-4, -2, 0, 2$

EXAMPLE 2 Write the consecutive numbers in mathematical terms.

		Answers
Add 1 to get the next integer.	Four integers, beginning with n	$n, n+1, n+2, n+3$
Add 2 to get the next odd integer.	Three odd integers, beginning with the odd integer x	$x, x+2, x+4$
Add 2 to get the next even integer.	Four even integers, beginning with the even integer t	$t, t+2, t+4, t+6$

EXAMPLE 3 Let n be the first of three consecutive integers. Write the expressions in mathematical terms.

Add 1 to get the next integer.

Let n = first consecutive integer
$n+1$ = second consecutive integer
$n+2$ = third consecutive integer

The *sum* of the first and third integers

$$n + (n + 2)$$

The *product* of the second and third integers

$$(n + 1)(n + 2)$$

The *square* of the third integer ⟶ $(n + 2)^2$

EXAMPLE 4 Let x be the first of three consecutive odd integers. Write the expression in mathematical terms.

x is an odd integer. → Let x, $x + 2$, and $x + 4$ be the three consecutive odd integers.

Twice the third integer, decreased by the second

$$2(x + 4) - (x + 2)$$

EXAMPLE 5 Let t be the first of three consecutive even integers. Write the expression in mathematical terms.

Three consecutive even integers, beginning with t are t, $t + 2$, $t + 4$.

The sum of the squares of the first two integers

$$t^2 + (t + 2)^2$$

EXAMPLE 6 Find three consecutive odd integers whose sum is 81.

Let n be the first odd; add 2 to get next odd.

Let n, $n + 2$, $n + 4$ be the three consecutive odd integers.

Write an equation. Their sum is 81.

$$n + (n + 2) + (n + 4) = 81$$

Solve the equation.

$$\begin{aligned} 3n + 6 &= 81 \\ 3n &= 75 \\ n &= 25 \\ n + 2 &= 27 \\ n + 4 &= 29 \end{aligned}$$

Check. $25 + 27 + 29 = 52 + 29$
$= 81$

Thus, the three consecutive odd integers are 25, 27, and 29.

EXAMPLE 7 Find three consecutive integers so that the first times the third, decreased by the second, is equal to one more than 10 times the third.

Let x be the first integer; add 1 to get next *integer*.

Let x, $x + 1$, $x + 2$ be the three consecutive integers.

1st times 3rd, minus 2nd, is 1 more than 10 times 3rd.

Write an equation. →

$$x(x + 2) - (x + 1) = 10(x + 2) + 1$$

$-(x + 1) = -1(x + 1) = -x - 1$ →

$$x^2 + 2x - x - 1 = 10x + 20 + 1$$

Put in standard form.

$$x^2 + x - 1 = 10x + 21$$

One side must be zero. →

$$x^2 - 9x - 22 = 0$$

$$(x - 11)(x + 2) = 0$$

$$x - 11 = 0 \quad \text{or} \quad x + 2 = 0$$

$$x = 11 \qquad\qquad x = -2$$

Two sets of solutions. →

$$x + 1 = 12 \qquad\qquad x + 1 = -1$$

$$x + 2 = 13 \qquad\qquad x + 2 = 0$$

x , $x + 1$, $x + 2$
↑ ↑ ↑
first second third

Thus, the three consecutive integers are 11, 12, and 13, or -2, -1, and 0.

EXAMPLE 8 Find three consecutive even integers so that the square of the first, increased by the square of the third, is 136.

The second integer is not used in the equation.

Let n, $n+2$, $n+4$ be the three consecutive even integers.

(first)² + (third)² = 136 → $n^2 + (n+4)^2 = 136$

$(n+4)^2 = (n+4)(n+4)$ → $n^2 + (n^2 + 8n + 16) = 136$ Quadratic equation

$2n^2 + 8n + 16 = 136$ ← Right side is not 0.

$2n^2 + 8n - 120 = 0$ ← Right side is 0.

2 is GCF of terms. → $2(n^2 + 4n - 60) = 0$

Divide each side by 2. → $n^2 + 4n - 60 = 0$

$(n-6)(n+10) = 0$

$n - 6 = 0$ or $n + 10 = 0$

$n = 6$ $n = -10$

$n + 2 = 8$ $n + 2 = -8$

$n + 4 = 10$ $n + 4 = -6$

There are two sets of solutions. Check both sets of solutions.

(1st)² + (3rd)²	is 136.
$6^2 + 10^2$	136
$36 + 100$	
136	

(1st)² + (3rd)²	is 136.
$(-10)^2 + (-6)^2$	136
$100 + 36$	
136	

Thus, the integers are 6, 8, 10, or −10, −8, −6.

ORAL EXERCISES

Give four consecutive integers, beginning with the given number.

1. 14 *14, 15, 16, 17*
2. −1 *−1, 0, 1, 2*
3. −7 *−7, −6, −5, −4*
4. x *$x, x+1, x+2, x+3$*

Give four consecutive odd integers, beginning with the given number.

5. 15 *15, 17, 19, 21*
6. −1 *−1, 1, 3, 5*
7. −7 *−7, −5, −3, −1*
8. n *$n, n+2, n+4, n+6$*

Give four consecutive even integers, beginning with the given number.

9. 14 *14, 16, 18, 20*
10. −2 *−2, 0, 2, 4*
11. −8 *−8, −6, −4, −2*
12. n *$n, n+2, n+4, n+6$*

Let x be the first of three consecutive odd integers. Give the expression in mathematical terms.

13. The sum of the first and second odd integers *$x + (x+2)$*

14. The product of the first and third odd integers *$x(x+4)$*

15. Four more than the square of the second odd integer *$(x+2)^2 + 4$*

16. Twice the first odd integer, decreased by 5 times the third *$2x - 5(x+4)$*

Let n be the first of three consecutive integers. Give the expression in mathematical terms.

17. The sum of the squares of the three integers *$n^2 + (n+1)^2 + (n+2)^2$*

18. Four times the second integer, increased by 6 times the third *$4(n+1) + 6(n+2)$*

19. Two less than the product of the first two integers *$n(n+1) - 2$*

20. The square of the third integer, decreased by the square of the first *$(n+2)^2 - n^2$*

EXERCISES

PART A

1. Find three consecutive integers whose sum is 174. *57, 58, 59*
2. Find three consecutive even integers whose sum is 78. *24, 26, 28*
3. Find three consecutive integers so that the sum of the first two is 47. *23, 24, 25*
4. Find three consecutive odd integers so that the sum of the first and third is 94.
5. Find three consecutive integers so that 8 more than 4 times the second is 176.
6. Find three consecutive even integers so that 6 less than twice the third is 70.
7. Find three consecutive integers so that the second, increased by the third, is 67.
8. Find three consecutive odd integers so the first, increased by the third, is 106.
9. Find three consecutive integers so that 5 times the first, increased by 3 times the second, is 515. *64, 65, 66*
10. Find three consecutive even integers so that twice the second, increased by 4 times the third, is 380. *60, 62, 64*
11. Find three consecutive integers so that 8 times the first, decreased by the third, is 488. *70, 71, 72*
12. Find three consecutive odd integers so that 6 times the second, decreased by the third, is 363. *71, 73, 75*
13. Find three consecutive integers so that the second times the third is 42.
14. Find three consecutive even integers so that the first times the second is 24.
15. Find three consecutive integers so that the square of the first, increased by the square of the third, is 100. $6, 7, 8; -8, -7, -6$
16. Find three consecutive odd integers so that the sum of the squares of the first two integers is 130. $7, 9, 11; -9, -7, -5$

PART B

17. Find three consecutive integers so that 25 more than the second is equal to 5 times the third. *4, 5, 6*
18. Find three consecutive even integers so that 7 times the third is equal to 20 less than the first. $-8, -6, -4$
19. Find three consecutive integers so that the first times the second is 47 more than the third. $7, 8, 9; -7, -6, -5$
20. Find three consecutive odd integers so that the second times the third is 13 more than 10 times the first.
21. Find three consecutive integers so that the square of the third is 39 more than the square of the second. *18, 19, 20*
22. Find three consecutive even integers so that the square of the third is 76 more than the square of the second. *16, 18, 20*
23. Find three consecutive integers so that the sum of their squares is 245.
24. Find three consecutive odd integers so that the sum of their squares is 83.
25. Find three consecutive integers so that 4 times the square of the third, decreased by 3 times the square of the first, is 41 more than twice the square of the second. *3, 4, 5; 9, 10, 11*
26. Find three consecutive even integers so that twice the square of the second, increased by the square of the third, is 40 less than 6 times the square of the first. *8, 10, 12*

PART C

27. Find three consecutive integers so that 3 less than twice the third is the sum of the first and second.
28. Find three consecutive odd integers so that the square of the second is equal to the first times the third.

Factoring Completely

OBJECTIVE

- To factor polynomials completely, like $12x^3 - 2x^2 - 24x = 2x(3x+4)(2x-3)$

REVIEW CAPSULE

FACTORING

GCF	$5x^4 + 15x^2 - 10x = 5x(x^3 + 3x - 2)$
Two Binomials	$6x^2 + 7x - 3 = (3x-1)(2x+3)$
Difference of Squares	$9x^2 - 4 = (3x+2)(3x-2)$
Perfect Square	$x^2 + 10x + 25 = (x+5)^2$

EXAMPLE 1 Factor $4x^2 + 8x - 12$ completely.

$$4x^2 + 8x - 12$$

First factor out the GCF. 4 is the GCF. → $4(x^2 + 2x - 3)$

Factor: $x^2 + 2x - 3 = (x+3)(x-1)$. → $4(x+3)(x-1)$

Thus, $4x^2 + 8x - 12 = 4(x+3)(x-1)$.

Factoring completely →

> To factor a polynomial completely:
> First, factor out the GCF, if any.
> Then, factor the remaining polynomial, if possible.

EXAMPLE 2 Factor $12x^3 + 14x^2 - 10x$ completely.

$$12x^3 + 14x^2 - 10x$$

$$2x(6x^2) + 2x(7x) + 2x(-5)$$

Factor out the GCF, $2x$. → $2x(6x^2 + 7x - 5)$

Factor $(6x^2 + 7x - 5)$. → $2x(3x+5)(2x-1)$

Thus, $12x^3 + 14x^2 - 10x = 2x(3x+5)(2x-1)$.

EXAMPLE 3 Factor $2x^2 - x - 3$ completely.

$$2x^2 - x - 3$$

The GCF is 1. → $1(2x^2 - x - 3)$

Factor the polynomial. → $(2x-3)(x+1)$

Thus, $2x^2 - x - 3 = (2x-3)(x+1)$.

EXAMPLE 4 Factor $2x^3 - 32xy^2$ completely.

Factor out the GCF. ⟶ $2x^3 - 32xy^2 = 2x(x^2 - 16y^2)$

Difference of Squares ⟶ $= 2x(x + 4y)(x - 4y)$

Thus, $2x^3 - 32xy^2 = 2x(x + 4y)(x - 4y)$.

EXAMPLE 5 Factor $45x^2 + 30x + 5$ completely.

5 is the GCF. ⟶ $45x^2 + 30x + 5 = 5(9x^2 + 6x + 1)$

Perfect square trinomial ⟶ $= 5(3x + 1)^2$

Thus, $45x^2 + 30x + 5 = 5(3x + 1)^2$.

ORAL EXERCISES

Tell if the expression is factored completely.

1. $6(2a + 8b)$ *no*
2. $4c(5c + d)$ *yes*
3. $3(x^2 + xy)$ *no*
4. $7(x^2 - 2x + 3)$ *yes*
5. $3y(y^2 - 7y + 12)$ *no*
6. $-5(9x^2 - 1)$ *no*

EXERCISES

PART A

Factor completely.

1. $3x^2 + 21x + 36$ *$3(x + 3)(x + 4)$*
2. $12y^2 - 3$ *$3(2y + 1)(2y - 1)$*
3. $2x^2 + 16x + 32$ *$2(x + 4)^2$*
4. $2n^2 + 4n - 30$
5. $18x^2 - 50$
6. $3n^2 + 30n + 75$
7. $x^3 - 6x^2 + 8x$
8. $4x^2y - 9y$
9. $4a^3 + 4a^2b + ab^2$
10. $x^2 - 6x$
11. $x^4 - x^2y^2$
12. $25ax^2 - 20ax + 4a$
13. $6x^2 - 14x + 4$
14. $30x^2 + 2xy - 4y^2$
15. $100x^2 + 100xy + 25y^2$
16. $8a^3 - 4a^2 - 40a$ *$4a(2a - 5)(a + 2)$*
17. $3a^3 - 75ab^2$ *$3a(a + 5b)(a - 5b)$*
18. $2x^3 - 4x^2 + 2x$ *$2x(x - 1)^2$*

PART B

EXAMPLE Factor $-6x^2 + 5x + 4$ completely.

$-6x^2 + 5x + 4$

Factor out -1. ⟶ $-1(6x^2 - 5x - 4)$

Factor the polynomial. ⟶ $-1(3x - 4)(2x + 1)$

Factor completely.

19. $-4x^2 + 4x + 3$ *$-1(2x - 3)(2x + 1)$*
20. $-y^2 - 9y - 14$ *$-1(y + 7)(y + 2)$*
21. $-n^2 + 2n - 1$ *$-1(n - 1)^2$*
22. $-x^2 + 8x - 16$ *$-1(x - 4)^2$*
23. $-3x^2 + 27$ *$-3(x + 3)(x - 3)$*
24. $-40x^2 - 14x + 6$ *$-2(5x + 3)(4x - 1)$*
25. $-9n^2 - 30n - 25$ *$-1(3n + 5)^2$*
26. $-4c^3 + c$ *$-c(2c + 1)(2c - 1)$*
27. $-4ac^2 - 4ac - a$ *$-a(2c + 1)^2$*
28. $-x^3 + 6x^2y - 9xy^2$ *$-x(x - 3y)^2$*
29. $ax^2 - 3a^2x - 10a^3$ *$a(x - 5a)(x + 2a)$*
30. $\pi r^2 - r$ *$r(\pi r - 1)$*

Chapter Three Review

Simplify. [p. 45]

1. $3x^2 \cdot 5x^3$
2. $-2x \cdot 4x^4$
3. $(a^3b^2)(ab^3)$
4. $-2xy(5x^2y^3)$
5. $(x^3)^4$
6. $(-x^4)^3$
7. $(5c)^2$
8. $(3ab^4)^3$

Evaluate if $x = 3$ and $y = -2$. [p. 45]

9. $2x^2y^3$
10. $(2xy^2)^2$
11. $2xy + 3xy^2 + 4$
12. $2x^2 + 3y^2$

Which expressions are polynomials? [p. 49]

13. $x^2 + \frac{3x}{y} - 4$
14. $x^2 + \sqrt{x}$
15. $2x^2y + 3xy + xy^2$

16. Give the degree of the polynomial $4x^3 + 2x^2y^2 + 5x^4y$. [p. 49]

Classify the polynomial. [p. 49]

17. $x^3 + x - 4$
18. $-2x^2y$
19. $x^2 - 9y^2$
20. $2x^3 + x^2 + 7x - 5$

Add. [p. 49]

21. $(4x^3 + x - x^2) + (3x^2 - 6x + 4)$
22. $(2x^2y + 3xy + xy^2) + (4xy^2 - xy)$

Subtract. [p. 49]

23. $(5x^2 - 7x + 3) - (3 + 2x - x^2)$
24. $(x^3 - 4x^2 + 5 - 6x) - (-4x^2 + 7 - 2x)$

Multiply. [p. 53]

25. $2x(x^2 + y^2 + 3xy)$
26. $(3x - 5)(2x - 7)$
27. $(2x + 3y)(2x - 3y)$
28. $(3x + 4)(x^2 + 3x - 5)$
29. $(x - y)(4x - 3y - 6)$

Factor out the GCF. [p. 56]

30. $6x^2 + 9x - 3$
31. $3n^3 - 2n^2 + n$
32. $8x^3 - 12x^2$
33. Factor -1 from $-x^2 + 7x - 3$. [p. 56]
34. Square the binomial $(4x + 3y)^2$. [p. 64]

Factor completely. [p. 59, 62, 64, 80]

35. $x^2 - 5x - 14$
36. $6n^2 + 11n + 5$
37. $9y^2 - 18y + 8$
38. $9x^2 - 16y^2$
39. $9n^2 + 12n + 4$
40. $2x^2 - 11xy + 12y^2$
41. $4x^2 + 8x + 4$
42. $12y^2 - 27$
43. $x^3 - 2x^2 - 15x$
44. $18x^3 - 15x^2 + 3x$
45. $-x^2 + 10x - 25$
46. $-2x^2 + 3x + 14$

Find the solution set. [p. 69]

47. $3x^2 + 10x - 8 = 0$
48. $2x^2 + 15 = 11x$
49. $x^2 + 7x = 0$

Find and graph the solution set. [p. 73]

50. $x^2 + 2x - 8 < 0$
51. $x^2 - 2x - 15 \geq 0$
52. $-x^2 + 25 > 0$

53. Find three consecutive odd integers whose sum is 201. [p. 76]
54. Find three consecutive integers so that the second integer times the third is 30. [p. 76]

Chapter Three Test

Simplify.

1. $(2a^2b)(3a^2b^3)$
2. $(2n)^4$
3. $(5x^2y^3)^2$

Evaluate if $x = 2$ and $y = -1$.

4. $-4x^3y^2$
5. $(2x^2y)^3$
6. $8x^2y - 2xy^2$

Which expressions are polynomials?

7. $2x^2 + xy + y^2$
8. $\sqrt{x - 2}$
9. $\frac{4x}{y}$
10. Give the degree of the polynomial $5x^2 + 2xy - 3xy^2$.

Classify the polynomial.

11. $3x - 4$
12. $x^2 + 7x^5 + 12$
13. $-7x$

Add.

14. $(4x^2 - 7 + 3x) + (2x^3 + 4x - 3)$
15. $(3xy^2 + 4xy - x^2y) + (5xy + 3x^2y - 6xy^2)$

Subtract.

16. $(2x^2 - 4x + 5) - (5x^2 - 2x - 3)$
17. $(-8y^2 - 7y - 3) - (y^2 + 7y - 6)$

Multiply.

18. $3x(2x^2 - x + 1)$
19. $(5x - 2y)(2x + 3y)$
20. $(2x + 3)(x^2 - 4x - 5)$

Factor out the GCF.

21. $5x^2 - 20x + 10$
22. $2n^3 - 6n^2 + 3n$
23. $6y^3 - 6y^2 + 9y$
24. Factor -1 from $-x^2 + 9x - 4$.
25. Square the binomial $(2x - 5)^2$.

Factor completely.

26. $15x^2 + x - 2$
27. $x^2 + 12x + 36$
28. $6x^2 - 7xy + 2y^2$
29. $16x^2 - y^2$
30. $5n^2 - 45$
31. $2xy^2 - 2xy - 12x$

Find the solution set.

32. $x^2 - 10x + 21 = 0$
33. $2x^2 + 5x - 12 = 0$
34. $x^2 + 2x = 24$

Find and graph the solution set.

35. $x^2 - 3x - 10 < 0$
36. $x^2 - 2x - 3 \geq 0$
37. Find three consecutive integers whose sum is 75.
38. Find three consecutive even integers so that the second times the third is 48.

4 RATIONAL EXPRESSIONS

Maria Gaetana Agnesi

Maria Gaetana Agnesi
(1718–1799)

Born into a wealthy family of Milan, Maria Gaetana Agnesi was educated privately and showed outstanding ability at a very early age. When Agnesi was nine, she published a paper defending the right of women to have a liberal arts education. She knew seven languages by the age of thirteen and had had a very broad educational background.

During Agnesi's teenage years her father, a professor of mathematics at the University of Bologna, introduced her to philosophy. She was especially interested in the philosophy of Sir Isaac Newton. Her early adult life was spent studying mathematics and teaching her younger brothers. The first important surviving work written by a woman was written by Agnesi to help a younger brother with his studies. This book, *Analytical Institutions,* consists of two volumes. The first deals with geometry and algebra, the second with differential and integral calculus. Because of the book's clarity and completeness, it became widely used as a textbook.

Agnesi was appointed honorary lecturer at the University of Bologna by Pope Benedict XIV. When her father died shortly thereafter, she retired to a solitary life of study and service to the poor and sick. Agnesi is best remembered for her work in mathematics and philosophy.

Identifying Rational Expressions

OBJECTIVES

- To identify and evaluate rational expressions
- To determine when a rational expression is undefined

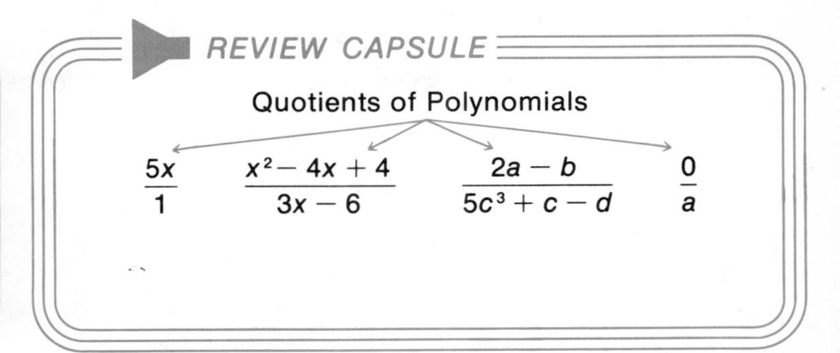

Definition of rational expression →

A *rational expression* is a polynomial or a quotient of polynomials.

EXAMPLE 1 Which are rational expressions?

		Answers
Polynomial →	$3a - b$	Yes
Not a polynomial →	$\sqrt{x^2 - 4}$	No
Quotient of polynomials →	$\frac{x^2 - 9}{4x + 12}$	Yes

EXAMPLE 2 Evaluate $\frac{2x}{x - 4}$, if $x = 8$.

Substitute 8 for x. → $\frac{2x}{x - 4} = \frac{2(8)}{8 - 4}$

Value is 4, if $x = 8$. → $= \frac{16}{4}$, or 4

EXAMPLE 3 Find the quotient, if it exists.

$\frac{10}{5} = 2$ because $5 \cdot 2 = 10$. → $\frac{10}{5} = 2$

$\frac{0}{6} = 0$ because $6 \cdot 0 = 0$. → $\frac{0}{6} = 0$

$\frac{5}{0}$ is undefined because no number multiplied by 0 equals 5. → $\frac{5}{0}$ Quotient does not exist.

$\frac{0}{7} = 0$ because $7 \cdot 0 = 0$. → $\frac{4 - 2^2}{7} = \frac{0}{7}$, or 0

$\frac{5}{12 - (3)(4)} = \frac{5}{12 - 12}$, undefined → $\frac{5}{12 - (3)(4)} = \frac{5}{0}$ Quotient does not exist.

A quotient does not exist if the denominator is 0. →

The denominator of a rational expression cannot be zero.

EXAMPLE 4 Evaluate $\frac{2x}{x-4}$, if $x = 4$.

Substitute 4 for x. → $\frac{2x}{x-4} = \frac{2(4)}{4-4}$

The denominator is 0. → $= \frac{8}{0}$

Thus, the value does *not* exist if $x = 4$.

A rational expression is *undefined* for values of the variable which make the denominator zero.

EXAMPLE 5 For what values of x is $\frac{7}{x^2-9}$ undefined?

Undefined when the denominator is equal to 0 →

$$x^2 - 9 = 0$$
$$(x+3)(x-3) = 0$$
$$x + 3 = 0 \quad \text{or} \quad x - 3 = 0$$
$$x = -3 \qquad\qquad x = 3$$

Check.

$\frac{7}{(-3)^2-9} = \frac{7}{9-9} = \frac{7}{0}$

$\frac{7}{(3)^2-9} = \frac{7}{9-9} = \frac{7}{0}$

Thus, $\frac{7}{x^2-9}$ is undefined if $x = -3$ or $x = 3$.

EXERCISES

Which are rational expressions?

1. $\frac{x^2-5x+6}{2x-4}$ *yes* **2.** $4a^2 - 25b^2$ *yes* **3.** $\frac{4x+8}{0}$ *no* **4.** $\frac{\sqrt{y-3}}{2x+6}$ *no*

Find the quotient, if it exists.

5. $\frac{-8}{2}$ *−4* **6.** $\frac{0}{4}$ *0* **7.** $\frac{7-7}{5+5}$ *0* **8.** $\frac{3}{0}$ *does not exist* **9.** $\frac{5+5}{6-6}$ *does not exist* **10.** $\frac{2+4}{2^2-4}$ *does not exist*

Evaluate if $x = 2$.

11. $\frac{x^2+2x+1}{x+1}$ *3* **12.** $\frac{2x-4}{3x+6}$ *0* **13.** $\frac{x+5}{4x}$ *$\frac{7}{8}$* **14.** $\frac{3x}{x-2}$ *does not exist* **15.** $\frac{x+3}{x^2-4}$ *does not exist*

Evaluate if $x = -3$.

16. $\frac{x^2-x-2}{x+5}$ *5* **17.** $\frac{2x+6}{3x-1}$ *0* **18.** $\frac{x+4}{2x}$ *$-\frac{1}{6}$* **19.** $\frac{3x}{x+3}$ *does not exist* **20.** $\frac{x+5}{x^2-9}$ *does not exist*

For what values of the variable is the expression undefined?

21. $\frac{2}{5x}$ *0* **22.** $\frac{a-3}{a+2}$ *−2* **23.** $\frac{5}{3-y}$ *3* **24.** $\frac{2x}{x^2-25}$ *−5; 5* **25.** $\frac{4}{b^2-7b+12}$ *3; 4*

Simplifying Rational Expressions

OBJECTIVE

- To simplify rational expressions like $\frac{x^2 - 5x + 6}{2x^2 - 3x - 2}$

REVIEW CAPSULE

Factor $-x^2 - 3x + 10$ completely.

$-x^2 - 3x + 10$

Factor out -1. ⟶ $-1(x^2 + 3x - 10)$

Factor the trinomial. ⟶ $-1(x + 5)(x - 2)$

EXAMPLE 1 Simplify $\frac{30}{45}$.

Factor numerator and denominator. ⟶ $\frac{30}{45} = \frac{5 \cdot 3 \cdot 2}{5 \cdot 3 \cdot 3}$

$= 1 \cdot 1 \cdot \frac{2}{3}$, or $\frac{2}{3}$

Short form

$\frac{30}{45} = \frac{\overset{1}{\cancel{5}} \cdot \overset{1}{\cancel{3}} \cdot 2}{\underset{1}{\cancel{5}} \cdot \underset{1}{\cancel{3}} \cdot 3}$

$= \frac{2}{3}$

Simplifying a rational expression ⟶

To simplify a rational expression:
First, factor numerator and denominator.
Second, divide out common factors.
Third, multiply remaining factors.

EXAMPLE 2 Simplify $\frac{6a + 9b}{6c}$.

Factor numerator and denominator. ⟶ $\frac{6a + 9b}{6c} = \frac{3(2a + 3b)}{3 \cdot 2 \cdot c}$

Divide out common factors. ⟶ $= \frac{\overset{1}{\cancel{3}}(2a + 3b)}{\underset{1}{\cancel{3}} \cdot 2 \cdot c}$

The fraction is in simplest form. $= \frac{2a + 3b}{2c}$

EXAMPLE 3 Simplify $\frac{2x - 8}{3x - 12}$.

Factor. ⟶ $\frac{2x - 8}{3x - 12} = \frac{2(x - 4)}{3(x - 4)}$

Divide out common factors. ⟶ $= \frac{2\overset{1}{\cancel{(x - 4)}}}{3\underset{1}{\cancel{(x - 4)}}}$, or $\frac{2}{3}$

EXAMPLE 4 Simplify $\frac{x^2 - 5x + 6}{9 - x^2}$. Simplify $\frac{6x^2 + 5x - 4}{4x^2 + 8x - 5}$.

Arrange denominator in descending order. ⟶ $\frac{x^2 - 5x + 6}{-x^2 + 9}$

$-x^2 + 9 = -1(x^2 - 9) = -1(x + 3)(x - 3)$

Factor and divide out common factors. ⟶ $\frac{(x - 2)\cancel{(x - 3)}^{1}}{-1(x + 3)\cancel{(x - 3)}_{1}}$ | $\frac{(3x + 4)\cancel{(2x - 1)}^{1}}{\cancel{(2x - 1)}_{1}(2x + 5)}$

The fraction is in simplest form. $\frac{x - 2}{-1(x + 3)}$ | $\frac{3x + 4}{2x + 5}$

EXAMPLE 5 Simplify $\frac{b^6}{b^2}$. Simplify $\frac{c^3}{c^5}$. Simplify $\frac{t^4}{t^4}$.

Factor. $\frac{b \cdot b \cdot b \cdot b \cdot b \cdot b}{b \cdot b}$ | $\frac{c \cdot c \cdot c}{c \cdot c \cdot c \cdot c \cdot c}$ | $\frac{t \cdot t \cdot t \cdot t}{t \cdot t \cdot t \cdot t}$

Divide out common factors. $\frac{\cancel{b}^1 \cdot \cancel{b}^1 \cdot b \cdot b \cdot b \cdot b}{\cancel{b}_1 \cdot \cancel{b}_1}$ | $\frac{\cancel{c}^1 \cdot \cancel{c}^1 \cdot \cancel{c}^1}{\cancel{c}_1 \cdot \cancel{c}_1 \cdot \cancel{c}_1 \cdot c \cdot c}$ | $\frac{\cancel{t}^1 \cdot \cancel{t}^1 \cdot \cancel{t}^1 \cdot \cancel{t}^1}{\cancel{t}_1 \cdot \cancel{t}_1 \cdot \cancel{t}_1 \cdot \cancel{t}_1}$

Multiply remaining factors. b^4 | $\frac{1}{c^2}$ | 1

Thus, $\frac{b^6}{b^2} = b^{6-2}$, or b^4; $\frac{c^3}{c^5} = \frac{1}{c^{5-3}}$, or $\frac{1}{c^2}$; $\frac{t^4}{t^4} = 1$.

Example 5 suggests this. ⟶

Quotient of Powers

$\frac{x^m}{x^n} = x^{m-n}$ if $m > n$.

$\frac{x^m}{x^n} = \frac{1}{x^{n-m}}$ if $n > m$.

$\frac{x^m}{x^n} = 1$ if $m = n$.

EXAMPLE 6 Simplify $\frac{x^4y^2z^5}{x^7y^2z^3}$.

$\frac{x^4 \cdot y^2 \cdot z^5}{x^7 \cdot y^2 \cdot z^3}$

Use quotient of powers to simplify. ⟶ $\frac{1 \cdot 1 \cdot z^2}{x^3}$

Multiply remaining factors. $\frac{z^2}{x^3}$

Short form

$\frac{\cancel{x^4}^{1} \cdot \cancel{y^2}^{1} \cdot \cancel{z^5}^{z^2}}{\cancel{x^7}_{x^3} \cdot \cancel{y^2}_{1} \cdot \cancel{z^3}_{1}}$

$\frac{z^2}{x^3}$

EXAMPLE 7 Simplify $\frac{-9x^2y}{12xy^3}$.

Simplify; multiply remaining factors. $\frac{\cancel{-9}^{-3} \cdot \cancel{x^2}^{x} \cdot \cancel{y}^{1}}{\cancel{12}_{4} \cdot \cancel{x}_{1} \cdot \cancel{y^3}_{y^2}} = \frac{-3x}{4y^2}$

EXAMPLE 8 Simplify $\frac{x^2 + 2x - 8}{6 - 3x}$.

Arrange denominator in descending order.

Factor and simplify.

$$\frac{x^2 + 2x - 8}{-3x + 6} = \frac{(x + 4)\overset{1}{\cancel{(x - 2)}}}{-3\underset{1}{\cancel{(x - 2)}}}, \text{ or } \frac{x + 4}{-3}$$

EXERCISES

PART A

Simplify.

1. $\frac{x^3yz^3}{xy^5z^3}$ $\frac{x^2}{y^4}$
2. $\frac{10a^2b^6}{15a^6b^2}$ $\frac{2b^4}{3a^4}$
3. $\frac{a^2bc^5}{a^4bc^2}$ $\frac{c^3}{a^2}$
4. $\frac{-4x^4y^2}{6xy^5}$ $\frac{-2x^3}{3y^3}$
5. $\frac{-9ab^2}{-6ab^8c^3}$ $\frac{3}{2b^6c^3}$
6. $\frac{8x^4y^2z}{12x^3z^2}$ $\frac{2xy^2}{3z}$
7. $\frac{6x + 8}{4}$ $\frac{3x + 4}{2}$
8. $\frac{6}{6a - 3}$ $\frac{2}{2a - 1}$
9. $\frac{10x - 15}{5a}$ $\frac{2x - 3}{a}$
10. $\frac{3x + 9}{5x + 15}$ $\frac{3}{5}$
11. $\frac{5a + 10}{5a - 15}$ $\frac{a + 2}{a - 3}$
12. $\frac{4x - 20y}{6x - 30y}$ $\frac{2}{3}$
13. $\frac{2a - 6}{3 - a}$ -2
14. $\frac{3x - 15}{10 - 2x}$ $\frac{3}{-2}$
15. $\frac{3x - 6}{12 - 6x}$ $\frac{1}{-2}$
16. $\frac{n^2 + 7n + 10}{3n + 6}$ $\frac{n + 5}{3}$
17. $\frac{x^2 - 4}{5x - 10}$ $\frac{x + 2}{5}$
18. $\frac{a^2 + a - 12}{2a - 6}$ $\frac{a + 4}{2}$
19. $\frac{3n - 6}{3n^2 - 5n - 2}$ $\frac{3}{3n + 1}$
20. $\frac{2x^2 + 5x - 12}{4x - 6}$ $\frac{x + 4}{2}$
21. $\frac{5a - 5}{4a^2 - 9a + 5}$ $\frac{5}{4a - 5}$
22. $\frac{x^2 - 2x - 15}{x^2 + 2x - 3}$ $\frac{x - 5}{x - 1}$
23. $\frac{a^2 + 7a + 12}{a^2 - 2a - 24}$ $\frac{a + 3}{a - 6}$
24. $\frac{c^2 - 5c + 4}{c^2 - 10c + 24}$ $\frac{c - 1}{c - 6}$

PART B

Simplify.

25. $\frac{5x + 20}{16 - x^2}$ $\frac{5}{-1(x - 4)}$
26. $\frac{a^2 - 9}{6 - a - a^2}$ $\frac{a - 3}{-1(a - 2)}$
27. $\frac{c^2 + 4c - 12}{7c - c^2 - 10}$ $\frac{c + 6}{-1(c - 5)}$
28. $\frac{-2x^2 + 4x + 16}{5x + 10}$ $\frac{-2(x - 4)}{5}$
29. $\frac{3a^2 - 9a + 6}{a^2 + 2a - 3}$ $\frac{3(a - 2)}{a + 3}$
30. $\frac{n^2 + 3n - 10}{-2n^2 + 50}$ $\frac{n - 2}{-2(n - 5)}$
31. $\frac{a^2 + 7a - 8}{4a^2 - 8a + 4}$ $\frac{a + 8}{4(a - 1)}$
32. $\frac{-5x^2 - 30x - 45}{x^2 - 4x - 21}$ $\frac{-5(x + 3)}{x - 7}$
33. $\frac{3c^2 - 12c + 12}{c^2 + 4c - 12}$ $\frac{3(c - 2)}{c + 6}$

PART C

Simplify.

34. $\frac{x^{6a}y^{3b}}{x^{2a}y^b}$
35. $\frac{x^{3c + 2}y^{5d - 1}}{x^{c - 1}y^{2d + 3}}$
36. $\frac{x^{4a + 3b}y^{2a - 4b}}{x^{a + b}y^{a - 3b}}$
37. $\frac{x^3 - 8}{3x - 6}$
38. $\frac{x^3 + 64}{x^2 - 4x + 16}$
39. $\frac{x^3 - 27}{4x - 12}$

Multiplying and Dividing

OBJECTIVES

- To multiply rational expressions like $\frac{2x}{3x^2 - 5x - 2} \cdot \frac{x^2 + x - 6}{4x + 12}$
- To find the reciprocal of a rational expression
- To divide rational expressions

REVIEW CAPSULE

Multiplying Fractions

$$\frac{2}{3} \cdot \frac{5}{7} = \frac{2 \cdot 5}{3 \cdot 7}, \text{ or } \frac{10}{21}$$

$$\frac{a}{b} \cdot \frac{c}{d} = \frac{a \cdot c}{b \cdot d}$$

$\frac{3}{4}$ $\frac{4}{3}$

Reciprocals

EXAMPLE 1 Multiply $\frac{3}{4} \cdot \frac{10}{9}$.

Use $\frac{a}{b} \cdot \frac{c}{d} = \frac{a \cdot c}{b \cdot d}$. $\longrightarrow$ $\frac{3}{4} \cdot \frac{10}{9} = \frac{3 \cdot 10}{4 \cdot 9}$

Factor. Divide out common factors. $\longrightarrow$ $= \frac{\overset{1}{\cancel{3}} \cdot \overset{1}{\cancel{2}} \cdot 5}{\underset{1}{\cancel{2}} \cdot 2 \cdot \underset{1}{\cancel{3}} \cdot 3}$

$= \frac{5}{6}$

EXAMPLE 2 Multiply $\frac{3x^2}{2a} \cdot \frac{5ab}{6x}$.

Multiply. Divide out common factors. $\longrightarrow$ $\frac{3x^2}{2a} \cdot \frac{5ab}{6x} = \frac{\overset{1}{\cancel{3}}\overset{x}{\cancel{x^2}} \cdot 5\overset{1}{\cancel{a}}b}{2\underset{1}{\cancel{a}} \cdot \underset{2\ 1}{\cancel{6}\cancel{x}}}$

Multiply remaining factors. $\longrightarrow$ $= \frac{5bx}{4}$

EXAMPLE 3 Multiply $\frac{4x}{2x + 6} \cdot \frac{x^2 + 4x + 3}{3xy}$.

Factor. $\longrightarrow$ $\frac{4x(x + 3)(x + 1)}{2(x + 3) \cdot 3 \cdot x \cdot y}$

Divide out common factors. $\longrightarrow$ $\frac{\overset{2}{\cancel{4}}\overset{1}{\cancel{x}}\overset{1}{\cancel{(x + 3)}}(x + 1)}{\underset{1}{\cancel{2}}\underset{1}{\cancel{(x + 3)}} \cdot 3 \cdot \underset{1}{\cancel{x}} \cdot y}$

$\frac{2(x + 1)}{3y}$

EXAMPLE 4 Multiply $\frac{2x^2 - 18}{2x^2 + 5x - 3} \cdot \frac{2x^2 - 9x + 4}{6x - 24}$.

$$\frac{(2x^2 - 18) \cdot (2x^2 - 9x + 4)}{(2x^2 + 5x - 3) \cdot (6x - 24)}$$

$2x^2 - 18 = 2(x^2 - 9) = 2(x + 3)(x - 3)$ ⟶

$$\frac{\overset{1}{2}\overset{1}{\cancel{(x + 3)}}(x - 3)\overset{1}{\cancel{(2x - 1)}}\overset{1}{\cancel{(x - 4)}}}{\underset{1}{\cancel{(2x - 1)}}\underset{1}{\cancel{(x + 3)}} \cdot \underset{3}{\cancel{6}}\underset{1}{\cancel{(x - 4)}}}$$

$$\frac{x - 3}{3}$$

EXAMPLE 5 Multiply $(x + 5) \cdot \frac{x}{x - 4}$.

Rewrite $x + 5$ as $\frac{x + 5}{1}$. ⟶

$$\frac{(x + 5)}{1} \cdot \frac{x}{x - 4} = \frac{x(x + 5)}{x - 4}$$

EXAMPLE 6 Multiply $\frac{6 - 3a}{5} \cdot \frac{a}{a^2 - 4}$.

$6 - 3a = -3a + 6 = -3(a - 2)$ ⟶

$$\frac{(-3a + 6) \cdot a}{5(a^2 - 4)} = \frac{-3\overset{1}{\cancel{(a - 2)}} \cdot a}{5(a + 2)\underset{1}{\cancel{(a - 2)}}} = \frac{-3a}{5(a + 2)}$$

EXAMPLE 7 Multiply.

Each problem is of the form $\frac{a}{b} \cdot \frac{b}{a}$.

$\frac{x - 5}{x + 3} \cdot \frac{x + 3}{x - 5}$	$\frac{2}{7} \cdot \frac{7}{2}$	$\frac{-4n}{3y} \cdot \frac{3y}{-4n}$
$\frac{\overset{1}{\cancel{(x - 5)}}\overset{1}{\cancel{(x + 3)}}}{\underset{1}{\cancel{(x + 3)}}\underset{1}{\cancel{(x - 5)}}}$	$\frac{\overset{1}{\cancel{2}} \cdot \overset{1}{\cancel{7}}}{\underset{1}{\cancel{7}} \cdot \underset{1}{\cancel{2}}}$	$\frac{\overset{1}{\cancel{(-4n)}} \cdot \overset{1}{\cancel{(3y)}}}{\underset{1}{\cancel{(3y)}} \cdot \underset{1}{\cancel{(-4n)}}}$
1	1	1

Each of the products is equal to 1. ⟶

$$\frac{a}{b} \cdot \frac{b}{a} = 1$$

$\frac{a}{b}$ and $\frac{b}{a}$ are reciprocals of each other if $a \neq 0$ and $b \neq 0$.

Reciprocals are also called *multiplicative inverses.*

EXAMPLE 8 Find the reciprocal of the expression.

	Expression	*Reciprocal*	*Expression*	*Reciprocal*
$x \cdot \frac{1}{x} = 1$ ⟶	x	$\frac{1}{x}$	2	$\frac{1}{2}$
$\frac{n + 2}{a - c} \cdot \frac{a - c}{n + 2} = 1$ ⟶	$\frac{n + 2}{a - c}$	$\frac{a - c}{n + 2}$	$\frac{-7x^2}{5y}$	$\frac{5y}{-7x^2}$
$\frac{3c + d}{1} \cdot \frac{1}{3c + d} = 1$ ⟶	$3c + d$	$\frac{1}{3c + d}$	$\frac{3x}{x^2 - 4}$	$\frac{x^2 - 4}{3x}$

EXAMPLE 9 Divide $\frac{8}{9} \div \frac{2}{3}$.

Multiply by the reciprocal.

$\frac{3}{2}$ is the reciprocal of $\frac{2}{3}$. ⟶

$$\frac{8}{9} \div \frac{2}{3} = \frac{8}{9} \cdot \frac{3}{2} = \frac{\overset{4}{\cancel{8}} \cdot \overset{1}{\cancel{3}}}{\underset{3}{\cancel{9}} \cdot \underset{1}{\cancel{2}}} = \frac{4}{3}$$

Dividing by a rational expression ⟶

$$\frac{a}{b} \div \frac{c}{d} = \frac{a}{b} \cdot \frac{d}{c}$$

To divide by a rational expression, multiply by its reciprocal.

EXAMPLE 10 Divide $\frac{x^2y^4}{6a^2} \div \frac{x^3y}{10a^2}$.

$\frac{10a^2}{x^3y}$ is the reciprocal of $\frac{x^3y}{10a^2}$. ⟶

$$\frac{x^2y^4}{6a^2} \div \frac{x^3y}{10a^2} = \frac{x^2y^4}{6a^2} \cdot \frac{10a^2}{x^3y}$$

Divide out common factors. ⟶

$$= \frac{\overset{1}{\cancel{x^2}} \cdot \overset{y^3}{\cancel{y^4}} \cdot \overset{5}{\cancel{10}} \cdot \overset{1}{\cancel{a^2}}}{\underset{3}{\cancel{6}} \cdot \underset{1}{\cancel{a^2}} \cdot \underset{x}{\cancel{x^3}} \cdot \underset{1}{\cancel{y}}} = \frac{5y^3}{3x}$$

EXAMPLE 11 Divide $\frac{4a+8}{5} \div \frac{a^2-3a-10}{a}$.

Reciprocal is $\frac{a}{a^2-3a-10}$. ⟶

Factor and simplify.

$$\frac{4a+8}{5} \cdot \frac{a}{a^2-3a-10} = \frac{4\overset{1}{\cancel{(a+2)}} \cdot a}{5 \cdot \underset{1}{\cancel{(a+2)}}(a-5)} = \frac{4a}{5(a-5)}$$

ORAL EXERCISES

Multiply.

1. $\frac{3}{5} \cdot \frac{2}{11}$ *$\frac{6}{55}$*
2. $\frac{3}{7} \cdot \frac{5}{6}$ *$\frac{5}{14}$*
3. $\frac{x}{2} \cdot \frac{y}{3x}$ *$\frac{y}{6}$*
4. $3 \cdot \frac{2}{7}$ *$\frac{6}{7}$*
5. $x \cdot \frac{3}{y}$ *$\frac{3x}{y}$*
6. $\frac{3x}{2a} \cdot \frac{2a}{3x}$ *1*

Find the reciprocal of the expression.

7. $\frac{3}{4}$ *$\frac{4}{3}$*
8. 7 *$\frac{1}{7}$*
9. $\frac{-2x}{5y}$
10. $\frac{x+2}{x^2-9}$
11. $\frac{1}{x-y}$ *$x - y$*
12. $3x+6$
13. 0 *no reciprocal*
14. -1 *-1*
15. 1 *1*

EXERCISES

PART A

Multiply.

1. $\frac{xy^4}{8} \cdot \frac{6}{x^3y^2}$ $\quad \frac{3y^2}{4x^2}$
2. $\frac{-xy}{6a^3} \cdot \frac{-4a}{xy^2z}$ $\quad \frac{2}{3a^2yz}$
3. $5c \cdot \frac{3b}{10c^2}$ $\quad \frac{3b}{2c}$
4. $\frac{x-2}{4} \cdot \frac{-2}{3x-6}$ $\quad -\frac{1}{6}$
5. $\frac{9c}{x+4} \cdot \frac{2x+8}{-3c^2}$ $\quad -\frac{6}{c}$
6. $\frac{5n+15}{2d-8} \cdot \frac{12-3d}{4n+12}$ $\quad -\frac{15}{8}$
7. $\frac{3y-15}{2x+6} \cdot \frac{x^2-9}{y-5}$ $\quad \frac{3(x-3)}{2}$
8. $\frac{3}{5-c} \cdot \frac{c^2-25}{2}$ $\quad \frac{3(c+5)}{-2}$
9. $(x+4) \cdot \frac{3c}{x^2-16}$ $\quad \frac{3c}{x-4}$

Divide.

10. $\frac{x^2y^3}{10} \div \frac{x^5y}{12}$ $\quad \frac{6y^2}{5x^3}$
11. $\frac{-10a^2}{b^3} \div \frac{15ab}{2c}$ $\quad \frac{-4ac}{3b^4}$
12. $\frac{3x+15}{10n} \div \frac{2x+10}{5}$ $\quad \frac{3}{4n}$
13. $\frac{3x^4}{2c-8} \div \frac{-9x}{5c-20}$ $\quad \frac{5x^3}{-6}$
14. $\frac{-4a+8}{-2x-10} \div \frac{6-3a}{x^2-25}$ $\quad \frac{2(x-5)}{-3}$
15. $\frac{3x+12}{5x} \div (x^2-16)$ $\quad \frac{3}{5x(x-4)}$

PART B

Multiply.

16. $\frac{x^2-5x+6}{3x+12} \cdot \frac{2x+8}{x^2+3x-10}$ $\quad \frac{2(x-3)}{3(x+5)}$
17. $\frac{n^2+6n}{n^2+2n-8} \cdot \frac{10-5n}{-10n}$ $\quad \frac{n+6}{2(n+4)}$
18. $\frac{3c^2-5c-2}{6c^2} \cdot \frac{4c-8}{c^2-4c+4}$ $\quad \frac{2(3c+1)}{3c^2}$
19. $\frac{x^2-3x-10}{2x^2+7x+6} \cdot \frac{6x+6}{15-3x}$ $\quad \frac{2(x+1)}{-1(2x+3)}$
20. $\frac{a^3-4a}{6a} \cdot \frac{2a^2-14a+12}{a^2-4a-12}$ $\quad \frac{(a-2)(a-1)}{3}$
21. $\frac{3x^2+30x+75}{-2x-10} \cdot \frac{8x}{3x^2-75}$ $\quad \frac{4x}{-1(x-5)}$
22. $(x^2-9y^2) \cdot \frac{2x-10y}{x^2-2xy-15y^2}$ $\quad 2(x-3y)$
23. $\frac{4x}{6x+12y} \cdot (x^2+4xy+4y^2)$ $\quad \frac{2x(x+2y)}{3}$

Divide.

24. $\frac{6a}{2n^2+3n-9} \div \frac{8a^4}{4n^2-9}$ $\quad \frac{3(2n+3)}{4a^3(n+3)}$
25. $\frac{3x^2-7x+2}{4a^5b^2} \div \frac{2x^2-3x-2}{8a^4b^2}$ $\quad \frac{2(3x-1)}{a(2x+1)}$
26. $\frac{x^2-5xy+6y^2}{10x+20y} \div \frac{4x-12y}{5}$ $\quad \frac{x-2y}{8(x+2y)}$
27. $\frac{c^2+2cd+d^2}{a^2-b^2} \div \frac{8c+8d}{-4a+4b}$ $\quad \frac{-1(c+d)}{2(a+b)}$
28. $(n^2-2n+1) \div \frac{6n-6}{12n^2}$ $\quad 2n^2(n-1)$
29. $\frac{3x^2+5x-2}{-2+6x} \div (5x^2+6x-8)$ $\quad \frac{1}{2(5x-4)}$

PART C

Multiply.

30. $\frac{x^{7a}}{y^{6a}} \cdot \frac{y^a}{x^{5a}}$
31. $\frac{x^3-8}{x^2-4x+16} \cdot \frac{x^3+64}{3x-6}$
32. $\frac{x^{2a}-7x^a+12}{2x^{2a}+6} \cdot \frac{x^{4a}-9}{3x^a-12}$
33. $\frac{x^{2a}-y^{2b}}{4x^{2c}-20x^c+25} \cdot \frac{8x^c-20}{3x^a+3y^b}$

Adding: Same Denominators

OBJECTIVE

■ To add rational expressions like $\frac{2x+3}{x^2+3x-10} + \frac{x-9}{x^2+3x-10}$

REVIEW CAPSULE

$$\frac{2}{7} + \frac{4}{7} = \frac{2+4}{7}, \text{ or } \frac{6}{7}$$

Same denominator

EXAMPLE 1 Add $\frac{x}{5} + \frac{3}{5}$.

Denominators are alike. Add numerators. → $\frac{x}{5} + \frac{3}{5} = \frac{x+3}{5}$

Add $\frac{3}{5y} + \frac{7n}{5y}$.

$$\frac{3}{5y} + \frac{7n}{5y} = \frac{3+7n}{5y}$$

$$\frac{a}{c} + \frac{b}{c} = \frac{a+b}{c}$$

To add fractions with like denominators, add the numerators.

EXAMPLE 2 Add $\frac{7}{3x^2y} + \frac{5}{3x^2y}$.

$\frac{a}{c} + \frac{b}{c} = \frac{a+b}{c}$ → $\frac{7+5}{3x^2y} = \frac{12}{3x^2y}$

Simplify. → $= \frac{\overset{4}{\cancel{12}}}{\underset{1}{\cancel{3}} \cdot x^2 \cdot y}$

$$= \frac{4}{x^2y}$$

EXAMPLE 3 Add $\frac{5x}{2n-6} + \frac{-9x}{2n-6}$.

$\frac{a}{c} + \frac{b}{c} = \frac{a+b}{c}$ → $\frac{5x + (-9x)}{2n-6} = \frac{-4x}{2n-6}$

Combine: $5x + (-9x) = -4x$.

Factor and simplify. $2n - 6 = 2(n-3)$ → $= \frac{\overset{-2}{\cancel{-4}} \cdot x}{\underset{1}{\cancel{2}}(n-3)}$

$$= \frac{-2x}{n-3}$$

EXAMPLE 4 Add $\frac{x-3}{xy}+\frac{3x+7}{xy}+\frac{x-4}{xy}$.

$\frac{a}{c}+\frac{b}{c}+\frac{d}{c}=\frac{a+b+d}{c}$ ⟶ $\frac{(x-3)+(3x+7)+(x-4)}{xy}=\frac{5x}{xy}$

$\frac{5\cdot \overset{1}{\cancel{x}}}{\underset{1}{\cancel{x}}\cdot y}=\frac{5}{y}$ ⟶ $=\frac{5}{y}$

EXAMPLE 5 Add $\frac{x+2}{x^2-7x+12}+\frac{x-8}{x^2-7x+12}$.

$\frac{a}{c}+\frac{b}{c}=\frac{a+b}{c}$ ⟶ $\frac{(x+2)+(x-8)}{x^2-7x+12}=\frac{2x-6}{x^2-7x+12}$

Factor and simplify. ⟶ $=\frac{2\overset{1}{\cancel{(x-3)}}}{\underset{1}{\cancel{(x-3)}}(x-4)}$, or $\frac{2}{x-4}$

EXERCISES

PART A

Add.

1. $\frac{3a}{8}+\frac{a}{8}$ $\frac{a}{2}$
2. $\frac{5x}{6}+\frac{7x}{6}$ $2x$
3. $\frac{5a}{4}+\frac{-3}{4}$ $\frac{5a-3}{4}$
4. $\frac{2}{3y}+\frac{7}{3y}$ $\frac{3}{y}$
5. $\frac{a-4}{ab}+\frac{a+4}{ab}$ $\frac{2}{b}$
6. $\frac{11c}{2x^2y}+\frac{-5c}{2x^2y}$ $\frac{3c}{x^2y}$
7. $\frac{5x}{12}+\frac{-7x}{12}+\frac{11x}{12}$ $\frac{3x}{4}$
8. $\frac{12}{5a}+\frac{7}{5a}+\frac{-2}{5a}$ $\frac{17}{5a}$
9. $\frac{x-3}{xy}+\frac{x+2}{xy}+\frac{x+1}{xy}$ $\frac{3}{y}$
10. $\frac{7}{3n-12}+\frac{2}{3n-12}$ $\frac{3}{n-4}$
11. $\frac{x}{2x+6}+\frac{3}{2x+6}$ $\frac{1}{2}$
12. $\frac{2c}{4c-20}+\frac{-10}{4c-20}$ $\frac{1}{2}$
13. $\frac{2x+5}{5x+10}+\frac{x+1}{5x+10}$ $\frac{3}{5}$
14. $\frac{5c-15}{3c-9}+\frac{3-c}{3c-9}$ $\frac{4}{3}$
15. $\frac{4n+3}{2n+8}+\frac{2n+3}{2n+8}$ $\frac{3(n+1)}{n+4}$
16. $\frac{x-5}{x^2-6x+8}+\frac{x+1}{x^2-6x+8}$ $\frac{2}{x-4}$
17. $\frac{3x}{x^2-25}+\frac{15}{x^2-25}$ $\frac{3}{x-5}$
18. $\frac{2a-5}{a^2+2a-15}+\frac{2a-7}{a^2+2a-15}$ $\frac{4}{a+5}$
19. $\frac{n^2+2n}{n^2+6n+9}+\frac{2n+3}{n^2+6n+9}$ $\frac{n+1}{n+3}$

PART B

EXAMPLE Add $\frac{x}{x^3-9x}+\frac{3}{x^3-9x}$.

x is the GCF of x^3-9x.
Factor completely.

$\frac{x+3}{x^3-9x}=\frac{x+3}{x(x^2-9)}=\frac{\overset{1}{\cancel{x+3}}}{x\underset{1}{\cancel{(x+3)}}(x-3)}=\frac{1}{x(x-3)}$

Add.

20. $\frac{n}{n^3-16n}+\frac{4}{n^3-16n}$ $\frac{1}{n(n-4)}$
21. $\frac{3x+7}{2x^2+10x+12}+\frac{x+5}{2x^2+10x+12}$ $\frac{2}{x+2}$
22. $\frac{2c^2-7}{c^3-10c^2+25c}+\frac{7-10c}{c^3-10c^2+25c}$ $\frac{2}{c-5}$
23. $\frac{a-4}{3a^2+12a-36}+\frac{2}{3a^2+12a-36}$ $\frac{1}{3(a+6)}$

Sequence Suspense

Choose a fraction, say $\frac{2}{3}$. Form a second fraction by placing 3 in the numerator and the sum 2 + 3 in the denominator.

Form a third fraction by placing 5 in the numerator and the sum 3 + 5 in the denominator.

$$\frac{2}{3}, \frac{3}{5}, \frac{5}{8}$$

← denominator of previous fraction (numerator)

← sum of numerator and denominator of previous fraction (denominator)

Continue the process until you have a sequence of 10 fractions.

$$\frac{2}{3}, \frac{3}{5}, \frac{5}{8}, \frac{8}{13}, \frac{13}{21}, \ldots \frac{?}{?}$$ tenth fraction

Convert the tenth fraction to a decimal with three places.

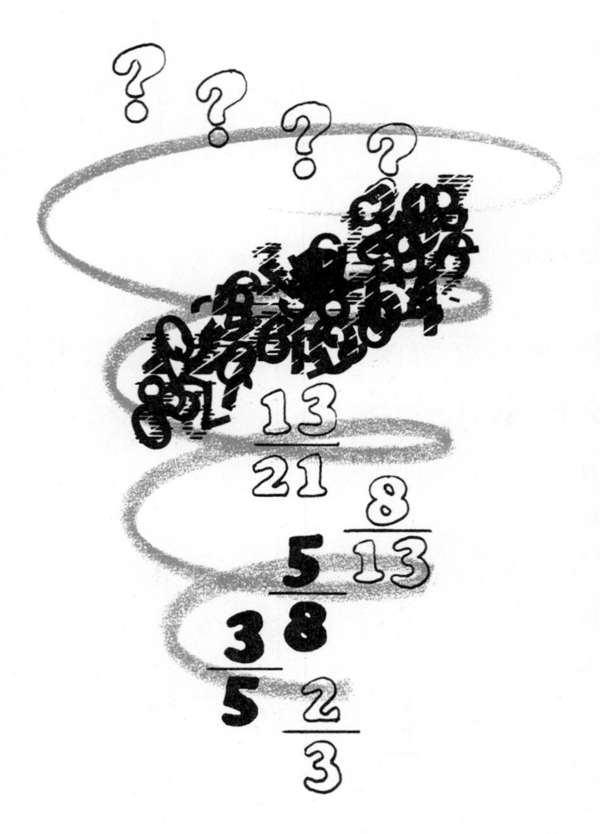

PROJECT

Choose a fraction (other than $\frac{2}{3}$) and form a sequence in the same manner as above. Carry your sequence as far as you like beyond the tenth fraction. Convert your last fraction to a decimal with three places.

Form as many sequences as you wish, but start with a different fraction each time. Convert your last fraction to a decimal with three places. What can you conclude?

Follow this flow chart to find the tenth fraction in a sequence.

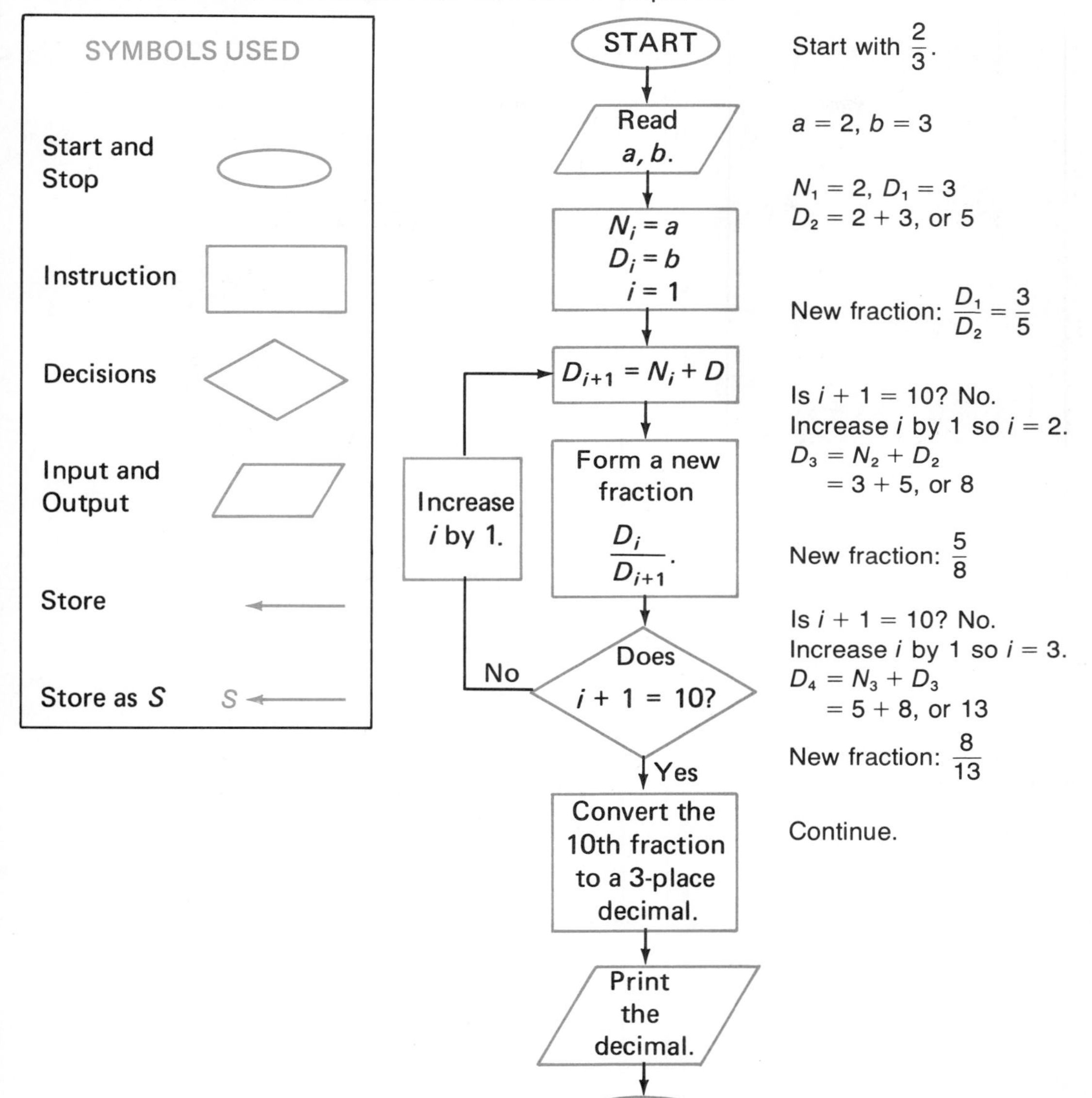

PROJECT Use the flow chart to find the tenth fraction in a sequence. Start with $\frac{1}{3}$.

Adding: Different Denominators

OBJECTIVE

- To add rational expressions like $\frac{5x}{x^2 - 7x + 12} + \frac{3}{2x - 6}$

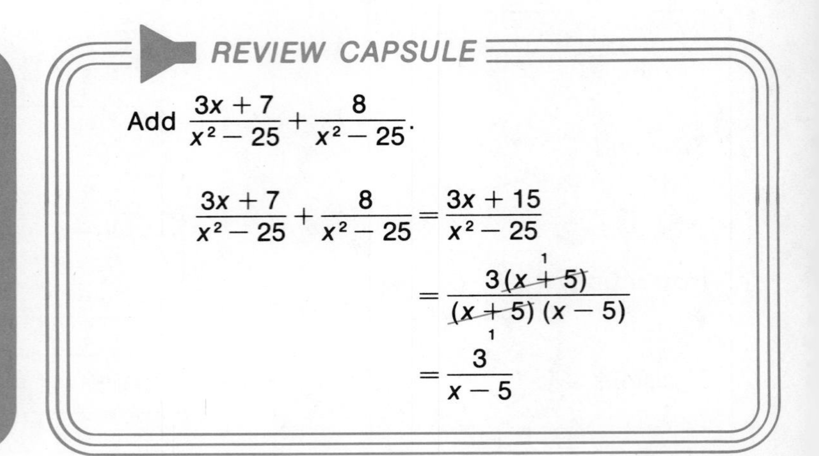

REVIEW CAPSULE

Add $\frac{3x+7}{x^2-25} + \frac{8}{x^2-25}$.

$$\frac{3x+7}{x^2-25} + \frac{8}{x^2-25} = \frac{3x+15}{x^2-25}$$

$$= \frac{3\cancel{(x+5)}^{1}}{\cancel{(x+5)}_{1}(x-5)}$$

$$= \frac{3}{x-5}$$

EXAMPLE 1 Find the least common denominator (LCD) of $\frac{7}{20}$ and $\frac{4}{15}$. Rewrite each fraction in terms of the LCD.

Factor denominators. →

$$\frac{7}{20} = \frac{7}{2 \cdot 2 \cdot 5} \qquad \frac{4}{15} = \frac{4}{3 \cdot 5}$$

LCD has all factors from both denominators. The common factor, 5 in this case, is included once.

LCD is $2 \cdot 2 \cdot 5 \cdot 3$.

$\frac{3}{3} = 1$, $\frac{2 \cdot 2}{2 \cdot 2} = 1$.

$$\frac{7}{2 \cdot 2 \cdot 5}\left(\frac{3}{3}\right) = \frac{21}{60} \qquad \frac{4}{3 \cdot 5}\left(\frac{2 \cdot 2}{2 \cdot 2}\right) = \frac{16}{60}$$

Multiply by 1.

Common denominators →

Thus, the LCD is 60 and $\frac{7}{20} = \frac{21}{60}$, $\frac{4}{15} = \frac{16}{60}$.

EXAMPLE 2 Add $\frac{x}{5} + \frac{3}{25} + \frac{2x+1}{10}$.

Factor denominators. Find the LCD.

$$\frac{x}{5} + \frac{3}{5 \cdot 5} + \frac{2x+1}{2 \cdot 5} \leftarrow \text{LCD is } 2 \cdot 5 \cdot 5.$$

Rewrite each fraction in terms of the LCD. →

$$\frac{x(5 \cdot 2)}{5(5 \cdot 2)} + \frac{3(2)}{5 \cdot 5(2)} + \frac{(2x+1)(5)}{2 \cdot 5(5)}$$

$$\frac{10x}{50} + \frac{6}{50} + \frac{10x+5}{50}$$

The denominators are the same. Add the numerators. →

$$\frac{10x + 6 + 10x + 5}{50}$$

Combine like terms. →

$$\frac{20x+11}{50}$$

EXAMPLE 3 Add $\frac{2}{ac} + \frac{3}{ab^2} + \frac{5}{c^3}$.

Find the LCD. ⟶ $\frac{2}{ac} + \frac{3}{abb} + \frac{5}{ccc}$

Multiply numerator and denominator by factors needed. Rewrite. ⟶

$$\frac{2(bbcc)}{ac(bbcc)} + \frac{3(ccc)}{abb(ccc)} + \frac{5(abb)}{ccc(abb)} \leftarrow \text{LCD is } ab^2c^3.$$

$$\frac{2b^2c^2}{ab^2c^3} + \frac{3c^3}{ab^2c^3} + \frac{5ab^2}{ab^2c^3}$$

Add. ⟶ $\frac{2b^2c^2 + 3c^3 + 5ab^2}{ab^2c^3}$

EXAMPLE 4 Add $\frac{5}{3x+9} + \frac{2x}{x^2 - 2x - 15}$.

Factor denominators. ⟶ $\frac{5}{3(x+3)} + \frac{2x}{(x+3)(x-5)}$

LCD is $3(x+3)(x-5)$.

Multiply by factors needed.

$$\frac{5(x-5)}{3(x+3)(x-5)} + \frac{(3)2x}{(3)(x+3)(x-5)}$$

$5(x-5) = 5x - 25$; $(3)2x = 6x$ ⟶ $\frac{5x - 25 + 6x}{3(x+3)(x-5)}$

Combine like terms.

The fraction is in simplest form.

$$\frac{11x - 25}{3(x+3)(x-5)}$$

EXAMPLE 5 Add $3c + 1 + \frac{5}{2c}$.

$3c + 1 = \frac{3c+1}{1}$ ⟶ $\frac{3c+1}{1} + \frac{5}{2c}$

LCD is 2c.

$$\frac{2c(3c+1)}{2c \cdot 1} + \frac{5}{2c}$$

The fraction is in simplest form.

$$\frac{6c^2 + 2c + 5}{2c}$$

EXAMPLE 6 Add $\frac{5x+12}{x^2+3x} + \frac{x+2}{x+3}$. | Add $\frac{x-1}{x} + \frac{x+2}{2x}$.

Factor the denominators.

$x^2 + 3x = x(x+3)$ ⟶

LCD is $x(x+3)$.

$$\frac{5x+12}{x(x+3)} + \frac{x+2}{x+3} \qquad \Bigg| \qquad \frac{x-1}{x} + \frac{x+2}{2x}$$

$$\frac{5x+12}{x(x+3)} + \frac{x(x+2)}{x(x+3)} \qquad \Bigg| \qquad \frac{2(x-1)}{2x} + \frac{x+2}{2x}$$

$$\frac{5x + 12 + x^2 + 2x}{x(x+3)} \qquad \Bigg| \qquad \frac{2x - 2 + x + 2}{2x}$$

Combine like terms and arrange in descending order.

$$\frac{x^2 + 7x + 12}{x(x+3)} \qquad \Bigg| \qquad \frac{3x}{2x}$$

Factor the numerator and simplify. ⟶

$$\frac{(x+4)\cancel{(x+3)}^{1}}{x\cancel{(x+3)}_{1}} \qquad \Bigg| \qquad \frac{3\cancel{x}^{1}}{2\cancel{x}_{1}}$$

$$\frac{x+4}{x} \qquad \Bigg| \qquad \frac{3}{2}$$

ORAL EXERCISES

Find the LCD.

1. $\frac{5a}{6} + \frac{3b}{8} + \frac{7c}{12}$ $\quad 24$
2. $\frac{2}{x^2y} + \frac{4}{xy} + \frac{5}{xy^3}$ $\quad x^2y^3$
3. $\frac{c}{4(x+2)} + \frac{d}{6(x+3)}$ $\quad 12(x+2)(x+3)$
4. $\frac{x-2}{(x+3)(x+3)} + \frac{3}{2(x+3)}$ $\quad 2(x+3)(x+3)$
5. $2a + 3 + \frac{4}{a-2}$ $\quad a-2$
6. $\frac{6x}{(x+2)(x-2)} + \frac{2}{3(x-2)}$ $\quad 3(x+2)(x-2)$

EXERCISES

PART A

Add.

1. $\frac{2c}{3} + \frac{7c}{6}$ $\quad \frac{11c}{6}$
2. $\frac{3x}{8} + \frac{5y}{12}$ $\quad \frac{9x+10y}{24}$
3. $\frac{n+2}{5} + \frac{n-3}{7}$ $\quad \frac{12n-1}{35}$
4. $\frac{3b}{a} + \frac{-2}{ab}$ $\quad \frac{3b^2-2}{ab}$
5. $\frac{7}{x^2} + \frac{4}{xy}$ $\quad \frac{7y+4x}{x^2y}$
6. $\frac{3}{4xy^2} + \frac{1}{6x^2y}$ $\quad \frac{9x+2y}{12x^2y^2}$
7. $\frac{3x}{10} + \frac{2x-3}{15} + \frac{x}{5}$ $\quad \frac{19x-6}{30}$
8. $\frac{4}{ab} + \frac{-2}{a^2} + \frac{3}{ab^3}$
9. $\frac{y}{2x} + \frac{4}{xy^2} + \frac{x}{6y}$
10. $\frac{4x}{3x+12} + \frac{2x}{5x+20}$
11. $\frac{a-2}{2a-6} + \frac{a}{2a-8}$
12. $\frac{n}{3n-2} + \frac{-3}{4n+1}$
13. $x - 3 + \frac{1}{4x}$
14. $\frac{3}{a+2} + 2a$
15. $2n + 1 + \frac{5n}{n-3}$ $\quad \frac{2n^2-3}{n-3}$
16. $\frac{2x}{x^2-4} + \frac{4}{x+2}$ $\quad \frac{6x-8}{(x+2)(x-2)}$
17. $\frac{5}{2a-6} + \frac{3a-1}{a^2-9}$ $\quad \frac{11a+13}{2(a+3)(a-3)}$
18. $\frac{2n+3}{n^2-16} + \frac{4n}{3n+12}$ $\quad \frac{4n^2-10n+9}{3(n+4)(n-4)}$

Add.

19. $\frac{x+2}{x} + \frac{x-6}{3x}$ $\quad \frac{4}{3}$
20. $\frac{a+12}{4a} + \frac{a-3}{a}$ $\quad \frac{5}{4}$
21. $\frac{2x}{3} + \frac{5x}{6}$ $\quad \frac{3x}{2}$
22. $\frac{20-x}{x^2+4x} + \frac{x+10}{x+4}$ $\quad \frac{x+5}{x}$
23. $\frac{a-7}{a-5} + \frac{10}{a^2-5a}$ $\quad \frac{a-2}{a}$
24. $\frac{2n-6}{n^2+6n} + \frac{n+3}{n+6}$ $\quad \frac{n-1}{n}$

PART B

Add.

25. $\frac{x+3}{x^2-6x+8} + \frac{2}{3x-6}$ $\quad \frac{5x+1}{3(x-2)(x-4)}$
26. $\frac{3x-1}{x^2+8x+15} + \frac{5}{2x+10}$ $\quad \frac{11x+13}{2(x+3)(x+5)}$
27. $\frac{2x-1}{x^2+2x-3} + \frac{x+2}{x^2-9}$ $\quad \frac{3x^2-6x+1}{(x+3)(x-1)(x-3)}$
28. $\frac{2x+1}{x^2+2x-24} + \frac{x-1}{x^2-16}$
29. $\frac{3x}{2x^2-x-10} + \frac{-2x}{2x^2-7x+5}$
30. $\frac{x+1}{3x^2-2x-1} + \frac{x-1}{3x^2+4x+1}$
31. $\frac{4}{n+1} + \frac{n+2}{n^2-1} + \frac{3}{n-1}$
32. $\frac{5}{a-2} + \frac{a+3}{a^2-4} + \frac{-2}{a+2}$
33. $\frac{2a+b}{a^2-b^2} + \frac{2}{5a+5b}$ $\quad \frac{12a+3b}{5(a+b)(a-b)}$
34. $\frac{3a}{a+b} + \frac{2b}{2a-3b}$ $\quad \frac{6a^2-7ab+2b^2}{(a+b)(2a-3b)}$
35. $\frac{x+2y}{x^2-2xy+y^2} + \frac{4}{3x-3y}$ $\quad \frac{7x+2y}{3(x-y)(x-y)}$
36. $\frac{2x-y}{9x^2-4y^2} + \frac{3}{6x+4y}$ $\quad \frac{13x-8y}{2(3x+2y)(3x-2y)}$

Rational Expressions: $\frac{a}{b} - \frac{c}{d}$

OBJECTIVE

- To simplify rational expressions like $\frac{4x}{x^2-9} - \frac{3}{2x+6}$

REVIEW CAPSULE

Factor -1 from $4 - x$.

$$4 - x = -x + 4$$
$$= -1x + 4$$
$$= -1(x - 4)$$

EXAMPLE 1 Simplify $-\frac{8}{2}$. Simplify $\frac{-1(8)}{2}$. Simplify $\frac{8}{-1(2)}$.

$\frac{-1(8)}{2} = \frac{-8}{2} = -4$

$\frac{8}{-1(2)} = \frac{8}{-2} = -4$

$-\frac{8}{2}$	$\frac{-1(8)}{2}$	$\frac{8}{-1(2)}$
-4	-4	-4

Each quotient is equal to -4. ⟶ **Thus,** $-\frac{8}{2} = \frac{-1(8)}{2} = \frac{8}{-1(2)}$.

Example 1 suggests this. ⟶

$$-\frac{a}{b} = \frac{-1(a)}{b} = \frac{a}{-1(b)}$$

EXAMPLE 2 Rewrite $-\frac{3x-2}{x^2-4}$.

$-\frac{a}{b} = \frac{-1(a)}{b}$ ⟶ $-\frac{3x-2}{x^2-4} = \frac{-1(3x-2)}{x^2-4}$, or $\frac{-3x+2}{x^2-4}$

EXAMPLE 3 Simplify $\frac{n}{3} - \frac{5n}{3}$.

$$\frac{n}{3} - \frac{5n}{3}$$

$-\frac{5n}{3} = \frac{-1(5n)}{3}$ ⟶ $\frac{n}{3} + \frac{-1(5n)}{3}$

$$\frac{n}{3} + \frac{-5n}{3}$$

$\frac{a}{c} + \frac{b}{c} = \frac{a+b}{c}$ ⟶ $\frac{n-5n}{3}$

Combine like terms. $1n - 5n = -4n$ } $\frac{-4n}{3}$

EXAMPLE 4 Simplify $\frac{2x}{x+2} - \frac{3}{4}$.

$-\frac{3}{4} = \frac{-1(3)}{4}$ ⟶ $\frac{2x}{x+2} - \frac{3}{4} = \frac{2x}{x+2} + \frac{-1(3)}{4}$

LCD is $4(x+2)$. $= \frac{4 \cdot 2x}{4(x+2)} + \frac{-3(x+2)}{4(x+2)}$

$\frac{a}{c} + \frac{b}{c} = \frac{a+b}{c}$

$-3(x+2) = -3x - 6$ $= \frac{8x - 3x - 6}{4(x+2)}$

Combine like terms. $= \frac{5x-6}{4(x+2)}$

EXAMPLE 5 Simplify $\frac{3n}{n^2-25} - \frac{15}{n^2-25}$.

$-\frac{15}{n^2-25} = \frac{-1(15)}{n^2-25}$ ⟶ $\frac{3n}{n^2-25} + \frac{-1(15)}{n^2-25}$

$\frac{a}{c} + \frac{b}{c} = \frac{a+b}{c}$ $\frac{3n-15}{n^2-25}$

Factor the numerator and denominator. ⟶ $\frac{3(n-5)}{(n+5)(n-5)}$

Write in simplest form. ⟶ $\frac{3}{n+5}$

EXAMPLE 6 Simplify $\frac{7}{x-3} - \frac{4x-5}{x^2-9}$.

$-\frac{4x-5}{x^2-9} = \frac{-1(4x-5)}{x^2-9}$ ⟶ $\frac{7}{x-3} + \frac{-1(4x-5)}{x^2-9}$

$-1(4x-5) = -4x+5$ ⟶ $\frac{7}{x-3} + \frac{-4x+5}{(x-3)(x+3)}$

Factor: $x^2 - 9 = (x-3)(x+3)$. ⟶

LCD is $(x-3)(x+3)$. $\frac{7(x+3)}{(x-3)(x+3)} + \frac{-4x+5}{(x-3)(x+3)}$

$\frac{a}{c} + \frac{b}{c} = \frac{a+b}{c}$ ⟶ $\frac{7x+21-4x+5}{(x-3)(x+3)}$

Combine like terms. $\frac{3x+26}{(x-3)(x+3)}$

EXAMPLE 7 Simplify $\frac{7}{5-n} + \frac{4}{n-5}$.

Arrange the denominator of $\frac{7}{5-n}$ in descending order. ⟶ $\frac{7}{-n+5} + \frac{4}{n-5}$

Factor -1 from $-n+5$. ⟶ $\frac{7}{-1(n-5)} + \frac{4}{n-5}$

$\frac{a}{-1(b)} = \frac{-1(a)}{b}$ ⟶ $\frac{-1(7)}{n-5} + \frac{4}{n-5}$

$\frac{-7+4}{n-5}$

$\frac{-3}{n-5}$

EXAMPLE 8 Simplify $\frac{5x}{x^2-9}+\frac{2}{3-x}$.

Rewrite $\frac{2}{3-x}$ so that the denominator contains $x-3$. → $\frac{5x}{x^2-9}+\frac{2}{-x+3}$

$\frac{5x}{x^2-9}+\frac{2}{-1(x-3)}$

$\frac{a}{-1(b)}=\frac{-1(a)}{b}$ → $\frac{5x}{x^2-9}+\frac{-1(2)}{x-3}$

Factor x^2-9. → $\frac{5x}{(x-3)(x+3)}+\frac{-2}{x-3}$

LCD is $(x-3)(x+3)$. → $\frac{5x}{(x-3)(x+3)}+\frac{-2(x+3)}{(x-3)(x+3)}$

$-2(x+3)=-2x-6$ → $\frac{5x-2x-6}{(x-3)(x+3)}$

Combine like terms. $\frac{3x-6}{(x-3)(x+3)}$

EXAMPLE 9 Simplify $\frac{5x}{x^2-9}-\frac{2}{3-x}$.

$-\frac{2}{3-x}=\frac{2}{-1(3-x)}$ → $\frac{5x}{x^2-9}+\frac{2}{-1(3-x)}$

$-1(3-x)=-3+x$ → $\frac{5x}{x^2-9}+\frac{2}{-3+x}$

$-3+x=x-3$ → $\frac{5x}{x^2-9}+\frac{2}{x-3}$

$\frac{5x}{(x-3)(x+3)}+\frac{2(x+3)}{(x-3)(x+3)}$

$\frac{5x+2x+6}{(x-3)(x+3)}$

Combine like terms. $\frac{7x+6}{(x-3)(x+3)}$

ORAL EXERCISES

True or false?

1. $-\frac{7}{4}=\frac{7}{-4}$ *T*

2. $-\frac{8}{3}=\frac{-8}{3}$ *T*

3. $-\frac{2}{5}=\frac{-2}{-5}$ *F*

4. $\frac{10}{-2}=\frac{-10}{2}$ *T*

5. $-\frac{x+3}{x-2}=\frac{-1(x+3)}{x-2}$ *T*

6. $\frac{x-5}{-1(x+4)}=\frac{-1(x-5)}{x+4}$ *T*

7. $6-x=x-6$ *F*

8. $7-n=-n+7$ *T*

9. $-x^2+9=-1(x^2-9)$ *T*

10. $-\frac{c+3}{c^2-1}=\frac{-1(c+3)}{-1(c^2-1)}$ *F*

11. $-\frac{x+y}{3-x}=\frac{x+y}{-1(3-x)}$ *T*

12. $\frac{2n+3}{-1(n^2-4)}=\frac{-1(2n+3)}{n^2-4}$ *T*

EXERCISES

PART A

Simplify.

1. $\frac{5x}{6} - \frac{x}{6}$ $\frac{2x}{3}$

2. $\frac{c+d}{cd} - \frac{c-d}{cd}$ $\frac{2}{c}$

3. $\frac{9}{4x+12} - \frac{1}{4x+12}$ $\frac{2}{x+3}$

4. $\frac{3a}{5a+10} - \frac{-6}{5a+10}$ $\frac{3}{5}$

5. $\frac{7a-10}{3a-12} - \frac{a+8}{3a-12}$ $\frac{2(a-3)}{a-4}$

6. $\frac{6n+5}{2n+6} - \frac{2n-7}{2n+6}$ 2

7. $\frac{7n}{12} - \frac{3n}{8}$ $\frac{5n}{24}$

8. $\frac{3}{4c^2d^2} - \frac{-5}{6cd}$ $\frac{9+10cd}{12c^2d^2}$

9. $\frac{5x}{2x-8} - \frac{3x}{2x-6}$ $\frac{2x^2-3x}{2(x-4)(x-3)}$

10. $\frac{x}{x+3} - \frac{2}{x-3}$ $\frac{x^2-5x-6}{(x+3)(x-3)}$

11. $\frac{7}{x-5} + \frac{3}{5-x}$ $\frac{4}{x-5}$

12. $\frac{2a}{a+3} + \frac{-6}{-a-3}$ 2

13. $\frac{8}{x-4} - \frac{5}{4-x}$ $\frac{13}{x-4}$

14. $\frac{4a}{a+2} - \frac{8}{-a-2}$ 4

PART B

Simplify.

15. $\frac{4a}{a^2-16} - \frac{a-12}{a^2-16}$ $\frac{3}{a-4}$

16. $\frac{7n-15}{n^2-36} - \frac{2n+15}{n^2-36}$ $\frac{5}{n+6}$

17. $\frac{2n-1}{n^2-4} - \frac{-4n}{3n+6}$ $\frac{4n^2-2n-3}{3(n+2)(n-2)}$

18. $\frac{7}{2a-10} - \frac{3a-2}{a^2-25}$ $\frac{a+39}{2(a+5)(a-5)}$

19. $\frac{4x-5}{x^2-7x+12} - \frac{x+7}{x^2-7x+12}$ $\frac{3}{x-3}$

20. $\frac{7n+5}{n^2-3n-10} - \frac{2n-5}{n^2-3n-10}$ $\frac{5}{n-5}$

21. $\frac{2a+3}{a^2+2a-3} - \frac{3}{2a+6}$ $\frac{a+9}{2(a+3)(a-1)}$

22. $\frac{c-3}{c^2-2c-8} - \frac{-2}{3c-12}$ $\frac{5c-5}{3(c+2)(c-4)}$

23. $\frac{3x+17}{x^2-25} - \frac{x+3}{25-x^2}$ $\frac{4}{x-5}$

24. $\frac{2x+5}{x^2-9} + \frac{4-x}{9-x^2}$ $\frac{3x+1}{(x+3)(x-3)}$

PART C

Simplify.

25. $\frac{10x-3}{2x-4} + \frac{2x-5}{4-2x} - \frac{4x+10}{2x-4}$

26. $\frac{3x}{x-5} + \frac{x}{5-x} - \frac{2x}{5-x}$

27. $\frac{7}{x} - \frac{3}{y} - \frac{5}{z}$

28. $\frac{8}{x+4} - \frac{3}{x} - \frac{2}{x+2}$

29. $\frac{5x}{x^2-9} - \frac{7}{2x-6} + \frac{3}{4x+12}$

30. $\frac{3x}{4-x^2} + \frac{5}{6-3x} - \frac{4}{5x+10}$

31. $\frac{5}{a^2+4a+4} - \frac{4}{a^2+5a+6}$

32. $\frac{2c}{3c^2+5c-2} - \frac{c}{3c^2-7c+2}$

Fractured Fractions

PROBLEM

Find the two rational expressions whose sum is $\frac{7x+9}{(x+1)(x+2)}$.

$$\frac{7x+9}{(x+1)(x+2)} = \frac{A}{x+1} + \frac{B}{x+2}$$
$$= \frac{A}{x+1}\left(\frac{x+2}{x+2}\right) + \frac{B}{x+2}\left(\frac{x+1}{x+1}\right)$$
$$= \frac{Ax+2A+Bx+B}{(x+1)(x+2)}$$
$$\frac{7x+9}{(x+1)(x+2)} = \frac{(A+B)x+(2A+B)}{(x+1)(x+2)}$$

Equal fractions with equal denominators have equal numerators.

$$7x+9 = \underbrace{(A+B)}_{7}x + \underbrace{(2A+B)}_{9}$$

Since $A+B=7$
$A = 7-B$

Since $2A+B=9$
$2(7-B)+B=9$
$14-2B+B=9$
$-B=-5$
$B=5$
$A=7-5$, or 2

So, $\frac{7x+9}{(x+1)(x+2)} = \frac{2}{x+1} + \frac{5}{x+2}$.

Check.

$\frac{7x+9}{(x+1)(x+2)}$	$\frac{2}{x+1} + \frac{5}{x+2}$
	$\frac{2(x+2)}{(x+1)(x+2)} + \frac{5(x+1)}{(x+1)(x+2)}$
	$\frac{2x+4+5x+5}{(x+1)(x+2)}$
	$\frac{7x+9}{(x+1)(x+2)}$

PROJECT

Find the two rational expressions whose sum is given below.

1. $\frac{2x+8}{(x+5)(x+3)}$ 2. $\frac{3x+15}{(x+1)(x+7)}$ 3. $\frac{4x+9}{(x+1)(x+5)}$

Chapter Four Review

Which are rational expressions? [p. 85]

1. $\frac{3a+2}{x^2-4x+4}$
2. $\frac{2n-4}{0}$
3. $16a^2-b^2$

Evaluate if $x=3$. [p. 85]

4. $\frac{x^2-4x+4}{x+3}$
5. $\frac{3x}{x^2+5x+6}$
6. $\frac{4x}{x^2-9}$

For what value(s) of the variable is the expression undefined? [p. 85]

7. $\frac{3}{4n}$
8. $\frac{a}{4a-12}$
9. $\frac{4c}{c^2-5c+6}$

Simplify. [p. 87]

10. $\frac{-8a^5b^3cd}{12a^2b^5c}$
11. $\frac{8x-10}{6}$
12. $\frac{3x+6}{5x+10}$
13. $\frac{x^2-3x-10}{2x^2+3x-2}$
14. $\frac{2a^2-50}{6a+30}$
15. $\frac{a-2}{10-5a}$

Find the reciprocal of the expression. [p. 90]

16. $\frac{2a-1}{a+4}$
17. x^2-9
18. $\frac{-3x}{y}$

Multiply. [p. 90]

19. $\frac{10a^2}{3b}\cdot\frac{b^4c^2}{-15a^3}$
20. $\frac{4x^2-4}{3x+12}\cdot\frac{x^2-16}{x^2-5x+4}$
21. $(3c^2+c-2)\cdot\frac{6}{8c+8}$

Divide. [p. 90]

22. $\frac{9x}{2y}\div\frac{-12x^3}{y^3z}$
23. $\frac{5c-10}{3x+3y}\div\frac{2c^2-8}{x^2+2xy+y^2}$
24. $\frac{n^2-2n-8}{n^2+6n+8}\div(2n-8)$

Add. [p. 94, 98]

25. $\frac{11c}{4xy}+\frac{-3c}{4xy}$
26. $\frac{2a-9}{5a-10}+\frac{a+3}{5a-10}$
27. $\frac{2x+5}{x^2-9}+\frac{3x+10}{x^2-9}$
28. $\frac{5}{6xy}+\frac{2y}{9x^2}$
29. $\frac{2a}{3a-6}+\frac{4a}{3a-12}$
30. $n+2+\frac{4n}{n+3}$
31. $\frac{2x+7}{x^2-x-12}+\frac{2x+5}{x^2-x-12}$
32. $\frac{3n+2}{2n^2+3n-5}+\frac{4n-9}{2n^2+3n-5}$
33. $\frac{3x-1}{x^2+2x-15}+\frac{2}{3x-9}$
34. $\frac{2n}{n^2+4n+4}+\frac{3n}{n^2+7n+10}$

Simplify. [p. 101]

35. $\frac{7x+5}{3x+6}-\frac{2x-5}{3x+6}$
36. $\frac{2}{x-4}-\frac{9}{4-x}$
37. $\frac{3}{2x-3}+\frac{-5}{3-2x}$
38. $\frac{5}{x+2}-\frac{3}{x-4}$
39. $\frac{4}{x-5}-\frac{-2}{25-x^2}$
40. $\frac{2}{3x-12}+\frac{3}{16-x^2}$

Chapter Four Test

Which are rational expressions?

1. $5x^2 - 10x$
2. $\frac{x+7}{n^2+5n+3}$
3. $\frac{x-2}{0}$

Evaluate if $x = 5$.

4. $\frac{x+3}{x-3}$
5. $\frac{x^2+2x}{3x}$
6. $\frac{2x}{x^2-25}$

For what value(s) of the variable is the expression undefined?

7. $\frac{-6}{x}$
8. $\frac{3a}{5a-10}$
9. $\frac{n+3}{n^2-36}$

Simplify.

10. $\frac{5a^3b^2c}{10ab^5c}$
11. $\frac{x^2-25}{2x^2-9x-5}$
12. $\frac{4x-12}{3-x}$

Find the reciprocal of the expression.

13. $\frac{x-5}{3y+2}$
14. $x^2 - 25$

Multiply.

15. $\frac{4ac}{b^2} \cdot \frac{3b^5}{8a^3c}$
16. $\frac{5x+15}{4x-8} \cdot \frac{2x-4}{3x+9}$
17. $\frac{x^2-9}{c^2-10c+25} \cdot \frac{6c-30}{4x+12}$
18. $(n^2-25) \cdot \frac{2}{3n+15}$

Divide.

19. $\frac{n^2-4}{2x-3y} \div \frac{5n+10}{4x-6y}$
20. $\frac{a^2-8a+15}{4} \div (a-3)$

Add.

21. $\frac{4a+21}{3a+15} + \frac{a+4}{3a+15}$
22. $\frac{4x+3}{x^2-3x-10} + \frac{3-x}{x^2-3x-10}$
23. $\frac{3}{4a^2b} + \frac{5}{6ab^2}$
24. $\frac{3}{2a-6} + \frac{3}{2a+8}$
25. $x - 3 + \frac{6}{x+2}$
26. $\frac{4}{c^2-5c+6} + \frac{3}{2c-6}$

Simplify.

27. $\frac{7x-2}{2x-1} - \frac{x+1}{2x-1}$
28. $\frac{4}{x-7} + \frac{2}{7-x}$
29. $\frac{3}{x-1} - \frac{4}{1-x^2}$

5 USING RATIONAL EXPRESSIONS

Is 2 Ever Equal to 1?

PROBLEM

Follow the steps in this proof.

	1. Let $a = b$
Multiply each side by a.	2. Then $a \cdot a = a \cdot b$
$a \cdot a = a^2$	3. $a^2 = ab$
Add $-b^2$ to each side.	4. $a^2 - b^2 = ab - b^2$
Factor each side.	5. $(a + b)(a - b) = b(a - b)$
Divide each side by $(a - b)$.	6. $\frac{(a + b)(a - b)}{a - b} = \frac{b(a - b)}{a - b}$
Simplify both fractions. Divide out common factors.	7. $\frac{(a + b)\overset{1}{\cancel{(a - b)}}}{\underset{1}{\cancel{(a - b)}}} = \frac{b\overset{1}{\cancel{(a - b)}}}{\underset{1}{\cancel{(a - b)}}}$
	8. $a + b = b$
From step 1, $a = b$, substitute b for a.	9. $b + b = b$
$b + b = 2b$	10. Thus, $2b = b$
Divide each side by b.	11. $\frac{2b}{b} = \frac{b}{b}$
Simplify both fractions.	12. $\frac{2\overset{1}{\cancel{b}}}{\underset{1}{\cancel{b}}} = \frac{\overset{1}{\cancel{b}}}{\underset{1}{\cancel{b}}}$
$\frac{2b}{b} = 2$ and $\frac{b}{b} = 1$	13. Thus, $2 = 1$

PROJECT

Since $2 \neq 1$, there must be an error in the proof above.

Can you find the error?

Fractional Equations

OBJECTIVES

- To solve equations like $\frac{3}{x+1} + \frac{x-2}{3} = \frac{13}{3x+3}$ and $\frac{x-2}{2x+3} = \frac{-5}{4}$
- To identify extraneous solutions

REVIEW CAPSULE

Solve and check $\frac{2}{3}x = 4$.

$$3\left(\frac{2}{3}x\right) = 3(4)$$
$$2x = 12$$
$$x = 6$$

Check.

$\frac{2}{3}x$	4
$\frac{2}{3}(6)$	4
4	

Thus, the solution is 6.

EXAMPLE 1 Solve $\frac{x}{4} + \frac{7}{12} = \frac{x+1}{3}$

Find the LCD. → $2 \cdot 2 \quad 2 \cdot 2 \cdot 3 \quad 3$ ← LCD is $2 \cdot 2 \cdot 3$ or 12.

Multiply each side by LCD. → $12\left(\frac{x}{4} + \frac{7}{12}\right) = 12\left(\frac{x+1}{3}\right)$

Distribute the 12. → $\left(12 \cdot \frac{x}{4}\right) + \left(12 \cdot \frac{7}{12}\right) = 12 \cdot \frac{x+1}{3}$

Simplify each product. → $\left(\overset{3}{\cancel{12}} \cdot \frac{x}{\underset{1}{\cancel{4}}}\right) + \left(\overset{1}{\cancel{12}} \cdot \frac{7}{\underset{1}{\cancel{12}}}\right) = \overset{4}{\cancel{12}} \cdot \frac{x+1}{\underset{1}{\cancel{3}}}$

The new equation has no fractions. Solve the new equation. →

$$3x + 7 = 4(x+1)$$
$$3x + 7 = 4x + 4$$
$$-x + 7 = 4$$
$$-x = -3$$
$$x = 3$$

Thus, the solution is 3.

EXAMPLE 2 Find the solution set of $\frac{3}{4a} - \frac{5}{6a^2} = \frac{1}{3a}$.

Rewrite $-\frac{5}{6a^2}$ as $\frac{-5}{6a^2}$. → $\frac{3}{4a} + \frac{-5}{6a^2} = \frac{1}{3a}$

The LCD is $12a^2$.
Multiply each side by $12a^2$.
Distribute and simplify.

$$\overset{3a}{\cancel{12a^2}} \cdot \frac{3}{\underset{1\,1}{\cancel{4a}}} + \overset{2\,1}{\cancel{12a^2}} \cdot \frac{-5}{\underset{1\,1}{\cancel{6a^2}}} = \overset{4a}{\cancel{12a^2}} \cdot \frac{1}{\underset{1\,1}{\cancel{3a}}}$$

The new equation has no fractions. →

$$3a(3) + 2(-5) = 4a(1)$$
$$9a - 10 = 4a$$
$$5a = 10$$
$$a = 2$$

Thus, the solution set is $\{2\}$.

EXAMPLE 3 Solve $\frac{2x-9}{x-7}+\frac{x}{2}=\frac{5}{x-7}$ and check.

The LCD is $2(x-7)$. Multiply each side by $2(x-7)$. ⟶

$$2(x-7)\left[\frac{2x-9}{x-7}+\frac{x}{2}\right]=2(x-7)\left[\frac{5}{x-7}\right]$$

Distribute $2(x-7)$ and simplify. ⟶

$$\left[2(x-7)\cdot\frac{2x-9}{x-7}\right]+\left[2(x-7)\cdot\frac{x}{2}\right]=2(x-7)\cdot\frac{5}{x-7}$$

$$2(2x-9)+x(x-7)=2(5)$$

Quadratic equation ⟶

$$4x-18+x^2-7x=10$$

$$x^2-3x-18=10$$

Write in standard form. ⟶

$$x^2-3x-28=0$$

Factor.

$$(x-7)(x+4)=0$$

Set each factor equal to 0.

$$x-7=0 \quad \text{or} \quad x+4=0$$

$$x=7 \qquad\qquad x=-4$$

Check both solutions. ⟶

7 and -4 are solutions of the derived equation, $x^2-3x-28=0$.

$\frac{2x-9}{x-7}+\frac{x}{2}$	$\frac{5}{x-7}$
$\frac{2(7)-9}{(7)-7}+\frac{7}{2}$	$\frac{5}{(7)-7}$
$\frac{14-9}{0}+\frac{7}{2}$	$\frac{5}{0}$

No denominator may be 0.

So, 7 is *not* a solution.

$\frac{2x-9}{x-7}+\frac{x}{2}$	$\frac{5}{x-7}$
$\frac{2(-4)-9}{(-4)-7}+\frac{-4}{2}$	$\frac{5}{(-4)-7}$
$\frac{-17}{-11}+(-2)$	
$\frac{17}{11}-\frac{22}{11}$, or $\frac{-5}{11}$	$\frac{-5}{11}$

So, -4 *is* a solution.

Only -4 is a solution of the original equation.

Thus, -4 is the only solution of the original equation.

7 is an extraneous solution in Example 3.

An *extraneous* solution of an equation is a solution of a derived equation which does not check in the *original* equation.

EXAMPLE 4 Solve $\frac{5}{x-3}-\frac{6}{x^2-9}=\frac{4}{x+3}$.

$-\frac{6}{x^2-9}=\frac{-1(6)}{x^2-9}$ ⟶

$$\frac{5}{x-3}+\frac{-1(6)}{x^2-9}=\frac{4}{x+3}$$

Factor denominators. ⟶

$$\frac{5}{x-3}+\frac{-6}{(x-3)(x+3)}=\frac{4}{x+3}$$

The LCD is $(x-3)(x+3)$. Multiply each side by LCD. Distribute and simplify.

$$\left[(x-3)(x+3)\cdot\frac{5}{x-3}\right]+\left[(x-3)(x+3)\cdot\frac{-6}{(x-3)(x+3)}\right]$$

$$=(x-3)(x+3)\cdot\frac{4}{x+3}$$

$$5x+15-6=4x-12$$

$$5x+9=4x-12$$

$$x=-21$$

Check to see if -21 makes any denominator equal to 0. ⟶ **Thus,** the solution is -21.

A *proportion* is an equation that sets *two* fractions equal.

$$\frac{a}{b} = \frac{c}{d}$$

a and *d* are the *extremes*; *b* and *c* are the *means*.

EXAMPLE 5 Solve the proportion $\frac{3}{x+1} = \frac{2}{x}$.

First Method

$$x(x+1) \cdot \frac{3}{x+1} = x(x+1) \cdot \frac{2}{x}$$

$$3x = 2(x+1)$$
$$3x = 2x + 2$$
$$x = 2$$

Second Method

$$\frac{3}{x+1} \times \frac{2}{x}$$

$$3x = 2(x+1)$$
$$3x = 2x + 2$$
$$x = 2$$

Thus, the solution is 2.

In a proportion, the product of the extremes equals the product of the means.

If $\frac{a}{b} = \frac{c}{d}$, then $ad = bc$.

EXAMPLE 6 Solve the proportion $\frac{x+1}{2} = \frac{3}{x}$.

If $\frac{a}{b} = \frac{c}{d}$ → $\frac{x+1}{2} \times \frac{3}{x}$

then $ad = bc$. → $x(x+1) = 2(3)$

Quadratic equation → $x^2 + x = 6$

Write in standard form. → $x^2 + x - 6 = 0$

$$(x+3)(x-2) = 0$$

$x + 3 = 0$ or $x - 2 = 0$

$x = -3$ $\quad$ $x = 2$

Neither −3 nor 2 is an extraneous solution. → **Thus,** −3 and 2 are the solutions.

EXAMPLE 7 Solve $.72n + .3 = 1.5 - .03n$.

Add $.03n$ to each side. → $.75n + .3 = 1.5$

Add $-.3$ to each side. → $.75n = 1.2$

First Method

$$\frac{75n}{100} = \frac{12}{10}$$

$$\frac{75n}{100} \times \frac{12}{10}$$

$$750n = 1200$$

$$n = 1.6$$

$750\overline{)1200.0}$ = 1.6 →

Second Method

In the second method, to multiply a decimal by 100, move the decimal point two places to the right.

$$.75n = 1.2$$
$$.75n = 1.20$$
$$75n = 120$$
$$n = 1.6$$

Thus, the solution is 1.6.

EXERCISES

PART A

Solve.

1. $\frac{2a}{5} + \frac{2a+1}{5} = 1$ *1*
2. $\frac{5-2x}{4} - \frac{x+7}{4} = \frac{x+6}{4}$ *−2*
3. $\frac{3}{4} + \frac{n-4}{8} = \frac{2n-3}{2}$ *2*
4. $\frac{a-1}{3} + \frac{a+2}{6} = 2$ *4*
5. $\frac{1}{4a^2} + \frac{5}{6a} = \frac{2}{3a^2}$ $\frac{1}{2}$
6. $\frac{3}{2x^2} - \frac{2}{5x} = \frac{7}{4x^2}$ $-\frac{5}{8}$

Find the solution set.

7. $\frac{2}{n-3} + \frac{2}{n} = 1$ $\{1, 6\}$
8. $\frac{3}{a} + \frac{2}{a+2} = 2$ $\left\{-\frac{3}{2}, 2\right\}$
9. $\frac{10}{x+4} - \frac{1}{x} = 1$ $\{1, 4\}$
10. $\frac{3n-7}{n-5} + \frac{n}{2} = \frac{8}{n-5}$ $\{-6\}$
11. $\frac{3a+2}{a+3} + \frac{a}{4} = \frac{a-4}{a+3}$ $\{-8\}$
12. $\frac{4x-3}{x-4} - \frac{2x}{3} = \frac{2x+5}{x-4}$ $\{3\}$

Solve the proportion.

13. $\frac{5}{n+4} = \frac{3}{n-2}$ *11*
14. $\frac{a-5}{2} = \frac{a-6}{4}$ *4*
15. $\frac{x-4}{x+1} = \frac{-2}{3}$ *2*
16. $\frac{3}{2n+1} = \frac{2}{3n-2}$ $\frac{8}{5}$
17. $\frac{4a+3}{3} = \frac{2a+5}{4}$ $\frac{3}{10}$
18. $\frac{2x+3}{3x-5} = \frac{1}{-2}$ $-\frac{1}{7}$
19. $\frac{x-3}{2} = \frac{1}{x-4}$ *2; 5*
20. $\frac{a+1}{8} = \frac{2}{a+1}$ *−5; 3*
21. $\frac{2a-3}{2a+3} = \frac{a-1}{2a+3}$ *2*

Solve.

22. $.25n = 75$ *300*
23. $.4a + .1 = 4.9$ *12*
24. $.03x + .6 = 7.8$ *240*
25. $.12n - .3 = .10n + .5$ *40*
26. $.08x - 2.73 = .02x + 3$ *95.5*
27. $x - 4.6 = .2x + 1$ *7*

PART B

Find the solution set.

28. $\frac{3}{n-2} + \frac{6}{n^2-5n+6} = \frac{4}{n-3}$ $\{5\}$
29. $\frac{2}{a+4} + \frac{2a+1}{a^2+2a-8} = \frac{1}{a-2}$ $\{\frac{7}{3}\}$
30. $\frac{7x-9}{x^2-25} - \frac{3}{x-5} = \frac{2}{x+5}$ $\{7\}$
31. $\frac{3}{x-2} - \frac{2x}{x^2-4} = \frac{5}{x+2}$ $\{4\}$
32. $\frac{n}{2n-6} - \frac{3}{n^2-6n+9} = \frac{n-2}{3n-9}$ $\{-6, 5\}$
33. $\frac{1}{2a} - \frac{9}{a^2+6a} = \frac{2-a}{2a+12}$ $\{-3, 4\}$

PART C

Solve.

34. $\frac{4}{x^2+2x-15} + \frac{5}{x^2-x-6} = \frac{3}{x^2+7x+10}$
35. $\frac{6}{x^2-9} + \frac{2}{x^2+5x+6} = \frac{4}{x^2-x-6}$
36. $\frac{5}{n-7} = \frac{5}{7-n}$
37. $\frac{x^2+5x+4}{x+4} = \frac{x^2-2x-3}{x-3}$

Work Problems

OBJECTIVE

- To find the time needed for a job when people work together at different rates

REVIEW CAPSULE

Solve $\frac{3}{5} + \frac{3}{x} = 1$.

$$5x\left(\frac{3}{5} + \frac{3}{x}\right) = 5x(1)$$
$$5x \cdot \frac{3}{5} + 5x \cdot \frac{3}{x} = 5x$$
$$3x + 15 = 5x$$
$$15 = 2x$$
$$7\tfrac{1}{2} = x$$

Thus, the solution is $7\frac{1}{2}$.

EXAMPLE 1 Al can paint a house in 6 days. Write in mathematical terms the part completed in 1, 2, 4, x, and 6 days.

Given number of days →

In mathematical terms the part of the job done →

$\frac{6}{6}$ job = 1 job →

Number of days	1	2	4	x	6
Part done	$\frac{1}{6}$	$\frac{2}{6}$	$\frac{4}{6}$	$\frac{x}{6}$	$\frac{6}{6}$, or 1

Part done in x days

1 means the whole job done.

EXAMPLE 2 Carol can clean the house in 5 hours. David can do it in 8 hours. Write in mathematical terms the parts done in 1 hour, 2 hours, 3 hours, and x hours, if Carol and David work together.

Add the parts done by Carol and David alone to find the part done *together*.

	Carol	David	Together
Part done in 1 hr	$\frac{1}{5}$	$\frac{1}{8}$	$\frac{1}{5} + \frac{1}{8}$
Part done in 2 hr	$\frac{2}{5}$	$\frac{2}{8}$	$\frac{2}{5} + \frac{2}{8}$
Part done in 3 hr	$\frac{3}{5}$	$\frac{3}{8}$	$\frac{3}{5} + \frac{3}{8}$
Part done in x hr	$\frac{x}{5}$	$\frac{x}{8}$	$\frac{x}{5} + \frac{x}{8}$

EXAMPLE 3 Carson's crew can do the cement work for a new building in 6 days. Green's crew would need 8 days. How many days will it take if the crews work together?

Let x = number of days if crews work together

	Carson	Green
Part done in 1 day	$\frac{1}{6}$	$\frac{1}{8}$
Part done in x days	$\frac{x}{6}$	$\frac{x}{8}$

1 means total job done.
Write an equation. ⟶ Part done together in x days

$$\frac{x}{6} + \frac{x}{8} = 1$$

The LCD is 24. Multiply each side by 24.

$$24\left(\frac{x}{6} + \frac{x}{8}\right) = 24(1)$$

$\overset{4}{\cancel{24}} \cdot \frac{x}{\underset{1}{\cancel{6}}} + \overset{3}{\cancel{24}} \cdot \frac{x}{\underset{1}{\cancel{8}}} = 4x + 3x$

$$4x + 3x = 24$$
$$7x = 24$$
$$x = \frac{24}{7}, \text{ or } 3\frac{3}{7}$$

Thus, it will take $3\frac{3}{7}$ days if the crews work together.

EXAMPLE 4 Phil can paint a house in x days. Write in mathematical terms the part he does in 1, 2, 3, and 4 days.

x days = 1 job

$\frac{x \text{ days}}{x} = \frac{1 \text{ job}}{x}$

$1 \text{ day} = \frac{1}{x} \text{ job}$

Number of days	1	2	3	4
Part done	$\frac{1}{x}$	$\frac{2}{x}$	$\frac{3}{x}$	$\frac{4}{x}$

EXAMPLE 5 Tom can mow his lawn in 5 hours. If Julia helps Tom, the job is done in 2 hours. How many hours would it take Julia working alone?

Let x = number of hours for Julia working alone

	Tom	Julia
Part mowed in 1 hour	$\frac{1}{5}$	$\frac{1}{x}$
Part mowed in 2 hours	$\frac{2}{5}$	$\frac{2}{x}$

1 is completed job.
Write an equation. ⟶ Part mowed together in 2 hours

$$\frac{2}{5} + \frac{2}{x} = 1$$

LCD is $5x$. Multiply each side by $5x$.

$$5x\left(\frac{2}{5} + \frac{2}{x}\right) = 5x(1)$$

$\overset{1}{\cancel{5}}x \cdot \frac{2}{\underset{1}{\cancel{5}}} + 5\overset{1}{\cancel{x}} \cdot \frac{2}{\underset{1}{\cancel{x}}} = 2x + 10$

$$2x + 10 = 5x$$
$$10 = 3x$$
$$3\tfrac{1}{3} = x$$

Thus, it would take Julia $3\frac{1}{3}$ hours working alone.

EXERCISES

PART A

1. A farmer can harvest crops in 15 days. Express in mathematical terms the part done in 1 day; in 4 days; in x days. $\frac{1}{15}$; $\frac{4}{15}$; $\frac{x}{15}$

2. A machine can do a job in x hours. Express in mathematical terms the part done in 1 hour; in 4 hours. $\frac{1}{x}$; $\frac{4}{x}$

3. Paul can plant his wheat crop in 10 days. His daughter can do it in 15 days. How many days will it take if they work together? *6*

4. Mrs. Smith can paint her house in 8 hours. Her son can do it in 12 hours. How many hours will it take if they work together? $4\frac{4}{5}$

5. Samuel can carpet a floor in 10 hours. If Irene helps him, the job is done in 6 hours. How many hours would it take Irene working alone? *15*

6. Mike can clean the house in 6 hours. If Diane helps him, the job is done in 4 hours. How many hours would it take Diane working alone? *12*

7. Machine A can do a job in 6 hours. Machine B can do the job in 8 hours. How many hours will it take if both machines are working? $3\frac{3}{7}$

8. Machine C can do a job in 15 hours. If machine D works at the same time, the job can be done in 10 hours. How many hours would it take machine D alone? *30*

PART B

EXAMPLE Machine A can do a job in 10 hours. Machine B can do the job in 8 hours. If B starts 3 hours after A has begun, how many hours will it take to do the job?

A works alone for 3 hours. Then A and B work together for x hours. ⟶

$$\left(\begin{array}{c}\text{Part done by } A \text{ alone}\\ \text{in first 3 hr}\end{array}\right) + \left(\begin{array}{c}\text{Part done by } A \text{ and } B\\ \text{together in } x \text{ hr}\end{array}\right) = 1$$

A alone: $\frac{1}{10}$ in 1 hr; $\frac{3}{10}$ in 3 hr.
A and B together:
$\frac{1}{10} + \frac{1}{8}$ in 1 hr; $\frac{x}{10} + \frac{x}{8}$ in x hr.

$$\frac{3}{10} + \left(\frac{x}{10} + \frac{x}{8}\right) = 1$$

The LCD is 40. Multiply each side by 40.
Solve the equation.

$$12 + 4x + 5x = 40$$
$$x = 3\tfrac{1}{9}$$

A alone, 3 hr, then together, $3\frac{1}{9}$ hr. ⟶ **Thus,** the job will take $3 + 3\frac{1}{9}$, or $6\frac{1}{9}$ hours.

9. Nita can mow a lawn in 90 minutes. Pedro can do it in 60 minutes. If Pedro starts 10 minutes after Nita has begun, how many minutes are needed to mow the lawn? *42*

10. Charles can paint the garage in 9 hours. Rita can do it in 6 hours. If Charles helps Rita after she has painted alone for 1 hour, how many hours will it take to paint the garage? *4*

11. Machine A can produce 1,000 items in 12 hours. Machine B can do this in 10 hours. If machine B starts 2 hours after A has begun, how many hours will it take to produce 1,000 items? $6\frac{6}{11}$

12. Machine C can do a job in 20 hours. Machine D can do this in 15 hours. If C works alone for 2 hours and then D works alone for 3 hours, how many additional hours are needed to complete the job with both machines at work? *6*

Complex Rational Expressions

OBJECTIVE

- To simplify expressions like $\dfrac{2+\dfrac{4n}{a-2}}{\dfrac{3n}{a^2-4}-\dfrac{5}{a+2}}$

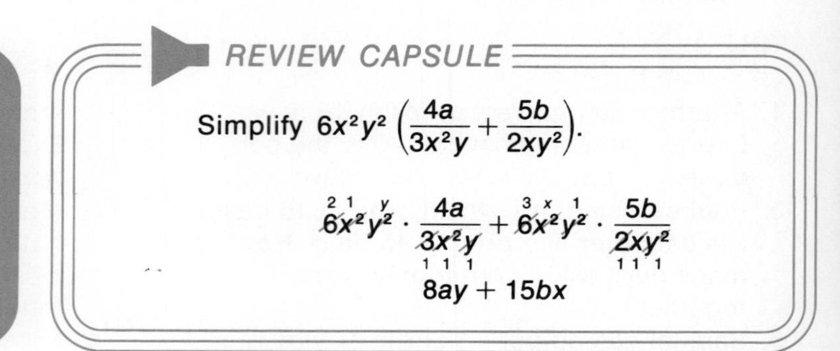

REVIEW CAPSULE

Simplify $6x^2y^2\left(\dfrac{4a}{3x^2y}+\dfrac{5b}{2xy^2}\right)$.

$$\overset{2\ 1\ y}{\cancel{6}\cancel{x^2}\cancel{y^2}}\cdot\frac{4a}{\underset{1\ 1\ 1}{\cancel{3}\cancel{x^2}\cancel{y}}}+\overset{3\ x\ 1}{\cancel{6}\cancel{x^2}\cancel{y^2}}\cdot\frac{5b}{\underset{1\ 1\ 1}{\cancel{2}\cancel{x}\cancel{y^2}}}$$

$$8ay+15bx$$

EXAMPLE 1 Simplify $\dfrac{\dfrac{3}{8}+\dfrac{7}{12}}{\dfrac{5}{6}-\dfrac{1}{4}}$.

The LCD of $\frac{3}{8}, \frac{7}{12}, \frac{5}{6}, \frac{1}{4}$ is 24.

Multiply numerator and denominator by 24.

$-\frac{1}{4}=\frac{-1}{4}$

$$\frac{24\left(\frac{3}{8}+\frac{7}{12}\right)}{24\left(\frac{5}{6}+\frac{-1}{4}\right)}=\frac{\overset{3}{\cancel{24}}\cdot\frac{3}{\underset{1}{\cancel{8}}}+\overset{2}{\cancel{24}}\cdot\frac{7}{\underset{1}{\cancel{12}}}}{\overset{4}{\cancel{24}}\cdot\frac{5}{\underset{1}{\cancel{6}}}+\overset{6}{\cancel{24}}\cdot\frac{-1}{\underset{1}{\cancel{4}}}}$$

$$=\frac{9+14}{20-6},\text{ or }\frac{23}{14}$$

Complex fractions have at least one fraction in the numerator or denominator.

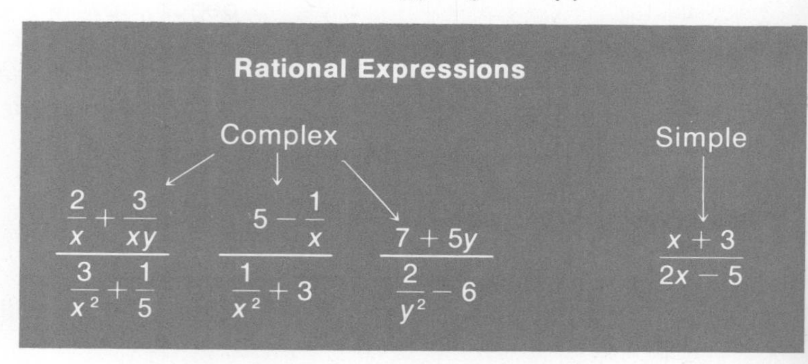

EXAMPLE 2 Simplify $\dfrac{5a-\dfrac{1}{2a}}{\dfrac{1}{3b}+4b}$.

The LCD of $\frac{1}{2a}$ and $\frac{1}{3b}$ is $6ab$.

Multiply numerator and denominator by $6ab$.

$-\frac{1}{2a}=\frac{-1}{2a}$

$$\frac{6ab\left(5a+\frac{-1}{2a}\right)}{6ab\left(\frac{1}{3b}+4b\right)}=\frac{6ab\cdot 5a+\overset{3\,1}{\cancel{6}\cancel{a}b}\cdot\frac{-1}{\underset{1\,1}{\cancel{2}\cancel{a}}}}{\overset{2\,1}{\cancel{6}a\cancel{b}}\cdot\frac{1}{\underset{1\,1}{\cancel{3}\cancel{b}}}+6ab\cdot 4b}$$

$$=\frac{30a^2b-3b}{2a+24ab^2}$$

EXAMPLE 3 Simplify $\dfrac{\dfrac{2}{x}+\dfrac{3}{xy}}{\dfrac{3}{x^2}+\dfrac{1}{5}}$.

The LCD of the four fractions is $5x^2y$.

Multiply numerator and denominator by $5x^2y$.

$$\frac{5x^2y\left(\dfrac{2}{x}+\dfrac{3}{xy}\right)}{5x^2y\left(\dfrac{3}{x^2}+\dfrac{1}{5}\right)} = \frac{5x^2y\cdot\dfrac{2}{x}+5x^2y\cdot\dfrac{3}{xy}}{5x^2y\cdot\dfrac{3}{x^2}+5x^2y\cdot\dfrac{1}{5}}$$

$$= \frac{10xy+15x}{15y+x^2y}$$

EXAMPLE 4 Simplify $\dfrac{\dfrac{2x}{x^2-25}+\dfrac{5}{x-5}}{\dfrac{4}{x+5}+\dfrac{3}{x-5}}$.

First, factor all denominators.

$$\frac{\dfrac{2x}{(x+5)(x-5)}+\dfrac{5}{x-5}}{\dfrac{4}{x+5}+\dfrac{3}{x-5}}$$

The LCD is $(x+5)(x-5)$. Multiply numerator and denominator by $(x+5)(x-5)$.

$$\frac{(x+5)(x-5)\left(\dfrac{2x}{(x+5)(x-5)}+\dfrac{5}{x-5}\right)}{(x+5)(x-5)\left(\dfrac{4}{x+5}+\dfrac{3}{x-5}\right)}$$

Distribute the $(x+5)(x-5)$ and simplify.

$$\frac{(x+5)(x-5)\cdot\dfrac{2x}{(x+5)(x-5)}+(x+5)(x-5)\cdot\dfrac{5}{x-5}}{(x+5)(x-5)\cdot\dfrac{4}{x+5}+(x+5)(x-5)\cdot\dfrac{3}{x-5}}$$

$$\frac{2x+5(x+5)}{4(x-5)+3(x+5)}$$

$5(x+5) = 5x+25$; $4(x-5) = 4x-20$

Combine like terms. ⟶

$$\frac{2x+5x+25}{4x-20+3x+15}, \text{ or } \frac{7x+25}{7x-5}$$

EXAMPLE 5 Simplify $\dfrac{1-\dfrac{6}{x}+\dfrac{5}{x^2}}{1-\dfrac{3}{x}-\dfrac{10}{x^2}}$.

The LCD is x^2. Multiply numerator and denominator by x^2.

$-\dfrac{6}{x} = \dfrac{-1(6)}{x} = \dfrac{-6}{x}$

$$\frac{x^2\left(1+\dfrac{-6}{x}+\dfrac{5}{x^2}\right)}{x^2\left(1+\dfrac{-3}{x}+\dfrac{-10}{x^2}\right)} = \frac{x^2\cdot 1+x^2\cdot\dfrac{-6}{x}+x^2\cdot\dfrac{5}{x^2}}{x^2\cdot 1+x^2\cdot\dfrac{-3}{x}+x^2\cdot\dfrac{-10}{x^2}}$$

$$= \frac{x^2-6x+5}{x^2-3x-10}$$

Factor numerator and denominator. Then simplify. ⟶

$$= \frac{(x-1)(x-5)}{(x+2)(x-5)}, \text{ or } \frac{x-1}{x+2}$$

EXERCISES

PART A

Simplify.

1. $\dfrac{\frac{2}{5}+\frac{1}{4}}{\frac{3}{2}+\frac{7}{10}}$ $\frac{13}{44}$

2. $\dfrac{\frac{9}{4}-\frac{2}{3}}{\frac{11}{6}-\frac{1}{2}}$ $\frac{19}{16}$

3. $\dfrac{\frac{1}{4}+4}{3-\frac{1}{3}}$ $\frac{51}{32}$

4. $\dfrac{\frac{5}{x}+\frac{2}{y}}{\frac{3}{x}+\frac{4}{y}}$ $\frac{5y+2x}{3y+4x}$

5. $\dfrac{\frac{1}{a}-\frac{1}{b}}{\frac{1}{a}+\frac{1}{b}}$ $\frac{b-a}{b+a}$

6. $\dfrac{c-\frac{1}{c}}{\frac{1}{d}+d}$ $\frac{c^2d-d}{c+cd^2}$

7. $\dfrac{\frac{4}{xy}-\frac{6}{y^2}}{\frac{7}{x^2}+\frac{2}{xy}}$ $\frac{4xy-6x^2}{7y^2+2xy}$

8. $\dfrac{\frac{5}{3a}+\frac{3}{2b}}{\frac{1}{6a}-\frac{3}{4b}}$ $\frac{20b+18a}{2b-9a}$

9. $\dfrac{\frac{2}{5x}+2}{1-\frac{3}{10y}}$ $\frac{4y+20xy}{10xy-3x}$

10. $\dfrac{\frac{4}{3x^2}+\frac{3}{2x}}{\frac{5}{6x}+\frac{1}{x^2}}$ $\frac{8+9x}{5x+6}$

11. $\dfrac{\frac{1}{5a^2}-\frac{2}{b}}{\frac{7}{10a}+\frac{3}{2b^2}}$ $\frac{2b^2-20a^2b}{7ab^2+15a^2}$

12. $\dfrac{a-\frac{1}{4c^2}}{\frac{1}{6c}-3a}$ $\frac{12ac^2-3}{2c-36ac^2}$

13. $\dfrac{\frac{4}{x-2}+\frac{3}{x}}{\frac{5}{x}+\frac{3}{x-2}}$ $\frac{7x-6}{8x-10}$

14. $\dfrac{\frac{6}{a+3}-\frac{4}{a-4}}{\frac{2}{a-4}+\frac{5}{a+3}}$ $\frac{2a-36}{7a-14}$

15. $\dfrac{1-\frac{9}{x^2}}{\frac{1}{x}-\frac{3}{x^2}}$ $x+3$

PART B

Simplify.

16. $\dfrac{\frac{3x}{x^2-4}+\frac{2}{x-2}}{\frac{4}{x+2}+\frac{5}{x-2}}$ $\frac{5x+4}{9x+2}$

17. $\dfrac{\frac{4}{a-5}+\frac{2}{a+2}}{\frac{2a}{a^2-3a-10}+\frac{3}{a-5}}$

18. $\dfrac{\frac{4c}{2c^2-c-1}-\frac{4}{c-1}}{\frac{1}{c-1}+\frac{6}{2c+1}}$

19. $\dfrac{\frac{x-7}{x^2-5x+6}}{\frac{5}{x-2}-\frac{4}{x-3}}$ 1

20. $\dfrac{\frac{7a}{a^2+6a+8}}{\frac{3}{a+4}+\frac{5}{a+2}}$ $\frac{7a}{8a+26}$

21. $\dfrac{\frac{7}{c+5}-\frac{4}{c-5}}{\frac{7c}{c^2-25}}$ $\frac{3c-55}{7c}$

22. $\dfrac{1+\frac{7}{x}+\frac{10}{x^2}}{1+\frac{6}{x}+\frac{8}{x^2}}$ $\frac{x+5}{x+4}$

23. $\dfrac{1-\frac{2}{a}-\frac{8}{a^2}}{1-\frac{7}{a}+\frac{12}{a^2}}$ $\frac{a+2}{a-3}$

24. $\dfrac{2+\frac{5}{c}+\frac{3}{c^2}}{3+\frac{5}{c}+\frac{2}{c^2}}$ $\frac{2c+3}{3c+2}$

25. $\dfrac{\frac{x}{y}+3-\frac{4y}{x}}{\frac{x}{y}-\frac{y}{x}}$ $\frac{x+4y}{x+y}$

26. $\dfrac{\frac{a}{b}-1-\frac{6b}{a}}{\frac{a}{b}+4+\frac{4b}{a}}$ $\frac{a-3b}{a+2b}$

27. $\dfrac{\frac{3}{y}+\frac{3}{x}-\frac{6}{xy}}{\frac{4}{y}+\frac{4}{x}-\frac{8}{xy}}$ $\frac{3}{4}$

Literal Equations and Formulas

OBJECTIVES

- To solve equations like $ax + b = c - dx$ for x
- To solve a formula like $A = \frac{h}{2}(b + c)$ for any one of its variables

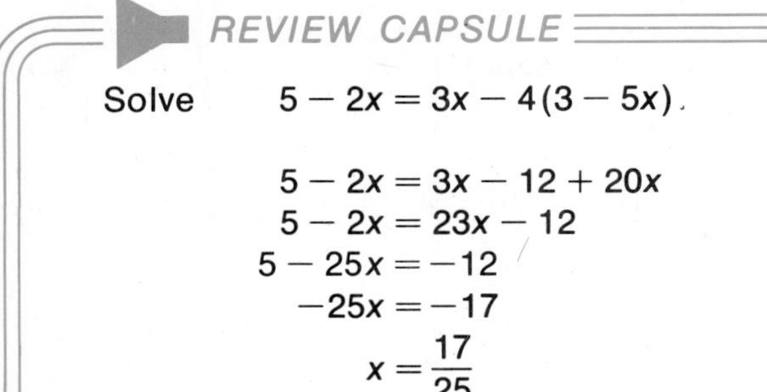

REVIEW CAPSULE

Solve $5 - 2x = 3x - 4(3 - 5x)$.

$$5 - 2x = 3x - 12 + 20x$$
$$5 - 2x = 23x - 12$$
$$5 - 25x = -12$$
$$-25x = -17$$
$$x = \frac{17}{25}$$

EXAMPLE 1 Solve $ax + b = cx$ for x.

	$ax + b = cx$
Add $-b$ to each side. →	$ax = cx - b$
Add $-cx$ to each side. →	$ax - cx = -b$
Factor out x. →	$(a - c)x = -b$
Divide each side by $(a - c)$. →	$\frac{(a-c)x}{a-c} = \frac{-b}{a-c}$
$\frac{-b}{a-c} \cdot \frac{-1}{-1} = \frac{b}{-a+c} = \frac{b}{c-a}$ →	$x = \frac{-b}{a-c}$, or $\frac{b}{c-a}$

EXAMPLE 2 Solve $a(b - nx) = c$ for x.

x is inside parentheses. →	$a(b - nx) = c$
Distribute the a.	$ab - anx = c$
Add $-ab$ to each side. →	$-anx = c - ab$
Divide each side by $-an$. →	$\frac{-anx}{-an} = \frac{c - ab}{-an}$
	$x = \frac{c - ab}{-an}$, or $\frac{ab - c}{an}$

EXAMPLE 3 Solve $a = \frac{x}{b} - c$ for x.

Multiply each side by the LCD, b. →	$b(a) = b\left(\frac{x}{b} - c\right)$
	$ab = b \cdot \frac{x}{b} - bc$
$\overset{1}{\cancel{b}} \cdot \frac{x}{\underset{1}{\cancel{b}}} = x$ →	$ab = x - bc$
Add bc to each side. →	$ab + bc = x$
Reverse the sides. →	**Thus,** $x = ab + bc$.

EXAMPLE 4 Solve $\frac{1}{a} = \frac{1}{x} + \frac{1}{b}$ for x.

x appears in a denominator. ⟶ $\frac{1}{a} = \frac{1}{x} + \frac{1}{b}$

The LCD is abx. Multiply each side by abx. $abx\left(\frac{1}{a}\right) = abx\left(\frac{1}{x} + \frac{1}{b}\right)$

Distribute the abx. $abx \cdot \frac{1}{a} = abx \cdot \frac{1}{x} + abx \cdot \frac{1}{b}$

$bx = ab + ax$

Add $-ax$ to each side. ⟶ $bx - ax = ab$

Factor out x. ⟶ $(b - a)x = ab$

$x = \frac{ab}{b - a}$

EXAMPLE 5 Solve $a = \frac{bx}{c + 2x}$ for x.

Write as a proportion: $a = \frac{a}{1}$. ⟶ $\frac{a}{1} = \frac{bx}{c + 2x}$

Solve the proportion. $ac + 2ax = bx$

$ac = bx - 2ax$

Factor out x. ⟶ $ac = (b - 2a)x$

Divide each side by $(b - 2a)$. ⟶ $\frac{ac}{b - 2a} = x$

Reverse the sides. ⟶ **Thus,** $x = \frac{ac}{b - 2a}$.

EXAMPLE 6 Solve $A = p + prt$ for p.
Then find p if $A = 6{,}600$, $r = .08$, and $t = 4$.

p appears in two terms. ⟶ $A = p + prt$

$p + prt = 1p + prt = p(1 + rt)$ ⟶ $A = p(1 + rt)$

Divide each side by $1 + rt$. ⟶ $\frac{A}{1 + rt} = p$

Reverse the sides. ⟶ or $p = \frac{A}{1 + rt}$

Substitute for A, r, and t. ⟶ $p = \frac{6{,}600}{1 + .08(4)}$

$= \frac{6{,}600}{1 + .32}$

$= \frac{6{,}600}{1.32}$

$1.32\overline{)6{,}600.00}$ = 5,000. ⟶ $= 5{,}000$

Thus, $p = \frac{A}{1 + rt}$
and $p = 5{,}000$, if $A = 6{,}600$, $r = .08$, $t = 4$.

EXERCISES

PART A

Solve for x.

1. $ax - b = c$ $x = \frac{b+c}{a}$
2. $a + bx = c$ $x = \frac{c-a}{b}$
3. $a(x - b) = c$ $x = \frac{c+ab}{a}$
4. $\frac{x}{a} = b$ $x = ab$
5. $1 = \frac{a}{x}$ $x = a$
6. $\frac{a}{b} \cdot x = c$ $x = \frac{bc}{a}$
7. $ax = b - cx$ $\frac{b}{a+c}$
8. $ax - b = 3x + 2$ $x = \frac{b+2}{a-3}$
9. $2ax + b = 3b - ax$ $x = \frac{2b}{3a}$
10. $ax = b - x$ $\frac{b}{a+1}$
11. $ax = ab$
12. $ax - 3 = bx + c$ $\frac{c+3}{a-b}$
13. $a(bx - c) = d$ $\frac{ac+d}{ab}$
14. $a(x - b) = c(x + d)$ $\frac{ab+cd}{a-c}$
15. $a(2x + b) = c(nx - 1)$ $\frac{-ab-c}{2a-cn}$
16. $a = \frac{bx}{c} + d$ $\frac{ac-cd}{b}$
17. $\frac{a}{b} = \frac{x}{b} + c$ $a - bc$
18. $ax = \frac{bx-c}{d}$ $\frac{-c}{ad-b}$
19. $\frac{3}{a} = \frac{b}{x}$ $\frac{ab}{3}$
20. $\frac{1}{x} = \frac{1}{a} + \frac{1}{b}$ $\frac{ab}{a+b}$
21. $\frac{2}{a} = \frac{3}{x} - \frac{4}{b}$ $\frac{3ab}{4a+2b}$
22. $\frac{a+x}{b-x} = c$ $\frac{bc-a}{1+c}$
23. $a = \frac{2x}{x+3}$ $x = \frac{3a}{2-a}$
24. $a = \frac{b+cx}{dx}$ $\frac{b}{ad-c}$

25. Solve $i = prt$ for p. Then find p if $i = 360$, $r = .06$, and $t = 10$. *600*
26. Solve $p = 2(l + w)$ for w. Then find w if $p = 100$ *and* $l = 33$. *17*
27. Solve $A = \frac{1}{2}bh$ for h. Then find h if $A = 32.8$ and $b = 16$. $h = \frac{2A}{b}$; *4.1*
28. Solve $F = \frac{9}{5}C + 32$ for C. Then find C if $F = 68$. $C = \frac{5F-160}{9}$; *20*

PART B

Solve the proportion for x.

29. $\frac{x}{a} = \frac{x+b}{c}$ $\frac{ab}{c-a}$
30. $\frac{x-a}{x-b} = \frac{c}{d}$ $\frac{ad-bc}{d-c}$
31. $\frac{a}{bx-c} = \frac{2}{x-a}$ $\frac{a^2-2c}{a-2b}$

32. Solve $A = ah + bh + ch$ for h. Then find h if $A = 135$, $a = 3$, $b = 5$, and $c = 7$.
33. Solve $V = \frac{1}{3}Bh$ for h. Then find h if $V = 154$ and $B = 42$. *11*
34. Solve $\frac{1}{a} = \frac{1}{b} + \frac{1}{c}$ for c. Then find c if $a = 6$ and $b = 8$. $c = \frac{ab}{b-a}$; *24*
35. Solve $T = mg - mf$ for m. Then find m if $T = 70$, $g = 32$, and $f = 22$. *7*

PART C

Solve for x.

36. $\frac{x}{4} + \frac{a+4}{12} = \frac{a+1}{3}$
37. $\frac{x-4}{8} + \frac{3}{4} = \frac{2a-3}{2}$
38. $ax + b(x - c) = d(x + e)$
39. $(x - c)(x + d) = 0$
40. $x^2 - 5ax + 6a^2 = 0$ [Hint: Factor into two binomials.]
41. $x^2 - 25a^2 = 0$
42. $x^2 + 6ax + 9a^2 = 0$
43. $x^2 + (a + b)x + ab = 0$
44. $x^2 - (c + d)^2 = 0$
45. $x^2 - (a^2 - 2ab + b^2) = 0$

Wet Mixture Problems *(optional)*

OBJECTIVE

- To solve problems about wet mixtures

REVIEW CAPSULE

A 40-liter solution of pesticide and water is 8% pesticide. How many liters are pesticide?

8% of 40 liters is pesticide.
8% of 40 = $.08(40) = 3.20$

Thus, there are 3.2 liters of pesticide.

EXAMPLE 1 How many liters of water must be evaporated from 60 liters of a 12% alcohol solution to make it a 36% solution?

Decrease the total solution by x. → Let x = number of liters (L) of water to be evaporated

Make a chart. →
12% alcohol means 88% water. →
36% alcohol means 64% water. →

	Total L	L of alcohol	L of water
12% solution	60	$.12(60)$	$.88(60)$
36% solution	$60 - x$	$.36(60 - x)$	$.64(60 - x)$

Since only water is decreased, amount of alcohol in each stays the same.
Write an equation. →

Alcohol in 12% solution = Alcohol in 36% solution

$$.12(60) = .36(60 - x)$$
$$7.20 = 21.60 - .36x$$

Multiply each side by 100. →

$$720 = 2160 - 36x$$
$$-1440 = -36x$$
$$40 = x$$

Thus, 40 liters of water must be evaporated.

EXAMPLE 2 How many liters of water must be added to 40 liters of a 30% salt solution to reduce it to a 20% solution?

Increase the total solution by x. → Let x = number of liters of water to be added

Make a chart. →
30% salt means 70% water. →
20% salt means 80% water. →

	Total L	L of salt	L of water
30% solution	40	$.30(40)$	$.70(40)$
20% solution	$40 + x$	$.20(40 + x)$	$.80(40 + x)$

Since only water is added, amount of salt in each stays the same.
Write an equation. →

Salt in 30% solution = Salt in 20% solution

$$.30(40) = .20(40 + x)$$
$$12.00 = 8.00 + .20x$$

Multiply each side by 100. →

$$1200 = 800 + 20x$$
$$400 = 20x$$
$$20 = x$$

Thus, 20 liters of water must be added.

EXAMPLE 3 A pharmacist has one solution that is 16% iodine and a more concentrated solution that is 80% iodine. How much of the more concentrated solution must be added to 20 cubic centimeters of the original solution to obtain a solution that is 30% iodine?

Increase the total solution by x. → Let $x =$ number of cubic centimeters (cm^3) of 80% solution to be added

Make a chart. →

	Total cm^3	cm^3 of iodine
16% solution	20	$.16(20)$
80% solution	x	$.80(x)$
30% solution	$20 + x$	$.30(20 + x)$

Amount of iodine in 30% solution is the same as the total of the other two solutions.

$$\begin{pmatrix}\text{Iodine in}\\ \text{16\% solution}\end{pmatrix} + \begin{pmatrix}\text{Iodine in}\\ \text{80\% solution}\end{pmatrix} = \begin{pmatrix}\text{Iodine in}\\ \text{30\% solution}\end{pmatrix}$$

Write an equation. →

$$.16(20) + .80(x) = .30(20 + x)$$

$$3.20 + .80x = 6.00 + .30x$$

Multiply each side by 100. →

$$320 + 80x = 600 + 30x$$

$$50x = 280$$

$$x = 5.6$$

Thus, 5.6 cm^3 of the 80% solution must be added.

EXERCISES

1. How many cubic centimeters of water must be evaporated from a 160 cubic centimeters of a 20% iodine solution to to make it a 40% solution? *80 cm^3*

2. A chemist has 120 cubic centimeters of a 15% alcohol solution. How much water must be evaporated in order to make a 20% alcohol solution? *30 cm^3*

3. A farmer has 600 liters of milk that tests 3.5% butterfat. How many liters of skimmed milk must be added to obtain 2% butterfat? *450 L*

4. How many deciliters of water must be added to 15 deciliters of a 28% solution of hydrochloric acid to reduce it to a 12% solution? *20 dL*

5. How many cubic centimeters of a solution that is 80% sulfuric acid must be added to 60 cubic centimeters of a 30% sulfuric acid solution to make a 50% solution? *40 cm^3*

6. A nurse has one solution that is 24% medicine and another solution that is 52% medicine. How much of the 52% solution must be added to 30 milliliters of the 24% solution to obtain a solution that is 40% medicine? *40 mL*

Rational Numbers and Decimals

OBJECTIVES

- To identify rational numbers
- To write repeating decimals in fraction form

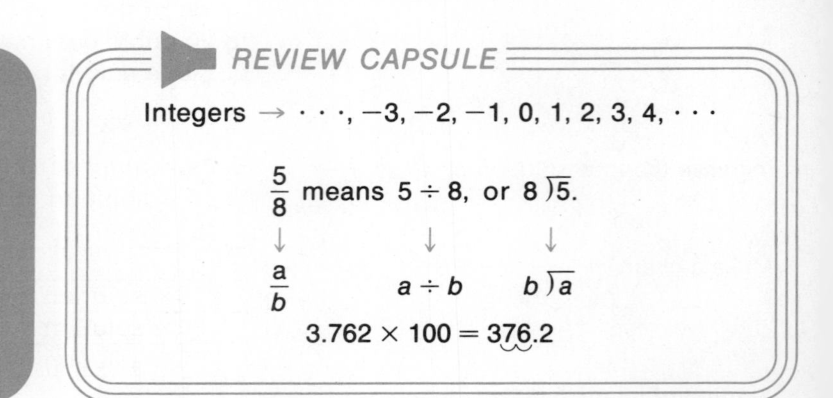

EXAMPLE 1 Solve $2x = 7$. Solve $4x = -3$. Solve $5x = 10$.

Equations of the form $bx = a$, where a and b are integers →

Solve $bx = a$:

$\frac{bx}{b} = \frac{a}{b}$, or $x = \frac{a}{b}$.

$2x = 7$	$4x = -3$	$5x = 10$
$\frac{2x}{2} = \frac{7}{2}$	$\frac{4x}{4} = \frac{-3}{4}$	$\frac{5x}{5} = \frac{10}{5}$
$x = \frac{7}{2}$	$x = \frac{-3}{4}$	$x = \frac{2}{1}$
Solution is $\frac{7}{2}$.	Solution is $\frac{-3}{4}$.	Solution is 2.

Solutions are of the form $\frac{a}{b}$, where a and b are integers.

Definition of rational number →

$b \neq 0$: $\frac{8}{0}$ is undefined.

A *rational* number is one which can be written in the form $\frac{a}{b}$, where a and b are *integers*, $b \neq 0$.

EXAMPLE 2 Show that -4, 0, and $2\frac{1}{3}$ are rational numbers.

Write them in the form $\frac{a}{b}$, where a and b are integers.

$-4 = \frac{-4}{1}$	$0 = \frac{0}{5}$	$2\frac{1}{3} = \frac{7}{3}$
-4 and 1 are integers.	0 and 5 are integers.	7 and 3 are integers.

EXAMPLE 3 Show that .7 and 5.31 are rational numbers.

Write them in the form $\frac{a}{b}$, where a and b are integers.

$.7 = \frac{7}{10}$	$5.31 = \frac{531}{100}$
7 and 10 are integers.	531 and 100 are integers.

EXAMPLE 4 Find decimals for $\frac{3}{4}$ and $\frac{1}{3}$.

Divide the denominator into the numerator.

$$\begin{array}{r} .75 \\ 4\overline{)3.00} \\ \underline{2\,8} \\ 20 \\ \underline{20} \\ 0 \end{array} \qquad \begin{array}{r} .333\cdots \\ 3\overline{)1.000\cdots} \\ \underline{9} \\ 10 \\ \underline{9} \\ 10 \\ \underline{9} \\ 1 \end{array}$$

$.3\overline{3}$ ← Bar above $\overline{3}$ means that 3 continues to repeat.

Thus, $\frac{3}{4} = .75$ — Terminating decimal

Thus, $\frac{1}{3} = .333\cdots$, or $.3\overline{3}$ — Repeating decimal

> Every rational number has a decimal numeral which either *terminates* or *repeats.*

EXAMPLE 5 Give decimals for $\frac{4}{11}$ and $\frac{3}{8}$ and tell if they terminate or repeat.

Terminating decimals have a 0 remainder.

Repeating decimals never have a 0 remainder.

$$\begin{array}{r} .363636\cdots \\ 11\overline{)4.000000} \\ \underline{3\,3} \\ 70 \\ \underline{66} \\ 40 \\ \underline{33} \\ 70 \\ \underline{66} \\ 40 \\ \underline{33} \\ 70 \\ \underline{66} \\ 4 \end{array} \qquad \begin{array}{r} .375 \\ 8\overline{)3.0} \\ \underline{2\,4} \\ 60 \\ \underline{56} \\ 40 \\ \underline{40} \\ 0 \end{array}$$

Bar above $.36\overline{36}$ means 36 continues to repeat.

Thus, $\frac{4}{11} = .36\overline{36}$ and $.36\overline{36}$ is a repeating decimal.

Thus, $\frac{3}{8} = .375$ and .375 is a terminating decimal.

EXAMPLE 6 Write $.5\overline{5}$ in the form $\frac{a}{b}$, where a and b are integers.

$.5\overline{5}$ means $.555\cdots$

Let $n = .555\cdots$

Multiply each side by 10; move decimal point 1 place.

Then $10n = 5.555\cdots$
and $n = .555\cdots$

Subtract. →

$9n = 5$

$n = \frac{5}{9}$

$\frac{5}{9}$ is a rational number because 5 and 9 are integers.

Thus, $.5\overline{5} = \frac{5}{9}$.

Example 6 suggests this. ⟶ **Repeating decimals name rational numbers.**

EXAMPLE 7 Show that $.62\overline{2}$ and $.71\overline{71}$ are rational numbers.

If 1 digit repeats, use $10(n)$.
If 2 digits repeat, use $100(n)$.

Subtract.
5.6 *is not* an integer.
Multiply numerator and denominator by 10: $\frac{5.6(10)}{9(10)} = \frac{56}{90}$.

$$\begin{aligned} \text{Let } n &= .6222\cdots \\ \text{Then } 10n &= 6.2222\cdots \\ \text{and } n &= .6222\cdots \\ \hline 9n &= 5.6 \\ n &= \frac{5.6}{9} \\ n &= \frac{56}{90} \end{aligned}$$

$$\begin{aligned} \text{Let } n &= .717171\cdots \\ \text{Then } 100n &= 71.717171\cdots \\ \text{and } n &= .717171\cdots \\ \hline 99n &= 71 \\ n &= \frac{71}{99} \end{aligned}$$

ORAL EXERCISES

Show that each is a rational number.

1. 5 *$\frac{5}{1}$* **2.** .03 *$\frac{3}{100}$* **3.** $2\frac{2}{5}$ *$\frac{12}{5}$* **4.** −3 *$\frac{-3}{1}$* **5.** −4.3 *$\frac{-43}{10}$* **6.** 0 *$\frac{0}{1}$* **7.** $-3\frac{1}{2}$ *$\frac{-7}{2}$*

EXERCISES

PART A

Give a decimal for the fraction and tell if it terminates or repeats.

1. $\frac{1}{4}$ *.25; term.* **2.** $\frac{2}{3}$ *$.\overline{6}6$; rep.* **3.** $\frac{5}{8}$ *.625; term.* **4.** $\frac{4}{9}$ *$.4\overline{4}$; rep.* **5.** $\frac{8}{5}$ *1.6; term.* **6.** $\frac{11}{18}$ *$.61\overline{1}$; rep.*

Show that the decimal is a rational number.

7. $.7\overline{7}$ *$\frac{7}{9}$* **8.** $.32\overline{32}$ *$\frac{32}{99}$* **9.** $.8\overline{8}$ *$\frac{8}{9}$* **10.** $.47\overline{47}$ *$\frac{47}{99}$* **11.** $.08\overline{08}$ *$\frac{8}{99}$*

PART B

Show that the decimal is a rational number.

12. $.34\overline{4}$ *$\frac{31}{90}$* **13.** $.72\overline{2}$ *$\frac{65}{90}$* **14.** $.28\overline{8}$ *$\frac{26}{90}$* **15.** $.53\overline{3}$ *$\frac{48}{90}$* **16.** $.07\overline{7}$ *$\frac{7}{90}$*
17. $8.2\overline{2}$ *$\frac{74}{9}$* **18.** $-.43\overline{43}$ *$\frac{-43}{99}$* **19.** $-.4\overline{4}$ *$\frac{-4}{9}$* **20.** $2.69\overline{69}$ *$\frac{267}{99}$* **21.** $4.36\overline{6}$ *$\frac{393}{90}$*
22. $.9\overline{9}$ *1* **23.** $2.9\overline{9}$ *3* **24.** $-4.9\overline{9}$ *−5* **25.** $.49\overline{9}$ *$\frac{1}{2}$* **26.** $-2.49\overline{9}$ *$\frac{-225}{90}$*

PART C

Show that the decimal is a rational number.

27. $.361\overline{361}$ **28.** $-.438\overline{438}$ **29.** $.349\overline{49}$ **30.** $4.562\overline{62}$

Dividing Polynomials

OBJECTIVE

- To divide polynomials like $(3x^3 + 2x^2 - 6x - 4) \div (x - 2)$

REVIEW CAPSULE

$$3x^2 - 7x^2 = -4x^2$$

$$(15x^3) \div (3x) = \frac{15x^3}{3x} = 5x^2$$

$$\begin{array}{rrl} & 2 & \leftarrow \text{Quotient (whole number)} \\ \text{Divisor} \rightarrow 5 & \overline{)14} & \leftarrow \text{Dividend} \\ & \underline{10} & \\ & 4 & \leftarrow \text{Remainder} \end{array}$$

Check.
$(2)(5) + 4 = 14$

EXAMPLE 1 Divide $679 \div 32$.

Divide: $600 \div 30 = 20$.
Multiply: $20(30 + 2) = 600 + 40$.
Subtract.
Divide: $30 \div 30 = 1$.
Multiply: $1(30 + 2) = 30 + 2$.
Subtract. →

Ordinary Form	*Expanded Form*
$$\begin{array}{r} 21 \\ 32\overline{)679} \\ \underline{64} \\ 39 \\ \underline{32} \\ 7 \end{array}$$	$$\begin{array}{r} 20 + 1 \\ 30 + 2\overline{)600 + 70 + 9} \\ \underline{600 + 40} \\ 30 + 9 \\ \underline{30 + 2} \\ 7 \end{array}$$

Thus, the quotient is 21 and the remainder is 7.

EXAMPLE 2 Divide $(3x^2 - 5x - 30) \div (x - 4)$. Check by multiplying.

Divide: $3x^2 \div x = 3x$. →
Multiply: $3x(x - 4) = 3x^2 - 12x$. →
Subtract. →
$-5x - (-12x) = 7x$

Step 1

$$\begin{array}{r} 3x \\ x - 4\overline{)3x^2 - 5x - 30} \\ \underline{3x^2 - 12x} \\ 7x - 30 \end{array}$$

Divide: $7x \div x = 7$. →
Multiply: $7(x - 4) = 7x - 28$. →
Subtract. →

Step 2

$$\begin{array}{r} 3x + 7 \\ x - 4\overline{)3x^2 - 5x - 30} \\ \underline{3x^2 - 12x} \\ 7x - 30 \\ \underline{7x - 28} \\ -2 \end{array}$$

Thus, the quotient is $3x + 7$ and the remainder is -2.

$(3x + 7)(x - 4) = 3x^2 - 5x - 28$;
$3x^2 - 5x - 28 + (-2) = 3x^2 - 5x - 30$ →

Check by multiplying:
$$(3x + 7)(x - 4) + (-2) = 3x^2 - 5x - 30.$$
(Quotient)(Divisor) + Remainder = Dividend

EXAMPLE 3 Divide $51 \div 3$. Is 3 a factor of 51?

$$\begin{array}{r} 17 \\ 3\overline{)51} \\ \underline{3} \\ 21 \\ \underline{21} \\ 0 \end{array}$$

Check.

$(17)(3) + 0$	51
$51 + 0$	51
51	

Remainder is 0. →

$(17)(3) = 51$ → **Thus,** $51 \div 3 = 17$ and $(17)(3) = 51$.
So, 3 is a factor of 51.

Divisor ↘ $3\overline{)51}$ ← Dividend (quotient 17)

In division, if the remainder is 0, then the divisor is a factor of the dividend.

EXAMPLE 4 Divide $(6x^3 - 2x^2 + 30 - 2x) \div (3x + 5)$. Is $3x + 5$ a factor of $6x^3 - 2x^2 + 30 - 2x$?

Divide: $6x^3 \div (3x) = 2x^2$. →
Multiply: $2x^2(3x + 5) = 6x^3 + 10x^2$. →
Subtract. →

Step 1

$$\begin{array}{r} 2x^2 \\ 3x + 5\overline{)6x^3 - 2x^2 - 2x + 30} \\ \underline{6x^3 + 10x^2 } \\ -12x^2 - 2x \end{array}$$

← Arrange dividend in descending order.

$-2x^2 - (+10x^2) = -12x^2$
Divide: $-12x^2 \div (3x) = -4x$. →
Multiply: $-4x(3x + 5) = -12x^2 - 20x$. →
Subtract. →
Continue: divide, multiply, subtract.

Steps 2 and 3

$$\begin{array}{r} 2x^2 - 4x + 6 \\ 3x + 5\overline{)6x^3 - 2x^2 - 2x + 30} \\ \underline{6x^3 + 10x^2 } \\ -12x^2 - 2x \\ \underline{-12x^2 - 20x } \\ 18x + 30 \\ \underline{18x + 30} \\ 0 \end{array}$$

Remainder is 0.

Thus, the quotient is $2x^2 - 4x + 6$ and the remainder is 0. So, $3x + 5$ is a factor of $6x^3 - 2x^2 + 30 - 2x$.

EXAMPLE 5 Is $2x + 4$ a factor of $10x^3 - 34x + 8$?

Divide: $(10x^3 - 34x + 8) \div (2x + 4)$.
Write dividend so that all terms appear.
Use $0 \cdot x^2$ for x^2-term.

$$\begin{array}{r} 5x^2 - 10x + 3 \\ 2x + 4\overline{)10x^3 + 0x^2 - 34x + 8} \\ \underline{10x^3 + 20x^2 } \\ -20x^2 - 34x \\ \underline{-20x^2 - 40x } \\ 6x + 8 \\ \underline{6x + 12} \\ -4 \end{array}$$

← x^2-term was missing in dividend.

The remainder is *not* 0. → **Thus,** $2x + 4$ is *not* a factor of $10x^3 - 34x + 8$.

ORAL EXERCISES

1. Is 2 a factor of 14? *yes*
2. Is 7 a factor of 15? *no*
3. Is x a factor of $3x^3$? *yes*
4. Is n a factor of $2n^2 - 5n$? *yes*
5. If $(x^2 - 7x + 12) \div (x - 3)$ has a remainder of 0, is $x - 3$ a factor of $x^2 - 7x + 12$? *yes*
6. If $(x^2 - 3x - 12) \div (x + 2)$ has a remainder of -2, is $x + 2$ a factor of $x^2 - 3x - 12$? *no*

EXERCISES

PART A

Divide. Check by multiplying.

1. $(5x^2 - 4x - 12) \div (x - 2)$ *$5x + 6$*
2. $(3y^2 + 7y - 4) \div (y + 3)$ *$3y - 2$, rem. 2*
3. $(2n^2 + 19n + 35) \div (2n + 5)$ *$n + 7$*
4. $(6a^2 - 13a + 2) \div (3a - 2)$ *$2a - 3$, rem. -4*

Divide.

5. $(3n^4 + 13n^3 + 4n^2 - 2n - 8) \div (n + 4)$
6. $(2x^3 - x^2 - 17x + 9) \div (x - 3)$
7. $(-14n + 8n^2 + 5) \div (2n - 1)$ *$4n - 5$*
8. $(14x^2 - 3x + 3x^3 + 7) \div (x + 5)$
9. $(a^3 - 13a - 12) \div (a + 3)$ *$a^2 - 3a - 4$*
10. $(9y^4 + 5y^2 - 6) \div (3y - 2)$ *$3y^3 + 2y^2 + 3y + 2$, rem. -2*

Is the given binomial a factor of the given polynomial?

11. $x + 6$ and $3x^3 + 17x^2 - 8x - 12$ *yes*
12. $n - 3$ and $2n^3 - 11n^2 + 18n - 12$ *no*
13. $3n - 2$ and $6n^3 - n^2 - 5n + 4$ *no*
14. $a + 1$ and $a^3 + 1$ *yes*

PART B

EXAMPLE Divide $(6x^3 + 15x^2 - 4x - 5) \div (3x^2 - 2)$.

$$
\begin{array}{r}
2x + 5 \\
3x^2 - 2 \overline{\big)\, 6x^3 + 15x^2 - 4x - 5} \\
\underline{6x^3 \qquad\qquad - 4x \phantom{{}-5}} \\
15x^2 + 0x - 5 \\
\underline{15x^2 \qquad\quad - 10} \\
5
\end{array}
$$

Subtract. ⟶

Divide.

15. $(2n^4 - 4n^3 + 7n^2 - 12n + 9) \div (n^2 + 3)$ *$2n^2 - 4n + 1$, rem. 6*
16. $(3x^4 + 2x^3 - 8x - 48) \div (x^2 - 4)$ *$3x^2 + 2x + 12$*
17. $(6x^3 + 4x^2 - 3x - 5) \div (2x^2 - 1)$ *$3x + 2$, rem. -3*
18. $(9a^6 - 3a^5 + a^2 - 4) \div (3a^3 - 1)$ *$3a^3 - a^2 + 1$, rem. -3*

PART C

19. Divide $(3x^3 - x^2 - 3x - 8) \div (x - 2)$. Find the remainder. Find a value of $3x^3 - x^2 - 3x - 8$ by substituting 2 for x. Is this value equal to the remainder?
20. Divide $(2x^4 - 2x^2 - 20x - 300) \div (x - 4)$. Find the remainder. Find a value of the dividend by substituting 4 for x. Is this value equal to the remainder?

Chapter Five Review

Solve. [p. 109]

1. $\frac{a-3}{4}+\frac{5}{4}=\frac{3a}{4}$
2. $\frac{5x}{3}-\frac{2x+9}{3}=1$
3. $\frac{n+3}{4}+\frac{5}{2}=\frac{3n-1}{8}$
4. $\frac{a-1}{3}-\frac{a-3}{2}=\frac{a-2}{4}$
5. $\frac{5}{6x^2}+\frac{2}{3x}=\frac{1}{2x}$
6. $\frac{5}{4a^2}-\frac{1}{5a}=\frac{3}{2a^2}$
7. $.15x=45$
8. $n-.6=.7n+1.2$
9. $.10a-3=.06a+.2$

Solve the proportion. [p. 109]

10. $\frac{2}{3x-2}=\frac{3}{x+4}$
11. $\frac{n-2}{6}=\frac{2}{n-1}$
12. $\frac{a-2}{a-2}=\frac{a}{2a-1}$

Find the solution set. [p. 109]

13. $\frac{3}{x+2}+\frac{2}{x}=1$
14. $\frac{2a-1}{a-3}+\frac{a}{4}=\frac{a+2}{a-3}$
15. $\frac{1}{x-5}+\frac{x-3}{x^2-7x+10}=\frac{7}{x-2}$
16. $\frac{x}{x+5}-\frac{5}{x^2-25}=\frac{x-6}{2x-10}$

17. Nita can clean the house in 5 hours. Nick can do it in 3 hours. How many hours will it take if they work together? [p. 113]

18. Marita can carpet a floor in 9 hours. If Breda helps her, the job is done in 6 hours. How many hours would it take Breda alone? [p. 113]

Simplify. [p. 116]

19. $\dfrac{\frac{1}{3x^2}+\frac{3}{2y}}{\frac{5}{6x}-\frac{2}{y^2}}$
20. $\dfrac{2a-\frac{1}{3a^2}}{\frac{1}{6b}+3b}$
21. $\dfrac{\frac{4n}{n^2-5n+6}-\frac{2}{n-3}}{\frac{5}{n-3}+\frac{3}{n-2}}$

Solve for x. [p. 119]

22. $ax-b=cx+d$
23. $a(x-b)=c(2x+3)$
24. Solve $\frac{1}{a}=\frac{1}{b}+\frac{1}{c}$ for b. Then find b if $a=3$ and $c=6$.
25. Solve $p=2(l+w)$ for l. Then find l if $p=24$ and $w=5$.

Show that each is a rational number. [p. 124]

26. $.444\cdots$
27. $.72\overline{72}$
28. $.42\overline{2}$
29. $-3\frac{1}{3}$
30. $.07$

Divide. [p. 127]

31. $(6n^3-22n+1)\div(2n-4)$
32. $(4x-3x^2+2x^3-6)\div(x^2+2)$

Is the given binomial a factor of the given polynomial? [p. 127]

33. $x+5$ and $3x^3+17x^2+8x-10$
34. $2n-3$ and $2n^3-n^2-n-9$

Chapter Five Test

Solve.

1. $\frac{n+2}{5}+\frac{2n-1}{5}=2$

2. $\frac{a}{2}-\frac{a-1}{3}=\frac{2a-3}{6}$

3. $\frac{3}{5x}+\frac{7}{10x^2}=\frac{1}{2x}$

4. $.3a-2=1.5-.2a$

Solve the proportion.

5. $\frac{2x-3}{x+3}=\frac{5}{4}$

6. $\frac{n}{4}=\frac{3}{n+1}$

Find the solution set.

7. $\frac{3}{x}+\frac{2}{x-2}=1$

8. $\frac{3}{x-4}+\frac{x-6}{x^2-2x-8}=\frac{2}{x+2}$

9. Machine *A* can do a job in 10 hours. Machine *B* can do the same job in 8 hours. How many hours will the job take if both machines work at the same time?

10. Sandra can decorate for a party in 6 hours. If Louis helps her, the work is done in 4 hours. How many hours would it take Louis if he decorated alone?

Simplify.

11. $$\frac{3+\frac{1}{4a^2}}{\frac{1}{6a}-2}$$

12. $$\frac{\frac{2x}{x^2-9}+\frac{5}{x+3}}{\frac{4}{x+3}+\frac{2}{x-3}}$$

Solve for *x*.

13. $2ax-b=cx+3d$

14. $a(bx-c)=dx$

15. Solve $A=\frac{h}{2}(b+c)$ for c. Then find c if $A=26$, $h=4$, and $b=5$.

16. Solve $T=2ac+2bc+2ab$ for c. Then find c if $T=94$, $a=4$, and $b=5$.

Show that each is a rational number.

17. $.59\overline{59}$

18. $.37\overline{7}$

19. $-7\frac{1}{2}$

20. 4.51

21. Divide $(3n^3-9n^2+9n-22)\div(n^2+3)$.

22. Is $2x-6$ a factor of $2x^3-22x+12$?

REAL NUMBERS

Fields

A set is a *field* for the operations of addition and multiplication if the following properties exist.

I. ADDITION
- Closure
- Associative
- Identity
- Inverse
- Commutative

II. MULTIPLICATION
- Closure
- Associative
- Identity
- Inverse
- Commutative

III. Distributive Property of Multiplication Over Addition

Does the set I form a field for the operations of addition and multiplication?

$$I = \{\ldots, -3, -2, -1, 0, 1, 2, 3, \ldots\}$$

I. Addition

Closed: Is the sum of two integers an integer?

$$-7 + 3 = -4$$

Integers

Yes

Associative: Does $(a + b) + c = a + (b + c)$?

$$(-7 + -3) + 6 = -7 + (-3 + 6)$$
$$-10 + 6 = -7 + 3$$
$$-4 = -4$$

Same

Yes

Identity: Does $a + 0 = a$?	Inverse: Does $a + -a = 0$?	Commutative: Does $a + b = b + a$?

$2 + 0 = 2$
Same
Yes

$-5 + 5 = 0 \quad 3 + -3 = 0$
Inverses
Yes

$-3 + 4 = 4 + -3$
$1 = 1$
Same
Yes

Thus, all addition properties hold.

II. MULTIPLICATION

Closed: Is the product of two integers an integer?	Associative: Does $(ab)c = a(bc)$?
Yes	Yes

Identity: Does $a \cdot 1 = a$?	Inverse: Is there an integer b such that $a \cdot b = 1$?
Yes	No integer b such that $3 \cdot b = 1$ No

Thus, the set of integers, $I = \{\ldots -3, -2, -1, 0, 1, 2, 3 \ldots\}$, does not form a field.

PROJECT

1. Does the set of rational numbers form a field under the operations of addition and multiplication?
2. Does the set of whole numbers, $W = \{0, 1, 2, 3, 4, \ldots\}$, form a field under the operations of addition and multiplication?
3. Look at the addition and multiplication tables mod 5 on page 27. Does the set of numbers modulo 5 form a field?
4. Would the set of numbers modulo 4 form a field for addition and multiplication?

Solving $x^2 = k$

OBJECTIVES

- To solve equations like $x^2 = 7$
- To simplify expressions like $\sqrt{25}$ and $-\sqrt{16}$
- To locate a number like $\sqrt{20}$ between consecutive integers

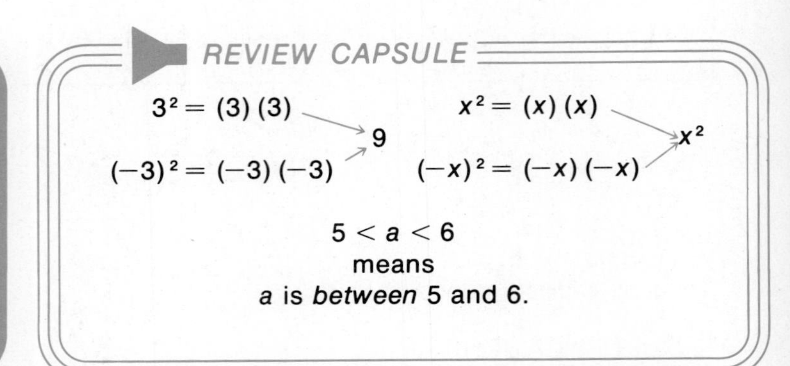

EXAMPLE 1 Find the solution set of $x^2 = 9$.

First Method	Second Method
$x^2 = 9$	$x^2 = 9$
$x^2 - 9 = 0$	
$(x-3)(x+3) = 0$	$x = \sqrt{9}$ or $x = -\sqrt{9}$
$x - 3 = 0$ or $x + 3 = 0$	$\downarrow \qquad \downarrow$
$x = 3 \qquad x = -3$	$x = 3$ or $x = -3$

Check:

Let $x = 3$.

$$\begin{array}{c|c} x^2 & 9 \\ \hline (3)^2 & 9 \\ 9 & \end{array}$$

Let $x = -3$.

$$\begin{array}{c|c} x^2 & 9 \\ \hline (-3)^2 & 9 \\ 9 & \end{array}$$

Thus, the solution set is $\{3, -3\}$.

The second method shows this. → The solutions of $x^2 = 9$ are the square roots of 9.

We say. → The positive square root of 9 is 3.

We write. → $\sqrt{9} = 3$

The negative square root of 9 is -3.

$-\sqrt{9} = -3$

EXAMPLE 2 Find the solution set of $x^2 = 16$.

The square roots of 16 are $\sqrt{16}$ and $-\sqrt{16}$, or 4 and -4. →

$4^2 = 16$ and $(-4)^2 = 16$

$$x^2 = 16$$

$x = \sqrt{16}$, or 4 $\qquad x = -\sqrt{16}$, or -4

Thus, the solution set is $\{4, -4\}$.

If $x^2 = k$, then $x = \sqrt{k}$ or $x = -\sqrt{k}$.
$\sqrt{k}$ is the *principal*, or *positive*, square root of k.

EXAMPLE 3 Find the solution set of $y^2 = 10$.

If $x^2 = k$, →

$$y^2 = 10$$

then $x = \sqrt{k}$ or $x = -\sqrt{k}$. →

$$y = \sqrt{10} \text{ or } y = -\sqrt{10}$$

$\sqrt{10}$ is the *principal*, or *positive*, square root of 10.

$\sqrt{10}$ cannot be simplified.

Thus, the solution set is $\{\sqrt{10}, -\sqrt{10}\}$.

$\sqrt{10}$: positive number whose square is 10. →

$$(\sqrt{10})^2 = \sqrt{10} \cdot \sqrt{10} = 10$$
$$(-\sqrt{10})^2 = (-\sqrt{10})(-\sqrt{10}) = 10$$

Definition of $\sqrt{x}$ →

$$(\sqrt{x})^2 = \sqrt{x} \cdot \sqrt{x} = x$$
$$(-\sqrt{x})^2 = (-\sqrt{x})(-\sqrt{x}) = x$$

EXAMPLE 4 Find the solution set of $x^2 = -9$.

$$x^2 = -9$$
$$x = \sqrt{-9} \text{ or } x = -\sqrt{-9}$$

$(+)^2 = (+)(+) = +$
$(-)^2 = (-)(-) = +$
$0^2 = 0 \cdot 0 = 0$
Numbers on our number line are positive, negative, or zero. →
$\sqrt{-9}$ is defined in *another* number system.

There is no *positive* number whose square is -9.
There is no *negative* number whose square is -9.
The square of 0 is not -9.

Thus, $\sqrt{-9}$ and $-\sqrt{-9}$ are undefined in the real number system. So, the solution set of $x^2 = -9$ is the empty set, ϕ, in the set of real numbers.

25 and 49 are perfect squares.
A *perfect square* is the square of an integer.

Perfect Squares

$25 = 5^2$ $\quad 49 = 7^2$

EXAMPLE 5 Determine if the integer is a perfect square.

Integer	*Answer*
1	Yes: $1 = 1^2$
9	Yes: $9 = 3^2$
10	No
36	Yes: $36 = 6^2$
50	No

No integer squared is 10. → (10)

No integer squared is 50. → (50)

EXAMPLE 6 Locate 14 between consecutive perfect squares. Then locate $\sqrt{14}$ between consecutive integers.

$9 = 3^2$ and $16 = 4^2$ ⟶

$$9 < 14 < 16$$
$$\sqrt{9} < \sqrt{14} < \sqrt{16}$$

3 and 4 are consecutive integers.

$$3 < \sqrt{14} < 4$$

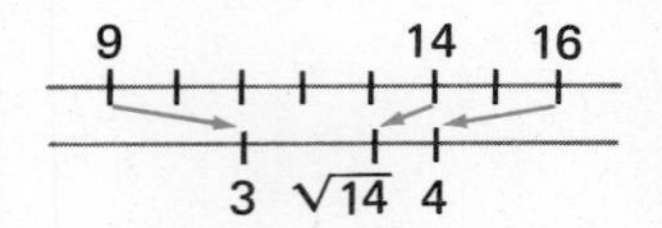

EXAMPLE 7 Locate $\sqrt{53}$ between consecutive integers.

$49 = 7^2$ and $64 = 8^2$

$$49 < 53 < 64$$
$$\sqrt{49} < \sqrt{53} < \sqrt{64}$$

7 and 8 are consecutive integers.

$$7 < \sqrt{53} < 8$$

EXAMPLE 8 Simplify, if possible.

100 and 49 are perfect squares.
5 is not a perfect square.

$\sqrt{100}$	$\sqrt{5}$	$-\sqrt{49}$
10	Not possible	-7

EXERCISES

PART A

Find the solution set.

1. $x^2 = 36$ *$\{-6, 6\}$* **2.** $x^2 = 7$ *$\{-\sqrt{7}, \sqrt{7}\}$* **3.** $x^2 = 100$ *$\{-10, 10\}$* **4.** $x^2 = 2$ *$\{-\sqrt{2}, \sqrt{2}\}$*

Determine if the integer is a perfect square.

5. 4 *yes; 2^2* **6.** 12 *no* **7.** 81 *yes; 9^2* **8.** 48 *no*

Simplify, if possible.

9. $\sqrt{25}$ *5* **10.** $\sqrt{22}$ **11.** $-\sqrt{4}$ *-2* **12.** $-\sqrt{17}$

Locate each square root between consecutive integers.

13. $\sqrt{8}$ *$2 < \sqrt{8} < 3$* **14.** $\sqrt{3}$ *$1 < \sqrt{3} < 2$* **15.** $\sqrt{20}$ *$4 < \sqrt{20} < 5$* **16.** $\sqrt{15}$ *$3 < \sqrt{15} < 4$*

PART B

Find the solution set.

17. $x^2 = 400$ *$\{-20, 20\}$* **18.** $x^2 = 35$ *$\{-\sqrt{35}, \sqrt{35}\}$* **19.** $144 = x^2$ *$\{-12, 12\}$* **20.** $x^2 = 70$ *$\{-\sqrt{70}, \sqrt{70}\}$*

Simplify, if possible.

21. $-\sqrt{121}$ *-11* **22.** $\sqrt{-25}$ **23.** $-\sqrt{900}$ *-30* **24.** $-\sqrt{-16}$

Locate each square root between consecutive integers.

25. $\sqrt{35}$ *$5 < \sqrt{35} < 6$* **26.** $\sqrt{61}$ *$7 < \sqrt{61} < 8$* **27.** $\sqrt{39}$ *$6 < \sqrt{39} < 7$* **28.** $\sqrt{97}$ *$9 < \sqrt{97} < 10$*

Irrational Numbers

OBJECTIVES

- To tell if a number is rational or irrational
- To approximate a square root like $\sqrt{5}$ as a decimal to the nearest tenth

REVIEW CAPSULE

Rational Numbers

Terminating Decimals	Repeating Decimals	Mixed Numerals	Integers
$2.67 = \frac{267}{100}$	$.777 \cdots = .7\overline{7} = \frac{7}{9}$	$2\frac{3}{4} = \frac{11}{4}$	$-4 = \frac{-4}{1}$

$.763 \doteq .76$ to the nearest hundredth, 2 decimal places.

$.763 \doteq .8$ to the nearest tenth, 1 decimal place.

$2.444 \cdots$ is rational. $2.444 \cdots = \frac{22}{9}$

$2.444 \cdots$ — Repeating decimal — Can be written with a bar: $2.4\overline{4}$.

$2.4343343334 \cdots$ — *Nonrepeating* decimal — *Cannot* be written with a bar.

It does not terminate or repeat.

A number like $2.4343343334 \cdots$ is *not rational.* It is an *irrational* number.

> *Irrational* numbers have decimal numerals which are *nonrepeating* and nonterminating.

EXAMPLE 1 Tell if the number is rational or irrational.

	Number	*Answer*
Repeating decimal $8.3\overline{3}$	$8.333 \cdots$	Rational
Nonrepeating, nonterminating	$8.3131131113 \cdots$	Irrational
Repeating $-2.76\overline{76}$	$-2.767676 \cdots$	Rational
Nonrepeating	$.11211221122211 \cdots$	Irrational
Repeating $.2838383 \cdots$	$.28\overline{383}$	Rational
Nonrepeating	$-.78798081 \cdots$	Irrational
Terminating decimal	5.347	Rational
Integer $\sqrt{16} = 4$	$\sqrt{16}$	Rational
Mixed numeral $4\frac{2}{3} = \frac{14}{3}$	$4\frac{2}{3}$	Rational

EXAMPLE 2 Approximate $\sqrt{35}$ to the nearest tenth.

Step 1 Locate $\sqrt{35}$ between consecutive integers. Then average the two integers.

To average two numbers, divide their sum by 2. $(5 + 6) \div 2 = 5.5$

$5 < \sqrt{35} < 6$ Average of 5 and 6 is 5.5.

Step 2 Divide 35 by the average. Divide to hundredths. Do not round the quotient.

Division steps are not shown.

$5.5_{\wedge}\overline{)35.0_{\wedge}00}$ with quotient 6.36 So, $5.5 < \sqrt{35} < 6.36$.

Step 3 Average the divisor and the quotient. Round the average to the nearest tenth.

To average two numbers, find their sum and divide by 2.

$$\begin{array}{r} 5.5 \\ +\ 6.36 \\ \hline 11.86 \end{array} \qquad 2\overline{)11.86} = 5.93 \qquad 5.93 \doteq 5.9 \text{ to the nearest tenth.}$$

Check to see if the new average is the same as the previous divisor.

$5.9 \neq 5.5$ So, we must continue.

Step 4 Divide 35 by the new average. Divide to hundredths. Do not round the quotient.

Division steps are not shown.

$5.9_{\wedge}\overline{)35.0_{\wedge}00}$ with quotient 5.93 So, $5.9 < \sqrt{35} < 5.93$.

Check to see if the divisor and quotient are the same in the tenths place.

If they were not the same, we would repeat Step 3.

Divisor → 5.9 5.93 ← Quotient

9 and 9 in the tenths place are the *same*.

$5.90 < \sqrt{35} < 5.93$ →

Thus, $\sqrt{35} \doteq 5.9$ to the nearest tenth.

The decimal for $\sqrt{35}$ is nonrepeating and nonterminating: $\sqrt{35} \doteq 5.9160797830996$, to 13 decimal places. Numbers like $\sqrt{35}$ are irrational numbers.

$\sqrt{35}$ is irrational. But, $\sqrt{36}$ is rational; $\sqrt{36} = 6$.

If the square root of a whole number is not a whole number, then the square root is irrational.

EXAMPLE 3 Approximate $\sqrt{15}$ to the nearest tenth.

Step 1 Locate $\sqrt{15}$ between consecutive integers. Then, average the two integers.

Average of 3 and 4: $(3 + 4) \div 2 = 7 \div 2 = 3.5$. →

$3 < \sqrt{15} < 4$ Average of 3 and 4 is 3.5.

Step 2 Divide 15 by the average. Divide to hundredths. Do not round the quotient.

Division steps are not shown.

$$3.5\overline{)15.000}\ = 4.28$$

So, $3.5 < \sqrt{15} < 4.28$.

To average two numbers, find their sum and divide by 2.

Step 3 Average the divisor and quotient. Round the average to the nearest tenth.

$$\begin{array}{r} 3.5 \\ +4.28 \\ \hline 7.78 \end{array} \qquad 2\overline{)7.78} = 3.89$$

$3.89 \doteq 3.9$ to the nearest tenth.

Check to see if the new average is the same as the previous divisor.

$3.9 \neq 3.5$ So, we must continue.

Step 4 Divide 15 by the new average. Divide to hundredths. Do not round the quotient.

Division steps are not shown.

$$3.9\overline{)15.000}\ = 3.84$$

So, $3.84 < \sqrt{15} < 3.9$.

Check to see if the divisor and quotient are the same in the tenths place.

Divisor → 3.9 3.84 ← Quotient

So, we continue by repeating Step 3.

9 and 8 in the tenths place are *not* the same.

Average divisor and quotient; round average to nearest tenth.

Repeat Step 3.

$$\begin{array}{r} 3.9 \\ +3.84 \\ \hline 7.74 \end{array} \qquad 2\overline{)7.74} = 3.87$$

$3.87 \doteq 3.9$ to the nearest tenth.

Check to see if the average is the same as the previous divisor.

Average is 3.9; previous divisor was 3.9; they are the same. →

$3.9 = 3.9$

$\sqrt{15}$ is irrational.

Thus, $\sqrt{15} \doteq 3.9$ to the nearest tenth.

EXAMPLE 4 Tell if the square root of the whole number is rational or irrational.

7 is a whole number. But, $\sqrt{7}$ is not a whole number. ⟶

$\sqrt{7}$	$-\sqrt{25}$
$\sqrt{7}$ is irrational.	$-\sqrt{25}$ is rational; $-\sqrt{25}=-5$.

{rationals} ∪ {irrationals} = {real numbers}

Real Numbers

All rational numbers together with all irrational numbers are called the real numbers.

ORAL EXERCISES

Approximate each number to the nearest tenth.

1. 5.38 *5.4* **2.** 7.42 *7.4* **3.** 2.79 *2.8* **4.** 4.03 *4.0*

EXERCISES

PART A

Approximate the square root to the nearest tenth.

1. $\sqrt{17}$ *4.1* **2.** $\sqrt{10}$ *3.2* **3.** $\sqrt{45}$ *6.7* **4.** $\sqrt{30}$ *5.5*

Determine if the number is rational or irrational.

5. 4.7373373337 · · · *irrat.* **6.** .44144114411144 · · · *irrat.* **7.** .623623623 · · · *rat.*

8. $\sqrt{21}$ *irrat.* **9.** $-\sqrt{100}$ *rat.* **10.** $-\sqrt{12}$ *irrat.*

PART B

Approximate the square root to the nearest tenth.

11. $\sqrt{40}$ *6.3* **12.** $\sqrt{59}$ *7.7* **13.** $\sqrt{76}$ *8.7* **14.** $\sqrt{88}$ *9.4*

Determine if the number is rational or irrational.

15. $5.74\overline{646}$ *rat.* **16.** −39.404142 · · · *irrat.* **17.** .242242242 · · · *rat.*

18. $-\sqrt{35}$ *irrat.* **19.** $\sqrt{144}$ *rat.* **20.** $\sqrt{99}$ *irrat.*

PART C

21. Approximate $\sqrt{40}$ to the nearest hundredth.

22. Approximate $\sqrt{59}$ to the nearest hundredth.

23. Approximate $\sqrt{76}$ to the nearest hundredth.

24. Approximate $\sqrt{88}$ to the nearest hundredth.

25. Approximate $\sqrt{44}$ to the nearest thousandth.

26. Approximate $\sqrt{54}$ to the nearest thousandth.

Multiplying and Simplifying Square Roots

OBJECTIVES

- To multiply expressions like $7\sqrt{2} \cdot 5\sqrt{3}$ and $\sqrt{x} \cdot \sqrt{x^5}$
- To simplify square roots like $\sqrt{18}$ and $\sqrt{x^6 y^5}$

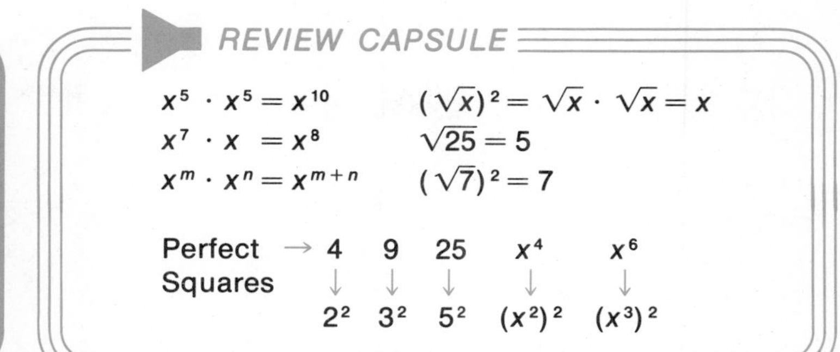

REVIEW CAPSULE

$x^5 \cdot x^5 = x^{10}$ $\qquad (\sqrt{x})^2 = \sqrt{x} \cdot \sqrt{x} = x$

$x^7 \cdot x = x^8$ $\qquad \sqrt{25} = 5$

$x^m \cdot x^n = x^{m+n}$ $\qquad (\sqrt{7})^2 = 7$

Perfect Squares → 4, 9, 25, x^4, x^6

↓ 2^2, 3^2, 5^2, $(x^2)^2$, $(x^3)^2$

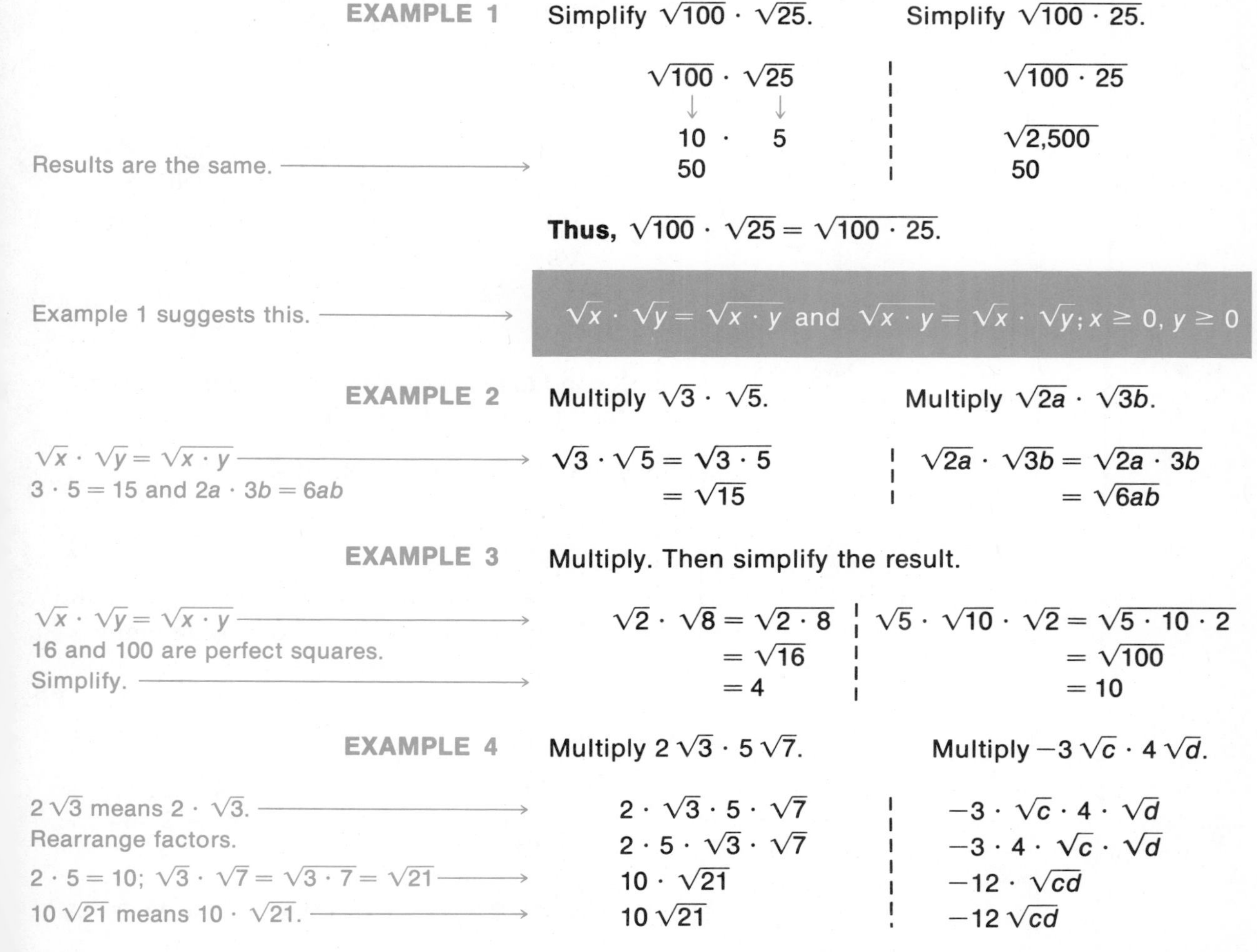

EXAMPLE 1 Simplify $\sqrt{100} \cdot \sqrt{25}$. $\qquad$ Simplify $\sqrt{100 \cdot 25}$.

$\sqrt{100} \cdot \sqrt{25}$ → $10 \cdot 5$ → 50 $\qquad$ | $\qquad \sqrt{100 \cdot 25}$ → $\sqrt{2{,}500}$ → 50

Results are the same. →

Thus, $\sqrt{100} \cdot \sqrt{25} = \sqrt{100 \cdot 25}$.

Example 1 suggests this. →

$\sqrt{x} \cdot \sqrt{y} = \sqrt{x \cdot y}$ and $\sqrt{x \cdot y} = \sqrt{x} \cdot \sqrt{y}; x \geq 0, y \geq 0$

EXAMPLE 2 Multiply $\sqrt{3} \cdot \sqrt{5}$. $\qquad$ Multiply $\sqrt{2a} \cdot \sqrt{3b}$.

$\sqrt{x} \cdot \sqrt{y} = \sqrt{x \cdot y}$ →
$3 \cdot 5 = 15$ and $2a \cdot 3b = 6ab$

$\sqrt{3} \cdot \sqrt{5} = \sqrt{3 \cdot 5} = \sqrt{15}$ $\qquad$ | $\qquad \sqrt{2a} \cdot \sqrt{3b} = \sqrt{2a \cdot 3b} = \sqrt{6ab}$

EXAMPLE 3 Multiply. Then simplify the result.

$\sqrt{x} \cdot \sqrt{y} = \sqrt{x \cdot y}$ →
16 and 100 are perfect squares.
Simplify. →

$\sqrt{2} \cdot \sqrt{8} = \sqrt{2 \cdot 8} = \sqrt{16} = 4$ $\qquad$ | $\qquad \sqrt{5} \cdot \sqrt{10} \cdot \sqrt{2} = \sqrt{5 \cdot 10 \cdot 2} = \sqrt{100} = 10$

EXAMPLE 4 Multiply $2\sqrt{3} \cdot 5\sqrt{7}$. $\qquad$ Multiply $-3\sqrt{c} \cdot 4\sqrt{d}$.

$2\sqrt{3}$ means $2 \cdot \sqrt{3}$. →
Rearrange factors.
$2 \cdot 5 = 10$; $\sqrt{3} \cdot \sqrt{7} = \sqrt{3 \cdot 7} = \sqrt{21}$ →
$10\sqrt{21}$ means $10 \cdot \sqrt{21}$. →

$2 \cdot \sqrt{3} \cdot 5 \cdot \sqrt{7}$	$-3 \cdot \sqrt{c} \cdot 4 \cdot \sqrt{d}$
$2 \cdot 5 \cdot \sqrt{3} \cdot \sqrt{7}$	$-3 \cdot 4 \cdot \sqrt{c} \cdot \sqrt{d}$
$10 \cdot \sqrt{21}$	$-12 \cdot \sqrt{cd}$
$10\sqrt{21}$	$-12\sqrt{cd}$

EXAMPLE 5 Simplify $\sqrt{9 \cdot 5}$.

$\sqrt{x \cdot y} = \sqrt{x} \cdot \sqrt{y}$ →

$$\begin{aligned}\sqrt{9 \cdot 5} &= \sqrt{9} \cdot \sqrt{5} \\ &= 3 \cdot \sqrt{5} \\ &= 3\sqrt{5}\end{aligned}$$

EXAMPLE 6 Simplify $\sqrt{18}$.

$\sqrt{18}$

Factor 18. Use a perfect square factor. → $\sqrt{9 \cdot 2}$ ← 9 is a perfect square.

$\sqrt{x \cdot y} = \sqrt{x} \cdot \sqrt{y}$ → $\sqrt{9} \cdot \sqrt{2}$

$\sqrt{9} = 3$ → $3\sqrt{2}$

EXAMPLE 7 Simplify $-3\sqrt{75}$. Simplify $-3\sqrt{20}$.

25 and 4 are perfect squares.
$\sqrt{x \cdot y} = \sqrt{x} \cdot \sqrt{y}$

$$\begin{aligned}-3\sqrt{75} &= -3\sqrt{25 \cdot 3} \\ &= -3 \cdot \sqrt{25} \cdot \sqrt{3} \\ &= -3 \cdot 5 \cdot \sqrt{3} \\ &= -15\sqrt{3}\end{aligned}$$

$$\begin{aligned}-3\sqrt{20} &= -3\sqrt{4 \cdot 5} \\ &= -3 \cdot \sqrt{4} \cdot \sqrt{5} \\ &= -3 \cdot 2 \cdot \sqrt{5} \\ &= -6\sqrt{5}\end{aligned}$$

EXAMPLE 8 Simplify $\sqrt{72}$.

9 is a perfect square factor of 72.
4 is a perfect square factor of 8.

First Method

$$\begin{aligned}\sqrt{72} &= \sqrt{9 \cdot 8} \\ &= \sqrt{9 \cdot 4 \cdot 2} \\ &= \sqrt{9} \cdot \sqrt{4} \cdot \sqrt{2} \\ &= 3 \cdot 2 \cdot \sqrt{2} \\ &= 6\sqrt{2}\end{aligned}$$

Second Method

$$\begin{aligned}\sqrt{72} &= \sqrt{36 \cdot 2} \\ &= \sqrt{36} \cdot \sqrt{2} \\ &\downarrow \\ &= 6\sqrt{2}\end{aligned}$$

Results are the same: $6\sqrt{2}$. →

In the Second Method we used the *greatest* perfect square factor of 72, which is 36.

EXAMPLE 9 Multiply $\sqrt{8} \cdot \sqrt{3}$. Then simplify the result.

$$\sqrt{8} \cdot \sqrt{3} = \sqrt{24} = \sqrt{4 \cdot 6} = 2\sqrt{6}$$

EXAMPLE 10 Given $\sqrt{2} \doteq 1.414$ and $\sqrt{3} \doteq 1.732$. Simplify $\sqrt{50}$ and $\sqrt{12}$. Then approximate each to the nearest tenth.

$$\sqrt{50} = \sqrt{25 \cdot 2} = \sqrt{25} \cdot \sqrt{2} \qquad \sqrt{12} = \sqrt{4 \cdot 3} = \sqrt{4} \cdot \sqrt{3}$$

Simplified → $= 5\sqrt{2}$ | $= 2\sqrt{3}$

Substitute approximations. → $5\sqrt{2} \doteq 5(1.414)$ | $2\sqrt{3} \doteq 2(1.732)$

Multiply. → $\doteq 7.070$ | $\doteq 3.464$

To nearest tenth → $\doteq 7.1$ | $\doteq 3.5$

Thus, $\sqrt{50} = 5\sqrt{2} \doteq 7.1$. **Thus,** $\sqrt{12} = 2\sqrt{3} \doteq 3.5$.

For the rest of this chapter, all variables represent positive numbers, unless otherwise stated.

EXAMPLE 11 Simplify $\sqrt{a^7}$.

a^6 is a perfect square.

$a^6 = a^3 \cdot a^3$ →

$\sqrt{x} \cdot \sqrt{x} = x$ →

$$\sqrt{a^6 \cdot a^1}$$
$$\sqrt{a^3 \cdot a^3 \cdot a^1}$$
$$\sqrt{a^3} \cdot \sqrt{a^3} \cdot \sqrt{a}$$
$$a^3 \cdot \sqrt{a}$$

Thus, $\sqrt{a^7} = a^3\sqrt{a}$.

Simplify $\sqrt{a^6}$.

$$\sqrt{a^6}$$
$$\sqrt{a^3 \cdot a^3}$$
$$\sqrt{a^3} \cdot \sqrt{a^3}$$
$$a^3$$

Thus, $\sqrt{a^6} = a^3$.

EXAMPLE 12 Simplify $\sqrt{18c^5}$.

9 and c^4 are perfect squares. →
Write perfect squares *first*. →
Simplify.
Write $\sqrt{\ }$ sign last. →

$$\sqrt{18c^5} = \sqrt{9 \cdot 2 \cdot c^4 \cdot c^1}$$
$$= \sqrt{9 \cdot c^4 \cdot 2 \cdot c}$$
$$= \sqrt{9} \cdot \sqrt{c^4} \cdot \sqrt{2 \cdot c}$$
$$= 3c^2\sqrt{2c}$$

Simplify $\sqrt{25d^8}$.

$$\sqrt{25d^8} = \sqrt{25} \cdot \sqrt{d^8}$$
$$= 5d^4$$

EXAMPLE 13 Simplify $\sqrt{25a^8b^3}$.

25, a^8, and b^2 are perfect squares. →

$\sqrt{a^8} = a^4$; $\sqrt{b^2} = b$ →

$$\sqrt{25a^8b^3} = \sqrt{25 \cdot a^8 \cdot b^2 \cdot b^1}$$
$$= \sqrt{25} \cdot \sqrt{a^8} \cdot \sqrt{b^2} \cdot \sqrt{b}$$
$$= 5a^4b\sqrt{b}$$

EXAMPLE 14 Simplify $\sqrt{20x^5y^7}$.

4, x^4, y^6 are perfect squares. →
Write perfect squares *first*. →

Write $\sqrt{\ }$ sign last. →

$$\sqrt{20x^5y^7} = \sqrt{4 \cdot 5 \cdot x^4 \cdot x^1 \cdot y^6 \cdot y^1}$$
$$= \sqrt{4 \cdot x^4 \cdot y^6 \cdot 5 \cdot x \cdot y}$$
$$= \sqrt{4} \cdot \sqrt{x^4} \cdot \sqrt{y^6} \cdot \sqrt{5 \cdot x \cdot y}$$
$$= 2x^2y^3\sqrt{5xy}$$

EXAMPLE 15 Multiply. Then simplify the result.

$\sqrt{x} \cdot \sqrt{y} = \sqrt{x \cdot y}$ →
Simplify.

$\sqrt{4} = 2$; $\sqrt{a^4} = a^2$ →

$$\sqrt{6a} \cdot \sqrt{2a^3} = \sqrt{12a^4}$$
$$= \sqrt{4 \cdot 3 \cdot a^4}$$
$$= \sqrt{4 \cdot a^4 \cdot 3}$$
$$= 2a^2\sqrt{3}$$

$$-2\sqrt{c} \cdot 5\sqrt{c}$$
$$-2 \cdot 5 \cdot \sqrt{c} \cdot \sqrt{c}$$
$$-10 \cdot c$$
$$-10c$$

EXAMPLE 16 Square $(-5\sqrt{2})^2$.

$(y)^2$ means $(y)(y)$. →
Rearrange factors. →
$\sqrt{x} \cdot \sqrt{x} = x$; $\sqrt{2} \cdot \sqrt{2} = 2$ →
and $\sqrt{a-4} \cdot \sqrt{a-4} = a - 4$

$$(-5\sqrt{2})^2$$
$$(-5\sqrt{2})(-5\sqrt{2})$$
$$(-5)(-5)(\sqrt{2} \cdot \sqrt{2})$$
$$25 \cdot 2$$
$$50$$

Square $(3\sqrt{a-4})^2$.

$$(3\sqrt{a-4})^2$$
$$(3\sqrt{a-4})(3\sqrt{a-4})$$
$$3 \cdot 3 \cdot (\sqrt{a-4})(\sqrt{a-4})$$
$$9 \cdot (a-4)$$
$$9a - 36$$

ORAL EXERCISES

Simplify.

1. $\sqrt{9} \cdot \sqrt{16}$ *12*
2. $\sqrt{4 \cdot 25}$ *10*
3. $\sqrt{9 \cdot 7}$ *$3\sqrt{7}$*
4. $\sqrt{11} \cdot \sqrt{11}$ *11*
5. $\sqrt{c} \cdot \sqrt{c}$ *c*
6. $\sqrt{x^2}$ *x*
7. $\sqrt{x^{10}}$ *x^5*
8. $\sqrt{x^8 \cdot x}$ *$x^4\sqrt{x}$*

EXERCISES

PART A

Simplify.

1. $\sqrt{27}$ *$3\sqrt{3}$*
2. $-\sqrt{45}$ *$-3\sqrt{5}$*
3. $-\sqrt{24}$ *$-2\sqrt{6}$*
4. $\sqrt{90}$ *$3\sqrt{10}$*
5. $\sqrt{32}$ *$4\sqrt{2}$*
6. $-\sqrt{80}$ *$-4\sqrt{5}$*
7. $-\sqrt{360}$
8. $\sqrt{96}$ *$4\sqrt{6}$*
9. $4\sqrt{20}$ *$8\sqrt{5}$*
10. $-2\sqrt{150}$
11. $-5\sqrt{63}$
12. $6\sqrt{125}$ *$30\sqrt{5}$*

Given $\sqrt{2} \doteq 1.414$, $\sqrt{3} \doteq 1.732$, and $\sqrt{5} \doteq 2.236$. Simplify the square root. Then approximate it to the nearest tenth.

13. $\sqrt{18}$ *$3\sqrt{2}$; 4.2*
14. $\sqrt{20}$ *$2\sqrt{5}$; 4.5*
15. $\sqrt{48}$ *$4\sqrt{3}$; 6.9*
16. $\sqrt{45}$ *$3\sqrt{5}$; 6.7*

Multiply.

17. $\sqrt{11} \cdot \sqrt{7}$ *$\sqrt{77}$*
18. $(-2\sqrt{7})(-3\sqrt{3})$
19. $\sqrt{5a} \cdot \sqrt{2b}$ *$\sqrt{10ab}$*
20. $-3\sqrt{5m} \cdot 4\sqrt{n}$

Multiply. Then simplify the result.

21. $\sqrt{2} \cdot \sqrt{32}$ *8*
22. $\sqrt{3} \cdot \sqrt{6} \cdot \sqrt{2}$ *6*
23. $\sqrt{6} \cdot \sqrt{2}$ *$2\sqrt{3}$*
24. $\sqrt{5} \cdot \sqrt{10}$ *$5\sqrt{2}$*

Simplify.

25. $\sqrt{9y^6}$ *$3y^3$*
26. $\sqrt{7a^4}$ *$a^2\sqrt{7}$*
27. $\sqrt{12b^2}$ *$2b\sqrt{3}$*
28. $\sqrt{24x^{10}}$
29. $\sqrt{x^9}$ *$x^4\sqrt{x}$*
30. $\sqrt{16y^7}$ *$4y^3\sqrt{y}$*
31. $\sqrt{5a^5}$ *$a^2\sqrt{5a}$*
32. $\sqrt{8b^3}$ *$2b\sqrt{2b}$*

PART B

Simplify.

33. $\sqrt{x^4y^6}$ *x^2y^3*
34. $\sqrt{36a^2b^8}$ *$6ab^4$*
35. $\sqrt{24c^{10}d^4}$ *$2c^5d^2\sqrt{6}$*
36. $\sqrt{x^6y^5}$ *$x^3y^2\sqrt{y}$*
37. $\sqrt{a^3b^2}$ *$ab\sqrt{a}$*
38. $\sqrt{18m^8n^9}$
39. $\sqrt{x^3y^7}$ *$xy^3\sqrt{xy}$*
40. $\sqrt{12a^5b^3}$ *$2a^2b\sqrt{3ab}$*

Multiply. Then simplify the result.

41. $\sqrt{12x^3} \cdot \sqrt{3x}$ *$6x^2$*
42. $\sqrt{6y} \cdot \sqrt{3y^7}$ *$3y^4\sqrt{2}$*
43. $2\sqrt{3x^3} \cdot 5\sqrt{2x}$ *$10x^2\sqrt{6}$*
44. $\sqrt{3x^3y^3} \cdot \sqrt{8xy}$ *$2x^2y^2\sqrt{6}$*

Square the expression.

45. $(2\sqrt{7})^2$ *28*
46. $(a\sqrt{b})^2$ *a^2b*
47. $(-3c\sqrt{d})^2$ *$9c^2d$*
48. $(2x\sqrt{3y})^2$ *$12x^2y$*
49. $(\sqrt{x+5})^2$ *$x+5$*
50. $(-\sqrt{a-3})^2$ *$a-3$*
51. $(2\sqrt{x+1})^2$ *$4x+4$*
52. $(3\sqrt{2a+5})^2$ *$18a+45$*

PART C

Simplify.

53. $\sqrt{x^{2m}}$
54. $\sqrt{x^{2m+1}}$
55. $\sqrt{x^{4m}}$
56. $\sqrt{x^{6m+1}}$
57. $\sqrt{x^{3m}} \cdot \sqrt{x^m}$
58. $\sqrt{x^{m+3}} \cdot \sqrt{x^{5m-1}}$
59. $\sqrt{x^{4m}y^{2n+1}}$
60. $\sqrt{x^{6m+4}y^{4n+3}}$

Adding and Multiplying Square Roots

OBJECTIVES

- To simplify expressions like $7\sqrt{2} + \sqrt{8} - 3\sqrt{2}$
- To multiply expressions like $(3\sqrt{2} + \sqrt{3})(\sqrt{2} - 5\sqrt{3})$

REVIEW CAPSULE

Combine like terms.

$2x + 5y + 3x - 7y$

$2x + 3x + 5y - 7y$

$5x - 2y$

Multiply.

$$\begin{array}{r} 3x \quad + 5 \\ 2x \quad - 4 \\ \hline 6x^2 + 10x \qquad \\ - 12x - 20 \\ \hline 6x^2 - \ \ 2x - 20 \end{array}$$

EXAMPLE 1 Simplify $3\sqrt{10} + 5\sqrt{10} - 2\sqrt{10}$.

$\sqrt{10}$ is a factor of each term. ⟶

$$3\sqrt{10} + 5\sqrt{10} - 2\sqrt{10} = (3 + 5 - 2)\sqrt{10} = 6\sqrt{10}$$

Like square roots may be combined just as *like terms* are combined.

$3\sqrt{10}$, $5\sqrt{10}$, and $-2\sqrt{10}$ are *like square roots.*

EXAMPLE 2 Simplify $2\sqrt{3} + 7\sqrt{5}$.

$\sqrt{3}$ is *not* a factor of *both* terms. ⟶

$$2\sqrt{3} + 7\sqrt{5}$$

$2\sqrt{3}$ and $7\sqrt{5}$ are *not* like square roots. ⟶ *Not* like square roots

Thus, $2\sqrt{3}$ and $7\sqrt{5}$ cannot be combined.

EXAMPLE 3 Simplify $8\sqrt{7} + 3\sqrt{5} - 2\sqrt{7} + \sqrt{5}$.

Group like square roots; $\sqrt{5} = 1\sqrt{5}$. ⟶

$$8\sqrt{7} + 3\sqrt{5} - 2\sqrt{7} + \sqrt{5} = \underbrace{8\sqrt{7} - 2\sqrt{7}} + \underbrace{3\sqrt{5} + 1\sqrt{5}}$$

Combine like square roots. ⟶

$$= \quad 6\sqrt{7} \quad + \quad 4\sqrt{5}$$

EXAMPLE 4 Simplify $\sqrt{8} + \sqrt{18} - \sqrt{2}$.

Simplify each square root.

$-\sqrt{2} = -1\sqrt{2}$; combine like roots.

$$\sqrt{8} + \sqrt{18} - \sqrt{2} = \sqrt{4 \cdot 2} + \sqrt{9 \cdot 2} - \sqrt{2}$$
$$= 2\sqrt{2} + 3\sqrt{2} - 1\sqrt{2} = 4\sqrt{2}$$

EXAMPLE 5 Simplify.

Simplify each square root.

$7\sqrt{a} + \sqrt{4a} + \sqrt{a}$	$\sqrt{25cd^3} + 3d\sqrt{cd} - \sqrt{cd^3}$
$7\sqrt{a} + \sqrt{4} \cdot \sqrt{a} + \sqrt{a}$	$\sqrt{25d^2cd} + 3d\sqrt{cd} - \sqrt{d^2cd}$
$7\sqrt{a} + 2\sqrt{a} + 1\sqrt{a}$	$5d\sqrt{cd} + 3d\sqrt{cd} - d\sqrt{cd}$
$10\sqrt{a}$	$7d\sqrt{cd}$

EXAMPLE 6 Multiply $5(\sqrt{3} - 2\sqrt{7})$.

Distribute the 5. → $5(\sqrt{3} - 2\sqrt{7}) = 5 \cdot \sqrt{3} + 5(-2\sqrt{7})$

$5(-2\sqrt{7}) = 5(-2) \cdot \sqrt{7} = -10\sqrt{7}$ → $= 5\sqrt{3} - 10\sqrt{7}$

EXAMPLE 7 Multiply $3\sqrt{2}(\sqrt{6} + 5\sqrt{2})$.

Distribute $3\sqrt{2}$. → $3\sqrt{2}(\sqrt{6} + 5\sqrt{2}) = 3\sqrt{2} \cdot \sqrt{6} + 3\sqrt{2} \cdot 5\sqrt{2}$

Multiply. → $= 3\sqrt{12} + 15\sqrt{4}$

Simplify the square roots.

$$= 3 \cdot \sqrt{4 \cdot 3} + 15 \cdot 2$$
$$= 3 \cdot 2 \cdot \sqrt{3} + 15 \cdot 2$$
$$= 6\sqrt{3} + 30$$

EXAMPLE 8 Multiply $(2\sqrt{3} + \sqrt{2})(4\sqrt{3} - 5\sqrt{2})$.

Write one expression below the other.

$$\begin{array}{rrr} 2\sqrt{3} & & +\sqrt{2} \\ 4\sqrt{3} & & -5\sqrt{2} \\ \hline 8 \cdot 3 & +4\sqrt{6} & \\ & -10\sqrt{6} & -5 \cdot 2 \\ \hline 24 & -6\sqrt{6} & -10 \end{array}$$

$4\sqrt{3}(2\sqrt{3} + \sqrt{2}) = 8 \cdot 3 + 4\sqrt{6}$ →

$-5\sqrt{2}(2\sqrt{3} + \sqrt{2}) = -10\sqrt{6} - 5 \cdot 2$ →

Combine like terms.

$24 - 10 = 14$ → **Thus,** $(2\sqrt{3} + \sqrt{2})(4\sqrt{3} - 5\sqrt{2}) = 14 - 6\sqrt{6}$.

EXAMPLE 9 Multiply $(2\sqrt{5} - \sqrt{3})(2\sqrt{5} + \sqrt{3})$.

$2\sqrt{5} - \sqrt{3}$ and $2\sqrt{5} + \sqrt{3}$ are a pair of *conjugates*. Their product will not contain a square root.

$$\begin{array}{rrr} 2\sqrt{5} & & -\sqrt{3} \\ 2\sqrt{5} & & +\sqrt{3} \\ \hline 4 \cdot 5 & -2\sqrt{15} & \\ & +2\sqrt{15} & -3 \\ \hline 20 & +\ 0 & -3 \end{array}$$

Middle term is zero. →

$20 + 0 - 3 = 17$ → **Thus,** $(2\sqrt{5} - \sqrt{3})(2\sqrt{5} + \sqrt{3}) = 17$.

EXAMPLE 10 Square $(3\sqrt{x} - 2\sqrt{y})^2$.

$(a)^2$ means $(a)(a)$. Write the expression twice.

Multiply.

$$\begin{array}{rrr} 3\sqrt{x} & & -2\sqrt{y} \\ 3\sqrt{x} & & -2\sqrt{y} \\ \hline 9 \cdot x & -6\sqrt{xy} & \\ & -6\sqrt{xy} & +4 \cdot y \\ \hline 9x & -12\sqrt{xy} & +4y \end{array}$$

Thus, $(3\sqrt{x} - 2\sqrt{y})^2 = 9x - 12\sqrt{xy} + 4y$.

EXERCISES

PART A

Simplify.

1. $2\sqrt{10} + 8\sqrt{5} - 6\sqrt{10} - 3\sqrt{5}$ *$-4\sqrt{10} + 5\sqrt{5}$*

2. $3\sqrt{5} - 7\sqrt{2} - \sqrt{5} + \sqrt{2}$ *$2\sqrt{5} - 6\sqrt{2}$*

3. $\sqrt{45} - \sqrt{20} + 4\sqrt{5}$ *$5\sqrt{5}$*

4. $\sqrt{50} + 2\sqrt{18} + \sqrt{32}$ *$15\sqrt{2}$*

5. $\sqrt{16n} - 3\sqrt{n} + \sqrt{4n}$ *$3\sqrt{n}$*

6. $\sqrt{c} - \sqrt{25c} + 2\sqrt{c}$ *$-2\sqrt{c}$*

Multiply.

7. $3(2\sqrt{3} - \sqrt{5})$ *$6\sqrt{3} - 3\sqrt{5}$*

8. $-2(4 + 3\sqrt{2})$ *$-8 - 6\sqrt{2}$*

9. $\sqrt{3}(4 + \sqrt{3})$

10. $-\sqrt{5}(2\sqrt{5} - \sqrt{2})$

11. $3\sqrt{2}(\sqrt{2} - 2\sqrt{5})$

12. $-2\sqrt{5}(2\sqrt{5} + 3)$

13. $\sqrt{2}(\sqrt{8} - \sqrt{6})$ *$4 - 2\sqrt{3}$*

14. $-5\sqrt{2}(\sqrt{10} - 2\sqrt{2})$ *$-10\sqrt{5} + 20$*

Multiply.

15. $(4\sqrt{3} + 5)(\sqrt{3} - 2)$ *$2 - 3\sqrt{3}$*

16. $(2\sqrt{7} - 3\sqrt{5})(\sqrt{7} - \sqrt{5})$ *$29 - 5\sqrt{35}$*

17. $(2\sqrt{7} - 4\sqrt{2})(3\sqrt{7} + \sqrt{2})$

18. $(\sqrt{10} - 2\sqrt{3})(4\sqrt{10} - \sqrt{3})$

19. $(3\sqrt{2} - 4)(3\sqrt{2} + 4)$ *2*

20. $(4\sqrt{5} + 3\sqrt{3})(4\sqrt{5} - 3\sqrt{3})$

Square the expression.

21. $(\sqrt{5} + \sqrt{3})^2$ *$8 + 2\sqrt{15}$*

22. $(3\sqrt{2} - \sqrt{7})^2$ *$25 - 6\sqrt{14}$*

23. $(\sqrt{10} - 2\sqrt{3})^2$ *$22 - 4\sqrt{30}$*

24. $(2\sqrt{5} + 4\sqrt{3})^2$ *$68 + 16\sqrt{15}$*

PART B

Simplify.

25. $\sqrt{8} + \sqrt{12} + \sqrt{18} + \sqrt{27}$ *$5\sqrt{2} + 5\sqrt{3}$*

26. $\sqrt{50} - \sqrt{20} + 3\sqrt{2} - \sqrt{45}$ *$8\sqrt{2} - 5\sqrt{5}$*

27. $\sqrt{48} + \sqrt{5} + \sqrt{80} - \sqrt{3}$ *$3\sqrt{3} + 5\sqrt{5}$*

28. $2\sqrt{c^2d} + 3c\sqrt{d}$ *$5c\sqrt{d}$*

29. $3b\sqrt{ab} + \sqrt{4ab^3} + \sqrt{ab^3}$ *$6b\sqrt{ab}$*

30. $5\sqrt{m^2n^3} + 2mn\sqrt{n}$ *$7mn\sqrt{n}$*

Multiply.

31. $(4\sqrt{c} + 5\sqrt{d})(\sqrt{c} - \sqrt{d})$

32. $(5\sqrt{a} - \sqrt{b})(3\sqrt{a} - \sqrt{b})$

33. $(\sqrt{c} - 3\sqrt{d})(2\sqrt{c} + \sqrt{d})$

34. $(\sqrt{a} + \sqrt{b})(\sqrt{a} - \sqrt{b})$ *$a - b$*

35. $(3\sqrt{c} - 2\sqrt{d})(3\sqrt{c} + 2\sqrt{d})$ *$9c - 4d$*

36. $(5\sqrt{x} + y)(5\sqrt{x} - y)$ *$25x - y^2$*

Square the expression.

37. $(\sqrt{a} + \sqrt{b})^2$ *$a + 2\sqrt{ab} + b$*

38. $(3\sqrt{x} - 2y)^2$ *$9x - 12y\sqrt{x} + 4y^2$*

39. $(2\sqrt{c} + 3\sqrt{d})^2$ *$4c + 12\sqrt{cd} + 9d$*

PART C

Multiply.

40. $(\sqrt{x+3} + 2)(\sqrt{x+3} + 5)$

41. $(\sqrt{2x-5} - 3)(\sqrt{2x-5} + 3)$

42. $(\sqrt{x+y} + \sqrt{x})(\sqrt{x+y} - \sqrt{x})$

43. $(4 - \sqrt{x-3})^2$

44. $(\sqrt{x+2} - \sqrt{x})^2$

45. $(3\sqrt{x-9} + 2\sqrt{x})^2$

Dividing Square Roots

OBJECTIVES

- To divide square roots like $\frac{\sqrt{30}}{\sqrt{6}}$ and $\frac{\sqrt{18x^3}}{\sqrt{2x}}$
- To change the denominators of expressions like $\frac{2}{\sqrt{5}}$ and $\frac{4}{\sqrt{7}-\sqrt{2}}$ to integers

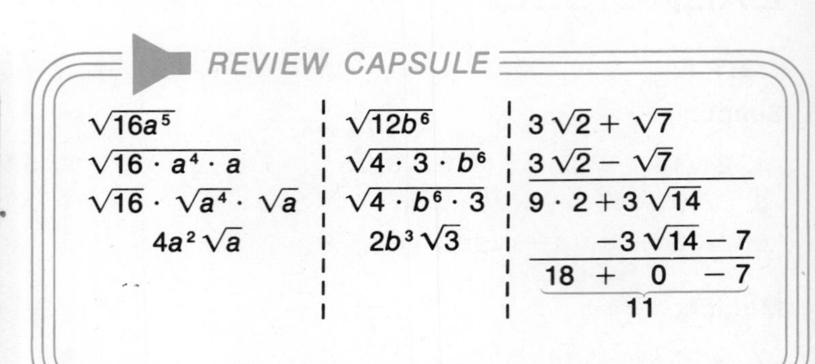

EXAMPLE 1 Simplify $\frac{\sqrt{36}}{\sqrt{4}}$ and $\sqrt{\frac{36}{4}}$. Compare the results.

$$\frac{\sqrt{36}}{\sqrt{4}} = \frac{6}{2} = 3 \text{ and } \sqrt{\frac{36}{4}} = \sqrt{9} = 3$$

Results are the same: 3. → **Thus,** $\frac{\sqrt{36}}{\sqrt{4}} = \sqrt{\frac{36}{4}}$.

Example 1 suggests this. $\frac{\sqrt{36}}{\sqrt{4}} = \sqrt{\frac{36}{4}}$ →

$$\frac{\sqrt{x}}{\sqrt{y}} = \sqrt{\frac{x}{y}} \text{ and } \sqrt{\frac{x}{y}} = \frac{\sqrt{x}}{\sqrt{y}}$$

EXAMPLE 2 Divide. Then simplify the result.

$\frac{\sqrt{x}}{\sqrt{y}} = \sqrt{\frac{x}{y}}$ →

$\frac{22}{11} = 2$ →

$\sqrt{9} = 3$

$\sqrt{12} = 2\sqrt{3}$

$$\frac{\sqrt{22}}{\sqrt{11}} = \sqrt{\frac{22}{11}} = \sqrt{2}$$

$$\frac{\sqrt{45}}{\sqrt{5}} = \sqrt{\frac{45}{5}} = \sqrt{9} = 3$$

$$\frac{\sqrt{24}}{\sqrt{2}} = \sqrt{\frac{24}{2}} = \sqrt{12} = \sqrt{4 \cdot 3} = 2\sqrt{3}$$

EXAMPLE 3 Divide $\frac{\sqrt{20a^5}}{\sqrt{5a}}$. Divide $\frac{\sqrt{15b^4}}{\sqrt{5b}}$.

$\frac{\sqrt{x}}{\sqrt{y}} = \sqrt{\frac{x}{y}}$ →

$\frac{20a^5}{5a} = 4a^4$ →

$\sqrt{4a^4} = 2a^2$ →

$\sqrt{3b^3} = b\sqrt{3b}$

$$\frac{\sqrt{20a^5}}{\sqrt{5a}} = \sqrt{\frac{20a^5}{5a}} = \sqrt{4a^4} = \sqrt{4} \cdot \sqrt{a^4} = 2a^2$$

$$\frac{\sqrt{15b^4}}{\sqrt{5b}} = \sqrt{\frac{15b^4}{5b}} = \sqrt{3b^3} = \sqrt{3 \cdot b^2 \cdot b} = \sqrt{b^2 \cdot 3 \cdot b} = b\sqrt{3b}$$

EXAMPLE 4 Given $\sqrt{3} \doteq 1.732$. Approximate $\frac{2}{\sqrt{3}}$ to the nearest tenth.

First Method

$$\frac{2}{\sqrt{3}} \doteq \frac{2}{1.732}$$

$$\begin{array}{r} 1.154 \\ 1.732_\wedge\overline{)2.000_\wedge000} \\ \underline{1\,732} \\ 268\,0 \\ \underline{173\,2} \\ 94\,80 \\ \underline{86\,60} \\ 8\,200 \\ \underline{6\,928} \\ 1\,272 \end{array}$$

Second Method

$$\frac{2}{\sqrt{3}} = \frac{2}{\sqrt{3}} \cdot 1 = \frac{2}{\sqrt{3}} \cdot \frac{\sqrt{3}}{\sqrt{3}}$$

$$= \frac{2\sqrt{3}}{3}$$

$$\doteq \frac{2(1.732)}{3}$$

$$\doteq \frac{3.464}{3}$$

$$\begin{array}{r} 1.154 \\ 3\overline{)3.464} \end{array}$$

It is *easier* to divide by 3 (Second Method) than by 1.732 (First Method).

$1.154 \doteq 1.2$ to nearest tenth.

Thus, $\frac{2}{\sqrt{3}} \doteq 1.2$ to the nearest tenth.

$\frac{2}{\sqrt{3}} = \frac{2}{\sqrt{3}} \cdot \frac{\sqrt{3}}{\sqrt{3}} = \frac{2\sqrt{3}}{3}$ ⟶

In the Second Method, $\frac{2}{\sqrt{3}}$ was changed to $\frac{2\sqrt{3}}{3}$.

Denominator is irrational. Denominator is rational.

> Changing a denominator from an irrational to a rational number is called *rationalizing the denominator.*

EXAMPLE 5 Rationalize the denominator.

Multiply by 1: $\frac{\sqrt{7}}{\sqrt{7}} = 1$; $\frac{\sqrt{5}}{\sqrt{5}} = 1$. ⟶

Denominators, 7 and 15, are rational numbers. ⟶

$$\frac{3}{\sqrt{7}} = \frac{3}{\sqrt{7}} \cdot \frac{\sqrt{7}}{\sqrt{7}}$$

$$= \frac{3\sqrt{7}}{7}$$

$$\frac{-4\sqrt{2}}{3\sqrt{5}} = \frac{-4\sqrt{2}}{3\sqrt{5}} \cdot \frac{\sqrt{5}}{\sqrt{5}}$$

$$= \frac{-4\sqrt{10}}{15}$$

EXAMPLE 6 Rationalize the denominator of $\frac{3}{\sqrt{8}}$.

Second Method: First, simplify the denominator ($\sqrt{8} = 2\sqrt{2}$).

First Method: Simplify the resulting fraction $\left(\frac{6}{8} = \frac{3}{4}\right)$.

First Method

$$\frac{3}{\sqrt{8}} \cdot \frac{\sqrt{8}}{\sqrt{8}} = \frac{3\sqrt{8}}{8}$$

$$= \frac{3 \cdot 2\sqrt{2}}{8}$$

$$= \frac{6\sqrt{2}}{8}$$

$$= \frac{3\sqrt{2}}{4}$$

Second Method

$$\frac{3}{\sqrt{8}} = \frac{3}{2\sqrt{2}}$$

$$= \frac{3}{2\sqrt{2}} \cdot \frac{\sqrt{2}}{\sqrt{2}}$$

$$= \frac{3\sqrt{2}}{4}$$

Thus, $\frac{3}{\sqrt{8}} = \frac{3\sqrt{2}}{4}$.

EXAMPLE 7 Rationalize the denominator of $\frac{2}{\sqrt{x^3}}$.

First, simplify $\sqrt{x^3}$: $\sqrt{x^2 \cdot x} = x\sqrt{x}$. ⟶

$$\frac{2}{\sqrt{x^3}} = \frac{2}{x\sqrt{x}}$$

Multiply by 1: $\frac{\sqrt{x}}{\sqrt{x}} = 1$.

$x \cdot \sqrt{x} \cdot \sqrt{x} = x \cdot x = x^2$ ⟶

$$= \frac{2}{x\sqrt{x}} \cdot \frac{\sqrt{x}}{\sqrt{x}} = \frac{2\sqrt{x}}{x^2}$$

EXAMPLE 8 Simplify $\sqrt{\frac{3}{5}}$.

$\sqrt{\frac{x}{y}} = \frac{\sqrt{x}}{\sqrt{y}}$ ⟶

$$\sqrt{\frac{3}{5}} = \frac{\sqrt{3}}{\sqrt{5}}$$

Rationalize the denominator.

$$= \frac{\sqrt{3}}{\sqrt{5}} \cdot \frac{\sqrt{5}}{\sqrt{5}} = \frac{\sqrt{15}}{5}$$

EXAMPLE 9 Simplify $\sqrt{\frac{3a}{2b^3}}$.

$\sqrt{\frac{x}{y}} = \frac{\sqrt{x}}{\sqrt{y}}$ ⟶

$$\sqrt{\frac{3a}{2b^3}} = \frac{\sqrt{3a}}{\sqrt{2b^3}}$$

Rationalize the denominator.

First, simplify $\sqrt{2b^3}$. ⟶

$$= \frac{\sqrt{3a}}{b\sqrt{2b}}$$

$$= \frac{\sqrt{3a}}{b\sqrt{2b}} \cdot \frac{\sqrt{2b}}{\sqrt{2b}}$$

$b \cdot \sqrt{2b} \cdot \sqrt{2b} = b \cdot 2b = 2b^2$ ⟶

$$= \frac{\sqrt{6ab}}{2b^2}$$

Simplify $\sqrt{\frac{2c-3}{5}}$.

$$\sqrt{\frac{2c-3}{5}} = \frac{\sqrt{2c-3}}{\sqrt{5}}$$

$$= \frac{\sqrt{2c-3}}{\sqrt{5}} \cdot \frac{\sqrt{5}}{\sqrt{5}}$$

$$= \frac{\sqrt{5(2c-3)}}{5}$$

$$= \frac{\sqrt{10c-15}}{5}$$

EXAMPLE 10 Rationalize the denominator of $\frac{5}{3-\sqrt{2}}$.

Multiply by 1: $\frac{3+\sqrt{2}}{3+\sqrt{2}} = 1$. ⟶

$$\frac{5}{3-\sqrt{2}} = \frac{5}{3-\sqrt{2}} \cdot \frac{3+\sqrt{2}}{3+\sqrt{2}}$$

$3+\sqrt{2}$ is the *conjugate* of $3-\sqrt{2}$.

$$\begin{array}{l} 3 - \sqrt{2} \\ \underline{3 + \sqrt{2}} \\ 9 - 3\sqrt{2} \\ \underline{\quad + 3\sqrt{2} - 2} \\ 9 + \;\; 0 \;\; - 2 = 7 \end{array}$$ ⟶

$$= \frac{5(3+\sqrt{2})}{(3-\sqrt{2})(3+\sqrt{2})}$$

$$= \frac{15+5\sqrt{2}}{9-2}$$

$$= \frac{15+5\sqrt{2}}{7}$$

EXAMPLE 11 Rationalize the denominator of $\frac{\sqrt{2}}{2\sqrt{3}+\sqrt{5}}$.

$2\sqrt{3}-\sqrt{5}$ is the *conjugate* of $2\sqrt{3}+\sqrt{5}$.

$$\begin{array}{l} 2\sqrt{3} + \sqrt{5} \\ \underline{2\sqrt{3} - \sqrt{5}} \\ 4 \cdot 3 + 2\sqrt{15} \\ \underline{\quad - 2\sqrt{15} - 5} \\ 12 \; + \;\; 0 \;\; - 5 = 7 \end{array}$$ ⟶

$$\frac{\sqrt{2}}{2\sqrt{3}+\sqrt{5}} \cdot \frac{2\sqrt{3}-\sqrt{5}}{2\sqrt{3}-\sqrt{5}} = \frac{\sqrt{2}(2\sqrt{3}-\sqrt{5})}{(2\sqrt{3}+\sqrt{5})(2\sqrt{3}-\sqrt{5})}$$

$$= \frac{2\sqrt{6}-\sqrt{10}}{12-5}$$

$$= \frac{2\sqrt{6}-\sqrt{10}}{7}$$

EXERCISES

PART A

Divide. Then simplify the result.

1. $\frac{\sqrt{30}}{\sqrt{10}}$ $\sqrt{3}$
2. $\frac{\sqrt{40}}{\sqrt{10}}$ 2
3. $\frac{\sqrt{80}}{\sqrt{10}}$ $2\sqrt{2}$
4. $\frac{\sqrt{15}}{\sqrt{3}}$ $\sqrt{5}$
5. $\frac{\sqrt{60}}{\sqrt{3}}$ $2\sqrt{5}$
6. $\frac{\sqrt{a^9}}{\sqrt{a^3}}$ a^3
7. $\frac{\sqrt{12b^3}}{\sqrt{3b}}$ $2b$
8. $\frac{\sqrt{6c^6}}{\sqrt{3c}}$ $c^2\sqrt{2c}$
9. $\frac{\sqrt{18n^8}}{\sqrt{2n^3}}$ $3n^2\sqrt{n}$
10. $\frac{\sqrt{24x^7}}{\sqrt{2x^3}}$ $2x^2\sqrt{3}$

Rationalize the denominator.

11. $\frac{1}{\sqrt{6}}$ $\frac{\sqrt{6}}{6}$
12. $\frac{2}{\sqrt{5}}$ $\frac{2\sqrt{5}}{5}$
13. $\frac{\sqrt{3}}{\sqrt{2}}$ $\frac{\sqrt{6}}{2}$
14. $\frac{-5}{2\sqrt{7}}$ $\frac{-5\sqrt{7}}{14}$
15. $\frac{-3\sqrt{5}}{4\sqrt{3}}$ $\frac{-\sqrt{15}}{4}$
16. $\frac{1}{\sqrt{8}}$ $\frac{\sqrt{2}}{4}$
17. $\frac{5}{\sqrt{12}}$ $\frac{5\sqrt{3}}{6}$
18. $\frac{\sqrt{7}}{\sqrt{18}}$ $\frac{\sqrt{14}}{6}$
19. $\frac{-3}{\sqrt{32}}$ $\frac{-3\sqrt{2}}{8}$
20. $\frac{-7\sqrt{3}}{\sqrt{20}}$ $\frac{-7\sqrt{15}}{10}$
21. $\frac{5}{\sqrt{a}}$ $\frac{5\sqrt{a}}{a}$
22. $\frac{3}{\sqrt{b^3}}$ $\frac{3\sqrt{b}}{b^2}$
23. $\frac{3a}{5\sqrt{c}}$ $\frac{3a\sqrt{c}}{5c}$
24. $\frac{-4}{7\sqrt{d^3}}$ $\frac{-4\sqrt{d}}{7d^2}$
25. $\frac{1}{\sqrt{xy^3}}$ $\frac{\sqrt{xy}}{xy^2}$
26. $\frac{3}{3-\sqrt{5}}$
27. $\frac{-2}{3+\sqrt{2}}$
28. $\frac{\sqrt{3}}{\sqrt{10}-2}$
29. $\frac{-\sqrt{2}}{\sqrt{7}+2}$
30. $\frac{3\sqrt{2}}{\sqrt{5}-\sqrt{3}}$

Given $\sqrt{2} \doteq 1.414$ and $\sqrt{3} \doteq 1.732$. Rationalize the denominator. Then approximate to the nearest tenth.

31. $\frac{3}{\sqrt{2}}$ $\frac{3\sqrt{2}}{2}$; 2.1
32. $\frac{10}{\sqrt{3}}$ $\frac{10\sqrt{3}}{3}$; 5.8
33. $\frac{11}{5\sqrt{2}}$
34. $\frac{5}{2\sqrt{3}}$ $\frac{5\sqrt{3}}{6}$; 1.4
35. $\frac{5}{\sqrt{8}}$ $\frac{5\sqrt{2}}{4}$; 1.8

Simplify.

36. $\sqrt{\frac{2}{3}}$ $\frac{\sqrt{6}}{3}$
37. $\sqrt{\frac{5}{7}}$ $\frac{\sqrt{35}}{7}$
38. $\sqrt{\frac{c}{d}}$ $\frac{\sqrt{cd}}{d}$
39. $\sqrt{\frac{2a}{5b}}$ $\frac{\sqrt{10ab}}{5b}$
40. $\sqrt{\frac{x-2}{6}}$ $\frac{\sqrt{6x-12}}{6}$

PART B

Rationalize the denominator.

41. $\frac{5}{3\sqrt{2}-\sqrt{7}}$
42. $\frac{\sqrt{2}}{\sqrt{15}+2\sqrt{3}}$
43. $\frac{3\sqrt{3}}{2\sqrt{5}-3\sqrt{2}}$
44. $\frac{2}{3\sqrt{x}+\sqrt{y}}$
45. $\frac{1}{2\sqrt{a}-b}$

Simplify.

46. $\sqrt{\frac{5a^2}{2b}}$
47. $\sqrt{\frac{4x}{3y^3}}$
48. $\sqrt{\frac{5m}{8n}}$
49. $\sqrt{\frac{a^3}{7b^5}}$
50. $\sqrt{\frac{3a-2}{3}}$
51. $\sqrt{\frac{3}{ab^3}}$
52. $\sqrt{\frac{8}{c^2d^3}}$
53. $\sqrt{\frac{4c}{x^4y^5}}$
54. $\sqrt{\frac{5a}{m^3n^5}}$
55. $\sqrt{\frac{5c+4}{2}}$

A Method for Finding Square Roots

Square roots of numbers can be found by following these steps.

Step 1 $\sqrt{6_{\wedge}60_{\wedge}49.}$ Begin at the decimal point and move left, marking off groups of two digits.

Step 2 The first digit of the answer is the whole number whose square is equal to or is less than the left-hand group. ($2^2 < 6$)

$$\begin{array}{r} 2 \\ \sqrt{6_{\wedge}60_{\wedge}49} \\ 4 \\ \hline 2 \end{array}$$

Step 3 Bring down the next group and divide by the partial answer multiplied by 20. ($2 \times 20 = 40$)

$$\begin{array}{r} 2 \\ \sqrt{6_{\wedge}60_{\wedge}49} \\ 4 \\ 40\overline{)260} \end{array}$$

Step 4 Estimate the answer and replace 0 with this digit. Multiply. (Note product is greater than 260. Use next smaller digit.)

$$\begin{array}{r} 2\ 6 \\ \sqrt{6_{\wedge}60_{\wedge}49} \\ 4 \\ 46\overline{)260} \\ 276 \end{array}$$

Step 5 Bring down the next group and divide by the partial answer multiplied by 20. ($25 \times 20 = 500$) Repeat step 4.

$$\begin{array}{r} 2\ 5\ 7 \\ \sqrt{6_{\wedge}60_{\wedge}49} \\ 4 \\ 45\overline{)260} \\ 225 \\ 507\overline{)3549} \\ 3549 \\ \hline 0 \end{array}$$

Thus, $\sqrt{66{,}049} = 257$.

To find the square root of a decimal begin at the decimal point and move to the right, marking off groups of two digits. Place the decimal point in the answer as in division and repeat steps 2-6.

PROJECT

Follow the steps given above to find each of the following.

1. $\sqrt{12{,}769}$ **2.** $\sqrt{455{,}625}$ **3.** $\sqrt{9{,}604}$ **4.** $\sqrt{141{,}376}$

5. $\sqrt{.2601}$ **6.** $\sqrt{.8281}$ **7.** $\sqrt{202.4929}$ **8.** $\sqrt{325.0809}$

Cube Roots and Fourth Roots

OBJECTIVE

- To simplify expressions which have cube roots or fourth roots

REVIEW CAPSULE

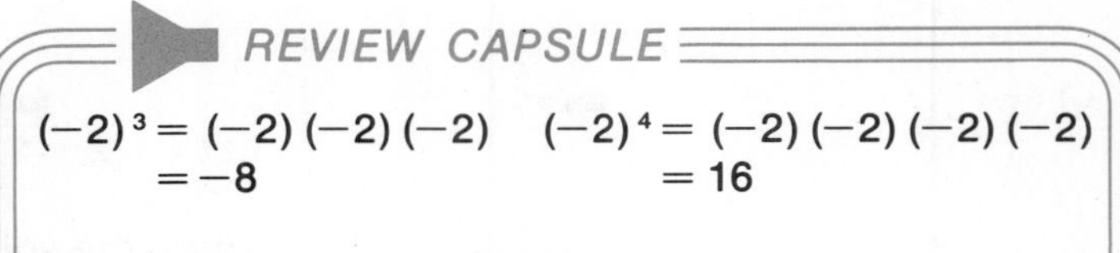

$(-2)^3 = (-2)(-2)(-2)$
$= -8$

$(-2)^4 = (-2)(-2)(-2)(-2)$
$= 16$

Third powers (cubes)	8	27	64	125	x^6	x^9	x^{12}
	↓	↓	↓	↓	↓	↓	↓
	2^3	3^3	4^3	5^3	$(x^2)^3$	$(x^3)^3$	$(x^4)^3$

EXAMPLE 1 Solve $x^3 = -8$.

$x^3 = -8$
↓ ↓
$x = -2$

$(-2)^3 = (-2)(-2)(-2) = -8$ ⟶

Check.

x^3	-8
$(-2)^3$	-8
-8	

Thus, the solution is -2.

The solution of $x^3 = -8$ is the *cube root* of -8.

We say ⟶ The cube root of -8 is -2.

We write ⟶ $\sqrt[3]{-8} = -2$

$\sqrt[3]{}$ is the cube root sign.

$\sqrt[3]{k}$ is the cube root of k.

If $x^3 = k$, then $x = \sqrt[3]{k}$.

EXAMPLE 2 Simplify $\sqrt[3]{27}$. Simplify $\sqrt[3]{-1{,}000}$.

$3 \cdot 3 \cdot 3 = 27$
$(-10)(-10)(-10) = -1{,}000$

$\sqrt[3]{27} = 3$

$\sqrt[3]{-1{,}000} = -10$

EXAMPLE 3 Simplify $\sqrt[3]{x^6}$. Simplify $\sqrt[3]{x^{15}}$. Simplify $\sqrt[3]{x^3}$.

$\sqrt[3]{x^6} = x^2$

Check.
$(x^2)^3 = x^{2 \cdot 3}$
$= x^6$

Thus, $\sqrt[3]{x^6} = x^2$.

$\sqrt[3]{x^{15}} = x^5$

Check.
$(x^5)^3 = x^{5 \cdot 3}$
$= x^{15}$

Thus, $\sqrt[3]{x^{15}} = x^5$.

$\sqrt[3]{x^3} = x$

Check.
$(x^1)^3 = x^{1 \cdot 3}$
$= x^3$

Thus, $\sqrt[3]{x^3} = x$.

$(x^m)^n = x^{m \cdot n}$ ⟶

EXAMPLE 4 Find the solution set of $x^4 = 16$.

$2^4 = (2)(2)(2)(2) = 16$;
$(-2)^4 = (-2)(-2)(-2)(-2) = 16$ →
Check. →

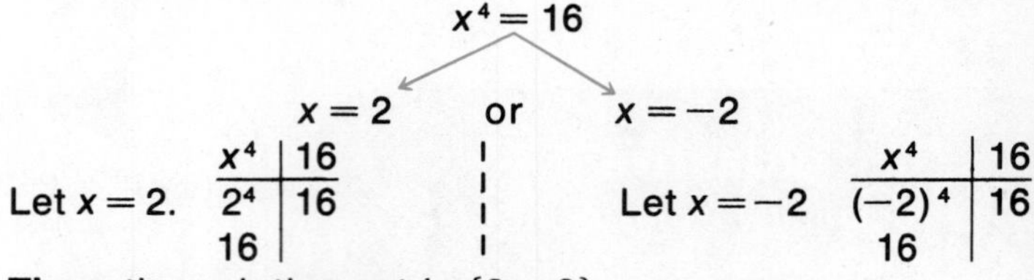

Thus, the solution set is $\{2, -2\}$.

We say →

We write →
$\sqrt[4]{}$ is the fourth root sign.

The solutions of $x^4 = 16$ are *fourth roots* of 16.

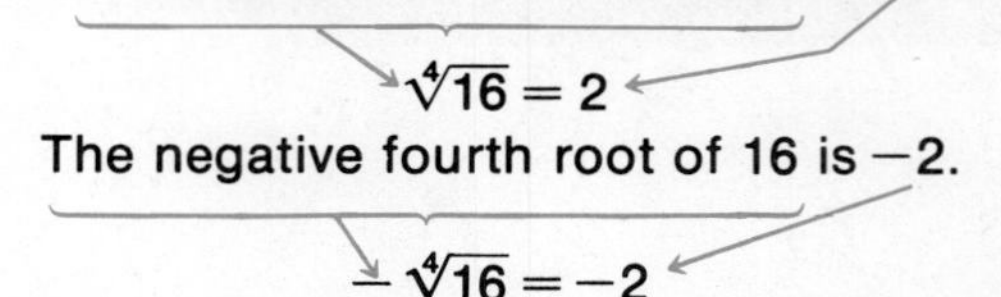

$\sqrt[4]{k}$ is the positive fourth root of k.
$-\sqrt[4]{k}$ is the negative fourth root of k.

> If $x^4 = k$, then $x = \sqrt[4]{k}$ or $x = -\sqrt[4]{k}$.

EXAMPLE 5 Simplify $\sqrt[4]{81}$. | Simplify $-\sqrt[4]{81}$.

$3^4 = 3 \cdot 3 \cdot 3 \cdot 3 = 81$ →

$\sqrt[4]{81} = 3$

$-\sqrt[4]{81} = -1\sqrt[4]{81}$
$= -3$

EXAMPLE 6 Simplify $\sqrt[4]{x^8}$. | Simplify $\sqrt[4]{x^{12}}$.

$(x^2)^4 = x^{2 \cdot 4} = x^8$ →
$(x^3)^4 = x^{12}$

$\sqrt[4]{x^8} = x^2$

$\sqrt[4]{x^{12}} = x^3$

EXAMPLE 7 Simplify. Compare the results.

$\sqrt[3]{8 \cdot 125}$	$\sqrt[3]{8} \cdot \sqrt[3]{125}$	$\sqrt[4]{16 \cdot 81}$	$\sqrt[4]{16} \cdot \sqrt[4]{81}$
$\sqrt[3]{1{,}000}$	$2 \cdot 5$	$\sqrt[4]{1{,}296}$	$2 \cdot 3$
10	10	6	6

Same   Same

Simplifying cube roots and fourth roots →

> $\sqrt[3]{x \cdot y} = \sqrt[3]{x} \cdot \sqrt[3]{y}$ | $\sqrt[4]{x \cdot y} = \sqrt[4]{x} \cdot \sqrt[4]{y}$

EXAMPLE 8 Simplify $\sqrt[3]{-54}$. | Simplify $\sqrt[4]{48}$.

$-27 = (-3)^3$; $16 = 2^4$
$\sqrt[3]{x \cdot y} = \sqrt[3]{x} \cdot \sqrt[3]{y}$ →
$\sqrt[4]{x \cdot y} = \sqrt[4]{x} \cdot \sqrt[4]{y}$

$\sqrt[3]{-54} = \sqrt[3]{-27 \cdot 2}$
$= \sqrt[3]{-27} \cdot \sqrt[3]{2}$
$= -3\sqrt[3]{2}$

$\sqrt[4]{48} = \sqrt[4]{16 \cdot 3}$
$= \sqrt[4]{16} \cdot \sqrt[4]{3}$
$= 2\sqrt[4]{3}$

EXAMPLE 9 Simplify $\sqrt[3]{5a} \cdot \sqrt[3]{2a}$. Simplify $\sqrt[4]{3b^2} \cdot \sqrt[4]{4b}$.

$\sqrt[3]{x} \cdot \sqrt[3]{y} = \sqrt[3]{x \cdot y}$
$\sqrt[4]{x} \cdot \sqrt[4]{y} = \sqrt[4]{x \cdot y}$

$$\sqrt[3]{5a} \cdot \sqrt[3]{2a} = \sqrt[3]{5a \cdot 2a} = \sqrt[3]{10a^2}$$

$$\sqrt[4]{3b^2} \cdot \sqrt[4]{4b} = \sqrt[4]{3b^2 \cdot 4b} = \sqrt[4]{12b^3}$$

EXAMPLE 10 Simplify $\sqrt[3]{c^7}$. Simplify $\sqrt[4]{d^{15}}$.

$c^6 = (c^2)^3$ and $d^{12} = (d^3)^4$
$\sqrt[3]{x \cdot y} = \sqrt[3]{x} \cdot \sqrt[3]{y}$
$\sqrt[4]{x \cdot y} = \sqrt[4]{x} \cdot \sqrt[4]{y}$

$$\sqrt[3]{c^7} = \sqrt[3]{c^6 \cdot c^1} = \sqrt[3]{c^6} \cdot \sqrt[3]{c} = c^2 \sqrt[3]{c}$$

$$\sqrt[4]{d^{15}} = \sqrt[4]{d^{12} \cdot d^3} = \sqrt[4]{d^{12}} \cdot \sqrt[4]{d^3} = d^3 \sqrt[4]{d^3}$$

EXAMPLE 11 Simplify $\sqrt[3]{24} + 5\sqrt[3]{3}$. Simplify $\sqrt[4]{8d^5} \cdot \sqrt[4]{2d^2}$.

$\sqrt[3]{24} + 5\sqrt[3]{3}$
$\sqrt[3]{8 \cdot 3} + 5\sqrt[3]{3}$
$\sqrt[3]{8 \cdot 3} = \sqrt[3]{8} \cdot \sqrt[3]{3} = 2\sqrt[3]{3}$ ⟶ $2\sqrt[3]{3} + 5\sqrt[3]{3}$
Combine like cube roots. ⟶ $7\sqrt[3]{3}$

$\sqrt[4]{8d^5} \cdot \sqrt[4]{2d^2}$
$\sqrt[4]{16d^7}$
$\sqrt[4]{16 \cdot d^4 \cdot d^3}$
$2d\sqrt[4]{d^3}$

EXERCISES

PART A

Simplify.

1. $\sqrt[3]{8}$ *2* **2.** $\sqrt[3]{-27}$ *−3* **3.** $\sqrt[3]{64}$ *4* **4.** $\sqrt[3]{-125}$ *−5*
5. $\sqrt[4]{625}$ *5* **6.** $-\sqrt[4]{10{,}000}$ *−10* **7.** $\sqrt[4]{1}$ *1* **8.** $-\sqrt[4]{625}$ *−5*
9. $\sqrt[3]{-64}$ *−4* **10.** $\sqrt[3]{1000}$ *10* **11.** $\sqrt[3]{-1}$ *−1* **12.** $\sqrt[3]{125}$ *5*
13. $\sqrt[3]{x^{12}}$ x^4 **14.** $\sqrt[4]{x^4}$ x **15.** $\sqrt[3]{x^9}$ x^3 **16.** $\sqrt[4]{x^{20}}$ x^5

Simplify.

17. $\sqrt[3]{5} \cdot \sqrt[3]{4}$ $\sqrt[3]{20}$ **18.** $\sqrt[3]{3a} \cdot \sqrt[3]{2b}$ $\sqrt[3]{6ab}$ **19.** $\sqrt[3]{7c} \cdot \sqrt[3]{2c}$ $\sqrt[3]{14c^2}$ **20.** $\sqrt[3]{4x^2} \cdot \sqrt[3]{3y^2}$ $\sqrt[3]{12x^2y^2}$
21. $\sqrt[4]{3} \cdot \sqrt[4]{5}$ $\sqrt[4]{15}$ **22.** $\sqrt[4]{2a} \cdot \sqrt[4]{4b}$ $\sqrt[4]{8ab}$ **23.** $\sqrt[4]{6c^2} \cdot \sqrt[4]{5c}$ $\sqrt[4]{30c^3}$ **24.** $\sqrt[4]{7x^2} \cdot \sqrt[4]{3y^3}$ $\sqrt[4]{21x^2y^3}$

Simplify.

25. $5\sqrt[3]{4} + 2\sqrt[3]{4}$ $7\sqrt[3]{4}$ **26.** $3\sqrt[4]{7} + 6\sqrt[4]{7}$ $9\sqrt[4]{7}$ **27.** $2\sqrt[3]{c} + 4\sqrt[3]{c}$ $6\sqrt[3]{c}$ **28.** $8\sqrt[4]{x} - 3\sqrt[4]{x}$ $5\sqrt[4]{x}$

PART B

Simplify.

29. $\sqrt[3]{16}$ $2\sqrt[3]{2}$ **30.** $\sqrt[4]{32}$ $2\sqrt[4]{2}$ **31.** $\sqrt[3]{-24}$ $-2\sqrt[3]{3}$ **32.** $-\sqrt[4]{48}$ $-2\sqrt[4]{3}$
33. $\sqrt[3]{a^5}$ $a\sqrt[3]{a^2}$ **34.** $\sqrt[4]{b^7}$ $b\sqrt[4]{b^3}$ **35.** $\sqrt[3]{x^{10}}$ $x^3\sqrt[3]{x}$ **36.** $\sqrt[4]{y^9}$ $y^2\sqrt[4]{y}$

Simplify.

37. $\sqrt[3]{16} + 4\sqrt[3]{2}$ $6\sqrt[3]{2}$ **38.** $\sqrt[4]{32} + 3\sqrt[4]{2}$ $5\sqrt[4]{2}$ **39.** $\sqrt[3]{54} + \sqrt[3]{2}$ $4\sqrt[3]{2}$
40. $\sqrt[3]{x^5} \cdot \sqrt[3]{x^2}$ $x^2\sqrt[3]{x}$ **41.** $\sqrt[4]{y^2} \cdot \sqrt[4]{y^7}$ $y^2\sqrt[4]{y}$ **42.** $\sqrt[3]{a^4} \cdot \sqrt[3]{a^7}$ $a^3\sqrt[3]{a^2}$
43. $\sqrt[3]{9a^2} \cdot \sqrt[3]{3a^2}$ $3a\sqrt[3]{a}$ **44.** $\sqrt[4]{25c^3} \cdot \sqrt[4]{25c^6}$ $5c^2\sqrt[4]{c}$ **45.** $\sqrt[3]{5x^4} \cdot \sqrt[3]{25x^4}$ $5x^2\sqrt[3]{x^2}$
46. $\sqrt[3]{8a^5} + \sqrt[3]{27a^5}$ $5a\sqrt[3]{a^2}$ **47.** $\sqrt[4]{16c^7} + \sqrt[4]{81c^7}$ $5c\sqrt[4]{c^3}$ **48.** $\sqrt[3]{2x^4} + \sqrt[3]{16x^4}$ $3x\sqrt[3]{2x}$

Zero and Negative Exponents

OBJECTIVE

- To simplify expressions which have zero exponents or negative exponents

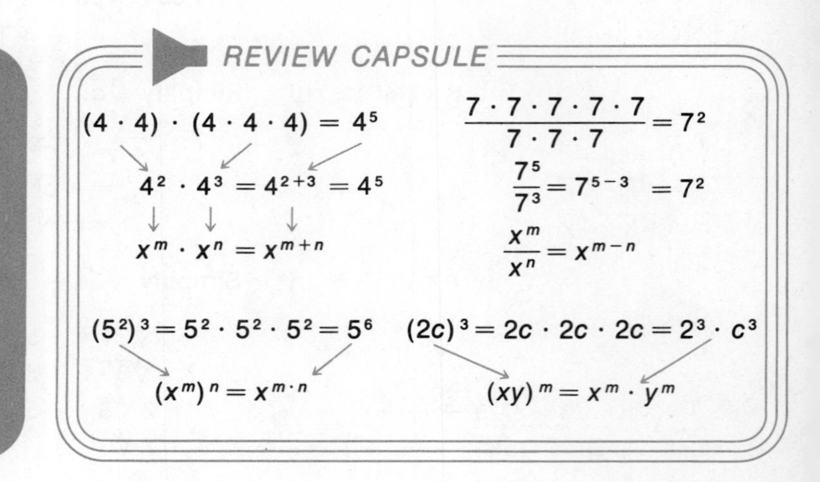

EXAMPLE 1 Find the value of 3^0.

Exponents are decreasing 4, 3, 2, 1, 0. Divide by 3 to get the next value.

$$\begin{aligned} 3^4 &= 81 \\ 3^3 &= 27 && \leftarrow 81 \div 3 = 27 \\ 3^2 &= 9 && \leftarrow 27 \div 3 = 9 \\ 3^1 &= 3 && \leftarrow 9 \div 3 = 3 \\ 3^0 &= 1 && \leftarrow 3 \div 3 = 1 \end{aligned}$$

$3^0 = 1$, if the pattern continues.

Thus, $3^0 = 1$.

Definition of zero exponent ⟶ $x^0 = 1$, for each $x \neq 0$.

EXAMPLE 2 Find the value of 3^{-2} and 3^{-3}.

Exponents are decreasing 2, 1, 0, −1, −2, −3.

Divide by 3 to get the next value.

$\frac{1}{9} \div 3 = \frac{1}{9} \cdot \frac{1}{3} = \frac{1}{27}$

$$\begin{aligned} 3^2 &= 9 \\ 3^1 &= 3 && \leftarrow 9 \div 3 = 3 \\ 3^0 &= 1 && \leftarrow 3 \div 3 = 1 \\ 3^{-1} &= \tfrac{1}{3} && \leftarrow 1 \div 3 = \tfrac{1}{3} \\ 3^{-2} &= \tfrac{1}{9} && \leftarrow \tfrac{1}{3} \div 3 = \tfrac{1}{9} \\ 3^{-3} &= \tfrac{1}{27} && \leftarrow \tfrac{1}{9} \div 3 = \tfrac{1}{27} \end{aligned}$$

Thus, $3^{-2} = \frac{1}{9} = \frac{1}{3^2}$ and $3^{-3} = \frac{1}{27} = \frac{1}{3^3}$.

Definition of negative exponents ⟶ $x^{-n} = \frac{1}{x^n}$, for each $x \neq 0$.

EXAMPLE 3 Find the value of 4^{-2} and $2 \cdot 5^{-3}$.

$x^{-n} = \frac{1}{x^n}$ →

$$4^{-2} = \frac{1}{4^2} = \frac{1}{16}$$

$$2 \cdot 5^{-3} = 2 \cdot \frac{1}{5^3} = 2 \cdot \frac{1}{125}, \text{ or } \frac{2}{125}$$

EXAMPLE 4 Simplify $5c^{-2}$ and $(5c)^{-2}$. Write with positive exponents.

In $5c^{-2}$ the base is c, not $5c$.

$x^{-n} = \frac{1}{x^n}$ →

$$5c^{-2} = 5(c^{-2}) = 5 \cdot \frac{1}{c^2} = \frac{5}{c^2}$$

$$(5c)^{-2} = \frac{1}{(5c)^2} = \frac{1}{25c^2}$$

EXAMPLE 5 Write $\frac{1}{x^{-3}}$ with a positive exponent.

$x^{-3} = \frac{1}{x^3}$; $1 \div \frac{1}{x^3} = 1 \cdot \frac{x^3}{1} = x^3$

$$\frac{1}{x^{-3}} = \frac{1}{\frac{1}{x^3}} = x^3$$

Negative exponent in denominator. →

$$\frac{1}{x^{-n}} = x^n$$

EXAMPLE 6 Find the value of $\frac{1}{4^{-2}}$ and $\frac{1}{2 \cdot 5^{-3}}$.

$\frac{1}{x^{-n}} = x^n$ →

$4^2 = 16$ and $5^3 = 125$ →

$$\frac{1}{4^{-2}} = 4^2 = 16$$

$$\frac{1}{2 \cdot 5^{-3}} = \frac{1}{2} \cdot \frac{1}{5^{-3}} = \frac{1}{2} \cdot \frac{5^3}{1} = \frac{125}{2}$$

EXAMPLE 7 Simplify $\frac{1}{(3a)^{-2}}$ and $\frac{1}{3a^{-2}}$. Write with positive exponents.

$\frac{1}{x^{-n}} = x^n$ →

In $3a^{-2}$ the base is a, not $3a$.

$$\frac{1}{(3a)^{-2}} = (3a)^2 = 9a^2$$

$$\frac{1}{3a^{-2}} = \frac{1}{3} \cdot \frac{1}{a^{-2}} = \frac{1}{3} \cdot \frac{a^2}{1} = \frac{a^2}{3}$$

EXAMPLE 8 Write $\frac{3a^2b^{-3}c^{-2}}{5x^{-2}yz^{-3}}$ with positive exponents.

$$\frac{3}{5} \cdot \frac{a^2}{1} \cdot \frac{b^{-3}}{1} \cdot \frac{c^{-2}}{1} \cdot \frac{1}{x^{-2}} \cdot \frac{1}{y} \cdot \frac{1}{z^{-3}}$$

$x^{-n} = \frac{1}{x^n}$ and $\frac{1}{x^{-n}} = x^n$ →

$$\frac{3}{5} \cdot \frac{a^2}{1} \cdot \frac{1}{b^3} \cdot \frac{1}{c^2} \cdot \frac{x^2}{1} \cdot \frac{1}{y} \cdot \frac{z^3}{1}$$

Multiply. →

$$\frac{3a^2x^2z^3}{5b^3c^2y}$$

EXAMPLE 9 Simplify $x^4 \cdot x^{-7}$. Write with a positive exponent.

First method:
$x^m \cdot x^n = x^{m+n}$; $4 + (-7) = -3$ →

$\frac{x^4}{x^7} = \frac{x^4 \cdot 1}{x^4 \cdot x^3} = \frac{1}{x^3}$

First Method

$$x^4 \cdot x^{-7} = x^{-3} = \frac{1}{x^3}$$

Second Method

$$x^4 \cdot x^{-7} = x^4 \cdot \frac{1}{x^7} = \frac{x^4}{x^7} = \frac{1}{x^3}$$

Thus, $x^4 \cdot x^{-7} = \frac{1}{x^3}$.

EXAMPLE 10 Simplify $-3a^{-4} \cdot 5b^{-2} \cdot 2a^0 \cdot b^{-3} \cdot 4a^6$.

Regroup factors. → $(-3 \cdot 5 \cdot 2 \cdot 4) \cdot (a^{-4} \cdot a^0 \cdot a^6) \cdot (b^{-2} \cdot b^{-3})$

Add exponents: $x^m \cdot x^n = x^{m+n}$. → $-120 \cdot a^2 \cdot b^{-5}$

$b^{-5} = \frac{1}{b^5}$ → $\frac{-120a^2}{b^5}$

EXAMPLE 11 Simplify $\frac{x^5}{x^{-3}}$. Write with a positive exponent.

First method:
$\frac{x^m}{x^n} = x^{m-n}$; $5 - (-3) = 8$ →

First Method

$$\frac{x^5}{x^{-3}} = x^8$$

Second Method

$$\frac{x^5}{x^{-3}} = \frac{x^5}{1} \cdot \frac{1}{x^{-3}} = x^5 \cdot x^3 = x^8$$

Thus, $\frac{x^5}{x^{-3}} = x^8$.

EXAMPLE 12 Simplify $\frac{6a^{-2}b^{-4}c^{-2}}{9a^3b^{-1}c^{-5}}$. Write with positive exponents.

Subtract exponents.

$$\frac{6}{9} \cdot \frac{a^{-2}}{a^3} \cdot \frac{b^{-4}}{b^{-1}} \cdot \frac{c^{-2}}{c^{-5}}$$

$\frac{x^m}{x^n} = x^{m-n}$ → $\frac{2}{3} \cdot a^{-5} \cdot b^{-3} \cdot c^3$

$x^{-n} = \frac{1}{x^n}$ → $\frac{2}{3} \cdot \frac{1}{a^5} \cdot \frac{1}{b^3} \cdot \frac{c^3}{1}$

Multiply.

$$\frac{2c^3}{3a^5b^3}$$

EXAMPLE 13 Simplify $(x^{-4})^3$. Write with a positive exponent.

$(x^m)^n = x^{m \cdot n}$ →

$$(x^{-4})^3 = x^{-4 \cdot 3} = x^{-12} = \frac{1}{x^{12}}$$

Thus, $(x^{-4})^3 = \frac{1}{x^{12}}$.

EXAMPLE 14 Simplify and write with positive exponents.

$(x^m)^n = x^{mn}; -3(-2) = 6$ ⟶

$(c^{-3})^{-2} = c^6$ | $(y^3)^{-5} = y^{-15} = \frac{1}{y^{15}}$

EXAMPLE 15 Simplify and write with positive exponents.

$(xy)^m = x^m \cdot y^m$ ⟶
$(x^m)^n = x^{mn}$ ⟶
$c^{-6} = \frac{1}{c^6}$ ⟶

$$(-5c^{-2})^3 = (-5)^3(c^{-2})^3 = -125 \cdot c^{-6} = \frac{-125}{c^6}$$

$$(b^{-4}d^3)^{-2} = (b^{-4})^{-2}(d^3)^{-2} = b^8 \cdot d^{-6} = \frac{b^8}{d^6}$$

EXAMPLE 16 Simplify $\left(\frac{2}{5}\right)^3$. Simplify $\left(\frac{a}{b}\right)^4$.

$x^3 = x \cdot x \cdot x$
$x^4 = x \cdot x \cdot x \cdot x$

$$\left(\frac{2}{5}\right)^3 = \frac{2}{5} \cdot \frac{2}{5} \cdot \frac{2}{5} = \frac{8}{125}$$

$$\left(\frac{a}{b}\right)^4 = \frac{a}{b} \cdot \frac{a}{b} \cdot \frac{a}{b} \cdot \frac{a}{b} = \frac{a^4}{b^4}$$

Thus, $\left(\frac{2}{5}\right)^3 = \frac{2^3}{5^3} = \frac{8}{125}$ and $\left(\frac{a}{b}\right)^4 = \frac{a^4}{b^4}$.

Power of a Quotient ⟶

$$\left(\frac{x}{y}\right)^m = \frac{x^m}{y^m}, \text{ for each } x \text{ and } y \neq 0.$$

EXAMPLE 17 Simplify and write with positive exponents.

$\left(\frac{x}{y}\right)^m = \frac{x^m}{y^m}$ ⟶
$(xy)^m = x^m \cdot y^m$ ⟶
$(x^m)^n = x^{mn}$ ⟶

$$\left(\frac{2a^3}{-3b^2}\right)^4 = \frac{(2a^3)^4}{(-3b^2)^4} = \frac{2^4(a^3)^4}{(-3)^4(b^2)^4} = \frac{16a^{12}}{81b^8}$$

EXAMPLE 18 Simplify and write with positive exponents.

$\left(\frac{x}{y}\right)^m = \frac{x^m}{y^m}$ ⟶
$(xy)^m = x^m \cdot y^m$ ⟶
$(x^m)^n = x^{mn}$ ⟶
$a^{-6} = \frac{1}{a^6}$
$\frac{1}{c^{-12}} = c^{12}$

$$\left(\frac{5a^{-2}}{b^3c^{-4}}\right)^3 = \frac{(5a^{-2})^3}{(b^3c^{-4})^3} = \frac{5^3(a^{-2})^3}{(b^3)^3(c^{-4})^3} = \frac{125a^{-6}}{b^9c^{-12}} = \frac{125c^{12}}{b^9a^6}$$

ORAL EXERCISES

Evaluate if $x = 2$.

1. x^0 *1* **2.** $5x^0$ *5* **3.** $(5x)^0$ *1* **4.** x^{-1} $\frac{1}{2}$ **5.** $3x^{-1}$ $\frac{3}{2}$ **6.** $(3x)^{-1}$ $\frac{1}{6}$

EXERCISES

PART A

Find the value.

1. 7^0 *1* **2.** $4 \cdot 5^0$ *4* **3.** $(3 \cdot 4)^0$ *1* **4.** 5^{-3} **5.** 2^{-4} **6.** $3 \cdot 4^{-2}$

7. $(3 \cdot 4)^{-2}$ $\frac{1}{144}$ **8.** $\frac{1}{5^{-3}}$ *125* **9.** $\frac{10}{2^{-4}}$ *160* **10.** $\frac{1}{3 \cdot 4^{-2}}$ $\frac{16}{3}$ **11.** $\frac{1}{(3 \cdot 4)^{-2}}$ **12.** $\frac{2^3}{3^{-2}}$ *72*

13. $\frac{2^{-3}}{3^2}$ $\frac{1}{72}$ **14.** $\frac{2^{-3}}{3^{-2}}$ $\frac{9}{8}$ **15.** $\frac{5^{-2}}{4^{-3}}$ $\frac{64}{25}$ **16.** $7^3 \cdot 7^{-5}$ $\frac{1}{49}$ **17.** $7^{-3} \cdot 7^5$ *49* **18.** $5^2 \cdot 5^0$ *25*

Simplify and write with positive exponents.

19. $8x^{-3}$ $\frac{8}{x^3}$ **20.** $(2x)^{-4}$ $\frac{1}{16x^4}$ **21.** $-10c^{-3}$ $-\frac{10}{c^3}$ **22.** $(-2c)^{-3}$ $-\frac{1}{8c^3}$

23. $\frac{1}{4x^{-3}}$ $\frac{x^3}{4}$ **24.** $\frac{1}{(4x)^{-3}}$ $64x^3$ **25.** $\frac{3}{5a^{-2}}$ $\frac{3a^2}{5}$ **26.** $\frac{4}{(5a)^{-2}}$ $100a^2$

27. $\frac{a^4b^{-3}}{c^2d^{-5}}$ $\frac{a^4d^5}{c^2b^3}$ **28.** $\frac{x^2y^{-4}}{z^{-3}}$ $\frac{x^2z^3}{y^4}$ **29.** $\frac{2a^{-2}b^5}{3c^3d^{-4}}$ $\frac{2b^5d^4}{3c^3a^2}$ **30.** $\frac{3x^{-5}}{4y^{-2}z^{-1}}$ $\frac{3y^2z}{4x^5}$

31. $x^{-2} \cdot x^5$ x^3 **32.** $a^3 \cdot a^{-6}$ $\frac{1}{a^3}$ **33.** $n^{-4} \cdot n^{-2}$ $\frac{1}{n^6}$ **34.** $y^{-3} \cdot y^0$ $\frac{1}{y^3}$

35. $\frac{x^4}{x^{-2}}$ x^6 **36.** $\frac{a^{-3}}{a^2}$ $\frac{1}{a^5}$ **37.** $\frac{6n^{-2}}{8n^{-5}}$ $\frac{3n^3}{4}$ **38.** $\frac{-2c^{-6}}{4c^{-4}}$ $-\frac{1}{2c^2}$

39. $(x^{-2})^3$ $\frac{1}{x^6}$ **40.** $(a^4)^{-3}$ $\frac{1}{a^{12}}$ **41.** $(n^{-3})^{-4}$ n^{12} **42.** $5(y^{-2})^4$ $\frac{5}{y^8}$

43. $(2x^{-3})^4$ $\frac{16}{x^{12}}$ **44.** $(3a^3)^{-2}$ $\frac{1}{9a^6}$ **45.** $(4n^{-4})^{-2}$ $\frac{n^8}{16}$ **46.** $(-5y^{-2})^3$ $-\frac{125}{y^6}$

47. $\left(\frac{x^4}{5}\right)^3$ $\frac{x^{12}}{125}$ **48.** $\left(\frac{a^3}{b^{-2}}\right)^2$ a^6b^4 **49.** $\left(\frac{c^{-2}}{d^3}\right)^4$ $\frac{1}{c^8d^{12}}$ **50.** $\left(\frac{x^{-3}}{y^{-2}}\right)^5$ $\frac{y^{10}}{x^{15}}$

PART B

Simplify and write with positive exponents.

51. $2x^3 \cdot 5y^{-2} \cdot 3x^{-5} \cdot 4y^6$ $\frac{120y^4}{x^2}$ **52.** $3a^{-3} \cdot 2b^4 \cdot a^5 \cdot 10b^{-6}$ $\frac{60a^2}{b^2}$

53. $4x^{-2} \cdot y^3 \cdot 5x^{-3} \cdot 2y^4$ $\frac{40y^7}{x^5}$ **54.** $-5a^4 \cdot 3b^2 \cdot 4a^0 \cdot b^{-5}$ $-\frac{60a^4}{b^3}$

55. $\frac{x^3y^{-4}}{x^{-2}y^2}$ $\frac{x^5}{y^6}$ **56.** $\frac{a^2b^{-3}}{a^5b^{-5}}$ $\frac{b^2}{a^3}$ **57.** $\frac{10c^{-6}d^4}{15c^{-2}d^{-2}}$ $\frac{2d^6}{3c^4}$

58. $(x^3y^{-2})^3$ $\frac{x^9}{y^6}$ **59.** $(a^{-3}b^{-4})^2$ $\frac{1}{a^6b^8}$ **60.** $(-2c^{-2}d^3)^4$ $\frac{16d^{12}}{c^8}$

61. $(x^4y^{-3})^{-2}$ $\frac{y^6}{x^8}$ **62.** $(a^{-2}b^4)^{-3}$ $\frac{a^6}{b^{12}}$ **63.** $(-3c^{-3}d^{-2})^{-3}$ $\frac{c^9d^6}{-27}$

64. $\left(\frac{3a^2}{-5b^3}\right)^3$ $-\frac{27a^6}{125b^9}$ **65.** $\left(\frac{4x^{-3}}{y^3z^{-2}}\right)^2$ $\frac{16z^4}{x^6y^6}$ **66.** $\left(\frac{a^{-2}b^3}{2c^{-1}}\right)^4$ $\frac{b^{12}c^4}{16a^8}$

67. $\left(\frac{a^3b^{-4}}{c^{-2}d^5}\right)^2$ $\frac{a^6c^4}{b^8d^{10}}$ **68.** $\left(\frac{-2x^{-2}y}{3z^{-3}w}\right)^3$ $-\frac{8y^3z^9}{27x^6w^3}$ **69.** $\left(\frac{ab^{-2}c^2}{de^3f^{-3}}\right)^4$ $\frac{a^4c^8f^{12}}{b^8d^4e^{12}}$

Fractional Exponents

OBJECTIVES

- To write a number like $\sqrt[3]{4}$ in exponent form
- To find the value of expressions like $27^{\frac{2}{3}}$ and $9^{-\frac{3}{2}}$

REVIEW CAPSULE

$$3^2 \cdot 3^5 = 3^7 \rightarrow x^m \cdot x^n = x^{m+n}$$

$$(5^2)^3 = 5^6 \rightarrow (x^m)^n = x^{mn}$$

$$\sqrt{x} \cdot \sqrt{x} = x \qquad \sqrt[3]{x} \cdot \sqrt[3]{x} \cdot \sqrt[3]{x} = x$$

$$5^{-2} = \frac{1}{5^2} \leftarrow x^{-n} = \frac{1}{x^n}$$

EXAMPLE 1 Show that $3^{\frac{1}{2}} = \sqrt{3}$.

$x^m \cdot x^n = x^{m+n}$: $\frac{1}{2}+\frac{1}{2}=1$ ⟶ $3^{\frac{1}{2}} \cdot 3^{\frac{1}{2}} = 3^1$

$\sqrt{x} \cdot \sqrt{x} = x$ ⟶ $\sqrt{3} \cdot \sqrt{3} = 3$

$(3^{\frac{1}{2}})^2 = (\sqrt{3})^2$ ⟶ **Thus,** $3^{\frac{1}{2}} = \sqrt{3}$.

EXAMPLE 2 Show that $5^{\frac{1}{3}} = \sqrt[3]{5}$. | Show that $2^{\frac{1}{4}} = \sqrt[4]{2}$.

Add exponents. ⟶ $5^{\frac{1}{3}} \cdot 5^{\frac{1}{3}} \cdot 5^{\frac{1}{3}} = 5^1$ | $2^{\frac{1}{4}} \cdot 2^{\frac{1}{4}} \cdot 2^{\frac{1}{4}} \cdot 2^{\frac{1}{4}} = 2^1$

$\sqrt[3]{x} \cdot \sqrt[3]{x} \cdot \sqrt[3]{x} = x$ ⟶ $\sqrt[3]{5} \cdot \sqrt[3]{5} \cdot \sqrt[3]{5} = 5$ | $\sqrt[4]{2} \cdot \sqrt[4]{2} \cdot \sqrt[4]{2} \cdot \sqrt[4]{2} = 2$

$(5^{\frac{1}{3}})^3 = (\sqrt[3]{5})^3$ ⟶ **Thus,** $5^{\frac{1}{3}} = \sqrt[3]{5}$. | **Thus,** $2^{\frac{1}{4}} = \sqrt[4]{2}$.

Definition of $x^{\frac{1}{n}}$ ⟶ $x^{\frac{1}{n}} = \sqrt[n]{x}$, where n is a positive integer.

EXAMPLE 3 Write $\sqrt{7}$ and $\sqrt[3]{15}$ in exponent form.

$\sqrt[n]{x} = x^{\frac{1}{n}}$ ⟶ $\sqrt{7} = 7^{\frac{1}{2}}$ | $\sqrt[3]{15} = 15^{\frac{1}{3}}$

EXAMPLE 4 Find the value of $25^{\frac{1}{2}}$ and $81^{\frac{1}{4}}$.

$x^{\frac{1}{n}} = \sqrt[n]{x}$ ⟶ $25^{\frac{1}{2}} = \sqrt{25} = 5$ | $81^{\frac{1}{4}} = \sqrt[4]{81} = 3$

EXAMPLE 5 Write $4c^{\frac{1}{3}}$ and $(4c)^{\frac{1}{3}}$ with a cube root sign.

$x^{\frac{1}{n}} = \sqrt[n]{x}$; in $4c^{\frac{1}{3}}$ the base is c, not $4c$. $4c^{\frac{1}{3}} = 4(c^{\frac{1}{3}}) = 4\sqrt[3]{c}$ | $(4c)^{\frac{1}{3}} = \sqrt[3]{4c}$

EXAMPLE 6 Simplify $(\sqrt[3]{8})^2$ and $\sqrt[3]{8^2}$.

	$(\sqrt[3]{8})^2$	$\sqrt[3]{8^2}$
$\sqrt[3]{8}=2$	$(2)^2$	
$\sqrt[3]{64}=4$		$\sqrt[3]{64}$
Results are the same. →	4	4

> $(\sqrt[n]{x})^m = \sqrt[n]{x^m}$, where m and n are positive integers.

EXAMPLE 7 Write $(\sqrt[3]{8})^2$ and $\sqrt[3]{8^2}$ in exponent form.

	$(\sqrt[3]{8})^2$	$\sqrt[3]{8^2}$
$\sqrt[n]{x} = x^{\frac{1}{n}}$ →	$(8^{\frac{1}{3}})^2$	$(8^2)^{\frac{1}{3}}$
Multiply exponents: $(x^m)^n = x^{mn}$ →	$8^{\frac{2}{3}}$	$8^{\frac{2}{3}}$

Results are the same.

Thus, $8^{\frac{2}{3}} = (\sqrt[3]{8})^2$ and $8^{\frac{2}{3}} = \sqrt[3]{8^2}$.

Property of fractional exponents →

> $x^{\frac{m}{n}} = (\sqrt[n]{x})^m = \sqrt[n]{x^m}$, where m and n are positive integers.

EXAMPLE 8 Find the value of $27^{\frac{2}{3}}$.

$x^{\frac{m}{n}} = (\sqrt[n]{x})^m$ or $\sqrt[n]{x^m}$

$729 = 9 \cdot 9 \cdot 9 = 9^3$

Results are the same. →

First Method	*Second Method*
$27^{\frac{2}{3}} = (\sqrt[3]{27})^2$	$27^{\frac{2}{3}} = \sqrt[3]{27^2}$
$= (3)^2$	$= \sqrt[3]{729}$
$= 9$	$= 9$

Thus, $27^{\frac{2}{3}} = 9$.

EXAMPLE 9 Find the value of $8^{-\frac{2}{3}}$.

$x^{-n} = \frac{1}{x^n}$ → $8^{-\frac{2}{3}} = \frac{1}{8^{\frac{2}{3}}}$

$x^{\frac{m}{n}} = (\sqrt[n]{x})^m$ → $= \frac{1}{(\sqrt[3]{8})^2}$

$\sqrt[3]{8} = 2$ → $= \frac{1}{(2)^2}$

$= \frac{1}{4}$

Thus, $8^{-\frac{2}{3}} = \frac{1}{4}$.

EXAMPLE 10 Write $2a^{\frac{3}{4}}$ and $(2a)^{\frac{3}{4}}$ with a fourth root sign.

$x^{\frac{m}{n}} = \sqrt[n]{x^m}$

$2a^{\frac{3}{4}} = 2(a^{\frac{3}{4}}) = 2\sqrt[4]{a^3}$ | $(2a)^{\frac{3}{4}} = \sqrt[4]{(2a)^3} = \sqrt[4]{8a^3}$

EXAMPLE 11 Write $\sqrt[3]{x^2}$ and $5\sqrt[4]{a}$ in exponent form.

$\sqrt[n]{x^m} = x^{\frac{m}{n}}$

$\sqrt[3]{x^2} = x^{\frac{2}{3}}$ | $5\sqrt[4]{a^1} = 5a^{\frac{1}{4}}$

EXERCISES

PART A

Find the value.

1. $9^{\frac{1}{2}}$ *3* **2.** $8^{\frac{1}{3}}$ *2* **3.** $16^{\frac{1}{4}}$ *2* **4.** $25^{\frac{3}{2}}$ *125* **5.** $8^{\frac{2}{3}}$ *4* **6.** $16^{\frac{3}{4}}$ *8*

7. $16^{\frac{1}{2}}$ *4* **8.** $27^{\frac{1}{3}}$ *3* **9.** $64^{\frac{1}{3}}$ *4* **10.** $4^{\frac{5}{2}}$ *32* **11.** $8^{\frac{4}{3}}$ *16* **12.** $16^{\frac{5}{4}}$ *32*

Write in exponent form.

13. $\sqrt{6}$ $6^{\frac{1}{2}}$ **14.** $\sqrt[3]{20}$ $20^{\frac{1}{3}}$ **15.** $\sqrt[4]{10}$ $10^{\frac{1}{4}}$ **16.** $4\sqrt[3]{a}$ $4a^{\frac{1}{3}}$ **17.** $5\sqrt{b}$ $5b^{\frac{1}{2}}$ **18.** $3\sqrt[4]{b}$ $3b^{\frac{1}{4}}$

Write with a square root sign.

19. $10^{\frac{1}{2}}$ $\sqrt{10}$ **20.** $a^{\frac{1}{2}}$ $\sqrt{a}$ **21.** $5a^{\frac{1}{2}}$ $5\sqrt{a}$ **22.** $(5a)^{\frac{1}{2}}$ $\sqrt{5a}$ **23.** $3b^{\frac{1}{2}}$ $3\sqrt{b}$ **24.** $(3b)^{\frac{1}{2}}$ $\sqrt{3b}$

Write with a cube root or a fourth root sign.

25. $7^{\frac{1}{3}}$ $\sqrt[3]{7}$ **26.** $a^{\frac{1}{4}}$ $\sqrt[4]{a}$ **27.** $6c^{\frac{1}{3}}$ $6\sqrt[3]{c}$ **28.** $(6c)^{\frac{1}{3}}$ $\sqrt[3]{6c}$ **29.** $2n^{\frac{1}{4}}$ $2\sqrt[4]{n}$ **30.** $(2n)^{\frac{1}{4}}$ $\sqrt[4]{2n}$

Write in exponent form.

31. $\sqrt[3]{c^2}$ $c^{\frac{2}{3}}$ **32.** $(\sqrt[3]{9})^4$ $9^{\frac{4}{3}}$ **33.** $\sqrt[4]{x^3}$ $x^{\frac{3}{4}}$ **34.** $(\sqrt[4]{7})^3$ $7^{\frac{3}{4}}$ **35.** $\sqrt{a^5}$ $a^{\frac{5}{2}}$ **36.** $(\sqrt{10})^3$ $10^{\frac{3}{2}}$

PART B

EXAMPLE Write $\sqrt[3]{25}$ in exponent form.

$25 = 5^2$

$\sqrt[n]{x^m} = x^{\frac{m}{n}}$ ⟶

$\sqrt[3]{25} = \sqrt[3]{5^2} = 5^{\frac{2}{3}}$

Write in exponent form.

37. $\sqrt[3]{49}$ $7^{\frac{2}{3}}$ **38.** $\sqrt[3]{16}$ $2^{\frac{4}{3}}$ **39.** $\sqrt{8}$ $2^{\frac{3}{2}}$ **40.** $\sqrt{27}$ $3^{\frac{3}{2}}$ **41.** $\sqrt[4]{27}$ $3^{\frac{3}{4}}$ **42.** $\sqrt[4]{32}$ $2^{\frac{5}{4}}$

Write with a square root, cube root, or fourth root sign.

43. $3^{\frac{3}{2}}$ $\sqrt{27}$ **44.** $5^{\frac{3}{4}}$ $\sqrt[4]{125}$ **45.** $4x^{\frac{2}{3}}$ $4\sqrt[3]{x^2}$ **46.** $(4x)^{\frac{2}{3}}$ $\sqrt[3]{16x^2}$ **47.** $5x^{\frac{3}{4}}$ $5\sqrt[4]{x^3}$ **48.** $(5x)^{\frac{3}{4}}$ $\sqrt[4]{125x^3}$

Find the value.

49. $25^{-\frac{1}{2}}$ $\frac{1}{5}$ **50.** $8^{-\frac{1}{3}}$ $\frac{1}{2}$ **51.** $16^{-\frac{1}{4}}$ $\frac{1}{2}$ **52.** $9^{-\frac{1}{2}}$ $\frac{1}{3}$ **53.** $27^{-\frac{1}{3}}$ $\frac{1}{3}$ **54.** $625^{-\frac{1}{4}}$ $\frac{1}{5}$

55. $25^{-\frac{3}{2}}$ $\frac{1}{125}$ **56.** $27^{-\frac{2}{3}}$ $\frac{1}{9}$ **57.** $625^{-\frac{3}{4}}$ $\frac{1}{125}$ **58.** $4^{-\frac{5}{2}}$ $\frac{1}{32}$ **59.** $8^{-\frac{4}{3}}$ $\frac{1}{16}$ **60.** $16^{-\frac{5}{4}}$ $\frac{1}{32}$

Using Fractional Exponents

OBJECTIVE

- To simplify expressions which have fractional exponents

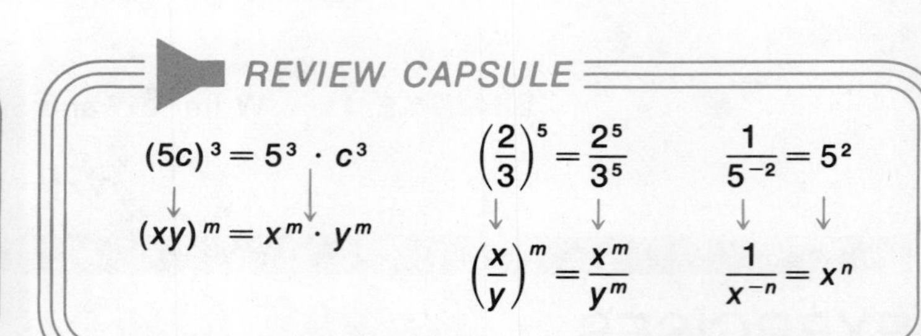

EXAMPLE 1 Multiply $5^{\frac{2}{3}} \cdot 5^{\frac{1}{2}}$. Multiply $a^{\frac{4}{5}} \cdot a^{-\frac{3}{5}}$.

Add exponents: $x^m \cdot x^n = x^{m+n}$.

$$5^{\frac{2}{3}} \cdot 5^{\frac{1}{2}} = 5^{\frac{2}{3}+\frac{1}{2}} = 5^{\frac{7}{6}}$$

$$a^{\frac{4}{5}} \cdot a^{-\frac{3}{5}} = a^{\frac{4}{5}+(-\frac{3}{5})} = a^{\frac{1}{5}}$$

EXAMPLE 2 Simplify $(y^{\frac{2}{3}})^6$. Simplify $(x^9)^{\frac{4}{3}}$.

Multiply exponents: $(x^m)^n = x^{mn}$.

$$(y^{\frac{2}{3}})^6 = y^{\frac{2}{3}\cdot\frac{6}{1}} = y^4$$

$$(x^9)^{\frac{4}{3}} = x^{\frac{9}{1}\cdot\frac{4}{3}} = x^{12}$$

EXAMPLE 3 Divide $\frac{x^3}{x^{\frac{3}{4}}}$. Divide $\frac{b^{\frac{2}{3}}}{b^{\frac{1}{2}}}$.

Subtract exponents: $\frac{x^m}{x^n} = x^{m-n}$.

$$\frac{x^3}{x^{\frac{3}{4}}} = x^{3-\frac{3}{4}} = x^{\frac{9}{4}}$$

$$\frac{b^{\frac{2}{3}}}{b^{\frac{1}{2}}} = b^{\frac{2}{3}-\frac{1}{2}} = b^{\frac{1}{6}}$$

EXAMPLE 4 Simplify $(r^{\frac{1}{2}}t^{\frac{2}{3}})^6$.

$(xy)^m = x^m \cdot y^m$ ⟶ $(r^{\frac{1}{2}}t^{\frac{2}{3}})^6 = (r^{\frac{1}{2}})^6(t^{\frac{2}{3}})^6$

$(x^m)^n = x^{mn}$ ⟶ $= r^3t^4$

EXAMPLE 5 Simplify $(27c^6d^3)^{\frac{2}{3}}$.

$(xyz)^m = x^m \cdot y^m \cdot z^m$ ⟶ $(27c^6d^3)^{\frac{2}{3}} = 27^{\frac{2}{3}}(c^6)^{\frac{2}{3}}(d^3)^{\frac{2}{3}}$

$x^{\frac{m}{n}} = (\sqrt[n]{x})^m$: $27^{\frac{2}{3}} = (\sqrt[3]{27})^2$ ⟶ $= (\sqrt[3]{27})^2 \cdot c^{\frac{12}{3}} \cdot d^{\frac{6}{3}}$

Simplify. ⟶ $= 9c^4d^2$

EXAMPLE 6 Simplify $\left(\frac{c^{\frac{2}{3}}}{d^{\frac{1}{3}}}\right)^6$. Simplify $\left(\frac{27a^{12}}{8b^9}\right)^{\frac{1}{3}}$.

$\left(\frac{x}{y}\right)^m = \frac{x^m}{y^m}$ → $\frac{(c^{\frac{2}{3}})^6}{(d^{\frac{1}{3}})^6}$ | $\frac{(27a^{12})^{\frac{1}{3}}}{(8b^9)^{\frac{1}{3}}}$

$(x^m)^n = x^{mn}$ → $\frac{c^{\frac{12}{3}}}{d^{\frac{6}{3}}}$ | $\frac{27^{\frac{1}{3}}(a^{12})^{\frac{1}{3}}}{8^{\frac{1}{3}}(b^9)^{\frac{1}{3}}}$

$27^{\frac{1}{3}} = \sqrt[3]{27} = 3$

Simplify. → $\frac{c^4}{d^2}$ | $\frac{3a^4}{2b^3}$

EXAMPLE 7 Find the value of $\frac{9^{-\frac{1}{2}}}{16^{-\frac{3}{4}}}$ and $\left(\frac{25}{16}\right)^{-\frac{1}{2}}$.

$\frac{9^{-\frac{1}{2}}}{1} \cdot \frac{1}{16^{-\frac{3}{4}}}$ | $\frac{25^{-\frac{1}{2}}}{16^{-\frac{1}{2}}}$

$x^{-n} = \frac{1}{x^n}$; $\frac{1}{x^{-n}} = x^n$ → $\frac{1}{9^{\frac{1}{2}}} \cdot \frac{16^{\frac{3}{4}}}{1}$ | $\frac{25^{-\frac{1}{2}}}{1} \cdot \frac{1}{16^{-\frac{1}{2}}}$

$16^{\frac{3}{4}} = (\sqrt[4]{16})^3 = (2)^3 = 8$ → $\frac{1}{3} \cdot \frac{8}{1}$ | $\frac{1}{25^{\frac{1}{2}}} \cdot \frac{16^{\frac{1}{2}}}{1}$

$9^{\frac{1}{2}} = \sqrt{9} = 3$

$\frac{8}{3}$ | $\frac{1}{5} \cdot \frac{4}{1}$ or $\frac{4}{5}$

EXAMPLE 8 Simplify and write with positive exponents.

$x^{-n} = \frac{1}{x^n}$ →

$a^{-\frac{2}{5}} \cdot a^{-\frac{1}{5}} = a^{-\frac{3}{5}}$
$= \frac{1}{a^{\frac{3}{5}}}$

$(d^{-4})^{\frac{1}{2}} = d^{-2}$
$= \frac{1}{d^2}$

EXAMPLE 9 Simplify and write with positive exponents.

$x^{-n} = \frac{1}{x^n}$

$\frac{1}{x^{-n}} = x^n$

$\frac{c^{\frac{1}{4}}}{c^{\frac{1}{2}}} = c^{\frac{1}{4}-\frac{1}{2}} = c^{-\frac{1}{4}}$
$= \frac{1}{c^{\frac{1}{4}}}$

$\frac{a^{-\frac{2}{3}}}{b^{-\frac{1}{2}}} = \frac{a^{-\frac{2}{3}}}{1} \cdot \frac{1}{b^{-\frac{1}{2}}}$
$= \frac{b^{\frac{1}{2}}}{a^{\frac{2}{3}}}$

ORAL EXERCISES

Simplify.

1. $a^{\frac{1}{3}} \cdot a^{\frac{1}{3}}$ $a^{\frac{2}{3}}$ **2.** $(b^6)^{\frac{1}{2}}$ b^3 **3.** $(c^{\frac{1}{2}})^6$ c^3 **4.** $\frac{x^{\frac{3}{5}}}{x^{\frac{1}{5}}}$ $x^{\frac{2}{5}}$ **5.** $\frac{1}{c^{-\frac{2}{3}}}$ $c^{\frac{2}{3}}$ **6.** $c^{-\frac{2}{3}}$ $\frac{1}{c^{\frac{2}{3}}}$

EXERCISES

PART A

Simplify and write with positive exponents.

1. $7^{\frac{3}{5}} \cdot 7^{\frac{1}{5}}$ $7^{\frac{4}{5}}$ **2.** $5^{\frac{4}{3}} \cdot 5^{-\frac{2}{3}}$ $5^{\frac{2}{3}}$ **3.** $3^{-\frac{2}{7}} \cdot 3^{\frac{5}{7}}$ $3^{\frac{3}{7}}$ **4.** $2^{3} \cdot 2^{-\frac{1}{2}}$ $2^{\frac{5}{2}}$

5. $x^{\frac{2}{3}} \cdot x^{\frac{5}{3}}$ $x^{\frac{7}{3}}$ **6.** $a^{\frac{5}{7}} \cdot a^{-\frac{3}{7}}$ $a^{\frac{2}{7}}$ **7.** $n^{-\frac{1}{5}} \cdot n^{\frac{4}{5}}$ $n^{\frac{3}{5}}$ **8.** $b^{2} \cdot b^{-\frac{3}{4}}$ $b^{\frac{5}{4}}$

9. $(x^{\frac{3}{4}})^{4}$ x^{3} **10.** $(a^{8})^{\frac{3}{4}}$ a^{6} **11.** $(n^{10})^{\frac{2}{5}}$ n^{4} **12.** $(b^{-\frac{2}{3}})^{-6}$ b^{4}

13. $\frac{x^{\frac{4}{5}}}{x^{\frac{1}{5}}}$ $x^{\frac{3}{5}}$ **14.** $\frac{a^{\frac{4}{7}}}{a^{-\frac{2}{7}}}$ $a^{\frac{6}{7}}$ **15.** $\frac{n^{-\frac{2}{3}}}{n^{-\frac{4}{3}}}$ $n^{\frac{2}{3}}$ **16.** $\frac{b^{2}}{b^{\frac{1}{2}}}$ $b^{\frac{3}{2}}$

17. $(x^{\frac{1}{3}}y^{\frac{2}{3}})^{3}$ xy^{2} **18.** $(a^{4}b^{6})^{\frac{1}{2}}$ $a^{2}b^{3}$ **19.** $(3m^{\frac{1}{2}}n^{\frac{5}{2}})^{2}$ $9mn^{5}$ **20.** $(16c^{4}d^{8})^{\frac{3}{4}}$ $8c^{3}d^{6}$

21. $\left(\frac{x^{\frac{1}{2}}}{y^{\frac{3}{4}}}\right)^{4}$ $\frac{x^{2}}{y^{3}}$ **22.** $\left(\frac{a^{\frac{3}{4}}}{b^{\frac{2}{3}}}\right)^{12}$ $\frac{a^{9}}{b^{8}}$ **23.** $\left(\frac{25m^{4}}{9n^{6}}\right)^{\frac{1}{2}}$ $\frac{5m^{2}}{3n^{3}}$ **24.** $\left(\frac{8c^{3}}{27d^{9}}\right)^{\frac{2}{3}}$ $\frac{4c^{2}}{9d^{6}}$

Find the value.

25. $\frac{1}{9^{-\frac{1}{2}}}$ 3 **26.** $\frac{1}{8^{-\frac{2}{3}}}$ 4 **27.** $\frac{5}{16^{-\frac{1}{4}}}$ 10 **28.** $\frac{-10}{25^{-\frac{3}{2}}}$ -1250

PART B

Simplify and write with positive exponents.

29. $x^{\frac{2}{5}} \cdot x^{-\frac{4}{5}}$ **30.** $a^{\frac{1}{3}} \cdot a^{-\frac{1}{2}}$ **31.** $n^{-\frac{3}{4}} \cdot n^{-\frac{1}{2}}$ **32.** $b^{-2} \cdot b^{\frac{1}{2}}$

33. $(x^{\frac{2}{3}})^{-6}$ **34.** $(a^{-\frac{3}{4}})^{8}$ **35.** $(n^{-9})^{\frac{1}{3}}$ **36.** $(b^{6})^{-\frac{1}{3}}$

37. $\frac{x^{\frac{1}{3}}}{x^{\frac{2}{3}}}$ $\frac{1}{x^{\frac{1}{3}}}$ **38.** $\frac{a^{\frac{1}{6}}}{a^{\frac{1}{3}}}$ $\frac{1}{a^{\frac{1}{6}}}$ **39.** $\frac{n^{-\frac{2}{3}}}{n^{-\frac{1}{3}}}$ $\frac{1}{n^{\frac{1}{3}}}$ **40.** $\frac{b^{-\frac{1}{2}}}{b^{\frac{3}{4}}}$ $\frac{1}{b^{\frac{5}{4}}}$

41. $\frac{x^{-\frac{1}{2}}}{y^{-\frac{1}{3}}}$ $\frac{y^{\frac{1}{3}}}{x^{\frac{1}{2}}}$ **42.** $\frac{a^{-\frac{2}{3}}}{b^{-\frac{3}{4}}}$ $\frac{b^{\frac{3}{4}}}{a^{\frac{2}{3}}}$ **43.** $\frac{m^{\frac{1}{3}}}{n^{-\frac{1}{4}}}$ $m^{\frac{1}{3}}n^{\frac{1}{4}}$ **44.** $\frac{c^{-\frac{3}{5}}}{d^{\frac{2}{5}}}$ $\frac{1}{c^{\frac{3}{5}}d^{\frac{2}{5}}}$

45. $\left(\frac{x^{-\frac{2}{3}}}{y^{-\frac{1}{3}}}\right)^{3}$ $\frac{y}{x^{2}}$ **46.** $\left(\frac{a^{-\frac{1}{2}}}{b^{-\frac{1}{4}}}\right)^{4}$ $\frac{b}{a^{2}}$ **47.** $\left(\frac{27m^{-6}}{8n^{-3}}\right)^{\frac{1}{3}}$ $\frac{3n}{2m^{2}}$ **48.** $\left(\frac{25c^{-4}}{4d^{-6}}\right)^{\frac{3}{2}}$ $\frac{125d^{9}}{8c^{6}}$

Find the value.

49. $\frac{16^{-\frac{1}{2}}}{27^{-\frac{1}{3}}}$ $\frac{3}{4}$ **50.** $\frac{8^{-\frac{2}{3}}}{16^{-\frac{1}{4}}}$ $\frac{1}{2}$ **51.** $\frac{25^{-\frac{3}{2}}}{100^{\frac{1}{2}}}$ $\frac{1}{1250}$ **52.** $\left(\frac{27}{8}\right)^{-\frac{1}{3}}$ $\frac{2}{3}$ **53.** $\left(\frac{4}{9}\right)^{-\frac{3}{2}}$ $\frac{27}{8}$ **54.** $\left(\frac{81}{16}\right)^{-\frac{3}{4}}$ $\frac{8}{27}$

PART C

Find the real number solution set.

55. $x^{5} = 32$ **56.** $x^{5} = -32$ **57.** $x^{6} = 64$ **58.** $x^{6} = -64$ **59.** $x^{3} = 7$ **60.** $x^{5} = 7$

Flow Chart: Dividing Polynomials

Follow the flow chart for dividing $ax^n + bx^{n-1} + cx^{n-2} + \ldots$ by $x + t$.

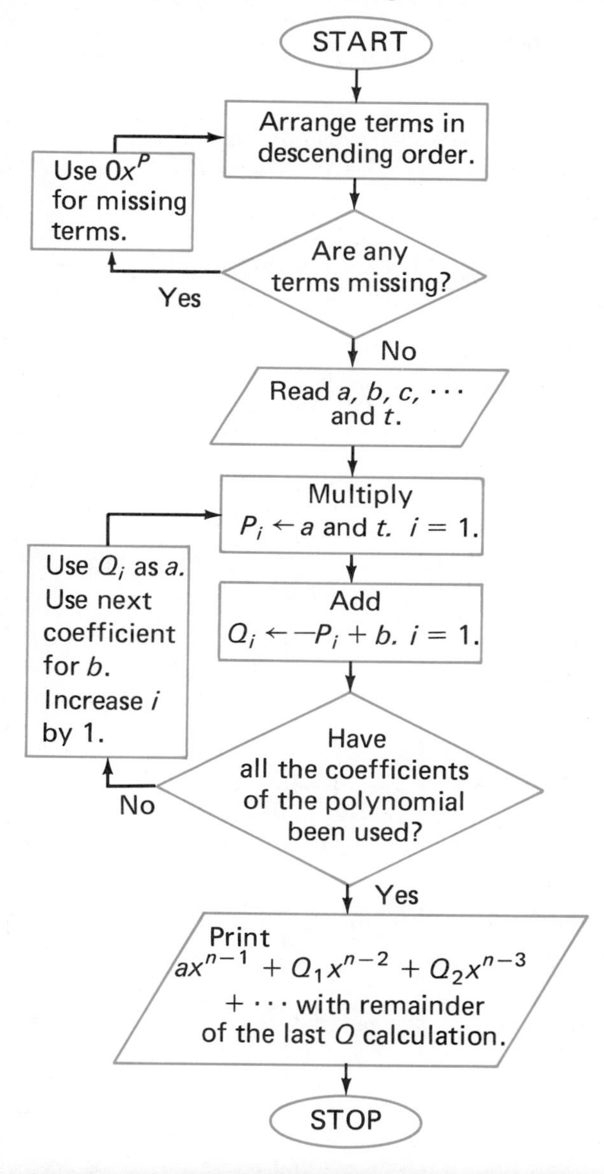

Divide $2x^3 - 5x + 12 + 4x^4$ by $x - 2$.

Arrange terms in descending order.
$4x^4 + 2x^3 - 5x + 12$

Any terms missing? Yes, kx^2.
Use $0x^2$.
Arrange terms again.
$4x^4 + 2x^3 + 0x^2 - 5x + 12$
Any terms missing? No.

Read coefficients.
$a = 4$, $b = 2$, $c = 0$, $d = -5$, $e = 12$, $t = -2$

$at = 4(-2)$ or -8
$P_1 = -8$

$-P_i + b = -(-8) + 2$ or 10
$Q_1 = 10$

Are all the coefficients used? No.
Let $Q_1 = a$. Use c for b. $i = 2$

$(Q_1)t = 10(-2)$ or -20
$P_2 = -20$

$-P_2 + c = -(-20) + 0 = 20$
$Q_2 = 20$

(Repeat until all coefficients are used.)

Print
$4x^3 + 10x^2 + 20x + 35$ remainder 82.

PROJECT Use the flow chart to divide $2x^3 + x - 3x^2 + 2$ by $x - 1$.

Chapter Six Review

Find the solution set. [p. 134]

1. $x^2 = 100$
2. $x^2 = 14$
3. $x^2 = 64$
4. $x^2 = 17$
5. Locate $\sqrt{21}$ between consecutive integers. [p. 134]
6. Approximate $\sqrt{55}$ to the nearest tenth. [p. 137]

Determine if the number is rational or irrational. [p. 137]

7. 2.696969 . . .
8. 20.21222324 . . .
9. $\sqrt{27}$
10. $-\sqrt{9}$

Simplify. [p. 141, 145]

11. $-\sqrt{28}$
12. $-2\sqrt{50}$
13. $\sqrt{4a^6b^5}$
14. $\sqrt{c^3d^5}$
15. $6\sqrt{7} - 3\sqrt{5} - \sqrt{7} + \sqrt{5}$
16. $\sqrt{8} + \sqrt{50} - 3\sqrt{2}$
17. $\sqrt{25a} - 2\sqrt{a} + \sqrt{9a}$
18. $4a\sqrt{b} + \sqrt{a^2b}$

Multiply. Then simplify the result [p. 141, 145]

19. $3\sqrt{5} \cdot 2\sqrt{10}$
20. $\sqrt{2x^5} \cdot \sqrt{8x}$
21. $(-3\sqrt{2})^2$
22. $(2\sqrt{3} - \sqrt{5})(3\sqrt{3} + 2\sqrt{5})$
23. $(4 + 2\sqrt{3})(4 - 2\sqrt{3})$
24. $(\sqrt{5} - 3\sqrt{3})^2$
25. $(\sqrt{x} + 5\sqrt{y})^2$

Rationalize the denominator [p. 148]

26. $\frac{3}{\sqrt{7}}$
27. $\frac{-\sqrt{5}}{2\sqrt{a}}$
28. $\frac{5}{\sqrt{18x^3}}$
29. $\frac{3}{2\sqrt{3} - \sqrt{10}}$

Divide. Then simplify the result. [p. 148]

30. $\frac{\sqrt{24}}{\sqrt{3}}$
31. $\frac{\sqrt{10c^7}}{\sqrt{5c^3}}$
32. $\frac{\sqrt{27a^6}}{\sqrt{3a^3}}$
33. $\frac{\sqrt{40x^9}}{\sqrt{5x^3}}$

Simplify. [p. 148]

34. $\sqrt{\frac{3}{5}}$
35. $\sqrt{\frac{2c}{3d}}$
36. $\sqrt{\frac{7}{12bc^3}}$
37. $\sqrt{\frac{2n-1}{3}}$

Simplify. [p. 153]

38. $\sqrt[3]{-8}$
39. $\sqrt[4]{81x^{12}}$
40. $\sqrt[4]{9b^3} \cdot \sqrt[4]{9b^8}$
41. $\sqrt[3]{16} + \sqrt[3]{2}$

Find the value. [p. 156, 161, 164]

42. $3 \cdot 2^0$
43. $\frac{3^{-2}}{2^{-3}}$
44. $36^{-\frac{1}{2}}$
45. $125^{\frac{2}{3}}$
46. $\frac{4^{-\frac{1}{2}}}{27^{-\frac{1}{3}}}$
47. $\left(\frac{81}{16}\right)^{-\frac{1}{4}}$

Simplify and write with positive exponents. [p. 156, 164]

48. $5x^{-3}$
49. $3x^{-5} \cdot 2x^3$
50. $(-2x^{-2}y^3)^{-4}$
51. $a^{\frac{2}{7}} \cdot a^{-\frac{5}{7}}$
52. $(a^{-9})^{-\frac{1}{3}}$
53. $\frac{a^3b^{-2}}{c^{-4}d^5}$
54. $\frac{6x^{-3}}{8y^{-2}}$
55. $\left(\frac{3x^{-2}}{y^{-3}}\right)^2$
56. $\frac{c^{\frac{1}{4}}}{c^{-\frac{1}{2}}}$
57. $(8x^3y^6)^{\frac{2}{3}}$

Write in exponent form. [p. 161]

58. $\sqrt{3}$
59. $\sqrt[3]{a^2}$
60. $(\sqrt[4]{5})^3$

Write with a square root, cube root, or fourth root sign. [p. 161]

61. $6^{\frac{1}{2}}$
62. $5x^{\frac{1}{3}}$
63. $(2a)^{\frac{3}{4}}$

Chapter Six Test

Find the solution set.

1. $x^2 = 16$
2. $x^2 = 15$
3. Locate $\sqrt{38}$ between consecutive integers.
4. Approximate $\sqrt{11}$ to the nearest tenth.

Determine if the number is rational or irrational.

5. 2.373737 . . .
6. 3.212212221 . . .
7. $\sqrt{2}$
8. $\sqrt{16}$

Simplify.

9. $2\sqrt{18}$
10. $\sqrt{8a^6}$
11. $\sqrt{x^4y^3}$
12. $5\sqrt{7} - 2\sqrt{3} + \sqrt{7} - 3\sqrt{3}$
13. $\sqrt{12} + \sqrt{27} - \sqrt{3}$
14. $5x\sqrt{y} + \sqrt{x^2y}$

Multiply. Simplify the result.

15. $2\sqrt{6} \cdot 5\sqrt{2}$
16. $\sqrt{3x} \cdot \sqrt{12x^3}$
17. $(-2\sqrt{3a})^2$
18. $(3\sqrt{2} + \sqrt{3})(\sqrt{2} - 2\sqrt{3})$
19. $(2\sqrt{5} - 3)(2\sqrt{5} + 3)$
20. $(3\sqrt{x} + \sqrt{y})^2$

Rationalize the denominator.

21. $\dfrac{1}{\sqrt{3}}$
22. $\dfrac{-5}{2\sqrt{c^3}}$
23. $\dfrac{2}{3\sqrt{2} + \sqrt{7}}$

Divide. Simplify the result.

24. $\dfrac{\sqrt{40}}{\sqrt{5}}$
25. $\dfrac{\sqrt{24a^6}}{\sqrt{2a^3}}$

Simplify.

26. $\sqrt{\dfrac{5}{6}}$
27. $\sqrt{\dfrac{3}{8x^3}}$

Simplify.

28. $\sqrt[3]{-27}$
29. $\sqrt[4]{16x^8}$
30. $\sqrt[3]{3} + \sqrt[3]{24}$
31. $\sqrt[4]{4x^3} \cdot \sqrt[4]{4x^2}$

Find the value.

32. $6 \cdot 7^0$
33. $\dfrac{4^{-2}}{3^{-3}}$
34. $9^{\frac{3}{2}}$
35. $\left(\dfrac{9}{25}\right)^{-\frac{1}{2}}$

Simplify and write with positive exponents.

36. $2a^{-6} \cdot 3a^2$
37. $(x^3y^{-4})^{-2}$
38. $a^{\frac{1}{5}} \cdot a^{-\frac{4}{5}}$
39. $(x^{-4})^{\frac{1}{2}}$
40. $\dfrac{x^4y^{-2}}{x^{-2}y^3}$
41. $\left(\dfrac{2c^{-2}}{d^{-3}}\right)^3$
42. $\dfrac{x^{\frac{3}{5}}}{x^{-\frac{1}{5}}}$
43. $(25a^4b^6)^{\frac{1}{2}}$

44. Write $\sqrt[4]{x^3}$ in exponent form.
45. Write $4x^{\frac{2}{3}}$ with a cube root sign.

7 QUADRATIC EQUATIONS

Fun for Philatelists

A philatelist (fi-lat′e-list) is a stamp collector. Philatelists who also like mathematics will be especially interested in a group of ten stamps issued by Nicaragua in 1971. The stamps show the ten equations "that changed the face of the earth." Each equation represents a major turning point in mathematics or science. On the back of each stamp, a brief history of the equation is written in Spanish. The comments below have been adapted from the backs of the stamps.

$$V = V_e \ln \frac{m_0}{m_1}$$

This equation gives the changing speed of a rocket as it burns away the weight of the fuel it is carrying.

Interstellar space travel would be almost impossible without using this equation to plan the fuel consumption.

$$e^{\ln N} = N$$

This formula is one way to express a natural, or Napier logarithm. Logarithms provided a powerful tool for multiplying and dividing large numbers. By simplifying these processes to addition and subtraction, much effort could be saved. The impact of logarithms on the fields of navigation and astronomy in the 17th century can be compared with the effect of computers today.

Designed and printed by the Thos De la Rue & Co. Ltd. Each stamp is 48 x 32 mm in sheets of 50 stamps.

Area Problems

OBJECTIVE

- To solve problems about area

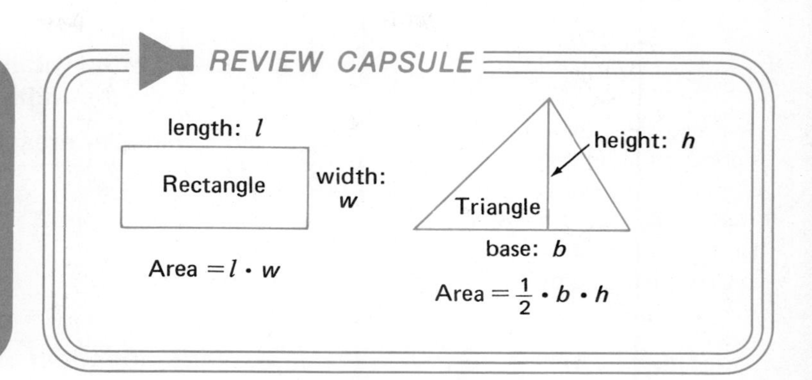

EXAMPLE 1 The length of a rectangle is 3 cm more than twice the width. The area is 44 cm².
Find the length and the width

w is the width in centimeters (cm). → Let $w =$ the width

3 more than twice w → $2w + 3 =$ the length

Write an equation: (width)(length) = area. → $w(2w + 3) = 44$

Multiply: $w(2w + 3) = 2w^2 + 3w$. → $2w^2 + 3w = 44$

Add -44 to each side. → $2w^2 + 3w - 44 = 0$

Factor. → $(2w + 11)(w - 4) = 0$

$2w + 11 = 0$ $\quad$ $w - 4 = 0$

Cannot use $-5\frac{1}{2}$ for w. Width cannot be negative. → $w = -5\frac{1}{2}$ $\quad$ $w = 4$

$2w + 3 = 2(4) + 3 = 11$ → **Thus,** the width is 4 cm and the length is 11 cm.

EXAMPLE 2 A triangle's height is 2 m less than the length of its base. The area is 12 m².
Find the length of the base and the height

Let $b =$ the length of the base

2 less than b → $b - 2 =$ the height

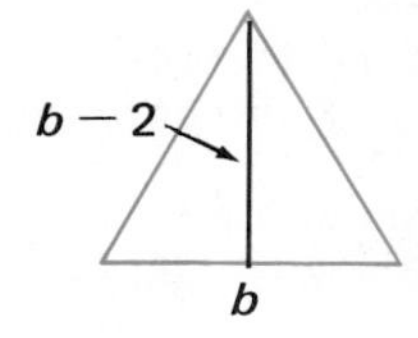

For a triangle, $(\frac{1}{2})$(length of base)(height) = area. → $\frac{1}{2}b(b - 2) = 12$

Multiply each side by 2, LCD. → $b(b - 2) = 24$

$$b^2 - 2b = 24$$

$$b^2 - 2b - 24 = 0$$

Factor. → $(b - 6)(b + 4) = 0$

$b - 6 = 0$ or $b + 4 = 0$ → $b = 6$ or $b = -4$

Cannot use -4 for base, b.
$b - 2 = 6 - 2 = 4$ → **Thus,** The length of the base is 6 m and the height is 4 m.

EXAMPLE 3 The width of one square is 3 times the width of a second square. The sum of their areas is 160 cm^2. Find the width of each square.

Let s = the width of the smaller square
$3s$ = the width of the larger square

For a square, area is (length of side) (length of side).

$$\underbrace{(\text{smaller area})} + \underbrace{(\text{larger area})} = 160$$

Sum of areas is 160. → $s^2 + (3s)^2 = 160$

$(3s)^2 = 3s \cdot 3s = 9s^2$ → $s^2 + 9s^2 = 160$

Combine like terms: $s^2 = 1s^2$. → $10s^2 = 160$

Divide each side by 10. → $s^2 = 16$

If $x^2 = k$, then $x = \sqrt{k}$ or $x = -\sqrt{k}$. → $s = 4$ or $s = -4$

$3s = 3(4) = 12$ → **Thus,** the width of the smaller square is 4 cm and the width of the larger square is 12 cm.

EXAMPLE 4 A triangle's height and a rectangle's width are the same. The rectangle's length is 3 times its width. The triangle's base measures the same as the rectangle's length. The rectangle's area decreased by the triangle's area is 24 dm^2. Find the area of each figure.

Always draw a sketch for problems about geometric figures.

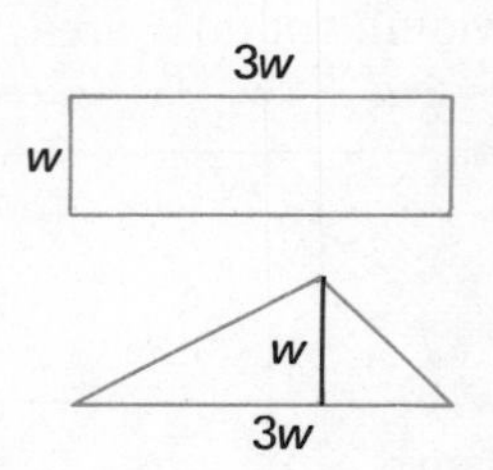

Let w = the width of the rectangle
and w = the height of the triangle
Then $3w$ = the length of the rectangle
and $3w$ = the length of the base of the triangle

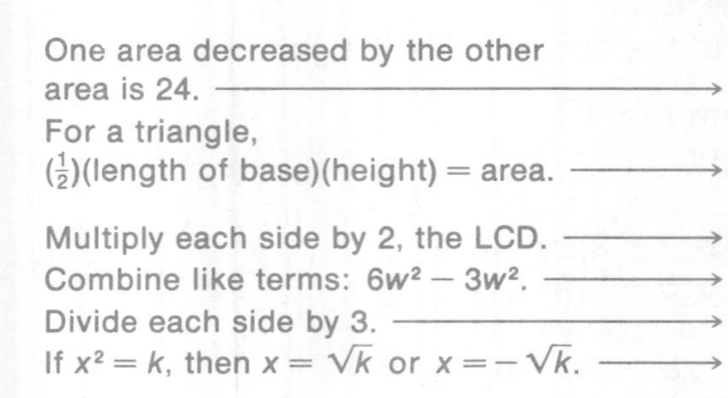

One area decreased by the other area is 24. → $(\text{area}) - (\text{area}) = 24$

For a triangle, $(\frac{1}{2})$(length of base)(height) = area. → $3w \cdot w - \frac{1}{2} \cdot 3w \cdot w = 24$

$$3w^2 - \frac{1}{2} \cdot 3w^2 = 24$$

Multiply each side by 2, the LCD. → $6w^2 - 3w^2 = 48$

Combine like terms: $6w^2 - 3w^2$. → $3w^2 = 48$

Divide each side by 3. → $w^2 = 16$

If $x^2 = k$, then $x = \sqrt{k}$ or $x = -\sqrt{k}$. → $w = 4$ or $w = -4$

Rectangle: Width, w is 4 dm; length, $3w$ is 12 dm.
Triangle: Length of base, $3w$ is 12 dm; height, w is 4 dm.

Area $= l \cdot w = 12 \cdot 4 = 48$ → **Thus,** the area of the rectangle is 48 dm^2

Area $= \frac{1}{2}bh = \frac{1}{2} \cdot 12 \cdot 4 = 24$ → and the area of the triangle is 24 dm^2.

ORAL EXERCISES

Find the area. *40 m²* *25 cm²* *15 dm²*

1.

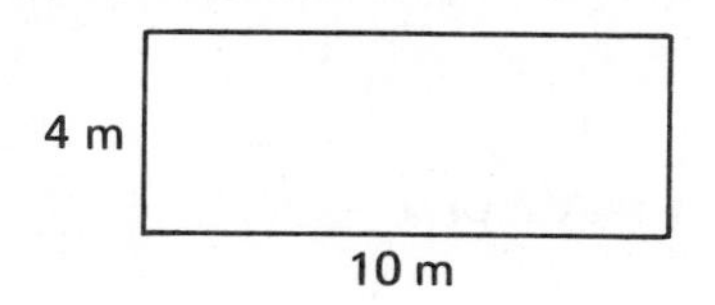

2.

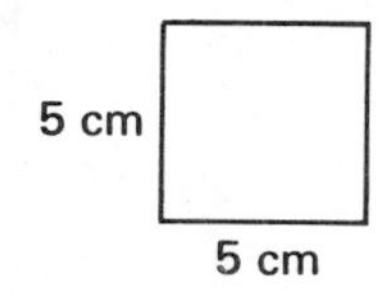

3.

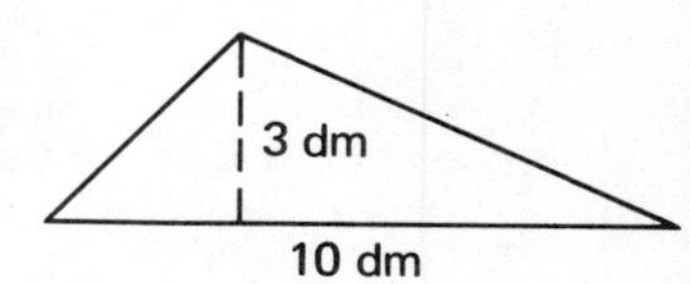

EXERCISES

PART A

1. The length of a rectangle is 3 times its width. The area is 75 mm^2. Find the length and the width. *15 mm, 5 mm*
2. The length of a rectangle is 4 cm more than its width. The area is 45 cm^2. Find the length and the width.
3. A triangle's height is twice the length of its base. The area is 16 m^2. Find the length of the base and the height. *4 m, 8 m*
4. A triangle's height is 6 times the length of its base. The area is 12 mm^2. Find the length of the base and the height.
5. The area of a square is 36 dm^2. Find the length of a side. *6 dm*
6. The area of a square is 144 cm^2. Find the length of a side. *12 cm*
7. The length of a rectangle is 3 cm less than twice its width. The area is 20 cm^2. Find the length and the width. *5 cm, 4 cm*
8. The length of a rectangle is 5 m more than 3 times its width. The area is 42 m^2. Find the length and the width. *14 m, 3 m*

PART B

9. A triangle's height is 4 dm less than the length of its base. The area is 6 dm^2. Find the length of the base and the height. *6 dm, 2 dm*
10. A triangle's height is 3 cm more than the length of its base. The area is 44 cm^2. Find the length of the base and the height. *8 cm, 11 cm*
11. The width of one square is twice the width of a second square. The sum of their areas is 125 mm^2. Find the width of each square.
12. The width of one square is 3 times the width of a second square. One area decreased by the other is 32 m^2. Find each width.
13. A square and a rectangle have the same width. The rectangle's length is 4 times its width. The sum of their areas is 45 mm^2. Find the area of each figure. *9 mm², 36 mm²*
14. A square and a rectangle have the same width. The rectangle's length is 5 cm more than twice its width. One area decreased by the other is 36 cm^2. Find the area of each figure.
15. A triangle's base measures the same as a rectangle's width. The rectangle's length is twice its width. The triangle's height is twice the length of its base. The sum of their areas is 108 m^2. Find the area of each figure. *36 m², 72 m²*
16. A triangle's base measures 2 m more than a rectangle's width. The rectangle's length is 4 times its width. The triangle's height is the same as the rectangle's length. The rectangle's area decreased by the triangle's area is 30 dm^2. Find each area. *70 dm², 100 dm²*

Solving $(x - a)^2 = k$

OBJECTIVE

- To find solution sets of equations like $(2x - 5)^2 = 7$

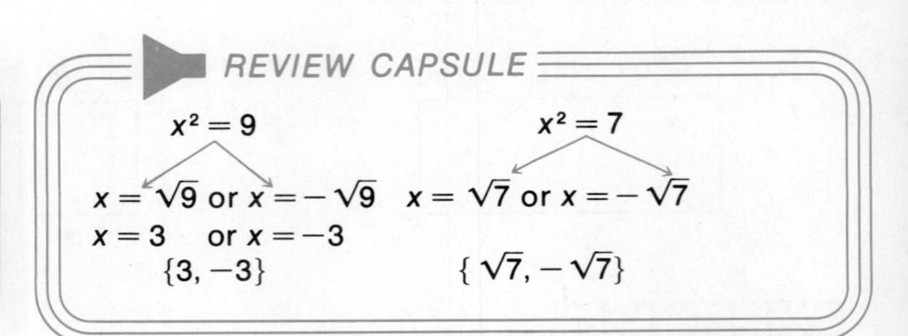

EXAMPLE 1 Find the solution set of $4x^2 = 5$.

$$4x^2 = 5$$

Divide each side by 4. →

$$x^2 = \frac{5}{4}$$

If $x^2 = k$, then $x = \sqrt{k}$ or $x = -\sqrt{k}$. →

$$x = \sqrt{\frac{5}{4}} \quad \text{or} \quad x = -\sqrt{\frac{5}{4}}$$

$\sqrt{\frac{a}{b}} = \frac{\sqrt{a}}{\sqrt{b}}$ →

$$x = \frac{\sqrt{5}}{\sqrt{4}} \quad \text{or} \quad x = -\frac{\sqrt{5}}{\sqrt{4}}$$

$$x = \frac{\sqrt{5}}{2} \quad \text{or} \quad x = -\frac{\sqrt{5}}{2}$$

Both solutions are written together.
Read: $x =$ positive or negative $\frac{\sqrt{5}}{2}$. →

$$x = \pm\frac{\sqrt{5}}{2}$$

Check both solutions in the original equation, $4x^2 = 5$. → **Thus,** $\left\{\frac{\sqrt{5}}{2}, -\frac{\sqrt{5}}{2}\right\}$ is the solution set.

EXAMPLE 2 Find the solution set of $(n - 3)^2 = 16$.

If $x^2 = k$, →

$$(n - 3)^2 = 16$$

then $x = \sqrt{k}$ or $x = -\sqrt{k}$. →

$$n - 3 = \sqrt{16} \qquad n - 3 = -\sqrt{16}$$

$\sqrt{16} = 4$ →

$$n - 3 = 4 \qquad n - 3 = -4$$

Add 3 to each side. →

$$\begin{array}{rr} +3 & +3 \\ \hline n & = 7 \end{array} \qquad \begin{array}{rr} +3 & +3 \\ \hline n & = -1 \end{array}$$

Check. →
Substitute. →

Let $n = 7$.

$(n - 3)^2$	16
$(7 - 3)^2$	16
4^2	
16	
True	

Let $n = -1$.

$(n - 3)^2$	16
$(-1 - 3)^2$	16
$(-4)^2$	
16	
True	

Thus, $\{7, -1\}$ is the solution set.

EXAMPLE 3 Find the solution set of $(2y + 3)^2 = 8$.

$$(2y+3)^2 = 8$$

If $x^2 = k$, then $x = \sqrt{k}$ or $x = -\sqrt{k}$. → $2y + 3 = \sqrt{8}$ or $2y + 3 = -\sqrt{8}$

Simplify: $\sqrt{8} = \sqrt{4 \cdot 2} = 2\sqrt{2}$. → $2y + 3 = 2\sqrt{2}$ | $2y + 3 = -2\sqrt{2}$

Add -3 to each side. → $2y = -3 + 2\sqrt{2}$ | $2y = -3 - 2\sqrt{2}$

Divide each side by 2. → $y = \frac{-3 + 2\sqrt{2}}{2}$ | $y = \frac{-3 - 2\sqrt{2}}{2}$

Write both solutions together. → $y = \frac{-3 \pm 2\sqrt{2}}{2}$

Thus, $\left\{\frac{-3 + 2\sqrt{2}}{2}, \frac{-3 - 2\sqrt{2}}{2}\right\}$ is the solution set.

EXAMPLE 4 Find the solution set of $2(3x + 4)^2 - 5 = 19$.

Add 5 to each side. → $2(3x + 4)^2 = 24$

Divide each side by 2. → $(3x + 4)^2 = 12$

$3x + 4 = \sqrt{12}$ or $3x + 4 = -\sqrt{12}$

Add -4 to each side. → $3x = -4 + \sqrt{12}$ | $3x = -4 - \sqrt{12}$

Simplify: $\sqrt{12} = \sqrt{4 \cdot 3} = 2\sqrt{3}$. → $3x = -4 + 2\sqrt{3}$ | $3x = -4 - 2\sqrt{3}$

Divide each side by 3. → $x = \frac{-4 + 2\sqrt{3}}{3}$ | $x = \frac{-4 - 2\sqrt{3}}{3}$

Thus, $\left\{\frac{-4 + 2\sqrt{3}}{3}, \frac{-4 - 2\sqrt{3}}{3}\right\}$ is the solution set.

ORAL EXERCISES

Determine the two numbers.

1. $\frac{2 \pm 3}{4}$ $\frac{5}{4}, -\frac{1}{4}$
2. $\frac{-5 \pm 3}{2}$ $-1, -4$
3. $\frac{4 \pm 4}{12}$ $\frac{2}{3}, 0$
4. $\frac{-3 \pm 7}{5}$ $\frac{4}{5}, -2$
5. $\frac{1 \pm 3\sqrt{2}}{2}$ $\frac{1 + 3\sqrt{2}}{2}, \frac{1 - 3\sqrt{2}}{2}$

EXERCISES

PART A

Find the solution set.

1. $9a^2 = 5$ $\{\pm\frac{\sqrt{5}}{3}\}$
2. $3y^2 = 24$ $\{\pm 2\sqrt{2}\}$
3. $(y - 3)^2 = 16$ $\{7, -1\}$
4. $(n + 5)^2 = 12$
5. $(2x - 5)^2 = 9$ $\{4, 1\}$
6. $(3n + 4)^2 = 25$ $\{\frac{1}{3}, -3\}$
7. $(3x - 2)^2 = 10$
8. $(4a + 5)^2 = 3$

PART B

Find the solution set.

9. $(5n + 3)^2 = 8$
10. $(2y - 7)^2 = 12$
11. $(x - 2\sqrt{5})^2 = 5$ $\{3\sqrt{5}, \sqrt{5}\}$
12. $(y + 3\sqrt{2})^2 = 2$ $\{-2\sqrt{2}, -4\sqrt{2}\}$
13. $(n - 1)^2 - 5 = 0$ $\{1 \pm \sqrt{5}\}$
14. $2(a - 4)^2 - 3 = 13$ $\{4 \pm 2\sqrt{2}\}$
15. $2(y + 1)^2 - 18 = 0$ $\{2, -4\}$
16. $9x^2 - 7 = 0$ $\{\pm\frac{\sqrt{7}}{3}\}$

Solving by Completing the Square

OBJECTIVE

- To solve equations like $3x^2 + 4x - 2 = 0$ by using the form $(x - a)^2 = k$

REVIEW CAPSULE

$(x + 4)^2 = x^2 + 2 \cdot x \cdot 4 + 4^2 = x^2 + 8x + 16$

$(a + b)^2 = a^2 + 2 \cdot a \cdot b + b^2$

Square of a Binomial — Perfect Square Trinomial

EXAMPLE 1 Find a number so that $x^2 + 14x + \underline{?}$ is a perfect square trinomial.

14 is the coefficient of x.

$$x^2 + (14)x + \underline{\quad}$$

Add the square of $\frac{1}{2}(14)$.

$$\left(\frac{14}{2}\right)^2 = 7^2, \text{ or } 49$$

Check by factoring. ⟶ $x^2 + 14x + 49 = (x + 7)(x + 7)$, or $(x + 7)^2$

A perfect square trinomial is the square of a binomial.

Thus, $x^2 + 14x + 49$ is a perfect square trinomial.

The procedure in Example 1 is called *completing the square.*

Completing the square ⟶

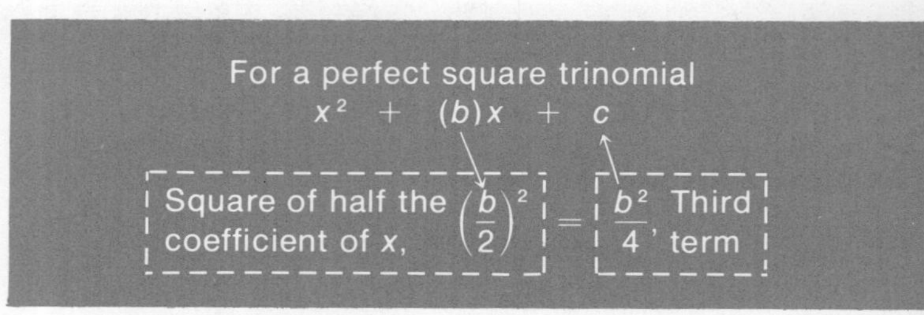

EXAMPLE 2 Complete the square of $x^2 - 5x + \underline{\quad}$.

-5 is the coefficient of x. ⟶ $x^2 + (-5)x + \underline{\quad}$

Half of the coefficient: $-\frac{5}{2}$

Square the half: $(-\frac{5}{2})^2 = \frac{25}{4}$.

Add the square.

$$\left(\frac{-5}{2}\right)^2 = \frac{25}{4}$$

Check by factoring. ⟶ $x^2 - 5x + \frac{25}{4} = \left(x - \frac{5}{2}\right)\left(x - \frac{5}{2}\right)$, or $\left(x - \frac{5}{2}\right)^2$

$x^2 - 5x + \frac{25}{4} = \left(x - \frac{5}{2}\right)^2$ ⟶ **Thus,** $x^2 - 5x + \frac{25}{4}$ is a perfect square trinomial.

EXAMPLE 3 Complete the square on the left side of the equation $n^2 - 8n = 3$. Solve the equation.

$\left(\frac{-8}{2}\right)^2 = (-4)^2 = 16$; Add 16 to each side. →

$$n^2 - 8n + \_\_\_\_ = 3 + \_\_\_\_$$
$$n^2 - 8n + 16 = 3 + 16$$

Factor the perfect square trinomial.

$$(n - 4)^2 = 19$$

If $x^2 = k$, then $x = \sqrt{k}$ or $x = -\sqrt{k}$. →

$$n - 4 = \sqrt{19} \quad \text{or} \quad n - 4 = -\sqrt{19}$$

Add 4 to each side. →

$$n = 4 + \sqrt{19} \qquad n = 4 - \sqrt{19}$$

Solutions may be written together: $4 \pm \sqrt{19}$. →

Thus, the solutions are $4 + \sqrt{19}$ and $4 - \sqrt{19}$.

EXAMPLE 4 Find the solution set of $x^2 + 10x + 4 = 0$ by completing the square.

Get the constant, 4, alone. Add -4 to each side. →

$$x^2 + 10x + 4 = 0$$
$$x^2 + 10x + \_\_\_\_ = -4 + \_\_\_\_$$

Complete the square: $(\frac{10}{2})^2 = 25$. Add 25 to each side. →

$$x^2 + 10x + 25 = -4 + 25$$
$$(x + 5)^2 = 21$$

Solve.

$$x + 5 = \sqrt{21} \quad \text{or} \quad x + 5 = -\sqrt{21}$$
$$x = -5 + \sqrt{21} \qquad x = -5 - \sqrt{21}$$

Thus, $\{-5 + \sqrt{21}, -5 - \sqrt{21}\}$ is the solution set.

SUMMARY

To solve a quadratic equation by completing the square:
1. **Get the constant term alone on one side.**
2. **Complete the square. Add to each side.**
3. **Solve the new equation.**

EXAMPLE 5 Find the solution set of $x^2 - 3x - 9 = 0$ by completing the square.

Add 9 to each side. →

$$x^2 - 3x + \_\_\_\_ = 9 + \_\_\_\_$$

Complete the square. Add to each side. $\left(-\frac{3}{2}\right)^2 = \frac{9}{4}$ →

$$x^2 - 3x + \frac{9}{4} = 9 + \frac{9}{4}$$

Factor the perfect square trinomial.

Solve the new equation

$$\left(x - \frac{3}{2}\right)^2 = \frac{45}{4}$$

$$x - \frac{3}{2} = \sqrt{\frac{45}{4}} \quad \text{or} \quad x - \frac{3}{2} = -\sqrt{\frac{45}{4}}$$

$\sqrt{\frac{45}{4}} = \frac{\sqrt{45}}{\sqrt{4}} = \frac{\sqrt{45}}{2}$ →

$$x = \frac{3}{2} + \frac{\sqrt{45}}{2} \qquad x = \frac{3}{2} - \frac{\sqrt{45}}{2}$$

Add: same denominators. →

$$x = \frac{3 + \sqrt{45}}{2} \qquad x = \frac{3 - \sqrt{45}}{2}$$

Simplify: $\sqrt{45} = \sqrt{9 \cdot 5} = 3\sqrt{5}$. →

$$x = \frac{3 + 3\sqrt{5}}{2} \qquad x = \frac{3 - 3\sqrt{5}}{2}$$

Thus, $\left\{\frac{3 + 3\sqrt{5}}{2}, \frac{3 - 3\sqrt{5}}{2}\right\}$ is the solution set.

EXAMPLE 6 Solve $3x^2 + 7x + 1 = 0$ by completing the square.

$$3x^2 + 7x = -1$$

Coefficient of x^2 is not 1. Coefficient of x^2 must be 1 to complete the square. Divide each side by 3.

$$x^2 + \frac{7}{3}x + \_\_\_\_ = \frac{-1}{3} + \_\_\_\_$$

$\left(\frac{1}{2} \cdot \frac{7}{3}\right)^2 = \left(\frac{7}{6}\right)^2 = \frac{49}{36}$

Add $\frac{49}{36}$ to complete the square.

$$x^2 + \frac{7}{3}x + \frac{49}{36} = \frac{-1}{3} + \frac{49}{36}$$

$$\left(x + \frac{7}{6}\right)^2 = \frac{37}{36}$$

$$x + \frac{7}{6} = \sqrt{\frac{37}{36}} \quad \text{or} \quad x + \frac{7}{6} = -\sqrt{\frac{37}{36}}$$

Add $\frac{-7}{6}$ to each side. $\sqrt{\frac{37}{36}} = \frac{\sqrt{37}}{\sqrt{36}} = \frac{\sqrt{37}}{6}$

$$x = \frac{-7}{6} + \frac{\sqrt{37}}{6} \qquad x = \frac{-7}{6} - \frac{\sqrt{37}}{6}$$

$$x = \frac{-7 + \sqrt{37}}{6} \qquad x = \frac{-7 - \sqrt{37}}{6}$$

Cannot simplify $\sqrt{37}$.

Thus, the solutions are $\frac{-7 + \sqrt{37}}{6}$ and $\frac{-7 - \sqrt{37}}{6}$.

ORAL EXERCISES

Complete the square to form a perfect square trinomial.

1. $a^2 + 12a$ $a^2 + 12a + 36$
2. $n^2 - 6n$ $n^2 - 6n + 9$
3. $x^2 + 3x$ $x^2 + 3x + \frac{9}{4}$
4. $b^2 - b$ $b^2 - b + \frac{1}{4}$
5. $c^2 + \frac{3}{2}c$ $c^2 + \frac{3}{2}c + \frac{9}{16}$
6. $x^2 - \frac{2}{3}x$ $x^2 - \frac{2}{3}x + \frac{1}{9}$

EXERCISES

PART A

Find the solution set by completing the square.

1. $x^2 + 6x = 16$ $\{2, -8\}$
2. $x^2 + 3x = 10$ $\{2, -5\}$
3. $x^2 - 8x = -15$ $\{5, 3\}$
4. $n^2 - 6n = 5$ $\{3 \pm \sqrt{14}\}$
5. $n^2 + 8n = -9$ $\{-4 \pm \sqrt{7}\}$
6. $n^2 - 4n = 7$ $\{2 \pm \sqrt{11}\}$
7. $y^2 + 3y - 3 = 0$
8. $y^2 - 5y + 2 = 0$
9. $y^2 + y - 5 = 0$
10. $x^2 + 8x - 3 = 0$
11. $x^2 - 10x + 15 = 0$
12. $x^2 + 12x - 2 = 0$
13. $n^2 - 5n - 5 = 0$
14. $n^2 + 7n + 1 = 0$
15. $n^2 - n - 11 = 0$

PART B

Solve by completing the square.

16. $5x^2 + 3x - 2 = 0$ $\frac{2}{5}, -1$
17. $2x^2 - 7x - 4 = 0$ $4, -\frac{1}{2}$
18. $3x^2 + 10x + 3 = 0$ $-\frac{1}{3}, -3$
19. $2n^2 + 5n - 5 = 0$
20. $5n^2 - 7n + 1 = 0$
21. $3n^2 + 4n - 3 = 0$
22. $y^2 - 6y - 3 = 0$ $3 \pm 2\sqrt{3}$
23. $y^2 + 10y + 7 = 0$ $-5 \pm 3\sqrt{2}$
24. $y^2 - 8y - 4 = 0$ $4 \pm 2\sqrt{5}$

PART C

Solve by completing the square.

25. $6x = x^2 - 6$
26. $\frac{1}{2}x^2 + 3x = 7$
27. $36x^2 - 60x + 5 = 0$
28. $-3x^2 - 5x + 4 = 0$
29. $\frac{1}{3}x^2 - \frac{1}{2}x - \frac{1}{6} = 0$
30. $8x^2 + 8x - 7 = 0$

Quadratic Formula

OBJECTIVE

- To solve equations like $3x^2 + 4x - 2 = 0$ by using a formula

REVIEW CAPSULE

Solve $3x^2 + 4x - 2 = 0$ by completing the square.

$$3x^2 + 4x = 2$$
$$x^2 + \frac{4}{3}x = \frac{2}{3}$$
$$x^2 + \frac{4}{3}x + \frac{4}{9} = \frac{2}{3} + \frac{4}{9}$$
$$\left(x + \frac{2}{3}\right)^2 = \frac{10}{9}$$
$$x + \frac{2}{3} = \frac{\sqrt{10}}{3} \quad \text{or} \quad x + \frac{2}{3} = \frac{-\sqrt{10}}{3}$$

The solutions are $\frac{-2 + \sqrt{10}}{3}$ and $\frac{-2 - \sqrt{10}}{3}$.

EXAMPLE 1 Solve $ax^2 + bx + c = 0$ for x by completing the square.

$$ax^2 + bx + c = 0$$

Add $-c$ to each side. → $ax^2 + bx = -c$

Divide each side by a. → $x^2 + \frac{b}{a}x = \frac{-c}{a}$

Complete the square: $\left(\frac{1}{2} \cdot \frac{b}{a}\right)^2 = \left(\frac{b}{2a}\right)^2 = \frac{b^2}{4a^2}$. Add to each side. → $x^2 + \frac{b}{a}x + \frac{b^2}{4a^2} = \frac{-c}{a} + \frac{b^2}{4a^2}$

$4a^2$ is LCD of the right side. → $x^2 + \frac{b}{a}x + \frac{b^2}{4a^2} = \frac{-4ac}{4a^2} + \frac{b^2}{4a^2}$

Factor the left side. Simplify the right side. → $\left(x + \frac{b}{2a}\right)^2 = \frac{b^2 - 4ac}{4a^2}$

If $x^2 = k$, then $x = \sqrt{k}$ or $x = -\sqrt{k}$. → $x + \frac{b}{2a} = \sqrt{\frac{b^2 - 4ac}{4a^2}}$ or $x + \frac{b}{2a} = -\sqrt{\frac{b^2 - 4ac}{4a^2}}$

$\sqrt{\frac{b^2 - 4ac}{4a^2}} = \frac{\sqrt{b^2 - 4ac}}{\sqrt{4a^2}} = \frac{\sqrt{b^2 - 4ac}}{2a}$ → $x + \frac{b}{2a} = \frac{\sqrt{b^2 - 4ac}}{2a}$ $\quad$ $x + \frac{b}{2a} = -\frac{\sqrt{b^2 - 4ac}}{2a}$

Add $-\frac{-b}{2a}$ to each side. → $x = \frac{-b}{2a} + \frac{\sqrt{b^2 - 4ac}}{2a}$ $\quad$ $x = \frac{-b}{2a} - \frac{\sqrt{b^2 - 4ac}}{2a}$

Add: same denominator. → $x = \frac{-b + \sqrt{b^2 - 4ac}}{2a}$ $\quad$ $x = \frac{-b - \sqrt{b^2 - 4ac}}{2a}$

Write solutions together. → $x = \frac{-b \pm \sqrt{b^2 - 4ac}}{2a}$

Quadratic Formula

If $ax^2 + bx + c = 0$, then $x = \frac{-b \pm \sqrt{b^2 - 4ac}}{2a}$.

a, b, and c may be any numbers, $a \neq 0$.

EXAMPLE 2 Solve $3x^2 + 4x - 2 = 0$ by using the quadratic formula.

Quadratic formula →
$$x = \frac{-b \pm \sqrt{b^2 - 4ac}}{2a}$$

Determine values of a, b, and c. →
$$3x^2 + 4x + (-2) = 0$$
$$a = 3 \quad b = 4 \quad c = -2$$

Substitute in the formula: 3 for a, 4 for b, -2 for c. →
$$x = \frac{-(4) \pm \sqrt{4^2 - 4(3)(-2)}}{2(3)}$$

$-4(3)(-2) = +24$ →
$$= \frac{-4 \pm \sqrt{16 + 24}}{6}$$
$$= \frac{-4 \pm \sqrt{40}}{6}$$

Simplify: $\sqrt{40} = \sqrt{4 \cdot 10} = 2\sqrt{10}$. →
$$= \frac{-4 \pm 2\sqrt{10}}{6}$$

Factor numerator and denominator. →
$$= \frac{2(-2 \pm \sqrt{10})}{2(3)}$$

Simplify. →
$$= \frac{-2 \pm \sqrt{10}}{3}$$

Thus, the solutions are $\frac{-2 + \sqrt{10}}{3}$ and $\frac{-2 - \sqrt{10}}{3}$.

EXAMPLE 3 Rewrite the equation in standard form, $ax^2 + bx + c = 0$. Determine a, b, and c.

	Equation	*Standard Form*	*a, b, c*
$-x = -1x$ →	$3x^2 + 4 = x$	$3x^2 - x + 4 = 0$	3, −1, 4
$x^2 = 1x^2$ →	$8x + x^2 = 2$	$x^2 + 8x - 2 = 0$	1, 8, −2
	$-4x = 3 - 5x^2$	$5x^2 - 4x - 3 = 0$	5, −4, −3
Write 0 for the missing term. →	$x = -3x^2$	$3x^2 + x + 0 = 0$	3, 1, 0

EXAMPLE 4 Find the solution set of $3x^2 + 1 = 6x$.

Write in standard form. →
$$3x^2 - 6x + 1 = 0$$

$a = 3$, $b = -6$, $c = 1$. Use the quadratic formula. →
$$x = \frac{-(-6) \pm \sqrt{(-6)^2 - 4(3)(1)}}{2(3)}$$

$-(-6) = 6$ →
$$= \frac{6 \pm \sqrt{36 - 12}}{6}$$
$$= \frac{6 \pm \sqrt{24}}{6}$$

Simplify: $\sqrt{24} = \sqrt{4 \cdot 6} = 2\sqrt{6}$ →
$$= \frac{6 \pm 2\sqrt{6}}{6}$$

Factor numerator and denominator. Simplify. →
$$= \frac{2(3 \pm \sqrt{6})}{2(3)}, \text{ or } \frac{3 \pm \sqrt{6}}{3}$$

Thus, $\left\{\frac{3 + \sqrt{6}}{3}, \frac{3 - \sqrt{6}}{3}\right\}$ is the solution set.

ORAL EXERCISES

Determine *a*, *b*, and *c* in the equation.

1. $5x^2 + 2x - 3 = 0$ — *$a = 5, b = 2, c = -3$*
2. $x^2 - 5x + 7 = 0$ — *$a = 1, b = -5, c = 7$*
3. $-2x^2 - x = 0$ — *$a = 2, b = 1, c = 0$*
4. $x^2 - 7 = 0$ — *$a = 1, b = 0, c = -7$*

Give the equation in standard form, $ax^2 + bx + c = 0$.

5. $x^2 - 8x = 7$ — *$x^2 - 8x - 7 = 0$*
6. $4 + 5x^2 = -x$ — *$5x^2 + x + 4 = 0$*
7. $-2 = x - 4x^2$ — *$4x^2 - x - 2 = 0$*
8. $3x = -2x^2$ — *$2x^2 + 3x + 0 = 0$*

EXERCISES

PART A

Simplify.

1. $\dfrac{-3 \pm \sqrt{8}}{2}$ — *$\frac{-3 \pm 2\sqrt{2}}{2}$*
2. $\dfrac{4 \pm \sqrt{20}}{6}$ — *$\frac{2 \pm \sqrt{5}}{3}$*
3. $\dfrac{-6 \pm \sqrt{32}}{2}$
4. $\dfrac{-2 \pm \sqrt{12}}{2}$
5. $\dfrac{6 \pm \sqrt{9 - 9}}{4}$ — *$\frac{3}{2}$*

Solve by using the quadratic formula.

6. $2x^2 + 5x - 2 = 0$
7. $4x^2 + x - 2 = 0$
8. $x^2 - 3x + 1 = 0$
9. $x^2 + 7x + 11 = 0$
10. $3x^2 + 7x + 3 = 0$
11. $2x^2 - x - 2 = 0$
12. $2x^2 - 4x + 1 = 0$
13. $x^2 + 4x - 1 = 0$
14. $2x^2 + 6x + 3 = 0$
15. $4x^2 - 12x + 7 = 0$
16. $4x^2 + 8x + 1 = 0$
17. $9x^2 - 6x - 7 = 0$
18. $x^2 + 5x = 3$
19. $5x^2 = 2 - x$
20. $2x^2 = 5 + x$
21. $3x^2 + 6x = 2$
22. $2x^2 = 6x - 1$
23. $5x^2 = 1 - 2x$
24. $x^2 - 6x + 9 = 0$ *3*
25. $9x^2 + 12x + 4 = 0$
26. $4x^2 + 25 = 20x$ *$\frac{5}{2}$*
27. $x^2 - 5x = 0$ *$5, 0$*
28. $3x^2 + 2x = 0$ *$0, -\frac{2}{3}$*
29. $2x^2 = 7x$ *$\frac{7}{2}, 0$*
30. $x^2 - 7 = 0$ *$\pm\sqrt{7}$*
31. $x^2 - 12 = 0$ *$\pm 2\sqrt{3}$*
32. $2x^2 - 11 = 0$ *$\pm \frac{\sqrt{22}}{2}$*

PART B

EXAMPLE Find the solution set of $\frac{3}{2}x^2 + \frac{1}{2}x - 3 = 0$.

Multiply each side by the LCD, 2. Now, coefficients are integers. — $3x^2 + 1x - 6 = 0$

Use the quadratic formula. — $x = \dfrac{-(1) \pm \sqrt{1^2 - 4(3)(-6)}}{2(3)}$

$1^2 - 4(3)(-6) = 1 + 72 = 73$ ⟶ $= \dfrac{-1 \pm \sqrt{73}}{6}$

Thus, $\left\{\dfrac{-1 + \sqrt{73}}{6}, \dfrac{-1 - \sqrt{73}}{6}\right\}$ is the solution set.

Find the solution set.

33. $\frac{1}{2}x^2 + \frac{1}{2}x - 2 = 0$
34. $\frac{1}{2}x^2 - \frac{3}{2}x + \frac{5}{6} = 0$
35. $\frac{1}{2}x^2 - x = \frac{3}{4}$
36. $\frac{1}{15}x^2 + \frac{5}{3} = \frac{2}{3}x$ *$\{5\}$*
37. $3 = 7x - x^2$
38. $5x = 1 - 2x^2$
39. $2x = 3 - 2x^2$
40. $2 = 6x - 3x^2$
41. $5x^2 - 2x + 8 = 3 + 4x^2 + 5x$ *$\left\{\frac{7 \pm \sqrt{29}}{2}\right\}$*
42. $2x^2 + 11x + 4 = 3x - x^2 + 1$ *$\left\{\frac{-4 \pm \sqrt{7}}{3}\right\}$*

PART C

Find the solution set. Use the quadratic formula.

43. $x^2 - (2\sqrt{2})x - 6 = 0$
44. $(\sqrt{6})x^2 - 4x - 2\sqrt{6} = 0$
45. $5x^2 + (2\sqrt{10})x - 1 = 0$

The Discriminant

OBJECTIVE

- To determine whether an equation like $3x^2 + 4x - 2 = 0$ has two, one, or no real number solutions

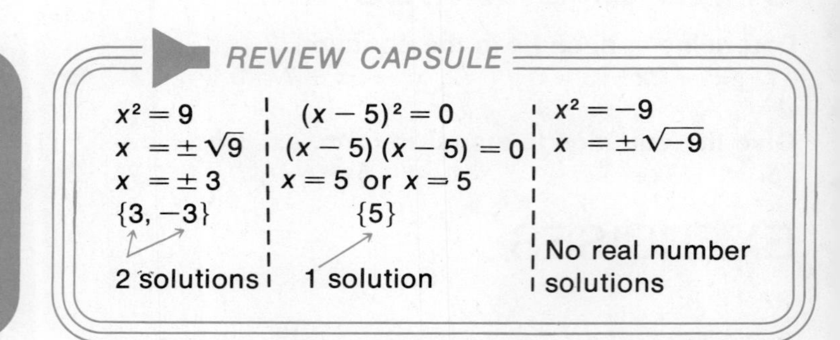

EXAMPLE 1 Solve $x^2 + 6x + 9 = 0$. Then tell the number of solutions.

Use the quadratic formula. ⟶

$$x = \frac{-(6) \pm \sqrt{6^2 - 4(1)(9)}}{2(1)}$$

$$= \frac{-6 \pm \sqrt{36 - 36}}{2}$$

$$= \frac{-6 \pm \sqrt{0}}{2} \quad \longleftarrow \quad b^2 - 4ac = 0$$

$$= \frac{-6 \pm 0}{2}, \text{ or } -3$$

There is only 1 solution. ⟶ **Thus,** the only solution is -3.

EXAMPLE 2 Solve $2x^2 + 3x + 4 = 0$. Then tell the number of solutions.

Use the quadratic formula.

$$x = \frac{-(3) \pm \sqrt{3^2 - 4(2)(4)}}{2(2)}$$

$$= \frac{-3 \pm \sqrt{9 - 32}}{4}$$

$$= \frac{-3 \pm \sqrt{-23}}{4} \quad \longleftarrow \quad b^2 - 4ac < 0$$

$\sqrt{-23}$ is neither positive, nor negative, nor 0: $(+)(+) = +$; $(-)(-) = +$; $0 \cdot 0 = 0$.

$\sqrt{-23}$ is not a real number. ⟶ **Thus,** there is no real number solution.

EXAMPLE 3 Solve $x^2 + 3x - 5 = 0$. Then tell the number of solutions.

Use the quadratic formula.

$$x = \frac{-(3) \pm \sqrt{3^2 - 4(1)(-5)}}{2(1)}$$

$$x = \frac{-3 \pm \sqrt{29}}{2} \quad \longleftarrow \quad b^2 - 4ac > 0$$

Thus, there are two solutions, $\frac{-3 + \sqrt{29}}{2}$ and $\frac{-3 - \sqrt{29}}{2}$.

See Example 3. →
See Example 1. →
See Example 2. →

If $ax^2 + bx + c = 0$, then $x = \dfrac{-b \pm \sqrt{b^2 - 4ac}}{2a}$

and, if $b^2 - 4ac > 0$, there are 2 solutions
if $b^2 - 4ac = 0$, there is 1 solution
if $b^2 - 4ac < 0$, there are no solutions.
$b^2 - 4ac$ is the *discriminant* of the equation.

EXAMPLE 4 Compute the discriminant. Then determine the number of solutions.

	$3x^2 + 2 = 4x$	$x^2 + 5x - 3 = 0$	$4x^2 - 20x + 25 = 0$
Write the equation in standard form. →	$3x^2 - 4x + 2 = 0$		
$b^2 - 4ac$ is the discriminant. →	$b^2 - 4ac$	$b^2 - 4ac$	$b^2 - 4ac$
Find the value of $b^2 - 4ac$.	$(-4)^2 - 4(3)(2)$	$5^2 - 4(1)(-3)$	$(-20)^2 - 4(4)(25)$
	$16 - 24$	$25 + 12$	$400 - 400$
	-8	37	0
	Thus,	**Thus,**	**Thus,**
	$b^2 - 4ac < 0$.	$b^2 - 4ac > 0$.	$b^2 - 4ac = 0$.
No solution means no real number solution. →	No solution	Two solutions	One solution

EXERCISES

PART A

Compute the discriminant. Determine the number of solutions.

1. $x^2 + 4x - 5 = 0$
2. $x^2 - 8x + 16 = 0$
3. $x^2 + 2x + 3 = 0$
4. $x^2 - 4x = 0$ *16; 2*
5. $9x^2 + 6x + 1 = 0$
6. $x^2 - 3x - 5 = 0$
7. $2x^2 - 3x + 2 = 0$
8. $2x^2 - 4x + 1 = 0$
9. $3x^2 + 7x + 3 = 0$ *13; 2*
10. $4x^2 - 12x + 9 = 0$ *0; 1*
11. $3x^2 - 9 = 0$ *108; 2*
12. $3x^2 + 5x + 3 = 0$ *−11; none*

PART B

Compute the discriminant. Determine the number of solutions.

13. $x^2 + 12 = 7x$ *1; 2*
14. $x^2 + 25 = 10x$ *0; 1*
15. $x^2 + 5x = 2$ *33; 2*
16. $4x^2 + x = 2$ *33; 2*
17. $x^2 + 4 = 3x$ *−7; none*
18. $3x^2 + 5x = 2$ *49; 2*
19. $3x^2 = 2 - 6x$ *60; 2*
20. $25x^2 = 20x - 4$ *0; 1*
21. $5x^2 = 3x + 3$ *69; 2*
22. $2x^2 = 5x + 12$ *121; 2*
23. $5x^2 = x - 1$ *−19; none*
24. $4x^2 = 12x - 7$ *32; 2*

PART C

Compute the discriminant. Determine the number of solutions.

25. $(x - 5)(x + 2) = 8$
26. $(x - 10)(x + 2) = -36$
27. $(x - 4)(x + 2) = -10$
28. $x^2 + (2\sqrt{2})x - 6 = 0$
29. $x^2 - (2\sqrt{3})x + 3 = 0$
30. $x^2 + (3\sqrt{2})x + 5 = 0$
31. $-2x^2 - 5x + 12 = 0$
32. $-4x^2 + 4x - 1 = 0$
33. $-3x^2 - 2x - 1 = 0$

Synthetic Division

There is an easy method for dividing a polynomial by a binomial of the form $x - a$.

The long way:

$$
\begin{array}{r|l}
 & 3x^3 - 4x^2 + 5x - 6 \text{ remainder: } -4 \\
x - 2 & 3x^4 - 10x^3 + 13x^2 - 16x + 8 \\
 & \underline{3x^4 - 6x^3} \\
 & -4x^3 + 13x^2 \\
 & \underline{-4x^3 + 8x^2} \\
 & 5x^2 - 16x \\
 & \underline{5x^2 - 10x} \\
 & -6x + 8 \\
 & \underline{-6x + 12} \\
 & -4
\end{array}
$$

The easy way is called *synthetic division.*

Divide $x - 2$ into $3x^4 - 10x^3 + 13x^2 - 16x + 8$

Coefficients of dividend

$x - a = x - 2$; so, $a = 2$
Bring down the 3.
2(3) gives the 6.
$-10 + (6)$ gives the -4.
Continue: multiply by 2 and then add.
Answer.

$$
\begin{array}{r|rrrrr}
2 & 3 & -10 & 13 & -16 & 8 \\
 & & 6 & -8 & 10 & -12 \\
\hline
 & 3 & -4 & 5 & -6 & -4 \\
 & 3x^3 & -4x^2 & +5x & -6 &
\end{array}
$$

remainder: -4

PROBLEM

Divide $2x^4 - 3x^3 - 5x^2 - 14x + 10$ by $x - 3$. Use synthetic division.

$$
\begin{array}{r|rrrrr}
3 & 2 & -3 & -5 & -14 & 10 \\
 & & 6 & 9 & 12 & -6 \\
\hline
 & 2 & 3 & 4 & -2 & 4
\end{array}
$$

Answer: $2x^3 + 3x^2 + 4x - 2$ remainder: 4

PROJECT

Divide the following. Use synthetic division.

1. $5x^4 - 3x^3 + 2x^2 - 2x - 40$ by $x - 2$
2. $2x^4 - 8x^3 + 5x^2 + 6x - 10$ by $x - 3$
3. $3x^3 + 14x^2 - 7x - 8$ by $x + 5$ [Hint: $x - a = x + 5$; so, $a = -5$]

Sum and Product of Solutions

OBJECTIVES

- To write a quadratic equation given its solution set
- To find the sum and the product of the solutions of an equation like $2x^2 + 3x - 5 = 0$

REVIEW CAPSULE

Solve by factoring.

$$x^2 + 2x - 15 = 0$$
$$(x + 5)(x - 3) = 0$$

$x + 5 = 0$ or $x - 3 = 0$

$x = -5$ $\qquad$ $x = 3$

Thus, the solutions are -5 and 3.

EXAMPLE 1 Write a quadratic equation whose solutions are -1 and 4.

Solutions are -1 and 4. → $x = -1$ $\qquad$ $x = 4$

$+1 \quad +1$ $\qquad$ $-4 \quad -4$

Make right side 0. → $x + 1 = 0$ $\qquad$ $x - 4 = 0$

If $a = 0$ or $b = 0$, then $(a)(b) = 0$. → $(x + 1)(x - 4) = 0$

Multiply. → $x^2 - 3x - 4 = 0$

Thus, $x^2 - 3x - 4 = 0$ is a quadratic equation whose solutions are -1 and 4.

EXAMPLE 2 Solve $3x^2 + 7x + 2 = 0$. Find the sum and product of the solutions.

Solve by factoring.

$$(3x + 1)(x + 2) = 0$$

$3x + 1 = 0$ or $x + 2 = 0$

$3x = -1$

$x = -\frac{1}{3}$ $\qquad$ $x = -2$

The solutions are $-\frac{1}{3}$ and -2.

Add and multiply to find the sum and product.

Sum of Solutions	*Product of Solutions*
$-\frac{1}{3} + (-2)$	$-\frac{1}{3}(-2)$
$-\frac{7}{3}$	$\frac{2}{3}$

From Example 2 → $3x^2 + 7x + 2 = 0$

Divide each side by 3 to make the coefficient of $x^2 = 1$. → $x^2 + \left(\frac{7}{3}\right)x + \frac{2}{3} = 0$

Coefficient of x is $\frac{7}{3}$, same as *−1 times* sum of solutions.

sum of solutions $= -\left(\frac{7}{3}\right)$ $\qquad$ $\frac{2}{3} =$ product of solutions

---

Standard form →

Divide each side by a. →

If $ax^2 + bx + c = 0$,

then $x^2 + \left(\frac{b}{a}\right)x + \frac{c}{a} = 0$.

$-\left(\frac{b}{a}\right)$ is the sum of the solutions.

$\frac{c}{a}$ is the product of the solutions.

EXAMPLE 3 Find the sum and product of the solutions without solving the equation.

Find $-\left(\frac{b}{a}\right)$ and $\frac{c}{a}$.

Divide each side by 4. →

$\frac{b}{a}$, the coefficient of x, is $-\frac{7}{4}$.

$$4x^2 - 7x + 3 = 0$$
$$x^2 - \frac{7}{4}x + \frac{3}{4} = 0$$
Sum: $-\left(\frac{b}{a}\right) = -\left(-\frac{7}{4}\right) = \frac{7}{4}$

Product: $\frac{c}{a} = \frac{3}{4}$

$$2x^2 + 3x - 5 = 0$$
$$x^2 + \frac{3}{2}x - \frac{5}{2} = 0$$
Sum: $-\left(\frac{b}{a}\right) = -\frac{3}{2}$

Product: $\frac{c}{a} = -\frac{5}{2}$

EXAMPLE 4 Write a quadratic equation whose solution set is $\{\frac{2}{3}, \frac{-3}{4}\}$.

Find $\frac{b}{a}$ and $\frac{c}{a}$.

Sum of Solutions

$\frac{2}{3} + \frac{-3}{4} = -\frac{1}{12}$

If $-\frac{b}{a} = -\frac{1}{12}$, then $\frac{b}{a} = \frac{1}{12}$. →

So, $-\left(\frac{b}{a}\right) = -\frac{1}{12}$ and $\frac{b}{a} = \frac{1}{12}$.

Product of Solutions

$\frac{2}{3} \cdot \frac{-3}{4} = \frac{-6}{12} = -\frac{1}{2}$

So, $\frac{c}{a} = -\frac{1}{2}$.

Divide each side of $ax^2 + bx + c = 0$ by a. →

$$x^2 + \frac{b}{a}x + \frac{c}{a} = 0$$

Substitute for $\frac{b}{a}$ and $\frac{c}{a}$. →

$$x^2 + \frac{1}{12}x - \frac{1}{2} = 0$$

Multiply each side by the LCD, 12. →

$$12x^2 + 1x - 6 = 0$$

Solution set is $\{\frac{2}{3}, \frac{-3}{4}\}$. →

Thus, $12x^2 + x - 6 = 0$ is the quadratic equation desired.

EXAMPLE 5 Write a quadratic equation whose solution set is $\{\frac{3}{4}\}$.

Write the only solution twice: $\{\frac{3}{4}\} = \{\frac{3}{4}, \frac{3}{4}\}$.

Solution set: $\{\frac{3}{4}, \frac{3}{4}\}$

Find $\frac{b}{a}$ and $\frac{c}{a}$.

Sum of Solutions

$\frac{3}{4} + \frac{3}{4} = \frac{6}{4} = \frac{3}{2}$

So, $-\left(\frac{b}{a}\right) = \frac{3}{2}$

If $-\frac{b}{a} = \frac{3}{2}$, then $\frac{b}{a} = -\frac{3}{2}$. →

$\frac{b}{a} = -\frac{3}{2}$.

Product of Solutions

$\frac{3}{4} \cdot \frac{3}{4} = \frac{9}{16}$

So, $\frac{c}{a} = \frac{9}{16}$.

$$x^2 + \frac{b}{a}x + \frac{c}{a} = 0$$

Substitute for $\frac{b}{a}$ and $\frac{c}{a}$. →

$$x^2 + \left(-\frac{3}{2}x\right) + \frac{9}{16} = 0$$

Multiply each side by the LCD, 16. →

$$16x^2 + (-24x) + 9 = 0$$

Solution set is $\{\frac{3}{4}\}$. →

Thus, $16x^2 - 24x + 9 = 0$ is the quadratic equation desired.

ORAL EXERCISES

Find the sum and product of the solutions without solving the equation.

1. $2x^2 - 7x + 4 = 0$
2. $3x^2 + 2x - 2 = 0$
3. $4x^2 - 8x - 3 = 0$
4. $x^2 + 8x + 12 = 0$
5. $5x^2 + 3x = 0$ *$-\frac{3}{5}$; 0*
6. $3x^2 - 6 = 0$ *0; −2*
7. $5x^2 - 2x - 3 = 0$
8. $4x^2 + 7x - 1 = 0$

EXERCISES

PART A

Write a quadratic equation which has the given solution set.

1. $\{2, 3\}$
2. $\{-5, 2\}$
3. $\{-2, 5\}$
4. $\{-3, -4\}$
5. $\{3\}$
6. $\{-4\}$
7. $\{5\}$
8. $\{-2\}$
9. $\{\frac{1}{2}, \frac{1}{4}\}$
10. $\{\frac{1}{3}, \frac{-2}{3}\}$
11. $\{\frac{3}{4}, \frac{-1}{4}\}$
12. $\{\frac{-2}{5}, \frac{-3}{5}\}$
13. $\{\frac{1}{2}\}$
14. $\{\frac{-1}{3}\}$
15. $\{\frac{2}{5}\}$
16. $\{\frac{-3}{4}\}$
17. $\{2, \frac{1}{2}\}$
18. $\{-1, \frac{1}{4}\}$
19. $\{2, \frac{-1}{3}\}$
20. $\{-3, \frac{-2}{3}\}$

$2x^2 - 5x + 2 = 0$ *$4x^2 + 3x - 1 = 0$* *$3x^2 - 5x - 2 = 0$* *$3x^2 + 11x + 6 = 0$*

PART B

EXAMPLE Write a quadratic equation whose solution set is $\left\{\frac{4 \pm 2\sqrt{3}}{3}\right\}$.

Find $\frac{b}{a}$ and $\frac{c}{a}$.

$4 + 2\sqrt{3}$
$4 - 2\sqrt{3}$
$16 + 8\sqrt{3}$
$\quad -8\sqrt{3} - 4 \cdot 3$
$16 + 0 - 12 \; = 4$

Sum of Solutions

$$\frac{4 + 2\sqrt{3}}{3} + \frac{4 - 2\sqrt{3}}{3} = \frac{8}{3}$$

$$-\left(\frac{b}{a}\right) = \frac{8}{3} \text{ and } \frac{b}{a} = -\frac{8}{3}$$

Product of Solutions

$$\frac{4 + 2\sqrt{3}}{3} \cdot \frac{4 - 2\sqrt{3}}{3} = \frac{16 - 12}{9} = \frac{4}{9}$$

$$\frac{c}{a} = \frac{4}{9}$$

$$x^2 + \left(-\frac{8}{3}\right)x + \frac{4}{9} = 0$$

Multiply each side by LCD, 9. ⟶

$$9x^2 - 24x + 4 = 0$$

Write a quadratic equation which has the given solution set.

21. $\{3 \pm \sqrt{5}\}$
22. $\{-2 \pm \sqrt{3}\}$
23. $\{2 \pm \sqrt{7}\}$
24. $\{-3 \pm \sqrt{10}\}$
25. $\left\{\frac{2 \pm \sqrt{3}}{2}\right\}$
26. $\left\{\frac{-3 \pm \sqrt{2}}{3}\right\}$
27. $\left\{\frac{1 \pm \sqrt{5}}{4}\right\}$
28. $\left\{\frac{-2 \pm \sqrt{7}}{3}\right\}$
29. $\{5 \pm 2\sqrt{3}\}$
30. $\{-5 \pm 3\sqrt{2}\}$
31. $\{3 \pm 2\sqrt{5}\}$
32. $\{-6 \pm 2\sqrt{10}\}$
33. $\left\{\frac{1 \pm 2\sqrt{2}}{2}\right\}$
34. $\left\{\frac{-3 \pm 2\sqrt{3}}{3}\right\}$
35. $\left\{\frac{3 \pm 2\sqrt{2}}{2}\right\}$
36. $\left\{\frac{-4 \pm 2\sqrt{5}}{3}\right\}$
37. $\{5, 10\}$
38. $\{-20, 15\}$
39. $\{-20, 25\}$
40. $\{-10, -20\}$
41. $\{2.5, .4\}$
42. $\{-.2, .5\}$
43. $\{-1.2, .5\}$
44. $\{-.7, -.3\}$

$10x^2 - 29x + 10 = 0$ *$10x^2 - 3x - 1 = 0$* *$10x^2 + 7x - 6 = 0$* *$100x^2 + 100x + 21 = 0$*

PART C

45. Add $\frac{-b \pm \sqrt{b^2 - 4ac}}{2a}$ to show that $-\frac{b}{a}$ is the sum of the solutions of $ax^2 + bx + c = 0$.

46. Multiply $\frac{-b \pm \sqrt{b^2 - 4ac}}{2a}$ to show that $\frac{c}{a}$ is the product of the solutions of $ax^2 + bx + c = 0$.

Word Problems and Quadratic Equations

OBJECTIVE

- To solve word problems leading to quadratic equations

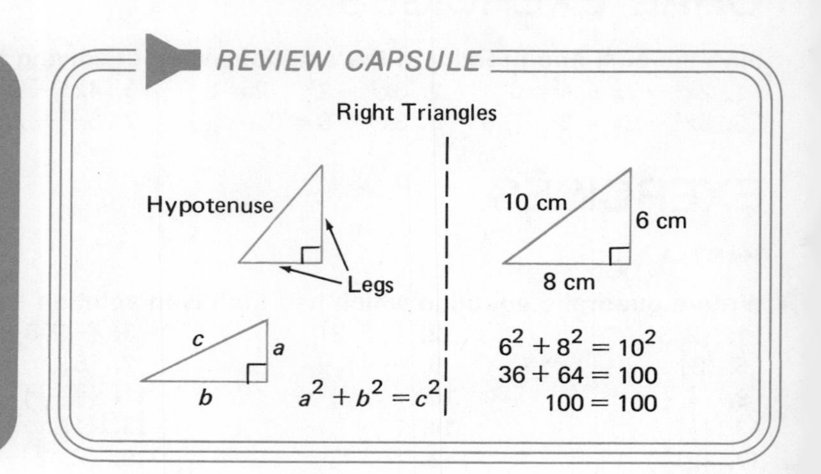

EXAMPLE 1 One leg of a right triangle is 7 m longer than the other leg. The hypotenuse is 13 m long. Find the length of each leg.

x is the length in meters (m). → Let x = length of shorter leg

7 more than x → $x + 7$ = length of longer leg

For a right triangle, $a^2 + b^2 = c^2$. → $x^2 + (x + 7)^2 = 13^2$

$(x + 7)^2 = x^2 + 14x + 49$ — $x^2 + (x^2 + 14x + 49) = 169$

Write in standard form. → $2x^2 + 14x - 120 = 0$

2 is a common factor. → $2(x^2 + 7x - 60) = 0$

Divide each side by 2. → $x^2 + 7x - 60 = 0$

Solve by factoring. — $(x + 12)(x - 5) = 0$

$x + 12 = 0$ or $x - 5 = 0$

$x = -12$ $\qquad$ $x = 5$

Cannot use -12 for x: length cannot be *negative*.

Thus, the shorter leg is 5 m long; the other is 12 m long.

EXAMPLE 2 One leg of a right triangle is twice the length of the other leg. The hypotenuse is 10 cm long. Find the length of each leg.

Let x = length of shorter leg

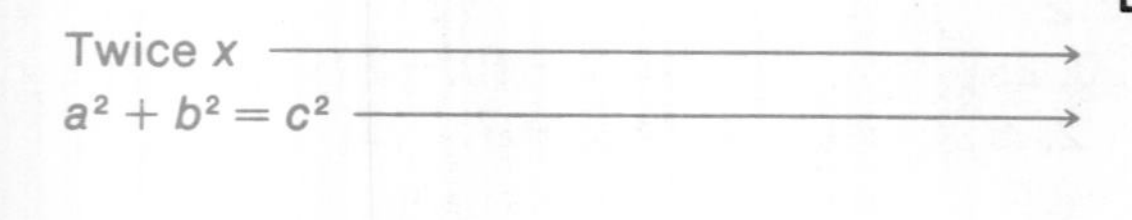

Twice x → $2x$ = length of longer leg

$a^2 + b^2 = c^2$ → $x^2 + (2x)^2 = 10^2$

$$x^2 + 4x^2 = 100$$

$$5x^2 = 100$$

$$x^2 = 20$$

$x = \sqrt{20}$ or $x = -\sqrt{20}$

$\sqrt{20} = \sqrt{4 \cdot 5}$, or $2\sqrt{5}$ → **Thus,** the shorter leg is $2\sqrt{5}$ cm long; the other is $4\sqrt{5}$ cm long.

EXAMPLE 3 Sixty light bulbs are placed in rows on a tray. The number of bulbs in each row is 4 less than the number of rows. Find the number of bulbs in one row.

Let r = number of rows

4 less than r → $r - 4$ = number of bulbs in one row

Number of rows × number in one row = total number of bulbs.

$$r(r-4) = 60$$
$$r^2 - 4r = 60$$

Write in standard form. →

$$r^2 - 4r - 60 = 0$$

Solve by factoring.

$$(r+6)(r-10) = 0$$
$$r + 6 = 0 \quad \text{or} \quad r - 10 = 0$$
$$r = -6 \qquad\qquad r = 10$$

Cannot use -6 for r, the number of rows.
Number of rows, r, is 10. → **Thus,** the number of bulbs in one row is $10 - 4$ or 6.

EXAMPLE 4 A rectangle's length is twice its width. A diagonal's length is 3 m more than the rectangle's length. Find the rectangle's length and width and the diagonal's length.

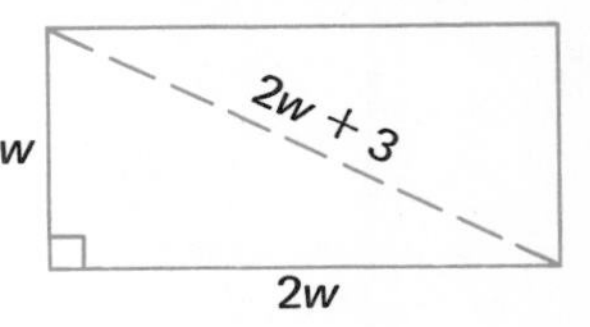

Let w = the width

Twice w → $2w$ = the length

3 more than $2w$ → $2w + 3$ = a diagonal's length

A right triangle is formed.
$a^2 + b^2 = c^2$

$$w^2 + (2w)^2 = (2w+3)^2$$
$$w^2 + 4w^2 = 4w^2 + 12w + 9$$

Write in standard form. →

$$w^2 - 12w - 9 = 0$$

Left side cannot be factored. Use formula. →

$$w = \frac{-(-12) \pm \sqrt{(-12)^2 - 4(1)(-9)}}{2(1)}$$

$(-12)^2 - 4 \cdot 1(-9) = 144 + 36 = 180$ →

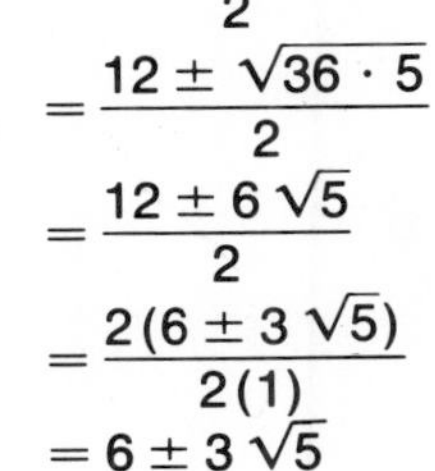

$$= \frac{12 \pm \sqrt{180}}{2}$$
$$= \frac{12 \pm \sqrt{36 \cdot 5}}{2}$$
$$= \frac{12 \pm 6\sqrt{5}}{2}$$

2 is a common factor of numerator and denominator. Write in simplest form. →

$$= \frac{2(6 \pm 3\sqrt{5})}{2(1)}$$

$6 - 3\sqrt{5}$ is a *negative* number. It cannot be used for w, the width. →

$$= 6 \pm 3\sqrt{5}$$

Rectangle's width, $w = (6 + 3\sqrt{5})$ m

$2(6 + 3\sqrt{5}) = 12 + 6\sqrt{5}$ → Rectangle's length, $2w = (12 + 6\sqrt{5})$ m

$(12 + 6\sqrt{5}) + 3 = 15 + 6\sqrt{5}$ → Diagonal's length, $2w + 3 = (15 + 6\sqrt{5})$ m

---

Examples 1, 2, 3, and 4 show cases where one solution of the equation cannot be a solution for the problem.

EXAMPLE 5 A second number is 5 less than twice the first. If the second number is multiplied by 3 more than the first, the result is 21. Find the numbers.

Let $f =$ first number

5 less than twice f → $2f - 5 =$ second number

Second number times 3 more than first is 21.

$$(2f - 5)(f + 3) = 21$$

Multiply.

$$2f^2 + f - 15 = 21$$

Add -21 to each side.

$$2f^2 + f - 36 = 0$$

Solve by factoring.

$$(2f + 9)(f - 4) = 0$$

$$2f + 9 = 0 \quad \text{or} \quad f - 4 = 0$$

$$f = -4\tfrac{1}{2} \qquad\qquad f = 4$$

The negative solution, $-4\frac{1}{2}$, *can* be used for f, the first number. → There will be *two* pairs of numbers.

$2f - 5 = 2(-4\frac{1}{2}) - 5 = -9 - 5 = -14$ →

First and second numbers,	First and second numbers,
$f = -4\frac{1}{2}$ and $2f - 5 = -14$	$f = 4$ and $2f - 5 = 3$

EXAMPLE 6 The length of a rectangle is 2 dm more than twice the width. The area is 5 dm^2. Find the length and the width.

w is the width in decimeters (dm).

Let $w =$ the width

2 more than twice w → $2w + 2 =$ the length

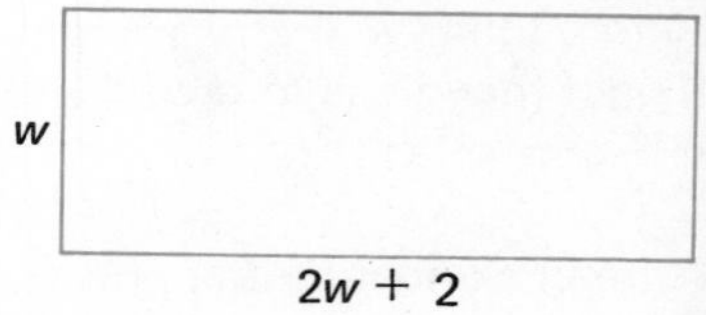

(width)(length) = area →

$$w(2w + 2) = 5$$

$$2w^2 + 2w = 5$$

Write in standard form. →

$$2w^2 + 2w - 5 = 0$$

Left side cannot be factored. Use quadratic formula. →

$$w = \frac{-(2) \pm \sqrt{2^2 - 4(2)(-5)}}{2(2)}$$

$$= \frac{-2 \pm \sqrt{44}}{4}$$

$$= \frac{-2 \pm 2\sqrt{11}}{4}$$

$\frac{-2 \pm 2\sqrt{11}}{4} = \frac{2(-1 \pm \sqrt{11})}{2(2)}$ Simplify. →

$$= \frac{-1 \pm \sqrt{11}}{2}$$

$-1 - \sqrt{11}$ is a negative number; width cannot be negative.

The width w is $\frac{-1 + \sqrt{11}}{2}$.

The length $2w + 2$ is equal to $2\left(\frac{-1 + \sqrt{11}}{2}\right) + 2$:

$$\overset{1}{\cancel{2}}\left(\frac{-1 + \sqrt{11}}{\underset{1}{\cancel{2}}}\right) + 2 = -1 + \sqrt{11} + 2 = 1 + \sqrt{11}.$$

Thus, the width is $\left(\frac{-1 + \sqrt{11}}{2}\right)$ dm and the length is $(1 + \sqrt{11})$ dm.

EXERCISES

PART A

Find the lengths of the two legs and the hypotenuse of the right triangle. Every length is given in meters (m).

1. 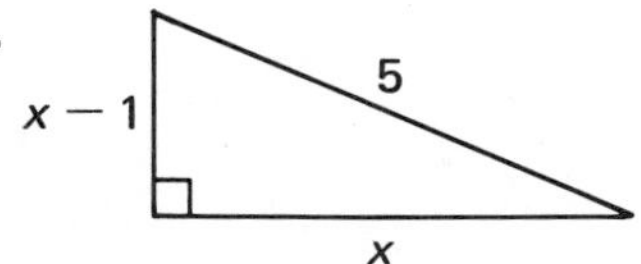

2. 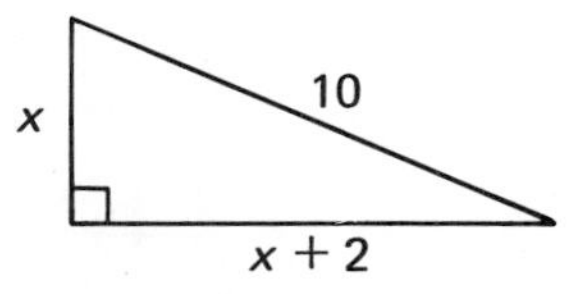

3. 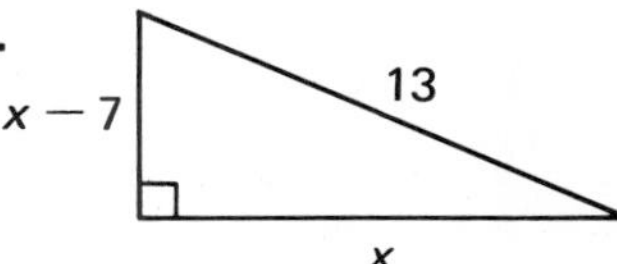

4. 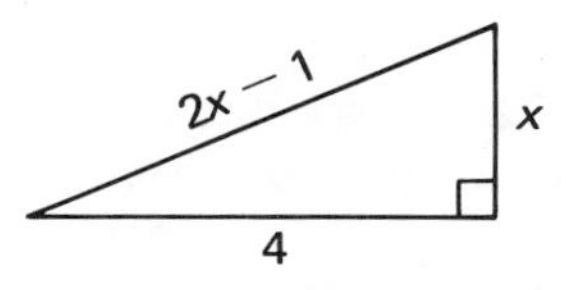

5. 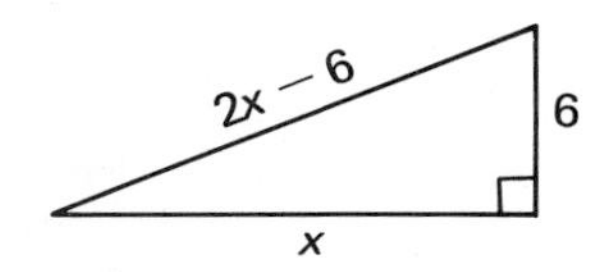

6. 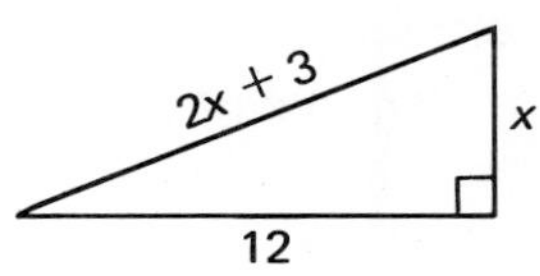

7. 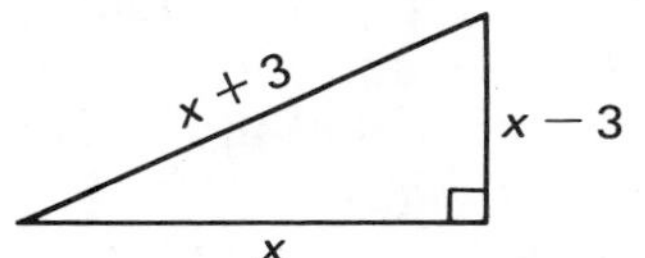

8. 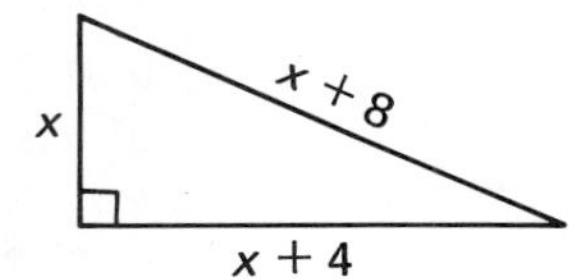

9. 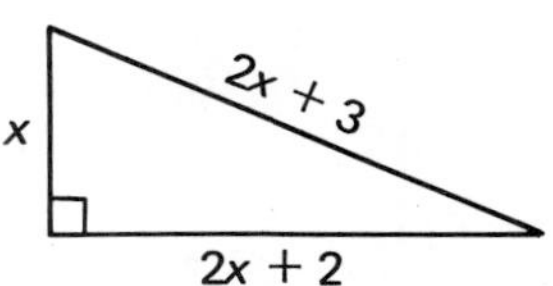

Solve these problems.

10. Forty chairs are placed in rows. The number of chairs in each row is 3 less than the number of rows. Find the number of chairs in one row. *5*

11. Forty-four cookies are placed in rows on a tray. The number of cookies in each row is 7 more than the number of rows. Find the number of cookies in one row. *11*

12. Forty-eight trees are planted in rows in an orchard. The number of rows is 8 less than the number of trees in each row. Find the number of rows. *4*

13. Thirty-two holes are drilled in rows on a metal block. The number of rows is 4 more than the number of holes in each row. Find the number of rows. *8*

14. A rectangle's length is 1 m more than its width. A diagonal's length is 1 m more than the rectangle's length. Find the diagonal's length. *5 m*

15. A rectangle's length is 4 cm more than 4 times its width. A diagonal's length is 1 cm more than the rectangle's length. Find the diagonal's length. *41 cm*

16. A second number is 4 less than the first. Their product is 21. Find the numbers. $7, 3; -3, -7$

17. A second number is 2 more than 3 times the first. Their product is 33. Find the numbers. $3, 11; -3\frac{2}{3}, -9$

18. A second number is 5 more than the first. If the second number is multiplied by 2 less than the first, the result is 8. Find the numbers. $3, 8; -6, -1$

19. A second number is 3 less than twice the first. If the second number is multiplied by 3 less than the first, the result is 5. Find the numbers. $4, 5; \frac{1}{2}, -2$

PART B

Find the lengths of the two legs and the hypotenuse of the right triangle. Every length is given in decimeters (dm).

20.

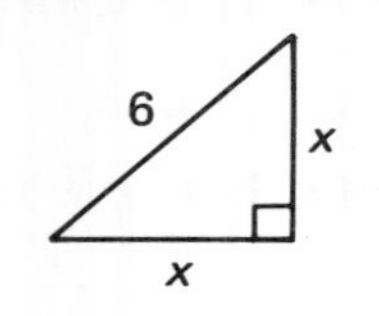

21.

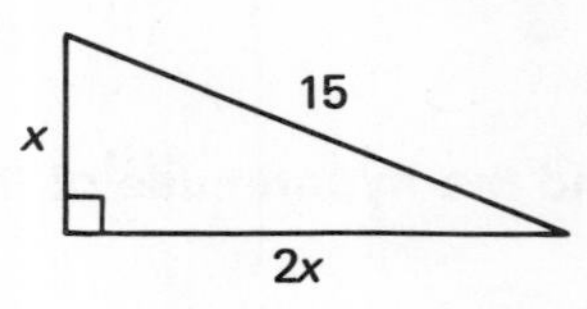

22.

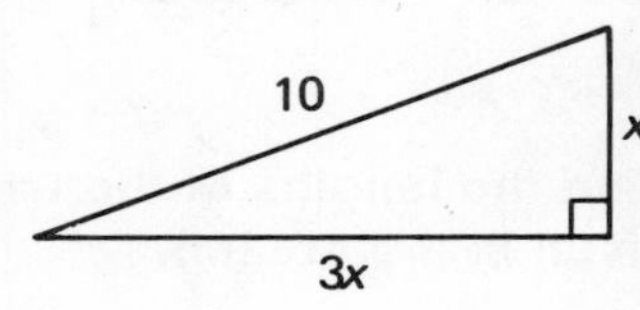

23.

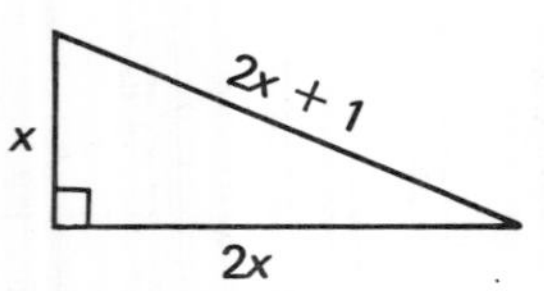

24.

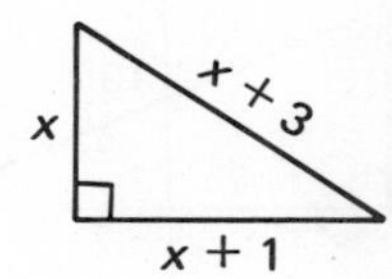

25.

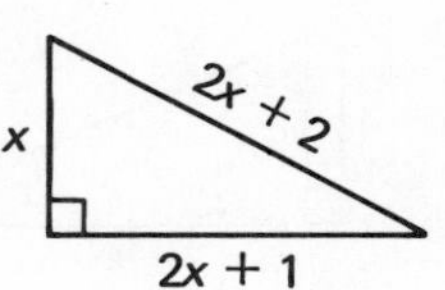

Solve these problems.

26. A diagonal of a square is 4 cm long. Find the width of the square. *$2\sqrt{2}$ cm*

27. A diagonal of a square is 8 mm long. Find the width of the square. *$4\sqrt{2}$ mm*

28. A diagonal of a square is 2 m longer than the length of a side. Find the lengths of a side and a diagonal of the square.
$2 + 2\sqrt{2}$ m; $4 + 2\sqrt{2}$ m

29. A diagonal of a square is 3 dm longer than the length of a side. Find the lengths of a side and a diagonal of the square.
$3 + 3\sqrt{2}$ dm; $6 + 3\sqrt{2}$ dm

30. The length of a rectangle is 2 cm more than the width. The area is 4 cm². Find the length and the width.
$w = -1 + \sqrt{5}$ cm, $l = 1 + \sqrt{5}$ cm

31. The length of a rectangle is 4 m more than the width. The area is 6 m². Find the length and the width.
$w = -2 + \sqrt{10}$ m, $l = 2 + \sqrt{10}$ m

32. A second number is 2 less than the first. Their product is 4. Find the numbers.
$1 + \sqrt{5}, -1 + \sqrt{5}; 1 - \sqrt{5}, -1 - \sqrt{5}$

33. A second number is 4 more than the first. Their product is 1. Find the numbers.
$-2 + \sqrt{5}, 2 + \sqrt{5}; -2 - \sqrt{5}, 2 - \sqrt{5}$

34. A second number is 2 less than twice the first. Their product is 3. Find the numbers.

35. A second number is 4 more than twice the first. Their product is 3. Find the numbers.

36. The length of a rectangle is 2 dm more than twice the width. The area is 3 dm². Find the length and the width.

37. The length of a rectangle is 4 m more than twice the width. The area is 7 m². Find the length and the width.

38. A rectangle's length is twice its width. A diagonal's length is 2 cm more than the rectangle's length. Find the rectangle's length and width and the diagonal's length.
$w = 4 + 2\sqrt{5}, l = 8 + 4\sqrt{5}, d = 10 + 4\sqrt{5}$

39. A rectangle's length is 3 times its width. A diagonal's length is 2 m more than the rectangle's length. Find the rectangle's length and width and the diagonal's length.
$w = 6 + 2\sqrt{10}, l = 18 + 6\sqrt{10}, d = 20 + 6\sqrt{10}$

40. The length of a rectangle is 2 dm more than the width. A diagonal's length is 1 dm more than the rectangle's length. Find the rectangle's length and width and the diagonal's length.
$w = 1 + \sqrt{6}, l = 3 + \sqrt{6}, d = 4 + \sqrt{6}$

41. A rectangle's length is 1 mm more than 3 times the width. A diagonal's length is 1 mm more than the rectangle's length. Find the rectangle's length and width and the diagonal's length.
$w = 3 + 2\sqrt{3}, l = 10 + 6\sqrt{3}, d = 11 + 6\sqrt{3}$

Radical Equations

OBJECTIVE

■ To solve equations like $1 + \sqrt{2x+1} = 2\sqrt{x}$

REVIEW CAPSULE

$(3\sqrt{2x-1})^2$	$(4-\sqrt{3x})^2$
$(3\sqrt{2x-1})(3\sqrt{2x-1})$	$(4-\sqrt{3x})(4-\sqrt{3x})$
$3 \cdot 3 \cdot (\sqrt{2x-1})(\sqrt{2x-1})$	$16 - 8\sqrt{3x} + 3x$
$9 \cdot (2x-1)$	

EXAMPLE 1 Find the solution set of $x - 2 = \sqrt{x}$.

An equation like $x - 2 = \sqrt{x}$ is a *radical equation.*

$$x - 2 = \sqrt{x} \quad \leftarrow \text{Original equation}$$

Square each side. ⟶

$$(x-2)^2 = (\sqrt{x})^2$$

$(\sqrt{x})^2 = \sqrt{x} \cdot \sqrt{x} = x$

$$x^2 - 4x + 4 = x \quad \leftarrow \text{Squared equation}$$

Add $-x$ to each side.

$$x^2 - 5x + 4 = 0$$

Solve by factoring.

$$(x-1)(x-4) = 0$$

$$x - 1 = 0 \text{ or } x - 4 = 0$$

$$x = 1 \qquad x = 4$$

The solution set of the *squared* equation is $\{1, 4\}$. Check in the *original* equation.

Substitute. ⟶

Let $x = 1$.

$x-2$	$\sqrt{x}$
$1-2$	$\sqrt{1}$
-1	1

$-1 \neq 1$

1 is *not* a solution of $x - 2 = \sqrt{x}$.

Let $x = 4$.

$x-2$	$\sqrt{x}$
$4-2$	$\sqrt{4}$
2	2

$2 = 2$

4 *is* a solution of $x - 2 = \sqrt{x}$.

One solution of the squared equation does not check in the original equation.

Thus, $\{4\}$ is the solution set of the original equation.

In Example 1, 1 is an extra solution. ⟶

Squaring each side of an equation may introduce an *extra solution.*

EXAMPLE 2 Solve and check $\sqrt{5x-7} = 2\sqrt{x}$.

Square each side. ⟶

$$(\sqrt{5x-7})^2 = (2\sqrt{x})^2$$

$(2\sqrt{x})^2 = 2\sqrt{x} \cdot 2\sqrt{x} = 4x$ ⟶

$$5x - 7 = 4x$$

$$x = 7$$

Check.

Let $x = 7$.

$\sqrt{5x-7}$	$2\sqrt{x}$
$\sqrt{5(7)-7}$	$2\sqrt{7}$
$\sqrt{28}$	
$2\sqrt{7}$	

7 checks in the original equation. ⟶ **Thus,** the solution is 7.

EXAMPLE 3 Find the solution set of $3 + \sqrt{3x+1} = x$.

$$3 + \sqrt{3x+1} = x$$

Get the $\sqrt{\ }$ alone. →

$$\sqrt{3x+1} = x - 3$$

Square each side. →

$$(\sqrt{3x+1})^2 = (x-3)^2$$

Solve the squared equation.

$$3x + 1 = x^2 - 6x + 9$$
$$0 = x^2 - 9x + 8$$
$$0 = (x-1)(x-8)$$
$$x - 1 = 0 \quad \text{or} \quad x - 8 = 0$$
$$x = 1 \qquad\qquad x = 8$$

Solutions of the *squared* equation are 1 and 8.
Check 1 and 8 in the *original* equation.

Let $x = 1$.

$3 + \sqrt{3x+1}$	x
$3 + \sqrt{3 \cdot 1 + 1}$	1
$3 + \sqrt{4}$	
$3 + 2$	
5	

$5 \neq 1$

1 is an extra solution. →

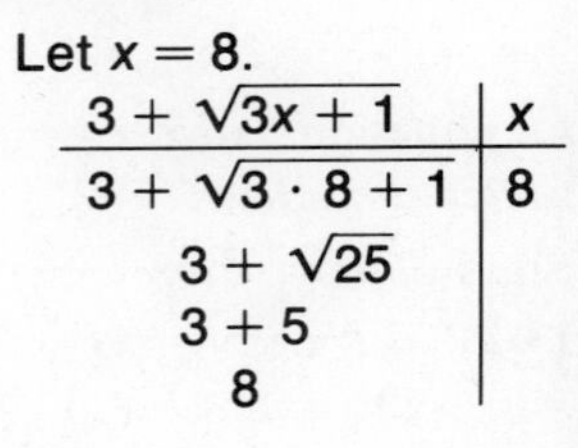

Let $x = 8$.

$3 + \sqrt{3x+1}$	x
$3 + \sqrt{3 \cdot 8 + 1}$	8
$3 + \sqrt{25}$	
$3 + 5$	
8	

1 does not check in the original equation. →

Thus, $\{8\}$ is the solution set.

EXAMPLE 4 Solve $\sqrt{5x+1} - \sqrt{3x-5} = 2$.

$$\sqrt{5x+1} - \sqrt{3x-5} = 2$$

Get one of the two square roots alone. →

$$\sqrt{5x+1} = 2 + \sqrt{3x-5}$$

Square each side. →

$$(\sqrt{5x+1})^2 = (2 + \sqrt{3x-5})^2$$

$(2 + \sqrt{3x-5})^2 = (2 + \sqrt{3x-5})(2 + \sqrt{3x-5})$ →

$$5x + 1 = 4 + 4\sqrt{3x-5} + (3x-5)$$
$$5x + 1 = 4\sqrt{3x-5} + 3x - 1$$

Get the $\sqrt{\ }$ alone. →

$$2x + 2 = 4\sqrt{3x-5}$$

Divide each side by 2. →

$$x + 1 = 2\sqrt{3x-5}$$

Square each side. →

$$(x+1)^2 = (2\sqrt{3x-5})^2$$
$$x^2 + 2x + 1 = 4(3x-5)$$

Solve the squared equation.

$$x^2 + 2x + 1 = 12x - 20$$
$$x^2 - 10x + 21 = 0$$
$$(x-3)(x-7) = 0$$
$$x - 3 = 0 \quad \text{or} \quad x - 7 = 0$$

$x = 3$ or $x = 7$ → Solutions of the *squared* equation are 3 and 7.

Check 3 and 7 in the *original* equation.

Let $x = 3$.

$\sqrt{5x+1} - \sqrt{3x-5}$	2
$\sqrt{5 \cdot 3 + 1} - \sqrt{3 \cdot 3 - 5}$	2
$\sqrt{16} - \sqrt{4}$	
$4 - 2$	
2	

Let $x = 7$.

$\sqrt{5x+1} - \sqrt{3x-5}$	2
$\sqrt{5 \cdot 7 + 1} - \sqrt{3 \cdot 7 - 5}$	2
$\sqrt{36} - \sqrt{16}$	
$6 - 4$	
2	

There is no extra solution.

Thus, 3 and 7 are the solutions.

ORAL EXERCISES

Check 3 and 5 in each equation. Then state whether the solution set is {3}, {5}, {3, 5}, or ϕ.

1. $\sqrt{x-1}=2$ *{5}*
2. $\sqrt{x+1}=-2$ ϕ
3. $\sqrt{x-1}=x-3$ *{5}*
4. $\sqrt{x+1}=x-1$ *{3}*
5. $\sqrt{2x-6}=x-3$ *{3, 5}*

EXERCISES

PART A

Solve.

1. $\sqrt{x}=5$ *25*
2. $\sqrt{x-2}=3$ *11*
3. $\sqrt{2x}=-4$ *no sol.*
4. $\sqrt{x+3}=6$ *33*
5. $\sqrt{2x}=x-4$ *8*
6. $\sqrt{x-3}=x-5$ *7*
7. $\sqrt{x+2}=x+2$
8. $\sqrt{2x-3}=x-3$ *6*
9. $\sqrt{3x}=\sqrt{x+6}$ *3*
10. $\sqrt{5x-3}=\sqrt{2x+3}$ *2*
11. $\sqrt{6x-1}=\sqrt{2x+1}$ *$\frac{1}{2}$*
12. $\sqrt{2x+5}=\sqrt{x+3}$ *−2*

Find the solution set.

13. $\sqrt{3x-2}-5=0$
14. $\sqrt{x-4}+2=0$ ϕ
15. $-2+\sqrt{4x+3}=0$
16. $3+\sqrt{2x+5}=0$ ϕ
17. $6+\sqrt{3x}=x$ *{12}*
18. $2+\sqrt{4x-3}=x$
19. $\sqrt{3x+10}=x+4$
20. $2+\sqrt{5x+6}=3x$
21. $2\sqrt{x}=\sqrt{3x+5}$ *{5}*
22. $\sqrt{12x+3}=3\sqrt{2x}$ *{$\frac{1}{2}$}*
23. $2\sqrt{x-1}=\sqrt{3x-1}$ *{3}*
24. $\sqrt{x+10}=3\sqrt{2x+3}$ *{−1}*

PART B

Find the solution set.

25. $4+2\sqrt{x-1}=x$ *{10}*
26. $2\sqrt{3x-2}=x+2$ *{2, 6}*
27. $3\sqrt{2x+3}=2x+5$ *{−1, $\frac{1}{2}$}*

Solve.

28. $\sqrt{4x-3}=2+\sqrt{2x-5}$ *3, 7*
29. $\sqrt{3x-5}=2-\sqrt{x-1}$ *2*
30. $\sqrt{3x}=2+\sqrt{x-2}$ *3*
31. $\sqrt{3x-3}-\sqrt{x}=1$ *4*
32. $\sqrt{2x+3}-\sqrt{x+1}=1$
33. $\sqrt{3x-5}-\sqrt{x-2}=1$ *2, 3*
34. $\sqrt{3x+1}-\sqrt{x-4}=3$ *5, 8*
35. $\sqrt{4x+1}-\sqrt{x-2}=3$ *2, 6*
36. $\sqrt{2x}+\sqrt{x-1}=3$ *2*
37. $\sqrt{4x}-\sqrt{2x+7}=1$ *9*
38. $\sqrt{2x+1}-\sqrt{x}=1$ *0, 4*
39. $\sqrt{x+15}-\sqrt{2x+7}=1$ *1*

PART C

EXAMPLE Solve and check $\sqrt[3]{7x-6}=4$.

A cube root: find *third* power of each side. ⟶

$$(\sqrt[3]{7x-6})^3=(4)^3$$
$$7x-6=64$$
$$7x=70$$
$$x=10$$

Check. Let $x=10$.

$\sqrt[3]{7x-6}$	4
$\sqrt[3]{7\cdot 10-6}$	4
$\sqrt[3]{64}$	
4	

Thus, 10 is the solution.

Solve.

40. $\sqrt[3]{2x}=2$
41. $\sqrt[3]{7x-1}=\sqrt[3]{5x+7}$
42. $\sqrt[4]{x}=3$
43. $\sqrt[4]{5x+1}=\sqrt[4]{6x-2}$
44. $\sqrt{\frac{x-3}{x+2}}=\frac{2}{3}$
45. $\sqrt{\frac{2}{x+3}}=\sqrt{\frac{3}{2x+2}}$
46. $\frac{\sqrt{x-2}}{x-2}=\frac{x-5}{\sqrt{x-2}}$
47. $\frac{x-3}{\sqrt{x-3}}=\frac{4\sqrt{x-3}}{3}$

Complex Numbers

OBJECTIVES

- To simplify expressions like $\sqrt{-9}$ and $2\sqrt{-5}$
- To solve equations like $x^2 = -12$
- To add numbers like $(3-2i) + (-5+6i)$

REVIEW CAPSULE

Real Numbers

Rational Numbers

-3 $\quad 7.4 \quad -2\frac{1}{3} \quad 0$

$.666\ldots \quad -\sqrt{16} \quad \sqrt[3]{-8}$

Irrational Numbers

$\sqrt[4]{5} \quad 3-\sqrt{2} \quad .4242242224\ldots$

EXAMPLE 1 Solve $x^2 = -1$ and find $\sqrt{-1}$.

$$x^2 = -1$$

If $x^2 = k$, then $x = \sqrt{k}$ or $x = -\sqrt{k}$. →

$$x = \sqrt{-1} \quad \text{or} \quad x = -\sqrt{-1}$$

$(+)^2 = +;\ (-)^2 = +;\ 0^2 = 0$ → There is no real number, rational or irrational, whose square is -1.

i is not on our number line. To solve this problem, mathematicians invented a number, i.

Definition of i. →
i is not a real number.

$$i = \sqrt{-1} \text{ and } i^2 = -1$$

EXAMPLE 2 Write $\sqrt{-4}$ and $\sqrt{-7}$ in terms of i.

$(2i)^2 = 2i \cdot 2i = 2 \cdot 2 \cdot i \cdot i$ →

$i^2 = -1$ →

$$(2i)^2 = 2 \cdot 2 \cdot i \cdot i = 4 \cdot i^2 = 4(-1) = -4$$

Thus, $\sqrt{-4} = 2i$.

$$(i\sqrt{7})^2 = i \cdot i \cdot \sqrt{7} \cdot \sqrt{7} = i^2 \cdot 7 = -1 \cdot 7 = -7$$

Thus, $\sqrt{-7} = i\sqrt{7}$.

Example 2 suggests this. →

$$\sqrt{-x} = i\sqrt{x}, \text{ for each positive number } x.$$

EXAMPLE 3 Simplify $\sqrt{-16}$ and $-\sqrt{-3}$.

$\sqrt{-x} = i\sqrt{x}$ →

$$\sqrt{-16} = i\sqrt{16} = i \cdot 4 = 4i \qquad -\sqrt{-3} = -i\sqrt{3}$$

EXAMPLE 4 Simplify $-2 + \sqrt{-25}$ and $3 - \sqrt{-2}$.

$\sqrt{-x} = i\sqrt{x}$ →

$-2 + \sqrt{-25} = -2 + i\sqrt{25} = -2 + 5i$ | $3 - \sqrt{-2} = 3 - i\sqrt{2}$

$-2 + 5i$ and $3 - i\sqrt{2}$ → Complex numbers

Definition of complex number →

> $a + bi$ is a *complex number* if a and b are real numbers.

EXAMPLE 5 Show that 7, $-3i$, 0, $-\sqrt{2}$, and $3 - 4i$ are complex numbers.

7	$-3i$	0	$-\sqrt{2}$	$3-4i$
↓	↓	↓	↓	↓
$7 + 0i$	$0 + (-3i)$	$0 + 0i$	$-\sqrt{2} + 0i$	$3 + (-4i)$

Each can be written as $a + bi$, where a and b are real numbers.

Example 5 suggests this. →

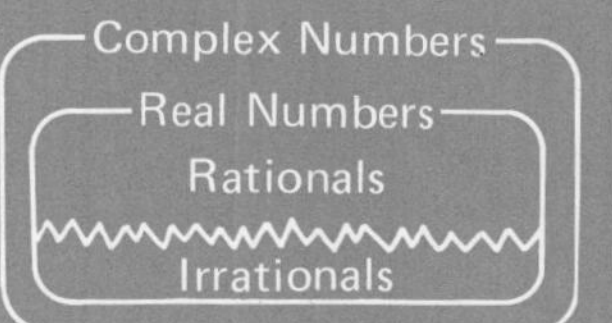

Every real number, rational or irrational, is also a complex number.

EXAMPLE 6 Solve and check $x^2 = -25$.

If $x^2 = k$, → $x^2 = -25$

then $x = \sqrt{k}$ or $x = -\sqrt{k}$. → $x = \sqrt{-25}$ or $x = -\sqrt{-25}$

$\sqrt{-a} = i\sqrt{a}$ → $= i\sqrt{25}$ | $= -i\sqrt{25}$

$i\sqrt{25} = i \cdot 5 = 5i$ → $= 5i$ | $= -5i$

Check. Substitute.

Let $x = 5i$.

x^2	-25
$(5i)^2$	-25
$5i \cdot 5i$	
$25i^2$	
$25(-1)$	
-25	

Let $x = -5i$.

x^2	-25
$(-5i)^2$	-25
$(-5i)(-5i)$	
$25i^2$	
$25(-1)$	
-25	

Thus, $5i$ and $-5i$ are the solutions.

EXAMPLE 7 Solve $x^2 = -8$. Solve $x^2 + 10 = 0$.

If $x^2 = k$, then $x = \sqrt{k}$ or $x = -\sqrt{k}$. →

$$x^2 = -8$$
$$x = \sqrt{-8} \quad \text{or} \quad x = -\sqrt{-8}$$

$\sqrt{8} = \sqrt{4 \cdot 2} = 2\sqrt{2}$ →

$$= i\sqrt{8} \qquad = -i\sqrt{8}$$
$$= i \cdot 2\sqrt{2} \qquad = -i \cdot 2\sqrt{2}$$

Write i next to $\sqrt{\ }$ sign. →

$$= 2i\sqrt{2} \qquad = -2i\sqrt{2}$$

$$x^2 + 10 = 0$$
$$x^2 = -10$$
$$x = \sqrt{-10} \quad \text{or} \quad x = -\sqrt{-10}$$
$$= i\sqrt{10} \qquad = -i\sqrt{10}$$

We will assume the commutative, associative, and distributive properties for complex numbers.

EXAMPLE 8 Simplify $(3 + 8i) + (5 + 2i)$.

Regroup. →
i is a common factor of $8i$ and $2i$.

$$\begin{aligned}(3 + 8i) + (5 + 2i) &= (3 + 5) + (8i + 2i) \\ &= (3 + 5) + (8 + 2)i \\ &= 8 + 10i\end{aligned}$$

Thus, $(3 + 8i) + (5 + 2i) = (3 + 5) + (8i + 2i)$, or $8 + 10i$.

Example 8 suggests this.
To add complex numbers, regroup and combine like terms.

$$(a + bi) + (c + di) = (a + c) + (b + d)i$$

EXAMPLE 9 Simplify $(-7 + \sqrt{-18}) + (4 - \sqrt{-2})$.

$$(-7 + \sqrt{-18}) + (4 - \sqrt{-2})$$

$\sqrt{18} = \sqrt{9 \cdot 2} = 3\sqrt{2}$; Write in terms of i. →

$$(-7 + 3i\sqrt{2}) + (4 - i\sqrt{2})$$

Regroup. →

$$(-7 + 4) + (3i\sqrt{2} - 1i\sqrt{2})$$

Combine like terms. →

$$-3 + 2i\sqrt{2}$$

EXAMPLE 10 Simplify $(8 + 3i) - (2 - 5i)$.

$-x = -1x$ →

$$(8 + 3i) - 1(2 - 5i)$$

Distribute the -1. →

$$(8 + 3i) + (-2 + 5i)$$

Combine like terms. →

$$6 + 8i$$

ORAL EXERCISES

Show that the number is a complex number by giving it in the form $a + bi$, where a and b are real numbers.

1. $2i$ $0 + 2i$
2. 6 $6 + 0i$
3. -4 $-4 + 0i$
4. i $0 + 1i$
5. $-4i$ $0 + (-4i)$
6. -2 $-2 + 0i$
7. 0 $0 + 0i$
8. -3.5 $-3.5 + 0i$
9. $\sqrt{-9}$ $0 + 3i$
10. $-\sqrt{9}$ $-3 + 0i$
11. $\sqrt{-2}$ $0 + i\sqrt{2}$
12. $\sqrt{5}$ $\sqrt{5} + 0i$

EXERCISES

PART A

Simplify.

1. $\sqrt{-25}$ $5i$
2. $-\sqrt{-16}$ $-4i$
3. $\sqrt{-64}$ $8i$
4. $\sqrt{-5}$ $i\sqrt{5}$
5. $-\sqrt{-2}$ $-i\sqrt{2}$
6. $\sqrt{-21}$ $i\sqrt{21}$
7. $\sqrt{-8}$ $2i\sqrt{2}$
8. $-\sqrt{-27}$ $-3i\sqrt{3}$
9. $\sqrt{-45}$ $3i\sqrt{5}$
10. $-3-\sqrt{-4}$ $-3-2i$
11. $4+\sqrt{-2}$ $4+i\sqrt{2}$
12. $7-\sqrt{-15}$ $7-i\sqrt{15}$
13. $3+\sqrt{-18}$ $3+3i\sqrt{2}$
14. $-2-\sqrt{-12}$ $-2-2i\sqrt{3}$
15. $6-\sqrt{-50}$ $6-5i\sqrt{2}$

Solve.

16. $x^2=-1$ $-i, i$
17. $x^2=-2$ $-i\sqrt{2}, i\sqrt{2}$
18. $x^2=-12$ $-2i\sqrt{3}, 2i\sqrt{3}$
19. $x^2=-20$ $-2i\sqrt{5}, 2i\sqrt{5}$
20. $x^2=-30$ $-i\sqrt{30}, i\sqrt{30}$
21. $x^2=-36$ $-6i, 6i$

Simplify.

22. $(2+5i)+(3+2i)$ $5+7i$
23. $(5+6i)-(3+2i)$ $2+4i$
24. $(-2-4i)+(6+7i)$ $4+3i$
25. $(3+i)+(-4-3i)$ $-1-2i$
26. $(4+5i)-(-2+3i)$ $6+2i$
27. $(4-i)+(-6+5i)$ $-2+4i$
28. $(-8+2i)+(6-2i)$ -2
29. $(-5+3i)-(2-i)$ $-7+4i$
30. $(2+i)-(-3-4i)$ $5+5i$

PART B

EXAMPLE Simplify $-5\sqrt{-8}$.

Write in terms of i. Simplify $\sqrt{8}$.
Rearrange factors.

$$-5\sqrt{-8}=-5\cdot i\sqrt{8}=-5\cdot i\cdot 2\sqrt{2}$$
$$=-5\cdot 2\cdot i\cdot\sqrt{2}, \text{ or } -10i\sqrt{2}$$

Simplify.

31. $3\sqrt{-4}$ $6i$
32. $-5\sqrt{-20}$ $-10i\sqrt{5}$
33. $4\sqrt{-12}$ $8i\sqrt{3}$
34. $5+2\sqrt{-9}$ $5+6i$
35. $-2-3\sqrt{-25}$ $-2-15i$
36. $2+5\sqrt{-8}$ $2+10i\sqrt{2}$
37. $7-3\sqrt{-12}$ $7-6i\sqrt{3}$
38. $-1+4\sqrt{-18}$ $-1+12i\sqrt{2}$
39. $-4-2\sqrt{-45}$ $-4-6i\sqrt{5}$

Solve.

40. $x^2+25=0$ $-5i, 5i$
41. $x^2+45=0$ $-3i\sqrt{5}, 3i\sqrt{5}$
42. $x^2+35=0$ $-i\sqrt{35}, i\sqrt{35}$

Simplify.

43. $(3+4\sqrt{-2})+(5+3\sqrt{-2})$ $8+7i\sqrt{2}$
44. $(2+\sqrt{-25})-(5+\sqrt{-4})$ $-3+3i$
45. $(6+5\sqrt{-3})-(1-2\sqrt{-3})$ $5+7i\sqrt{3}$
46. $(-4+6\sqrt{-5})+(7-2\sqrt{-5})$ $3+4i\sqrt{5}$
47. $(6+\sqrt{-18})+(-4+\sqrt{-2})$ $2+4i\sqrt{2}$
48. $(7-\sqrt{-8})-(3-\sqrt{-50})$ $4+3i\sqrt{2}$

PART C

Solve for $x+yi$. Then answer the question for complex numbers.

49. $(a+bi)+(x+yi)=a+bi$. Which complex number is the additive identity?

50. $(a+bi)+(x+yi)=0+0i$. What is the additive inverse of $a+bi$?

Powers of i

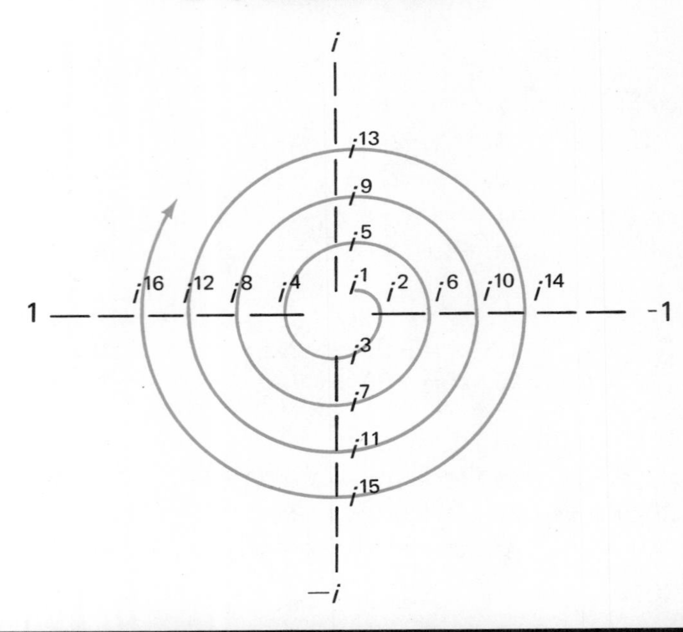

i, -1, $-i$, 1 occur in a cycle for consecutive powers of i.

PROBLEM

Simplify i^{21} and i^{39} as i, -1, $-i$, or 1.

$$\begin{aligned} i^{21} &= i^{20} \cdot i \\ &= (i^4)^5 \cdot i \\ &= (1)^5 \cdot i \\ &= i \end{aligned}$$

$$\begin{aligned} i^{39} &= i^{38} \cdot i \\ &= (i^2)^{19} \cdot i \\ &= (-1)^{19} \cdot i \\ &= -i \end{aligned}$$

PROJECT

Simplify each to i, -1, $-i$, or 1.

1. i^{19} **2.** i^{49} **3.** i^{30}

4. i^{52} **5.** i^{142} **6.** i^{59}

Multiplying and Dividing

OBJECTIVES

- To multiply numbers like $\sqrt{-5} \cdot \sqrt{-3}$ and $(2+5i)(3-4i)$
- To divide numbers like $(5-2i) \div (4+3i)$

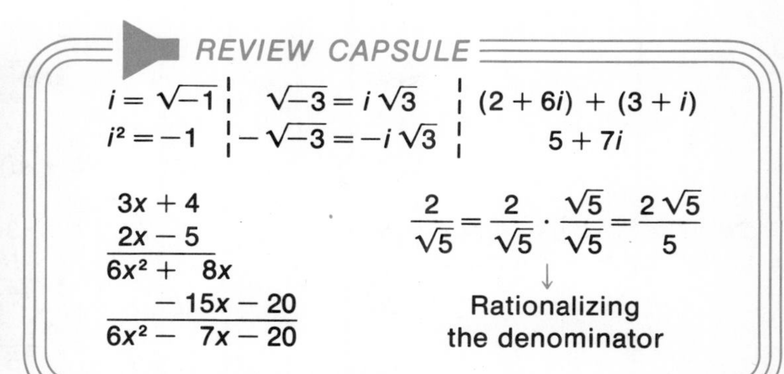

EXAMPLE 1 Multiply the complex numbers $5i$ and $3i$.

$5i \cdot 3i = 5 \cdot 3 \cdot i \cdot i = 15i^2$ →
$i^2 = -1$ →

$$5i \cdot 3i = 15i^2 = 15(-1), \text{ or } -15$$

Thus, $5i \cdot 3i = -15$.

EXAMPLE 2 Multiply $\sqrt{-2} \cdot \sqrt{-5}$.

First, write in terms of i. →
Rearrange factors.
$i^2 = -1$; $\sqrt{2} \cdot \sqrt{5} = \sqrt{10}$ →

$$\sqrt{-2} \cdot \sqrt{-5} = i\sqrt{2} \cdot i\sqrt{5}$$
$$= i^2 \cdot \sqrt{2} \cdot \sqrt{5}$$
$$= -1\sqrt{10}, \text{ or } -\sqrt{10}$$

Thus, $\sqrt{-2} \cdot \sqrt{-5} = -\sqrt{10}$.

EXAMPLE 3 Does $\sqrt{-4} \cdot \sqrt{-9} = \sqrt{(-4)(-9)}$?

Write $\sqrt{-4}$ and $\sqrt{-9}$ in terms of i. →
$6i^2 = 6(-1) = -6$ →
Results are *different*. →

$\sqrt{-4} \cdot \sqrt{-9}$	$\sqrt{(-4)(-9)}$
$i\sqrt{4} \cdot i\sqrt{9}$	$\sqrt{36}$
$2i \cdot 3i$	↓
$6i^2$	
-6	6

$$-6 \neq 6$$

Thus, $\sqrt{-4} \cdot \sqrt{-9} \neq \sqrt{(-4)(-9)}$.

SUMMARY

To multiply numbers like $\sqrt{-4} \cdot \sqrt{-9}$:
First, write in terms of i. → $2i \cdot 3i$
Then, multiply. → $6i^2$, **or** -6

EXAMPLE 4 Multiply $-\sqrt{-6} \cdot \sqrt{-2}$. Multiply $2\sqrt{5} \cdot \sqrt{-3}$.

First, write in terms of i.
$i^2 = -1; -i^2 = -(-1)$;
$\sqrt{12} = \sqrt{4 \cdot 3} = 2\sqrt{3}$ ⟶

$$-\sqrt{-6} \cdot \sqrt{-2} = -i\sqrt{6} \cdot i\sqrt{2} = -i^2\sqrt{12} = -(-1)(2\sqrt{3}) = 2\sqrt{3}$$

$$2\sqrt{5} \cdot \sqrt{-3} = 2\sqrt{5} \cdot i\sqrt{3} = 2i\sqrt{15}$$

EXAMPLE 5 Multiply $(3 + 4i)(5 + 2i)$.

Complex numbers $a + bi$ and $c + di$ are multiplied as two binomials.

$8i^2 = 8(-1) = -8$ ⟶

$$\begin{array}{ll} 3 & +4i \\ 5 & +2i \\ \hline 15 & +20i \\ & +\ 6i + 8i^2 \\ \hline 15 & +26i + 8i^2 \\ 15 & +26i - 8, \text{ or } 7 + 26i \end{array}$$

Thus, $(3 + 4i)(5 + 2i) = 7 + 26i$.

EXAMPLE 6 Multiply $(-5 + 3i\sqrt{2})(2 - i\sqrt{2})$.

Multiply as binomials.

$(-1i\sqrt{2})(3i\sqrt{2}) = (-1 \cdot 3)(i \cdot i)(\sqrt{2} \cdot \sqrt{2})$ ⟶

$-3i^2(2) = -3(-1)(2) = +6$
Combine: $-10 + 6 = -4$. ⟶

$$\begin{array}{l} -5 + \ 3i\sqrt{2} \\ \ \ 2 - \ 1i\sqrt{2} \\ \hline -10 + \ 6i\sqrt{2} \\ \quad + \ 5i\sqrt{2} - 3i^2(2) \\ \hline -10 + 11i\sqrt{2} + 6, \text{ or } -4 + 11i\sqrt{2} \end{array}$$

Thus, $(-5 + 3i\sqrt{2})(2 - i\sqrt{2}) = -4 + 11i\sqrt{2}$.

EXAMPLE 7 Multiply $(2 + 3i)(2 - 3i)$.

$2 + 3i$ and $2 - 3i$ are a pair of conjugates.

The product of conjugate complex numbers is a real number.

$$\begin{array}{l} 2 + 3i \\ 2 - 3i \\ \hline 4 + 6i \\ \quad - 6i - 9i^2 \\ \hline 4 + 0i + 9, \text{ or } 13 \end{array}$$

$\sqrt{7} + i\sqrt{2}$ and $\sqrt{7} - i\sqrt{2}$ are a pair of conjugates.

> $a + bi$ and $a - bi$ are a pair of *conjugate* complex numbers.

EXAMPLE 8 Find the conjugate of $-5 - 2i$. Multiply the pair of conjugates.

Conjugate of $a - bi$ is $a + bi$. ⟶ The conjugate of $-5 - 2i$ is $-5 + 2i$.

$-4i^2 = -4(-1) = +4$;
combine: $25 + 0 + 4 = 29$. ⟶

$$\begin{array}{l} -5 - \ 2i \\ -5 + \ 2i \\ \hline 25 + 10i \\ \quad - 10i - 4i^2 \\ \hline 25 + \ 0i + 4, \text{ or } 29 \end{array}$$

EXAMPLE 9 Divide $-5 \div (3i)$ and rationalize the denominator.

$x \div y = \frac{x}{y}$ →
$$-5 \div (3i) = \frac{-5}{3i}$$

Multiply by 1: $\frac{i}{i} = 1$. →
$$= \frac{-5}{3i} \cdot \frac{i}{i}$$

Multiply fractions. →
$$= \frac{-5i}{3i^2}$$

$i^2 = -1$ →
$$= \frac{-5i}{-3}$$

Simplify. →
$$= \frac{5i}{3}$$

Thus, $-5 \div (3i) = \frac{5i}{3}$.

EXAMPLE 10 Divide $(4 - 2i) \div (3 + 5i)$ and rationalize the denominator.

$x \div y = \frac{x}{y}$ →
$$(4 - 2i) \div (3 + 5i) = \frac{4 - 2i}{3 + 5i}$$

Multiply by 1: $\frac{3 - 5i}{3 - 5i} = 1$ →
$$= \frac{4 - 2i}{3 + 5i} \cdot \frac{3 - 5i}{3 - 5i}$$

$3 - 5i$ is the conjugate of $3 + 5i$.
$$= \frac{(4 - 2i)(3 - 5i)}{(3 + 5i)(3 - 5i)}$$

$(4 - 2i)(3 - 5i) = 2 - 26i$ →
$(3 + 5i)(3 - 5i) = 34$ →
$$= \frac{2 - 26i}{34}$$

Simplify: $\frac{2(1 - 13i)}{2 \cdot 17} = \frac{1 - 13i}{17}$ →
$$= \frac{1 - 13i}{17}$$

EXAMPLE 11 Multiply $(2i)^7$.

$(xy)^m = x^m \cdot y^m$ →
$$(2i)^7 = 2^7 \cdot i^7$$

Write i^2 as a factor as many times as possible.
$$= 2^7 \cdot i^2 \cdot i^2 \cdot i^2 \cdot i$$
$$\downarrow \quad \downarrow \quad \downarrow$$
$$= 2^7(-1)(-1)(-1)i$$

$2^7 = 128$ →
$$= -128i$$

Thus, $(2i)^7 = -128i$.

ORAL EXERCISES

Multiply.

1. $i \cdot i$ *-1* **2.** $i^2 \cdot i$ *$-i$* **3.** $i^2 \cdot i^2$ *1*
4. $i^2 \cdot i^2 \cdot i$ *i* **5.** $i^2 \cdot i^2 \cdot i^2$ *-1* **6.** $i^2 \cdot i^2 \cdot i^2 \cdot i^2 \cdot i$ *i*

Give the conjugate of the number.

7. $4 - 3i$ *$4 + 3i$* **8.** $1 + 5i$ *$1 - 5i$* **9.** $-7 - 4i$ *$-7 + 4i$* **10.** $-5 + 2i\sqrt{3}$ *$-5 - 2i\sqrt{3}$*

EXERCISES

PART A

Multiply.

1. $2i \cdot 3i$ -6
2. $-4i \cdot 5i$ 20
3. $i\sqrt{2} \cdot i\sqrt{3}$ $-\sqrt{6}$
4. $\sqrt{-3} \cdot \sqrt{-5}$ $-\sqrt{15}$
5. $(-\sqrt{-3})(\sqrt{-7})$ $\sqrt{21}$
6. $\sqrt{-2} \cdot \sqrt{-7}$ $-\sqrt{14}$
7. $3\sqrt{-2} \cdot 2\sqrt{-5}$ $-6\sqrt{10}$
8. $(-3\sqrt{-2})(4\sqrt{-11})$ $12\sqrt{22}$
9. $4\sqrt{-3} \cdot 5\sqrt{-10}$ $-20\sqrt{30}$
10. $2\sqrt{3} \cdot \sqrt{-2}$ $2i\sqrt{6}$
11. $3\sqrt{-2} \cdot \sqrt{5}$ $3i\sqrt{10}$
12. $3i(-2)$ $-6i$
13. $(3i)^2$ -9
14. $(-4i)^2$ -16
15. $(i\sqrt{5})^2$ -5
16. $(2+3i)(1+2i)$ $-4+7i$
17. $(3-4i)(2+2i)$ $14-2i$
18. $(2-5i)(1-2i)$ $-8-9i$
19. $(-3+4i)(-2-2i)$ $14-2i$
20. $(2+i)(5+i)$ $9+7i$
21. $(3-i)(-2+i)$ $-5+5i$
22. $(3+4i)^2$ $-7+24i$
23. $(5-2i)^2$ $21-20i$
24. $(-1+i)^2$ $-2i$
25. $(3+2i)(3-2i)$ 13
26. $(4-5i)(4+5i)$ 41
27. $(-3+4i)(-3-4i)$ 25
28. $(2-i)(2+i)$ 5
29. $(1+3i)(1-3i)$ 10
30. $(-5-6i)(-5+6i)$ 61

Divide and rationalize the denominator.

31. $7 \div (2i)$ $\frac{7i}{-2}$
32. $-3 \div (4i)$ $\frac{3i}{4}$
33. $5 \div (-3i)$ $\frac{5i}{3}$
34. $\frac{-2}{7i}$ $\frac{2i}{7}$
35. $\frac{1}{-6i}$ $\frac{i}{6}$
36. $\frac{8}{i}$ $-8i$

PART B

Divide and rationalize the denominator.

37. $(1+i) \div (3+2i)$ $\frac{5+i}{13}$
38. $(3-4i) \div (1-i)$ $\frac{7-i}{2}$
39. $(3+2i) \div (2-3i)$ i
40. $\frac{4}{2+3i}$ $\frac{8-12i}{13}$
41. $\frac{-3}{4-2i}$ $\frac{-6-3i}{10}$
42. $\frac{5i}{3+i}$ $\frac{1+3i}{2}$
43. $\frac{3+5i}{1+i}$ $4+i$
44. $\frac{3+2i}{4-3i}$ $\frac{6+17i}{25}$
45. $\frac{1-3i}{2-4i}$ $\frac{7-i}{10}$

Multiply.

46. $\sqrt{-2} \cdot \sqrt{-12}$ $-2\sqrt{6}$
47. $-3\sqrt{-2} \cdot \sqrt{-20}$ $6\sqrt{10}$
48. $2\sqrt{-2} \cdot \sqrt{-14}$ $-4\sqrt{7}$
49. $(2i)^3$ $-8i$
50. $(2i)^5$ $32i$
51. i^{11} $-i$
52. $(4+2i\sqrt{2})(3+2i\sqrt{2})$ $4+14i\sqrt{2}$
53. $(3-2i\sqrt{3})(3+2i\sqrt{3})$ 21
54. $(2+3i\sqrt{2})^2$ $-14+12i\sqrt{2}$
55. $(3-i\sqrt{5})^2$ $4-6i\sqrt{5}$
56. $(2-3i\sqrt{5})(1-2i\sqrt{5})$ $-28-7i\sqrt{5}$
57. $(-2+i\sqrt{7})(-2-i\sqrt{7})$ 11

PART C

Solve for $x+yi$. Then answer the question for complex numbers.

58. $(a+bi)(1+0i) = x+yi$. Which complex number is the multiplicative identity?
59. $(a+bi)\left(\frac{a-bi}{a^2+b^2}\right) = x+yi$. What is the multiplicative inverse of $a+bi$?

Let $a+bi$ and $a-bi$ represent conjugate complex numbers; a and b are real numbers.

60. Prove that the product of conjugate complex numbers is a real number.
61. Prove that the sum of conjugate complex numbers is a real number.

Complex Number Solutions of Equations

OBJECTIVES

- To solve equations like $3x^2 - 4x + 2 = 0$
- To determine if a quadratic equation has complex number solutions by using the discriminant

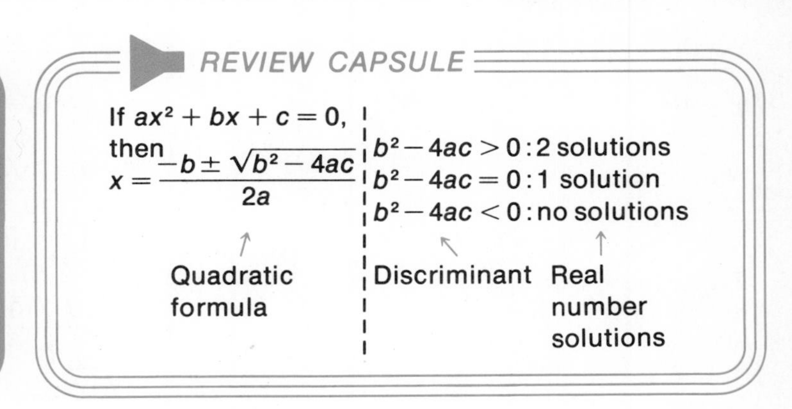

EXAMPLE 1 Find the solution set of $4x^2 + 3x + 1 = 0$.

Use the quadratic formula. $a = 4,\ b = 3,\ c = 1$ →

$$x = \frac{-(3) \pm \sqrt{3^2 - 4(4)(1)}}{2(4)}$$

$3^2 - 4(4)(1) = 9 - 16 = -7$

$$= \frac{-3 \pm \sqrt{-7}}{8} \quad \leftarrow b^2 - 4ac < 0$$

Write in terms of i. $\sqrt{-a} = i\sqrt{a}$

$$= \frac{-3 \pm i\sqrt{7}}{8} \quad \leftarrow \text{Complex numbers}$$

Solutions may be written together: $\left\{\frac{-3 \pm i\sqrt{7}}{8}\right\}$. →

Thus, $\left\{\frac{-3 + i\sqrt{7}}{8}, \frac{-3 - i\sqrt{7}}{8}\right\}$ is the solution set.

EXAMPLE 2 Solve $3x^2 + 2 = 2x$.

Write in standard form. Add $-2x$ to each side. → $3x^2 - 2x + 2 = 0$

$a = 3,\ b = -2,\ c = 2$ →

$$x = \frac{-(-2) \pm \sqrt{(-2)^2 - 4(3)(2)}}{2(3)}$$

$(-2)^2 - 4(3)(2) = 4 - 24 = -20$

$$= \frac{2 \pm \sqrt{-20}}{6} \quad \leftarrow b^2 - 4ac < 0$$

Write in terms of i.

$$= \frac{2 \pm i\sqrt{20}}{6} \quad \leftarrow \text{Complex numbers}$$

Simplify: $\sqrt{20} = \sqrt{4 \cdot 5} = 2\sqrt{5}$. →

$$= \frac{2 \pm 2i\sqrt{5}}{6}$$

2 is a common factor of numerator and denominator. →

$$= \frac{2(1 \pm i\sqrt{5})}{2(3)}$$

Simplify. →

$$= \frac{1 \pm i\sqrt{5}}{3}$$

Solutions may be written together.

Thus, $\frac{1 + i\sqrt{5}}{3}$ and $\frac{1 - i\sqrt{5}}{3}$ are the solutions.

EXAMPLE 3 Solve $5x^2 + 2 = 0$.

Write in standard form. Write $0x$ for missing term. → $5x^2 + 0x + 2 = 0$

$a = 5,\ b = 0,\ c = 2$ → $x = \frac{-(0) \pm \sqrt{0^2 - 4(5)(2)}}{2(5)}$

$0^2 - 4(5)(2) = 0 - 40 = -40$

$= \frac{0 \pm \sqrt{-40}}{10}$ ← $b^2 - 4ac < 0$

Write in terms of i.

$= \frac{0 \pm i\sqrt{40}}{10}$ ← Complex numbers

$\sqrt{40} = \sqrt{4 \cdot 10} = 2\sqrt{10}$

$= \frac{\pm 2i\sqrt{10}}{2(5)}$

2 is a common factor of numerator and denominator. Simplify.

$= \frac{\pm i\sqrt{10}}{5}$

Thus, $\frac{i\sqrt{10}}{5}$ and $\frac{-i\sqrt{10}}{5}$ are the solutions.

Examples 1, 2, and 3 suggest this. →

$b^2 - 4ac$ is the *discriminant* of the equation.

> If $ax^2 + bx + c = 0$, then $x = \frac{-b \pm \sqrt{b^2 - 4ac}}{2a}$
> and, if $b^2 - 4ac < 0$,
> there are *two complex number solutions* which are not real numbers.

EXAMPLE 4 Compute the discriminant of $3x^2 + 4x = 2$. Determine if the solutions are complex numbers which are not real numbers.

$3x^2 + 4x = 2$

Write in standard form. Add -2 to each side. → $3x^2 + 4x - 2 = 0$

$a = 3,\ b = 4,\ c = -2$ → $b^2 - 4ac = 4^2 - 4(3)(-2)$

$= 16 + 24$

Discriminant is *positive*. → $= 40$

$b^2 - 4ac > 0$

Thus, there are 2 solutions which are real numbers.

EXAMPLE 5 Compute the discriminant. Determine if the solutions are complex numbers which are not real numbers.

$4x^2 + 2x + 3 = 0$	$x^2 - 10x + 25 = 0$
$b^2 - 4ac = 2^2 - 4(4)(3)$	$b^2 - 4ac = (-10)^2 - 4(1)(25)$
$= -44$	$= 0$
Thus, there are 2 complex number solutions which are not real numbers.	**Thus,** there is 1 solution, which is a real number.

Discriminant is *negative*. → (left column)

ORAL EXERCISES

The discriminant of a quadratic equation is given. Tell if the solutions are complex numbers which are not real numbers.

1. -6 *yes* **2.** 12 *no* **3.** -18 *yes* **4.** 25 *no* **5.** 0 *no*

EXERCISES

PART A

Simplify.

1. $\frac{3 \pm \sqrt{-36}}{6}$ $\frac{1 \pm 2i}{2}$ **2.** $\frac{-4 \pm \sqrt{-20}}{4}$ $\frac{-2 \pm i\sqrt{5}}{2}$ **3.** $\frac{6 \pm \sqrt{-8}}{2}$ $3 \pm i\sqrt{2}$

Find the solution set.

4. $x^2 + 3x + 3 = 0$ **5.** $x^2 - 3x + 4 = 0$ $\left\{\frac{3 \pm i\sqrt{7}}{2}\right\}$ **6.** $x^2 + x + 2 = 0$ $\left\{\frac{-1 \pm i\sqrt{7}}{2}\right\}$

7. $x^2 + 4x + 5 = 0$ $\{-2 \pm i\}$ **8.** $x^2 - 2x + 2 = 0$ $\{1 \pm i\}$ **9.** $x^2 - 8x + 20 = 0$ $\{4 \pm 2i\}$

10. $2x^2 + 3x + 3 = 0$ **11.** $2x^2 - x + 1 = 0$ **12.** $2x^2 + 5x + 4 = 0$

13. $2x^2 + 2x + 1 = 0$ **14.** $2x^2 - 2x + 5 = 0$ **15.** $2x^2 - 2x + 13 = 0$

16. $3x^2 + 3x + 1 = 0$ **17.** $3x^2 - x + 1 = 0$ **18.** $3x^2 + x + 2 = 0$

19. $x^2 + 6x + 11 = 0$ $\{-3 \pm i\sqrt{2}\}$ **20.** $2x^2 - 4x + 3 = 0$ **21.** $2x^2 - 6x + 7 = 0$

22. $3x^2 + 4x + 2 = 0$ **23.** $3x^2 - 2x + 1 = 0$ **24.** $4x^2 + 2x + 1 = 0$

25. $2x^2 + 5 = 0$ $\left\{\pm\frac{i\sqrt{10}}{2}\right\}$ **26.** $4x^2 + 3 = 0$ $\left\{\pm\frac{i\sqrt{3}}{2}\right\}$ **27.** $5x^2 + 3 = 0$ $\left\{\pm\frac{i\sqrt{15}}{5}\right\}$

28. $2x^2 + 9 = 0$ $\left\{\pm\frac{3i\sqrt{2}}{2}\right\}$ **29.** $3x^2 + 4 = 0$ $\left\{\pm\frac{2i\sqrt{3}}{3}\right\}$ **30.** $4x^2 + 1 = 0$ $\left\{\pm\frac{i}{2}\right\}$

PART B

Compute the discriminant. Determine if the solutions are complex numbers which are not real numbers.

31. $2x^2 - 5x + 4 = 0$ *-7; yes* **32.** $2x^2 - 5x + 2 = 0$ *9; no* **33.** $2x^2 - 5x + 6 = 0$ *-23; yes*

34. $4x^2 + 2 = 4x$ *-16; yes* **35.** $4x^2 = 4x + 1$ *32; no* **36.** $4x^2 = 4x - 1$ *0; no*

Solve.

37. $x^2 + 13 = 7x$ $\frac{7 \pm i\sqrt{3}}{2}$ **38.** $2 + 3x^2 = 3x$ $\frac{3 \pm i\sqrt{15}}{6}$ **39.** $x - 1 = 4x^2$ $\frac{1 \pm i\sqrt{15}}{8}$

40. $2x^2 = 6x - 5$ $\frac{3 \pm i}{2}$ **41.** $10 + x^2 = 6x$ $3 \pm i$ **42.** $4x - 3 = 4x^2$ $\frac{1 \pm i\sqrt{2}}{2}$

PART C

Find the solution set.

43. $x^2 + 2ix + 3 = 0$ **44.** $x^2 - 6ix + 8 = 0$ **45.** $2ix^2 + 5x - 2i = 0$

46. $\sqrt{x} = 3i$ **47.** $\sqrt{x - 1} = 2i$ **48.** $\sqrt{x + 8} = 3 - i$

49. $x^3 = 8$ [Hint: $x^3 - 8 = 0$; factor $x^3 - 8$ (see page 67); set each factor equal to zero.]

50. $x^3 = 27$ **51.** $x^3 = -64$ **52.** $x^4 = 16$

53. Show that $(-1 + i\sqrt{3})^3 = 8$. **54.** Show that $\left(\frac{-3 - 3i\sqrt{3}}{2}\right)^3 = 27$.

Chapter Seven Review

Find the solution set. [p. 174, 176, 179, 193]

1. $2x^2 = 24$
2. $(n+2)^2 = 36$
3. $3x^2 + 2x = 2$
4. $x^2 = 4x + 2$
5. $\sqrt{5x-3} = 2\sqrt{x}$
6. $\sqrt{x+5} = x - 1$

7. Solve $x^2 - 3x = -1$ by completing the square. [p. 176]

Compute the discriminant. Determine the number of solutions. [p. 182]

8. $2x^2 + 5x - 2 = 0$
9. $9x^2 = 12x - 4$

10. Find the sum and product of the solutions of $4x^2 + 3x - 3 = 0$ without solving the equation. [p. 185]

11. Write a quadratic equation whose solution set is $\{-3 + \sqrt{7}, -3 - \sqrt{7}\}$. [p. 185]

Find the lengths of the two legs and hypotenuse of the right triangle. [p. 188]

12.

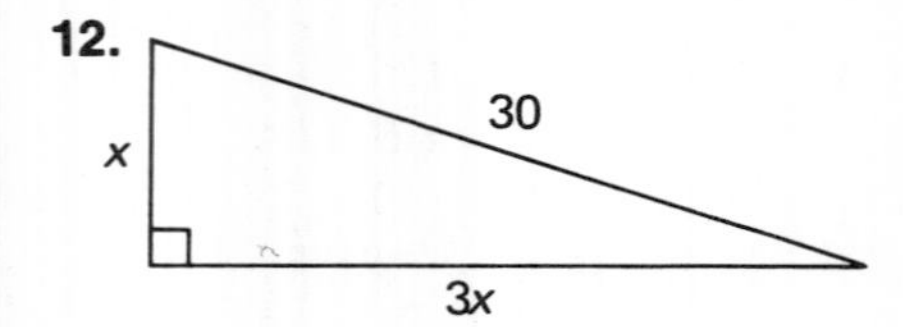

13.

x
x + 4
x + 1

14. A diagonal of a square is 4 cm longer than the length of a side. Find the lengths of a side and a diagonal of the square. [p. 188]

15. Fifty-four cans are placed in rows on a shelf. The number of cans in each row is 3 less than the number of rows. Find the number of cans in one row. [p. 188]

16. A second number is 1 less than 3 times the first. If the second number is multiplied by 2 more than the first, the result is 10. Find the numbers. [p. 188]

17. The length of a rectangle is 5 m more than twice the width. The area is 33 m^2. Find the length and the width. [p. 171]

Simplify. [p. 196, 201]

18. $-4\sqrt{-7}$
19. $6 + 5\sqrt{-12}$
20. $-4 - 3\sqrt{-8}$
21. $(6 - 3i) + (-2 + 5i)$
22. $(3 + 4i) - (5 - i)$
23. $(2i)^9$

Multiply. [p. 201]

24. $5i \cdot 4i$
25. $-2\sqrt{-5} \cdot 4\sqrt{-3}$
26. $(i\sqrt{3})^2$
27. $(3 - 2i)(5 + 3i)$
28. $(4 + 3i)(4 - 3i)$
29. $(3 - 5i)^2$

Divide and rationalize the denominator. [p. 201]

30. $\frac{-2}{3i}$
31. $\frac{3 + 2i}{-2 - i}$

Solve. Which solutions are complex numbers which are not real numbers? [p. 205]

32. $x^2 = -16$
33. $x^2 + 36 = 0$
34. $x^2 + 3x + 4 = 0$
35. $x^2 + 4x + 6 = 0$
36. $2x^2 - 3x + 2 = 0$

Chapter Seven Test

1. Solve $x^2 - 8x = 5$ by completing the square.

Find the solution set.

2. $5x^2 = 15$
3. $(n-3)^2 = 25$
4. $2x^2 + 3x - 3 = 0$
5. $x^2 - 4x + 1 = 0$
6. $\sqrt{2x-2} = 2\sqrt{x-3}$
7. $\sqrt{2x-1} = x - 2$

Compute the discriminant. Determine the number of solutions.

8. $x^2 - 10x + 25 = 0$
9. $2x^2 + 3x = 4$

10. Find the sum and product of the solutions of $2x^2 + 6x - 1 = 0$ without solving the equation.
11. Write a quadratic equation whose solution set is $\{3, 4\}$.

Find the lengths of the two legs and hypotenuse of the right triangle.

12.
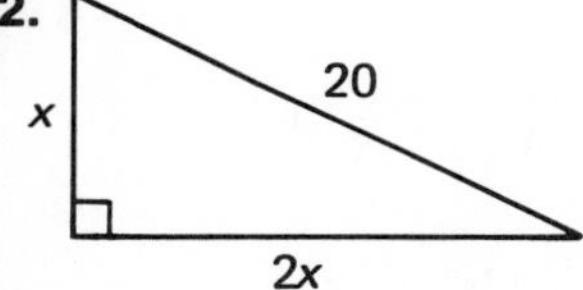

13.
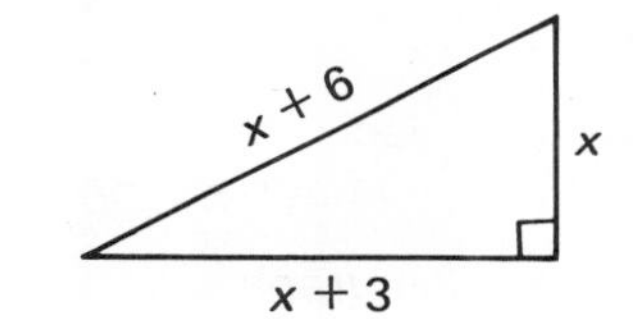

14. The length of a rectangle is 1 dm more than twice the width. The area is 21 dm². Find the length and the width.
15. Thirty-six chairs are placed in rows. The number of chairs in each row is 5 less than the number of rows. Find the number of chairs in one row.
16. A diagonal of a square is 1 m longer than the length of a side. Find the lengths of a diagonal and a side of the square.
17. A second number is 6 less than the first. The second number multiplied by 3 more than the first is 10. Find the numbers.

Simplify.

18. $\sqrt{-6}$
19. $-3\sqrt{-16}$
20. $5 - 3\sqrt{-20}$
21. $(5-4i) + (1+i)$
22. $(7+2i) - (3-4i)$

Multiply.

23. $3i \cdot 6i$
24. $-\sqrt{-2} \cdot \sqrt{-5}$
25. $2\sqrt{-3} \cdot \sqrt{-12}$
26. $(i\sqrt{6})^2$
27. $(2+5i)(1+2i)$
28. $(3+i\sqrt{2})(3-i\sqrt{2})$
29. $(5-2i)^2$
30. $(2i)^5$

Divide and rationalize the denominator.

31. $\frac{5}{2i}$
32. $\frac{1+2i}{2-3i}$

Solve. Which solutions are complex numbers which are not real numbers?

33. $x^2 = -6$
34. $x^2 + 9 = 0$
35. $2x^2 + x + 1 = 0$
36. $x^2 - 6x + 10 = 0$

LINEAR SENTENCES

Graphs of Complex Numbers

To graph a complex number $a + bi$:

(1) Plot the real value a along the horizontal (real) axis.

(2) Plot the value of b along the vertical (imaginary) axis.

(3) Find the point (a, b).

The point (a, b) represents the complex number $a + bi$.

The ray from the origin also represents the complex number $a + bi$.

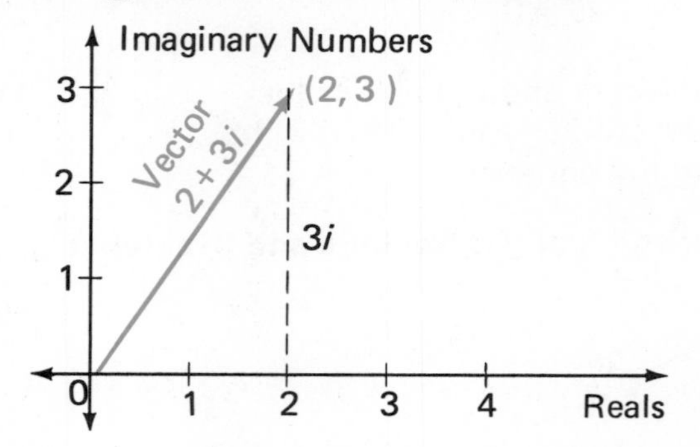

To add $(2 + 3i) + (1 - 2i)$ graphically.

(1) Draw the vectors representing $2 + 3i$ and $1 - 2i$ in the plane.

(2) Complete a parallelogram and draw its diagonal.

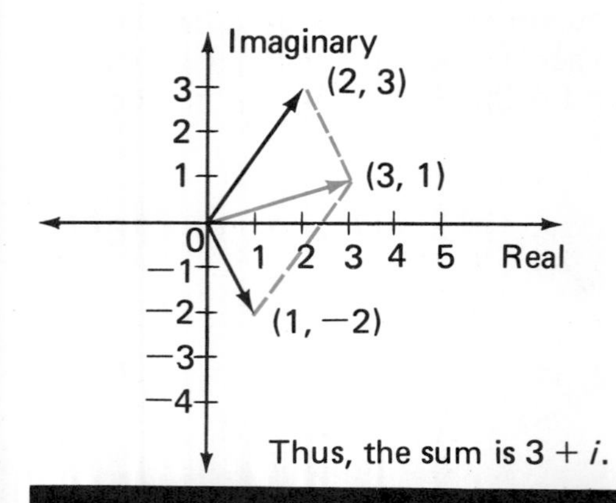

Thus, the sum is $3 + i$.

To subtract $(2 - 3i) - (3 + 2i)$ graphically.

(1) Draw the vectors $2 - 3i$ and $3 + 2i$.

(2) Draw the vector $-(3 + 2i) = -3 - 2i$.

(3) Complete a parallelogram and draw its diagonal.

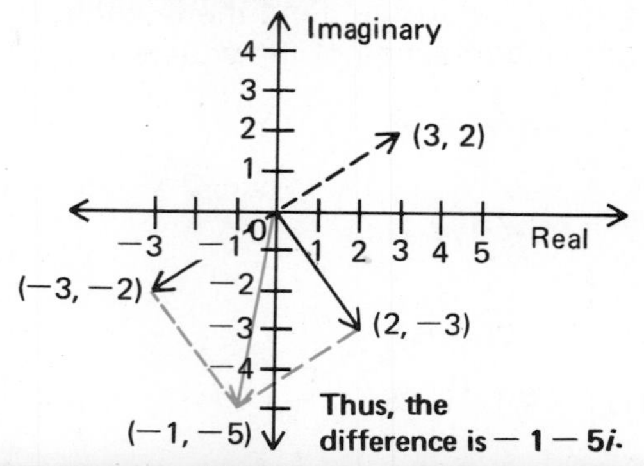

Thus, the difference is $-1 - 5i$.

PROJECT

Add graphically.

1. $(1 + 2i) + (4 - 3i)$

2. $(6 - i) + (3 - 2i)$

Subtract graphically.

3. $(3 + 5i) - (2 - i)$

4. $(-2 + i) - (-2 - i)$

Vertical and Horizontal Lines

OBJECTIVES

- To locate points in the Cartesian coordinate plane
- To find the length between two points on a horizontal or vertical line

REVIEW CAPSULE

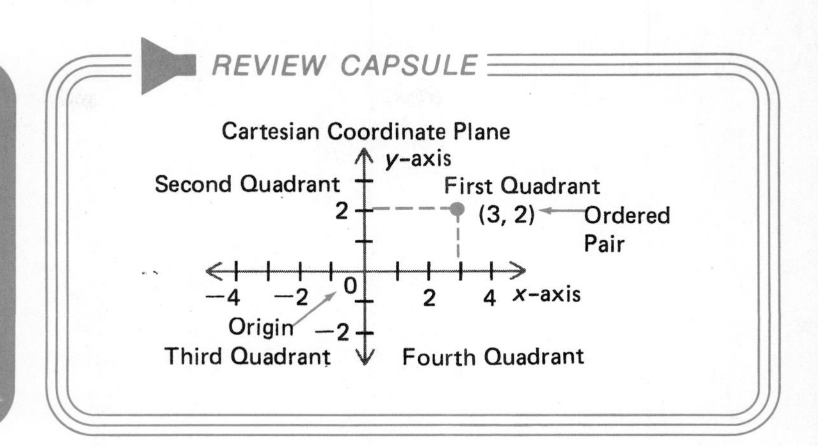

EXAMPLE 1 Plot the points (5, 2), (0, 3), (−2, 4), (−3, −2), (2, −3), (3, 0).

Ordered pair
(5, 2)
↓ ↓
(x, y)
(first coordinate, second coordinate)
↓ ↓
abscissa ordinate

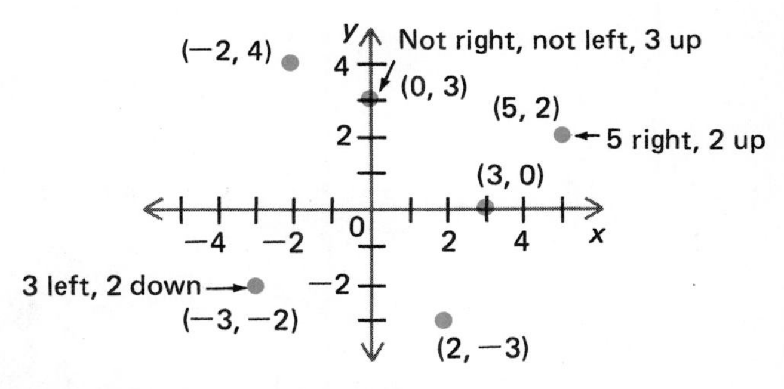

EXAMPLE 2 Give the coordinates of each point.

$A(3, 4)$ means point A with coordinates (3, 4).

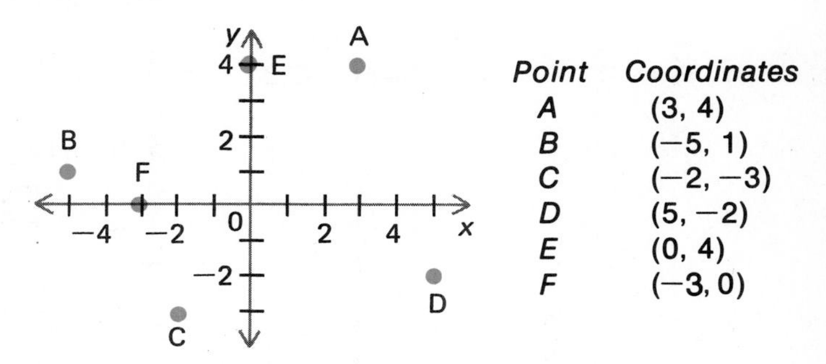

Point	*Coordinates*
A	(3, 4)
B	(−5, 1)
C	(−2, −3)
D	(5, −2)
E	(0, 4)
F	(−3, 0)

EXAMPLE 3 In which quadrant does the point lie?

$A(3, -2)$: right 3, down 2 | $B(-4, -5)$: left 4, down 5

Thus, A is in the 4th quadrant; B is in the 3rd quadrant.

EXAMPLE 4 Locate the points (3, 4), (0, 6), (−3, −4). Find the distance of each point from the x-axis and y-axis.

The first coordinate tells the distance and direction from the y-axis

The second coordinate tells the distance and the direction from the x-axis.

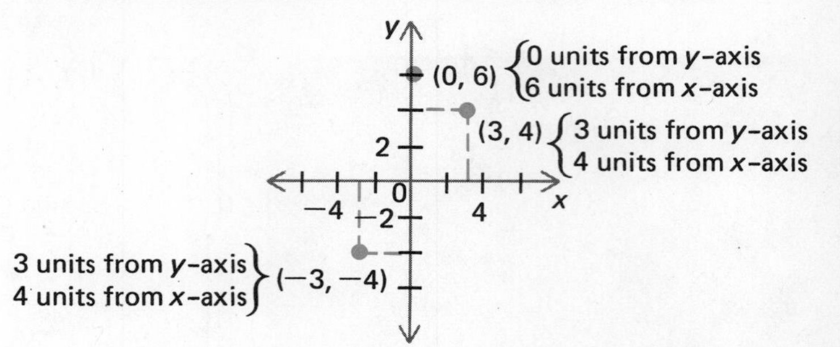

EXAMPLE 5 Find the length of line segment $\overline{AB}$.

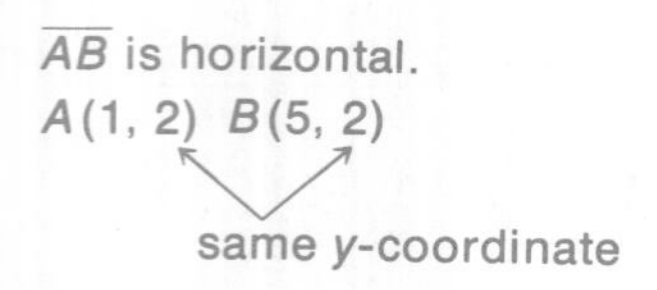

Length of $\overline{AB}$ is the *absolute value* of the difference of x-coordinates.

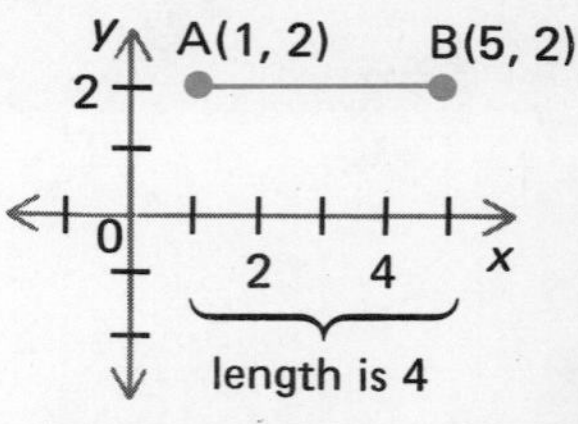

|x-coord. B − x-coord. A|
$|5-1|$
$|4|$

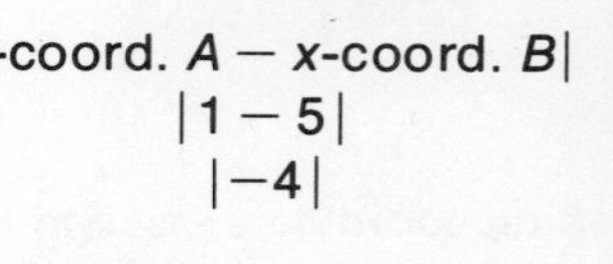

$|4| = 4$; $|-4| = 4$ → **Thus,** the length of $\overline{AB}$ is 4.

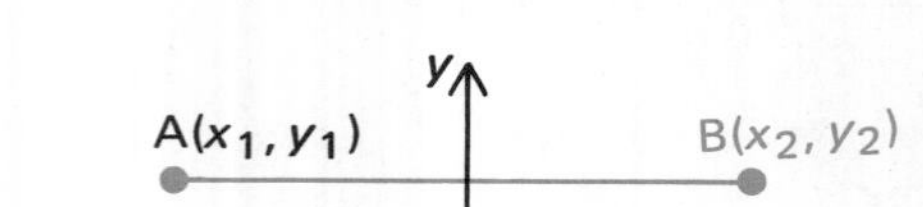
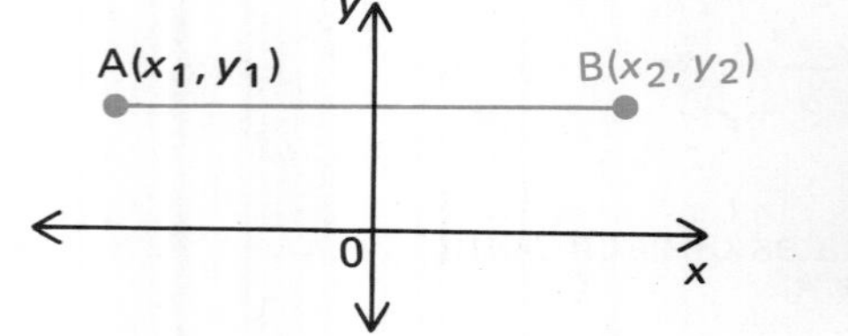

If $\overline{AB}$ is horizontal, then the length of $\overline{AB}$ is $|x_2 - x_1|$, or $|x_1 - x_2|$.

x-coordinate of B x-coordinate of A

EXAMPLE 6 Find the length of $\overline{MN}$.

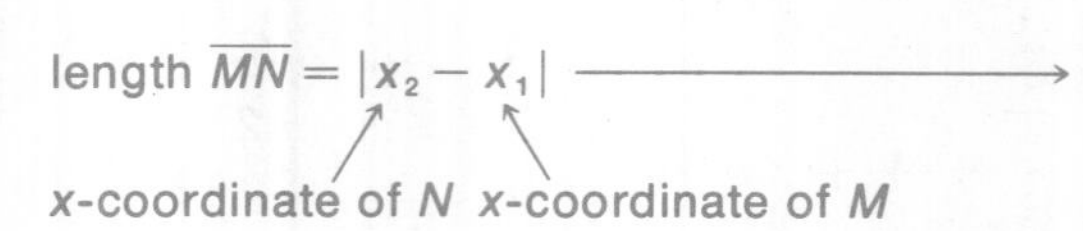

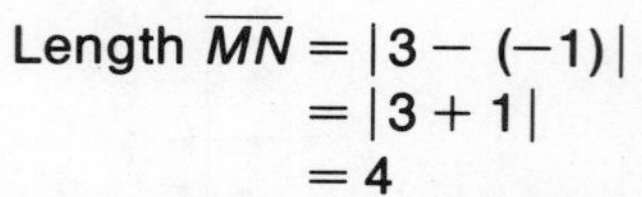

$$\text{Length } \overline{MN} = |3-(-1)| = |3+1| = 4$$

Thus, the length of $\overline{MN}$ is 4.

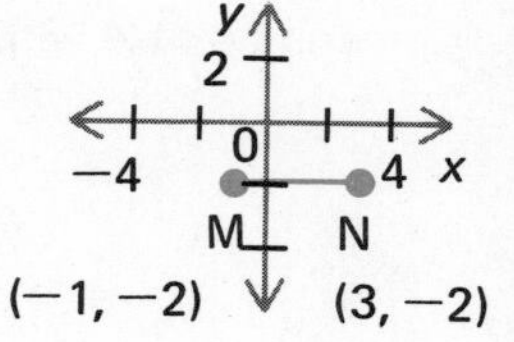

EXAMPLE 7 What is the length of $\overline{AB}$?

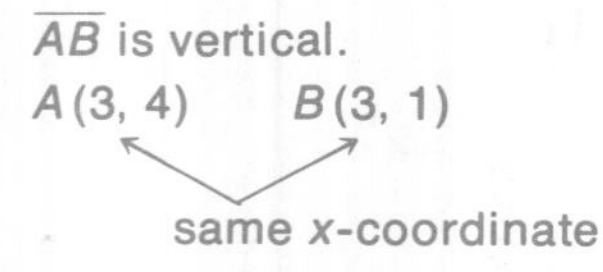

$$\text{Length } \overline{AB} = |4-1| = |3| = 3$$

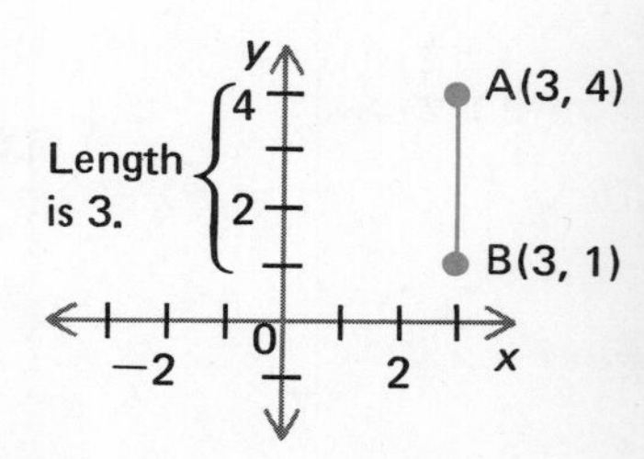

Thus, the length of $\overline{AB}$ is 3.

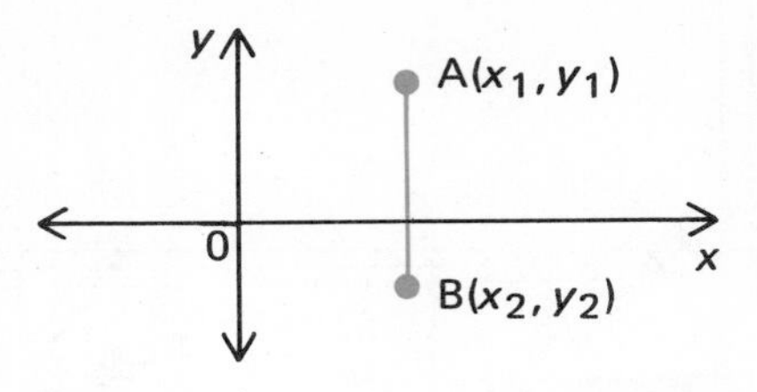

If $\overline{AB}$ is vertical, then the length of $\overline{AB} = |y_2 - y_1|$.

y-coordinate of B   y-coordinate of A

EXAMPLE 8

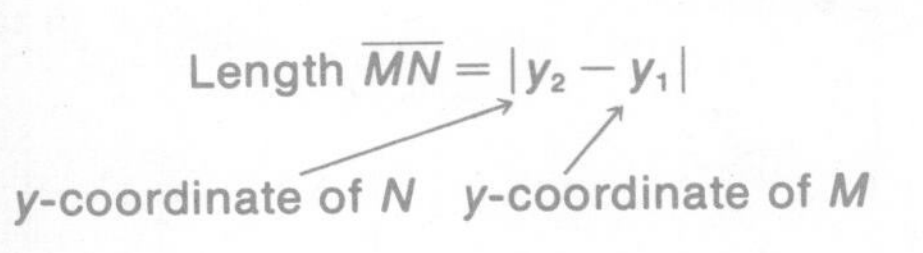

Find the length of $\overline{MN}$.

Length $\overline{MN} = |-1 - 2|$
$= |-3|$
$= 3$

Thus, the length of $\overline{MN}$ is 3.

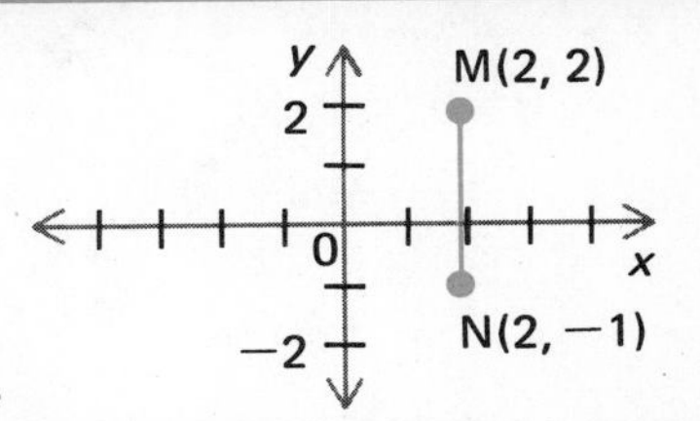

ORAL EXERCISES

Tell in which quadrant each point lies.

1. $(-4, 3)$ *2* **2.** $(5, 4)$ *1* **3.** $(-2, -2)$ *3* **4.** $(-2, 7)$ *2* **5.** $(6, -7)$ *4*

EXERCISES

PART A

Plot the points.

1. $A(3, 2)$ **2.** $B(-2, 3)$ **3.** $C(-1, -4)$ **4.** $D(2, -3)$ **5.** $E(2, 3)$

6. Give the coordinates of each point.

7. Find the distances of each point from the vertical and horizontal axes.

8. Find the lengths of $\overline{AF}$, $\overline{JK}$, $\overline{DG}$, $\overline{BC}$, and $\overline{BE}$.

9. Find the lengths of $\overline{FE}$, $\overline{GH}$, $\overline{JI}$, $\overline{AB}$, and $\overline{DC}$.

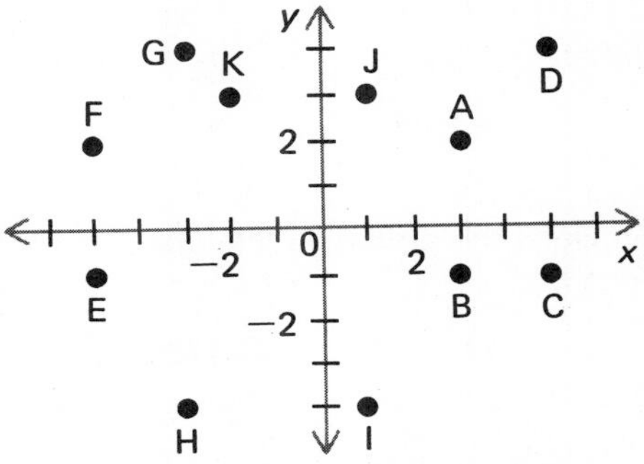

Find the length of $\overline{AB}$.

10. $A(3, 2)$ and $B(3, -3)$ *5* **11.** $A(1, -4)$ and $B(-2, -4)$ *3* **12.** $A(2, 3)$ and $B(-3, 3)$ *5*
13. $A(-4, 2)$ and $B(-4, -2)$ *4* **14.** $A(-2, -1)$ and $B(-2, 3)$ *4* **15.** $A(1, 5)$ and $B(2, 5)$ *1*

PART B

Name the quadrant in which the point (x, y) lies.

16. $x > 0$ and $y > 0$ *1* **17.** $x < 0$ and $y > 0$ *2* **18.** $x > 0$ and $y < 0$ *4* **19.** $x < 0$ and $y < 0$ *3*

PART C

Find the length of $\overline{PQ}$. Plot P and Q.

20. $P(a_1, b_1)$ and $Q(a_2, b_1)$ **21.** $P(a_1, b_1)$ and $Q(a_1, b_2)$

Directed Distances

OBJECTIVE

- To find directed distances on vertical and horizontal lines

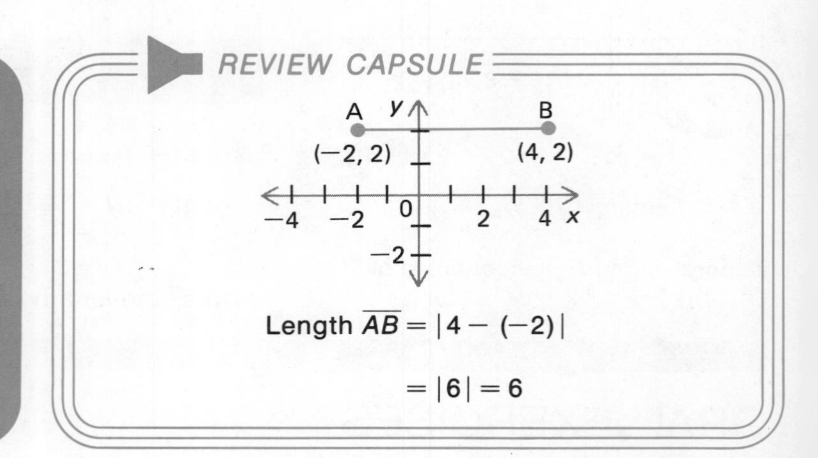

Length $\overline{AB} = |4 - (-2)|$

$= |6| = 6$

EXAMPLE 1 Find the directed distance from P to Q and from Q to P.

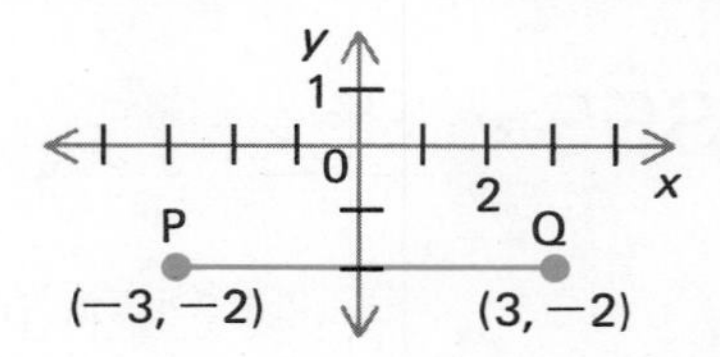

Find the length of $\overline{PQ}$ and $\overline{QP}$. Length is the same.

x-coord. of Q x-coord. of P	x-coord. of P x-coord. of Q
Length of $\overline{PQ} = \|3 - (-3)\|$	Length of $\overline{QP} = \|-3 - 3\|$
$= 6$	$= 6$
Move right from P to Q. The directed distance, P to Q, is 6.	Move left from Q to P. The directed distance, Q to P, is -6.

Right is +. Left is −.
$\vec{d}(PQ)$ means the directed distance from P to Q.

Thus, $\vec{d}(PQ) = 6$ and $\vec{d}(QP) = -6$.

EXAMPLE 2 Find the directed distance from A to B and from B to A.

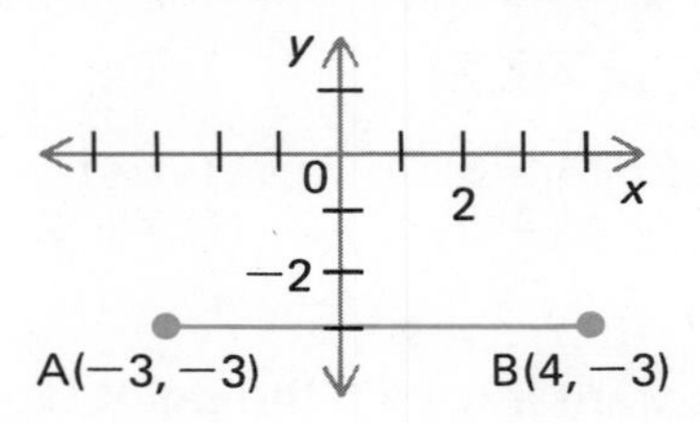

Find the length of $\overline{AB}$ and $\overline{BA}$.
Move *right* from A to B.
Move *left* from B to A.

x-coord. of B x-coord. of A	x-coord. of A x-coord. of B
Length of $\overline{AB} = \|4 - (-3)\|$	Length of $\overline{BA} = \|-3 - 4\|$
$= 7$	$= 7$
The directed distance is 7.	The directed distance is -7.

Thus, $\vec{d}(AB) = 7$ and $\vec{d}(BA) = -7$.

Directed distance from left to right is positive.

Directed distance from right to left is negative.

If $\overline{PQ}$ is horizontal, then
$\vec{d}(PQ) = x_Q - x_P$.
directed distance PQ (↑ $\vec{d}(PQ)$); x-coordinate of Q (↑ x_Q); x-coordinate of P (↖ x_P)

If $\overline{PQ}$ is horizontal, then $\vec{d}(QP) = x_P - x_Q$.

EXAMPLE 3 Find $\vec{d}(PQ)$ and $\vec{d}(QP)$.

$P(-3, 2)$ ← x-coord. of P $\qquad Q(4, 2)$ ← x-coord. Q

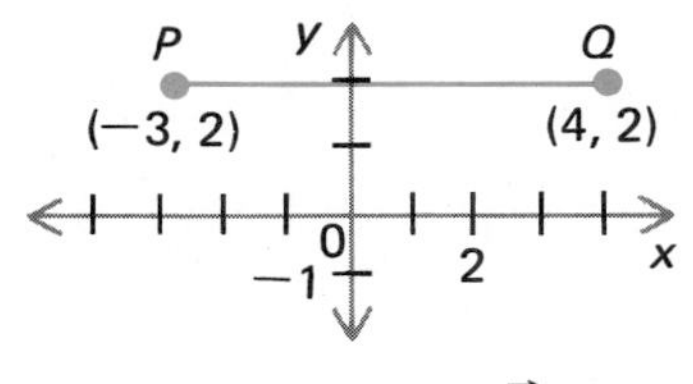

$$\vec{d}(PQ) = x_Q - x_P = 4 - (-3) = 7$$

$$\vec{d}(QP) = x_P - x_Q = -3 - 4 = -7$$

Thus, $\vec{d}(PQ) = 7$ and $\vec{d}(QP) = -7$.

EXAMPLE 4 Find the directed distance from P to Q and from Q to P.

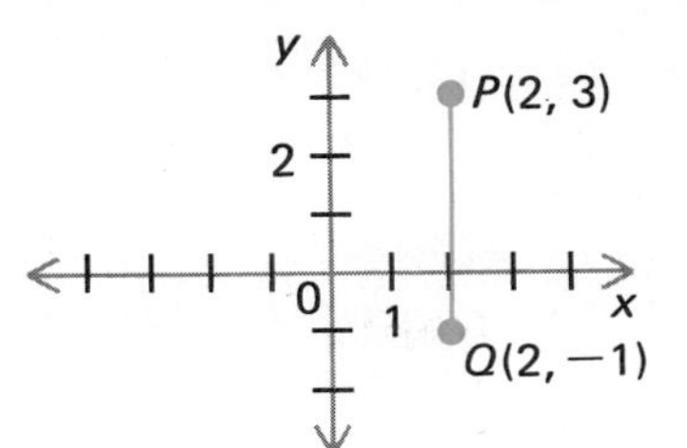

$P(2, 3)$ ← y-coord. of P $\qquad Q(2, -1)$ ← y-coord. of Q

y-coord. Q y-coord. P	y-coord. P y-coord. Q
length of $\overline{PQ} = \lvert -1 - 3 \rvert = 4$	length of $\overline{QP} = \lvert 3 - (-1) \rvert = 4$
The directed distance is -4.	The directed distance is 4.
(Move *down* from P to Q)	(Move *up* from Q to P)

Thus, $\vec{d}(PQ) = -4$ and $\vec{d}(QP) = 4$.

Directed distance down is negative.

Directed distance up is positive.

If $\overline{PQ}$ is vertical, then
$\vec{d}(PQ) = y_Q - y_P$.
directed distance PQ (↑ $\vec{d}(PQ)$); y-coordinate of Q (↑ y_Q); y-coordinate of P (↖ y_P)

If $\overline{PQ}$ is vertical, then $\vec{d}(QP) = y_P - y_Q$.

EXAMPLE 5 Find $\vec{d}(PQ)$ and $\vec{d}(QP)$.

$P(-3, 3)$ — y-coord. of P $\quad$ $Q(-3, -2)$ — y-coord. of Q

(−3, 3)
P
y
2
0
2
x
Q
(−3, −2)

$\vec{d}(PQ) = y_Q - y_P$
$= -2 - 3$
$= -5$

$\vec{d}(QP) = y_P - y_Q$
$= 3 - (-2)$
$= 5$

Thus, $\vec{d}(PQ) = -5$ and $\vec{d}(QP) = 5$.

EXAMPLE 6

$P(4, 5)$ and $Q(-2, 5)$ are on a horizontal line.

$M(-6, -4)$ and $N(-6, -3)$ are on a vertical line.

Find $\vec{d}(PQ)$ for $P(4, 5)$ and $Q(-2, 5)$.

$\vec{d}(PQ) = -2 - 4$
$= -6$

Find $\vec{d}(MN)$ for $M(-6, -4)$ and $N(-6, -3)$.

$\vec{d}(MN) = -3 - (-4)$
$= -3 + 4$
$= 1$

ORAL EXERCISES

Is $\vec{d}(AB)$ positive or negative?

1. $A(2, 3)$, $B(-4, 3)$ *negative*
2. $A(-2, 4)$, $B(-2, -3)$ *negative*
3. $A(1, -7)$, $B(1, -5)$ *positive*
4. $A(-1, -5)$, $B(-1, -2)$ *positive*

EXERCISES

PART A

Find $\vec{d}(PQ)$.

1. $P(3, 4)$, $Q(-2, 4)$ *−5*
2. $P(5, 1)$, $Q(1, 1)$ *−4*
3. $P(2, 3)$, $Q(-2, 3)$ *−4*
4. $P(1, 6)$, $Q(1, -4)$ *−10*
5. $P(7, -3)$, $Q(-2, -3)$ *−9*
6. $P(-1, -3)$, $Q(2, -3)$ *3*

Find $\vec{d}(QP)$.

7. $P(3, 5)$, $Q(3, -2)$ *7*
8. $P(2, -3)$, $Q(2, 7)$ *−10*
9. $P(2, -4)$, $Q(6, -4)$ *−4*
10. $P(5, -2)$, $Q(-3, -2)$ *8*
11. $P(-4, -5)$, $Q(-2, -5)$ *−2*
12. $P(-5, -2)$, $Q(-1, -2)$ *−4*

PART B

Find $\vec{d}(MN)$.

13. $M(-3\frac{1}{2}, 6)$, $N(4\frac{1}{4}, 6)$ *$7\frac{3}{4}$*
14. $M(6, -2\frac{1}{4})$, $N(6, 5\frac{1}{2})$ *$7\frac{3}{4}$*

Find $\vec{d}(NM)$.

15. $M(-5, 4\frac{1}{2})$, $N(-5, -3\frac{1}{4})$ *$7\frac{3}{4}$*
16. $M(-3\frac{1}{5}, 7)$, $N(4\frac{1}{2}, 7)$ *$-7\frac{7}{10}$*

PART C

Find $\vec{d}(AB)$.

17. $A(4a, 2b)$, $B(7a, 2b)$
18. $A(5x, 3y)$, $B(-6x, 3y)$
19. $A(-3x, 7y)$, $B(-3x, -5y)$
20. $A(2a, 3b)$, $B(2a, 8b)$

Find $\vec{d}(BA)$.

21. $A(6a, -2b)$, $B(6a, 5b)$
22. $A(2a, 3b)$, $B(7a, 3b)$
23. $A(-5x, 6y)$, $B(-5x, -8y)$
24. $A(10x, 7y)$, $B(-3x, 7y)$

Mathematics in Drafting

Pictured above, a draftswoman does structural analyses on nuclear submarines. A draftsperson makes mechanical drawings which show several views of an object.

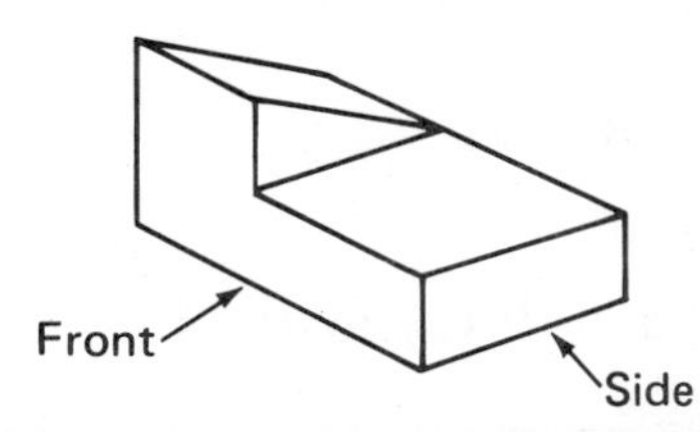

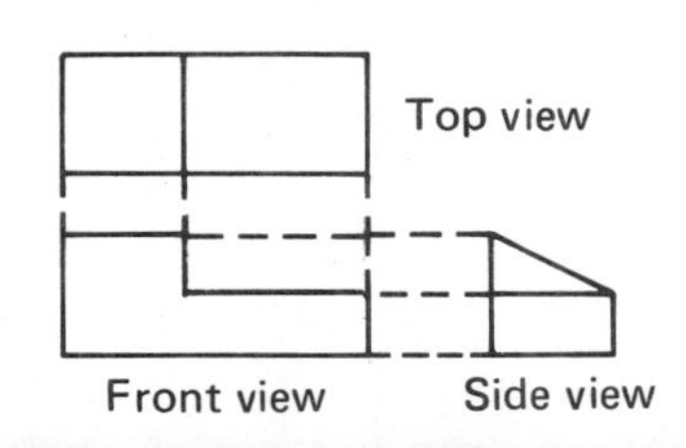

PROJECT

Draw the top, front, and side views of the object shown below.

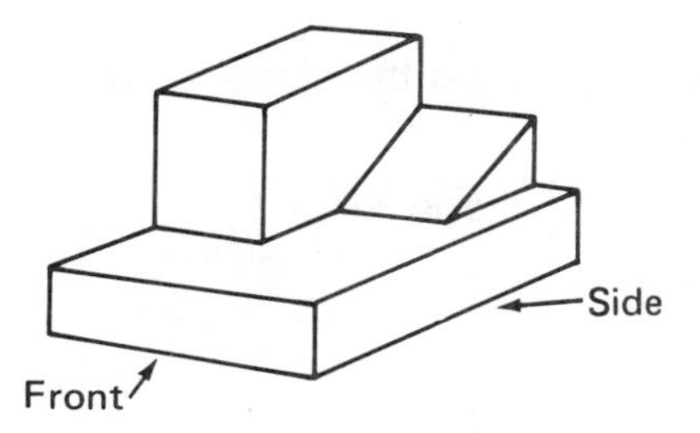

Slope of a Line

OBJECTIVES

- To find the slope of a line given any two points on it
- To classify the slope of a line as positive, negative, zero, or undefined

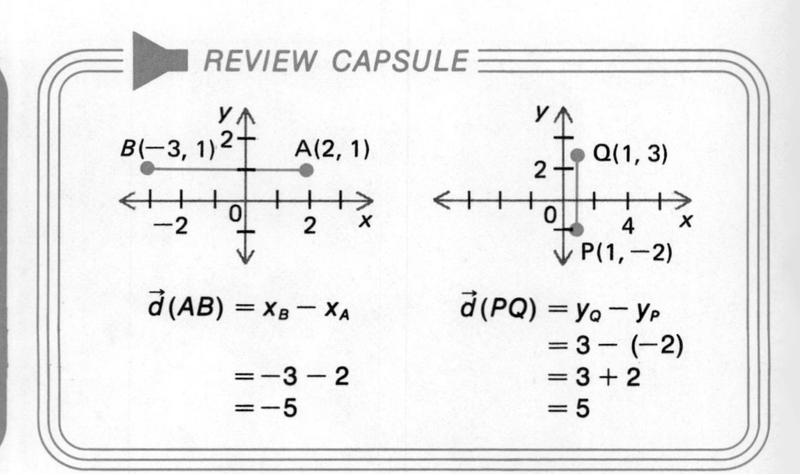

EXAMPLE 1 Find the ratio of $\vec{d}(AC)$ to $\vec{d}(CB)$.

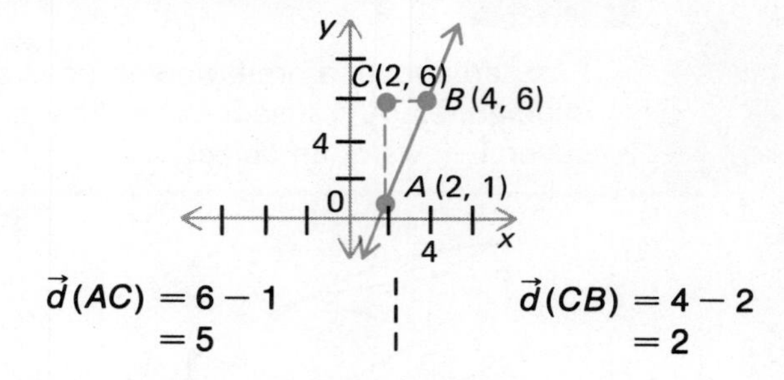

$\overline{AC}$ is vertical. Use $y_C - y_A$.
$\overline{CB}$ is horizontal. Use $x_B - x_C$.

$\vec{d}(AC) = 6 - 1$
$= 5$

$\vec{d}(CB) = 4 - 2$
$= 2$

Write the ratio of $\vec{d}(AC)$ to $\vec{d}(CB)$.

$\frac{\text{Difference in } y\text{-coordinates}}{\text{Difference in } x\text{-coordinates}}$

$$\frac{\vec{d}(AC)}{\vec{d}(CB)} = \frac{6-1}{4-2}, \text{ or } \frac{5}{2}$$

Thus, the ratio of $\vec{d}(AC)$ to $\vec{d}(CB)$ is $\frac{5}{2}$.

The slope of $\overleftrightarrow{AB}$ is $\frac{5}{2}$. ⟶ This ratio is the *slope* of $\overleftrightarrow{AB}$.

EXAMPLE 2 Find the slope of $\overleftrightarrow{AB}$.

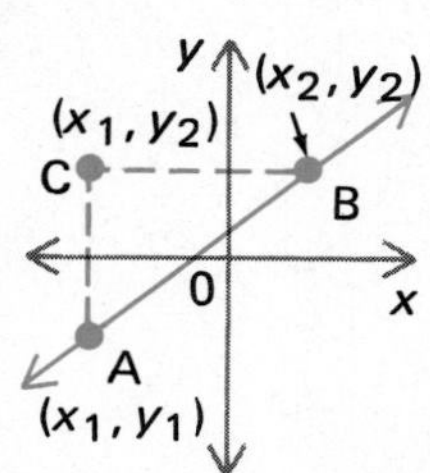

$$\text{slope } \overleftrightarrow{AB} = \frac{\vec{d}(AC)}{\vec{d}(CB)}$$
$$= \frac{y_2 - y_1}{x_2 - x_1}$$

$\frac{y\text{-coordinate } B - y\text{-coordinate } A}{x\text{-coordinate } B - x\text{-coordinate } A}$

Thus, slope of $\overleftrightarrow{AB}$ is $\frac{y_2 - y_1}{x_2 - x_1}$.

Definition of slope

$\dfrac{\text{Difference in } y\text{-coordinates}}{\text{Difference in } x\text{-coordinates}}$

For $A(x_1, y_1)$ and $B(x_2, y_2)$, the *slope* of $\overleftrightarrow{AB}$ is $\dfrac{y_2 - y_1}{x_2 - x_1}$.

EXAMPLE 3 Find the slope of $\overleftrightarrow{AB}$. Find the slope of $\overleftrightarrow{PQ}$.

Line slants up, slope positive.
Line slants down, slope negative.

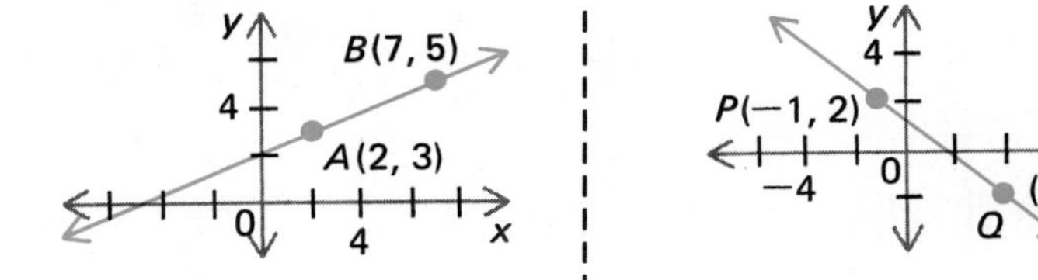

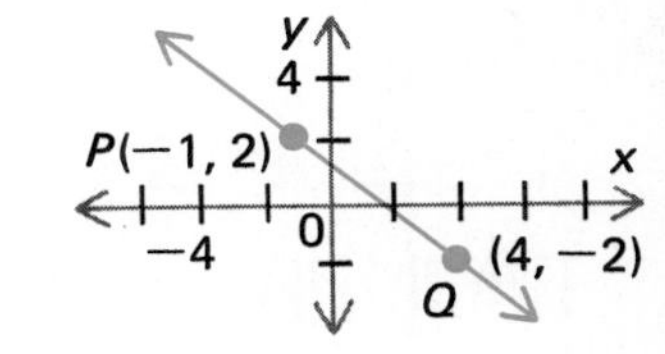

$\dfrac{y_B - y_A}{x_B - x_A}$ or $\dfrac{y_A - y_B}{x_A - x_B}$

Changing the order of the points does *not* change the slope.

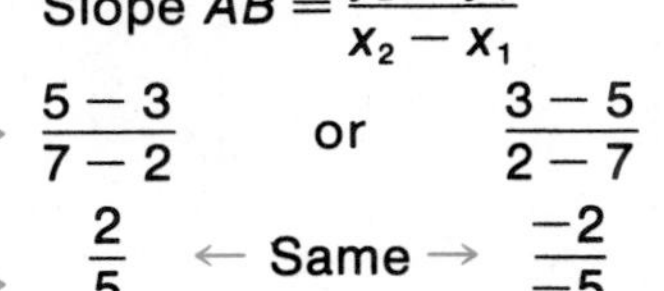

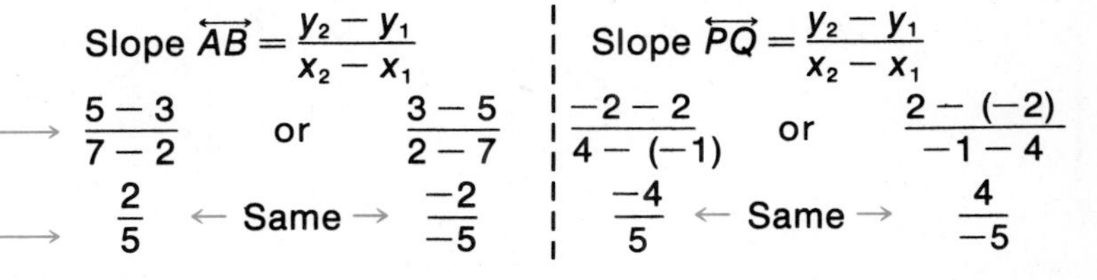

Slope $\overleftrightarrow{AB} = \dfrac{y_2 - y_1}{x_2 - x_1}$

$\dfrac{5-3}{7-2}$ or $\dfrac{3-5}{2-7}$

$\dfrac{2}{5}$ ← Same → $\dfrac{-2}{-5}$

Slope $\overleftrightarrow{PQ} = \dfrac{y_2 - y_1}{x_2 - x_1}$

$\dfrac{-2-2}{4-(-1)}$ or $\dfrac{2-(-2)}{-1-4}$

$\dfrac{-4}{5}$ ← Same → $\dfrac{4}{-5}$

EXAMPLE 4 Find the slope of $\overleftrightarrow{AB}$, $\overleftrightarrow{BC}$, and $\overleftrightarrow{CD}$.

Slope $= \dfrac{y_2 - y_1}{x_2 - x_1}$.

Slope $\overleftrightarrow{AB} = \dfrac{3-(-3)}{2-(-2)} = \dfrac{6}{4} = \dfrac{3}{2}$

Slope $\overleftrightarrow{BC} = \dfrac{6-3}{4-2} = \dfrac{3}{2}$

Slope $\overleftrightarrow{CD} = \dfrac{9-6}{6-4} = \dfrac{3}{2}$

y
D(6, 9)
C(4, 6)
4
B(2, 3)
0
4
x
A(−2, −3)

Points A, B, C and D are all on the *same* line.

Thus, slope $\overleftrightarrow{AB}$ = slope $\overleftrightarrow{BC}$ = slope $\overleftrightarrow{CD} = \frac{3}{2}$.

The slope of a line is the same no matter which two points on it are used to compute the slope.

EXAMPLE 5 Find the slope of $\overleftrightarrow{AB}$.

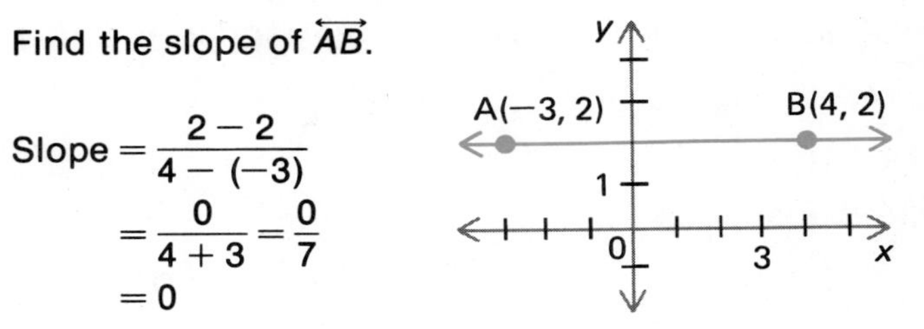

$$\text{Slope} = \frac{2-2}{4-(-3)} = \frac{0}{4+3} = \frac{0}{7} = 0$$

A horizontal line does have a slope. It is zero.

The slope of a horizontal line is zero.

EXAMPLE 6 Find the slope of $\overleftrightarrow{RS}$.

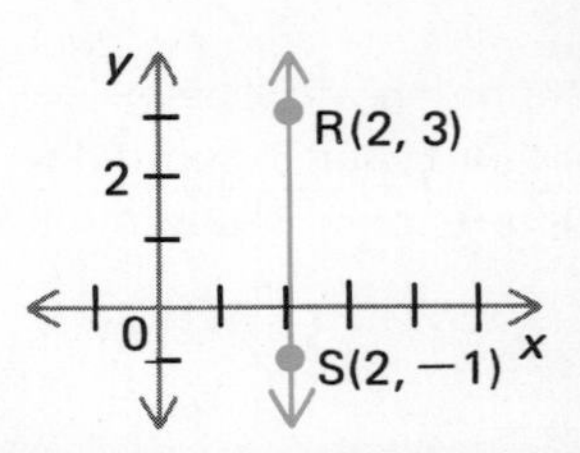

$$\text{Slope} = \frac{-1-3}{2-2}$$
$$= \frac{-4}{0}$$

$\frac{a}{0}$ is undefined. ⟶ The slope of $\overleftrightarrow{RS}$ is undefined.

The slope of a vertical line is undefined.

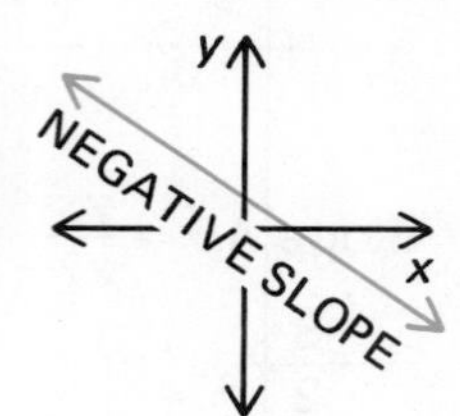

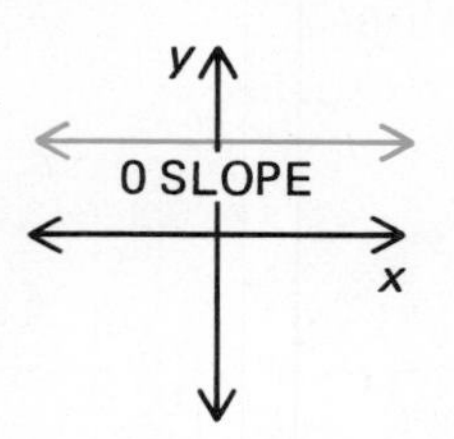

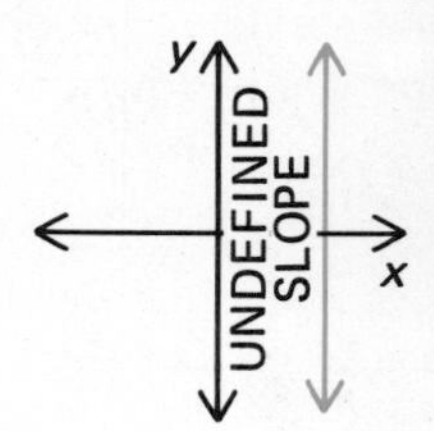

EXERCISES

PART A

Classify the slope of each line as positive, negative, zero, or undefined.

1.
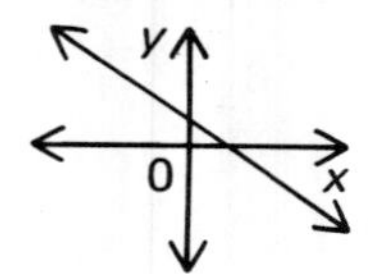

2.
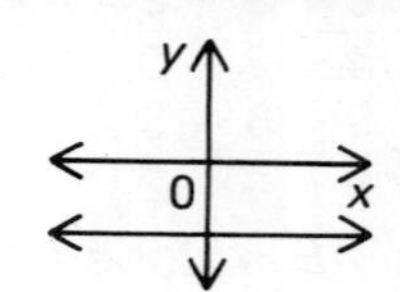

3.
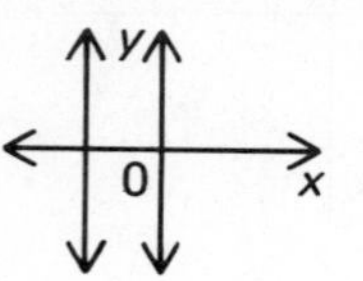

4.
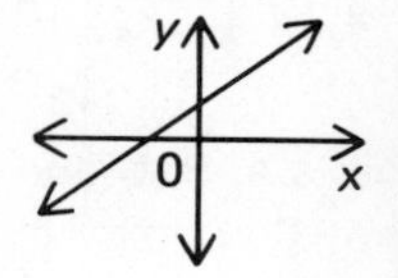

Find the slope of $\overleftrightarrow{AB}$.

5. $A(3, 2), B(4, 1)$ *−1*
6. $A(-3, -2), B(-1, 4)$ *3*
7. $A(6, 2), B(-1, -4)$ *$\frac{6}{7}$*
8. $A(3, 2), B(-3, 2)$ *0*
9. $A(-4, 1), B(-2, 1)$ *0*
10. $A(4, 1), B(4, -3)$
11. $A(-1, 2), B(1, 1)$ *$-\frac{1}{2}$*
12. $A(4, 2), B(6, 1)$ *$-\frac{1}{2}$*
13. $A(-2, 1), B(-2, 4)$
14. $A(-4, -1), B(-2, 4)$ *$\frac{5}{2}$*
15. $A(1, 2), B(5, 6)$ *1*
16. $A(-3, -1), B(-6, -4)$

Verify that for the line containing *A, B, C,* slope $\overleftrightarrow{AB}$ = slope $\overleftrightarrow{BC}$ = slope $\overleftrightarrow{AC}$.

17. $A(-2, -3), B(0, 0), C(2, 3)$ *slopes =, $\frac{3}{2}$*
18. $A(-6, -4), B(-4, -3), C(-2, -2)$ *slopes =, $\frac{1}{2}$*
19. $A(3, 6), B(0, 4), C(-3, 2)$ *slopes =, $\frac{2}{3}$*

PART B

Find the slope of $\overleftrightarrow{PQ}$.

20. $P(2, -3), Q(\frac{1}{2}, -\frac{1}{3})$ *$-\frac{16}{9}$*
21. $P(-\frac{1}{3}, \frac{1}{2}), Q(\frac{5}{6}, 4)$ *3*
22. $P(\frac{1}{5}, \frac{1}{4}), Q(\frac{3}{5}, \frac{3}{4})$ *$\frac{5}{4}$*

PART C

Find the slope of $\overleftrightarrow{MN}$.

23. $M(a, -b), N(-a, b)$
24. $M(4x, 5y), N(-2x, -7y)$
25. $M(x_1, y_1), N(x_2, y_2)$

Equation of a Line: Slope-Intercept

OBJECTIVES

- To write an equation of a line given a point and y-intercept
- To write an equation of a line given its slope and y-intercept
- To determine the slope and y-intercept of a line given its equation

REVIEW CAPSULE

Solve the proportion $\frac{4}{x+2} = \frac{3}{x}$.

First Method

$$x(x+2)\frac{4}{x+2} = x(x+2)\cdot\frac{3}{x}$$

$$4x = 3(x+2)$$
$$4x = 3x + 6$$
$$x = 6$$

Second Method

$$\frac{4}{x+2} = \frac{3}{x}$$
$$4x = 3(x+2)$$
$$4x = 3x + 6$$
$$x = 6$$

Thus, the solution is 6.

EXAMPLE 1 Write an equation of the line passing through $P(0, 2)$ and $Q(2, 3)$. Solve for y.

The y-intercept is the y-coordinate of the point at which the line intersects the y-axis.

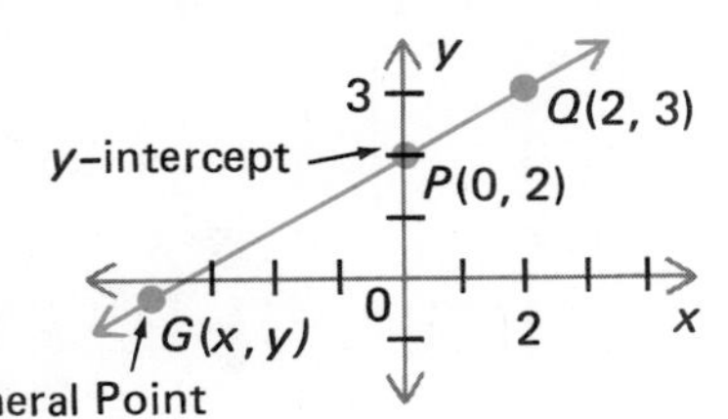

Choose $G(x, y)$ as a general point.

Find the slope of $\overleftrightarrow{PQ}$. → Slope $\overleftrightarrow{PQ} = \frac{3-2}{2-0}$, or $\frac{1}{2}$

Find the slope of $\overleftrightarrow{GP}$. → Slope $\overleftrightarrow{GP} = \frac{y-2}{x-0}$, or $\frac{y-2}{x}$

G, P, and Q are on the same line. → Slope $\overleftrightarrow{GP}$ = slope $\overleftrightarrow{PQ}$

An equation of the line → $\frac{y-2}{x} = \frac{1}{2}$

Solve for y.

In the second method, use prod. of extremes = prod. of means.

First Method

$$\frac{y-2}{x} = \frac{1}{2}$$

Multiply each side by the LCD $2x$. → $2x\frac{(y-2)}{x} = 2x\left(\frac{1}{2}\right)$

$$2y - 4 = x$$

Add 4 to each side. → $2y = x + 4$

Divide each side by 2. → $y = \frac{1}{2}x + 2$

slope ($\frac{1}{2}$) y-intercept (2)

Second Method

$$\frac{y-2}{x} = \frac{1}{2}$$
$$2(y-2) = 1x$$
$$2y - 4 = x$$
$$2y = x + 4$$
$$y = \tfrac{1}{2}x + 2$$

Thus, an equation of the line is $y = \frac{1}{2}x + 2$.

EXAMPLE 2 Write an equation of the line whose y-intercept is b and whose slope is m.

$(0, b)$ is any point on the y-axis.
(x, y) is any point on the line. →

y
P(0, b) ← y-intercept
General Point →
0
x
G(x, y)

$$\text{Slope } \overleftrightarrow{GP} = \frac{y_2 - y_1}{x_2 - x_1}$$
$$= \frac{y - b}{x - 0}, \text{ or } \frac{y - b}{x}$$

The slope of the line is m.

$$\frac{y - b}{x} = m$$

Solve for y. →

$$x\left(\frac{y - b}{x}\right) = xm$$
$$y - b = mx$$
$$y = mx + b$$

(↑ slope, ↑ y-intercept)

Thus, an equation of the line is $y = mx + b$.

Slope-intercept form of the equation of a line →

> $y = mx + b$ is an equation of the line whose slope is m and whose y-intercept is b.

EXAMPLE 3 Write an equation of the line whose slope is $\frac{1}{3}$ and whose y-intercept is -2.

Slope-intercept form of the equation →
Substitute: $m = \frac{1}{3}$, $b = -2$. →

$$y = mx + b$$
$$y = \tfrac{1}{3}x + (-2)$$

Thus, an equation is $y = \frac{1}{3}x - 2$.

EXAMPLE 4 Find the slope and the y-intercept of the line whose equation is $3x + 2y = 10$.

Solve for y.
Add $-3x$ to each side. →

Divide each side by 2. →

$\frac{-3x + 10}{2} = \frac{-3x}{2} + \frac{10}{2} = -\frac{3}{2}x + 5$ →

$$3x + 2y = 10$$
$$2y = -3x + 10$$
$$\frac{2y}{2} = \frac{-3x + 10}{2}$$
$$y = -\tfrac{3}{2}x + 5$$

(↗ m, ↖ b)

Thus, the slope is $-\frac{3}{2}$ and the y-intercept is 5.

ORAL EXERCISES

Give the slope and the y-intercept.

1. $y = 2x + 3$ — $m = 2, b = 3$
2. $y = -\frac{1}{2}x - 3$ — $m = -\frac{1}{2}, b = -3$
3. $y = -4x + 5$ — $m = -4, b = 5$
4. $y = x - 3$ — $m = 1, b = -3$
5. $y = 4x - \frac{1}{2}$ — $m = 4, b = -\frac{1}{2}$

EXERCISES

PART A

Write an equation of line $\overleftrightarrow{AB}$ in the form $y = mx + b$.

1. $A(0, -2)$, $B(1, 2)$ — $y = 4x - 2$
2. $A(0, -4)$, $B(-1, 3)$ — $y = -7x - 4$
3. $A(0, -2)$, $B(2, -3)$ — $y = -\frac{1}{2}x - 2$
4. $A(-1, 1)$, $B(0, 1)$ — $y = 1$
5. $A(1, -2)$, $B(0, -1)$ — $y = -x - 1$
6. $A(-2, 4)$, $B(0, -4)$ — $y = -4x - 4$
7. $A(0, 2)$, $B(3, 2)$ — $y = 2$
8. $A(0, -3)$, $B(-1, -2)$ — $y = -x - 3$
9. $A(0, 4)$, $B(2, 4)$ — $y = 4$

Write an equation of the line.

10. $m = 2, b = 3$ — $y = 2x + 3$
11. $m = -\frac{1}{2}, b = -2$ — $y = -\frac{1}{2}x - 2$
12. $m = -2, b = -\frac{1}{2}$
13. $m = \frac{1}{2}, b = -\frac{1}{3}$ — $y = \frac{1}{2}x - \frac{1}{3}$
14. $m = -\frac{1}{3}, b = 2$ — $y = -\frac{1}{3}x + 2$
15. $m = -4, b = 4$
16. $m = 2, b = -1$ — $y = 2x - 1$
17. $m = -2, b = 3$ — $y = -2x + 3$
18. $m = -\frac{1}{2}, b = -\frac{1}{2}$

Find the slope and the y-intercept of the line whose equation is given.

19. $y = \frac{1}{2}x + 2$
20. $y = 3x - \frac{1}{2}$
21. $y = -4x + 7$
22. $y = -7x - 3$
23. $2x + 3y = 9$ — $m = -\frac{2}{3}, b = 3$
24. $3x - 2y = 10$ — $m = \frac{3}{2}, b = -5$
25. $4x + 6y - 9 = 0$ — $m = -\frac{2}{3}, b = \frac{3}{2}$
26. $2y + 3x = 7$ — $m = -\frac{3}{2}, b = \frac{7}{2}$

PART B

EXAMPLE Find the slope and the y-intercept of the line whose equation is $4x - (2y - 6) = 0$.

Solve for y.	$4x - (2y - 6) = 0$
$4x - (2y - 6) = 4x - (1)(2y - 6)$ →	$4x - 2y + 6 = 0$
Add -6 to each side. →	$4x - 2y = -6$
Add $-4x$ to each side. →	$-2y = -4x - 6$
Divide each side by -2. →	$y = \frac{-4x - 6}{-2}$
Rewrite. →	$y = 2x + 3$

slope ↗ (2) ↖ y-intercept (3)

Find the slope and the y-intercept of the line whose equation is given.

27. $3y - (4 + 6x) = 0$
28. $7x - 3(4 - y) = 0$
29. $5y - 2(5 - 2x) = 0$
30. $2x - (2y + 4) = 0$
31. $6y - 4(-2 - 3x) = 0$
32. $15 - 3(-2x - 3y) = 8$

PART C

***$Ax + By + C = 0$*, where *A*, *B*, and *C* are non-zero constants, is the equation of a line.**

33. Show that the slope of the line is $-\frac{A}{B}$.
34. Show that its y-intercept is $-\frac{C}{B}$.

Permutations

PROBLEM 1.

A truck driver wants to travel from New York City to Hartford, Connecticut and then to Bar Harbor, Maine. How many possible routes are there?

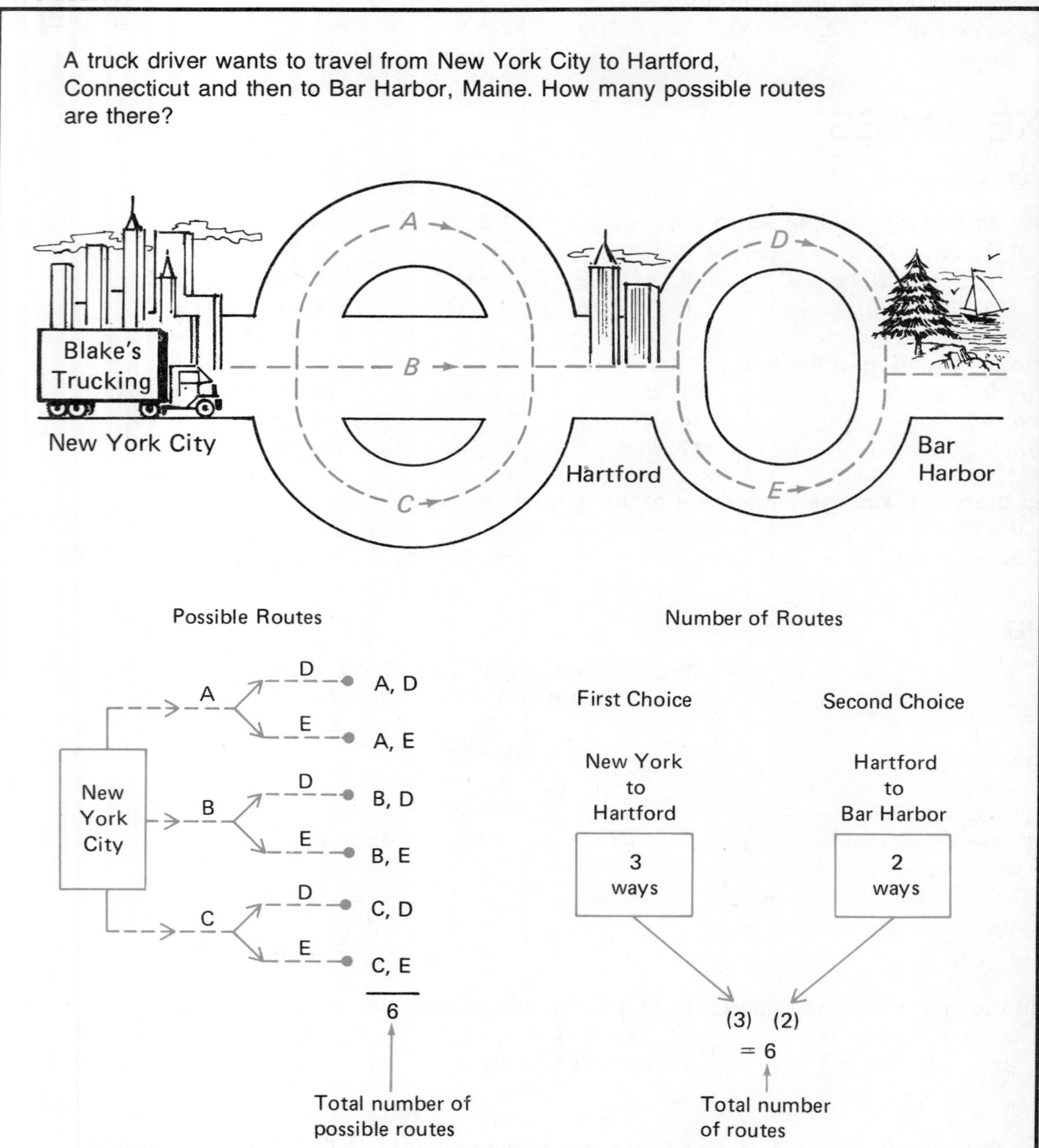

An arrangement of objects in a definite order is a *permutation* of the objects.

The number of permutations of n things taken r at a time can be found by using this formula.

$$nPr = n(n-1)(n-2)\ldots(n-r+1)$$

PROBLEM 2.

Find the number of permutations of 6 things taken 4 at a time.

$$\begin{aligned} nPr &= n(n-1)(n-2)\ldots(n-r+1) \\ {}_6P_4 &= 6(5)(4)(3) \qquad \leftarrow n = 6,\ r = 4,\ (n-r+1) = (6-4+1) = 3 \\ &= 360 \end{aligned}$$

Thus, there are 360 permutations of 6 things taken 4 at a time.

PROBLEM 3.

How many 3-digit numbers can be formed from the digits 1 through 7, if no digit is repeated in a number?

$$\begin{aligned} nPr &= n(n-1)(n-2)\ldots(n-r+1) \\ {}_7P_3 &= 7(6)(5) \qquad \leftarrow (n-r+1) = (7-3+1) = 5 \\ &= 210 \end{aligned}$$

Thus, 210 three-digit numbers can be formed from the 7 digits if no digit is repeated in a number.

PROJECT

Evaluate.

1. ${}_6P_2$ **2.** ${}_7P_5$ **3.** ${}_3P_2$ **4.** ${}_9P_4$ **5.** ${}_3P_3$

6. How many different 5-flag signals can be formed in a straight line if 8 different flags are available?

7. How many different 4-digit numbers can be formed from the digits 1 through 9 if no digit is repeated in a number?

Equation of a Line: Point-Slope

OBJECTIVES

- To write an equation of a line given its slope and a point on it
- To write an equation of a line given any two points on it

REVIEW CAPSULE

Write an equation of the line whose slope is -2 and which passes through the point $(0, 4)$.

$m = -2 \qquad b = 4$

$y = mx + b$ ← slope-intercept form

$y = -2x + 4$

EXAMPLE 1 Write an equation of the line which passes through the points $P(2, 3)$ and $Q(4, 6)$.

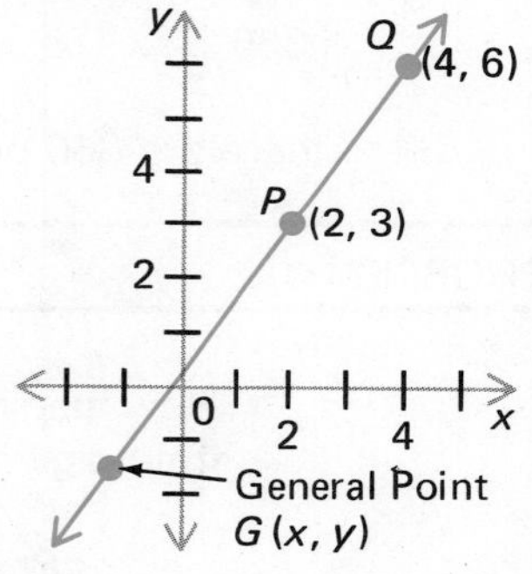

Slope $= \frac{y_2 - y_1}{x_2 - x_1}$. → Slope $\overleftrightarrow{PQ} = \frac{6-3}{4-2} = \frac{3}{2}$

Slope $\overleftrightarrow{GP} = \frac{y-3}{x-2}$

Slope $\overleftrightarrow{GP}$ = slope $\overleftrightarrow{PQ}$. → $\frac{y-3}{x-2} = \frac{3}{2}$

Prod. of extremes = prod. of means. → $2(y-3) = 3(x-2)$

Divide each side by 2. → $y - 3 = \frac{3}{2}(x-2)$

Thus, an equation of the line is $y - 3 = \frac{3}{2}(x-2)$.

EXAMPLE 2 Write an equation of the line which passes through the point $P(x_1, y_1)$ and whose slope is m.

Slope $\overleftrightarrow{GP} = \frac{y - y_1}{x - x_1}$

Slope $\overleftrightarrow{GP} = m$. → $\frac{y - y_1}{x - x_1} = m$

Multiply each side by the LCD $(x - x_1)$. → $(x - x_1)\left(\frac{y - y_1}{x - x_1}\right) = (x - x_1)m$

$y - y_1 = m(x - x_1)$

Thus, an equation of the line is $y - y_1 = m(x - x_1)$.

Point-slope form of the equation of a line →

$$y - y_1 = m(x - x_1)$$

is an equation of the line whose slope is m and which passes through the point (x_1, y_1).

EXAMPLE 3 Write an equation of the line passing through the point $(2, -3)$ and whose slope is 2. Solve for y.

Point-slope form of the equation → $y - y_1 = m(x - x_1)$

$y - (-3) = 2(x - 2)$

$-(-3) = -1(-3) = 3$ → $y + 3 = 2(x - 2)$

Solve for y.

$y + 3 = 2(x - 2)$

$2(x - 2) = 2x - 4$ — $y + 3 = 2x - 4$

Add -3 to each side. → $y = 2x - 7$

Thus, an equation of the line is $y = 2x - 7$.

EXAMPLE 4 Write an equation of the line passing through the points $(3, 4)$ and $(-2, -3)$. Solve for y.

First find the slope. → $m = \frac{4 - (-3)}{3 - (-2)} = \frac{4 + 3}{3 + 2} = \frac{7}{5}$

Use the point-slope form. → $y - y_1 = m(x - x_1)$

Use either point for (x_1, y_1). → $y - 4 = \frac{7}{5}(x - 3)$ or $y - (-3) = \frac{7}{5}[x - (-2)]$

Multiply each side by 5. → $5(y - 4) = 7(x - 3)$ $\quad$ $5(y + 3) = 7(x + 2)$

Solve for y.

$5y - 20 = 7x - 21$ $\quad$ $5y + 15 = 7x + 14$

$5y = 7x - 1$ $\quad$ $5y = 7x - 1$

$y = \frac{7x - 1}{5}$ $\quad$ $y = \frac{7x - 1}{5}$

$\frac{7x - 1}{5} = \frac{7}{5}x + \left(-\frac{1}{5}\right)$ → $y = \frac{7}{5}x - \frac{1}{5}$ $\quad$ $y = \frac{7}{5}x - \frac{1}{5}$

Thus, $y = \frac{7}{5}x - \frac{1}{5}$ is an equation of the line.

EXERCISES

PART A

Write an equation in the form $y - y_1 = m(x - x_1)$. Then solve for y.

1. $A(2, 3)$, $B(1, -2)$ **2.** $P(-2, -3)$, $Q(3, 5)$ **3.** $P(4, 1)$, $Q(-2, 3)$ **4.** $M(4, 2)$, $N(-2, 1)$

5. $R(8, 4)$, $S(2, 1)$ $\quad y - 4 = \frac{1}{2}(x - 8)$

6. $A(6, 3)$, $B(2, 2)$ $\quad y - 3 = \frac{1}{4}(x - 6)$

7. $A(-5, -6)$, $B(2, 3)$ $\quad y + 6 = \frac{9}{7}(x + 5)$

8. $A(-4, -5)$, $B(2, 5)$ $\quad y + 5 = \frac{5}{3}(x + 4)$

9. $A(1, 5)$, $m = 3$ **10.** $B(4, -3)$, $m = -2$ **11.** $P(-2, -3)$, $m = -\frac{1}{2}$ **12.** $P(2, 4)$, $m = -4$

13. $P(-2, 4)$, $m = -4$ **14.** $A(-5, 1)$, $m = 0$ **15.** $M(-4, 2)$, $m = -3$ **16.** $A(-1, 3)$, $m = -\frac{1}{2}$

PART B

Write an equation in the form $y - y_1 = m(x - x_1)$. Then solve for y.

17. $R(-3, -\frac{1}{2})$, $m = -\frac{1}{3}$ **18.** $P(-\frac{1}{3}, 2)$, $m = -2$ **19.** $M(\frac{1}{4}, \frac{7}{4})$, $N(\frac{5}{3}, 6)$

PART C

20. A line contains the points $A(a, 0)$ and $B(0, b)$, where $a \neq 0$ and $b \neq 0$. Show that $\frac{x}{a} + \frac{y}{b} = 1$ is an equation of the line.

Graphing Lines

OBJECTIVES

- To sketch the graph of a line given its equation
- To determine whether a point lies on a line
- To graph horizontal and vertical lines by recognizing their equations

REVIEW CAPSULE

Identify the slope and the y-intercept of the line with equation $3y - 2x - 9 = 0$.

$$3y = 2x + 9$$

$$y = \frac{2x + 9}{3}$$

Slope-intercept form $\rightarrow y = \frac{2}{3}x + 3$

slope ($\frac{2}{3}$) $\quad$ y-intercept (3)

EXAMPLE 1 Graph the line with slope $\frac{2}{3}$ and passing through the point $(0, 2)$.

Plot the point $A(0, 2)$.
Use slope to find point B.

$m = \frac{2}{3}$ $\begin{matrix}\leftarrow \text{up } 2 \\ \leftarrow \text{right } 3\end{matrix}$

Plot (0, 2).

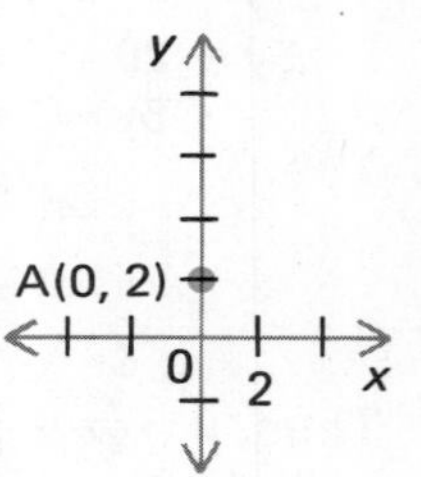

Use the slope $\frac{2}{3}$.

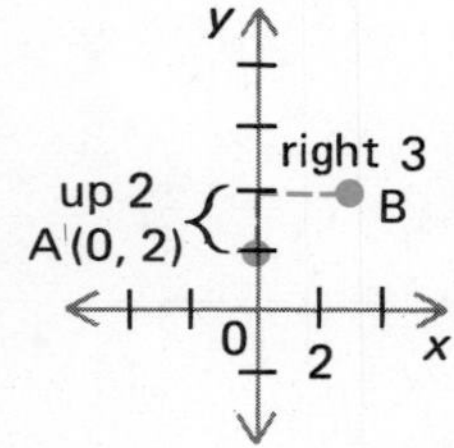

Draw the line $\overleftrightarrow{AB}$.

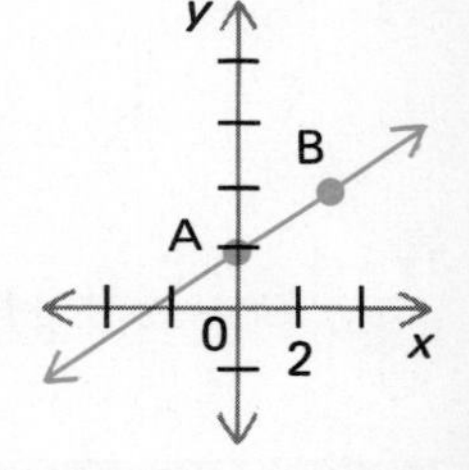

EXAMPLE 2 Sketch the graph of the line with the equation $y = \frac{2}{3}x + 2$.

$$y = \frac{2}{3}x + 2$$

Identify the slope and y-intercept. ⟶ Slope is $\frac{2}{3}$. $\quad$ y-intercept is 2.

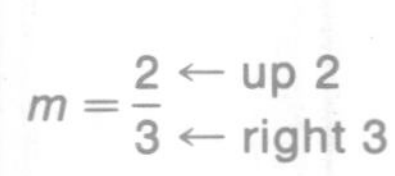

Plot (0, 2). $\quad$ Use the slope $\frac{2}{3}$. $\quad$ Draw the line.

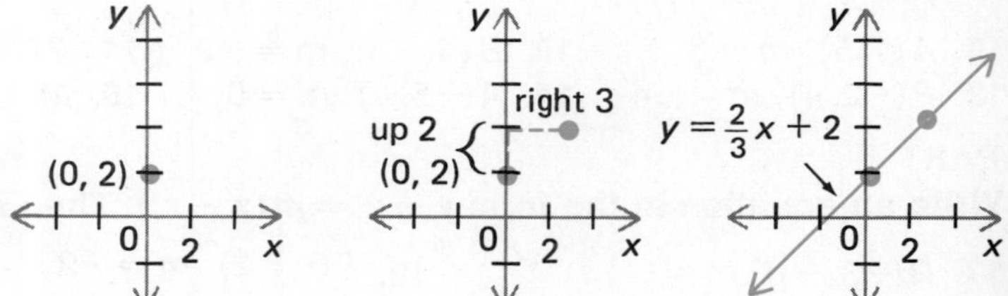

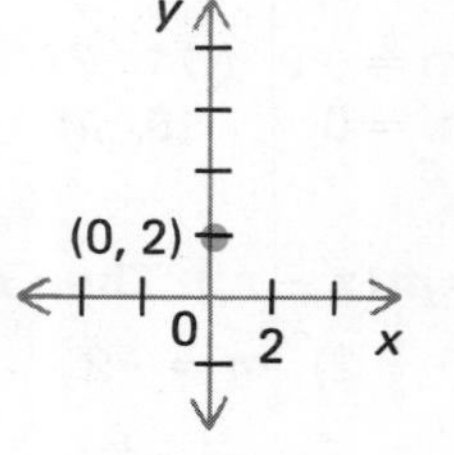

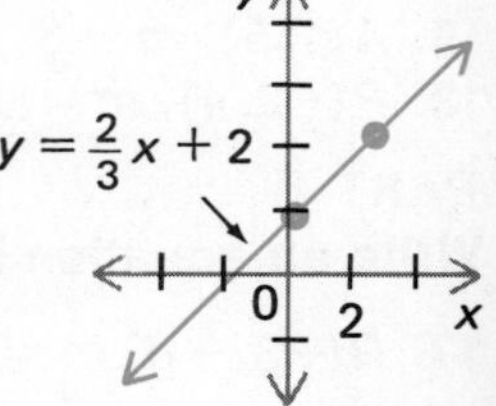

The graph is the *set of points* (x, y) *such that* $y = \frac{2}{3}x + 2$.

A line is a set of points. $\quad$ $\{(x, y) \mid y = \frac{2}{3}x + 2\}$

EXAMPLE 3 Graph $\{(x, y) \mid -4x - 2y = 6\}$.

Solve for y.

$$-4x - 2y = 6$$

Add $-4x$ to each side. →

$$-2y = 4x + 6$$

Divide each side by -2. →

$$y = \frac{4x + 6}{-2}$$

$$y = -2x - 3$$

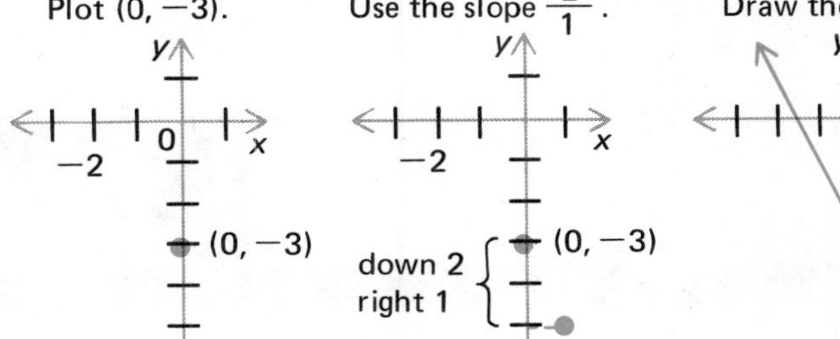

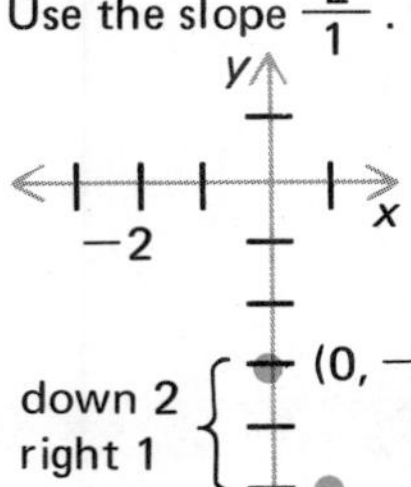

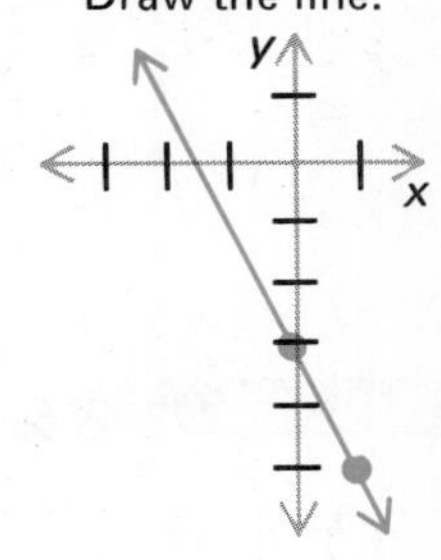

EXAMPLE 4 Is the point (2, 6) on the line with equation $2y - 3x - 6 = 0$?

(2, 6)

$x = 2$ $y = 6$

Check in the equation.

$2y - 3x - 6$	0
$2(6) - 3(2) - 6$	0
$12 - 6 - 6$	
0	

The graph of $2y - 3x - 6 = 0$ passes through (2, 6). → (2, 6) satisfies the equation.

Thus, (2, 6) is a point on the line.

> Every point on a line must satisfy the equation. Each point (x, y) which satisfies the equation is on the line.

EXAMPLE 5 Graph the line with equation $y = 6$.

Use any x-coordinate. (1, 6) and (5, 6) satisfy the equation. → (1, 6) and (5, 6) are on the line.

Plot (1, 6) and (5, 6). Draw the line.

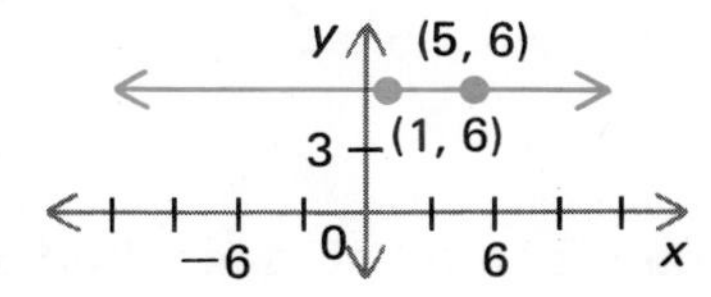

$y = 6$ for all values of x. → The graph of $y = 6$ is a *horizontal* line.

> The graph of the equation $y = b$, where b is any number, is a *horizontal* line.

EXAMPLE 6 Graph $\{(x, y) \mid x = -2\}$.

Use any y-coordinate. → $(-2, 0)$ and $(-2, 4)$ are on the line.

$(-2, 0)$ and $(-2, 4)$ satisfy the equation.

Plot $(-2, 0)$ and $(-2, 4)$. Draw the line.

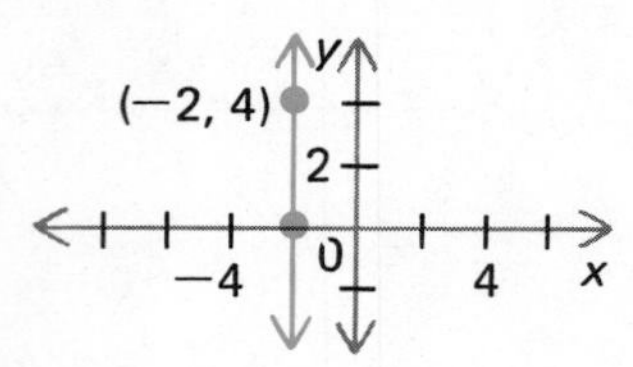

$x = -2$ for all values of y. → The graph of $\{(x, y) \mid x = -2\}$ is a vertical line.

The graph of the equation $x = k$ is a *vertical* line.

ORAL EXERCISES

Which sets of points are vertical lines? Which are horizontal lines?

1. $\{(x, y) \mid y = -3\}$ *h*
2. $\{(x, y) \mid x = 5\}$ *v*
3. $\{(x, y) \mid x = 2\}$ *v*
4. $\{(x, y) \mid x = -7\}$ *v*
5. $\{(x, y) \mid y = -2\}$ *h*
6. $\{(x, y) \mid y = -6\}$ *h*

EXERCISES

PART A

Sketch the graph of the line described by each equation.

1. $y = \frac{1}{2}x + 3$
2. $y = -\frac{2}{3}x - 2$
3. $y = \frac{1}{3}x + 1$
4. $y = 2x + 4$
5. $y = -\frac{1}{2}x - 1$
6. $y = -2x + 1$
7. $y = -3x - 2$
8. $y = \frac{2}{3}x + 3$

Graph.

9. $\{(x, y) \mid y = -2x + 4\}$
10. $\{(x, y) \mid y = -3x - 1\}$
11. $\{(x, y) \mid y = 4x + 3\}$
12. $\{(x, y) \mid y = \frac{1}{2}x - 5\}$
13. $\{(x, y) \mid y = \frac{1}{3}x + 7\}$
14. $\{(x, y) \mid x = 3\}$
15. $\{(x, y) \mid x = -1\}$
16. $\{(x, y) \mid y = -3\}$
17. $\{(x, y) \mid y = 2\}$

Is the given point on the line described by the equation?

18. $(1, 2)$, $2y + 3x = 7$ *yes*
19. $(2, 3)$, $x - 3y = -7$ *yes*
20. $(-1, 3)$, $x + 2y = 7$ *no*
21. $(-2, -1)$, $2x - y = -5$ *no*
22. $(-1, -2)$, $-2x + y = -4$ *no*
23. $(2, -3)$, $-2x + y = -7$ *yes*

PART B

Sketch the graph of the line described by each equation.

24. $2y + 3x = 4$
25. $x - 3y - 6 = 0$
26. $2x + 4y = 8$
27. $3x - y + 5 = 0$
28. $y - 3 = 0$
29. $x + 7 = 0$
30. $x - y = 4$
31. $2x + 7y = 14$

PART C

32. Find the coordinates of a point on the graph of $x + 4y = 18$ such that the ordinate of the point is twice the abscissa. Do not graph.

Combinations

$7! = 7 \times 6 \times 5 \times 4 \times 3 \times 2 \times 1 = 5{,}040$

Read: 7 factorial

$$n! = n(n-1)(n-2)\ldots \times 3 \times 2 \times 1$$

PROBLEM 1.

Simplify $\frac{8!}{3!(8-3)!}$.

$$\frac{8!}{3!(8-3)!} = \frac{8!}{3!(5!)} = \frac{8 \times 7 \times \overset{1}{\cancel{6}} \times \cancel{5 \times 4 \times 3 \times 2 \times 1}}{\underset{1}{\cancel{3}} \times \underset{1}{\cancel{2}} \times 1 \times (\cancel{5 \times 4 \times 3 \times 2 \times 1})} = 56$$

An arrangement of objects not in a definite order is a *combination.*

The number of combinations of n things taken r at a time

$$\binom{n}{r} = \frac{n!}{r!(n-r)!}$$

PROBLEM 2.

How many ways can a committee of 3 be selected from 18 people?

$$\binom{n}{r} = \frac{n!}{r!(n-r)!}$$

$$\binom{18}{3} = \frac{18!}{3!(18-3)!} = \frac{18!}{3!(15!)}$$

$$= \frac{\overset{3}{\cancel{18}} \times 17 \times 16 \times \overset{1}{\cancel{15!}}}{\cancel{3} \times \cancel{2} \times 1 \times \underset{1}{\cancel{15!}}} = 816$$

PROJECT

Simplify.

1. $\frac{7!}{5!}$ **2.** $\frac{10!}{7!}$ **3.** $\frac{7!}{3!(7-3)!}$ **4.** $\frac{9!}{7!(9-7)!}$

5. How many ways can a committee of 4 be selected from 18 people?

6. How many ways can a tennis team of 4 be selected from 17 players?

Chapter Eight Review

Plot the points. [p. 211]

1. $(-1, 3)$ **2.** $(4, 7)$ **3.** $(-5, -3)$ **4.** $(2, -4)$

Find the length of the line segment. [p. 211]

5. $M(3, 5)$, $N(3, -6)$ **6.** $A(1, -3)$, $B(6, -3)$ **7.** $P(4, -1)$, $Q(4, 5)$

Name the quadrant in which the point (*x*, *y*) lies. [p. 211]

8. $x < 0$ and $y > 0$ **9.** $x > 0$ and $y > 0$ **10.** $x < 0$ and $y < 0$ **11.** $x > 0$ and $y < 0$

Find $\vec{d}(PQ)$. [p. 214]

12. $P(-3, 7)$, $Q(6, 7)$ **13.** $P(6, -4)$, $Q(7, -4)$ **14.** $P(-3, 10)$, $Q(-3, -2)$

Find the slope of $\overleftrightarrow{AB}$. [p. 218]

15. $A(7, -3)$, $B(-2, -5)$ **16.** $A(-3, 4)$, $B(7, -1)$ **17.** $A(-1, -1)$, $B(-1, -2)$

Write an equation of the line. [p. 221, 226]

18. $P(-3, 5)$, $Q(0, -2)$ **19.** $A(3, 6)$, $m = 8$ **20.** $m = -\frac{1}{3}$, $b = 5$

Find the slope and the *y*-intercept of the line whose equation is given. [p. 221]

21. $y = \frac{2}{3}x - 1$ **22.** $4x - y = 7$ **23.** $y - 2(3 - 6x) = 8$

Sketch the graph of the line described by each equation. [p. 228]

24. $y = 5x - 1$ **25.** $2y - 3x = 8$

Graph. [p. 228]

26. $\{(x, y) \mid y = 7x - 1\}$ **27.** $\{(x, y) \mid y = -3\}$ **28.** $\{(x, y) \mid x = 7\}$

Is the given point on the line described by the equation? [p. 228]

29. $(1, -2)$, $3x - 2y = 7$ **30.** $(-1, -1)$, $-x - y = 2$

Chapter Eight Test

Plot the points.

1. $(7, -3)$
2. $(0, 6)$
3. $(-6, 7)$
4. $(-1, -5)$

Find the length of the line segment.

5. $A(-1, 7)$, $B(3, 7)$
6. $M(10, -1)$, $N(-2, -1)$
7. $P(7, 6)$, $Q(7, 10)$

Name the quadrant in which the point (x, y) lies.

8. $x < 0$ and $y < 0$
9. $x > 0$ and $y > 0$
10. $x < 0$ and $y > 0$

Find $\vec{d}(PQ)$.

11. $P(-1, 10)$, $Q(-1, 6)$
12. $P(6, -3)$, $Q(10, -3)$
13. $P(-3, -2)$, $Q(7, -2)$

Find the slope of $\overleftrightarrow{AB}$.

14. $A(-1, -3)$, $B(7, -5)$
15. $A(6, -3)$, $B(-1, -2)$
16. $A(-10, -3)$, $B(7, -3)$
17. $A(-2, 6)$, $B(-2, 8)$

Write an equation of the line.

18. $A(0, -3)$, $B(6, -1)$
19. $P(6, -2)$, $Q(7, 2)$
20. $P(6, -2)$, $m = 4$
21. $m = -8$, $b = 3$

Find the slope and the y-intercept of the line whose equation is given.

22. $y = -\frac{3}{5}x + 7$
23. $3x + 2y = 8$
24. $y - 2(x - 3) = 0$

Sketch the graph of the line described by each equation.

25. $y = -2x - 3$
26. $2y + 3x = 8$

Graph.

27. $\left\{(x, y) \mid y = \frac{2}{3}x - 1\right\}$
28. $\{(x, y) \mid y = -1\}$
29. $\{(x, y) \mid x = 4\}$

Is the given point on the line described by the equation?

30. $(2, 3)$, $2x - 2y = -2$
31. $(-2, -1)$, $x - 3y = -5$

9 ANALYTIC GEOMETRY

Mathematics and the Electric Company

This is a picture of the control room of a large power control center that coordinates the interchange of power among eight major power systems. Dispatchers can easily spot and correct problems immediately by using sophisticated telemetering, computers, and telephone equipment.

Pythagorean Theorem and its Converse

OBJECTIVES

- To find the length of the hypotenuse or a leg of a right triangle
- To identify right triangles using the converse of the Pythagorean theorem
- To solve problems using the Pythagorean theorem

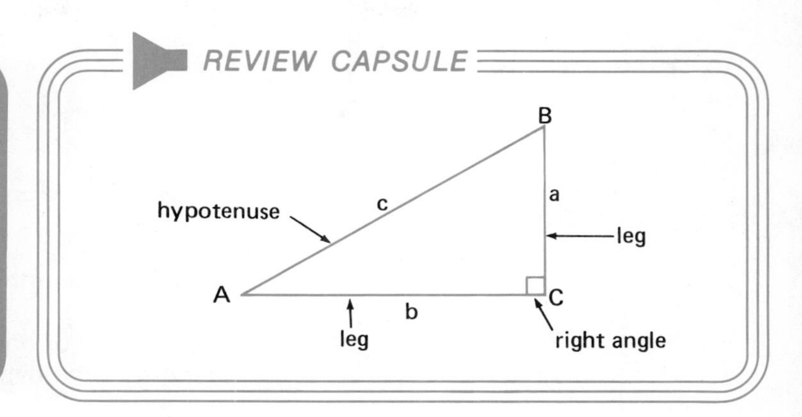

EXAMPLE 1 Find the area of each of the three squares. What is the relationship among the areas?

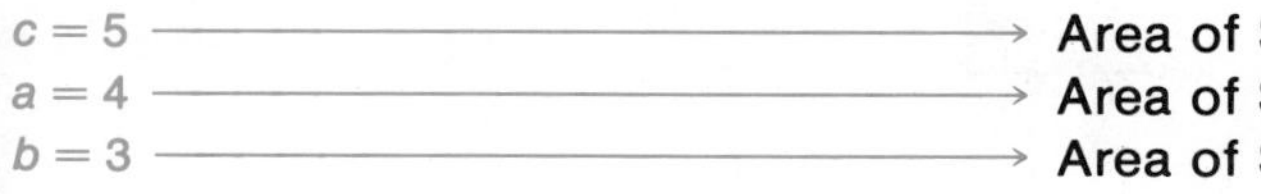

Area of Square I = 5(5) = 25
Area of Square II = 4(4) = 16
Area of Square III = 3(3) = 9

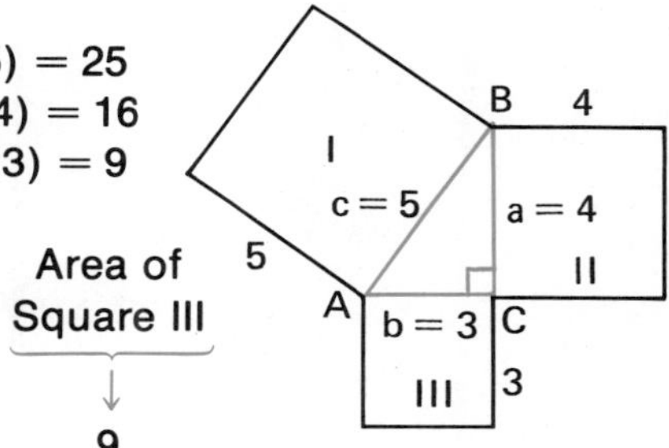

Area of Square I = Area of Square II + Area of Square III

$$25 = 16 + 9$$

The relationship among the areas of the squares gives a relationship among the sides of right $\triangle ABC$. → $c^2 = a^2 + b^2$

Pythagorean theorem →

If $\triangle ABC$ is a right triangle, then $c^2 = a^2 + b^2$. The square of the length of the hypotenuse is equal to the sum of the squares of the lengths of the two legs.

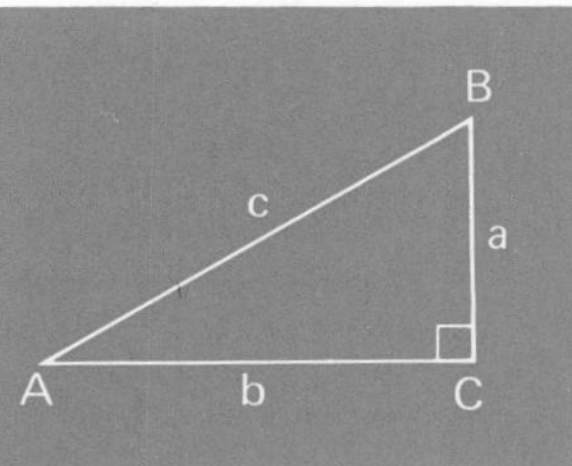

EXAMPLE 2 Find the length of the hypotenuse.

B
c = ?
a = 15
A
C
b = 8

Use the Pythagorean theorem. → $c^2 = a^2 + b^2$

$a = 15$ and $b = 8$ → $c^2 = (15)^2 + (8)^2$

$c^2 = 225 + 64$

$c = \sqrt{289}$; $289 = 17 \cdot 17$ → $c = \sqrt{289}$, or $c = 17$

Thus, the length of the hypotenuse is 17.

EXAMPLE 3 The length of one leg of a right triangle is 5 and the length of the hypotenuse is 13. Find the length of the other leg.

Draw a diagram. →

Use the Pythagorean theorem. →

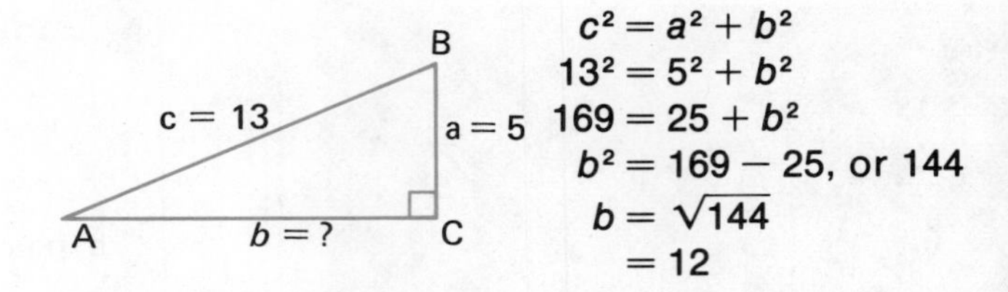

$$c^2 = a^2 + b^2$$
$$13^2 = 5^2 + b^2$$
$$169 = 25 + b^2$$
$$b^2 = 169 - 25, \text{ or } 144$$
$$b = \sqrt{144}$$
$$= 12$$

Thus, the length of the other leg is 12.

EXAMPLE 4 The length of one leg of a right triangle is 5 and the length of the hypotenuse is 10. Find the length of the other leg.

B
c = 10
a = ?
A
C
b = 5

$$c^2 = a^2 + b^2$$
$$10^2 = a^2 + 5^2$$
$$100 = a^2 + 25$$
$$a^2 = 75$$
$$a = \sqrt{75}$$
$$= \sqrt{25 \cdot 3}$$
$$= 5\sqrt{3}$$

$\sqrt{75} = \sqrt{25 \cdot 3}$
$= \sqrt{25} \cdot \sqrt{3}$, or $5\sqrt{3}$

Thus, the length of the other leg is $5\sqrt{3}$.

EXAMPLE 5 A triangle has sides measuring 2, 3, and $\sqrt{13}$. Show that the measures satisfy the equation $c^2 = a^2 + b^2$.

c is the length of the longest side.
$c = \sqrt{13}$, $a = 2$, $b = 3$

$2^2 = 4$, $3^2 = 9$, $(\sqrt{13})^2 = 13$
So, $\sqrt{13} > 2$
and $\sqrt{13} > 3$;
$c = \sqrt{13}$

c^2	$a^2 + b^2$
$(\sqrt{13})^2$	$2^2 + 3^2$
13	$4 + 9$
	13

The triangle with sides measuring 2, 3, and $\sqrt{13}$ is a right triangle.

Thus, the measures satisfy $c^2 = a^2 + b^2$.

Converse of the Pythagorean theorem →

If $c^2 = a^2 + b^2$,
then
$\triangle ABC$ is a
right triangle.

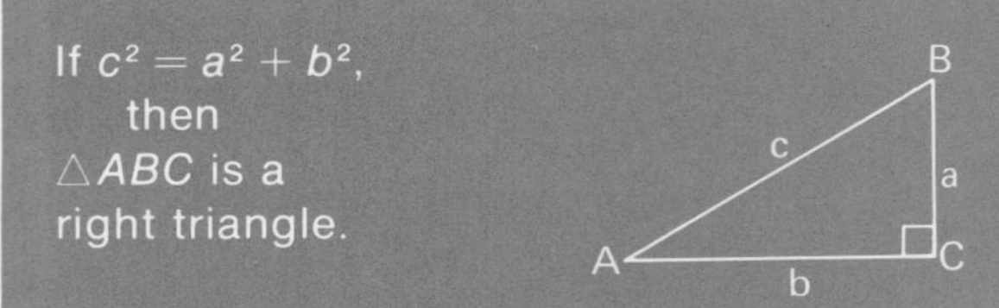

If the square of the length of the longest side of a $\triangle$ equals the sum of the squares of the lengths of the other two sides, then the $\triangle$ is a right $\triangle$.

EXAMPLE 6 If the lengths of the sides of a triangle are $5, \sqrt{7}, 4\sqrt{2}$, is the triangle a right triangle?

$5^2 = 25$, $(\sqrt{7})^2 = 7$,
$(4\sqrt{2})^2 = 4^2(\sqrt{2})^2 = 16(2) = 32$
So, $4\sqrt{2}$ is the length of the longest side c.

c^2	$a^2 + b^2$
$(4\sqrt{2})^2$	$(5)^2 + (\sqrt{7})^2$
$16(2)$	$25 + 7$
32	32

Use the converse of the Pythagorean theorem. → **Thus,** the triangle is a right triangle.

ORAL EXERCISES

Is the triangle a right triangle?

1. 8, 15, 17 *yes* **2.** $\sqrt{5}, \sqrt{2}, 3$ *no* **3.** $\sqrt{6}, 2, \sqrt{10}$ *yes* **4.** 5, 12, 13 *yes* **5.** $\sqrt{9}, \sqrt{18}, 3$ *yes*

EXERCISES

PART A

Find the length of the missing side of each right triangle. (c is the hypotenuse.)

1. $c = 11, b = 5$ **2.** $c = 15, a = 8$ **3.** $a = 4, b = 7$ **4.** $c = 14, a = 7$ **5.** $a = 6, b = 5$
6. $a = 3, b = \sqrt{3}$ *$2\sqrt{3}$* **7.** $a = \sqrt{8}, b = 9$ *$\sqrt{89}$* **8.** $c = 10, a = 6$ *8* **9.** $a = \sqrt{2}, b = \sqrt{3}$ *$\sqrt{5}$* **10.** $c = 4\sqrt{2}, a = 4$ *4*

Is the triangle a right triangle?

11. 6, 8, 11 *no* **12.** 15, 12, 9 *yes* **13.** 10, 24, 26 **14.** 12, 20, 16 **15.** 4, 4, 8 *no*
16. $\sqrt{5}, 2, 1$ *yes* **17.** $2\sqrt{3}, 2\sqrt{6}, 6$ *yes* **18.** $\sqrt{2}, \sqrt{3}, \sqrt{5}$ *yes* **19.** $8, \sqrt{19}, 3\sqrt{5}$ *yes* **20.** $4\sqrt{3}, 2\sqrt{3}, 6$ *yes*

PART B

EXAMPLE A baseball diamond is a square with a distance of 90 feet between bases. Find the distance from home plate to 2nd base to the nearest tenth.

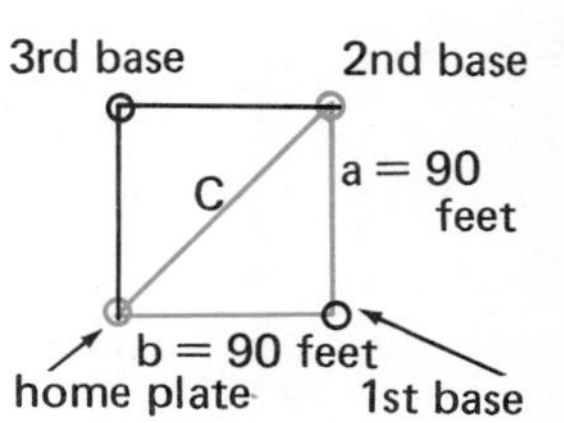

$c^2 = a^2 + b^2$
$c^2 = 90^2 + 90^2 = 8100 + 8100$
$c = \sqrt{16200}$, or $90\sqrt{2}$
$c \doteq 90(1.41)$, or 126.9 to the nearest tenth
Thus, the distance is 126.9 ft to the nearest tenth.

21. A ladder 13 m long leans against a building. The foot of the ladder is 5 m from the base of the building. At what height does it touch the building? *12 m*

22. The length of the diagonal of a TV screen is 50 cm. One side is 40 cm long. Find the area of the screen. *1,200 cm²*

The Distance Between Two Points

OBJECTIVE

- To determine the distance between any two points in a plane

REVIEW CAPSULE

Horizontal Distance	Vertical Distance
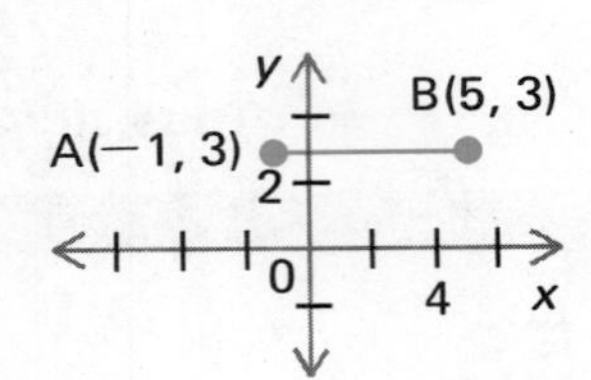	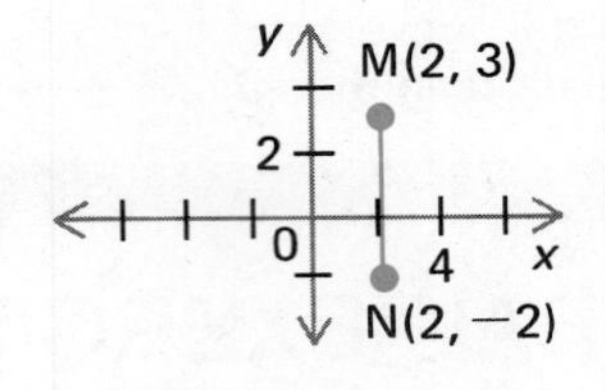
$AB = \lvert 5 - (-1) \rvert$, or 6	$MN = \lvert -2 - 3 \rvert$, or 5
$A(x_1, y)$, $B(x_2, y)$	$M(x, y_1)$, $N(x, y_2)$
$AB = \lvert x_2 - x_1 \rvert$	$MN = \lvert y_2 - y_1 \rvert$

EXAMPLE 1 Right triangle ABC has coordinates $A(2, 2)$, $B(5, 6)$ and $C(5, 2)$. Find the length of each side.

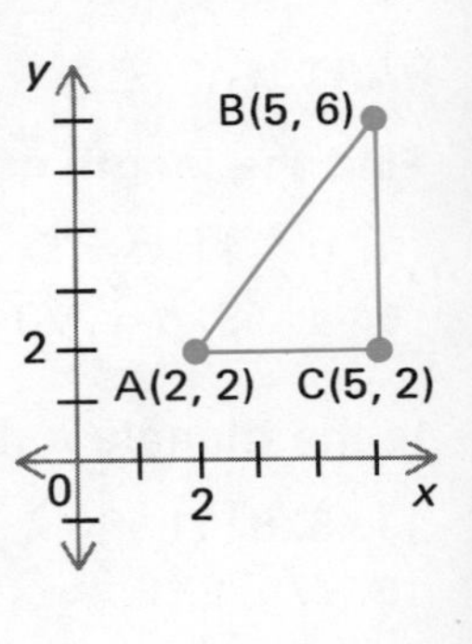

$AC = |5 - 2| = 3$
$BC = |6 - 2| = 4$

Use the Pythagorean theorem. ⟶ $(AB)^2 = 3^2 + 4^2$
$(AB)^2 = 9 + 16$, or 25
$AB = 5$

Thus, $AC = 3$, $BC = 4$, and $AB = 5$.

EXAMPLE 2 Right $\triangle ABC$ has coordinates $A(2, 1)$ and $B(5, 6)$. Find the length of $\overline{AB}$.

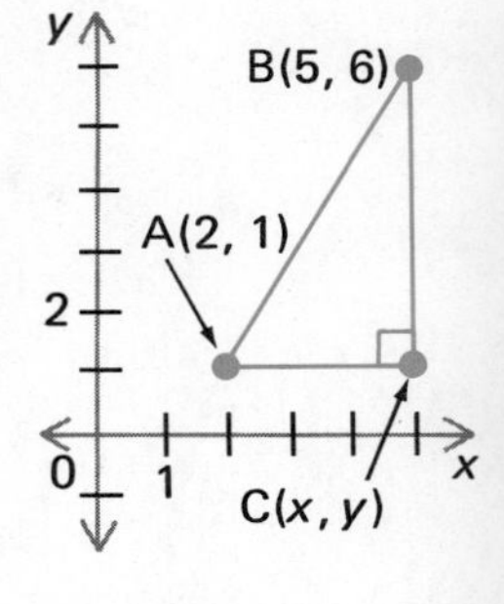

Find the coordinates of C. ⟶ ($\overline{CB}$ is vertical.) Same x-coord. as $B(5, 6)$. ($\overline{AC}$ is horizontal.) Same y-coord. as $A(2, 1)$.

$C(5, 1)$

Points ⟶ $A(2, 1)$, $B(5, 6)$, $C(5, 1)$

Pythagorean theorem ⟶ $(AB)^2 = (AC)^2 + (CB)^2$
$= |5 - 2|^2 + |6 - 1|^2$

$3^2 + 5^2 = 9 + 25$ or 34 ⟶ $= 3^2 + 5^2$
$(AB)^2 = 34$

Take the square root of each side. ⟶ $AB = \sqrt{34}$

Thus, the length of $\overline{AB}$ is $\sqrt{34}$.

EXAMPLE 3 Right $\triangle ABC$ has coordinates $A(x_1, y_1)$ and $B(x_2, y_2)$. Find the length of $\overline{AB}$.

$C(x_2, y_1)$

x-coord. of B ↑ ↑ y-coord. of A

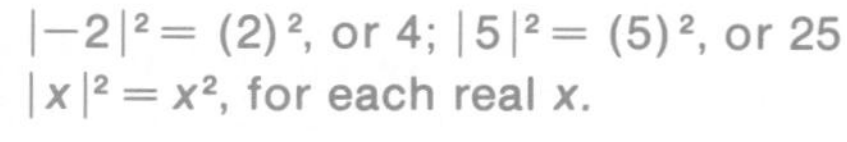

$|-2|^2 = (2)^2$, or 4; $|5|^2 = (5)^2$, or 25

$|x|^2 = x^2$, for each real x.

$$(AB)^2 = (AC)^2 + (CB)^2$$
$$= |x_2 - x_1|^2 + |y_2 - y_1|^2$$
$$(AB)^2 = (x_2 - x_1)^2 + (y_2 - y_1)^2$$
$$AB = \sqrt{(x_2 - x_1)^2 + (y_2 - y_1)^2}$$

Thus, the length of $\overline{AB}$ is $\sqrt{(x_2 - x_1)^2 + (y_2 - y_1)^2}$.

Distance formula →

> The distance d between $A(x_1, y_1)$ and $B(x_2, y_2)$ is given by this formula.
> $$d = \sqrt{(x_2 - x_1)^2 + (y_2 - y_1)^2}$$

EXAMPLE 4 Find the distance between $A(2, 1)$ and $B(6, 8)$.

$A(2, 1)$ → (x_1, y_1) $B(6, 8)$ → (x_2, y_2)

$$d = \sqrt{(x_2 - x_1)^2 + (y_2 - y_1)^2}$$
$$= \sqrt{(6-2)^2 + (8-1)^2}$$
$$= \sqrt{4^2 + 7^2}$$
$$= \sqrt{16 + 49}$$
$$= \sqrt{65}$$

Thus, the distance is $\sqrt{65}$.

EXERCISES

PART A

Find the distance between the given points. Answer may be left in simplified radical form.

1. $A(2, 4), B(6, 6)$
2. $R(-2, -3), S(-7, 5)$
3. $P(-2, 4), Q(7, 8)$
4. $M(7, 8), N(-2, -4)$ *15*
5. $A(6, -2), B(-3, 4)$
6. $A(5, 1), B(-2, 4)$
7. $M(8, 4), N(-2, -3)$
8. $A(0, 3), B(4, 0)$ *5*
9. $M(0, -4), N(5, 0)$
10. $M(9, 1), N(-2, -1)$
11. $A(5, 1), B(7, -3)$
12. $P(10, 2), Q(-1, -3)$
13. $R(-5, -4), S(-2, -3)$
14. $P(-1, 4), Q(5, -3)$
15. $R(-2, -2), S(-4, -4)$
16. $R(-10, 12), S(2, -5)$
17. $P(10, 0), Q(4, 6)$ *$6\sqrt{2}$*
18. $A(7, 2), B(-2, -1)$ *$3\sqrt{10}$*
19. $A(-7, -2), B(6, 1)$ *$\sqrt{178}$*
20. $M(12, -1), N(-3, 5)$ *$3\sqrt{29}$*

PART B

Find the distance between the given points. Answer may be left in simplified radical form.

21. $P(1, -3), Q(\frac{2}{3}, -\frac{1}{2})$ *$\frac{\sqrt{229}}{6}$*
22. $A(\frac{1}{2}, \frac{1}{4}), B(-3, 2)$ *$\frac{7\sqrt{5}}{4}$*
23. $A(1, 2), B(\frac{1}{6}, -\frac{1}{3})$ *$\frac{\sqrt{221}}{6}$*
24. $A(\frac{1}{3}, \frac{1}{4}), B(-\frac{2}{3}, \frac{3}{4})$ *$\frac{\sqrt{5}}{2}$*
25. $P(\frac{1}{5}, -\frac{1}{4}), Q(\frac{3}{5}, \frac{1}{2})$ *$\frac{17}{20}$*
26. $A(\frac{2}{3}, \frac{1}{3}), B(\frac{1}{3}, \frac{2}{3})$ *$\frac{\sqrt{2}}{3}$*
27. $M(a, b), N(3a, 2b)$
28. $M(2x, 3y), N(-3x, y)$
29. $M(s, p), N(-3s, -2p)$

Cost of Electricity

A 100-watt light bulb burns for 1 hour. Therefore, 100 watt-hours of electricity are used. A 300-watt light bulb burns for 4 hours. Therefore, 1,200 watt-hours of electricity are used. The watt-hour is too small a unit for measuring the cost of electricity. The kilowatt (1,000 watts) is used.

$$8{,}760 \text{ watts} = \frac{8{,}760}{1{,}000} = 8.76 \text{ kilowatts}$$

Cost of Electricity (for Residential Use) per Month

1st 12 kilowatt hours	$2.35 per KWH
Next 43 kilowatt hours	.0541 per KWH
Next 45 kilowatt hours	.0437 per KWH
Next 400 kilowatt hours	.0341 per KWH

PROBLEM

Find the cost of electricity for a 30-day month if a family uses, on the average, 16 kilowatt hours per day.
The family uses 30(16), or 480 kilowatt hours per month.

The first 12 kilowatt hours cost 12(2.35), or $28.20.
The next 43 kilowatt hours cost 43(.0541), or $2.33.
The next 45 kilowatt hours cost 45(.0437), or $1.97.
The next 380 kilowatt hours cost 380(.0341), or $12.96.

Thus, the cost of electricity is $45.46.

PROJECT

Find the cost of electricity for a 30-day month if the Johnson family's average use each evening is the following.

1	220-watt television	4.5	hours
4	100-watt light bulbs	6	hours
5	150-watt light bulbs	3	hours
1	250-watt stereo	2.25	hours
1	1,000-watt hairdryer	1.75	hours

Midpoint Formula

OBJECTIVES

- To determine the coordinates of the midpoint of a segment
- To determine the coordinates of an endpoint of a segment given the coordinates of the other endpoint and the midpoint

REVIEW CAPSULE

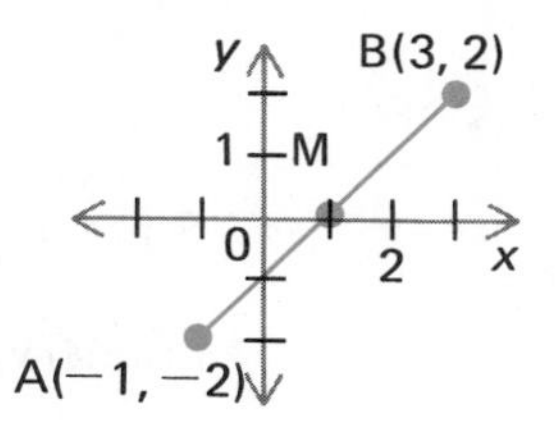

M is the midpoint of $\overline{AB}$ if $AM = MB$.

A midpoint divides a segment into two segments of equal length.

EXAMPLE 1 Determine the coordinates (x, y) of M, the midpoint of the line segment joining $P(3, 2)$ and $Q(7, 2)$.

Use the definition of midpoint. $\overline{PM}$ and $\overline{MQ}$ are horizontal.

$$PM = MQ$$

From the diagram, $x > 3$ so $|x - 3| = x - 3$. $7 > x$ so $|7 - x| = 7 - x$.

$$|x - 3| = |7 - x|$$
$$x - 3 = 7 - x$$
$$2x = 10$$
$$x = 5$$

Every point on a horizontal line has the same y-coordinate. and $y = 2$

Thus, $M(5, 2)$ is the midpoint of $\overline{PQ}$.

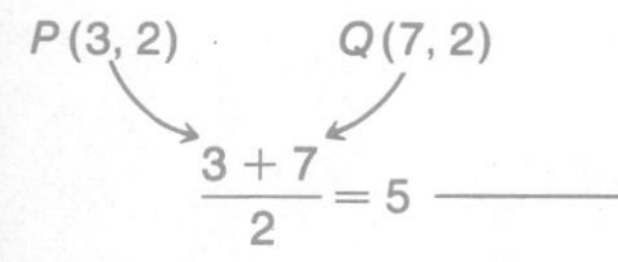

$\frac{3+7}{2} = 5$ → The x-coordinate of the midpoint M is the average of the x-coordinates of P and Q.

EXAMPLE 2 Determine the coordinates (x, y) of M, the midpoint of the line segment joining $P(4, 1)$ and $Q(4, 5)$.

$\overline{QM}$ and $\overline{MP}$ are vertical.

$$QM = MP$$

From the diagram, $5 > y$ so $|5 - y| = 5 - y$. $y > 1$ so $|y - 1| = y - 1$.

$$|5 - y| = |y - 1|$$
$$5 - y = y - 1$$
$$6 = 2y$$
$$y = 3$$

Every point on a vertical line has the same x-coordinate. and $x = 4$

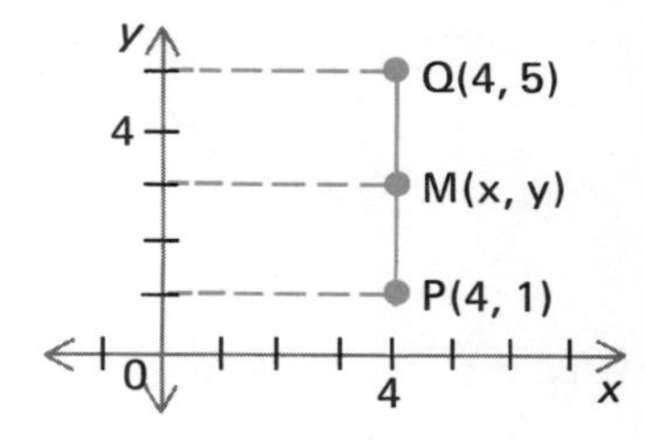

Thus, $M(4, 3)$ is the midpoint of $\overline{PQ}$.

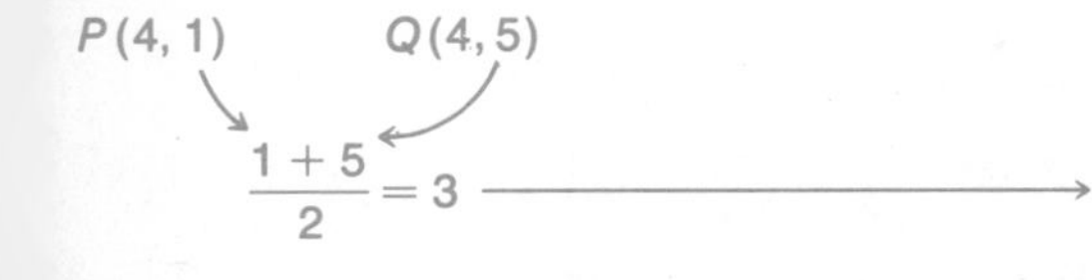

$\frac{1+5}{2} = 3$ → The y-coordinate of the midpoint M is the average of the y-coordinates of P and Q.

EXAMPLE 3 Determine the coordinates (x, y) of M, the midpoint of the line segment joining $P(2, 1)$ and $Q(6, 5)$.

Draw a vertical line from Q and a horizontal line from P.

$\overline{PS}$ is horizontal.
$P(2, 1)$ $S(x, y)$
$1 = y$
$\overline{QS}$ is vertical.
$Q(6, 5)$ $S(x, y)$
$6 = x$

The coordinates of S are $(6, 1)$.

Let $M(x, y)$ be the midpoint of $\overline{PQ}$.

Use the property of the midpoint of a horizontal line.

$T(x, 1)$ is the midpoint of $\overline{PS}$.
$P(2, 1)$ $S(6, 1)$

Use the observation following Example 1.

$$x = \frac{2+6}{2}, \text{ or } 4$$

Use the property of the midpoint of a vertical line.

$R(6, y)$ is the midpoint of $\overline{QS}$.
$Q(6, 5)$ $S(6, 1)$

Use the observation following Example 2.

$$y = \frac{5+1}{2}, \text{ or } 3$$

Thus, $M(4, 3)$ is the midpoint of $\overline{PQ}$.

Midpoint Formula.

If $\overline{PQ}$ has endpoints $P(x_1, y_1)$ and $Q(x_2, y_2)$, then the midpoint M has coordinates $\left(\frac{x_1 + x_2}{2}, \frac{y_1 + y_2}{2}\right)$.

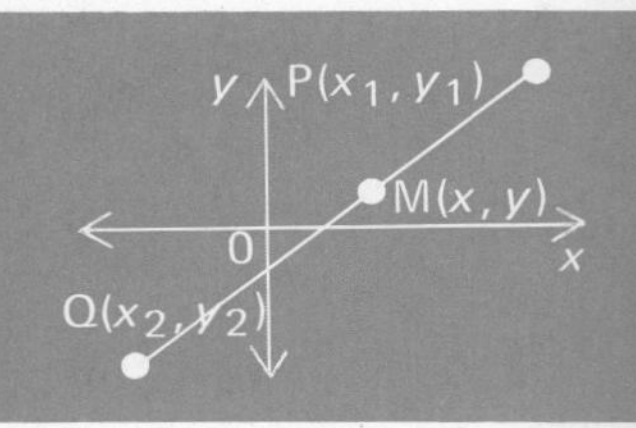

EXAMPLE 4 Determine the coordinates (x, y) of M, the midpoint of the line segment joining $P(4, -2)$ and $Q(6, 4)$.

$P(4, -2)$ $Q(6, 4)$
(x_1, y_1) (x_2, y_2)

Use the midpoint formula.

$$x = \frac{x_1 + x_2}{2} = \frac{4+6}{2} = 5 \qquad y = \frac{y_1 + y_2}{2} = \frac{-2+4}{2} = 1$$

Thus, $M(5, 1)$ is the midpoint of $\overline{PQ}$.

EXERCISES

PART A

Determine the coordinates (*x*, *y*) of *M*, the midpoint of the segment joining *P* and *Q*.

1. $P(3, 1)$, $Q(5, 3)$ *(4, 2)*
2. $P(-2, 1)$, $Q(8, 7)$ *(3, 4)*
3. $P(6, 2)$, $Q(-2, -2)$ *(2, 0)*
4. $P(7, 1)$, $Q(3, 2)$ *(5, $\frac{3}{2}$)*
5. $P(5, 2)$, $Q(3, 8)$ *(4, 5)*
6. $P(-6, 1)$, $Q(4, 5)$ *(−1, 3)*
7. $P(5, -4)$, $Q(-1, -2)$
8. $P(9, -2)$, $Q(1, 6)$ *(5, 2)*
9. $P(-3, 5)$, $Q(7, -3)$ *(2, 1)*
10. $P(5, 1)$, $Q(3, 5)$ *(4, 3)*
11. $P(8, 2)$, $Q(4, 6)$ *(6, 4)*
12. $P(-3, -1)$, $Q(5, 7)$ *(1, 3)*
13. $P(0, 4)$, $Q(6, 0)$ *(3, 2)*
14. $P(8, 0)$, $Q(0, -10)$ *(4, −5)*
15. $P(0, -6)$, $Q(-8, 0)$ *(−4, −3)*
16. $P(-10, 0)$, $Q(0, -12)$
17. $P(-6, -3)$, $Q(-4, -7)$
18. $P(-7, -5)$, $Q(-3, -1)$
19. $P(-8, -2)$, $Q(-2, -4)$ *(−5, −3)*
20. $P(-9, -3)$, $Q(-1, -7)$ *(−5, −5)*
21. $P(-11, -5)$, $Q(-2, -8)$ *($-\frac{13}{2}$, $-\frac{13}{2}$)*

PART B

EXAMPLE Find the coordinates of the endpoint Q of $\overline{PQ}$ for $P(2, 3)$ and its midpoint $M(4, 7)$.

$P(2, 3)$ (x_1, y_1) $M(4, 7)$ (x, y) $Q(x_2, y_2)$

Use the midpoint formula. ⟶ $x = \frac{x_1 + x_2}{2}$ $y = \frac{y_1 + y_2}{2}$

Substitute known values. ⟶ $4 = \frac{2 + x_2}{2}$ $7 = \frac{3 + y_2}{2}$

Solve: Multiply each side by 2. ⟶ $8 = 2 + x_2$ $14 = 3 + y_2$

$6 = x_2$ $11 = y_2$

Thus, $Q(6, 11)$ is the other endpoint.

Find the coordinates of the endpoint *Q* of $\overline{PQ}$, given endpoint *P* and midpoint *M*.

22. $P(4, 3)$, $M(6, 5)$ *(8, 7)*
23. $P(-1, -2)$, $M(4, 3)$ *(9, 8)*
24. $P(8, -1)$, $M(4, 6)$ *(0, 13)*
25. $P(9, 2)$, $M(4, 0)$ *(−1, −2)*
26. $P(-3, -2)$, $M(1, 2)$ *(5, 6)*
27. $P(-8, -6)$, $M(-2, -4)$
28. $P(8, -2)$, $M(-2, -1)$
29. $P(4, 3)$, $M(6, 2)$ *(8, 1)*
30. $P(3, 4)$, $M(4, 8)$ *(5, 12)*
31. $P(-6, -3)$, $M(2, 0)$ *(10, 3)*
32. $P(-4, -2)$, $M(-2, -2)$
33. $P(-1, -1)$, $M(4, 4)$ *(9, 9)*
34. $P(0, 6)$, $M(2, -7)$ *(4, −20)*
35. $P(0, 8)$, $M(4, -2)$ *(8, −12)*
36. $P(0, 8)$, $M(0, 6)$ *(0, 4)*
37. $P(5, 8)$, $M(-2, -3)$
38. $P(4, -1)$, $M(6, 4)$ *(8, 9)*
39. $P(-4, -7)$, $M(0, 0)$ *(4, 7)*
40. $P(8, -2)$, $M(-3, -7)$ *(−14, −12)*
41. $P(-5, -3)$, $M(2, -3)$ *(9, −3)*
42. $P(-7, 3)$, $M(2, 8)$ *(11, 13)*

PART C

43. Given $P(-1, 2)$, $Q(1, -1)$, $R(5, 1)$, and $S(1, 3)$. *M*, *N*, *T*, *V* are midpoints of $\overline{PQ}$, $\overline{QR}$, $\overline{RS}$, and $\overline{SP}$ respectively. Graph and find *M*, *N*, *T*, and *V*. Show that $MN = VT$ and $MV = NT$.

44. Given $P(a, 0)$, $Q(b, c)$, $R(d, e)$, and $S(f, g)$. *M*, *N*, *T*, *V* are midpoints of $\overline{PQ}$, $\overline{QR}$, $\overline{RS}$, and $\overline{SP}$ respectively. Graph and find *M*, *N*, *T*, and *V*. Show that $MN = VT$ and $MV = NT$.

Parallels and Perpendiculars

OBJECTIVES

- To determine whether two lines are parallel, or perpendicular, from their slopes
- To find the slope or write the equation of a line parallel, or perpendicular, to a given line

REVIEW CAPSULE

Write an equation of $\overleftrightarrow{AB}$ for $A(4, -1)$ and $B(2, 7)$.

$m = \frac{7-(-1)}{2-4}$, or -4 ← Find the slope.

$y - 7 = -4(x-2)$ ← Use the point-slope form.

EXAMPLE 1 $\overleftrightarrow{PQ} \parallel \overleftrightarrow{MN} \parallel \overleftrightarrow{RS}$. Find the slope of $\overleftrightarrow{PQ}$, $\overleftrightarrow{MN}$, and $\overleftrightarrow{RS}$.

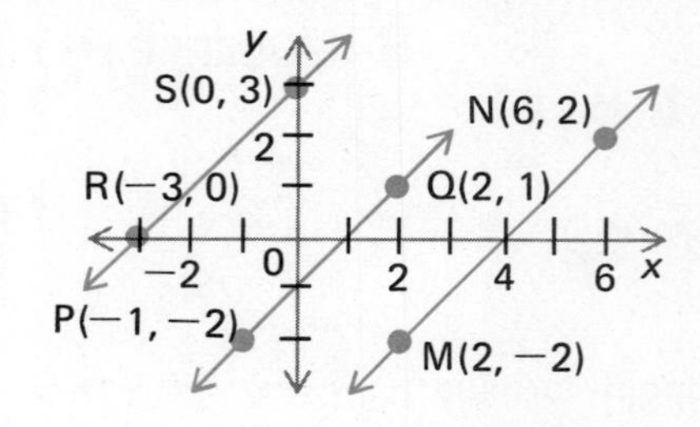

$m = \frac{\text{diff. of } y\text{-coordinates}}{\text{diff. of } x\text{-coordinates}}$

$$m\ (\overleftrightarrow{PQ}) = \frac{1-(-2)}{2-(-1)} = \frac{1+2}{2+1} = 1$$

$$m\ (\overleftrightarrow{MN}) = \frac{2-(-2)}{6-2} = \frac{2+2}{4} = 1$$

$$m\ (\overleftrightarrow{RS}) = \frac{3-0}{0-(-3)} = \frac{3}{3} = 1$$

Thus, the slope of each is 1.

Two vertical lines are parallel, but their slopes are undefined.

> Parallel nonvertical lines have the same slope.
> If two lines have the same slope, they are parallel.

EXAMPLE 2 Find the slope of a line which is parallel to the line with equation $2x + y = 8$.

Write $2x + y = 8$ in $y = mx + b$ form.

$$y = -2x + 8$$

↑ m ↑ b

-2 is the slope.

Two parallel lines have the same slope.

Thus, the slope of a line parallel to the line with equation $2x + y = 8$ is -2.

EXAMPLE 3 Write an equation of the line passing through the point (2, 3) and which is parallel to the line with equation $3x - 2y = 8$.

$$3x - 2y = 8$$
$$-2y = -3x + 8$$
$$y = \frac{3}{2}x - 4$$

Parallel lines have same slope. $\downarrow$ $m = \frac{3}{2}$ | Point $\swarrow$ $(x_1, y_1) = (2, 3)$

Write the equation. (Use the point-slope form.)

$$y - y_1 = m(x - x_1)$$
$$y - 3 = \frac{3}{2}(x - 2)$$

Use the point-slope form. $\longrightarrow$ **Thus,** $y - 3 = \frac{3}{2}(x - 2)$ is an equation of the line.

EXAMPLE 4 $\overleftrightarrow{PQ} \perp \overleftrightarrow{RS}$. Find the slope of $\overleftrightarrow{PQ}$ and $\overleftrightarrow{RS}$.

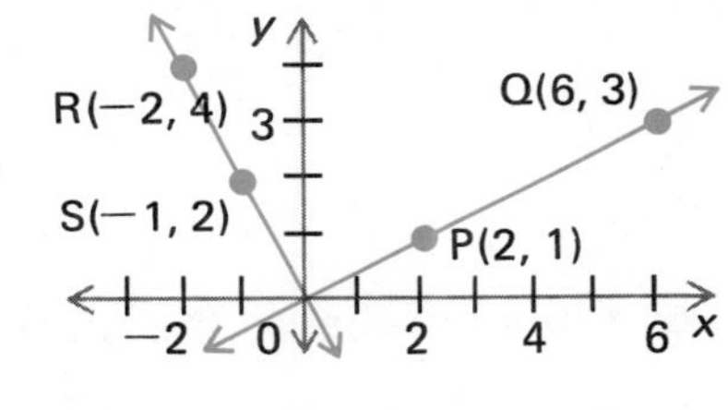

$$m(\overleftrightarrow{PQ}) = \frac{3-1}{6-2} = \frac{2}{4} = \frac{1}{2}$$
$$m(\overleftrightarrow{RS}) = \frac{4-2}{-2-(-1)} = \frac{2}{-2+1} = -2$$

$m(\overleftrightarrow{PQ}) \cdot m(\overleftrightarrow{RS}) = -2 \cdot \frac{1}{2} = -1$

Two numbers whose product is -1 are negative reciprocals.

Thus, the slope of $\overleftrightarrow{PQ}$ is $\frac{1}{2}$ and the slope of $\overleftrightarrow{RS}$ is -2.

For lines which are neither vertical nor horizontal

> Perpendicular lines have slopes which are negative reciprocals.
> If two lines have slopes which are negative reciprocals, then they are perpendicular.

EXAMPLE 5 Write an equation of the line passing through the point (2, 3) and perpendicular to the line with equation $3y - 2x = 4$.

Write in $y = mx + b$ form.

$$3y = 2x + 4 \quad \text{or} \quad y = \frac{2}{3}x + \frac{4}{3}$$

($\frac{2}{3}$ is m)

negative reciprocals

$$\left(\frac{2}{3}\right)\left(-\frac{3}{2}\right) = -1$$

The slope of a perpendicular line is $-\frac{3}{2}$.

Use $y - y_1 = m(x - x_1)$. $\longrightarrow$ **Thus,** the equation is $y - 3 = -\frac{3}{2}(x - 2)$.

EXAMPLE 6 Write an equation of the line perpendicular to the one joining $A(2,-3)$ and $B(2,5)$ and passing through B.

$$m(\overleftrightarrow{AB}) = \frac{5-(-3)}{2-2} \text{ is undefined.}$$

Horizontal and vertical lines are perpendicular. ⟶

So, $\overleftrightarrow{AB}$ is a vertical line.
A line perpendicular to $\overleftrightarrow{AB}$ is horizontal.
So, $m = 0$. A point on the line is $(2, 5)$.

$y - 5 = 0(x - 2)$ ⟶ **Thus,** $y = 5$ is an equation of the line.

EXAMPLE 7 Write an equation of the line perpendicular to the line with equation $y = 3$ and passing through $(-1, 4)$.

$y = 3$ represents a horizontal line. A line perpendicular to this line is vertical.

A vertical line has equation $x = k$. ⟶ **Thus,** $x = -1$ is an equation of the line.

EXERCISES

PART A

Find the slope of a line parallel to the given line.

1. $2x - 3y = 7$ $\frac{2}{3}$ **2.** $-6x + 5y = 4$ $\frac{6}{5}$ **3.** $x - 2y = 16$ $\frac{1}{2}$ **4.** $3x + 4y = -8$

Find the slope of a line perpendicular to the given line.

5. $-3x + 5y = 8$ $-\frac{5}{3}$ **6.** $7x + 2y = -10$ $\frac{2}{7}$ **7.** $5x - 6y = 4$ $-\frac{6}{5}$ **8.** $-x + 2y = 2$

Write an equation of the line which is parallel to the given line and passes through the given point.

9. $3x - 2y = 10$, $(-2, 1)$ **10.** $2x - 3y = 9$, $(0, 0)$ $y = \frac{2}{3}x$ **11.** $-2x + 4y = 8$, $(2, 3)$
12. $2x + 4y = 8$, $(2, 4)$ **13.** $y = 7$, $(0, 2)$ $y = 2$ **14.** $-3y = 9$, $(10, 6)$ $y = 6$

Write an equation of the line which is perpendicular to the given line and passes through the given point.

15. $2x + 4y = 12$, $(4, 3)$ **16.** $-3x + 2y = 10$, $(1, -3)$ **17.** $4y - 3x = 16$, $(-1, 4)$
18. $7y = 14$, $(-3, 2)$ $x = -3$ **19.** $2x - 3y = 9$, $(-1, 4)$ **20.** $-3x + 4y = 8$, $(2, -1)$

PART B

Determine whether $\overleftrightarrow{PQ} \parallel \overleftrightarrow{RS}$, $\overleftrightarrow{PQ} \perp \overleftrightarrow{RS}$, or neither.

21. $P(2, 3)$, $Q(6, 5)$, $R(8, 1)$, $S(10, 5)$ N **22.** $P(-1, 2)$, $Q(9, 6)$, $R(-4, 7)$, $S(-2, 2)$ $\perp$

PART C

Write an equation of the line which is the perpendicular bisector of the line joining points P and Q.

23. $P(-2, 3)$, $Q(4, 7)$ **24.** $P(1, 5)$, $Q(7, -3)$ **25.** $P(6, -3)$, $Q(4, 7)$

Applications of Formulas

OBJECTIVE

- To apply slope, distance, and midpoint formulas

REVIEW CAPSULE

To show $ABCD$ is parallelogram, show $\overline{AB} \parallel \overline{DC}$ and $\overline{AD} \parallel \overline{BC}$, or show $AB = DC$ and $AD = BC$.

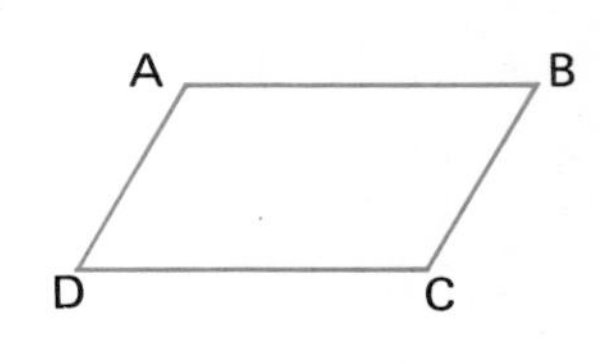

EXAMPLE 1 Show that the points $A(-2, 1)$, $B(5, 2)$, and $C(2, 5)$ form a right triangle. Use the slope formula.

Draw a sketch.
Find the slopes of the segments that appear $\perp$.

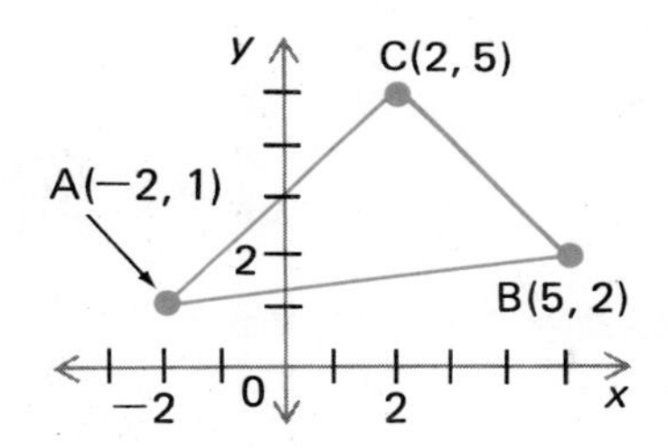

$$m(\overline{AC}) = \frac{5-1}{2-(-2)} = \frac{4}{4} = 1$$

$$m(\overline{BC}) = \frac{5-2}{2-5} = \frac{3}{-3} = -1$$

Slopes are negative reciprocals. ⟶ $\overline{AC}$ is perpendicular to $\overline{BC}$ ($\overline{AC} \perp \overline{BC}$).

$\triangle ABC$ has a right angle. ⟶ **Thus,** $\triangle ABC$ is a right triangle.

EXAMPLE 2 Show that the triangle in Example 1 is a right triangle by using the distance formula.

$d = \sqrt{(x_2 - x_1)^2 + (y_2 - y_1)^2}$ ⟶

$$AB = \sqrt{[5-(-2)]^2 + (2-1)^2} = \sqrt{49+1} = \sqrt{50}$$
$$AC = \sqrt{[2-(-2)]^2 + (5-1)^2} = \sqrt{16+16} = \sqrt{32}$$
$$BC = \sqrt{(5-2)^2 + (2-5)^2} = \sqrt{9+9} = \sqrt{18}$$

Check if $c^2 = a^2 + b^2$. ⟶

$(AB)^2$	$(AC)^2 + (BC)^2$
$(\sqrt{50})^2$	$(\sqrt{32})^2 + (\sqrt{18})^2$
50	32 + 18
	50

If $(AB)^2 = (AC)^2 + (BC)^2$, then $\triangle ABC$ is a right $\triangle$. ⟶ **Thus,** $\triangle ABC$ is a right triangle.

EXAMPLE 3 Do the points $A(-3, -2)$, $B(0, 0)$, and $C(6, 4)$ belong to the same line?

Find the slopes of $\overleftrightarrow{AB}$ and $\overleftrightarrow{BC}$.

$$m(\overleftrightarrow{AB}) = \frac{-2-0}{-3-0} = \frac{-2}{-3} = \frac{2}{3} \qquad m(\overleftrightarrow{BC}) = \frac{4-0}{6-0} = \frac{4}{6} = \frac{2}{3}$$

If $m(\overleftrightarrow{AB}) = m(\overleftrightarrow{BC})$, then the points are on the same line. ⟶ **Thus,** A, B, and C are on the same line.

EXAMPLE 4 Find the coordinates of the midpoints D and E of $\overline{AB}$ and $\overline{BC}$. Show that $\overline{DE} \parallel \overline{AC}$.

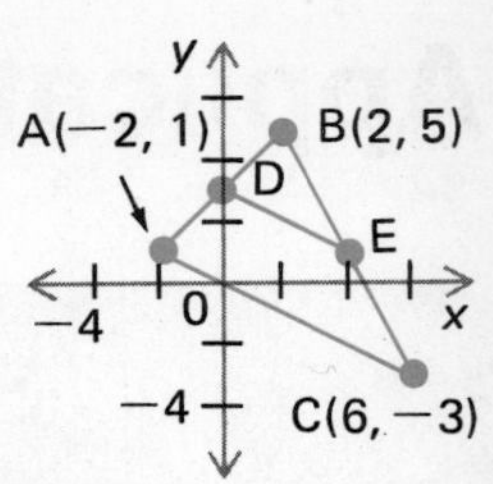

Use the midpoint formula.

x-coord. $= \dfrac{x_1 + x_2}{2}$

y-coord. $= \dfrac{y_1 + y_2}{2}$

$A(-2, 1)$, $B(2, 5)$

$$x = \frac{-2+2}{2}, \text{ or } 0$$

$$y = \frac{1+5}{2}, \text{ or } 3$$

The midpoint of $\overline{AB}$ is $D(0, 3)$.

$B(2, 5)$, $C(6, -3)$

$$x = \frac{2+6}{2}, \text{ or } 4$$

$$y = \frac{5+(-3)}{2}, \text{ or } 1$$

The midpoint of $\overline{BC}$ is $E(4, 1)$.

Use the slope formula. ⟶

$$m(\overline{DE}) = \frac{3-1}{0-4}$$

$\overline{AC}$ and $\overline{DE}$ have the same slope. ⟶

$$= \frac{2}{-4}, \text{ or } -\frac{1}{2}$$

$$m(\overline{AC}) = \frac{-3-1}{6-(-2)} = \frac{-4}{8}, \text{ or } -\frac{1}{2}$$

Thus, $\overline{DE} \parallel \overline{AC}$.

EXAMPLE 5 Show $DE = \frac{1}{2}(AC)$ in Example 4.

Use the distance formula. ⟶

$D(0, 3)$, $E(4, 1)$

Find DE.

$\sqrt{(0-4)^2 + (3-1)^2}$

$\sqrt{(-4)^2 + 2^2}$

$\sqrt{16+4}$

$\sqrt{20}$

$A(-2, 1)$, $C(6, -3)$

Find AC.

$\sqrt{(-2-6)^2 + [1-(-3)]^2}$

$\sqrt{(-8)^2 + 4^2}$

$\sqrt{64+16} = \sqrt{80}$

$\sqrt{80} = 2\sqrt{20}$

$\sqrt{80} = \sqrt{4}\sqrt{20} = 2\sqrt{20}$

Thus, $DE = \frac{1}{2}(AC)$.

EXAMPLE 6 The vertices of a quadrilateral are $A(0, 0)$, $B(6, 1)$, $C(5, 4)$, and $D(-1, 3)$. Show that $ABCD$ is a parallelogram.

Draw a sketch.

Use slopes to show $\overline{AB} \parallel \overline{DC}$.

$$m(\overline{AB}) = \frac{1-0}{6-0} = \frac{1}{6}$$

$$m(\overline{DC}) = \frac{4-3}{5-(-1)} = \frac{1}{6}$$

The slopes are the same. ⟶ So, $\overline{AB} \parallel \overline{DC}$.

Use slopes to show $\overline{AD} \parallel \overline{BC}$.

$$m(\overline{AD}) = \frac{3-0}{-1-0} = -3$$

$$m(\overline{BC}) = \frac{4-1}{5-6} = \frac{3}{-1} = -3$$

The slopes are the same. ⟶ So, $\overline{AD} \parallel \overline{BC}$.

Both pairs of opposite sides are parallel. ⟶ **Thus,** $ABCD$ is a parallelogram.

EXERCISES

PART A

Use the slope formula to show that the given points form right triangles.

1. $A(1, 2)$, $B(2, 5)$, $C(-1, 6)$
2. $A(-1, -2)$, $B(5, 1)$, $C(3, 5)$
   $m(\overline{AB}) = \frac{1}{2}$, $m(\overline{BC}) = -2$

Use the distance formula to show that the given points form right triangles.

3. $A(5, -5)$, $B(7, 3)$, $C(-1, 5)$
4. $A(2, 3)$, $B(4, -1)$, $C(8, 1)$

Determine whether the three points belong to the same line.

5. $A(0, 0)$, $B(2, 3)$, $C(4, 6)$ *yes*
6. $A(4, 2)$, $B(8, 4)$, $C(-2, -1)$ *yes*
7. $R(-1, -2)$, $S(1, 2)$, $T(4, 8)$ *yes*
8. $R(-3, -1)$, $S(6, 2)$, $T(9, 3)$ *yes*
9. $M(-1, -3)$, $N(0, 0)$, $P(1, 4)$ *no*
10. $M(0, 0)$, $N(5, 2)$, $P(8, 4)$ *no*

11. The vertices of a quadrilateral are $A(0, 0)$, $B(1, 2)$, $C(6, 5)$, and $D(5, 3)$. Show that $ABCD$ is a parallelogram.
    $m(\overline{AB}) = m(\overline{DC}) = 2$; $m(\overline{AD}) = m(\overline{BC}) = \frac{3}{5}$
12. The vertices of a triangle are $A(2, 1)$, $B(7, 4)$, and $C(2, 7)$. Show that $\triangle ABC$ is isosceles. *$AB = BC = \sqrt{34}$*
13. Find the coordinates of the midpoints of sides $\overline{AB}$ and $\overline{BC}$ of triangle $A(2, 1)$, $B(8, 3)$, and $C(4, 5)$. Show that the length of the line segment joining the midpoints is $\frac{1}{2}$ the length of the third side $\overline{AC}$.
    midpt. $\overline{AB}(5, 2)$, midpt. $\overline{BC}(6, 4)$
14. Find the coordinates of the midpoints of sides $\overline{BC}$ and $\overline{AC}$ of triangle $A(-1, -2)$, $B(7, 6)$, and $C(3, 8)$. Show that the line segment joining the midpoints is parallel to the third side $\overline{AB}$.
    midpt. $\overline{BC}(5, 7)$, midpt. $\overline{AC}(1, 3)$

PART B

Given quadrilateral $R(0, 0)$, $S(6, 0)$, $T(8, 3)$, $V(2, 3)$.

15. Show that the opposite sides are parallel.
16. Is $RT = VS$? *no*
17. Show that the opposite sides have the same length.
18. Show that $\overline{RT}$ bisects $\overline{VS}$.

19. For quadrilateral $R(0, 0)$, $S(6, 1)$, $T(5, 4)$, $V(-2, 5)$, find the midpoint of each side. Connect the midpoints in order as shown. Show $\overline{AB} \parallel \overline{DC}$ and $\overline{AD} \parallel \overline{BC}$. Show $AB = DC$ and $AD = BC$. What kind of figure is $ABCD$? *parallelogram*

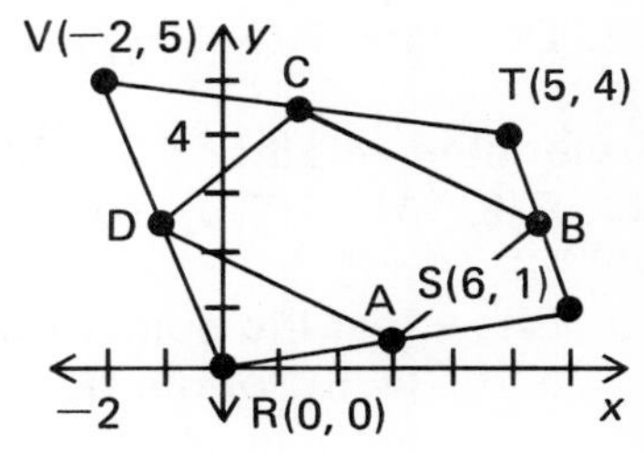

PART C

20. For quadrilateral $A(0, 0)$, $B(a, b)$, $C(c, d)$, $D(e, f)$, find the midpoint of each side. Connect the midpoints in order as shown. Show $\overline{EF} \parallel \overline{GH}$ and $\overline{EH} \parallel \overline{FG}$. Show $EF = HG$ and $EH = FG$. What kind of figure is $GHEF$?

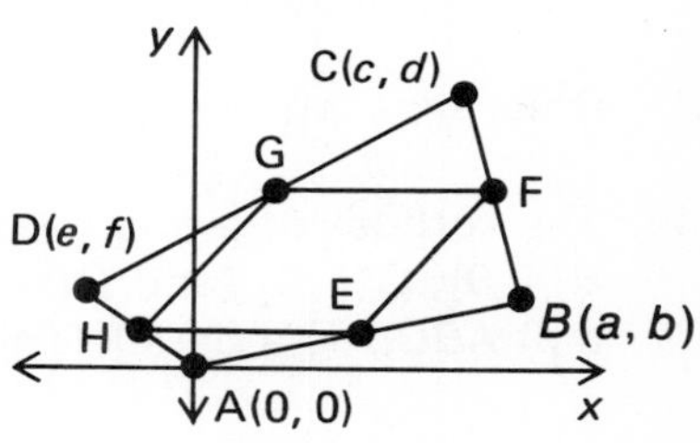

Chapter Nine Review

Find the length of the missing side of each right triangle. (c is the hypotenuse.) [p. 235]

1. $c = 13, a = 5$ **2.** $a = \sqrt{3}, b = \sqrt{6}$ **3.** $c = 14, b = 6$ **4.** $a = 2\sqrt{5}, b = 4\sqrt{2}$ **5.** $c = 5, b = 3$

Is the triangle a right triangle? [p. 235]

6. 7, 4, 11 **7.** 6, 8, 10 **8.** $\sqrt{5}, \sqrt{12}, \sqrt{17}$ **9.** $3\sqrt{5}, 2\sqrt{2}, \sqrt{37}$ **10.** 6, 6, 12

11. Each side of a square field is 15 m long. How long is a diagonal connecting the opposite corners? [p. 235]

12. If you drive 30 km due west and 40 km due north, how far are you from your starting point? [p. 235]

Find the distance between the given points. [p. 238]

13. $P(-1, 2), Q(6, 4)$ **14.** $A(7, -1), B(8, -4)$ **15.** $M(3, 0), N(10, 8)$ **16.** $R(-1, -8), S(6, 12)$

Find the midpoint of $\overline{PQ}$. [p. 241]

17. $P(4, -3), Q(7, 4)$ **18.** $P(-1, -5), Q(6, -3)$ **19.** $P(4, 4), Q(7, 8)$ **20.** $P(10, -1), Q(6, 8)$

Find the endpoint Q of $\overline{PQ}$, given point P and midpoint M. [p. 241]

21. $P(-3, 6), M(1, 2)$ **22.** $P(6, -3), M(4, -1)$ **23.** $P(-4, -1), M(4, 3)$ **24.** $P(7, -1), M(-1, -2)$

Find the slope of a line parallel to the given line. [p. 244]

25. $3x - 4y = 6$ **26.** $-8x + 5y = 7$ **27.** $x - 3y = 1$ **28.** $5x - 10y = -9$

Find the slope of a line perpendicular to the given line. [p. 244]

29. $-5x + 6y = 2$ **30.** $4x + 2y = -3$ **31.** $-2x - 3y = 9$ **32.** $4x + 5y = 10$

Write an equation of the line parallel to the given line passing through the given point. [p. 244]

33. $3x - 5y = 15$, (1, 2) **34.** $-2x - 3y = 10$, (4, −1) **35.** $7x - 3y = 12$, (1, 6)

Write an equation of the line perpendicular to the given line passing through the given point. [p. 244]

36. $2x + 5y = 7$, (−1, −3) **37.** $3y + 6x = 8$, (2, −7) **38.** $-6x + 4y = 9$, (−1, 3)

Determine whether $\overleftrightarrow{PQ} \parallel \overleftrightarrow{RS}$, $\overleftrightarrow{PQ} \perp \overleftrightarrow{RS}$, or neither. [p. 244]

39. $P(2, -1), Q(5, 0), R(8, -3), S(11, -2)$ **40.** $P(-1, 7), Q(3, 8), R(7, -6), S(8, -10)$

41. Show that the points $A(2, 3)$, $B(3, 6)$, and $C(1, 5)$ form a right triangle. Use the slope formula. [p. 247]

42. Show that the points $A(-1, 2)$, $B(12, -3)$, and $C(8, 6)$ form a right triangle. Use the distance formula. [p. 247]

43. Determine whether the points $A(-1, 2)$, $B(0, 0)$, and $C(1, -2)$ belong to the same line. [p. 247]

44. The vertices of a quadrilateral are $A(-5, 3)$, $B(4, 0)$, $C(6, 6)$, and $D(-3, 9)$. Show that $ABCD$ is a parallelogram. [p. 247]

45. The vertices of a triangle are $A(-2, -3)$, $B(6, 2)$, and $C(3, 6)$. Find the midpoints M of $\overline{AC}$ and N of $\overline{BC}$. Show that $\overline{MN} \parallel \overline{AB}$. Show that $MN = \frac{1}{2}(AB)$. [p. 247]

Chapter Nine Test

Find the length of the missing side of each right triangle. (c is the hypotenuse.)

1. $c = 8$, $b = 7$

2. $a = 3\sqrt{2}$, $b = 4\sqrt{3}$

Is the triangle a right triangle?

3. 9, 12, 15

4. $2\sqrt{5}$, $3\sqrt{2}$, $\sqrt{38}$

5. A football field is 100 yards long by 50 yards wide. How long is a diagonal connecting the opposite corners?

Find the distance between the given points.

6. $P(3, -8)$, $Q(6, -10)$

7. $P(-1, -3)$, $Q(-2, -5)$

Find the midpoint of $\overline{PQ}$.

8. $P(-3, 6)$, $Q(4, -8)$

Find the endpoint Q of $\overline{PQ}$ given point P and midpoint M.

9. $P(-2, -4)$, $M(6, 3)$

Find the slope of a line parallel to the given line.

10. $2x - 5y = 5$

11. $x + 3y = 6$

Find the slope of a line perpendicular to the given line.

12. $3x - 2y = 6$

13. $-x + 4y = 3$

Write an equation of the line parallel to the line with the given equation and passing through the given point.

14. $2x + 4y = 9$, $(2, -1)$

15. $3y + 5x = -12$, $(-3, -1)$

Write an equation of the line perpendicular to the line with the given equation and passing through the given point.

16. $-2x - 3y = 14$, $(-3, -4)$

17. $6x + 4y = 11$, $(5, -1)$

Determine whether $\overleftrightarrow{PQ} \parallel \overleftrightarrow{RS}$, $\overleftrightarrow{PQ} \perp \overleftrightarrow{RS}$, or neither.

18. $P(-1, 3)$, $Q(6, 4)$, $R(8, 2)$, $S(15, 3)$

19. $P(4, -3)$, $Q(8, 10)$, $R(-3, 8)$, $S(10, 4)$

20. Show that the points $A(6, -10)$, $B(7, -3)$, and $C(10, -7)$ form a right triangle. Use the slope formula.

21. Show that the points $A(6, -10)$, $B(7, -3)$, and $C(10, -7)$ form a right triangle. Use the distance formula.

22. Determine whether the points $A(5, 3)$, $B(0, 5)$, and $C(-5, 7)$ belong to the same line.

23. The vertices of a quadrilateral are $A(3, 0)$, $B(9, 4)$, $C(6, 9)$, and $D(0, 5)$. Show that $ABCD$ is a parallelogram.

24. The vertices of a triangle are $A(-1, -4)$, $B(4, 3)$, and $C(2, 8)$. Find the midpoints M of $\overline{AC}$ and N of $\overline{BC}$. Show that $\overline{MN} \parallel \overline{AB}$. Show that $MN = \frac{1}{2}(AB)$.

10 FUNCTIONS

Sonya Kovalevsky

Sonya Kovalevsky
(1850–1891)

Sonya Kovalevsky (also Sophie Korvin-Krukovsky) was born in Moscow. Her family belonged to the Russian nobility. Her early interest in mathematics was stimulated by her Uncle Pyotr.

When she was 17, Sonya studied calculus at the naval school in St. Petersburg. Her parents opposed this study so Sonya followed a practice of other Russian girls who wanted to study abroad. They would contract a marriage with a young man who shared their views. The wife would then be free to study abroad. Sonya's sister introduced her to Vladimir Kovalevsky. He agreed to the plan and they were married.

They moved to Heidelberg, Germany and Sonya became the private pupil of the famous mathematician Karl Weierstrass. In 1874, she was awarded a Doctor of Philosophy degree from the University of Gottingham. She and Vladimir returned to Russia where Sonya continued her research.

Nine years later Vladimir died under very tragic circumstances. Grieving the loss of her husband, Sonya accepted the opportunity to lecture at Stockholm University in Sweden. She was very popular at the university and was named a professor for life.

Although Sonya's scientific career was brief, it can only be regarded as brilliant. She published several papers which were highly regarded by other mathematicians. She was awarded the "Prix Bordin" by the Institut de France. Sonya was the first woman to belong to the Russian Academy of Sciences.

Relations and Functions

OBJECTIVES

- To graph a relation
- To determine the domain and range of a relation
- To determine whether a relation is a function

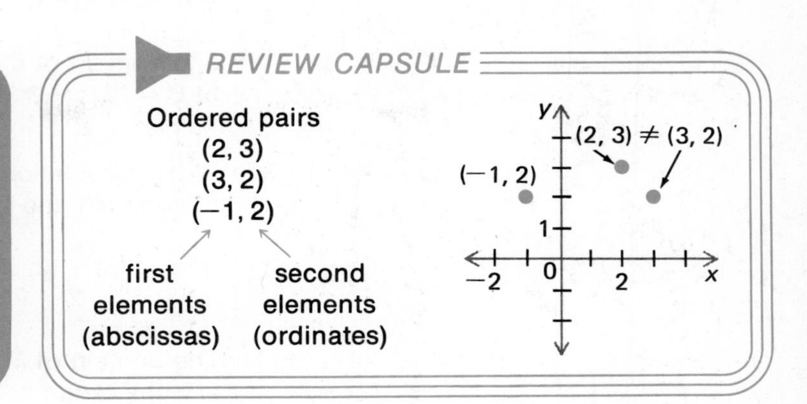

EXAMPLE 1 Graph the set of ordered pairs.
$A = \{(-1, 2), (2, 2), (2, 3), (4, -1)\}$.

A is a set of 4 ordered pairs.

$\{-1, 2, 4\}$ Set of first elements

$\{2, 3, -1\}$ Set of second elements

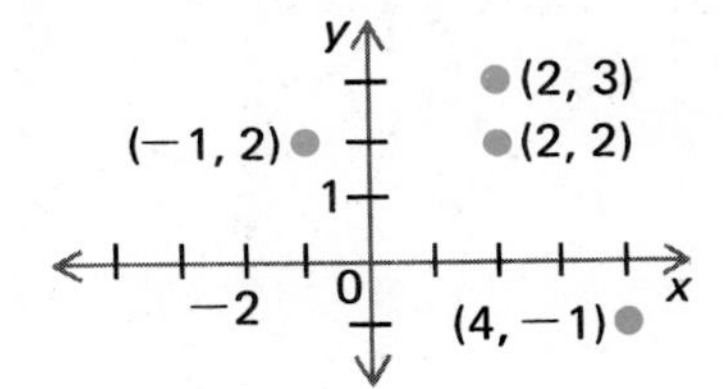

Definition of relation →

A *relation* is any set of ordered pairs.

EXAMPLE 2 Graph relation M. $M = \{(2, 3), (3, 2), (-2, -1), (3, -2)\}$.
Write the set of first elements of the ordered pairs of M.
Write the set of second elements.

Plot the four points. →
The point (2, 3) is not the same as (3, 2).

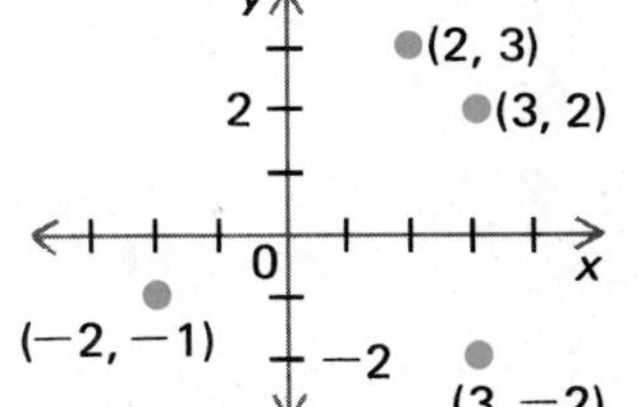

The set of first elements is $\{2, 3, -2\}$.

The set of second elements is $\{3, 2, -1, -2\}$.

Definition of domain →

Definition of range →

The set of all first elements of the ordered pairs of a relation is the *domain* of the relation.
The set of all second elements of the ordered pairs of a relation is the *range* of the relation.

EXAMPLE 3 Determine the domain and the range for the relation C.

$$C = \{(-1, 2), (0, 1), (2, -3), (4, -2), (5, 1)\}.$$

Set of first elements → The domain of $C = \{-1, 0, 2, 4, 5\}$ and the range
Set of second elements → of $C = \{2, 1, -3, -2\}$.

EXAMPLE 4 Write the relation M whose graph is shown. Give the domain and the range of M.

Find the coordinates of the plotted points. → $M = \{(3, 1), (2, 3), (-1, 2), (-2, 1)\}$.

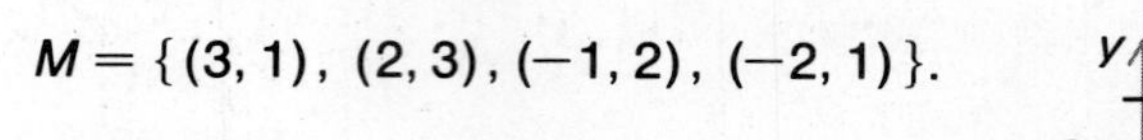

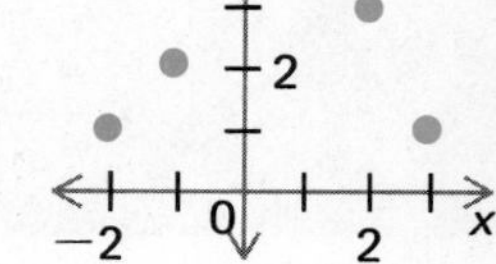

The domain of $M = \{3, 2, -1, -2\}$ and the range of $M = \{1, 3, 2\}$.

This is not true for all relations. → For the pairs in M, no two first elements are the same.

Definition of function →

> A *function* is a relation in which no two ordered pairs have the same first element.

EXAMPLE 5 Determine whether the set is a function.

$A = \{(-1, 2), (2, 2), (3, -1)\}$ $B = \{(2, 3), (3, 4), (2, 6)\}$ (same)

No first element is repeated. → Set A is a function. B is not a function.

EXAMPLE 6 Give the ordered pairs in relation M. Is M a function?

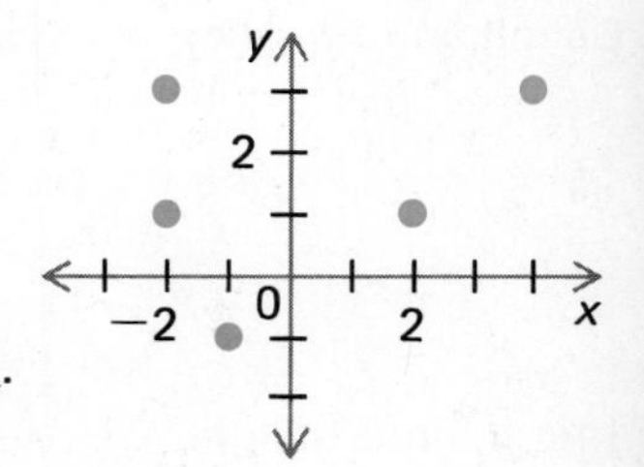

$M = \{(-2, 1), (-2, 3), (-1, -1), (2, 1), (4, 3)\}$.

A first element, −2, is repeated. **Thus,** M is *not* a function.

EXAMPLE 7 Use the graph of relation S to give the domain and range of S. Is S a function?

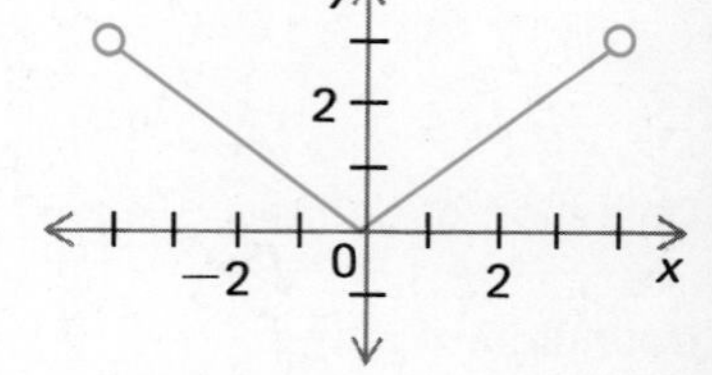

The x-values go from −4 to 4. → Domain of S is $\{x \mid -4 < x < 4\}$.

The y-values go from 0 to 3, including 0. → Range of S is $\{y \mid 0 \le y < 3\}$.

No two ordered pairs have the same first element. → **Thus,** S is a function.

We cannot list all the ordered pairs. The function S is an infinite set of ordered pairs.

ORAL EXERCISES

Give the domain and range of each relation.

1. $\{(3,5),\ (-1,3),\ (3,-4)\}$
2. $\{(-2,-3),\ (6,1),\ (7,-3)\}$
3. $\{(8,-1),\ (-7,6),\ (8,-6)\}$ *$D=\{8,-7\}$ $R=\{-1,6,-6\}$*

EXERCISES

PART A

Graph each relation. Determine the domain and range. Is the relation a function?

1. $\{(0,0),\ (2,3),\ (-1,2),\ (3,2),\ (4,3)\}$
2. $\{(8,0),\ (7,0),\ (-1,6),\ (-7,-1),\ (0,7)\}$ *yes*
3. $\{(-1,-2),\ (-2,-1),\ (-1,-3),\ (-3,-1)\}$
4. $\{(3,-2),\ (4,-2),\ (-4,2),\ (2,-3)\}$ *yes*
5. $\{(4,-1),\ (6,-1),\ (-2,3),\ (5,3),\ (0,0)\}$
6. $\{(4,0),\ (-4,0),\ (0,-4),\ (0,4)\}$ *no*
7. $\{(-1,4),\ (-2,4),\ (-3,4),\ (4,4)\}$ *yes*
8. $\{(-1,2),\ (-1,-2),\ (-1,-3),\ (-1,-4)\}$ *no*
9. $\{(2,-3),\ (3,-3),\ (-2,-3),\ (-3,-3)\}$
10. $\{(5,1),\ (4,2),\ (-4,-2),\ (-1,-5),\ (-5,1)\}$

9. $D=\{2,3,-2,-3\}$ $R=\{-3\}$; yes
10. $D=\{5,4,-1,-4,-5\}$ $R=\{1,2,-2,-5\}$; yes

Write the relation whose graph is shown. Give the domain and range of each. Is the relation a function?

11.

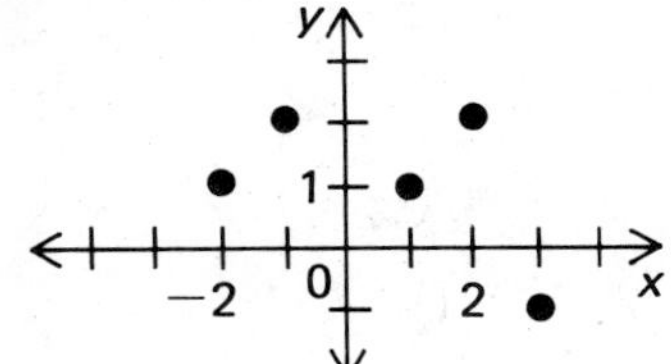

12.

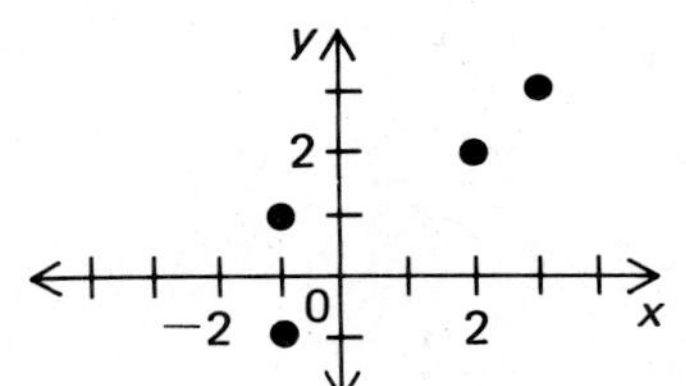

13.

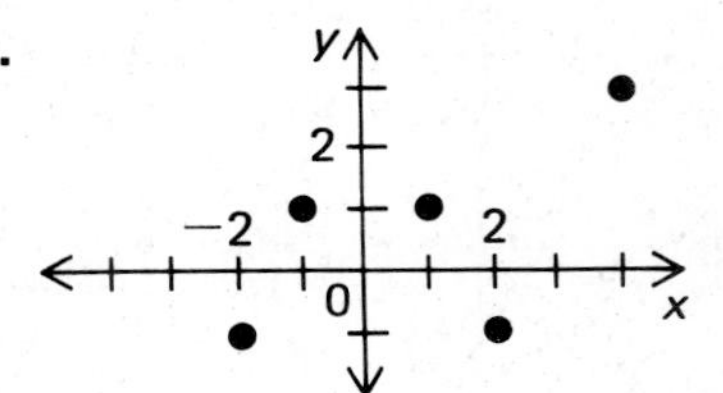

PART B

Give the domain and range of each relation. Is the relation a function?

14.

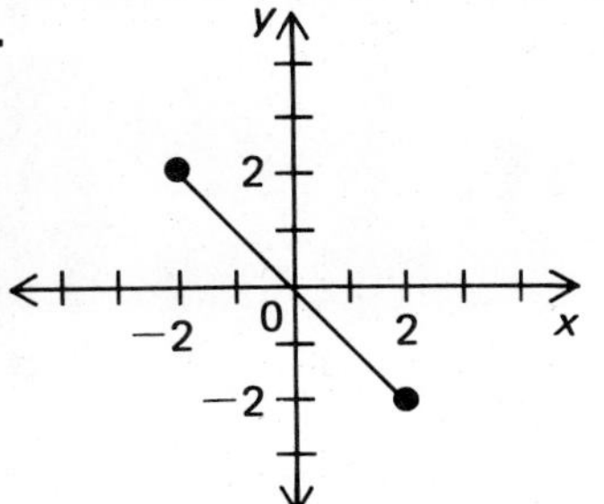

15.

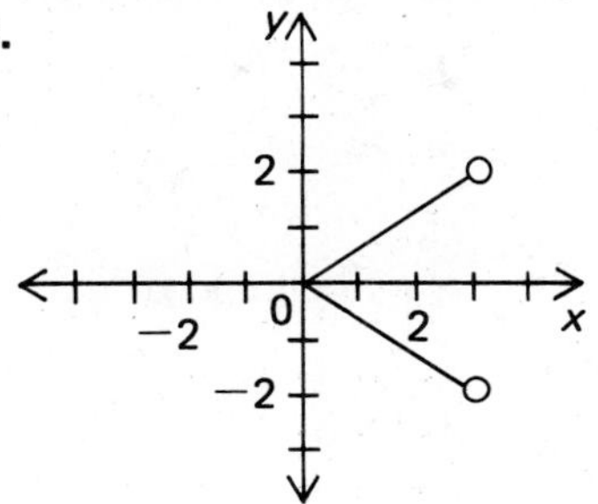

16.

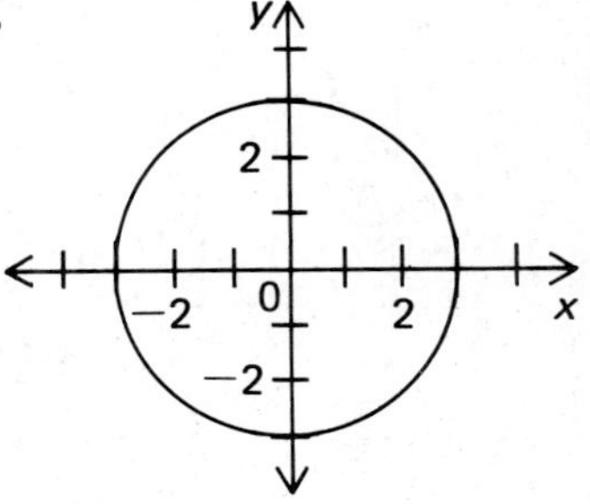

PART C

Give the domain and range of each relation. Is the relation a function?

17.

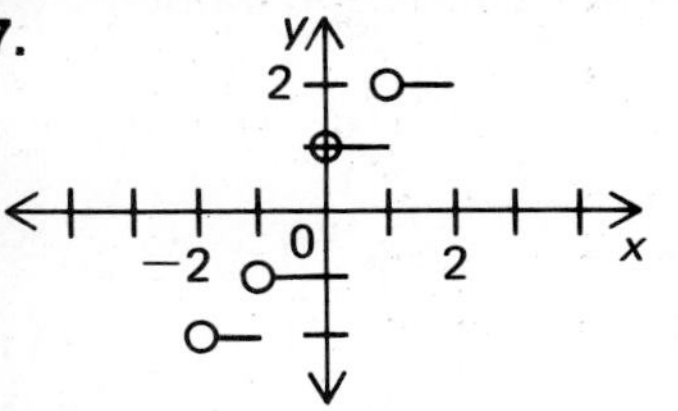

18.

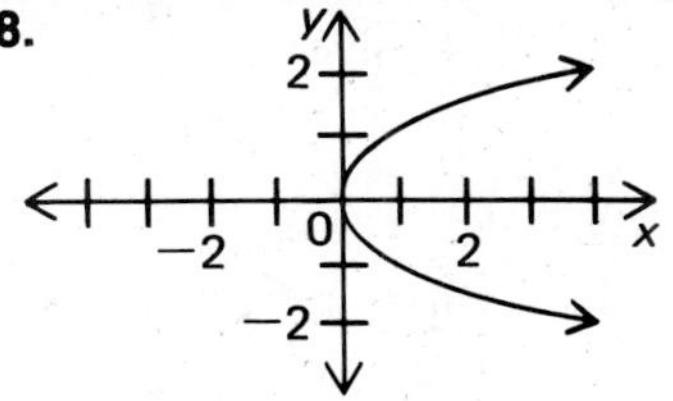

19.

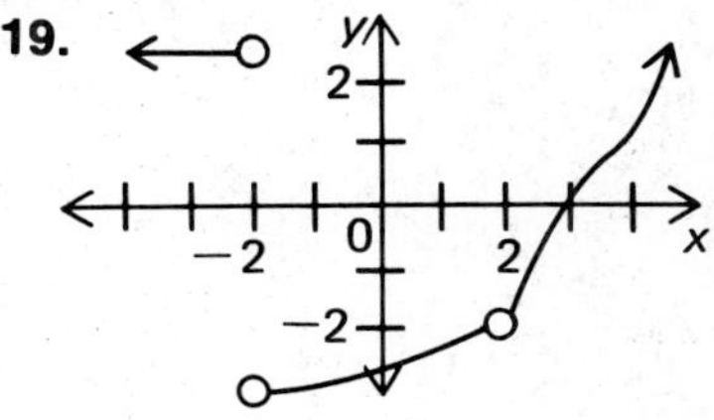

Function Values

OBJECTIVES

- To determine values of functions
- To find the range of a function given its domain

REVIEW CAPSULE

Graph the function f.
$f = \{(3, 1), (-1, -2), (0, 2)\}$.

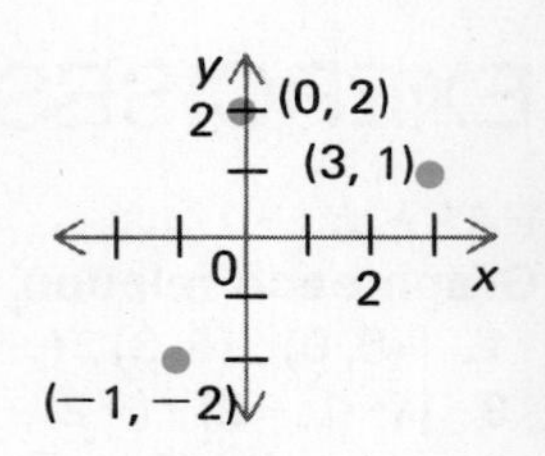

Domain of f is $\{3, -1, 0\}$.
Range of f is $\{1, -2, 2\}$.

EXAMPLE 1 Graph the function f. Find the range value corresponding to each number in the domain.

$$f = \{(1, 2), (3, 2), (4, -1), (-2, 3)\}$$

Functions are usually labeled with small letters. →

1st coordinates → Domain of $f = \{1, 3, 4, -2\}$.

2nd coordinates → Range of $f = \{2, 2, -1, 3\}$.

For $(1, 2)$, $f(1) = 2$ → The value of f at $x = 1$ is 2.
For $(3, 2)$, $f(3) = 2$ → The value of f at $x = 3$ is 2.
For $(4, -1)$, $f(4) = -1$ → The value of f at $x = 4$ is -1.
For $(-2, 3)$, $f(-2) = 3$ → The value of f at $x = -2$ is 3.

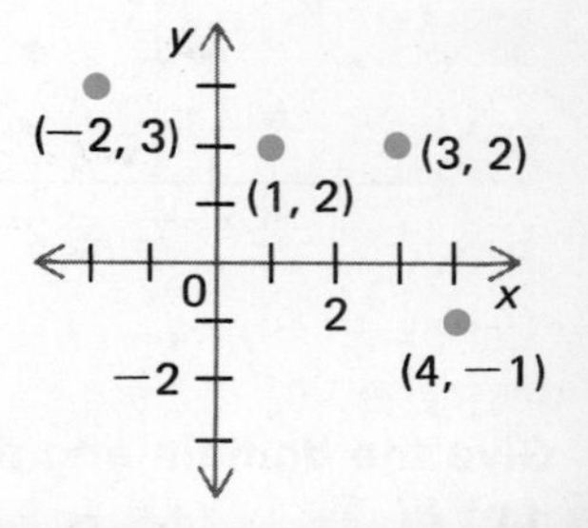

Definition of function value →

> For (x, y), an ordered pair of a function, f,
> $$f(x) = y.$$
> (The value of f at x is y.)

EXAMPLE 2 Let $g = \{(-1, -2), (0, 3), (2, 3), (3, -1)\}$.
Find $g(-1)$, $g(0)$, $g(2)$, $g(3)$.

For $(-1, -2)$, $g(-1) = -2$ → **Thus,** $g(-1) = -2$, $g(0) = 3$, $g(2) = 3$, and $g(3) = -1$.

EXAMPLE 3 Write the function h if each of its function values are given.

$$h(-2) = 5 \qquad h(0) = 4 \qquad h(-1) = 2$$

$h(-2) = 5$: -2 is x, 5 is y → $(-2, 5)$ → **Thus,** $h = \{(-2, 5), (0, 4), (-1, 2)\}$.

EXAMPLE 4 Let $f(x) = 3x - 1$.
Find $f(1)$, $f(2)$ and $f(-1)$.

$y = f(x)$ →
Substitute for x. →

For $x = 1$	For $x = 2$	For $x = -1$
$f(x) = 3x - 1$	$f(x) = 3x - 1$	$f(x) = 3x - 1$
$f(1) = 3(1) - 1$	$f(2) = 3(2) - 1$	$f(-1) = 3(-1) - 1$
$= 3 - 1$	$= 6 - 1$	$= -3 - 1$
$= 2$	$= 5$	$= -4$

Thus, $f(1) = 2$, $f(2) = 5$, and $f(-1) = -4$.

EXAMPLE 5 Let $g(x) = x^2 + 1$. If the domain of g is $D = \{-1, 0, 2\}$, find the range of g.

The domain is the set of all x-values. →
Substitute for x. →

For $x = -1$	For $x = 0$	For $x = 2$
$g(x) = x^2 + 1$	$g(x) = x^2 + 1$	$g(x) = x^2 + 1$
$g(-1) = (-1)^2 + 1$	$g(0) = 0^2 + 1$	$g(2) = 2^2 + 1$
$= 1 + 1$	$= 0 + 1$	$= 4 + 1$
$= 2$	$= 1$	$= 5$

The range is the set of all values of the function. → **Thus,** the range of g is $\{2, 1, 5\}$.

EXERCISES

PART A

Let $f = \{(1, 6), (-1, 3), (2, -7), (-3, 8), (0, 1), (-2, 4)\}$. Find the function values.

1. $f(1)$ *6*
2. $f(-1)$ *3*
3. $f(2)$ *−7*
4. $f(-3)$ *8*
5. $f(0)$ *1*
6. $f(-2)$ *4*

Let $h(x) = x^2 - 2x + 5$. Find the function values.

7. $h(-1)$ *8*
8. $h(1)$ *4*
9. $h(2)$ *5*
10. $h(-2)$ *13*
11. $h(0)$ *5*
12. $h(5)$ *20*

Find the range. D is the domain.

13. $f(x) = 3x + 5$, $D = \{0, 2, 5\}$ *$\{5, 11, 20\}$*
14. $h(x) = -5x + 3$, $D = \{-1, 0, 2\}$ *$\{8, 3, -7\}$*
15. $g(x) = 6x^2 + 2$, $D = \{-2, 1, 4\}$ *$\{26, 8, 98\}$*
16. $m(x) = 2x^2 - 7x + 3$, $D = \{3, -1\}$ *$\{0, 12\}$*

PART B

Determine the range. D is the domain.

17. $h(x) = (x - 2)^2$, $D = \{-1, 0, 3\}$ *$\{9, 4, 1\}$*
18. $f(x) = \dfrac{-2x^2 + 3}{5}$, $D = \{-2, 0, 4, \frac{1}{4}\}$
19. $f(x) = |x|$, $D = \{-4, -1, 0, 2, 4\}$ *$\{4, 1, 0, 2\}$*
20. $h(x) = \sqrt{x}$, $D = \{0, 2, 4, 9\}$ *$\{0, \sqrt{2}, 2, 3\}$*

PART C

Let $f(x) = 2x - 3$ and $g(x) = x^2$. Find the value of the expression.

21. $f(2) + g(3)$
22. $g(-1) - f(-2)$
23. $f(-3) - g(0)$
24. $\dfrac{g(a) - g(t)}{a - t}$
25. $f(a) + g(b)$
26. $f(a + h)$
27. $g(a + h) - g(a)$
28. $f[g(3)]$

Identifying and Graphing Functions

OBJECTIVES

- To identify functions from their graphs by using the vertical line test
- To graph certain functions and classify them

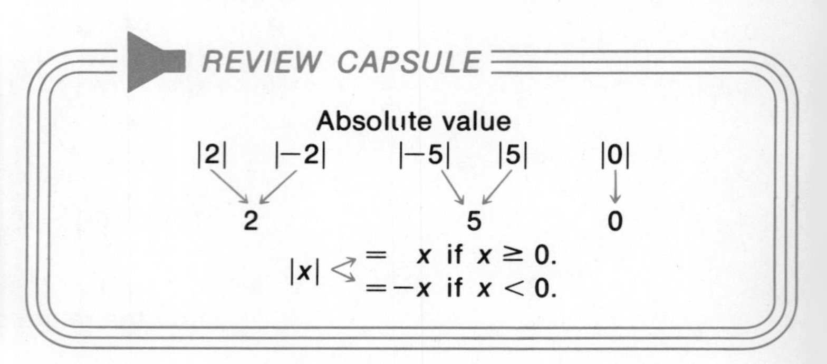

EXAMPLE 1 Is the following graph the graph of a function?

Draw vertical lines.
Check if any cross the graph in more than one point.

A vertical line crosses the graph in two points.

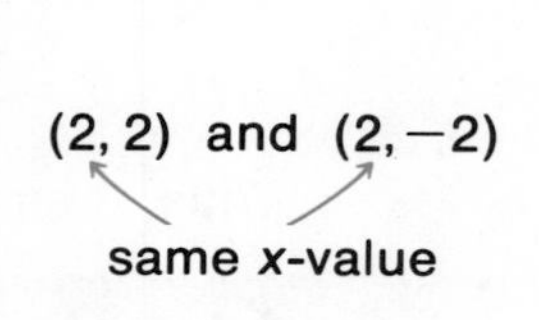

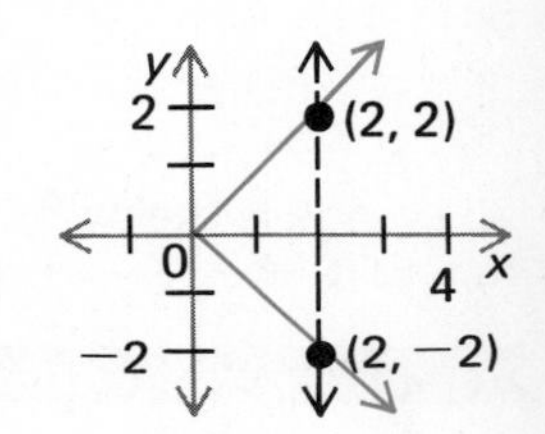

Two different ordered pairs have the same x-value. → **Thus,** the graph is not the graph of a function.

EXAMPLE 2 Determine which are graphs of functions and which are not.

Draw vertical lines.
See if any intersect the graph in more than one point.

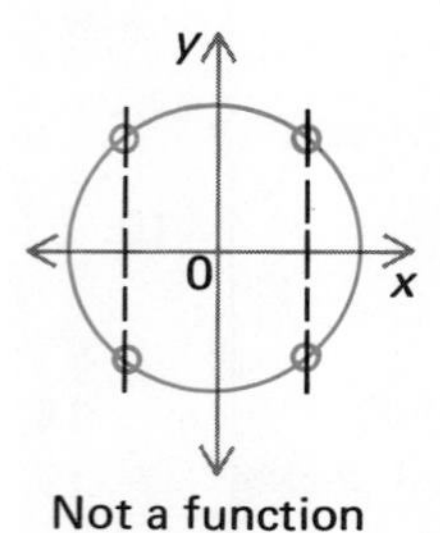

Not a function

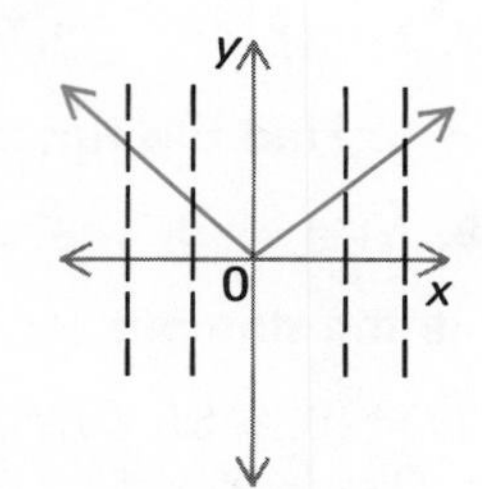

Function

Not a function

A vertical line is not the graph of a function since every ordered pair has the same x-value.

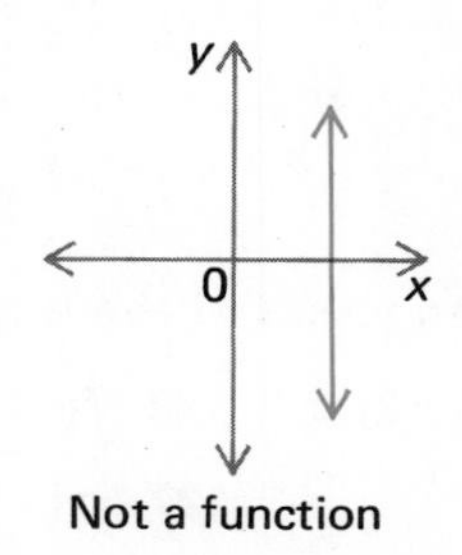

Not a function

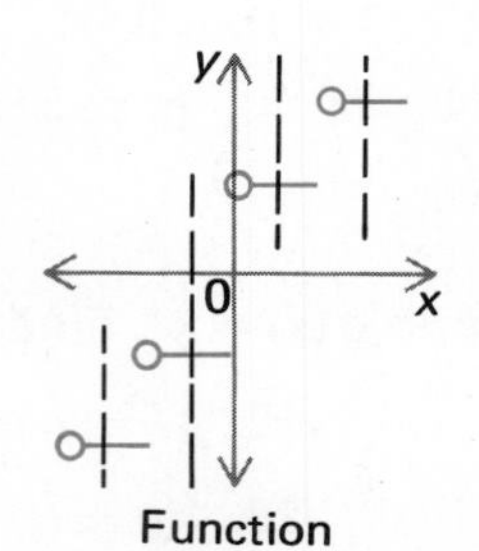

Function

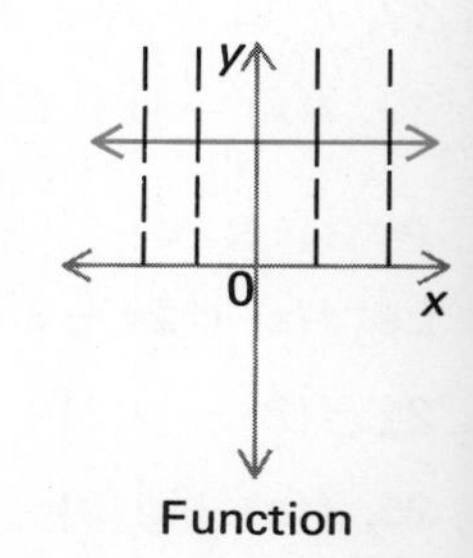

Function

EXAMPLE 3 Graph $\{(x, y) \mid y = |x|\}$. Is $\{(x, y) \mid y = |x|\}$ a function?

Find and plot ordered pairs. Connect the points.

x	$\lvert x \rvert$	y
1	$\lvert 1 \rvert$	1
2	$\lvert 2 \rvert$	2
−1	$\lvert -1 \rvert$	1
−2	$\lvert -2 \rvert$	2
0	$\lvert 0 \rvert$	0

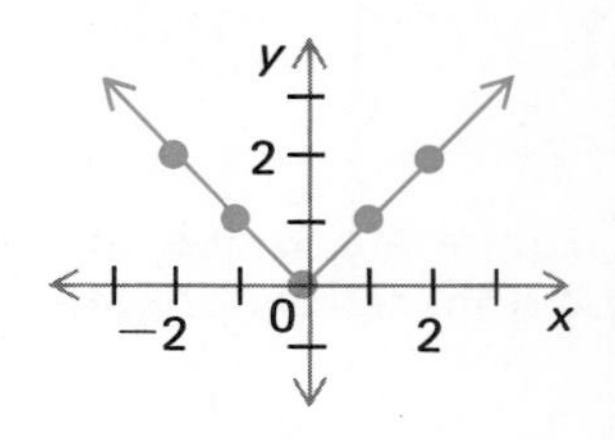

Use the vertical line test. ⟶ **Thus,** $\{(x, y) \mid y = |x|\}$ is a function.

EXAMPLE 4 Graph $\{(x, y) \mid y - 2x + 1 = 0\}$. Is $\{(x, y) \mid y - 2x + 1 = 0\}$ a function?

Write in $y = mx + b$ form. ⟶

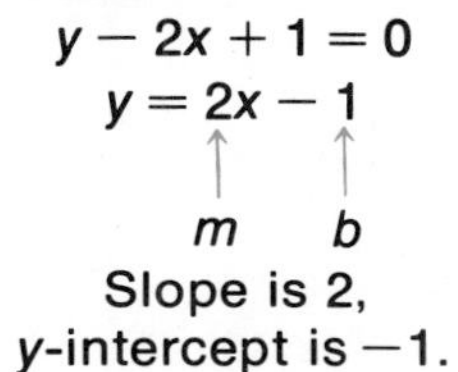

$$y - 2x + 1 = 0$$
$$y = 2x - 1$$

m b

Graph is a straight line.

Slope is 2,
y-intercept is −1.

(2, 3)
(1, 1)
(0, −1)

Use the vertical line test. **Thus,** $\{(x, y) \mid y - 2x + 1 = 0\}$ is a function.

$y = mx + b$, $m \neq 0$, describes a linear function.

> A *linear* function is a function whose graph is a nonvertical, nonhorizontal line or line segment.

EXAMPLE 5 Graph $\{(x, y) \mid y = 4\}$. Is $\{(x, y) \mid y = 4\}$ a function?

$y = 4$ represents a horizontal line.

All values of the function are the same.

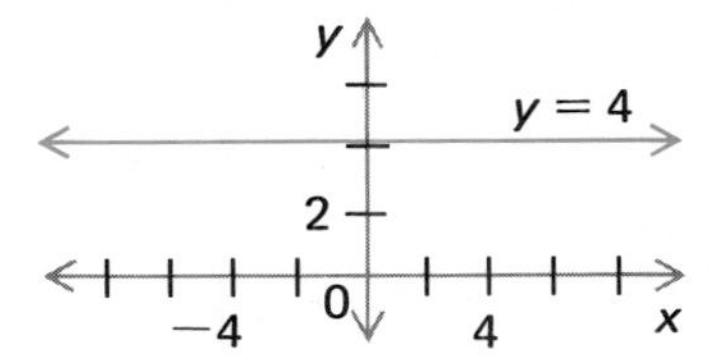

Use vertical line test. ⟶ **Thus,** $\{(x, y) \mid y = 4\}$ is a function.

$y = b$ describes a constant function.

> A *constant* function is a function whose graph is a horizontal line or line segment.

EXAMPLE 6 Graph $\{(x, y) \mid x = 2\}$. Is $\{(x, y) \mid x = 2\}$ a constant function?

$x = 2$ represents a vertical line.

$\{(x, y) \mid x = 2\}$ is not a function.

Thus, $\{(x, y) \mid x = 2\}$ is not a constant function.

EXERCISES

PART A

Determine which are graphs of functions and which are not. Are there any linear functions? Are there any constant functions?

1.

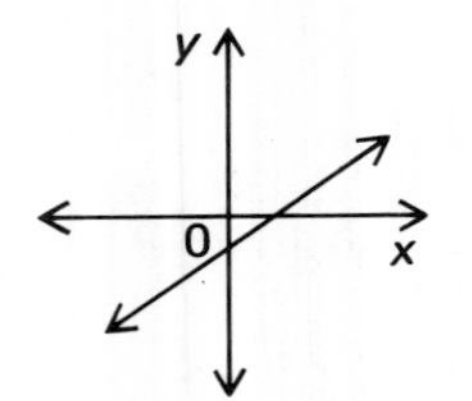

2.

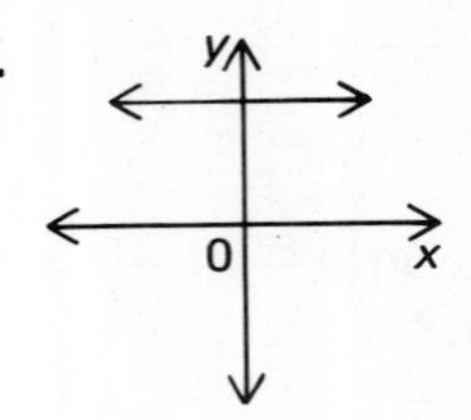

3.

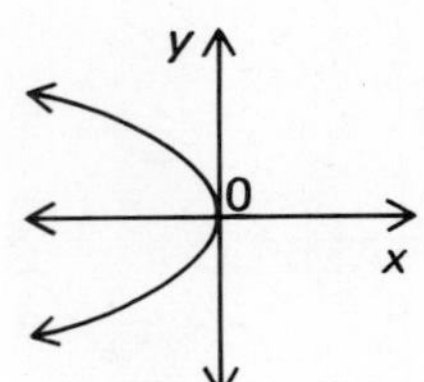

4.

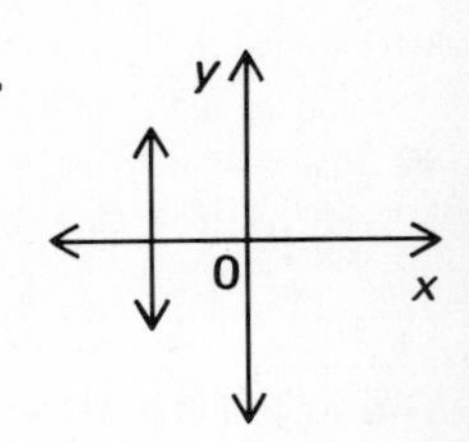

Graph each relation. Which are functions? Which are linear functions? Which are constant functions?

5. $\{(x, y) \mid y = 2x\}$ *LF*
6. $\{(x, y) \mid y = 4x\}$ *LF*
7. $\{(x, y) \mid y = -3x\}$ *LF*
8. $\{(x, y) \mid y = 3x - 2\}$ *LF*
9. $\{(x, y) \mid y = -2x + 3\}$ *LF*
10. $\{(x, y) \mid y = -2x - 5\}$ *LF*
11. $\{(x, y) \mid x = -6\}$ *no*
12. $\{(x, y) \mid y = 4\}$ *CF*
13. $\{(x, y) \mid y = 0\}$ *CF*

PART B

Graph each relation. Which are functions? Which are linear functions? Which are constant functions?

14. $\{(x, y) \mid 2x + 3y = 9\}$ *LF*
15. $\{(x, y) \mid 3x - 2y = 7\}$
16. $\{(x, y) \mid 3x - 2y = 6\}$
17. $\{(x, y) \mid y = -|x|\}$ *Function*
18. $\{(x, y) \mid y = |2x|\}$
19. $\{(x, y) \mid y = |x| + 2\}$
20. $\{(x, y) \mid y = |x - 2|\}$
21. $\{(x, y) \mid y = |x + 2|\}$
22. $\{(x, y) \mid |x| + |y| = 4\}$

PART C

EXAMPLE Graph $\{(x, y) \mid y = [x]\}$. Is $\{(x, y) \mid y = [x]\}$ a function?

$[x]$ means the greatest *integer* which is less than or equal to x.

Use the definition of $[x]$.
$0 \le \frac{1}{3}$ →
$1 \le 1\frac{1}{2}$ →
$-1 \le -\frac{1}{3}$ →

x	$[x]$	y
0	$[0]$	0
$\frac{1}{3}$	$[\frac{1}{3}]$	0
1	$[1]$	1
$1\frac{1}{2}$	$[1\frac{1}{2}]$	1
$-\frac{1}{3}$	$[-\frac{1}{3}]$	-1
$-1\frac{1}{2}$	$[-1\frac{1}{2}]$	-2

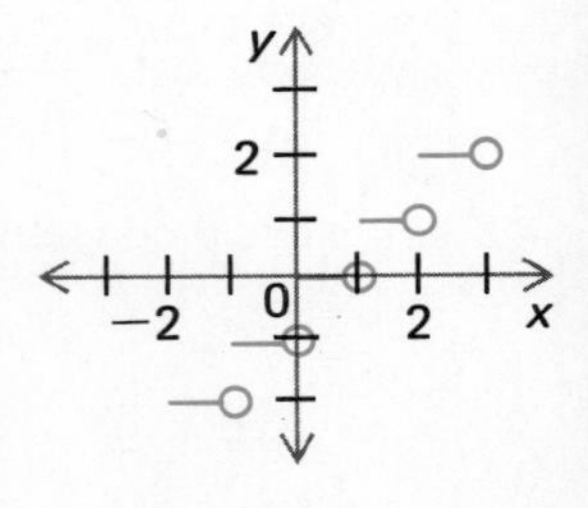

It is called the greatest integer function. → **Thus,** $\{(x, y) \mid y = [x]\}$ is a function.

Graph each relation. Which are functions?

23. $\{(x, y) \mid y = -[x]\}$
24. $y = 2 + [x]$
25. $y = [x] - 2$
26. $\{(x, y) \mid y = x + [x]\}$
27. $\{(x, y) \mid y = [x + 2]\}$
28. $\{(x, y) \mid y = [|x|]\}$

Direct Variation

OBJECTIVES

- To determine whether a relation is a direct variation
- To apply direct variations

REVIEW CAPSULE

The ratio of two numbers, x to y is $\frac{x}{y}$.

Ratio of 6 to 3: $\frac{6}{3}$, or 2

Ratio of 4 to 12: $\frac{4}{12}$, or $\frac{1}{3}$

EXAMPLE 1 Write an equation which describes the ratio y to x.

y	8	12	16	32	64	128	256
x	2	3	4	8	16	32	64

In each case, the ratio y to x is 4. → **Thus,** $\frac{y}{x} = 4$, or $y = 4x$.

Definition of direct variation →

A *direct variation* is a function in which the *ratio* of the two variables is always the same.

$\frac{y}{x} = k$, or $y = kx$, where k is the constant of proportionality.

y varies directly as x.
y is directly proportional to x.

EXAMPLE 2 Does the equation $y = 2x$ express a direct variation?

$$y = 2x$$
$$\frac{y}{x} = 2 \qquad (x \neq 0)$$

Ratio $\frac{y}{x}$ is always 2. → **Thus,** $y = 2x$ expresses a direct variation.

EXAMPLE 3 Does the equation $y = 3x + 4$ express a direct variation?

Test the ratio y to x.

x	$y = 3x + 4$	$\frac{y}{x}$
1	7	$\frac{7}{1}$, or 7
2	10	$\frac{10}{2}$, or 5

$7 \neq 5$

The ratio $\frac{y}{x}$ is not always the same. → **Thus,** $y = 3x + 4$ does not express a direct variation.

EXAMPLE 4 From the table, determine if y varies directly as x. If so, find the constant of proportionality.

x	y	$\frac{y}{x}$
1	−6	$\frac{-6}{1}$, or −6
−2	12	$\frac{12}{-2}$, or −6
3	−18	$\frac{-18}{3}$, or −6

The ratio $\frac{y}{x}$ is always −6.

Thus, y varies directly as x and the constant of proportionality is −6.

EXAMPLE 5 C varies directly as d. If the constant of proportionality is 5, find C when $d = 10$.

The ratio of C to d is 5. ⟶ $\frac{C}{d} = 5$

Substitute: $d = 10$. ⟶ $\frac{C}{10} = 5$

Multiply each side by 10. ⟶ $C = 50$

Thus, C is 50 when d is 10.

EXAMPLE 6 y varies directly as x and $y = 16$ when $x = 4$. Find y when $x = 7$.

Let (x_1, y_1) and (x_2, y_2) be any two pairs of corresponding variables.

The ratios are always the same, k. ⟶ $\frac{y_1}{x_1} = k$ and $\frac{y_2}{x_2} = k$

So, $\frac{y_1}{x_1} = \frac{y_2}{x_2}$

Substitute. ⟶ $y = 16$ when $x = 4$; $y = ?$ when $x = 7$. $\frac{16}{4} = \frac{y}{7}$

Solve the proportion. ⟶

$$4y = 112$$
$$y = 28$$

Thus, y is 28 when x is 7.

EXAMPLE 7 y varies directly as the square of x. $y = 36$ when $x = 2$. Find y when $x = 10$.

$\frac{y_1}{x_1^2} = \frac{y_2}{x_2^2}$

$$\frac{36}{2^2} = \frac{y}{10^2}$$
$$\frac{36}{4} = \frac{y}{100}$$
$$4y = 3600$$
$$y = 900$$

Thus, $y = 900$ when $x = 10$.

ORAL EXERCISES

Which express direct variations? For each direct variation, determine the constant of proportionality.

1. $y = 15x$ *k = 15* **2.** $y = 3x + 7$ *No* **3.** $m = \frac{1}{4}n^2$ *$k = \frac{1}{4}$* **4.** $c = 6d$ *k = 6* **5.** $-4x = y$ *k = −4*

6. $c = 3.14d$ *k = 3.14* **7.** $A = 2r + 9$ *No* **8.** $l = 6w$ *k = 6* **9.** $p = r + s$ *No* **10.** $\frac{c}{d^2} = 7$ *k = 7*

EXERCISES

PART A

Which tables express direct variation? For each direct variation, determine the constant of proportionality.

1.

x	y
−2	1
−4	2
2	−1
4	−2

2.

c	d
5	2
7	3
11	5
15	7

3. *No*

m	n
−1	1
3	−1
−3	2
7	−3

4. *k = 5*

A	r
15	3
10	2
20	4
25	5

5.

R	S
−1	1
−2	2
3	−3
4	−4

***y* varies directly as *x*.**

6. $y = 16$ when $x = -4$. Find y when $x = 8$.

7. $y = -12$ when $x = -3$. Find y when $x = -8$. *−32*

8. $y = -81$ when $x = 9$. Find y when $x = 7$.

9. $y = 27$ when $x = 27$. Find y when $x = -6$.

10. $y = -3$ when $x = -15$. Find x when $y = 2$.

11. $y = 5$ when $x = 25$. Find y when $x = 30$.

12. C varies directly as d. If the constant of proportionality is 12, find C when $d = 3$.

13. V varies directly as T. If the constant of proportionality is 20, find V when $T = 2$. *40*

PART B

14. y varies directly as the square of x. $y = 25$ when $x = 3$. Find y when $x = 6$. *100*

15. y varies directly as x^2. $y = 30$ when $x = 5$. Find y when $x = 10$. *120*

16. r varies directly as the positive square root of s. $r = 15$ when $s = 25$. Find r when $s = 36$.

17. m varies directly as $\sqrt{n}$. $m = 12$ when $n = 4$. Find m when $n = 16$. *24*

PART C

18. $y = 4x$ expresses a direct variation. Plot points and graph $y = 4x$. What is the graph of this direct variation?

19. In a direct variation $\frac{y}{x} = k$, $y > 0$. If y remains constant, what happens to k as x increases? as x decreases?

20. $y = -2x$ expresses a direct variation. Plot points and graph $y = -2x$. What is the graph of this direct variation?

21. In a direct variation $\frac{y}{x} = k$, $y < 0$. If y remains constant, what happens to k as x increases? As x decreases?

22. The circumference c of a circle varies directly as the length of a diameter d. If $c = 9.42$ when $d = 3$, find c when $d = 7$.

23. With a constant rate of speed, the distance d varies directly as the time t. If $d = 120$ when $t = 2$, find d when $t = 6$.

24. The perimeter p of an equilateral triangle varies directly as the length of a side s. If $p = 12$ when $s = 4$, find p when $s = 6$.

25. If a boat travels 240 km in 12 hours, how far can it travel in 18 hours traveling at the same rate of speed?

Inverse Variation

OBJECTIVES

- To determine whether a relation is an inverse variation
- To apply inverse variations

REVIEW CAPSULE

If y varies directly as x and $y = 20$ when $x = 4$, find the constant of proportionality.

$$\frac{y}{x} = k \rightarrow \frac{20}{4} = 5$$

The constant of proportionality is 5.

In a direct variation, the *ratio* of the variables is constant.

EXAMPLE 1 Write an equation which describes the relationship between x and y.

x	-16	-2	$\frac{1}{4}$	$\frac{1}{2}$	2	4	8	32	128
y	-4	-32	256	128	32	16	8	2	$\frac{1}{2}$

In each case, the product xy is 64. ⟶ **Thus,** $xy = 64$

Definition of inverse variation ⟶

An *inverse variation* is a function in which the *product* of the two variables is always the same.

$xy = k$, or $y = \frac{k}{x}$. k is the constant of variation.

y varies inversely as x.
y is inversely proportional to x.

EXAMPLE 2 From the table, determine if y varies inversely as x. If so, find the constant of variation.

y	4	6	2	-1	$-\frac{1}{2}$
x	3	2	6	-12	-24

Check each product. ⟶

$4 \cdot 3$	$6 \cdot 2$	$2 \cdot 6$	$(-1)(-12)$	$(-\frac{1}{2})(-24)$
12	12	12	12	12

$xy = 12$ for each pair x and y.

Thus, y varies inversely as x, and the constant of variation is 12.

EXAMPLE 3 Does the equation $y = 5x$ express an inverse variation?

Find some valves of x and y in $y = 5x$.

x	y	xy
2	10	20
4	20	80
−1	−5	5

$x \cdot y$ is not the same for each pair.

Thus, $y = 5x$ does not express an inverse variation.

EXAMPLE 4 y varies inversely as x. $y = 12$ when $x = 3$. Find y when $x = 9$.

Let (x_1, y_1) and (x_2, y_2) be any two pairs of corresponding variables.

The products are the same, k. ⟶

$$x_1y_1 = k \quad \text{and} \quad x_2y_2 = k$$

So,

$x = 3$ when $y = 12$.
$x = 9$ when $y = ?$. ⟶

$$x_1y_1 = x_2y_2$$
$$3 \cdot 12 = 9y$$
$$36 = 9y$$
$$4 = y$$

Thus, y is 4 when x is 9.

EXAMPLE 5 C is inversely proportional to R. $C = 12$ when $R = \frac{1}{4}$. Find C when $R = 9$.

The products are the same. ⟶

$$\tfrac{1}{4} \cdot 12 = 9C$$
$$3 = 9C$$
$$\tfrac{1}{3} = C$$

Thus, $C = \frac{1}{3}$ when $R = 9$.

EXAMPLE 6 If x varies inversely as y^2 and if $x = 2$ when $y = 6$, find x when $y = 12$.

x varies inversely as the square of y. ⟶

$$6^2 \cdot 2 = 12^2 \cdot x$$
$$36 \cdot 2 = 144x$$
$$72 = 144x$$
$$\tfrac{1}{2} = x$$

Thus, x is $\frac{1}{2}$ when y is 12.

ORAL EXERCISES

Which equations express inverse variations? For each inverse variation, determine the constant of variation.

1. $xy = 12$ *k = 12*
2. $y = \frac{16}{x}$ *k = 16*
3. $\frac{y}{x} = 10$ *no*
4. $12 = lw$ *k = 12*
5. $c = 6d$
6. $x = \frac{12}{y}$ *k = 12*
7. $a \cdot b = 2$ *k = 2*
8. $-7m = n$ *no*
9. $\frac{1}{4}r = s$ *no*
10. $n = \frac{m}{7}$ *no*

EXERCISES

PART A

Which tables express inverse variations? For each inverse variation, determine the constant of variation.

1.

x	y
3	15
5	9
15	3

2. *no*

c	d
−2	3
6	1
−3	2

3. *no*

R	S
−2	−2
−1	−1
2	2

4.

l	w
10	10
−20	−5
$\frac{1}{2}$	200

5.

m	n
1	1
$\frac{3}{2}$	$\frac{2}{3}$
$\frac{4}{5}$	$\frac{5}{4}$

***y* varies inversely as *x*.**

6. $y = 12$ when $x = 3$. Find y when $x = 6$.
7. $y = 9$ when $x = 5$. Find y when $x = 15$. *3*
8. $y = 7$ when $x = -4$. Find y when $x = 14$.
9. $y = 81$ when $x = \frac{1}{9}$. Find x when $y = 3$.
10. $y = -3$ when $x = -12$. Find x when $y = 12$.
11. $y = 27$ when $x = \frac{1}{3}$. Find y when $x = 9$. *1*

PART B

12. If x varies inversely as y^2 and if $x = 3$ when $y = 4$, find x when $y = 2$.
13. If R varies inversely as the square of d and $R = 4$ when $d = 2$, find R when $d = 3$.
14. If x varies inversely as $\sqrt{y}$ and if $x = 6$ when $y = 9$, find x when $y = 16$. *$4\frac{1}{2}$*
15. If S varies inversely as the positive square root of t and $S = 7$ when $t = 25$, find S when $t = 36$. *$\frac{35}{6}$*

PART C

16. $y = \frac{9}{x}$ expresses an inverse variation. Plot points and graph $y = \frac{9}{x}$.
17. In an inverse variation $y = \frac{k}{x}$, $k > 0$, what happens to y as x increases? as x decreases?
18. Plot points and graph $x \cdot y = 16$.
19. If $xy = k$, $k < 0$, what happens to x as y increases?
20. If the distance remains constant, the rate r varies inversely as the time t. If $r = 50$ when $t = 4$, find r when $t = 2$.
21. If the area of a rectangle remains constant, the length l varies inversely as the width w. If $l = 12$ when $w = 3$, find l when $w = 9$.
22. The frequency f of a radio wave is inversely proportional to its wave length w. If a wave 300 meters long has a frequency of 1200 kilocycles/sec, what is the length of a wave with a frequency of 900 kilocycles/sec?
23. For a fixed sum of money, the number of articles n that can be purchased varies inversely as the cost c of the article. If 6 articles can be purchased for \$2.00 per article, how many articles can be purchased for \$1.00 per article?

Combined Variations

OBJECTIVES

- To apply joint variation
- To solve variation problems

REVIEW CAPSULE

Direct variation	Inverse variation
$\frac{y}{x} = k$	$xy = k$
The ratio is constant.	The product is constant.

EXAMPLE 1 y varies directly as the product of x and z. If the constant of proportionality is 2, write an equation which expresses the relationship between x, y, and z.

The ratio of y to the product of x and z is always 2. →

Thus, $\frac{y}{xz} = 2$ expresses the relationship.

Definition of joint variation. →

A *joint variation* is a function in which one variable varies directly as the product of two others.

$$\frac{y}{xz} = k, \text{ or } y = kxz$$

y varies jointly as x and z.

EXAMPLE 2 Does the equation $\frac{xz}{y} = 3$ represent a joint variation?

Put the equation in $\frac{y}{xz} = k$ form.

$$\frac{xz}{y} = 3$$

$$xz = 3y$$

$$1 = \frac{3y}{xz}$$

The constant of proportionality is $\frac{1}{3}$. →

$$\frac{1}{3} = \frac{y}{xz}$$

Thus, $\frac{xz}{y} = 3$ expresses a joint variation.

EXAMPLE 3 y varies jointly as x and z. $y = 24$ when $x = 2$ and $z = 3$. Find y when $x = 3$ and $z = 4$.

$\frac{y}{xz}$ is always the same. →

$$\frac{y_1}{x_1z_1} = \frac{y_2}{x_2z_2}$$

Write the proportion. →

$$\frac{24}{2 \cdot 3} = \frac{y}{3 \cdot 4}$$

Solve the proportion.

$$6y = 288, \text{ or } y = 48$$

Thus, $y = 48$ when $x = 3$ and $z = 4$.

EXAMPLE 4 y varies jointly as the square of x and the positive square root of z. If $y = 54$ when $x = 3$ and $z = 9$, find y when $x = 2$ and $z = 4$.

$\frac{y}{x^2\sqrt{z}}$ is always the same. ⟶

$$\frac{54}{3^2\sqrt{9}} = \frac{y}{2^2\sqrt{4}}$$

$3^2 = 9 \quad \sqrt{9} = 3$, $2^2 = 4 \quad \sqrt{4} = 2$ ⟶

$$\frac{54}{9 \cdot 3} = \frac{y}{4 \cdot 2}, \text{ or } \frac{54}{27} = \frac{y}{8}$$

Solve the proportion.

$$27y = 432$$
$$y = 16$$

Thus, y is 16 when $x = 2$ and $z = 4$.

> *Combined variations* are variations which are combinations of direct and inverse variations.

EXAMPLE 5 The pressure P of a gas varies directly as the temperature T and inversely as the volume V. If $P = 50$ when $T = 25$ and $V = 2$, find P when $T = 40$ and $V = 4$.

P varies inversely as V. P varies directly as T. ⟶

$$\frac{P_1V_1}{T_1} = \frac{P_2V_2}{T_2}$$

Substitute given values. ⟶

$$\frac{50 \cdot 2}{25} = \frac{P \cdot 4}{40}$$

The product of the means equals the product of the extremes. ⟶

$$100P = 4{,}000$$
$$P = 40$$

Thus, $P = 40$ when $T = 40$ and $V = 4$.

EXAMPLE 6 y varies jointly as x and z and inversely as w. $y = 12$ when $x = 2$, $z = 6$ and $w = 3$. Find y when $x = 5$, $z = 7$, and $w = 21$.

y varies jointly as x and z. y varies inversely as w. ⟶

$$\frac{y_1w_1}{x_1z_1} = \frac{y_2w_2}{x_2z_2}$$

Substitute given values. ⟶

$$\frac{12(3)}{2(6)} = \frac{y(21)}{5(7)}$$
$$\frac{36}{12} = \frac{21y}{35}$$
$$252y = 1{,}260$$
$$y = 5$$

Thus, $y = 5$ when $x = 5$, $z = 7$, and $w = 21$.

ORAL EXERCISES

Identify each of the following as direct variation, inverse variation, or joint variation. Determine the constant of proportionality, or the constant of variation.

1. $xy = 2$ *I* **2.** $x = 3y$ *D* **3.** $y = \frac{6}{x}$ *I* **4.** $t = 4sr$ *J* **5.** $a = \frac{b}{6}$ *D* **6.** $kx = \frac{y}{z}$

EXERCISES

PART A

Which of the following equations express joint variations?

1. $\frac{ab}{c} = 5$ **2.** $y = \frac{3x^2}{z}$ **3.** $xy = 15$ *No* **4.** $s = \frac{16\sqrt{t}}{r}$ **5.** $y = 150x$ *No*

***y* varies jointly as *x* and *z*.**

6. $y = 24$ when $x = 2$ and $z = 3$. Find y when $x = 4$ and $z = 7$. *112*
7. $y = 30$ when $x = 3$ and $z = 15$. Find y when $x = 5$ and $z = 2$. $\frac{20}{3}$
8. $y = 72$ when $x = 3$ and $z = 8$. Find y when $x = -2$ and $z = -3$. *18*
9. $y = 100$ when $x = 5$ and $z = 10$. Find y when $x = 16$ and $z = 4$. *128*
10. $y = 12$ when $x = 3$ and $z = 2$. Find y when $x = 5$ and $z = 6$. *60*
11. $y = 60$ when $x = 2$ and $z = 5$. Find y when $x = 6$ and $z = 4$. *144*
12. $y = 18$ when $x = 1$ and $z = 6$. Find y when $x = 4$ and $z = 10$. *120*

13. y varies directly as the square of x and inversely as z. $y = 2$ when $x = 5$ and $z = 100$. Find y when $x = 3$ and $z = 4$. *18*

14. y varies jointly as x and z and inversely as the positive square root of w. $y = 2$ when $x = 16$, $z = 4$ and $w = 16$. Find y when $x = 7$, $z = 8$ and $w = 25$. $\frac{7}{5}$

15. y varies directly as the positive square root of x and inversely as the square of z. $y = 8$ when $x = 4$ and $z = 5$. Find y when $x = 16$ and $z = 3$. $\frac{400}{9}$

16. y varies jointly as x and the positive square root of z and inversely as w. $y = 3$ when $x = 2$, $z = 4$ and $w = 16$. Find y, when $x = 1$, $z = 25$ and $w = 5$. *12*

PART B

17. The area A of a parallelogram varies jointly as the lengths of its base b and altitude h. $A = 16$ when $b = 2$ and $h = 8$. Find A when $b = 8$ and $h = 16$. *128*

18. The distance D traveled at a uniform rate varies jointly as the rate r and the time t. $D = 120$ when $r = 60$ and $t = 2$. Find D when $r = 80$ and $t = 3$. *240*

PART C

19. The force of attraction F between two particles varies jointly as their masses m_1 and m_2 and inversely as the square of the distance r between their centers. When $m_1 = 1$ g, $m_2 = 500$ g, and $r = 5$ cm, $F = 1.33 \times 10^{-6}$ dynes. Find F when $m_1 = 2$ g, $m_2 = 600$ g, and $r = 6$ cm.

20. The load l that a beam of fixed length can support varies jointly as its width w and the square of its depth d. A beam whose width $w = 4$ and whose depth $d = 2$ can support a load of 1,760 kg. Find l when $w = 4$ and $d = 8$.

Chapter Ten Review

Graph each relation. Determine the domain and range. Is the relation a function? [p. 253]

1. $\{(4, 3), (-1, 2), (3, 3), (-5, 2)\}$ **2.** $\{(0, 3), (5, 0), (-3, 6), (7, 0)\}$ **3.** $\{(-1, 2), (-2, 3), (-3, 3)\}$

Give the domain and range of each relation. Is the relation a function? [p. 253]

4.

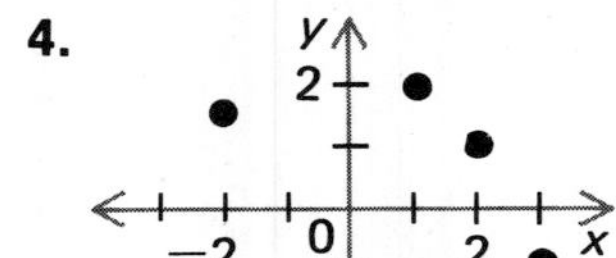

5.

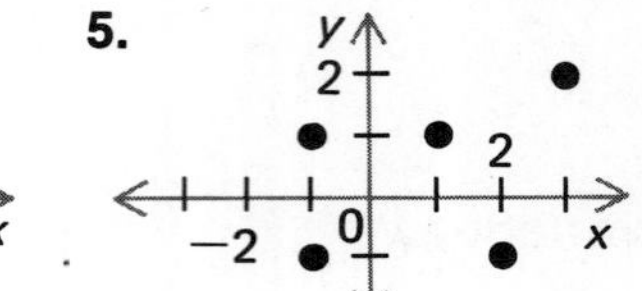

6.

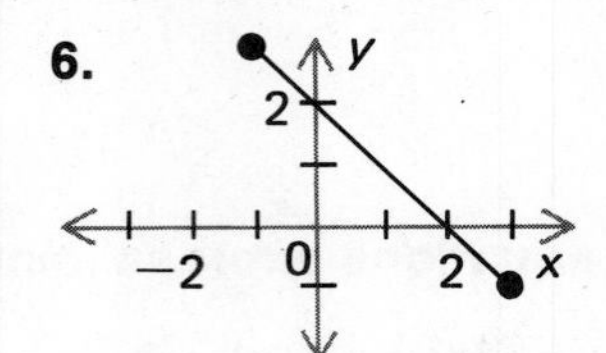

7.

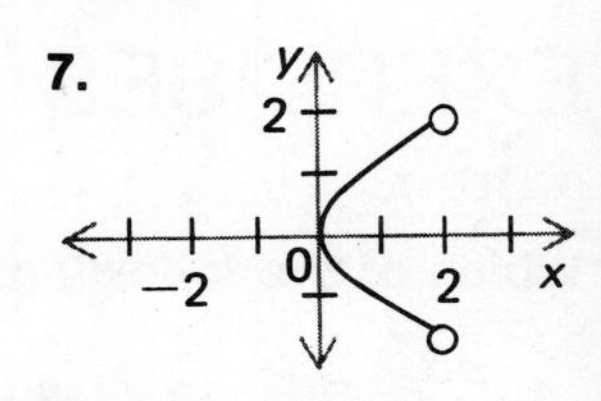

Let $h = \{(2, 3), (-3, 1), (4, 0), (-6, 7)\}$. Find. [p. 256] **Let $f(x) = x^2 + 1$. Find.** [p. 256]

8. $h(2)$ **9.** $h(-3)$ **10.** $h(4)$ **11.** $h(-6)$ **12.** $f(1)$ **13.** $f(0)$ **14.** $f(-2)$ **15.** $f(-3)$

Determine the range. *D* is the domain. [p. 256]

16. $g(x) = 3x - 1, D = \{-1, 0, 3\}$ **17.** $h(x) = x^2 - 1, D = \{0, -1, 2\}$ **18.** $f(x) = (x - 1)^2, D = \{-1, 0, \frac{1}{2}\}$

Determine which are graphs of functions and which are not. Are there any linear functions? Are there any constant functions? [p. 258]

19. **20.**

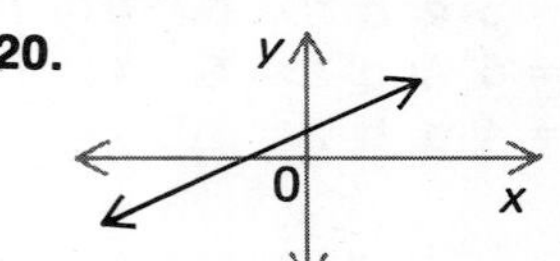

21.

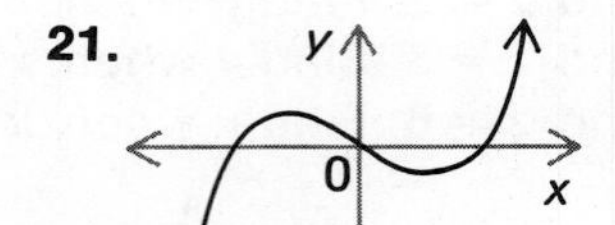

22.

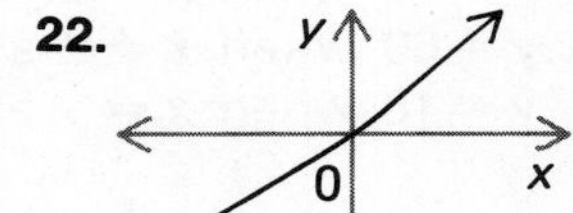

Graph each relation. Which are functions? Which are linear functions? Which are constant functions? [p. 258]

23. $\{(x, y) \mid y = |x| + 3\}$ **24.** $\{(x, y) \mid 2x - 3y = 6\}$ **25.** $\{(x, y) \mid y = -5\}$

Which tables express direct variation? Inverse variation? For each variation, determine the constant of proportionality, or the constant of variation. [pp. 261, 264]

26.

x	y
3	18
1	6
−2	−12

27.

s	t
12	1
6	2
−4	−3

28.

m	n
8	3
12	2
−6	−4

29.

c	d
−6	2
−3	1
6	−2

30.

c	d
7	−2
−28	$\frac{1}{2}$
14	−1

***y* varies directly as *x*.** [p. 261]

31. $y = 18$ when $x = 3$. Find y when $x = 5$.

32. $y = -8$ when $x = 2$. Find y when $x = -4$.

***y* varies inversely as *x*.** [p. 264]

33. $y = 12$ when $x = 2$. Find y when $x = 12$.

34. $y = -3$ when $x = 5$. Find y when $x = -3$.

35. y varies directly as the square of x. $y = 80$ when $x = 4$, find y when $x = 6$. [p. 261]

36. If s varies inversely as the square of t and $s = 4$ when $t = 3$, find s when $t = 2$. [p

37. y varies jointly as x and z. $y = -12$ when $x = 3$ and $z = -1$. Find y when $x = 5$ and $z = 6$. [p. 267]

38. y varies directly as the square of x and inversely as the positive square root of z. $y = 9$ when $x = 3$ and $z = 16$. Find y when $x = -1$ and $z = 4$. [p. 267]

Chapter Ten Test

Graph each relation. Determine the domain and range. Is the relation a function?

1. $\{(-2, 1), (-3, 1), (4, -3), (6, 1)\}$

2. $\{(0, 3), (3, 0), (5, 0), (1, 3)\}$

Give the domain and range of each relation. Is the relation a function?

3.

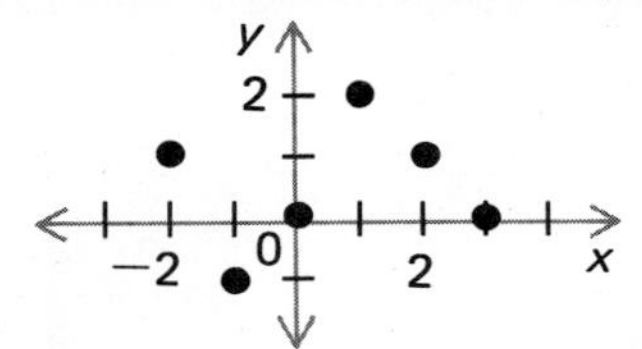

4.

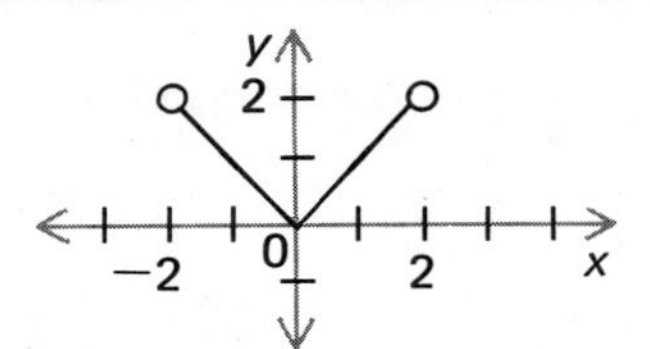

5.

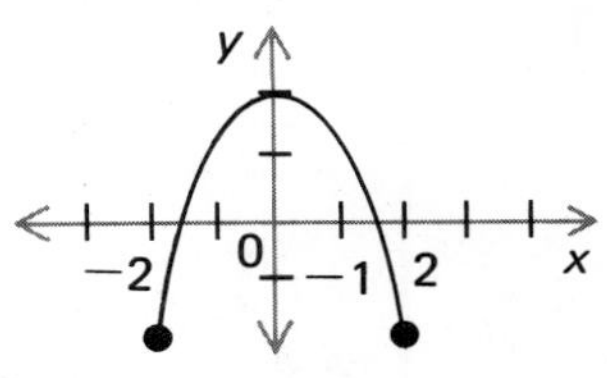

Let $f = \{(1, -1), (2, -3), (-2, 4), (6, 0)\}$. Find.

6. $f(-2)$ **7.** $f(2)$ **8.** $f(6)$

Let $h(x) = 2x + 3$. Find.

9. $h(-1)$ **10.** $h(0)$ **11.** $h(3)$

Determine the range. D is the domain.

12. $g(x) = x^2 + 3, D = \{-1, 0, 3\}$

13. $f(x) = 4x + 7, D = \{-1, 3, 4\}$

Determine which are graphs of functions and which are not. Which are linear functions? Which are constant functions?

14.

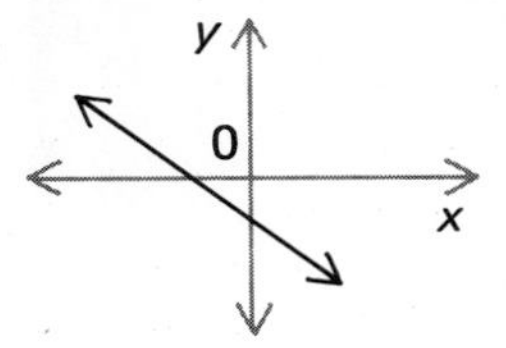

15.

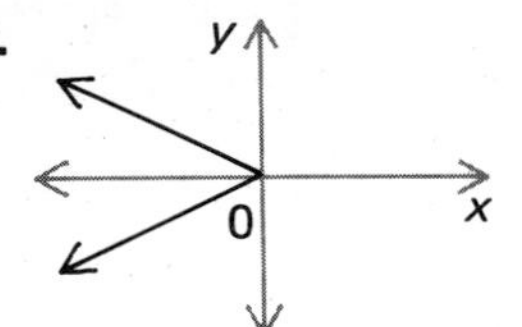

16.

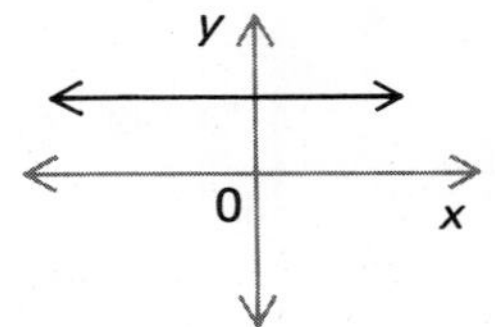

Graph each relation. Which are functions? Which are linear functions? Which are constant functions?

17. $\{(x, y) \mid 3x - 5y = 15\}$

18. $\{(x, y) \mid y = 8\}$

19. $\{(x, y) \mid y = |x| - 1\}$

Which tables express direct variation? inverse variation? For each variation, determine the constant of proportionality, or the constant of variation.

20.

x	y
3	8
−4	−6
−2	−12

21.

r	s
5	1
−5	−1
−10	−2

22.

m	n
−6	−3
9	2
−18	−1

23. y varies directly as x. $y = 10$ when $x = 2$. Find y when $x = 6$.

24. y varies inversely as x. $y = 5$ when $x = 6$. Find y when $x = 15$.

25. r varies jointly as s and t. $r = 24$ *when* $s = 4$ and $t = 2$. Find r when $s = -2$ and $t = -3$.

26. a varies directly as the positive square root of b and inversely as the square of c. $a = 2$ when $b = 4$ and $c = 5$. Find a when $b = 9$ and $c = 6$.

11 CONIC SECTIONS

Karl Friedrich Gauss

Karl Friedrich Gauss
(1777–1855)

Karl Friedrich Gauss is considered to be one of the greatest mathematicians of the nineteenth century. He has been linked with Archimedes and Newton as one of the three greatest mathematicians of all time.

Gauss was an infant prodigy. It is told that at the age of three he detected an error in his father's bookkeeping. When Gauss was in elementary school, he was asked to find the sum of all whole numbers from 1 to 100. To the teacher's chagrin, Gauss had the answer in a few minutes. He added the numbers in pairs.

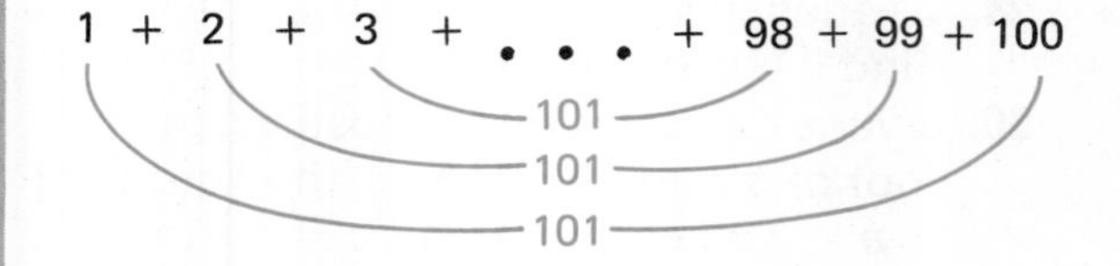

He discovered that the sum was 50 pairs of 101, or 5,050.

He discovered at age nineteen that a regular polygon of 17 sides can be constructed with a straightedge and compass. At age twenty, he received his doctorate for a dissertation proving the Fundamental Theorem of Algebra.

Gauss was a pioneer in Non-Euclidean geometry and one of the first to show the significance and use of complex numbers.

He was the first to apply probability theory to work in astronomy. The normal probability curve or the "curve of error" was used by Gauss to represent the distribution of errors in astronomical experiments. The bell shaped curve is sometimes called the graph of "Gaussian Law."

The Parabola

OBJECTIVES

- To graph a parabola
- To determine the coordinates of the maximum or minimum point and to write the equation of the axis of symmetry
- To estimate roots from the graph

REVIEW CAPSULE

Find the values of x that make $f(x) = 0$.

$$f(x) = x^2 - 5x + 6$$

Let $f(x) = 0$. $\rightarrow$ $x^2 - 5x + 6 = 0$

Solve the equation $\rightarrow$ $(x - 3)(x - 2) = 0$

$x - 3 = 0$ or $x - 2 = 0$

$x = 3$ $\qquad$ $x = 2$

Thus, $f(x) = 0$ when x is 3 or 2.

EXAMPLE 1 Graph the function $\{(x, y) \mid y = x^2\}$.

To graph a function:
1. Find some points and plot them.
2. Draw a smooth curve.

x	x^2	y
0	0^2	0
1	1^2	1
2	2^2	4
−1	$(-1)^2$	1
−2	$(-2)^2$	4

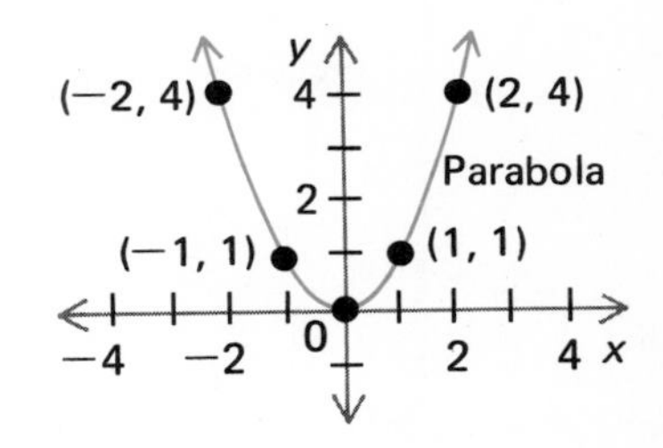

EXAMPLE 2 Graph the function $\{(x, y) \mid y = x^2 - 2x - 3\}$. Find the coordinates of the lowest point on the graph.

Plot as many points as you need to see the shape of the curve.

x	$x^2 - 2x - 3$	y
−2	$(-2)^2 - 2(-2) - 3$	5
−1	$(-1)^2 - 2(-1) - 3$	0
0	$0^2 - 2(0) - 3$	−3
1	$1^2 - 2(1) - 3$	−4
2	$2^2 - 2(2) - 3$	−3
3	$3^2 - 2(3) - 3$	0
4	$4^2 - 2(4) - 3$	5

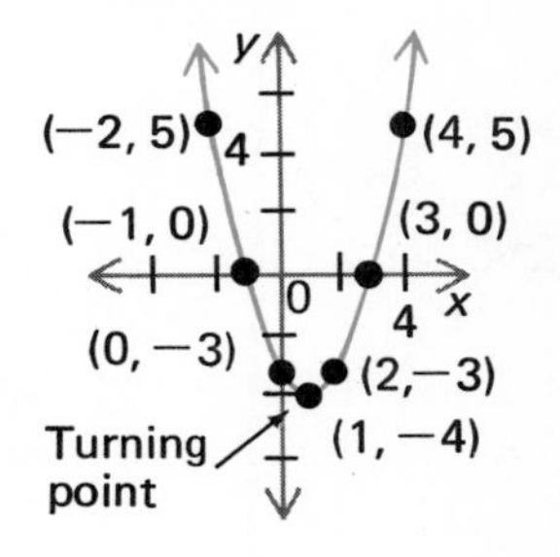

(1, −4) is the minimum point. ⟶ **Thus,** the lowest point is $(1, -4)$.

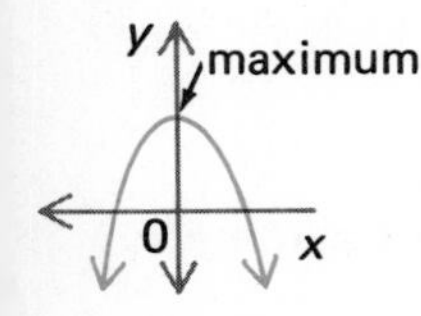

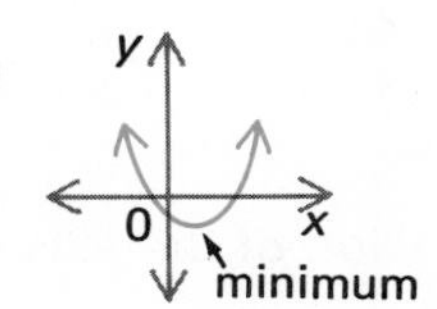

Equation of a Parabola

$y = ax^2 + bx + c$, where a, b, and c are real $[a \neq 0]$.
The turning point is a maximum or minimum point.

EXAMPLE 3 For the graph of $y = -x^2 - 2x + 8$, label the turning point and draw the axis of symmetry. Write the coordinates of the turning point and the equation of the axis of symmetry.

$(-1, 9)$ is the turning point.

The axis of symmetry is a vertical line drawn through the turning point.

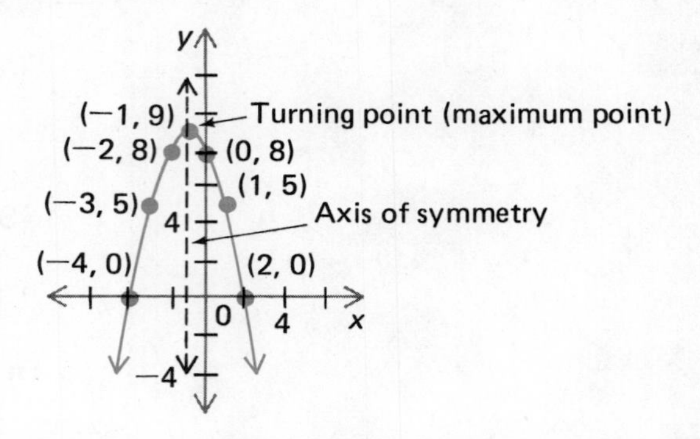

Thus, $(-1, 9)$ is the turning point. $x = -1$ is the equation of the axis of symmetry.

EXAMPLE 4 Use the graph of $y = -x^2 - 2x + 8$ to find the roots.

Find the values of x where $y = 0$.
The parabola intersects the x-axis at $(-4, 0)$ and $(2, 0)$.

Thus, the roots are -4 and 2.

Examples 3 and 4 show this. → The axis of symmetry passes halfway between the roots.

An equation of the axis of symmetry is

$$x = \frac{r_1 + r_2}{2},$$

where r_1 and r_2 are the roots.

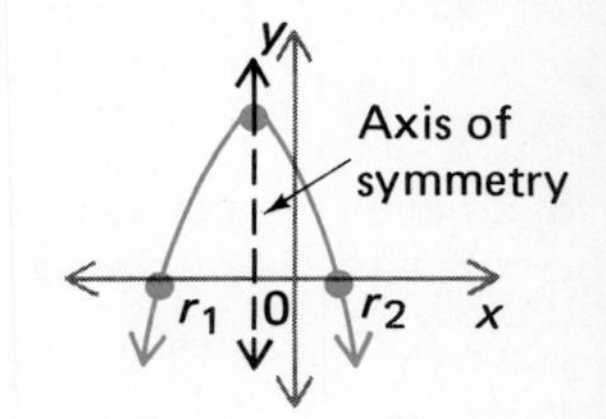

EXAMPLE 5 Show that the equation of the axis of symmetry for the parabola with equation $y = ax^2 + bx + c$ is $x = \frac{-b}{2a}$.

The sum of the roots of $ax^2 + bx + c = 0$ is $\frac{-b}{a}$. → Let r_1 and r_2 be the roots. $r_1 + r_2 = \frac{-b}{a}$.

Substitute $\frac{-b}{a}$ for $r_1 + r_2$. →

$$x = \frac{r_1 + r_2}{2} \text{ or } x = \frac{-b}{2a}$$

Thus, $x = \frac{-b}{2a}$ is the equation of the axis of symmetry.

The turning point is on the axis of symmetry.

For a parabola with equation $y = ax^2 + bx + c$, the equation of the axis of symmetry is $x = \frac{-b}{2a}$.

EXAMPLE 6 Find the equation of the axis of symmetry for the parabola with equation $y = x^2 - 3x + 1$. Then determine the coordinates of the turning point.

$$y = 1x^2 - 3x + 1$$

$$a = 1 \quad b = -3 \quad c = 1$$

$$x = \frac{-b}{2a}$$

$-b = -1(-3)$ ⟶

$$x = \frac{-1(-3)}{2(1)}$$

$x = \frac{3}{2}$ is the equation of the axis of symmetry.

The axis of symmetry passes through the turning point. Use the equation to find the y-coordinate.

$\left(\frac{3}{2}, y\right)$ are the coordinates of the turning point.

$$y = x^2 - 3x + 1$$

$$y = \left(\frac{3}{2}\right)^2 - 3\left(\frac{3}{2}\right) + 1$$

$$y = -\frac{5}{4}$$

Thus, $\left(\frac{3}{2}, -\frac{5}{4}\right)$ is the turning point.

EXAMPLE 7 Graph the function defined by $y = x^2 - 3x + 1$. Estimate the roots to the nearest tenth from the graph.

From Example 6 ⟶ $\left(\frac{3}{2}, -\frac{5}{4}\right)$ is the turning point.

Plot points. ⟶ Use 3 values of x on each side of the turning point.

3 values less than $\frac{3}{2}$

Turning point ⟶

3 values greater than $\frac{3}{2}$

x	$x^2 - 3x + 1$	y
-1	$(-1)^2 - 3(-1) + 1$	5
0	$(0)^2 - 3(0) + 1$	1
1	$1^2 - 3(1) + 1$	-1
$\frac{3}{2}$	$(\frac{3}{2})^2 - 3(\frac{3}{2}) + 1$	$-\frac{5}{4}$
2	$(2)^2 - 3(2) + 1$	-1
3	$(3)^2 - 3(3) + 1$	1
4	$(4)^2 - 3(4) + 1$	5

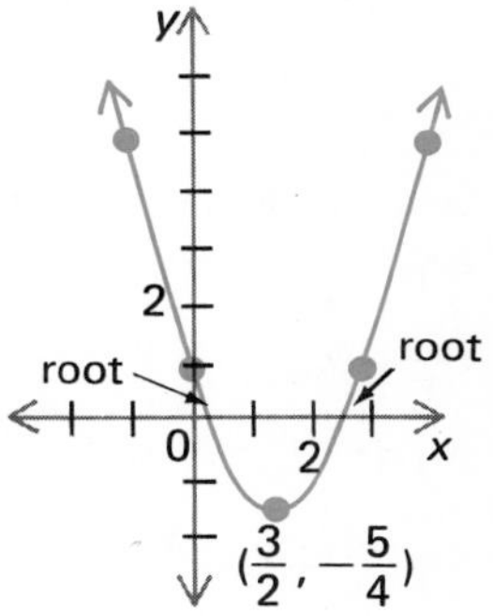

Determine where the graph intersects the x-axis. ⟶ **Thus,** the roots are approximately .4 and 2.5.

EXERCISES

PART A

Find the equation of the axis of symmetry for the parabola with the given equation. Then determine the coordinates of the turning point.

1. $y = 2x^2 + 6x + 9$
2. $y = -3x^2 - 6x + 5$
3. $y = 4x^2 - 6x - 10$
4. $y = -x^2 - 6x - 6$ *$x = -3, (-3, 3)$*
5. $y = x^2 + 4x - 3$ *$x = -2, (-2, -7)$*
6. $y = -x^2 + 6x + 4$ *$x = 3; (3, 13)$*

Graph each function. Draw the axis of symmetry and label the coordinates of the turning point.

7. $y = x^2 - 4x + 3$ *$(2, -1)$*
8. $y = \frac{1}{2}x^2$ *$(0, 0)$*
9. $y = -2x^2$ *$(0, 0)$*
10. $y = x^2 + 5$ *$(0, 5)$*
11. $y = -2x^2 + 4x + 1$ *$(1, 3)$*
12. $y = x^2 + 8x + 3$ *$(-4, -13)$*

Graph each function. Estimate the roots to the nearest tenth from the graph.

13. $y = 2x^2 + 7x - 3$ *.4; −3.9*
14. $y = x^2 - 2x - 2$ *2.7; −.7*
15. $y = -x^2 + 4x + 1$ *4.2; −.2*
16. $y = -3x^2 + 9x + 5$ *3.5; −.5*
17. $y = -x^2 + 4x + 2$ *4.4; −.4*
18. $y = 3x^2 + 5x - 2$ *−2; .3*

PART B

EXAMPLE Graph $y = \frac{1}{2}x^2$, $y = x^2$, and $y = 2x^2$ on the same set of axes. Write the equation of the axis of symmetry and the coordinates of the turning point for each.

x	x^2	$\frac{1}{2}x^2$	$2x^2$
−2	4	2	8
−1	1	$\frac{1}{2}$	2
0	0	0	0
1	1	$\frac{1}{2}$	2
2	4	2	8

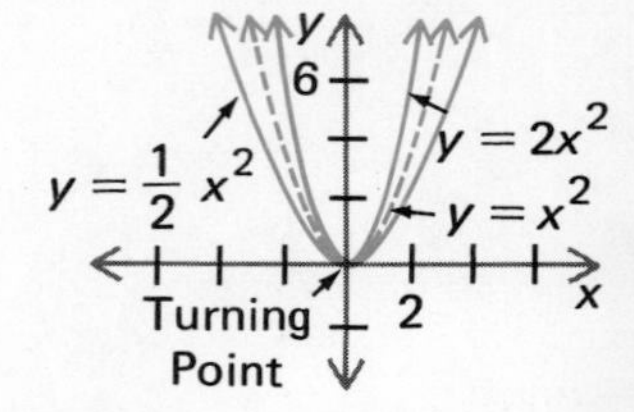

$x = \frac{-b}{2a}$; substitute 0 for b. → The equation of the axis of symmetry is $x = 0$ for each. Each parabola has the same turning point $(0, 0)$.

A set of parabolas which has the same turning point and same axis of symmetry forms a family of parabolas.

Graph the family of parabolas. Determine the turning point and the equation of the axis of symmetry.

19. $y = -\frac{1}{4}x^2$, $y = -x^2$, $y = -4x^2$ *$(0, 0)$, $(x = 0)$*
20. $y = \frac{1}{4}x^2 - 2$, $y = x^2 - 2$, $y = 4x^2 - 2$ *$(0, -2)$; $x = 0$*

PART C

21. Show that the coordinates of the turning point of the parabola with equation $y = ax^2 + bx + c$ are $\left(-\frac{b}{2a}, \frac{4ac - b^2}{4a}\right)$.

The Circle

OBJECTIVES

- To write the equation of a circle whose center is at the origin
- To graph a circle
- To determine the radius of a circle from the equation

REVIEW CAPSULE

Find the radius r.

$d = \sqrt{(x_2 - x_1)^2 + (y_2 - y_1)^2}$
$r = \sqrt{(2-0)^2 + (3-0)^2}$
$r = \sqrt{4+9}$
$r = \sqrt{13}$

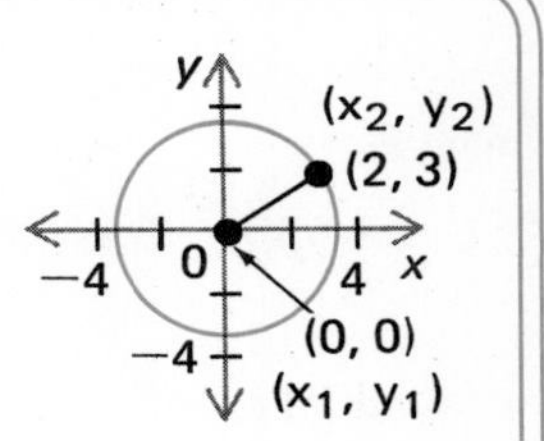

EXAMPLE 1 Write an equation of the circle whose center is at the origin and with a radius of 4.

Draw a diagram.
The distance is the radius. → Let $P(x, y)$ be any point on the circle. The distance between $P(x, y)$ and the origin is 4.

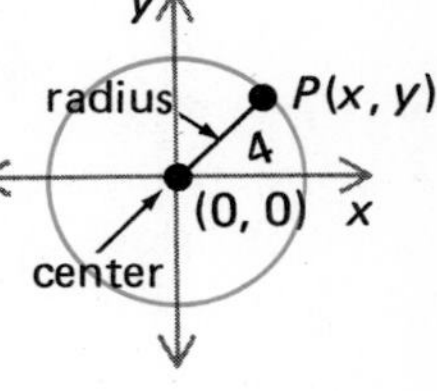

Use the distance formula. → $\sqrt{(x-0)^2 + (y-0)^2} = 4$

$\sqrt{x^2 + y^2} = 4$

Square each side. → $x^2 + y^2 = 4^2$

$P(x, y)$ is any point 4 units away from the origin.

Thus, $x^2 + y^2 = 16$ is an equation of the circle whose center is at the origin and with a radius of 4.

Definition of a circle →

A *circle* is the set of all points in a plane which are at a fixed distance, the radius, from a fixed point, the center. Standard form of the equation of a circle with center (0, 0) and radius r is $x^2 + y^2 = r^2$.

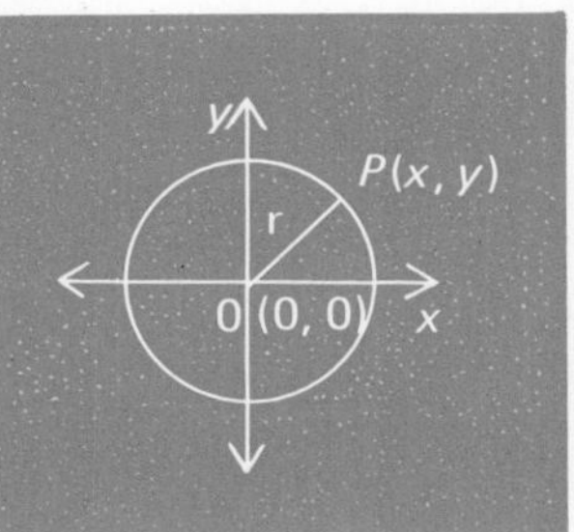

EXAMPLE 2 Write an equation of the circle whose center is at the origin and with radius 5.

Standard form → $x^2 + y^2 = r^2$
$r = 5$ $\quad x^2 + y^2 = 25$

Thus, $x^2 + y^2 = 25$ is the equation.

EXAMPLE 3 An equation of a circle is $x^2 + y^2 = 36$. Find the radius.

$x^2 + y^2 = r^2$ →

$$x^2 + y^2 = 36$$

So, $r^2 = 36$

Use only the positive square root. →

$$r = \sqrt{36}, \text{ or } 6$$

Thus, the radius is 6.

EXAMPLE 4 Write an equation of the circle whose center is at the origin and which passes through the point (3, 4).

The equation of a circle →

$$x^2 + y^2 = r^2$$

Use the point (3, 4) to find r^2. →
↑ ↑
x y

$$3^2 + 4^2 = r^2$$
$$9 + 16 = r^2$$
$$r^2 = 25$$

Thus, $x^2 + y^2 = 25$ is the equation of the circle centered at the origin and passing through (3, 4).

EXAMPLE 5 Sketch the graph of $x^2 + y^2 = 4$.

$$x^2 + y^2 = 4$$

$x^2 + y^2 = r^2$ → So, $r^2 = 4$

$$r = 2$$

Plot points on the axes which are 2 units from the origin. →

(2, 0)
(−2, 0)
(0, 2)
(0, −2)

are some points on the graph.

Draw a smooth curve.

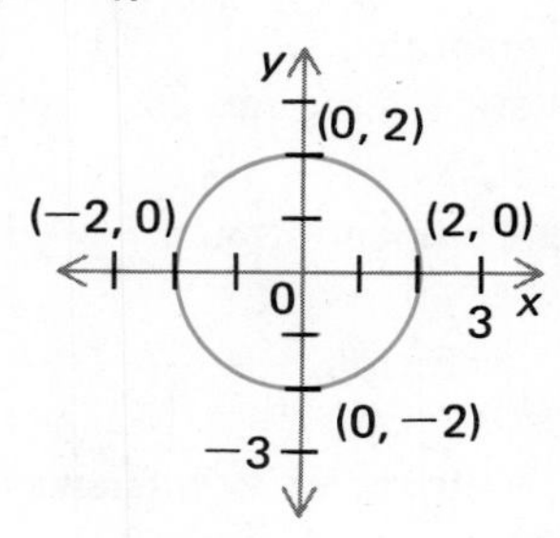

EXAMPLE 6 Sketch the graph of $x^2 + y^2 = 12$.

$$x^2 + y^2 = r^2$$

So, $r^2 = 12$

$$r = \sqrt{12}$$

$$r = 3.5$$

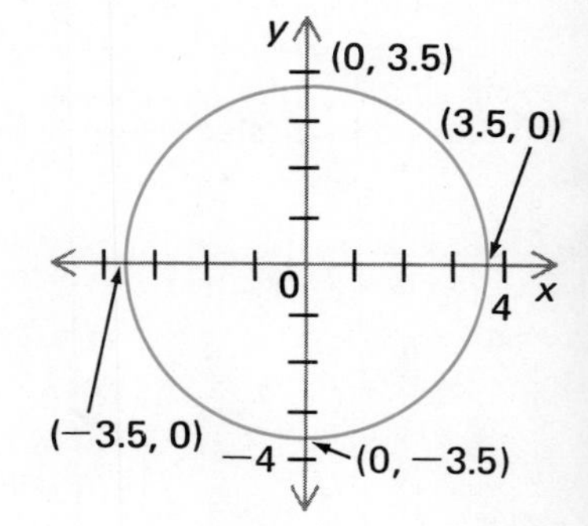

ORAL EXERCISES

For the circle whose equation is given, find the radius.

1. $x^2 + y^2 = 25$ *5* **2.** $x^2 + y^2 = 64$ *8* **3.** $x^2 + y^2 = 16$ *4* **4.** $x^2 + y^2 = 49$ *7* **5.** $x^2 + y^2 = 15$ *$\sqrt{15}$*

EXERCISES

PART A

Write an equation of the circle whose center is at the origin and with radius as given.

1. 8 **2.** 5 **3.** 11 **4.** 1 **5.** 13 **6.** 20
7. 4 **8.** 3 **9.** 2 **10.** 6 **11.** 9 **12.** 7
13. 10 $x^2 + y^2 = 100$ **14.** 12 $x^2 + y^2 = 144$ **15.** 15 $x^2 + y^2 = 225$ **16.** $\sqrt{6}$ $x^2 + y^2 = 6$ **17.** $\sqrt{7}$ $x^2 + y^2 = 7$ **18.** $\sqrt{3}$ $x^2 + y^2 = 3$

Write an equation of the circle whose center is at the origin and which passes through the given point.

19. $(5, 12)$ **20.** $(3, 5)$ **21.** $(-1, 2)$ **22.** $(-3, -4)$ $x^2 + y^2 = 25$ **23.** $(-3, 4)$ $x^2 + y^2 = 25$
24. $(0, 4)$ $x^2 + y^2 = 16$ **25.** $(2, -3)$ $x^2 + y^2 = 13$ **26.** $(6, -8)$ $x^2 + y^2 = 100$ **27.** $(0, -3)$ $x^2 + y^2 = 9$ **28.** $(5, 4)$ $x^2 + y^2 = 41$

Determine the radius of the circle whose equation is given. Then sketch the graph.

29. $x^2 + y^2 = 16$ *4* **30.** $x^2 + y^2 = 100$ **31.** $x^2 + y^2 = 4$ *2* **32.** $x^2 + y^2 = 9$ *3* **33.** $x^2 + y^2 = 36$
34. $x^2 + y^2 = 81$ *9* **35.** $x^2 + y^2 = 17$ $\sqrt{17}$ **36.** $x^2 + y^2 = 12$ $2\sqrt{3}$ **37.** $x^2 + y^2 = 6$ $\sqrt{6}$ **38.** $x^2 + y^2 = 3$ $\sqrt{3}$

PART B

EXAMPLE Show that $7x^2 + 7y^2 = 30$ is an equation of a circle.

Write the equation in the form $x^2 + y^2 = r^2$. →

$$7x^2 + 7y^2 = 30$$
$$x^2 + y^2 = \frac{30}{7}$$
$$r^2 = \frac{30}{7}$$

$r^2 = \frac{30}{7}$ → **Thus,** $7x^2 + 7y^2 = 30$ is an equation of the circle with radius $\sqrt{\frac{30}{7}}$ and center at the origin.

Determine which of the following are equations of circles. Determine the radius for each circle.

39. $6x^2 + 6y^2 = 36$ $\sqrt{6}$ **40.** $5x^2 + 4y^2 = 20$ *no* **41.** $7x^2 + 7y^2 = 49$ $\sqrt{7}$
42. $7x^2 - 7y^2 = 42$ *no* **43.** $8x^2 + 8y^2 = 70$ **44.** $x^2 - y^2 = 25$ *no*
45. $4x^2 + 4y^2 = 17$ **46.** $3x^2 - 3y^2 = 18$ *no* **47.** $3x^2 + 3y^2 = 12$ *2*
48. $-3x^2 - 3y^2 = -16$ **49.** $4x^2 + 4y^2 = 8$ $\sqrt{2}$ **50.** $2x^2 + 2y^2 = 16$ $2\sqrt{2}$
51. $16x^2 + 16y^2 = 1$ $\frac{1}{4}$ **52.** $-12x^2 - 12y^2 = -2$ **53.** $10x^2 - 10y^2 = 100$ *no*

PART C

54. Write an equation of the circle whose center is at the point $(3, 2)$ and with radius 6.

55. Write an equation of the circle whose center is at the point $(-1, 2)$ and with radius 4.

56. Write an equation of the circle whose center is at the point $(4, 1)$ and which passes through the point $(8, 4)$.

57. Write an equation of the circle whose center is at the point $(-1, -2)$ and which passes through the point $(4, 3)$.

The Ellipse

OBJECTIVES

- To write the equation of an ellipse whose center is at the origin
- To determine the intercepts of an ellipse
- To graph an ellipse

REVIEW CAPSULE

Solve $\sqrt{x-3}+\sqrt{x}=3$.

$\sqrt{x-3}=3-\sqrt{x}$

$x-3=9-6\sqrt{x}+x$ ← Square each side.

$-12=-6\sqrt{x}$ ← Isolate the radical.

$2=\sqrt{x}$

$4=x$ ← Square each side again. Check the solution.

Thus, 4 is the solution.

EXAMPLE 1 Write an equation of the set of points the sum of whose distances from $F_1(-3, 0)$ and $F_2(3, 0)$ is 10.

F_1 and F_2 are fixed points.

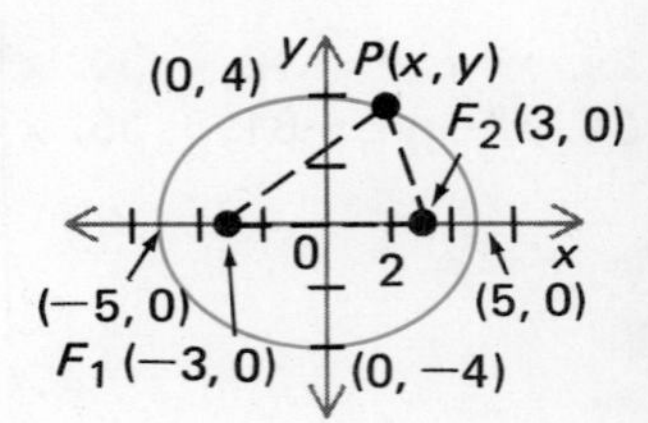

Let $P(x, y)$ be a general point.
d_1 is the distance between $P(x, y)$ and $F_1(-3, 0)$.
d_2 is the distance between $P(x, y)$ and $F_2(3, 0)$.

$$d_1+d_2=10$$

Use the distance formula to represent d_1 and d_2. →

$d_1=\sqrt{(x-(-3))^2+(y-0)^2}$, or $\sqrt{(x+3)^2+y^2}$

$d_2=\sqrt{(x-3)^2+(y-0)^2}$, or $\sqrt{(x-3)^2+y^2}$

$d_1+d_2=10$ →

Thus, $\sqrt{(x+3)^2+y^2}+\sqrt{(x-3)^2+y^2}=10$ is the equation.

The equation in Example 1 can be simplified to $\frac{x^2}{25}+\frac{y^2}{16}=1$.

Definition of an ellipse →

An *ellipse* is a set of points the sum of whose distances from two fixed points (the foci), is constant.

$$\frac{x^2}{a^2}+\frac{y^2}{b^2}=1$$

is the equation in standard form of an ellipse with center (0, 0), x-intercepts $\pm a$, and y-intercepts $\pm b$.

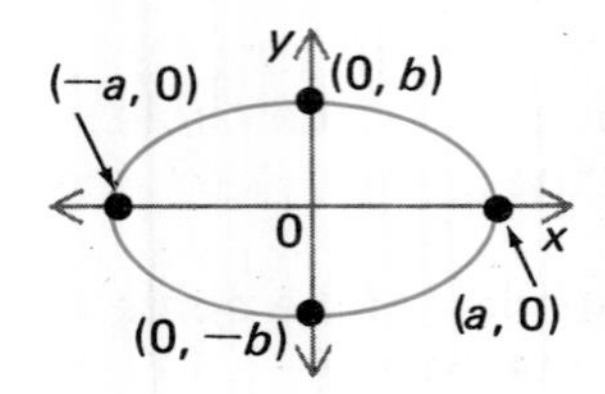

EXAMPLE 2 Write an equation in standard form of the ellipse whose center is at the origin and whose x-intercepts are ± 6 and y-intercepts are ± 5.

$\pm a$ are x-intercepts.
$\pm b$ are y-intercepts.

$$\frac{x^2}{a^2} + \frac{y^2}{b^2} = 1$$

$$\frac{x^2}{6^2} + \frac{y^2}{5^2} = 1$$

Thus, $\frac{x^2}{36} + \frac{y^2}{25} = 1$ is the equation.

EXAMPLE 3 Sketch the graph of the ellipse whose equation is given.

Find a^2 and b^2.

At $\pm a$, $y = 0$.
At $\pm b$, $x = 0$.

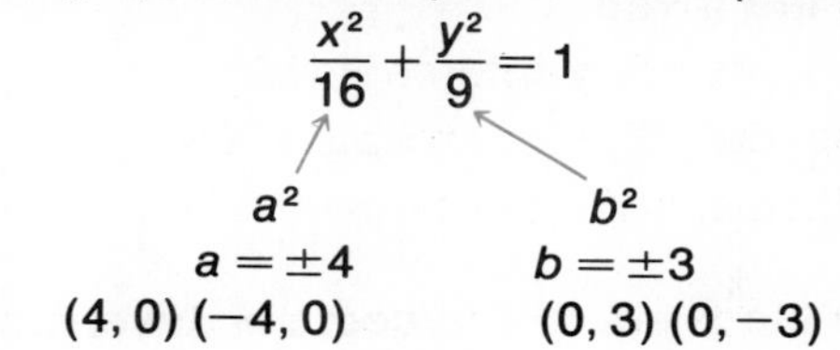

Plot the points.
Draw a smooth curve.

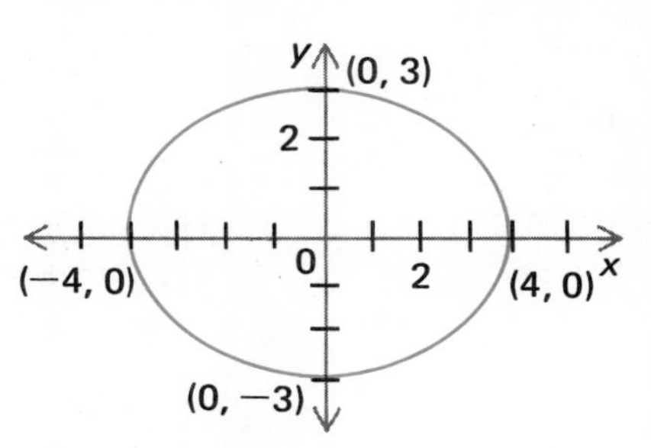

EXAMPLE 4 Sketch the graph of the ellipse whose equation is $4x^2 + 9y^2 = 36$.

Put in standard form $\frac{x^2}{a^2} + \frac{y^2}{b^2} = 1$.

$$4x^2 + 9y^2 = 36$$

$$\frac{4x^2}{36} + \frac{9y^2}{36} = \frac{36}{36}$$

$$\frac{x^2}{9} + \frac{y^2}{4} = 1$$

Write the points of intersection with the axes.

Plot the points.
Draw a smooth curve.

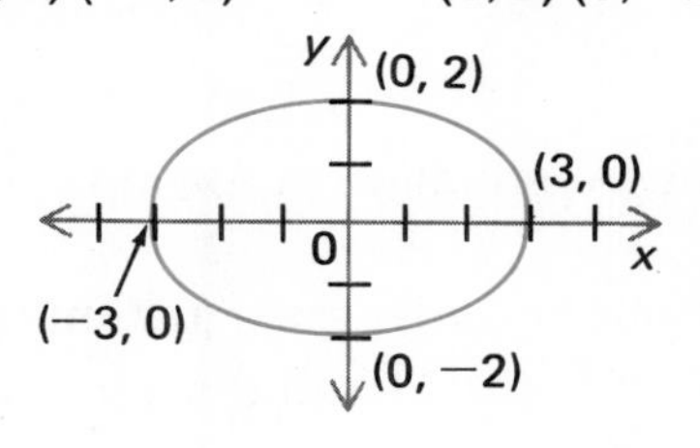

ORAL EXERCISES

For the ellipse whose equation is given, determine the *x*- and *y*-intercepts.

1. $\frac{x^2}{16}+\frac{y^2}{4}=1$
2. $\frac{x^2}{81}+\frac{y^2}{16}=1$
3. $\frac{x^2}{9}+\frac{y^2}{25}=1$
4. $3x^2+y^2=75$ $x=\pm 5;\ y=\pm 5\sqrt{3}$
5. $x^2+4y^2=100$ $x=\pm 10;\ y=\pm 5$
6. $4x^2+6y^2=36$ $x=\pm 3;\ y=\pm\sqrt{6}.$

EXERCISES

PART A

Write an equation in standard form of the ellipse whose center is at the origin and whose intercepts are given.

1. *x*-intercepts ± 5; *y*-intercepts ± 3
2. *x*-intercepts ± 7; *y*-intercepts ± 8
3. *x*-intercepts ± 2; *y*-intercepts $\pm\sqrt{3}$
4. *x*-intercepts ± 8; *y*-intercepts $\pm\sqrt{3}$
5. *x*-intercepts $\pm\sqrt{8}$; *y*-intercepts ± 6 $\frac{x^2}{8}+\frac{y^2}{36}=1$
6. *x*-intercepts $\pm\sqrt{5}$; *y*-intercepts $\pm\sqrt{7}$ $\frac{x^2}{5}+\frac{y^2}{7}=1$

Determine the *x*- and *y*-intercepts of the ellipse whose equation is given. Then sketch the graph.

7. $\frac{x^2}{25}+\frac{y^2}{4}=1$ $x=\pm 5;\ y=\pm 2$
8. $\frac{x^2}{16}+\frac{y^2}{81}=1$ $x=\pm 4;\ y=\pm 9$
9. $\frac{x^2}{25}+\frac{y^2}{9}=1$ $x=\pm 5;\ y=\pm 3$
10. $\frac{x^2}{100}+\frac{y^2}{25}=1$
11. $\frac{x^2}{16}+\frac{y^2}{4}=1$
12. $\frac{x^2}{9}+\frac{y^2}{100}=1$
13. $4x^2+y^2=100$
14. $5x^2+25y^2=25$
15. $3x^2+y^2=75$
16. $x^2+9y^2=36$ $x=\pm 6;\ y=\pm 2$
17. $2x^2+y^2=32$ $x=\pm 4;\ y=\pm 4\sqrt{2}$
18. $3x^2+2y^2=48$ $x=\pm 4;\ y=\pm 2\sqrt{6}$

PART B

EXAMPLE Show that $7x^2+8y^2=30$ is an equation of an ellipse.

Put in standard form $\frac{x^2}{a^2}+\frac{y^2}{b^2}=1.$

$$7x^2+8y^2=30$$

$$\frac{7x^2}{30}+\frac{8y^2}{30}=1$$

$$\frac{x^2}{\frac{30}{7}}+\frac{y^2}{\frac{30}{8}}=1$$

x-intercepts $\pm\sqrt{\frac{30}{7}}$

y-intercepts $\pm\sqrt{\frac{30}{8}}$

Thus, $7x^2+8y^2=30$ is an equation of an ellipse.

Show that each is an equation of an ellipse.

19. $3x^2+2y^2=30$
20. $4x^2+5y^2=16$
21. $16x^2+4y^2=14$

PART C

22. Show that the equation in Example 1 simplifies to $\frac{x^2}{25}+\frac{y^2}{16}=1.$

Using the definition, write an equation in standard form for each ellipse.

23. Foci $(-3,0)\ (1,0)$; sum of distances is 8.
24. Foci $(-2,0)$, $(2,0)$; sum of distances is 12.

The Hyperbola

OBJECTIVES

- To write the equation of a hyperbola whose center is at the origin
- To determine the intercepts of a hyperbola
- To graph a hyperbola

REVIEW CAPSULE

Find the distance between $A(-4, 0)$ and $B(6, 3)$.

$d = \sqrt{(x_2 - x_1)^2 + (y_2 - y_1)^2}$ ← Distance formula

$= \sqrt{(6 - (-4))^2 + (3 - 0)^2}$ ← Substitute.

$= \sqrt{10^2 + 3^2}$ ← Simplify.

$= \sqrt{100 + 9}$

$= \sqrt{109}$

EXAMPLE 1 Write an equation of the set of points the difference of whose distances from $F_1(-5, 0)$ and $F_2(5, 0)$ is 8.

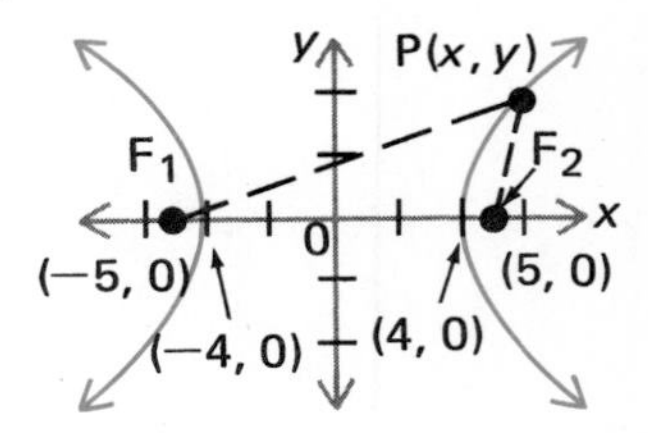

Let $P(x, y)$ be a general point.
d_1 is the distance between $P(x, y)$ and $F_1(-5, 0)$.
d_2 is the distance between $P(x, y)$ and $F_2(5, 0)$.

$$d_1 - d_2 = 8$$

Use the distance formula to represent d_1 and d_2.

$d_1 = \sqrt{(x - (-5))^2 + (y - 0)^2}$, or $\sqrt{(x + 5)^2 + y^2}$

$d_2 = \sqrt{(x - 5)^2 + (y - 0)^2}$, or $\sqrt{(x - 5)^2 + y^2}$

$d_1 - d_2 = 8$ → **Thus,** $\sqrt{(x + 5)^2 + y^2} - \sqrt{(x - 5)^2 + y^2} = 8$ is the equation.

The equation in Example 1 can be simplified to $\frac{x^2}{16} - \frac{y^2}{9} = 1$.

Definition of a hyperbola →

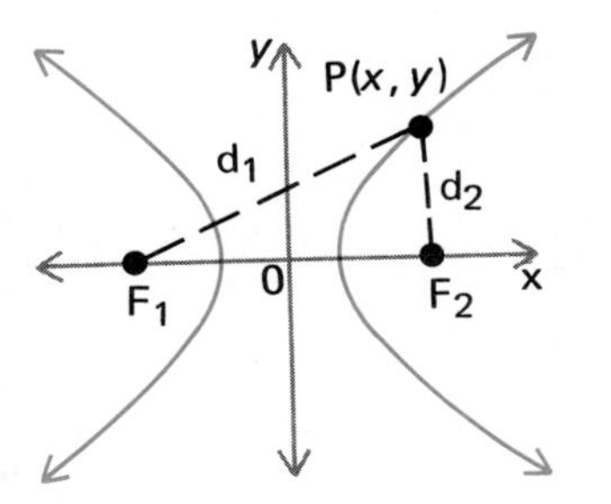

A *hyperbola* is a set of points the difference of whose distances from two fixed points (the foci) is constant.

$$\frac{x^2}{a^2} - \frac{y^2}{b^2} = 1$$

is the standard form of an equation of a hyperbola with center (0, 0) and x-intercepts $\pm a$.

EXAMPLE 2 Sketch the graph of the hyperbola with the given equation.

$$\frac{x^2}{16} - \frac{y^2}{4} = 1$$

$\frac{x^2}{a^2} - \frac{y^2}{b^2} = 1$ ⟶

$$a^2 = 16 \qquad b^2 = 4$$
$$a = \pm 4 \qquad b = \pm 2$$

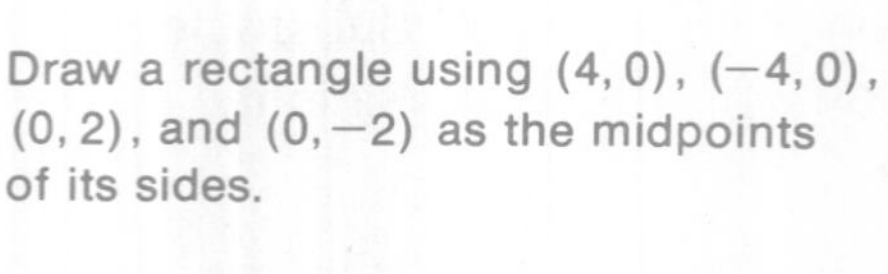

Draw a rectangle using (4, 0), (−4, 0), (0, 2), and (0, −2) as the midpoints of its sides.

Use the diagonals of the rectangle and the x-intercepts to sketch the hyperbola.

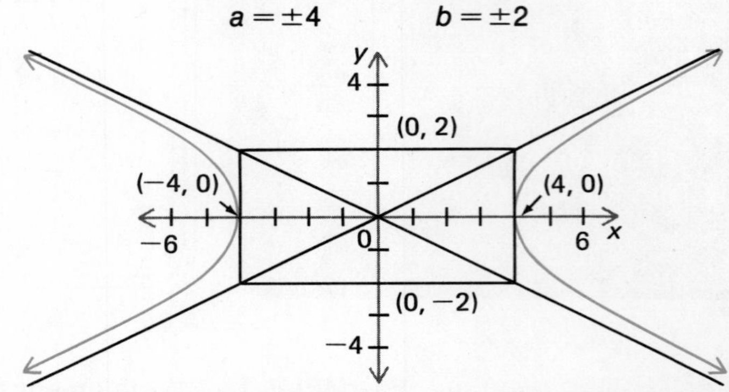

The diagonals are called *asymptotes*. The hyperbola will get closer and closer to the asymptotes, but never touch them.

EXAMPLE 3 Sketch the graph of the hyperbola whose equation is $4x^2 - y^2 = 16$.

Write in standard form.
$\frac{x^2}{a^2} - \frac{y^2}{b^2} = 1$

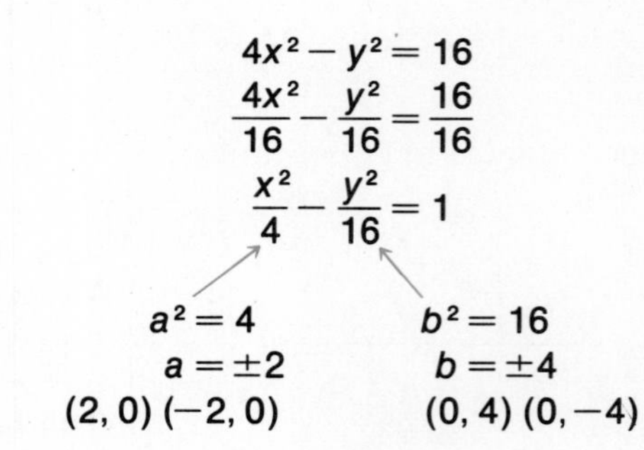

$$4x^2 - y^2 = 16$$
$$\frac{4x^2}{16} - \frac{y^2}{16} = \frac{16}{16}$$
$$\frac{x^2}{4} - \frac{y^2}{16} = 1$$

Find a and b. ⟶

$$a^2 = 4 \qquad b^2 = 16$$
$$a = \pm 2 \qquad b = \pm 4$$
$$(2, 0)\ (-2, 0) \qquad (0, 4)\ (0, -4)$$

Draw a rectangle using (2, 0), (−2, 0), (0, 4), and (0, −4) as the midpoints of its sides.

Use the diagonals of the rectangle (asymptotes) and the x-intercepts to sketch the hyperbola.

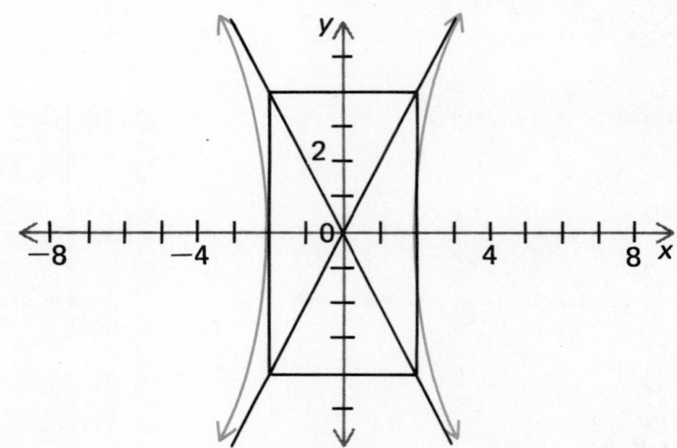

EXAMPLE 4 Sketch the graph of $xy = 6$.

Solve for y. → $y = \frac{6}{x}$

Write a table. →

x	$\frac{6}{x}$	y
1	$\frac{6}{1}$	6
−1	$\frac{6}{-1}$	−6
2	$\frac{6}{2}$	3
−2	$\frac{6}{-2}$	−3
3	$\frac{6}{3}$	2
−3	$\frac{6}{-3}$	−2

Plot the 6 points.
Draw a smooth curve.
The graph of $xy = 6$ is the graph of an inverse variation.

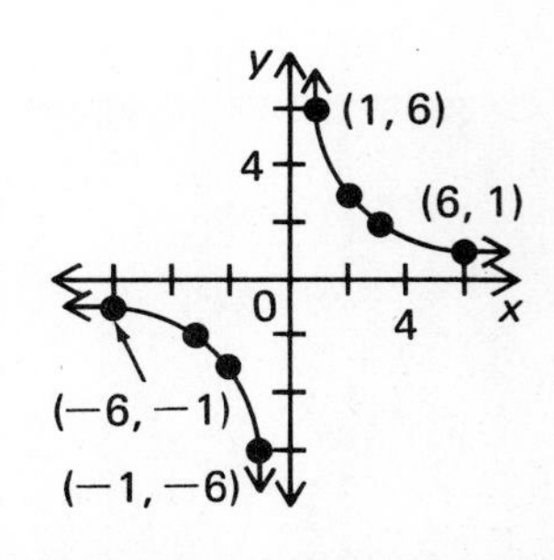

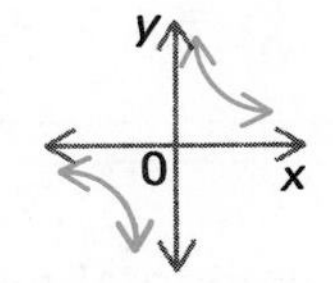

$xy = k$ is an equation of a hyperbola which does not intersect the x-axis or the y-axis.

ORAL EXERCISES

For the hyperbola whose equation is given, determine the x-intercepts.

1. $\frac{x^2}{4} - \frac{y^2}{16} = 1$ *±2* **2.** $\frac{x^2}{36} - \frac{y^2}{9} = 1$ *±6* **3.** $5x^2 - 4y^2 = 80$ *±4* **4.** $2x^2 - 3y^2 = 18$ *±3*

EXERCISES

PART A

Sketch the graph of the hyperbola whose equation is given.

1. $\frac{x^2}{4} - \frac{y^2}{16} = 1$ **2.** $\frac{x^2}{36} - \frac{y^2}{9} = 1$ **3.** $\frac{x^2}{9} - \frac{y^2}{16} = 1$ **4.** $\frac{x^2}{25} - \frac{y^2}{36} = 1$

5. $\frac{x^2}{100} - \frac{y^2}{25} = 1$ **6.** $\frac{x^2}{81} - \frac{y^2}{16} = 1$ **7.** $\frac{x^2}{25} - \frac{y^2}{4} = 1$ **8.** $\frac{x^2}{49} - \frac{y^2}{4} = 1$

9. $4x^2 - y^2 = 36$ **10.** $4x^2 - 9y^2 = 36$ **11.** $16x^2 - 9y^2 = 144$ **12.** $12x^2 - 3y^2 = 48$

13. $xy = 4$ **14.** $xy = 9$ **15.** $xy = 1$ **16.** $xy = 3$

PART B

Sketch the graph of the hyperbola described by the equation.

17. $4x^2 - 16y^2 = 32$ **18.** $4x^2 - 5y^2 = 80$ **19.** $xy = -4$ **20.** $xy = -8$

PART C

21. Write an equation in terms of a and b for the asymptotes of the hyperbola with equation $\frac{x^2}{a^2} - \frac{y^2}{b^2} = 1$.

22.–33. Write equations for the asymptotes of each of the hyperbolas whose equations are given in Exercises 1–12.

Identifying Conics

OBJECTIVE

- To identify an equation of a parabola, circle, ellipse, or hyperbola

REVIEW CAPSULE

Equations of Conic Sections

Parabola: $y = ax^2 + bx + c$

Circle: $x^2 + y^2 = r^2$

Ellipse: $\frac{x^2}{a^2} + \frac{y^2}{b^2} = 1$

Hyperbola: $\frac{x^2}{a^2} - \frac{y^2}{b^2} = 1$, or $xy = k$

EXAMPLE 1 Identify $4x^2 - 9y^2 = 36$ as an equation of a parabola, a circle, an ellipse, or a hyperbola.

Since both variables are squared, divide each side by the constant. →

$$4x^2 - 9y^2 = 36$$
$$\frac{4x^2}{36} - \frac{9y^2}{36} = 1$$
$$\frac{x^2}{9} - \frac{y^2}{4} = 1$$

$\frac{x^2}{9} - \frac{y^2}{4} = 1$ describes a hyperbola. → **Thus,** $4x^2 - 9y^2 = 36$ is an equation of a hyperbola.

EXAMPLE 2 Identify $x^2 = 16 - 4y^2$ as an equation of a parabola, a circle, an ellipse, or a hyperbola.

Since both variables are squared, add $4y^2$ to each side. →

Divide each side by 16. →

$$x^2 = 16 - 4y^2$$
$$x^2 + 4y^2 = 16$$
$$\frac{x^2}{16} + \frac{4y^2}{16} = \frac{16}{16}$$
$$\frac{x^2}{16} + \frac{y^2}{4} = 1$$

$\frac{x^2}{16} + \frac{y^2}{4} = 1$ describes an ellipse. → **Thus,** $x^2 = 16 - 4y^2$ is an equation of an ellipse.

EXAMPLE 3 Identify $x^2 + 12 = y - 7x$ as an equation of a parabola, a circle, an ellipse, or a hyperbola.

Since only one variable is squared, solve for y. →

$$x^2 + 12 = y - 7x$$
$$x^2 + 7x + 12 = y$$

$y = x^2 + 7x + 12$ is an equation of a parabola. → **Thus,** $x^2 + 12 = y - 7x$ is an equation of a parabola.

EXAMPLE 4 Identify $4x^2 + 4y^2 = 16$ as an equation of a parabola, a circle, an ellipse, or a hyperbola.

Since both variables are squared, divide each side by 16. →

$$4x^2 + 4y^2 = 16$$
$$\frac{4x^2}{16} + \frac{4y^2}{16} = \frac{16}{16}$$
$$\frac{x^2}{4} + \frac{y^2}{4} = 1, \text{ or } x^2 + y^2 = 4$$

$x^2 + y^2 = 4$ is an equation of a circle. → **Thus,** $4x^2 + 4y^2 = 16$ is an equation of a circle.

EXAMPLE 5 Identify $x = \frac{6}{y}$ as an equation of a parabola, a circle, an ellipse, or a hyperbola.

$$x = \frac{6}{y}$$
$$xy = 6$$

$xy = 6$ is an equation of a hyperbola. → **Thus,** $x = \frac{6}{y}$ is an equation of a hyperbola.

EXERCISES

PART A

Identify the conic section whose equation is given.

1. $\frac{x^2}{9} + \frac{y^2}{4} = 1$ *E*
2. $\frac{x^2}{4} + \frac{y^2}{4} = 1$ *C*
3. $\frac{x^2}{16} - \frac{y^2}{16} = 1$ *H*
4. $\frac{x^2}{4} - \frac{y^2}{16} = 1$ *H*
5. $\frac{x^2}{10} + \frac{y^2}{15} = 1$ *E*
6. $\frac{x^2}{12} - \frac{y^2}{3} = 1$ *H*
7. $\frac{x^2}{3} + \frac{y^2}{5} = 2$ *E*
8. $\frac{x^2}{4} + \frac{y^2}{4} = 5$ *C*
9. $\frac{x^2}{15} - \frac{y^2}{12} = 2$ *H*
10. $8x^2 + 8y^2 = 4$ *C*
11. $x^2 - y = 4$ *P*
12. $6x^2 - 5y^2 = 12$ *H*
13. $y = \frac{16}{x}$ *H*
14. $10x^2 + 10y^2 = 100$ *C*
15. $4x^2 + 3y^2 = 15$ *E*
16. $9x^2 - 4y^2 = 6$ *H*
17. $xy = 15$ *H*
18. $2x^2 = 4 - 2y^2$ *C*
19. $y = x^2 + 4$ *P*
20. $-y - x^2 - 5 = 0$ *P*
21. $x^2 - 2y^2 = 8$ *H*

PART B

Identify the conic section whose equation is given.

22. $\frac{3x^2}{18} + \frac{4y^2}{16} = 1$ *E*
23. $\frac{7x^2}{15} - \frac{8y^2}{25} = 1$ *H*
24. $\frac{4x^2}{15} + \frac{4y^2}{15} = 1$ *C*
25. $\frac{2x^2}{13} + \frac{3y^2}{15} = 4$ *E*
26. $\frac{6x^2}{15} + \frac{5y^2}{14} = 3$ *E*
27. $\frac{6x^2}{14} + \frac{6y^2}{14} = 12$ *C*
28. $16x^2 - 16y^2 = 5$ *H*
29. $12x^2 + 16y^2 = 17$ *E*
30. $9y^2 + 9x^2 = 5$ *C*
31. $6x^2 + 6y^2 = 8$ *C*
32. $15x^2 - 12y^2 = 6$ *H*
33. $10x^2 + 20y^2 = 20$ *E*

Translations

OBJECTIVE

■ To graph an equation like $y = |x + 1|$ or $y = x^2 - 2$ by moving the basic graphs $y = |x|$ and $y = x^2$

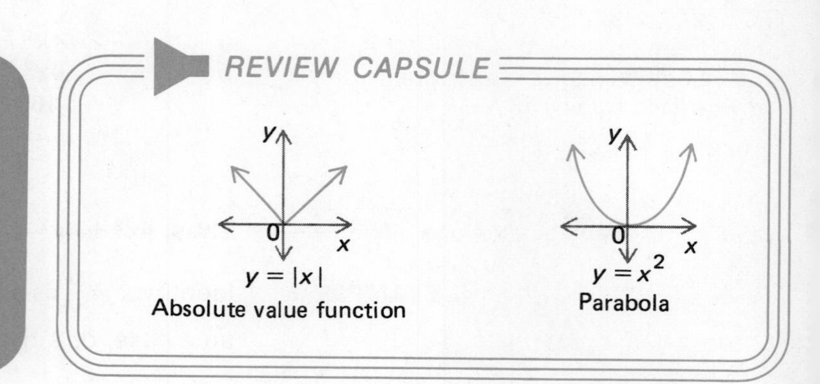

EXAMPLE 1 Graph $y = |x - 1|$ and $y = |x + 1|$. Compare with $y = |x|$.

x	-2	-1	0	1	2
$\|x\|$	$\|-2\|$	$\|-1\|$	$\|0\|$	$\|1\|$	$\|2\|$
y	2	1	0	1	2

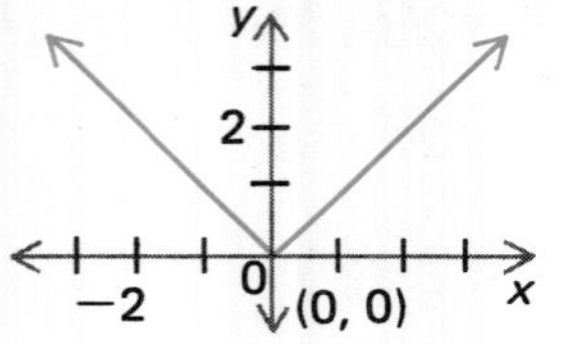

$y = |x|$
is a basic graph.

x	-2	-1	0	1	2
$\|x - 1\|$	$\|-3\|$	$\|-2\|$	$\|-1\|$	$\|0\|$	$\|1\|$
y	3	2	1	0	1

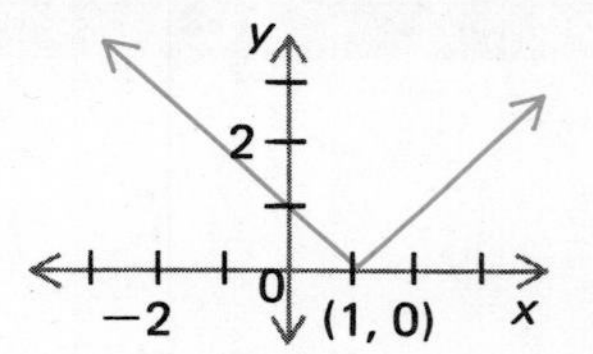

$y = |x - 1|$
Move the basic graph $y = |x|$ one unit *right*.

x	-2	-1	0	1	2
$\|x + 1\|$	$\|-1\|$	$\|0\|$	$\|1\|$	$\|2\|$	$\|3\|$
y	1	0	1	2	3

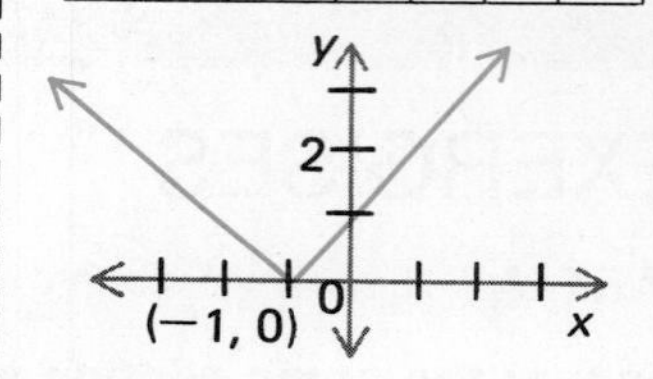

$y = |x + 1|$
Move the basic graph $y = |x|$ one unit *left*.

EXAMPLE 2 Graph $y = (x - 1)^2$ and $y = (x + 1)^2$. Compare with $y = x^2$.

x	-2	-1	0	1	2
x^2	$(-2)^2$	$(-1)^2$	0^2	1^2	2^2
y	4	1	0	1	4

$y = x^2$
is a basic graph.

x	-1	0	1	2	3
$(x - 1)^2$	$(-2)^2$	$(-1)^2$	0^2	1^2	2^2
y	4	1	0	1	4

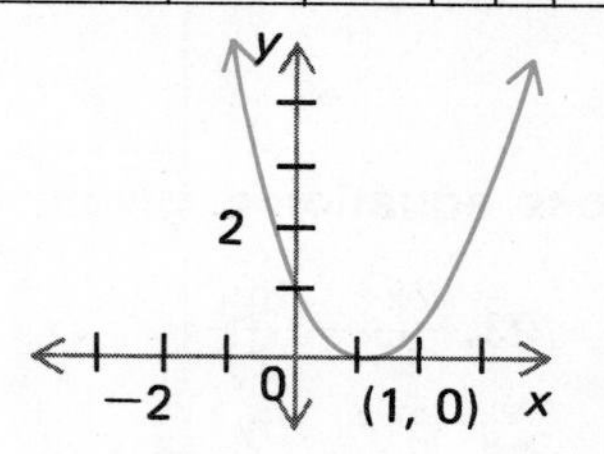

$y = (x - 1)^2$
Move the basic graph $y = x^2$ one unit *right*.

x	-3	-2	-1	0	1
$(x + 1)^2$	$(-2)^2$	$(-1)^2$	0^2	1^2	2^2
y	4	1	0	1	4

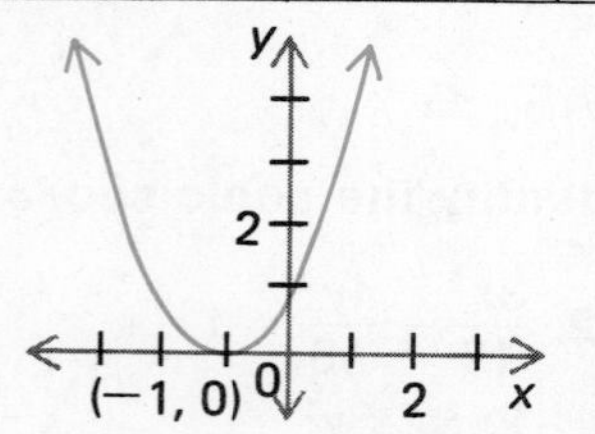

$y = (x + 1)^2$
Move the basic graph $y = x^2$ one unit *left*.

Examples 1 and 2 suggest this. ⟶

k is a positive number. ⟶

$y = f(x - k)$ Move the graph of $y = f(x)$ k units to the right.	$y = f(x + k)$ Move the graph of $y = f(x)$ k units to the left.

EXAMPLE 3 Graph $y = |x| + 1$ and $y = |x| - 1$. Compare with $y = |x|$.

x	-2	-1	0	1	2
$\lvert x\rvert$	$\lvert -2\rvert$	$\lvert -1\rvert$	$\lvert 0\rvert$	$\lvert 1\rvert$	$\lvert 2\rvert$
y	2	1	0	1	2

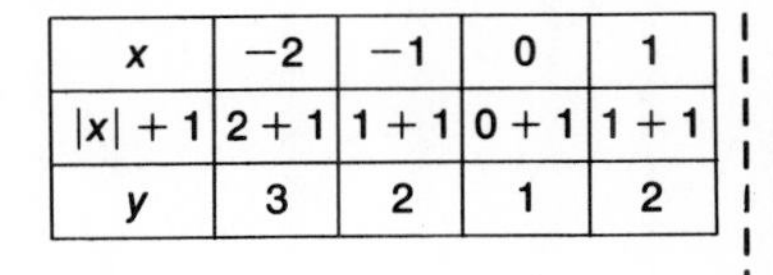

x	-2	-1	0	1
$\lvert x\rvert + 1$	$2 + 1$	$1 + 1$	$0 + 1$	$1 + 1$
y	3	2	1	2

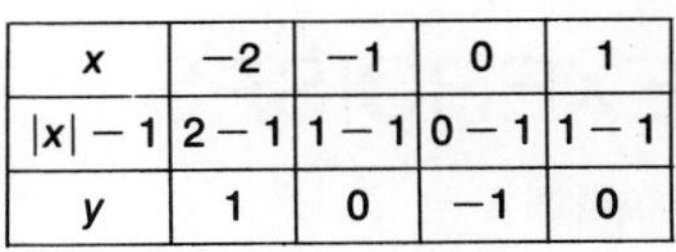

x	-2	-1	0	1
$\lvert x\rvert - 1$	$2 - 1$	$1 - 1$	$0 - 1$	$1 - 1$
y	1	0	-1	0

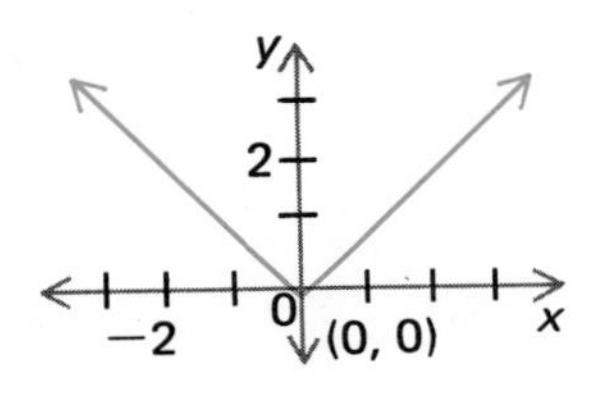

$y = |x|$ is a basic graph.

$y = |x| + 1$
Move the basic graph $y = |x|$ one unit *up*.

$y = |x| - 1$
Move the basic graph $y = |x|$ one unit *down*.

Example 3 suggests this. ⟶

k is a positive number. ⟶

$y = f(x) + k$ Move the graph of $y = f(x)$ k units up.	$y = f(x) - k$ Move the graph of $y = f(x)$ k units down.

EXAMPLE 4 Graph $y = (x - 1)^2 + 2$.

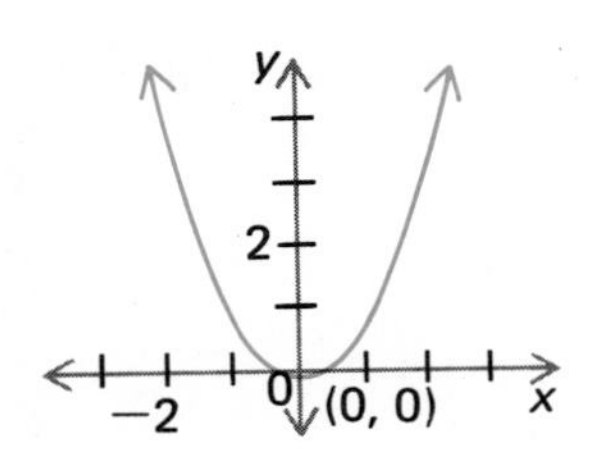

$y = x^2$ is a basic graph.

Move the basic graph 1 unit right.

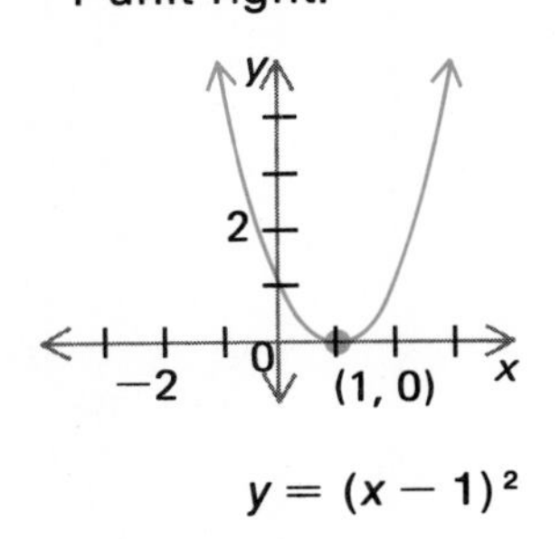

$y = (x - 1)^2$

Move the basic graph 2 units up.

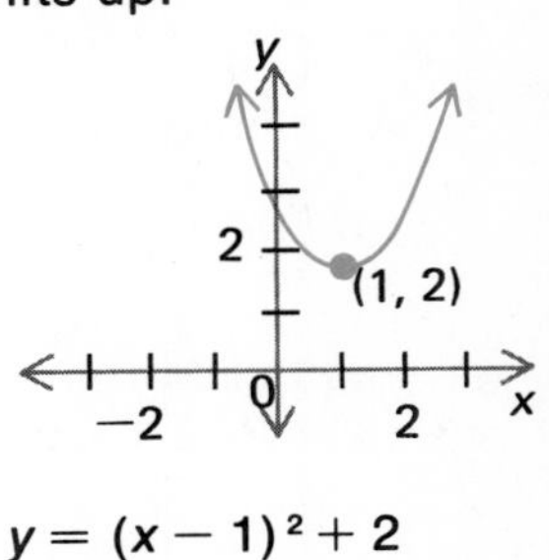

$y = (x - 1)^2 + 2$

EXAMPLE 5 Explain how to graph $y = x^2 - 4x + 4$.

Rewrite: $x^2 - 4x + 4 = (x - 2)^2$. ⟶ $y = (x - 2)^2$

1. Draw the graph of $y = x^2$.
2. Move the graph 2 units right.

ORAL EXERCISES

Describe how the graph of each function compares with the basic graph.

1. $y = x^2$ **2.** $y = (x - 5)^2$ **3.** $y = x^2 + 1$ **4.** $y = (x + 3)^2 - 2$
5. $y = |x|$ **6.** $y = |x + 3|$ **7.** $y = |x| - 2$ **8.** $y = -3 + |x - 1|$

EXERCISES

PART A

Graph each of the following.

1. $y = |x + 2|$ **2.** $y = |x - 2|$ **3.** $y = |x + 3| + 1$ **4.** $y = |x - 3| - 2$
5. $y = (x - 2)^2$ **6.** $y = (x + 2)^2$ **7.** $y = (x + 3)^2 - 3$ **8.** $y = (x - 3)^2 + 1$
9. $y = 2 + |x|$ **10.** $y = -2 + (x + 1)^2$ **11.** $y = -3 + |x - 4|$ **12.** $y = 4 + |x + 2|$

PART B

EXAMPLE Graph $y = x^2 + 8x + 6$.

Complete the square.

$$y - 6 = x^2 + 8x$$
$$y - 6 + 16 = x^2 + 8x + 16$$
$$y + 10 = (x + 4)^2$$
$$y = (x + 4)^2 - 10$$

Draw $y = x^2$.
Move the graph 4 units left.
Move the graph 10 units down.

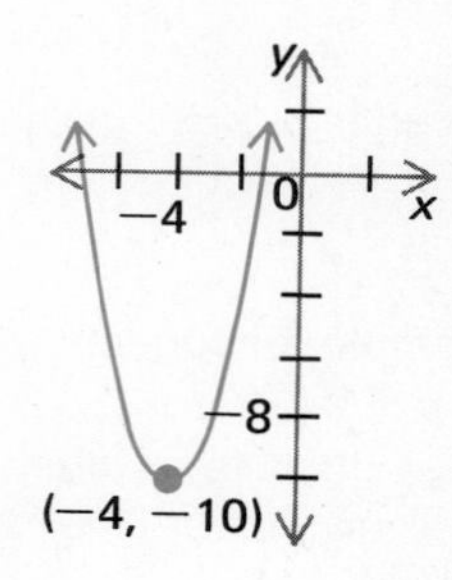

Complete the square and graph each of the following.

13. $y = x^2 + 8x + 4$ **14.** $y = x^2 + 4x - 1$ **15.** $y = x^2 + 6x - 4$
16. $y = x^2 + 2x - 5$ **17.** $y = x^2 + 10x + 22$ **18.** $y - 3 = x^2 + 4x - 1$
$y = (x + 1)^2 - 6$ *$y = (x + 5)^2 - 3$* *$y = (x + 2)^2 - 2$*

PART C

EXAMPLE Graph $(x + 2)^2 + (y - 1)^2 = 10$.

The center moves 2 units left and 1 unit up. Radius remains the same.

Circle $x^2 + y^2 = 10$
has center $(0, 0)$
and radius $\sqrt{10}$.
Circle $(x + 2)^2 + (y - 1)^2 = 10$
has center $(-2, 1)$
and radius $\sqrt{10}$.

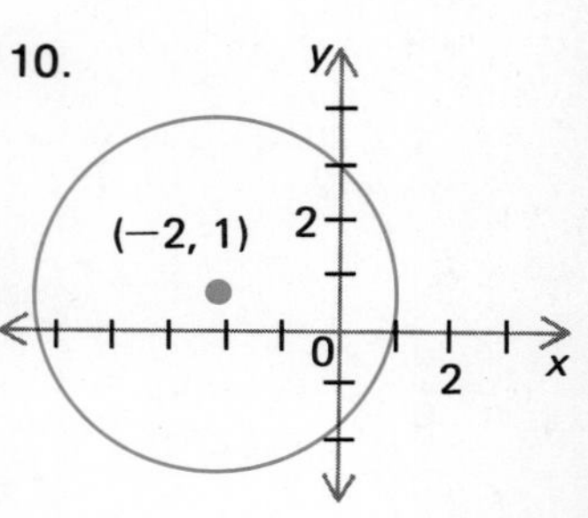

Graph each of the following.

19. $(x - 1)^2 + (y + 1)^2 = 16$ **20.** $(x + 2)^2 + (y - 1)^2 = 12$ **21.** $(x + 1)^2 + (y - 2)^2 = 1$

22. $\dfrac{(x + 2)^2}{16} + \dfrac{(y - 1)^2}{4} = 1$ **23.** $\dfrac{(x + 1)^2}{25} - \dfrac{(y + 2)^2}{4} = 1$ **24.** $9(x - 3)^2 + 4(y + 2) = 36$

Conic Sections

Conic sections can be constructed by folding paper.

PARABOLA

Draw a straight line and mark a dot on a piece of paper.

Fold the paper so that the line is on the dot.

Make 20 to 30 different folds, each with the line on the dot. The curve formed is a parabola.

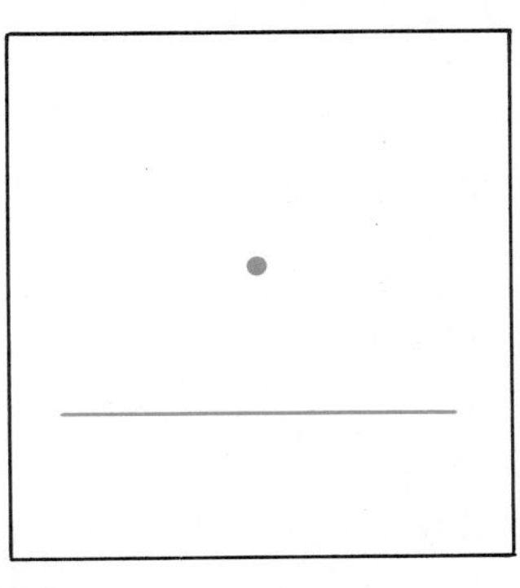

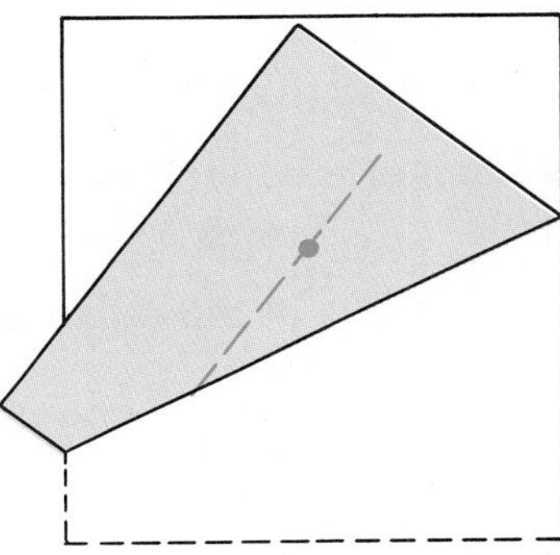

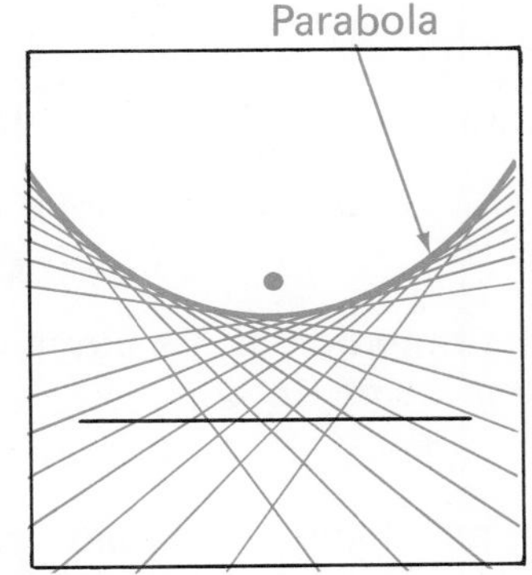

ELLIPSE

Draw a circle on a piece of paper. Mark a point inside the circle but not at the center.

Fold the paper so that the dot is on the circle. Do this 30 to 40 places along the circle. The straight lines form a curve which is an ellipse.

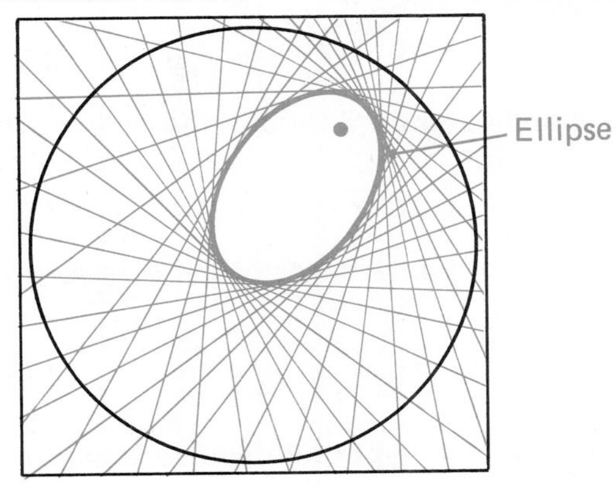

PROJECT

1. Draw a circle on a piece of paper. This time mark a dot outside the circle. Fold the paper so that the dot is on the circle. What conic section is formed after 30 to 40 such folds?

2. What figure is formed using a dot in the center of a circle?

Chapter Eleven Review

Find the equation of the axis of symmetry for the parabola with the given equation. Then determine the coordinates of the turning point. [p. 273]

1. $y = 3x^2 - 6x + 8$ **2.** $y = -4x^2 + 8x - 7$ **3.** $y = x^2 + 4x - 3$

Graph each function. Draw the axis of symmetry and label the coordinates of the turning point. [p. 273]

4. $y = x^2 - 6x + 12$ **5.** $y = -2x^2 + 8x + 2$ **6.** $y = -x^2 - 12x - 30$

Graph each function. Estimate the roots to the nearest tenth from the graph. [p. 273]

7. $y = 3x^2 + 8x - 2$ **8.** $y = -2x^2 + 7x + 1$ **9.** $y = x^2 + 4x - 3$

10. Graph the family of parabolas described by $y = x^2$, $y = 2x^2$, and $y = 4x^2$. [p. 273]

Write an equation of the circle whose center is at the origin and with radius as given. [p. 277]

11. 13 **12.** $\sqrt{15}$ **13.** 17 **14.** $\sqrt{8}$ **15.** $\sqrt{10}$ **16.** 25

Write an equation of the circle whose center is at the origin and which passes through the given point. [p. 277]

17. $(-2, -1)$ **18.** $(3, -2)$ **19.** $(7, -4)$ **20.** $(0, -4)$ **21.** $(-5, -3)$

Determine which of the following are equations of circles. Determine the radius of each circle. Then sketch the graph. [p. 277]

22. $x^2 + y^2 = 49$ **23.** $4x^2 + 4y^2 = 16$ **24.** $8x^2 + 8y^2 = 16$ **25.** $4x^2 + 6y^2 = 36$

Write an equation in standard form of the ellipse whose center is at the origin and whose intercepts are given. [p. 280]

26. $x = \pm 4, y = \pm 2$ **27.** $x = \pm 7, y = \pm 3$ **28.** $x = \pm 5, y = \pm 5$ **29.** $x = \pm 2, y = \pm 8$

Determine the *x*- and *y*-intercepts of the ellipse whose equation is given. Then sketch the graph. [p. 280]

30. $\frac{x^2}{36} + \frac{y^2}{4} = 1$ **31.** $\frac{x^2}{49} + \frac{y^2}{16} = 1$ **32.** $x^2 + 4y^2 = 16$ **33.** $5x^2 + 25y^2 = 125$

Sketch the graph of the hyperbola whose equation is given. [p. 283]

34. $\frac{x^2}{36} - \frac{y^2}{4} = 1$ **35.** $xy = -6$ **36.** $xy = 12$ **37.** $16x^2 - 4y^2 = 64$

Identify the conic section whose equation is given. [p. 286]

38. $\frac{x^2}{16} - \frac{y^2}{16} = 1$ **39.** $\frac{x^2}{16} + \frac{y^2}{36} = 1$ **40.** $4x^2 + 4y^2 = 12$ **41.** $x^2 - y = 8$

Graph each of the following. [p. 288]

42. $y = |x| - 4$ **43.** $y = (x - 2)^2 + 3$ **44.** $y = 2 + |x + 6|$ **45.** $y = -3 + x^2$

Complete the square and graph each of the following. [p. 288]

46. $y = x^2 + 6x + 4$ **47.** $y = x^2 - 10x + 20$ **48.** $y = x^2 + 12x + 30$

Chapter Eleven Test

Find the equation of the axis of symmetry for the parabola with the given equation. Then determine the coordinates of the turning point.

1. $y = x^2 + 8x - 2$
2. $y = -3x^2 + 1$

Graph each function. Draw the axis of symmetry and label the coordinates of the turning point.

3. $y = x^2 + 3x - 4$
4. $y = x^2 - 6x + 5$

Graph each function. Estimate the roots to the nearest tenth from the graph.

5. $y = 5x^2 + 9x - 1$
6. $y = -3x^2 + 5x + 1$
7. Graph the family of parabolas described by $y = -x^2$, $y = -2x^2$, and $y = -\frac{1}{2}x^2$.

Write an equation of the circle whose center is at the origin and with radius as given.

8. 10
9. $\sqrt{13}$
10. 30

Write an equation of the circle whose center is at the origin and which passes through the given point.

11. $(-3, -4)$
12. $(6, -1)$

Determine which of the following are equations of circles. Determine the radius of each circle. Then sketch the graph.

13. $x^2 + y^2 = 25$
14. $6x^2 + 6y^2 = 24$
15. $4x^2 + 10y^2 = 20$

Write an equation in standard form of the ellipse whose center is at the origin and whose intercepts are given.

16. $x = \pm 3, y = \pm 2$
17. $x = \pm 6, y = \pm 2$

Determine the *x*- and *y*-intercepts of the ellipse whose equation is given. Then sketch the graph.

18. $\frac{x^2}{64} + \frac{y^2}{36} = 1$
19. $\frac{x^2}{4} + \frac{y^2}{81} = 1$
20. $x^2 + 4y^2 = 36$

Sketch the graph of the hyperbola whose equation is given.

21. $\frac{x^2}{64} - \frac{y^2}{16} = 1$
22. $xy = 9$
23. $9x^2 - y^2 = 81$

Identify the conic section whose equation is given.

24. $\frac{x^2}{4} + \frac{y^2}{16} = 1$
25. $\frac{x^2}{2} - \frac{y^2}{6} = 1$
26. $\frac{x^2}{18} + \frac{y^2}{18} = 1$
27. $xy = 8$

Graph each of the following.

28. $y = |x - 6|$
29. $y = (x - 3)^2 + 1$
30. $y = 3 + (x - 1)^2$

Complete the square and graph.

31. $y = x^2 - 8x + 11$

12 SYSTEMS OF EQUATIONS AND INEQUALITIES

How the U.S. Uses Its Energy

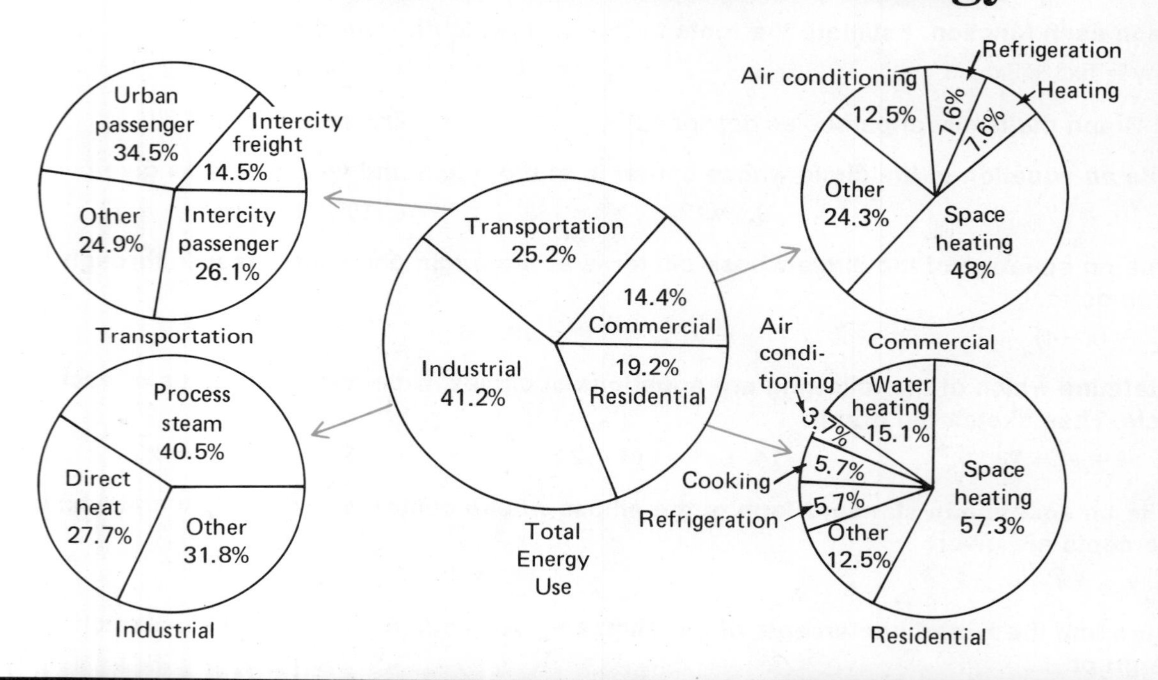

PROJECT

1. What percent of total energy use is Intercity Passenger?
2. What percent of total energy use is residential space heating?
3. How much will a 10% reduction in residential water heating reduce the total energy use?
4. How much will a 20% reduction in urban passenger reduce the total energy use?
5. How much will a 15% reduction in commercial space heating reduce the total energy use?
6. How much would residential space heating, water heating, and air conditioning have to be reduced in order to prevent a 2% reduction in energy use by industry?

Source: Federal Energy Administration, Washington D.C.

Solving Linear Systems By Graphing

OBJECTIVES

- To find the solution of a system of two linear equations graphically
- To determine whether a system is consistent or inconsistent

REVIEW CAPSULE

Graph $\{(x, y) \mid y = 2x + 1\}$.

$y = 2x + 1$ ← y-intercept

Slope $= \frac{2}{1}$

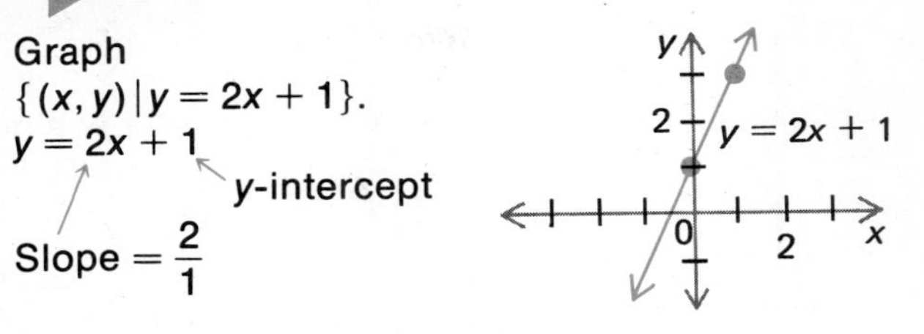

Every point on the line is a solution of $y = 2x + 1$. Every solution of $y = 2x + 1$ is a point on the line.

EXAMPLE 1 Graph $\{(x, y) \mid y = -2x + 4\}$ and $\{(x, y) \mid y = 3x - 1\}$ on the same set of axes. Find their common solution.

Plot the y-intercept. Use the slope to find other points on the line.

$y = -2x + 4$

slope $= -\frac{2}{1}$ $\quad$ y-intercept $= 4$

$y = 3x - 1$

slope $= \frac{3}{1}$ $\quad$ y-intercept $= -1$

The common solution is the point of intersection.

From the graph $(1, 2)$ is a point common to each line.
Thus, $(1, 2)$ is the common solution.

EXAMPLE 2 Solve the system of equations $\begin{array}{l} x - y = 1 \\ 2x + y = -4 \end{array}$ graphically.

Write in $y = mx + b$ form. →

$x - y = 1$
$y = x - 1$

Graph using slope-intercept method.

slope $= \frac{1}{1}$

$2x + y = -4$
$y = -2x - 4$

slope $= -\frac{2}{1}$

$(-1, -2)$ is a point common to each. → **Thus,** $(-1, -2)$ is the common solution.

> To solve a system of two linear equations graphically:
> 1. Graph both lines on the same set of axes.
> 2. Find their point of intersection.

EXAMPLE 3 Show algebraically that $(-1, -2)$ satisfies both equations of Example 2, $x - y = 1$ and $2x + y = -4$.

Substitute in each equation: $x = -1$ and $y = -2$. ⟶

$x - y$	1
$(-1) - (-2)$	1
$-1 + 2$	
1	

$2x + y$	-4
$2(-1) + (-2)$	-4
$-2 - 2$	
-4	

Thus, $(-1, -2)$ satisfies both equations.

EXAMPLE 4 Solve the system of equations graphically.

$$3x - 2y = 12$$
$$x + 3y = -7$$

Check the solution.

Write each equation in $y = mx + b$ form. (slope: m; y-intercept: b)

$$3x - 2y = 12$$
$$-2y = -3x + 12$$
$$y = \frac{3}{2}x - 6$$

slope $= \frac{3}{2}$

$$x + 3y = -7$$
$$3y = -x - 7$$
$$y = -\frac{1}{3}x - \frac{7}{3}$$

slope $= -\frac{1}{3}$

Sketch the graph of each line on the same set of axes.

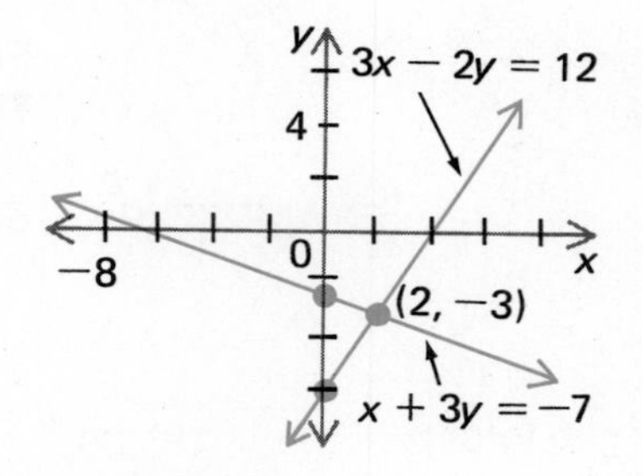

The two lines intersect at $(2, -3)$. ⟶ From the graph $(2, -3)$ is a point common to each line.

Check in each equation. ⟶ Substitute in each equation: $x = 2$ and $y = -3$.

$3x - 2y$	12
$3(2) - 2(-3)$	12
$6 + 6$	
12	

$x + 3y$	-7
$2 + 3(-3)$	-7
$2 - 9$	
-7	

Thus, $(2, -3)$ is the solution of the system.

ORAL EXERCISES

Give the coordinates of the point of intersection.

1.

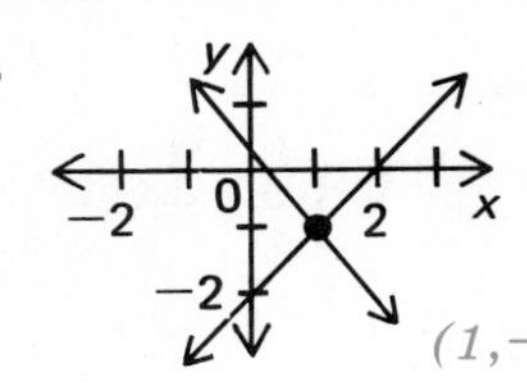

(1, −1)

2.

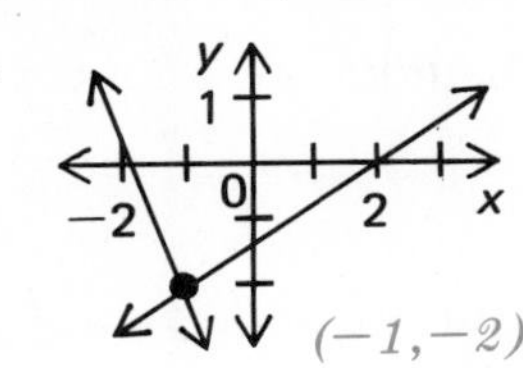

(−1, −2)

3.

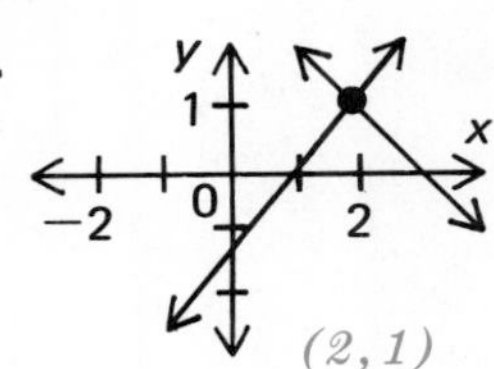

(2, 1)

4.

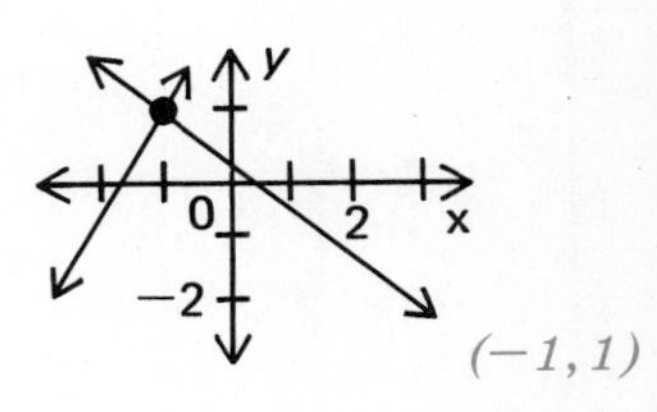

(−1, 1)

EXERCISES

PART A

Solve each system graphically.

1. $x + y = 4$
$x - y = 2$ *(3, 1)*

2. $3x + y = 5$
$x - y = 7$ *(3, −4)*

3. $4x - y = 9$
$x - 3y = 16$ *(1, −5)*

4. $4x - y = 5$ *(2.5, 5)*
$2x - y = 0$

5. $3x - y = -1$
$2x + y = 1$ *(0, 1)*

6. $3x + 4y = 18$
$5x - y = 7$ *(2, 3)*

7. $2x + 3y = 1$
$2x - y = 1$ *(.5, 0)*

8. $x + 2y = 6$
$3x - 2y = 6$ *(3, 1.5)*

9. $2x + y = 5$
$x - y = 1$
(2, 1)

10. $x - y = -1$
$2x - 3y = 0$
(−3, −2)

11. $5x - 2y = 19$
$7x + 3y = 15$
(3, −2)

12. $x + 2y = 11$
$x - y = 5$ *(7, 2)*

PART B

EXAMPLE Solve the system of equations graphically.

$$y = 2x - 1$$
$$y = 2x + 2$$

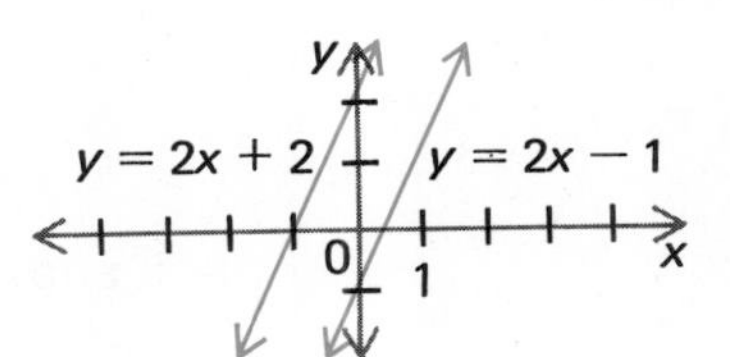

The two lines are parallel. There is no common point.

The two lines do not intersect.
Thus, there is no solution.

A system of equations which has no solution is said to be an *inconsistent* system.
A *consistent* system is one which has a solution.

Which systems are consistent? Which are inconsistent?

13. $2x + 3y = 9$
$8x + 12y = 1$ *I*

14. $3x - 2y = 7$
$2x - 3y = 9$ *C*

15. $y = 2x - 3$
$2y = 4x + 12$ *I*

16. $x + y = 7$
$x - y = 6$ *C*

PART C

Write a system of two linear equations for the given number of solutions.

17. no solution

18. one solution

19. an infinite number of solutions

Solving Linear Systems Algebraically

OBJECTIVES

- To solve a system of two linear equations by substitution
- To solve a system of two linear equations by addition

REVIEW CAPSULE

Substitute $2x - 5$ for y in $x + y = 7$.

$$x + y = 7$$

Replace y with $2x - 5$. → $x + (2x - 5) = 7$

EXAMPLE 1 Solve the system $x + y = 6$, $y = 2x + 3$ by substitution.

$$y = 2x + 3$$

Replace the y in $x + y = 6$ with $2x + 3$.

$$x + y = 6$$

New equation has one variable. Solve. ⟶

$$\begin{aligned} x + (2x + 3) &= 6 \\ x + 2x + 3 &= 6 \\ 3x + 3 &= 6 \\ 3x &= 3 \\ x &= 1 \end{aligned}$$

Substitute $x = 1$ in either equation to find y.
Both equations give the same y-value.

$$\begin{aligned} y &= 2x + 3 \\ y &= 2(1) + 3 \\ y &= 5 \end{aligned} \qquad \text{or} \qquad \begin{aligned} x + y &= 6 \\ 1 + y &= 6 \\ y &= 5 \end{aligned}$$

So, $x = 1$ and $y = 5$.

Check in both equations.
Substitute in each equation.
$x = 1$ and $y = 5$

y	$2x + 3$
5	$2(1) + 3$
	$2 + 3$
	5

$x + y$	6
$1 + 5$	6
6	

$(1, 5)$ means $x = 1$ and $y = 5$.

Thus, $(1, 5)$ is the solution of the system.

To solve a system of two linear equations by substitution:
1. Solve one equation for either variable.
2. Substitute in the other equation.

EXAMPLE 2 Solve the system $2x + 3y = 9$ by substitution.
$-2x - 5y = 1$

Solve for x. ⟶

$$2x + 3y = 9$$
$$x = \frac{9 - 3y}{2}$$

Substitute for x in the second equation. $2\left(\frac{9 - 3y}{2}\right) = 2x$

$$-2x - 5y = 1$$
$$-(9 - 3y) - 5y = 1$$

Combine like terms. ⟶

$$-2y = 10$$
$$y = -5$$

To find x let $y = -5$ in either equation.

$$-2x - 5y = 1$$
$$-2x - 5(-5) = 1$$
$$-2x = -24$$
$$x = 12$$

Check in both equations:
Let $x = 12$ and $y = -5$. ⟶

$2x + 3y$	9
$2(12) + 3(-5)$	9
9	

$-2x - 5y$	1
$-2(12) - 5(-5)$	1
1	

Thus, the solution of the system is $(12, -5)$.

EXAMPLE 3 Add the two equations $2x + 3y = 9$ from Example 2.
$-2x - 5y = 1$
Then find the solution of the system.

Add the equations.
New equation has one variable.

$$2x + 3y = 9$$
$$\underline{-2x - 5y = 1}$$
$$0x - 2y = 10$$
$$-2y = 10$$
$$y = -5$$

To find x let $y = -5$ in either equation.

$$2x + 3y = 9$$
$$2x + 3(-5) = 9$$
$$x = 12$$

Check in both equations.

$2x + 3y$	9
$2(12) + 3(-5)$	9
9	

$-2x - 5y$	1
$-2(12) - 5(-5)$	1
1	

Thus, the solution of the system is $(12, -5)$.

The method used to solve the system in Example 3 is the method of addition.

EXAMPLE 4 Solve the system $\begin{aligned} 3x + y &= -1 \\ 2x - 3y &= -8 \end{aligned}$ by addition.

Eliminate the y-terms.
Multiply first equation by 3 →

$$3x + y = -1$$
$$3(3x + y) = 3(-1) \quad \text{or} \quad 9x + 3y = -3$$

y-terms are additive inverses. →

$$\begin{aligned} 9x + 3y &= -3 \\ 2x - 3y &= -8 \\ \hline 11x \qquad &= -11 \end{aligned}$$

Add the equations.
Solve for x.

$$x = -1$$

To find y, substitute $x = -1$ in either equation.

$$\begin{aligned} 3x + y &= -1 \\ 3(-1) + y &= -1 \\ y &= 2 \end{aligned}$$

Thus, $(-1, 2)$ is the solution of the system.

> To solve a system of two linear equations by addition:
> 1. Rewrite the equations so that the coefficients of one variable become additive inverses.
> 2. Add the equations.

EXAMPLE 5 Solve the system $\begin{aligned} 3x + 2y &= 5 \\ 4x + 5y &= 2 \end{aligned}$ by addition.

Multiply each equation so that the coefficients of one variable are additive inverses.

First Method

$$\begin{aligned} 5(3x + 2y) &= 5(5) \\ -2(4x + 5y) &= -2(2) \end{aligned}$$

$$\begin{aligned} 15x + 10y &= 25 \\ -8x - 10y &= -4 \\ \hline 7x \qquad &= 21 \\ x &= 3 \end{aligned}$$

Find y by letting $x = 3$ in the first equation.

$$\begin{aligned} 3x + 2y &= 5 \\ 3(3) + 2y &= 5 \\ 9 + 2y &= 5 \\ 2y &= -4 \\ y &= -2 \end{aligned}$$

Second Method

$$\begin{aligned} 4(3x + 2y) &= 4(5) \\ -3(4x + 5y) &= -3(2) \end{aligned}$$

$$\begin{aligned} 12x + 8y &= 20 \\ -12x - 15y &= -6 \\ \hline -7y &= 14 \\ y &= -2 \end{aligned}$$

Find x by letting $y = -2$ in the second equation.

$$\begin{aligned} 4x + 5y &= 2 \\ 4x + 5(-2) &= 2 \\ 4x - 10 &= 2 \\ 4x &= 12 \\ x &= 3 \end{aligned}$$

Add the equations. →
Solve. →

Check in both equations.

$3x + 2y$	5
$3(3) + 2(-2)$	
5	

$4x + 5y$	2
$4(3) + 5(-2)$	
2	

Each method gives the same solution.

Thus, $(3, -2)$ is the solution of the system.

ORAL EXERCISES

Determine multipliers for each equation so that the sum of the two equations is an equation in one variable.

1. $2x + 3y = 9$ *(2 and 3) or*
   $5x - 2y = 7$ *(5 and −2)*
2. $5x - 3y = 6$ *(7 and 3) or*
   $2x + 7y = 10$ *(2 and −5)*
3. $4x + 5y = 9$ *(2 and 5) or*
   $7x - 2y = 4$ *(7 and −4)*
4. $6x - 3y = 7$ *(−2 and 3) or*
   $5x - 2y = 6$ *(−5 and 6)*
5. $5x + 9y = 7$ *(−4 and 9) or*
   $3x + 4y = 10$ *(−3 and 5)*
6. $6x - 9y = 2$ *(−2 and 1) or*
   $7x - 18y = 5$ *(−7 and 6)*

EXERCISES

PART A

Solve by substitution.

1. $2x - 4y = -6$
   $-x + y = 1$
   (1, 2)
2. $4x - 3y = 2$
   $2x + y = -4$
   (−1, −2)
3. $3x - 6y = 12$
   $-2x + y = -5$
   (2, −1)
4. $-x + y = -5$
   $5x + 6y = 14$
   (4, −1)
5. $-x + 2y = -2$
   $2x + 3y = 11$
   (4, 1)
6. $-7x + 14y = 7$
   $4x - y = -11$
   (−3, −1)
7. $-x + 4y = -3$
   $2x - 3y = 11$
   (7, 1)
8. $4x + 12y = 4$
   $5x - y = -11$
   (−2, 1)

Solve by addition.

9. $2x + y = 6$
   $x - y = 3$
   (3, 0)
10. $x - 2y = 6$
    $-x + 3y = -4$
    (10, 2)
11. $3x + 4y = -15$
    $3x + 6y = -21$
    (−1, −3)
12. $-3x + 2y = 7$
    $4x + 2y = -14$
    (−3, −1)
13. $x - 2y = 8$
    $3x + y = 3$
    (2, −3)
14. $4x + 5y = 6$
    $5x + 2y = -1$
    (−1, 2)
15. $7x - 3y = 5$
    $-4x + 7y = 13$
    (2, 3)
16. $5x - 7y = -16$
    $2x + 8y = 26$
    (1, 3)

PART B

Solve. Use whichever method is more convenient.

17. $2x + 3y = 8$
    $3x + 4y = 11$
    (1, 2)
18. $3x - 5y = -13$
    $4x + y = -2$
    (−1, 2)
19. $5x + 2y = 11$
    $3x - 4y = 17$
    (3, −2)
20. $2x + 7y = -12$
    $3x - y = 5$
    (1, −2)
21. $7x - 3y = 14$
    $3x + y = 6$
    (2, 0)
22. $9x + 7y = 21$
    $x + 12y = 36$
    (0, 3)
23. $3x - y = b$
    $x - y = c$
24. $2x + 3y = a$
    $-x + y = b$

PART C

Solve.

25. $\frac{x}{2} + \frac{y}{3} = 4$
    $\frac{x}{4} + \frac{y}{3} = 3$
26. $\frac{x}{5} - \frac{y}{3} = -2$
    $\frac{x}{2} + \frac{y}{6} = 7$
27. $\frac{2x}{3} + \frac{y}{7} = -11$
    $\frac{x}{7} - \frac{y}{3} = -10$
28. $\frac{x}{4} - \frac{y}{7} = 7$
    $\frac{x}{5} - \frac{y}{3} = 1$
29. $-\frac{x}{2} + \frac{y}{3} = 2$
    $\frac{x}{4} + \frac{y}{9} = 0$
30. $\frac{3x}{5} + \frac{y}{7} = 4$
    $\frac{x}{5} + \frac{y}{7} = 6$

2 by 2 Matrices

$\begin{bmatrix} a & b \\ c & d \end{bmatrix}$ is a 2 by 2 *matrix*.

Read: matrix *abcd*. 2 rows 2 columns

Each matrix has a corresponding *determinant*. Notice how its value is found.

$$\begin{vmatrix} 5 & 4 \\ 2 & 3 \end{vmatrix} = 5(3) - 2(4) = 15 - 8, \text{ or } 7$$

$$\begin{vmatrix} -1 & -5 \\ 6 & 7 \end{vmatrix} = (-1)(7) - (6)(-5) = -7 + 30, \text{ or } 23$$

$$\begin{vmatrix} a & b \\ c & d \end{vmatrix} = ad - cb$$

Read: determinant *abcd*.

PROBLEM

Find the value of the determinant $\begin{vmatrix} 2 & -3 \\ 4 & -1 \end{vmatrix}$.

$$\begin{vmatrix} 2 & -3 \\ 4 & -1 \end{vmatrix} = 2(-1) - 4(-3)$$
$$= -2 + 12$$
$$= 10$$

PROJECT

Find the value of the determinant.

1. $\begin{vmatrix} 1 & 2 \\ 3 & 9 \end{vmatrix}$ **2.** $\begin{vmatrix} 5 & -3 \\ -4 & 2 \end{vmatrix}$ **3.** $\begin{vmatrix} 6 & 4 \\ -5 & -3 \end{vmatrix}$ **4.** $\begin{vmatrix} -2 & 0 \\ 4 & 2 \end{vmatrix}$

Determinants can be used to solve a system of equations like

$5x + 2y = 4$
$2x - 3y = 13$ for x and y.

$$\begin{array}{cccc} 5\,x & + & 2\,y & = 4 \\ 2\,x & + & -3\,y & = 13 \end{array}$$

$$x = \frac{\begin{vmatrix} 4 & 2 \\ 13 & -3 \end{vmatrix}}{\begin{vmatrix} 5 & 2 \\ 2 & -3 \end{vmatrix}} = \frac{\begin{vmatrix} 4 & 2 \\ 13 & -3 \end{vmatrix}}{\begin{vmatrix} 5 & 2 \\ 2 & -3 \end{vmatrix}} = \frac{(4)\,(-3) - (13)\,(2)}{5\,(-3) - (2)\,(2)}$$

$$= \frac{-12 - 26}{-15 - 4}\text{, or } 2$$

$$y = \frac{\begin{vmatrix} 5 & 4 \\ 2 & 13 \end{vmatrix}}{\begin{vmatrix} 5 & 2 \\ 2 & -3 \end{vmatrix}} = \frac{\begin{vmatrix} 5 & 4 \\ 2 & 13 \end{vmatrix}}{\begin{vmatrix} 5 & 2 \\ 2 & -3 \end{vmatrix}} = \frac{5\,(13) - 2\,(4)}{5\,(-3) - 2\,(2)}$$

$$= \frac{65 - 8}{-15 - 4}\text{, or } -3$$

Thus, $x = 2$ and $y = -3$.

To solve the system $\begin{matrix} ax + by = c \\ dx + ey = f \end{matrix}$ for x and y, use

$$x = \frac{\begin{vmatrix} c & b \\ f & e \end{vmatrix}}{\begin{vmatrix} a & b \\ d & e \end{vmatrix}} \text{ and } y = \frac{\begin{vmatrix} a & c \\ d & f \end{vmatrix}}{\begin{vmatrix} a & b \\ d & e \end{vmatrix}}.$$

PROJECT

Solve for x and y.

5. $2x + 3y = 7$
$x + 2y = 3$

6. $5x - 12y = 4$
$4x - 7y = -2$

7. $3x - 4y = 7$
$4x + 6y = 15$

Systems of Three Linear Equations

OBJECTIVE

- To find solutions of systems of three linear equations

REVIEW CAPSULE

Solve the system $x - y = 3$.
$x + 2y = 9$

Multiply by 2. → $2x - 2y = 6$
$x + 2y = 9$
Add. → $3x = 15$
$x = 5$

$x - y = 3$
$5 - y = 3$
$-y = -2$
$y = 2$

Thus, the solution is $(5, 2)$.

EXAMPLE 1 Solve the system
$$-3x + y + 2z = 2$$
$$x + 2y - z = 1$$
$$-2x + y + 3z = 5$$

Choose any two equations. Eliminate one variable by adding. →

$-3x + y + 2z = 2$ → $-3x + y + 2z = 2$
$x + 2y - z = 1$ Multiply by 2. → $2x + 4y - 2z = 2$
$-x + 5y = 4$

z is eliminated. Use a different pair of equations. Elimininate z by adding. →

$x + 2y - z = 1$ Multiply by 3. → $3x + 6y - 3z = 3$
$-2x + y + 3z = 5$ → $-2x + y + 3z = 5$
$x + 7y = 8$

Use the two new equations. Solve for x and y by addition. →

$$-x + 5y = 4$$
$$x + 7y = 8$$
$$12y = 12$$
$$\boxed{y = 1}$$

To find x, substitute $y = 1$ in either new equation.

$$-x + 5y = 4$$
$$-x + 5(1) = 4$$
$$\boxed{x = 1}$$

To find z, substitute $x = 1$ and $y = 1$ in any of the original equations.

$$x + 2y - z = 1$$
$$1 + 2(1) - z = 1$$
$$-z = -2$$
$$\boxed{z = 2}$$

Check in the original equations: $x = 1, y = 1, z = 2$.

$-3x + y + 2z$	2	$x + 2y - z$	1	$-2x + y + 3z$	5
$-3 + 1 + 4$	2	$1 + 2 - 2$	1	$-2 + 1 + 6$	5
2		1		5	

$(1, 1, 2)$ means $x = 1$, $y = 1$, and $z = 2$.

Thus, $(1, 1, 2)$ is the solution.

EXAMPLE 2 Solve the system.

$$\begin{aligned} x - 2y + 3z &= 9 \\ 3x - y \quad &= -10 \\ 5y + 4z &= 2 \end{aligned}$$

Eliminate z from first and third equations. →

$x - 2y + 3z = 9$ Multiply by -4. → $-4x + 8y - 12z = -36$

$5y + 4z = 2$ Multiply by 3. → $15y + 12z = 6$

$-4x + 23y = -30$

Second equation → $3x - y = -10$ Multiply by 23. → $69x - 23y = -230$

New equation → $-4x + 23y = -30$ → $-4x + 23y = -30$

$65x = -260$

$\boxed{x = -4}$

To find y substitute $x = -4$ in $3x - y = -10$.

$$\begin{aligned} 3(-4) - y &= -10 \\ -12 - y &= -10 \\ y &= -2 \end{aligned}$$

$$\begin{aligned} 5y + 4z &= 2 \\ 5(-2) + 4z &= 2 \\ 4z &= 12 \\ z &= 3 \end{aligned}$$

Check in all three equations.

Thus, $(-4, -2, 3)$ is the solution.

EXERCISES

PART A

Solve.

1. $2x + 3y + 4z = 4$
$5x + 7y + 8z = 9$
$3x - 2y - 6z = 7$
(3, −2, 1)

2. $x + y - 3z = 8$
$2x - 3y + z = -6$
$3x + 4y - 2z = 20$
(2, 3, −1)

3. $2x + y + 2z = 3$
$x + y + z = 2$
$3x - y + z = 4$
(2, 1, −1)

4. $a + 2b + 3c = 9$
$-3a + 5b - 4c = -7$
$3a - b + 2c = -1$
(−3, 0, 4)

5. $3a + 2b - 2c = 3$
$-a + 3b + c = -10$
$2a - b - 4c = -7$
(7, −3, 6)

6. $a + 15b - 9c = 52$
$4a - 5b + 6c = 13$
$7a + b + 3c = 52$
(7, 3, 0)

PART B

Solve.

7. $3x + 2y = -5$
$-x - 2z = 5$
$y + z = -3$
(−1, −1, −2)

8. $2x + 3y = -5$
$4y - 5z = -32$
$3x + 2z = 14$
(2, −3, 4)

9. $x + 2y = -6$
$y + 2z = 11$
$2x + z = 16$
(4, −5, 8)

3 by 3 Matrices

$$\begin{bmatrix} 3 & -2 & 1 \\ -1 & 4 & 10 \\ 2 & -3 & 5 \end{bmatrix}$$ is a 3 by 3 matrix.

3 rows 3 columns

PROBLEM 1.

Find the value of the determinant of the matrix above.

- Repeat columns 1 and 2.

$$\begin{vmatrix} 3 & -2 & 1 \\ -1 & 4 & 10 \\ 2 & -3 & 5 \end{vmatrix} \begin{matrix} 3 & -2 \\ -1 & 4 \\ 2 & -3 \end{matrix}$$

- Multiply on the down diagonals.
- Multiply on the up diagonals.

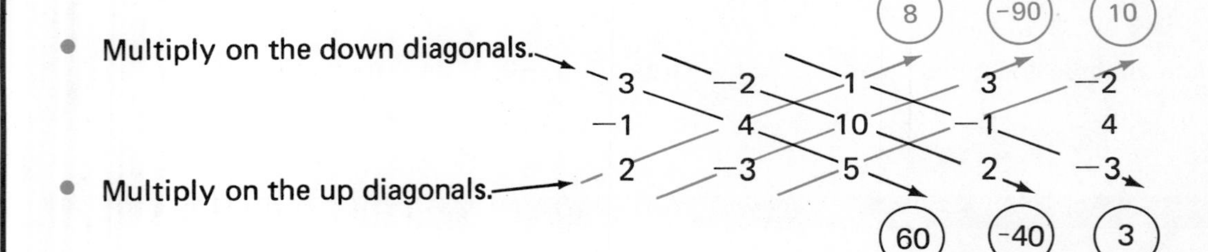

- Subtract the products.

$(60 + -40 + 3) - (8 + -90 + 10) = 95$

down diagonal products up diagonal products

Thus, the value of the determinant is 95.

PROJECT

Find the value of the determinant.

1. $\begin{vmatrix} 2 & 3 & 5 \\ 4 & 2 & 1 \\ -1 & -3 & 2 \end{vmatrix}$

2. $\begin{vmatrix} -3 & 1 & 5 \\ -2 & 0 & 2 \\ 6 & 3 & 4 \end{vmatrix}$

3. $\begin{vmatrix} 1 & 3 & -2 \\ 1 & -4 & 5 \\ 1 & 2 & 3 \end{vmatrix}$

To solve the system $\begin{aligned} ax + by + cz &= d \\ ex + fy + gz &= h \\ ix + jy + kz &= l \end{aligned}$ for x, y, and z use

$$x = \frac{\begin{vmatrix} d & b & c \\ h & f & g \\ l & j & k \end{vmatrix}}{\begin{vmatrix} a & b & c \\ e & f & g \\ i & j & k \end{vmatrix}}, \quad y = \frac{\begin{vmatrix} a & d & c \\ e & h & g \\ i & l & k \end{vmatrix}}{\begin{vmatrix} a & b & c \\ e & f & g \\ i & j & k \end{vmatrix}}, \text{ and } z = \frac{\begin{vmatrix} a & b & d \\ e & f & h \\ i & j & l \end{vmatrix}}{\begin{vmatrix} a & b & c \\ e & f & g \\ i & j & k \end{vmatrix}}$$

PROBLEM 2.

Solve the system $\begin{aligned} 2x + 3y + 4z &= 4 \\ 5x + 7y + 8z &= 9 \\ 3x - 2y - 6z &= 7 \end{aligned}$ for x by using determinants.

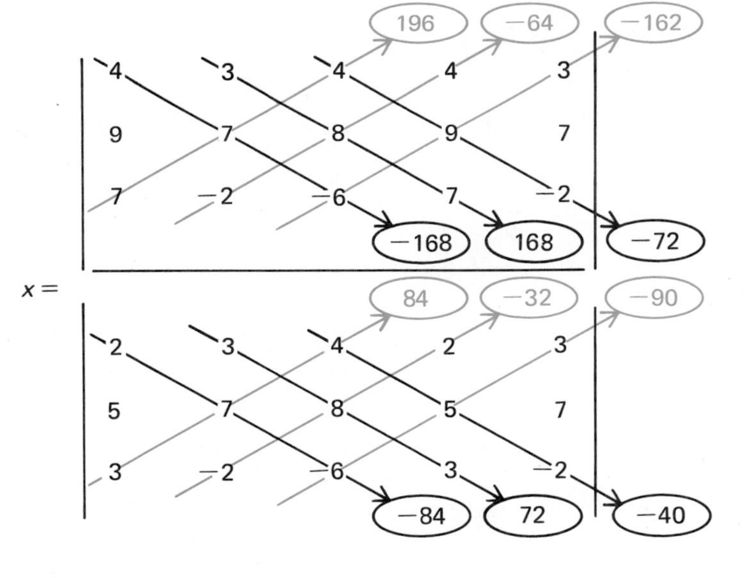

$$x = \frac{(-168 + 168 - 72) - (196 - 64 - 162)}{(-84 + 72 - 40) - (84 - 32 - 90)} = \frac{-42}{-14} = 3$$

PROJECT

4. Solve the system above for y and z.

Linear-Quadratic Systems

OBJECTIVE

- To find the real solutions of systems of one quadratic and one linear equation

REVIEW CAPSULE

Equations of Conic Sections

Parabola	*Circle*
$y = ax^2 + bx + c$	$x^2 + y^2 = r^2$
Ellipse	*Hyperbola*
$\frac{x^2}{a^2} + \frac{y^2}{b^2} = 1$	$\frac{x^2}{a^2} - \frac{y^2}{b^2} = 1$ or $xy = k$

EXAMPLE 1 Give the coordinates of the points of intersection.

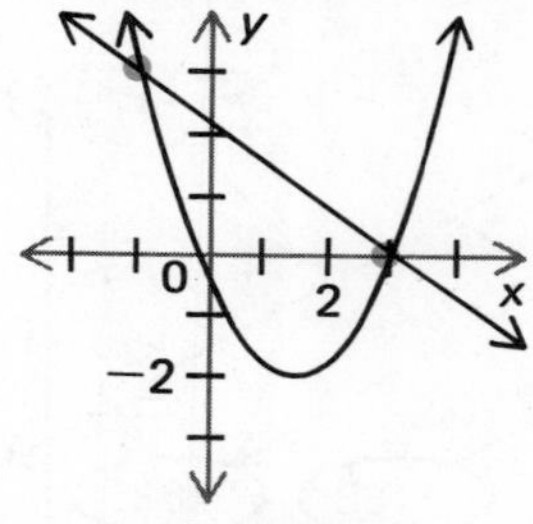

Points of intersection ⟶ (−1, 3) and (3, 0)

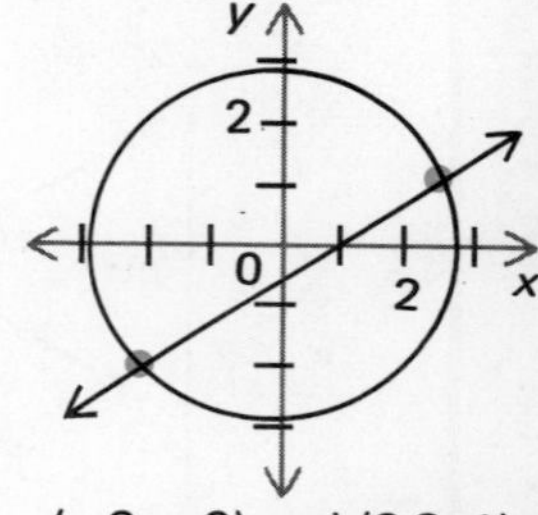

(−2, −2) and (2.8, 1)

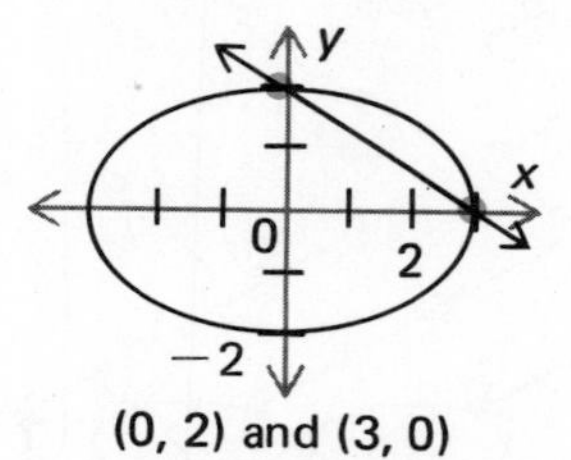

(0, 2) and (3, 0)

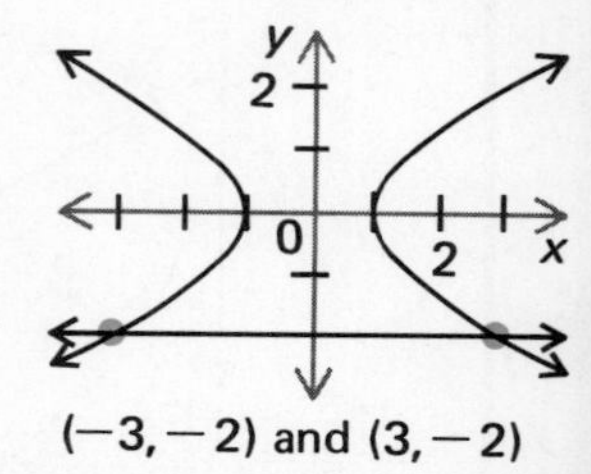

(−3, −2) and (3, −2)

EXAMPLE 2 How many points of intersection are there in each of the following?

The points of intersection are common solutions.

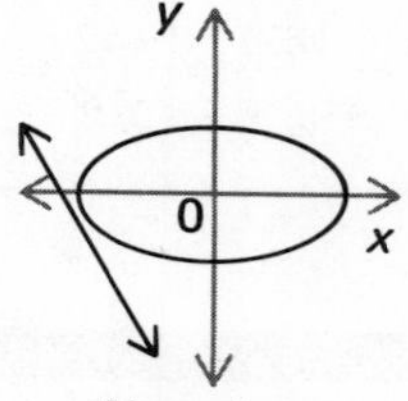

No points

Two points

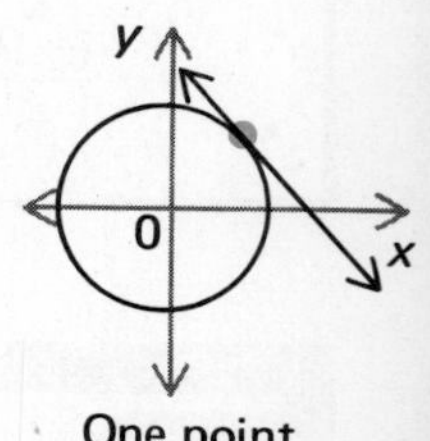

One point

One quadratic equation and one linear equation may have no, one, or two common solutions.

EXAMPLE 3 Solve the system $y = x^2 + 2x - 3$ graphically.
$x + 2y = 4$
Estimate to the nearest tenth if the roots are not integers.

x	y
-4	5
-3	0
-2	-3
-1	-4
0	-3
1	0
2	5

$(-3.8, 3.9)$ $(1.4, 1.3)$ $x + 2y = 4$ $y = x^2 + 2x - 3$

$$x + 2y = 4$$
$$2y = -x + 4$$
$$y = -\frac{1}{2}x + 2$$

slope ($-\frac{1}{2}$), y-intercept (2)

Thus, $(-3.8, 3.9)$ and $(1.4, 1.3)$ are the solutions of the system.

EXAMPLE 4 Solve the system $\frac{x^2}{9} + \frac{y^2}{4} = 1$ graphically.
$x + 3y = -10$

$\frac{x^2}{9} + \frac{y^2}{4} = 1$ is the equation of an ellipse.
x-intercepts $(3, 0)$, $(-3, 0)$.
y-intercepts $(0, 2)$, $(0, -2)$.

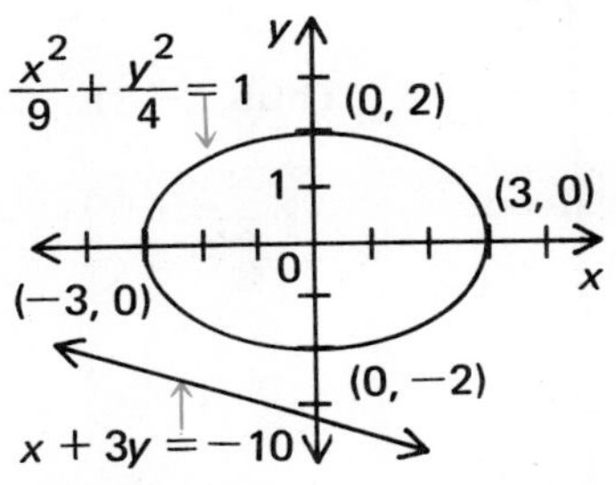

$$x + 3y = -10$$
$$3y = -x - 10$$
$$y = -\frac{1}{3}x - \frac{10}{3}$$

$m = -\frac{1}{3}$, $b = -\frac{10}{3}$

There are no points of intersection. → **Thus,** there are no real solutions of the system.

EXAMPLE 5 Solve the system $x^2 + y^2 = 1$ algebraically.
$x + y = 1$

Solve linear equation for y. →
$$x + y = 1$$
$$y = -x + 1$$

Substitute. →
$$x^2 + (y)^2 = 1$$
$$x^2 + (-x + 1)^2 = 1$$

$(-x + 1)(-x + 1) = x^2 - 2x + 1$ →
$$x^2 + x^2 - 2x + 1 = 1$$
$$2x^2 - 2x + 1 = 1$$
$$2x^2 - 2x = 0$$

Factor: $2x^2 - 2x = 2x(x - 1)$ →
$$2x(x - 1) = 0$$

Set each factor equal to 0. →
$$2x = 0 \qquad x - 1 = 0$$
$$\boxed{x = 0} \qquad \boxed{x = 1}$$

Find y. →
$$y = -x + 1 \qquad y = -x + 1$$
$$y = -0 + 1 \qquad y = -1 + 1$$
$$\boxed{y = 1} \qquad \boxed{y = 0}$$

The graphs intersect in two points. → **Thus,** $(0, 1)$ and $(1, 0)$ are the solutions.

EXAMPLE 6 Solve the system $4x^2 - 3y^2 = 36$ algebraically.
$x - y = 3$

Solve for y. ⟶ $x - y = 3$
$y = x - 3$

Substitute.
Let $y = x - 3$. ⟶ $4x^2 - 3(y)^2 = 36$
$4x^2 - 3(x-3)^2 = 36$
$(x-3)^2 = x^2 - 6x + 9$ ⟶ $4x^2 - 3(x^2 - 6x + 9) = 36$
Simplify. $4x^2 - 3x^2 + 18x - 27 = 36$
$x^2 + 18x - 63 = 0$

Factor: $x^2 + 18x - 63 = (x+21)(x-3)$. ⟶ $(x+21)(x-3) = 0$

Solve for x. ⟶ $x + 21 = 0$, $\boxed{x = -21}$ $\qquad x - 3 = 0$, $\boxed{x = 3}$

Find y. ⟶ $y = x - 3$, $y = -21 - 3$, $\boxed{y = -24}$ $\qquad y = x - 3$, $y = 3 - 3$, $\boxed{y = 0}$

The graphs of $4x^2 - 3y^2 = 36$ and $x - y = 3$ intersect in two points. ⟶ **Thus,** $(-21, -24)$ and $(3, 0)$ are the solutions.

EXERCISES

PART A

Solve graphically. Estimate to the nearest tenth if the roots are not integers.

1. $y = x^2 - 6x + 9$
$x - y = 3$

2. $x^2 + y^2 = 4$
$y = 4x - 5$

3. $x^2 - y^2 = 36$
$x + 2y = 2$

4. $x^2 + 4y^2 = 32$
$x + 2y = 8$

5. $\frac{x^2}{9} + \frac{y^2}{4} = 1$
$3x - 2y = 6$

6. $\frac{x^2}{16} - \frac{y^2}{4} = 1$
$2x + 3y = 9$

7. $xy = 6$
$3x + 4y = 9$

8. $\frac{x^2}{4} + \frac{y^2}{4} = 1$
$2x - 6y = 12$

Solve algebraically.

9. $4x^2 + 8y^2 = 36$
$y = 2x$
(1, 2), (−1, −2)

10. $xy = 16$
$y = x$
(4, 4), (−4, −4)

11. $x^2 + y^2 = 41$
$x + y = 9$
(5, 4), (4, 5)

12. $2x^2 - y^2 = 31$
$2x - y = 9$
(4, −1), (14, 19)

13. $y = x^2 - 6x + 8$
$x - y = 2$
(5, 3), (2, 0)

14. $3x^2 + y^2 = 12$
$y = 1 - 4x$
$(1, -3), (-\frac{11}{19}, \frac{63}{19})$

15. $x^2 + y^2 = 9$
$x - y = 3$
(0, −3), (3, 0)

16. $xy = -6$
$x - y = 5$
(3, −2), (2, −3)

PART B

Solve algebraically and check by sketching the graph.

17. $x^2 - y = 3$
$2x - y = 3$

18. $x^2 + y^2 = 18$
$2x + y = 3$
$(3, -3), (-\frac{3}{5}, \frac{21}{5})$

19. $3x^2 + 2y^2 = 5$
$2x + y = 1$
$(1, -1), (-\frac{3}{11}, \frac{17}{11})$

20. $5x^2 - 2y^2 = 2$
$x + y = -1$
$(2, -3), (-\frac{2}{3}, -\frac{1}{3})$

Quadratic Systems

OBJECTIVE

- To find the real solutions of systems of two quadratic equations

REVIEW CAPSULE

Solve $\begin{aligned} 2x + 3y &= 9 \\ x + y &= 3 \end{aligned}$ by addition.

$$\begin{aligned} & 2x + 3y = 9 \\ \text{Multiply by } -2. \rightarrow \quad & \underline{-2x - 2y = -6} \\ \text{Add.} \rightarrow \quad & y = 3 \end{aligned} \qquad \begin{aligned} x + y &= 3 \\ x + 3 &= 3 \\ x &= 0 \end{aligned}$$

Thus, (0, 3) is the solution.

EXAMPLE 1 How many points of intersection are there in each of the following?

Each is the intersection of two conic sections.

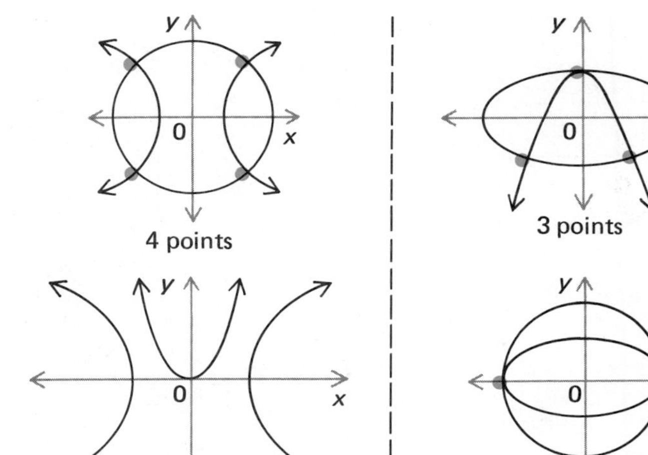

Points of intersection are common solutions.

Two quadratic equations may have no common solutions, one, two, three, or four common solutions.

EXAMPLE 2 Solve $\begin{aligned} x^2 + y^2 &= 16 \\ \frac{x^2}{16} + \frac{y^2}{4} &= 1 \end{aligned}$ by graphing.

Ellipse with
x-intercepts $(a, 0)$, $(-a, 0)$
y-intercepts $(0, b)$, $(0, -b)$

$x^2 + y^2 = 16$

Circle with
$r = 4$

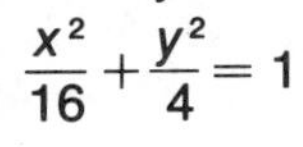

Ellipse

$a^2 = 16$, $b^2 = 4$ ⟶ $a = \pm 4 \qquad b = \pm 2$

(0, 2) (0, 4) (−4, 0) x y 0 (4, 0) (0, −2) (0, −4)

There are two points of intersection.

Thus, the solutions are $(-4, 0)$ and $(4, 0)$.

EXAMPLE 3 Solve the system $x^2 + y^2 = 25$ by graphing.
$x^2 - y^2 = 7$

$x^2 + y^2 = 25$
Circle with $r = 5$.

Find the asymptotes of the hyperbola. →

$x^2 - y^2 = 7$

$\frac{x^2}{7} - \frac{y^2}{7} = 1$

$a = \pm\sqrt{7} \qquad b = \pm\sqrt{7}$

$\sqrt{7} \doteq 2.6$

Hyperbola with x-intercepts $(\sqrt{7}, 0)$ $(-\sqrt{7}, 0)$.

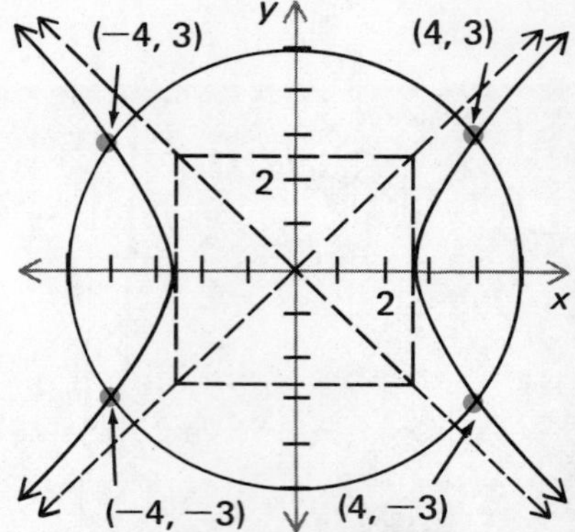

Thus, $(-4, 3)$, $(-4, -3)$, $(4, 3)$, and $(4, -3)$ are the solutions.

EXAMPLE 4 Solve the system $x^2 + y^2 = 25$ algebraically.
$x^2 - y^2 = 7$

Adding the equations will eliminate the y^2 term.

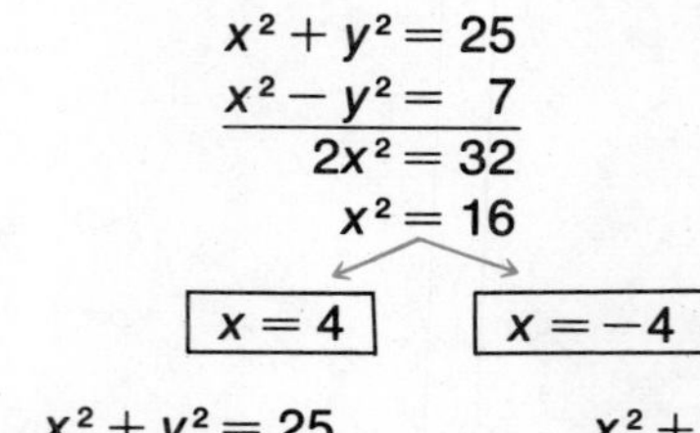

Add. →

$$\begin{aligned} x^2 + y^2 &= 25 \\ x^2 - y^2 &= 7 \\ \hline 2x^2 &= 32 \end{aligned}$$

Divide by 2. →

$$x^2 = 16$$

$x = \pm\sqrt{16}$ →

$$\boxed{x = 4} \qquad \boxed{x = -4}$$

Solve for y.
Use either equation.

$$\begin{aligned} x^2 + y^2 &= 25 \\ (4)^2 + y^2 &= 25 \\ 16 + y^2 &= 25 \\ y^2 &= 9 \\ \boxed{y = \pm 3} \end{aligned} \qquad \begin{aligned} x^2 + y^2 &= 25 \\ (-4)^2 + y^2 &= 25 \\ 16 + y^2 &= 25 \\ y^2 &= 9 \\ \boxed{y = \pm 3} \end{aligned}$$

$x = 4$ and $y = \pm 3$ give ordered pairs $(4, 3)$ and $(4, -3)$.
$x = -4$ and $y = \pm 3$ give ordered pairs $(-4, 3)$ and $(-4, -3)$. →

Thus, $(4, 3)$, $(4, -3)$, $(-4, 3)$, and $(-4, -3)$ are the solutions.

EXAMPLE 5 Solve the system $x^2 + y^2 = 16$ algebraically.
$\frac{x^2}{16} + \frac{y^2}{4} = 1$

Write $\frac{x^2}{16} + \frac{y^2}{4} = 1$ without fractions. →

$x^2 + y^2 = 16$ — Multiply by -1. → $-x^2 - y^2 = -16$
$x^2 + 4y^2 = 16$ → $x^2 + 4y^2 = 16$
Add. → $3y^2 = 0$

$$\boxed{y = 0}$$

Solve for x. →

$$\begin{aligned} x^2 + y^2 &= 16 \\ x^2 + (0)^2 &= 16 \\ x^2 &= 16 \\ \boxed{x = \pm 4} \end{aligned}$$

$x = \pm 4$ and $y = 0$ give the ordered pairs $(4, 0)$ and $(-4, 0)$.

Graphs intersect in two points.

Thus, the solutions are $(4, 0)$ and $(-4, 0)$.

ORAL EXERCISES

For each of the following tell how many solutions there are. Identify the curves. Estimate the solutions of each system.

1.

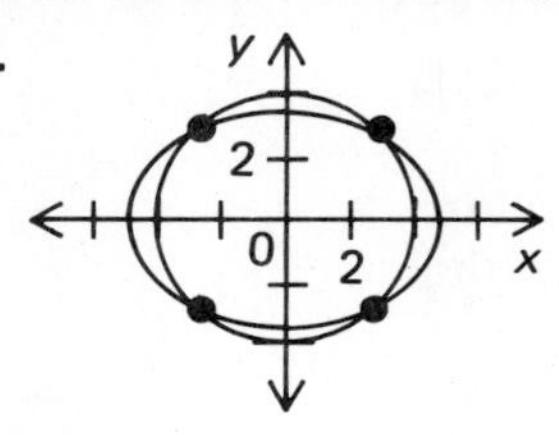

2.

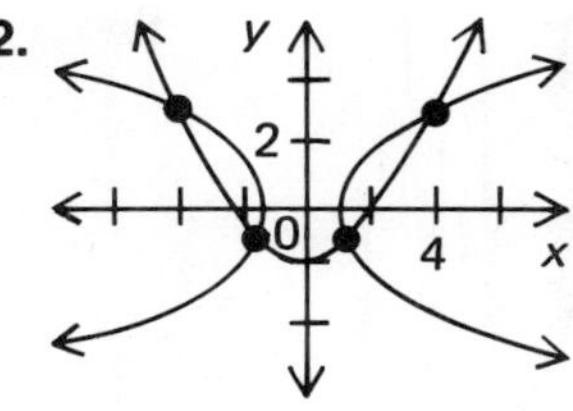

3.

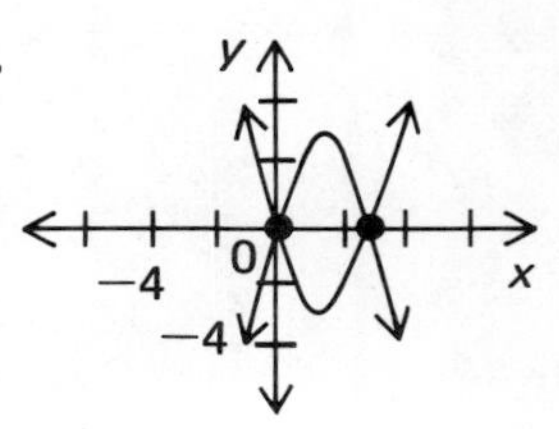

EXERCISES

PART A

Solve each system graphically. Then solve algebraically.

1. $x^2 - y^2 = 16$
   $x^2 + y^2 = 16$
   (4, 0), (−4, 0)
2. $x^2 + y^2 = 7$
   $2x^2 + 3y^2 = 18$
   $(\sqrt{3}, 2)(-\sqrt{3}, 2), (\sqrt{3}, -2), (-\sqrt{3}, -2)$
3. $8x^2 + 4y^2 = 16$
   $8x^2 + 3y^2 = 12$
4. $x^2 + 5y^2 = 25$
   $x^2 + y^2 = 9$
5. $x^2 + y^2 = 13$
   $x^2 - y^2 = 5$
   (3, 2), (−3, 2), (3, −2), (−3, −2)
6. $3x^2 + 2y^2 = 33$
   $3x^2 + y^2 = 17$
7. $x^2 + y^2 = 25$
   $\frac{x^2}{25} + \frac{y^2}{5} = 1$
8. $x^2 + 2y^2 = 27$
   $x^2 - 2y^2 = 23$

PART B

EXAMPLE Solve the system $x^2 + y^2 = 10$ algebraically.
$xy = 3$

$$xy = 3$$

Solve for y. ⟶ $$y = \frac{3}{x}$$

$$x^2 + y^2 = 10$$

Substitute $\frac{3}{x}$ for y. ⟶ $$x^2 + \left(\frac{3}{x}\right)^2 = 10$$

$$x^2 + \frac{9}{x^2} = 10$$

Multiply each side by x^2. ⟶ $$x^4 + 9 = 10x^2$$

$$x^4 - 10x^2 + 9 = 0$$

Factor. ⟶ $$(x^2 - 9)(x^2 - 1) = 0$$

Solve for x. Find $y = \frac{3}{x}$.

$x^2 - 9 = 0$ or $x^2 - 1 = 0$
$x = 3$ or -3 or $x = 1$ or -1
$y = \frac{3}{3}$ or $\frac{3}{-3}$ or $y = \frac{3}{1}$ or $\frac{3}{-1}$
$y = 1$ or -1 or $y = 3$ or -3

The graphs of $x^2 + y^2 = 10$ and $xy = 3$ intersect in four points. ⟶ **Thus,** (3, 1), (−3, −1), (1, 3), and (−1, −3) are the solutions.

Solve each system algebraically.

9. $x^2 + y^2 = 25$
   $xy = 12$
   (4, 3), (3, 4), (−4, −3), (−3, −4)
10. $x^2 - 2y = 8$
    $x^2 + 4y^2 = 10$
11. $x^2 + y^2 = 26$
    $xy = 5$
12. $x^2 - y^2 = 3$
    $y = x^2 - 3$

Digit Problems

OBJECTIVE

- To solve digit problems using systems of equations

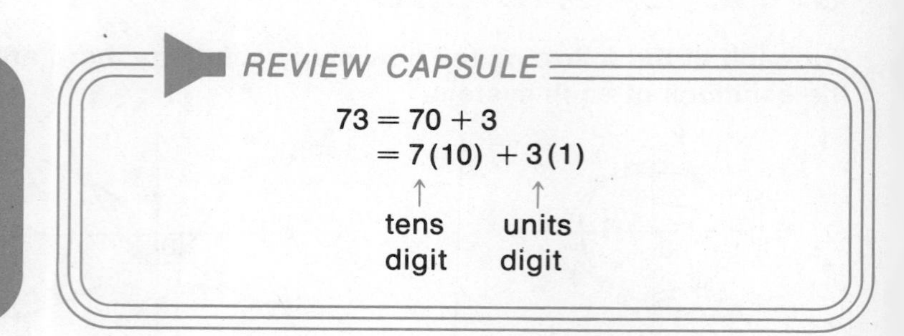

EXAMPLE 1 Identify the tens digit and the units digit.

$65 = 60 + 5$
$= 6(10) + 5(1)$
(6: tens digit; 5: units digit)

$49 = 40 + 9$
$= 4(10) + 9(1)$
(4: tens digit; 9: units digit)

EXAMPLE 2 Write a two-digit number whose tens digit is represented by t and whose units digit is represented by u.

tens digit units digit
$65 = 60 + 5$
$= 6(10) + 5(1)$ ⟶

$$t(10) + u(1)$$
$$10t + u$$

$10t + u$ represents any two-digit number.

t represents the tens digit. u represents the units digit.

EXAMPLE 3 Reverse the digits of each number. Identify the tens digit and the units digit of the new number.

Original number ⟶	$34 = 3(10) + 4$	$49 = 4(10) + 9$
Number with the digits reversed. ⟶	$43 = 4(10) + 3$	$94 = 9(10) + 4$

(4 and 9: tens digit; 3 and 4: units digit)

EXAMPLE 4 If $10t + u$ represents a two-digit number, represent the number with the digits reversed.

Original number → $10t + u$

Number with digits reversed → $10u + t$

EXAMPLE 5 Find the sum of the digits of each.

$65 = 6(10) + 5$	$49 = 4(10) + 9$
tens digit(t) ↑, units digit(u) ↑	tens digit(t) ↑, units digit(u) ↑
$t + u = 6 + 5$	$t + u = 4 + 9$
$= 11$	$= 13$

$t + u$ is the sum of digits. → (above)

$10t + u$ Original number	$10u + t$ Number with digits reversed	$t + u$ Sum of digits

EXAMPLE 6 The tens digit of a two-digit number is 4 more than twice the units digit. The sum of the digits is 10. Find the number.

Let $t =$ the tens digit
$u =$ the units digit
$10t + u =$ the number

The tens digit is 4 more than twice the units digit.

Write an equation. → $t = 2u + 4$

The sum of the digits is 10.

Write another equation. → $t + u = 10$

Solve the system: $t = 2u + 4$, $t + u = 10$

$t = 2u + 4$

Use substitution. → $t + u = 10$

Replace t with $2u + 4$. → $(2u + 4) + u = 10$

$3u + 4 = 10$

$3u = 6$

$\boxed{u = 2}$

$t + 2 = 10$

Find t. → $\boxed{t = 8}$

$10t + u = 10(8) + 2 = 82$ → **Thus,** the number is 82.

EXAMPLE 7 The tens digit of a certain two-digit number is 6 more than the units digit. The sum of the digits of the number and the number obtained when the digits are reversed is 51. Find the original number.

Let $t =$ tens digit
$u =$ units digit
$10t + u =$ the number
$10u + t =$ the number with the digits reversed

The tens digit is 6 more than the units digit.

First equation. → $t = u + 6$

The *sum* of the digits of the number and the number with digits reversed is 51.

Second equation. → $(t + u) + 10u + t = 51$

Combine like terms. → $2t + 11u = 51$

Solve the system.

$t = u + 6$
$2t + 11u = 51$

Substitute $u + 6$ for t in the second equation. →

$$2(u + 6) + 11u = 51$$
$$2u + 12 + 11u = 51$$
$$13u = 39$$
$$u = 3$$

Find t. →

$$t = u + 6$$
$$= 3 + 6$$
$$= 9$$

Check

$(t + u) + (10u + t)$	51
$9 + 3 + 10(3) + 9$	51
$12 + 30 + 9$	
51	

Original number: $10t + u = 10(9) + 3$
$= 90 + 3$
$= 93$

Thus, the original number is 93.

ORAL EXERCISES

Identify the tens digit and the units digit.

1. 72 **2.** 36 **3.** 91 **4.** 86 **5.** 43 **6.** 12

Give each of the numbers with the digits reversed. Then identify the tens and units digit.

7. 36 **8.** 21 **9.** 67 **10.** 94 **11.** 13 **12.** 70

Find the sum of the digits.

13. 61 *7* **14.** 12 *3* **15.** 73 *10* **16.** 19 *10* **17.** 83 *11* **18.** 98 *17*

EXERCISES

PART A

1. The tens digit of a two-digit number is 3 more than the units digit. The sum of the digits is 7. Find the number. *52*

2. The tens digit of a two-digit number is 2 more than the units digit. The sum of the digits is 14. Find the number. *86*

3. The units digit of a two-digit number is 7 less than twice the tens digit. The sum of the digits is 11. Find the number. *65*

4. The tens digit of a two-digit number is 4 more than twice the units digit. The sum of the digits is 10. Find the number. *82*

5. The tens digit of a two-digit number is 3 more than twice the units digit. If the digits are reversed, the resulting number is 45 less than the original number. Find the number. *72*

6. The units digit of a two-digit number is 2 less than three times the tens digit. If the digits are reversed, the resulting number is 18 more than the original number. Find the number. *24*

7. The sum of the digits of a two-digit number is 8. If the digits are reversed, the new number is 18 more than the original number. Find the number. *35*

8. The sum of the digits of a two-digit number is 6. If the digits are reversed, the new number is 18 less than the original number. Find the number. *42*

9. The tens digit of a two digit number exceeds the units digit by 5. The sum of the digits is 9. Find the number. *72*

10. The units digit of a two-digit number exceeds the tens digit by 1. The sum of the digits is 9. Find the number. *45*

PART B

11. A two-digit number is 9 less than 4 times the sum of its digits. If the digits are reversed, the new number is 54 more than the original number. Find the original number. *39*

12. A two-digit number is 2 less than 5 times the sum of its digits. If the digits are reversed, the new number is 9 more than the original number. Find the original number. *23*

13. The sum of the digits of a two-digit number is 9. The number is 3 times the sum of the digits. Find the number. *27*

14. The sum of the digits of a two-digit number is 12. The number is 4 times the sum of the digits. Find the number. *48*

15. A two-digit number is 18 more than the number obtained by reversing the digits. The number is also 14 more than 5 times the sum of the digits. Find the number. *64*

16. A two-digit number is 27 more than the number obtained by reversing the digits. The number is also 38 more than twice the sum of the digits. Find the number. *52*

Motion Problems

OBJECTIVE

■ To solve problems about motion

A car traveled 90 kilometers per hour for 3 hours. How far did it travel?

$$\begin{aligned} \text{distance} &= \text{rate} \times \text{time} \\ &= 90 \times 3 \\ &= 270 \end{aligned}$$

Thus, the car traveled 270 kilometers.

EXAMPLE 1 A car traveled 240 kilometers at an average speed of 80 kilometers per hour. How long did it travel?

distance = rate × time

$$\frac{\text{distance}}{\text{rate}} = \frac{\text{rate} \times \text{time}}{\text{rate}}$$

$$\frac{\text{distance}}{\text{rate}} = \text{time}$$

distance = 240 km
rate = 80 km/h

$$\begin{aligned} \text{time} &= \frac{\text{distance}}{\text{rate}} \\ &= \frac{240}{80} \\ &= 3 \end{aligned}$$

Thus, the car traveled for 3 hours.

EXAMPLE 2 A car traveled 420 kilometers in 7 hours. What was its average speed?

distance = rate × time

$$\frac{\text{distance}}{\text{time}} = \frac{\text{rate} \times \text{time}}{\text{time}}$$

$$\frac{\text{distance}}{\text{time}} = \text{rate}$$

distance = 420 km
time = 7 hr

$$\begin{aligned} \text{rate} &= \frac{\text{distance}}{\text{time}} \\ &= \frac{420}{7} \\ &= 60 \end{aligned}$$

Thus, the car traveled at an average speed of 60 km/h.

Motion formulas
d = distance
r = rate
t = time

$d = rt$ | $t = \frac{d}{r}$ | $r = \frac{d}{t}$

EXAMPLE 3 Two cars started from the same point at the same time and traveled in opposite directions. The faster of the two cars traveled 80 km/h. The slower car traveled at 70 km/h. In how many hours were they 375 km apart?

Draw a diagram. The cars traveled away from each other. The sum of the distances is the total distance.

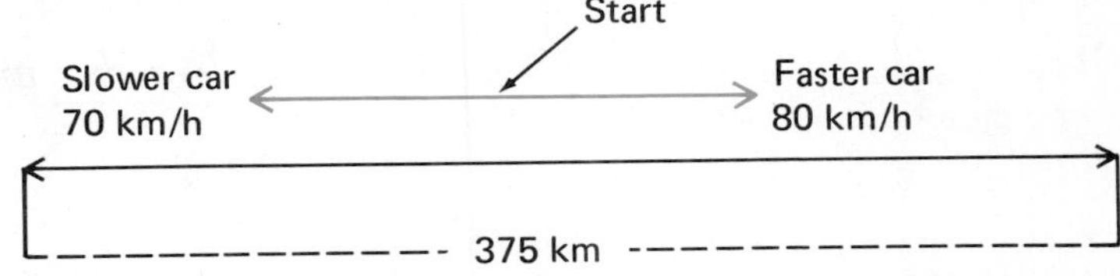

Use $d = rt$. Both travel the same number of hours (t).

	Rate (km/h)	Time (hr)	Distance (km)
Slower car	70	t	$70t$
Faster car	80	t	$80t$

Use the diagram. →

$$\begin{pmatrix}\text{distance}\\ \text{of slower car}\end{pmatrix} + \begin{pmatrix}\text{distance}\\ \text{of faster car}\end{pmatrix} = \text{total distance}$$

$$70t + 80t = 375$$

$70t + 80t = 150t$ →

$$150t = 375$$

$$t = \frac{375}{150}, \text{ or } 2\frac{1}{2}$$

Thus, the cars were 375 km apart after $2\frac{1}{2}$ hr.

EXAMPLE 4 Two trains started toward each other at the same time from towns 732 km apart. One train traveled at 148 km/h; the other at 96 km/h. In how many hours did they meet?

Draw a diagram. The trains traveled toward each other. The sum of the distances is the total distance.

Slower train 96 km/h

Meet

Faster train 148 km/h

732 km

Use $d = rt$. Both travel the same number of hours (t).

	Rate (km/h)	Time (hr)	Distance (km)
Slower train	96	t	$96t$
Faster train	148	t	$148t$

When the trains met,

Use the diagram. →

$$\begin{pmatrix}\text{distance}\\ \text{of slower train}\end{pmatrix} + \begin{pmatrix}\text{distance}\\ \text{of faster train}\end{pmatrix} = \text{total distance}$$

$$96t + 148t = 732$$

$$244t = 732$$

Divide each side by 244. →

$$t = 3$$

Check. →

$\begin{pmatrix}\text{distance}\\ \text{of slower train}\end{pmatrix} + \begin{pmatrix}\text{distance}\\ \text{of faster train}\end{pmatrix}$	total distance
$96(3) + 148(3)$	732
$288 + 444$	
732	

Thus, the trains met in 3 hours.

EXAMPLE 5 An aircraft carrier left a port traveling 30 km/h. Two hours later, a destroyer left the same port and traveled in the same direction at 40 km/h. How many hours did the destroyer travel before it overtook the carrier?

Draw a diagram. The ships travel in the same direction.

Carrier 30 km/h

Destroyer 40 km/h

Start

The aircraft carrier travels 2 hours more than the destroyer since it left 2 hours earlier.

	Rate (km/h)	Time	Distance (km)
Aircraft carrier	30	$t+2$	$30(t+2)$
Destroyer	40	t	$40t$

When the destroyer catches up with the carrier, they will both have traveled the same distance.

$$30(t+2) = 40t$$
$$30t + 60 = 40t$$
$$60 = 10t$$
$$6 = t$$

Thus, the destroyer traveled for 6 hours before it overtook the carrier.

EXAMPLE 6 A family drove into the country at the rate of 80 km/h. They returned over the same road at 70 km/h. If they arrived home in 6 hours, how far did they drive into the country?

Draw a diagram.

$t = \frac{d}{r}$

	Rate (km/h)	Time (hr)	Distance (km)
Trip out	80	$\frac{x}{80}$	x
Trip back	70	$\frac{x}{70}$	x

The total trip took 6 hr. → $\frac{x}{80} + \frac{x}{70} = 6$

Multiply by the LCD, 560. → $7x + 8x = 3360$

$7x + 8x = 15x$ → $15x = 3360$

Divide each side by 15. → $x = 224$

Check. →

(time of trip out) + (time of trip back)	total time
$\frac{224}{80} + \frac{224}{70}$	6
$2\frac{4}{5} + 3\frac{1}{5}$	
6	

Thus, they drove 224 km into the country.

EXAMPLE 7 A boat can travel 20 km/h in still water. If the rate of the current is 4 km/h, how fast does the boat travel upstream (against the current)?
How fast does it travel downstream (with the current)?

Going upstream the current slows the boat down.

Going downstream the current speeds the boat up.

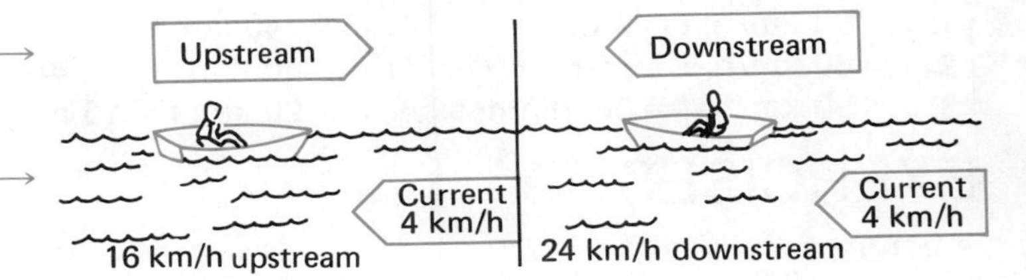

Upstream-downstream motion problems

> Let s = rate of the boat in still water (in km/h)
> c = rate of the current (in km/h)
> then the rate of the boat: upstream $= s - c$; downstream $= s + c$.

EXAMPLE 8 A boat can travel 20 km/h in still water. It can travel 36 km downstream in the same time that it can travel 24 km upstream. What is the rate of the current?

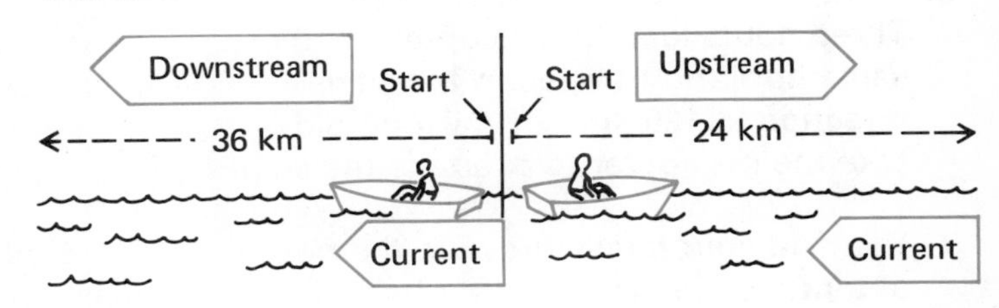

Let c = rate of the current

	Rate (km/h)	Time (hr)	Distance (km)
Downstream (with current)	$20 + c$	$\frac{36}{20 + c}$	36
Upstream (against current)	$20 - c$	$\frac{24}{20 - c}$	24

$\text{Time} = \frac{\text{distance}}{\text{rate}}$

The time for the downstream trip is the same as the time for the upstream trip. Solve the proportion.

$$\frac{36}{20 + c} = \frac{24}{20 - c}$$
$$36(20 - c) = 24(20 + c)$$
$$720 - 36c = 480 + 24c$$
$$240 - 36c = 24c$$
$$240 = 60c$$
$$4 = c$$

Thus, the rate of the current is 4 km/h.

ORAL EXERCISES

Find the distance.

1. $r = 90$ km/h, $t = 3$ hr
2. $r = 60$ km/h, $t = \frac{1}{2}$ hr
3. $r = 90$ km/h, $t = 30$ minutes *45*

Find the rate.

4. $d = 140$ km, $t = 2$ hr
5. $d = 40$ km, $t = \frac{1}{2}$ hr
6. $d = 20$ km, $t = 15$ minutes *80*

Find the time.

7. $d = 200$ km, $r = 40$ km/h
8. $d = 180$ km, $r = 30$ km/h
9. $d = 400$ km, $r = 50$ km/h *8*

EXERCISES

PART A

1. Two cars started from the same point at the same time and traveled in opposite directions. The first car traveled 80 km/h. The second car traveled 70 km/h. In how many hours were they 600 km apart? *4*

2. Two cars started from the same point at the same time and traveled in opposite directions. The faster car traveled 85 km/h. The slower car traveled 75 km/h. In how many hours were they 640 km apart? *4*

3. Two trains started toward each other from towns 870 km apart at the same time. The passenger train traveled 80 km/h. The freight train traveled 65 km/h. In how many hours did they meet? *6*

4. Two trains started toward each other from towns 800 km apart at the same time. The passenger train traveled 3 times as fast as the freight train. They met in 4 hr. Find the rate of each. *freight, 50 km/h; passenger 150 km/h*

5. A ship left port traveling at 30 km/h. Three hours later, a helicopter left the same port and traveled in the same direction at 120 km/h. How long did it take the helicopter to overtake the ship? *1 hr*

6. A jet plane and an airplane left the same terminal at the same time and flew in the same direction. After 6 hr they were 2,100 km apart. The jet flew at 1,000 km/h. Find the rate of the airplane. *650 km/h*

7. You and your family drove to the beach at a rate of 60 km/h. After 2 hr at the beach you returned home at 90 km/h. If the entire trip took 7 hr, how far was the beach from your house? *180 km*

8. Janet, Harvey, and Bob hiked along the road at 5 km/h. Sara met them with a truck and they returned at a rate of 40 km/h. If the total trip took $4\frac{1}{2}$ hr, find the distance hiked. *20 km*

9. A scout in her canoe traveled 12 km/h in still water. She traveled 30 km downstream in the same time that she traveled 15 km upstream. What was the rate of the current? *4 km/h*

10. A camper rowed 36 km downstream in 4 hr. It took him 12 hr to return upstream. Find his rate in still water. Find the rate of the current. *rate = 6 km/h, rate of current = 3 km/h*

PART B

11. For 3 hr a fishing boat traveled at the same rate. For the next hour, the fishing boat traveled at one half that rate. If the boat traveled a total distance of 161 km what was the initial rate? *46 km/h*

12. Two students started toward each other at the same time from opposite ends of a hall 210 meters long. The first student walked at 2 meters per second and the other at $1\frac{1}{2}$ meters per second. In how many seconds did they meet? *60 sec*

Systems of Linear Inequalities

OBJECTIVES

- To graph linear inequalities in the plane
- To find the solutions of systems of two linear inequalities graphically

REVIEW CAPSULE

Graph the solution set of $-2x + 3 > 7$.

$$-2x + 3 > 7$$

Add -3 to each side. ⟶ $-2x > 4$

Divide by -2. Reverse the order of the inequality. $x < -2$

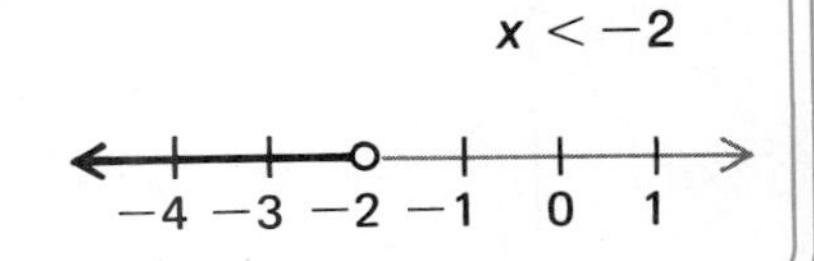

EXAMPLE 1 Graph $\{(x, y) \mid y > 2x - 1\}$.

First graph $y = 2x - 1$.
Use slope-intercept method.

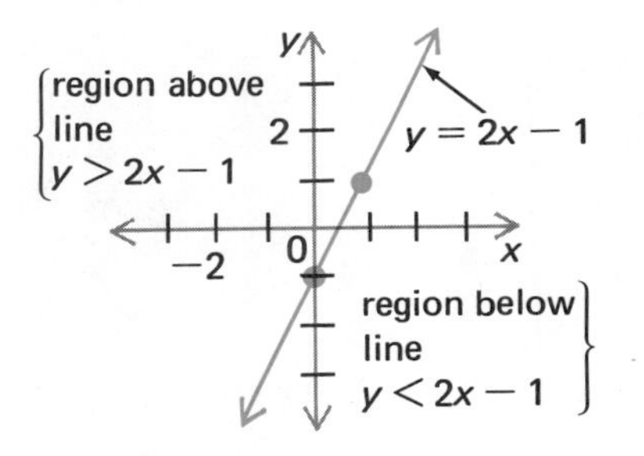

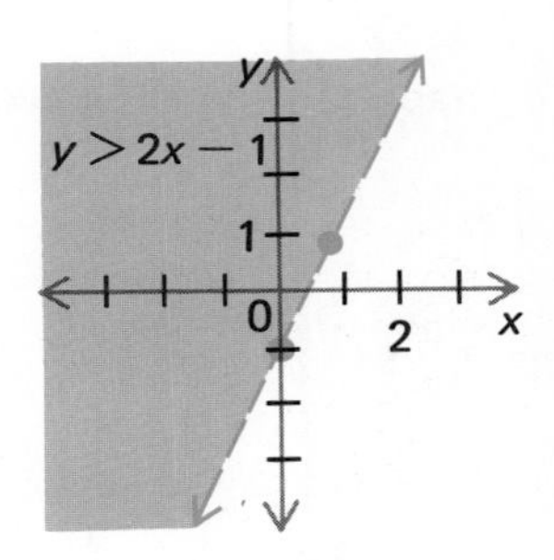

Check. Pick a point in the shaded area. $(-1, 2)$.

y	$2x - 1$
2	$2(-1) - 1$
	-3

$2 > -3$

The shaded region does not include the line. Use a dashed line.

Thus, the graph of $\{(x, y) \mid y > 2x - 1\}$ is the shaded region above the line but not the line.

The graph of $y = mx + b$ is a line.
The graph of $y > mx + b$ is above the line.
The graph of $y < mx + b$ is below the line.

EXAMPLE 2 Graph $\{(x, y) \mid y \leq -2x + 1\}$.

$y \leq -2x + 1$ means
$y = -2x + 1$ or $y < -2x + 1$.
Graph $y = 2x + 1$ with a solid line.
$y < -2x + 1$ is below the line.

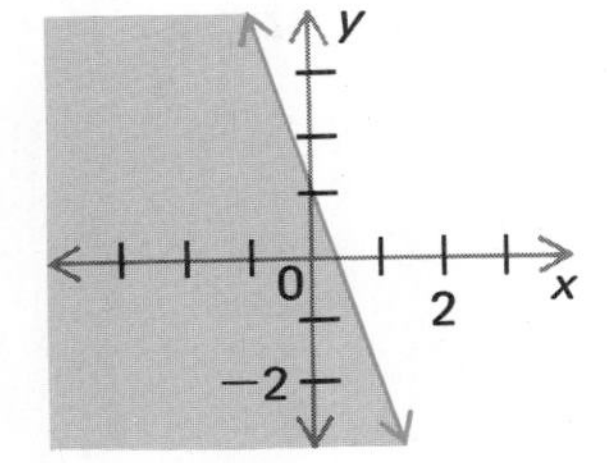

Check a point. $(-2, -3)$

y	$-2x + 1$
-3	$-2(-2) + 1$
	5

$-3 \leq 5$

Line is solid because it is included in the graph.

Thus, the graph of $\{(x, y) \mid y \leq -2x + 1\}$ is the line and the shaded region below the line.

EXAMPLE 3 Graph $\{(x, y) \mid x - 2y > 8\}$.

Write in $y = mx + b$ form.

$x - 2y > 8$

Add $-x$ to each side. → $-2y > -x + 8$

Divide each side by -2. Reverse the order. → $y < \frac{1}{2}x - 4$

Graph the line $y = \frac{1}{2}x - 4$. Use a dashed line.

Check a point in the shaded region.
Test $(5, -4)$.

$x - 2y$	8
$5 - 2(-4)$	8
$5 + 8$	
13	

$13 > 8$

The graph of $y < \frac{1}{2}x - 4$ is below the line. Shade below the line.

Thus, the graph of $\{(x, y) \mid x - 2y > 8\}$ is the shaded region below the line.

EXAMPLE 4 Graph $y \geq x$.

Graph the line $y = x$ with a solid line.

$y > x$ is above the line. Shade above the line.

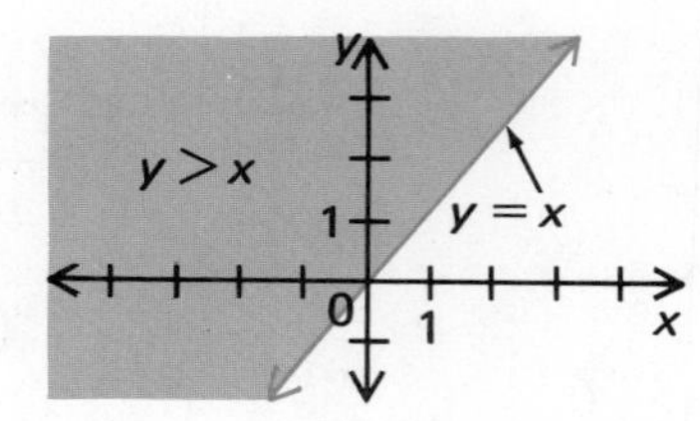

Check a point in the shaded region.
Test $(-3, 2)$.

y	x
2	-3

$2 \geq -3$

Thus, the graph of $y \geq x$ is the line and the shaded region above the line.

EXAMPLE 5 Graph $y > 2$.

Graph the horizontal line $y = 2$ with a dashed line.

$y > 2$ is above the line. Shade above the line.

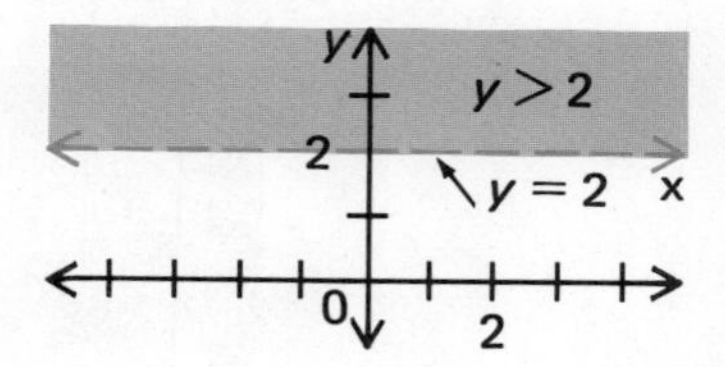

Check a point in the shaded region.
Test $(1, 3)$.

y	2
3	2

$3 > 2$

Thus, the graph of $y > 2$ is the region above the line.

EXAMPLE 6 Graph $x \leq 1$.

Graph the vertical line $x = 1$ with a solid line.

$x < 1$ is to the left of the line. Shade the region to the left of the line.

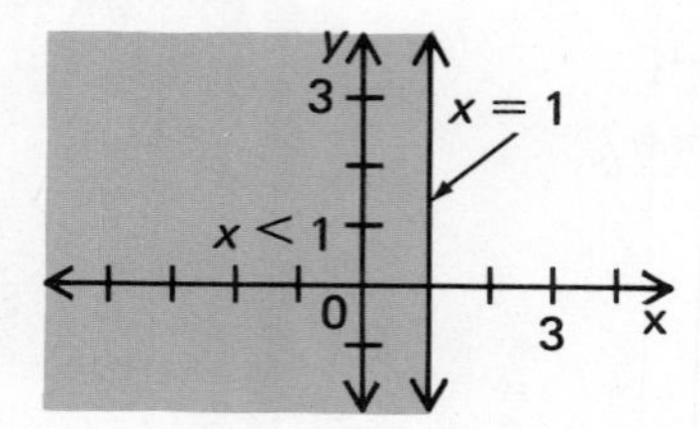

Check a point in the shaded region.
Test $(-1, 2)$.

x	1
-1	1

$-1 \leq 1$

Thus, the graph of $x \leq 1$ is the line and the shaded region to the left of the line.

EXAMPLE 7 Solve the system of inequalities $y > -x - 1$ graphically.
$y > 2x + 1$

Graph $y > -x - 1$.
1. Graph $y = -x - 1$. Use a dashed line.
2. Shade above the line.

Graph $y > 2x + 1$.
1. Graph $y = 2x + 1$. Use a dashed line.
2. Shade above the line with a different color.

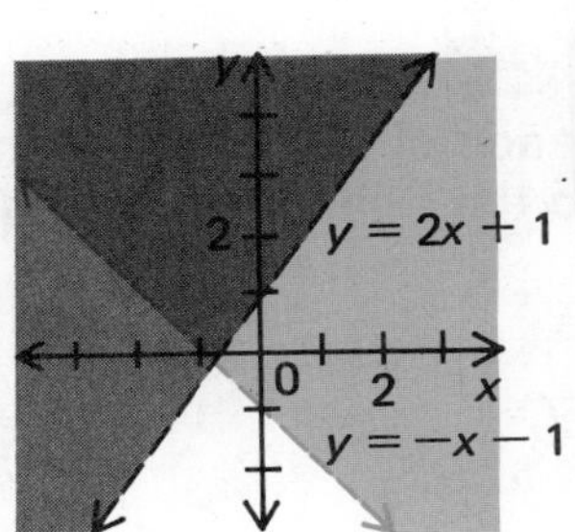

Check a test point in the double-shaded region in *both* inequalities.

Test $(-1, 3)$.

y	$-x - 1$
3	$-(-1) - 1$
3	$1 - 1$
3	0

$3 > 0$

y	$2x + 1$
3	$2(-1) + 1$
3	$-2 + 1$
3	-1

$3 > -1$

The graph of the system is the intersection of the two shadings. →

Thus, the solution is the double-shaded region. Every point in this region is a solution of the system.

EXAMPLE 8 Solve the system $y \geq 2x + 1$ graphically.
$y \geq 2x + 3$

Graph $y \geq 2x + 1$.
1. Graph $y = 2x + 1$. Use a solid line.
2. Shade above the line.

Graph $y \geq 2x + 3$.
1. Graph $y = 2x + 3$. Use a solid line.
2. Shade above the line with a different color.

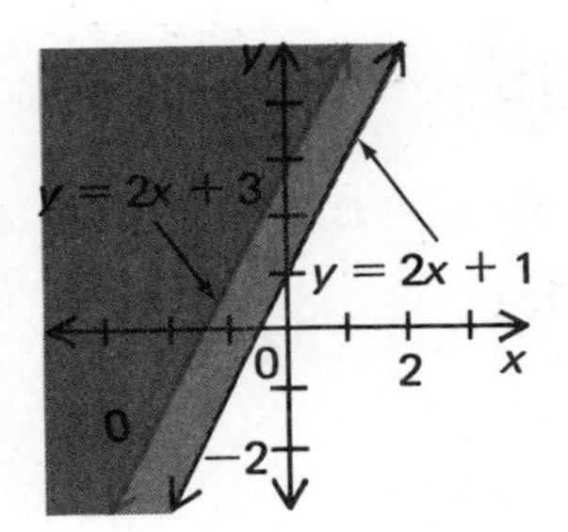

Check a test point in the double-shaded region in *both* inequalities.

Test $(-3, 1)$.

y	$2x + 1$
1	$2(-3) + 1$
1	$-6 + 1$
1	-5

$1 \geq -5$

y	$2x + 3$
1	$2(-3) + 3$
1	$-6 + 3$
1	-3

$1 \geq -3$

Thus, the solution is the double-shaded region including the line $y = 2x + 3$. Every point in this region is a solution of the system.

EXAMPLE 9 Solve the system $y \geq 1$ graphically.
$x \leq 2$

Graph $y \geq 1$.
1. Graph $y = 1$. Use a solid line.
2. Shade above the line.

Graph $x \leq 2$.
1. Graph $x = 2$. Use a solid line.
2. Shade to the left of the line with a different color.

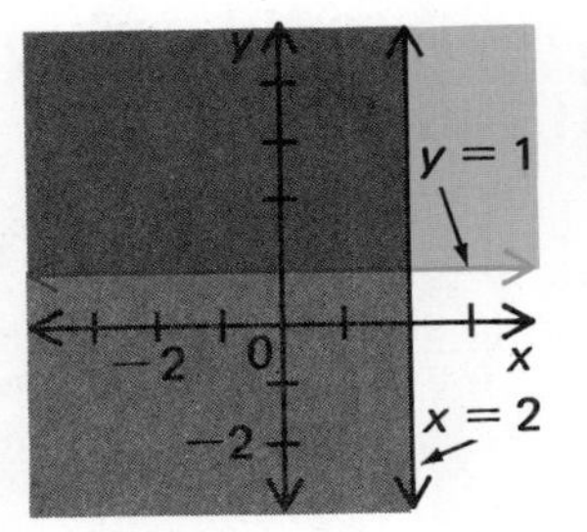

Check a test point in the double-shaded region in *both* inequalities.

Test $(-2, 3)$.

y	1
3	1

$3 \geq 1$

x	2
-2	2

$-2 \leq 2$

Thus, the solution is the double-shaded region including parts of both lines $y = 1$ and $x = 2$. Every point in this region is a solution of the system.

ORAL EXERCISES

Determine whether the graph of each of the following includes the line. Tell whether the shaded region is above, below, to the right, or to the left of the line.

1. $x \leq 3$
2. $y \leq 2x + 3$
3. $y > 3x + 5$
4. $y > -1$
5. $y \leq -2x + 1$
6. $x > -1$
7. $y \leq -2$
8. $y < x + 9$
9. $y > -4$ *no, above*
10. $y \leq -7$ *yes, below*
11. $x \leq -4$ *yes, left*
12. $x \geq -7$ *yes, right*

EXERCISES

PART A

Graph each of the following and check with a test point.

1. $\{(x, y) \mid y \leq 3x - 1\}$
2. $\{(x, y) \mid y \geq 3x + 4\}$
3. $\{(x, y) \mid y < 2x + 1\}$
4. $\{(x, y) \mid y > -2x + 3\}$
5. $\{(x, y) \mid y < -x - 4\}$
6. $\{(x, y) \mid y \geq x - 3\}$
7. $\{(x, y) \mid y \leq 3x + 4\}$
8. $\{(x, y) \mid y \leq -5x + 7\}$
9. $\{(x, y) \mid y \leq -3x + 6\}$
10. $y \leq -1$
11. $y > 3$
12. $y \leq 4$
13. $x \geq 2$
14. $x < -3$
15. $y \geq 2x - 2$
16. $y \leq -x - 3$
17. $y < -2x + 3$

Solve the following systems graphically. Check with a test point.

18. $y < -x + 3$
    $y > 2x - 2$
19. $y > -3x + 4$
    $y < 2x - 3$
20. $y > x - 3$
    $y < -4x - 1$
21. $y \geq 2x + 6$
    $y \leq -3x - 1$
22. $y \leq 2x + 8$
    $y \leq 2x + 3$
23. $y \geq 3x + 4$
    $y \geq 3x - 1$

PART B

Graph each of the following and check with a test point.

24. $2x - 3y \geq 5$
25. $-x + 2y \leq 8$
26. $-2x - 3y \leq 18$
27. $-2x + 4y < 8$
28. $-3x - 3y > 12$
29. $5x + 2y > 10$

Solve the following systems graphically. Check with a test point.

30. $2x + y < 3$
    $y \geq x$
31. $-3x - y \geq 6$
    $y - 2x > 1$
32. $2x - 3y \geq 9$
    $2y + x > 6$
33. $-2x - 3y \geq 6$
    $2y - x > -4$
34. $y \leq 2$
    $x \leq 4$
35. $x - 2y > 9$
    $x + 2y \leq 4$

PART C

Solve the following systems graphically. Check with a test point.

36. $x + y \leq 3$
    $4x - 5y \geq 1$
    $-x + y > 2$
37. $2x - y \geq 8$
    $x + 3y < 7$
    $x + 2y \geq 4$
38. $-x + 3y < 5$
    $3x - 2y \geq 6$
    $-4x + y < 3$

Systems of Quadratic Inequalities

OBJECTIVES

- To graph quadratic inequalities
- To find solutions of systems of one linear and one quadratic inequality graphically

REVIEW CAPSULE

Solve $x + y = 3$
$x^2 + y^2 = 9$.

Graph each equation. Find the points of intersection. →

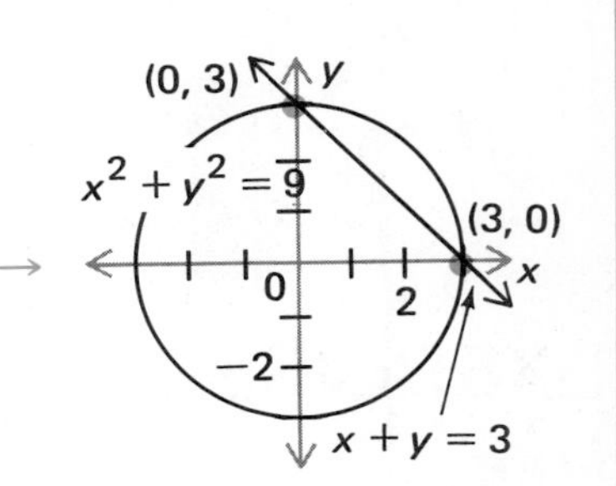

Thus, the solutions are $(0, 3)$ and $(3, 0)$.

EXAMPLE 1 Graph $x^2 + y^2 < 9$.

Graph $x^2 + y^2 = 9$.
(A circle with center at origin and radius 3.)
Use a dashed curve.

$x^2 + y^2 < 9$ is inside the circle.
(Check by testing a point.)

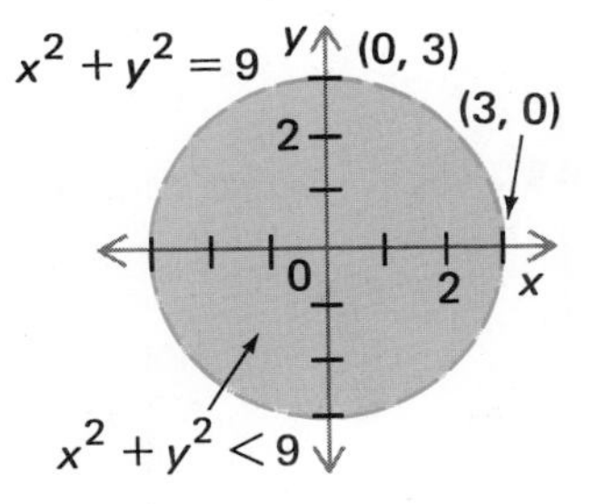

Check a test point in the shaded region.
It must satisfy $x^2 + y^2 < 9$.
Test $(1, 2)$.

$x^2 + y^2$	9
$(1)^2 + (2)^2$	9
$1 + 4$	
5	

$5 < 9$

Shade inside the circle.

Thus, the graph of $x^2 + y^2 < 9$ includes the set of all points (x, y) within the circle.

EXAMPLE 2 Graph $xy \geq 6$.

Graph $xy = 6$

$$y = \frac{6}{x}$$

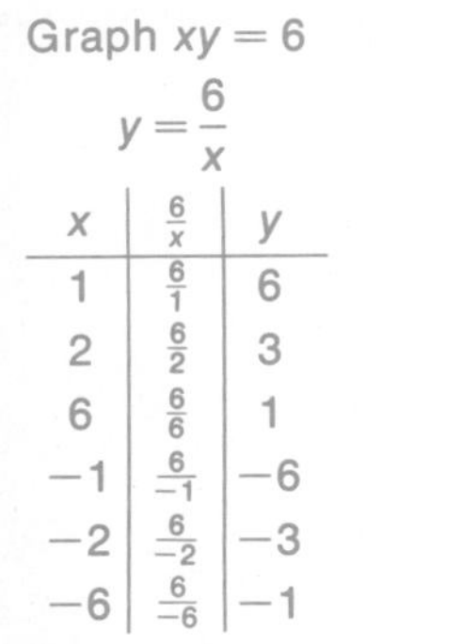

x	$\frac{6}{x}$	y
1	$\frac{6}{1}$	6
2	$\frac{6}{2}$	3
6	$\frac{6}{6}$	1
−1	$\frac{6}{-1}$	−6
−2	$\frac{6}{-2}$	−3
−6	$\frac{6}{-6}$	−1

Use a solid curve.
$xy > 6$ is outside the branches.
Shade outside the branches.

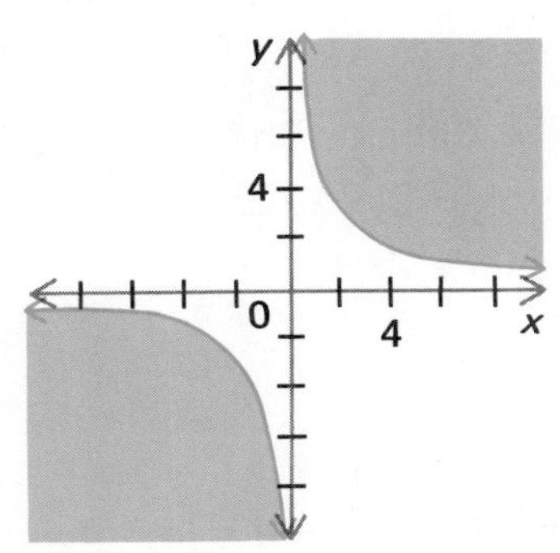

Check a test point in the shaded region.
Test $(-3, -4)$.

xy	6
$(-3)(-4)$	6
12	

$12 \geq 6$

Test $(3, 4)$.

xy	6
$(3)(4)$	6
12	

$12 \geq 6$

Thus, the graph of $xy \geq 6$ includes the points in the shaded region and the points on the hyperbola.

ORAL EXERCISES

State the inequality which describes the shaded region.

1.

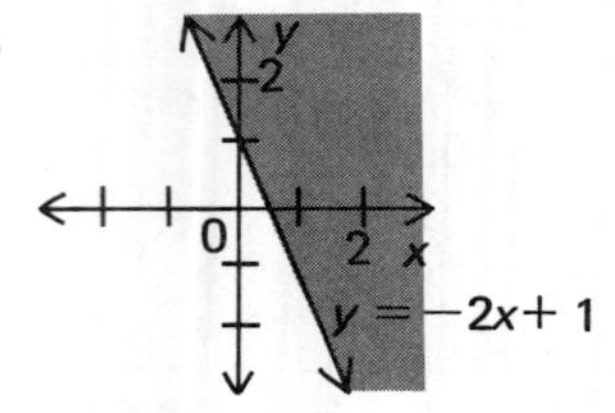

2.

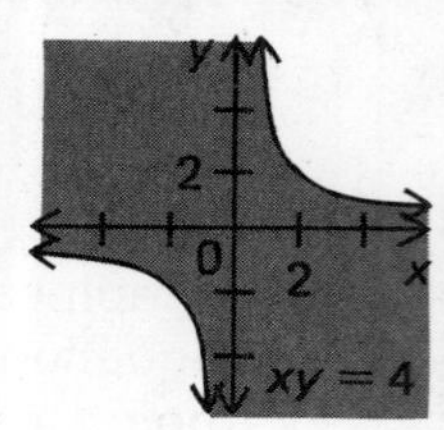

3.

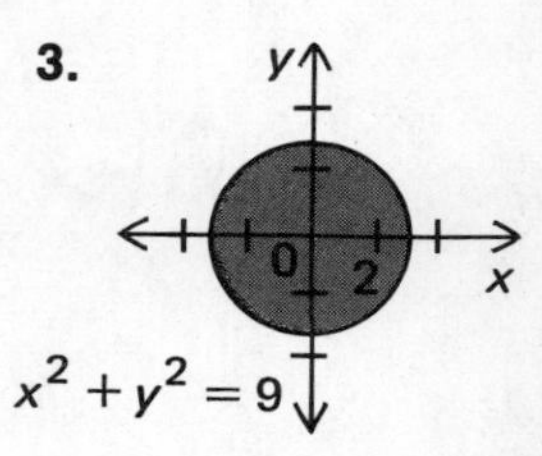

EXERCISES

PART A

Graph each of the following and check with a test point.

1. $x^2 + y^2 < 4$
2. $xy \leq 4$
3. $x^2 + y^2 \geq 16$
4. $2x^2 + 3y^2 \geq 18$
5. $x^2 - y^2 > 8$
6. $4x^2 + 16y^2 < 64$
7. $xy < 5$
8. $x^2 + y^2 \leq 4$
9. $9x^2 - 4y^2 \leq 36$
10. $x^2 - 4y^2 \geq 6$
11. $3x^2 + 9y^2 < 27$
12. $xy < 1$

PART B

EXAMPLE Solve the system $x^2 + y^2 \leq 16$ graphically.
$y \geq -2x - 1$

Graph $x^2 + y^2 \leq 16$
1. Graph $x^2 + y^2 = 16$
(circle centered at origin with radius 4.)
Use a solid curve.
2. Shade inside the circle.

Graph $y \geq -2x - 1$
1. Graph $y = -2x - 1$
Use a solid line.
2. Shade above the line.

Check a test point in the double-shaded region in both inequalities.
Test (3, 2).

$x^2 + y^2$	16
$3^2 + 2^2$	16
$9 + 4$	16
$13 \leq 16$	

y	$-2x - 1$
2	$-2(3) - 1$
	$-6 - 1$
$2 \geq -7$	

Thus, the solution is the double-shaded region including part of the line $y = -2x - 1$ and part of the circle.

Solve each system graphically. Check with a test point.

13. $x^2 + y^2 \leq 9$
$y \leq -x + 3$

14. $4x^2 + 9y^2 \leq 36$
$2x + 3y \geq 9$

15. $xy \leq 4$
$x + 2y < 8$

16. $2x^2 - 4y^2 \leq 16$
$x + 3y < 12$

17. $x^2 + y^2 \leq 4$
$x - 2y < 4$

18. $4x^2 + 16y^2 \leq 64$
$-2x - 3y > 9$

PART C

Solve each system graphically. Check with a test point.

19. $x^2 + y^2 \leq 16$
$xy \leq 4$

20. $x^2 - y^2 \leq 4$
$x^2 + y^2 \leq 9$

21. $4x^2 + 9y^2 < 36$
$x^2 + y^2 \leq 36$

Crossnumber Puzzle

1	2	3	4	5	6		7
8					9	10	
11				12			
					13	14	15
16	17	18			19		20
21			22	23		24	
25		26		27	28		
		29	30			31	

Across

1. The consecutive prime numbers between 40 and 50.
7. The solution of $\frac{y^2 - 1}{y + 1} = 3$.
8. The solution of $2x - 1492 = x$.
9. $\sqrt[3]{1{,}000{,}000}$
11. $7 \times (2 \times 10^3 + 1)$
13. The first prime number after 100.
16. Find the height of a cylinder if its volume equals $5{,}516{,}160\pi$ cm^3 and r = 104 cm.
19. $\sqrt[4]{81}$
20. Solve this system for x $\frac{x + y}{3} = 2$ and $\frac{x - y}{2} = -1$.
21. The prime number closest to 80.
22. The odd solution of $50 - 27x = -x^2$.
24. The even solution of $x^2 - 15x = -36$.
25. Solve this system for x $x + y = 0$ and $x - y = 1332$.
27. 2^{10}
29. $(2)(3)(17)$
31. The solution of $\frac{-10}{1 - 3x} = 5$.

Down

1. The next number greater than 410 divisible by 3.
2. The square of 12.
3. The solution of $(x - 490)^2 = 0$.
4. $16 \cdot 20$
5. The positive solution of $x^2 + 7x = 44$.
6. The first prime number greater than 70.
7. The x-value for the system $x + 2y = 160$ and $2x - y = 20$.
10. The smallest solution of $32x^2 = 17x$.
12. The solution of $\frac{2}{x + 3} = \frac{3}{2x + 1}$.
13. The odd solution of $x^2 - 19x = -78$.
14. The largest solution of $(x + 2)^2 = 4$.
15. $2^3(2^7 + 2^4 + 2^3 + 2^0)$
16. 24^2
17. 14^2
18. The smallest whole number that is not prime.
22. The solution of $\sqrt{x + 1} = \sqrt{2x - 1}$.
23. 2^9
24. 11^2
26. The positive solution of $x^2 - 60x = 61$.
28. The solution of $5(m + 2) = 10$.
30. The solution of $121^x = 1$.

Chapter Twelve Review

Solve each system graphically and check. [p. 295]

1. $2x - 4y = 2$
   $x + 2y = 9$
2. $x + 2y = 11$
   $2x - y = -13$
3. $3x - y = 13$
   $2x - 3y = 11$

Solve by substitution. [p. 298]

4. $-2x + 4y = -1$
   $8x - 2y = 4$
5. $4x - y = 3$
   $5x + 2y = 7$

Solve by addition. [p. 298]

6. $2x - y = 3$
   $5x + y = 11$
7. $5x - 3y = 7$
   $2x + y = 5$

Solve. [p. 304]

8. $-2x + 3y - z = 5$
   $4x + 2y + 5z = 15$
   $x + 3y + 2z = 11$
9. $2x - 3y + 2z = 3$
   $-3x + 4y - 2z = -2$
   $5x - 3y + 3z = -1$

Solve graphically. [p. 308]

10. $x^2 + y^2 = 36$
    $x + 2y = 10$
11. $x^2 - y^2 = 9$
    $x - 2y = -3$

Solve algebraically. [p. 308]

12. $x^2 + y^2 = 16$
    $x - y = 4$
13. $2x^2 + y^2 = 18$
    $x + y = 3$

Solve each system graphically. Then solve algebraically. [p. 311]

14. $x^2 + y^2 = 12$
    $x^2 - y^2 = 4$
15. $x^2 + y^2 = 13$
    $xy = 6$
16. $x^2 + 3y^2 = 10$
    $x^2 - y^2 = 2$

17. The tens digit of a two-digit number is 3 more than twice the units digit. The sum of the digits is 12. Find the number. [p. 314]

18. A two-digit number is 3 less than 4 times the sum of its digits. If the digits are reversed, the new number is 36 more than the original number. Find the original number. [p. 314]

19. Two cars started toward each other at the same time from towns 672 km apart. The first car traveled 80 km/h. The second car traveled 88 km/h. In how many hours did they meet? [p. 318]

20. A family in their row boat traveled 5 km/h in still water. They traveled 30 km downstream in the same time that they traveled 20 km upstream. What was the rate of the current? [p. 318]

Graph and check with a test point. [p. 323]

21. $\{(x, y) \mid y \leq -2x + 1\}$
22. $\{(x, y) \mid y > 5x + 4\}$
23. $2x + 4y < 8$
24. $-3x - 3y \geq 14$

Solve graphically and check with a test point. [p. 323]

25. $y < 2x + 1$
    $y > -x - 5$
26. $y \leq -3x + 6$
    $y \geq 2x - 3$
27. $2x - 3y < 3$
    $y \geq x$
28. $-3x + 2y \geq 8$
    $y - 2x < 6$

Graph and check with a test point. [p. 327]

29. $x^2 + y^2 \leq 16$
30. $xy \leq 3$
31. $4x^2 + 9y^2 \geq 36$

Solve graphically and check with a test point. [p. 327]

32. $x^2 + y^2 \leq 4$ and $x + 2y > 8$
33. $xy \geq 6$ and $2x - 3y < 12$

Chapter Twelve Test

Solve each system graphically and check.

1. $6x - 3y = 15$
$2x + y = 15$

2. $-2x - 3y = 3$
$4x + 2y = -10$

Solve by substitution.

3. $4x - 3y = 3$
$x - 2y = -8$

Solve by addition.

4. $-2x + 4y = -22$
$4x - 6y = 44$

Solve.

5. $2x - 3y + z = -1$
$-x + 4y + 3z = 5$
$4x - 2y - 2z = -14$

Solve graphically.

6. $x^2 - y^2 = 9$
$x - 2y = -3$

7. $x^2 + y^2 = 25$
$xy = 12$

Solve algebraically.

8. $x^2 + y^2 = 25$
$3x - y = 5$

9. $x^2 - y^2 = 7$
$x^2 + y^2 = 25$

10. The units digit of a two-digit number is 2 less than 3 times the tens digit. The sum of the digits is 6. Find the number.

11. Two trains started from the same point at the same time and traveled in opposite directions. The faster train traveled 190 km/h. The slower train traveled 80 km/h. In how many hours were they 1,080 km apart?

Graph and check with a test point.

12. $\{(x, y) \mid y > -3x + 6\}$

13. $3x - 6y \leq 12$

Solve graphically and check with a test point.

14. $y \geq -2x + 3$
$y < 3x - 2$

15. $2x - 3y \geq 6$
$-4x + 7y < 10$

Graph and check with a test point.

16. $x^2 + y^2 < 36$

17. $3x^2 - 4y^2 \geq 12$

Solve graphically and check with a test point.

18. $6x^2 + 3y^2 \geq 12$
$-2x + y < 8$

13 PROGRESSIONS AND SERIES

Fermat's Last Theorem

Pierre–Simon de Fermat
(1601–1665)

Fermat, a lawyer by profession, was an amateur mathematician. His notebooks and correspondence with mathematicians of his day contained many brilliant contributions to geometry, calculus, probability, and the theory of numbers.

Fermat's Last Theorem
For each positive integer exponent $n \geq 3$, $x^n + y^n = z^n$ has no positive integer solutions x, y, z.

For example, if $n = 1$ or $n = 2$, $x^n + y^n = z^n$ has infinitely many solutions.

$x^1 + y^1 = z^1$	$x^2 + y^2 = z^2$
$x = 1 \quad y = 2 \quad z = 3$	$x = 3 \quad y = 4 \quad z = 5$
$x = 1 \quad y = 1 \quad z = 2$	$x = 5 \quad y = 12 \quad z = 13$
$x = 3 \quad y = 5 \quad z = 8$, etc.	$x = 8 \quad y = 15 \quad z = 17$, etc.

Fermat's theorem is one of the most famous unsolved problems of mathematics. He wrote in the margin of one of his volumes that he had discovered a wonderful proof but that the margin was too small to contain the proof. Unfortunately, he never recorded his proof anywhere.

For 300 years mathematicians have been trying to prove the theorem but no proof has been found. Selfridge and Pollack have shown that no solutions exist for $3 \leq n < 25{,}000$.

Arithmetic Progressions

OBJECTIVES

- To write arithmetic progressions like 3, 7, 11, 15, · · ·
- To find a given term in an arithmetic progression

$3 + (2 + 2 + 2 + 2)$
$= 3 + 4(2)$

5, 10, 15, 20, . . . ← Dots mean the pattern continues.

EXAMPLE 1 Find the number that is added to each term to get the next term. Then give the next three numbers.

	2	7	12	17	22	· · ·
Add 5 to each term. →	2	$2 + 5$	$7 + 5$	$12 + 5$	$17 + 5$	· · ·

$22 + 5 = 27$, $27 + 5 = 32$, $32 + 5 = 37$ → **Thus,** the next three numbers are 27, 32, 37.

A progression is any list of numbers. → 2, 7, 12, 17, 22, . . . is a *progression.*

The numbers in a progression are called terms. → 2 is the 1st *term*; 17 is the 4th *term*

$7 - 2 = 5$ $12 - 7 = 5$ $17 - 12 = 5$ → 5 is the *common difference* of this progression.

Add the common difference to get the next term.

> 2, 7, 12, 17, 22, · · · is an *arithmetic* progression. The same number is added to any term to get the next term.

EXAMPLE 2 Write the first 4 terms of the arithmetic progression whose first term is 7 and whose common difference is 3.

First term is 7.	7	$7 + 3$	$(7 + 3) + 3$	$(7 + 3 + 3) + 3$
Add 3 to get next term.	7	10	13	16

Thus, the first 4 terms are 7, 10, 13, 16.

EXAMPLE 3 Write the first 4 terms of the arithmetic progression with the given first term a and common difference d.

$a = 5, d = -3$	$a = -8, d = 4$
$5 + (-3) = 2$ $2 + (-3) = -1$ $-1 + (-3) = -4$	$-8 + 4 = -4$ $-4 + 4 = 0$ $0 + 4 = 4$
Terms are $5, 2, -1, -4$.	Terms are $-8, -4, 0, 4$.

EXAMPLE 4 If the progression is arithmetic, find the common difference, d.

	Progression	*Answer*
$7 - 5 = 2$; $10 - 7 = 3$; but $2 \neq 3$.	5, 7, 10, 14, . . .	Not arithmetic
$16 - 10 = 22 - 16 = 28 - 22 = 6$	10, 16, 22, 28, . . .	$6 = d$: arithmetic
$5 - 15 = -5 - 5 = -15 - (-5) = -10$	15, 5, −5, −15, . . .	$-10 = d$: arithmetic
$6 - 3 \neq 12 - 6$.	3, 6, 12, 24, . . .	Not arithmetic
$3.5 - 2 = 5 - 3.5 = 6.5 - 5 = 1.5$	2, 3.5, 5, 6.5, . . .	$1.5 = d$: arithmetic
Add $-\sqrt{2}$ to get next term.	$4, 4 - \sqrt{2}, 4 - 2\sqrt{2}, 4 - 3\sqrt{2}, \ldots$	$-\sqrt{2} = d$: arithmetic

EXAMPLE 5 Find the 26th term of 2, 7, 12, 17, 22 . . . .

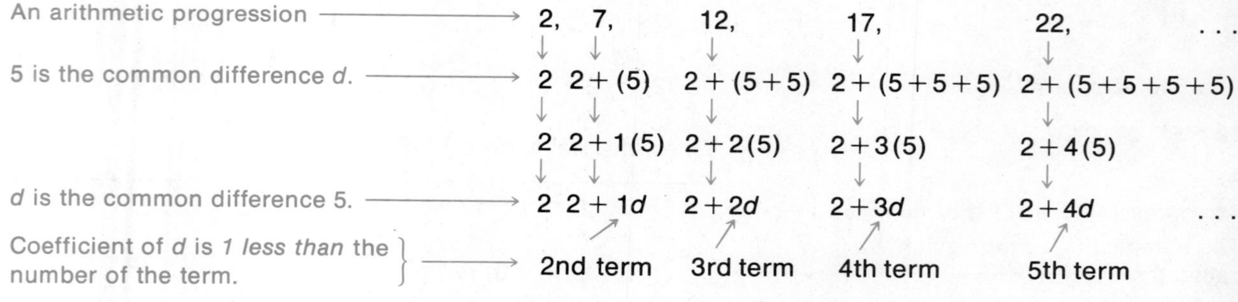

Thus, the 26th term is $2 + 25(5) = 2 + (26 - 1)d$, or 127.

(2 = first term; 25 = 26 − 1 = 1 less than number of term; 5 = d = common difference)

Formula for nth term of arithmetic progression. $(n - 1)$ is 1 less than the number of the term.

> a is first term; d is common difference
>
> $a, a + d, a + 2d, a + 3d, \cdots$ is an arithmetic progression. The nth term is $a + (n - 1)d$.

EXAMPLE 6 Find the 21st term of 8, 3, −2, −7, ...

a is the first term.	$a = 8$	$n\text{th term} = a + (n - 1)d$ ← Formula
n is the number of the term.	$n = 21$	$21\text{st term} = 8 + (21 - 1)(-5)$
d is the common difference.	$d = -5$	$= 8 + 20(-5)$, or -92

Thus, -92 is the 21st term.

ORAL EXERCISES

Give the next three numbers in the progression.

1. 3, 7, 11, 15, . . . *19, 23, 27*
2. 5, 3, 1, −1, . . . *−3, −5, −7*
3. −7, −4, −1, 2, . . . *5, 8, 11*
4. 3, $3\frac{1}{2}$, 4, $4\frac{1}{2}$, . . . *5, $5\frac{1}{2}$, 6*

State the first 4 terms of the arithmetic progression with the given first term *a* and common difference *d*.

5. $a = 2$, $d = 4$ *2, 6, 10, 14*
6. $a = -4$, $d = -2$ *−4, −6, −8, −10*
7. $a = -10$, $d = 5$ *−10, −5, 0, 5*
8. $a = 20$, $d = -10$ *20, 10, 0, −10*

If the progression is arithmetic, find the common difference *d*.

9. 3, 5, 7, 9, . . . *2*
10. 26, 23, 20, 17, . . . *−3*
11. 2, 4, 8, 16, . . . *not arith.*
12. −10, −8, −6, −4, . . .

EXERCISES

PART A

Write the next three numbers in the progression.

1. 3, 3.5, 4, 4.5, . . .
2. 2, 3.2, 4.4, 5.6, . . .
3. 11, 9.5, 8, 6.5, . . .
4. 1, $1\frac{1}{3}$, $1\frac{2}{3}$, 2, . . .
5. −4, $-3\frac{1}{2}$, −3, $-2\frac{1}{2}$, . . .
6. 6, $5\frac{3}{4}$, $5\frac{1}{2}$, $5\frac{1}{4}$, . . .
7. 2, $2 + \sqrt{5}$, $2 + 2\sqrt{5}$, . . . *$2 + 3\sqrt{5}$, $2 + 4\sqrt{5}$, $2 + 5\sqrt{5}$*
8. 1, $2 + \sqrt{3}$, $3 + 2\sqrt{3}$, . . . *$4 + 3\sqrt{3}$, $5 + 4\sqrt{3}$, $6 + 5\sqrt{3}$*
9. 3, $5 - \sqrt{2}$, $7 - 2\sqrt{2}$, . . . *$9 - 3\sqrt{2}$, $11 - 4\sqrt{2}$, $13 - 5\sqrt{2}$*

Write the first 4 terms of the arithmetic progression with the given first term *a* and common difference *d*.

10. $a = 8$, $d = 7$
11. $a = 9$, $d = -6$ *9, 3, −3, −9*
12. $a = -7$, $d = 4$
13. $a = 5$, $d = 2.4$
14. $a = -3$, $d = 2\frac{1}{3}$
15. $a = -6$, $d = -.3$
16. $a = -2$, $d = 2 + \sqrt{5}$ *-2, $\sqrt{5}$, $2 + 2\sqrt{5}$, $4 + 3\sqrt{5}$*
17. $a = 3$, $d = 1 - \sqrt{3}$ *3, $4 - \sqrt{3}$, $5 - 2\sqrt{3}$, $6 - 3\sqrt{3}$*
18. $a = i$, $d = 2 + i$ *i, $2 + 2i$, $4 + 3i$, $6 + 4i$*

If the progression is arithmetic, find the common difference *d*.

19. −9, −5, −1, 3, . . . *4*
20. 28, 19, 10, 1, . . . *−9*
21. 2, −4, 6, −8, . . .
22. 4, 6.6, 9.2, 11.8, . . . *2.6*
23. 5, 6.7, 7.4, 9.1, . . .
24. 6, $5\frac{1}{3}$, $4\frac{2}{3}$, 4, . . .
25. 3, $3 + \sqrt{2}$, $3 + 2\sqrt{2}$, . . . *$\sqrt{2}$*
26. 5, $3 + \sqrt{3}$, $1 + 2\sqrt{3}$, . . .
27. $1 + i$, 3, $5 - i$, . . .
28. $\sqrt{2}$, $4 + \sqrt{2}$, $8 + \sqrt{2}$, . . . *4*
29. $\sqrt{17}$, $\sqrt{19}$, $\sqrt{21}$, $\sqrt{23}$, . . . *not arith.*
30. $\sqrt{2}$, $\sqrt{8}$, $\sqrt{18}$, $\sqrt{32}$, . . . *$\sqrt{2}$*

Find the 21st term of the progression.

31. 7, 11, 15, 19, . . . *87*
32. 10, 4, −2, −8, . . . *−110*
33. −8, −5, −2, 1, . . .
34. 2, 2.4, 2.8, 3.2, . . . *10*
35. −1, −2.3, −3.6, −4.9, . . .
36. 7, $7\frac{1}{2}$, 8, $8\frac{1}{2}$, . . . *17*

Find the 26th term of the progression.

37. 30, 26, 22, 18, . . . *−70*
38. 3, 11, 19, 27, . . . *203*
39. −10, −13, −16, −19, . . .
40. 5, 5.2, 5.4, 5.6, . . . *10*
41. −3, −3.4, −3.8, −4.2, . . .
42. 4, $4\frac{2}{5}$, $4\frac{4}{5}$, $5\frac{1}{5}$, . . . *14*
43. 19, 25, 31, 37, . . . *169*
44. −4, −11, −18, −25, . . .
45. −27, −22, −17, −12, . . .

PART B

EXAMPLE Given an arithmetic progression and one of its terms, find n the number of the term.

$-4, -2, 0, 2, \ldots; 72$ is the nth term.

72 is a term of the progression. ⟶ $-4, -2, 0, 2, \ldots, 72, \ldots$ (nth term)

Formula for the nth term ⟶ $n\text{th term} = a + (n-1)d$

Substitute. $72 = -4 + (n-1)(2)$

Add 4 to each side. (Do not distribute 2.) $76 = (n-1)(2)$

Divide each side by 2. (Do not distribute 2.) $38 = n - 1$

Add 1 to each side. $39 = n$

Thus, 72 is the 39th term.

Given an arithmetic progression and one of its terms, find n the number of the term.

46. $-5, -2, 1, 4, \ldots; 61$ is nth term. *23*

47. $3, 7, 11, 15, \ldots; 127$ is nth term.

48. $11, 6, 1, -4, \ldots; -54$ is nth term. *14*

49. $-26, -30, -34, -38, \ldots; -106$ is nth term.

50. $7, 12, 17, 22, \ldots; 227$ is nth term. *45*

51. $-7, -10, -13, -16, \ldots; -103$ is nth term.

52. $1, 1.2, 1.4, 1.6, \ldots; 7$ is nth term. *31*

53. $-2, -1.6, -1.2, -.8, \ldots; 8$ is nth term.

54. $-1, -.2, .6, 1.4, \ldots; 23$ is nth term. *31*

55. $2, .9, -.2, -1.3, \ldots; -9$ is nth term.

56. $-2, -1\frac{1}{2}, -1, -\frac{1}{2}, \ldots; 28$ is nth term. *61*

57. $3, 3\frac{1}{5}, 3\frac{2}{5}, 3\frac{3}{5}, \ldots; 12$ is nth term.

Find the 21st term of the progression.

58. $4, 4 + 2\sqrt{3}, 4 + 4\sqrt{3}, 4 + 6\sqrt{3}, \ldots$

59. $2, 3 + \sqrt{5}, 4 + 2\sqrt{5}, 5 + 3\sqrt{5}, \ldots$

60. $3, 4 - \sqrt{2}, 5 - 2\sqrt{2}, 6 - 3\sqrt{2}, \ldots$ *$23 - 20\sqrt{2}$*

61. $3 + 2i, 4 + i, 5, 6 - i, \ldots$

Find the 26th term of the progression.

62. $-4, -3 - \sqrt{2}, -2 - 2\sqrt{2}, -1 - 3\sqrt{2}, \ldots$

63. $-2, \sqrt{3}, 2 + 2\sqrt{3}, 4 + 3\sqrt{3}, \ldots$

64. $6 - 4\sqrt{5}, 3 - 2\sqrt{5}, 0, -3 + 2\sqrt{5}, \ldots$ *$-69 + 46\sqrt{5}$*

65. $-3 - 5i, -2 - 3i, -1 - i, i, \ldots$

PART C

66. Mrs. Adams deposited $400 in her savings account on January 15. Then each month she deposited $45 more than on the previous month. What was her deposit on Dec. 15 of the same year?

67. A mechanic's salary in his 12th year of work was $13,400. His salary increases were $400 each year. What was his salary for the first year?

68. Find x so that $(x + 3), (4x + 1), (8x - 3), \ldots$ is an arithmetic progression. [Hint: common difference $d =$ (2nd term) $-$ (1st term) $=$ (3rd term) $-$ (2nd term).]

69. Find 2 values of x so that $\frac{1}{x}, 1, \frac{10}{x+3}, \ldots$ is an arithmetic progression.

70. Find the 1st term a of the arithmetic progession whose 16th term is 110 and and common difference d is 7.

71. Find the 1st term a of the arithmetic progression whose 41st term is -124 and common difference d is -3.

Arithmetic Means

OBJECTIVE

- To find arithmetic means

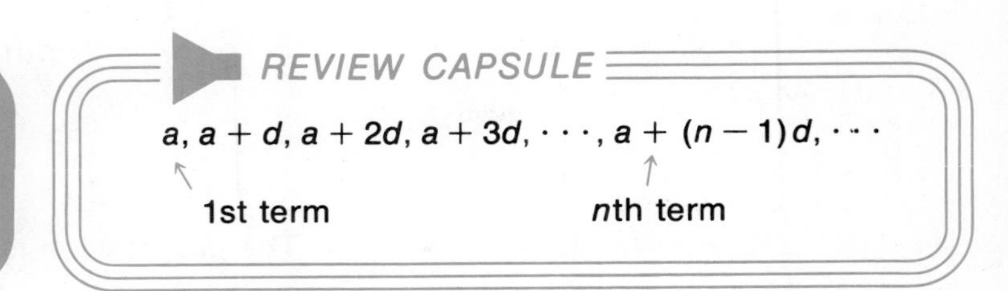

EXAMPLE 1 Find b so that 5, b, 17 forms an arithmetic progression.

Formula for the nth term → $n\text{th term} = a + (n-1)d$

17 is the 3rd term, $n = 3$. Solve the equation. →

$$17 = 5 + (3-1)d$$
$$17 = 5 + 2d$$
$$12 = 2d$$

Common difference d is 6. → $6 = d$

$5 + 6 = 11$; $11 + 6 = 17$ → The progression is 5, 11, 17.

Thus, b is 11.

Means are the terms *between* two given terms in a progression.

Arithmetic Progressions

5, 11, 17	5, 11, 17, 23, 29, 35
the one arithmetic *mean* between 5 and 17	the four arithmetic *means* between 5 and 35

EXAMPLE 2 Find the two arithmetic means between 29 and 8.

29, ____, ____, 8

(29: 1st term; 8: 4th term)

$n\text{th term} = a + (n-1)d$

8 is the 4th term, $n = 4$. →

$$8 = 29 + (4-1)d$$
$$8 = 29 + 3d$$
$$-21 = 3d$$

Common difference d is -7. → $-7 = d$

$29 + (-7) = 22$; $22 + (-7) = 15$; $15 + (-7) = 8$ → The progression is 29, 22, 15, 8.

29, 22, 15, 8 — 22, 15: the 2 arithmetic means

Thus, 22 and 15 are the two arithmetic means between 29 and 8.

EXAMPLE 3 Find the four arithmetic means between 20 and 24.

A 1st term (20), a last term (24), and 4 means is a total of 6 terms. → 20, ___, ___, ___, ___, 24 (1st term ↗, ↖ 6th term)

nth term $= a + (n-1)d$

24 is the 6th term, $n = 6$. → $24 = 20 + (6-1)d$

$24 = 20 + 5d$

$4 = 5d$

$\frac{4}{5} = .8$ → $.8 = d$

Add .8 to get next term. → The progression is 20, 20.8, 21.6, 22.4, 23.2, 24.

The 4 terms between 20 and 24. **Thus,** the 4 arithmetic means are 20.8, 21.6, 22.4, and 23.2.

ORAL EXERCISES

If the progression is arithmetic, state the means between the first and last terms listed.

1. 9, 6, 3 *6*
2. 1, 2, 4, 8 *not arith.*
3. 4, 1, −2, −5 *1, −2*
4. −4, −2, 2, 4 *not arith.*
5. −5, −3, −1, 1, 3, 5 *−3, −1, 1, 3*
6. 2, 2.5, 3, 3.5, 4, 4.5 *2.5, 3, 3.5, 4*

EXERCISES

PART A

Find the one arithmetic mean between the numbers.

1. 5 and 11 *8*
2. 20 and 12 *16*
3. 3 and 8 *5.5*
4. 4 and −6 *−1*

Find the two arithmetic means between the numbers.

5. 11 and 32 *18, 25*
6. 30 and 12 *24, 18*
7. −14 and 10 *−6, 2*
8. 2 and 6.5 *3.5, 5*

Find the four arithmetic means between the numbers.

9. 3 and 33
10. 40 and 5
11. −2 and 18
12. 15 and 0
13. 10 and 12
14. 30 and 33 *30.6, 31.2, 31.8, 32.4*
15. 4 and 15 *6.2, 8.4, 10.6, 12.8*
16. 9 and 0

PART B

Find the one arithmetic mean between the numbers.

17. 2.1 and 3.5 *2.8*
18. 4.2 and 10.8 *7.5*
19. 6 and 9.8 *7.9*
20. 7.8 and 12 *9.9*
21. $\sqrt{3}$ and $5\sqrt{3}$ *$3\sqrt{3}$*
22. 2 and $2 + 2\sqrt{5}$ *$2 + \sqrt{5}$*
23. $3 + \sqrt{2}$ and $5 + 3\sqrt{2}$ *$4 + 2\sqrt{2}$*
24. x and y *$\frac{x+y}{2}$*

25. Is the one arithmetic mean between two numbers equal to the *average* of the numbers? *yes*

Find the two arithmetic means between the numbers.

26. 3.4 and 10.3 *5.7, 8*
27. .1 and 5.8 *2, 3.9*
28. 4 and 15.1 *7.7, 11.4*
29. $2\sqrt{5}$ and $5\sqrt{5}$
30. 3 and $9 + 3\sqrt{2}$
31. $1 + \sqrt{3}$ and $7 - 5\sqrt{3}$

Arithmetic Series

OBJECTIVES

- To write an arithmetic progression as an arithmetic series
- To find the sum of an arithmetic series

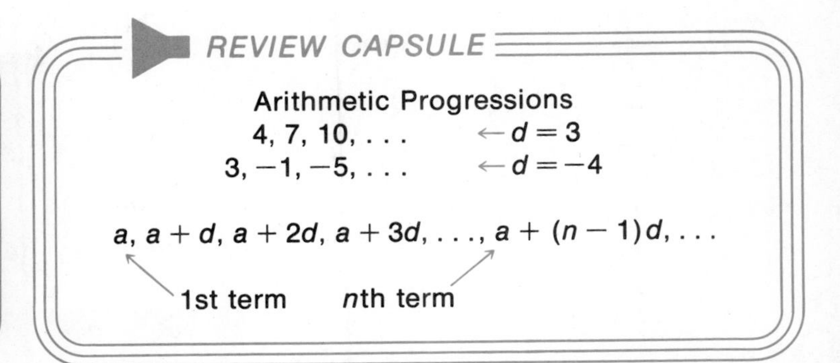

EXAMPLE 1 Indicate the sum of the progression 6, 8, 10, 12, ...

Progression → $6, 8, 10, 12, \ldots$

Replace commas with + signs. → ↓

Series → $6 + 8 + 10 + 12 + \cdots$

A series is an indicated sum.

> $6 + 8 + 10 + 12 + \cdots$ is an arithmetic *series*, if
> $6, 8, 10, 12, \cdots$ is an arithmetic progression.

EXAMPLE 2 Write the progression 8, 3, −2, −7, . . . as a series.

$8, 3, -2, -7, \ldots$

Replace commas with + signs.
$8 + 3 + (-2) + (-7) + \cdots$
$= 8 + 3 - 2 - 7 - \cdots$ → $8 + 3 - 2 - 7 - \cdots$

EXAMPLE 3 If the series is arithmetic, find the common difference d.

$5 + 7 + 10 + 14 + \cdots$	$15 + 5 - 5 - 15 - \cdots$
Not arithmetic	$-10 = d$: arithmetic

$7 - 5 = 2$; $10 - 7 = 3$; but $2 \neq 3$.
15, 5, −5, −15, . . . is an arithmetic progression.

EXAMPLE 4 Find the sum S of the first 20 terms of the arithmetic series $3 + 6 + 9 + 12 + \cdots$.

$a = 3$, $d = 3$, 20th term $= 60$. → $S = 3 + 6 + 9 + \cdots + 54 + 57 + 60$

Write terms in reverse order. → $S = 60 + 57 + 54 + \cdots + 9 + 6 + 3$

Add each column. → $2 \cdot S = 63 + 63 + 63 + \cdots + 63 + 63 + 63$

20 terms in the series → $2 \cdot S = 20(63)$

Sum of the first 20 terms → $S = 630$

Example 4 suggests a method to find a formula for the sum S of the first n terms of an arithmetic series.

$l = a + (n-1)d$ → Let l represent the nth term.

$l - d$ is term before l.

$$\begin{aligned} S &= a + (a+d) + (a+2d) + \cdots + (l-2d) + (l-d) + l \\ S &= l + (l-d) + (l-2d) + \cdots + (a+2d) + (a+d) + a \\ 2S &= \underbrace{(a+l) + (a+l) + (a+l) + \cdots + (a+l) + (a+l) + (a+l)}_{n \text{ binomials}} \end{aligned}$$

$$2S = n(a+l) \text{ or } S = \frac{n}{2}(a+l).$$

Formula →

If l is given, use this formula. →

The sum S of the first n terms of an arithmetic series

$$S = \frac{n}{2}(a+l)$$

1st term (a), nth term (l)

EXAMPLE 5 Find the sum of the first n terms of the arithmetic series given the 1st term a, the nth term l, and n.

$n = 35$, $a = 12$, $l = 114$

Formula → $S = \frac{n}{2}(a+l)$

Substitute: $n = 35$, $a = 12$, $l = 114$. → $= \frac{35}{2}(12 + 114)$

$\frac{35}{2}(126) = \frac{35}{\cancel{2}_1}(\cancel{126}^{63})$ → $= \frac{35}{2}(126)$, or 2,205

Thus, the sum of the first 35 terms is 2,205.

A second formula for the sum of an arithmetic series can be found by substitution.

The first sum formula and the nth term formula. → $S = \frac{n}{2}[a + l]$ and $l = a + (n-1)d$

Substitute $a + (n-1)d$ for l. → $S = \frac{n}{2}[a + a + (n-1)d]$

Simplify: $a + a = 2a$. → $S = \frac{n}{2}[2a + (n-1)d]$

If l is not given, use this formula. →

The sum S of the first n terms of an arithmetic series

$$S = \frac{n}{2}[2a + (n-1)d]$$

EXAMPLE 6 Find the sum of the first n terms of the arithmetic series given n and the series.

$n = 30$; $2.7 + 2.4 + 2.1 + 1.8 + \cdots$

If l is not given, use this formula. → $S = \frac{n}{2}[2a + (n-1)d]$

Substitute: $n = 30$, $a = 2.7$, $d = -.3$. → $= \frac{30}{2}[2(2.7) + (30-1)(-.3)]$

$2(2.7) = 5.4$; $29(-.3) = -8.7$

$= 15[5.4 - 8.7]$

$= 15(-3.3)$, or -49.5

EXERCISES

PART A

Write the progression as a series.

1. $5, 8, 11, 14, \cdots$ *$5 + 8 + 11 + 14 + \cdots$*
2. $4, 1, -2, -5, \cdots$ *$4 + 1 - 2 - 5 - \cdots$*
3. $-1, -2, -3, -4, \cdots$ *$-1 - 2 - 3 - 4 - \cdots$*

If the series is arithmetic, find the common difference d.

4. $2 + 5 + 8 + 11 + \cdots$ *3*
5. $3 + 1 - 1 - 3 - \cdots$ *-2*
6. $1 - 1 + 1 - 1 + \cdots$ *not arith.*
7. $1 + 1.4 + 1.8 + 2.2 + \cdots$ *.4*
8. $-.7 - .2 + .3 + .8 + \cdots$ *.5*
9. $\sqrt{2} + 3\sqrt{2} + 5\sqrt{2} + 7\sqrt{2} + \cdots$ *$2\sqrt{2}$*

Find the sum of the first n terms of the arithmetic series given the 1st term a, the nth term l, and n.

10. $n = 20$, $a = 1$, $l = 134$ *1,350*
11. $n = 40$, $a = 2$, $l = -115$ *$-2,260$*
12. $n = 25$, $a = 4$, $l = 196$ *2,500*
13. $n = 45$, $a = 3$, $l = -217$
14. $n = 35$, $a = -4$, $l = 64$
15. $n = 50$, $a = -5$, $l = -54$
16. $n = 40$, $a = .7$, $l = 20.2$ *418*
17. $n = 60$, $a = 2.3$, $l = -62.6$
18. $n = 55$, $a = 1.6$, $l = 125.8$ *3503.5*

Find the sum of the first n terms of the arithmetic series given n and the series.

19. $n = 20$; $5 + 15 + 25 + 35 + \cdots$ *2,000*
20. $n = 40$; $6 + 2 - 2 - 6 - \cdots$ *$-2,880$*
21. $n = 21$; $-4 - 1 + 2 + 5 + \cdots$ *546*
22. $n = 46$; $15 + 5 - 5 - 15 - \cdots$ *$-9,660$*
23. $n = 60$; $7 + 7.2 + 7.4 + 7.6 + \cdots$ *774*
24. $n = 45$; $-2 - .9 + .2 + 1.3 + \cdots$ *999*

PART B

Find the sum of the first n terms of the arithmetic series given n and the series.

25. $n = 26$; $.3 + .8 + 1.3 + 1.8 + \cdots$ *170.3*
26. $n = 41$; $-1.3 - .1 + 1.1 + 2.3 + \cdots$ *930.7*
27. $n = 20$; $\frac{1}{4} + 2\frac{1}{4} + 4\frac{1}{4} + 6\frac{1}{4} + \cdots$ *385*
28. $n = 30$; $1\frac{1}{3} + 1\frac{2}{3} + 2 + 2\frac{1}{3} + \cdots$ *185*
29. $n = 10$; $7\frac{1}{2} + 4\frac{1}{2} + 1\frac{1}{2} - 1\frac{1}{2} - \cdots$ *-60*
30. $n = 26$; $4 + 3\frac{3}{5} + 3\frac{1}{5} + 2\frac{4}{5} + \cdots$ *-26*
31. $n = 41$; $\sqrt{3} + 4\sqrt{3} + 7\sqrt{3} + 10\sqrt{3} + \cdots$ *$2501\sqrt{3}$*
32. $n = 16$; $3\sqrt{5} + \sqrt{5} - \sqrt{5} - 3\sqrt{5} - \cdots$ *$-192\sqrt{5}$*

PART C

33. Cans are stacked in 20 rows with 1 can in the top row, 2 in the next row below, 3 in the next row below, and so on. How many cans are in the stack?
34. Joan saves 1 dime the 1st day, 2 dimes the 2nd day, 3 dimes the 3rd day, and so on. How much money will she have saved at the end of 30 days?

Geometric Progressions

OBJECTIVES

- To write geometric progressions like 1, 3, 9, 27, · · ·
- To find a given term in a geometric progression

REVIEW CAPSULE

$3 \cdot x \cdot x \cdot x \cdot x = 3 \cdot x^4$

$(-1)^5 = -1 \longrightarrow (-1)^n = -1$, if n is odd

$(-1)^6 = 1 \longrightarrow (-1)^n = 1$, if n is even

EXAMPLE 1 Find the number that each term is multiplied by to get the next term. Then give the next three numbers.

Multiply each term by 2.

5	10	20	40	· · ·
↓	↓	↓	↓	
5	5 · 2	10 · 2	20 · 2	· · ·

$40 \cdot 2 = 80$, $80 \cdot 2 = 160$, $160 \cdot 2 = 320$

Thus, the next three numbers are 80, 160, and 320.

A progression is any list of numbers.

5, 10, 20, 40, . . . is a *progression*.
(1st *term*: 5; 3rd *term*: 20)

$10 \div 5 = 2$ $\quad 20 \div 10 = 2$ $\quad 40 \div 20 = 2$

2 is the *common ratio* of this progression.

Multiply by the common ratio to get the next term.

5, 10, 20, 40, · · · is a *geometric* progression.
The same number multiplies any term to get the next term.

EXAMPLE 2 Write the first 4 terms of the geometric progression whose first term is $\frac{1}{3}$ and whose common ratio is 3.

First term is $\frac{1}{3}$.

Multiply by 3 to get next term.

$$\frac{1}{3} \qquad \frac{1}{3} \cdot 3 \qquad \left(\frac{1}{3} \cdot 3\right) \cdot 3 \qquad \left(\frac{1}{3} \cdot 3 \cdot 3\right) \cdot 3$$

$$\downarrow \qquad \downarrow \qquad \downarrow \qquad \downarrow$$

$$\frac{1}{3} \qquad 1 \qquad 3 \qquad 9$$

Thus, the first 4 terms are $\frac{1}{3}$, 1, 3, 9.

EXAMPLE 3 Write the first 4 terms of the geometric progression with the given first term a and common ratio r.

$4 \cdot \frac{1}{2} = 2$ | $2(-5) = -10$
$2 \cdot \frac{1}{2} = 1$ | $-10(-5) = 50$
$1 \cdot \frac{1}{2} = \frac{1}{2}$ | $50(-5) = -250$

$a = 4$, $r = \frac{1}{2}$	$a = 2$, $r = -5$
Terms are 4, 2, 1, $\frac{1}{2}$.	Terms are 2, −10, 50, −250.

EXAMPLE 4 If the progression is geometric, find the common ratio r.

	Progression	*Answer*
$6 \div 3 = 2$; $9 \div 6 = 1.5$; *but* $2 \neq 1.5$	3, 6, 9, 12, . . .	Not geometric
$-8 \div (-2) = -32 \div (-8) = -128 \div (-32) = 4$	−2, −8, −32, −128, . . .	$4 = r$: geometric
$-16 \div 64 = 4 \div (-16) = -1 \div 4 = -\frac{1}{4}$	64, −16, 4, −1, . . .	$-\frac{1}{4} = r$: geometric
$0 \div 6 = 0$; *but* $0 \div 0$ is undefined	6, 0, 0, 0, . . .	Not geometric
$.7 \div 7 = .07 \div .7 = .007 \div .07 = .1$	7, .7, .07, .007, . . .	$.1 = r$: geometric
Multiply by $\sqrt{2}$ to get next term.	$5, 5\sqrt{2}, 10, 10\sqrt{2}, \ldots$	$\sqrt{2} = r$: geometric
Multiply by i; $i^2 = -1$.	$3, 3i, -3, -3i, \ldots$	$i = r$: geometric

EXAMPLE 5 Find the 10th term of 5, 10, 20, 40, 80, ...

A geometric progression → 5, 10, 20, 40, 80, . . .

2 is the common ratio r. → 5, $5(2)$, $5(2 \cdot 2)$, $5(2 \cdot 2 \cdot 2)$, $5(2 \cdot 2 \cdot 2 \cdot 2)$

5, $5 \cdot 2^1$, $5 \cdot 2^2$, $5 \cdot 2^3$, $5 \cdot 2^4$

r is the common ratio 2. → 5, $5 \cdot r^1$, $5 \cdot r^2$, $5 \cdot r^3$, $5 \cdot r^4$, . . .

Exponent with r is *1 less than* the number of the term. → 2nd term, 3rd term, 4th term, 5th term

Thus, the 10th term is $5 \cdot 2^9 = 5 \cdot 2^{10-1}$, or 2,560.

(5: first term; 2: common ratio; exponent is 1 less than number of term)

Formula for nth term of geometric progression

$(n - 1)$ is one less than the number of the term. →

> $a \neq 0$ is first term $r \neq 0$ is common ratio
>
> $a, a \cdot r, a \cdot r^2, a \cdot r^3, a \cdot r^4, \cdots$ is a geometric progression. The nth term is $a \cdot r^{n-1}$.

EXAMPLE 6 Find the 10th term of the geometric progression $\frac{1}{2}, -1, 2, -4, \ldots$.

a is the first term. → $a = \frac{1}{2}$

r is the common ratio. → $r = -2$

n is the number of the term. → $n = 10$

nth term $= a \cdot r^{n-1}$ ← Formula

10th term $= \frac{1}{2}(-2)^{10-1}$

$= \frac{1}{2}(-2)^9$

$= \frac{1}{2}(-512)$, or -256

Thus, −256 is the 10th term.

EXAMPLE 7 Find the specified term of the geometric progression with the given first term a and common ratio r.

5th term; $a = 12$, $r = .1$

Formula for nth term → $n\text{th term} = a \cdot r^{n-1}$

Substitute for a, r, and n. → $5\text{th term} = 12(.1)^{5-1}$

$= 12(.1)^4$

$= 12(.0001)$

$= .0012$

7th term; $a = -8$, $r = -\frac{1}{2}$.

$n\text{th term} = a \cdot r^{n-1}$

$7\text{th term} = -8(-\frac{1}{2})^{7-1}$

$= -8(-\frac{1}{2})^6$

$= -8(\frac{1}{64})$

$= -\frac{1}{8}$

$-8 \cdot \frac{1}{64} = \overset{-1}{\cancel{-8}} \cdot \frac{1}{\underset{8}{\cancel{64}}} = -\frac{1}{8}$

ORAL EXERCISES

If the progression is geometric, find the common ratio r.

1. 1, 3, 9, 27, . . . *3*
2. 1, −2, 4, −8, . . . *−2*
3. 8, 4, 2, 1, . . . *$\frac{1}{2}$*
4. 5, 10, 15, 20, . . . *not geom.*
5. −9, 3, −1, $\frac{1}{3}$, . . . *$-\frac{1}{3}$*
6. 3, .3, .03, .003, . . . *.1*
7. −5, −10, −20, −40, . . . *2*
8. 2, −2, 2, −2, . . . *−1*

EXERCISES

PART A

Give the next three numbers in the geometric progression.

1. 3, 6, 12, . . .
2. 64, 16, 4, . . . *1, $\frac{1}{4}$, $\frac{1}{16}$*
3. 1, .2, .04, . . .
4. $-\frac{1}{3}$, 1, −3, . . . *9, −27, 81*
5. 32, −16, 8, . . . *−4, 2, −1*
6. −1, −2, −4, . . .

Write the first 4 terms of the geometric progression with the given first term a and common ratio r.

7. $a = 10$, $r = 2$
8. $a = -\frac{1}{4}$, $r = 4$ *$-\frac{1}{4}$, −1, −4, −16*
9. $a = .2$, $r = 5$
10. $a = 16$, $r = \frac{1}{2}$
11. $a = -9$, $r = \frac{1}{3}$
12. $a = 5$, $r = .1$
13. $a = \frac{1}{2}$, $r = -2$
14. $a = 4$, $r = -\frac{1}{2}$ *4, −2, 1, $-\frac{1}{2}$*
15. $a = -2$, $r = -3$

Find the 10th term of the geometric progression.

16. 3, 6, 12, 24, . . . *1,536*
17. 10, −20, 40, −80, . . . *−5,120*
18. $-\frac{1}{2}$, −1, −2, −4, . . .
19. 4, 2, 1, $\frac{1}{2}$, . . . *$\frac{1}{128}$*
20. 16, −8, 4, −2, . . . *$-\frac{1}{32}$*
21. −2, −1, $-\frac{1}{2}$, $-\frac{1}{4}$, . . .

Find the 11th term of the geometric progression.

22. −4, −8, −16, −32, . . .
23. −5, 10, −20, 40, . . . *−5,120*
24. $\frac{1}{4}$, $\frac{1}{2}$, 1, 2, . . .
25. −8, −4, −2, −1, . . . *$-\frac{1}{128}$*
26. −4, 2, −1, $\frac{1}{2}$, . . . *$-\frac{1}{256}$*
27. 32, 16, 8, 4, . . .

Find the specified term of the geometric progression with the given first term a and common ratio r.

28. 5th term; $a = -10$, $r = 3$
29. 4th term; $a = 3$, $r = -5$ *−375*
30. 7th term; $a = 4$, $r = 10$
31. 6th term; $a = 9$, $r = .1$
32. 5th term; $a = -1$, $r = .2$
33. 4th term; $a = 7$, $r = -.1$
34. 5th term; $a = -5$, $r = \frac{1}{3}$
35. 4th term; $a = 8$, $r = -\frac{1}{4}$
36. 5th term; $a = 25$, $r = \frac{2}{5}$

PART B

EXAMPLE Find the specified term of the geometric progression with the given first term a and common ratio r.

6th term; $a = 3$, $r = \sqrt{2}$

Formula for nth term → $n\text{th term} = a \cdot r^{n-1}$

Substitute for a, r, and n. → $6\text{th term} = 3(\sqrt{2})^{6-1}$

$= 3(\sqrt{2})^5$

Write $(\sqrt{2} \cdot \sqrt{2})$ as a factor as many times as possible. → $= 3(\sqrt{2} \cdot \sqrt{2})(\sqrt{2} \cdot \sqrt{2})\sqrt{2}$

$(\sqrt{2} \cdot \sqrt{2}) = 2$ → $= 3 \cdot 2 \cdot 2 \cdot \sqrt{2}$

The specified term → $= 12\sqrt{2}$

8th term; $a = 2$, $r = i$

$n\text{th term} = a \cdot r^{n-1}$

$8\text{th term} = 2 \cdot i^{8-1}$

$= 2 \cdot i^7$

$= 2 \cdot i^2 \cdot i^2 \cdot i^2 \cdot i$

$= 2(-1)(-1)(-1)i$

$= -2i$

Find the specified term of the geometric progression with the given first term *a* and common ratio *r*.

37. 7th term; $a = 2$, $r = \sqrt{3}$ *54*

38. 6th term; $a = -4$, $r = \sqrt{5}$ $-100\sqrt{5}$

39. 9th term; $a = 5$, $r = -\sqrt{2}$ *80*

40. 8th term; $a = -2$, $r = -\sqrt{3}$ $54\sqrt{3}$

41. 5th term; $a = \sqrt{2}$, $r = 3\sqrt{2}$ $324\sqrt{2}$

42. 4th term; $a = -\sqrt{5}$, $r = 2\sqrt{5}$ -200

43. 8th term; $a = 3$, $r = i$ $-3i$

44. 7th term; $a = 4$, $r = i$ -4

45. 6th term; $a = 5$, $r = 2i$ $160i$

46. 5th term; $a = i$, $r = 3i$ $81i$

PART C

47. A tennis ball dropped from a height of 128 dm rebounds on each bounce $\frac{1}{2}$ of the distance from which it fell. How high does it go on its 9th rebound?

48. A tank has 4,000 liters of water. Each day $\frac{1}{2}$ of the water is removed. How much water is in the tank at the end of the 8th day?

49. A golf ball dropped from a height of 81 meters rebounds on each bounce $\frac{2}{3}$ of the distance from which it fell. How far does it fall on its 6th descent?

50. A tank contains 243 liters of gas. Each time a valve is operated, $\frac{1}{3}$ of the gas is released. How much gas is in the tank after the valve is operated 6 times?

Find two values of *x* so that the three terms form a geometric progression.

[Hint: $\frac{\text{2nd term}}{\text{1st term}} = \frac{\text{3rd term}}{\text{2nd term}}$, since each fraction is the common ratio *r*.]

51. $\frac{1}{2}$, x, 8

52. $\frac{1}{3}$, x, $\frac{1}{12}$

53. $\frac{1}{3}$, $x - 1$, $4x$

54. $2i$, x, $8i$

55. Which term of $-5, 10, -20, \ldots$ is $-5,120$? [Hint: Substitute in the formula for the nth term and solve for n.]

56. Which term of $3, -6, 12, \ldots$ is -384?

57. Which term of $\frac{1}{8}, \frac{1}{4}, \frac{1}{2}, \ldots$ is 128?

Geometric Means

OBJECTIVE

■ To find geometric means

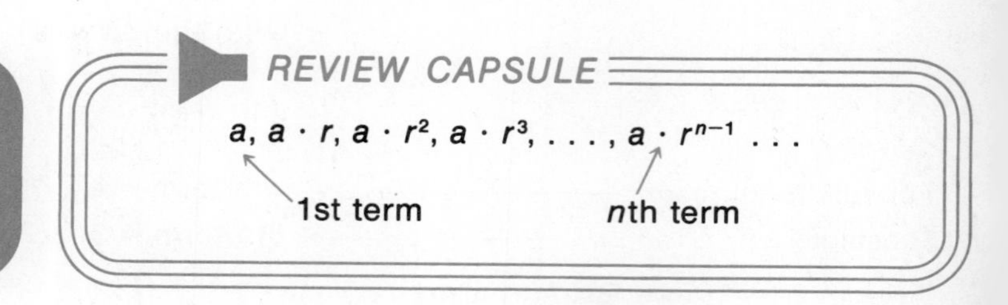

EXAMPLE 1 Find b so that 4, b, 16 forms a geometric progression.

Formula for the nth term → $n\text{th term} = a \cdot r^{n-1}$

16 is the 3rd term, $n = 3$. → $16 = 4 \cdot r^{3-1}$

Solve the equation.

$$16 = 4 \cdot r^2$$
$$4 = r^2$$

Two common ratios, 2 and -2 → $r = 2$ or $r = -2$

If $r = 2$, $4 \cdot 2 = 8$; $8 \cdot 2 = 16$.
If $r = -2$, $4(-2) = -8$; $-8(-2) = 16$. → There are *two* progressions: 4, 8, 16 and 4, -8, 16

Thus, the two values of b are 8 and -8.

Means are the terms *between* two given terms in a progression.

4, -8, 16
↑
the *negative* geometric mean between 4 and 16

Geometric Progressions

4, 8, 16	4, 8, 16, 32
the *positive* geometric *mean* between 4 and 16	two geometric *means* between 4 and 32

EXAMPLE 2 Find two geometric means between 10 and $-\frac{2}{25}$.

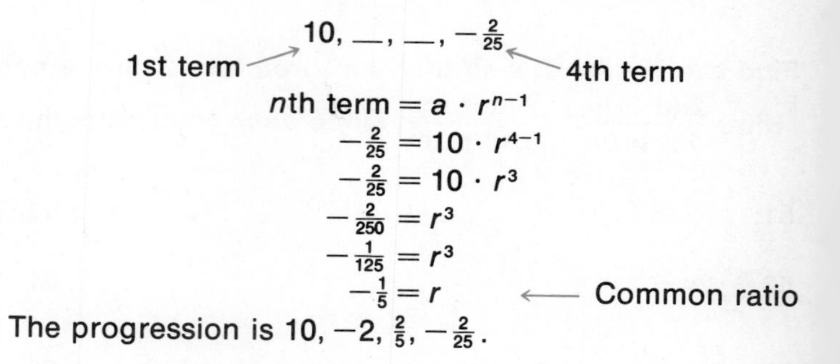

$10, \_\_, \_\_, -\frac{2}{25}$ (1st term: 10; 4th term: $-\frac{2}{25}$)

$$n\text{th term} = a \cdot r^{n-1}$$

$-\frac{2}{25}$ is the 4th term, $n = 4$. → $-\frac{2}{25} = 10 \cdot r^{4-1}$

$$-\frac{2}{25} = 10 \cdot r^3$$

Divide each side by 10. → $-\frac{2}{250} = r^3$

Simplify. → $-\frac{1}{125} = r^3$

$(-\frac{1}{5})^3 = -\frac{1}{125}$

$$-\frac{1}{5} = r \quad \leftarrow \text{Common ratio}$$

$10 \cdot \frac{-1}{5} = -2$; $-2 \cdot \frac{-1}{5} = \frac{2}{5}$; $\frac{2}{5} \cdot \frac{-1}{5} = -\frac{2}{25}$

The progression is $10, -2, \frac{2}{5}, -\frac{2}{25}$.

Thus, -2 and $\frac{2}{5}$ are two geometric means between 10 and $-\frac{2}{25}$.

EXAMPLE 3 Find two geometric means between $3\sqrt{2}$ and 12.

$3\sqrt{2}$, ___, ___, 12

1st term → $3\sqrt{2}$; 12 ← 4th term

nth term $= a \cdot r^{n-1}$

12 is the 4th term, $n = 4$ → $12 = 3\sqrt{2} \cdot r^{4-1}$

Divide each side by $3\sqrt{2}$. → $\frac{12}{3\sqrt{2}} = r^3$

$\frac{12}{3\sqrt{2}} = \frac{4}{\sqrt{2}}$; $\frac{4}{\sqrt{2}} \cdot \frac{\sqrt{2}}{\sqrt{2}} = \frac{4\sqrt{2}}{2} = \frac{2\sqrt{2}}{1}$ → $2\sqrt{2} = r^3$

$(\sqrt{2})^3 = \sqrt{2} \cdot \sqrt{2} \cdot \sqrt{2} = 2\sqrt{2}$

$\sqrt{2} = r$ ← Common ratio

$3\sqrt{2} \cdot \sqrt{2} = 3 \cdot 2 = 6$; $6 \cdot \sqrt{2} = 6\sqrt{2}$; $6\sqrt{2} \cdot \sqrt{2} = 6 \cdot 2 = 12$ → The progression is $3\sqrt{2}$, 6, $6\sqrt{2}$, 12.

Thus, 6 and $6\sqrt{2}$ are two geometric means between $3\sqrt{2}$ and 12.

EXERCISES

PART A

Find the *positive* geometric mean between the numbers.

1. 3 and 12 *6*
2. -4 and -36 *12*
3. $\frac{1}{2}$ and 8 *2*
4. $-\frac{2}{5}$ and -10
5. 8 and 2 *4*
6. -9 and -1 *3*
7. 4 and $\frac{1}{4}$ *1*
8. -15 and $-\frac{3}{5}$
9. 4 and .04 *.4*
10. 3 and .12 *.6*
11. 1 and .09 *.3*
12. .1 and .016

Find two geometric means between the numbers.

13. 2 and 54
14. -3 and 24 *6, −12*
15. -2 and -128
16. $\frac{1}{5}$ and -25
17. 9 and $\frac{1}{3}$
18. 8 and -1 *−4, 2*
19. 50 and $\frac{2}{5}$ *10, 2*
20. -8 and $\frac{1}{8}$
21. 6 and .006 *.6, .06*
22. 4 and .032 *.8, .16*
23. 1 and .027 *.3, .09*
24. .2 and .0128

PART B

Find the *positive* geometric mean between the numbers.

25. 3 and 6 *$3\sqrt{2}$*
26. $5\sqrt{2}$ and $10\sqrt{2}$ *10*
27. -4 and -12 *$4\sqrt{3}$*
28. $-3\sqrt{5}$ and $-15\sqrt{5}$
29. 1 and 8 *$2\sqrt{2}$*
30. $\sqrt{3}$ and $12\sqrt{3}$ *6*
31. 2 and 36 *$6\sqrt{2}$*
32. $-\sqrt{5}$ and $-20\sqrt{5}$ *10*

Find two geometric means between the numbers.

33. 2 and $6\sqrt{3}$
34. 5 and $10\sqrt{2}$
35. 3 and $21\sqrt{7}$
36. 4 and $20\sqrt{5}$
37. $\sqrt{2}$ and 4 *2, $2\sqrt{2}$*
38. $2\sqrt{3}$ and 18 *6, $6\sqrt{3}$*
39. $\sqrt{7}$ and 49 *7, $7\sqrt{7}$*
40. $3\sqrt{5}$ and 75 *15, $15\sqrt{5}$*

PART C

Find one geometric mean between the numbers. (Each exercise has two answers.)

41. i and $4i$
42. i and $-9i$
43. 1 and -4
44. -2 and 18
45. i and $2i$
46. i and $-5i$
47. 2 and -10
48. -3 and 6
49. $1 + i$ and $4 + 4i$
50. $1 - i$ and $9 - 9i$
51. x and y

Sums of Even and Odd Numbers

Each rectangle below is composed of the sum of even numbers.

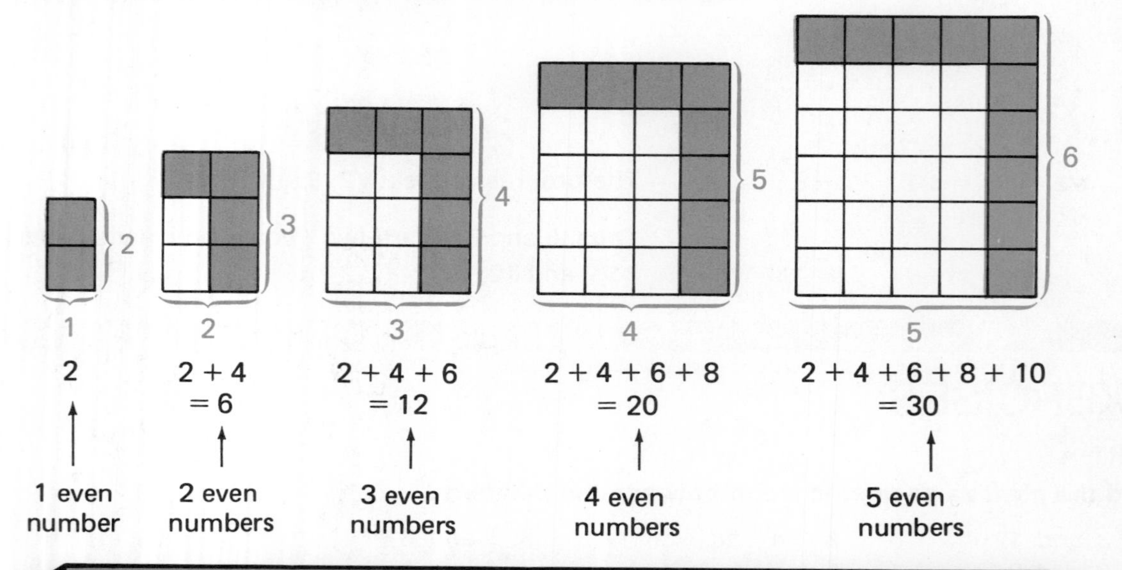

The sum of the first k even numbers is $k(k + 1)$.

PROBLEM

Find the sum of the first 7 even numbers, of the first 14 even numbers.

$$\text{Sum} = k(k+1) = 7(7+1) = 7(8) = 56$$

$$\text{Sum} = k(k+1) = 14(14+1) = 14(15) = 210$$

PROJECT Find the sum.

1. The first 10 even numbers
2. The first 16 even numbers
3. The first 29 even numbers
4. The first 100 even numbers

Each figure below is composed of the sum of odd numbers.

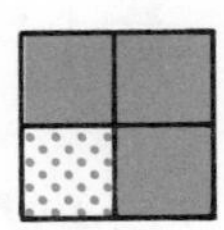
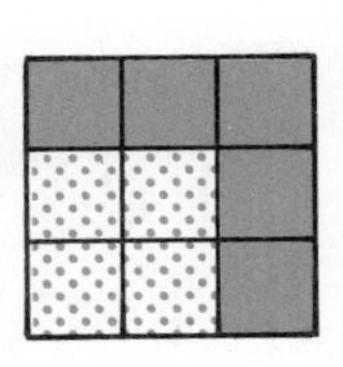
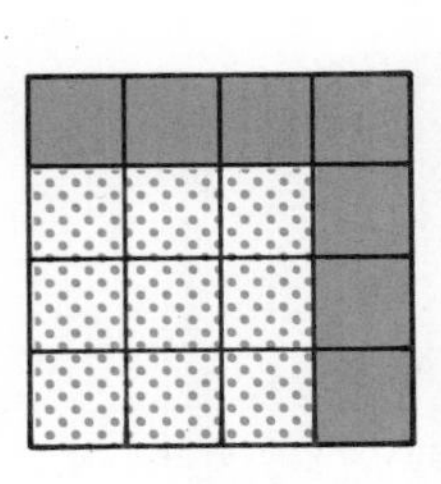
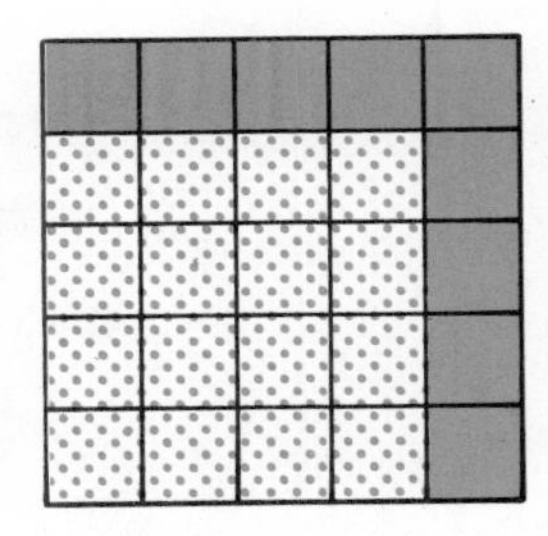
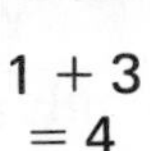

1	$1 + 3$ $= 4$	$1 + 3 + 5$ $= 9$	$1 + 3 + 5 + 7$ $= 16$	$1 + 3 + 5 + 7 + 9$ $= 25$

The sum of the first k odd numbers is k^2.

PROBLEM

Find the sum of the first 8 odd numbers; of the first 15 odd numbers.

$$\text{Sum} = k^2 = 8^2 = 64$$

$$\text{Sum} = k^2 = 15^2 = 225$$

PROJECT Find the sum.

5. The first 10 odd numbers
6. The first 12 odd numbers
7. The first 9 odd numbers
8. The first 20 odd numbers
9. The first 100 odd numbers
10. The first 40 odd numbers

Here are the first ten counting numbers: 1, 2, 3, 4, 5, 6, 7, 8, 9, 10.

11. How many odd numbers are there from 1 through 10?
12. How many even numbers are there from 1 through 10?
13. Find the sum of the first 10 counting numbers.
14. Find the sum of the first 100 counting numbers.

Geometric Series

OBJECTIVES

- To find the sum of a finite geometric series
- To find the sum of an infinite geometric series

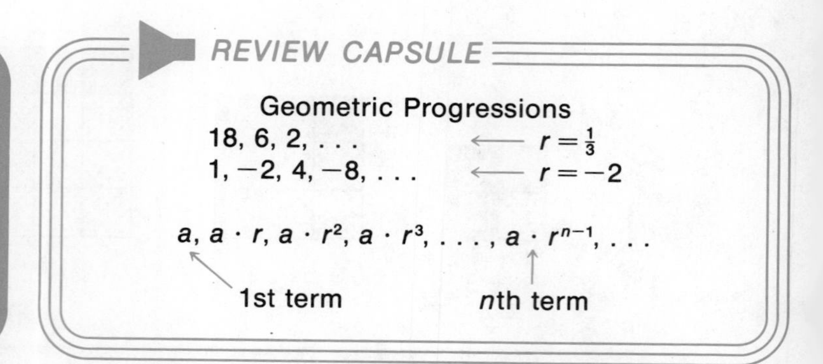

EXAMPLE 1 Indicate the sum of the progression 2, 10, 50, 250, ...

Progression → 2 , 10 , 50 , 250 , . . .

Replace commas with + signs. →

Series → $2 + 10 + 50 + 250 + \cdots$

A series is an indicated sum.

$2 + 10 + 50 + 250 + \cdots$ is a geometric *series*, if 2, 10, 50, 250, $\cdots$ is a geometric progression.

EXAMPLE 2 Write the geometric progression 2, −6, 18, −54, . . . as a series.

Replace commas with + signs.
$2 + (-6) + 18 + (-54) + \cdots$
$= 2 - 6 + 18 - 54 + \cdots$ →

2 , −6 , 18 , −54 , . . .

$2 - 6 + 18 - 54 + \cdots$

EXAMPLE 3 Find the sum S of the first 4 terms of the geometric series $5 + 5 \cdot 3 + 5 \cdot 3^2 + 5 \cdot 3^3 + \cdots$.

$r = 3$; multiply first 4 terms by 3. → $3 \cdot S = \quad 5 \cdot 3 + 5 \cdot 3^2 + 5 \cdot 3^3 + 5 \cdot 3^4$

Indicated sum of 1st 4 terms → $S = 5 + 5 \cdot 3 + 5 \cdot 3^2 + 5 \cdot 3^3$

Subtract in each column. → $2S = -5 + 0 + 0 + 0 + 5 \cdot 3^4$

Rewrite. → $2S = 5 \cdot 3^4 - 5 \cdot 1$

Factor 5 from right side. → $2S = 5(3^4 - 1)$

Divide each side by 2. $3^4 = 81$ → $S = \frac{5(81-1)}{2}$

$S = \frac{5 \cdot 80}{2}$, or 200

Thus, the sum is 200.

Example 3 suggests a method to find a formula for the sum S of the first n terms of a geometric series.

The first n terms of a geometric series →

$$S = a + ar + ar^2 + ar^3 + \cdots + ar^{n-1}$$

r is the common ratio; multiply the first n terms by r. →
Sum of the first n terms →
Subtract in each column. →
Rewrite. →
Factor each side. →
Divide each side by $r - 1$. →

$$\begin{aligned} r \cdot S &= \quad ar + ar^2 + ar^3 + \cdots + ar^{n-1} + ar^n \\ S &= a + ar + ar^2 + ar^3 + \cdots + ar^{n-1} \\ \hline rS - S &= -a + 0 + 0 + 0 + \cdots + 0 + ar^n \\ rS - 1S &= ar^n - 1a \\ (r-1)S &= a(r^n - 1) \\ S &= \frac{a(r^n - 1)}{r - 1} \end{aligned}$$

Formula →

> The sum S of the first n terms of a geometric series can be found by
>
> $$S = \frac{a(r^n - 1)}{r - 1},$$
>
> where a is the first term and r is the common ratio of the series.

EXAMPLE 4 Find the sum of the first n terms of the geometric series given the first term a, the common ratio r, and n.

$$a = \tfrac{1}{5},\ r = -5,\ n = 7$$

Series described by a, r, n: $\frac{1}{5} - 1 + 5 - 25 + \cdots$ →

Substitute in formula: $a = \frac{1}{5}$, $r = -5$, $n = 7$. →

Shortcut for $(-5)^7$: $(-5)^7 = (-5)^3 \cdot 5^4 = -125 \cdot 625 = -78{,}125$ →

$-78{,}126 \div (-6) = 13{,}021$; $13{,}021 \div 5 = 2{,}604\frac{1}{5}$ →

$$\begin{aligned} S = \frac{a(r^n - 1)}{r - 1} &= \frac{\frac{1}{5}[(-5)^7 - 1]}{(-5) - 1} \\ &= \frac{\frac{1}{5}(-78{,}125 - 1)}{-6} \\ &= \frac{1}{5} \cdot \frac{-78{,}126}{-6} \\ &= \tfrac{1}{5}(13{,}021), \text{ or } 2{,}604\tfrac{1}{5} \end{aligned}$$

Thus, the sum of the first 7 terms is $2{,}604\frac{1}{5}$.

EXAMPLE 5 Find the sum of the first n terms of the geometric series given n and the series.

$$n = 10;\ 2 + 1 + \tfrac{1}{2} + \tfrac{1}{4} + \cdots$$

$r = 1 \div 2 = \frac{1}{2}$, $a = 2$, $n = 10$; substitute. →

Shortcut for $(\frac{1}{2})^{10}$: $2^{10} = 2^5 \cdot 2^5 = 32 \cdot 32 = 1{,}024$ →

$\frac{2}{-\frac{1}{2}} = 2 \div \left(-\frac{1}{2}\right) = 2 \cdot \frac{-2}{1}$ →

$$\begin{aligned} S = \frac{a(r^n - 1)}{r - 1} &= \frac{2[(\frac{1}{2})^{10} - 1]}{(\frac{1}{2}) - 1} \\ &= \frac{2\left(\frac{1}{1{,}024} - \frac{1{,}024}{1{,}024}\right)}{-\frac{1}{2}} \\ &= 2 \cdot \frac{-2}{1} \cdot \frac{-1{,}023}{1{,}024} = \frac{1{,}023}{256}, \text{ or } 3\tfrac{255}{256} \end{aligned}$$

Thus, the sum of the first 10 terms is $3\frac{255}{256}$.

If $n = 12$, then $r^n = (\frac{1}{2})^{12} = \frac{1}{4{,}096}$.
$12 > 10$, but $(\frac{1}{2})^{12} < (\frac{1}{2})^{10}$.

In Example 5, observe that r^n is very small.
If n were larger than 10,
r^n would be smaller than $\frac{1}{1{,}024}$.

$$S = \frac{a[\,r^n - 1]}{r-1} = \frac{a[\,(\frac{1}{2})^{10} - 1]}{r-1} = \frac{a[\,\frac{1}{1{,}024} - 1]}{r-1}$$

In Example 5, $r = \frac{1}{2}$; $-1 < \frac{1}{2} < 1$.

If $-1 < r < 1$, then
r^n gets closer and closer to 0
as n gets larger and larger

Substiture 0 for r^n. ⟶

and $S = \frac{a(r^n - 1)}{r-1}$ gets closer and closer to $\frac{a(0-1)}{r-1}$

$$= \frac{a(-1)}{r-1}$$

$\frac{a(-1)}{r-1} = \frac{-1(a)}{-1+r} = \frac{-1(a)}{-1(1-r)} = \frac{a}{1-r}$ ⟶

$$= \frac{a}{1-r}.$$

A *finite* series has a *last* term.

In an *infinite* series, n gets larger and larger.

Geometric Series: $\frac{1}{3} + 1 + 3 + 9 + \cdots$

8th term and *last term*
$\frac{1}{3} + 1 + 3 + \cdots + 729$
Finite series

7th term but *no last term*
$\frac{1}{3} + 1 + 3 + \cdots + 243 + \cdots$
Infinite series

The sum S of an *infinite* geometric series can be found by
$S = \frac{a}{1-r}$, if $-1 < r < 1$ and a is the 1st term.

EXAMPLE 6

Find the sum of the infinite geometric series.

$$2 + 1 + \tfrac{1}{2} + \tfrac{1}{4} + \cdots$$

$r = \frac{1}{2}$: $-1 < \frac{1}{2} < 1$, $-1 < r < 1$ — Use the formula. ⟶

$$S = \frac{a}{1-r} = \frac{2}{1-\frac{1}{2}}$$

$\frac{2}{\frac{1}{2}} = 2 \div \frac{1}{2} = 2 \cdot \frac{2}{1} = 4$ ⟶

$$= \frac{2}{\frac{1}{2}}, \text{ or } 4$$

Compare this answer with that of Example 5. ⟶

Thus, the sum is 4.

EXAMPLE 7

Find the sum of the infinite geometric series.

$$\tfrac{1}{4} + \tfrac{1}{2} + 1 + 2 + \cdots$$

$$r = 2$$

An infinite geometric series has a sum, only if $-1 < r < 1$; 2 is *not* between -1 and 1.

So, r is *not between* -1 and 1.

Thus, this infinite series does *not* have a sum.

EXAMPLE 8 Find the sum of the infinite geometric series: $8 - 2 + \frac{1}{2} - \frac{1}{8} + \cdots$.

$r = -\frac{1}{4}$; $-1 < -\frac{1}{4} < 1$. $-1 < r < 1$ } Use the formula. ⟶

$$S = \frac{a}{1-r} = \frac{8}{1-(-\frac{1}{4})}$$
$$= \frac{8}{\frac{5}{4}}$$
$$= \frac{32}{5}, \text{ or } 6\frac{2}{5}$$

$\frac{8}{\frac{5}{4}} = 8 \div \frac{5}{4} = 8 \cdot \frac{4}{5} = \frac{32}{5} = 6\frac{2}{5}$

Thus, $6\frac{2}{5}$ is the sum.

EXAMPLE 9 Write the repeating decimal $.434343 \cdots$ as an infinite geometric series. Then find a and r, the first term and common ratio of the series.

$$\begin{array}{r} .43 \\ .0043 \\ +.000043 \\ \hline .434343 \end{array}$$ ⟶

$$.434343 \cdots = .43 + .0043 + .000043 + \cdots$$

$r = .0043 \div .43 = .01$ ⟶ **Thus,** $a = .43$ and $r = .01$.

EXAMPLE 10 Write the repeating decimal $.37\overline{37}$ in fraction form $\frac{a}{b}$, where a and b are integers.

Write $.37\overline{37}$ as an infinite geometric series. ⟶

$$.37\overline{37} = .37 + .0037 + .000037 + \cdots$$

$r = .01$; $-1 < .01 < 1$. $-1 < r < 1$ } Use the formula. ⟶

$$S = \frac{a}{1-r} = \frac{.37}{1-.01}$$

.37 and .99 are *not* integers. ⟶

$$= \frac{.37}{.99}$$

$\frac{.37}{.99} \cdot \frac{100}{100} = \frac{37}{99}$ ⟶

$$= \frac{37}{99}$$

37 and 99 are integers. **Thus,** $.37\overline{37} = \frac{37}{99}$.

ORAL EXERCISES

Give the geometric progression as a series.

1. 1, 3, 9, 27, . . . *$1 + 3 + 9 + 27 + \cdots$*
2. $3, -1, \frac{1}{3}, -\frac{1}{9}, \ldots$ *$3 - 1 + \frac{1}{3} - \frac{1}{9} + \cdots$*
3. $4, 2, 1, \frac{1}{2}, \ldots$ *$4 + 2 + 1 + \frac{1}{2} + \cdots$*
4. $-\frac{1}{5}, 1, -5, 25, \ldots$ *$-\frac{1}{5} + 1 - 5 + 25 - \cdots$*

If the series is geometric, find the common ratio *r*.

5. $\frac{1}{3} + 1 + 3 + 9 + \cdots$ *3*
6. $-1 - 2 - 4 - 8 - \cdots$ *2*
7. $4 + 8 + 12 + 16 + \cdots$ *not geom*
8. $2 - 1 + \frac{1}{2} - \frac{1}{4} + \cdots$ *$-\frac{1}{2}$*
9. $-\frac{1}{4} + 1 - 4 + 16 - \cdots$ *-4*
10. $25 + 5 + 1 + \frac{1}{5} + \cdots$ *$\frac{1}{5}$*
11. $6 + .6 + .06 + .006 + \cdots$ *.1*
12. $14 + .14 + .0014 + \cdots$ *.01*
13. $.02 + .002 + .0002 + \cdots$ *.1*

EXERCISES

PART A

Find the sum of the first n terms of the geometric series given n and the series.

1. $n = 10$; $5 + 10 + 20 + 40 + \cdots$ *5115*
2. $n = 9$; $-3 - 6 - 12 - 24 - \cdots$ *−1533*
3. $n = 9$; $1 - 2 + 4 - 8 + \cdots$ *171*
4. $n = 10$; $-4 + 8 - 16 + 32 - \cdots$ *1364*
5. $n = 7$; $5 + 15 + 45 + 135 + \cdots$ *5465*
6. $n = 10$; $-\frac{1}{2} - 1 - 2 - 4 - \cdots$ *$-511\frac{1}{2}$*
7. $n = 8$; $16 + 8 + 4 + 2 + \cdots$ *$31\frac{7}{8}$*
8. $n = 8$; $32 - 16 + 8 - 4 + \cdots$ *$21\frac{1}{4}$*
9. $n = 7$; $.3 + 3 + 30 + 300 + \cdots$ *333,333.3*
10. $n = 6$; $-1 + 10 - 100 + 1{,}000 - \cdots$ *90,909*

Find the sum of the first n terms of the geometric series given the first term a, the common ratio r, and n.

11. $a = 6$, $r = 10$, $n = 7$ *6,666,666*
12. $a = -2$, $r = 3$, $n = 6$ *−728*
13. $a = 70$, $r = .1$, $n = 5$ *77.777*
14. $a = 25$, $r = .2$, $n = 3$ *31*
15. $a = 20$, $r = .2$, $n = 3$ *24.8*
16. $a = 20$, $r = .1$, $n = 5$ *22.222*
17. $a = -3$, $r = .1$, $n = 5$ *−3.3333*
18. $a = 3$, $r = .2$, $n = 3$ *3.72*

Find the sum of the infinite geometric series.

19. $8 + 4 + 2 + 1 + \cdots$ *16*
20. $-2 - 1 - \frac{1}{2} - \frac{1}{4} - \cdots$ *−4*
21. $9 + 6 + 4 + \frac{8}{3} + \cdots$ *27*
22. $9 + 3 + 1 + \frac{1}{3} + \cdots$ *$13\frac{1}{2}$*
23. $\frac{1}{2} + \frac{1}{4} + \frac{1}{8} + \frac{1}{16} + \cdots$ *1*
24. $\frac{1}{4} - \frac{1}{8} + \frac{1}{16} - \frac{1}{32} + \cdots$ *$\frac{1}{6}$*
25. $2 + .2 + .02 + .002 + \cdots$ *$2\frac{2}{9}$*
26. $7 + .7 + .07 + .007 + \cdots$ *$7\frac{7}{9}$*
27. $1 - \frac{1}{3} + \frac{1}{9} - \frac{1}{27} + \cdots$ *$\frac{3}{4}$*
28. $.3 + .03 + .003 + .0003 + \cdots$ *$\frac{1}{3}$*

Write the repeating decimal as an infinite geometric series. Then find a and r.

29. $.555 \cdots$ *$.5 + .05 + .005 + \cdots$; $a = .5$; $r = .1$*
30. $.3\overline{3}$
31. $6.6\overline{6}$
32. $4.444 \cdots$
33. $.0222 \cdots$ *$.02 + .002 + .0002 + \cdots$; $a = .02$; $r = .1$*
34. $22.2\overline{2}$ *$20 + 2 + .2 + \cdots$; $a = 20$; $r = .1$*

Write the repeating decimal in fraction form $\frac{a}{b}$, where a and b are integers.

35. $.222 \cdots$ *$\frac{2}{9}$*
36. $.7\overline{7}$ *$\frac{7}{9}$*
37. $.05\overline{5}$ *$\frac{5}{90}$*
38. $.0444 \cdots$ *$\frac{4}{90}$*
39. $7.777 \cdots$ *$\frac{70}{9}$*
40. $3.3\overline{3}$ *$\frac{30}{9}$*

PART B

Find the sum of the first n terms of the geometric series given n and the series.

41. $n = 9$; $-32 + 16 - 8 + 4 - \cdots$ *$-21\frac{3}{8}$*
42. $n = 7$; $625 + 125 + 25 + 5 + \cdots$ *$781\frac{6}{25}$*
43. $n = 6$; $\frac{1}{2} + 2 + 8 + 32 + \cdots$ *$682\frac{1}{2}$*
44. $n = 6$; $-2 - 8 - 32 - 128 - \cdots$ *−2730*
45. $n = 7$; $\frac{2}{3} + 2 + 6 + 18 + \cdots$ *$728\frac{2}{3}$*
46. $n = 7$; $\frac{2}{5} + 2 + 10 + 50 + \cdots$ *$7812\frac{2}{5}$*
47. $n = 7$; $1 - 5 + 25 - 125 + \cdots$ *13,021*
48. $n = 6$; $.2 + 1 + 5 + 25 + \cdots$ *781.2*

Find the sum of the first *n* terms of the geometric series given the first term *a*, the common ratio *r*, and *n*.

49. $a = -243$, $r = \frac{1}{3}$, $n = 6$ *−364*
50. $a = 27$, $r = \frac{1}{3}$, $n = 6$ $40\frac{4}{9}$
51. $a = 200$, $r = \frac{1}{2}$, $n = 6$ $393\frac{3}{4}$
52. $a = -128$, $r = -\frac{1}{4}$, $n = 6$ $-102\frac{3}{8}$
53. $a = 64$, $r = -\frac{1}{4}$, $n = 4$ *51*
54. $a = 500$, $r = -\frac{1}{5}$, $n = 4$ *416*
55. $a = 5$, $r = 4$, $n = 6$ *6,825*
56. $a = -3$, $r = 5$, $n = 6$ *−11,718*
57. $a = -1$, $r = -3$, $n = 7$ *−547*
58. $a = -1$, $r = -4$, $n = 5$ *−205*

Find the sum of the infinite geometric series.

59. $\frac{1}{5} + 1 + 5 + 25 + \cdots$ *no sum*
60. $4 - 2 + 1 - \frac{1}{2} + \cdots$ $2\frac{2}{3}$
61. $-64 - 16 - 4 - 1 - \cdots$ $-85\frac{1}{3}$
62. $-25 - 5 - 1 - \frac{1}{5} - \cdots$ $-31\frac{1}{4}$
63. $-27 + 9 - 3 + 1 - \cdots$ $-20\frac{1}{4}$
64. $5 - 1 + \frac{1}{5} - \frac{1}{25} + \cdots$ $4\frac{1}{6}$
65. $-16 + 12 - 9 + \frac{27}{4} - \cdots$ $-9\frac{1}{7}$
66. $25 - 10 + 4 - \frac{8}{5} + \cdots$ $17\frac{6}{7}$
67. $8\sqrt{5} + 4\sqrt{5} + 2\sqrt{5} + \sqrt{5} + \cdots$ $16\sqrt{5}$
68. $2\sqrt{3} + \sqrt{3} + \frac{1}{2}\sqrt{3} + \frac{1}{4}\sqrt{3} + \cdots$ $4\sqrt{3}$
69. $27\sqrt{2} + 18\sqrt{2} + 12\sqrt{2} + 8\sqrt{2} + \cdots$ $81\sqrt{2}$
70. $64\sqrt{7} + 48\sqrt{7} + 36\sqrt{7} + 27\sqrt{7} + \cdots$ $256\sqrt{7}$

Write the repeating decimal as an infinite geometric series. Then find *a* and *r*.

71. $.383838 \cdots$ *a = .38; r = .01*
72. $.6\overline{565}$ *a = .65; r = .01*
73. $53.5\overline{353}$ *a = 53; r = .01*
74. $74.7474 \cdots$ *a = 74; r = .01*
75. $.0262626 \cdots$ *a = .026; r = .01*
76. $.02\overline{929}$ *a = .029; r = .01*

Write the repeating decimal in fraction form $\frac{a}{b}$, where *a* and *b* are integers.

77. $.353535 \cdots$ $\frac{35}{99}$
78. $.2\overline{828}$ $\frac{28}{99}$
79. $17.\overline{1717}$ $\frac{1700}{9}$
80. $62.6262 \cdots$ $\frac{6200}{99}$
81. $.0474747 \cdots$ $\frac{47}{990}$
82. $.0\overline{3232}$ $\frac{32}{990}$

PART C

A steel ball dropped from a height of 128 meters rebounds on each bounce $\frac{1}{2}$ of the distance from which it fell.

83. When the ball hits the ground the 8th time, how far has it traveled since it was first dropped?
84. How many meters will the ball travel before coming to rest?

A golf ball dropped from a height of 81 meters rebounds on each bounce $\frac{2}{3}$ of the distance from which it fell.

85. When the ball hits the ground the 6th time, how far has it traveled since it was first dropped?
86. How many meters will the ball travel before coming to rest?

Chapter Thirteen Review

Write the first 4 terms of the arithmetic progression with the given first term *a* and common difference *d*. [p. 333]

1. $a = -3$, $d = 5$
2. $a = 6$, $d = -4$
3. $a = 4$, $d = 1.6$
4. $a = -3$, $d = 2 - \sqrt{3}$

Find the 36th term of the arithmetic progression. [p. 333]

5. $6, 3, 0, -3, \ldots$
6. $-20, -22, -24, -26, \ldots$
7. $2, 2\frac{3}{5}, 3\frac{1}{5}, 3\frac{4}{5}, \ldots$
8. Find the one arithmetic mean between 7 and 23. [p. 337]
9. Find the four arithmetic means between 5 and 7. [p. 337]
10. Write the progression $5, 2, -1, -4, \ldots$ as a series. [p. 339]

Find the sum of the first *n* terms of the arithmetic series given the first term *a*, the *n*th term *l*, and *n*. [p. 339]

11. $n = 20$, $a = 2.4$, $l = 8.1$
12. $n = 25$, $a = -5$, $l = 45$

Find the sum of the first *n* terms of the arithmetic series given *n* and the series. [p. 339]

13. $n = 21$; $4 + 1 - 2 - 5 - \cdots$
14. $n = 26$; $3 + 3\frac{3}{5} + 4\frac{1}{5} + 4\frac{4}{5} + \cdots$

Write the first 4 terms of the geometric progression with the given first term *a* and common ratio *r*. [p. 342]

15. $a = 5$, $r = -2$
16. $a = \frac{1}{2}$, $r = 4$
17. $a = -16$, $r = \frac{1}{2}$
18. $a = 2$, $r = \sqrt{3}$

Find the 11th term of the geometric progression. [p. 342]

19. $3, 6, 12, 24, \ldots$
20. $8, 4, 2, 1, \ldots$
21. $\frac{1}{2}, -1, 2, -4, \ldots$

Find the specified term of the geometric progression with the given first term *a* and common ratio *r*. [p. 342]

22. 5th term; $a = 4$, $r = -3$
23. 7th term; $a = -2$, $r = \sqrt{3}$
24. Find the *positive* geometric mean between -2 and -18. [p. 346]
25. Find two geometric means between 2 and -16. [p. 346]

Find the sum of the first *n* terms of the geometric series given *n* and the series. [p. 350]

26. $n = 7$; $\frac{1}{5} + 2 + 20 + 200 + \cdots$
27. $n = 8$; $16 - 8 + 4 - 2 + \cdots$

Find the sum of the first *n* terms of the geometric series given the first term *a*, the common ratio *r*, and *n*. [p. 350]

28. $a = -4$, $r = .1$, $n = 5$
29. $a = 4$, $r = -2$, $n = 6$

Find the sum of the infinite geometric series. [p. 350]

30. $-4 - 2 - 1 - \frac{1}{2} - \frac{1}{4} - \cdots$
31. $32 - 8 + 2 - \frac{1}{2} + \cdots$
32. $5 + .5 + .05 + \cdots$
33. Write the repeating decimal $.353535\ldots$ as an infinite geometric series. Then find *a* and *r*. [p. 350]
34. Write the repeating decimal $.2\overline{2}$ in fraction form $\frac{a}{b}$, where *a* and *b* are integers. [p. 350]

Chapter Thirteen Test

Write the first 4 terms of the arithmetic progression with the given first term *a* and common difference *d*.

1. $a = 3, d = -3$

2. $a = 2, d = 1.5$

Find the 31st term of the arithmetic progression.

3. $-1, -4, -7, -10, \ldots$

4. $2, 2\frac{1}{3}, 2\frac{2}{3}, 3, \ldots$

5. Find the four arithmetic means between 7 and 8.

6. Write the progression $-3, -1, 1, 3, \ldots$ as a series.

Find the sum of the first *n* terms of the arithmetic series given the first term *a*, the *n*th term *l*, and *n*.

7. $n = 25, a = .6, l = 29.8$

8. $n = 20, a = 27, l = -6$

Find the sum of the first *n* terms of the arithmetic series given *n* and the series.

9. $n = 21; 3 + 3\frac{1}{2} + 4 + 4\frac{1}{2} + \cdots$

10. $n = 40; -5 - 15 - 25 - 35 - \cdots$

Write the first 4 terms of the geometric progression with the given first term *a* and common ratio *r*.

11. $a = 3, r = -2$

12. $a = 9, r = \frac{1}{3}$

Find the 10th term of the geometric progression.

13. $\frac{1}{2}, 1, 2, 4, \ldots$

14. $2, 1, \frac{1}{2}, \frac{1}{4}, \ldots$

Find the specified term of the geometric progression with the given first term *a* and common ratio *r*.

15. 6th term; $a = 2, r = -3$

16. 5th term; $a = -3, r = \sqrt{2}$

17. Find two geometric means between 3 and 24.

Find the sum of the first *n* terms of the geometric series given *n* and the series.

18. $n = 10; \frac{1}{2} + 1 + 2 + 4 + 8 + \cdots$

19. $n = 6; 100 - 50 + 25 - 12\frac{1}{2} + \cdots$

Find the sum of the first *n* terms of the geometric series given the first term *a*, the common ratio *r*, and *n*.

20. $a = 3, r = .1, n = 4$

21. $a = 5, r = -10, n = 4$

Find the sum of the infinite geometric series.

22. $-2 - 1 - \frac{1}{2} - \frac{1}{4} - \cdots$

23. $16 - 4 + 1 - \frac{1}{4} + \cdots$

24. Write the repeating decimal .272727 . . . as an infinite geometric series. Then find *a* and *r*.

25. Write the repeating decimal $.4\overline{4}$ in fraction form $\frac{a}{b}$, where *a* and *b* are integers.

14 LOGARITHMS

Mathematics in Business

Organizing data into graphs and tables is very important to a business. They help executives to study trends in sales, profits, cost, supply, demand, etc. Statistics in the form of graphs are very helpful to inform stockholders and employees of details about a company. They are also a vital part of analyzing and solving problems within a business.

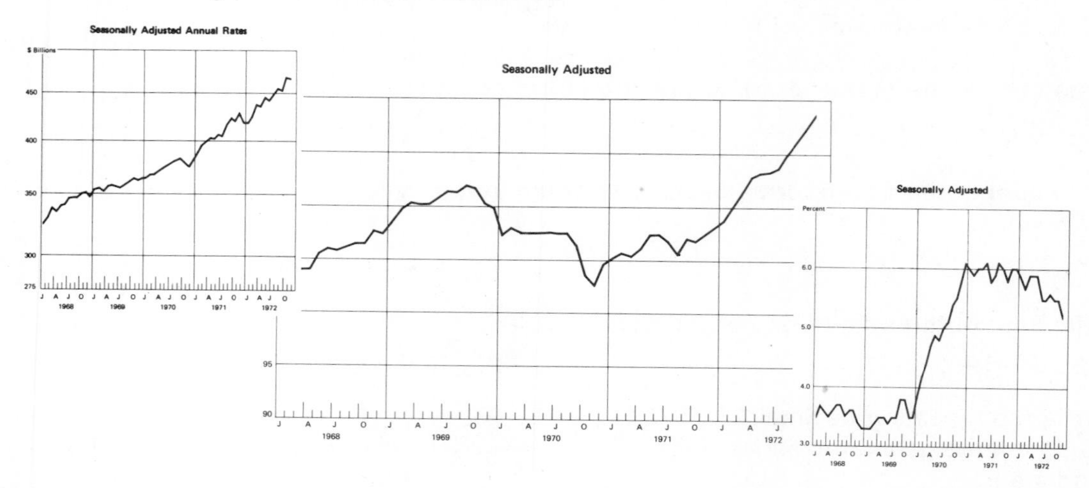

Price of n items at \$$p$ per item = np.

PROBLEM

Ms. Gray Eagle bought some stock for \$2,880. When the price dropped \$2 per share, she purchased 6 more than the original number of shares of stock for the same amount. What was the original purchase price of each share?

Let n = original number of shares
p = original price of each share
$n + 6$ = number of shares in 2nd purchase
$p - 2$ = price per share in 2nd purchase

1st purchase: $np = 2{,}880$ → $n = \dfrac{2{,}880}{p}$

2nd purchase: $(n + 6)(p - 2) = 2{,}880$ → $np + 6p - 2n - 12 = 2{,}880$

Substitute for n. → $\dfrac{2{,}880}{p} \cdot p + 6p - 2\left(\dfrac{2{,}880}{p}\right) - 12 = 2{,}880$

Simplify. → $2{,}880 + 6p - \dfrac{5{,}760}{p} - 12 = 2{,}880$

Multiply by p. → $2{,}880p + 6p^2 - 5{,}760 - 12p = 2{,}880p$

Simplify. → $6p^2 - 12p - 5{,}760 = 0$

Divide by 6. → $p^2 - 2p - 960 = 0$

Factor. → $(p - 32)(p + 30) = 0$

Set each factor = 0. → $p - 32 = 0$ or $p + 30 = 0$

Solve for p. → $p = 32$ or $p = -30$

-30 is extraneous.

Thus, the original price of each share was \$32.

PROJECT

1. A high school group chartered a bus for \$120. If 10 more students had gone on the trip, the cost to each student would have been \$1 less. How many students went on the trip and how much did each have to pay?

2. The student store paid \$80 for a number of school jerseys. Two of the jerseys were torn, but the others were sold for a profit of \$1 each. The store made \$10 on the total transaction. How many jerseys were originally purchased? How much did the store pay for each?

Numbers as Powers of 10

OBJECTIVES

- To show numbers as powers of 10 like $47.6 = 10^{1.6776}$
- To find the value of a power of 10 like $10^{3.8407} = 6{,}930$

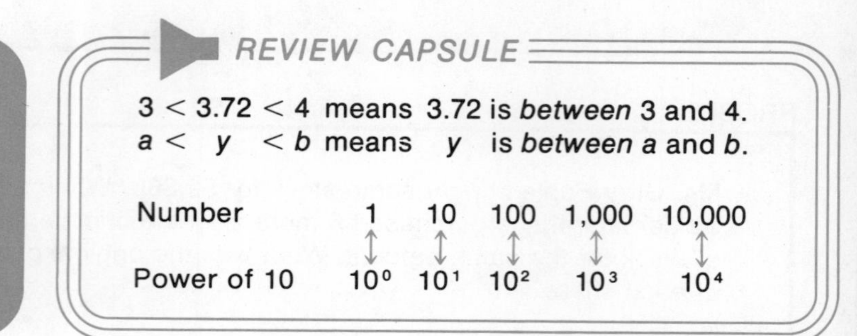

REVIEW CAPSULE

$3 < 3.72 < 4$ means 3.72 is *between* 3 and 4.
$a < y < b$ means y is *between* a and b.

Number	1	10	100	1,000	10,000
Power of 10	10^0	10^1	10^2	10^3	10^4

EXAMPLE 1 Show 6,320 as a power of 10.

1,000 and 10,000 are consecutive whole-number powers of 10.

$1{,}000 < 6{,}320 < 10{,}000$ ← Locate 6,320 between powers of 10.

Change to exponential form. → $10^3 < 10^y < 10^4$ ← Let $6{,}320 = 10^y$.

Compare the exponents: $3 < 3 + (\text{decimal}) < 4$. → $10^3 < 10^{3+(\text{decimal})} < 10^4$ ← $3 < y < 4$

The decimal .8007 is found in a table. → $10^3 < 10^{3+.8007} < 10^4$ ← $10^y = 10^{3+.8007}$ (whole-number part, decimal part)

$6{,}320 = 10^y = 10^{3+.8007} = 10^{3.8007}$ → **Thus,** $6{,}320 = 10^{3.8007}$.

In Example 1 the *decimal* part (.8007) is found in a table like this.

Part of the table is shown.
The complete table is in the back of the book.
Decimal points are not shown in the table.

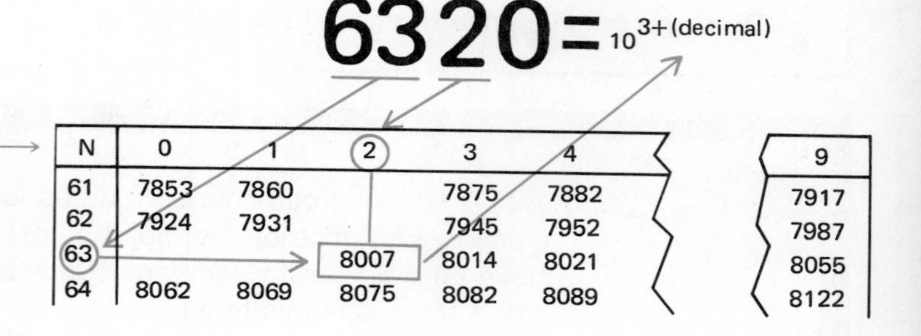

$6320 = 10^{3+(\text{decimal})}$

N	0	1	2	3	4	9
61	7853	7860		7875	7882	7917
62	7924	7931		7945	7952	7987
63			8007	8014	8021	8055
64	8062	8069	8075	8082	8089	8122

EXAMPLE 2 Show 6.32 and 632 as powers of 10.

Locate 6.32 between powers of 10. → $1 < 6.32 < 10$ | $100 < 632 < 1{,}000$

Change to exponential form. → $10^0 < 10^{0+\text{decimal}} < 10^1$ | $10^2 < 10^{2+\text{decimal}} < 10^3$

Use the table to find the decimal part in 63 row, under 2. → $10^0 < 10^{0+.8007} < 10^1$ | $10^2 < 10^{2+.8007} < 10^3$

$0.8007 = .8007$ → **Thus,** $6.32 = 10^{0.8007}$, or $10^{.8007}$ and $632 = 10^{2.8007}$.

Sandwich the number between powers of 10. →

> To show a number as a power of 10:
> 1. Find the whole-number part of the exponent.
> 2. Use the table to find the decimal part.

EXAMPLE 3 Show 6.21, 62.1, and 621 as powers of 10.

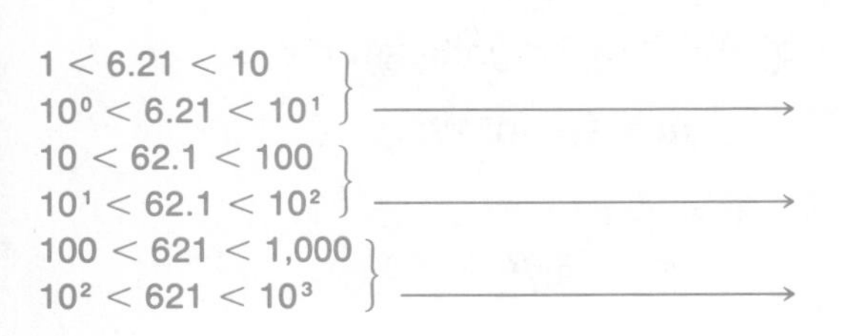

$1 < 6.21 < 10$, $10^0 < 6.21 < 10^1$ →

$10 < 62.1 < 100$, $10^1 < 62.1 < 10^2$ →

$100 < 621 < 1{,}000$, $10^2 < 621 < 10^3$ →

Number	*As a Power of 10*
1	10^0
6.21	$10^{0 + \text{decimal}}$
10	10^1
62.1	$10^{1 + \text{decimal}}$
100	10^2
621	$10^{2 + \text{decimal}}$
1,000	10^3

Look in the *62* row, under *1*. → From the table, 621 gives the decimal part .7931.

The decimal parts of the exponents are the same for all 3 numbers. The numbers have the same digits. →

Thus, $6.21 = 10^{0+.7931}$, or $10^{0.7931}$,
$62.1 = 10^{1+.7931}$, or $10^{1.7931}$,
$621 = 10^{2+.7931}$, or $10^{2.7931}$.

EXAMPLE 4 Find the value of $10^{3.8082}$

$$10^{3.8082} = 10^{3+.8082}$$

Locate .8082 in the table to find three digits in the value.

N	0	1	2	(3)	4	…	9
62	7924	7931	7938		7952		7987
63	7993	8000	8007		8021		8055
(64)				8082	8089		8122

.8082 is in the *64* row, under *3*. → The three digits are 643.

Sandwich the number between powers of 10. →

$$10^3 < 10^{3+.8082} < 10^4$$
$$1{,}000 < 10^{3+.8082} < 10{,}000$$
$$1{,}000 < 6{,}430 < 10{,}000$$

The answer is correct with 3-digit accuracy since the table gives only 3 digits. → **Thus,** $10^{3.8082} = 6{,}430$, accurate to 3 significant digits.

EXAMPLE 5 Find the value of $10^{1.7952}$ to 3 significant digits.

Look for .7952 in the table. The digits are 624.
Place the decimal point. →

$$10^1 < 10^{1+.7952} < 10^2$$
$$10 < 10^{1+.7952} < 100$$
$$10 < 62.4 < 100$$

The table gives 3 digits. → **Thus,** $10^{1.7952} = 62.4$ to 3 significant digits.

> To find the value of a power of 10 to 3 significant digits:
> 1. Use the decimal part of the exponent in the table to find the digits.
> 2. Use the whole-number part of the exponent to place the decimal point.

EXAMPLE 6 Find the value of $10^{.8021}$ to 3 significant digits.

0 is the whole-number part of the exponent. → $10^{.8021} = 10^{0.8021}$

The digits are 634; place the decimal point. → $10^0 < 10^{0+.8021} < 10^1$; $1 < 6.34 < 10$

To 3 significant digits → **Thus,** $10^{.8021} = 6.34$.

EXERCISES

PART A

Show the number as a power of 10.

1. 100 *10^2* **2.** 10,000 *10^4* **3.** 10 *10^1* **4.** 1 *10^0*
5. 54.6 *$10^{1.7372}$* **6.** 5,460 *$10^{3.7372}$* **7.** 5.46 *$10^{0.7372}$* **8.** 546 *$10^{2.7372}$*
9. 213 *$10^{2.3284}$* **10.** 3.48 *$10^{0.5416}$* **11.** 46.1 *$10^{1.6637}$* **12.** 1,750 *$10^{3.2430}$*
13. 30.7 *$10^{1.4871}$* **14.** 702 *$10^{2.8463}$* **15.** 5.06 *$10^{0.7042}$* **16.** 4,080 *$10^{3.6107}$*
17. 204 *$10^{2.3096}$* **18.** 6.03 *$10^{0.7803}$* **19.** 80.1 *$10^{1.9036}$* **20.** 10,900 *$10^{4.0374}$*
21. 6.52 *$10^{0.8142}$* **22.** 739 *$10^{2.8686}$* **23.** 82,400 *$10^{4.9159}$* **24.** 98.7 *$10^{1.9943}$*

Find the value to 3 significant digits.

25. $10^{0.9201}$ *8.32* **26.** $10^{1.9201}$ *83.2* **27.** $10^{2.9201}$ *832* **28.** $10^{3.9201}$ *8,320*
29. $10^{1.4425}$ *27.7* **30.** $10^{0.8727}$ *7.46* **31.** $10^{2.5705}$ *372* **32.** $10^{3.8331}$ *6,810*
33. $10^{2.6425}$ *439* **34.** $10^{4.9652}$ *92,300* **35.** $10^{0.2253}$ *1.68* **36.** $10^{1.7356}$ *54.4*
37. $10^{3.9112}$ *8,150* **38.** $10^{1.5119}$ *32.5* **39.** $10^{2.8882}$ *773* **40.** $10^{.6637}$ *4.61*
41. $10^{.9894}$ *9.76* **42.** $10^{4.3692}$ *23,400* **43.** $10^{1.0864}$ *12.2* **44.** $10^{2.7612}$ *577*

PART B

Show the number as a power of 10.

45. 43 *$10^{1.6335}$* **46.** 7.6 *$10^{0.8808}$* **47.** 580 *$10^{2.7634}$* **48.** 2,100 *$10^{3.3222}$*
49. 6.2 *$10^{0.7924}$* **50.** 19 *$10^{1.2788}$* **51.** 8,400 *$10^{3.9243}$* **52.** 350 *$10^{2.5441}$*
53. 5 *$10^{0.6990}$* **54.** 30 *$10^{1.4771}$* **55.** 2,000 *$10^{3.3010}$* **56.** 900 *$10^{2.9542}$*

Find the value to 3 significant digits.

57. $10^{2.3802}$ *240* **58.** $10^{3.8751}$ *7,500* **59.** $10^{2.7482}$ *560* **60.** $10^{2.6021}$ *400*
61. $10^{3.2304}$ *1,700* **62.** $10^{2.5798}$ *380* **63.** $10^{4.9191}$ *83,000* **64.** $10^{3.7782}$ *6,000*

Multiplying and Dividing Powers of 10

OBJECTIVE

- To multiply and divide numbers by using powers of 10

REVIEW CAPSULE

$10^5 \cdot 10^3 = 10^8 \longrightarrow x^m \cdot x^n = x^{m+n}$

$\frac{10^7}{10^2} = 10^5 \longrightarrow \frac{x^m}{x^n} = x^{m-n}$

EXAMPLE 1 Multiply 54.6×113 to 3 significant digits. Use powers of 10.

54.6×113

Show the numbers as powers of 10. → $10^{1.7372} \times 10^{2.0531}$

$10^m \cdot 10^n = 10^{m+n}$; Add exponents. → $10^{(1.7372+2.0531)}$

$10^{3.7903}$

Find the value of $10^{3.7903}$. → 6,170 to 3 significant digits

To 3 significant digits → **Thus,** $54.6 \times 113 = 6{,}170$.

EXAMPLE 2 Multiply $2.68 \times 5{,}720$ to 3 significant digits.

$2.68 \times 5{,}720$

Show the numbers as powers of 10. → $10^{0.4281} \times 10^{3.7574}$

Add exponents. → $10^{4.1855}$

$10^{4+(.1855)}$ ← .1855 is NOT in the table.

$10^{4+(.1847)}$ ← .1847 is in the table.

15,300

.1855 is *between* .1847 and .1875 in the table. Use the *nearest* decimal part (.1847) in the table. Find the value of $10^{4.1847}$.

Thus, $2.68 \times 5{,}720 = 15{,}300$ to 3 significant digits.

EXAMPLE 3 Divide $\frac{8{,}230}{34.6}$ to 3 significant digits. Use powers of 10.

Show the numbers as powers of 10. → $\frac{8{,}230}{34.6} = \frac{10^{3.9154}}{10^{1.5391}}$

$\frac{10^m}{10^n} = 10^{m-n}$; Subtract exponents. → $= 10^{(3.9154-1.5391)}$

$= 10^{2.3763}$

.3766 is nearest decimal part in the table. → $\doteq 10^{2.3766}$

Find the value of $10^{2.3766}$. → $\doteq 238$

Thus, $8{,}230 \div 34.6 = 238$ to 3 significant digits.

ORAL EXERCISES

Find each product or quotient.

1. $10^{1.5} \times 10^{0.5}$ *100*
2. $10^{2.8} \times 10^{0.2}$ *1,000*
3. $\frac{10^{2.5}}{10^{0.5}}$ *100*
4. $\frac{10^{4.6}}{10^{1.6}}$ *1,000*

EXERCISES

PART A

Multiply to 3 significant digits. Use powers of 10.

1. 23.7×3.16 *74.9*
2. 42.8×19.3 *826*
3. 2.15×446 *959*
4. 52.4×693 *36,300*
5. 7.77×3.04 *23.6*
6. $8.24 \times 9{,}320$ *76,800*
7. $70.8 \times 5{,}430$ *384,000*
8. $2.56 \times 18{,}600$ *47,600*
9. $58.2 \times 30{,}600$ *1,780,000*
10. 5.7×690 *3,930*
11. $79 \times 3{,}800$ *300,000*
12. $460 \times 9{,}600$ *4,420,000*

Divide to 3 significant digits. Use powers of 10.

13. $\frac{56.4}{3.27}$ *17.3*
14. $\frac{632}{4.08}$ *155*
15. $\frac{782}{53.6}$ *14.6*
16. $\frac{4{,}070}{29.4}$ *138*
17. $\frac{34.6}{5.83}$ *5.93*
18. $\frac{434}{62.1}$ *6.99*
19. $\frac{5{,}020}{73.2}$ *68.6*
20. $\frac{18{,}400}{809}$ *22.7*

PART B

EXAMPLE Multiply $42.6 \times 2.8 \times 519$ to 3 significant digits.

Three factors →	$42.6 \times 2.8 \times 519$
Show numbers as powers of 10. →	$10^{1.6294} \times 10^{0.4472} \times 10^{2.7152}$
Add exponents. $10^m \cdot 10^n \cdot 10^p = 10^{m+n+p}$ →	$10^{4.7918}$
.7917 is nearest decimal in table. →	$10^{4.7917}$
Find the value of $10^{4.7917}$. →	61,900

Thus, $42.6 \times 2.8 \times 519 = 61{,}900$ to 3 significant digits.

Multiply to 3 significant digits. Use powers of 10.

21. $38.4 \times 2.51 \times 603$ *58,100*
22. $2.76 \times 86.3 \times 5.04$ *1,200*
23. $39.5 \times 4.17 \times 63.2$ *10,400*
24. $47 \times 6.92 \times 5{,}800$ *1,890,000*
25. $53 \times 33.2 \times 7.9$ *13,900*
26. $4.3 \times 18.4 \times 70.2$ *5,550*

PART C

Divide to 3 significant digits. Use powers of 10.

27. $\frac{53.4 \times 296}{87.1}$
28. $\frac{6{,}120}{72.4 \times 3.8}$
29. $\frac{4.35 \times 3{,}980}{564}$
30. $\frac{7{,}840}{36.3 \times 57}$
31. $\frac{39 \times 72}{4.3 \times 68}$
32. $\frac{530 \times 26}{34 \times 77}$

Find the power to 3 significant digits. Use powers of 10. [Hint: $x^3 = x \cdot x \cdot x$]

33. $(28.7)^2$
34. $(5.64)^3$
35. $(173)^4$
36. $(6.8)^5$

Logarithms

OBJECTIVES

- To find the logarithm of a number
- To find a number when given its logarithm

Show 68.4 as a power of 10.

$10^1 < 68.4 < 10^2$ ← Sandwich number between powers of 10.

$68.4 = 10^{1 + (\text{decimal})}$

Decimal part is .8351 ← Found in table in *68* row, under *4*

Thus, $68.4 = 10^{1.8351}$.

EXAMPLE 1 Show 1,000 and 75.3 as powers of 10.

1,000	75.3
↓	↓
10^3	$10^{1.8768}$

Thus, $1{,}000 = 10^3$ and $75.3 = 10^{1.8768}$.

Read: the log of 1,000 is 3. → We write $\log 1{,}000 = 3$ $\log 75.3 = 1.8768$.

log is the abbreviation of *logarithm*.

$\log x = y$ means $x = 10^y$ — The log of x is the *exponent* when x is shown as a power of 10.

EXAMPLE 2 Find log 637.

Show 637 as a power of 10. → $637 = 10^{2.8041}$

$\log x = y$ if $x = 10^y$. → **Thus,** $\log 637 = 2.8041$.

EXAMPLE 3 Find log 45.8. | Find log 4,580.

Show the number as a power of 10. $45.8 = 10^{1.6609}$ | $4{,}580 = 10^{3.6609}$

$\log x = y$, if $x = 10^y$ **Thus,** $\log 45.8 = 1.6609$, or $1 + .6609$
and $\log 4{,}580 = 3.6609$, or $3 + .6609$.

A logarithm has an integer part and a decimal part. → Integer part — Decimal part

Each part has a special name. → *Characteristic* — *Mantissa*

Definition of characteristic and mantissa

The *characteristic* of a log is the integer part.	The *mantissa* of a log is the decimal part.

EXAMPLE 4 Find the characteristic and mantissa of the logarithm.

$7.23 = 10^{0.8591}$

$0.8591 = 0 + .8591$

Characteristic is integer part.

Mantissa is decimal part.

log 7.23	log 531
$\log 7.23 = 0.8591$	$\log 531 = 2.7251$
Characteristic = 0	Characteristic = 2
Mantissa = .8591	Mantissa = .7251

EXAMPLE 5 Find x if $\log x = 4.5391$.

$\log x = y$

if $x = 10^y$

Mantissa (.5391), in *34* row under *6*, gives digits 346.

Characteristic is 4: $10^4 < x < 10^5$.

$$\log x = 4.5391$$
$$x = 10^{4.5391}, \text{ or } 10^{4+.5391}$$
$$x = 34{,}600$$

Thus, $x = 34{,}600$ if $\log x = 4.5391$.

34,600 is the *antilog* of 4.5391.

EXAMPLE 6 Find the antilog of 1.7875.

The antilog of 1.7875 is the number x whose log is 1.7875.

Mantissa (.7875) gives digits 613.

Characteristic is 1: $10^1 < x < 10^2$.

$1.7875 = \log 61.3$

$$\log x = 1.7875$$
$$x = 10^{1.7875}$$
$$x = 61.3$$

Thus, the antilog of 1.7875 is 61.3.

EXAMPLE 7 Find the antilog.

The antilog of 0.5832 is the number x whose log is 0.5832.

Mantissa (.5832) gives digits 383.

Characteristic is 0: $10^0 < x < 10^1$.

0.5832	3.6702
$\log x = 0.5832$	$\log x = 3.6702$
$x = 10^{0.5832}$	$x = 10^{3.6702}$
$x = 3.83$	$x = 4{,}680$

Thus, the antilog of 0.5832 is 3.83,
and the antilog of 3.6702 is 4,680.

SUMMARY

The mantissa of log *x* gives the digits of *x*.
The characteristic of log *x* places the decimal point in *x*.

ORAL EXERCISES

Give the logarithm.

1. log 100 *2*
2. log 10 *1*
3. log 10,000 *4*
4. log 1 *0*
5. log $10^{2.7661}$ *2.7661*

Find the antilog.

6. 2 *100*
7. 3 *1,000*
8. 1 *10*
9. 0 *1*
10. 4 *10,000*

Give the characteristic of the logarithm.

11. log 716 *2*
12. log 3,860 *3*
13. log 54.3 *1*
14. log 2.94 *0*
15. log 67 *1*
16. log 430 *2*
17. log 4.7 *0*
18. log 23,000 *4*

Identify the characteristic and the mantissa of the logarithm.

19. log 49.2 = 1.6920
20. log 2,840 = 3.4533
21. log 7.48 = 0.8739

EXERCISES

PART A

Find the logarithm.

1. log 3.52 *0.5465*
2. log 35.2 *1.5465*
3. log 3,520 *3.5465*
4. log 463 *2.6656*
5. log 2.79 *0.4456*
6. log 17.8 *1.2504*
7. log 8,030 *3.9047*
8. log 50.6 *1.7042*
9. log 6.04 *0.7810*

Find the antilog.

10. 1.3711 *23.5*
11. 0.3711 *2.35*
12. 2.3711 *235*
13. 0.7619 *5.78*
14. 2.5132 *326*
15. 3.9899 *9,770*
16. 2.0899 *123*
17. 1.9206 *83.3*
18. 4.2227 *16,700*
19. 1.8887 *77.4*
20. 0.7987 *6.29*
21. 3.6646 *4,620*

Find the characteristic and the mantissa of the logarithm.

22. log 472
23. log 2.16
24. log 70.4
25. log 9.27
26. log 6,030
27. log 32,600

PART B

Find the logarithm.

28. log 53 *1.7243*
29. log 1.9 *0.2788*
30. log 720 *2.8573*
31. log 9.4 *0.9731*
32. log 38 *1.5798*
33. log 6,400 *3.8062*
34. log 26,400 *4.4216*
35. log 80,700 *4.9069*
36. log 433,000 *5.6365*

Find the antilog.

37. 1.5315 *34*
38. 0.2553 *1.8*
39. 2.8633 *730*
40. 0.7559 *5.7*
41. 1.9345 *86*
42. 2.4771 *300*
43. 2.6335 *430*
44. 3.8129 *6,500*
45. 3.8451 *7,000*
46. 4.7093 *51,200*
47. 4.3483 *22,300*
48. 5.9638 *920,000*

Logarithm of a Product

OBJECTIVE

- To multiply numbers by using logarithms

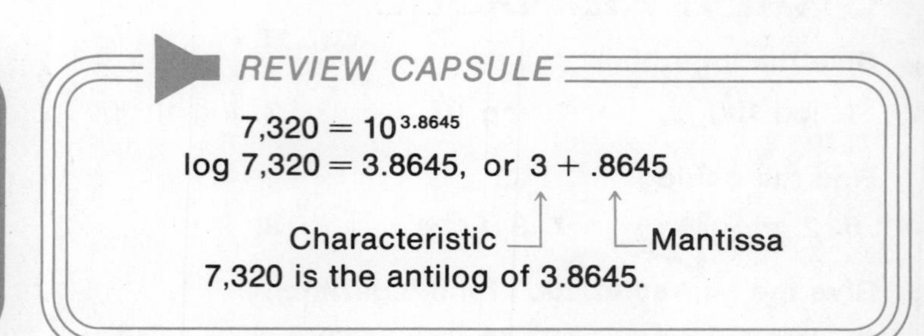

EXAMPLE 1 Find $\log(100 \times 1{,}000)$. Find $\log 100 + \log 1{,}000$.

$\log(100 \times 1{,}000)$	$\log 100 + \log 1{,}000$
$\log(100{,}000)$	$\log 10^2 + \log 10^3$
$\log(10^5)$, or 5	$2 + 3$, or 5

Results are the same: 5. →

Thus, $\log(100 \times 1{,}000) = \log 100 + \log 1{,}000$.

Log of a product →

> $\log(a \cdot b) = \log a + \log b$, for all positive numbers a and b.

EXAMPLE 2 Multiply $3.42 \times 5{,}170$ to 3 significant digits. Use logarithms.

$\log(a \cdot b) = \log a + \log b$ → $\log(3.42 \times 5{,}170) = \log 3.42 + \log 5{,}170$

Let $x = 3.42 \times 5{,}170$. Find log 3.42 and log 5,170. → $\log x = 0.5340 + 3.7135$

Add: $0.5340 + 3.7135 = 4.2475$. $\log x = 4.2475$ ← .2475 is NOT in the table.

.2475 is *between* .2455 and .2480 in the table. Use the nearest mantissa (.2480). $\log x \doteq 4.2480$ ← .2480 is in the table.

$x = 10^y$ if $\log x = y$. → So, $x \doteq 10^{4.2480}$.

Find the antilog of 4.2480. → $x \doteq 17{,}700$

Thus, $3.42 \times 5{,}170 = 17{,}700$ to 3 significant digits.

EXAMPLE 3 Multiply $37 \times 845 \times 2.09$ to 3 significant digits.

$\log(a \cdot b \cdot c) = \log a + \log b + \log c$ → $\log(37 \times 845 \times 2.09) = \log 37 + \log 845 + \log 2.09$

Let $x = 37 \times 845 \times 2.09$ → $\log x = 1.5682 + 2.9269 + 0.3201$

$= 4.8152$

.8149 is nearest mantissa to .8152 in the table. → $\log x \doteq 4.8149$

Antilog of 4.8149 is 65,300. → So, $x \doteq 10^{4.8149}$, or 65,300.

Thus, $37 \times 845 \times 2.09 = 65{,}300$ to 3 significant digits.

ORAL EXERCISES

True or false?

1. $\log(5 \times 7) = \log 5 + \log 7$ *T*
2. $\log 12 = \log 4 + \log 3$ *T*
3. $\log 8 = \log 3 + \log 5$ *F*
4. $\log 6 = \log 2 + \log 3$ *T*
5. $\log 30 = \log 2 + \log 3 + \log 5$ *T*
6. $\log 15 = \log 5 + \log 5 + \log 5$ *F*

EXERCISES

PART A

Multiply to 3 significant digits. Use logarithms.

1. 27.6×31.3 *864*
2. 1.94×5.03 *9.76*
3. 60.7×428 *26,000*
4. 5.37×806 *4,330*
5. $3.9 \times 53,800$ *210,000*
6. $37.6 \times 5,400$ *203,000*
7. $6.23 \times 53.7 \times 420$ *141,000*
8. $32.6 \times 4.69 \times 57$ *8,720*
9. $304 \times 21.6 \times 5.4$ *35,500*
10. $6,740 \times 8.3 \times 49.2$ *2,750,000*
11. $1.86 \times 40.2 \times 76.3$ *5,700*
12. $503 \times 9.1 \times 54.6$ *250,000*

PART B

EXAMPLE The volume of a rectangular solid is given by a formula, $V = lwh$. Show $\log V$ as a *sum* of logarithms.

Formula → $V = lwh$

Take the log of each side. → $\log V = \log\ (l \cdot w \cdot h)$

$\log(a \cdot b \cdot c) = \log a + \log b + \log c$ → $= \log l + \log w + \log h$

$\log V$ shown as a *sum* of logs → **Thus,** $\log V = \log l + \log w + \log h$.

Given a formula, show the indicated log as a sum of logarithms.

13. $A = lw$; $\log A$
14. $V = Bh$; $\log V$
15. $i = prt$; $\log i$
16. $L = \pi rs$; $\log L$
17. $C = 2\pi r$; $\log C$ ↘ *$\log C = \log 2 + \log \pi + \log r$*
18. $L = 2\pi rh$; $\log L$ ↘ *$\log L = \log 2 + \log \pi + \log r + \log h$*

Multiply to 3 significant digits. Use logarithms.

19. $4.27 \times 38 \times 8.6 \times 57.4$ *80,100*
20. $3.12 \times 74.6 \times 208 \times 17.6$ *852,000*
21. $87 \times 56 \times 29 \times 32$ *4,520,000*
22. $2.7 \times 3.4 \times 4.6 \times 5.3$ *224*

PART C

Prove that each of the following is true.

23. $\log\ (x^2 - y^2) = \log\ (x + y) + \log\ (x - y)$
24. $\log\ (x^2 + 2xy + y^2) = 2 \cdot \log\ (x + y)$

Scientific Notation

OBJECTIVES

- To change numbers from ordinary to scientific notation and vice-versa
- To find logs and antilogs by using scientific notation

REVIEW CAPSULE

$$10^{-2} = \frac{1}{10^2} \rightarrow x^{-n} = \frac{1}{x^n}$$

$$\frac{1}{10^3} = 10^{-3} \rightarrow \frac{1}{x^n} = x^{-n}$$

$6.43 \times 1{,}000 = 6{,}430.$
$6.43 \div 100 = .0643$
} Multiplying or dividing by a power of 10 moves the decimal point.

EXAMPLE 1 Show the number as 6.29 times a power of 10.

$6{,}290 = 6.29 \times 1{,}000 = 6.29 \times 10^3$
$62.9 = 6.29 \times 10 = 6.29 \times 10^1$
$6.29 = 6.29 \times 1 = 6.29 \times 10^0$

Number	*Factorization*
6,290	6.29×10^3
62.9	6.29×10^1
6.29	6.29×10^0

Number between 1 and 10 (6.29) — Integer power of 10 (exponent)

In Example 1 the numbers are shown in *scientific notation.*

Scientific notation
A number *from* 1 to 10 times an *integer* power of 10.

$a \times 10^c$ is shown in *scientific notation*
if $1 \leq a < 10$ and c is an integer.

EXAMPLE 2 Show .00629 and .629 as 6.29 times a power of 10.

$a \div b = a \cdot \frac{1}{b}$

$\frac{1}{x^n} = x^{-n}$: $\frac{1}{10^3} = 10^{-3}$

$$.00629 = 6.29 \div 1{,}000 = 6.29 \times \frac{1}{1{,}000} = 6.29 \times \frac{1}{10^3} = 6.29 \times 10^{-3}$$

$$.629 = 6.29 \div 10 = 6.29 \times \frac{1}{10} = 6.29 \times \frac{1}{10^1} = 6.29 \times 10^{-1}$$

Thus, $.00629 = 6.29 \times 10^{-3}$ in scientific notation,
and $.629 = 6.29 \times 10^{-1}$ in scientific notation.

Integer power of 10

-3 and -1 are integers.

Examples 1 and 2 suggest a short method for showing numbers in scientific notation.

EXAMPLE 3 Show the number in scientific notation.

	73,200.	.00458	5.16
Move decimal point to get a number from 1 to 10: 7.32. →	7‸3 2 0 0	0 0 4‸5 8	5‸1 6
Count the moves that return the decimal point to original place. →	7‸3 2 0 0.	.0 0 4‸5 8	5‸1 6
Direction and number of moves that return the decimal point give the exponent on 10.	Move 4 places *right*.	Move 3 places *left*.	Move 0 places.
Scientific notation →	7.32×10^4	4.58×10^{-3}	5.16×10^0

EXAMPLE 4 Change 6.38×10^{-2} and 3.52×10^5 to ordinary notation.

$x^{-n} = \frac{1}{x^n}$: $10^{-2} = \frac{1}{10^2}$ →

$$6.38 \times 10^{-2} = 6.38 \times \frac{1}{10^2}$$

$a \div b = a \cdot \frac{1}{b}$ →

$$= 6.38 \times \frac{1}{100}$$
$$= 6.38 \div 100$$

Ordinary notation →

$$= .0638$$

$$3.52 \times 10^5 = 3.52 \times 100{,}000 = 352{,}000$$

Thus, $6.38 \times 10^{-2} = .0638$ and $3.52 \times 10^5 = 352{,}000$.

Example 4 suggests a short method for changing from scientific to ordinary notation.

Scientific notation →	6.38×10^{-2}	3.52×10^5
Exponent on 10 tells the direction and number of places to move decimal point.	Move 2 places *left*. .0 6‸3 8	Move 5 places *right*. 3‸5 2 0 0 0.
Ordinary notation →	.0 6 3 8	3 5 2,0 0 0

EXAMPLE 5 Use scientific notation to find log 6,320.

Show 6,320 in scientific notation. →
$$6{,}320 = 6.32 \times 10^3$$

Take log of each side. →
$$\log 6{,}320 = \log(6.32 \times 10^3)$$

$\log(a \cdot b) = \log a + \log b$ →
$$\log 6{,}320 = \log 6.32 + \log 10^3$$

.8007 is in the *63* row, under *2*.
$$\log 6{,}320 = .8007 + 3$$

Mantissa (.8007) Characteristic (3)

$.8007 + 3 = 3 + .8007$ → **Thus,** $\log 6{,}320 = 3.8007$.

$1 \le a < 10$, c is an integer.

If x is shown in scientific notation
$$x = a \times 10^c,$$
then $\log x = (\log a) + c$.
Mantissa ↑ (log a) — Characteristic ↑ (c)

EXAMPLE 6 Find log 268 and log 2.68. Use scientific notation.

Show 268 in scientific notation.
Characteristic 2 is exponent on 10.
Mantissa = log 2.68 = .4281.

$268 = 2.68 \times 10^2$	$2.68 = 2.68 \times 10^0$
$\log 268 = (\log 2.68) + 2$	$\log 2.68 = (\log 2.68) + 0$
$= .4281 + 2$	$= .4281 + 0$
Thus, log 268 = 2.4281.	**Thus,** log 2.68 = 0.4281.

EXAMPLE 7 Use scientific notation to find the antilog of 3.8998.

Find x if $\log x = 3.8998$. → $\log x = 3 + .8998$

$\log 10^3 = 3$; $\log 7.94 = .8998$; 7.94 is antilog of .8998. → $\log x = \log 10^3 + \log 7.94$

$\log a + \log b = \log (a \cdot b)$ → $\log x = \log(10^3 \times 7.94)$

$10^3 \times 7.94 = 7.94 \times 10^3 = 7940$ → So, $x = 10^3 \times 7.94$, or 7,940

Thus, the antilog of 3.8998 is 7,940.

c is an integer. $1 \le a < 10$

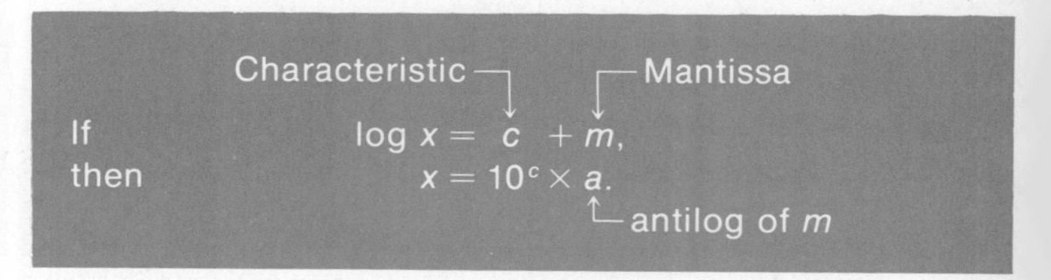

EXAMPLE 8 Find the antilog. Use scientific notation.

2.7505	0.7505
$\log x = 2 + .7505$	$\log x = 0 + .7505$
$x = 10^2 \times 5.63$	$x = 10^0 \times 5.63$
$= 563$	$= 5.63$

Find x if $\log x = 2.7505$.
5.63 is antilog of .7505.
Multiply by 10^2.

Thus, the antilog of 2.7505 is 563
and the antilog of 0.7505 is 5.63.

ORAL EXERCISES

Give the number in scientific notation.

1. 2,640 *2.64×10^3* **2.** 2.64 *2.64×10^0* **3.** .264 *2.64×10^{-1}* **4.** .00264 *2.64×10^{-3}*

Give the number in ordinary notation.

5. 4.73×10^4 *47,300* **6.** 4.73×10^0 *4.73* **7.** 4.73×10^{-2} *.0473* **8.** 4.73×10^{-1} *.473*

EXERCISES

PART A

Show the number in scientific notation.

1. 345 *3.45×10^2* **2.** 72.6 *7.26×10^1* **3.** 27,900 *2.79×10^4* **4.** 4.62 *4.62×10^0*
5. .0216 *2.16×10^{-2}* **6.** .513 *5.13×10^{-1}* **7.** .0378 *3.78×10^{-2}* **8.** .00735 *7.35×10^{-3}*
9. 23 *2.3×10^1* **10.** .0057 *5.7×10^{-3}* **11.** 7,600 *7.6×10^3* **12.** .48 *4.8×10^{-1}*

Write the number in ordinary notation.

13. 8.17×10^2 *817* **14.** 3.22×10^1 *32.2* **15.** 5.63×10^3 *5,630* **16.** 6.75×10^4 *67,500*
17. 6.24×10^{-1} *.624* **18.** 7.89×10^{-3} *.00789* **19.** 3.52×10^{-2} *.0352* **20.** 1.53×10^0 *1.53*
21. 2.9×10^4 *29,000* **22.** 3.6×10^{-1} *.36* **23.** 6.3×10^2 *630* **24.** 7.8×10^{-2} *.078*

Find the logarithm. Use scientific notation.

25. log 843 *2.9258* **26.** log 73.6 *1.8669* **27.** log 2,370 *3.3747* **28.** log 3.12 *0.4942*
29. log 6.3 *0.7993* **30.** log 68,400 *4.8351* **31.** log 52.7 *1.7218* **32.** log 4,300 *3.6335*

Find the antilog. Use scientific notation.

33. 2.7649 *582* **34.** 3.2742 *1,880* **35.** 1.8109 *64.7* **36.** 0.4346 *2.72*
37. 0.9289 *8.49* **38.** 1.8591 *72.3* **39.** 3.7160 *5,200* **40.** 4.6493 *44,600*

PART B

Find the logarithm. Use scientific notation.

41. log 5,240,000 *6.7193* **42.** log 372,000 *5.5705* **43.** log 8,600,000 *6.9345* **44.** log 28,000,000 *7.4472*

Find the antilog. Use scientific notation.

45. 5.6875 *487,000* **46.** 6.7993 *6,300,000* **47.** 5.7959 *625,000* **48.** 7.6532 *45,000,000*

PART C

Show the number in scientific notation.

49. 150,000,000 km; the distance between the earth and the sun

50. .000,000,000,2 cm; the wave length of gamma rays

51. 29,980,000,000 cm per second; the velocity of light

52. .000,000,000,000,000,000,02 erg; the photon energy of radar waves

Logarithm of a Quotient

OBJECTIVE

- To divide numbers by using logarithms

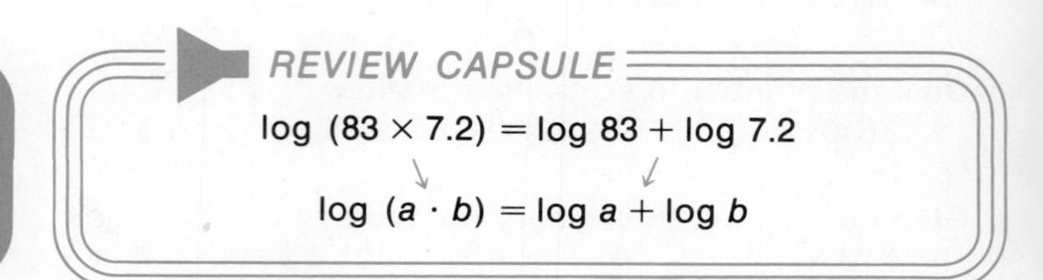

EXAMPLE 1 Find $\log \frac{100{,}000}{1{,}000}$. Find $\log 100{,}000 - \log 1{,}000$.

$\log \frac{100{,}000}{1{,}000}$ | $\log 100{,}000 - \log 1{,}000$

100,000 ÷ 1,000 = 100 — $\log 100$ | $\log 10^5 - \log 10^3$

Results are the same: 2. → $\log 10^2$, or 2 | $5 - 3$, or 2

Thus, $\log \frac{100{,}000}{1{,}000} = \log 100{,}000 - \log 1{,}000$.

Log of a quotient →

> $\log \frac{a}{b} = \log a - \log b$, for all positive numbers a and b.

EXAMPLE 2 Divide $\frac{3{,}420}{51.7}$ to 3 significant digits. Use logarithms.

$\log \frac{a}{b} = \log a - \log b$ → $\log \frac{3{,}420}{51.7} = \log 3{,}420 - \log 51.7$

Let $x = \frac{3{,}420}{51.7}$. → $\log x = 3.5340 - 1.7135$

$\log x = 1.8205$

.8202 is the nearest mantissa in the table. — $\log x \doteq 1.8202$

Find the antilog of 1.8202 → So, $x \doteq 10^1 \times 6.61$, or 66.1.

Thus, $\frac{3{,}420}{51.7} = 66.1$ to 3 significant digits.

EXAMPLE 3 Divide $\frac{3{,}700 \times 209}{84.5}$ to 3 significant digits.

$\log \frac{a \cdot b}{c} = \log (a \cdot b) - \log c$ → $\log \frac{3{,}700 \times 209}{84.5} = [\log (3{,}700 \times 209)] - \log 84.5$

Let $x = \frac{3{,}700 \times 209}{84.5}$. $\log (3{,}700 \times 209) = \log 3{,}700 + \log 209$

$$\log x = [\log 3{,}700 + \log 209] - \log 84.5$$
$$= [3.5682 + 2.3201] - 1.9269$$
$$= 3.9614$$

Find the antilog of 3.9614. → So, $x = 10^3 \times 9.15$, or 9,150.

To 3 significant digits → **Thus,** $\frac{3{,}700 \times 209}{84.5} = 9{,}150$.

EXAMPLE 4 Divide $\frac{48,700}{6.2 \times 51.4}$ to 3 significant digits.

$\log \frac{a}{b \cdot c} = \log a - \log (b \cdot c)$ ⟶ $\log \frac{48,700}{6.2 \times 51.4} = \log 48,700 - [\log (6.2 \times 51.4)]$

Let $x = \frac{48,700}{6.2 \times 51.4}$

$\log (6.2 \times 51.4) = \log 6.2 + \log 51.4$

$$\log x = \log 48,700 - [\log 6.2 + \log 51.4]$$

Add: 0.7924 + 1.7110 = 2.5034. ⟶

$$\begin{aligned} &= 4.6875 - [0.7924 + 1.7110] \\ &= 4.6875 - 2.5034 \\ &= 2.1841 \end{aligned}$$

.1847 is the nearest mantissa in the table.

$$\log x \doteq 2.1847$$

Find the antilog of 2.1847. ⟶ So, $x \doteq 10^2 \times 1.53$, or 153.

To 3 significant digits ⟶ **Thus,** $\frac{48,700}{6.2 \times 51.4} = 153$.

ORAL EXERCISES

True or false?

1. $\log \frac{14}{3} = \log 14 - \log 3$ *T*
2. $\log \frac{12}{5} = \log 12 \div \log 5$ *F*
3. $\log \frac{22 \times 3}{5} = \log 22 + \log 3 - \log 5$ *T*
4. $\log \frac{3 \times 5}{7} = \log 3 - \log 5 - \log 7$ *F*
5. $\log \frac{25}{3 \times 7} = \log 25 - (\log 3 + \log 7)$ *T*
6. $\log \frac{17}{2 \times 3} = \log 17 - \log 2 + \log 3$ *F*

EXERCISES

PART A

Divide to 3 significant digits. Use logarithms.

1. $\frac{51.7}{3.62}$ *14.3*
2. $\frac{643}{4.05}$ *159*
3. $\frac{776}{51.3}$ *15.1*
4. $\frac{3,080}{17.6}$ *175*
5. $\frac{37.5}{5.43}$ *6.91*
6. $\frac{427}{61.2}$ *6.98*
7. $\frac{4,030}{82.3}$ *49.0*
8. $\frac{28,400}{706}$ *40.2*

PART B

Divide to 3 significant digits. Use logarithms.

9. $\frac{52.6 \times 274}{83.2}$ *173*
10. $\frac{4.25 \times 3,750}{543}$ *29.4*
11. $\frac{34.6 \times 47.2}{7.18}$ *227*
12. $\frac{6,340}{71.6 \times 3.4}$ *26.0*
13. $\frac{72,500}{340 \times 47}$ *4.54*
14. $\frac{918}{2.3 \times 30.4}$ *13.1*

PART C

Divide to 3 significant digits.

15. $\frac{43 \times 78}{37 \times 59}$
16. $\frac{26.3 \times 704}{9.22 \times 8.14}$
17. $\frac{62 \times 3.4 \times 280}{53 \times 42}$

Logarithm of a Power

OBJECTIVE

- To find a power of a number by using logarithms

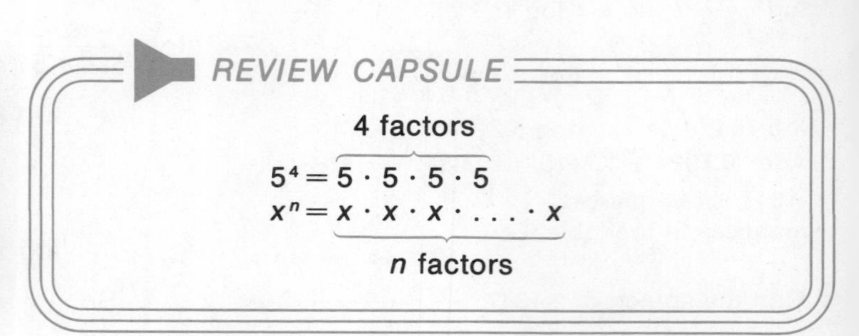

EXAMPLE 1 Find $\log 5^3$. Find $3 \cdot \log 5$.

$5^3 = 5 \cdot 5 \cdot 5$ ⟶ $\log 5^3 = \log (5 \times 5 \times 5)$

$= \log 5 + \log 5 + \log 5$

$\log 5 = 0.6990$ ⟶ $= .6990 + .6990 + .6990$

Results are the same: 2.0970. ⟶ $= 2.0970$

$3 \cdot \log 5$ ⟶ $3(.6990)$ ⟶ 2.0970

Thus, $\log 5^3 = 3 \cdot \log 5.$

Log of a power ⟶ $\log a^r = r \cdot \log a$

EXAMPLE 2 Find 31.7^4 to 3 significant digits. Use logarithms.

$\log a^r = r \cdot \log a$ ⟶ $\log 31.7^4 = 4 \cdot \log 31.7$

Let $x = 31.7^4$. ⟶ $\log x = 4(1.5011)$

.0043 is the nearest mantissa in the table. $\log x = 6.0044$

$\log x \doteq 6.0043$

Find the antilog of 6.0043. ⟶ So, $x \doteq 10^6 \times 1.01$, or 1,010,000.

To 3 significant digits ⟶ **Thus,** $31.7^4 = 1{,}010{,}000$.

EXAMPLE 3 Find $(28.5 \times 4.2)^2$ to 3 significant digits.

$\log a^r = r \cdot \log a$ ⟶ $\log (28.5 \times 4.2)^2 = 2 \cdot \log (28.5 \times 4.2)$

Let $x = (28.5 \times 4.2)^2$.
$\log (28.5 \times 4.2) = \log 28.5 + \log 4.2$ ⟶ $\log x = 2(\log 28.5 + \log 4.2)$

$= 2(1.4548 + 0.6232)$

$= 2(2.0780)$

$= 4.1560$

.1553 is the nearest mantissa in the table. $\log x \doteq 4.1553$

Find the antilog of 4.1553. ⟶ So, $x \doteq 10^4 \times 1.43$, or 14,300.

To 3 significant digits ⟶ **Thus,** $(28.5 \times 4.2)^2 = 14{,}300$.

EXAMPLE 4 Find $\left(\frac{576}{48.2}\right)^3$ to 3 significant digits.

$\log a^r = r \cdot \log a$ → $\log\left(\frac{576}{48.2}\right)^3 = 3 \cdot \log \frac{576}{48.2}$

Let $x = \left(\frac{576}{48.2}\right)^3$. $\log \frac{576}{48.2} = \log 576 - \log 48.2$ → $\log x = 3(\log 576 - \log 48.2)$

$= 3(2.7604 - 1.6830)$

$= 3(1.0774)$

$= 3.2322$

.2330 is the nearest mantissa in the table. $\log x \doteq 3.2330$

Find the antilog of 3.2330. → So, $x \doteq 10^3 \times 1.71$, or 1,710.

To 3 significant digits → **Thus,** $\left(\frac{576}{48.2}\right)^3 = 1{,}710$.

EXERCISES

PART A

Find the power to 3 significant digits. Use logarithms.

1. 56.3^2 *3,170*
2. 29.4^3 *25,400*
3. 4.26^4 *329*
4. 329^2 *108,000*
5. 7.05^3 *350*
6. 2.3^5 *64.3*
7. $(7.53 \times 56.8)^2$ *183,000*
8. $(30.6 \times 12.3)^2$ *142,000*
9. $(4.91 \times 6.3)^3$ *29,600*
10. $(32.4 \times 17.3)^2$ *314,000*
11. $(1.89 \times 4.4)^3$ *575*
12. $(3.7 \times 2.8)^4$ *11,500*
13. $\left(\frac{52.3}{3.46}\right)^2$ *228*
14. $\left(\frac{826}{6.7}\right)^2$ *15,200*
15. $\left(\frac{74.6}{5.32}\right)^3$ *2,760*
16. $\left(\frac{420}{61}\right)^3$ *326*
17. $\left(\frac{39}{7.6}\right)^4$ *694*
18. $\left(\frac{27.5}{8.4}\right)^4$ *115*

PART B

Find the power to 3 significant digits.

19. $(5.3 \times 27 \times 4.6)^2$ *434,000*
20. $(3.8 \times 62 \times 18)^2$ *18,000,000*
21. $(8.1 \times 3.6 \times 2.9)^3$ *605,000*
22. $(2.14 \times 3.56 \times 4.37)^3$ *36,900*
23. $\left(\frac{27 \times 41}{340}\right)^2$ *10.6*
24. $\left(\frac{410 \times 3.6}{57}\right)^3$ *17,400*
25. $\left(\frac{64}{3.7 \times 4.3}\right)^2$ *16.2*
26. $\left(\frac{732}{2.9 \times 74}\right)^3$ *39.7*

PART C

27. Show that $\log\left(\frac{ab}{c}\right)^r = r(\log a + \log b - \log c)$.
28. Show that $\log\left(\frac{a}{bc}\right)^r = r[\log a - (\log b + \log c)]$.

Logarithm of a Root

OBJECTIVE

- To find a root of a number like $\sqrt[3]{\frac{276}{4.51}}$ by using logarithms

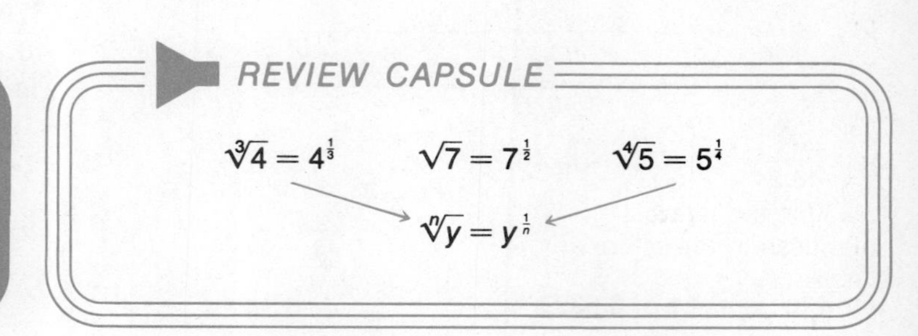

EXAMPLE 1 Find $\sqrt{7{,}850}$ to 3 significant digits. Use logarithms.

Write $\sqrt{7{,}850}$ as a power. $\sqrt{y} = y^{\frac{1}{2}}$ → $\sqrt{7{,}850} = 7{,}850^{\frac{1}{2}}$

$\log \sqrt{7{,}850} = \log (7{,}850^{\frac{1}{2}})$

Let $x = \sqrt{7{,}850}$.
$\log a^r = r \cdot \log a$; $\log a^{\frac{1}{2}} = \frac{1}{2} \cdot \log a$ → $\log x = \frac{1}{2} \cdot \log 7{,}850$

$\log 7{,}850 = 3.8949$ → $\log x = \frac{1}{2}(3.8949)$

$3.8949 \div 2 \doteq 1.9475$ → $\log x = 1.9475$

.9474 is the nearest mantissa in the table. → $\log x \doteq 1.9474$

Find the antilog of 1.9474. → So, $x \doteq 10^1 \times 8.86$, or 88.6.

Thus, $\sqrt{7{,}850} = 88.6$ to 3 significant digits.

EXAMPLE 2 Find $\sqrt[3]{27.4 \times 32}$ to 3 significant digits.

$\sqrt[3]{y} = y^{\frac{1}{3}}$ → $\log \sqrt[3]{27.4 \times 32} = \log (27.4 \times 32)^{\frac{1}{3}}$

$\log a^r = r \cdot \log a$; $\log a^{\frac{1}{3}} = \frac{1}{3} \cdot \log a$ → $\log x = \frac{1}{3} \cdot \log (27.4 \times 32)$

$\log (a \cdot b) = \log a + \log b$ → $= \frac{1}{3}(\log 27.4 + \log 32)$

$= \frac{1}{3}(1.4378 + 1.5051)$

$= \frac{1}{3}(2.9429)$

$2.9429 \div 3 \doteq .9810$ → $= 0.9810$

.9809 is nearest mantissa in table. → $\log x \doteq 0.9809$

Find the antilog of 0.9809; $10^0 = 1$. → So, $x \doteq 10^0 \times 9.57$, or 9.57.

Thus, $\sqrt[3]{27.4 \times 32} = 9.57$ to 3 significant digits.

EXAMPLE 3 Find $\sqrt[4]{\frac{67,000}{3.56}}$ to 3 significant digits.

$\sqrt[4]{y} = y^{\frac{1}{4}}$; $\log y^{\frac{1}{4}} = \frac{1}{4}(\log y)$ ⟶ $\log \sqrt[4]{\frac{67,300}{3.56}} = \log \left(\frac{67,300}{3.56}\right)^{\frac{1}{4}} = \frac{1}{4} \cdot \left(\log \frac{67,300}{3.56}\right)$

$\log \frac{a}{b} = \log a - \log b$ ⟶ $\log x = \frac{1}{4}(\log 67,300 - \log 3.56)$

$= \frac{1}{4}(4.8280 - 0.5514)$

$4.2766 \div 4 = 1.0692$ ⟶ $= \frac{1}{4}(4.2766)$, or 1.0692

Antilog of .0692 ≐ 1.17 ⟶ **Thus,** $\sqrt[4]{\frac{67,300}{3.56}} \doteq 10^1 \times 1.17$, or 11.7.

EXERCISES

PART A

Find each to 3 significant digits. Use logarithms.

1. $\sqrt{369}$ *19.2*
2. $\sqrt[3]{82.1}$ *4.35*
3. $\sqrt{5,120}$ *71.6*
4. $\sqrt[4]{19.6}$ *2.10*
5. $\sqrt{7.6 \times 48}$ *19.1*
6. $\sqrt[4]{56.2 \times 4.87}$ *4.07*
7. $\sqrt{38.2 \times 54.3}$ *45.5 or 45.6*
8. $\sqrt[3]{192 \times 37}$ *19.2*
9. $\sqrt{\frac{680}{4.7}}$ *12.0*
10. $\sqrt[3]{\frac{7,300}{3.4}}$ *12.9*
11. $\sqrt{\frac{51.6}{3.82}}$ *3.68*
12. $\sqrt[4]{\frac{94,200}{6.2}}$ *11.1*

PART B

EXAMPLE Find $\sqrt[3]{73.6^2}$ to 3 significant digits.

$\log \sqrt[3]{y} = \frac{1}{3}[\log y]$; $\log 73.6^2 = 2(\log 73.6)$ ⟶ $\log \sqrt[3]{73.6^2} = \frac{1}{3}[\log 73.6^2] = \frac{1}{3}[2(\log 73.6)]$

$\log 73.6 = 1.8669$ ⟶ $\log x = \frac{1}{3}[2(1.8669)]$

Multiply by 2; then divide by 3. ⟶ $= \frac{1}{3}[3.7338]$, or 1.2446

Antilog of .2446 ≐ 1.76 ⟶ **Thus,** $\sqrt[3]{73.6^2} \doteq 10^1 \times 1.76$, or 17.6.

Find each to 3 significant digits. Use logarithms.

13. $\sqrt{5.76^3}$ *13.8*
14. $\sqrt[3]{24.5^2}$ *8.44*
15. $\sqrt{1.93^5}$ *5.18*
16. $\sqrt[4]{3.81^3}$ *2.73*
17. $\sqrt{38.4^3}$ *238*
18. $\sqrt[3]{502^2}$ *63.2*
19. $\sqrt[3]{3.7^4}$ *5.72*
20. $\sqrt[4]{18.6^3}$ *8.96*

PART C

Find each to 3 significant digits. Use logarithms.

21. $\frac{\sqrt{37.6}}{\sqrt[3]{23.8}}$
22. $\frac{2.74^3}{\sqrt{42.3}}$
23. $\frac{68 \times 3.02}{\sqrt{8,140}}$
24. $\sqrt[4]{\frac{734^2}{25.3^3}}$
25. $\sqrt[3]{\frac{24.8^2}{5.9 \times 63}}$
26. $\frac{47^3 \times \sqrt{69.2}}{28^2 \times \sqrt[3]{3,120}}$

Chapter Fourteen Review

Show the number as a power of 10. [p. 360]

1. 1,000
2. 820
3. 5.02

Find the value to 3 significant digits. [p. 360]

4. $10^{1.2253}$
5. $10^{3.8882}$
6. $10^{0.5705}$

7. Multiply 3.24×17.5 to 3 significant digits. Use powers of 10. [p. 363]
8. Divide $243 \div 12.1$ to 3 significant digits. Use powers of 10. [p. 363]

Find the logarithm. [p. 365]

9. log 5.63
10. log 30,800
11. log 27

Find the antilog. [p. 365]

12. 2.7619
13. 0.2227
14. 4.8451

Find the characteristic and the mantissa of the logarithm. [p. 365]

15. log 4,920 = 3.6920
16. log 74.8 = 1.8739
17. log 2.84 = 0.4533

Show the number in scientific notation. [p. 370]

18. 8,430
19. 46.8
20. .207
21. .0062

Write the number in ordinary notation. [p. 370]

22. 3.15×10^4
23. 5.03×10^2
24. 7.8×10^{-2}
25. 9.14×10^{-1}

Find each to 3 significant digits. Use logarithms. [p. 368, 374, 376, 378]

26. $3{,}280 \times 5.43$
27. $46 \times 2.7 \times 19 \times 8.3$
28. $\dfrac{7{,}820}{21.3}$
29. $\dfrac{5{,}400}{3.7 \times 24.1}$
30. 174^2
31. 26.3^3
32. $(207 \times 4.36)^3$
33. $(3.8 \times 21 \times 4.6)^2$
34. $\left(\dfrac{34.6}{8.7}\right)^3$
35. $\left(\dfrac{520 \times 3.4}{46}\right)^2$
36. $\sqrt{6{,}380}$
37. $\sqrt[4]{9.02}$
38. $\sqrt[3]{253 \times 8.7}$
39. $\sqrt{\dfrac{930}{6.2}}$
40. $\sqrt{6.16^3}$
41. $\sqrt[4]{22.7^3}$

42. Given the formula $i = 500\,rt$, show log i as a sum of logarithms. [p. 368]
43. Given the formula $V = 2wh$, show log V as a sum of logarithms. [p. 368]

Chapter Fourteen Test

1. Show 52.4 as a power of 10.

2. Find the value of $10^{2.4425}$ to 3 significant digits.

3. Multiply 2.17×32.4 to 3 significant digits. Use powers of 10.

4. Divide $98.4 \div 1.26$ to 3 significant digits. Use powers of 10.

5. Find log 8,400.

6. Find the antilog of 1.7818.

Find the characteristic and the mantissa of the logarithm.

7. $\log 4{,}370 = 3.6405$

8. $\log 3.64 = 0.5611$

Show the number in scientific notation.

9. 6,250

10. .0372

Write the number in ordinary notation.

11. 4.62×10^2

12. 2.08×10^{-1}

Find each to 3 significant digits. Use logarithms.

13. $7.2 \times 203 \times 32$

14. $(27.6 \times 4.9)^2$

15. $\dfrac{8{,}420}{35.7}$

16. $\dfrac{324 \times 8.6}{47}$

17. $\left(\dfrac{612}{22.3}\right)^2$

18. $\sqrt{\dfrac{821}{3.47}}$

19. $\sqrt[3]{7{,}640}$

20. $\sqrt[3]{4.35^2}$

21. Given the formula $i = 7pr$, show $\log i$ as a sum of logarithms.

15 USING LOGARITHMS

Fibonacci Sequence

Some of the outstanding work in mathematics done in the 13th century is attributed to a man known by many names: Leonardo de Pisa, Leonardo Fibonacci, Leonardo Pisano, or just Fibonacci (1170–1250).

Fibonacci Sequence

0, 1, 1, 2, 3, 5, 8, 13, ...

PROJECT

1. Write the next 4 terms of the sequence.
2. How is this sequence formed?
3. Write a formula for the nth term y_n of this sequence.

The Fibonacci sequence is also found in botany. This diagram shows the double spiraling of the sunflower. Opposite sets of rotating spirals are formed by the arrangement of the florets in the head. Counting the spirals, you find 21 in the clockwise and 34 in the counter-clockwise direction. This 21:34 ratio is composed of two adjacent terms in the Fibonacci sequence. This spiraling can also be found in pine cones, pine-apples and other plants with spiral leaf growth patterns. The pine cone scales number 5 one way and 8 the other, for the Fibonacci ratio 5:8. The bumps on a pineapple number 8 and 13, for the ratio 8:13.

Logarithms of Decimals

OBJECTIVES

- To find the log of a number like .00374
- To find the antilog of a logarithm like 8.7348 − 10

REVIEW CAPSULE

Show .00854 in scientific notation.

$.00854 = 8.54 \times 10^{-3}$

Show 4.76×10^{-2} in ordinary notation.

$4.76 \times 10^{-2} = .0476$

EXAMPLE 1 Show each number as a power of 10. Then find its log.

$.001 = \frac{1}{1000} = \frac{1}{10^3} = 10^{-3}$

Number:	.001	.01	.1	1
	↓	↓	↓	↓
Power of 10:	10^{-3}	10^{-2}	10^{-1}	10^0
	↓	↓	↓	↓
log:	−3	−2	−1	0

If $x = 10^y$, then $\log x = y$. ⟶

$\log .01 = \log 10^{-2} = -2$

If $0 < x < 1$, then $\log x$ is negative.

EXAMPLE 2 Show .0421 in scientific notation. Then find log .0421.

.0421 in scientific notation ⟶ $.0421 = 4.21 \times 10^{-2}$

If $a = b$, then $\log a = \log b$. ⟶ $\log .0421 = \log (4.21 \times 10^{-2})$

$\log (a \cdot b) = \log a + \log b$ ⟶ $= \log 4.21 + \log 10^{-2}$

$\log 4.21 = .6243$; $\log 10^{-2} = -2$

$= .6243 + (-2)$

$= -2 + .6243$

Characteristic (negative): −2; Mantissa (positive): .6243

$\left\{ \begin{array}{l} .0421 = 10^{-2} \times 4.21 \\ \log .0421 = -2 + .6243 \end{array} \right\}$

Thus, $\log .0421 = -2 + .6243$.

We rewrite this logarithm in a different form.

To keep the mantissa (.6243) *positive*, do *not* add −2 and .6243.

$\log .0421 = -2 + .6243$

Use 8 − 10 for −2. ⟶ $\log .0421 = (8 - 10) + .6243$

$(8 - 10) + .6243 = 8 + .6243 - 10$
$= 8.6243 - 10$

$\log .0421 = 8.6243 - 10$

The mantissa remains positive. ⟶ Characteristic is 8 − 10, or −2. Mantissa is .6243.

$1 \le a < 10$ and $-c$ is an integer.

> If x is written in scientific notation,
> $x = a \times 10^{-c}$
> then $\log x = (\log a) + (-c)$.

EXAMPLE 3 Find log .00264. Then find the characteristic.

Show .00264 in scientific notation. → $.00264 = 2.64 \times 10^{-3}$

$\log 10^{-3} = -3$ → $\log .00264 = (\log 2.64) + (-3)$

$\log 2.64 = .4216$ → $= -3 + .4216$

Use $7 - 10$ for -3. → $= (7 - 10) + .4216$

$= 7.4216 - 10$

Thus, $\log .00264 = 7.4216 - 10$ and the characteristic is $7 - 10$, or -3.

EXAMPLE 4 Find the antilog of $8.7875 - 10$.

Let x be the antilog. → Find x if $\log x = 8.7875 - 10$.

$\log x = (8 - 10) + .7875$

Characteristic is $8 - 10$ or -2. → $\log x = -2 + .7875$

$\log 10^{-2} = -2$; $\log 6.13 = .7875$; 6.13 is antilog of .7875. → $\log x = \log 10^{-2} + \log 6.13$

$\log a + \log b = \log (a \cdot b)$ → $\log x = \log (10^{-2} \times 6.13)$

So, $x = 10^{-2} \times 6.13$

Move decimal point 2 places *left*. → $= .0613$

Thus, the antilog of $8.7875 - 10$ is .0613.

To find the antilog of a log →

> Characteristic — Mantissa
> If $\log x = -c + m$
> then $x = 10^{-c} \times$ antilog of m.

EXAMPLE 5 Find the antilog of $9.6758 - 10$.

Find x. → $\log x = 9.6758 - 10$

Characteristic is $9 - 10$ or -1. → $= -1 + .6758$

4.74 is antilog of .6758. → So, $x = 10^{-1} \times 4.74$

Move decimal point 1 place *left*. → $= .474$

Thus, the antilog of $9.6758 - 10$ is .474.

ORAL EXERCISES

Give the logarithm.

1. log .01 *−2*
2. log .001 *−3*
3. log .1 *−1*
4. log .0001 *−4*
5. log 1 *0*

Find the antilog.

6. −2 *.01*
7. −1 *.1*
8. 0 *1*
9. −3 *.001*
10. −4 *.0001*

Give the characteristic of the logarithm in two ways.

11. log .064 = 8.8062 − 10 *8 − 10; −2*
12. log .64 = 9.8062 − 10 *9 − 10; −1*
13. log .0064 = 7.8062 − 10 *7 − 10; −3*
14. log .00064 = 6.8062 − 10 *6 − 10; −4*

Give the number in scientific notation.

15. .0372 *3.72×10^{-2}*
16. .214 *2.14×10^{-1}*
17. .0076 *7.6×10^{-3}*
18. .00053 *5.3×10^{-4}*

Give the characteristic of the logarithm.

19. log .00293 *−3 or 7 − 10*
20. log .735 *−1 or 9 − 10*
21. log .0507 *−2 or 8 − 10*
22. log .017 *−2 or 8 − 10*
23. log .00082 *−4 or 6 − 10*
24. log 3.24 *0*

Give the number in ordinary notation.

25. $10^{-1} \times 5.63$ *.563*
26. $10^{-3} \times 8.7$ *.0087*
27. $10^{-2} \times 2.06$ *.0206*
28. $10^{-4} \times 4.2$ *.00042*

EXERCISES

PART A

Find the logarithm.

1. log .354
2. log .0354 *8.5490 − 10*
3. log .000354 *6.5490 − 10*
4. log .00613
5. log .0167
6. log .00822
7. log .207 *9.3160 − 10*
8. log .0405 *8.6075 − 10*
9. log .506 *9.7042 − 10*

Find the antilog.

10. 8.5132 − 10 *.0326*
11. 9.5132 − 10 *.326*
12. 7.5132 − 10 *.00326*
13. 9.3711 − 10 *.235*
14. 8.7619 − 10 *.0578*
15. 6.2227 − 10 *.000167*
16. 8.6646 − 10 *.0462*
17. 9.0899 − 10 *.123*
18. 7.9206 − 10 *.00833*
19. 9.7987 − 10 *.629*
20. 8.9899 − 10 *.0977*
21. 6.8887 − 10 *.000774*

PART B

Find the logarithm.

22. log .73 *9.8633 − 10*
23. log .006 *7.7782 − 10*
24. log .031 *8.4914 − 10*
25. log .00054 *6.7324 − 10*
26. log .0008 *6.9031 − 10*
27. log .000024 *5.3802 − 10*

Find the antilog.

28. 9.6335 − 10 *.43*
29. 8.5315 − 10 *.034*
30. 7.7559 − 10 *.0057*
31. 6.2553 − 10 *.00018*
32. 7.8633 − 10 *.0073*
33. 6.9345 − 10 *.00086*
34. 5.4771 − 10 *.00003*
35. 6.8451 − 10 *.0007*
36. 5.3010 − 10 *.00002*

Using Logarithms of Decimals

OBJECTIVE

- To compute by using logarithms

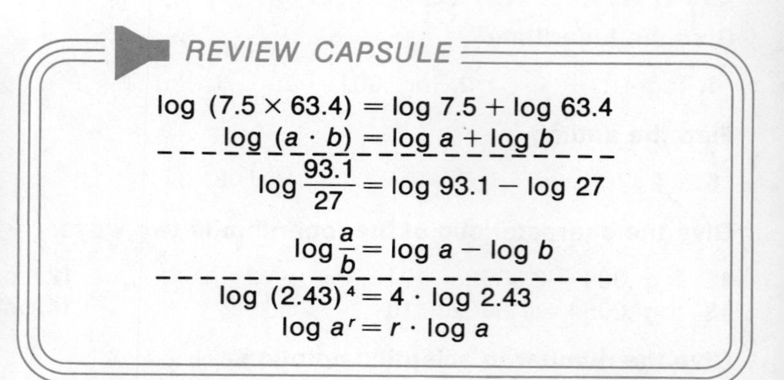

EXAMPLE 1 Find $.0845 \times 3.7$ to 3 significant digits.

$\log(a \cdot b) = \log a + \log b$ →	$\log(.0845 \times 3.7) = \log .0845 + \log 3.7$
Let $x = .0845 \times 3.7$. Find the logarithms. →	$\log x = (8.9269 - 10) + (0.5682)$
Group the positive numbers. →	$\log x = (8.9269 + 0.5682) - 10$
Add. →	$\log x = 9.4951 - 10$
.4955 is nearest mantissa. →	$\log x \doteq 9.4955 - 10$
Characteristic: $9 - 10 = -1$. →	$\log x \doteq -1 + .4955$
Find the antilog of $-1 + .4955$. →	So, $x \doteq 10^{-1} \times 3.13$, or .313.

Thus, $.0845 \times 3.7 = .313$ to 3 significant digits.

EXAMPLE 2 Find $\frac{.487}{51.4}$ to 3 significant digits.

$\log \frac{a}{b} = \log a - \log b$ →	$\log \frac{.487}{51.4} = \log .487 - \log 51.4$
Let $x = \frac{.487}{51.4}$.	$\log x = (9.6875 - 10) - (1.7110)$
Rearrange. →	$= (9.6875 - 1.7110) - 10$
Subtract. →	$= 7.9765 - 10$
.9763 is nearest mantissa. →	$\log x \doteq 7.9763 - 10$
Characteristic: $7 - 10 = -3$. Find antilog of $-3 + .9763$. →	So, $x \doteq 10^{-3} \times 9.47$, or .00947.
To 3 significant digits →	**Thus,** $\frac{.487}{51.4} = .00947$.

EXAMPLE 3 Find $.074^3$ to 3 significant digits.

Reason	Step
$\log a^r = r \cdot \log a$ →	$\log(.074^3) = 3 \cdot \log .074$
Let $x = .074^3$. →	$\log x = 3(8.8692 - 10)$
Distribute the 3. →	$= 26.6076 - 30$
.6075 is nearest mantissa.	$\log x \doteq 26.6075 - 30$
Characteristic: $26 - 30 = -4$. Find antilog of $-4 + .6075$. →	So, $x \doteq 10^{-4} \times 4.05$, or .000405.
To 3 significant digits →	**Thus,** $.074^3 = .000405$.

EXAMPLE 4 Find $\frac{61.7}{.0913}$ to 3 significant digits.

Reason	Step
$\log \frac{a}{b} = \log a - \log b$ →	$\log \frac{61.7}{.0913} = \log 61.7 - \log .0913$
	$\log x = (1.7903) - (8.9605 - 10)$
Rewrite the characteristic 1 as $11 - 10$. →	$= (11.7903 - 10) - (8.9605 - 10)$
Subtract: $11.7903 - 10$, $8.9605 - 10$, giving 2.8298 →	$= 2.8298$
.8299 is nearest mantissa.	$\log x \doteq 2.8299$
Find the antilog of 2.8299. →	So, $x \doteq 10^2 \times 6.76$, or 676.
To 3 significant digits →	**Thus,** $\frac{61.7}{.0913} = 676$.

EXAMPLE 5 Find $\sqrt[3]{.0532}$ to 3 significant digits.

Reason	Step
$\sqrt[3]{y} = y^{\frac{1}{3}}$ →	$\sqrt[3]{.0532} = .0532^{\frac{1}{3}}$
	$\log \sqrt[3]{.0532} = \log(.0532^{\frac{1}{3}})$
$\log a^r = r \cdot \log a$; $\log a^{\frac{1}{3}} = \frac{1}{3} \cdot \log a$ →	$\log x = \frac{1}{3}(\log .0532)$
	$= \frac{1}{3}(8.7259 - 10)$ ← Add $(20 - 20)$.
Rewrite the characteristic: -2 as $8 - 10$ (*Not* divisible by 3) or $28 - 30$ (Divisible by 3) →	$= \frac{1}{3}(28.7259 - 30)$
$(28.7259 - 30) \div 3$ →	$= 9.5753 - 10$
.5752 is nearest mantissa.	$\log x \doteq 9.5752 - 10$
Find antilog of $-1 + .5752$. →	So, $x \doteq 10^{-1} \times 3.76$, or .376.
To 3 significant digits →	**Thus,** $\sqrt[3]{.0532} = .376$.

EXERCISES

PART A

Find each to 3 significant digits.

1. $.062 \times 4.23$ *.262*
2. $.235 \times 66$ *15.5*
3. $.034 \times 2.17$ *.0738*
4. $.0058 \times 7.26$ *.0421*
5. $.417 \times 5.9$ *2.46*
6. $.0028 \times 64.3$ *.180*
7. $\frac{.629}{27}$ *.0233*
8. $\frac{.43}{8.74}$ *.0492*
9. $\frac{.036}{5.32}$ *.00677*
10. $.532^3$ *.151*
11. $.087^2$ *.00757*
12. $.816^4$ *.443*
13. $\frac{54.6}{.23}$ *237*
14. $\frac{78.2}{.029}$ *2,700*
15. $\frac{3.6}{.0793}$ *45.4*
16. $\sqrt[3]{.0736}$ *.419*
17. $\sqrt{.0059}$ *.0768*
18. $\sqrt[4]{.836}$ *.956*
19. $\sqrt{.637}$ *.798*
20. $\sqrt[4]{.072}$ *.518*
21. $\sqrt[3]{.0068}$ *.189*

PART B

EXAMPLE Find $\frac{.00939}{.277}$.

$\log \frac{.00939}{.277}$

$\log .00939 - \log .277$

$(7.9727 - 10) - (9.4425 - 10)$

(7.9727 − 9.4425) would give a *negative* mantissa. Rewrite the characteristic: use 17 − 20 for 7 − 10.

$17.9727 - 20$
$9.4425 - 10$
$8.5302 - 10$

Subtract. Mantissa is *positive*. →

Find antilog of −2 + .5302. → **Thus,** $\frac{.00939}{.277} = 10^{-2} \times 3.39$, or .0339.

Find $\frac{46.8}{983}$.

$\log \frac{46.8}{983}$

$\log 46.8 - \log 983$

$1.6702 - 2.9926$

$11.6702 - 10$
2.9926
$8.6776 - 10$

Thus, $\frac{46.8}{983} = 10^{-2} \times 4.76$, or .0476.

Find each to 3 significant digits.

22. $\frac{.0754}{.312}$ *.242*
23. $\frac{.0236}{.761}$ *.0310*
24. $\frac{.00833}{.269}$ *.0310*
25. $\frac{.00372}{.604}$ *.00616*
26. $\frac{63.5}{361}$ *.176*
27. $\frac{234}{783}$ *.299*
28. $\frac{804}{27{,}300}$ *.0295*
29. $\frac{4.52}{63.8}$ *.0709*
30. $\frac{.734}{.321}$ *2.29*
31. $\frac{.278}{.629}$ *.442*
32. $\frac{.652}{.023}$ *28.3*
33. $\frac{.043}{.076}$ *.566*
34. $.622 \times .423$ *.263*
35. $.057 \times .394$ *.0225*
36. $.026 \times .345$ *.00897*
37. $.0063 \times .047$ *.000296*

PART C

Write each as the log of a single expression. [Hint: $\log a - 2 \cdot \log b = \log a - \log b^2 = \log \frac{a}{b^2}$]

38. $\log 7 + \log m - \log n$
39. $2 \cdot \log t - \log v - \log w$
40. $\log 3 + 2 \cdot \log f + \log h$
41. $3(\log m + \log n - \log p)$
42. $\log 7 + \frac{1}{2} \cdot \log t - \log y$
43. $\frac{1}{3}(\log x + \log y - \log z)$

Computations with Logarithms

OBJECTIVE

- To do computations like $\sqrt[3]{\frac{.0069}{5.23}}$ by using logarithms

REVIEW CAPSULE

$$\log \frac{721}{34 \times 56} = \log 721 - \log (34 \times 56) = \log 721 - (\log 34 + \log 56)$$

$$\log \frac{a}{b \cdot c} = \log a - \log (b \cdot c) = \log a - (\log b + \log c)$$

EXAMPLE 1 Find $.038 \times 2780 \times .0063$ to 3 significant digits.

$\log (a \cdot b \cdot c) = \log a + \log b + \log c \longrightarrow$

Let $x = .038 \times 2780 \times .0063$

$$\begin{aligned} \log x &= \log .038 + \log 2{,}780 + \log .0063 \\ &= (8.5798 - 10) + (3.4440) + (7.7993 - 10) \\ &= 19.8231 - 20 \\ \log x &\doteq 19.8228 - 20 \end{aligned}$$

So, $x \doteq 10^{-1} \times 6.65$, or .665.

.8228 is nearest mantissa.
Characteristic: $19 - 20 = -1$.
Find antilog of $-1 + .8228$.

Thus, $.038 \times 2780 \times .0063 = .665$.

EXAMPLE 2 Find $\frac{.076}{3.24 \times 27}$ to 3 significant digits.

$\log \frac{a}{b \cdot c} = \log a - \log (b \cdot c) = \log a - (\log b + \log c) \longrightarrow$

$$\begin{aligned} \log \frac{.076}{3.24 \times 27} &= \log .076 - (\log 3.24 + \log 27) \\ \log x &= (8.8808 - 10) - (0.5105 + 1.4314) \\ &= (8.8808 - 10) - (1.9419) \\ &= 6.9389 - 10 \\ \log x &\doteq 6.9390 - 10 \end{aligned}$$

So, $x \doteq 10^{-4} \times 8.69$, or .000869.

.9390 is nearest mantissa.
Characteristic: $6 - 10 = -4$.
Find antilog of $-4 + .9390$.

Thus, $\frac{.076}{3.24 \times 27} = .000869$.

EXAMPLE 3 Find $(.00047 \times 629)^3$ to 3 significant digits.

$\log a^r = r \cdot \log a \longrightarrow$
$\log (a \cdot b) = \log a + \log b \longrightarrow$

$$\begin{aligned} \log (.00047 \times 629)^3 &= 3 \cdot \log (.00047 \times 629) \\ \log x &= 3(\log .00047 + \log 629) \\ &= 3[(6.6721 - 10) + 2.7987] \\ &= 3[9.4708 - 10] \\ &= 28.4124 - 30 \\ \log x &\doteq 28.4116 - 30 \end{aligned}$$

So, $x \doteq 10^{-2} \times 2.58$, or .0258.

Distribute the 3. $\longrightarrow$
.4116 is nearest mantissa.
Find antilog of $-2 + .4116$. $\longrightarrow$

Thus, $(.00047 \times 629)^3 = .0258$.

EXAMPLE 4 Find $\sqrt[3]{\frac{.094}{22.7}}$ to 3 significant digits.

$\sqrt[3]{y} = y^{\frac{1}{3}}$ →	$\log \sqrt[3]{\frac{.094}{22.7}} = \log \left(\frac{.094}{22.7}\right)^{\frac{1}{3}}$
$\log a^r = r \cdot \log a$; $\log a^{\frac{1}{3}} = \frac{1}{3} \cdot \log a$ →	$\log x = \frac{1}{3} \cdot \log \frac{.094}{22.7}$
$\log \frac{a}{b} = \log a - \log b$ →	$= \frac{1}{3}[\log .094 - \log 22.7]$
	$= \frac{1}{3}[(8.9731 - 10) - (1.3560)]$
	$= \frac{1}{3}[7.6171 - 10]$
Add (20 − 20) to (7 − 10). →	$= \frac{1}{3}[27.6171 - 30]$
(27.6171 − 30) ÷ 3 →	$= 9.2057 - 10$
.2068 is nearest mantissa.	$\log x \doteq 9.2068 - 10$
Find antilog of −1 + .2068. →	So, $x \doteq 10^{-1} \times 1.61$, or .161.
To 3 significant digits →	**Thus,** $\sqrt[3]{\frac{.094}{22.7}} = .161$.

EXAMPLE 5 Find $\log \frac{2t^3}{mn}$.

$\log \frac{a}{b} = \log a - \log b$ →	$\log \frac{2t^3}{mn} = \log (2t^3) - \log (mn)$
$\log (a \cdot b) = \log a + \log b$ →	$= \log 2 + \log t^3 - (\log m + \log n)$
$\log a^r = r \cdot \log a$; $-(x + y) = -1(x + y) = -x - y$ →	$= \log 2 + 3 \cdot \log t - \log m - \log n$

Thus, $\log \frac{2t^3}{mn} = \log 2 + 3 \cdot \log t - \log m - \log n$

EXAMPLE 6 Find $\log \sqrt[3]{\frac{5w}{d}}$.

$\sqrt[3]{y} = y^{\frac{1}{3}}$ →	$\log \sqrt[3]{\frac{5w}{d}} = \log \left(\frac{5w}{d}\right)^{\frac{1}{3}}$
$\log a^r = r \cdot \log a$ →	$= \frac{1}{3} \cdot \log \frac{5w}{d}$
$\log \frac{a}{b} = \log a - \log b$ →	$= \frac{1}{3}[\log (5w) - \log d]$
$\log (a \cdot b) = \log a + \log b$ →	$= \frac{1}{3}[\log 5 + \log w - \log d]$

Thus, $\log \sqrt[3]{\frac{5w}{d}} = \frac{1}{3}(\log 5 + \log w - \log d)$.

ORAL EXERCISES

Give the logarithm.

1. $\log 2a$ *$\log 2 + \log a$*

2. $\log a^2$ *$2 \cdot \log a$*

3. $\log \frac{7}{c}$ *$\log 7 - \log c$*

4. $\log \frac{a}{3}$ *$\log a - \log 3$*

5. $\log \sqrt{a}$ *$\frac{1}{2}(\log a)$*

6. $\log \sqrt[3]{a}$ *$\frac{1}{3}(\log a)$*

EXERCISES

PART A

Find each to 3 significant digits.

1. $.263 \times 42.7 \times .086$ *.966*
2. $.0073 \times 5.24 \times 360$ *13.8*
3. $.374 \times .298 \times .607$ *.0677*
4. $.067 \times .034 \times 742$ *1.69*
5. $3690 \times .048 \times .806$ *143*
6. $.0084 \times 692 \times 2.57$ *14.9*
7. $\frac{.437 \times 5.11}{34.6}$ *.0645*
8. $\frac{.083 \times .942}{2.78}$ *.0281*
9. $\frac{.043 \times 782}{507}$ *.0663*
10. $\frac{.312}{5.73 \times 26}$ *.00209*
11. $\frac{.087}{7.2 \times 3.06}$ *.00395*
12. $\frac{.643}{3.9 \times 4.2}$ *.0393*
13. $(.073 \times 26.8)^2$ *3.83*
14. $(.206 \times 3.12)^3$ *.266*
15. $(.00063 \times 742)^3$ *.102*
16. $\left(\frac{.309}{6.42}\right)^2$ *.00232*
17. $\left(\frac{.824}{23.7}\right)^2$ *.00121*
18. $\left(\frac{.756}{18.3}\right)^3$ *.0000705*
19. $\sqrt{.037 \times 69.2}$ *1.60*
20. $\sqrt{.263 \times 3.14}$ *.909*
21. $\sqrt[3]{.049 \times 7.24}$ *.708*
22. $\sqrt{\frac{.028}{43.2}}$ *.0255*
23. $\sqrt[3]{\frac{.813}{2.78}}$ *.664*
24. $\sqrt[3]{\frac{.091}{34.6}}$ *.138*

Find the logarithm.

25. $\log 3fg$
26. $\log m^2n$
27. $\log 2tv^3$
28. $\log x\sqrt{y}$
29. $\log \frac{5f}{g}$
30. $\log \frac{v^2}{2h}$
31. $\log \frac{3d}{t^2}$
32. $\log \frac{wv^3}{2g}$
33. $\log (7fw)^2$
34. $\log (t^2v)^3$
35. $\log \sqrt{2gh}$
36. $\log \sqrt[3]{5d^2}$
37. $\log \left(\frac{3d}{g}\right)^2$ *$2(\log 3 + \log d - \log g)$*
38. $\log \left(\frac{v^2w}{mn}\right)^3$
39. $\log \sqrt{\frac{2d}{g}}$ *$\frac{1}{2}(\log 2 + \log d - \log g)$*
40. $\log \sqrt[3]{\frac{tv^2}{gh}}$

PART B

Find each to 3 significant digits.

41. $\frac{.312 \times 27.8}{.083}$ *105*
42. $\left(\frac{2.34}{.469}\right)^2$ *24.9*
43. $\sqrt{\frac{68.5}{.072}}$ *30.8*
44. $\frac{.084}{.54 \times .073}$ *2.13*
45. $\left(\frac{.376}{.052}\right)^3$ *378*
46. $\sqrt[3]{\frac{.069}{.832}}$ *.436*

PART C

Find each to 3 significant digits.

47. $\frac{\sqrt{.0462}}{1.73^3}$
48. $\frac{\sqrt{72.3}}{\sqrt[3]{694}}$
49. $\frac{\sqrt[3]{93.2}}{.064 \times 27.3}$
50. $\sqrt[4]{\frac{2.36^2}{4.38^3}}$
51. $\sqrt[3]{\frac{.813^2}{283 \times .046}}$
52. $\frac{.476^2 \times \sqrt[3]{.038}}{.732^3 \times \sqrt{.542}}$

53. Can logarithms be used to do a computation like $3.71 + 2.75 - 3.21$?

54. List the kinds of computations that are convenient to do using logarithms.

Interpolation

OBJECTIVES

- To interpolate between logs and between antilogs
- To do computations by using interpolation

REVIEW CAPSULE

Solve the proportion $\frac{y}{10} = \frac{7}{11}$.

$\frac{y}{10} = \frac{7}{11}$ ⟵ $\frac{a}{b} = \frac{c}{d}$

$11y = 70$ ⟵ $ad = bc$

$y = \frac{70}{11}$, or $6\frac{4}{11}$

EXAMPLE 1 At a constant speed a car traveled 48 km in 30 minutes and 64 km in 40 minutes. How far did it travel in 33 minutes?

Subtract numbers at end of arrows.
$33 - 30 = 3$
$40 - 30 = 10$ and $64 - 48 = 16$

Write a proportion.

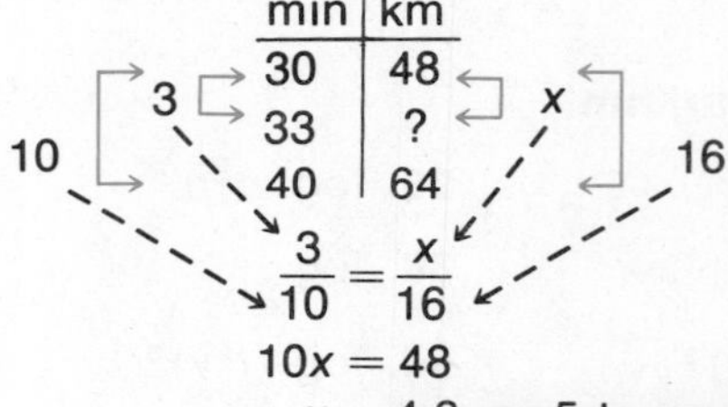

$$\frac{3}{10} = \frac{x}{16}$$

Solve the proportion.

$10x = 48$

$x = 4.8$, or 5 to nearest whole number

The number of km is $48 + x$: $48 + 5$, or 53.

Thus, the car traveled $48 + 5$, or 53 km in 33 minutes.

In Example 1 we used linear *interpolation* to find 53. We can interpolate to approximate the log of a 4-digit number.

EXAMPLE 2 Find log 35.26.

Use mantissas for log 352 and log 353.

The mantissas for log 3520 and log 3530 are in the table.

Omit all decimal points.

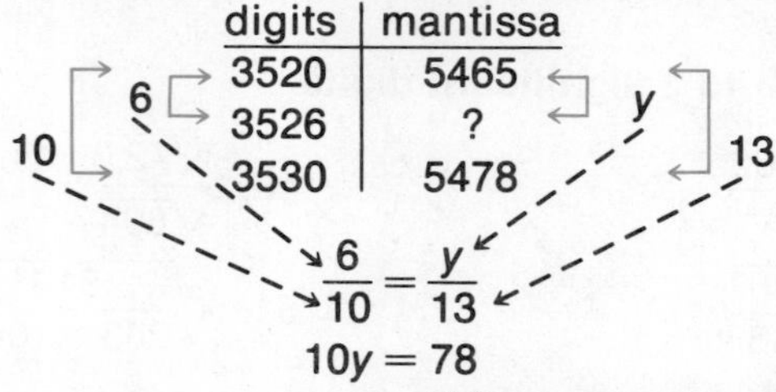

Subtract numbers at ends of arrows. Write a proportion.

$$\frac{6}{10} = \frac{y}{13}$$

Solve the proportion.

$10y = 78$

Round to nearest whole number.

$y = 7.8$, or 8 to nearest whole number

The mantissa .5473 is interpolated to 4 digits.

So, the mantissa is $5465 + 8$, or 5473.

The characteristic is 1.

Thus, $\log 35.26 = 1 + .5473$, or 1.5473.

EXAMPLE 3 Find the antilog of $8.4239 - 10$ interpolated to 4 digits.

The mantissa (.4239) is not in the table. Sandwich .4239 between 2 mantissas in the table.

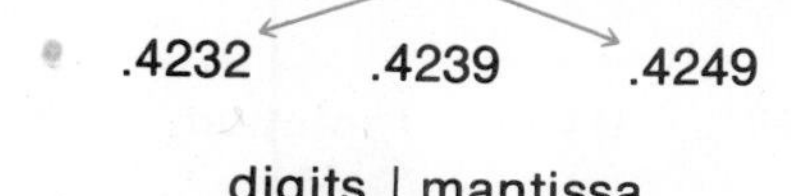

Find 4 digits for each mantissa.
2650 for 4232
2660 for 4249
Omit all decimal points.

Find the differences.
Write a proportion.
Solve the proportion.
Round to nearest whole number.

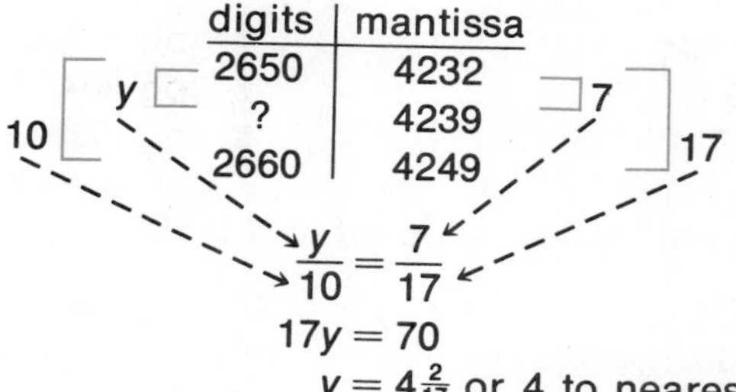

digits	mantissa
2650	4232
?	4239
2660	4249

$$\frac{y}{10} = \frac{7}{17}$$

$$17y = 70$$

$y = 4\frac{2}{17}$ or 4 to nearest whole number

2.654 is the antilog of .4239. → The four digits are $2650 + 4$, or 2654.

x is the antilog of $8.4239 - 10$. → $\log x \doteq 8.4239 - 10$

Characteristic: $8 - 10 = -2$
Find antilog of $-2 + .4239$. → So, $x \doteq 10^{-2} \times 2.654$, or .02654.

Interpolated to 4 digits → **Thus,** the antilog of $8.4239 - 10$ is .02654.

EXAMPLE 4 Find $589.3 \times .04716$ interpolated to 4 digits.

$\log (a \cdot b) = \log a + \log b$ → $\log (589.3 \times .04716) = \log 589.3 + \log .04716$

Use interpolation to find log 589.3 and log .04716.

log 589.3

digits	mantissa
5890	7701
5893	
5900	7709

(differences: 3 and 10 in digits; y and 8 in mantissas)

$$\frac{3}{10} = \frac{y}{8}$$

$$10y = 24$$

$$y = 2.4$$

$$y \doteq 2$$

So, $\log 589.3 = 2.7703$

log .04716

digits	mantissa
4710	6730
4716	
4720	6739

(differences: 6 and 10 in digits; y and 9 in mantissas)

$$\frac{6}{10} = \frac{y}{9}$$

$$10y = 54$$

$$y = 5.4$$

$$y \doteq 5$$

So, $\log .04716 = 8.6735 - 10$

$7701 + 2 = 7703$
$6730 + 5 = 6735$

Let $x = 589.3 \times .04716$. → $\log (589.3 \times .04716) = (2.7703) + (8.6735 - 10)$

$\log x = 1.4438$

Interpolate to find x the antilog of 1.4438.

antilog of 1.4438

digits	mantissa
2770	4425
	4438
2780	4440

(differences: y and 10 in digits; 13 and 15 in mantissas)

The nearest mantissas are .4425 and .4440.

$$\frac{y}{10} = \frac{13}{15}$$

$$15y = 130$$

$$y = 8\frac{2}{3}$$

$$y \doteq 9$$

The characteristic is 1. → The digits are 2779. So, $x \doteq 10^1 \times 2.779$, or 27.79.

Thus, $589.3 \times .04716 = 27.79$ interpolated to 4 digits.

EXERCISES

PART A

Find the logarithm. Use interpolation.

1. log 19.22
2. log 1,924
3. log .01927 *8.2849 − 10*
4. log 2.521
5. log 354.5
6. log 60.68
7. log .7347 *9.8661 − 10*
8. log .03979 *8.5998 − 10*
9. log .002073 *7.3166 − 10*

Find the antilog interpolated to 4 digits.

10. 1.2128 *16.32*
11. 0.4189 *2.624*
12. 2.6157 *412.7*
13. 3.6800 *4,787*
14. 1.2588 *18.15*
15. 2.9526 *896.6*
16. 1.7786 *60.06*
17. 0.6851 *4.843*
18. 3.3394 *2,185*
19. 9.2681 − 10 *.1854*
20. 8.5152 − 10 *.03275*
21. 7.8614 − 10 *.007268*

Find each interpolated to 4 digits.

22. 37.45×4.53 *169.7*
23. $6.202 \times 5{,}463$
24. $.2744 \times 34.6$ *9.495*
25. $\frac{89.27}{3.62}$ *24.67*
26. $\frac{276.6}{54.21}$ *5.103*
27. $\frac{.09343}{2.53}$ *.03693*
28. $(58.62)^2$ *3435*
29. $(.4736)^3$ *.1062*
30. $(2.403)^4$ *33.33*
31. $\sqrt{9{,}346}$ *96.68*
32. $\sqrt[3]{52.35}$ *3.741*
33. $\sqrt{.007024}$ *.08382*

PART B

Find each interpolated to 4 digits.

34. $25.37 \times 4.28 \times .853$ *92.62*
35. $.06814 \times 345 \times 7.42$ *174.4*
36. $\frac{4{,}264 \times 5.23}{37.2}$ *599.5*
37. $\frac{.8632}{6.24 \times 3.07}$ *.04506*
38. $(7.316 \times 2.58)^2$ *356.3*
39. $(.2075 \times 3.24)^3$ *.3039*
40. $\left(\frac{30.73}{6.32}\right)^2$ *23.65*
41. $\left(\frac{.7841}{1.92}\right)^3$ *.06813*
42. $\sqrt{31.46 \times 2.73}$ *9.267*
43. $\sqrt[3]{.04823 \times 73.6}$ *1.526*
44. $\sqrt{\frac{2{,}793}{43.6}}$ *8.004*
45. $\sqrt[3]{\frac{.8125}{2.58}}$ *.6805*
46. $\sqrt[3]{483.5^2}$ *61.60*
47. $\sqrt{7.074^3}$ *18.81*

PART C

The formula for compound interest is $A = P(1 + \frac{r}{n})^{nt}$, where P represents the principal invested at r percent annual interest compounded n times a year for t years. A is the amount accumulated. Solve the following problems using logs. [Hint: simplify the expression in parentheses first.]

48. A bond paying 6% compounded semiannually is bought for $3,000. How much will it be worth 10 years from now?
49. A savings bank pays 5% compounded quarterly. If $2,000 is deposited, how much will be in the account after 5 years?

Logarithms: Bases Other Than 10

OBJECTIVE

- To find logarithms of numbers by using a base other than 10

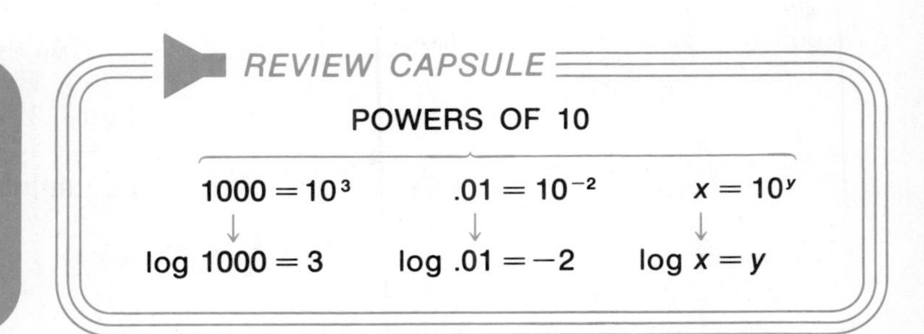

EXAMPLE 1 Show 8 as a power of 2. Show 32 as a power of 2.

$8 = 2 \cdot 2 \cdot 2 = 2^3$

$32 = 2 \cdot 2 \cdot 2 \cdot 2 \cdot 2 = 2^5$

$8 \to 2^3$ | $32 \to 2^5$

Thus, $8 = 2^3$ and $32 = 2^5$

Read: the log, base 2, of 8 is 3. → We write $\log_2 8 = 3$ and $\log_2 32 = 5$.

Definition of base 2 logarithm →

$\log_2 x = y$ means $x = 2^y$ — The base 2 logarithm of x is the *exponent* when x is shown as a power of 2.

EXAMPLE 2 Find $\log_2 16$. Find $\log_2 2$.

Show number as a power of 2. → $16 = 2^4$ | $2 = 2^1$

$\log_2 x = y$, if $x = 2^y$. → **Thus,** $\log_2 16 = 4$. | **Thus,** $\log_2 2 = 1$.

EXAMPLE 3 Write the *base 5* logarithm of 125.

Show 125 as a power of 5. → $125 = 5^3$

Use the exponent. → $\log_5 125 = 3$

Definition of base n logarithm (for $n > 0$ and $n \neq 1$) →

$\log_n x = y$ means $x = n^y$ — The base n logarithm of x is the *exponent* when x is shown as a power of n.

Logarithmic equations can be written as exponential equations by using the definition of base n log.

EQUATION

Logarithmic form		*Exponential form*
$\log_n x = y$	$\longrightarrow$	$x = n^y$
$\log_2 16 = 4$	$\longrightarrow$	$16 = 2^4$
$\log_5 125 = 3$	$\longrightarrow$	$125 = 5^3$

EXAMPLE 4 Write the equation in exponential form.

Logarithmic form: $\log_n x = y$ → $\log_5 625 = y$ | $\log_n \frac{1}{25} = -2$ | $\log_2 x = 7$

Exponential form: $x = n^y$ → $625 = 5^y$ | $\frac{1}{25} = n^{-2}$ | $x = 2^7$

EXAMPLE 5 Find $\log_5 625$.

Write a logarithmic equation. → $\log_5 625 = y$

Change to exponential form. → $625 = 5^y$

Show 625 as a power of 5. → $5^4 = 5^y$

Compare the exponents. (Bases are the same.) → $4 = y$

Thus, $\log_5 625 = 4$.

Find $\log_2 \frac{1}{8}$.

Write a logarithmic equation. → $\log_2 \frac{1}{8} = y$.

Change to exponential form. → $\frac{1}{8} = 2^y$

Show $\frac{1}{8}$ as a power of 2. → $\frac{1}{2^3} = 2^y$

$2^{-3} = 2^y$

Compare the exponents. (Bases are the same.) → $-3 = y$

Thus, $\log_2 \frac{1}{8} = -3$.

EXAMPLE 6 Solve $\log_2 x = 7$ for x.

Change to exponential form. → $x = 2^7$

$= 128$

Thus, $x = 128$.

Solve $\log_3 x = -4$ for x.

$x = 3^{-4}$

$= \frac{1}{3^4}$, or $\frac{1}{81}$

Thus, $x = \frac{1}{81}$.

EXAMPLE 7 Solve $\log_n 243 = 5$ for the base n.

Change to exponential form. → $243 = n^5$

Show 243 as a 5th power: $243 = 3^5$ → $3^5 = n^5$

Compare the bases. (Exponents are the same.) → $3 = n$

Thus, the base n is 3.

Solve $\log_n \frac{1}{25} = -2$ for n.

$\frac{1}{25} = n^{-2}$

$\frac{1}{5^2} = n^{-2}$

$5^{-2} = n^{-2}$

$5 = n$

Thus, the base n is 5.

EXAMPLE 8 Find $\log_6 \sqrt[3]{6}$.

Write a logarithmic equation. → $\log_6 \sqrt[3]{6} = y$

Change to exponential form. → $\sqrt[3]{6} = 6^y$

Show $\sqrt[3]{6}$ as a power of 6. → $6^{\frac{1}{3}} = 6^y$

Compare the exponents. (Bases are the same.) → $\frac{1}{3} = y$

Thus, $\log_6 \sqrt[3]{6} = \frac{1}{3}$.

ORAL EXERCISES

Find the logarithm.

1. $\log_6 36$ *2* **2.** $\log_7 7$ *1* **3.** $\log_4 1$ *0* **4.** $\log_3 \sqrt{3}$ *$\frac{1}{2}$* **5.** $\log_5 \frac{1}{5}$ *-1*

Give the equation in exponential form.

6. $\log_2 32 = y$ **7.** $\log_3 x = 5$ *$x = 3^5$* **8.** $\log_n 125 = 3$ *$125 = n^3$*

9. $\log_3 \frac{1}{9} = y$ **10.** $\log_5 x = -2$ *$x = 5^{-2}$* **11.** $\log_n \sqrt{7} = \frac{1}{2}$ *$\sqrt{7} = n^{\frac{1}{2}}$*

EXERCISES

PART A

Find the logarithm.

1. $\log_3 9$ *2* **2.** $\log_2 64$ *6* **3.** $\log_5 5$ *1*

4. $\log_5 25$ *2* **5.** $\log_4 64$ *3* **6.** $\log_2 1$ *0*

7. $\log_3 81$ *4* **8.** $\log_2 128$ *7* **9.** $\log_7 49$ *2*

10. $\log_5 \sqrt{5}$ *$\frac{1}{2}$* **11.** $\log_2 \sqrt[3]{2}$ *$\frac{1}{3}$* **12.** $\log_3 \sqrt[4]{3}$ *$\frac{1}{4}$*

13. $\log_2 \frac{1}{2}$ *-1* **14.** $\log_3 \frac{1}{9}$ *-2* **15.** $\log_5 \frac{1}{25}$ *-2*

16. $\log_2 \frac{1}{4}$ *-2* **17.** $\log_4 \frac{1}{16}$ *-2* **18.** $\log_2 \frac{1}{16}$ *-4*

Solve for x.

19. $\log_2 x = 4$ *16* **20.** $\log_3 x = 2$ *9* **21.** $\log_5 x = 3$ *125*

22. $\log_3 x = -1$ *$\frac{1}{3}$* **23.** $\log_5 x = -2$ *$\frac{1}{25}$* **24.** $\log_2 x = -3$ *$\frac{1}{8}$*

Solve for the base *n*.

25. $\log_n 8 = 3$ *2* **26.** $\log_n 81 = 4$ *3* **27.** $\log_n 5 = 1$ *5*

28. $\log_n \frac{1}{9} = -2$ *3* **29.** $\log_n \frac{1}{16} = -4$ *2* **30.** $\log_n \frac{1}{125} = -3$ *5*

PART B

EXAMPLE Find $\log_2 \sqrt[5]{8}$.

Write a logarithmic equation. →	$\log_2 \sqrt[5]{8} = y$
Change to exponential form. →	$\sqrt[5]{8} = 2^y$
Show 8 as a power of 2: $8 = 2^3$. →	$\sqrt[5]{2^3} = 2^y$
$\sqrt[n]{x^m} = x^{\frac{m}{n}}$; $\sqrt[5]{2^3} = 2^{\frac{3}{5}}$ →	$2^{\frac{3}{5}} = 2^y$
Compare the exponents. (Bases are the same.) →	$\frac{3}{5} = y$

Thus, $\log_2 \sqrt[5]{8} = \frac{3}{5}$.

Find the logarithm.

31. $\log_2 \sqrt[3]{4}$ *$\frac{2}{3}$* **32.** $\log_2 \sqrt[5]{16}$ *$\frac{4}{5}$* **33.** $\log_2 \sqrt[3]{32}$ *$\frac{5}{3}$* **34.** $\log_2 \sqrt{8}$ *$\frac{3}{2}$*

35. $\log_3 \sqrt[5]{9}$ *$\frac{2}{5}$* **36.** $\log_3 \sqrt[4]{27}$ *$\frac{3}{4}$* **37.** $\log_4 \sqrt[5]{16}$ *$\frac{2}{5}$* **38.** $\log_5 \sqrt[3]{25}$ *$\frac{2}{3}$*

39. $\log_{\frac{1}{3}} \frac{1}{9}$ *2* **40.** $\log_{\frac{1}{2}} \frac{1}{8}$ *3* **41.** $\log_{\frac{1}{4}} \frac{1}{4}$ *1* **42.** $\log_{\frac{2}{3}} \frac{4}{9}$ *2*

Exponential Equations

OBJECTIVE

- To solve an equation like $6^{3x-2} = 5$

REVIEW CAPSULE

$$1{,}000 = 10^3 \quad\quad \text{If } a = b$$

$$\log 1{,}000 = \log 10^3 \quad\quad \text{then } \log a = \log b$$

If two numbers are equal, then their logs are equal.

EXAMPLE 1 Solve $3^x = 17$ for x to three significant digits.

If $a = b$, then $\log a = \log b$. Take the log of each side. → $3^x = 17$; $\log 3^x = \log 17$

$\log a^r = r \cdot \log a$ → $x \cdot \log 3 = \log 17$

Divide each side by log 3. → $x = \dfrac{\log 17}{\log 3}$

$\log 17 = 1.2304$, $\log 3 = 0.4771$ → $x = \dfrac{1.2304}{0.4771}$

Divide $1.2304 \div .4771$ to 4 digits. → $x = 2.579$

Round the quotient to 3 digits. → $x \doteq 2.58$

Thus, $x = 2.58$ to 3 significant digits.

EXAMPLE 2 Solve $9^{2x-1} = 5$ for x to 3 significant digits.

Take the log of each side. → $\log 9^{2x-1} = \log 5$

$\log a^r = r \cdot \log a$ → $(2x - 1) \log 9 = \log 5$

Divide each side by log 9. → $2x - 1 = \dfrac{\log 5}{\log 9}$

$2x - 1 = \dfrac{0.6990}{0.9542}$

Divide to 4 digits. → $2x - 1 = .7326$

Add 1 to each side. → $2x = 1.7326$

Divide $1.7326 \div 2$ to 4 digits. → $x = .8663$

Round .8663 to 3 digits. → **Thus,** $x = .866$ to 3 significant digits.

EXERCISES

Solve for x to 3 significant digits.

1. $4^x = 21$ *2.20*
2. $4^{x-1} = 21$ *3.20*
3. $4^{x+1} = 21$ *1.20*
4. $3^x = 15$ *2.47*
5. $3^{2x} = 15$ *1.23*
6. $3^{x+2} = 15$ *.465*
7. $9^x = 6$ *.816*
8. $9^{x-2} = 6$ *2.82*
9. $9^{2x-1} = 6$ *.908*
10. $7^x = 5$ *.827*
11. $7^{2x} = 5$ *.414*
12. $7^{2x-2} = 5$ *1.41*

Binomial Expansions

OBJECTIVE

- To find a power of a binomial like $(3x - y)^5$

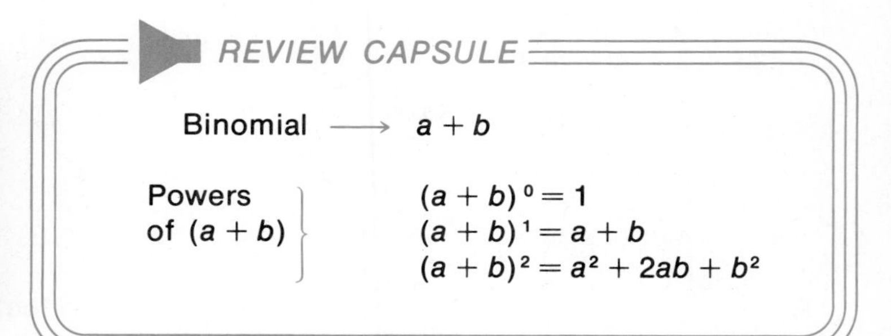

EXAMPLE 1 Find the 3rd, 4th, and 5th powers of $(a + b)$.

$(a + b)^2 (a + b) = (a^2 + 2ab + b^2)(a + b)$ ⟶ $(a + b)^3 = a^3 + 3a^2b^1 + 3a^1b^2 + b^3$

$(a + b)^4 = (a + b)^3(a + b)$ ⟶ $(a + b)^4 = a^4 + 4a^3b^1 + 6a^2b^2 + 4a^1b^3 + b^4$

$(a + b)^5 = (a + b)^4(a + b)$ ⟶ $(a + b)^5 = a^5 + 5a^4b^1 + 10a^3b^2 + 10a^2b^3 + 5a^1b^4 + b^5$

A pattern for the *exponents* of a and b is observed in $(a + b)^3$.

Coefficients are not written. ⟶ $(a + b)^3 = a^3 + \_\_ a^2b^1 + \_\_ a^1b^2 + b^3$

Exponents of a ⟶ 3 2 1

Exponents of b ⟶ 1 2 3

There is a similar pattern for the other powers of $(a + b)$.

The exponents of a *decrease* by 1: 3, 2, 1.
The exponents of b increase by 1: 1, 2, 3.

EXAMPLE 2 Write the product $(a + b)^6$ without its coefficients.

Use the pattern above.
Start with a^6; end with b^6. ⟶

$$(a + b)^6 = a^6 + \_\_ a^5b^1 + \_\_ a^4b^2 + \_\_ a^3b^3 + \_\_ a^2b^4 + \_\_ a^1b^5 + \_\_ b^6$$

Showing the *coefficients* of powers of $(a + b)$ in a triangular form reveals a pattern.

$(a + b)^0 = \underline{1}$ ⟶ 1

$(a + b)^1 = \underline{1}a + \underline{1}b$ ⟶ 1 1

$(a + b)^2 = \underline{1}a^2 + \underline{2}ab + \underline{1}b^2$ ⟶ 1 2 1

Find these coefficients in Example 1.
- $(a + b)^3$ ⟶ 1 3 3 1
- $(a + b)^4$ ⟶ 1 4 6 4 1
- $(a + b)^5$ ⟶ 1 5 10 10 5 1

Coefficients in any row are *symmetrical*. For example,
1 4 6 4 1

The coefficients form *Pascal's triangle.*

0 row ⟶ 1
1st row ⟶ 1 1
2nd row ⟶ 1 2 1

Each *row* in Pascal's triangle is formed by *adding* elements of the previous row.

3rd row of Pascal's triangle [Coefficients of $(a + b)^3$] ⟶

4th row of triangle; each row begins and ends with 1. ⟶

5th row of Pascal's triangle [Coefficients of $(a + b)^5$] ⟶

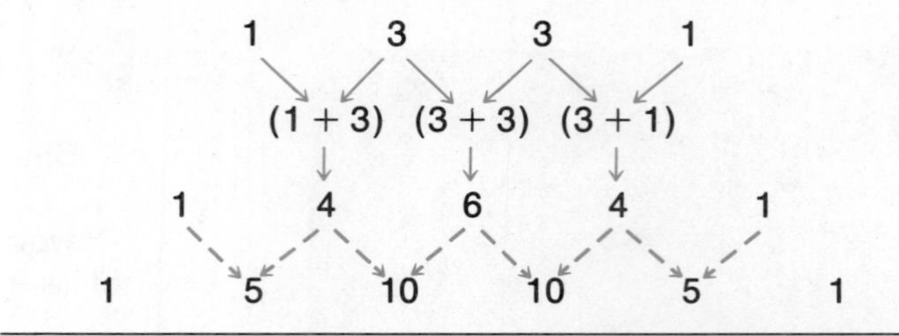

EXAMPLE 3 Write the product $(a + b)^6$ with its coefficients.

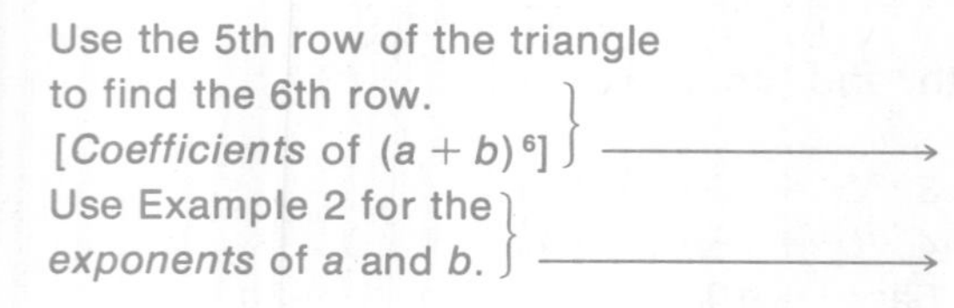

Use the 5th row of the triangle to find the 6th row. [*Coefficients* of $(a + b)^6$] ⟶

1 5 10 10 5 1
1 6 15 20 15 6 1

Use Example 2 for the *exponents* of a and b. ⟶

Thus, $(a + b)^6 = a^6 + 6a^5b^1 + 15a^4b^2 + 20a^3b^3 + 15a^2b^4 + 6a^1b^5 + b^6$.

EXAMPLE 4 Find the product $(4x + y)^3$. Simplify each term.

Use $(a + b)^3$ to get coefficients and exponents. ⟶ $(a + b)^3 = a^3 + 3a^2b^1 + 3a^1b^2 + b^3$

Substitute $(4x)$ for a, y for b. ⟶ $(4x + y)^3 = (4x)^3 + 3(4x)^2 \cdot y^1 + 3(4x)^1 \cdot y^2 + y^3$

Simplify the powers of $(4x)$. ⟶ $= 64x^3 + 3(16x^2) \cdot y + 3(4x) \cdot y^2 + y^3$

Simplify each term. ⟶ $= 64x^3 + 48x^2y + 12xy^2 + y^3$

Thus, $(4x + y)^3 = 64x^3 + 48x^2y + 12xy^2 + y^3$.

EXAMPLE 5 Find the product $(x^2 - 2y)^3$. Simplify each term.

Use $(a + b)^3$. ⟶ $a^3 + 3a^2b^1 + 3a^1b^2 + b^3$

Substitute (x^2) for a, $(-2y)$ for b. ⟶ $(x^2)^3 + 3(x^2)^2(-2y)^1 + 3(x^2)^1(-2y)^2 + (-2y)^3$

Simplify powers of (x^2) and $(-2y)$. ⟶ $x^6 + 3(x^4)(-2y) + 3(x^2)(4y^2) - 8y^3$

Simplify each term. ⟶ $x^6 - 6x^4y + 12x^2y^2 - 8y^3$

Thus, $(x^2 - 2y)^3 = x^6 - 6x^4y + 12x^2y^2 - 8y^3$.

EXAMPLE 6 Find the 5th term of the product $(2x^3 + \sqrt{3})^6$.

Use 6th row of the triangle to get coefficients of $(a+b)^6$. ⟶ 1 6 15 20 15 6 1

Exponents of a decrease by 1. Exponents of b increase by 1. ⟶ 5th term of $(a + b)^6$: $15a^2b^4$.

Substitute $(2x^3)$ for a, $(\sqrt{3})$ for b. Then simplify the 5th term.

5th term of $(2x^3 + \sqrt{3})^6$: $15(2x^3)^2(\sqrt{3})^4$, or $15(4x^6)(9)$

Thus, $540x^6$ is the 5th term of $(2x^3 + \sqrt{3})^6$.

EXERCISES

PART A

Find the product. Simplify each term.

1. $(a+b)^7$
2. $(a+b)^8$
3. $(a+b)^9$
4. $(x+2)^6$
5. $(x-3)^5$
6. $(x+1)^7$
7. $(x-y)^5$
8. $(m-2n)^5$
9. $(x+3y)^4$
10. $(2x+y)^4$
11. $(3x-y)^5$
12. $(2r+t)^6$
13. $(x^2+y)^4$
14. $(x^2-3)^5$
15. $(x+y^2)^5$
16. $(x+.1)^4$
17. $(x+.2)^4$ $x^4+.8x^3+.24x^2+.032x+.0016$
18. $(2x+.1)^5$

Find the indicated term of the product. Simplify the term.

19. 5th term of $(a+b)^7$
20. 3rd term of $(2x+y)^5$
21. 4th term of $(x-3)^6$
22. 4th term of $(c-3d)^5$
23. 5th term of $(x^2+y)^6$
24. 5th term of $(x^3-2)^7$

PART B

Find the product. Simplify each term.

25. $(2x+3)^5$
26. $(3x-2)^4$
27. $(2x-5)^5$
28. $(x^2+3y)^5$
29. $(x^3-2y)^4$
30. $(x^2+y^2)^5$
31. $(3x+2y)^5$
32. $(2x-5y)^4$
33. $(2x-2y)^5$
34. $(x+\sqrt{2})^6$
35. $(x-\sqrt{3})^5$
36. $(x-\sqrt{y})^4$
37. $\left(x+\frac{y}{2}\right)^4$
38. $\left(\frac{1}{x}+\frac{1}{y}\right)^5$ $\frac{1}{x^5}+\frac{5}{x^4y}+\frac{10}{x^3y^2}+\frac{10}{x^2y^3}+\frac{5}{xy^4}+\frac{1}{y^5}$
39. $\left(\frac{1}{m}-\frac{1}{n}\right)^6$

Find the indicated term of the product. Simplify the term.

40. 3rd term of $(2x+3y)^5$ $720x^3y^2$
41. 4th term of $(5x-2y)^6$
42. 4th term of $(x^3+2y)^5$ $80x^6y^3$
43. 5th term of $(3x^2-\sqrt{y})^6$
44. 5th term of $\left(4x^2+\frac{1}{2}\right)^6$ $15x^4$
45. 4th term of $\left(x^3-\frac{1}{x}\right)^5$

PART C

Binomial Theorem

$$(a+b)^n=a^n+\frac{n}{1}a^{n-1}b^1+\frac{n(n-1)}{1\cdot 2}a^{n-2}b^2+\frac{n(n-1)(n-2)}{1\cdot 2\cdot 3}a^{n-3}b^3+\cdots+b^n,$$

for each positive integer n.

EXAMPLE Find the *first four* terms of $(x+2y)^{12}$.

Use the binomial theorem. $n=12$, $a=x$, $b=2y$ → $x^{12}+\frac{12}{1}x^{11}(2y)^1+\frac{12\cdot 11}{1\cdot 2}x^{10}(2y)^2+\frac{12\cdot 11\cdot 10}{1\cdot 2\cdot 3}x^9(2y)^3+\cdots$

Divide out common factors. → $x^{12}+12x^{11}(2y)^1+\frac{\overset{6}{12}\cdot 11}{1\cdot \underset{1}{2}}x^{10}(4y^2)+\frac{\overset{2}{4}\cdot\overset{1}{3}\cdot 11\cdot 10}{1\cdot\underset{1}{2}\cdot\underset{1}{3}}x^9(8y^3)+\cdots$

First 4 terms of $(x+2y)^{12}$ → $x^{12}+24x^{11}y+264x^{10}y^2+1{,}760x^9y^3+\cdots$

Find the first four terms of the product. Use the binomial theorem.

46. $(x+y)^{11}$
47. $(x-2y)^{10}$
48. $(x^2+y^3)^{10}$
49. $(x+\frac{y}{2})^{12}$

Chapter Fifteen Review

Find the logarithm. [p. 383]

1. log .01
2. log .00346
3. log .805

Find the antilog. [p. 383]

4. 8.2227 − 10
5. 7.0899 − 10
6. 9.9899 − 10

Find each to 3 significant digits. [p. 386]

7. $.371 \times 584$
8. $.0762^3$
9. $\frac{.0076}{.345}$
10. $\frac{38.4}{703}$
11. $\sqrt{.0059}$
12. $\sqrt[3]{.0436}$

Find each to 3 significant digits. [p. 389]

13. $.89 \times 340 \times .023$
14. $\frac{.039 \times 786}{53.4}$
15. $\frac{.873}{2.6 \times 3.8}$
16. $\left(\frac{2.75}{.372}\right)^3$
17. $\sqrt{.331 \times 2.4}$
18. $\sqrt[3]{\frac{98.7}{.034}}$

Find the logarithm. [p. 389]

19. $\log 5t^2w$
20. $\log \frac{2d}{mn}$
21. $\log (7t^4)^3$
22. $\log \sqrt{\frac{3vw}{x}}$

Find the logarithm. Use interpolation. [p. 392]

23. log 59.32
24. log .02763

Find the antilog interpolated to 4 digits. [p. 392]

25. 2.6851
26. 8.3394 − 10

Find each interpolated to 4 digits. [p. 392]

27. 72.47×2.06
28. $\left(\frac{603.5}{28.4}\right)^2$
29. $\sqrt[3]{37.64^2}$

Find the logarithm. [p. 395]

30. $\log_3 81$
31. $\log_4 \frac{1}{16}$
32. $\log_5 \sqrt{5}$
33. $\log_2 \sqrt[3]{16}$

Solve for *x* to 3 significant digits. [p. 398]

34. $3^x = 26$
35. $9^{2x} = 4$
36. $5^{x-1} = 17$

Find the product. Simplify each term. [p. 399]

37. $(x + 2y)^6$
38. $(x^2 - 3)^4$
39. $(2x - \frac{1}{2}y)^5$

Find the indicated term of the product. Simplify the term. [p. 399]

40. 4th term of $(m + 3n)^6$
41. 5th term of $(x^2 - \sqrt{2})^6$

Chapter Fifteen Test

1. Find log .0374.

2. Find the antilog of $9.6656 - 10$.

Find each to 3 significant digits.

3. $.0514 \times 43.3$

4. $\frac{.875}{.024}$

5. $.602^3$

6. $(.632 \times .037)^2$

7. $\frac{.048 \times 730}{863}$

8. $\sqrt{\frac{5.91}{.217}}$

Find the logarithm.

9. $\log \frac{m^2 n}{p}$

10. $\log (5t^3)^2$

11. Find log 304.3. Use interpolation.

12. Find the antilog of $9.6800 - 10$ interpolated to 4 digits.

Find each interpolated to 4 digits.

13. 4.324×71.3

14. $\sqrt[3]{274.4^2}$

Find the logarithm.

15. $\log_5 25$

16. $\log_3 \frac{1}{9}$

17. $\log_2 \sqrt[5]{16}$

18. Solve $3^{x-1} = 14$ for x to 3 significant digits.

Find the product. Simplify each term.

19. $(x + 2y)^5$

20. $(x^2 - y)^4$

21. Find the 4th term of $(4x - \frac{1}{2}y)^6$. Simplify the term.

16 TRIGONOMETRIC FUNCTIONS

Fibonacci Lab

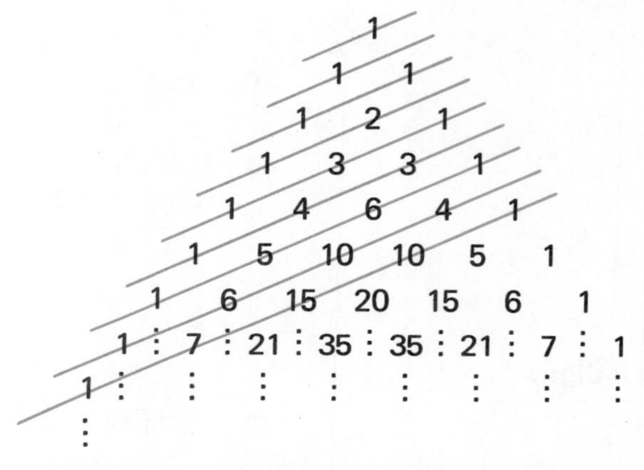

Notice that the Fibonacci sequence is generated by diagonal sums.

From the Fibonacci sequence,

0, 1, 1, 2, 3, 5, 8, 13, . . .

A sequence of fractions can be written

$$0, 1, \frac{1}{2}, \frac{2}{4}, \frac{3}{8}, \frac{5}{16}, \ldots$$

The fractional version of the Fibonacci sequence is found in the work of the geneticist Gregor Mendel (1822–1884). It is used in almost every facet of genetics concerned with genetic traits.

PROJECT

1. Write the next four terms of the fractional sequence.
2. How is the fractional sequence formed?
3. Write the *n*th term of the fractional sequence.

Angles of Rotation and Their Measures

OBJECTIVES

- To associate a directed degree measure with any angle of rotation
- To determine the position of the terminal side for a given rotation

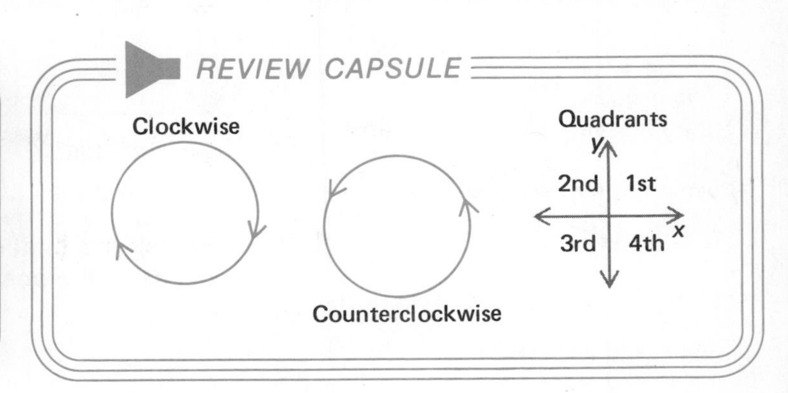

A rotating ray forms an *angle of rotation.*

Counterclockwise Rotations

Rotate the ray counterclockwise. →

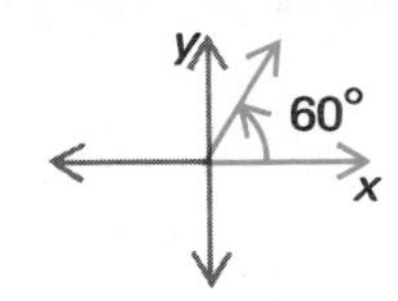

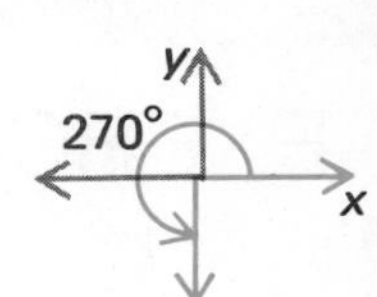

The measures of the angles are positive. →

Angles formed by counterclockwise rotations are positive.

EXAMPLE 1 Show angles with measures of $+60°$, $+135°$, $+270°$.

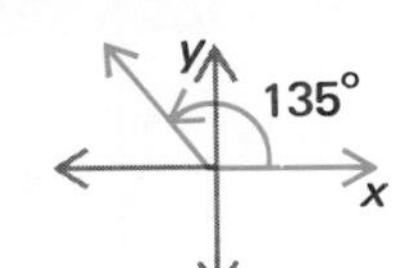

Clockwise Rotations

Rotate the ray clockwise. →

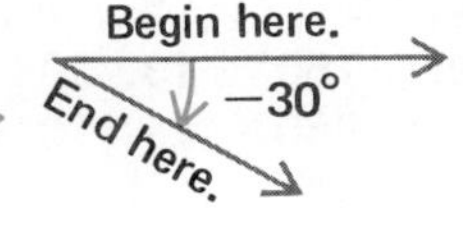

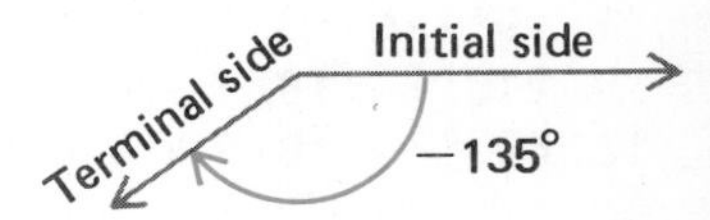

The measures of the angles are negative. →

Angles formed by clockwise rotations are negative.

EXAMPLE 2 Show angles with measures of $-45°$, $-135°$, $-330°$.

Start with ray pointing right and endpoint at the origin. →

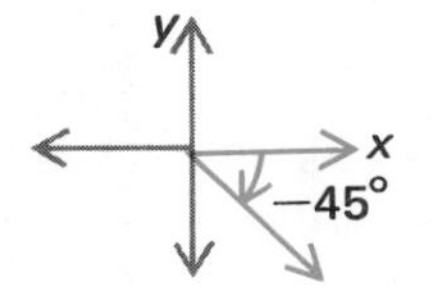

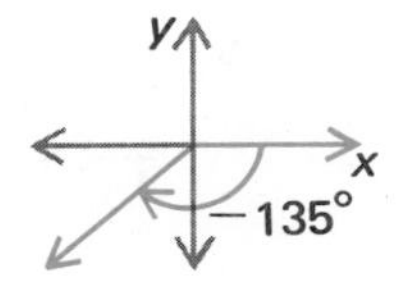

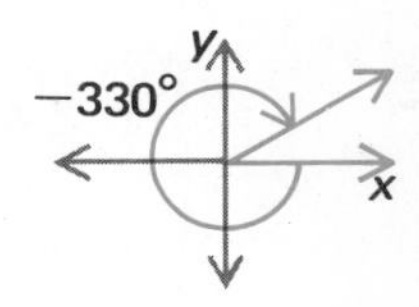

EXAMPLE 3 Label the initial side I and the terminal side T for the angle shown. Determine the quadrant in which the terminal side lies.

I: initial side
T: terminal side

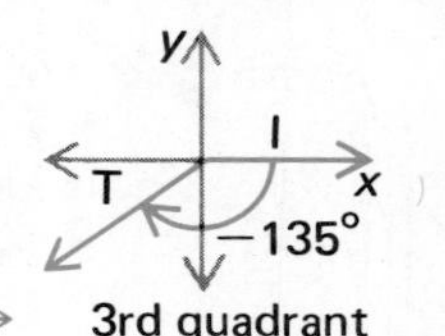

315°, I, T — y, x

−315°, T, I — y, x

Position of terminal side. ⟶ 3rd quadrant | 4th quadrant | 1st quadrant

EXAMPLE 4 Sketch the rotations: 360°, −540°, 450°.
Describe the position of the terminal side.

Angles with terminal sides on axes are called quadrantal angles. ⟶

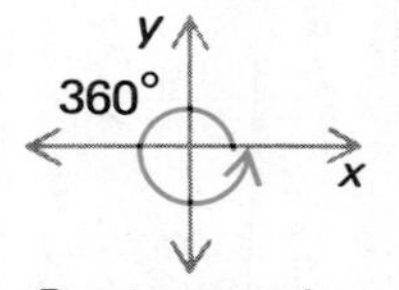

−360° + (−180°), −540°

360° + 90°, 450°

Position of terminal side. ⟶ Postive x-axis | Negative x-axis | Positive y-axis

EXAMPLE 5 State a positive and negative measure for angles which have the same terminal side as the one shown.

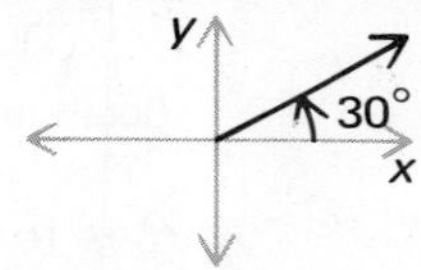

$360° + 30° = 390°$
$30° - 360° = -330°$

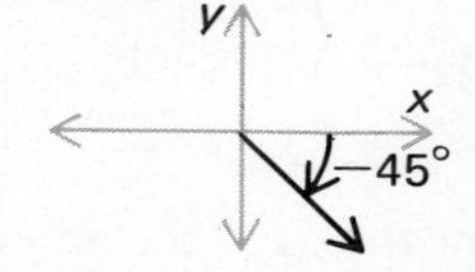

$360° + -45° = 315°$
$-45° + -360° = -405°$

EXERCISES

PART A

Sketch the angle of rotation. Label the initial side I and the terminal side T. Describe the position of the terminal side. *(Quadrant is given.)*

1. 45° *1st* **2.** −120° **3.** −45° *4th* **4.** 330° *4th* **5.** −135° **6.** 120° *2nd* **7.** 150° *2nd* **8.** −30°
9. 210° *3rd* **10.** −330° **11.** 135° *2nd* **12.** −180° **13.** 90° **14.** −225° **15.** 300° *4th* **16.** −90°
17. 400° *1st* **18.** 720° *positive x-axis* **19.** −130° *3rd* **20.** −650° *1st* **21.** 10° *1st* **22.** −190° *2nd* **23.** −350° *1st* **24.** 515° *2nd*

PART B

State two more measures, one positive and one negative, for angles which have the same terminal side as the one shown.

25.

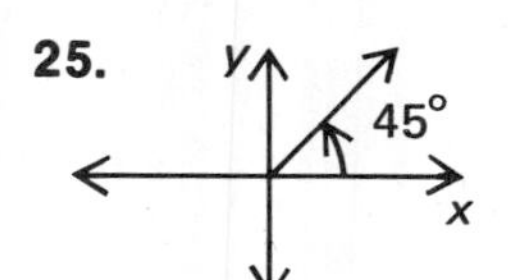

26.

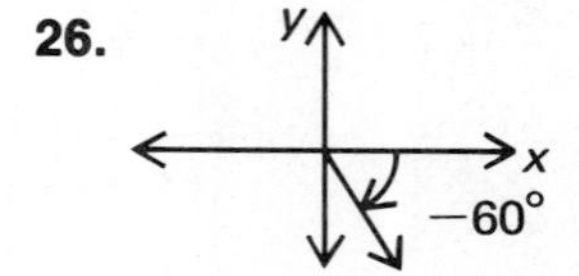

27.

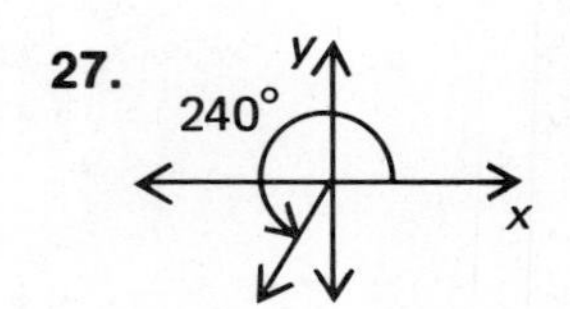

28.

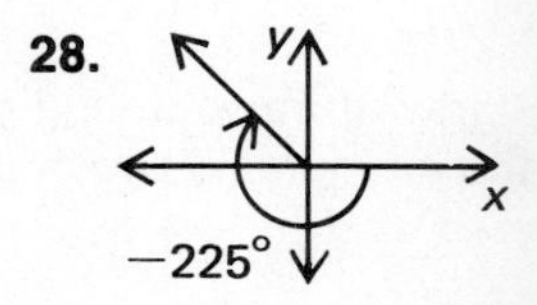

Reference Angles and Triangles

OBJECTIVE

- To locate a reference angle and reference triangle for any angle

REVIEW CAPSULE

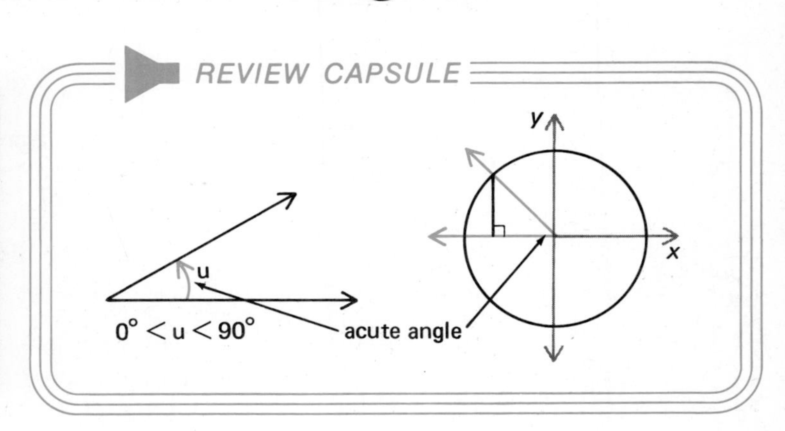

Each angle whose terminal side is not on an axis has a *reference angle.*

The terminal side is a side of a reference angle.

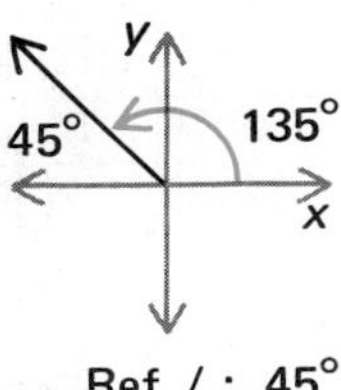

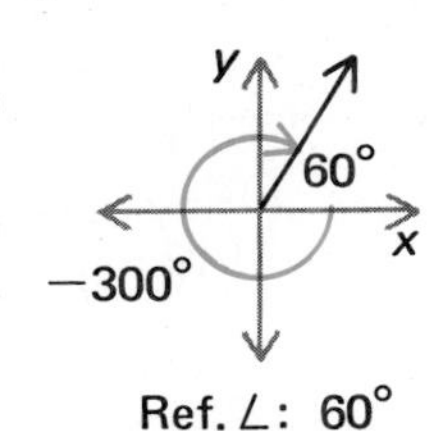

y
210°
30°
x

Use nondirected measures. ⟶

Ref. ∠: 45° | Ref. ∠: 60° | Ref. ∠: 30°

EXAMPLE 1 Show the angle of rotation: −250°, 300°, −140°. Identify the reference angle.

Sketch the rotation. Select the terminal side and x-axis.

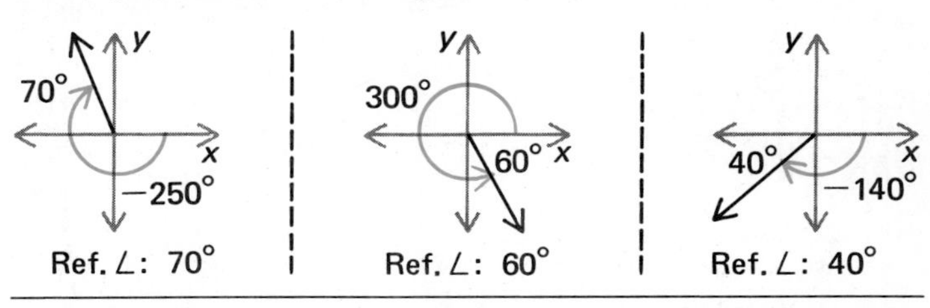

A reference triangle includes the reference angle.

Each angle that has a reference angle has a *reference triangle.*

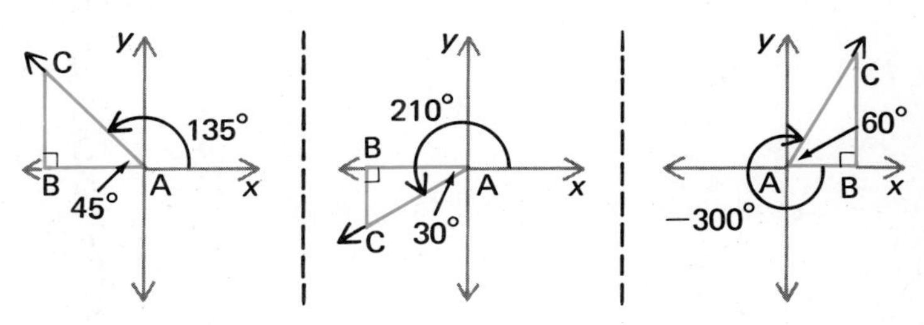

EXAMPLE 2 Sketch the angle of rotation: $-150°$, $45°$, $300°$. Identify the reference triangle.

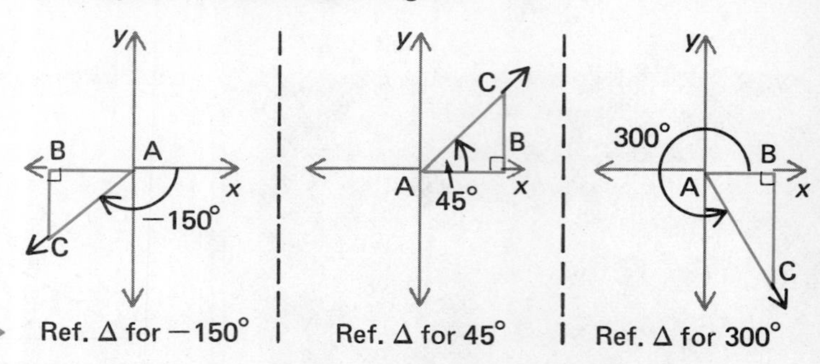

ABC is a reference △. ⟶

EXAMPLE 3 Sketch the angle of rotation: $45°$, $210°$, $-220°$. Use a circle with any radius. Determine the reference angle and triangle.

Place the vertex at the center of the circle.
Choose the point where the terminal side meets the circle.
Draw a ⊥ to the *x*-axis.

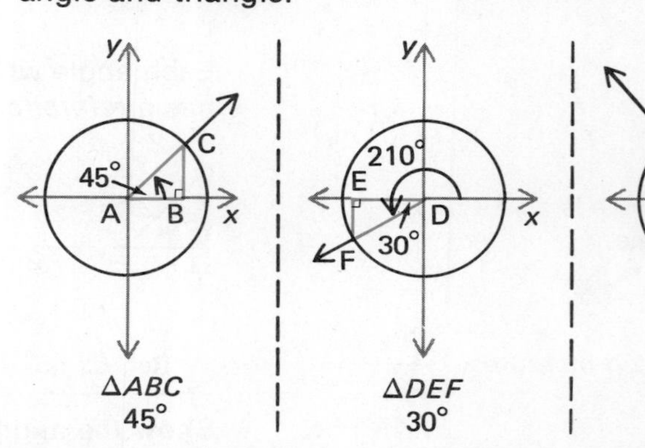

Reference triangle. ⟶ ΔABC | ΔDEF | ΔGHI

Reference angle. ⟶ 45° | 30° | 40°

EXERCISES

PART A

Sketch the angle of rotation. Identify the reference angle.

1. 160° *20°* **2.** −40° *40°* **3.** 225° *45°* **4.** −195° *15°* **5.** −150° *30°* **6.** 290° *70°*
7. −300° *60°* **8.** 190° *10°* **9.** −245° *65°* **10.** 330° *30°* **11.** 495° *45°* **12.** −780° *60°*

Sketch the angle of rotation. Identify the reference triangle.

13. 75° **14.** −220° **15.** 150° **16.** −75° **17.** −150° **18.** 300°
19. −330° **20.** 280° **21.** 250° **22.** −295° **23.** −510° **24.** 780°

PART B

For a circle with any radius, determine the reference angle and the reference triangle for the rotation.

25. 150° *30°* **26.** 135° *45°* **27.** −135° *45°* **28.** −210° *30°* **29.** 290° *70°* **30.** 315° *45°*
31. −315° *45°* **32.** 225° *45°* **33.** −75° *75°* **34.** 80° *80°* **35.** 345° *15°* **36.** 750° *30°*

Special Right Triangles

OBJECTIVE

- To find lengths of sides in a 30°-60° or 45°-45° right triangle

REVIEW CAPSULE

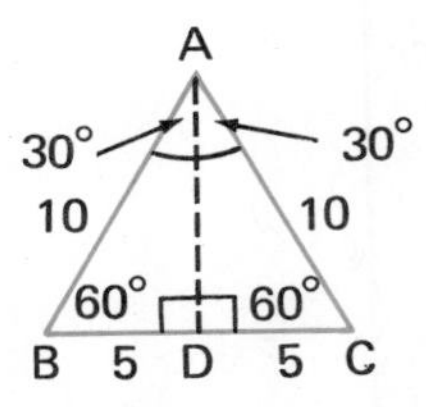

$\triangle ABC$ is equilateral.
$\overline{AD} \perp \overline{BC}$

In an equilateral $\triangle$, an altitude bisects the base and the vertex angle from which it is drawn.

EXAMPLE 1 $\triangle ABC$ is a 30°-60° right $\triangle$, $AB = 6$. Find BC.

Build an equilateral $\triangle$.

$BD = AB$ ⟶ $BD = 6$

$\overline{AC}$ bisects $\overline{BD}$ ⟶ $BC = \frac{1}{2} BD$

(See the Review Capsule.)

Thus, $BC = 3$.

A, 30°, 30°, 6, 60°, 60°, B, C, D

A, 30°, 6, 60°, B, C

EXAMPLE 2 For $\triangle ABC$ in Example 1, find AC.

Use the Pythagorean theorem. ⟶ $(BC)^2 + (AC)^2 = (AB)^2$

From Example 1, $BC = 3$. ⟶ $3^2 + (AC)^2 = 6^2$

$9 + (AC)^2 = 36$

$(AC)^2 = 27$

$\sqrt{27} = \sqrt{9} \cdot \sqrt{3} = 3\sqrt{3}$ ⟶ $AC = \sqrt{27}$, or $3\sqrt{3}$

Thus, $AC = 3\sqrt{3}$.

Examples 1 and 2 suggest this. ⟶

In a 30°-60° right $\triangle$:
1. the leg opposite the 30° angle has $\frac{1}{2}$ the measure of the hypotenuse
2. the leg opposite the 60° angle has $\frac{1}{2}$ the measure of the hypotenuse times $\sqrt{3}$.

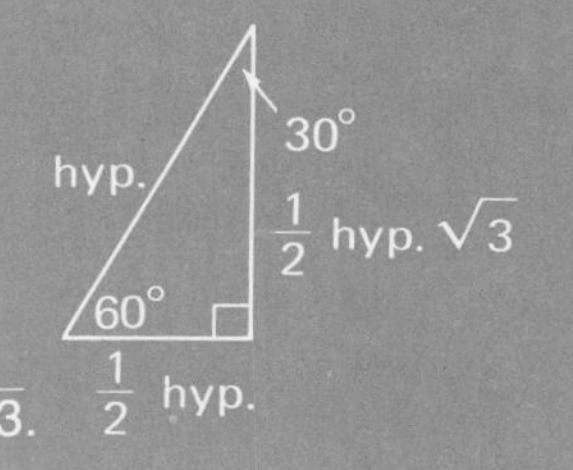

EXAMPLE 3 In right $\triangle ABC$, $AB = 4$. Find AC and BC.

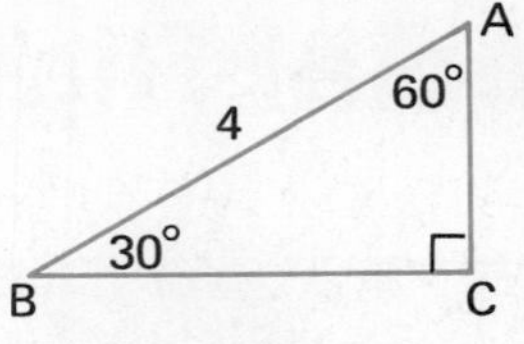

$\overline{AC}$ is opposite the 30° angle.
$\overline{BC}$ is opposite the 60° angle.

$$AC = \tfrac{1}{2}(AB) = \tfrac{1}{2}(4) = 2$$

$$BC = \tfrac{1}{2}(AB)(\sqrt{3}) = \tfrac{1}{2}(4)(\sqrt{3}) = 2\sqrt{3}$$

Thus, $AC = 2$ and $BC = 2\sqrt{3}$.

EXAMPLE 4 In right $\triangle XYZ$, $YZ = 5$. Find XY.

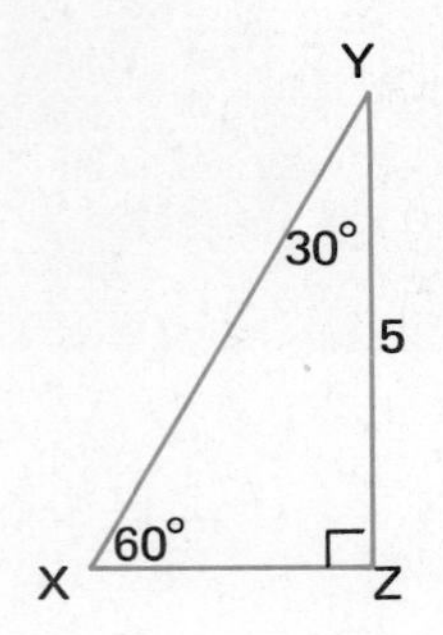

$\overline{YZ}$ is opposite the 60° angle.

$$YZ = \tfrac{1}{2}(XY)(\sqrt{3})$$

$YZ = 5$ ⟶ $5 = \tfrac{1}{2}(XY)(\sqrt{3})$

Multiply each side by 2. ⟶ $10 = (XY)(\sqrt{3})$

Divide each side by $\sqrt{3}$. ⟶ $\frac{10}{\sqrt{3}} = XY$

$\frac{10}{\sqrt{3}} = \frac{10}{\sqrt{3}} \cdot \frac{\sqrt{3}}{\sqrt{3}} = \frac{10\sqrt{3}}{3}$ ⟶ **Thus,** $XY = \frac{10}{\sqrt{3}}$, or $\frac{10\sqrt{3}}{3}$.

EXAMPLE 5 In right $\triangle DEF$, $FD = 12$. Find FE.

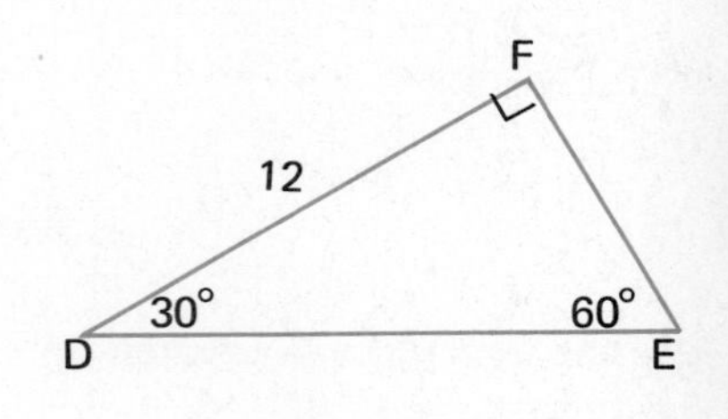

$\overline{FE}$ is opposite the 30° angle. $FE = \tfrac{1}{2}(DE)$. Find DE.

$$\tfrac{1}{2}(DE)(\sqrt{3}) = FD$$

$$\tfrac{1}{2}(DE)(\sqrt{3}) = 12$$

$$(DE)(\sqrt{3}) = 24$$

$\frac{24}{\sqrt{3}} = \frac{24 \cdot \sqrt{3}}{\sqrt{3} \cdot \sqrt{3}} = \frac{24\sqrt{3}}{3}$

$$DE = \frac{24}{\sqrt{3}}, \text{ or } 8\sqrt{3}$$

$FE = \tfrac{1}{2}(DE)$. ⟶ So, $FE = \tfrac{1}{2}(8\sqrt{3}) = 4\sqrt{3}$

Thus, $FE = 4\sqrt{3}$.

EXAMPLE 6 $\triangle ABC$ is a 45°-45° right triangle. $AB = 8$. Find AC and BC.

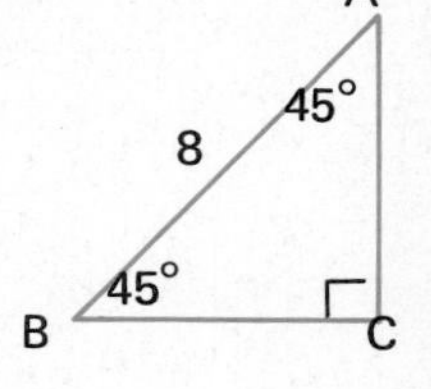

$\triangle ABC$ is isosceles. ⟶ $AC = BC$

Use the Pythagorean theorem. ⟶ $(AC)^2 + (BC)^2 = (AB)^2$

Let $AC = x$. ⟶ $x^2 + x^2 = 8^2$

$$2x^2 = 64$$

$$x^2 = 32$$

$\sqrt{32} = \sqrt{16} \cdot \sqrt{2} = 4\sqrt{2}$ ⟶ $x = \sqrt{32}$, or $4\sqrt{2}$

Thus, both $\overline{AC}$ and $\overline{BC}$ measure $4\sqrt{2}$.

Example 6 suggests this. →

> In a 45°-45° right triangle, each leg has $\frac{1}{2}$ the measure of the hypotenuse times $\sqrt{2}$.

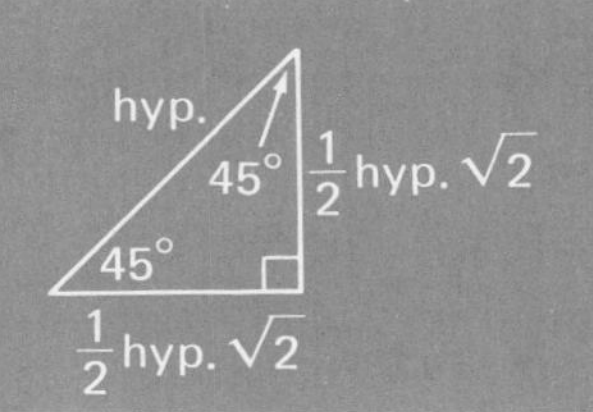

EXAMPLE 7 In a 45°-45° right triangle, one leg measures 6. Find the length of the hypotenuse.

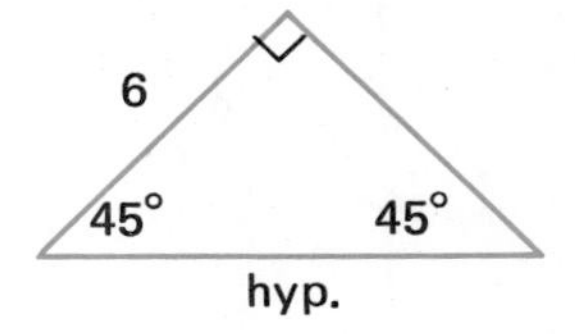

The leg measures 6. →

$$\tfrac{1}{2}\text{ hyp. }(\sqrt{2}) = 6$$

$$\text{hyp. }(\sqrt{2}) = 12$$

$\frac{12}{\sqrt{2}} = \frac{12}{\sqrt{2}} \cdot \frac{\sqrt{2}}{\sqrt{2}} = \frac{12\sqrt{2}}{2}$ →

$$\text{hyp.} = \frac{12}{\sqrt{2}}, \text{ or } 6\sqrt{2}$$

Thus, the length of the hypotenuse is $6\sqrt{2}$.

EXAMPLE 8 For a circle with a radius of 4 units and a reference triangle of 30°-60°-90° as shown, find the point (x, y).

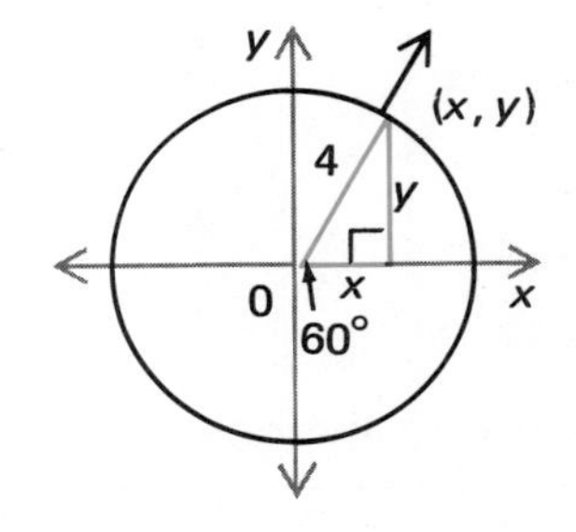

Use the 30°-60° right △. →

$$|x| = \tfrac{1}{2}(4)$$

$$|x| = 2$$

x is to the right *of the origin.* →

$$x = 2$$

Use the 30°-60° right △. →

$$|y| = \tfrac{1}{2}(4)\sqrt{3}$$

$$|y| = 2\sqrt{3}$$

y is above *the origin.* →

$$y = 2\sqrt{3}$$

(x, y) → **Thus,** the point is $(2, 2\sqrt{3})$.

EXAMPLE 9 For the following figure, determine the radius r and the point (x, y) if $y = 2$.

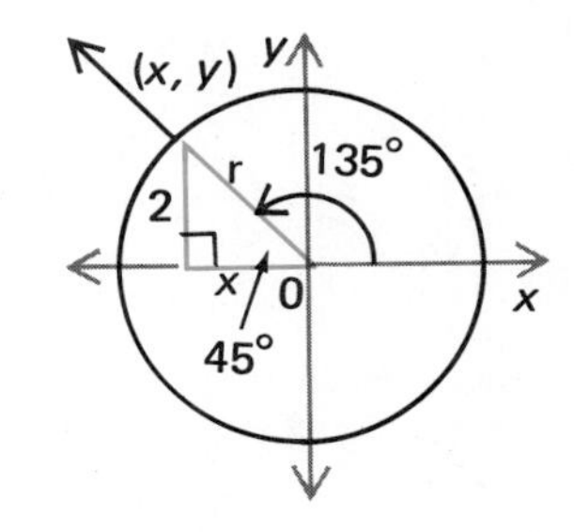

$$\tfrac{1}{2}r\sqrt{2} = y$$

$y = 2$ →

$$\tfrac{1}{2}r\sqrt{2} = 2$$

$$r\sqrt{2} = 4$$

$$r = \frac{4}{\sqrt{2}}, \text{ or } 2\sqrt{2}$$

The triangle is isosceles. $|x| = |y|$. →

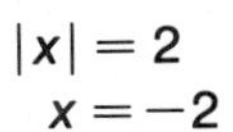

$$|x| = 2$$

x is to the left of the origin. →

$$x = -2$$

Thus, $r = 2\sqrt{2}$ and $(x, y) = (-2, 2)$.

EXERCISES

PART A

Determine the lengths of the indicated sides. Rationalize all denominators.

1.

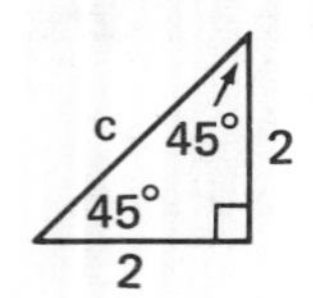

2.

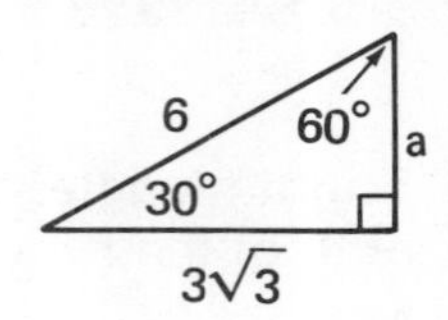

3.

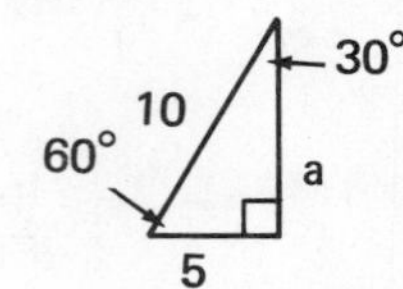

4.

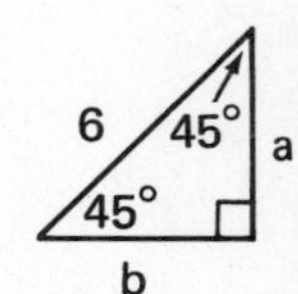

5.

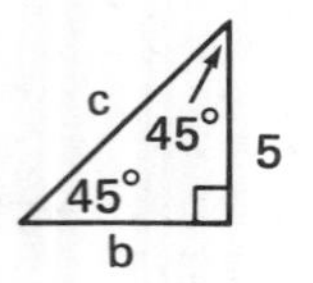

6.

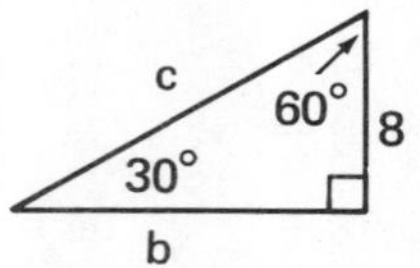

7.

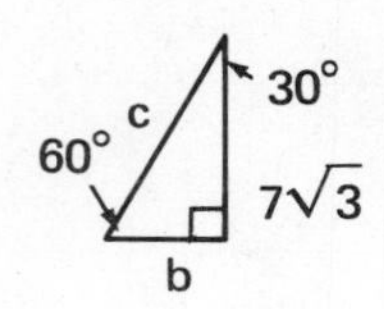

8.

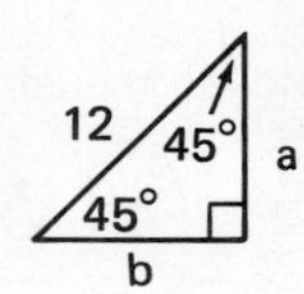

Determine the point (x, y) or the radius r.

9.

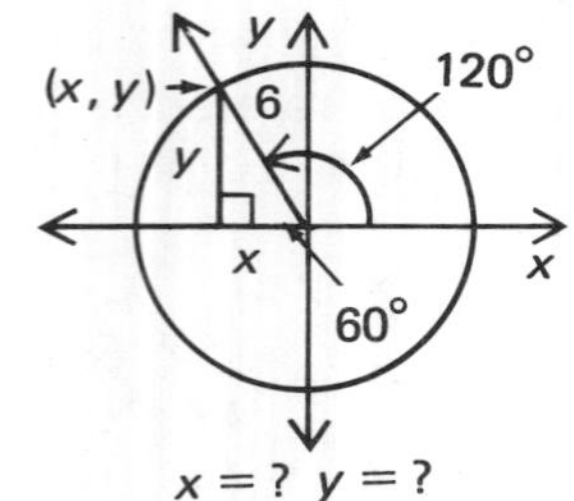

10.

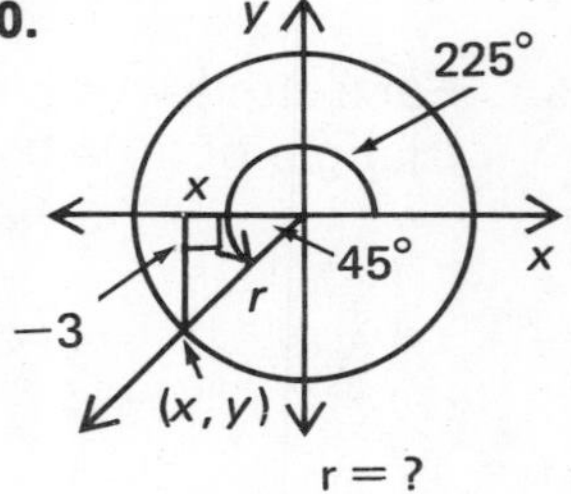

11.

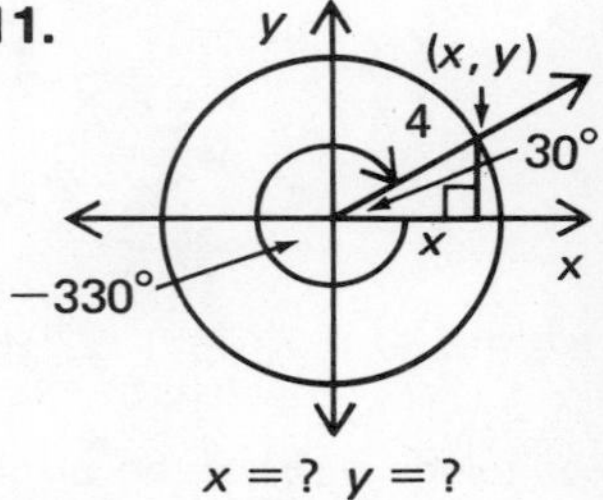

PART B

Complete the chart.

	Length of Hypotenuse	Length of Leg Opposite 30° ∠	Length of Leg Opposite 60° ∠
12.	6	? 3	? $3\sqrt{3}$
13.	? 4	2	? $2\sqrt{3}$
14.	? $\frac{8\sqrt{3}}{3}$	? $\frac{4\sqrt{3}}{3}$	4

	Length of Hypotenuse	Length of Leg Opposite 45° ∠	Length of Leg Opposite Other 45° ∠
15.	? $3\sqrt{2}$	? 3	3
16.	? $2\sqrt{2}$	2	? 2
17.	5	? $\frac{5\sqrt{2}}{2}$	? $\frac{5\sqrt{2}}{2}$

PART C

18. Use the methods of Examples 1 and 2 to show that in a 30°-60° rt. △, the length of the leg opposite the 30° angle is $\frac{1}{2}$ hyp.. Show that the length of the leg opposite the 60° angle is $\frac{1}{2}$ hyp. $(\sqrt{3})$.

19. Use the method of Example 6 to show that in a 45°-45° rt. △, the length of each leg is $\frac{1}{2}$ hyp. $(\sqrt{2})$.

Symmetric Points in a Plane

OBJECTIVE

- To determine pairs of points symmetric with respect to the x-axis, the y-axis, and the origin

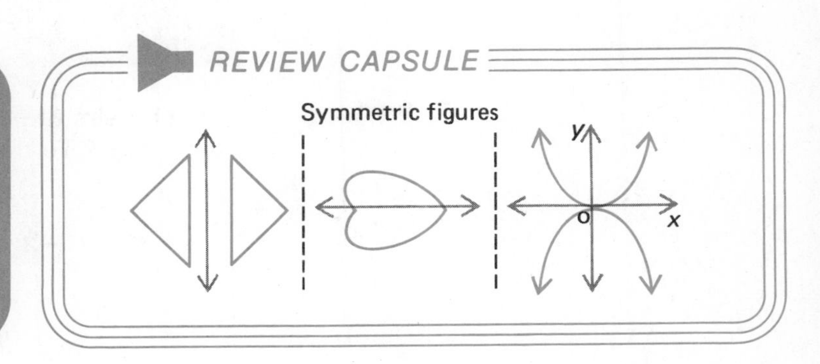

EXAMPLE 1 Determine which pairs of points are symmetric with respect to the x-axis.

$(-6, 3)$ and $(-6, -3)$ $(2, 1)$ and $(3, -1)$ $(5, 2)$ and $(5, -2)$

Plot the pairs of points.
Draw the segment between them.

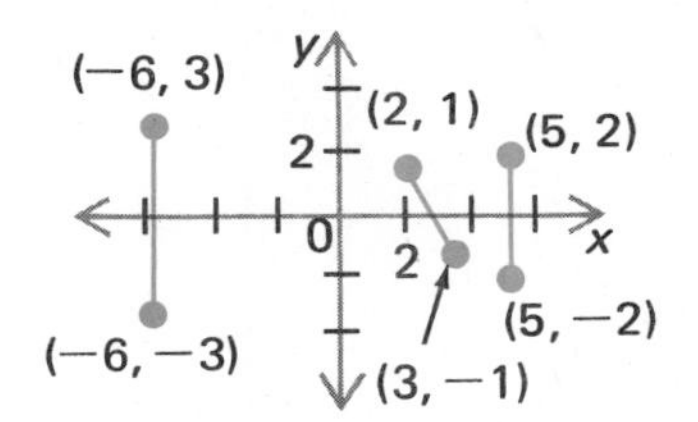

The x-axis is the $\perp$ bisector of the segment joining each pair of symmetric points.

$(-6, 3)$ and $(-6, -3)$	$(2, 1)$ and $(3, -1)$	$(5, 2)$ and $(5, -2)$
Symmetric	Not symmetric	Symmetric

(x, y) and $(x, -y)$ are symmetric with respect to the x-axis.

> Two points are symmetric with respect to the x-axis if their abscissas are the same and their ordinates are additive inverses.

EXAMPLE 2 Determine which pairs of points are symmetric with respect to the y-axis.
$(-2, 2)$ and $(2, 4)$ $(-3, 1)$ and $(3, 1)$ $(-3, -2)$ and $(3, -2)$

Plot the pairs of points.
Draw the segment between them.

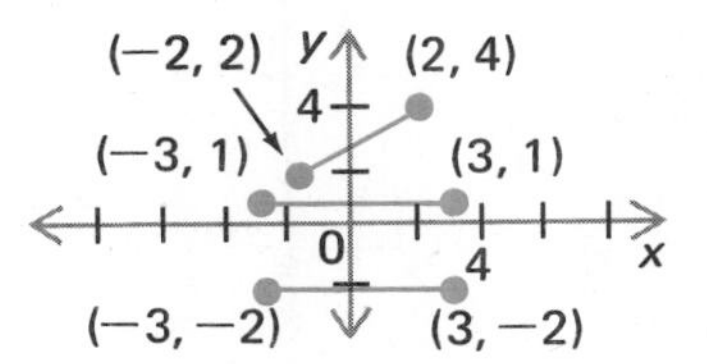

The y-axis is the $\perp$ bisector of the segment joining each pair of symmetric points.

$(-2, 2)$ and $(2, 4)$	$(-3, 1)$ and $(3, 1)$	$(-3, -2)$ and $(3, -2)$
Not symmetric	Symmetric	Symmetric

(x, y) and $(-x, y)$ are symmetric with respect to the y-axis.

Two points are symmetric with respect to the y-axis if their abscissas are additive inverses and their ordinates are the same.

EXAMPLE 3 Which pairs of points are symmetric with respect to the x-axis? To the y-axis?

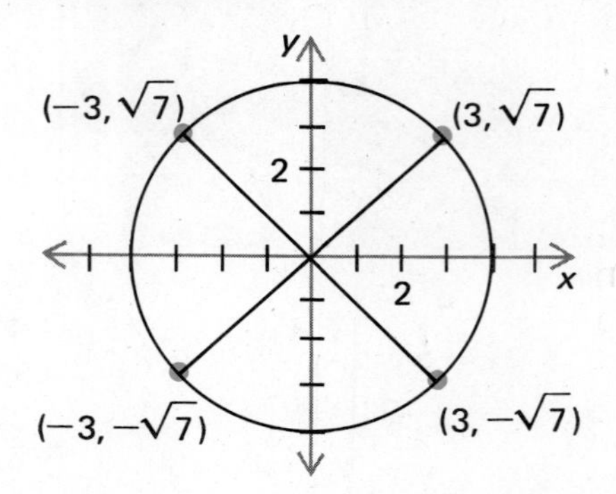

x-axis symmetry (x, y) and $(x, -y)$
y-axis symmetry (x, y) and $(-x, y)$

Symmetric to the x-axis	*Symmetric to the y-axis*
$(-3, \sqrt{7})$ and $(-3, -\sqrt{7})$	$(-3, \sqrt{7})$ and $(3, \sqrt{7})$
$(3, \sqrt{7})$ and $(3, -\sqrt{7})$	$(-3, -\sqrt{7})$ and $(3, -\sqrt{7})$

The two points are the endpoints of a diameter of the circle.

In the diagram for Example 3, two pairs of points:

$(-3, \sqrt{7})$ and $(3, -\sqrt{7})$ — additive inverses, additive inverses
$(3, \sqrt{7})$ and $(-3, -\sqrt{7})$ — additive inverses, additive inverses

are symmetric with respect to the origin.

(x, y) and $(-x, -y)$ are symmetric with respect to the origin.

Two points are symmetric with respect to the origin if their abscissas are additive inverses and their ordinates are additive inverses.

EXAMPLE 4 For $(-4, 6)$ find the symmetric points with respect to the x-axis, the y-axis, the origin.

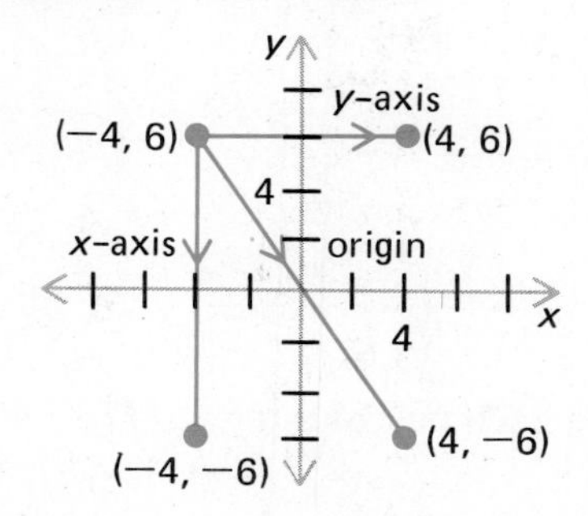

ORAL EXERCISES

Which pairs of points are symmetric with respect to the *x*-axis?

1. $(2, 3)$ and $(2, -3)$
2. $(-2, 3)$ and $(2, -3)$
3. $(-2, 3)$ and $(-2, -3)$

Which pairs of points are symmetric with respect to the *y*-axis?

4. $(5, 1)$ and $(-5, 1)$
5. $(0, 4)$ and $(0, -4)$
6. $(-4, 1)$ and $(4, 1)$

Which pairs of points are symmetric with respect to the origin?

7. $(7, 3)$ and $(-7, -3)$
8. $(8, 2)$ and $(8, -2)$
9. $(-6, 3)$ and $(6, -3)$

EXERCISES

PART A

Which pairs of points are symmetric with respect to the *x*-axis? the *y*-axis? the origin? Which pairs are not symmetrical?

1. $(2, -3)$ and $(2, 3)$
2. $(-6, 5)$ and $(6, 5)$ *y-axis*
3. $(-3, -1)$ and $(3, 1)$ *origin*
4. $(-7, 2)$ and $(7, -2)$
5. $(-1, 2)$ and $(1, 2)$ *y-axis*
6. $(9, 3)$ and $(8, -3)$
7. $(10, -1)$ and $(-10, 1)$
8. $(-3, 2)$ and $(3, -2)$ *origin*
9. $(0, 4)$ and $(0, -4)$

For each point, find the points symmetric with respect to the *x*-axis, the *y*-axis, and the origin.

10. $(2, 3)$
11. $(-3, 5)$
12. $(-1, -2)$
13. $(4, -3)$
14. $(-2, -3)$
15. $(-2, 1)$ *$(-2, -1), (2, 1), (2, -1)$*
16. $(2, -4)$ *$(2, 4), (-2, -4), (-2, 4)$*
17. $(4, 3)$ *$(4, -3), (-4, 3), (-4, -3)$*
18. $(-5, 2)$ *$(-5, -2), (5, 2), (5, -2)$*
19. $(4, 0)$ *$(4, 0), (-4, 0), (-4, 0)$*

PART B

For the circle and reference triangle shown, find (x, y) and determine the points symmetric with respect to the *x*-axis, the *y*-axis, and the origin.

20.

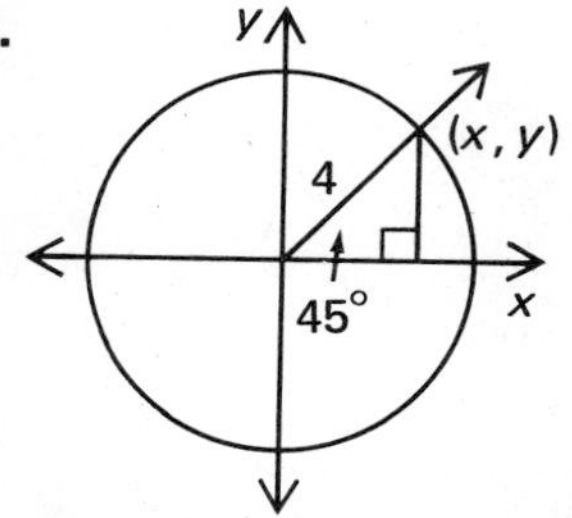

21.

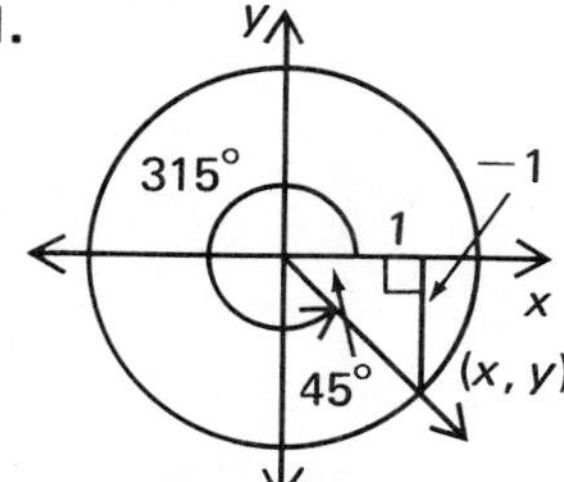

22.

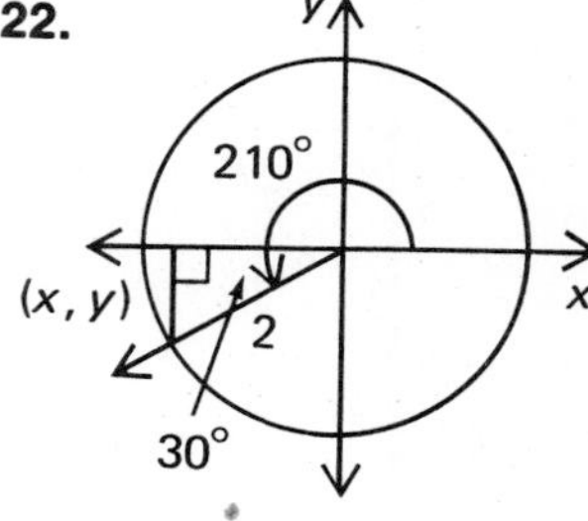

PART C

23. What does it mean for a *graph* to be symmetric with respect to the *x*-axis? to the *y*-axis? to the origin?

24. Draw a graph which is symmetric with respect to the *x*-axis; to the *y*-axis; to the origin.

A function is said to be *even* if its graph has *y*-axis symmetry and *odd* if it has origin symmetry. Which of the following are odd? even? neither?

25. $y = x^2$
26. $y = \frac{1}{x}$
27. $y = |x|$
28. $xy = 4$
29. $y = x^3$

Sine and Cosine

OBJECTIVES

- To find the sine and cosine of degree measures 0°, 30°, 45°, 60°, 90° and their multiples
- To determine if the sine or cosine is positive or negative for a given degree measure
- To solve equations like $\cos u = -\frac{1}{2}$

Find x and y.

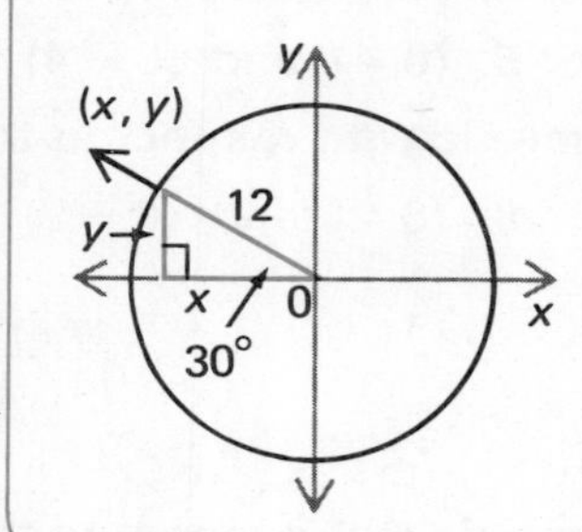

Use the 30°-60° right △.

$|x| = \frac{1}{2}r\sqrt{3}$ $|y| = \frac{1}{2}r$

$= \frac{1}{2}(12)\sqrt{3}$ $= \frac{1}{2}(12)$

$= 6\sqrt{3}$ $= 6$

So, $x = -6\sqrt{3}$.   So, $y = 6$.

EXAMPLE 1 Draw a circle with radius 6 on the coordinate plane. Sketch the terminal ray for a 30° angle. Draw the reference triangle and determine the ratios

$$\frac{y}{r} \text{ and } \frac{x}{r}.$$

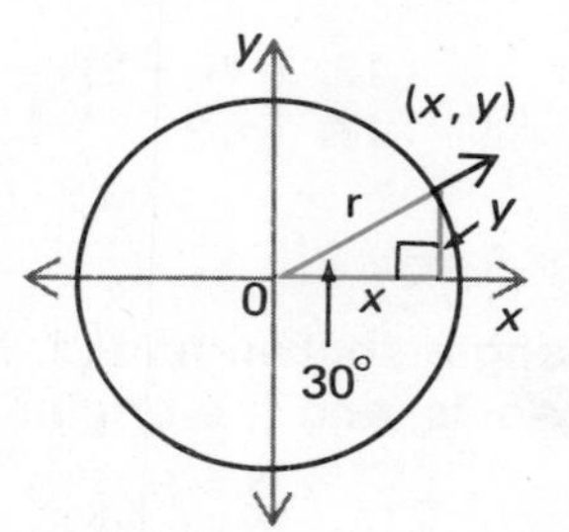

Find x and y for the reference triangle. (Use the formula for the 30°-60° right △.)

$r = 6$

$y = \frac{1}{2}(r)$ $x = \frac{1}{2}(r)\sqrt{3}$

$= \frac{1}{2}(6)$ $= \frac{1}{2}(6)\sqrt{3}$

$= 3$ $= 3\sqrt{3}$

$y = 3$, $x = 3\sqrt{3}$, $r = 6$ ⟶ **Thus,** for an angle of 30° with the given reference triangle, $\frac{y}{r}$ is $\frac{3}{6}$, or $\frac{1}{2}$; $\frac{x}{r}$ is $\frac{3\sqrt{3}}{6}$, or $\frac{\sqrt{3}}{2}$.

$\frac{y}{r}$ is the sine of 30°. $\sin 30° = \frac{1}{2}$ | $\frac{x}{r}$ is the cosine of 30°. $\cos 30° = \frac{\sqrt{3}}{2}$

Definition of sine (sin) and cosine (cos) ⟶

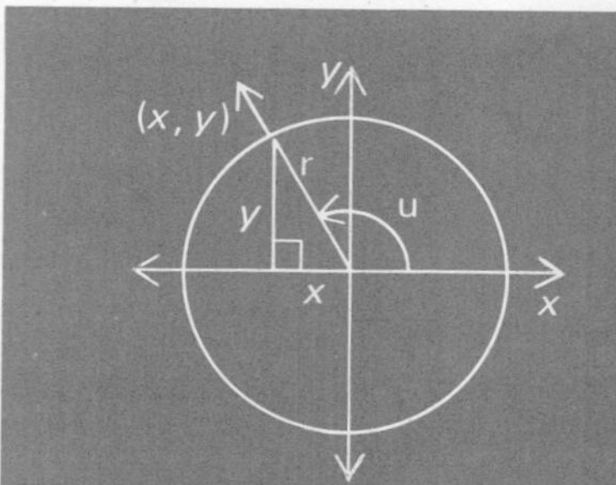

The sine of u is $\frac{y}{r}$.

$$\sin u = \frac{y}{r}$$

The cosine of u is $\frac{x}{r}$.

$$\cos u = \frac{x}{r}$$

For a given angle measure u, any convenient reference triangle may be used. Any two reference triangles will give the same value for $\sin u = \frac{y}{r}$ and $\cos u = \frac{x}{r}$.

The triangles will be similar.

EXAMPLE 2 Find sin 150° and cos 150°.

Sketch a circle with *any* radius and draw the reference △. (A radius of 2 is convenient.)

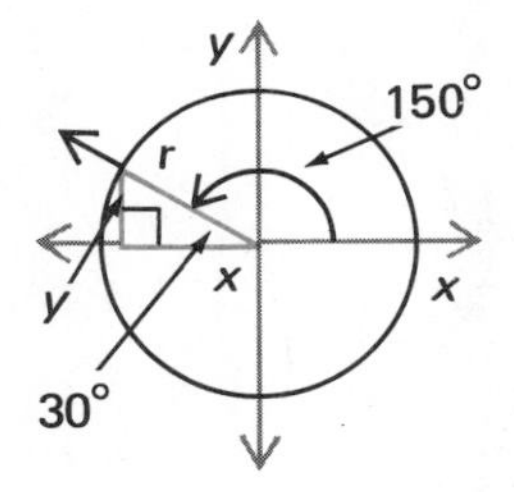

Reference ∠: 30°
Reference △: 30°-60° rt. △
$r = 2$

$|y| = \frac{1}{2}r$ $\quad$ $|x| = \frac{1}{2}(r)\sqrt{3}$

$|y| = \frac{1}{2}(2)$ $\quad$ $|x| = \frac{1}{2}(2)\sqrt{3}$

$|y| = 1$ $\quad$ $|x| = \sqrt{3}$

So, $y = 1$. $\quad$ So, $x = -\sqrt{3}$.

x is negative. y is positive. ⟶

$y = 1$, $x = -\sqrt{3}$, $r = 2$ ⟶ $\frac{y}{r} = \frac{1}{2}$ $\quad$ $\frac{x}{r} = -\frac{\sqrt{3}}{2}$

Use the definition.

Thus, $\sin 150° = \frac{1}{2}$ and $\cos 150° = -\frac{\sqrt{3}}{2}$.

EXAMPLE 3 Find sin 315° and cos 315°.

Draw a circle with radius 2. Sketch the reference △.

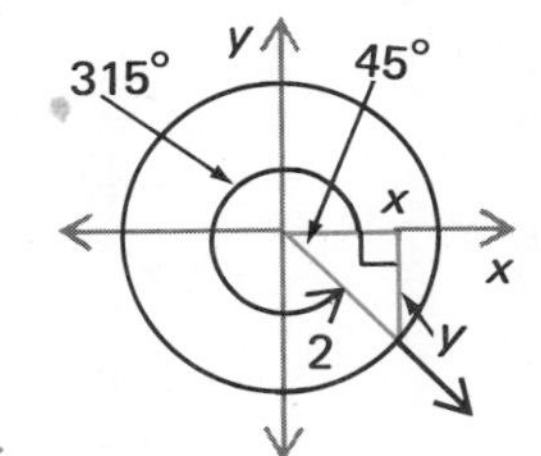

Reference ∠: 45°
Reference △: 45°-45° rt. △
$r = 2$

$|y| = \frac{1}{2}(r)\sqrt{2}$ $\quad$ $|x| = \frac{1}{2}(r)\sqrt{2}$

$|y| = \frac{1}{2}(2)\sqrt{2}$ $\quad$ $|x| = \frac{1}{2}(2)\sqrt{2}$

$|y| = \sqrt{2}$ $\quad$ $|x| = \sqrt{2}$

So, $y = -\sqrt{2}$. $\quad$ So, $x = \sqrt{2}$.

The △ is in the 4th quad.

x is positive. y is negative. ⟶

$y = -\sqrt{2}$, $x = \sqrt{2}$, $r = 2$ ⟶ $\frac{y}{r} = -\frac{\sqrt{2}}{2}$ $\quad$ $\frac{x}{r} = \frac{\sqrt{2}}{2}$

Thus, $\sin 315° = -\frac{\sqrt{2}}{2}$ and $\cos 315° = \frac{\sqrt{2}}{2}$.

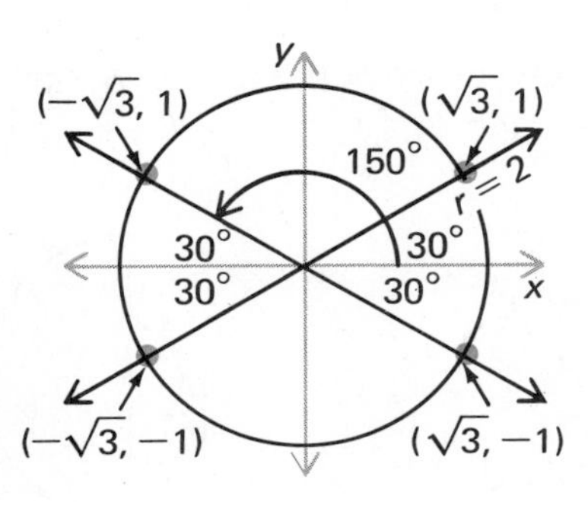

By observing the reference triangles in each quadrant, it can be determined whether sine and cosine will be positive (+) or negative (−) for angles in these quadrants.

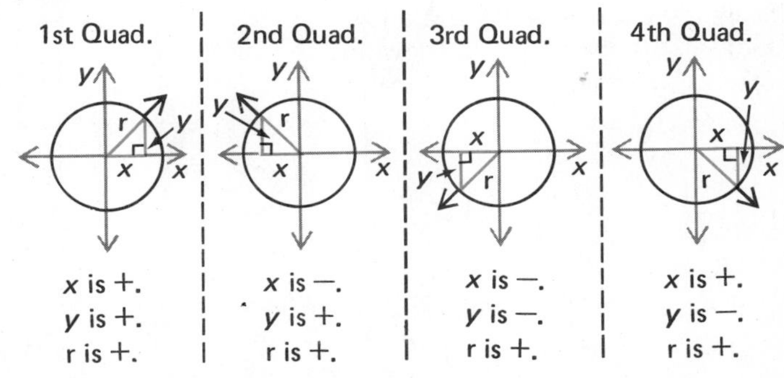

r is always positive. ⟶

EXAMPLE 4 Complete the chart.

$\sin = \frac{y}{r}$ $\cos = \frac{x}{r}$

Use the diagram on page 417.

$\frac{+}{+} = +$ $\frac{-}{+} = -$ $\frac{-}{-} = +$

	1st Quad.	2nd Quad.	3rd Quad.	4th Quad.
sine	+	+	−	−
cosine	+	−	−	+

EXAMPLE 5 Write each in terms of the same trigonometric function of an acute angle: cos 135° and cos 300°.

Make a sketch.
Draw the reference △.
Reference ∠ for 135° is 45°.
Reference ∠ for 300° is 60°.

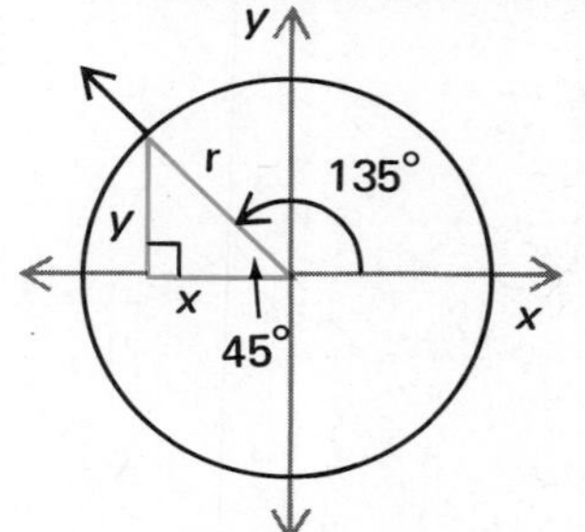

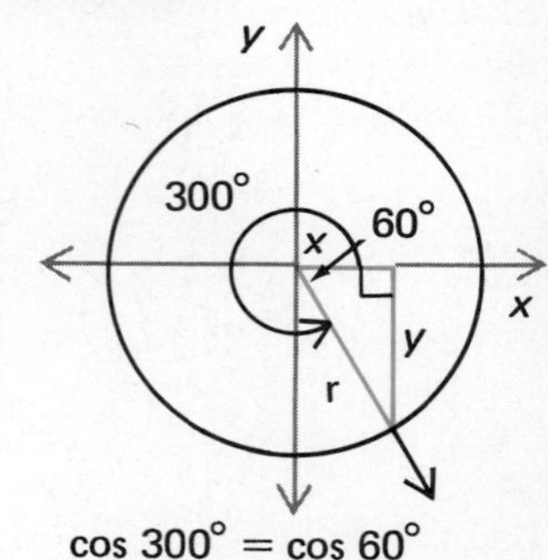

Use the table from Example 4 to get the correct sign.

Thus, cos 135° = −cos 45°

cos 300° = cos 60°

The sine and cosine of a quadrantal angle can be obtained by a similar method.

Quadrantal angles are 0°, 90°, 180°, and 270°.

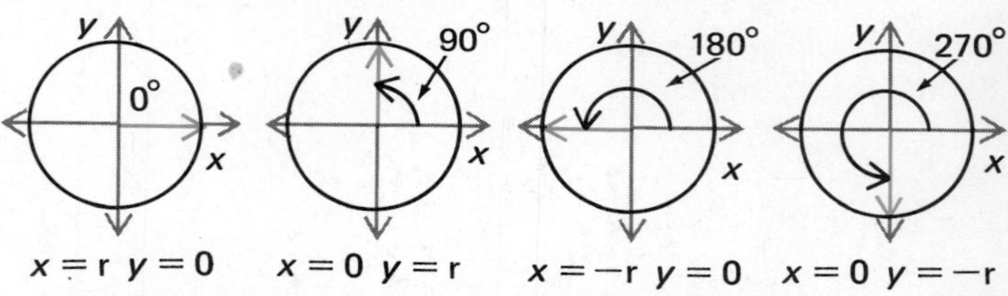

$x = r \;\; y = 0$ $x = 0 \;\; y = r$ $x = -r \;\; y = 0$ $x = 0 \;\; y = -r$

EXAMPLE 6 Complete the chart.

	0°	90°	180°	270°
sine	0	1	0	−1
cosine	1	0	−1	0

EXAMPLE 7 Find two values of u between 0° and 360° for which $\cos u = -\frac{1}{2}$.

Cosine is negative in the 2nd and 3rd quadrants. Draw diagrams showing the reference triangles.

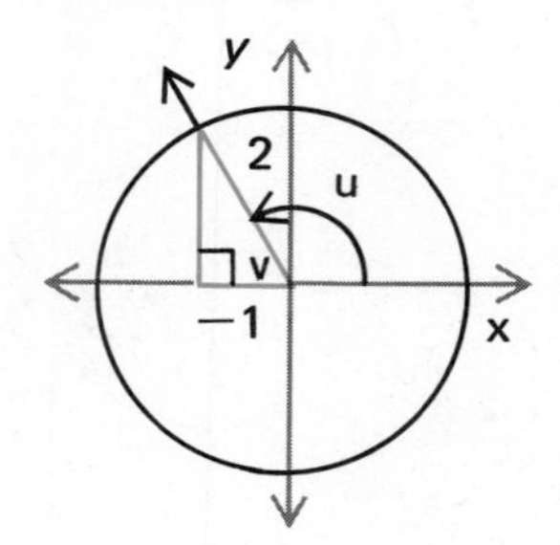

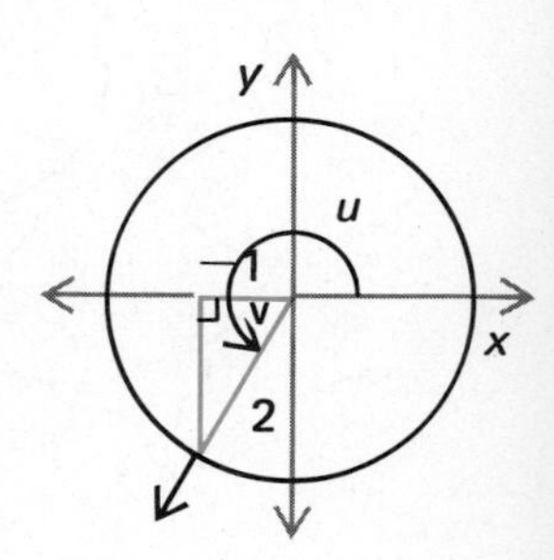

v is the measure of the reference angle.

$\cos 60° = \frac{1}{2}$ So, $v = 60°$.

$u = 180 - 60$ $u = 180 + 60$

Thus, $\cos u = -\frac{1}{2}$ for $u = 120°$ and $u = 240°$.

SUMMARY

To find the sine or cosine of an angle with measure u:

1. **Determine the quadrant and reference angle.**
2. **Determine the sign (+ or −) of the function in that quadrant.**
3. **Use the sine or cosine for the reference angle.**

ORAL EXERCISES

Tell whether the indicated function is positive or negative.

1. $\sin 45°$ *+* **2.** $\cos 135°$ *−* **3.** $\sin 210°$ *−* **4.** $\cos 300°$ *+* **5.** $\sin 150°$ *+* **6.** $\cos 225°$ *−*
7. $\sin 20°$ *+* **8.** $\cos 160°$ *−* **9.** $\sin 250°$ *−* **10.** $\cos 320°$ *+* **11.** $\sin 140°$ *+* **12.** $\cos 260°$ *−*

EXERCISES

PART A

Find each of the following. Rationalize any radical denominator.

1. $\sin 30°$ $\frac{1}{2}$ **2.** $\cos 30°$ $\frac{\sqrt{3}}{2}$ **3.** $\cos 45°$ $\frac{\sqrt{2}}{2}$ **4.** $\sin 45°$ $\frac{\sqrt{2}}{2}$ **5.** $\sin 0°$ *0* **6.** $\cos 90°$ *0*
7. $\sin 180°$ *0* **8.** $\cos 270°$ *0* **9.** $\sin 360°$ *0* **10.** $\cos 60°$ $\frac{1}{2}$ **11.** $\sin 60°$ $\frac{\sqrt{3}}{2}$ **12.** $\sin 150°$ $\frac{1}{2}$
13. $\cos 150°$ $-\frac{\sqrt{3}}{2}$ **14.** $\sin 135°$ $\frac{\sqrt{2}}{2}$ **15.** $\cos 210°$ **16.** $\sin 210°$ $-\frac{1}{2}$ **17.** $\cos 225°$ **18.** $\sin 330°$ $-\frac{1}{2}$
19. $\cos 330°$ $\frac{\sqrt{3}}{2}$ **20.** $\sin 315°$ **21.** $\sin 120°$ **22.** $\cos 240°$ $-\frac{1}{2}$ **23.** $\cos 300°$ $\frac{1}{2}$ **24.** $\cos 390°$

Write each in terms of the same trigonometric function of an acute angle.

25. $\sin 150°$ **26.** $\cos 150°$ **27.** $\cos 135°$ **28.** $\sin 135°$ **29.** $\sin 210°$ **30.** $\cos 210°$
31. $\cos 225°$ **32.** $\sin 225°$ **33.** $\sin 330°$ **34.** $\cos 330°$ **35.** $\cos 315°$ **36.** $\sin 315°$

In which quadrants is each of the following true?

37. $\sin u > 0$ *1, 2* **38.** $\sin u < 0$ *3, 4* **39.** $\cos u > 0$ *1, 4* **40.** $\cos u < 0$ *2, 3*

PART B

Find two values of u between 0° and 360° for which each is true.

41. $\sin u = \frac{1}{2}$ *30°, 150°* **42.** $\cos u = \frac{\sqrt{2}}{2}$ *45°, 315°* **43.** $\sin u = \frac{\sqrt{2}}{2}$ *45°, 135°* **44.** $\cos u = -\frac{1}{2}$ *120°, 240°*

45. $\cos u = -\frac{\sqrt{3}}{2}$ *150°, 210°* **46.** $\sin u = -\frac{\sqrt{3}}{2}$ *240°, 300°* **47.** $\cos u = -\frac{\sqrt{2}}{2}$ *135°, 225°* **48.** $\sin u = -\frac{1}{2}$ *210°, 330°*

PART C

Find two values of u between 0° and 720° for which each is true.

49. $\sin u = -\frac{1}{2}$ and $\cos u > 0$ **50.** $\cos u = -\frac{\sqrt{3}}{2}$ and $\sin u < 0$

Tangent and Cotangent

OBJECTIVES

- To find the tangent and cotangent of degree measures 0°, 30°, 45°, 60°, 90° and their multiples
- To find tangents and cotangents by using trigonometric relationships
- To solve equations like $\tan u = 1$

REVIEW CAPSULE

Find the measure of the reference angle.

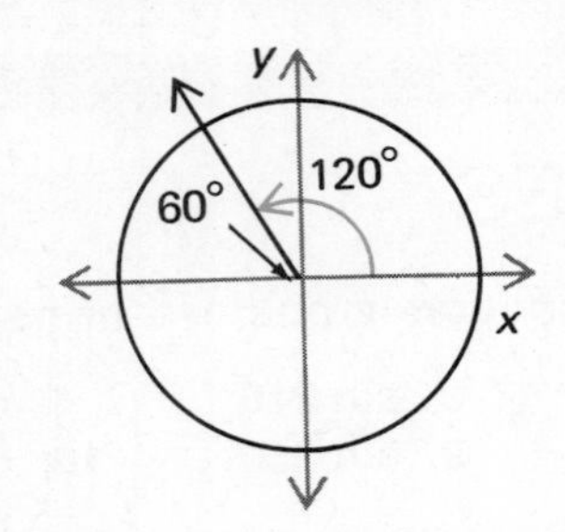

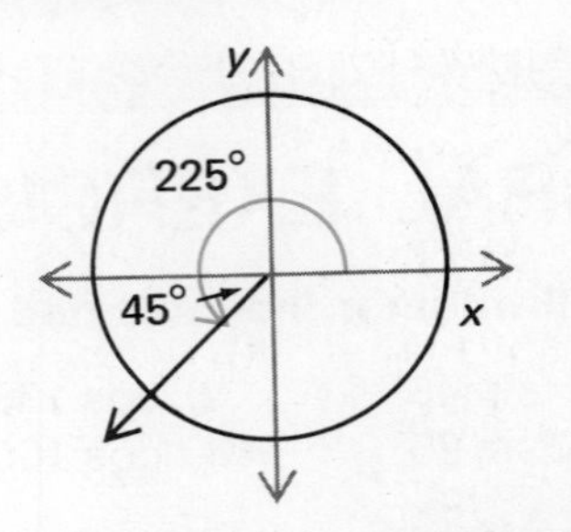

Reference ∠ measures 60°. Reference ∠ measures 45°.

EXAMPLE 1 For the given reference triangle find the ratios $\frac{y}{x}$ and $\frac{x}{y}$.

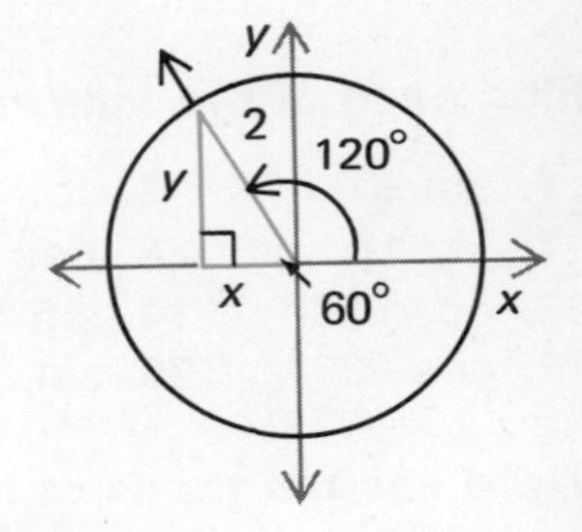

Use the 30°-60° rt. △ for a reference triangle.

$$|y| = \tfrac{1}{2}r\sqrt{3} \qquad |x| = \tfrac{1}{2}r$$
$$= \tfrac{1}{2}(2)\sqrt{3} \qquad = \tfrac{1}{2}(2)$$
$$= \sqrt{3} \qquad = 1$$

x is negative. *y* is positive. ——→ So, $y = \sqrt{3}$, and $x = -1$.

Thus, for an angle of 120°,

$$\frac{y}{x} \text{ is } \frac{\sqrt{3}}{-1}, \text{ or } -\sqrt{3} \text{ and } \frac{x}{y} = \frac{-1}{\sqrt{3}}, \text{ or } -\frac{\sqrt{3}}{3}.$$

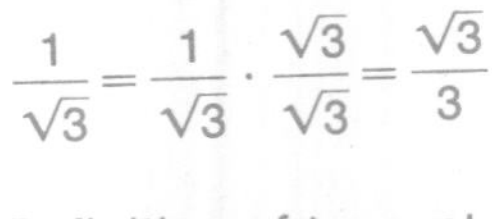

In Example 1,

$\frac{y}{x}$ is the tangent of 120° and $\frac{x}{y}$ is the cotangent of 120°.

$\tan 120° = -\sqrt{3}$ $\qquad$ $\cot 120° = -\frac{\sqrt{3}}{3}$

Definition of tangent and cotangent

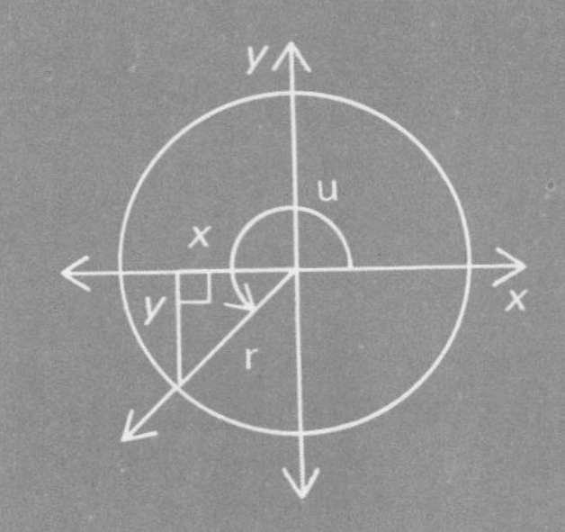

The tangent of $u = \frac{y}{x}$.

$\tan u = \frac{y}{x}$

The cotangent of $u = \frac{x}{y}$.

$\cot u = \frac{x}{y}$

$\tan u \cdot \cot u = \frac{y}{x} \cdot \frac{x}{y} = 1$ ⟶ **Tangent and cotangent are reciprocal functions.**

EXAMPLE 2 If $\tan x = -\frac{3}{4}$, find $\cot x$.

Find the reciprocal of $-\frac{3}{4}$. ⟶ **Thus,** $\cot x = -\frac{4}{3}$.

EXAMPLE 3 Show that $\frac{\sin u}{\cos u} = \tan u$.

$$\frac{\sin u}{\cos u} = \frac{\frac{y}{r}}{\frac{x}{r}} = \frac{y}{r} \cdot \frac{r}{x} = \frac{y}{x} = \tan u$$

Thus, $\frac{\sin u}{\cos u} = \tan u$.

Example 3 suggests this. ⟶

$$\tan u = \frac{\sin u}{\cos u} \qquad \cot u = \frac{\cos u}{\sin u}$$

EXAMPLE 4 If $\sin w = \frac{3}{5}$, $\cos w = \frac{4}{5}$, find $\tan w$ and $\cot w$.

$\tan w = \frac{\sin w}{\cos w}$, $\cot w = \frac{\cos w}{\sin w}$

$$\tan w = \frac{\frac{3}{5}}{\frac{4}{5}} = \frac{3}{4} \qquad \cot w = \frac{\frac{4}{5}}{\frac{3}{5}} = \frac{4}{3}$$

Thus, $\tan w = \frac{3}{4}$ and $\cot w = \frac{4}{3}$.

EXAMPLE 5 Complete the chart. Use + to indicate a positive value and − to indicate a negative value.

Use $\tan u = \frac{\sin u}{\cos u}$ and $\cot u = \frac{\cos u}{\sin u}$.

	1st Quad.	2nd Quad.	3rd Quad.	4th Quad.
Tangent	+	−	+	−
Cotangent	+	−	+	−

EXAMPLE 6 Find $\tan 225°$ and $\cot 225°$.

Draw the reference △.
Use the 45°-45° rt. △ to find tan 45°.

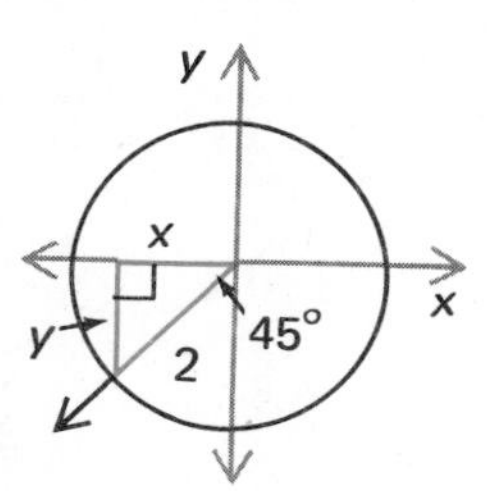

$|x| = \frac{1}{2}(2)\sqrt{2}$ $\qquad$ $|y| = \frac{1}{2}(2)\sqrt{2}$
$|x| = \sqrt{2}$ $\qquad$ $|y| = \sqrt{2}$
So, $x = -\sqrt{2}$ $\qquad$ So, $y = -\sqrt{2}$

$$\tan 45° = \frac{y}{x} = \frac{-\sqrt{2}}{-\sqrt{2}} = 1$$

The reference measure is 45°.
Tan and cot are positive in the 3rd quad.

$\tan 225° = 1$
$\cot 225° = 1$

Thus, $\tan 225° = 1$ and $\cot 225° = 1$.

EXAMPLE 7 Write cot 240° in terms of the same trigonometric function of an acute angle.

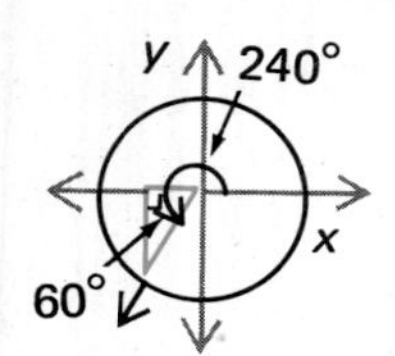

240° is in the third quad. Cotangent is positive in the 3rd quad.

Reference ∠ measures 60°.

Thus, $\cot 240° = \cot 60°$.

EXAMPLE 8 Find tan 90° and tan 270°.

Draw the angles. Determine x and y.

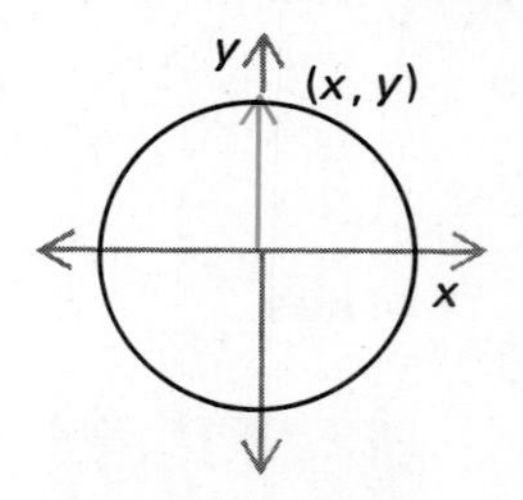

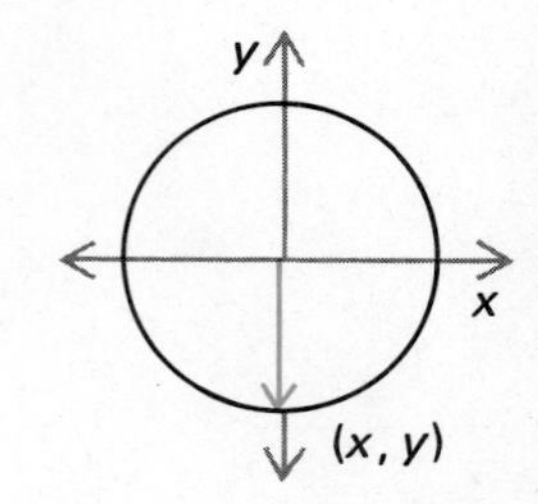

$x = 0,\ y = r$ $\qquad$ $x = 0,\ y = -r$

$\tan 90° = \dfrac{y}{x}$ $\qquad$ $\tan 270° = \dfrac{y}{x}$

Division by 0 is undefined. ⟶ $= \dfrac{r}{0}$ $\qquad$ $= \dfrac{-r}{0}$

Thus, tan 90° and tan 270° are undefined.

EXAMPLE 9 Complete the chart.

Use the diagrams above.

	0°	90°	180°	270°
Tangent	0	undefined	0	undefined
Cotangent	undefined	0	undefined	0

EXAMPLE 10 Find two values of u between 0° and 360° for which $\tan u = -1$.

Tangent is negative in the 2nd and 4th quadrants.

Draw diagrams showing the reference triangles.

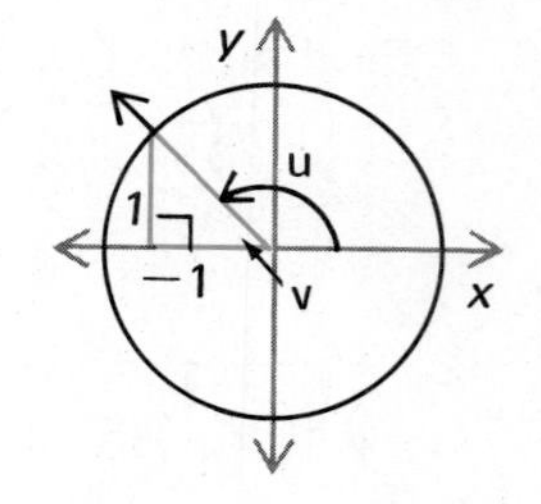

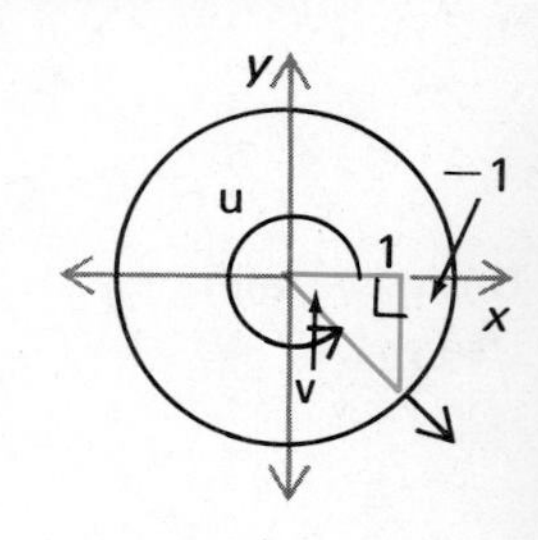

v is the measure of the reference ∠. ⟶

$\tan 45° = 1$ $\qquad$ So, $v = 45°$.

$u = 180° - 45°$ $\qquad$ $u = 360° - 45°$

$= 135°$ $\qquad$ $= 315°$

Thus, $\tan u = -1$ for $u = 135°$ and $u = 315°$.

ORAL EXERCISES

Tell whether the indicated value is positive or negative.

1. tan 45° *+* **2.** cot 135° *−* **3.** tan 315° *−* **4.** cot 225° *+* **5.** tan 150° *−* **6.** cot 300° *−*
7. tan 140° *−* **8.** tan 200° *+* **9.** cot 330° *−* **10.** cot 25° *+* **11.** tan 190° *+* **12.** cot 100° *−*

EXERCISES

PART A

Find each of the following. Rationalize any radical denominator.

1. tan 30° **2.** cot 30° *$\sqrt{3}$* **3.** cot 60° *$\frac{\sqrt{3}}{3}$* **4.** tan 60° *$\sqrt{3}$* **5.** tan 0° *0* **6.** cot 0° *undefined*
7. cot 90° *0* **8.** tan 90° **9.** tan 120° **10.** cot 120° **11.** cot 135° *−1* **12.** tan 135° *−1*
13. tan 150° **14.** cot 150° **15.** cot 180° **16.** tan 180° *0* **17.** tan 210° **18.** cot 210° *$\sqrt{3}$*
19. cot 225° *1* **20.** tan 225° *1* **21.** tan 240° **22.** cot 240° **23.** cot 270° **24.** tan 270°
25. tan 300° **26.** cot 315° *−1* **27.** cot 330° **28.** tan 360° *0* **29.** tan 495° *−1* **30.** cot 495° *−1*

Write each of the following in terms of the same trigonometric function of an acute angle.

31. tan 150° **32.** cot 150° **33.** cot 135° *−cot 45°* **34.** tan 135° *−tan 45°* **35.** tan 210° *tan 30°* **36.** cot 210° *cot 30°*
37. cot 225° **38.** tan 225° **39.** tan 330° **40.** cot 315° **41.** cot 120° **42.** tan 300° *−tan 60°*

43. If $\tan x = -\frac{1}{3}$, find $\cot x$. *-3*
44. If $\cot x = \frac{3}{5}$, find $\tan x$. *$\frac{5}{3}$*
45. If $\sin x = \frac{1}{2}$ and $\cos x = \frac{\sqrt{3}}{2}$, find $\tan x$ and $\cot x$. *$\tan x = \frac{\sqrt{3}}{3}$, $\cot x = \sqrt{3}$*
46. If $\sin x = -\frac{3}{5}$ and $\cos x = -\frac{4}{5}$, find $\tan x$ and $\cot x$. *$\tan x = \frac{3}{4}$, $\cot x = \frac{4}{3}$*
47. If $\sin x = \frac{5}{13}$ and $\cos x = -\frac{12}{13}$, find $\tan x$ and $\cot x$. *$\tan x = -\frac{5}{12}$, $\cot x = -\frac{12}{5}$*

In which quadrants is each of the following true?

48. $\tan u > 0$ *1, 3* **49.** $\tan u < 0$ *2, 4* **50.** $\cot u > 0$ *1, 3* **51.** $\cot u < 0$ *2, 4*

PART B

Find two values of u between 0° and 360° for which each is true.

52. $\tan u = 1$ *45°, 225°* **53.** $\cot u = -1$ *135°, 315°* **54.** $\tan u = \frac{\sqrt{3}}{3}$ *30°, 210°* **55.** $\cot u = -\frac{\sqrt{3}}{3}$ *120°, 300°*

56. $\cot u = \frac{\sqrt{3}}{3}$ *60°, 240°* **57.** $\tan u = -\frac{\sqrt{3}}{3}$ *150°, 330°* **58.** $\cot u = \sqrt{3}$ *30°, 210°* **59.** $\tan u = -1$ *135°, 315°*

PART C

Find two values of u between 0° and 720° for which each is true.

60. $\cot u = -\sqrt{3}$ and $\tan u < 0$ **61.** $\tan u = \sqrt{3}$ and $\sin u < 0$

62. $\tan u = -\frac{\sqrt{3}}{3}$ and $\cos u > 0$ **63.** $\cot u = 1$ and $\sin u > 0$

Secant and Cosecant

OBJECTIVES

- To find the secant and cosecant of degree measures 0°, 30°, 45°, 60°, 90° and their multiples
- To determine the quadrants in which the secant and cosecant are positive
- To solve equations like sec $u = 2$

REVIEW CAPSULE

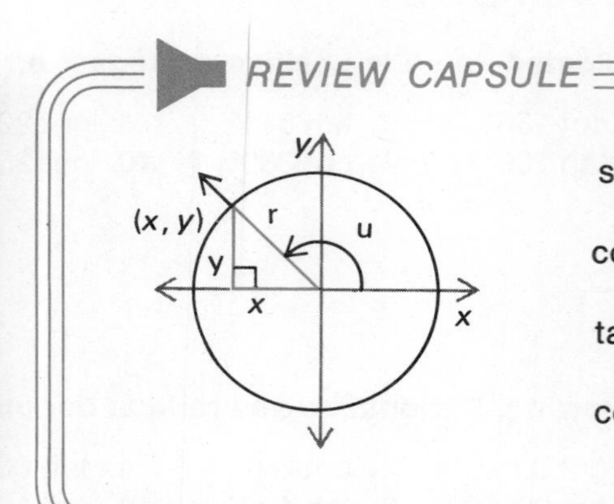

$\sin u = \frac{y}{r}$

$\cos u = \frac{x}{r}$

$\tan u = \frac{y}{x},\ x \neq 0$

$\cot u = \frac{x}{y},\ y \neq 0$

EXAMPLE 1 Write the reciprocals of sin u and cos u in terms of x, y, and r in the reference triangle.

Use the defintion. ⟶

$\sin u = \frac{y}{r}$	$\cos u = \frac{x}{r}$
Thus, $\frac{1}{\sin u} = \frac{r}{y}$.	**Thus,** $\frac{1}{\cos u} = \frac{r}{x}$.

Definition of secant and cosecant ⟶

The cosecant of $u = \frac{r}{y}$.

$\csc u = \frac{1}{\sin u}$

The secant of $u = \frac{r}{x}$.

$\sec u = \frac{1}{\cos u}$

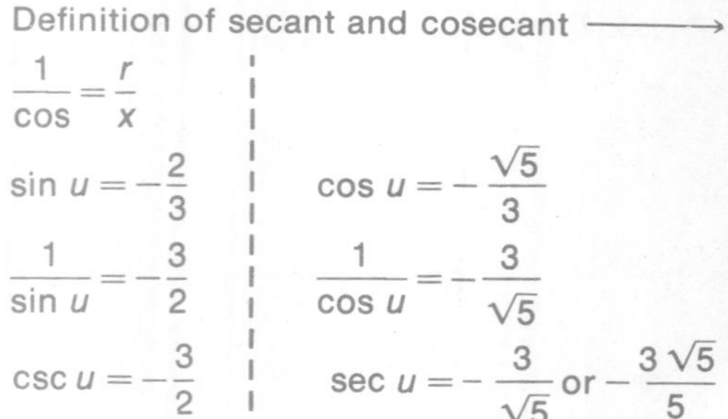

$\frac{1}{\cos} = \frac{r}{x}$

$\sin u = -\frac{2}{3}$	$\cos u = -\frac{\sqrt{5}}{3}$
$\frac{1}{\sin u} = -\frac{3}{2}$	$\frac{1}{\cos u} = -\frac{3}{\sqrt{5}}$
$\csc u = -\frac{3}{2}$	$\sec u = -\frac{3}{\sqrt{5}}$ or $-\frac{3\sqrt{5}}{5}$

EXAMPLE 2 Complete the chart.

Use $\sec u = \frac{1}{\cos u}$ and $\csc u = \frac{1}{\sin u}$.

	1st Quad.	2nd Quad.	3rd Quad.	4th Quad.
Secant	+	−	−	+
Cosecant	+	+	−	−

EXAMPLE 3 Find csc 150° and sec 150°.

Use reciprocals by definition.

Draw the reference triangle. The reference measure is 30°.

$\sin 150° = \frac{1}{2}$	$\cos 150° = -\frac{\sqrt{3}}{2}$

$\frac{2}{\sqrt{3}} = \frac{2}{\sqrt{3}} \cdot \frac{\sqrt{3}}{\sqrt{3}} = \frac{2\sqrt{3}}{3}$ ⟶ **Thus,** $\csc 150° = \frac{2}{1}$ or 2 and $\sec 150° = -\frac{2}{\sqrt{3}}$ or $-\frac{2\sqrt{3}}{3}$.

EXAMPLE 4 Write sec 315° and csc 210° as the same trigonometric function of an acute angle.

Find the reference ∠.

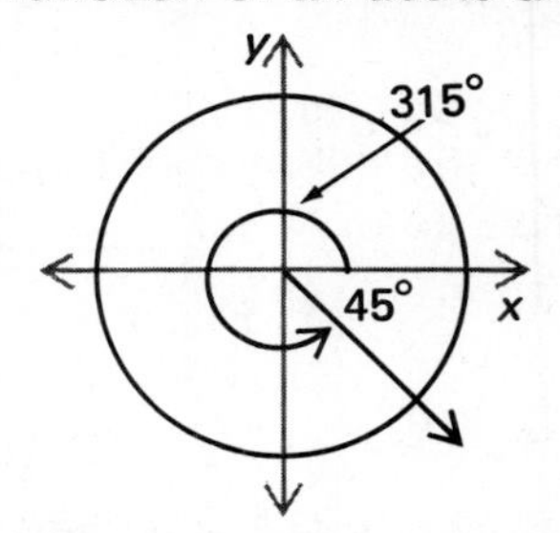

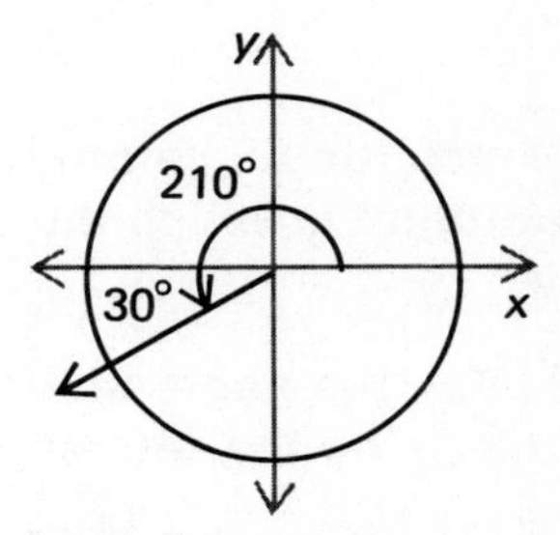

Use the correct sign. ⟶ **Thus,** sec 315° = sec 45° and csc 210° = −csc 30°.

EXAMPLE 5 Complete the chart.

sin 0° = 0; csc 0° = $\frac{1}{0}$ ⟶ undefined
sin 180° = 0; csc 180° = $\frac{1}{0}$ ⟶ undefined
cos 90° = 0; sec 90° = $\frac{1}{0}$ ⟶ undefined
cos 270° = 0; sec 270° = $\frac{1}{0}$ ⟶ undefined

	0°	90°	180°	270°
Secant	1	undefined	−1	undefined
Cosecant	undefined	1	undefined	−1

EXAMPLE 6 Find two values of u between 0° and 360° for which sec u = −2.

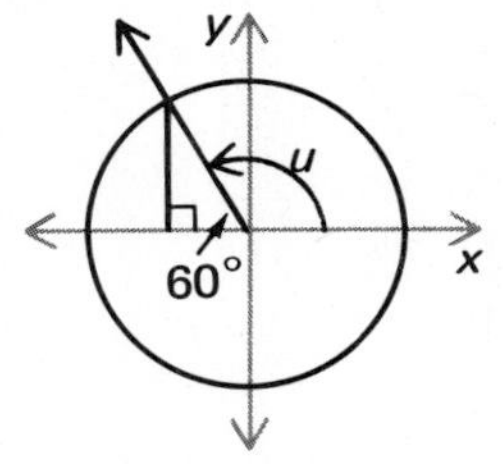

sec 60° = 2
u = 180° − 60°
= 120°

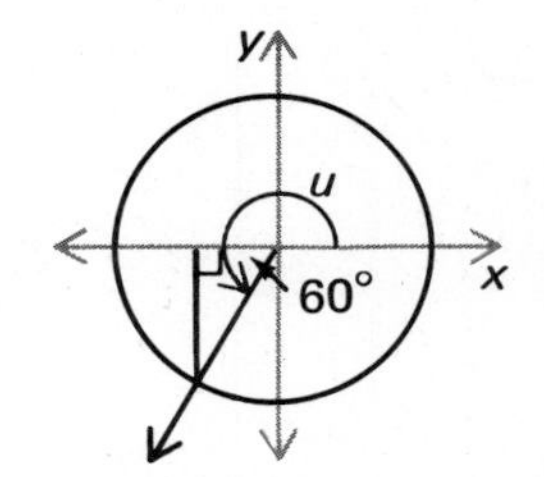

sec 60° = 2
u = 180° + 60°
= 240°

Thus, sec u = −2 for u = 120° and u = 240°.

EXERCISES

PART A

Find each of the following. Rationalize any radical denominator.

1. sec 30° $\frac{2\sqrt{3}}{3}$
2. csc 30° 2
3. csc 45° $\sqrt{2}$
4. sec 45° $\sqrt{2}$
5. sec 60° 2
6. csc 60° $\frac{2\sqrt{3}}{3}$
7. csc 135° $\sqrt{2}$
8. sec 135°
9. sec 150°
10. csc 150° 2
11. csc 225°
12. sec 225°
13. sec 240° −2
14. csc 240° $-\frac{2\sqrt{3}}{3}$
15. csc 300° $-\frac{2\sqrt{3}}{3}$
16. sec 300° 2
17. sec 315° $\sqrt{2}$
18. csc 315°

Write each of the following in terms of the same trigonometric function of an acute angle.

19. sec 150° −sec 30°
20. csc 190° −csc 10°
21. sec 220° −sec 40°
22. csc 320°
23. sec 130°

PART B

Find two values of u between and including 0° and 360° for which each is true.

24. sec u = 1 0°, 360°
25. csc u = −2 210°, 330°
26. sec u = $\frac{2\sqrt{3}}{3}$ 30°, 330°

Chapter Sixteen Review

Sketch the angle of rotation. Label the initial side I and terminal side T. Determine the quadrant in which the terminal side lies. [p. 405]

1. 60° **2.** −135° **3.** −30° **4.** 240° **5.** −300° **6.** 100° **7.** −200°

Sketch the angle of rotation. Determine the reference angle. [p. 407]

8. 150° **9.** −60° **10.** 320° **11.** −230°

Sketch the angle of rotation. Determine the reference triangle. [p. 407]

12. 60° **13.** −135° **14.** 120° **15.** −340°

For a circle with any radius, determine the reference angle and the reference triangle for the rotation. [p. 407]

16. 120° **17.** −60° **18.** −150° **19.** 310° **20.** 200° **21.** −255° **22.** 20°

Determine the length of the indicated side. [p. 409]

23.

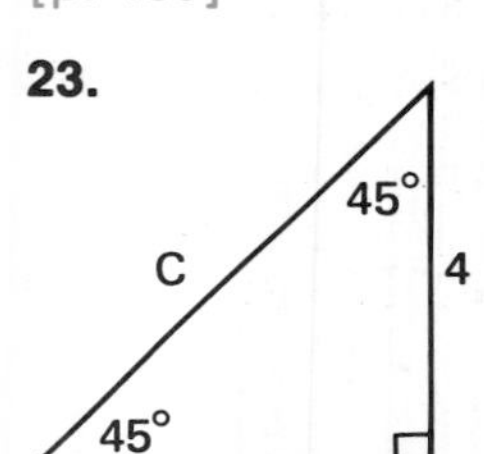

24.

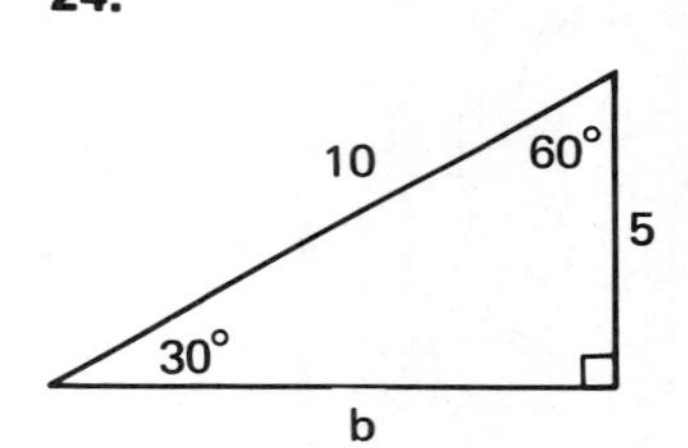

Determine the point (x, y) and the radius r. For (x, y), find the points symmetric with respect to the x-axis, the y-axis, and the origin. [p. 413]

25.

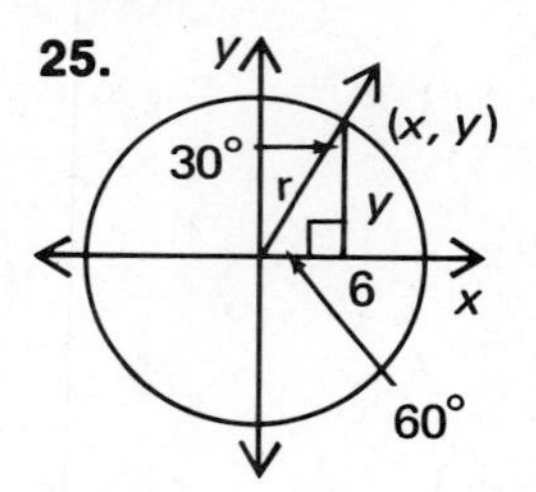

26.

Which pairs of points are symmetric with respect to the x-axis? the y-axis? the origin? not symmetrical? [p. 413]

27. (3, 8) and (−3, 8) **28.** (−6, 3) and (6, −3)

Find the symmetric point with respect to the x-axis, the y-axis, and the origin. [p. 413]

29. (6, −1) **30.** (−3, 1) **31.** (−2, −5)

Find each of the following. Rationalize any radical denominator. [pp. 416, 420, 424]

32. sin 30° **33.** tan 60° **34.** sec 120° **35.** cos 150° **36.** csc 240° **37.** cot 240°

Write each of the following in terms of the same trigonometric function of an acute angle. [pp. 416, 420, 424]

38. sin 120° **39.** tan 140° **40.** sec 210° **41.** cos 150° **42.** cot 315° **43.** csc 120°

Determine in which quadrants each of the following is true. [pp. 416, 420]

44. Sin is negative. **45.** Cos is positive. **46.** Tan is negative. **47.** Cot is positive.

Find two values of u between 0° and 360° for which each is true. [pp. 416, 420, 424]

48. $\sin u = \frac{\sqrt{3}}{2}$ **49.** $\cos u = \frac{1}{2}$ **50.** $\tan u = -1$ **51.** $\sin u = -\frac{\sqrt{2}}{2}$

52. $\csc u = 2$ **53.** $\sec u = -\sqrt{2}$ **54.** $\tan u = -\sqrt{3}$ **55.** $\cot u = \frac{\sqrt{3}}{3}$

Find each of the following. [pp. 416, 420, 424]

56. sin 0° **57.** sec 90° **58.** csc 270° **59.** cos 90° **60.** tan 180° **61.** cot 360°

Chapter Sixteen Test

Sketch the angle of rotation. Label the initial side I and the terminal side T. Determine in which quadrant the terminal side lies.

1. 30° **2.** −150° **3.** 210° **4.** 330° **5.** −100°

Sketch the angle of rotation. Determine the reference angle.

6. 120° **7.** −150° **8.** 310° **9.** −240°

Sketch the angle of rotation. Determine the reference triangle.

10. 45° **11.** −120° **12.** 300° **13.** 110°

For a circle with any radius, sketch the angle of rotation. Determine the reference angle and the reference triangle.

14. 150° **15.** −30° **16.** 210° **17.** 340°

Determine the lengths of the indicated sides.

18.

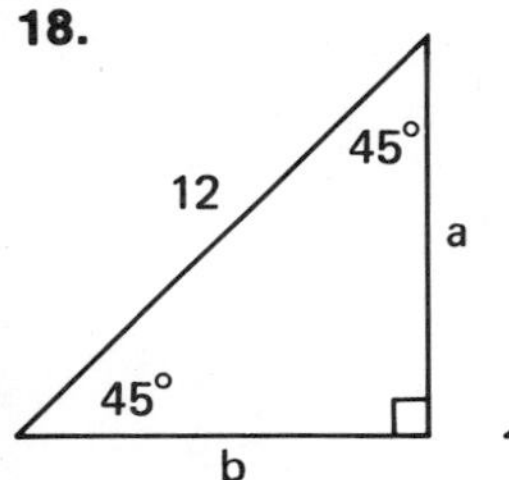

19.

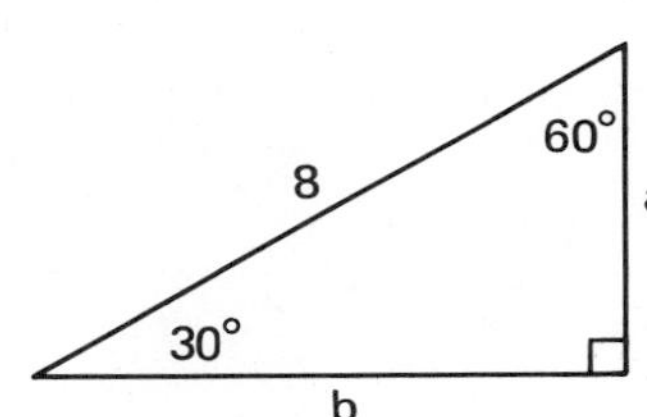

Determine the point (x, y) and the radius r. For (x, y), find the points symmetric with respect to the x-axis, the y-axis, and the origin.

20.

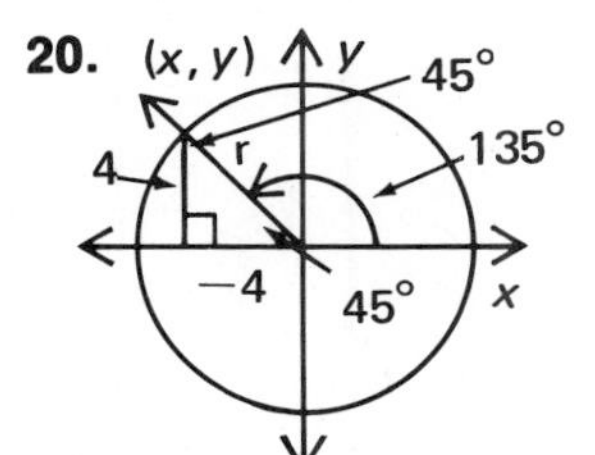

21.

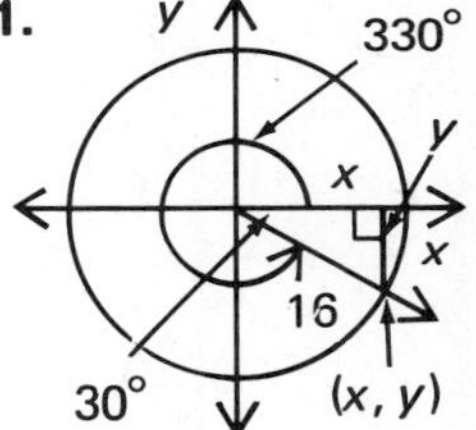

Which pairs of points are symmetric with respect to the x-axis? the y-axis? the origin? not symmetrical?

22. (3, −1) and (3, 1) **23.** (−1, 8) and (1, −8)

Find the points symmetric with respect to the x-axis, the y-axis, and the origin.

24. (3, 4) **25.** (−4, −5)

Find each of the following. Rationalize any radical denominator.

26. sin 45° **27.** cos 30° **28.** tan 135° **29.** cot 240° **30.** sec 300° **31.** csc 315°

Write each of the following in terms of the same trigonometric function of an acute angle.

32. cos 150° **33.** sin 300° **34.** tan 150° **35.** cot 225° **36.** sec 120° **37.** csc 330°

Determine in which quadrants each of the following is true.

38. Sin is positive. **39.** Cos is negative. **40.** Tan is positive. **41.** Cot is negative.

Find two values of *u* between 0° and 360° for which each is true.

42. $\sin u = -\frac{1}{2}$ **43.** $\cos u = \frac{\sqrt{3}}{2}$ **44.** $\tan u = 1$ **45.** $\cot u = -\frac{\sqrt{3}}{3}$ **46.** $\sec u = -\frac{2\sqrt{3}}{3}$

Find each of the following.

47. sin 90° **48.** cos 0° **49.** tan 270° **50.** cot 360° **51.** sec 90° **52.** csc 180°

17 USING TRIGONOMETRIC FUNCTIONS

Making a Sine Table

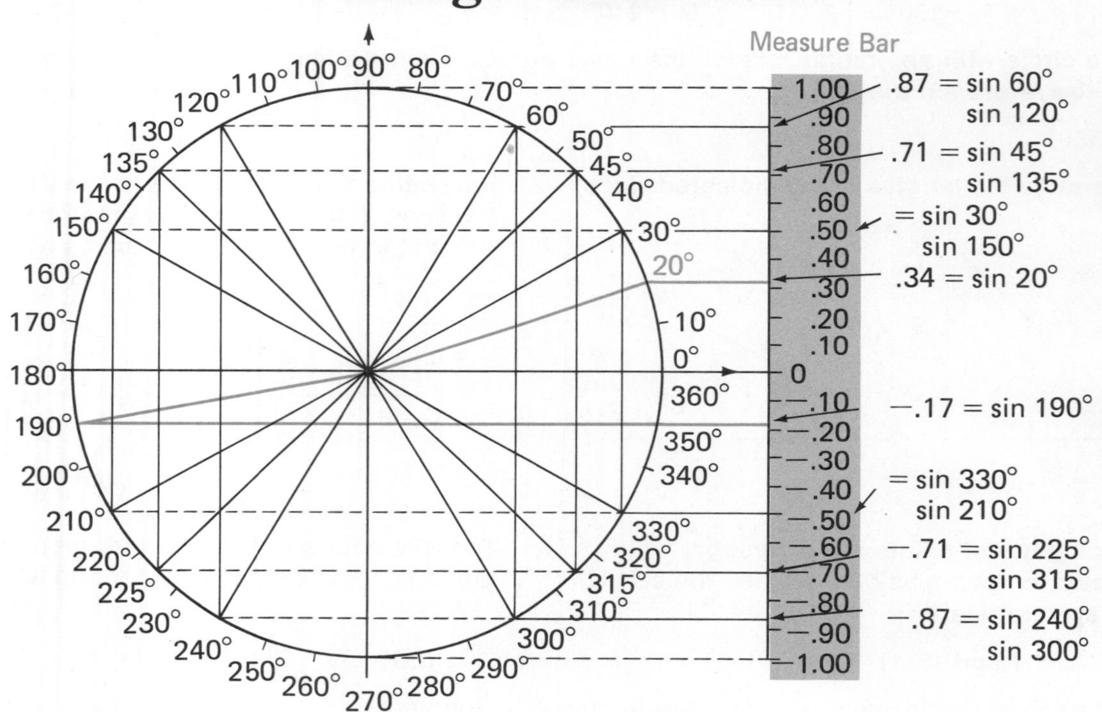

To find sin 20°:

- Draw a segment from the origin to 20° on the circle.
- Draw a horizontal segment to the measure bar.

Thus, sin 20° ≐ .34.

To find sin 190°:

- Draw a segment from the origin to 190° on the circle.
- Draw a horizontal segment to the measure bar.

Thus, sin 190° ≐ −.17.

PROJECT

Use the diagram to estimate sine values for each to 2 decimal places.

1. 50° **2.** 230° **3.** 120° **4.** 180° **5.** 310° **6.** 270°

7. Where would the measure bar be placed in order to approximate cosines?

Finding Trigonometric Values

OBJECTIVES

- To find the other five trigonometric functions of u given one of them
- To determine the quadrant in which the terminal side lies from information about trigonometric values

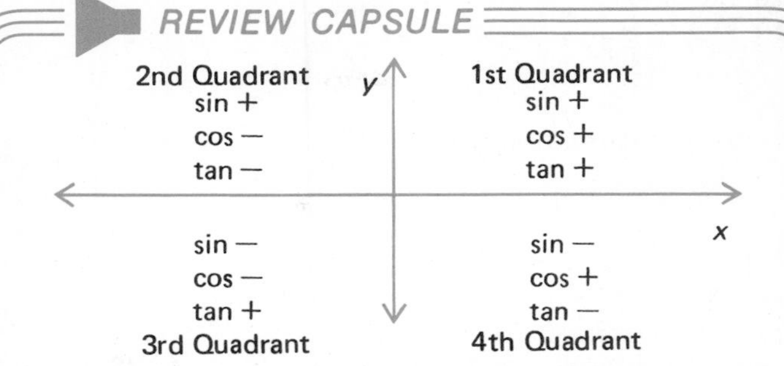

Cot, csc, and sec have the same signs as their reciprocals.

EXAMPLE 1 Find the value of each of the other five trigonometric functions of u if $\sin u = \frac{3}{5}$ and the terminal side is in the 2nd quadrant.

$$\sin u = \tfrac{3}{5}, \text{ so } \tfrac{y}{r} = \tfrac{3}{5}.$$

In the second quadrant, $y > 0$. ⟶ So, $y = 3$, $r = 5$.

Use the Pythagorean theorem.

$$x^2 + y^2 = r^2$$
$$x^2 + (3)^2 = (5)^2$$
$$x^2 = 16$$

In the second quadrant, $x < 0$. So, $x = -4$.

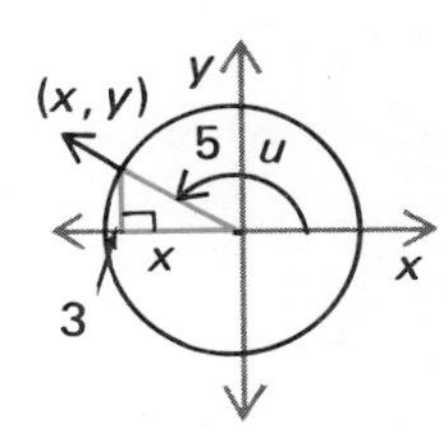

Use the definitions. ⟶ **Thus,** $\cos u = -\frac{4}{5}$, $\sec u = -\frac{5}{4}$, $\tan u = -\frac{3}{4}$
$\cot u = -\frac{4}{3}$, $\csc u = \frac{5}{3}$.

EXAMPLE 2 In which quadrants may the terminal side lie if $\cos u = -\frac{5}{12}$?

From the Review Capsule ⟶ Cosine is negative in the 2nd and 3rd quadrants.

Thus, the terminal side may lie in the 2nd or 3rd quadrant.

EXAMPLE 3 $\text{Tan } u = -\frac{2}{3}$ and $\sin u < 0$. In which quadrant does the terminal side lie?

Tan u is negative in the 2nd and 4th quadrants. Sin u is negative in the 3rd and 4th quadrants.

Both tan and sin are negative in the 4th quadrant only. ⟶ **Thus,** the terminal side lies in the 4th quadrant.

EXAMPLE 4 Find the value of each of the other five trigonometric functions of u if $\tan u = -\frac{2}{3}$ and $\sin u < 0$.

See Example 3. ⟶ The terminal side lies in the 4th quadrant.

$$\tan u = -\frac{2}{3}$$

$$\frac{y}{x} = -\frac{2}{3}$$

In the 4th quadrant, $y < 0$, $x > 0$. ⟶ So, $y = -2$, $x = 3$.

$$x^2 + y^2 = r^2$$

$$3^2 + (-2)^2 = r^2$$

$$13 = r^2$$

r is always > 0. ⟶ So, $r = \sqrt{13}$.

Draw the reference △.

Use the definitions.
Rationalize denominators.

Thus, $\sin u = -\frac{2}{\sqrt{13}}$ or $\frac{-2\sqrt{13}}{13}$, $\csc u = -\frac{\sqrt{13}}{2}$, $\cos u = \frac{3}{\sqrt{13}}$ or $\frac{3\sqrt{13}}{13}$, $\sec u = \frac{\sqrt{13}}{3}$, and $\cot u = -\frac{3}{2}$.

EXERCISES

PART A

State in which quadrants the terminal side of an angle may lie for each of the following.

1. $\sin u = -\frac{4}{5}$ *3; 4*
2. $\cos u = -\frac{3}{5}$ *2; 3*
3. $\cos u = \frac{5}{13}$ *1; 4*
4. $\sin u = \frac{2}{5}$ *1; 2*
5. $\tan u = \frac{5}{12}$ *1; 3*
6. $\cot u = -\frac{12}{5}$ *2; 4*
7. $\sec u = -\frac{15}{8}$ *2; 3*
8. $\csc u = 3$ *1; 2*
9. $\tan u = -12$
10. $\cot u = \sqrt{3}$
11. $\sin u < 0$ and $\cos u > 0$
12. $\tan u < 0$ and $\cot u > 0$

Find the value of each of the other five trigonometric functions of u.

13. $\tan u = -\frac{12}{5}$ and $\sin u > 0$
14. $\sin u = \frac{3}{5}$ and $\tan u < 0$
15. $\cot u = -1$ and $\cos u > 0$
16. $\csc u = -\frac{13}{5}$ and $\cos u < 0$

PART B

Find the value of each of the other five trigonometric functions of u.

17. $\tan u = -\sqrt{3}$ and $\cos u < 0$
18. $\tan u = \frac{\sqrt{3}}{3}$ and $\sin u < 0$
19. $\sec u = -\frac{2\sqrt{3}}{3}$ and $\tan u > 0$
20. $\cos u = -\frac{\sqrt{3}}{3}$ and $\csc u > 0$

PART C

Find the value(s) of each of the other trigonometric functions of u.

21. $\sin u = -\frac{1}{2}$ and $\tan u = \frac{\sqrt{3}}{3}$
22. $\cos u = \frac{\sqrt{2}}{2}$ and $\sec u = \sqrt{2}$
23. $\cos u = \frac{\sqrt{3}}{2}$ and $\csc u = 2$
24. $\tan u = -1$ and $\cot u = -1$

Using Trigonometric Tables

OBJECTIVES

- To determine the sin, cos, tan, and cot of a degree measure by reading a trigonometric table
- To find the degree measure when the sin, cos, tan, and cot of that degree measure are given
- To rewrite trigonometric functions using cofunctions

REVIEW CAPSULE

Find the complementary measure.

60°	14° 10′	35° 16′
$90° - 60° = 30°$	$90° - 14°10′ = 89°60′ - 14°10′ = 75°50′$	$90° - 35°16′ = 89°60′ - 35°16′ = 54°44′$

The sum of complementary measures is 90°.

EXAMPLE 1 Find sin 36° and cot 37° 30′

Functions at the *top* of the table must be used with degree measures in the left-hand column, (0° to 45°).

A portion of a trigonometric table is shown.

Angle	Sin	Cos	Tan	Cot	
36° 00′	.5878	.8090	.7265	1.3764	54° 00′
10	.5901	.8073	.7310	1.3680	50
20	.5925	.8056	.7355	1.3597	40
30	.5948	.8039	.7400	1.3514	30
40	.5972	.8021	.7445	1.3432	20
50	.5995	.8004	.7490	1.3351	10
37° 00′	.6018	.7986	.7536	1.3270	53° 00′
10	.6041	.7969	.7581	1.3190	50
20	.6065	.7951	.7627	1.3110	40
30	.6088	.7934	.7673	1.3032	30
40	.6111	.7916	.7720	1.2954	20
50	.6134	.7898	.7766	1.2876	10

36° → 36° 00′ → .5878

37° 30′ → 30 → 1.3032

Read from the table. → **Thus,** sin 36° = .5878, and cot 37° 30′ = 1.3032.

EXAMPLE 2 Find cos 45° 50′ and cot 46° 20′.

Angle measure is given in multiples of 10 minutes.

43° 00′	.6820	.7314	.9325	1.0724	47° 00′
10	.6841	.7294	.9380	1.0661	50
20	.6862	.7274	.9435	1.0599	40
30	.6884	.7254	.9490	1.0538	30
40	.6905	.7234	.9545	1.0477	20
50	.6926	.7214	.9601	1.0416	10
44° 00′	.6947	.7193	.9657	1.0355	46° 00′
10	.6967	.7173	.9713	1.0295	50
20	.6988	.7153	.9770	1.0235	40
30	.7009	.7133	.9827	1.0176	30
40	.7030	.7112	.9884	1.0117	20
50	.7050	.7092	.9942	1.0058	10
45° 00′	.7071	.7071	1.0000	1.0000	45° 00′
	Cos	Sin	Cot	Tan	Angle

46° 20′ → 20 → .9545

45° 50′ → 50 → .6967

Functions at the *bottom* of the table must be used with degree measures in the right-hand column, (45° to 90°).

Thus, cos 45° 50′ = .6967 and cot 46° 20′ = .9545.

EXAMPLE 3 Write each as the same trigonometric function of an acute angle. Then, find each value.

cos 216° | sin 175°

Sketch the angle of rotation and determine the reference angle.

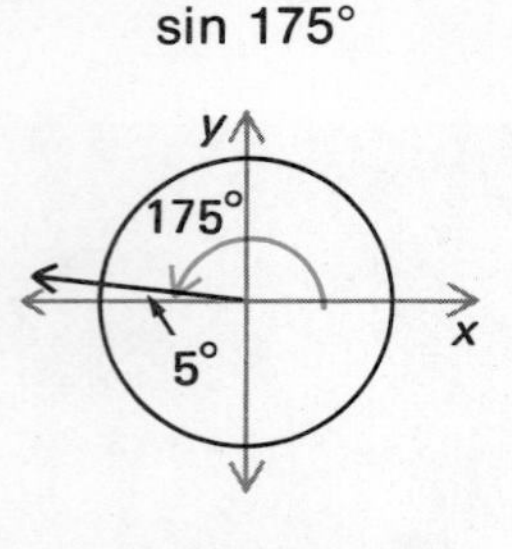

Find the reference measure.
Use the correct sign.

Look up the value in the table. ⟶

$\cos 216° = -\cos 36°$
$= -.8090.$

$\sin 175° = \sin 5°$
$= .0872.$

EXAMPLE 4 Find sin 124° 10′. | Find cot 310° 50′.

Sketch the angle of rotation and determine the reference angle.

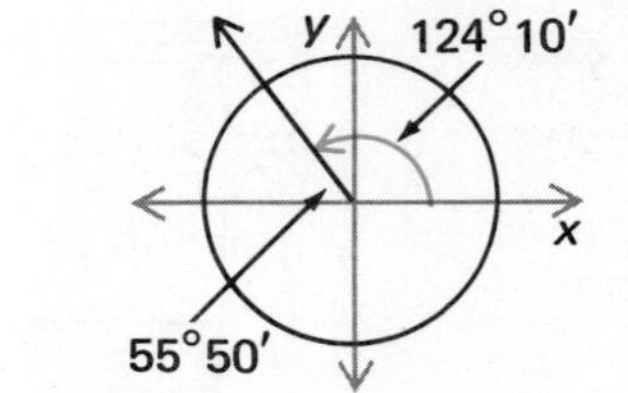

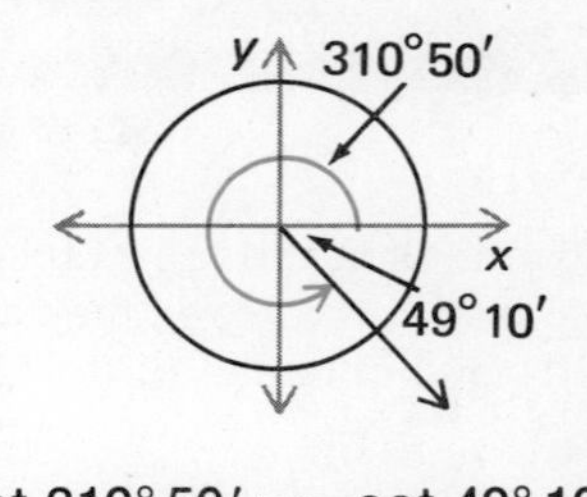

Write each in terms of the same trig function of the reference angle. ⟶
Look in the table. ⟶

$\sin 124° 10' = \sin 55° 50'$
$= .8274$

$\cot 310° 50' = -\cot 49° 10'$
$= -.8642$

EXAMPLE 5 Find the value for each pair from the table.

Complementary measures: 40° and 50°; 30° and 60°

sin 40° cos 50° ⟶ .6428

tan 30° cot 60° ⟶ .5774

They are also called complementary.

Sin and cos are cofunctions.
Tan and cot are cofunctions.

Complementary functions explain why the right-hand and left-hand columns of the trigonometric tables work.

$$\sin \theta = \cos (90° - \theta)$$
$$\cos \theta = \sin (90° - \theta)$$

$$\tan \theta = \cot (90° - \theta)$$
$$\cot \theta = \tan (90° - \theta)$$

EXAMPLE 6 Express each in terms of the cofunction.
cos 57°, sin 13°, cot 12° 40′

Use the cofunction and the complement.

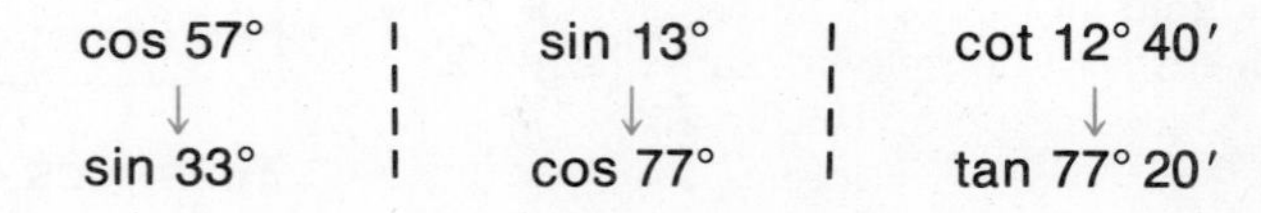

EXAMPLE 7 Cos $u = .7790$. Find u, $(u < 90°)$. | Cot $u = .8342$. Find u, $(u < 90°)$.

Look in the table to find the value.

38°50' → (row 38° 50) | ← 50°10' (row 50° 10)

Angle	Sin	Cos	Tan	Cot	
38° 00'	.6157	.7880	.7813	1.2799	52° 00'
10	.6180	.7862	.7860	1.2723	50
20	.6202	.7844	.7907	1.2647	40
30	.6225	.7826	.7954	1.2572	30
40	.6248	.7808	.8002	1.2497	20
50	.6271	.7790	.8050	1.2423	10
39° 00	.6293	.7771	.8098	1.2349	51° 00'
10	.6316	.7753	.8146	1.2276	50
20	.6338	.7735	.8195	1.2203	40
30	.6361	.7716	.8243	1.2131	30
40	.6383	.7698	.8292	1.2059	20
50	.6406	.7679	.8342	1.1988	10
40° 00'	.6428	.7660	.8391	1.1918	50° 00'
	Cos	Sin	Cot	Tan	Angle

Read the angle measure from the correct column.

$\cos u = .7790$
$u = 38°\,50'$

$\cot u = .8342$
$u = 50°\,10'$

EXAMPLE 8 Find all values of u, $0° < u < 360°$, for which $\sin u = -.6202$.

Read the table as in Example 7. (Use .6202.) → The measure of the reference ∠ is 38° 20'.

sin u is negative

3rd Quadrant | 4th Quadrant

Draw the reference ∠. →

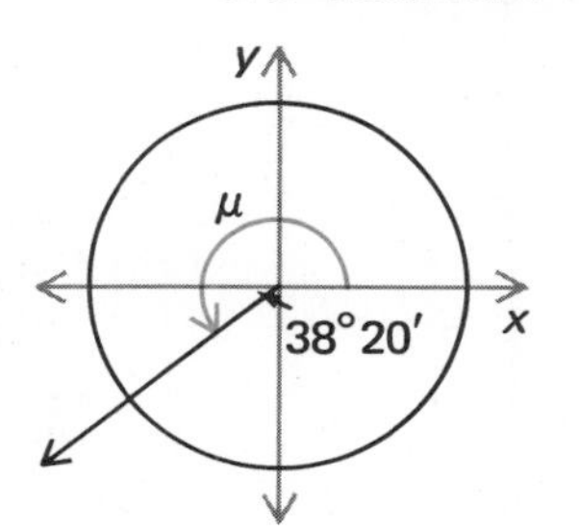

y
μ
x
38°20'

Find u. → **Thus,** $u = 218°\,20'$ or $u = 321°\,40'$.

ORAL EXERCISES

Use the table to find each of the following.

1. sin 33° *.5446* **2.** tan 18° **3.** cos 73° *.2924* **4.** cot 47°
5. tan 47° 10' **6.** sin 10° 30' **7.** cot 23° 20' **8.** cos 63° 30'
9. sin 70° 30' *.9426* **10.** tan 0° 20' *.0058* **11.** sin 44° 40' *.7030* **12.** tan 62° 30' *1.9210*

Use the table to find u, $(u < 90°)$.

13. $\sin u = .2079$ *12°* **14.** $\tan u = 1.4826$ *56°* **15.** $\cos u = .5150$ *59°*
16. $\cot u = 2.9042$ *19°* **17.** $\cos u = .3881$ *67° 10'* **18.** $\tan u = .2278$ *12° 50'*
19. $\tan u = 6.1970$ *80° 50'* **20.** $\sin u = .9932$ *83° 20'* **21.** $\cot u = .1673$ *80° 30'*

EXERCISES

PART A

Use the table to find *u*, ($u < 90°$).

1. $\sin u = .4540$ *27°* **2.** $\cos u = .4384$ *64°* **3.** $\tan u = .7536$ *37°*
4. $\cot u = 6.3138$ *9°* **5.** $\sin u = .9520$ *72° 10′* **6.** $\cos u = .9811$
7. $\tan u = 1.1988$ *50° 10′* **8.** $\tan u = 3.3759$ *73° 30′* **9.** $\cot u = 26.432$ *2° 10′*

Use the table to find each of the following.

10. sin 133° *.7314* **11.** tan 250° **12.** cos 335° **13.** cot 190°10′
14. tan 170°20′ *−.1703* **15.** sin 133° 30′ *.7254* **16.** cot 350°30′ *−5.9758* **17.** sin 300°10′ *−.8646*

Express each in terms of the cofunction.

18. sin 27° **19.** cos 46° **20.** cot 18° **21.** tan 73°
22. tan 63°30′ *cot 26° 30′* **23.** cot 79°10′ *tan 10° 50′* **24.** sin 40°50′ *cos 49° 10′* **25.** cos 88°40′ *sin 1° 20′*

Find all values of *u*, $0° < u < 360°$.

26. $\sin u = .1564$ *9°; 171°* **27.** $\tan u = .6249$ *32°; 212°* **28.** $\cos u = -.9511$
29. $\tan u = -1.1847$ *130°10′; 310°10′* **30.** $\cot u = 2.8502$ *19°20′; 199°20′* **31.** $\sin u = .3311$ *19°20′; 160°40′*

PART B

EXAMPLE Express as a function of an acute angle measure less than 45°.

sin 125°	tan 310°
sin 125° = sin 55°	tan 310° = −tan 50°
= cos 35°	= −cot 40°

Thus, sin 125° = cos 35° and tan 310° = −cot 40°.

Express each as a function of an acute angle measure less than 45°.

32. sin 210° *−sin 30°* **33.** cos 220° *−cos 40°* **34.** tan 305° **35.** cot 190°
36. cos 215° 10′ *−cos 35° 10′* **37.** sin 100° 30′ *cos 10° 30′* **38.** cos 320° 10′ *cos 39° 50′* **39.** tan 300° 30′ *−cot 30° 30′*

PART C

Express each as a reciprocal function of an acute angle. See Part B Example.

40. sin 125° **41.** tan 220° **42.** cos 82°50′ **43.** cot 100°10′

Find each of the following.

44. sec 12°10′ **45.** csc 15°30′

Find *u*, $u < 90°$.

46. $\sec u = 1.0932$ **47.** $\csc u = 1.2494$

Linear Trigonometric Equations

OBJECTIVE

- To solve trigonometric equations like $4 \sin \theta - 5 = 2 \sin \theta - 3$

	0°	30°	45°	60°	90°
sin	0	$\frac{1}{2}$	$\frac{\sqrt{2}}{2}$	$\frac{\sqrt{3}}{2}$	1
cos	1	$\frac{\sqrt{3}}{2}$	$\frac{\sqrt{2}}{2}$	$\frac{1}{2}$	0
tan	0	$\frac{\sqrt{3}}{3}$	1	$\sqrt{3}$	undefined

Values for other measures are found in the table.

EXAMPLE 1 Determine the values of u between 0° and 360° that make the equation true.

$$2 \cos u = \sqrt{3}$$

$\cos u = \frac{\sqrt{3}}{2}$ →

$$\cos u = \frac{\sqrt{3}}{2}$$

Reference angle: 30°

Cos is positive in the 1st and 4th quadrants. → **Thus,** $u = 30°$ and $u = 330°$.

EXAMPLE 2 Determine the values of u between 0° and 360° to the nearest ten minutes that make the equation true.

$$5 \sin u = 1$$

Solve for sin u.

$$\sin u = \frac{1}{5}, \text{ or } .2000$$

Use trigonometric table. Find the closest value. → Reference angle: 11° 30′

Sin is positive in the 1st and 2nd quadrants. → **Thus,** $u = 11° 30'$ and $u = 168° 30'$.

EXAMPLE 3 Determine the values of θ between 0° and 360° to the nearest ten minutes that make the equation true.

$$2 \tan \theta - 1 = -5$$

Solve for tan θ.

$$2 \tan \theta = -4$$
$$\tan \theta = -2$$

Use the trigonometric table. → Reference angle: 63° 30′

Tan is negative in the 2nd and 4th quadrants. → **Thus,** $\theta = 116° 30'$ and $\theta = 296° 30'$.

EXAMPLE 4 Determine the values of u between 0° and 360° which make the equation true.

$$4 \sin u - 5 = 2 \sin u - 3$$

Solve for $\sin u$.

Add 5 to each side. → $4 \sin u = 2 \sin u + 2$

Add $-2 \sin u$ to each side. → $2 \sin u = 2$

Divide each side by 2. → $\sin u = 1$

$\sin 90° = 1$. → **Thus,** $u = 90°$.

EXAMPLE 5 Determine the values of u between 0° and 360° which make the equation true.

$$2 \cos u = 4$$

$$\cos u = 2$$

From the trig table, $\cos u \leq 1.0000$. → **Thus,** there are no solutions.

EXERCISES

PART A

Determine the values of u between 0° and 360° which make the equation true.

1. $2 \sin u = 1$
2. $2 \cos u = -\sqrt{3}$
3. $\tan u = 1$ *45°; 225°*
4. $\cot u = \sqrt{3}$
5. $\cot u = -1$
6. $2 \sin u = -\sqrt{2}$
7. $\cos u = 0$ *90°; 270°*
8. $3 \tan u = -\sqrt{3}$
9. $2 \sin u + 1 = 0$ *210°; 330°*
10. $2 \cos u - \sqrt{3} = 0$ *30°; 330°*
11. $\sqrt{2} \cos u + 1 = 0$ *135°; 225°*
12. $2 \sin u - \sqrt{3} = 0$ *60°; 120°*

Determine the values of θ between 0° and 360° to the nearest ten minutes, which make the equation true.

13. $2 \sin \theta = -2$
14. $2 \cos \theta + 5 = -3 \cos \theta$
15. $5 \tan \theta + 3 = 2 - 3 \tan \theta$
16. $4 \cos \theta = 2$
17. $4 \sin \theta = 5 - 2 \sin \theta$
18. $-4 \cot \theta + 3 = 5 \cot \theta - 6$
19. $4 \sin \theta = 5$ *not possible*
20. $3 \cot \theta + 4 = 5$ *71° 30′; 251° 30′*
21. $5 \cos \theta - 2 = 3 \cos \theta$

PART B

Determine the values of θ between 0° and 720° to the nearest ten minutes, which make the equation true.

22. $2 \cos \theta = 3 - 2 \cos \theta$
23. $\sin \theta + 4 = 6 \sin \theta$
24. $4 \tan \theta = -3$
25. $\tan \theta + \sqrt{3} = 0$ *120°; 300°; 480°; 660°*
26. $2 \sin \theta = -\sqrt{3}$ *240°; 300°; 600°; 660°*
27. $-\cot \theta = \sqrt{3}$ *150°; 330°; 510°; 690°*

PART C

Determine the values of u between 0° and 360° which make the equation true.

28. $\sin u = \cos u$
29. $\sin u = -\cos u$
30. $\tan u = \cot u$

Trigonometric Interpolation

OBJECTIVES

- To determine the sin, cos, tan, and cot of a measure like 38° 34′
- To solve an equation like $\sin \theta = .2736$ for θ to the nearest minute

REVIEW CAPSULE

Find log 235.7.

	digits	mantissa
10 [7 [	2350	3711] n] 18
	2357	
	2360	3729

$$\frac{7}{10} = \frac{n}{18}$$

$$10n = 126$$

$$n = 12.6, \text{ or } n \doteq 13$$

Mantissa: 3711 + 13, or 3724

Characteristic: 2.

Thus, log 235.7 = 2.3724.

EXAMPLE 1 Find sin 27° 35′.

Interpolate: 27° 35′ lies between 27° 30′ and 27° 40′ in the table.

	measure	sine
10 [5 [	27° 30′	4617] n] 26
	27° 35′	
	27° 40′	4643

Find these values in the table.

Write the proportion. →

$$\frac{5}{10} = \frac{n}{26}$$

Solve. →

$$10n = 130$$

$$n = 13$$

4617 + 13 = 4630

Replace the decimal point.

Thus, $\sin 27° 35′ \doteq .4630$.

EXAMPLE 2 Find cos 38° 34′.

38° 34′ is between 38° 30′ and 38° 40′.

	measure	cosine
10 [4 [	38° 30′	7826] n] −18
	38° 34′	
	38° 40′	7808

Values of cos decrease as measure increases.

Write the proportion. →

$$\frac{4}{10} = \frac{n}{-18}$$

Solve. →

$$10n = -72$$

$$n = -7.2, \text{ or } n \doteq -7$$

7826 − 7 = 7819

Replace the decimal point.

Thus, $\cos 38° 34′ \doteq .7819$.

EXAMPLE 3 Sin $u = .5987$. Find u, $(0° < u < 90°)$ to the nearest minute.

.5972 and .5995 are the table entries in the sin column nearest .5987.

		measure	sine		
10	n	36° 40′	5972	15	23
			5987		
		36° 50′	5995		

$$\frac{n}{10} = \frac{15}{23}$$
$$23n = 150$$
$$n = \frac{150}{23}, \text{ or } n \doteq 7$$

$u = 36°\,40' + 7'$, or $36°\,47'$ ⟶ **Thus,** $u = 36°\,47'$.

EXAMPLE 4 Find all values of θ, $(0° < \theta < 360°)$ to the nearest minute which make the following equation true.

$$3 \tan \theta + 4 = 2$$

Solve for $\tan \theta$. ⟶

$$3 \tan \theta = -2$$
$$\tan \theta = \frac{-2}{3}$$
$$\tan \theta \doteq -.6667$$

Use .6667 to find the measure of the reference angle.

Interpolate. ⟶

.6661 and .6703 are the table entries in the tangent column nearest .6667.

		measure	tangent		
10	n	33° 40′	.6661	6	42
			.6667		
		33° 50′	.6703		

$$\frac{n}{10} = \frac{6}{42}$$
$$42n = 60$$
$$n = \frac{60}{42}$$
$$n \doteq 1$$

$33°\,40' + 1'$ ⟶ The measure of the reference angle is $33°\,41'$.

Tangent is negative in the 2nd and 4th quadrants.

$180° = 179°\,60'$
$360° = 359°\,60'$

2nd quadrant	4th quadrant
$179°\,60'$	$359°\,60'$
$-\ 33°\,41'$	$-\ 33°\,41'$
$146°\,19'$	$326°\,19'$

Thus, $\theta = 146°\,19'$ or $326°\,19'$.

EXERCISES

PART A

Find each of the following.

1. cos 11°43′ *.9791* **2.** tan 76°18′ **3.** sin 36°49′ *.5993* **4.** cot 62°57′
5. tan 42°12′ *.9068* **6.** sin 49°36′ **7.** cos 82°53′ *.1239* **8.** tan 8°13′
9 sin 10°12′ *.1771* **10.** cot 34°19′ *1.4650* **11.** tan 63°18′ *1.9883* **12.** cos 31°17′ *.8546*

Find u, ($0° < u < 90°$) to the nearest minute.

13. $\sin u = .4690$ **14.** $\cos u = .8577$ **15.** $\tan u = 1.6868$ *59°20′*
16. $\cot u = .6666$ **17.** $\sin u = .8430$ **18.** $\cos u = .0650$ *86°16′*
19. $\tan u = .3370$ *18°38′* **20.** $\tan u = 2.0053$ *63°30′* **21.** $\sin u = .1573$ *9°03′*

Find all values of θ, ($0° < \theta < 360°$) to the nearest minute.

22. $2 \cos \theta = -1$ *120°; 240°* **23.** $3 \sin \theta + 2 = -2 \sin \theta$ *203°35′; 336°25′* **24.** $4 \tan \theta - 1 = 4 - 2 \tan \theta$
25. $5 \cot \theta + 1 = 3 \cot \theta$ **26.** $4 \cos \theta - 1 = 4 - 2 \cos \theta$ *33°34′; 326°26′* **27.** $3 \sin \theta + 2 = 3 - \sin \theta$ *14°29′; 165°31′*

PART B

EXAMPLE Find sin 320°14′.

Rewrite using sine of an acute angle. sin 320°14′ = −sin 39°46′
Find sin 39°46′.

Interpolate. ⟶

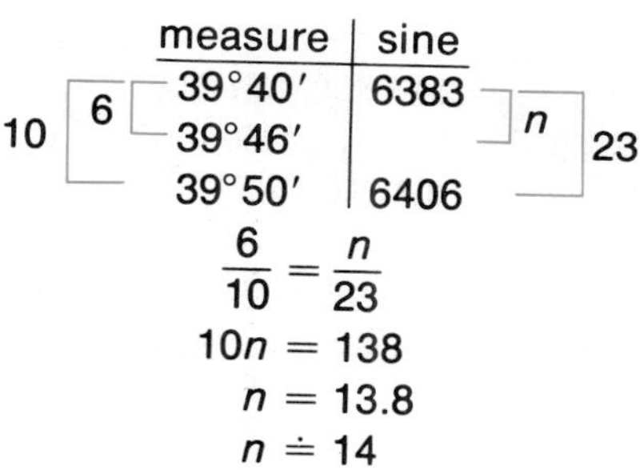

6383 + 14 = 6397
Replace decimal point.

So, sin 39°46′ = .6397.

Thus, sin 320°14′ = −.6397.

Find each of the following.

28. sin 135°36′ *.6996* **29.** cos 225°53′ *−.6961* **30.** tan 97°18′ *−7.8069* **31.** cot 195°48′ *3.5340*

Find u, ($0° < u < 360°$) to the nearest minute.

32. $\cos u = -.9177$ **33.** $\sin u = -.8221$ **34.** $\tan u = .4272$

PART C

Find all values of A, ($0° < A < 360°$).

35. $\frac{\sin A}{\cos A} = 1$ **36.** $\cos A \cdot \csc A = 2.3750$ **37.** $\cot A \cdot \tan A = 1$

Right Triangle Trigonometry

OBJECTIVES

- To find trigonometric functions of acute angles in a right triangle
- To simplify expressions like $\frac{\cos 60° + \sin 30°}{\tan 45°}$

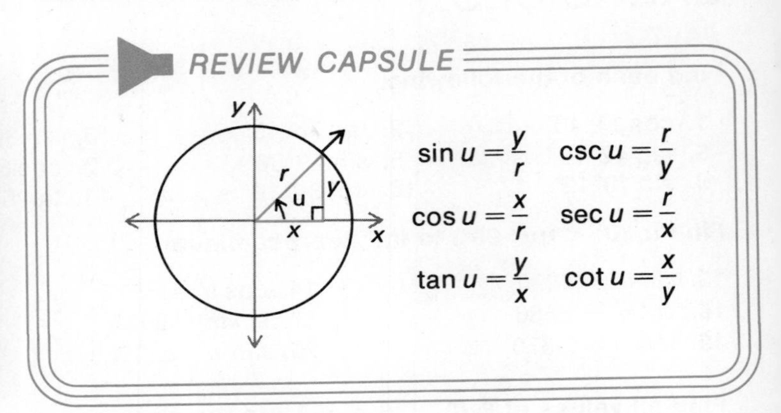

$\sin u = \frac{y}{r}$ $\csc u = \frac{r}{y}$

$\cos u = \frac{x}{r}$ $\sec u = \frac{r}{x}$

$\tan u = \frac{y}{x}$ $\cot u = \frac{x}{y}$

EXAMPLE 1 In triangle ABC, locate the side opposite $\angle A$, the side adjacent to angle A and the hypotenuse.

Label angles with capitals, sides with small letters.

Side a is opposite $\angle A$.
Side b is opposite $\angle B$.
Side c is opposite $\angle C$.

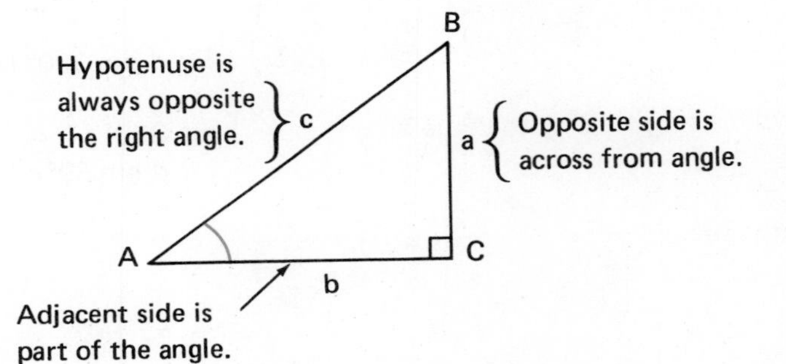

Thus, side a is opposite $\angle A$, side b is adjacent to $\angle A$, side c is the hypotenuse.

EXAMPLE 2 In triangle ABC locate the side opposite $\angle B$, the side adjacent to $\angle B$, and the hypotenuse.

Hypotenuse is opposite the right angle.

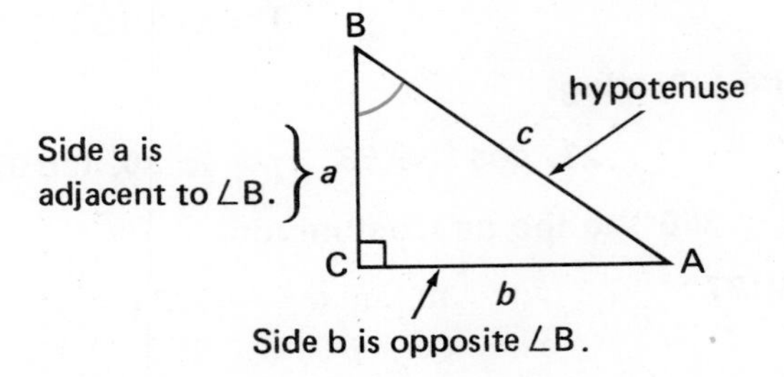

Thus, side b is opposite $\angle B$, side a is adjacent to $\angle B$, side c is the hypotenuse.

EXAMPLE 3 Write the definitions of $\sin A$, $\cos A$, $\tan A$, using right triangle ABC.

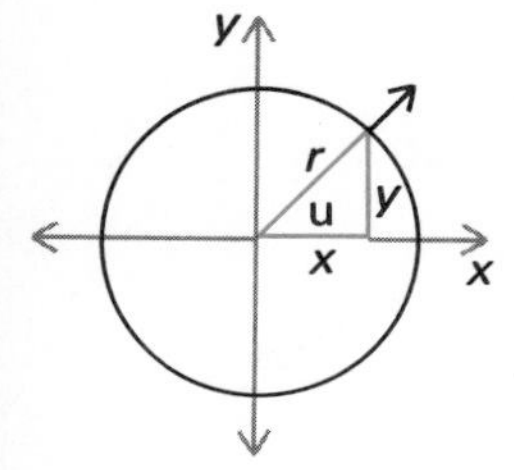

y is opposite side.
x is adjacent side.
r is hypotenuse.

$\sin u = \frac{y}{r}$ ⟶ $\sin A = \dfrac{\text{length of opposite side}}{\text{length of hypotenuse}}$

$\cos u = \frac{x}{r}$ ⟶ $\cos A = \dfrac{\text{length of adjacent side}}{\text{length of hypotenuse}}$

$\tan u = \frac{y}{x}$ ⟶ $\tan A = \dfrac{\text{length of opposite side}}{\text{length of adjacent side}}$

Thus, $\sin A = \frac{a}{c}$, $\cos A = \frac{b}{c}$, and $\tan A = \frac{a}{b}$.

EXAMPLE 4 For each right triangle, find $\sin 45°$, $\cos 45°$, and $\tan 45°$.

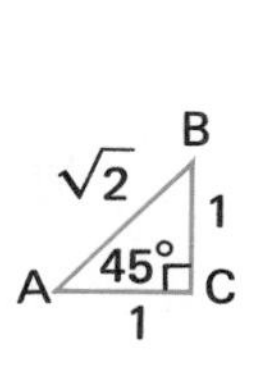

$\sin A = \frac{\text{opp.}}{\text{hyp.}}$ ⟶

$\cos A = \frac{\text{adj.}}{\text{hyp.}}$ ⟶

$\tan A = \frac{\text{opp.}}{\text{adj.}}$ ⟶

Value of each trig function is the same for any size right triangle.

$\sin 45° = \frac{1}{\sqrt{2}}$, or $\frac{\sqrt{2}}{2}$ | $\sin 45° = \frac{3}{3\sqrt{2}}$, or $\frac{\sqrt{2}}{2}$

$\cos 45° = \frac{1}{\sqrt{2}}$, or $\frac{\sqrt{2}}{2}$ | $\cos 45° = \frac{3}{3\sqrt{2}}$, or $\frac{\sqrt{2}}{2}$

$\tan 45° = \frac{1}{1}$, or 1 | $\tan 45° = \frac{3}{3}$, or 1

Thus, $\sin 45° = \frac{\sqrt{2}}{2}$, $\cos 45° = \frac{\sqrt{2}}{2}$, and $\tan 45° = 1$.

EXAMPLE 5 For the given triangle, find $\sin 30°$, $\cos 30°$, $\tan 30°$.

For 30°,
a is opposite side,
b is adjacent side,
c is hypotenuse.

$\sin 30° = \frac{1}{2}$

$\cos 30° = \frac{\sqrt{3}}{2}$

$\tan 30° = \frac{1}{\sqrt{3}}$, or $\frac{\sqrt{3}}{3}$

Thus, $\sin 30° = \frac{1}{2}$, $\cos 30° = \frac{\sqrt{3}}{2}$, and $\tan 30° = \frac{\sqrt{3}}{3}$.

EXAMPLE 6 For the given triangle, find sin 60°, cos 60°, tan 60°.

For 60°,
b is opposite side.
a is adjacent side.
c is hypotenuse.

$\sin 60° = \frac{\sqrt{3}}{2}$

$\cos 60° = \frac{1}{2}$

$\tan 60° = \sqrt{3}$

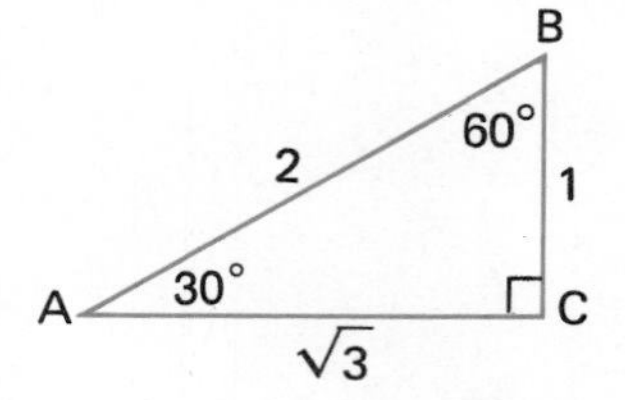

Thus, $\sin 60° = \frac{\sqrt{3}}{2}$, $\cos 60° = \frac{1}{2}$, and $\tan 60° = \sqrt{3}$.

From Examples 5 and 6 ⟶

$\sin 30° = \cos 60°$ $\cos 30° = \sin 60°$	$\sin A = \cos (90° - A)$ $\cos A = \sin (90° - A)$

EXAMPLE 7 Find u, $(0° < u < 90°)$ if $\sin 40° = \cos u$.

$\sin A = \cos (90° - A)$ ⟶

$\sin 40° = \cos u$
$\sin 40° = \cos (90° - 40°)$
$\sin 40° = \cos 50°$

Thus, $u = 50°$.

EXERCISES

PART A

Label the sides and find the sin, cos, and tan of the indicated acute angle.

1. 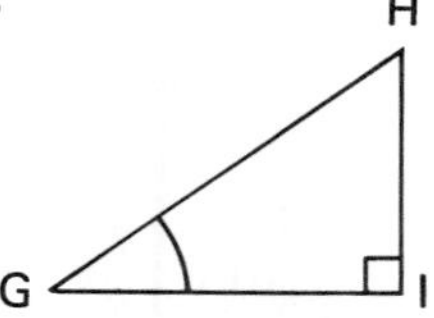

2. 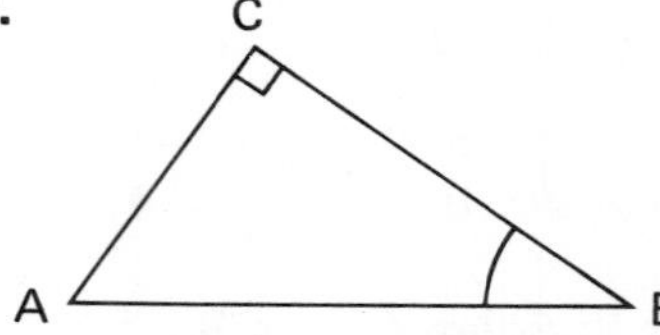

3. 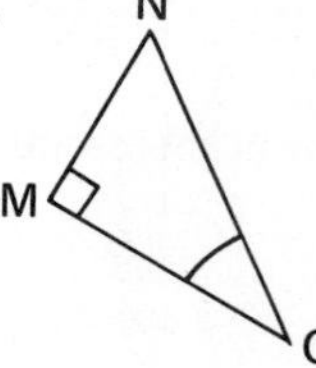

4. 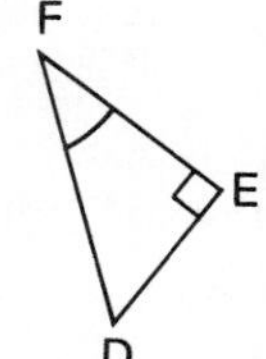

Find u, $0° < u < 90°$.

5. $\sin 19° = \cos u$ *71°*
6. $\cos 33° = \sin u$ *57°*
7. $\tan 20° = \cot u$ *70°*
8. $\cot 60° = \tan u$ *30°*

PART B

Evaluate and simplify. Use the results of Examples 4–6.

9. $\frac{\sin 30° + \cos 60°}{\tan 45°}$ *1*

10. $\frac{\sin 60° - \cos 30°}{\sin 60°}$ *0*

11. $\frac{\cos 60° + \tan 45°}{\cos 45°}$ *$\frac{3\sqrt{2}}{2}$*

Using Right Triangle Trigonometry

OBJECTIVES

- To find a length of a side of a right triangle given the length of one side and the measure of one acute angle
- To find the measure of an acute angle of a right triangle given the length of two sides
- To apply right triangle trigonometry

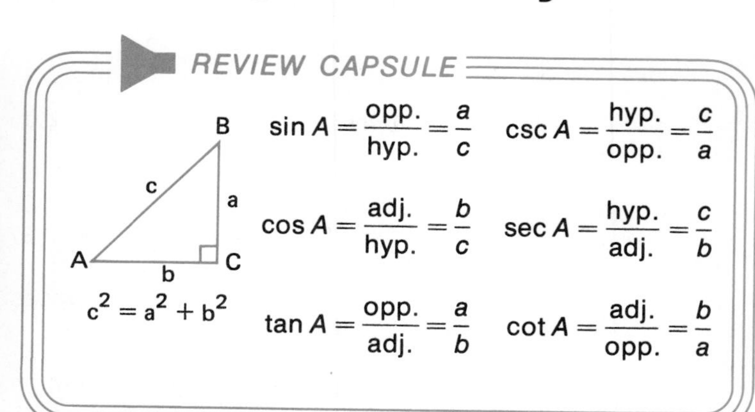

EXAMPLE 1 For each triangle, identify the trig. function needed to find x.

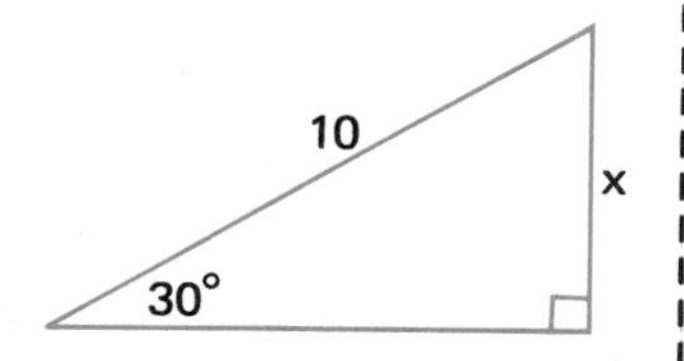

Given: hypotenuse
Find: opposite side

Use the function that relates the two sides. → Use $\sin A = \frac{\text{opp.}}{\text{hyp.}}$.

Thus, use sine.

Given: adjacent side
Find: opposite side

Use $\tan A = \frac{\text{opp.}}{\text{adj.}}$.

Thus, use tangent.

EXAMPLE 2 In right triangle ABC, $\angle A$ measures 58°, $b = 6.2$ m. Find a to the nearest meter.

Draw and label a diagram.

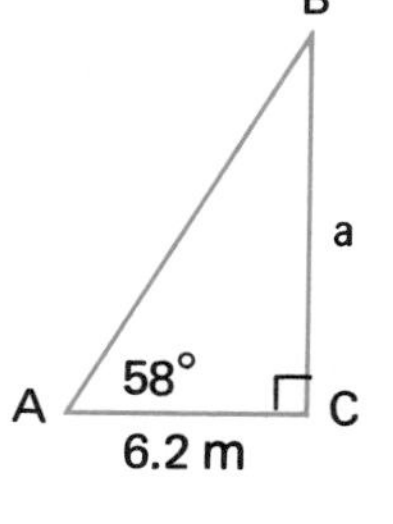

Given: adj. side
Find: opp. side

Use $\tan A = \frac{\text{opp.}}{\text{adj.}}$.

$m\angle A = 58$, $b = 6.2$ → $\tan 58° = \frac{a}{6.2}$

Multiply each side by 6.2. → $6.2(\tan 58°) = a$

Find tan 58° in trig table. → $6.2(1.6003) = a$

Multiply. → $9.92186 = a$

$9.92186 \doteq 10$ → **Thus,** $a = 10$ m to the nearest meter.

EXAMPLE 3 Find the measure of each acute angle in right triangle ABC if $a = 4$ and $c = 5$.

Given opp. side and hyp. → $\sin A = \frac{\text{opp.}}{\text{hyp.}}$

$a = 4$, $c = 5$ → $\sin A = \frac{4}{5} = .8000$

$m\angle A$ means degree measure of $\angle A$.
From the table, $\sin 53° = .7986$
$\angle A$ and $\angle B$ are complements.

$$m\angle A \doteq 53$$
$$53 + m\angle B = 90$$
$$m\angle B = 37$$

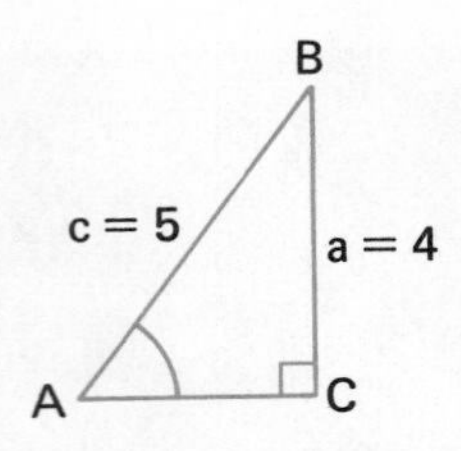

Thus, $\angle A$ measures 53° and $\angle B$ measures 37°.

EXAMPLE 4 In right triangle ABC, $m\angle B = 49$ and $a = 5$, find $m\angle A$. Find b to nearest whole number.

Find $m\angle A$.
$\angle A$ and $\angle B$ are complements. →

$$m\angle A + m\angle B = 90$$
$$m\angle A + 49 = 90$$
$$m\angle A = 41$$

Find b. → Given: $a = 5$
Find: b

$\tan x = \frac{\text{opp.}}{\text{adj.}}$

Use $\tan A = \frac{a}{b}$ or $\tan B = \frac{b}{a}$.

$$\tan 41° = \frac{5}{b}$$

In the first method, division by a 4-place decimal is needed. →

$$b = \frac{5}{\tan 41°} = \frac{5}{.8693} \doteq 6$$

or

$$\tan 49° = \frac{b}{5}$$
$$b = 5(\tan 49°) = 5(1.1504) = 5.7520 \doteq 6$$

Thus, $m\angle A = 41$ and $b \doteq 6$.

EXAMPLE 5 At a point on the ground 10 meters from the foot of a tree, the angle of elevation to the top of the tree is 38°. Find the height of the tree to the nearest meter.

Angle of elevation is the angle between the horizontal and the line of sight.
Given: adjacent
Find: opposite

$$\tan A = \frac{\text{opp.}}{\text{adj.}}$$
$$\tan 38° = \frac{a}{10}$$
$$10(\tan 38°) = a$$

From the table, $\tan 38° = .7813$. → $10(.7813) = a$

Multiply. → $7.813 = a$

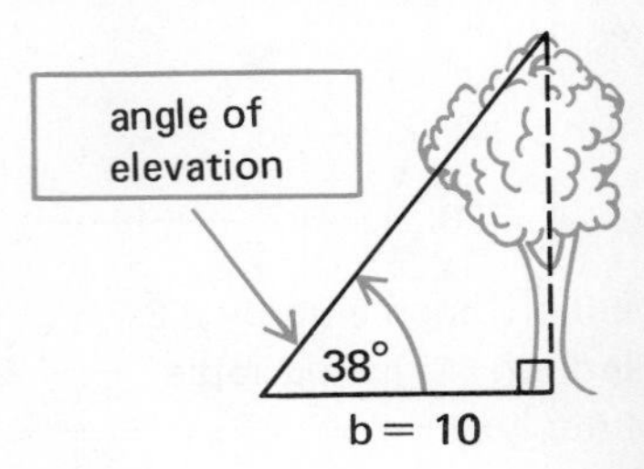

Thus, a is 8 m to the nearest meter.

EXAMPLE 6 A pilot flying at an altitude of 600 m observes that the angle of depression of an airport which he is approaching measures 62°. Find, to the nearest meter, the distance from the point on the ground directly below the plane to the airport.

Angle of depression is the angle between the horizontal and the line of sight.

Angle depression = angle elevation

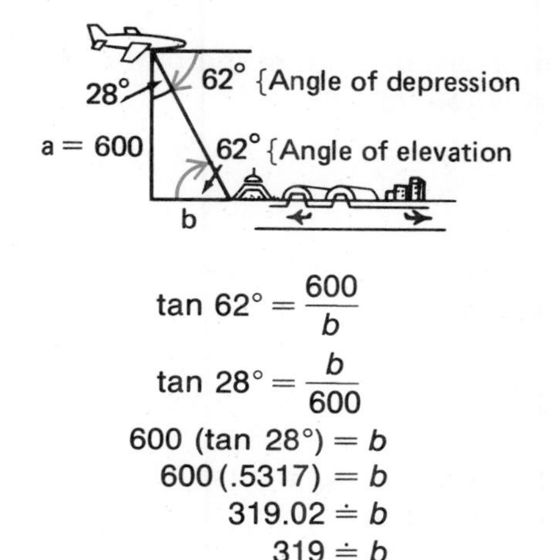

$$\tan 62° = \frac{600}{b}$$

Use the complement so that the unknown is in the numerator.

$$\tan 28° = \frac{b}{600}$$

$$600\,(\tan 28°) = b$$

From table, $\tan 28° = .5317$. → $600(.5317) = b$

Multiply. → $319.02 \doteq b$

Round to the nearest meter. $319 \doteq b$

Thus, b is 319 m.

EXAMPLE 7 A ladder leans against a building. The angle which the ladder makes with the ground measures 38°. The length of the ladder is 5 m. Find the distance from the foot of the ladder to the building, to the nearest meter.

Given: $m\angle A$ and hyp.
Find: adjacent side
Use $\cos A$.

$$\cos 38° = \frac{b}{5}$$

$$5(\cos 38°) = b$$

From the table, $\cos 38° = .7880$. $5(.7880) = b$

Multiply. → $3.9400 = b$

Round to the nearest meter. $b = 4$

Thus, b is 4 m.

EXERCISES

PART A

Find the measure of each of the other two sides of right triangle *ABC*, with $m\angle C = 90$, to the nearest whole number.

1. $m\angle A = 30$, $a = 7$
2. $m\angle B = 62$, $a = 12$
3. $m\angle B = 17$, $b = 6$
4. $m\angle A = 37$, $b = 9$
5. $m\angle B = 33$, $a = 6.1$
6. $m\angle A = 72$, $b = 7.3$

Find the measure of each acute angle of right triangle *ABC* with $m\angle C = 90$, to the nearest degree.

7. $c = 5$, $a = 3$ *37°; 53°*
8. $b = 12$, $a = 5$ *67°; 23°*
9. $a = 4$, $b = 3$ *53°; 37°*
10. $a = 6.2$, $b = 3.5$ *29°; 61°*
11. $c = 6.4$, $a = 4.3$ *42°; 48°*
12. $c = 7.3$, $a = 4.1$ *34°; 56°*

13. From the top of a lighthouse 70 meters high, the angle of depression of a boat at sea measures 36°. Find, to the nearest meter, the distance from the boat to the foot of the lighthouse. *96 m*

14. At a point on the ground 12 meters from the foot of a tree, the angle of elevation to the top of the tree measures 42°. Find the height of the tree to the nearest meter. *11 m*

15. A ladder leans against a building. The top of the ladder reaches a point on the building which is 11 meters above the ground. The foot of the ladder is 5 meters from the building. Find to the nearest degree the measure of the angle which the ladder makes with the level ground. *66°*

16. From an airplane flying 600 meters above sea level the angle of depression to a ship measures 29°. Find the distance, to the nearest meter, from the point at sea directly below the plane to the ship. *1,082 m*

PART B

Determine *BC* to the nearest tenth.

17. $m\angle A = 30$, $AD = 8$, $m\angle \theta = 45$ *10.9*
18. $m\angle A = 45$, $AD = 15$, $m\angle \theta = 60$ *35.5*
19. $m\angle A = 30$, $AD = 20$, $m\angle \theta = 60$ *17.3*

20. From point *B* on the ground, the angle of elevation to a balloon measures 45°. The balloon is attached to a rope 150 meters long. Find, to the nearest ten meters, the height of the balloon. *110 m*

PART C

21. From an airplane flying 1,000 meters above sea level the angles of depression to two ships due west measure 33° and 20°. Find the distance between the ships to the nearest meter.

Polar Form of Complex Numbers

$\cos\theta = \frac{a}{r}$ $\sin\theta = \frac{b}{r}$

$r\cos\theta = a$ $r\sin\theta = b$

$a + bi$

$r\cos\theta + (r\sin\theta)i$

$r(\cos\theta + i\sin\theta)$

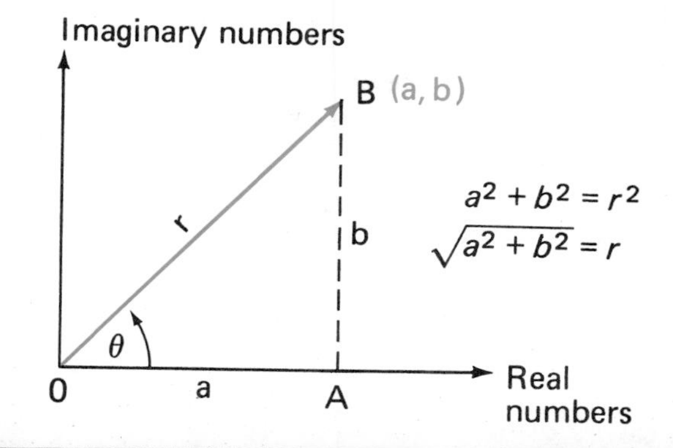

$a^2 + b^2 = r^2$

$\sqrt{a^2 + b^2} = r$

Polar Form of Complex Numbers

$a + bi = r(\cos\theta + i\sin\theta)$, where $r = \sqrt{a^2 + b^2}$

PROBLEMS

Write $1 + i$ in polar form.

$r = \sqrt{a^2 + b^2}$

$= \sqrt{1^2 + 1^2}$ or $\sqrt{2}$

$\cos\theta = \frac{a}{r}$

$= \frac{1}{\sqrt{2}}$ or $\frac{\sqrt{2}}{2}$

So $\theta = 45°$.

$\sin\theta = \frac{b}{r}$

$= \frac{1}{\sqrt{2}}$ or $\frac{\sqrt{2}}{2}$

So $\theta = 45°$.

$a + bi = r(\cos\theta + i\sin\theta)$

So $1 + i = \sqrt{2}(\cos 45° + i\sin 45°)$.

Change $2(\cos 60° + i\sin 60°)$ to $a + bi$ form.

$r(\cos\theta + i\sin\theta) = 2(\cos 60° + i\sin 60°)$

$a = r\cos\theta$

$= 2\cos 60°$

$= 2\left(\frac{1}{2}\right)$

$= 1$

$b = r\sin\theta$

$= 2\sin 60°$

$= 2\left(\frac{\sqrt{3}}{2}\right)$

$= \sqrt{3}$

So $a + bi = 1 + \sqrt{3}i$.

PROJECT

Change to polar form.

1. $\sqrt{3} + i$ **2.** $1 - i$ **3.** $\frac{1}{\sqrt{2}} + \frac{i}{\sqrt{2}}$

Change to $a + bi$ form.

4. $2(\cos 45° + i\sin 45°)$ **5.** $5(\cos 120° + i\sin 120°)$ **6.** $3(\cos 30° + i\sin 30°)$

Chapter Seventeen Review

State in which quadrants the terminal side of an angle may lie for each of the following. [p. 429]

1. $\cos u = -\frac{4}{5}$ 2. $\sin u = \frac{5}{12}$ 3. $\tan u = -1$ 4. $\cot u = \sqrt{3}$ 5. $\sin u = -\frac{7}{8}$

Find the value of each of the other five trigonometric functions of u. [p. 429]

6. $\tan u = \frac{5}{12}$, $\sin u < 0$ 7. $\cos u = -\frac{4}{5}$, $\tan u < 0$ 8. $\cot u = \frac{1}{2}$, $\cos u > 0$

Use the table to find each of the following. [p. 431]

9. $\sin 28°$ 10. $\tan 52°10'$ 11. $\cos 143°$ 12. $\cot 310°30'$

Find u, $u < 90°$. [p. 431]

13. $\cos u = .9827$ 14. $\sin u = .9907$

Find all values of u, $0° < u < 360°$. [p. 431]

15. $\tan u = -3.4874$ 16. $\cot u = 1.7556$

Express in terms of the cofunction. [p. 431]

17. $\tan 73°$ 18. $\sin 62°10'$

Express as a function of an acute angle measure less than 45°. [p. 431]

19. $\cos 300°$ 20. $\cot 100°$

Determine the values of θ between 0° and 360°, to the nearest ten minutes, which make the equation true. [p. 435]

21. $\cos\theta = -\frac{1}{2}$ 22. $2\tan\theta = -2$ 23. $5\cot\theta + 3 = 1$ 24. $\sin\theta + 3 = 5\sin\theta$

Find each of the following. [p. 437]

25. $\sin 37°18'$ 26. $\cot 120°12'$

Find all values of θ, $(0° < \theta < 360°)$, to the nearest minute, which make the equation true. [p. 437]

27. $\cos\theta = .7910$ 28. $\tan\theta = -1.0873$

Find all values of θ, $(0° < \theta < 360°)$, which make the equation true. [p. 437]

29. $5\sin\theta - 3 = -2\sin\theta + 4$ 30. $-3\tan\theta + 6 = 5 - \tan\theta$

Evaluate and simplify. [p. 440]

31. $\frac{\cos 60° + \sin 60°}{\sin 30°}$ 32. $\frac{\sec 45° - \tan 45°}{\cos 30°}$ 33. $\frac{\csc 45° + \cot 60°}{\tan 30°}$

Find the measure of each of the other two sides of right triangle ABC with $m\angle C = 90$ to the nearest whole number. [p. 443]

34. $m\angle A = 42$, $b = 3$ 35. $m\angle B = 65$, $a = 4.1$ 36. $m\angle A = 12$, $a = 9.7$

Find $m\angle A$ and $m\angle B$ of right triangle ABC with $m\angle C = 90$ to the nearest degree. [p. 443]

37. $a = 5$, $b = 7$ 38. $c = 13$, $a = 5$ 39. $b = 3.6$, $c = 10$

40. From the top of a lighthouse 90 meters high the angle of depression of a boat at sea measures 43°. Find, to the nearest meter, the distance from the boat to the foot of the lighthouse. [p. 443]

41. At a point on the ground 23 meters the foot of a building, the angle of elevation to the top of the building measures 38°. What is the height of the building to the nearest meter? [p. 443]

Chapter Seventeen Test

State in which quadrants the terminal side of an angle may lie for each of the following.

1. $\sin u = -\frac{7}{8}$ **2.** $\tan u = \frac{3}{4}$ **3.** $\cos u = -\frac{1}{2}$ **4.** $\cot u = -\sqrt{3}$

Find the value of each of the five other trigonometric functions of *u*.

5. $\cos u = -\frac{3}{5}$ and $\sin u < 0$ **6.** $\tan u = \frac{4}{3}$ and $\sin u > 0$

Use the table to find each of the following.

7. $\cos 49°$ **8.** $\sin 38°20'$ **9.** $\tan 310°20'$

Find *u*, $u < 90°$.

10. $\cos u = .7844$ **11.** $\tan u = 1.4460$

Find all values of *u*, $0° < u < 360°$

12. $\sin u = -.3118$ **13.** $\cot u = 1.6003$

Express in terms of the cofunction.

14. $\sin 18°$ **15.** $\cot 73°10'$

Express as a function of an acute angle measure less than 45°.

16. $\tan 310°$ **17.** $\cos 70°20'$

Determine the values of θ between 0° and 360°, to the nearest ten minutes, which make the equation true.

18. $4 \cot \theta + 4 = 0$ **19.** $\sqrt{3} \tan \theta + 1 = 0$ **20.** $4 \sin \theta - 5 = -2 \sin \theta$

Find each of the following.

21. $\cos 38°17'$ **22.** $\sin 320°14'$

Find all values of θ, $(0° < \theta < 360°)$, to the nearest minute, which make the equation true.

23. $\tan \theta = -.3568$

Find all values of θ, $(0° < \theta < 360°)$, to the nearest minute, which make the equation true.

24. $6 \cos \theta + 8 = -4 \cos \theta - 1$ **25.** $-5 \tan \theta - 6 = 3 - 2 \tan \theta$

Evaluate and simplify.

26. $\dfrac{\sin 30° + \cos 60°}{\tan 45°}$ **27.** $\dfrac{\sec 45° - \csc 30°}{\cot 60°}$

Find the measure of each of the other two sides of right triangle *ABC* with $m\angle C = 90$ to the nearest whole number.

28. $m\angle A = 34$, $b = 4$ **29.** $m\angle B = 72$, $a = 3.1$

Find $m\angle A$ and $m\angle B$ of right triangle *ABC* with $m\angle C = 90$ to the nearest degree.

30. $a = 3$, $b = 4$ **31.** $c = 8$, $a = 5$

32. From the top of a lighthouse 75 m high, the angle of depression of a boat measures 32°. Find to the nearest meter, the distance from the foot of the lighthouse to the boat.

33. At a point on the ground 30 m from the foot of a tree, the angle of elevation to the top of the tree is 42°. Find the height of the tree to the nearest meter.

18 GRAPHS AND IDENTITIES

Making a Tangent Table

We can use a unit circle to estimate tangent values.

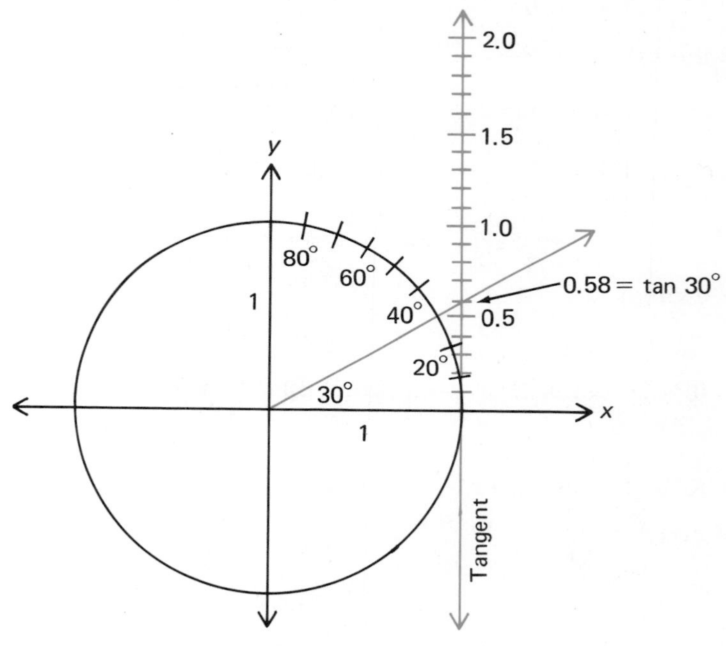

To find tan 30°:

- Draw a ray from the origin to 30° on the unit circle and extend it to the tangent line.
- Estimate the coordinate on the tangent number line at the intersection.

Thus, tan 30° = .58.

PROJECT Use the diagram to estimate tangent values for each to two decimal places.

1. 10° **2.** 45° **3.** 35° **4.** 50° **5.** 55° **6.** 90°

Functions of Negative Angles

OBJECTIVE

- To find the sine, cosine, tangent, cotangent, secant, and cosecant of angles of negative measure

REVIEW CAPSULE

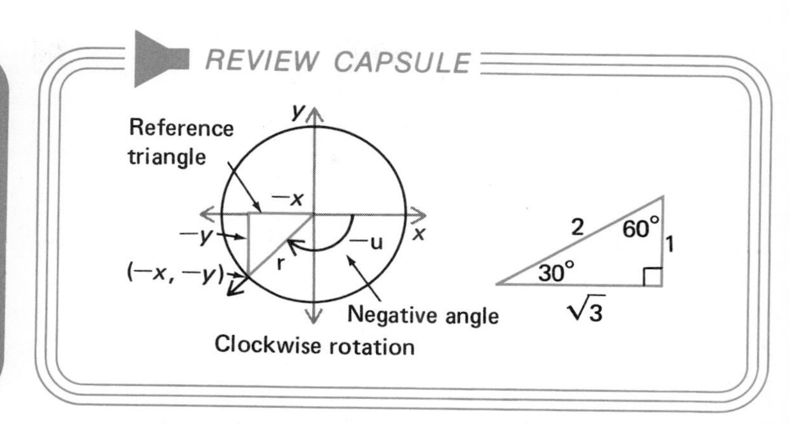

EXAMPLE 1 Find sin (−30°), cos (−30°), tan (−30°).

Draw the reference triangle.

Compare to the reference triangle for 30°.

$\sin 30° = \frac{1}{2}$

$-\sin 30° = -\frac{1}{2}$

$\sin(-30°) = -\frac{1}{2}$

$-\sin 30° = \sin(-30°)$

$\sin(-30°) = -\frac{1}{2}$

$\cos(-30°) = \frac{\sqrt{3}}{2}$

$\tan(-30°) = -\frac{\sqrt{3}}{3}$

EXAMPLE 2 Find sin (−150°), cos (−150°), tan (−150°).

Draw the reference △.

Compare it to the one for 150°.

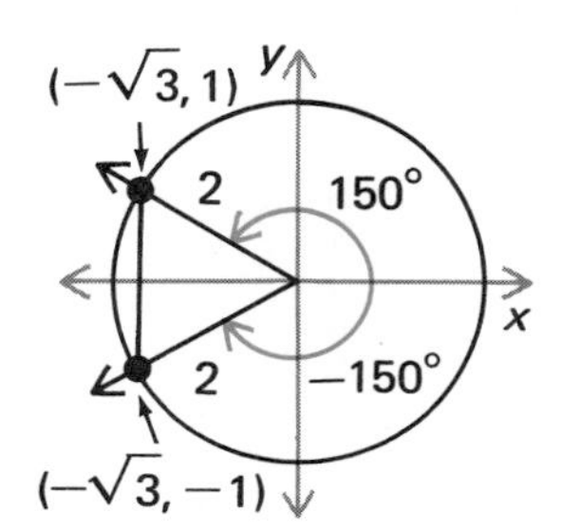

$\sin(-150°) = -\frac{1}{2}$

$\cos(-150°) = -\frac{\sqrt{3}}{2}$

$\tan(-150°) = \frac{\sqrt{3}}{3}$

Examples 1 and 2 suggest this. ⟶

The reference triangles for u and $-u$ are reflections in the x-axis.

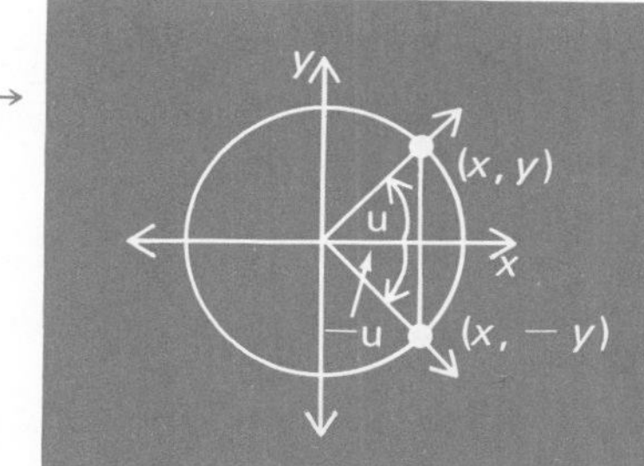

$\sin(-u) = -\sin u$

$\cos(-u) = \cos u$

$\tan(-u) = -\tan u$

EXAMPLE 3 Express sin (−225°) as a function of a positive acute angle.

$$\sin(-225°) = -\sin(225°) = -(-\sin 45°) = \sin 45°$$

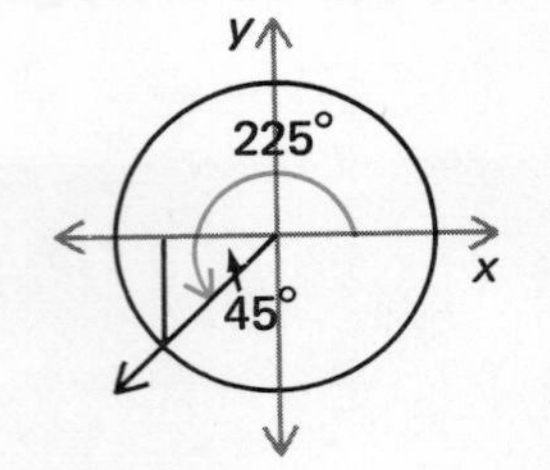

Draw reference triangles.
sec behaves like cos. sec (−u) = sec u
Csc behaves like sin. csc (−u) = −csc u

EXAMPLE 4 Express as a function of a positive acute angle.

sec (−160°)	sin (−300°)	csc (−45°)
sec (160°)	−sin (300°)	↓
−sec 20°	− (−sin 60°)	
	sin 60°	−csc 45°

ORAL EXERCISES

Express as a function of a positive acute angle.

1. sin (−20°) *−sin 20°* **2.** cos (−125°) **3.** tan (−15°) *−tan 15°* **4.** sin (−310°) *sin 50°*
5. cos (−340°) *cos 20°* **6.** tan (−350°) *tan 10°* **7.** sin (−240°) *sin 60°* **8.** cos (−170°) *−cos 10°*

EXERCISES

PART A

Express as a function of a positive acute angle.

1. tan (−10°) **2.** sin (−35°) **3.** cos (−20°) **4.** tan (−110°) *tan 70°* **5.** sin (−130°) *−sin 50°*
6. cos (−123°) **7.** cot (−18°) **8.** sin (−186°) **9.** csc (−38°) **10.** sec (−50°)
11. sin (−325°) *sin 35°* **12.** cos (−350°) *cos 10°* **13.** tan (−280°) *tan 80°* **14.** sin (−196° 18′) *sin 16° 18′* **15.** cos (−315° 33′) *cos 44° 27′*

PART B

EXAMPLE Express sin (−310°) as a function of a positive acute angle less than 45°.

$$\sin(-310°) = -\sin(310°) = -(-\sin 50°) = \sin 50° = \cos 40°$$

sin A = cos (90 − A) ⟶

Express as a function of a positive acute angle less than 45°.

16. sin (−100°) **17.** cos (−260°) **18.** tan (−190°) **19.** sin (−310°) **20.** cos (−80°)
21. sec (−35°) **22.** cot (−110°) *tan 20°* **23.** csc (−290°) *sec 20°* **24.** cos (260° 30′) *−sin 9° 30′* **25.** tan (−54° 23′)

Graphs of $y = \sin x$ and $y = \cos x$

OBJECTIVES

- To graph $y = \sin x$ and $y = \cos x$
- To identify their amplitude and period
- To locate the quadrants in which the sin increases or decreases and in which the cos increases or decreases

REVIEW CAPSULE

Sketch the graph of $y = x^2$.
Graph the set of (x, y) which satisfies $y = x^2$.

Domain: set of 1st coordinates
Range: set of 2nd coordinates

$(-2, 4)$ $(2, 4)$ $(-1, 1)$ $(1, 1)$

Function decreases. Function increases.

EXAMPLE 1 The graph of $y = \sin x$ for $-360° \leq x \leq 360°$ is shown. What is its maximum value?

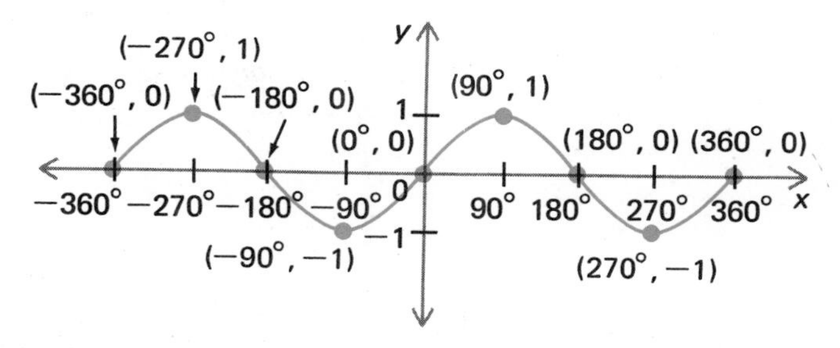

The maximum value of the function occurs for $x = 90°$, and $x = -270°$.

$\sin 90° = 1$, $\sin(-270°) = 1$ ⟶ **Thus,** the maximum value of $y = \sin x$ is 1.

Definition of amplitude ⟶

> The amplitude of the function $y = \sin x$ is its maximum value, 1.

From the graph in Example 1, notice that the function repeats itself every 360°.

> A periodic function repeats itself. $y = \sin x$ repeats every 360°. The period of $y = \sin x$ is 360°.

From the graph in Example 1

1st quadrant	2nd quadrant	3rd quadrant	4th quadrant
$0° < x < 90°$	$90° < x < 180°$	$180° < x < 270°$	$270° < x < 360°$
Function increases	decreases	decreases	increases

EXAMPLE 2 The graph of $y = \cos x$ for $-360° \leq x \leq 360°$ is shown. What is the amplitude of the function?

The maximum value of $y = \cos x$ occurs at $x = -360°$, $x = 0°$, $x = 360°$.

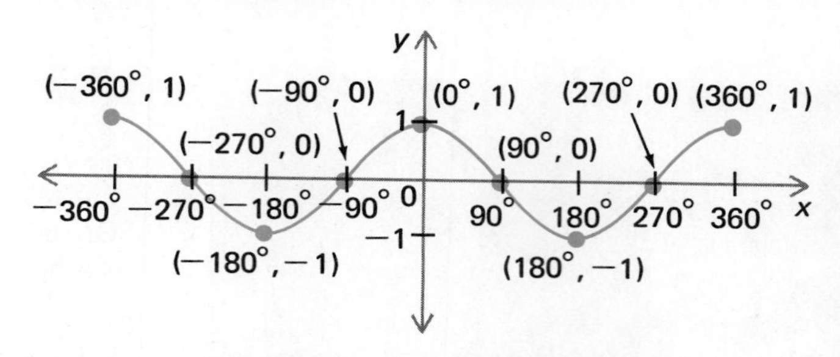

$\cos(-360°) = \cos 0° = \cos 360° = 1$ ⟶ **Thus,** the amplitude of $y = \cos x$ is 1.

EXAMPLE 3 From the graph of $y = \cos x$, shown in Example 2, determine the period of $y = \cos x$.

The graph repeats itself every 360°.

Thus, the period of $y = \cos x$ is 360°.

From the graph in Example 2.

1st quadrant	2nd quadrant	3rd quadrant	4th quadrant
$0° < x < 90°$	$90° < x < 180°$	$180° < x < 270°$	$270° < x < 360°$
Function decreases	decreases	increases	increases

EXAMPLE 4 Use the graph of $y = \cos x$ to estimate $\cos 45°$ to the nearest tenth.

From the graph in Example 2 ⟶

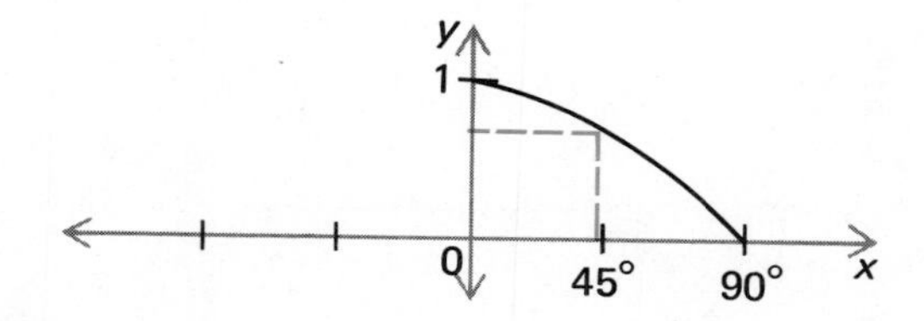

Read the value of y corresponding to $x = 45°$. ⟶ **Thus,** $\cos 45° \doteq .7$.

EXAMPLE 5 Sketch $y = \sin x$ for $0° \leq x \leq 90°$. Estimate $\sin 25°$ to the nearest tenth.

From the trigonometric tables ⟶

x	0°	30°	60°	90°
sin x	0	.5	.87	1

Read the value of y corresponding to $x = 25°$. ⟶

Thus, $\sin 25° \doteq .4$.

EXAMPLE 6 Sketch $y = \sin x$ and $y = \cos x$ for $-180° \leq x \leq 180°$ on the same set of axes. For which values of x does $\sin x = \cos x$?

Write a table of values for key points.

x	$-180°$	$-90°$	$0°$	$90°$	$180°$
$\sin x$	0	-1	0	1	0
$\cos x$	-1	0	1	0	-1

Plot the key points for each function and draw a smooth curve. →

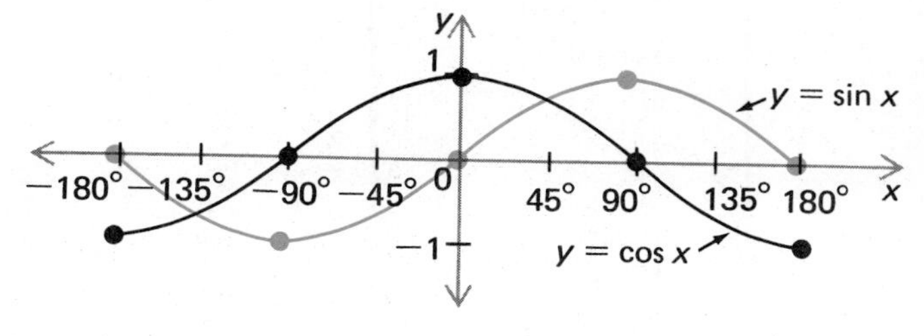

Read the values of x from the graph. → The values are the same where the graphs cross.

Thus, for $x = -135°$ and $x = 45°$, $\sin x = \cos x$.

EXERCISES

PART A

Complete the chart.

		$y = \sin x$	$y = \cos x$
1.	Amplitude	*1*	*1*
2.	Period	*360°*	*360°*
3.	Maximum	*1*	*1*
4.	Minimum	*−1*	*−1*
5.	Quadrants where function increases	*1st, 4th*	*3rd, 4th*
6.	Quadrants where function decreases	*2nd, 3rd*	*1st, 2nd*

Sketch $y = \cos x$ and $y = \sin x$. Estimate.

7. $\cos 30°$ *.9*
8. $\cos 60°$ *.5*
9. $\cos 75°$ *.3*
10. $\sin 45°$ *.7*
11. $\sin 60°$ *.9*
12. $\sin 15°$ *.3*

13. Sketch $y = \sin x$ and $y = \cos x$ for $0° \leq x \leq 360°$ on the same set of axes. For which values of x is $\sin x = \cos x$? *45°; 225°*

14. Sketch $y = \sin x$ and $y = \cos x$ for $-360° \leq x \leq 180°$ on the same set of axes. For which values of x is $\sin x = \cos x$? *−315°; −135°; 45°*

PART B

Sketch the graph for $0° \leq x \leq 360°$.

15. $y = -\sin x$
16. $y = -\cos x$
17. $y = \sin(-x)$
18. $y = \cos(-x)$

PART C

Sketch the graph for $0° \leq x \leq 360°$.

19. $y = \sec x$
20. $y = \csc x$
21. $y = \quad \sin x$
22. $y = \quad \cos(-x)$

Change of Amplitude

OBJECTIVES

- To determine the amplitude for functions like $y = 3 \sin x$ and $y = \frac{1}{2} \cos x$

REVIEW CAPSULE

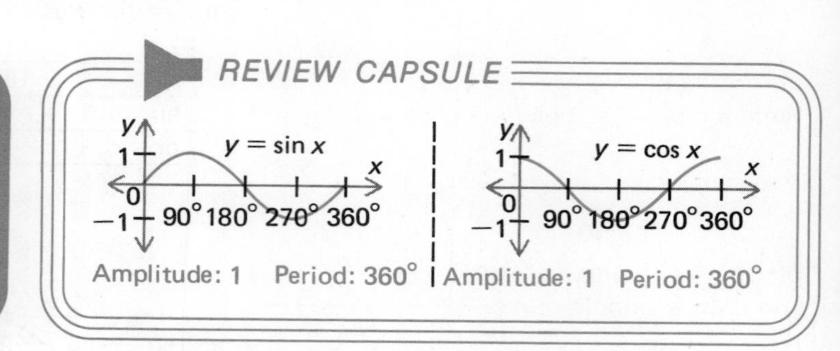

EXAMPLE 1 Sketch $y = 2 \sin x$ and $y = \sin x$ for $0° \le x \le 360°$ on the same set of axes. Determine the period and the amplitude for each.

Make a table of values.

Multiply $\sin x$ values by 2. ⟶

x	0°	90°	180°	270°	360°
$\sin x$	0	1	0	−1	0
$2 \sin x$	0	2	0	−2	0

Plot key points for each function and draw a smooth curve.

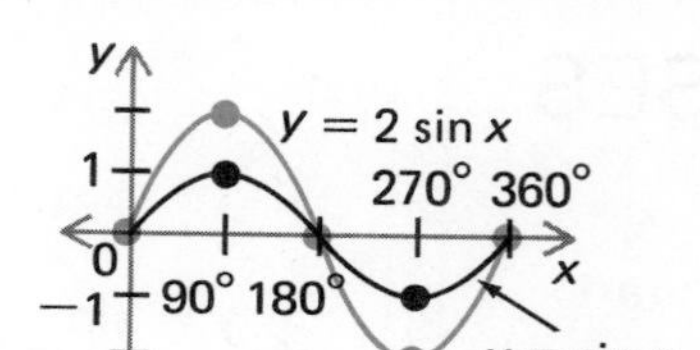

The period for both functions is 360°. The maximum of $y = 2 \sin x$ is 2.

Period of $y = \sin x$: 360° Amplitude of $y = \sin x$: 1
Period of $y = 2 \sin x$: 360° Amplitude of $y = 2 \sin x$: 2

EXAMPLE 2 Sketch $y = \frac{1}{2} \sin x$ and $y = \sin x$ for $0° \le x \le 360°$ on the same set of axes. Determine the period and the amplitude for each.

Make a table of values.

Multiply $\sin x$ values by $\frac{1}{2}$. ⟶

x	0°	90°	180°	270°	360°
$\sin x$	0	1	0	−1	0
$\frac{1}{2} \sin x$	0	$\frac{1}{2}$	0	$-\frac{1}{2}$	0

Plot key points for each function and draw a smooth curve.

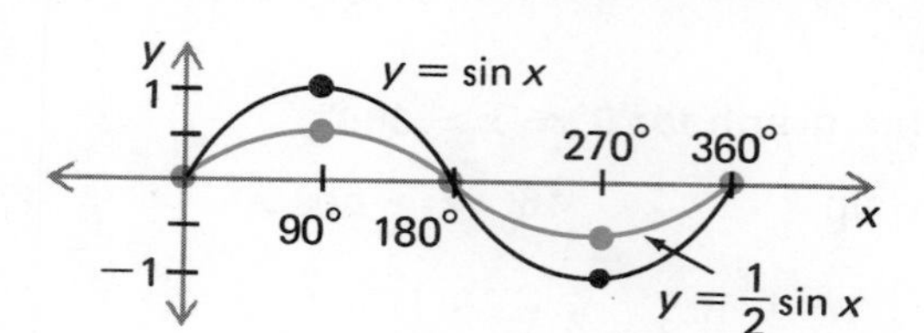

The period for both functions is 360°. The maximum of $y = \frac{1}{2} \sin x$ is $\frac{1}{2}$.

Period of $y = \sin x$: 360° Amplitude of $y = \sin x$: 1
Period of $y = \frac{1}{2} \sin x$: 360° Amplitude of $y = \frac{1}{2} \sin x$: $\frac{1}{2}$

EXAMPLE 3 Sketch $y = 3 \cos x$ and $y = \cos x$ for $0° \leq x \leq 360°$ on the same set of axes. Determine the period and the amplitude for each.

Multiply $\cos x$ by 3.
$3 \cos 0° = 3 (\cos 0°) = 3(1) = 3$ ⟶

x	0°	90°	180°	270°	360°
$\cos x$	1	0	−1	0	1
$3 \cos x$	3	0	−3	0	3

Plot key points for each function and draw a smooth curve.

The period for each function is 360°.

The maximum of $y = 3 \cos x$ is 3.

Period of $y = \cos x$: 360°   Amplitude of $y = \cos x$: 1
Period of $y = 3 \cos x$: 360°   Amplitude of $y = 3 \cos x$: 3

EXAMPLE 4 Sketch $y = \frac{1}{2} \cos x$ and $y = \cos x$ for $0° \leq x \leq 360°$ on the same set of axes. Determine the period and the amplitude for each.

Make a table.

$\frac{1}{2}(\cos 0°) = \frac{1}{2}(1) = \frac{1}{2}$ ⟶

x	0°	90°	180°	270°	360°
$\cos x$	1	0	−1	0	1
$\frac{1}{2} \cos x$	$\frac{1}{2}$	0	$-\frac{1}{2}$	0	$\frac{1}{2}$

Plot key points for each function and draw a smooth curve.

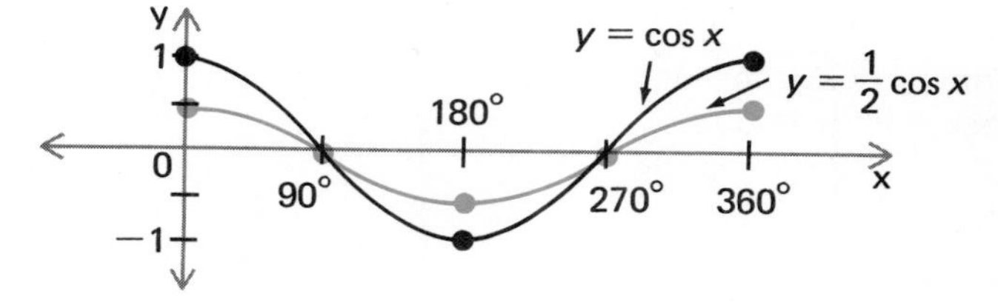

The period of each function is 360°.

The maximum of $y = \frac{1}{2} \cos x$ is $\frac{1}{2}$.

Period of $y = \cos x$: 360°   Amplitude of $y = \cos x$: 1
Period of $y = \frac{1}{2} \cos x$: 360°   Amplitude of $y = \frac{1}{2} \cos x$: $\frac{1}{2}$

If $y = a \sin x$, then the amplitude is a and the period is 360°, for each positive number a.

If $y = a \cos x$, then the amplitude is a and the period is 360°, for each positive number a.

EXAMPLE 5 Determine the ampitude.

$y = 5 \sin x$	$y = 6 \cos x$	$y = \frac{3}{4} \sin x$
Amplitude: 5	Amplitude: 6	Amplitude: $\frac{3}{4}$

ORAL EXERCISES

Determine the amplitude.

1. $y = 3 \sin x$ *3*
2. $y = \frac{1}{3} \cos x$ *$\frac{1}{3}$*
3. $y = \frac{1}{5} \sin x$ *$\frac{1}{5}$*
4. $y = 4 \cos x$ *4*
5. $y = \frac{1}{2} \cos x$ *$\frac{1}{2}$*
6. $y = 10 \sin x$ *10*
7. $y = \frac{1}{8} \cos x$ *$\frac{1}{8}$*
8. $y = 24 \cos x$ *24*

EXERCISES

PART A

Sketch the graph for $0° \le x \le 360'$.

1. $y = 2 \cos x$
2. $y = \frac{1}{2} \sin x$
3. $y = 2 \sin x$
4. $y = \frac{1}{2} \cos x$
5. $y = 3 \sin x$
6. $y = 3 \cos x$
7. $y = \frac{1}{3} \cos x$
8. $y = \frac{1}{3} \sin x$

9. Sketch $y = 2 \cos x$ and $y = \frac{1}{2} \cos x$ for $0° \le x \le 360°$ on the same set of axes. Determine the period and the amplitude for each. *360°, 2; 360°, $\frac{1}{2}$*

10. Sketch $y = 3 \sin x$ and $y = \frac{1}{2} \sin x$ for $0° \le x \le 360°$ on the same set of axes. Determine the period and the amplitude for each. *360°, 3; 360°, $\frac{1}{2}$*

PART B

EXAMPLE Sketch $y = -2 \sin x$ for $0° \le x \le 360°$. Determine the period and the amplitude.

Make a table of values.
Multiply values of $\sin x$ by -2.

x	0°	90°	180°	270°	360°
$-2 \sin x$	0	-2	0	2	0

Plot key points for the function and draw a smooth curve.

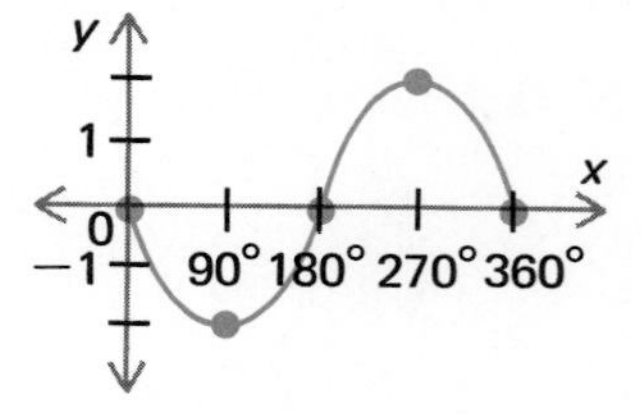

The maximum value for $y = -2 \sin x$ is 2. ⟶ **Thus,** the amplitude is 2 and the period is 360°.

11. Sketch $y = -\frac{1}{2} \cos x$ for $0° \le x \le 360°$. Determine the period and the amplitude.
12. Sketch $y = -2 \cos x$ for $0° \le x \le 360°$. Determine the period and the amplitude.
13. Sketch $y = -\frac{1}{2} \sin x$ for $0° \le x \le 360°$. Determine the period and the amplitude.
14. Sketch $y = -3 \sin x$ for $0° \le x \le 360°$. Determine the period and the amplitude.
15. Sketch $y = 2 \cos x$ and $y = \frac{1}{2} \sin x$ for $0° \le x \le 360°$ on the same set of axes. For how many values of x does $2 \cos x = \frac{1}{2} \sin x$? *2*
16. Sketch $y = -2 \cos x$ and $y = 2 \sin x$ for $0° \le x \le 360°$ on the same set of axes. For how many values of x does $-2 \cos x = 2 \sin x$? *2*

PART C

17. Sketch $y = 2 \csc x$ for $0° \le x \le 360°$.
18. Sketch $y = -\frac{1}{2} \csc x$ for $0° \le x \le 360°$.
19. Sketch $y = \frac{1}{2} \sec x$ for $0° \le x \le 360°$.
20. Sketch $y = -2 \sec x$ for $0° \le x \le 360°$.

Change of Period

OBJECTIVE

- To determine the period for functions like $y = \sin 3x$ and $y = \cos \frac{1}{2}x$

REVIEW CAPSULE

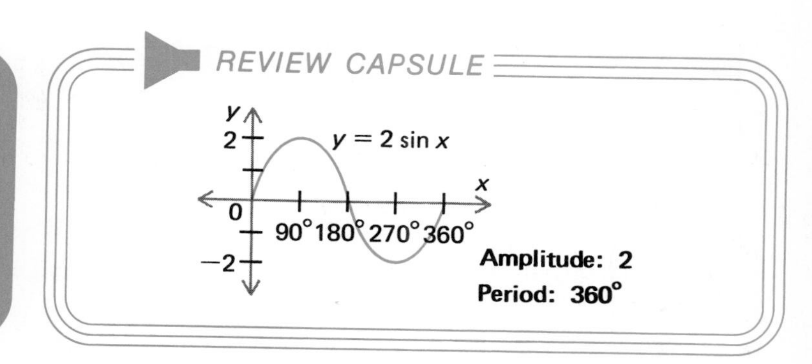

EXAMPLE 1 Sketch $y = \sin 2x$ and $y = \sin x$ for $0° \leq x \leq 360°$ on the same set of axes. Determine the period and the amplitude for each.

Make a table.

x	0°	45°	90°	135°	180°	225°	270°	315°	360°
$2x$	0°	90°	180°	270°	360°	450°	540°	630°	720°
$\sin 2x$	0	1	0	−1	0	1	0	−1	0

$y = \sin 2x$ repeats itself twice from 0° to 360°. ⟶

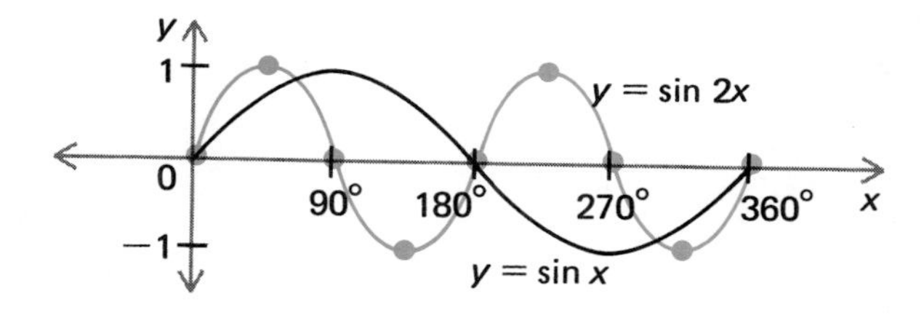

$y = \sin 2x$ completes one cycle from 0° to 180°.

Note: $\frac{360°}{2} = 180°$

Period of $y = \sin x$: 360°
Period of $y = \sin 2x$: 180°

Amplitude of $y = \sin x$: 1
Amplitude of $y = \sin 2x$: 1

EXAMPLE 2 Sketch $y = \sin \frac{1}{2}x$ and $y = \sin x$ for $0° \leq x \leq 720°$ on the same set of axes. Determine the period and the amplitude for each.

x	0°	180°	360°	540°	720°
$\frac{1}{2}x$	0°	90°	180°	270°	360°
$\sin \frac{1}{2}x$	0	1	0	−1	0

$y = \sin \frac{1}{2}x$ completes $\frac{1}{2}$ its cycle in 360°.

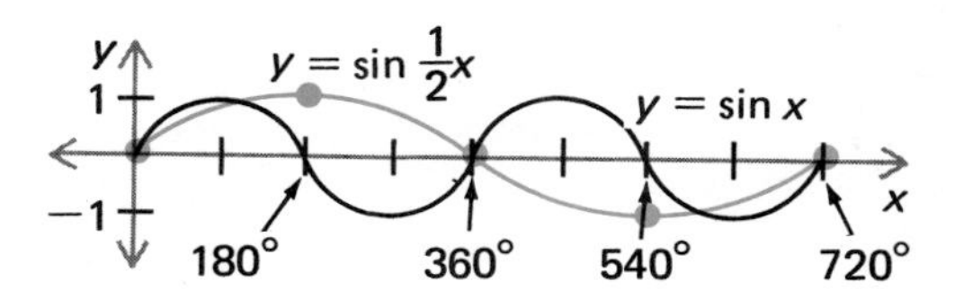

$y = \sin \frac{1}{2}x$ completes one cycle from 0° to 720°.

Note: $360° \div \frac{1}{2} = 720°$.

Period of $y = \sin x$: 360°
Period of $y = \sin \frac{1}{2}x$: 720°

Amplitude of $y = \sin x$: 1
Amplitude of $y = \sin \frac{1}{2}x$: 1

EXAMPLE 3 Sketch $y = \cos 3x$ and $y = \cos x$ for $0° \le x \le 360°$ on the same set of axes. Determine the period and the amplitude for each.

Make a table.

x	0°	30°	60°	90°	120°	180°	240°	300°	360°
$3x$	0°	90°	180°	270°	360°	540°	720°	900°	1,080°
$\cos 3x$	1	0	−1	0	1	−1	1	−1	1

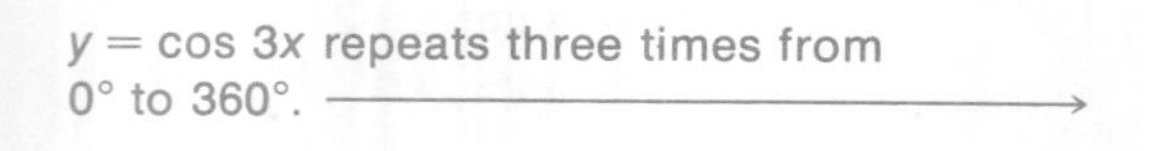

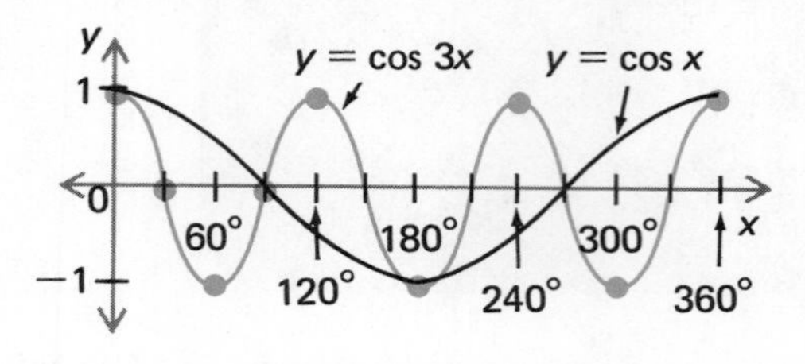

$y = \cos 3x$ completes one cycle from 0° to 120°.
Note: $360 \div 3 = 120°$.

Period of $y = \cos x$: 360° Amplitude of $y = \cos x$: 1
Period of $y = \cos 3x$: 120° Amplitude of $y = \cos 3x$: 1

> If $y = \sin bx$, then the period is $\frac{360°}{b}$ and the amplitude is 1, for each positive number b.
>
> If $y = \cos bx$, then the period is $\frac{360°}{b}$ and the amplitude is 1, for each positive number b.
>
> The function completes b cycles in 360°.

EXAMPLE 4 Determine the period.

Period $= \frac{360°}{b}$.

$y = \sin 4x$
Period $= \frac{360°}{4}$, or 90°

$y = \cos \frac{2}{3}x$
Period $= \frac{360°}{\frac{2}{3}}$, or 540°

EXAMPLE 5 Sketch $y = \sin 2x$ and $y = \cos \frac{1}{2}x$ for $0° \le x \le 360°$. For how many values of x does $\sin 2x = \cos \frac{1}{2}x$?

Plot key values: $y = \sin 2x$ completes 2 cycles in 360°.
$y = \cos \frac{1}{2}x$ completes only $\frac{1}{2}$ cycle in 360°.

The graphs intersect 5 times.

Thus, $\sin 2x = \cos \frac{1}{2}x$ for 5 values of x.

ORAL EXERCISES

Determine the number of cycles within 360° and the period for each.

1. $y = \sin 2x$ *2, 180°*
2. $y = \cos \frac{1}{3}x$ *$\frac{1}{3}$, 1,080°*
3. $y = \sin \frac{1}{2}x$ *$\frac{1}{2}$, 720°*
4. $y = \cos 3x$ *3, 120°*
5. $y = \cos \frac{1}{6}x$ *$\frac{1}{6}$, 2,160°*
6. $y = \sin 9x$ *9, 40°*
7. $y = \cos 6x$ *6, 60°*
8. $y = \sin \frac{1}{9}x$ *$\frac{1}{9}$, 3,240°*

EXERCISES

PART A

Determine the period.

1. $y = \sin x$ *360°*
2. $y = \cos x$ *360°*
3. $y = \cos 2x$ *180°*
4. $y = \sin 2x$ *180°*
5. $y = \cos \frac{1}{2}x$ *720°*
6. $y = \sin \frac{1}{2}x$ *720°*
7. $y = \sin 3x$ *120°*
8. $y = \cos 3x$ *120°*
9. $y = \sin 4x$ *90°*
10. $y = \cos 4x$ *90°*
11. $y = \cos \frac{1}{3}x$ *1,080°*
12. $y = \sin \frac{1}{3}x$ *1,080°*

Sketch the graph for $0° \le x \le 360°$.

13. $y = \sin 2x$
14. $y = \cos 2x$
15. $y = \cos \frac{1}{2}x$
16. $y = \sin \frac{1}{2}x$
17. $y = \cos \frac{1}{3}x$
18. $y = \sin \frac{1}{3}x$
19. $y = \sin 3x$
20. $y = \cos 3x$
21. $y = \sin 4x$
22. $y = \cos 4x$
23. $y = \cos \frac{1}{4}x$
24. $y = \sin \frac{1}{4}x$
25. Sketch $y = \cos \frac{1}{2}x$ and $y = \sin 2x$ for $0° \le x \le 360°$ on the same set of axes. For how many values of x does $\cos \frac{1}{2}x = \sin 2x$? *5*
26. Sketch $y = \sin \frac{1}{2}x$ and $y = \cos 2x$ for $0° \le x \le 360°$ on the same set of axes. For how many values of x does $\sin \frac{1}{2}x = \cos 2x$? *3*

PART B

EXAMPLE Sketch $y = \sin (-\frac{1}{2}x)$ in the interval 0° to 360°. Find the period.

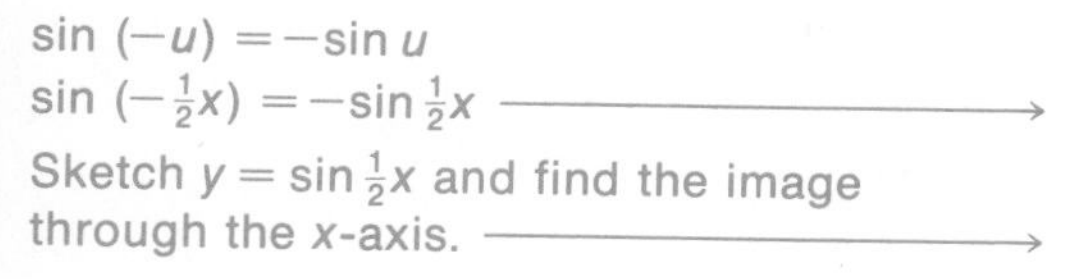

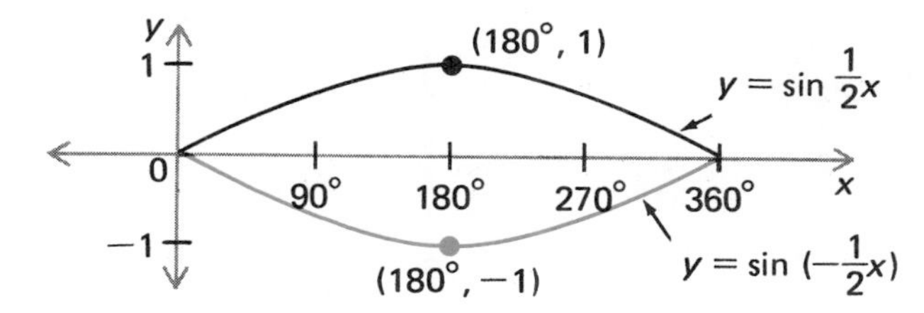

$$\sin\left(-\frac{1}{2}x\right) = -\sin\frac{1}{2}x\text{; period is } \frac{360°}{\frac{1}{2}}\text{, or } 720°.$$

Thus, the period of $y = \sin (-\frac{1}{2}x)$ is 720°.

27. Sketch $y = \sin (-2x)$ in the interval 0° to 360°. Determine the period. *180°*
28. Sketch $y = \cos (-\frac{1}{2}x)$ in the interval 0° to 360°. Determine the period. *720°*
29. Sketch $y = \cos (-2x)$ in the interval 0° to 360°. Determine the period. *180°*
30. Sketch $y = \cos (-2x)$ and $y = \sin (-\frac{1}{2}x)$ in the interval 0° to 360° on the same set of axes. For how many values of x does $\cos (-2x) = \sin (-\frac{1}{2}x)$? *3*

Change of Period and Amplitude

OBJECTIVE

- To determine changes in the period and amplitude for functions like $y = 3 \sin 2x$ and $y = 4 \cos \frac{1}{2}x$

REVIEW CAPSULE

$y = 3 \sin x$	$y = \cos \frac{1}{2}x$
Amplitude: 3	Amplitude: 1
Period: 360°	Period: 720°
$y = a \sin x$	$y = \cos bx$
Amplitude: a	Period: $\frac{360°}{b}$
Maximum value is a. Minimum value is $-a$.	The curve completes b cycles in 360°.

EXAMPLE 1 Sketch $y = \cos \frac{1}{2}x$ for $0° \le x \le 360°$.

The amplitude is 1.

Maximum is 1. Minimum is −1.

The period is $\frac{360}{\frac{1}{2}}$ or 720°.

The curve completes $\frac{1}{2}$ cycle in 360°.

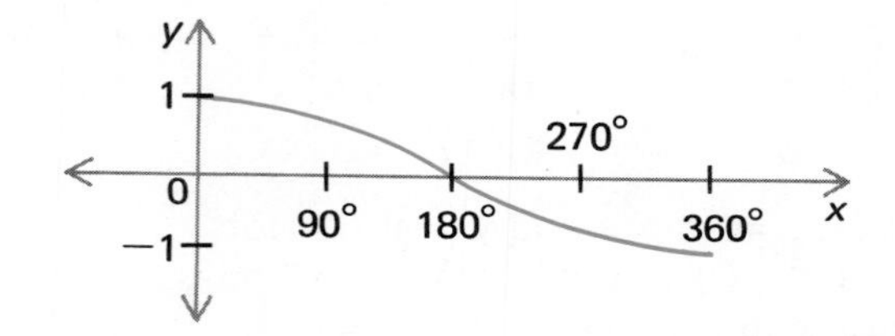

EXAMPLE 2 Sketch $y = 2 \cos x$ for $0° \le x \le 360°$.

The amplitude is 2.

Maximum is 2. Minimum is −2.

The period is $\frac{360}{1}$, or 360°.
The curve completes 1 cycle in 360°.

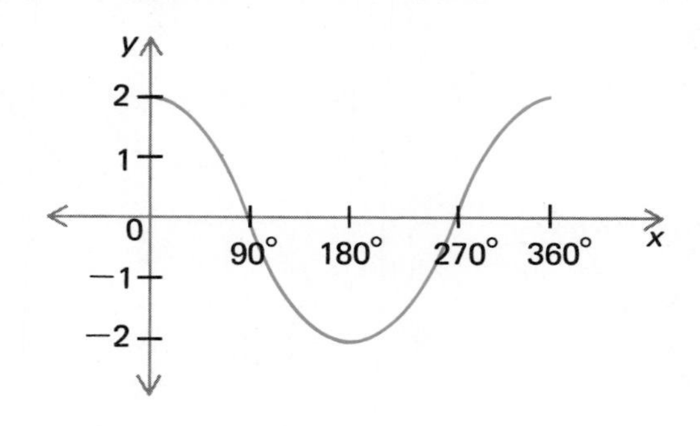

EXAMPLE 3 Sketch $y = 2 \cos \frac{1}{2}x$ for $0° \le x \le 360°$.

The curve completes $\frac{1}{2}$ cycle in 360° as in Example 1.

The amplitude is 2 as in Example 2.

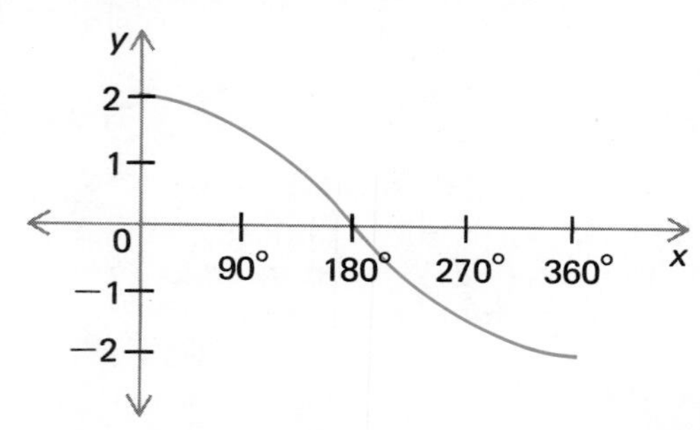

EXAMPLE 4 Sketch the graph of $y = \frac{1}{2} \sin 2x$ for $0° \le x \le 360°$.

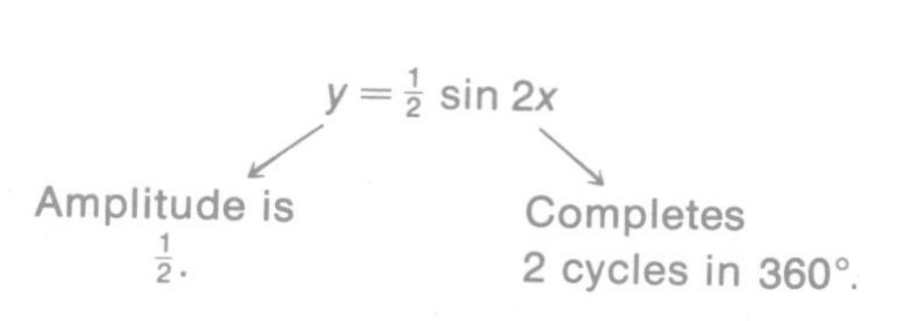

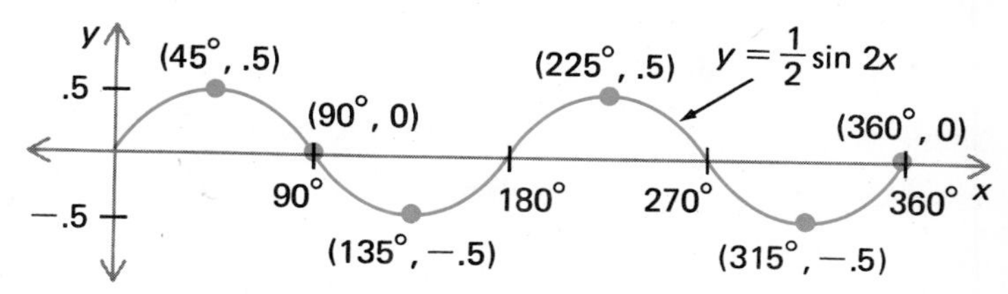

EXAMPLE 5 Sketch the graphs of $y = \cos \frac{1}{2}x$ and $y = 2 \sin x$ for $0° \le x \le 360°$ on the same set of axes. For how many values of x does $\cos \frac{1}{2}x = 2 \sin x$?

For each graph, use the amplitude and the period to draw a sketch.

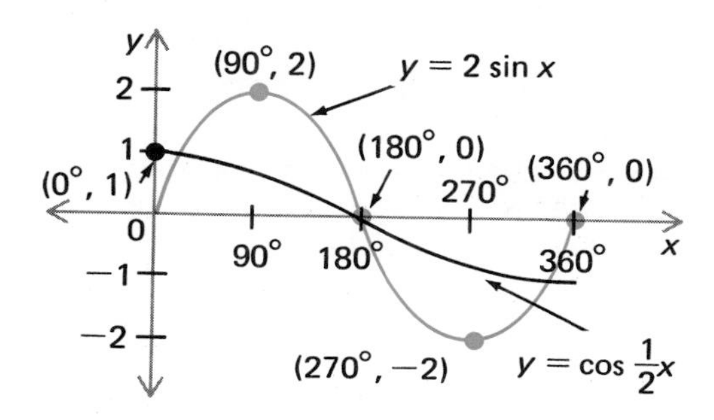

3 points of intersection → **Thus,** $\cos \frac{1}{2}x = 2 \sin x$ for 3 values of x.

EXAMPLE 6 Sketch the graphs of $y = \sin 2x$ and $y = 2 \sin x$ for $0° \le x \le 360°$ on the same set of axes. For which values of x does $\sin 2x = 2 \sin x$?

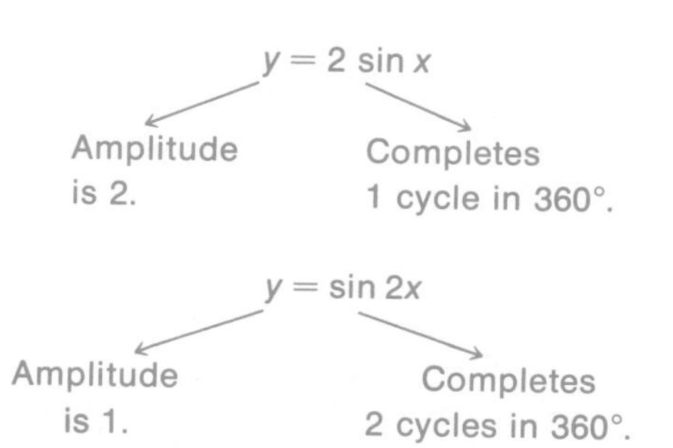

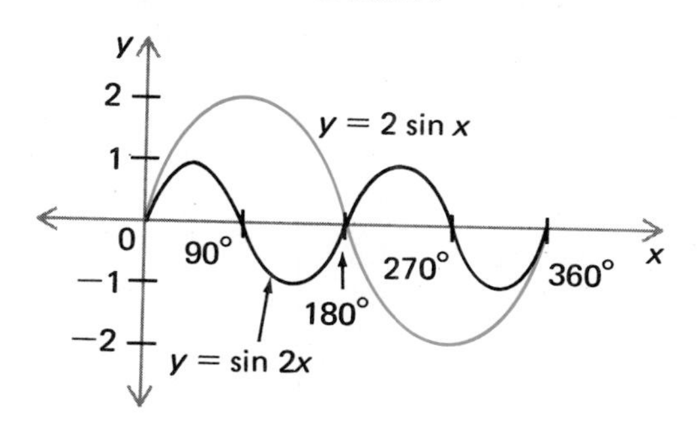

See where the graphs intersect. → **Thus,** $\sin 2x = 2 \sin x$ for $x = 0°$, 180°, and 360°.

ORAL EXERCISES

Determine the period and the amplitude.

1. $y = 3 \sin 2x$ *180°, 3*
2. $y = \frac{1}{2} \sin 2x$ *180°, $\frac{1}{2}$*
3. $y = 2 \cos \frac{1}{2}x$ *720°, 2*
4. $y = \frac{1}{2} \cos 3x$ *120°, $\frac{1}{2}$*
5. $y = \frac{1}{3} \sin 3x$ *120°, $\frac{1}{3}$*
6. $y = 3 \sin \frac{1}{3}x$ *1,080°, 3*
7. $y = 2 \cos 6x$ *60°, 2*
8. $y = 6 \sin \frac{1}{6}x$ *2,160°, 6*

EXERCISES

PART A

Determine the period and the amplitude for each of the following.

1,080°, 2

1. $y = 3 \sin \frac{1}{2}x$
2. $y = 2 \sin \frac{1}{2}x$
3. $y = \frac{1}{2} \cos \frac{1}{2}x$
4. $y = 2 \cos \frac{1}{3}x$
5. $y = \frac{1}{2} \cos 2x$
6. $y = 3 \cos \frac{1}{3}x$
7. $y = 2 \sin \frac{1}{2}x$
8. $y = \frac{1}{6} \sin 4x$ 90°, $\frac{1}{6}$
9. $y = \frac{1}{4} \sin 6x$ 60°, $\frac{1}{4}$
10. $y = 6 \cos \frac{1}{6}x$ 2,160°, 6
11. $y = 6 \cos 9x$ 40°, 6
12. $y = 4 \cos 12x$ 30°, 4

Sketch the graph for 0° ≤ x ≤ 360°. Then determine the period and the amplitude.

180°, $\frac{1}{2}$

13. $y = 2 \cos \frac{1}{2}x$
14. $y = 2 \sin \frac{1}{2}x$
15. $y = \frac{1}{2} \sin 2x$
16. $y = \frac{1}{2} \cos 2x$
17. $y = 3 \cos 3x$
18. $y = 3 \sin 3x$
19. $y = \frac{1}{3} \cos 3x$
20. $y = \frac{1}{3} \sin 3x$ 120°, $\frac{1}{3}$
21. $y = 3 \sin \frac{1}{3}x$ 1,080°, 3
22. $y = 3 \cos \frac{1}{3}x$ 1,080°, 3
23. $y = 2 \sin 6x$ 60°, 2
24. $y = \frac{1}{2} \cos 9x$ 40°, $\frac{1}{2}$
25. Sketch $y = 2 \cos x$ and $y = \sin \frac{1}{2}x$ for $0° \leq x \leq 360°$ on the same set of axes. For how many values of x does $2 \cos x = \sin \frac{1}{2}x$? 2
26. Sketch $y = \frac{1}{2} \cos x$ and $y = \sin \frac{1}{2}x$ for $0° \leq x \leq 360°$ on the same set of axes. For how many values of x does $\frac{1}{2} \cos x = \sin \frac{1}{2}x$? 2
27. Sketch $y = \cos 2x$ and $y = 2 \cos x$ for $0° \leq x \leq 360°$ on the same set of axes. For how many values of x does $\cos 2x = 2 \cos x$? 2
28. Sketch $y = \sin 3x$ and $y = 3 \sin x$ for $0° \leq x \leq 360°$ on the same set of axes. For how many values of x does $\sin 3x = 3 \sin x$? 3

PART B

29. Sketch $y = -2 \sin \frac{1}{2}x$ for $0° \leq x \leq 360°$. Determine the period and the amplitude. 720°, 2
30. Sketch $y = -\frac{1}{2} \cos 2x$ for $0° \leq x \leq 360°$. Determine the period and the amplitude. 180°, $\frac{1}{2}$
31. Sketch $y = -2 \sin (-\frac{1}{2}x)$ for $0° \leq x \leq 360°$. Determine the period and the amplitude. 720°, 2
32. Sketch $y = -\frac{1}{2} \cos (-2x)$ for $0° \leq x \leq 360°$. Determine the period and the amplitude. 180°, $\frac{1}{2}$
33. Sketch $y = -2 \cos \frac{1}{2}x$ for $-180° \leq x \leq 180°$. Determine the period and the amplitude. 720°, 2
34. Sketch $y = -\frac{1}{2} \sin 2x$ for $-180° \leq x \leq 180°$. Determine the period and the amplitude. 180°, $\frac{1}{2}$
35. Sketch $y = -2 \cos x$ and $y = \sin (-\frac{1}{2}x)$ in the interval 0° to 360° on the same set of axes. For how many values of x does $-2 \cos x = \sin (-\frac{1}{2}x)$? 2
36. Sketch $y = -\frac{1}{2} \sin x$ and $y = \cos (-2x)$ in the interval −180° to 180° on the same set of axes. For how many values of x does $-\frac{1}{2} \sin x = \cos (-2x)$? 4

PART C

37. Sketch the graph of $y = -2 \cos (-2x)$ and $y = -\frac{1}{2} \sin (-2x)$ in the interval 0° to 360° on the same set of axes. For how many values of x does $-2 \cos (-2x) = -\frac{1}{2} \sin (-2x)$?
38. Sketch the graph of $y = \frac{1}{2} \sec 2x$ and $y = 2 \sec \frac{1}{2}x$ for $0° \leq x \leq 360°$ on the same set of axes. For how many values of x does $\frac{1}{2} \sec 2x = 2 \sec \frac{1}{2}x$?

Graphs of $y = \tan x$ and $y = \cot x$

OBJECTIVES

- To graph $y = \tan x$ and $y = \cot x$
- To identify their amplitude and period
- To locate the quadrants in which they are increasing and decreasing

REVIEW CAPSULE

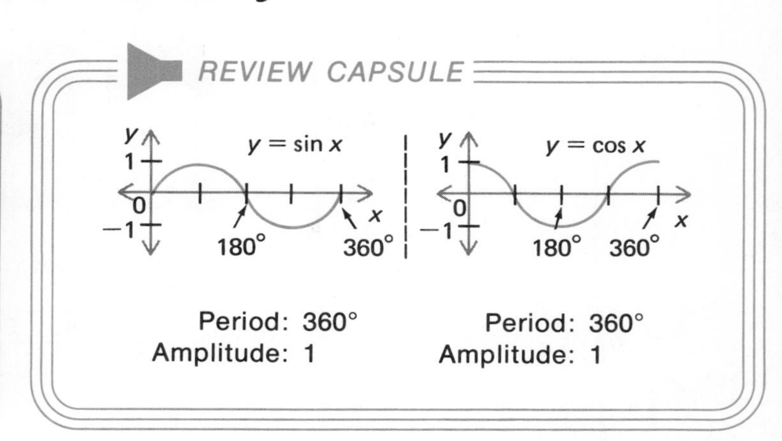

EXAMPLE 1 The graph of $y = \tan x$ for $0° \leq x \leq 360°$ is shown. In which intervals is the function increasing? decreasing? What is the amplitude? What is the period?

Tan 90° and tan 270° are not defined.

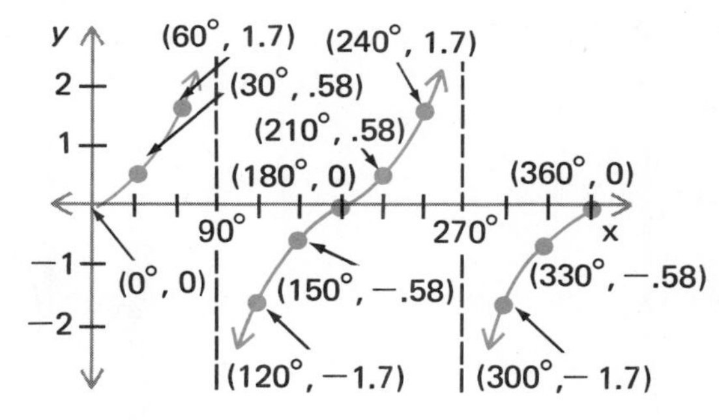

This function always increases.

$y = \tan x$ has no maximum value.

It completes 2 cycles in 360°.

Thus, $y = \tan x$ is always increasing.
It has no amplitude; its period is 180°.

EXAMPLE 2 Sketch $y = \tan 2x$ for $0° \leq x \leq 180°$.

Make a table.

x	0°	22.5°	45°	67.5°	90°	112.5°	135°	157.5°	180°
$2x$	0°	45°	90°	135°	180°	225°	270°	315°	360°
$\tan 2x$	0	1	—	−1	0	1	—	−1	0

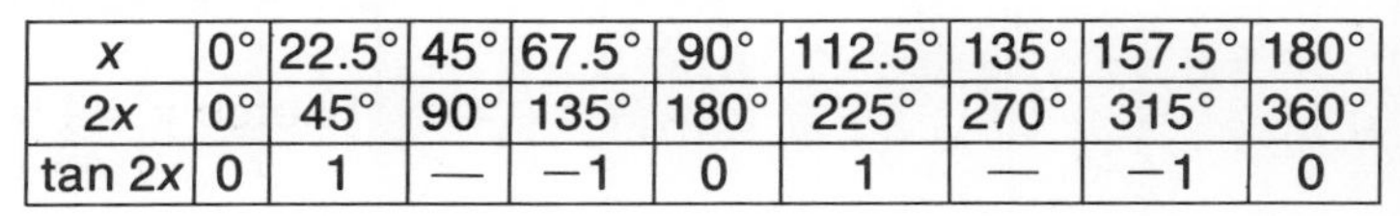

$y = \tan 2x$ completes 2 cycles in 180°.

Its period is $\frac{180°}{2}$, or 90°.

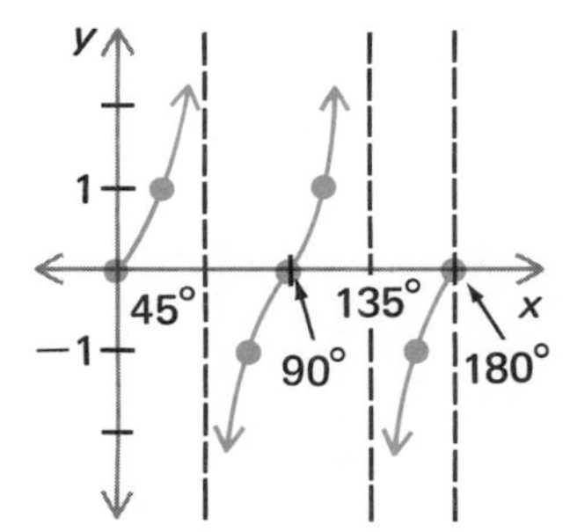

Examples 1 and 2 suggest this. →

$y = \tan x$

No amplitude Period: 180°

$y = \tan bx$

No amplitude Period: $\frac{180°}{b}$

The function completes b cycles in 180°.

EXAMPLE 3 Sketch $y = \cot x$ from 0° to 360°.

x	0°	45°	90°	135°	180°	225°	270°	315°	360°
$\cot x$	—	1	0	−1	—	1	0	−1	—

$y = \cot x$ has no maximum.
$y = \cot x$ repeats every 180°. →

$y = \cot x$ is always decreasing. →

$y = \cot x$ has no minimum. →

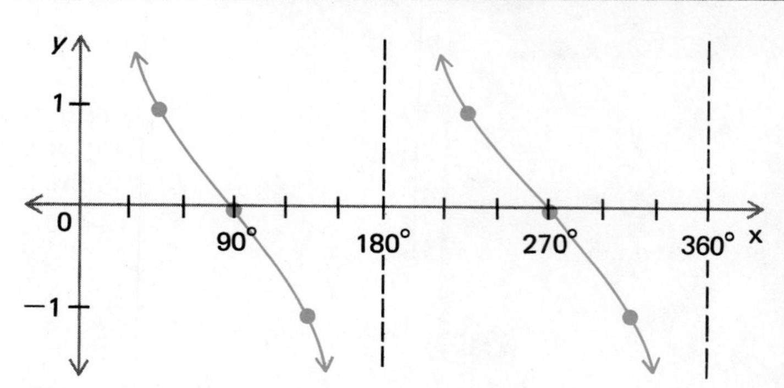

Example 3 and the similarity to $y = \tan x$ suggest this.

$y = \cot x$

No amplitude Period: 180°.

$y = \cot bx$

No amplitude Period: $\frac{180°}{b}$

The function completes b cycles in 180°.

EXAMPLE 4 Sketch $y = \frac{1}{2} \tan 2x$ for $-90° \leq x \leq 90°$.

$y = \frac{1}{2} \tan 2x$

Completes 2 cycles
Asymptotes are at
$x = -45°$ $x = 45°$

There is no amplitude.

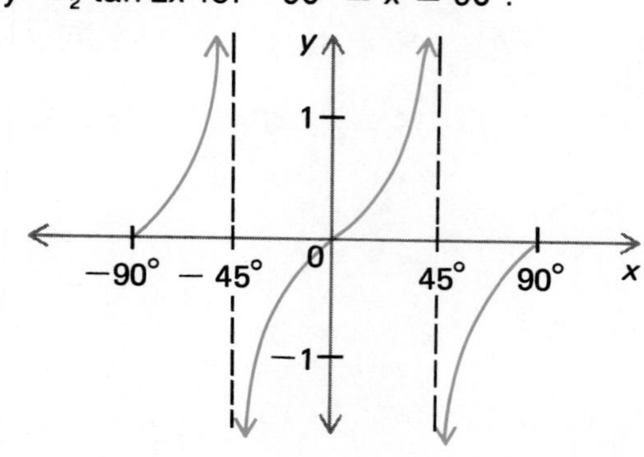

EXERCISES

PART A

Sketch the graph of $y = \tan x$ in the interval $-180°$ to $360°$.

1. In what intervals is the tan increasing? decreasing? *$-180°$ to $360°$; none*
2. What is the maximum value of the tan? minimum value?
3. What is the period of the tan? *180°*
4. What is the amplitude of the tan? *none*
5. Show on the graph that $\tan(-45°) = -\tan 45°$.

Sketch the graph of $y = \cot x$ in the interval $-180°$ to $360°$.

6. In what intervals is the cot increasing? decreasing? *$-180°$ to $360°$*
7. What is the maximum value of the cot? minimum value?
8. What is the period of the cot? *180°*
9. What is the amplitude of the cot? *none*
10. Show on the graph that $\cot(-60°) = -\cot 60°$.

Determine the period.

11. $y = 2 \tan \frac{1}{2}x$ *360°*
12. $y = \frac{1}{2} \cot 2x$ *90°*
13. $y = 3 \cot 6x$ *30°*
14. $y = \frac{1}{2} \tan 3x$ *60°*
15. $y = 6 \cot 12x$ *15°*
16. $y = 3 \tan 2x$ *90°*
17. $y = 6 \tan 6x$ *30°*
18. $y = 2 \cot 3x$ *60°*

Sketch the graph in the interval $0°$ to $360°$.

19. $y = \tan 2x$
20. $y = \tan \frac{1}{2}x$
21. $y = \cot 2x$
22. $y = \cot \frac{1}{2}x$
23. $y = 2 \cot \frac{1}{2}x$
24. $y = \frac{1}{2} \cot 2x$
25. $y = \frac{1}{2} \tan 2x$
26. $y = 2 \tan \frac{1}{2}x$
27. $y = 2 \tan 3x$
28. $y = 2 \cot 3x$
29. $y = 4 \tan \frac{1}{3}x$
30. $y = 4 \cot \frac{1}{3}x$

31. Sketch $y = \tan \frac{1}{2}x$ and $\cot 2x$ in the interval $0°$ to $360°$ on the same set of axes. For how many values of x does $\tan \frac{1}{2}x = \cot 2x$? *4*

32. Sketch $y = \frac{1}{2} \tan 2x$ and $y = 2 \cot 2x$ in the interval $0°$ to $360°$ on the same set of axes. For how many values of x does $\frac{1}{2} \tan 2x = 2 \cot 2x$? *8*

PART B

33. Sketch $y = \tan(-\frac{1}{2}x)$ for $0° \le x \le 360°$. What is the period? [Hint: $\tan(-u) = -\tan u$] *360°*

34. Sketch $y = \cot(-2x)$ for $0° \le x \le 360°$. What is the period? *90°* [Hint: $\cot(-u) = -\cot u$]

35. Sketch $y = -\frac{1}{2} \tan(-2x)$ in the interval $-180°$ to $180°$. What is the period? *90°*

36. Sketch $y = -2 \cot(-\frac{1}{2}x)$ in the interval $-180°$ to $180°$. What is the period? *360°*

PART C

37. Sketch $y = -2 \cot x$ and $y = \tan(-\frac{1}{2}x)$ in the interval $0°$ to $360°$ on the same set of axes. For how many values of x does $-2 \cot x = \tan(-\frac{1}{2}x)$? *2*

38. Sketch $y = -\frac{1}{2} \tan(2x)$ and $y = \cot(-2x)$ in the interval $0°$ to $360°$ on the same set of axes. For how many values of x does $-\frac{1}{2} \tan(2x) = \cot(-2x)$? *8*

Logarithms of Trigonometric Functions

OBJECTIVES

- To find logarithms of the sin, cos, tan and cot of degree measures using a log-trig. table
- To find the degree measure when the log-trig. value is given
- To interpolate using a log-trig. table

REVIEW CAPSULE

Write the log equation.

$$N = \sqrt[3]{\frac{2.67 \times 1.6}{(3.8)^4}}$$

$$\text{Log } N = \tfrac{1}{3}[(\log 2.67 + \log 1.6) - 4 \log 3.8]$$

$\log ab = \log a + \log b \qquad \log a^n = n \log a$

$\log \frac{a}{b} = \log a - \log b \qquad \log \sqrt[n]{a} = \frac{1}{n} \log a$

EXAMPLE 1 Find log sin 36° 20′.

Use the table of trigonometric values. ⟶

sin 36° 20′ = .5925
Now find log .5925.
Characteristic: 9. − 10

Interpolate for the mantissa. ⟶

		digits	mantissa		
10	5	5920	7723	n	8
		5925			
		5930	7731		

$$\frac{5}{10} = \frac{n}{8}$$

$$n = 4$$

Mantissa: 7723 + 4 = 7727 ⟶ So, log .5925 = 9.7727 − 10.

Thus, log sin 36° 20′ = 9.7727 − 10.

There are tables containing logarithms for trigonometric functions already worked out. ⟶

Logarithms of Trigonometric Functions*

Angle		L Sin	L Cos	L Tan	L Cot		
36°	00′	9.7692	9.9080	9.8613	10.1387	54°	00′
	10	9.7710	9.9070	9.8639	10.1361		50
	20	9.7727	9.9061	9.8666	10.1334		40
	30	9.7744	9.9052	9.8692	10.1308		30
	40	9.7761	9.9042	9.8718	10.1282		20
	50	9.7778	9.9033	9.8745	10.1255		10
37°	00′	9.7795	9.9023	9.8771	10.1229	53°	00′
	10	9.7811	9.9014	9.8797	10.1203		50

*These tables give the logarithms increased by 10. Hence in each case 10 should be subtracted.

Read across 36° 20′ under the L sin (log sin) column.

From the table, read 9.7727. Subtract 10.
So, log sin 36° 20′ = 9.7727 − 10.

EXAMPLE 2 Find log cos 45° 40′ and log cot 46° 20′.

Read up the table and use the bottom column heads for measures greater than or equal to 45°.

43°	00′	9.8338	9.8641	9.9697	10.0303	47°	00′
	10	9.8351	9.8629	9.9722	10.0278		50
	20	9.8685	9.8618	9.9747	10.0253		40
	30	9.8378	9.8606	9.9772	10.0228		30
	40	9.8391	9.8594	9.9798	10.0202		20
	50	9.8405	9.8582	9.9823	10.0177		10
44°	00′	9.8418	9.8569	9.9848	10.0152	46°	00′
	10	9.8431	9.8557	9.9874	10.0126		50
	20	9.8444	9.8545	9.9899	10.0101		40
	30	9.8457	9.8532	9.9924	10.0076		30
	40	9.8469	9.8520	9.9949	10.0051		20
	50	9.8482	9.8507	9.9975	10.0025		10
45°	00′	9.8495	9.8495	10.0000	10.0000	45°	00′
		L Cos	L Sin	L Cot	L Tan	Angle	

46° 20′

45° 40′

Thus, log cos 45° 40′ = 9.8444 − 10,
log cot 46° 20′ = 9.9798 − 10.

EXAMPLE 3 Find u, $u < 90°$, when log sin $u = 9.9950 - 10$.
when log tan $u = .8431$.

Log trig. tables give logs increased by 10.
Look up 9.9950 in log sin column, 10.8431 in log tan column.

8°	00′	9.1436	9.9958	9.1478	10.8522	82°	00′
	10	9.1525	9.9956	9.1569	10.8431		50
	20	9.1612	9.9954	9.1658	10.8342		40
	30	9.1697	9.9952	9.1745	10.8255		30
	40	9.1781	9.9950	9.1831	10.8169		20
	50	9.1863	9.9948	9.1915	10.8085		10
9°	00′	9.1943	9.9946	9.1997	10.8003	81°	00′
		L Cos	L Sin	L Cot	L Tan	Angle	

81° 50′

81° 20′

Thus, when log sin $u = 9.9950 - 10$, $u = 81° 20′$
and when log tan $u = .8431$, $u = 81° 50′$.

EXAMPLE 4 Find log cot 31° 34′.

Interpolate.
Write mantissa only.

		measure	log cotangent		
10	4	31° 30′	2127	n	−29
		31° 34′			
		31° 40′	2098		

$$\frac{4}{10} = \frac{n}{-29}$$

$$10n = -116$$

$$n = -11.6 \doteq -12$$

2127 + (−12) = 2115
10 .2115 − 10 = .2115

Thus, log cot 31° 34′ = .2115.

EXAMPLE 5 Find $u\,(0° < u < 90°)$ to the nearest minute if log cos $u = 9.8760 - 10$.

9.8756 and 9.8767 are the nearest values to 9.8760 in the log cos column.

		measure	log cosine		
10	n	41° 10′	9.8767	−7	−11
			9.8760		
		41° 20′	9.8756		

$$\frac{n}{10} = \frac{7}{11}$$

$$11n = 70$$

$$n = \frac{70}{11} \doteq 6$$

$$\frac{-7}{-11} = \frac{7}{11}$$

41° 10′ + 6′ = 41° 16′

Thus, if log cos $u = 9.8760 - 10$, $u = 41° 16′$.

EXAMPLE 6 Find the value of $\frac{\sin 75° \, (3.45)^2}{\sqrt[3]{0.876}}$ to the nearest tenth. Use logarithms.

Let S represent the expression.

$$S = \frac{\sin 75° \, (3.45)^2}{\sqrt[3]{0.876}}$$

Write the log equation. Substitute. Use log table and log-trig. table. ⟶

$$\log S = \log \sin 75° + 2 \log 3.45 - \frac{1}{3} \log .876$$

$$= (9.9849 - 10) + 2(.5378) - \frac{1}{3}(9.9425 - 10)$$

$\frac{1}{3}(9.9425 - 10) = \frac{1}{3}(29.9425 - 30)$
$= 9.9808 - 10$

$$= (9.9849 - 10) + 1.0756 - (9.9808 - 10)$$

$$= 9.9849 - 10 + 1.0756 - 9.9808 + 10$$

Combine the numbers.

$$= 11.0605 - 9.9808$$

$$= 1.0797$$

Look up .0797 in the table of mantissas. The nearest entry (0792) has significant digits 120.

Thus, $S = 12.0$ to the nearest tenth.

EXERCISES

PART A

Find.

1. $\log \tan 75°$
2. $\log \cos 36° 30'$ *9.9052 − 10*
3. $\log \sin 47° 50'$ *9.8699 − 10*
4. $\log \cot 87° 40'$ *8.6101 − 10*
5. $\log \cos 42° 10'$
6. $\log \tan 10° 10'$
7. $\log \cos 33° 13'$
8. $\log \sin 15° 19'$
9. $\log \sin 63° 32'$ *9.9225−10*
10. $\log \cot 17° 8'$ *.5111*
11. $\log \tan 35° 16'$ *9.8495 − 10*
12. $\log \cos 57° 12'$

Find u, $0° < u < 90°$.

13. $\log \sin u = 8.7188 - 10$ *3° 0′*
14. $\log \tan u = .2991$ *63°20′*
15. $\log \cot u = 8.7194 - 10$ *87° 0′*
16. $\log \cos u = 9.9752 - 10$
17. $\log \sin u = 9.9959 - 10$
18. $\log \sin u = 9.9547 - 10$
19. $\log \cot u = .7620$ *9°49′*
20. $\log \tan u = 9.5168 - 10$
21. $\log \cos u = 9.7375 - 10$

Find the value to the nearest tenth. Use logarithms.

22. $\frac{\tan 75°}{\sqrt{3.25}}$
23. $\frac{(2.35)^2 \sqrt{0.0876}}{\sin 43°}$ *2.4*
24. $3 \tan 60° \sqrt{\frac{36.3}{3.14}}$
25. $\frac{28.7 \tan 61° 40'}{38.6}$
26. $\cos 73° 30' \sqrt{\frac{143}{17.6}}$ *.8*
27. $\frac{(13.9)^3 \sqrt{13.2}}{\cot 18° 20'}$

PART B

Find the value to the nearest tenth. Use logarithms.

28. $\frac{151 \sin 49° 16'}{\sqrt[3]{192}}$
29. $\frac{(43.2)^3 (193.5)}{(\cot 18° 43')^2}$
30. $\frac{\cos 18° 49'}{\sqrt[3]{198.7}\,(16.3)^2}$

Quadratic Trigonometric Equations

OBJECTIVE

- To solve trigonometric equations like $3\cot^2\theta - 7\cot\theta + 2 = 0$

REVIEW CAPSULE

Solve $3x^2 - 7x + 2 = 0$.

$$x = \frac{-b \pm \sqrt{b^2 - 4ac}}{2a} \qquad a = 3,\ b = -7,\ c = 2$$

$$x = \frac{7 \pm \sqrt{(-7)^2 - 4(3)(2)}}{2(3)}, \text{ or } \frac{7 \pm \sqrt{25}}{6}$$

$$x = \frac{7+5}{6}, \text{ or } 2 \qquad x = \frac{7-5}{6}, \text{ or } \frac{1}{3}$$

Thus, the solutions are 2 and $\frac{1}{3}$.

EXAMPLE 1 Determine values of θ between 0° and 360° for which $3(\tan\theta)^2 - 1 = 0$ is true.

Think of $\tan\theta$ as x.

Solve for x: $3x^2 - 1 = 0$. → $3(\tan\theta)^2 - 1 = 0$

Add 1: $3x^2 = 1$. → $3(\tan\theta)^2 = 1$

Divide by 3: $x^2 = \frac{1}{3}$. → $(\tan\theta)^2 = \frac{1}{3}$

Find square root: $x = \pm\sqrt{\frac{1}{3}}$. → $\tan\theta = \pm\sqrt{\frac{1}{3}}$, or $\pm\frac{\sqrt{3}}{3}$

$\tan 30° = \frac{\sqrt{3}}{3}$ → Reference angle: 30°

Tan is (+) in 1st and 3rd quad.
Tan is (−) in 2nd and 4th quad.

$\tan\theta = \frac{\sqrt{3}}{3}$: 1st quad. 3rd quad.

$\tan\theta = -\frac{\sqrt{3}}{3}$: 2nd quad. 4th quad.

Use the reference triangles in all 4 quadrants.

Thus, $\theta = 30°$, 210°, 150°, and 330°.

$\tan^2\theta$ means $(\tan\theta)^2$.

The equation in Example 1 is usually written $3\tan^2\theta - 1 = 0$.

EXAMPLE 2 Determine the values of u, $0° \leq u < 360°$.

$$2\sin^2 u - \sin u = 0$$

Factor. → $\sin u(2\sin u - 1) = 0$

Set each factor equal to 0. →

$\sin u = 0$ $\qquad$ $2\sin u - 1 = 0$

$2\sin u = 1$

$\sin u = \frac{1}{2}$

Reference angle: 30°

1st Quad. 2nd Quad.

sin 0° = 0, sin 180° = 0
sin 30° = $\frac{1}{2}$
sin is (+) in 1st and 2nd quads.

$u = 0°$ $\quad u = 180°$ $\qquad u = 30°$ $\quad u = 150°$

Thus, $u = 0°$, 180°, 30° and 150°.

EXAMPLE 3 Determine values of u between 0° and 360° for which the equation is true.

$\cos^2 u$ means $(\cos u)^2$
Think: $2x^2 - 3x - 2 = 0$.
Factor: $(2x + 1)(x - 2) = 0$.
Set each factor equal to 0.
$2x + 1 = 0$ or $x - 2 = 0$

$2x = -1$ $x = 2$

$x = -\frac{1}{2}$

$$2 \cos^2 u - 3 \cos u - 2 = 0$$

$$(2 \cos u + 1)(\cos u - 2) = 0$$

$2 \cos u + 1 = 0$ or $\cos u - 2 = 0$

$2 \cos u = -1$ $\cos u = 2$

$\cos u = -\frac{1}{2}$

$\cos 60° = \frac{1}{2}$ → Reference angle: 60° — No solutions $(\cos u \leq 1)$

cos is neg. in quad. 2 and 3. → 2nd quad. 3rd quad.

Use the reference triangles. → **Thus,** $u = 120°$ and $240°$.

EXAMPLE 4 Determine values of θ to the nearest degree, between 0° and 360°, for which the equation is true.

Think: $3x^2 - 7x + 2 = 0$
(See the Review Capsule.)

$$3 \cot^2 \theta - 7 \cot \theta + 2 = 0$$

Use the quadratic formula $x = \frac{-b \pm \sqrt{b^2 - 4ac}}{2a}$.
$a = 3$, $b = -7$, $c = 2$

$$\cot \theta = \frac{-b \pm \sqrt{b^2 - 4ac}}{2a}$$

$$\cot \theta = \frac{-(-7) \pm \sqrt{(-7)^2 - 4(3)(2)}}{2(3)}$$

$$= \frac{7 \pm \sqrt{49 - 24}}{6}$$

$$= \frac{7 \pm \sqrt{25}}{6}$$

$$= \frac{7 \pm 5}{6}$$

$\frac{7 \pm 5}{6}$ → $\frac{7+5}{6}$, $\frac{7-5}{6}$

$\cot u = \frac{7 + 5}{6}$ or $\cot u = \frac{7 - 5}{6}$

$\cot u = 2$ or $\cot u = \frac{1}{3}$ or .3333

Look up the closest values to 2.0000 and .3333 in the cot column of the table. → Reference angle: 27° — 1st quad. 3rd quad.; Reference angle: 72° — 1st quad. 3rd quad.

Use the nearest degree. → **Thus,** $\theta = 27°, 207°, 72°$, and $252°$.

EXAMPLE 5 Determine values of u to the nearest degree, between 0° and 360°, for which the equation is true.

$$2\sin^2 u - 5\sin u + 1 = 0$$

Think: $2x^2 - 5x + 1 = 0$

Use the quadratic formula $x = \frac{-b \pm \sqrt{b^2 - 4ac}}{2a}$. → $\sin u = \frac{5 \pm \sqrt{25 - 8}}{4}$

$a = 2$, $b = -5$, $c = 1$

$$\sin u = \frac{5 \pm \sqrt{17}}{4}$$

$\sqrt{17} \doteq 4.123$
(See the table on p. 524.)

$$\sin u = \frac{5 + \sqrt{17}}{4} = \frac{5 + 4.123}{4} = \frac{9.123}{4} = 2.281$$

No solution ($\sin u \le 1$)

or

$$\sin u = \frac{5 - \sqrt{17}}{4} = \frac{5 - 4.123}{4} = \frac{.877}{4} = .219$$

Look up .2190 in the sin column.
Use the nearest degree.

Reference angle: 13°

Sin is pos. in quad. 1 and 2. →

1st quad. $u \doteq 13°$ 2nd quad. $u \doteq 167°$

Thus, $u \doteq 13°$ and 167°.

EXERCISES

PART A

Determine values to the nearest degree between 0° and 360° for which the equations are true.

1. $2\sin^2 u + 3 = 4$
2. $2\cos^2 u - 1 = 0$ *45°; 135°; 225°; 315°*
3. $\tan^2 u - 1 = 0$
4. $2\cos^2 u + \cos u = 0$
5. $\cot^2\theta + \cot\theta = 0$
6. $2\sin^2\theta - \sin\theta - 1 = 0$
7. $\cot^2\theta - 2\cot\theta + 1 = 0$ *45°; 225°*
8. $2\sin^2\theta + 3\sin\theta - 1 = 0$ *16°; 164°*
9. $\cos^2\theta - 11\cos\theta + 10 = 0$ *0°; 360°*

PART B

Determine values to the nearest minute between 0° and 360° for which the equation is true.

10. $2\cos^2 u - 5\cos u - 1 = 0$
11. $3\sin^2 u + 6\sin u + 1 = 0$ *190° 35′; 349° 25′*
12. $2\tan^2 u - 7\tan u + 1 = 0$
13. $4\sin^2\theta + 5\sin\theta - 2 = 0$ *18° 35′; 161° 25′*
14. $3\cos^2\theta + 6\cos\theta - 2 = 0$ *73° 5′; 282° 55′*
15. $6\sin^2\theta - 5 = 7\sin\theta$ *210°; 330°*

PART C

Solve for $\sin\theta$ or $\cos\theta$.

16. $3\sin^2\theta + 5\sin\theta - 2k\sin\theta = 2 + k\sin^2\theta$
17. $a\cos^2\theta - b(a + 1)\cos\theta + \cos^2\theta = 0$

Application of Sine Function

A ray of light passes through air and into water. The angle formed by the ray passing through the air is called the *angle of incidence.* The angle formed by the ray passing through the water is called the *angle of refraction.*

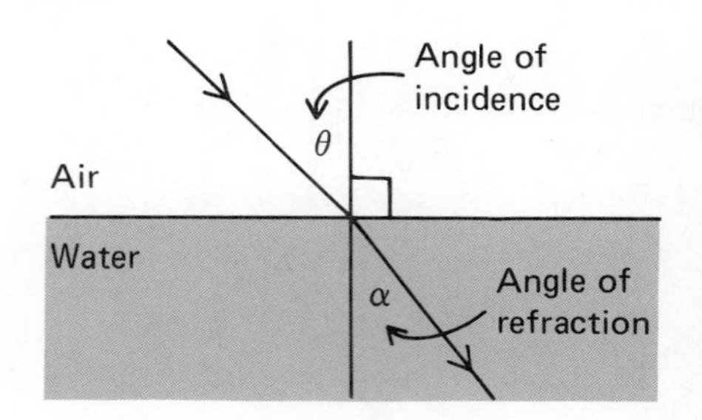

$\frac{\sin \theta}{\sin \alpha} = C_{\text{water}} \longleftarrow$ Constant for water

$\frac{\sin \theta}{\sin \alpha} = 1.33$ Refractive index for water = 1.33

PROBLEM

A ray of light passes through air into water at a 30° angle of incidence. Find the angle of refraction to the nearest 10 minutes.

$$\frac{\sin \theta}{\sin \alpha} = C_{\text{water}}$$

$$\frac{\sin 30^\circ}{\sin \alpha} = 1.33, \text{ or } 1.33 \sin \alpha = \sin 30^\circ$$

$$\sin \alpha = \frac{\frac{1}{2}}{1.33} = \frac{1}{2.66}$$

$$\sin \alpha = .3759$$

$$\alpha = 22^\circ 0' \text{ to the nearest 10 minutes.}$$

Thus, the angle of refraction is 22°0′.

Each medium has its own constant as light passes from air through that medium. The constants are called refractive indices.

$C_{\text{quartz crystal}} \doteq 1.54$ $C_{\text{ice}} \doteq 1.31$ $C_{\text{diamond}} \doteq 2.42$

PROJECT Find the angle of refraction to the nearest 10 minutes.

1. A ray of light passes through air into diamond at a 45° angle of incidence.
2. A ray of light passes through air into quartz crystal at a 30° angle of incidence.
3. A ray of light passes through air into ice at a 63° angle of incidence.

Verifying Trigonometric Identities

OBJECTIVE

- To verify basic trigonometric identities

REVIEW CAPSULE

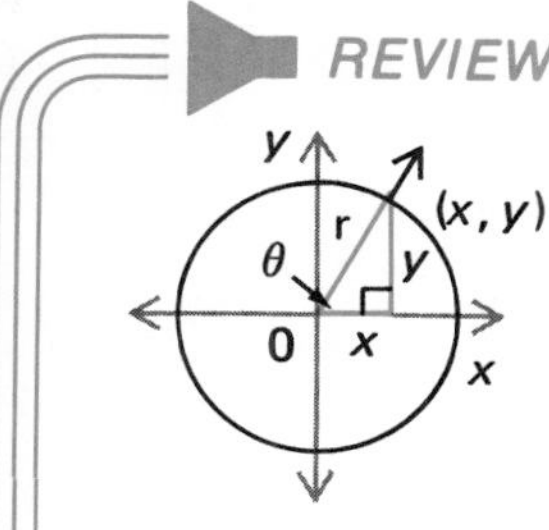

$\sin\theta = \frac{y}{r}$ $\quad \csc\theta = \frac{r}{y}$

$\cos\theta = \frac{x}{r}$ $\quad \sec\theta = \frac{r}{x}$

$\tan\theta = \frac{y}{x}$ $\quad \cot\theta = \frac{x}{y}$

Pythagorean theorem: $x^2 + y^2 = r^2$

EXAMPLE 1 For what values of x does $x + 4 = 4 + x$?

$x + 4 = 4 + x$ is a statement of the commutative property of addition.

Thus, $x + 4 = 4 + x$, for all real numbers.

Definition of identity →

> Equations which are true for all values are called *identities.*

EXAMPLE 2 Show that $(\sin 30°)(\csc 30°) = 1$.

By definition, $\csc 30° = \frac{1}{\sin 30°}$ →

$(\sin 30°)(\csc 30°)$	1
$\frac{1}{2} \cdot \frac{2}{1}$	1
1	

EXAMPLE 3 Show that for any degree measure, θ,
$(\sin\theta)(\csc\theta) = 1$, $\quad (\cos\theta)(\sec\theta) = 1$, $\quad (\tan\theta)(\cot\theta) = 1$.

Use the definitions. →

$\frac{y}{r} \cdot \frac{r}{y} = 1$, $\quad \frac{x}{r} \cdot \frac{r}{x} = 1$, $\quad \frac{y}{x} \cdot \frac{x}{y} = 1$.

For a given angle measure, their product is 1.

Sin and csc, cos and sec, tan and cot are reciprocal functions.

Reciprocal identities →

> $\sin A \csc A = 1$ or $\sin A = \frac{1}{\csc A}$ or $\csc A = \frac{1}{\sin A}$
>
> $\cos A \sec A = 1$ or $\cos A = \frac{1}{\sec A}$ or $\sec A = \frac{1}{\cos A}$
>
> $\tan A \cot A = 1$ or $\tan A = \frac{1}{\cot A}$ or $\cot A = \frac{1}{\tan A}$

EXAMPLE 4 Show that for any degree measure A,

$$\tan A = \frac{\sin A}{\cos A} \text{ and } \cot A = \frac{\cos A}{\sin A}$$

Use the definitions.
Show that both sides of the equation are the same.

$\tan A$	$\dfrac{\sin A}{\cos A}$
$\dfrac{y}{x}$	$\dfrac{\frac{y}{r}}{\frac{x}{r}}$
	$\dfrac{y}{x}$

$\cot A$	$\dfrac{\cos A}{\sin A}$
$\dfrac{x}{y}$	$\dfrac{\frac{x}{r}}{\frac{y}{r}}$
	$\dfrac{x}{y}$

Quotient identities ⟶

$$\tan A = \frac{\sin A}{\cos A} \qquad \cot A = \frac{\cos A}{\sin A}$$

EXAMPLE 5 Verify the quotient identities for $x = 150°$.

Reference angle: 30°
150° is in the 2nd quadrant.

$\tan 150°$	$\dfrac{\sin 150°}{\cos 150°}$
$-\dfrac{\sqrt{3}}{3}$	$\dfrac{\frac{1}{2}}{-\frac{\sqrt{3}}{2}}$
	$\dfrac{1}{2} \cdot \left(-\dfrac{2}{\sqrt{3}}\right)$
	$-\dfrac{1}{\sqrt{3}}$, or $-\dfrac{\sqrt{3}}{3}$

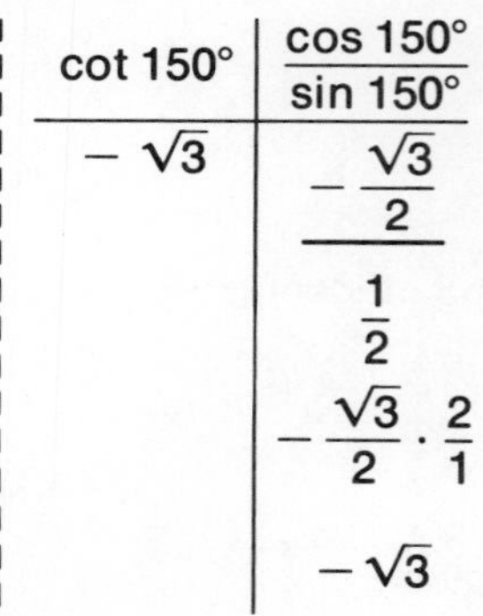

$\cot 150°$	$\dfrac{\cos 150°}{\sin 150°}$
$-\sqrt{3}$	$\dfrac{-\frac{\sqrt{3}}{2}}{\frac{1}{2}}$
	$-\dfrac{\sqrt{3}}{2} \cdot \dfrac{2}{1}$
	$-\sqrt{3}$

EXAMPLE 6 Show that for any degree B,

$$\sin^2 B + \cos^2 B = 1.$$

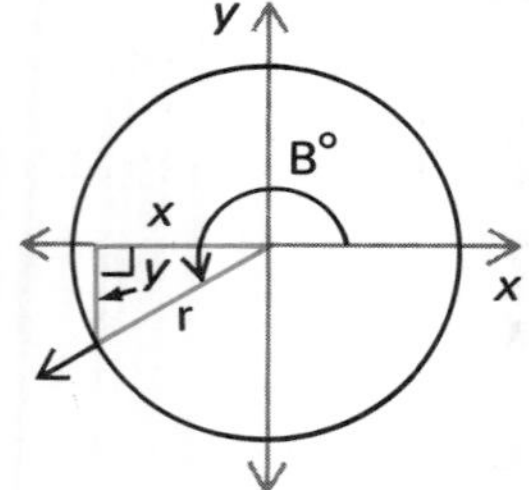

$\sin^2 B + \cos^2 B$	1
$\left(\dfrac{y}{r}\right)^2 + \left(\dfrac{x}{r}\right)^2$	1
$\dfrac{y^2 + x^2}{r^2}$	
1	

Thus, $\sin^2 B + \cos^2 B = 1$.

EXAMPLE 7 Verify $\sin^2 B + \cos^2 B = 1$ for $B = 240°$.

Reference angle: 60°
240° is in the 3rd quadrant.

$\sin^2 240° + \cos^2 240°$	1
$\left(-\dfrac{\sqrt{3}}{2}\right)^2 + \left(-\dfrac{1}{2}\right)^2$	1
$\dfrac{3}{4} + \dfrac{1}{4}$	
1	

Using $\sin^2 A + \cos^2 A = 1$, two other similar identities can be proved.

Pythagorean identities
These can also be proved directly from definitions.

$$\sin^2 A + \cos^2 A = 1$$
$$\tan^2 A + 1 = \sec^2 A$$
$$1 + \cot^2 A = \csc^2 A$$

EXERCISES

PART A

Express as sin, or cos, or both.

1. $\csc 30°$
2. $\cot 135°$
3. $\csc 320°$
4. $\sec (-75°)$
5. $\tan (-310°)$

Verify for $A = 60°$.

6. $\sin^2 A + \cos^2 A = 1$
7. $\tan^2 A + 1 = \sec^2 A$
8. $1 + \cot^2 A = \csc^2 A$
9. $\cos A = \dfrac{1}{\sec A}$

Verify for $x = 135°$.

10. $\sin^2 x = 1 - \cos^2 x$
11. $\sin x = \dfrac{1}{\csc x}$
12. $\csc^2 x - 1 = \cot^2 x$
13. $\tan x = \dfrac{\sin x}{\cos x}$

Verify for $B = 210°$.

14. $\sec^2 B = \tan^2 B + 1$
15. $\cos^2 B = 1 - \sin^2 B$
16. $\cot B = \dfrac{\cos B}{\sin B}$
17. $1 + \cot^2 B = \csc^2 B$

Verify for $x = 300°$.

18. $(\sin x)(\csc x) = 1$
19. $(\tan x)(\cot x) = 1$
20. $(\cos x)(\sec x) = 1$
21. $\sin x = \dfrac{1}{\csc x}$

PART B

EXAMPLE Write an equivalent expression for $\tan x$ using only $\sin x$.

$$\tan x = \frac{\sin x}{\cos x}$$

$\sin^2 x + \cos^2 x = 1$ So, $\cos x = \pm\sqrt{1 - \sin^2 x}$.

$$= \frac{\sin x}{\pm\sqrt{1 - \sin^2 x}}, \text{ or } \pm\frac{\sin x}{\sqrt{1 - \sin^2 x}}$$

Write an equivalent expression using only $\sin x$.

22. $\cos x$
23. $\cot x$
24. $\sec x$
25. $1 + \tan^2 x$

PART C

Prove the identity. Use the definitions for $\tan x$, $\cot x$, $\sec x$, and $\csc x$.

26. $\tan^2 x + 1 = \sec^2 x$
27. $1 + \cot^2 x = \csc^2 x$

Starting with $\sin^2 A + \cos^2 A = 1$, prove the identity.

28. $\tan^2 A + 1 = \sec^2 A$
29. $1 + \cot^2 A = \csc^2 A$

Proving Trigonometric Identities

OBJECTIVE

- To prove trigonometric identities

REVIEW CAPSULE

Reciprocal Identities

$(\sin A)(\csc A) = 1$
$(\cos A)(\sec A) = 1$
$(\tan A)(\cot A) = 1$

Quotient Identities

$\tan A = \dfrac{\sin A}{\cos A}$

$\cot A = \dfrac{\cos A}{\sin A}$

Pythagorean Identities

$\sin^2 A + \cos^2 A = 1 \qquad \tan^2 A + 1 = \sec^2 A$

$\cot^2 A + 1 = \csc^2 A$

EXAMPLE 1 Prove the identify $(\sin x)(\sec x) = \tan x$.

$\sin x \cdot \sec x$	$\tan x$	
$\sin x \cdot \dfrac{1}{\cos x}$	$\tan x$	Replace $\sec x$ with $\dfrac{1}{\cos x}$.
$\dfrac{\sin x}{\cos x}$		Multiply.
$\tan x$		Replace $\dfrac{\sin x}{\cos x}$ with $\tan x$.

Since each side of the equation is the same, the identity has been proved.

EXAMPLE 2 Prove the identity $\sec x = \cos x + \tan x \sin x$.

$\sec x$	$\cos x + \tan x \sin x$	
$\sec x$	$\cos x + \dfrac{\sin x}{\cos x} \cdot \sin x$	Replace $\tan x$ with $\dfrac{\sin x}{\cos x}$.
	$\dfrac{\cos^2 x + \sin^2 x}{\cos x}$	Add.
	$\dfrac{1}{\cos x}$	Replace $\sin^2 x + \cos^2 x$ with 1.
	$\sec x$	Replace $\dfrac{1}{\cos x}$ with $\sec x$.

Since each side of the equation is the same, the identity has been proved.

SUMMARY

To prove a trigonometric identity:

1. **Work with each side separately.**
2. **Substitute using basic trigonometric identities.**
3. **Do any addition, subtraction, multiplication or division necessary.**
4. **When each side is the same, the identity is proved.**

EXAMPLE 3 Prove the identity $\frac{1+\sec x}{\sin x} = \csc x + \csc x \sec x$.

$\frac{a+b}{c} = \frac{a}{c} + \frac{b}{c}$ →

Replace $\csc x$ with $\frac{1}{\sin x}$.

$\frac{1+\sec x}{\sin x}$	$\csc x + \csc x \sec x$
$\frac{1}{\sin x} + \frac{\sec x}{\sin x}$	$\frac{1}{\sin x} + \frac{1}{\sin x} \cdot \sec x$
	$\frac{1}{\sin x} + \frac{\sec x}{\sin x}$

EXAMPLE 4 Prove the identity $\sin^2 x - \cos^2 x = 2\sin^2 x - 1$.

By the Pythagorean identity, $\cos^2 x = 1 - \sin^2 x$ →

$\sin^2 x - \cos^2 x$	$2\sin^2 x - 1$
$\sin^2 x - (1 - \sin^2 x)$	$2\sin^2 x - 1$
$\sin^2 x - 1 + \sin^2 x$	
$2\sin^2 x - 1$	

EXERCISES

PART A

Prove the identity.

1. $\frac{\tan x}{\sin x} = \sec x$
2. $\tan x \cos x = \sin x$
3. $\sec x \cot x = \csc x$
4. $\sec x - \tan x \sin x = \cos x$
5. $\cos^2 x - \sin^2 x = 2\cos^2 x - 1$
6. $\cos^2 x(1 + \tan^2 x) = 1$

PART B

Prove the identity.

7. $\frac{1+\csc x}{\sec x} = \cos x + \cot x$
8. $\frac{\sin x}{1+\cos x} + \cot x = \csc x$
9. $\frac{1-\sin^2 x}{\sin x} \cdot \sec x = \cot x$
10. $\tan x + \frac{\cos x}{\sin x} = \sec x \cdot \csc x$
11. $\frac{\sin x}{1+\cos x} + \frac{\sin x}{1-\cos x} = 2\csc x$
12. $\frac{\sin x}{1-\cos x} - \frac{\cos x}{\sin x} = \csc x$
13. $\sec x \cos x = \tan x \cot x$
14. $\frac{\sec x + \csc x}{\cot x + \tan x} = \sin x + \cos x$

PART C

Prove the identity.

15. $\frac{2}{\sec^2 x} - 1 = \frac{1 - \frac{1}{\cot^2 x}}{1 + \frac{1}{\cot^2 x}}$
16. $\frac{\sec(180° - \theta)}{\sin(180° + \theta)} = \tan(180° + \theta) + \frac{1}{\cot(270° - \theta)}$

Radian Measure

OBJECTIVES

- To express radian measure in degrees
- To express degree measure in radians
- To evaluate expressions like $\cos \frac{\pi}{3} - \sin \frac{2\pi}{3}$

REVIEW CAPSULE

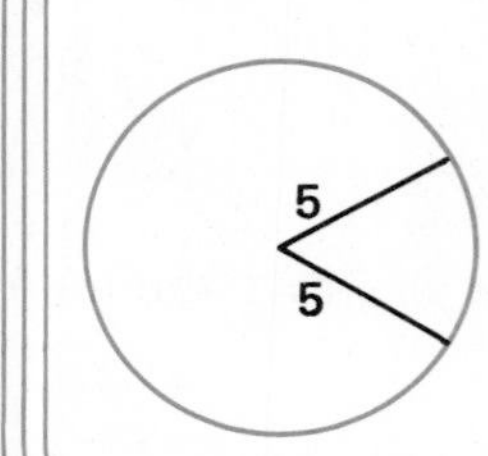

circumference = 2π (radius)

$c = 2\pi r$
$= 2\pi(5)$
$= 10\pi$

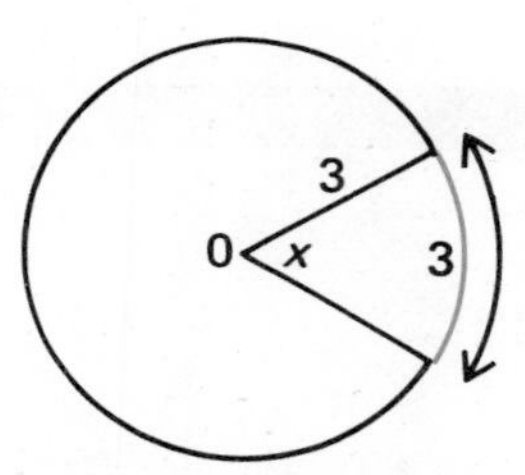

O intercepts an arc with the same length as a radius, *x* is 1 radian.

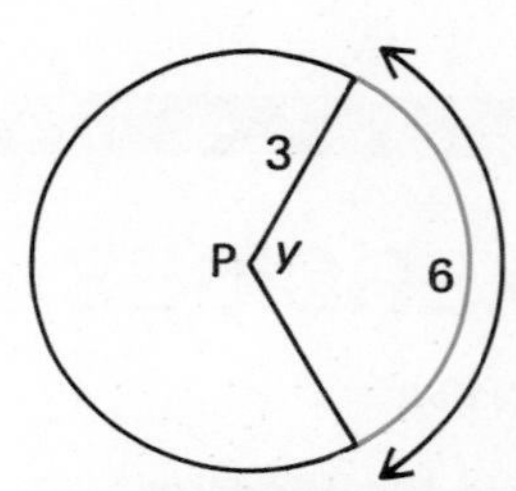

P intercepts an arc with length twice a radius, *y* is 2 radians.

Definition of a radian →

1 radian is the measure of a central angle which cuts off an arc whose length is equal to the length of a radius of the circle.

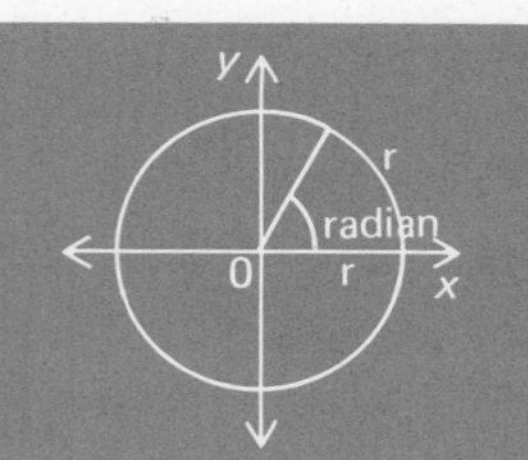

EXAMPLE 1 Find x in radians.

The radius is used as length of a radius.

x intercepts an arc whose length is 3 times the radius.

Thus, $x = 3$ radians.

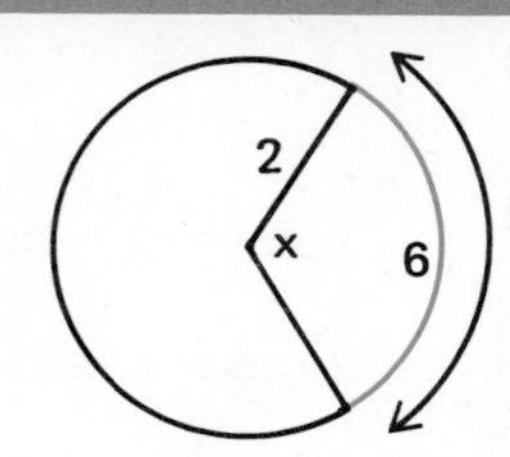

EXAMPLE 2 Express 180° in radian measure.

Circumference $= 2\pi r$ →

Length $\widehat{AB} = \frac{1}{2}(2\pi r)$
$= \pi r$

Thus, 180° is π radians.

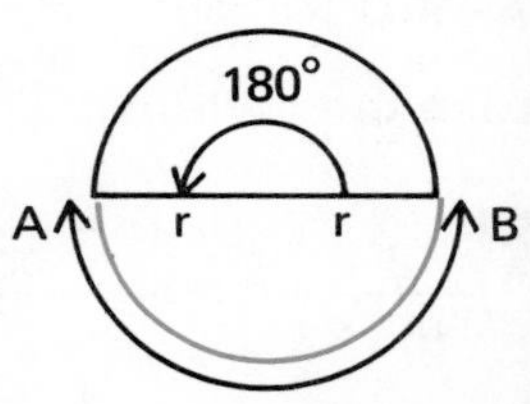

EXAMPLE 3 Express 90° in radian measure.

Set up a proportion. r is the radian measure. Solve the proportion.

$$\frac{90}{180} = \frac{r}{\pi}$$
$$180r = 90\pi$$
$$r = \frac{90\pi}{180}, \text{ or } \frac{\pi}{2}$$

Thus, 90° is $\frac{\pi}{2}$ radians.

EXAMPLE 4 Express 2π radians in degree measure.

Set up a proportion. d is the degree measure.

$$\frac{d}{180} = \frac{2\pi}{\pi}$$
$$\pi d = 180(2\pi)$$
$$d = \frac{360\pi}{\pi}, \text{ or } 360$$

Thus, 2π radians is 360°.

$$\frac{d}{180^\circ} = \frac{r}{\pi}$$

To change from degree measure to radian measure substitute degrees for d and solve for r.
To change from radian measure to degree measure, substitute radians for r and solve for d.

EXAMPLE 5 Express in radian measure.

Set up a proportion. Replace d with the given value.

45°

$$\frac{45}{180} = \frac{r}{\pi}$$
$$180r = 45\pi$$
$$r = \frac{45\pi}{180}, \text{ or } \frac{\pi}{4}$$

−225°

$$\frac{-225}{180} = \frac{r}{\pi}$$
$$180r = -225\pi$$
$$r = \frac{-225\pi}{180}, \text{ or } \frac{-5\pi}{4}$$

Thus, 45° is $\frac{\pi}{4}$ radians and −225° is $-\frac{5\pi}{4}$ radians.

EXAMPLE 6 Express in degree measure.

$\frac{3\pi}{5}$

Set up a proportion. →

$$\frac{d}{180} = \frac{\frac{3\pi}{5}}{\pi}$$

Solve.

$$\pi d = \frac{3\pi}{5}(180)$$
$$d = 108$$

$\frac{5\pi}{2}$

$$\frac{d}{180} = \frac{\frac{5\pi}{2}}{\pi}$$
$$\pi d = \frac{5\pi}{2}(180)$$
$$d = 450$$

Thus, $\frac{3\pi}{5}$ radians is 108° and $\frac{5\pi}{2}$ radians is 450°.

EXAMPLE 7 Evaluate.

$\frac{\pi}{3}$ is 60°, $\frac{\pi}{4}$ is 45°, $\frac{\pi}{6}$ is 30°

$\cos \frac{\pi}{3}$	$\tan \frac{\pi}{4}$	$\sin \frac{\pi}{6}$
$\cos 60°$	$\tan 45°$	$\sin 30°$
$\frac{1}{2}$	1	$\frac{1}{2}$

EXAMPLE 8 Evaluate $\dfrac{\sin \frac{5\pi}{6} - \cos \frac{3\pi}{2}}{\cot \frac{\pi}{4}}$

Change to degrees and write the value of each.

$$\frac{\sin 150° - \cos 270°}{\cot 45°}$$

$$\frac{\frac{1}{2} - (0)}{1}$$

$$\frac{1}{2}$$

Thus, the value of the expression is $\frac{1}{2}$.

ORAL EXERCISES

Express in radians.

1. 360° **2.** 90° **3.** 45° *$\frac{\pi}{4}$* **4.** 180° *π* **5.** 30° *$\frac{\pi}{6}$* **6.** 60° *$\frac{\pi}{3}$* **7.** 300° *$\frac{5\pi}{3}$*

Express in degrees.

8. $\frac{\pi}{2}$ **9.** $\frac{\pi}{4}$ **10.** $\frac{3\pi}{2}$ *270°* **11.** π *180°* **12.** $\frac{5\pi}{6}$ *150°* **13.** $\frac{\pi}{6}$ *30°* **14.** $\frac{7\pi}{6}$ *210°*

EXERCISES

PART A

Express in radians.

1. 15° **2.** 20° **3.** 80° *$\frac{4\pi}{9}$* **4.** 210° *$\frac{7\pi}{6}$* **5.** 135° *$\frac{3\pi}{4}$* **6.** 330° *$\frac{11\pi}{6}$* **7.** 315° *$\frac{7\pi}{4}$*
8. −45° **9.** −120° **10.** −135° *$-\frac{3\pi}{4}$* **11.** −210° **12.** −270° **13.** −300° **14.** −360°

Express in degrees.

15. $\frac{\pi}{3}$ **16.** $\frac{\pi}{9}$ **17.** $\frac{\pi}{4}$ **18.** $\frac{3\pi}{4}$ **19.** $\frac{4\pi}{3}$ **20.** $\frac{5\pi}{3}$ **21.** $\frac{9\pi}{10}$ *162°*
22. $\frac{\pi}{5}$ **23.** $-\frac{2\pi}{3}$ **24.** $-\frac{11\pi}{6}$ *−330°* **25.** $-\frac{3\pi}{2}$ **26.** $-\frac{\pi}{2}$ **27.** -2π **28.** $-\frac{2\pi}{9}$

PART B

Evaluate.

29. $\sin \frac{5\pi}{2}$ **30.** $\cos \frac{\pi}{4}$ *$\frac{\sqrt{2}}{2}$*
31. $\tan \frac{3\pi}{4}$ **32.** $\cos \frac{4\pi}{3}$
33. $\sin \frac{\pi}{6} \cos \pi + \tan \frac{3\pi}{4} \cos^2 \frac{3\pi}{4} + \cos \frac{\pi}{2}$

34. $\dfrac{\cos \frac{\pi}{4} + \sin \frac{5\pi}{6}}{\tan \frac{\pi}{4}}$ *$\frac{1 + \sqrt{2}}{2}$*

35. $\dfrac{\tan \frac{5\pi}{4} - \sin \frac{5\pi}{3}}{\cos \frac{11\pi}{6}}$

Applying Radian Measure

OBJECTIVES

- To sketch trigonometric functions and read the graphs if the x-axis is labeled in radians
- To find solutions in radians of trigonometric equations

REVIEW CAPSULE

Degrees	0°	30°	60°	90°	180°	270°	360°
Radians	0	$\frac{\pi}{6}$	$\frac{\pi}{3}$	$\frac{\pi}{2}$	π	$\frac{3\pi}{2}$	2π

Degrees	−45°	−60°	−90°	−150°	−180°	−270°
Radians	$-\frac{\pi}{4}$	$-\frac{\pi}{3}$	$-\frac{\pi}{2}$	$-\frac{5\pi}{6}$	$-\pi$	$-\frac{3\pi}{2}$

In graphing trigonometric functions, we often label the x-axis in radians instead of degrees.

EXAMPLE 1 Sketch the graph of $y = \sin x$ for $0 \le x \le 2\pi$. What is the period? For what value of x does $\sin x$ reach its maximum? its minimum?

The graph is exactly the same as when the x-axis was labeled in degrees.

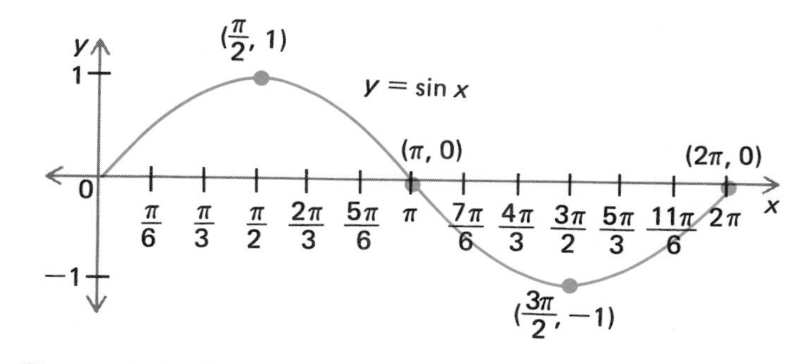

2π is 360°.
π is 180°.
$\frac{3\pi}{2}$ is 270°.

The period of $y = \sin x$ is 2π.
The maximum occurs at $x = \frac{\pi}{2}$.
The minimum occurs at $x = \frac{3\pi}{2}$.

EXAMPLE 2 Sketch $y = 2 \cos x$ for $-\pi \le x \le \pi$.

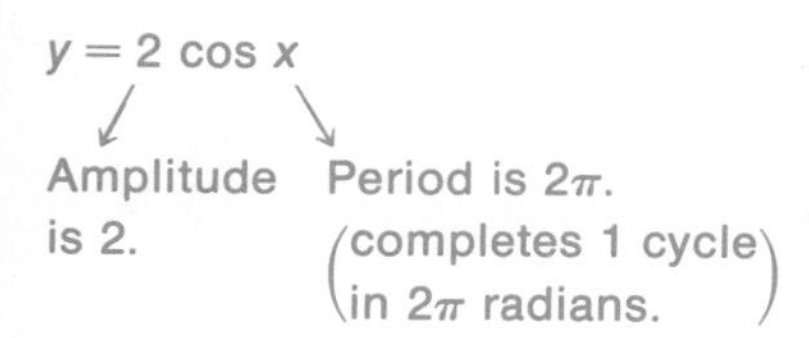

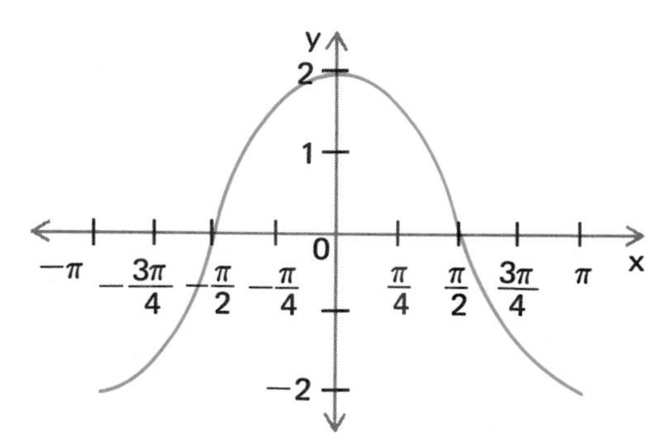

EXAMPLE 3 Sketch $y = \sin 2x$ for $0 \le x \le 2\pi$.

$y = \sin 2x$

Amplitude is 1. Period is $\frac{2\pi}{2}$ or π. (completes 2 cycles)

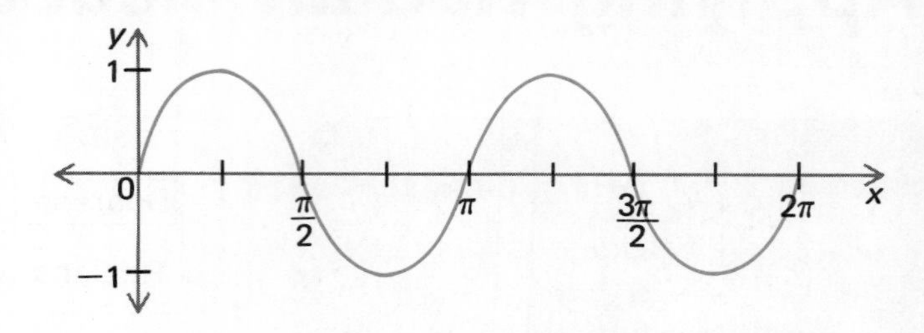

EXAMPLE 4 Sketch $y = \sin 2x$ and $y = 2 \cos x$ for $0 \le x \le \pi$ on the same set of axes. For what values of x does $\sin 2x = 2 \cos x$?

Use Example 3 to sketch $y = \sin 2x$.

$y = 2 \cos x$

Amplitude is 2. Period is 2π. (completes only $\frac{1}{2}$ cycle for $0 \le x \le \pi$.)

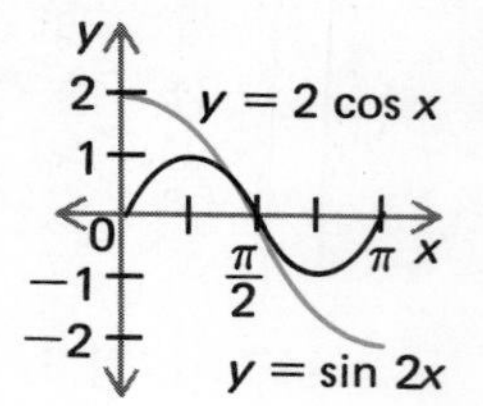

Thus, $\sin 2x = 2 \cos x$ for $x = \frac{\pi}{2}$.

EXAMPLE 5

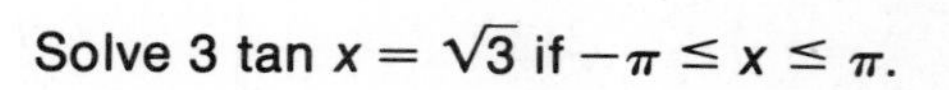

Solve $3 \tan x = \sqrt{3}$ if $-\pi \le x \le \pi$.

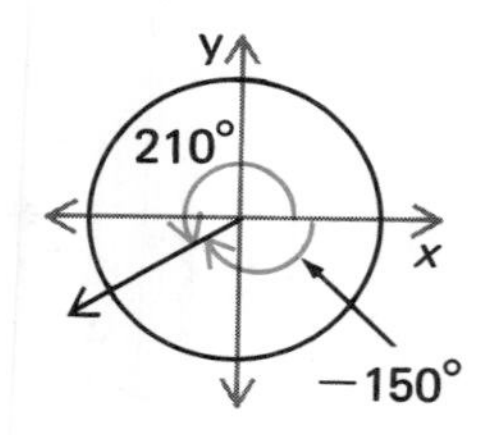

$$3 \tan x = \sqrt{3}$$
$$\tan x = \frac{\sqrt{3}}{3}$$

Reference angle: 30°

1st quad. 30° | 3rd quad. 210°, or −150°

Convert the degree measures to radians.

Thus, the solutions are $\frac{\pi}{6}$ and $-\frac{5\pi}{6}$.

EXAMPLE 6 Solve $2 \sin^2 x - 3 \sin x - 2 = 0$ if $0 \le x \le 2\pi$.

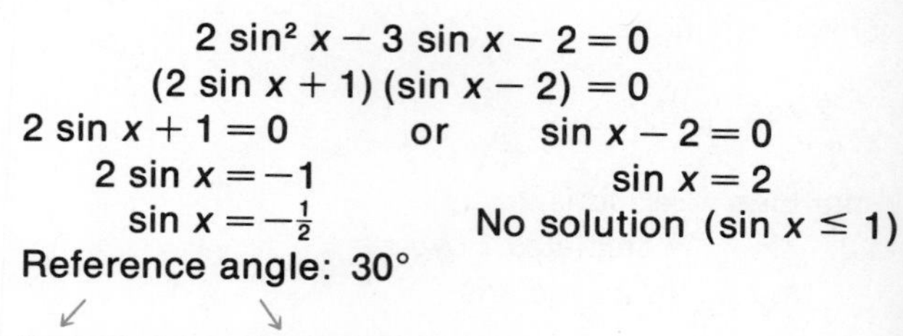

Factor →
Set each factor equal to zero →
Solve for $\sin x$. →

$$2 \sin^2 x - 3 \sin x - 2 = 0$$
$$(2 \sin x + 1)(\sin x - 2) = 0$$

$2 \sin x + 1 = 0$ or $\sin x - 2 = 0$

$2 \sin x = -1$ | $\sin x = 2$

$\sin x = -\frac{1}{2}$ | No solution ($\sin x \le 1$)

Reference angle: 30°

3rd quad. 210° | 4th quad. 330°

Convert 210° and 330° to radian measure. → **Thus,** the solutions are $\frac{7\pi}{6}$ and $\frac{11\pi}{6}$.

EXERCISES

PART A

Sketch $y = \cos 2x$ in the interval 0 to 2π.

1. Find the maximum and minimum values. *1; −1*
2. What is the period? *π*
3. What is the amplitude? *1*

Sketch $y = 2 \sin x$ for $-\pi \le x \le \pi$.

4. Find the maximum and minimum values. *2; −2*
5. What is the period? *2π*
6. What is the amplitude? *2*

Sketch the graph for $0 \le x \le 2\pi$. Determine the period and the amplitude.

7. $y = 3 \sin x$
8. $y = \cos \frac{1}{2}x$ *4π; 1*
9. $y = \frac{1}{3} \sin x$
10. $y = 4 \cos x$ *2π; 4*
11. $y = 2 \cos \frac{1}{2}x$
12. $y = \frac{1}{2} \sin 2x$ *π; $\frac{1}{2}$*
13. $y = 2 \cos 2x$
14. $y = \frac{1}{2} \sin \frac{1}{2}x$

15. Sketch $y = 2 \cos x$ and $y = \sin \frac{1}{2}x$ for $0 \le x \le 2\pi$ on the same set of axes. For how many values of x does $2 \cos x = \sin \frac{1}{2}x$? *2*

16. Sketch $y = \frac{1}{2} \cos x$ and $y = \sin 2x$ for $0 \le x \le 2\pi$ on the same set of axes. For how many values of x does $\frac{1}{2} \cos x = \sin 2x$? *4*

17. Sketch $y = \sin 2x$ and $y = 2 \cos x$ for $-\pi \le x \le \pi$ on the same set of axes. For what values of x does $\sin 2x = 2 \cos x$?

18. Sketch $y = 2 \sin x$ and $y = \sin \frac{1}{2}x$ for $-\pi \le x \le 2\pi$ on the same set of axes. For what values of x does $2 \sin x = \sin \frac{1}{2}x$?

Solve if $0 \le x \le 2\pi$.

19. $\sin x = \frac{1}{2}$ *$\frac{\pi}{6}$; $\frac{5\pi}{6}$*
20. $\cos x = -\frac{\sqrt{3}}{2}$ *$\frac{5\pi}{6}$; $\frac{7\pi}{6}$*
21. $\tan x = -1$
22. $2 \cos x + 1 = 0$
23. $3 \tan^2 x - 1 = 0$
24. $2 \sin^2 x + 3 = 4$
25. $2 \sin^2 x - \sin x = 0$
26. $2 \cos^2 x - 3 \cos x + 1 = 0$
27. $2 \sin^2 x - \sin x - 1 = 0$

PART B

Sketch the graph for $0 \le x \le 2\pi$.

28. $y = \tan x$
29. $y = \cot x$
30. $y = \sec x$
31. $y = \csc x$
32. $y = \tan 2x$
33. $y = \frac{1}{2} \cot x$

Sketch the graph for $-\pi \le x \le 2\pi$. Determine the period and the amplitude.

34. $y = -2 \sin x$ *2π; 2*
35. $y = -\frac{1}{2} \cos x$ *2π; $\frac{1}{2}$*
36. $y = \sin (-2x)$ *π; 1*
37. $y = \cos (-\frac{1}{2}x)$
38. $y = \tan (-x)$
39. $y = -2 \cot x$
40. $y = -\frac{1}{2} \sin 2x$
41. $y = 2 \cos (-2x)$ *π; 2*
42. $y = -\frac{1}{2} \cos 2x$

PART C

43. Sketch the graph of $y = -2 \sin \frac{1}{2}x$ and $y = -\frac{1}{2} \cos (-2x)$ for $0 \le x \le 2\pi$ on the same set of axes. For how many values of x does $-2 \sin \frac{1}{2}x = -\frac{1}{2} \cos (-2x)$?

44. Sketch the graph of $y = \tan (-2x)$ and $y = -2 \cos (\frac{1}{2}x)$ for $0 \le x \le 2\pi$ on the same set of axes. For how many values of x does $\tan (-2x) = -2 \cos (\frac{1}{2}x)$?

Inverses of Trigonometric Functions

OBJECTIVES

- To determine the inverse of a function
- To graph inverses of the sin, cos, and tan functions
- To apply inverse function notation like Arc cos $(\frac{3}{5})$

REVIEW CAPSULE

$\{(-3, 2), (2, 2), (-1, 6)\}$	$\{(-2, 3), (-2, 1), (1, 4)\}$
No two first elements are the same.	Two first elements are the same.
Is a function	Is not a function

EXAMPLE 1 $A = \{(1, 2), (2, -3), (5, 2)\}$ is a function. Form relation B by interchanging the two elements of each ordered pair of A. Is B a function?

A: domain, range
B: domain, range
A first element is repeated.

$$A = \{(1, 2), (2, -3), (5, 2)\}$$

$$B = \{(2, 1), (-3, 2), (2, 5)\}$$

Thus, B is not a function.

EXAMPLE 2 $A = \{(-1, 2), (-2, 2), (1, 5)\}$ is a function. Form relation B by interchanging the two elements of each ordered pair of A. Is B a function?

A is a function. → $A = \{(-1, 2), (-2, -2), (1, 5)\}$

B is a function. → $B = \{(2, -1), (-2, -2), (5, 1)\}$

No two first elements are the same.

Thus, B is a function.

Definition of inverse of a function →

> The inverse of a function is formed by interchanging the elements of each ordered pair of the function.

The inverse function is formed when the inverse of a function is a function.

EXAMPLE 3 Find the inverse of the function $y = 2x + 3$. Is the inverse an inverse function?

$$y = 2x + 3$$

Interchange x and y. → $x = 2y + 3$

Solve for y. → $y = \frac{x - 3}{2}$, or $y = \frac{1}{2}x - \frac{3}{2}$

$y = \frac{1}{2}x - \frac{3}{2}$ is a nonvertical line. → **Thus,** the inverse is an inverse function.

EXAMPLE 4 Find the inverse of $y = \sin x$.

$$y = \sin x$$

Interchange x and y. → $x = \sin y$

Thus, the inverse of $y = \sin x$ is $x = \sin y$.

$\sin y = x$ can be written in two other ways.

$y = \text{arc} \sin x$, or $y = \sin^{-1} x$
and read
y is the angle whose sin is x.

EXAMPLE 5 Find u if $u = \text{arc} \sin \frac{1}{2}$, $0° \leq u \leq 360°$.

u is the angle whose sin is $\frac{1}{2}$, i.e. $\sin u = \frac{1}{2}$. → Write $u = \text{arc} \sin \frac{1}{2}$ as $\sin u = \frac{1}{2}$.

Reference angle: 30°

1st quad. 30°   2nd quad. 150°

So, $\sin 30° = \frac{1}{2}$ and $\sin 150° = \frac{1}{2}$.

Thus, $u = 30°$ or $150°$.

EXAMPLE 6 Find θ if $\theta = \text{arc} \tan (-1)$, $0 \leq \theta \leq 2\pi$.

Write $\theta = \text{arc} \tan (-1)$ as $\tan \theta = (-1)$.

Reference angle: 45°

2nd quad. 135°   4th quad. 315°

So, $\tan 135° = -1$ and $\tan 315° = -1$

$135° = \frac{3\pi}{4}$; $315° = \frac{7\pi}{4}$ → **Thus,** $\theta = \frac{3\pi}{4}$ or $\frac{7\pi}{4}$.

EXAMPLE 7 Sketch $y = \text{arc sin } x$, $y = \text{arc cos } x$ and $y = \text{arc tan } x$.

Make tables for $y = \sin x$, $y = \cos x$, and $y = \tan x$ and interchange the elements of the ordered pairs.

$y = \text{arc sin } x$

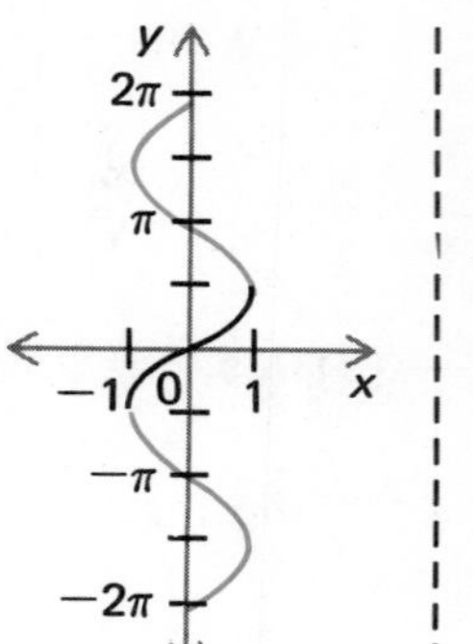

$y = \text{arc cos } x$

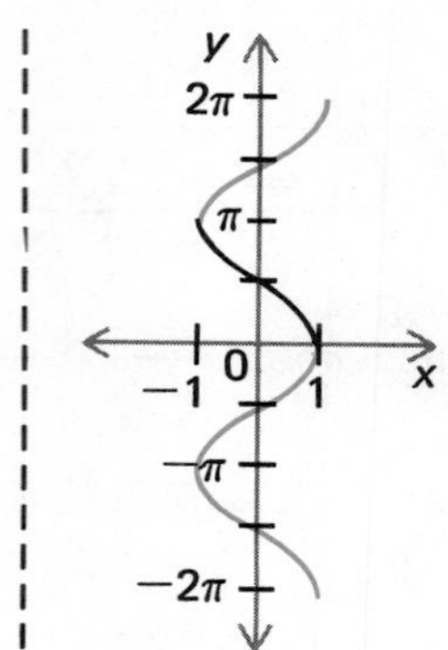

$y = \text{arc tan } x$

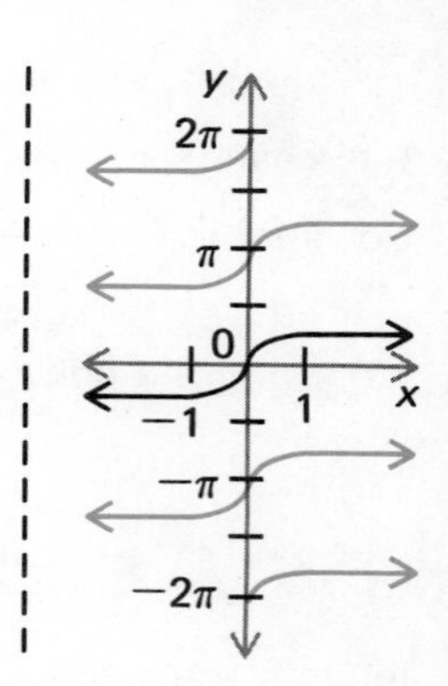

The graphs do not meet the vertical line test. → The inverses of $y = \sin x$, $y = \cos x$, and $y = \tan x$ are not functions (not inverse functions).

See the graphs in Example 7. →

To obtain inverse functions of $y = \sin x$, $y = \cos x$ and $y = \tan x$, restrict the range to values called principal values.

Capitalize the first letter of Arc sin x or $\text{Sin}^{-1} x$, etc., to indicate principal value.

$-\frac{\pi}{2} \leq \text{Arc sin } x \leq \frac{\pi}{2}$, or $-90° \leq \text{Arc sin } x \leq 90°$

$0 \leq \text{Arc cos } x \leq \pi$, or $0° \leq \text{Arc cos } x \leq 180°$

$-\frac{\pi}{2} \leq \text{Arc tan } x \leq \frac{\pi}{2}$, or $-90° \leq \text{Arc tan } x \leq 90°$

EXAMPLE 8 Find $\sin A$ if $A = \text{Arc cos } (\frac{-12}{13})$.

Arc means principal value $0° \leq A \leq 180°$ or $0 \leq A \leq \pi$

Write $A = \text{Arc cos } (-\frac{12}{13})$ as $\cos A = -\frac{12}{13}$.

$\cos A = -\frac{12}{13}$

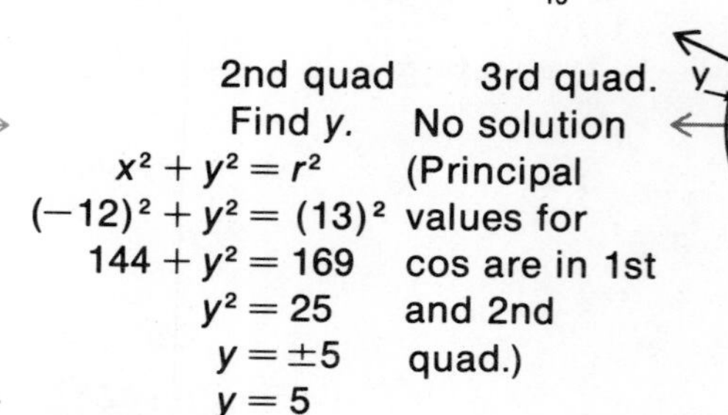

$x = -12$, $r = 13$ →

2nd quad	3rd quad.
Find y.	No solution
$x^2 + y^2 = r^2$	(Principal
$(-12)^2 + y^2 = (13)^2$	values for
$144 + y^2 = 169$	cos are in 1st
$y^2 = 25$	and 2nd
$y = \pm 5$	quad.)

In 2nd quadrant → $y = 5$

$\sin A = \frac{y}{r}$

Thus, $\sin A = \frac{5}{13}$.

EXAMPLE 9 Find $\tan(\text{Arc sin}(-\frac{3}{5}))$.

Principal values $-90° \leq A \leq 90°$

Let $A = \text{Arc sin}(-\frac{3}{5})$.
Write as $\sin A = -\frac{3}{5}$.

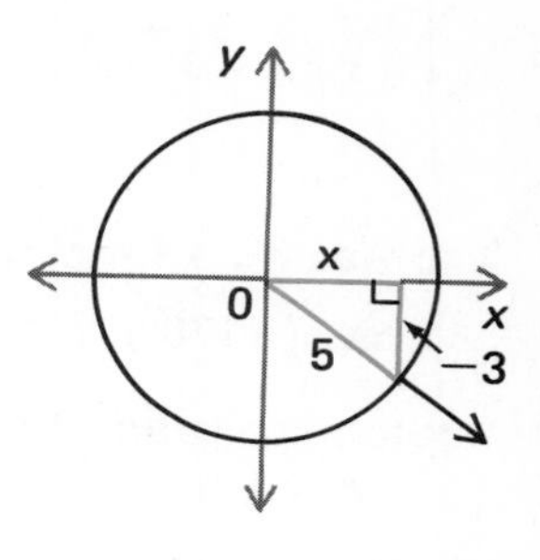

3rd quad.
No solution
(Principal values for sin are in 1st and 4th quad.)

4th quad.
Find x.
$x^2 + y^2 = r^2$
$x^2 + 9 = 25$
$x = 4$

$y = -3$, $r = 5$

$\tan A = \frac{y}{x}$ → **Thus,** $\tan A = -\frac{3}{4}$.

EXAMPLE 10 Find $\text{Sin}^{-1}\left(\tan \frac{5\pi}{6}\right)$.

$\tan 150° = -\tan 30° = -\frac{\sqrt{3}}{3}$ → $\tan \frac{5\pi}{6} = \tan 150°$
$= -\frac{\sqrt{3}}{3}$

Let $A = \text{Sin}^{-1}(-\frac{\sqrt{3}}{3})$. → Find $\text{Sin}^{-1}(-\frac{\sqrt{3}}{3})$.
Write as $\sin A = -\frac{\sqrt{3}}{3}$.

$1.732 \div 3 \doteq .5773$. Look up in the table to the nearest degree. → Reference angle: 35°

3rd quad.
No solution

4th quad.
325°

$\frac{325\pi}{180} = \frac{65\pi}{36}$ → **Thus,** $\text{Sin}^{-1}\left(\tan \frac{5\pi}{6}\right) = \frac{65\pi}{36}$.

EXERCISES

PART A

Find the inverse of each function. Is the inverse an inverse function?

1. $\{(-1, 6), (-2, 3), (5, 6)\}$
2. $\{(0, 7), (7, -1), (-8, 2)\}$ *{(7, 0), (−1, 7), (2, −8)}; yes*
3. $2x - 3y = 8$
4. $y = x^2$
5. $y = \sin 2x$

Find A.

6. $A = \text{Arc cos } \frac{1}{2}$
7. $A = \text{Arc tan } (-1)$ *$-\frac{\pi}{4}$ or $-45°$*
8. $A = \text{Arc sin } \frac{\sqrt{2}}{2}$
9. $A = \text{Arc tan } \frac{\sqrt{3}}{3}$

Find.

10. $\sin(\text{Arc cos}(-\frac{\sqrt{3}}{2}))$
11. $\text{Sin}^{-1}(\tan \frac{3\pi}{4})$
12. $\text{Arc cos}(\sin -\frac{\pi}{3})$
13. $\sin(\text{Arc cos } \frac{12}{13})$ *$\frac{5}{13}$*
14. $\cos(\text{Arc tan}(-\frac{3}{4}))$ *$\frac{4}{5}$*
15. $\tan(\text{Arc sin} \frac{-\sqrt{2}}{2})$ *-1*

PART B

Sketch the graph.

16. $y = \text{arc cot } x$
17. $y = \text{arc sin } 2x$
18. $y = 2 \text{ arc cos } 2x$

Chapter Eighteen Review

Express as a function of a positive acute angle less than 45°. [p. 451]

1. $\cos(-170°)$ **2.** $\tan(-330°)$ **3.** $\sin(-220°)$ **4.** $\cos(-100°)$ **5.** $\sin(-300°)$

Sketch the graph for $0° \le x \le 360°$. Determine the period and the amplitude. [pp. 453, 456, 459, 462, 465]

6. $y = 2 \sin x$ **7.** $y = \tan \frac{1}{2}x$ **8.** $y = \sin(-3x)$ **9.** $y = 3 \sin \frac{1}{2}x$

10. Sketch $y = \sin 2x$ and $y = \frac{1}{2}\cos x$ for $-180° \le x \le 180°$ on the same set of axes. For how many values of x does $\sin 2x = \frac{1}{2}\cos x$? [p. 462]

11. Sketch $y = -2\cos 2x$ and $y = \sin(-2x)$ for $-\pi \le x \le \pi$ on the same set of axes. For how many values of x does $-2\cos 2x = \sin(-2x)$? [p. 483]

Find. [p. 468]

12. $\log \tan 33° 18'$

Find u, $0° \le u \le 90°$. [p. 468]

13. $\log \cos u = 9.4870 - 10$

Find the value to the nearest tenth. Use logarithms. [p. 468]

14. $\dfrac{(\sin 35°)^2}{(49.2)^3}$ **15.** $\dfrac{\sqrt{4.34}\,(153)^2}{\tan 73°30'}$ **16.** $\cos 18° 40' \sqrt[3]{\dfrac{.087}{10.3}}$

Solve for θ to the nearest degree, $(0° \le \theta \le 360°)$. [p. 471]

17. $2\cos^2\theta + 5\cos\theta = 6$ **18.** $\sin^2\theta - 1 = 0$

Solve if $0 \le \theta \le 2\pi$. [p. 483]

19. $\tan^2\theta + \tan\theta = 0$ **20.** $2\sin^2\theta - \sin\theta = 1$

Express as sin, or cos, or both. [p. 475]

21. $\sec 45°$ **22.** $\tan 10°$ **23.** $\cot 80°$ **24.** $\csc(-320°)$

Verify for $A = 120°$. [p. 475]

25. $\sin^2 A + \cos^2 A = 1$

Write an equivalent expression using only sin x. [p. 475]

26. $(\sin x)(\csc x) = 1$ **27.** $\tan x$ **28.** $\cos x \cdot \csc x$

Prove the identity. [p. 478]

29. $\dfrac{\cos^2 x}{1 - \sin x} = 1 + \sin x$ **30.** $\csc x \sec x = \cot x + \tan x$ **31.** $\dfrac{\sec x + \csc x}{\csc x} = 1 + \tan x$

Express in radians. [p. 480]

32. $45°$ **33.** $-30°$ **34.** $300°$ **35.** $-210°$

Express in degrees. [p. 480]

36. $\frac{\pi}{9}$ **37.** $-\frac{5}{6}\pi$ **38.** $\frac{3\pi}{2}$ **39.** -4π

Evaluate. [p. 480]

40. $\tan \dfrac{11\pi}{6}$ **41.** $\cos\left(\dfrac{-7\pi}{6}\right)$ **42.** $\dfrac{\sin\frac{\pi}{6} + \cos\frac{\pi}{3}}{\tan\frac{\pi}{4}}$ **43.** $\dfrac{\tan(-\frac{7\pi}{6}) - \cot(-\frac{\pi}{4})}{\cos\frac{5\pi}{3}}$

Find the inverse of the function. Is the inverse an inverse function? [p. 486]

44. $\{(-3, 7), (7, -3), (6, 4)\}$ **45.** $\{(-1, 2), (-2, 3), (4, 2)\}$ **46.** $-2x + y = 6$ **47.** $y = \sin x$

Find A. [p. 486]

48. $A = \text{Arc}\cos\frac{\sqrt{3}}{2}$ **49.** $A = \text{Arc}\tan(-\frac{\sqrt{3}}{3})$

Find each. [p. 486]

50. $\cos(\text{Arc}\sin\frac{4}{5})$ **51.** $\sin(\text{Arc}\cos(-\frac{3}{5}))$

Chapter Eighteen Test

Express as a function of a positive acute angle less than 45°.

1. $\sin(-150°)$ **2.** $\tan(125°)$ **3.** $\cos(310°)$

Sketch the graph for $0° \le x \le 360°$. Determine the period and the amplitude.

4. $y = \frac{1}{2}\cos x$ **5.** $y = 2\sin 3x$ **6.** $y = \tan 2x$

7. Sketch $y = \cos 3x$ and $y = 2\sin x$ for $0° \le x \le 360°$ on the same set of axes. For how many values of x does $\cos 3x = 2\sin x$?

8. Sketch $y = \frac{1}{2}\sin x$ and $y = 2\cos 2x$ for $-\pi \le x \le \pi$ on the same set of axes. For how many values of x does $\frac{1}{2}\sin x = 2\cos 2x$?

Find.

9. $\log \cos 28° 40'$

10. $\log \tan 48° 24'$

Find u, $0° \le u \le 90°$.

11. $\log \tan u = 9.2819 - 10$

12. $\log \cos u = 9.9727 - 10$

Find the value to the nearest tenth. Use logarithms.

13. $\dfrac{(15.6)^2\sqrt{.87}}{\sin 38°}$ **14.** $\sqrt[3]{\dfrac{93.6}{876.3}}\cos 17° 40'$

Solve for θ to the nearest degree ($0° \le \theta \le 360°$).

15. $2\cos^2\theta - 1 = 0$ **16.** $3\sin^2\theta + 5\sin\theta - 2 = 0$

Solve if $0 \le \theta \le 2\pi$.

17. $\sin^2\theta - \sin\theta = 0$

Express as sin, or cos, or both.

18. $\csc 30°$ **19.** $\tan(-340°)$ **20.** $\cot 230°$

Verify for $A = 150°$.

21. $\sin^2 A + \cos^2 A = 1$

Write an equivalent expression using only sin x.

22. $\cot x$ **23.** $\cos x \cdot \cot x$

Prove the identity.

24. $\sec x = \dfrac{\sin x \csc x}{\cos x}$ **25.** $\csc x = \dfrac{\sin x}{1 + \cos x} + \cot x$

Express in radians.

26. $60°$ **27.** $-300°$

Express in degrees.

28. $\dfrac{5\pi}{6}$ **29.** $-\dfrac{2\pi}{3}$

Evaluate.

30. $\sin\left(-\dfrac{5\pi}{6}\right)$ **31.** $\dfrac{\cos\frac{\pi}{3} + \tan\frac{\pi}{6}}{\sin\frac{\pi}{4}}$

Find the inverse of the function. Is the inverse an inverse function?

32. $\{(-3, 2), (4, 6), (3, 2)\}$ **33.** $x - y = 3$ **34.** $y = \cos x$

Find A.

35. $A = \text{Arc sin}\ \dfrac{\sqrt{2}}{2}$

Find each.

36. $\tan\left(\text{Arc cos}\ \dfrac{3}{5}\right)$ **37.** $\text{Arc cos}\left(\sin\dfrac{2\pi}{3}\right)$

19 SOLVING TRIANGLES

Sin (t) Using Infinite Series

n	$n!$	$\frac{1}{n!}$
1	1	1
2	2	.500000
3	6	.166667
4	24	.041667
5	120	.008333
6	720	.001389
7	5,040	.000198
8	40,320	.000025

$3! = 3 \times 2 \times 1$
$5! = 5 \times 4 \times 3 \times 2 \times 1$
$n! = n \times (n-1) \times (n-2) \times \ldots \times 3 \times 2 \times 1$

$$\sin(t) = t - \frac{t^3}{3!} + \frac{t^5}{5!} - \frac{t^7}{7!} + \frac{t^9}{9!} - \ldots$$

where t is expressed in radians.

PROBLEM

Express sin (.2) correct to four decimal places.

$$\sin(.2) = .2 - (.2)^3\left(\frac{1}{3!}\right) + (.2)^5\left(\frac{1}{5!}\right) - (.2)^7\left(\frac{1}{7!}\right) + (.2)^9\left(\frac{1}{9!}\right) - \ldots$$

$$\doteq .2 - .008 \times .166667 + .00032 \times .008333 - \ldots$$

$$\doteq .2 - .001333336 + .00000266656 - \underbrace{\ldots}$$

$$\doteq .1987$$

(... will not change first four decimal places)

Thus, $\sin(.2) \doteq .1987$.

1. Find sin (.3) correct to four decimal places.
2. Find cos (.3) correct to four decimal places.

$$\left[\cos(t) = 1 - \frac{t^2}{2!} + \frac{t^4}{4!} - \frac{t^6}{6!} + \frac{t^8}{8!} - \ldots\right]$$

Law of Cosines

OBJECTIVES

- To derive the law of cosines
- To apply the law of cosines to find the length of a side

REVIEW CAPSULE

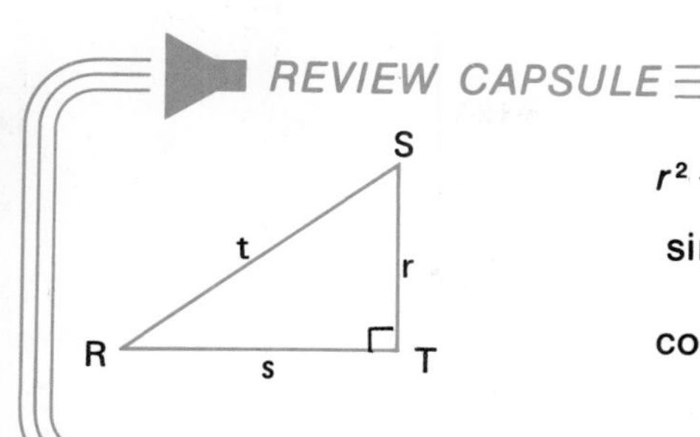

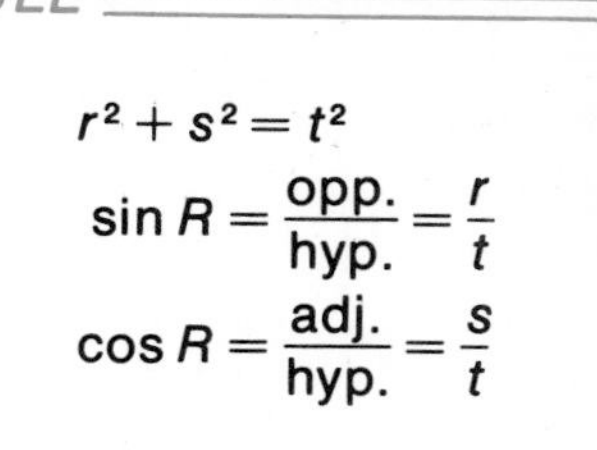

EXAMPLE 1 In $\triangle ABC$, find a to the nearest unit.

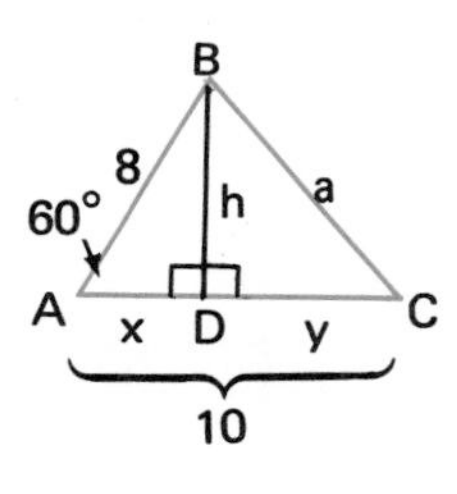

In right $\triangle CBD$

$a^2 = y^2 + h^2$

In right $\triangle ABD$, $x^2 + h^2 = 64$ or $h^2 = 64 - x^2$. → $= y^2 + (64 - x^2)$

$= (y^2 - x^2) + 64$

$= (y + x)(y - x) + 64$

Substitute: $AC = y + x = 10$ and $y = 10 - x$. → $= 10(10 - x - x) + 64$

$= 10(10 - 2x) + 64$

$= 100 - 20x + 64$

$= 164 - 20x$

In right $\triangle ABD$, $\cos 60° = \frac{x}{8}$, or $x = 8 \cos 60°$. → $= 164 - 20(8 \cos 60°)$

$\cos 60° = \frac{1}{2}$ → $= 164 - 20\left(8 \cdot \frac{1}{2}\right)$

$= 164 - 80$

$= 84$

$a = \sqrt{84}$ → **Thus,** $a = 9$ to the nearest unit.

EXAMPLE 2 In $\triangle ABC$, show that $a^2 = b^2 + c^2 - 2bc \cos A$.

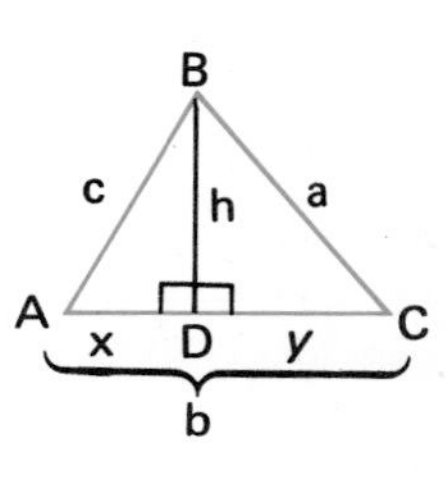

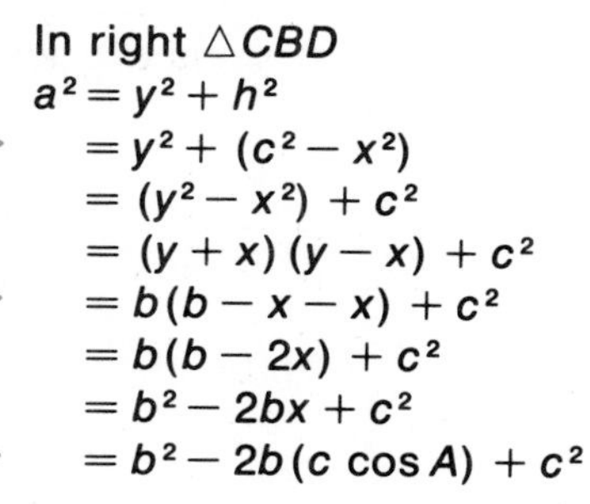

In right $\triangle CBD$

$a^2 = y^2 + h^2$

In right $\triangle ABD$, $x^2 + h^2 = c^2$ or $h^2 = c^2 - x^2$. → $= y^2 + (c^2 - x^2)$

$= (y^2 - x^2) + c^2$

$= (y + x)(y - x) + c^2$

Substitute: $y + x = b$ and $y = b - x$. → $= b(b - x - x) + c^2$

$= b(b - 2x) + c^2$

$= b^2 - 2bx + c^2$

In right $\triangle ABD$, $\cos A = \frac{x}{c}$, or $x = c \cos A$. → $= b^2 - 2b(c \cos A) + c^2$

Rearrange the terms. → **Thus,** $a^2 = b^2 + c^2 - 2bc \cos A$.

Example 2 suggests the law of cosines.

The statements for b^2 and c^2 can be derived by drawing the appropriate altitudes.

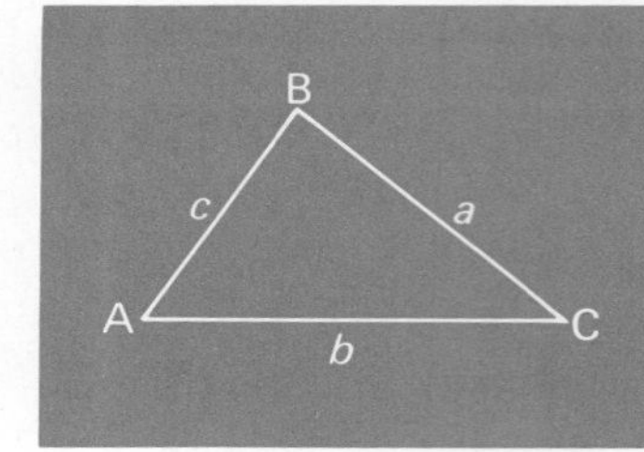

For any triangle *ABC*,

$a^2 = b^2 + c^2 - 2bc \cos A$

$b^2 = a^2 + c^2 - 2ac \cos B$

$c^2 = a^2 + b^2 - 2ab \cos C.$

EXAMPLE 3 In triangle *ABC*, $a = 4$, $b = 6$ and $m\angle C = 60$. Find c. (Answer may be left in radical form.)

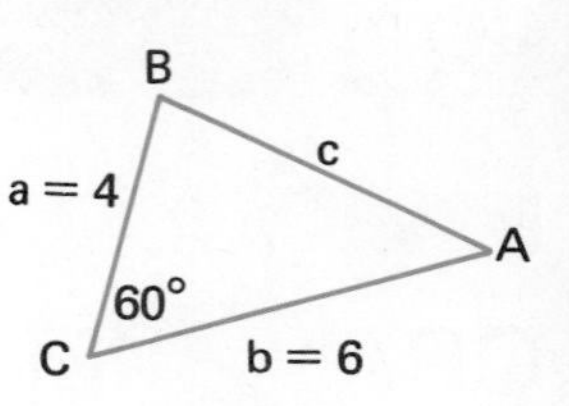

$c^2 = a^2 + b^2 - 2ab \cos C$

$a = 4$, $b = 6$, $m\angle C = 60$ ⟶ $c^2 = 4^2 + 6^2 - 2(4)(6) \cos 60°$

$= 16 + 36 - 48(\cos 60°)$

$\cos 60° = \frac{1}{2}$ ⟶ $= 52 - 48\left(\frac{1}{2}\right)$

$= 52 - 24$

$= 28$

Thus, $c = \sqrt{28}$, or $2\sqrt{7}$.

EXAMPLE 4 In triangle *ABC*, $b = 7$, $c = 5$, and $m\angle A = 150$. Find a to the nearest unit.

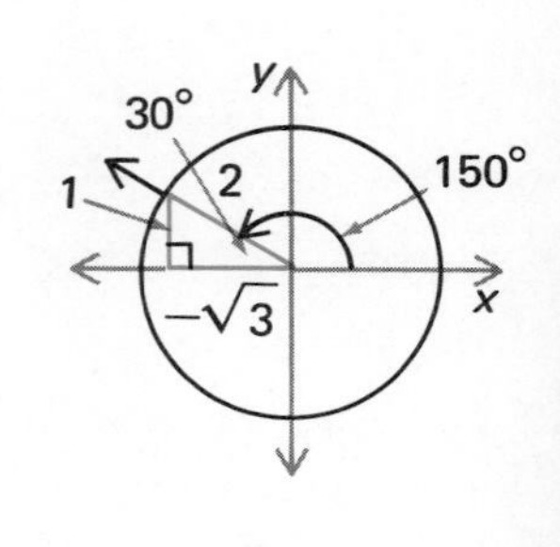

$a^2 = b^2 + c^2 - 2bc \cos A$

$b = 7$, $c = 5$, $m\angle A = 150$ ⟶ $a^2 = 7^2 + 5^2 - 2(7)(5) \cos 150°$

$\cos 150° = -\frac{\sqrt{3}}{2}$

$= 49 + 25 - 70\left(-\frac{\sqrt{3}}{2}\right)$

$= 74 + 35\sqrt{3}$

$= 74 + 35(1.732)$

$= 74 + 60.620$

$= 134.62$

$a = \sqrt{134.62}$ ⟶ **Thus,** $a = 12$ to the nearest unit.

EXAMPLE 5 In right triangle *ABC*, $m\angle C = 90$. Use the law of cosines to show that $c^2 = a^2 + b^2$.

Law of cosines ⟶ $c^2 = a^2 + b^2 - 2ab \cos C$

$m\angle C = 90$ ⟶ $c^2 = a^2 + b^2 - 2ab \cos 90°$

$\cos 90° = 0$ $\quad c^2 = a^2 + b^2 - 2ab(0)$

Pythagorean theorem ⟶ **Thus,** $c^2 = a^2 + b^2$.

For a right triangle, the law of cosines becomes the Pythagorean theorem.

EXAMPLE 6 Two straight roads RT and ST intersect at a town T and form with each other an acute angle of 67°. R is 35 kilometers from T and S is 50 kilometers from T. Find the distance, to the nearest kilometer, between towns R and S.

Draw a diagram.

Use the law of cosines. → $t^2 = r^2 + s^2 - 2rs \cos T$

$r = 50$, $s = 35$ → $t^2 = (50)^2 + (35)^2 - 2(50)(35) \cos 67°$

$\cos 67° = .3907$ → $= 2500 + 1225 - 3500\ (.3907)$

$= 2500 + 1225 - 1367.45$

$= 2357.55$

$t = \sqrt{2357.55}$

$\sqrt{2357.55} \doteq 49$ → **Thus,** R and S are 49 km apart to the nearest km.

ORAL EXERCISES

Give the form of the law of cosines needed to find x.

1.

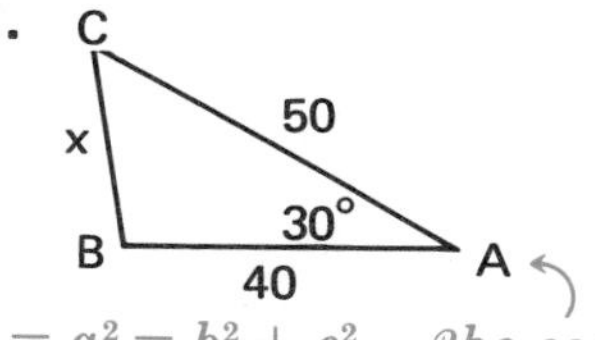

$x^2 = a^2 = b^2 + c^2 - 2bc \cos A$

$= (50)^2 + (40)^2 - 2(50)(40) \cos 30°$

2.

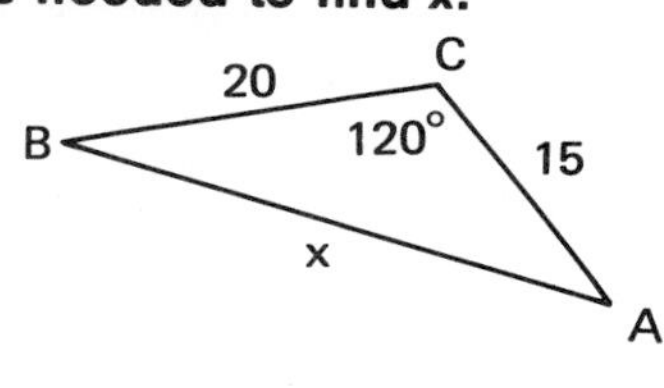

3.

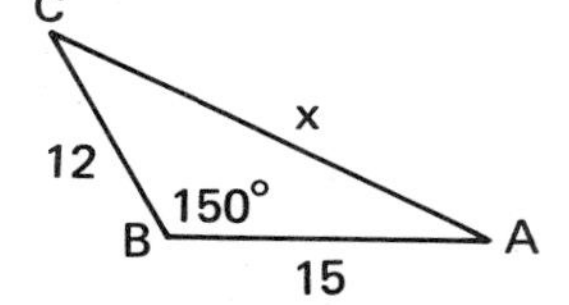

EXERCISES

PART A

Each refers to triangle *ABC*. Find the length of the indicated side to the nearest unit, or leave in radical form.

1. $a = 3$, $c = 5$, $m\angle B = 60$. Find b. *$\sqrt{19}$ or 4*

2. $c = 7$, $b = 6$, $m\angle A = 150$. Find a.

3. $a = 6$, $b = 7$, $m\angle C = 135$. Find c.

4. $a = 4$, $c = 7$, $m\angle B = 30$. Find b.

5. $c = 2$, $b = 3$, $m\angle A = 45$. Find a.

6. $a = 5$, $b = 8$, $m\angle C = 120$. Find c.

7. $b = 10$, $c = 12$, $\cos A = .896$. Find a.

8. $a = 10$, $c = 8$, $\cos B = -.75$. Find b.

9. $a = 6$, $b = 7$, $\cos C = -.56$. Find c. *$\sqrt{132}$ or 11*

10. $b = 7$, $c = 3$, $\cos A = .13$. Find a. *7*

PART B

11. Two ships which leave from the same point form an angle of 73°. Ship A travels 500 km and ship B 275 km. Both ships travel in a straight path. Find to the nearest kilometer the distance between them. *495 km*

12. To travel to city B from city A by bus it is necessary to go to city C first. City C is 300 km from city A. City B is 550 km from city C. Angle ACB is 90°. Find to the nearest kilometer the distance between A and B. *626 km*

13. Prove $b^2 = a^2 + c^2 - 2ac \cos B$.

14. Prove $c^2 = a^2 + b^2 - 2ab \cos C$.

Finding Angles by the Law of Cosines

OBJECTIVE

- To apply the law of cosines to find the measure of an angle

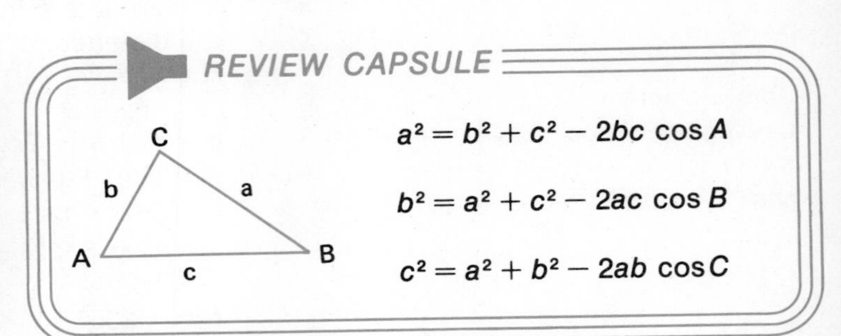

$a^2 = b^2 + c^2 - 2bc \cos A$

$b^2 = a^2 + c^2 - 2ac \cos B$

$c^2 = a^2 + b^2 - 2ab \cos C$

EXAMPLE 1 In $\triangle ABC$, $a = 3$, $b = 6$ and $c = 5$. Find $\cos A$.

$$a^2 = b^2 + c^2 - 2bc \cos A$$

$a = 3$, $b = 6$, $c = 5$ ⟶

$$3^2 = 6^2 + 5^2 - 2(6)(5) \cos A$$

$$9 = 36 + 25 - 60 \cos A$$

Add $60 \cos A$ and -9 to each side. ⟶

$$60 \cos A = 36 + 25 - 9$$

Divide each side by 60. ⟶

$$\cos A = \frac{36 + 25 - 9}{60}.$$

$$= \frac{52}{60}$$

$\frac{52}{60} \doteq .8667$ ⟶ **Thus,** $\cos A = .8667$.

EXAMPLE 2 In any triangle ABC, show that $\cos A = \frac{b^2 + c^2 - a^2}{2bc}$.

$$a^2 = b^2 + c^2 - 2bc \cos A$$

Add $2bc \cos A$ and $-a^2$ to each side. ⟶

$$2bc \cos A = b^2 + c^2 - a^2$$

Divide each side by $2bc$. ⟶ **Thus,** $\cos A = \frac{b^2 + c^2 - a^2}{2bc}$.

The statements for $\cos B$ and $\cos C$ can be derived by starting with the statements for b^2 and c^2. (See the Review Capsule.)

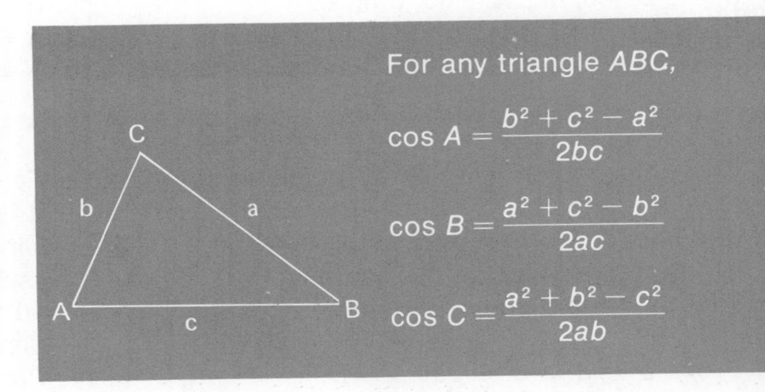

For any triangle ABC,

$$\cos A = \frac{b^2 + c^2 - a^2}{2bc}$$

$$\cos B = \frac{a^2 + c^2 - b^2}{2ac}$$

$$\cos C = \frac{a^2 + b^2 - c^2}{2ab}$$

EXAMPLE 3 In $\triangle ABC$ $a=7$, $b=8$, and $c=5$. Find the measure of $\angle B$ to the nearest degree.

$$\cos B = \frac{a^2+c^2-b^2}{2ac}$$
$$= \frac{7^2+5^2-8^2}{2(7)(5)}$$
$$= \frac{49+25-64}{70}$$
$$= \frac{10}{70}, \text{ or } .1429$$

Look up .1429 in the cos column of table. → **Thus,** the measure of $\angle B = 82°$ to the nearest degree.

EXAMPLE 4 In $\triangle ABC$, the lengths of the sides are 3, 5, and 6. Find the measure of the smallest angle to the nearest degree.

The angle opposite the shortest side has the least measure.

$$\cos C = \frac{a^2+b^2-c^2}{2ab}$$
$$= \frac{25+36-9}{2(5)(6)}$$

$\frac{25+36-9}{2(5)(6)} = \frac{52}{60}$ → $\cos C = .8667$

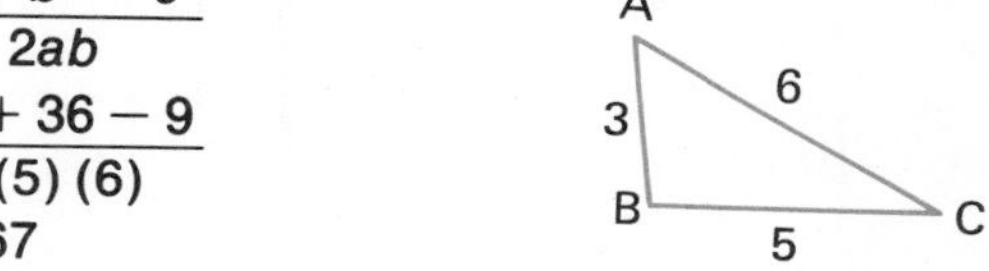

$m\angle C = 30$ → **Thus,** the measure of the smallest angle is 30°.

EXERCISES

PART A

1. $a=7$, $b=9$, $c=3$. Find $\cos A$. *.759*
2. $a=10$, $b=2$, $c=9$. Find $\cos B$. *.983*
3. $a=6$, $b=3$, $c=4$. Find $\cos C$. *.806*
4. $a=12$, $b=6$, $c=7$. Find $\cos A$. *−.702*

Find the measure of the angle to the nearest degree.

5. $a=3$, $b=8$, $c=7$. Find $m\angle A$. *22°*
6. $a=9$, $b=3$, $c=9$. Find $m\angle B$. *19°*
7. $a=13$, $b=12$, $c=5$. Find $m\angle B$. *67°*
8. $a=2$, $b=8$, $c=7$. Find $m\angle C$. *54°*

PART B

In $\triangle ABC$, find the measure of the smallest angle to the nearest degree.

9. $a=2$, $b=3$, $c=4$ *29°*
10. $a=7$, $b=10$, $c=9$ *43°*
11. In $\triangle PQR$, show that $\cos Q = \frac{p^2+r^2-q^2}{2pr}$.

Area of a Triangle

OBJECTIVES

- To find the area of a triangle given the measures of two sides and the included angle
- To apply the formulas

REVIEW CAPSULE

Find the area of the triangle.

$K = \frac{1}{2}bh$

$= \frac{1}{2}(16)(8)$

$= 64$ sq units

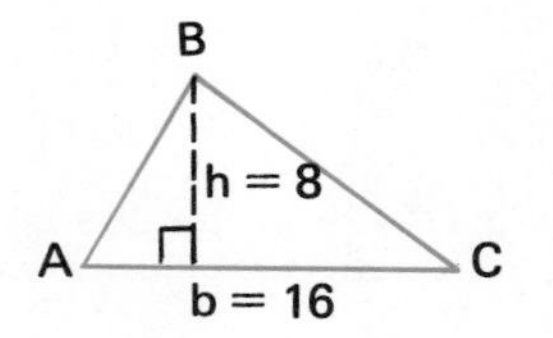

Thus, the area is 64 sq units.

EXAMPLE 1 Find the area to the nearest tenth of a square unit.

In right $\triangle ABD$, $\sin 60° = \frac{h}{8}$, or $h = 8 \sin 60°$.

$\sin 60° = \frac{\sqrt{3}}{2}$

$K = \frac{1}{2}bh$

$= \frac{1}{2}(12)h$

$= \frac{1}{2}(12)(8 \sin 60°)$

$= \frac{1}{2}(12)(8)\left(\frac{\sqrt{3}}{2}\right)$

$= 24\sqrt{3}$

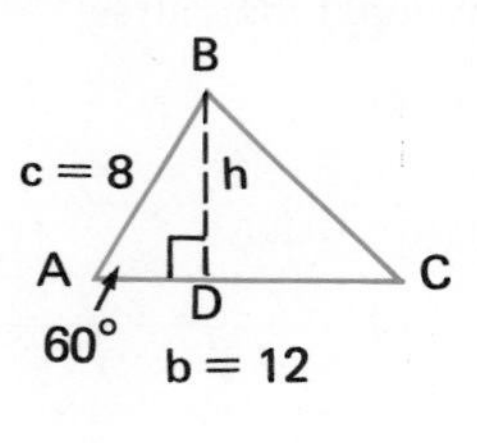

$\sqrt{3} \doteq 1.732$

$24(1.732) = 41.568$

Thus, the area of $\triangle ABC$ is 41.6 sq units.

EXAMPLE 2 Write a formula for finding the area of $\triangle ABC$ in terms of b, c, and $\angle A$.

In right $\triangle ABD$, $\sin A = \frac{h}{c}$, or $h = c \sin A$.

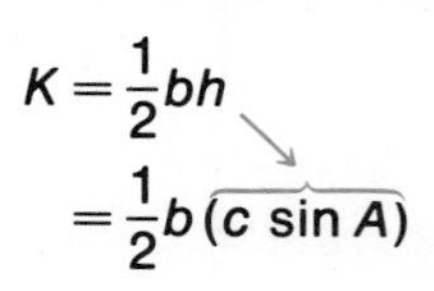

$K = \frac{1}{2}bh$

$= \frac{1}{2}b(c \sin A)$

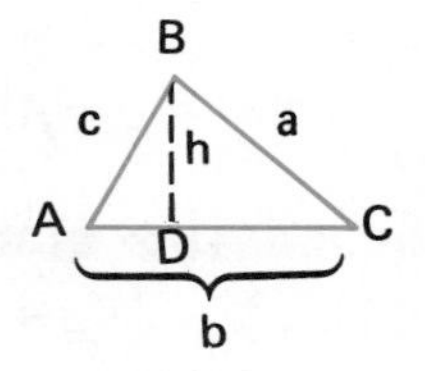

Note: sides b and c are adjacent to $\angle A$.

Thus, a formula for the area of $\triangle ABC$ is $K = \frac{1}{2}bc \sin A$.

The area of a triangle is one half the product of the lengths of any two sides and the sine of the included angle measure.

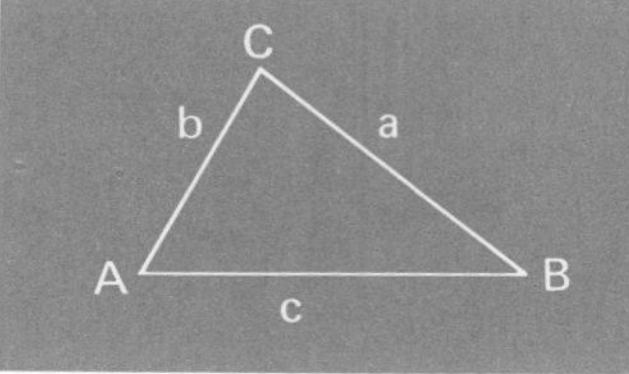

For any triangle ABC,

$K = \frac{1}{2}bc \sin A$

$K = \frac{1}{2}ab \sin C$

$K = \frac{1}{2}ac \sin B$.

EXAMPLE 3 In $\triangle ABC$, $a=6$, $b=8$, $m\angle C=30$. Find the area.

$\sin 30^\circ = \frac{1}{2}$

$$K = \tfrac{1}{2}ab \sin C$$
$$= \tfrac{1}{2}(6)(8) \sin 30^\circ$$
$$= \tfrac{1}{2}(6)(8)(\tfrac{1}{2})$$
$$= 12$$

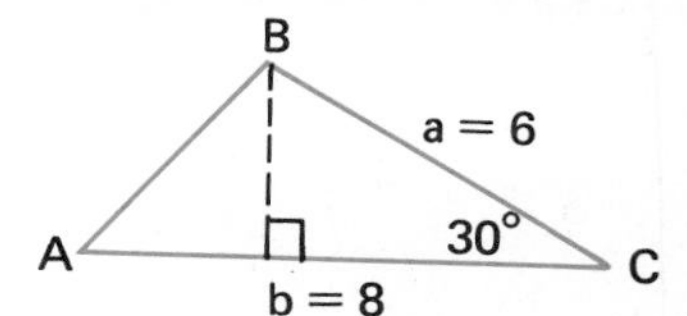

Thus, the area of $\triangle ABC$ is 12 sq units.

EXAMPLE 4 In triangle ABC, $c=4$, $b=6$, $m\angle A=140$. Find the area.

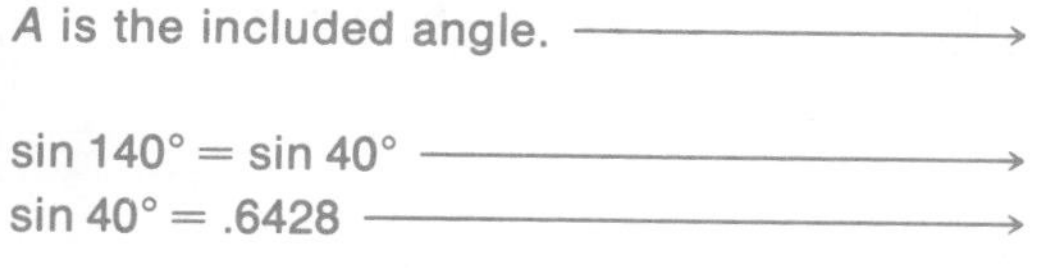

A is the included angle. → $K = \tfrac{1}{2}bc \sin A$

$= \tfrac{1}{2}(6)(4) \sin 140^\circ$

$\sin 140^\circ = \sin 40^\circ$ → $= 12 \sin 40^\circ$

$\sin 40^\circ = .6428$ → $= 12(.6428)$

$= 7.7136$

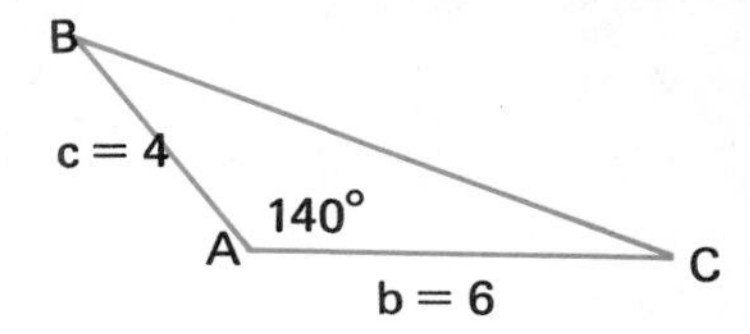

Thus, the area is 8 sq units to the nearest sq unit.

EXAMPLE 5 In $\triangle PQR$, $q=5$, $r=40$. Find the measure of $\angle P$ to the nearest degree if the area is 59 sq units.

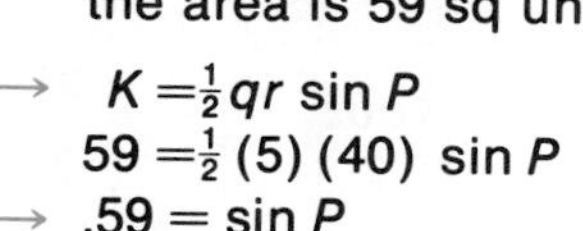

Use an area formula for *q*, *r*, and $\angle P$. → $K = \tfrac{1}{2}qr \sin P$

$59 = \tfrac{1}{2}(5)(40) \sin P$

$\frac{59}{100} = \sin P$ → $.59 = \sin P$

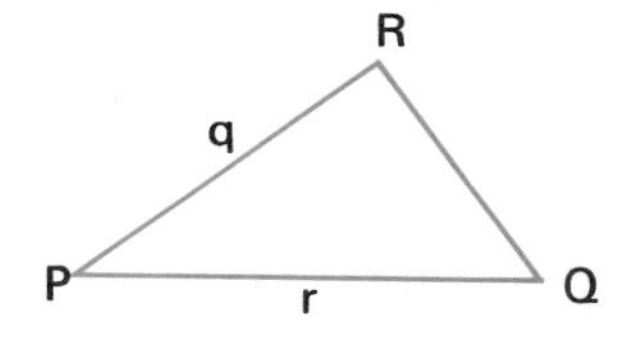

From the table, $m\angle P \doteq 36$. → **Thus,** the measure of $\angle P$ is 36° to the nearest degree.

EXERCISES

PART A

In $\triangle ABC$, find the area to the nearest square unit.

1. $a=4$, $b=8$, $m\angle C=50$ *12*
2. $a=5$, $c=8$, $m\angle B=70$ *19*
3. $b=7$, $c=10$, $m\angle A=40$ *22*
4. $a=6.2$, $b=4.4$, $m\angle C=30$ *7*

In $\triangle PQR$, find the measure of $\angle P$ to the nearest degree.

5. $q=10$, $r=20$, $K=47$ sq units *28°*
6. $q=30$, $r=5$, $K=36$ sq units *29°*
7. $q=18$, $r=7$, $K=27$ sq units *25°*
8. $q=63$, $r=12$, $K=100$ sq units *15°*

PART B

In $\triangle ABC$, find the area to the nearest square unit.

9. $a=1.7$, $b=8.6$, $\angle C=32^\circ 20'$ *4*
10. $b=9.3$, $c=7.6$, $\angle A=100^\circ 18'$ *35*

PART C

Prove.

11. $K=\tfrac{1}{2}ab \sin C$
12. $K=\tfrac{1}{2}ac \sin B$

Law of Sines

OBJECTIVE

- To apply the law of sines in finding unknown measures in a triangle

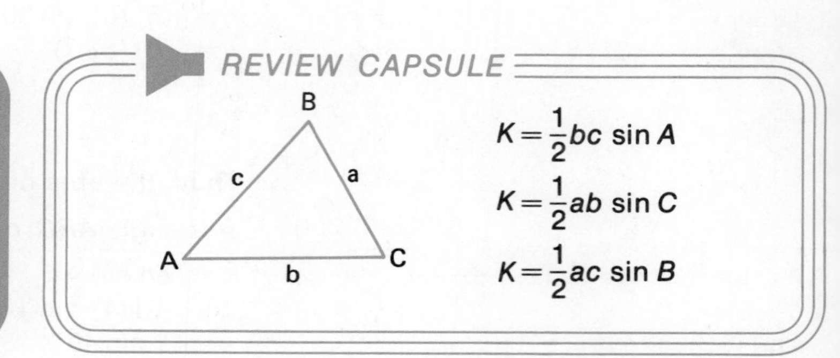

EXAMPLE 1 For $\triangle ABC$, show that $\frac{\sin A}{a}=\frac{\sin B}{b}=\frac{\sin C}{c}$.

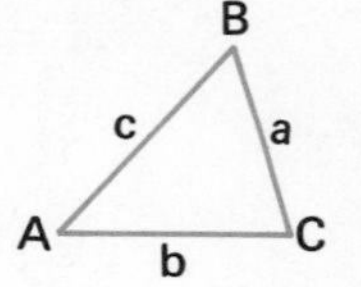

3 ways to write the area of $\triangle ABC$. → $\frac{1}{2}bc \sin A=\frac{1}{2}ab \sin C=\frac{1}{2}ac \sin B$

Multiply each side by 2. → $bc \sin A = ab \sin C = ac \sin B$

Divide each side by abc. → $\frac{bc \sin A}{abc}=\frac{ab \sin C}{abc}=\frac{ac \sin B}{abc}$

Divide out common factors. → **Thus,** $\frac{\sin A}{a}=\frac{\sin C}{c}=\frac{\sin B}{b}$.

Law of sines → In $\triangle ABC$, $\frac{\sin A}{a}=\frac{\sin B}{b}=\frac{\sin C}{c}$.

Can also be written as $\frac{a}{\sin A}=\frac{b}{\sin B}=\frac{c}{\sin C}$

EXAMPLE 2 For $\triangle ABC$, $a = 12$, $m\angle A = 50$, $m\angle C = 45$. Find c to the nearest unit.

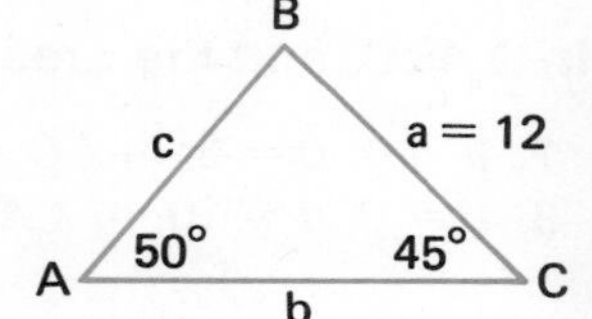

Given a, $m\angle A$, and $m\angle C$

Use the law of sines. → $\frac{\sin A}{a}=\frac{\sin C}{c}$

$a = 12$, $m\angle A = 50$, $m\angle C = 45$ → $\frac{\sin 50°}{12}=\frac{\sin 45°}{c}$

Solve the proportion. → $c \sin 50° = 12 \sin 45°$

Find sin 50° and sin 45°. → $c(.7660) = 12(.7071)$

Divide each side by .7660. → $c=\frac{8.4852}{.7660}$

$c \doteq 11.1$ → **Thus,** $c = 11$ to the nearest unit.

Given the measures of two angles and a side opposite one of the angles, use the law of sines to find the remaining measures of the triangle.

EXAMPLE 3 For $\triangle ABC$, $a = 40$, $b = 100$, $m\angle B = 60$. Find the measure of $\angle A$ to the nearest degree.

Given: a, $m\angle B$, b
Use the law of sines. → $\dfrac{\sin A}{a} = \dfrac{\sin B}{b}$

$m\angle B = 60$, $b = 100$, $a = 40$ → $\dfrac{\sin A}{40} = \dfrac{\sin 60^\circ}{100}$

Solve. → $\sin A = \dfrac{40(\sin 60^\circ)}{100}$

From the table, $\sin 60^\circ = .8660$. → $= \dfrac{40(.8660)}{100}$

$= .3464$

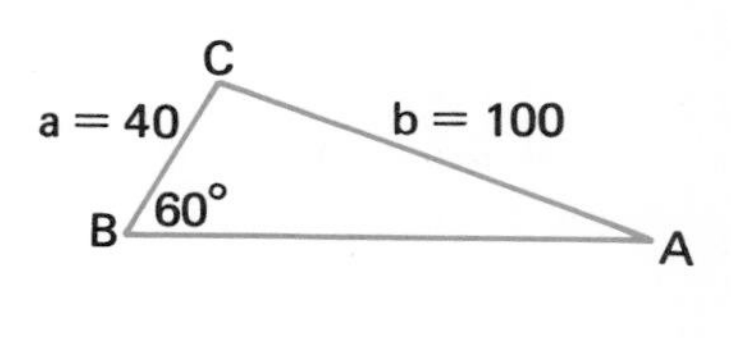

$\sin 20^\circ = .3420$ → **Thus,** the measure of $\angle A$ is 20° to the nearest degree.

EXAMPLE 4 For $\triangle ABC$, $b = 15$, $m\angle A = 60$, $m\angle C = 70$. Find c to the nearest unit.

The sum of the measures of the angles of a triangle is 180°. → First find $m\angle B$.
$m\angle B + 60 + 70 = 180$
$m\angle B = 50$

$\dfrac{\sin B}{b} = \dfrac{\sin C}{c}$ → $\dfrac{\sin 50^\circ}{15} = \dfrac{\sin 70^\circ}{c}$

$c \sin 50^\circ = 15 \sin 70^\circ$

$\sin 50^\circ = .7660$, $\sin 70^\circ = .9397$ → $c(.7660) = 15(.9397)$

$c = \dfrac{14.0955}{.7660}$

$= 18.4$

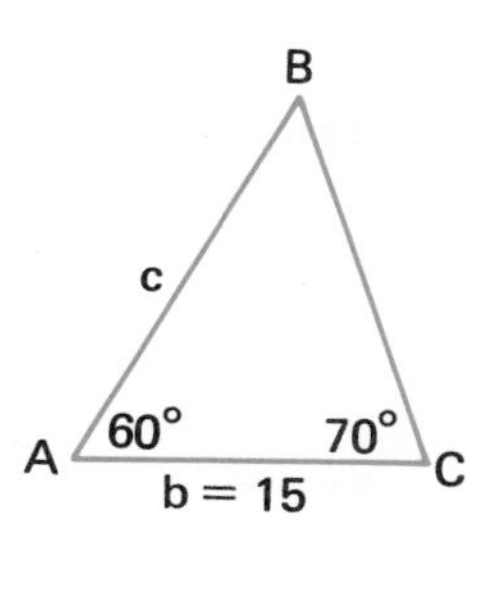

Thus, $c = 18$ to the nearest unit.

EXAMPLE 5 Two boats A and B are on the same side of a lake and are 300 kilometers apart. From boat A to a point C, the angle formed measures 120°20′. From boat B to C, the angle formed measures 28°30′. Find the distance from B to C to the nearest kilometer.

Find the measure of $\angle C$. → $180^\circ - (120^\circ 20' + 28^\circ 30')$
$180^\circ - 148^\circ 50' = 31^\circ 10'$

$\dfrac{\sin A}{a} = \dfrac{\sin C}{c}$ → $\dfrac{\sin 120^\circ 20'}{a} = \dfrac{\sin 31^\circ 10'}{300}$

$300 \sin 120^\circ 20' = a \sin 31^\circ 10'$

$\sin 120^\circ 20' = \sin 59^\circ 40'$, $\sin 59^\circ 40' = .8631$, $\sin 31^\circ 10' = .5175$ → $300(.8631) = a(.5175)$

$258.93 = a(.5175)$

$\dfrac{258.93}{.5175} = a$

Thus, $a = 500$ kilometers to the nearest kilometer.

EXAMPLE 6 For $\triangle ABC$, $m\angle A = 36$, $m\angle C = 48$, $a = 12$. Find c to the nearest unit.

$\frac{\sin A}{a} = \frac{\sin C}{c}$ ⟶ $\frac{\sin 36°}{12} = \frac{\sin 48°}{c}$

$c \sin 36° = 12 \sin 48°$

Divide each side by sin 36°. ⟶ $c = \frac{12 \sin 48°}{\sin 36°}$

Take the log of each side. ⟶ $\log c = \log 12 + \log \sin 48° - \log \sin 36°$

$= 1.0792 + (9.8711 - 10) - (9.7692 - 10)$

$\log 15.2 \doteq 1.811$ ⟶ $\log c \doteq 1.1811$

Thus, $c = 15$ to the nearest unit.

ORAL EXERCISES

Give the form of the law of sines that is needed to find the measure.

1.

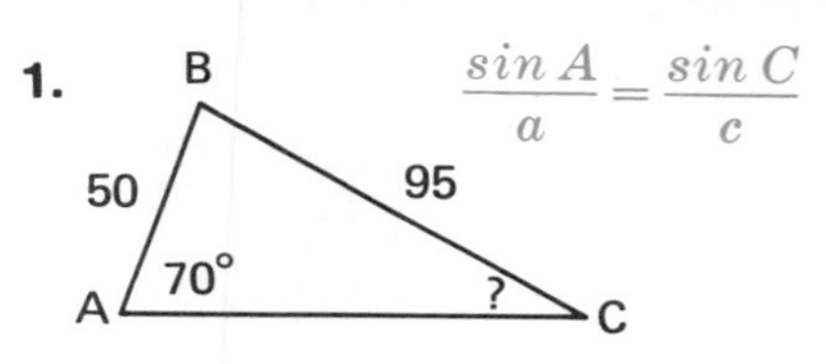

$\frac{\sin A}{a} = \frac{\sin C}{c}$

2.

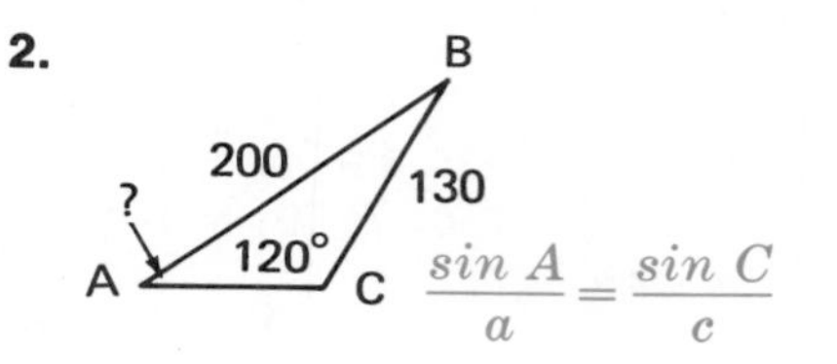

$\frac{\sin A}{a} = \frac{\sin C}{c}$

3.

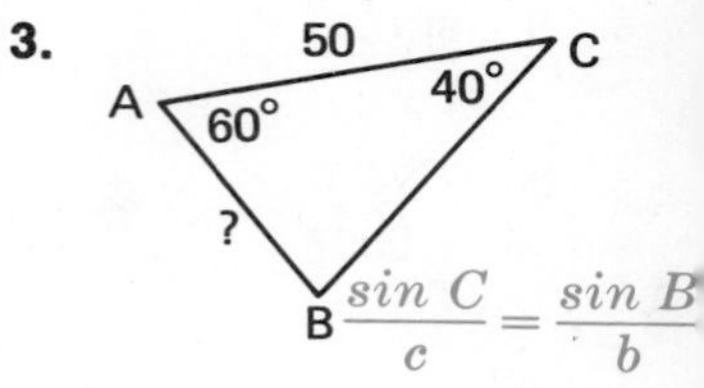

$\frac{\sin C}{c} = \frac{\sin B}{b}$

EXERCISES

PART A

1. For $\triangle ABC$, $a = 20$, $m\angle A = 70$, $m\angle B = 60$. Find b to the nearest unit. *18*
2. For $\triangle ABC$, $b = 35$, $m\angle C = 48$, $m\angle B = 62$. Find c to the nearest unit. *29*
3. For $\triangle ABC$, $a = 32$, $b = 63$, $m\angle B = 58$. Find $m\angle A$ to the nearest degree. *26°*
4. For $\triangle ABC$, $a = 38$, $c = 16$, $m\angle A = 73$. Find $m\angle C$ to the nearest degree.
5. For $\triangle PQR$, $q = 5$, $r = 7$, $m\angle Q = 40$. Find $\sin R$. *.8999*
6. For $\triangle ABC$, $a = 14$, $c = 10$, $m\angle C = 30$. Find $\sin A$. *.7*
7. For $\triangle PQR$, $p = 12$, $\sin P = \frac{1}{3}$ $\sin Q = \frac{3}{5}$. Find q. *22*
8. For $\triangle ABC$, $\sin A = .20$, $\sin B = .7$, $a = 12$. Find b. *42*

PART B

9. Two cars P and Q are parked on the same side of the street, 30 meters apart. From P to R the angle formed is 125°10′. From Q to R the angle formed is 20°40′. Find the distance from P to R to the nearest meter. *19 m*
10. For $\triangle ABC$, $m\angle B = 17$, $m\angle C = 83$, $b = 15$. Use logarithms to find c. *50.93*

PART C

11. For $\triangle RST$, show that $\frac{\sin R}{r} = \frac{\sin T}{t} = \frac{\sin S}{s}$.
12. For $\triangle QRS$, show that $\frac{q}{\sin Q} = \frac{r}{\sin R} = \frac{s}{\sin S}$.

The Ambiguous Case

OBJECTIVE

- To determine the number of triangles which can be constructed given the parts

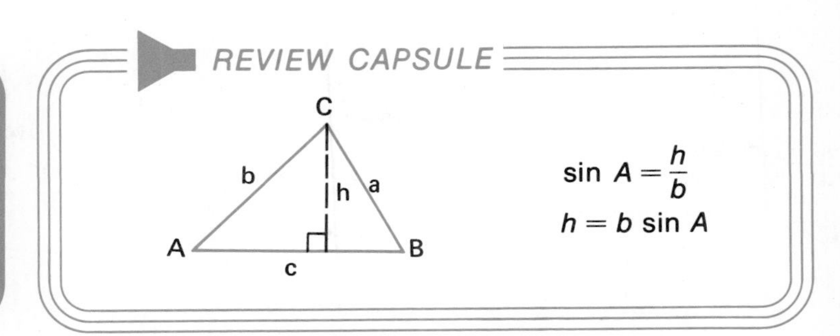

EXAMPLE 1 A is an acute angle and $a < b$ and $a < h$, $(a < b \sin A)$. Show a triangle cannot be constructed.

Check to see if side a touches the base of the $\triangle$. →

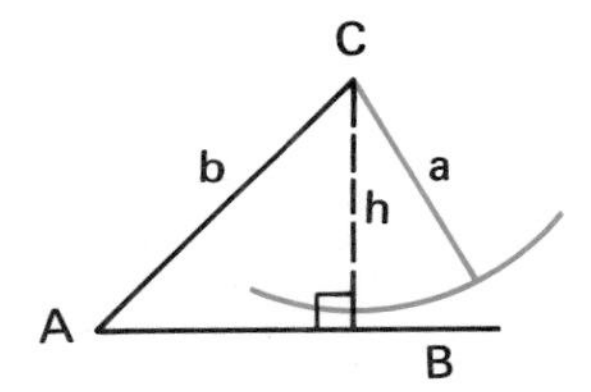

Side a cannot touch the base of the $\triangle$.

No triangle can be constructed. →

Thus, when A is an acute angle
and $a < b$
and $a < h$, $(a < b \sin A)$,
no triangle is formed.

EXAMPLE 2 A is an acute angle and $a < b$. How many triangles can be constructed if $a = h$, $(a = b \sin A)$? if $a > h$, $(a > b \sin A)$?

When $a = h$ and $a < b$, side a touches the base in only one point. One triangle is formed.

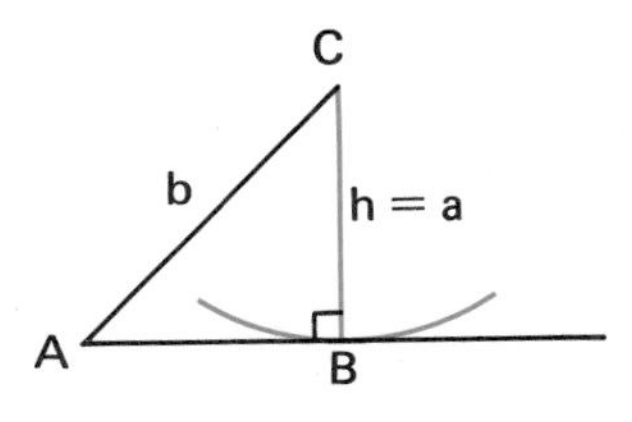

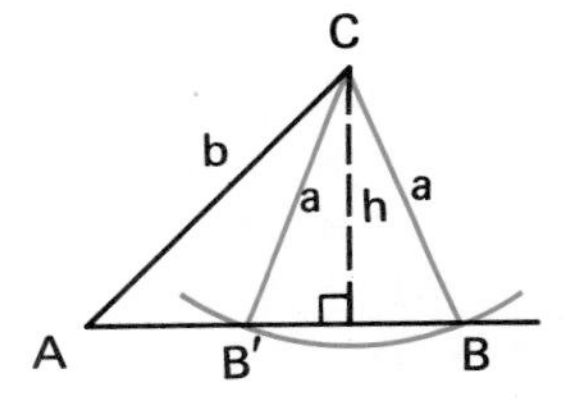

When $a > h$ and $a < b$, side a touches the base in two points. Two triangles are formed.

Thus, when A is an acute angle,

one triangle can be constructed if $a < b$ and $a = h$, $(a = b \sin A)$.	*two* triangles can be constructed if $a < b$ and $a > h$, $(a > b \sin A)$.

EXAMPLE 3 A is an acute angle. How many triangles can be constructed if $a = b$? if $a > b$?

A is an acute angle.
Side a = side b. An isosceles △ is formed.

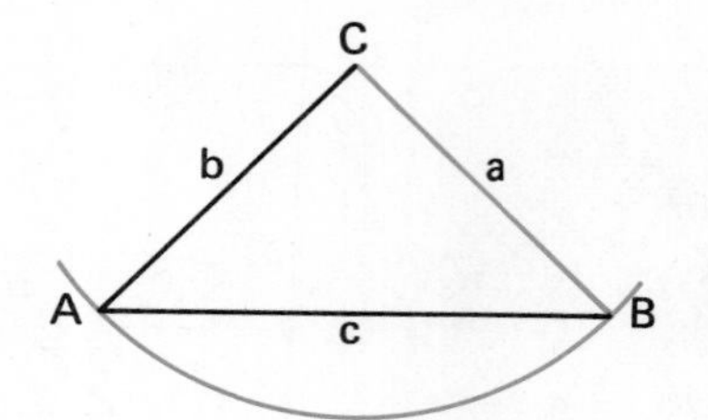

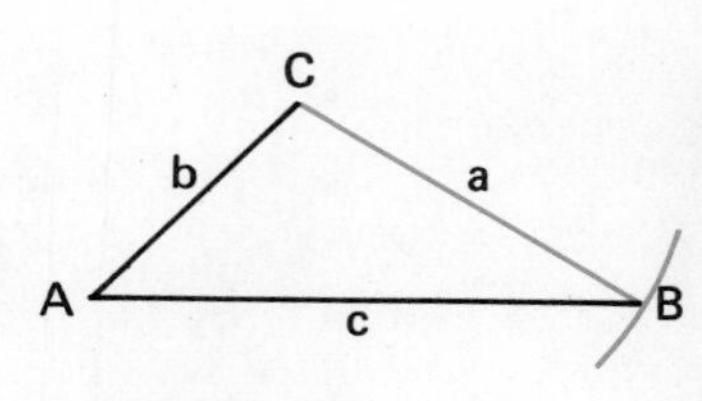

Side $a >$ side b. One △ is formed.

Thus, when A is an acute angle,

one isosceles triangle can be constructed if $a = b$.	one triangle can be constructed if $a > b$.

EXAMPLE 4 A is an obtuse angle or a right angle.
How many triangles can be constructed if $a > b$?

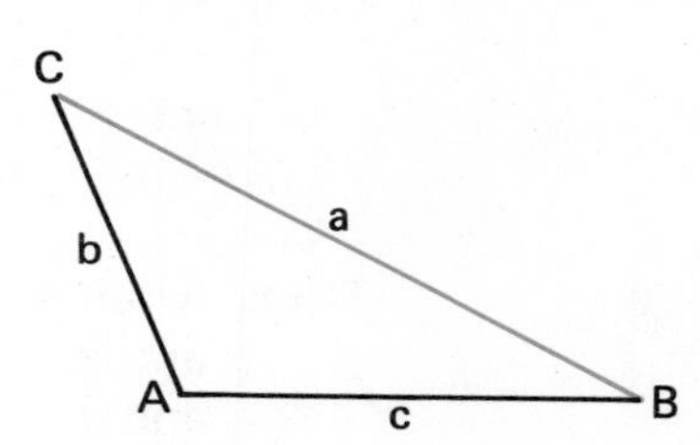

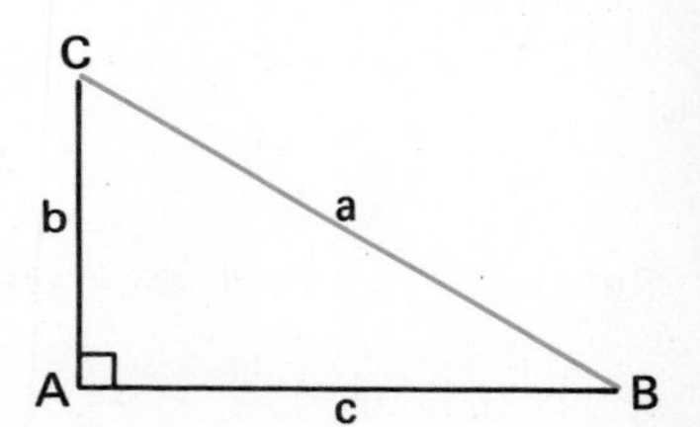

Thus, when A is an obtuse angle or a right angle, one triangle can be constructed if $a > b$.

EXAMPLE 5 A is an obtuse angle or a right angle.
How many triangles can be constructed if $a \leq b$?

A is an obtuse angle.
Side $a <$ side b. No triangle is formed.
A is a right angle.
Side a = side b. No triangle is formed.

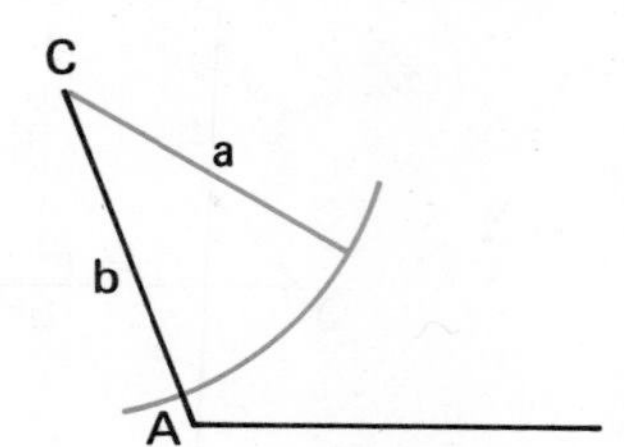

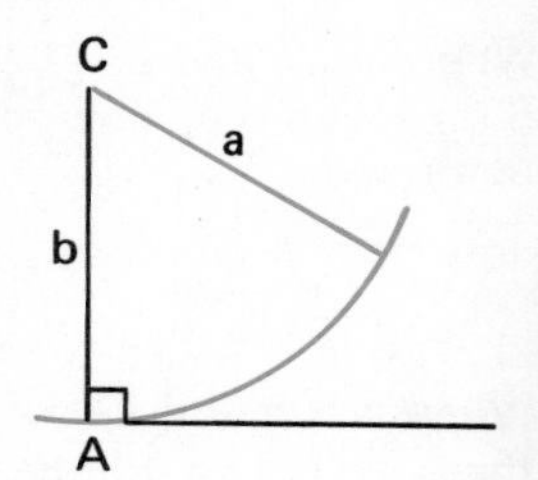

Thus, when A is an obtuse angle or a right angle, no triangle can be constructed if $a \leq b$.

SUMMARY

A is an acute angle and $a < b$:

1. **Two triangles can be constructed if $a > h$, ($a > b \sin A$).**
2. **One triangle can be constructed if $a = h$, ($a = b \sin A$).**
3. **No triangle can be constructed if $a < h$, ($a < b \sin A$).**

A is an acute angle:

1. **One triangle can be constructed if $a > b$.**
2. **One triangle can be constructed if $a = b$.**

A is an obtuse angle or a right angle:

1. **One triangle can be constructed if $a > b$.**
2. **No triangle can be constructed if $a = b$.**
3. **No triangle can be constructed if $a < b$.**

EXAMPLE 6 How many triangles can be constructed if $m\angle A = 120$, $a = 30$ and $b = 50$?

A is an obtuse angle and $a < b$.

Thus, no triangle can be constructed.

EXAMPLE 7 How many triangles can be constructed if $m\angle A = 40$, $a = 20$ and $b = 30$?

A is an acute angle and $a < b$.

Find h.

$$\begin{aligned} h &= b \sin A \\ &= 30 \sin 40 \\ &= 30(.6428) \\ h &= 19.284 \end{aligned}$$

Since $a > h$, side a touches the base in two points. → **Thus,** two triangles can be constructed.

EXERCISES

PART A

How many triangles can be constructed?

1. $m\angle A = 30$, $a = 15$, $b = 20$ *2*
2. $m\angle A = 150$, $a = 50$, $b = 20$ *1*
3. $m\angle A = 40$, $a = 20$, $b = 40$ *none*
4. $m\angle A = 100$, $a = 12$, $b = 6$ *1*
5. $m\angle A = 10$, $a = 5$, $b = 10$ *2*
6. $m\angle A = 123$, $a = 60$, $b = 70$ *none*

PART B

How many triangles can be constructed?

7. $m\angle A = 90$, $a = 10$, $b = 10$ *none*
8. $m\angle A = 90$, $a = 17$, $b = 30$ *none*
9. $m\angle A = 90$, $a = 35$, $b = 20$ *1*
10. $m\angle A = 90$, $a = 32$, $b = 32$ *none*
11. $m\angle B = 63$, $b = 17$, $c = 39$ *none*
12. $m\angle B = 150$, $b = 19$, $c = 19$ *none*

Sin $(x \pm y)$ and Sin $2x$

OBJECTIVE

- To apply the formulas for $\sin(x \pm y)$ and $\sin 2x$

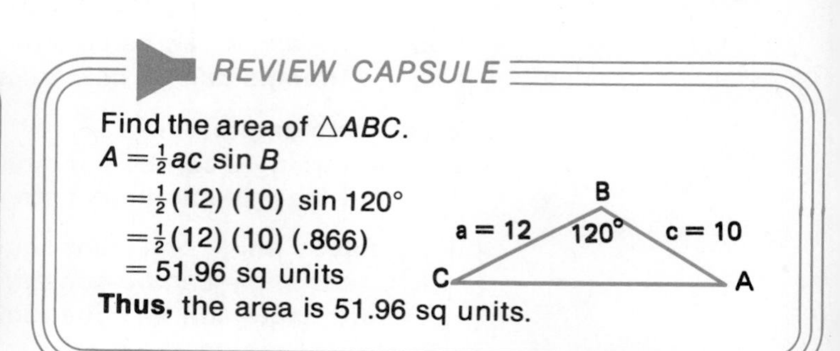

REVIEW CAPSULE

Find the area of $\triangle ABC$.

$A = \frac{1}{2}ac \sin B$

$= \frac{1}{2}(12)(10) \sin 120°$

$= \frac{1}{2}(12)(10)(.866)$

$= 51.96$ sq units

Thus, the area is 51.96 sq units.

EXAMPLE 1 Write the formulas for the areas of each of the triangles: $\triangle ABC$, $\triangle ABD$ and $\triangle CBD$.

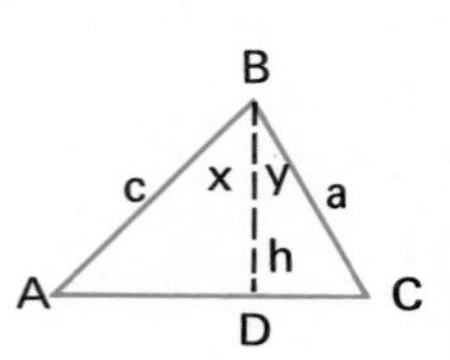

Use $\angle B$ and sides a and c. ⟶ Area $\triangle ABC$: $A = \frac{1}{2}ac \sin B$

Use $\angle x$ and sides c and h. ⟶ Area $\triangle ABD$: $A = \frac{1}{2}ch \sin x$

Use $\angle y$ and sides h and a. ⟶ Area $\triangle CBD$: $A = \frac{1}{2}ah \sin y$

EXAMPLE 2 Show that $\sin(x + y) = \sin x \cos y + \cos x \sin y$.

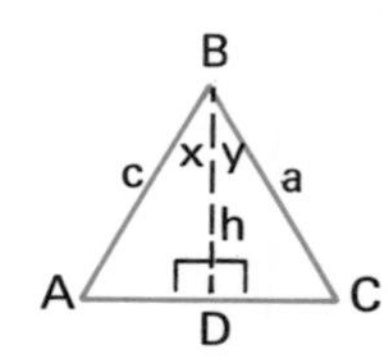

Area $\triangle ABC$ = area $\triangle ABD$ + area $\triangle CBD$

$m\angle B = m\angle x + m\angle y$ ⟶ $\frac{1}{2}ac \sin(x + y) = \frac{1}{2}ch \sin x + \frac{1}{2}ah \sin y$

Multiply each side by 2. Divide each side by ac.

$$\sin(x + y) = \frac{ch}{ac}\sin x + \frac{ah}{ac}\sin y$$

In rt. $\triangle CBD$, $\frac{h}{a} = \cos y$; In rt. $\triangle ABD$, $\frac{h}{c} = \cos x$ ⟶

$$= \frac{h}{a}\sin x + \frac{h}{c}\sin y$$

$$= \cos y \sin x + \cos x \sin y$$

$$= \sin x \cos y + \cos x \sin y$$

Thus, $\sin(x + y) = \sin x \cos y + \cos x \sin y$.

$$\sin(x + y) = \sin x \cos y + \cos x \sin y$$

EXAMPLE 3 Find $\sin 75°$. Use the formula for $\sin(x+y)$. Answer may be left in radical form.

Write 75° as 30° + 45°. ⟶

$$\sin(x+y) = \sin x \cos y + \cos x \sin y$$
$$\sin(30° + 45°) = \sin 30° \cos 45° + \cos 30° \sin 45°$$
$$= \frac{1}{2} \cdot \frac{\sqrt{2}}{2} + \frac{\sqrt{3}}{2} \cdot \frac{\sqrt{2}}{2}$$
$$= \frac{\sqrt{2} + \sqrt{6}}{4}$$

Thus, $\sin 75° = \frac{\sqrt{2}+\sqrt{6}}{4}$.

EXAMPLE 4 If $\sin A = \frac{12}{13}$ and $\sin B = \frac{3}{5}$, where A and B are acute angles, find the value of $\sin(A+B)$.

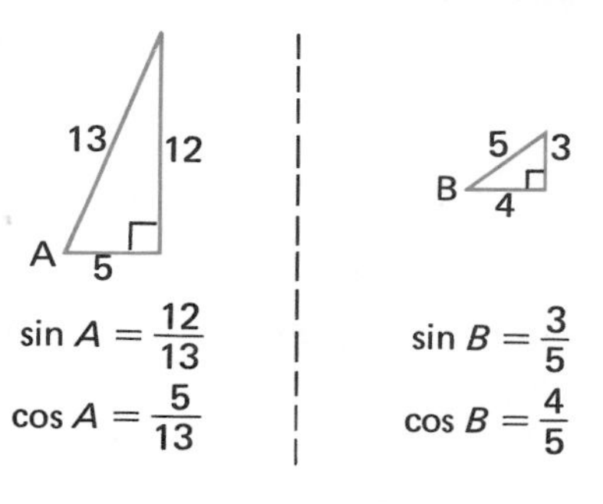

$\sin A = \frac{12}{13}$ $\quad$ $\sin B = \frac{3}{5}$

$\cos A = \frac{5}{13}$ $\quad$ $\cos B = \frac{4}{5}$

$$\sin(A+B) = \sin A \cos B + \cos A \sin B$$
$$= \left(\frac{12}{13}\right)\left(\frac{4}{5}\right) + \left(\frac{5}{13}\right)\left(\frac{3}{5}\right)$$
$$= \frac{48}{65} + \frac{15}{65}, \text{ or } \frac{63}{65}$$

Thus, $\sin(A+B) = \frac{63}{65}$.

EXAMPLE 5 Show that $\sin(x-y) = \sin x \cos y - \cos x \sin y$.

Rewrite $x - y$ as $x + (-y)$ ⟶

Use the formula for $\sin(x+y)$. ⟶

$\cos(-y) = \cos y$, $\sin(-y) = -\sin y$ ⟶

$$\sin(x-y) = \sin[x + (-y)]$$
$$= \sin x \cos(-y) + \cos x \sin(-y)$$
$$= \sin x \cos y + \cos x(-\sin y)$$
$$= \sin x \cos y - \cos x \sin y$$

Thus, $\sin(x-y) = \sin x \cos y - \cos x \sin y$.

$$\sin(x-y) = \sin x \cos y - \cos x \sin y$$

EXAMPLE 6 Find $\sin 15°$. Use the formula for $\sin(x-y)$. Answer may be left in radical form.

Write 15° as 45° − 30°. ⟶

$$\sin(45° - 30°) = \sin 45° \cos 30° - \cos 45° \sin 30°$$
$$= \frac{\sqrt{2}}{2} \cdot \frac{\sqrt{3}}{2} - \frac{\sqrt{2}}{2} \cdot \frac{1}{2}$$
$$= \frac{\sqrt{6} - \sqrt{2}}{4}$$

15° can also be written as 60° − 45°.

Thus, $\sin 15° = \frac{\sqrt{6}-\sqrt{2}}{4}$

EXAMPLE 7 If $\sin A = \frac{12}{13}$ and A lies in the 2nd quadrant and $\cos B = \frac{4}{5}$ and B lies in the 1st quadrant, find the value of $\sin (A - B)$.

Draw reference triangles for $\angle A$ and $\angle B$.

In the 2nd quadrant, cosine is negative.

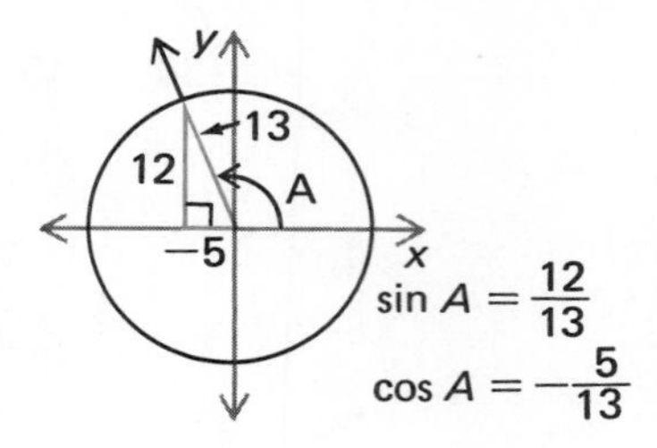

$\sin A = \frac{12}{13}$

$\cos A = -\frac{5}{13}$

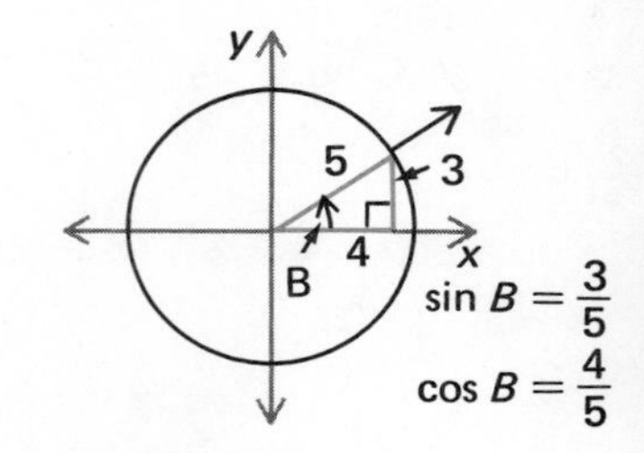

$\sin B = \frac{3}{5}$

$\cos B = \frac{4}{5}$

$$\sin (A - B) = \sin A \cos B - \cos A \sin B$$
$$= \left(\frac{12}{13}\right)\left(\frac{4}{5}\right) - \left(-\frac{5}{13}\right)\left(\frac{3}{5}\right) = \frac{48}{65} + \frac{15}{65}$$

Thus, $\sin (A - B) = \frac{63}{65}$.

EXAMPLE 8 Find $\sin 120°$. Use the formula for $\sin (x + y)$.

Write 120° as 60° + 60°. ⟶ $\sin (120°) = \sin (60° + 60°)$

Formula for $\sin (x + y)$ ⟶ $= \sin 60° \cos 60° + \cos 60° \sin 60°$

Combine terms. ⟶ $= 2 \sin 60° \cos 60°$

$\sin 60° = \frac{\sqrt{3}}{2}$; $\cos 60° = \frac{1}{2}$ ⟶ $= 2\left(\frac{\sqrt{3}}{2}\right)\left(\frac{1}{2}\right)$

$2\left(\frac{\sqrt{3}}{2}\right)\left(\frac{1}{2}\right) = \frac{\sqrt{3}}{2}$

Thus, $\sin 120° = \frac{\sqrt{3}}{2}$.

EXAMPLE 9 Show that $\sin 2A = 2 \sin A \cos A$.

Write $2A$ as $A + A$. ⟶ $\sin 2A = \sin (A + A)$

Formula for $\sin (x + y)$. ⟶ $= \sin A \cos A + \cos A \sin A$

$= 2 \sin A \cos A$

$$\sin 2x = 2 \sin x \cos x$$

EXAMPLE 10 If $\cos A = \frac{-12}{13}$ and A lies in the second quadrant, find $\sin 2A$.

Draw the reference triangle.

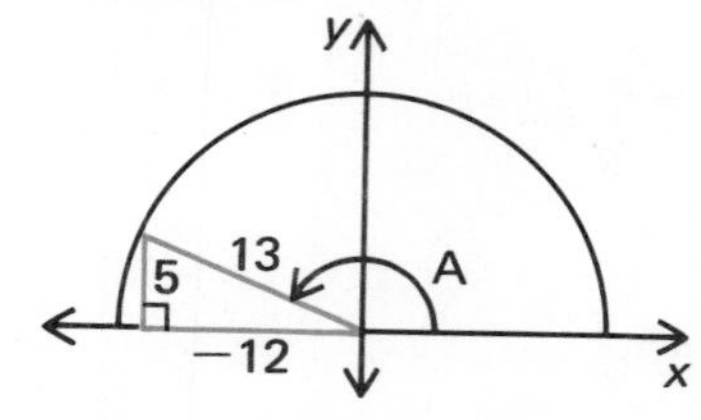

$\sin A = \frac{5}{13}$

$$\sin 2A = 2 \sin A \cos A$$
$$= 2\left(\frac{5}{13}\right)\left(\frac{-12}{13}\right)$$
$$= \frac{-120}{169}$$

Thus, $\sin 2A = -\frac{120}{169}$.

EXERCISES

PART A

Find each. Use the formula for sin $(x + y)$ or sin $(x - y)$. Answers may be left in radical form.

1. sin 105° **2.** sin 120° **3.** sin 75° **4.** sin 15° **5.** sin 60° $\frac{\sqrt{3}}{2}$
6. sin 135° $\frac{\sqrt{2}}{2}$ **7.** sin 150° $\frac{1}{2}$ **8.** sin 90° 1 **9.** sin 30° $\frac{1}{2}$ **10.** sin 180° 0

Find each. Use the formula for sin $2x$. Answers may be left in radical form.

11. sin 90° 1 **12.** sin 60° $\frac{\sqrt{3}}{2}$ **13.** sin 180° 0 **14.** sin 120° $\frac{\sqrt{3}}{2}$

Find the value of sin $(A + B)$ and sin $(A - B)$.

15. Sin $A = \frac{4}{5}$ and A lies in the 2nd quadrant, sin $B = \frac{5}{13}$ and B lies in the 1st quadrant.

16. Sin $A = \frac{4}{5}$ and cos $B = \frac{12}{13}$, A and B are both acute angles.

17. Sin $A = \frac{8}{17}$ and cos $B = \frac{3}{5}$. A and B are both acute angles.

18. Sin $A = \frac{12}{13}$ and cos $B = \frac{5}{13}$. A and B are both acute angles.

19. Sin $A = \frac{15}{17}$ and A lies in the 2nd quadrant. sin $B = \frac{12}{13}$ and B lies in the 1st quadrant.

20. Sin $A = \frac{3}{5}$ and A lies in the 2nd quadrant. cos $B = \frac{5}{13}$ and B lies in the 1st quadrant.

Find the value of sin $2A$.

21. Sin $A = -\frac{3}{5}$ and A lies in the 3rd quadrant.

22. Sin $A = \frac{2}{3}$ and A lies in the 2nd quadrant.

23. Sin $A = \frac{5}{7}$ and A is an acute angle.

24. Sin $A = -\frac{2}{\sqrt{13}}$ and A lies in the 4th quadrant.

PART B

Simplify each of the following for acute angle A.

25. $\frac{2 \sin A}{\sin 2A}$ $\sec A$ **26.** $\frac{\cos A}{\sin 2A}$ $\frac{1}{2} \csc A$ **27.** $\frac{\sin 2A}{\cos^2 A}$ $2 \tan A$

PART C

Prove the identity. Use the formulas for sin $(x \pm y)$.

28. $\sin (90° + u) = \cos u$ **29.** $\sin (90° - u) = \cos u$ **30.** $\sin (180° + u) = -\sin u$
31. $\sin (180° - u) = \sin u$ **32.** $\sin (270° - u) = -\cos u$ **33.** $\sin (360° - u) = -\sin u$

Cos $(x \pm y)$ and Cos $2x$

OBJECTIVE

- To apply the formulas for $\cos(x \pm y)$ and $\cos 2x$

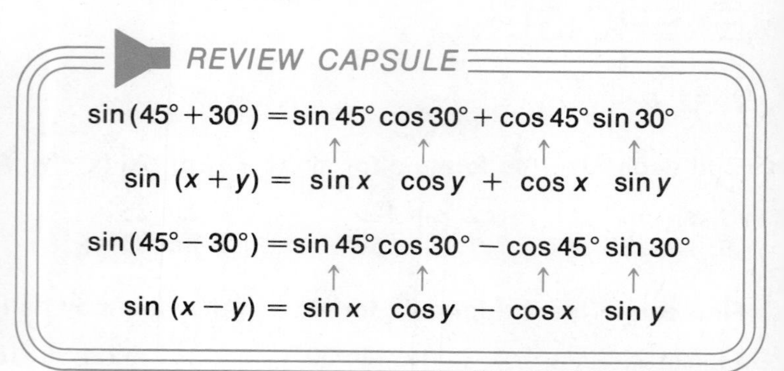

REVIEW CAPSULE

$$\sin(45° + 30°) = \sin 45° \cos 30° + \cos 45° \sin 30°$$

$$\sin(x + y) = \sin x \cos y + \cos x \sin y$$

$$\sin(45° - 30°) = \sin 45° \cos 30° - \cos 45° \sin 30°$$

$$\sin(x - y) = \sin x \cos y - \cos x \sin y$$

EXAMPLE 1 Show that $\cos(x + y) = \cos x \cos y - \sin x \sin y$.

$\cos u = \sin(90° - u)$ →
$$\cos(x + y) = \sin[90° - (x + y)]$$

$90 - (x + y) = 90 - x - y$ →
$$= \sin[(90° - x) - y]$$

Use the formula for $\sin(x - y)$. →
$$= \sin(90° - x)\cos y - \cos(90° - x)\sin y$$

$\cos(90° - x) = \sin x$ →
$$= \cos x \cos y - \sin x \sin y$$

Thus, $\cos(x + y) = \cos x \cos y - \sin x \sin y$.

$$\cos(x + y) = \cos x \cos y - \sin x \sin y$$

EXAMPLE 2 Find $\cos 105°$. Use the formula for $\cos(x + y)$. Answer may be left in radical form.

$$\cos(x + y) = \cos x \cos y - \sin x \sin y$$

Replace $105°$ by $60° + 45°$. →
$$\cos(60° + 45°) = \cos 60° \cos 45° - \sin 60° \sin 45°$$

$$= \left(\frac{1}{2}\right)\left(\frac{\sqrt{2}}{2}\right) - \left(\frac{\sqrt{3}}{2}\right)\left(\frac{\sqrt{2}}{2}\right)$$

$$= \frac{\sqrt{2} - \sqrt{6}}{4}$$

Thus, $\cos 105° = \dfrac{\sqrt{2} - \sqrt{6}}{4}$.

EXAMPLE 3 If $\cos A = \frac{-12}{13}$ and A is in the second quadrant and $\cos B = \frac{5}{13}$ and B is in the first quadrant, find the value of $\cos(A + B)$.

$$\sin A = \frac{5}{13} \qquad \sin B = \frac{12}{13}$$

$$\begin{aligned}\cos(A+B) &= \cos A \cos B - \sin A \sin B \\ &= \left(\frac{-12}{13}\right)\left(\frac{5}{13}\right) - \left(\frac{5}{13}\right)\left(\frac{12}{13}\right) \\ &= \frac{-60}{169} - \frac{60}{169} \\ &= \frac{-120}{169}\end{aligned}$$

Thus, $\cos(A + B) = -\frac{120}{169}$.

EXAMPLE 4 Show that $\cos(x - y) = \cos x \cos y + \sin x \sin y$.

Rewrite $x - y$ as $x + (-y)$.

$\cos(-y) = \cos y$
$\sin(-y) = -\sin y$

$$\begin{aligned}\cos(x-y) &= \cos[x + (-y)] \\ &= \cos x \cos(-y) - \sin x \sin(-y) \\ &= \cos x \quad \cos y \quad - \quad \sin x \quad (-\sin y) \\ &= \cos x \cos y + \sin x \sin y\end{aligned}$$

Thus, $\cos(x - y) = \cos x \cos y + \sin x \sin y$.

$$\cos(x - y) = \cos x \cos y + \sin x \sin y$$

EXAMPLE 5 If $\sin A = \frac{-8}{17}$ and A lies in the third quadrant and $\tan B = \frac{3}{4}$ and B lies in the first quadrant, find $\cos(A - B)$.

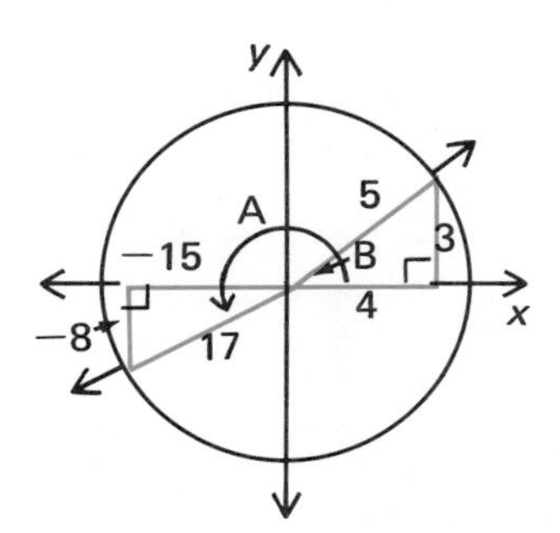

$$\sin A = \frac{-8}{17} \qquad \sin B = \frac{3}{5}$$

$$\cos A = \frac{-15}{17} \qquad \cos B = \frac{4}{5}$$

$$\begin{aligned}\cos(A-B) &= \cos A \cos B + \sin A \sin B \\ &= \left(\frac{-15}{17}\right)\left(\frac{4}{5}\right) + \left(\frac{-8}{17}\right)\left(\frac{3}{5}\right) \\ &= \frac{-60}{85} + \frac{-24}{85}, \text{ or } \frac{-84}{85}\end{aligned}$$

Thus, $\cos(A - B) = -\frac{84}{85}$.

EXAMPLE 6 Find $\cos 120°$. Use the formula for $\cos(A + B)$.

Write $120°$ as $60° + 60°$. → Formula for $\cos(x + y)$ → $\cos 60° = \frac{1}{2}$; $\sin 60° = \frac{\sqrt{3}}{2}$ →

$$\begin{aligned}\cos 120° &= \cos(60° + 60°)\\ &= \cos 60° \cos 60° - \sin 60° \sin 60°\\ &= \cos^2 60° - \sin^2 60°\\ &= \left(\frac{1}{2}\right)^2 - \left(\frac{\sqrt{3}}{2}\right)^2\\ &= \frac{1}{4} - \frac{3}{4}\end{aligned}$$

Thus, $\cos 120° = -\frac{1}{2}$.

EXAMPLE 7 Show that $\cos 2A = \cos^2 A - \sin^2 A$.

Write $2A$ as $A + A$. → Formula for $\cos(x + y)$ →

$$\begin{aligned}\cos 2A &= \cos(A + A)\\ &= \cos A \cos A - \sin A \sin A\\ &= \cos^2 A - \sin^2 A\end{aligned}$$

$$\cos 2x = \cos^2 x - \sin^2 x$$

EXAMPLE 8 If $\cos\theta = \frac{-3}{5}$ and θ lies in the second quadrant, find $\cos 2\theta$.

If $\cos\theta = \frac{-3}{5}$, $\sin\theta = \frac{4}{5}$. →

$$\begin{aligned}\cos 2\theta &= \cos^2\theta - \sin^2\theta\\ &= \left(\frac{-3}{5}\right)^2 - \left(\frac{4}{5}\right)^2\\ &= \frac{9}{25} - \frac{16}{25}\end{aligned}$$

Thus, $\cos 2\theta = -\frac{7}{25}$.

EXAMPLE 9 Write $\cos 2A = \cos^2 A - \sin^2 A$ in two other ways using the Pythagorean identity $\sin^2 A + \cos^2 A = 1$.

If $\sin^2 A + \cos^2 A = 1$, then $\cos^2 A = 1 - \sin^2 A$ and $\sin^2 A = 1 - \cos^2 A$.

$\cos^2 A - (1 - \cos^2 A) =$ $\cos^2 A - 1 + \cos^2 A$

Replace $\cos^2 A$ with $1 - \sin^2 A$.

$$\begin{aligned}\cos 2A &= \cos^2 A - \sin^2 A\\ &= (1 - \sin^2 A) - \sin^2 A\\ &= 1 - 2\sin^2 A\end{aligned}$$

Replace $\sin^2 A$ with $1 - \cos^2 A$.

$$\begin{aligned}\cos 2A &= \cos^2 A - \sin^2 A\\ &= \cos^2 A - (1 - \cos^2 A)\\ &= 2\cos^2 A - 1\end{aligned}$$

Three ways of expressing $\cos 2A$

$$\cos 2A = \cos^2 A - \sin^2 A$$
$$\cos 2A = 1 - 2\sin^2 A$$
$$\cos 2A = 2\cos^2 A - 1$$

EXERCISES

PART A

Find the value. Use the formula for cos $(x+y)$. Answers may be left in radical form.

1. $\cos(45° + 45°)$ *0*
2. $\cos 60°$ $\frac{1}{2}$
3. $\cos 120°$ $-\frac{1}{2}$
4. $\cos 105°$
5. $\cos 75°$
6. $\cos 135°$ $-\frac{\sqrt{2}}{2}$
7. $\cos 90°$ *0*
8. $\cos 180°$ *−1*

Find the value. Use the formula for cos $(x-y)$. Answers may be left in radical form.

9. $\cos 60°$ $\frac{1}{2}$
10. $\cos 30°$ $\frac{\sqrt{3}}{2}$
11. $\cos 120°$ $-\frac{1}{2}$
12. $\cos 0°$ *1*
13. $\cos 105°$
14. $\cos 75°$
15. $\cos 15°$ $\frac{\sqrt{6}+\sqrt{2}}{4}$
16. $\cos 90°$ *0*

Find each. Use the formula for cos 2x. Answers may be left in radical form.

17. $\cos 60°$ $\frac{1}{2}$
18. $\cos 90°$ *0*
19. $\cos 120°$ $-\frac{1}{2}$
20. $\cos 180°$ *−1*

Find the value of cos $(A+B)$ and cos $(A-B)$.

21. Cos $A = -\frac{4}{5}$ and A lies in the 2nd quadrant. Cos $B = \frac{12}{13}$ and B lies in the 1st quadrant.
22. Cos $A = -\frac{5}{13}$ and A lies in the 2nd quadrant. Cos $B = \frac{4}{5}$ and B lies in the 1st quadrant.
23. Cos $A = \frac{12}{13}$ and cos $B = \frac{5}{13}$. A and B are both acute angles.
24. Cos $A = \frac{11}{12}$ and cos $B = \frac{4}{7}$. A and B are both acute angles.
25. Cos $A = -\frac{12}{13}$ and A is in the 2nd quadrant. Cos $B = \frac{5}{13}$ and B lies in the 1st quadrant. $-\frac{120}{169}$; *0*
26. Cos $A = -\frac{12}{13}$ and A is in the 2nd quadrant. Cos $B = \frac{3}{5}$ and B lies in the 1st quadrant.

Find the value of cos 2A.

27. Cos $A = -\dfrac{12}{13}$ and A lies in the 2nd quadrant.
28. Cos $A = \dfrac{5}{13}$ and A lies in the 4th quadrant.
29. Cos $A = \dfrac{4}{5}$ and A is an acute angle. $\frac{7}{25}$
30. Cos $A = -\dfrac{3}{5}$ and A lies in the 3rd quadrant. $-\frac{7}{25}$

PART B

Evaluate.

31. $\dfrac{1-2\sin^2 A}{\cos 2A}$ *1*
32. $\dfrac{\cos^2 A - \sin^2 A}{2\cos^2 A - 1}$ *1*
33. $\dfrac{\cos 2A}{\sin^2 A - \cos^2 A}$ *−1*

PART C

Prove the identity. Use the formulas for cos $(x \pm y)$.

34. $\cos(90° + u) = -\sin u$
35. $\cos(180° - u) = -\cos u$
36. $\cos(90° - u) = \sin u$
37. $\cos(270° + u) = \sin u$
38. $\cos(180° + u) = -\cos u$
39. $\cos(360° - u) = \cos u$

Tan $(x \pm y)$ and Tan $2x$

OBJECTIVE

- To apply the formulas for $\tan(x \pm y)$ and $\tan 2x$

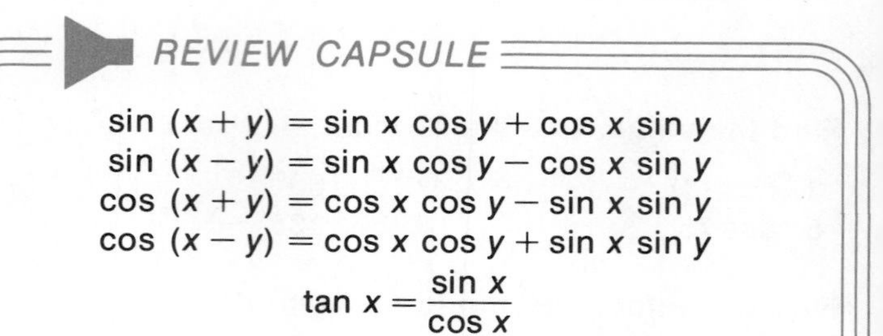

EXAMPLE 1 Show that $\tan(x+y) = \frac{\tan x + \tan y}{1 - \tan x \tan y}$.

$\tan A = \frac{\sin A}{\cos A}$ ⟶ $$\tan(x+y) = \frac{\sin(x+y)}{\cos(x+y)}$$

Use the formula for $\sin(x+y)$. ⟶
Use the formula for $\cos(x+y)$. ⟶ $$= \frac{\sin x \cos y + \cos x \sin y}{\cos x \cos y - \sin x \sin y}$$

Divide each term by $\cos x \cos y$ and simplify. ⟶ $$= \frac{\frac{\sin x \cos y}{\cos x \cos y} + \frac{\cos x \sin y}{\cos x \cos y}}{\frac{\cos x \cos y}{\cos x \cos y} - \frac{\sin x \sin y}{\cos x \cos y}}$$

$\tan x = \frac{\sin x}{\cos x}$; $\tan y = \frac{\sin y}{\cos y}$ $$= \frac{\tan x + \tan y}{1 - \tan x \tan y}$$

Thus, $\tan(x+y) = \frac{\tan x + \tan y}{1 - \tan x \tan y}$.

EXAMPLE 2 Show that $\tan(x-y) = \frac{\tan x - \tan y}{1 + \tan x \tan y}$.

$x - y = x + (-y)$ ⟶ $$\tan[x+(-y)] = \frac{\tan x + \tan(-y)}{1 - \tan x \tan(-y)}$$

$\tan(-y) = -\tan y$ $$= \frac{\tan x - \tan y}{1 + \tan x \tan y}$$

Thus, $\tan(x-y) = \frac{\tan x - \tan y}{1 + \tan x \tan y}$.

Tangent of the sum and the difference of two angles →

$$\tan(x+y) = \frac{\tan x + \tan y}{1 - \tan x \tan y}$$
$$\tan(x-y) = \frac{\tan x - \tan y}{1 + \tan x \tan y}$$

EXAMPLE 3 Find tan 105°. Use the formula for $\tan(x+y)$. Answer may be left in radical form.

Write 105° as 60° + 45°. Use formula for $\tan(x+y)$. →

$$\tan(60° + 45°) = \frac{\tan 60° + \tan 45°}{1 - \tan 60° \tan 45°}$$

$\tan 60° = \sqrt{3}$; $\tan 45° = 1$ →

$$= \frac{\sqrt{3}+1}{1-(\sqrt{3})(1)}$$

Thus, $\tan 105° = \frac{\sqrt{3}+1}{1-\sqrt{3}}$.

EXAMPLE 4 If $\tan A = \frac{3}{4}$ and $\tan B = \frac{1}{5}$, find $\tan(A-B)$.

Substitute given values in $\tan(x-y) = \frac{\tan x - \tan y}{1 + \tan x \tan y}$.

$$\tan(A-B) = \frac{\frac{3}{4} - \frac{1}{5}}{1 + \frac{3}{4}(\frac{1}{5})}, \text{ or } \frac{\frac{11}{20}}{\frac{23}{20}}$$

Thus, $\tan(A-B) = \frac{11}{23}$.

EXAMPLE 5 Find tan 120°. Use the formula for $\tan(x+y)$.

$120° = 60° + 60°$ →

$$\tan(60° + 60°) = \frac{\tan 60° + \tan 60°}{1 - \tan 60° \tan 60°}$$
$$= \frac{2 \tan 60°}{1 - \tan^2 60°}$$

$\tan 60° = \sqrt{3}$

$$= \frac{2\sqrt{3}}{1-(\sqrt{3})^2}, \text{ or } \frac{2\sqrt{3}}{-2}$$

$\frac{2\sqrt{3}}{-2} = -\sqrt{3}$

Thus, $\tan 120° = -\sqrt{3}$.

EXAMPLE 6 Show that $\tan 2A = \frac{2 \tan A}{1 - \tan^2 A}$.

Write $2A$ as $A + A$. →

$$\tan 2A = \tan(A+A)$$

Formula for $\tan(x+y)$ →

$$= \frac{\tan A + \tan A}{1 - \tan A \tan A}$$
$$= \frac{2 \tan A}{1 - \tan^2 A}$$

$$\tan 2x = \frac{2 \tan x}{1 - \tan^2 x}$$

EXAMPLE 7 If $\tan \theta = -\frac{4}{5}$, find $\tan 2\theta$.

$\tan 2\theta = \frac{2 \tan \theta}{1 - \tan^2 \theta}$ ⟶ $\tan 2\theta = \frac{2 \cdot (\frac{-4}{5})}{1 - (\frac{-4}{5})^2}$

$1 - \frac{16}{25} = \frac{25}{25} - \frac{16}{25} = \frac{9}{25}$ ⟶ $= \frac{\frac{-8}{5}}{\frac{9}{25}}$, or $\frac{-40}{9}$

Thus, $\tan 2\theta = -\frac{40}{9}$.

EXERCISES

PART A

Find the value. Use the formula for tan (x + y) or tan (x − y). Answers may be left in radical form.

1. $\tan 75°$ $\frac{3+\sqrt{3}}{3-\sqrt{3}}$
2. $\tan 90°$ *doesn't exist*
3. $\tan 15°$ $\frac{3-\sqrt{3}}{3+\sqrt{3}}$

Find tan (x + y).

4. $\tan x = \frac{3}{4}$ and $\tan y = -\frac{3}{4}$ 0
5. $\tan x = \sqrt{3}$ and $\tan y = \frac{4}{3}$
6. $\tan x = \frac{5}{12}$ and $\tan y = -\frac{4}{7}$ $-\frac{1}{8}$
7. $\tan x = -1$ and $\tan y = \frac{\sqrt{3}}{3}$ $\frac{-3+\sqrt{3}}{3+\sqrt{3}}$

Find tan (x − y).

8. $\tan x = -\frac{4}{3}$ and $\tan y = \frac{5}{12}$ $\frac{63}{16}$
9. $\tan x = -\frac{5}{12}$ and $\tan y = \frac{5}{7}$ $-\frac{95}{59}$
10. $\tan x = -\sqrt{3}$ and $\tan y = 1$
11. $\tan x = \frac{\sqrt{3}}{3}$ and $\tan y = \frac{12}{5}$

Find tan 2A.

12. $\tan A = -\frac{3}{4}$ $-\frac{24}{7}$
13. $\tan A = \frac{5}{12}$ $\frac{120}{119}$
14. $\tan A = -\sqrt{3}$ $\sqrt{3}$
15. $\tan A = 1$ *doesn't exist*
16. $\tan A = -\frac{4}{7}$ $-\frac{56}{33}$
17. $\tan A = \frac{4}{3}$ $-\frac{24}{7}$

PART B

18. If $\tan x = -\frac{3}{4}$ and $\cos y = -\frac{5}{13}$ and y lies in the 2nd quadrant,
    find $\tan (x + y)$.
    find $\tan (x - y)$.
    find $\tan 2x$. $\frac{-24}{7}$
    find $\tan 2y$. $\frac{120}{119}$

19. If $\tan x = \sqrt{3}$ and $\sin y = \frac{4}{5}$ and y lies in the 2nd quadrant,
    find $\tan (x + y)$. $\frac{3\sqrt{3}-4}{3+4\sqrt{3}}$
    find $\tan (x - y)$.
    find $\tan 2x$. $-\sqrt{3}$
    find $\tan 2y$. $\frac{24}{7}$

Sin $\frac{x}{2}$, Cos $\frac{x}{2}$, and Tan $\frac{x}{2}$

OBJECTIVE

- To apply the formulas for $\sin \frac{1}{2}x$ and $\cos \frac{1}{2}x$

REVIEW CAPSULE

$$\cos 2A = \cos^2 A - \sin^2 A$$
$$\cos 2A = 1 - 2 \sin^2 A$$
$$\cos 2A = 2 \cos^2 A - 1$$

EXAMPLE 1 Show that $\sin \frac{x}{2} = \pm\sqrt{\frac{1 - \cos x}{2}}$.

Formula for $\cos 2A$ → $\cos 2A = 1 - 2 \sin^2 A$

Replace A with $\frac{x}{2}$. → $\cos 2\left(\frac{x}{2}\right) = 1 - 2 \sin^2 \left(\frac{x}{2}\right)$

Simplify. → $\cos x = 1 - 2 \sin^2 \left(\frac{x}{2}\right)$

Add $2 \sin^2 \left(\frac{x}{2}\right)$ and $-\cos x$ to each side. } $2 \sin^2 \left(\frac{x}{2}\right) = 1 - \cos x$

Divide each side by 2. → $\sin^2 \left(\frac{x}{2}\right) = \frac{1 - \cos x}{2}$

Thus, $\sin \frac{x}{2} = \pm\sqrt{\frac{1 - \cos x}{2}}$

$$\sin \frac{x}{2} = \pm\sqrt{\frac{1 - \cos x}{2}}$$

EXAMPLE 2 Find $\sin 15°$. Use the formula for $\sin \frac{x}{2}$.

Answer may be left in radical form.

$$\sin \frac{x}{2} = \pm\sqrt{\frac{1 - \cos x}{2}}$$

Write $15°$ as $\frac{30°}{2}$. → $\sin \frac{30°}{2} = \pm\sqrt{\frac{1 - \cos 30°}{2}}$

$\cos 30° = \frac{\sqrt{3}}{2}$. → $= \sqrt{\frac{1 - \frac{\sqrt{3}}{2}}{2}}$

$\frac{1 - \frac{\sqrt{3}}{2}}{2} = \frac{\frac{2 - \sqrt{3}}{2}}{2} = \frac{2 - \sqrt{3}}{4}$ → $= \sqrt{\frac{2 - \sqrt{3}}{4}}$

Thus, $\sin 15° = \frac{\sqrt{2 - \sqrt{3}}}{2}$.

EXAMPLE 3 Show that $\cos \frac{x}{2} = \pm\sqrt{\frac{1 + \cos x}{2}}$.

Formula for cos 2A → $\cos 2A = 2\cos^2 A - 1$

Add +1 to each side. → $2\cos^2 A = \cos 2A + 1$

Divide each side by 2. → $\cos^2 A = \frac{\cos 2A + 1}{2}$

$$\cos A = \pm\sqrt{\frac{\cos 2A + 1}{2}}$$

Replace A by $\frac{x}{2}$. → **Thus,** $\cos \frac{x}{2} = \pm\sqrt{\frac{\cos x + 1}{2}}$.

$$\cos \frac{x}{2} = \pm\sqrt{\frac{1 + \cos x}{2}}$$

EXAMPLE 4 If $\cos x = -\frac{1}{2}$ and x lies in the third quadrant, find $\sin \frac{x}{2}$.

x lies in 3rd quadrant:
$180° < x < 270°$
$90° < \frac{x}{2} < 135°$

$$\sin \frac{x}{2} = \pm\sqrt{\frac{1 - \cos x}{2}}$$

Sin $\frac{x}{2}$ is positive.

$$= +\sqrt{\frac{1 - \left(-\frac{1}{2}\right)}{2}}$$

$$= +\sqrt{\frac{1 + \frac{1}{2}}{2}}$$

$\frac{1 + \frac{1}{2}}{2} = \frac{\frac{3}{2}}{2} = \frac{3}{4}$ →

$$= +\sqrt{\frac{3}{4}}$$

Thus, $\sin \frac{x}{2} = \frac{\sqrt{3}}{2}$.

ORAL EXERCISES

Find the value. Answers may be left in radical form.

1. If $\cos x = 1$, find $\cos \frac{1}{2}x$. ± 1
2. If $\cos x = 0$, find $\sin \frac{1}{2}x$. $\pm \frac{\sqrt{2}}{2}$
3. If $\cos x = -1$, find $\sin \frac{1}{2}x$. ± 1
4. If $\cos x = -1$, find $\cos \frac{1}{2}x$. 0
5. If $\cos x = \frac{1}{2}$, find $\cos \frac{1}{2}x$. $\pm \frac{\sqrt{3}}{2}$
6. If $\cos x = -\frac{1}{2}$, find $\sin \frac{1}{2}x$. $\pm \frac{\sqrt{3}}{2}$

EXERCISES

PART A

Find the value. Use the formula for sin $\frac{1}{2}x$ or cos $\frac{1}{2}x$. Answers may be left in radical form.

1. $\cos 15°$

2. $\sin 22\frac{1}{2}°$

3. $\cos 135°$ $-\frac{\sqrt{2}}{2}$

4. $\sin 45°$ $\frac{\sqrt{2}}{2}$

5. $\cos 22\frac{1}{2}°$

6. $\cos 67\frac{1}{2}°$

7. $\sin 67\frac{1}{2}°$

8. $\cos(-15°)$

Find the value of sin $\frac{x}{2}$ and cos $\frac{x}{2}$. Answers may be left in radical form.

9. Cos $x = \frac{3}{4}$ and x is in the 1st quadrant.

10. Cos $x = -\frac{\sqrt{3}}{2}$ and x is in the 2nd quadrant.

11. Cos $x = \frac{1}{3}$ and x is in the 1st quadrant.

12. Cos $x = \frac{1}{4}$ and x is in the 1st quadrant.

13. Cos $x = -\frac{1}{4}$ and x is in the 3rd quadrant.
$\frac{\sqrt{10}}{4}; -\frac{\sqrt{6}}{4}$

14. Cos $x = -a$ and x is in the 3rd quadrant.
$\frac{\sqrt{2+2a}}{2}; -\frac{\sqrt{2-2a}}{2}$

PART B

EXAMPLE Show that $\tan\frac{x}{2} = \pm\sqrt{\frac{1-\cos x}{1+\cos x}}$.

$$\tan\frac{x}{2} = \frac{\sin\frac{x}{2}}{\cos\frac{x}{2}}$$

$$= \frac{\pm\sqrt{\frac{1-\cos x}{2}}}{\pm\sqrt{\frac{1+\cos x}{2}}}$$

Thus, $\tan\frac{x}{2} = \pm\sqrt{\frac{1-\cos x}{1+\cos x}}$.

Find the value. Use the formula for tan $\frac{x}{2}$. Answers may be left in radical form.

15. Cos $x = \frac{\sqrt{3}}{2}$ and x is in the 4th quadrant.
$-\sqrt{7-4\sqrt{3}}$

16. Cos $x = -\frac{1}{7}$ and x is in the 3rd quadrant.
$-\frac{2\sqrt{3}}{3}$

PART C

17. Find $\tan\frac{x}{2}$ if $\sin x = \frac{1}{2}$ and $\cos x = \frac{\sqrt{3}}{2}$.

18. Find $\tan\frac{\theta}{2}$ if $\cos\theta = \frac{7}{25}$ and $\theta < 90°$.

Double and Half Angle Identities

OBJECTIVE

- To prove identities using double and half angle identities

REVIEW CAPSULE

$\cos 2A = \cos^2 A - \sin^2 A$ | $\sin 2A = 2 \sin A \cos A$

$\cos 2A = 1 - 2\sin^2 A$

$\cos 2A = 2\cos^2 A - 1$ | $\tan 2A = \dfrac{2 \tan A}{1 - \tan^2 A}$

EXAMPLE 1 Prove the identity $\dfrac{\sin 2A}{1 + \cos 2A} = \tan A$.

$\dfrac{\sin 2A}{1+\cos 2A}$	$\tan A$
$\dfrac{2 \sin A \cos A}{1 + (2\cos^2 A - 1)}$	$\tan A$
$\dfrac{\overset{1}{\cancel{2}} \sin A \cos A}{\underset{1}{\cancel{2}} \cos^2 A}$	
$\dfrac{\sin A}{\cos A}$	
$\tan A$	

Reasons:

$\sin 2A = 2 \sin A \cos A$ ⟶

$\cos 2A = 2\cos^2 A - 1$ ⟶

$1 + (2\cos^2 A - 1) = 2\cos^2 A$ ⟶

$\dfrac{\sin A \cos A}{\cos^2 A} = \dfrac{\sin A}{\cos A}$ ⟶

$\dfrac{\sin A}{\cos A} = \tan A$ ⟶

Since each side of the equation is the same, the identity has been proved.

EXAMPLE 2 Prove the identity $\tan 2\theta = \dfrac{\sin 2\theta}{1 - \sin 2\theta \tan \theta}$.

$\tan 2\theta$	$\dfrac{\sin 2\theta}{1 - \sin 2\theta \tan \theta}$
$\tan 2\theta$	$\dfrac{\sin 2\theta}{1 - 2 \sin \theta \overset{1}{\cancel{\cos \theta}} \dfrac{\sin \theta}{\underset{1}{\cancel{\cos \theta}}}}$
	$\dfrac{\sin 2\theta}{1 - 2\sin^2 \theta}$
	$\dfrac{\sin 2\theta}{\cos 2\theta}$
	$\tan 2\theta$

Reasons:

Sin $2\theta = 2 \sin \theta \cos \theta$, $\tan \theta = \dfrac{\sin \theta}{\cos \theta}$ ⟶

$\cos 2\theta = 1 - 2\sin^2 \theta$ ⟶

$\dfrac{\sin 2\theta}{\cos 2\theta} = \tan 2\theta$ ⟶

EXAMPLE 3 Prove the identity $\frac{\sec A - 1}{2 \sec A} = \sin^2 \frac{A}{2}$.

$\frac{\sec A - 1}{2 \sec A}$	$\sin^2 \frac{A}{2}$
$\frac{\frac{1}{\cos A} - 1}{2 \cdot \frac{1}{\cos A}}$	$\left(\pm\sqrt{\frac{1 - \cos A}{2}}\right)^2$
	$\frac{1 - \cos A}{2}$
$\frac{\frac{1 - \cos A}{\cos A}}{\frac{2}{\cos A}}$	
$\frac{1 - \cos A}{2}$	

$\text{Sec } A = \frac{1}{\cos A}$; $\sin \frac{A}{2} = \pm\sqrt{\frac{1 - \cos A}{2}}$ ⟶

$\frac{\frac{1 - \cos A}{\cos A}}{\frac{2}{\cos A}} = \frac{1 - \cos A}{\cancel{\cos A}_1} \cdot \frac{\cancel{\cos A}^1}{2}$

EXERCISES

PART A

Prove the identity.

1. $\sin 2A = \frac{2 \tan A}{1 + \tan^2 A}$

2. $1 - \sin 2x = (\sin x - \cos x)^2$

3. $\frac{\tan \theta \sin 2\theta}{2} = \sin^2 \theta$

4. $\frac{2}{\sin 2A} = \frac{\cos A}{\sin A} + \frac{\sin A}{\cos A}$

5. $\cos^2 \frac{A}{2} = \frac{\sec A + 1}{2 \sec A}$

6. $\tan \frac{\theta}{2} = \frac{\sin \theta}{1 + \cos \theta}$

PART B

Prove the identity.

7. $\tan 2A + \frac{1}{\cos 2A} = \frac{\cos A + \sin A}{\cos A - \sin A}$

8. $\tan 2A = \frac{2 \tan A}{\sec^2 A - 2 \tan^2 A}$

9. $\frac{\frac{1}{2} \sin^2 x}{\sin^2 \frac{1}{2} x} - 1 = \cos x$

10. $\left(\sin \frac{x}{2} + \cos \frac{x}{2}\right)^2 = 1 + \sin x$

11. $\frac{1 - \tan^2 x}{1 + \tan^2 x} = \cos 2x$

12. $2 \tan 2x = \frac{\sin x + \cos x}{\cos x - \sin x} + \frac{\sin x - \cos x}{\sin x + \cos x}$

Chapter Nineteen Review

Each refers to triangle *ABC*. Find the length of the indicated side to the nearest unit, or leave in radical form. [p. 493]

1. $a = 7$, $c = 4$, $m\angle B = 40$. Find b.

2. $c = 9$, $b = 7$, $m\angle A = 120$. Find a.

Each refers to triangle *ABC*. [p. 496]

3. $a = 6$, $b = 8$, $c = 5$. Find $\cos A$.

4. $a = 4$, $b = 5$, $c = 7$. Find $m\angle B$ to the nearest degree.

For $\triangle ABC$, find the area to the nearest square unit. [p. 498]

5. $a = 5$, $b = 7$, $m\angle C = 70$.

6. $a = 7.1$, $c = 4.3$, $\angle B$ measures $40°30'$.

For $\triangle PQR$, find $m\angle P$ to the nearest degree. [p. 498]

7. $q = 8$, $r = 12$, Area $= 27$ square units

8. $q = 10$, $r = 12$, Area $= 50$ square units

9. For $\triangle ABC$, $a = 40$, $m\angle A = 65$, $m\angle B = 57$. Find b to the nearest unit. [p. 500]

10. For $\triangle ABC$ $a = 14$, $c = 20$, $m\angle A = 43$. Find $m\angle C$ to the nearest degree.

How many triangles can be constructed? [p. 503]

11. $m\angle A = 40$, $a = 12$, $b = 16$

12. $m\angle A = 140$, $a = 30$, $b = 10$

Find the value. Use the formula for $\sin (x \pm y)$ or $\sin (\frac{x}{2})$. Answers may be left in radical form. [p. 506, 517]

13. $\sin 75°$

14. $\sin 60°$

15. $\sin 15°$

Find the value. Use the formula for $\cos (x \pm y)$ or $\cos (\frac{x}{2})$. Answers may be left in radical form. [p. 510, 517]

16. $\cos 15°$

17. $\cos 75°$

18. $\cos 105°$

Find $\sin (A + B)$ and $\sin (A - B)$.

19. Sin $A = \frac{-3}{5}$ and A lies in the 3rd quadrant. Sin $B = \frac{12}{13}$ and B lies in the 2nd quadrant. [p. 506]

Find $\cos (A + B)$ and $\cos (A - B)$.

20. Cos $A = \frac{-3}{5}$ and A is in the 2nd quadrant. Cos $B = \frac{4}{5}$ and B lies in the 1st quadrant. [p. 510]

21. If $\sin A = -\frac{5}{13}$ and A lies in the 3 rd quadrant, find $\sin 2A$. [p. 506]

22. If $\cos A = -\frac{1}{8}$ and A lies in the 2nd quadrant, find $\cos \frac{A}{2}$. [p. 517]

Find $\tan (x + y)$. [p. 514]

23. $\tan x = \frac{5}{12}$ and $\tan y = -\frac{3}{4}$

Find $\tan (x - y)$. [p. 514]

24. $\tan x = \frac{7}{9}$ and $\tan y = \frac{12}{5}$

25. $\tan A = -\frac{5}{12}$, find $\tan 2A$. [p. 514]

26. Find $\tan 75°$. Use the formula for $\tan (x + y)$. [p. 514]

Simplify. [p. 506]

27. $\dfrac{2 \cos A}{\sin 2A}$

Prove the identity. [p. 520]

28. $\cos^2 \dfrac{\theta}{2} = \dfrac{\sin \theta + \tan \theta}{2 \tan \theta}$

29. If $\tan x = \frac{3}{4}$, $\sin y = \frac{3}{5}$ and y lies in the 2nd quadrant, find $\tan (x \pm y)$ and $\tan 2x$. [p. 514]

Chapter Nineteen Test

Each refers to triangle *ABC*. Find the length of the indicated side to the nearest unit, or leave in radical form.

1. $a = 6$, $b = 10$, $m\angle C = 50$. Find c.
2. $c = 10$, $b = 6$, $m\angle A = 150$. Find a.

Each refers to triangle *ABC*.

3. $a = 7$, $b = 6$, $c = 10$. Find $\cos C$.
4. $a = 12$, $b = 4$, $c = 9$. Find $m\angle A$ to the nearest degree.

For $\triangle ABC$, find the area to the nearest square unit.

5. $b = 12$, $c = 10$, $m\angle A = 80$.
6. $a = 6.3$, $b = 8$, $\angle C$ measures $30°40'$.

For $\triangle PQR$, find $m\angle P$ to the nearest degree.

7. $q = 6$, $r = 9$, $A = 18$ sq units

8. For $\triangle ABC$, $a = 50$, $m\angle A = 70$, $m\angle C = 40$. Find c to the nearest unit.
9. For $\triangle ABC$, $a = 16$, $c = 10$, $m\angle A = 32$. Find $m\angle C$ to the nearest degree.

How many triangles can be constructed?

10. $m\angle A = 20$, $a = 6$, $b = 12$
11. $m\angle A = 120$, $a = 15$, $b = 8$

Find the value. Use the formula for $\sin(x \pm y)$ or $\sin\left(\frac{x}{2}\right)$. Answers may be left in radical form.

12. $\sin 105°$
13. $\sin 30°$
14. $\sin 22\frac{1}{2}°$

Find the value. Use the formula for $\cos(x + y)$ or $\cos(x - y)$. Answers may be left in radical form.

15. $\cos 135°$
16. $\cos 60°$

Find $\sin(A - B)$.

17. $\text{Sin } A = -\frac{5}{7}$ and A lies in the 3rd quadrant. $\text{Sin } B = \frac{7}{9}$ and B lies in the 2nd quadrant.

Find $\cos(A + B)$.

18. $\text{Cos } A = -\frac{5}{13}$ and A lies in the 2nd quadrant. $\text{Cos } B = \frac{3}{5}$ and B lies in the 4th quadrant.

19. If $\sin A = -\frac{12}{13}$ and A lies in the 4th quadrant, find $\sin 2A$.
20. If $\cos A = \frac{5}{13}$ and A lies in the 1st quadrant, find $\cos 2A$.

Find $\tan(x + y)$.

21. $\tan x = \frac{3}{7}$ and $\tan y = -\frac{4}{5}$

Find $\tan(x - y)$.

22. $\tan x = -\frac{12}{5}$ and $\tan y = -\frac{2}{5}$

23. If $\tan A = -\frac{3}{4}$, find $\tan 2A$.
24. Find $\tan 105°$. Use the formula for $\tan(x + y)$.

Simplify.

25. $\dfrac{4 \sin A \cos A}{\sin 2A}$

Prove the identity.

26. $\cos 2A = \dfrac{2 - \sec^2 A}{\sec^2 A}$

27. If $\tan x = -\frac{4}{3}$, $\cos y = -\frac{5}{13}$ and y lies in the 2nd quadrant, find $\tan(x - y)$.

Table of Roots and Powers

No.	Sq.	Sq. Root	Cube	Cu. Root	No.	Sq.	Sq. Root	Cube	Cu. Root
1	1	1.000	1	1.000	51	2,601	7.141	132,651	3.708
2	4	1.414	8	1.260	52	2,704	7.211	140,608	3.733
3	9	1.732	27	1.442	53	2,809	7.280	148,877	3.756
4	16	2.000	64	1.587	54	2,916	7.348	157,564	3.780
5	25	2.236	125	1.710	55	3,025	7.416	166,375	3.803
6	36	2.449	216	1.817	56	3,136	7.483	175,616	3.826
7	49	2.646	343	1.913	57	3,249	7.550	185,193	3.849
8	64	2.828	512	2.000	58	3,364	7.616	195,112	3.871
9	81	3.000	729	2.080	59	3,481	7.681	205,379	3.893
10	100	3.162	1,000	2.154	60	3,600	7.746	216,000	3.915
11	121	3.317	1,331	2.224	61	3,721	7.810	226,981	3.936
12	144	3.464	1,728	2.289	62	3,844	7.874	238,328	3.958
13	169	3.606	2,197	2.351	63	3,969	7.937	250,047	3.979
14	196	3.742	2,744	2.410	64	4,096	8.000	262,144	4.000
15	225	3.875	3,375	2.466	65	4,225	8.062	274,625	4.021
16	256	4.000	4,096	2.520	66	4,356	8.124	287,496	4.041
17	289	4.123	4,913	2.571	67	4,489	8.185	300,763	4.062
18	324	4.243	5,832	2.621	68	4,624	8.246	314,432	4.082
19	361	4.359	6,859	2.668	69	4,761	8.307	328,509	4.102
20	400	4.472	8,000	2.714	70	4,900	8.357	343,000	4.121
21	441	4.583	9,261	2.759	71	5,041	8.426	357,911	4.141
22	484	4.690	10,648	2.802	72	5,184	8.485	373,248	4.160
23	529	4.796	12,167	2.844	73	5,329	8.544	389,017	4.179
24	576	4.899	13,824	2.884	74	5,476	8.602	405,224	4.198
25	625	5.000	15,625	2.924	75	5,625	8.660	421,875	4.217
26	676	5.099	17,576	2.962	76	5,776	8.718	438,976	4.236
27	729	5.196	19,683	3.000	77	5,929	8.775	456,533	4.254
28	784	5.292	21,952	3.037	78	6,084	8.832	474,552	4.273
29	841	5.385	24,389	3.072	79	6,241	8.888	493,039	4.291
30	900	5.477	27,000	3.107	80	6,400	8.944	512,000	4.309
31	961	5.568	29,791	3.141	81	6,561	9.000	531,441	4.327
32	1,024	5.657	32,768	3.175	82	6,724	9.055	551,368	4.344
33	1,089	5.745	35,937	3.208	83	6,889	9.110	571,787	4.362
34	1,156	5.831	39,304	3.240	84	7,056	9.165	592,704	4.380
35	1,225	5.916	42,875	3.271	85	7,225	9.220	614,125	4.397
36	1,296	6.000	46,656	3.302	86	7,396	9.274	636,056	4.414
37	1,369	6.083	50,653	3.332	87	7,569	9.327	658,503	4.431
38	1,444	6.164	54,872	3.362	88	7,744	9.381	681,472	4.448
39	1,521	6.245	59,319	3.391	89	7,921	9.434	704,969	4.465
40	1,600	6.325	64,000	3.420	90	8,100	9.487	729,000	4.481
41	1,681	6.403	68,921	3.448	91	8,281	9.539	753,571	4.498
42	1,764	6.481	74,088	3.476	92	8,464	9.592	778,688	4.514
43	1,849	6.557	79,507	3.503	93	8,649	9.644	804,357	4.531
44	1,936	6.633	85,184	3.530	94	8,836	9.695	830,584	4.547
45	2,025	6.708	91,125	3.557	95	9,025	9.747	857,375	4.563
46	2,116	6.782	97,336	3.583	96	9,216	9.798	884,736	4.579
47	2,209	6.856	103,823	3.609	97	9,409	9.849	912,673	4.595
48	2,304	6.928	110,592	3.634	98	9,604	9.899	941,192	4.610
49	2,401	7.000	117,649	3.659	99	9,801	9.950	970,299	4.626
50	2,500	7.071	125,000	3.684	100	10,000	10.000	1,000,000	4.642

Trigonometric Functions

x	sin x	cos x	tan x	cot x	sec x	csc x	
0° 0′	.00000	1.0000	.00000		1.0000		**90° 0′**
10′	.00291	1.0000	.00291	343.77	1.0000	343.78	50′
20′	.00582	1.0000	.00582	171.88	1.0000	171.89	40′
30′	.00873	1.0000	.00873	114.59	1.0000	114.59	30′
40′	.01164	.9999	.01164	85.940	1.0001	85.946	20′
50′	.01454	.9999	.01455	68.750	1.0001	68.757	10′
1° 0′	.01745	.9998	.01746	57.290	1.0002	57.299	**89° 0′**
10′	.02036	.9998	.02036	49.104	1.0002	49.114	50′
20′	.02327	.9997	.02328	42.964	1.0003	42.976	40′
30′	.02618	.9997	.02619	38.188	1.0003	38.202	30′
40′	.02908	.9996	.02910	34.368	1.0004	34.382	20′
50′	.03199	.9995	.03201	31.242	1.0005	31.258	10′
2° 0′	.03490	.9994	.03492	28.6363	1.0006	28.654	**88° 0′**
10′	.03781	.9993	.03783	26.4316	1.0007	26.451	50′
20′	.04071	.9992	.04075	24.5418	1.0008	24.562	40′
30′	.04362	.9990	.04366	22.9038	1.0010	22.926	30′
40′	.04653	.9989	.04658	21.4704	1.0011	21.494	20′
50′	.04943	.9988	.04949	20.2056	1.0012	20.230	10′
3° 0′	.05234	.9986	.05241	19.0811	1.0014	19.107	**87° 0′**
10′	.05524	.9985	.05533	18.0750	1.0015	18.103	50′
20′	.05814	.9983	.05824	17.1693	1.0017	17.198	40′
30′	.06105	.9981	.06116	16.3499	1.0019	16.380	30′
40′	.06395	.9980	.06408	15.6048	1.0021	15.637	20′
50′	.06685	.9978	.06700	14.9244	1.0022	14.958	10′
4° 0′	.06976	.9976	.06993	14.3007	1.0024	14.336	**86° 0′**
10′	.07266	.9974	.07285	13.7267	1.0027	13.763	50′
20′	.07556	.9971	.07578	13.1969	1.0029	13.235	40′
30′	.07846	.9969	.07870	12.7062	1.0031	12.746	30′
40′	.08136	.9967	.08163	12.2505	1.0033	12.291	20′
50′	.08426	.9964	.08456	11.8262	1.0036	11.868	10′
5° 0′	.08716	.9962	.08749	11.4301	1.0038	11.474	**85° 0′**
10′	.09005	.9959	.09042	11.0594	1.0041	11.105	50′
20′	.09295	.9957	.09335	10.7119	1.0044	10.758	40′
30′	.09585	.9954	.09629	10.3854	1.0046	10.433	30′
40′	.09874	.9951	.09923	10.0780	1.0049	10.128	20′
50′	.10164	.9948	.10216	9.7882	1.0052	9.839	10′
6° 0′	.10453	.9945	.10510	9.5144	1.0055	9.5668	**84° 0′**
10′	.10742	.9942	.10805	9.2553	1.0058	9.3092	50′
20′	.11031	.9939	.11099	9.0098	1.0061	9.0652	40′
30′	.11320	.9936	.11394	8.7769	1.0065	8.8337	30′
40′	.11609	.9932	.11688	8.5555	1.0068	8.6138	20′
50′	.11898	.9929	.11983	8.3450	1.0072	8.4647	10′
7° 0′	.12187	.9925	.12278	8.1443	1.0075	8.2055	**83° 0′**
10′	.12476	.9922	.12574	7.9530	1.0079	8.0157	50′
20′	.12764	.9918	.12869	7.7704	1.0083	7.8344	40′
30′	.13053	.9914	.13165	7.5958	1.0086	7.6613	30′
	cos x	sin x	cot x	tan x	csc x	sec x	x

Trigonometric Functions

x	sin x	cos x	tan x	cot x	sec x	csc x	
30′	.1305	.9914	.1317	7.5958	1.0086	7.6613	30′
40′	.1334	.9911	.1346	7.4287	1.0090	7.4957	20′
50′	.1363	.9907	.1376	7.2687	1.0094	7.3372	10′
8° 0′	.1392	.9903	.1405	7.1154	1.0098	7.1853	**82° 0′**
10′	.1421	.9899	.1435	6.9682	1.0102	7.0396	50′
20′	.1449	.9894	.1465	6.8269	1.0107	6.8998	40′
30′	.1478	.9890	.1495	6.6912	1.0111	6.7655	30′
40′	.1507	.9886	.1524	6.5606	1.0116	6.6363	20′
50′	.1536	.9881	.1554	6.4348	1.0120	6.5121	10′
9° 0′	.1564	.9877	.1584	6.3138	1.0125	6.3925	**81° 0′**
10′	.1593	.9872	.1614	6.1970	1.0129	6.2772	50′
20′	.1622	.9868	.1644	6.0844	1.0134	6.1661	40′
30′	.1650	.9863	.1673	5.9758	1.0139	6.0589	30′
40′	.1679	.9858	.1703	5.8708	1.0144	5.9554	20′
50′	.1708	.9853	.1733	5.7694	1.0149	5.8554	10′
10° 0′	.1736	.9848	.1763	5.6713	1.0154	5.7588	**80° 0′**
10′	.1765	.9843	.1793	5.5764	1.0160	5.6653	50′
20′	.1794	.9838	.1823	5.4845	1.0165	5.5749	40′
30′	.1822	.9833	.1853	5.3955	1.0170	5.4874	30′
40′	.1851	.9827	.1883	5.3093	1.0176	5.4026	20′
50′	.1880	.9822	.1914	5.2257	1.0182	5.3205	10′
11° 0′	.1908	.9816	.1944	5.1446	1.0187	5.2408	**79° 0′**
10′	.1937	.9811	.1974	5.0658	1.0193	5.1636	50′
20′	.1965	.9805	.2004	4.9894	1.0199	5.0886	40′
30′	.1994	.9799	.2035	4.9152	1.0205	5.0159	30′
40′	.2022	.9793	.2065	4.8430	1.0211	4.9452	20′
50′	.2051	.9787	.2095	4.7729	1.0217	4.8765	10′
12° 0′	.2079	.9781	.2126	4.7046	1.0223	4.8097	**78° 0′**
10′	.2108	.9775	.2156	4.6382	1.0230	4.7448	50′
20′	.2136	.9769	.2186	4.5736	1.0236	4.6817	40′
30′	.2164	.9763	.2217	4.5107	1.0243	4.6202	30′
40′	.2193	.9757	.2247	4.4494	1.0249	4.5604	20′
50′	.2221	.9750	.2278	4.3897	1.0256	4.5022	10′
13° 0′	.2250	.9744	.2309	4.3315	1.0263	4.4454	**77° 0′**
10′	.2278	.9737	.2339	4.2747	1.0270	4.3901	50′
20′	.2306	.9730	.2370	4.2193	1.0277	4.3362	40′
30′	.2334	.9724	.2401	4.1653	1.0284	4.2837	30′
40′	.2363	.9717	.2432	4.1126	1.0291	4.2324	20′
50′	.2391	.9710	.2462	4.0611	1.0299	4.1824	10′
14° 0′	.2419	.9703	.2493	4.0108	1.0306	4.1336	**76° 0′**
10′	.2447	.9696	.2524	3.9617	1.0314	4.0859	50′
20′	.2476	.9689	.2555	3.9136	1.0321	4.0394	40′
30′	.2504	.9681	.2586	3.8667	1.0329	3.9939	30′
40′	.2532	.9674	.2617	3.8208	1.0337	3.9495	20′
50′	.2560	.9667	.2648	3.7760	1.0345	3.9061	10′
15° 0′	.2588	.9659	.2679	3.7321	1.0353	3.8637	**75° 0′**
	cos x	sin x	cot x	tan x	csc x	sec x	x

Trigonometric Functions

x	sin x	cos x	tan x	cot x	sec x	csc x	
15° 0′	.2588	.9659	.2679	3.7321	1.0353	3.8637	**75° 0′**
10′	.2616	.9652	.2711	3.6891	1.0361	3.8222	50′
20′	.2644	.9644	.2742	3.6470	1.0369	3.7817	40′
30′	.2672	.9636	.2773	3.6059	1.0377	3.7420	30′
40′	.2700	.9628	.2805	3.5656	1.0386	3.7032	20′
50′	.2728	.9621	.2836	3.5261	1.0394	3.6652	10′
16° 0′	.2756	.9613	.2867	3.4874	1.0403	3.6280	**74° 0′**
10′	.2784	.9605	.2899	3.4495	1.0412	3.5915	50′
20′	.2812	.9596	.2931	3.4124	1.0421	3.5559	40′
30′	.2840	.9588	.2962	3.3759	1.0430	3.5209	30′
40′	.2868	.9580	.2994	3.3402	1.0439	3.4867	20′
50′	.2896	.9572	.3026	3.3052	1.0448	3.4532	10′
17° 0′	.2924	.9563	.3057	3.2709	1.0457	3.4203	**73° 0′**
10′	.2952	.9555	.3089	3.2371	1.0466	3.3881	50′
20′	.2979	.9546	.3121	3.2041	1.0476	3.3565	40′
30′	.3007	.9537	.3153	3.1716	1.0485	3.3255	30′
40′	.3035	.9528	.3185	3.1397	1.0495	3.2951	20′
50′	.3062	.9520	.3217	3.1084	1.0505	3.2653	10′
18° 0′	.3090	.9511	.3249	3.0777	1.0515	3.2361	**72° 0′**
10′	.3118	.9502	.3281	3.0475	1.0525	3.2074	50′
20′	.3145	.9492	3314	3.0178	1.0535	3.1792	40′
30′	.3173	.9483	.3346	2.9887	1.0545	3.1516	30′
40′	.3201	.9474	.3378	2.9600	1.0555	3.1244	20′
50′	.3228	.9465	.3411	2.9319	1.0566	3.0977	10′
19° 0′	.3256	.9455	.3443	2.9042	1.0576	3.0716	**71° 0′**
10′	.3283	.9446	.3476	2.8770	1.0587	3.0458	50′
20′	.3311	.9436	.3508	2.8502	1.0598	3.0206	40′
30′	.3338	.9426	.3541	2.8239	1.0609	2.9957	30′
40′	.3365	.9417	.3574	2.7980	1.0620	2.9714	20′
50′	.3393	.9407	.3607	2.7725	1.0631	2.9474	10′
20° 0′	.3420	.9397	.3640	2.7475	1.0642	2.9238	**70° 0′**
10′	.3448	.9387	.3673	2.7228	1.0653	2.9006	50′
20′	.3475	.9377	.3706	2.6985	1.0665	2.8779	40′
30′	.3502	.9367	.3739	2.6746	1.0676	2.8555	30′
40′	.3529	.9356	.3772	2.6511	1.0688	2.8334	20′
50′	.3557	.9346	.3805	2.6279	1.0700	2.8118	10′
21° 0′	.3584	.9336	.3839	2.6051	1.0712	2.7904	**69° 0′**
10′	.3611	.9325	.3872	2.5826	1.0724	2.7695	50′
20′	.3638	.9315	.3906	2.5605	1.0736	2.7488	40′
30′	.3665	.9304	.3939	2.5386	1.0748	2.7285	30′
40′	.3692	.9293	.3973	2.5172	1.0760	2.7085	20′
50′	.3719	.9283	.4006	2.4960	1.0773	2.6888	10′
22° 0′	.3746	.9272	.4040	2.4751	1.0785	2.6695	**68° 0′**
10′	.3773	.9261	.4074	2.4545	1.0798	2.6504	50′
20′	.3800	.9250	.4108	2.4342	1.0811	2.6316	40′
30′	.3827	.9239	.4142	2.4142	1.0824	2 6131	30′
	cos x	sin x	cot x	tan x	csc x	sec x	x

Trigonometric Functions

x		sin x	cos x	tan x	cot x	sec x	csc x		
	30′	.3827	.9239	.4142	2.4142	1.0824	2.6131		30′
	40′	.3854	.9228	.4176	2.3945	1.0837	2.5949		20′
	50′	.3881	.9216	.4210	2.3750	1.0850	2.5770		10′
23°	**0′**	.3907	.9205	.4245	2.3559	1.0864	2.5593	**67°**	**0′**
	10′	.3934	.9194	.4279	2.3369	1.0877	2.5419		50′
	20′	.3961	.9182	.4314	2.3183	1.0891	2.5247		40′
	30′	.3987	.9171	.4348	2.2998	1.0904	2.5078		30′
	40′	.4014	.9159	.4383	2.2817	1.0918	2.4912		20′
	50′	.4041	.9147	.4417	2.2637	1.0932	2.4748		10′
24°	**0′**	.4067	.9135	.4452	2.2460	1.0946	2.4586	**66°**	**0′**
	10′	.4094	.9124	.4487	2.2286	1.0961	2.4426		50′
	20′	.4120	.9112	.4522	2.2113	1.0975	2.4269		40′
	30′	.4147	.9100	.4557	2.1943	1.0990	2.4114		30′
	40′	.4173	.9088	.4592	2.1775	1.1004	2.3961		20′
	50′	.4200	.9075	.4628	2.1609	1.1019	2.3811		10′
25°	**0′**	.4226	.9063	.4663	2.1445	1.1034	2.3662	**65°**	**0′**
	10′	.4253	.9051	.4699	2.1283	1.1049	2.3515		50′
	20′	.4279	.9038	.4734	2.1123	1.1064	2.3371		40′
	30′	.4305	.9026	.4770	2.0965	1.1079	2.3228		30′
	40′	.4331	.9013	.4806	2.0809	1.1095	2.3088		20′
	50′	.4358	.9001	.4841	2.0655	1.1110	2.2949		10′
26°	**0′**	.4384	.8988	.4877	2.0503	1.1126	2.2812	**64°**	**0′**
	10′	.4410	.8975	.4913	2.0353	1.1142	2.2677		50′
	20′	.4436	.8962	.4950	2.0204	1.1158	2.2543		40′
	30′	.4462	.8949	.4986	2.0057	1.1174	2.2412		30′
	40′	.4488	.8936	.5022	1.9912	1.1190	2.2282		20′
	50′	.4514	.8923	.5059	1.9768	1.1207	2.2154		10′
27°	**0′**	.4540	.8910	.5095	1.9626	1.1223	2.2027	**63°**	**0′**
	10′	.4566	.8897	.5132	1.9486	1.1240	2.1902		50′
	20′	.4592	.8884	.5169	1.9347	1.1257	2.1779		40′
	30′	.4617	.8870	.5206	1.9210	1.1274	2.1657		30′
	40′	.4643	.8857	.5243	1.9074	1.1291	2.1537		20′
	50′	.4669	.8843	.5280	1.8940	1.1308	2.1418		10′
28°	**0′**	.4695	.8829	.5317	1.8807	1.1326	2.1301	**62°**	**0′**
	10′	.4720	.8816	.5354	1.8676	1.1343	2.1185		50′
	20′	.4746	.8802	.5392	1.8546	1.1361	2.1070		40′
	30′	.4772	.8788	.5430	1.8418	1.1379	2.0957		30′
	40′	.4797	.8774	.5467	1.8291	1.1397	2.0846		20′
	50′	.4823	.8760	.5505	1.8165	1.1415	2.0736		10′
29°	**0′**	.4848	.8746	.5543	1.8040	1.1434	2.0627	**61°**	**0′**
	10′	.4874	.8732	.5581	1.7917	1.1452	2.0519		50′
	20′	.4899	.8718	.5619	1.7796	1.1471	2.0413		40′
	30′	.4924	.8704	.5658	1.7675	1.1490	2.0308		30′
	40′	.4950	.8689	.5696	1.7556	1.1509	2.0204		20′
	50′	.4975	.8675	.5735	1.7437	1.1528	2.0101		10′
30°	**0′**	.5000	.8660	.5774	1.7321	1.1547	2.0000	**60°**	**0′**
		cos x	sin x	cot x	tan x	csc x	sec x	x	

Trigonometric Functions

x	sin x	cos x	tan x	cot x	sec x	csc x	
30° 0′	.5000	.8660	.5774	1.7321	1.1547	2.0000	**60°** 0′
10′	.5025	.8646	.5812	1.7205	1.1567	1.9900	50′
20′	.5050	.8631	.5851	1.7090	1.1586	1.9801	40′
30′	.5075	.8616	.5890	1.6977	1.1606	1.9703	30′
40′	.5100	.8601	.5930	1.6864	1.1626	1.9606	20′
50′	.5125	.8587	.5969	1.6753	1.1646	1.9511	10′
31° 0′	.5150	.8572	.6009	1.6643	1.1666	1.9416	**59°** 0′
10′	.5175	.8557	.6048	1.6534	1.1687	1.9323	50′
20′	.5200	.8542	.6088	1.6426	1.1708	1.9230	40′
30′	.5225	.8526	.6128	1.6319	1.1728	1.9139	30′
40′	.5250	.8511	.6168	1.6212	1.1749	1.9049	20′
50′	.5275	.8496	.6208	1.6107	1.1770	1.8959	10′
32° 0′	.5299	.8480	.6249	1.6003	1.1792	1.8871	**58°** 0′
10′	.5324	.8465	.6289	1.5900	1.1813	1.8783	50′
20′	.5348	.8450	.6330	1.5798	1.1835	1.8699	40′
30′	.5373	.8434	.6371	1.5697	1.1857	1.8612	30′
40′	.5398	.8418	.6412	1.5597	1.1879	1.8527	20′
50′	.5422	.8403	.6453	1.5497	1.1901	1.8444	10′
33° 0′	.5446	.8387	.6494	1.5399	1.1924	1.8361	**57°** 0′
10′	.5471	.8371	.6536	1.5301	1.1946	1.8279	50′
20′	.5495	.8355	.6577	1.5204	1.1969	1.8198	40′
30′	.5519	.8339	.6619	1.5108	1.1992	1.8118	30′
40′	.5544	.8323	.6661	1.5013	1.2015	1.8039	20′
50′	.5568	.8307	.6703	1.4919	1.2039	1 7960	10′
34° 0′	.5592	.8290	.6745	1.4826	1.2062	1.7883	**56°** 0′
10′	.5616	.8274	.6787	1.4733	1.2086	1.7806	50′
20′	.5640	.8258	.6830	1.4641	1.2110	1.7730	40′
30′	.5664	.8241	.6873	1.4550	1.2134	1.7655	30′
40′	.5688	.8225	.6916	1.4460	1.2158	1.7581	20′
50′	.5712	.8208	.6959	1.4370	1.2183	1.7507	10′
35° 0′	.5736	.8192	.7002	1.4281	1.2208	1.7435	**55°** 0′
10′	.5760	.8175	.7046	1.4193	1.2233	1.7362	50′
20′	.5783	.8158	.7089	1.4106	1.2258	1.7291	40′
30′	.5807	.8141	.7133	1.4019	1.2283	1.7221	30′
40′	.5831	.8124	.7177	1.3934	1.2309	1.7151	20′
50′	.5854	.8107	.7221	1.3848	1.2335	1.7082	10′
36° 0′	.5878	.8090	.7265	1.3764	1.2361	1.7013	**54°** 0′
10′	.5901	.8073	.7310	1.3680	1.2387	1.6945	50′
20′	.5925	.8056	.7355	1.3597	1.2413	1.6878	40′
30′	.5948	.8039	.7400	1.3514	1.2440	1.6812	30′
40′	.5972	.8021	.7445	1.3432	1.2467	1.6746	20′
50′	.5995	.8004	.7490	1.3351	1.2494	1.6681	10′
37° 0′	.6018	.7986	.7536	1.3270	1.2521	1.6616	**53°** 0′
10′	.6041	.7969	.7581	1.3190	1.2549	1.6553	50′
20′	.6065	.7951	.7627	1.3111	1.2577	1.6489	40′
30′	.6088	.7934	.7673	1.3032	1.2605	1.6427	30′
	cos x	sin x	cot x	tan x	csc x	sec x	x

Trigonometric Functions

x	sin x	cos x	tan x	cot x	sec x	csc x	
30′	.6088	.7934	.7673	1.3032	1.2605	1.6427	30′
40′	.6111	.7916	.7720	1.2954	1.2633	1.6365	20′
50′	.6134	.7898	.7766	1.2876	1.2662	1.6304	10′
38° 0′	.6157	.7880	.7813	1.2799	1.2690	1.6243	**52° 0′**
10′	.6180	.7862	.7860	1.2723	1.2719	1.6183	50′
20′	.6202	.7844	.7907	1.2647	1.2748	1.6123	40′
30′	.6225	.7826	.7954	1.2572	1.2779	1.6064	30′
40′	.6248	.7808	.8002	1.2497	1.2808	1.6005	20′
50′	.6271	.7790	.8050	1.2423	1.2837	1.5948	10′
39° 0′	.6293	.7771	.8098	1.2349	1.2868	1.5890	**51° 0′**
10′	.6316	.7753	.8146	1.2276	1.2898	1.5833	50′
20′	.6338	.7735	.8195	1.2203	1.2929	1.5777	40′
30′	.6361	.7716	.8243	1.2131	1.2960	1.5721	30′
40′	.6383	.7698	.8292	1.2059	1.2991	1.5666	20′
50′	.6406	.7679	.8342	1.1988	1.3022	1.5611	10′
40° 0′	.6428	.7660	.8391	1.1918	1.3054	1.5557	**50° 0′**
10′	.6450	.7642	.8441	1.1847	1.3086	1.5504	50′
20′	.6472	.7623	.8491	1.1778	1.3118	1.5450	40′
30′	.6494	.7604	.8541	1.1708	1 3151	1.5398	30′
40′	.6517	.7585	.8591	1.1640	1.3184	1.5346	20′
50′	.6539	.7566	.8642	1.1571	1.3217	1.5294	10′
41° 0′	.6561	.7547	.8693	1.1504	1.3250	1.5243	**49° 0′**
10′	.6583	.7528	.8744	1.1436	1.3284	1.5192	50′
20′	.6604	.7509	.8796	1.1369	1.3318	1.5142	40′
30′	.6626	.7490	.8847	1.1303	1.3352	1.5092	30′
40′	.6648	.7470	.8899	1.1237	1.3386	1.5042	20′
50′	.6670	.7451	.8952	1.1171	1.3421	1.4993	10′
42° 0′	.6691	.7431	.9004	1.1106	1.3456	1.4945	**48° 0′**
10′	.6713	.7412	.9057	1.1041	1.3492	1.4897	50′
20′	.6734	.7392	.9110	1.0977	1.3527	1.4849	40′
30′	.6756	.7373	.9163	1.0913	1.3563	1.4802	30′
40′	.6777	.7353	.9217	1.0850	1.3600	1.4755	20′
50′	.6799	.7333	.9271	1.0786	1.3636	1.4709	10′
43° 0′	.6820	.7314	.9325	1.0724	1.3673	1.4663	**47° 0′**
10′	.6841	.7294	.9380	1.0661	1.3711	1.4617	50′
20′	.6862	.7274	.9435	1.0599	1.3748	1.4572	40′
30′	.6884	.7254	.9490	1.0538	1.3786	1.4527	30′
40′	.6905	.7234	.9545	1.0477	1.3824	1.4483	20′
50′	.6926	.7214	.9601	1.0416	1.3863	1.4439	10′
44° 0′	.6947	.7193	.9657	1.0355	1.3902	1.4396	**46° 0′**
10′	.6967	.7173	.9713	1.0295	1.3941	1.4352	50′
20′	.6988	.7153	.9770	1.0235	1.3980	1.4310	40′
30′	.7009	.7133	.9827	1.0176	1.4020	1.4267	30′
40′	.7030	.7112	.9884	1.0117	1.4061	1.4225	20′
50′	.7050	.7092	.9942	1.0058	1.4101	1.4184	10′
45° 0′	.7071	.7071	1.0000	1.0000	1.4142	1.4142	**45° 0′**
	cos x	sin x	cot x	tan x	csc x	sec x	x

Common Logarithms of Numbers

n	0	1	2	3	4	5	6	7	8	9
10	0000	0043	0086	0128	0170	0212	0253	0294	0334	0374
11	0414	0453	0492	0531	0569	0607	0645	0682	0719	0755
12	0792	0828	0864	0899	0934	0969	1004	1038	1072	1106
13	1139	1173	1206	1239	1271	1303	1335	1367	1399	1430
14	1461	1492	1523	1553	1584	1614	1644	1673	1703	1732
15	1761	1790	1818	1847	1875	1903	1931	1959	1987	2014
16	2041	2068	2095	2122	2148	2175	2201	2227	2253	2279
17	2304	2330	2355	2380	2405	2430	2455	2480	2504	2529
18	2553	2577	2601	2625	2648	2672	2695	2718	2742	2765
19	2788	2810	2833	2856	2878	2900	2923	2945	2967	2989
20	3010	3032	3054	3075	3096	3118	3139	3160	3181	3201
21	3222	3243	3263	3284	3304	3324	3345	3365	3385	3404
22	3424	3444	3464	3483	3502	3522	3541	3560	3579	3598
23	3617	3636	3655	3674	3692	3711	3729	3747	3766	3784
24	3802	3820	3838	3856	3874	3892	3909	3927	3945	3962
25	3979	3997	4014	4031	4048	4065	4082	4099	4116	4133
26	4150	4166	4183	4200	4216	4232	4249	4265	4281	4298
27	4314	4330	4346	4362	4378	4393	4409	4425	4440	4456
28	4472	4487	4502	4518	4533	4548	4564	4579	4594	4609
29	4624	4639	4654	4669	4683	4698	4713	4728	4742	4757
30	4771	4786	4800	4814	4829	4843	4857	4871	4886	4900
31	4914	4928	4942	4955	4969	4983	4997	5011	5024	5038
32	5051	5065	5079	5092	5105	5119	5132	5145	5159	5172
33	5185	5198	5211	5224	5237	5250	5263	5276	5289	5302
34	5315	5328	5340	5353	5366	5378	5391	5403	5416	5428
35	5441	5453	5465	5478	5490	5502	5514	5527	5539	5551
36	5563	5575	5587	5599	5611	5623	5635	5647	5658	5670
37	5682	5694	5705	5717	5729	5740	5752	5763	5775	5786
38	5798	5809	5821	5832	5843	5855	5866	5877	5888	5899
39	5911	5922	5933	5944	5955	5966	5977	5988	5999	6010
40	6021	6031	6042	6053	6064	6075	6085	6096	6107	6117
41	6128	6138	6149	6160	6170	6180	6191	6201	6212	6222
42	6232	6243	6253	6263	6274	6284	6294	6304	6314	6325
43	6335	6345	6355	6365	6375	6385	6395	6405	6415	6425
44	6435	6444	6454	6464	6474	6484	6493	6503	6513	6522
45	6532	6542	6551	6561	6571	6580	6590	6599	6609	6618
46	6628	6637	6646	6656	6665	6675	6684	6693	6702	6712
47	6721	6730	6739	6749	6758	6767	6776	6785	6794	6803
48	6812	6821	6830	6839	6848	6857	6866	6875	6884	6893
49	6902	6911	6920	6928	6937	6946	6955	6964	6972	6981
50	6990	6998	7007	7016	7024	7033	7042	7050	7059	7067
51	7076	7084	7093	7101	7110	7118	7126	7135	7143	7152
52	7160	7168	7177	7185	7193	7202	7210	7218	7226	7235
53	7243	7251	7259	7267	7275	7284	7292	7300	7308	7316
54	7324	7332	7340	7348	7356	7364	7372	7380	7388	7396

Common Logarithms of Numbers

n	0	1	2	3	4	5	6	7	8	9
55	7404	7412	7419	7427	7435	7443	7451	7459	7466	7474
56	7482	7490	7497	7505	7513	7520	7528	7536	7543	7551
57	7559	7566	7574	7582	7589	7597	7604	7612	7619	7627
58	7634	7642	7649	7657	7664	7672	7679	7686	7694	7701
59	7709	7716	7723	7731	7738	7745	7752	7760	7767	7774
60	7782	7789	7796	7803	7810	7818	7825	7832	7839	7846
61	7853	7860	7868	7875	7882	7889	7896	7903	7910	7917
62	7924	7931	7938	7945	7952	7959	7966	7973	7980	7987
63	7993	8000	8007	8014	8021	8028	8035	8041	8048	8055
64	8062	8069	8075	8082	8089	8096	8102	8109	8116	8122
65	8129	8136	8142	8149	8156	8162	8169	8176	8182	8189
66	8195	8202	8209	8215	8222	8228	8235	8241	8248	8254
67	8261	8267	8274	8280	8287	8293	8299	8306	8312	8319
68	8325	8331	8338	8344	8351	8357	8363	8370	8376	8382
69	8388	8395	8401	8407	8414	8420	8426	8432	8439	8445
70	8451	8457	8463	8470	8476	8482	8488	8494	8500	8506
71	8513	8519	8525	8531	8537	8543	8549	8555	8561	8567
72	8573	8579	8585	8591	8597	8603	8609	8615	8621	8627
73	8633	8639	8645	8651	8657	8663	8669	8675	8681	8686
74	8692	8698	8704	8710	8716	8722	8727	8733	8739	8745
75	8751	8756	8762	8768	8774	8779	8785	8791	8797	8802
76	8808	8814	8820	8825	8831	8837	8842	8848	8854	8859
77	8865	8871	8876	8882	8887	8893	8899	8904	8910	8915
78	8921	8927	8932	8938	8943	8949	8954	8960	8965	8971
79	8976	8982	8987	8993	8998	9004	9009	9015	9020	9025
80	9031	9036	9042	9047	9053	9058	9063	9069	9074	9079
81	9085	9090	9096	9101	9106	9112	9117	9122	9128	9133
82	9138	9143	9149	9154	9159	9165	9170	9175	9180	9186
83	9191	9196	9201	9206	9212	9217	9222	9227	9232	9238
84	9243	9248	9253	9258	9263	9269	9274	9279	9284	9289
85	9294	9299	9304	9309	9315	9320	9325	9330	9335	9340
86	9345	9350	9355	9360	9365	9370	9375	9380	9385	9390
87	9395	9400	9405	9410	9415	9420	9425	9430	9435	9440
88	9445	9450	9455	9460	9465	9469	9474	9479	9484	9489
89	9494	9499	9504	9509	9513	9518	9523	9528	9533	9538
90	9542	9547	9552	9557	9562	9566	9571	9576	9581	9586
91	9590	9595	9600	9605	9609	9614	9619	9624	9628	9633
92	9638	9643	9647	9652	9657	9661	9666	9671	9675	9680
93	9685	9689	9694	9699	9703	9708	9713	9717	9722	9727
94	9731	9736	9741	9745	9750	9754	9759	9763	9768	9773
95	9777	9782	9786	9791	9795	9800	9805	9809	9814	9818
96	9823	9827	9832	9836	9841	9845	9850	9854	9859	9863
97	9868	9872	9877	9881	9886	9890	9894	9899	9903	9908
98	9912	9917	9921	9926	9930	9934	9939	9943	9948	9952
99	9956	9961	9965	9969	9974	9978	9983	9987	9991	9996

Logarithms of Trigonometric Functions *

Angle	L Sin	L Tan	L Cot	L Cos	
0° 0′	——	——	——	10.0000	90° 0′
10′	7.4637	7.4637	12.5363	.0000	50′
20′	.7648	.7648	.2352	.0000	40′
30′	7.9408	7.9409	12.0591	.0000	30′
40′	8.0658	8.0658	11.9342	.0000	20′
50′	.1627	.1627	.8373	10.0000	10′
1° 0′	8.2419	8.2419	11.7581	9.9999	89° 0′
10′	.3088	.3089	.6911	.9999	50′
20′	.3668	.3669	.6331	.9999	40′
30′	.4179	.4181	.5819	.9999	30′
40′	.4637	.4638	.5362	.9998	20′
50′	.5050	.5053	.4947	.9998	10′
2° 0′	8.5428	8.5431	11.4569	9.9997	88° 0′
10′	.5776	.5779	.4221	.9997	50′
20′	.6097	.6101	.3899	.9996	40′
30′	.6397	.6401	.3599	.9996	30′
40′	.6677	.6682	.3318	.9995	20′
50′	.6940	.6945	.3055	.9995	10′
3° 0′	8.7188	8.7194	11.2806	9.9994	87° 0′
10′	.7423	.7429	.2571	.9993	50′
20′	.7645	.7652	.2348	.9993	40′
30′	.7857	.7865	.2135	.9992	30′
40′	.8059	.8067	.1933	.9991	20′
50′	.8251	.8261	.1739	.9990	10′
4° 0′	8.8436	8.8446	11.1554	9.9989	86° 0′
10′	.8613	.8624	.1376	.9989	50′
20′	.8783	.8795	.1205	.9988	40′
30′	.8946	.8960	.1040	.9987	30′
40′	.9104	.9118	.0882	.9986	20′
50′	.9256	.9272	.0728	.9985	10′
5° 0′	8.9403	8.9420	11.0580	9.9983	85° 0′
10′	.9545	.9563	.0437	.9982	50′
20′	.9682	.9701	.0299	.9981	40′
30′	.9816	.9836	.0164	.9980	30′
40′	8.9945	8.9966	11.0034	.9979	20′
50′	9.0070	9.0093	10.9907	.9977	10′
6° 0′	9.0192	9.0216	10.9784	9.9976	84° 0′
10′	.0311	.0336	.9664	.9975	50′
20′	.0426	.0453	.9547	.9973	40′
30′	.0539	.0567	.9433	.9972	30′
40′	.0648	.0678	.9322	.9971	20′
50′	.0755	.0786	.9214	.9969	10′
7° 0′	9.0859	9.0891	10.9109	9.9968	83° 0′
10′	.0961	.0995	.9005	.9966	50′
20′	.1060	.1096	.8904	.9964	40′
30′	.1157	.1194	.8806	.9963	30′
40′	.1252	.1291	.8709	.9961	20′
50′	.1345	.1385	.8615	.9959	10′
8° 0′	9.1436	9.1478	10.8522	9.9958	82° 0′
10′	.1525	.1569	.8431	.9956	50′
20′	.1612	.1658	.8342	.9954	40′
30′	.1697	.1745	.8255	.9952	30′
40′	.1781	.1831	.8169	.9950	20′
50′	.1863	.1915	.8085	.9948	10′
9° 0′	9.1943	9.1997	10.8003	9.9946	81° 0′
	L Cos	L Cot	L Tan	L Sin	Angle

* Subtract 10 from each entry in this table to obtain the proper logarithm of the indicated trigonometric function.

Logarithms of Trigonometric Functions

Angle	L Sin	L Tan	L Cot	L Cos	
9° 0′	9.1943	9.1997	10.8003	9.9946	**81° 0′**
10′	.2022	.2078	.7922	.9944	50′
20′	.2100	.2158	.7842	.9942	40′
30′	.2176	.2236	.7764	.9940	30′
40′	.2251	.2313	.7687	.9938	20′
50′	.2324	.2389	.7611	.9936	10′
10° 0′	9.2397	9.2463	10.7537	9.9934	**80° 0′**
10′	.2468	.2536	.7464	.9931	50′
20′	.2538	.2609	.7391	.9929	40′
30′	.2606	.2680	.7320	.9927	30′
40′	.2674	.2750	.7250	.9924	20′
50′	.2740	.2819	.7181	.9922	10′
11° 0′	9.2806	9.2887	10.7113	9.9919	**79° 0′**
10′	.2870	.2953	.7047	.9917	50′
20′	.2934	.3020	.6980	.9914	40′
30′	.2997	.3085	.6915	.9912	30′
40′	.3058	.3149	.6851	.9909	20′
50′	.3119	.3212	.6788	.9907	10′
12° 0′	9.3179	9.3275	10.6725	9.9904	**78° 0′**
10′	.3238	.3336	.6664	.9901	50′
20′	.3296	.3397	.6603	.9899	40′
30′	.3353	.3458	.6542	.9896	30′
40′	.3410	.3517	.6483	.9893	20′
50′	.3466	.3576	.6424	.9890	10′
13° 0′	9.3521	9.3634	10.6366	9.9887	**77° 0′**
10′	.3575	.3691	.6309	.9884	50′
20′	.3629	.3748	.6252	.9881	40′
30′	.3682	.3804	.6196	.9878	30′
40′	.3734	.3859	.6141	.9875	20′
50′	.3786	.3914	.6086	.9872	10′
14° 0′	9.3837	9.3968	10.6032	9.9869	**76° 0′**
10′	.3887	.4021	.5979	.9866	50′
20′	.3937	.4074	.5926	.9863	40′
30′	.3986	.4127	.5873	.9859	30′
40′	.4035	.4178	.5822	.9856	20′
50′	.4083	.4230	.5770	.9853	10′
15° 0′	9.4130	9.4281	10.5719	9.9849	**75° 0′**
10′	.4177	.4331	.5669	.9846	50′
20′	.4223	.4381	.5619	.9843	40′
30′	.4269	.4430	.5570	.9839	30′
40′	.4314	.4479	.5521	.9836	20′
50′	.4359	.4527	.5473	.9832	10′
16° 0′	9.4403	9.4575	10.5425	9.9828	**74° 0′**
10′	.4447	.4622	.5378	.9825	50′
20′	.4491	.4669	.5331	.9821	40′
30′	.4533	.4716	.5284	.9817	30′
40′	.4576	.4762	.5238	.9814	20′
50′	.4618	.4808	.5192	.9810	10′
17° 0′	9.4659	9.4853	10.5147	9.9806	**73° 0′**
10′	.4700	.4898	.5102	.9802	50′
20′	.4741	.4943	.5057	.9798	40′
30′	.4781	.4987	.5013	.9794	30′
40′	.4821	.5031	.4969	.9790	20′
50′	.4861	.5075	.4925	.9786	10′
18° 0′	9.4900	9.5118	10.4882	9.9782	**72° 0′**
	L Cos	L Cot	L Tan	L Sin	Angle

Logarithms of Trigonometric Functions

Angle	L Sin	L Tan	L Cot	L Cos	
18° 0′	9.4900	9.5118	10.4882	9.9782	**72° 0′**
10′	.4939	.5161	.4839	.9778	50′
20′	.4977	.5203	.4797	.9774	40′
30′	.5015	.5245	.4755	.9770	30′
40′	.5052	.5287	.4713	.9765	20′
50′	.5090	.5329	.4671	.9761	10′
19° 0′	9.5126	9.5370	10.4630	9.9757	**71° 0′**
10′	.5163	.5411	.4589	.9752	50′
20′	.5199	.5451	.4549	.9748	40′
30′	.5235	.5491	.4509	.9743	30′
40′	.5270	.5531	.4469	.9739	20′
50′	.5306	.5571	.4429	.9734	10′
20° 0′	9.5341	9.5611	10.4389	9.9730	**70° 0′**
10′	.5375	.5650	.4350	.9725	50′
20′	.5409	.5689	.4311	.9721	40′
30′	.5443	.5727	.4273	.9716	30′
40′	.5477	.5766	.4234	.9711	20′
50′	.5510	.5804	.4196	.9706	10′
21° 0′	9.5543	9.5842	10.4158	9.9702	**69° 0′**
10′	.5576	.5879	.4121	.9697	50′
20′	.5609	.5917	.4083	.9692	40′
30′	.5641	.5954	.4046	.9687	30′
40′	.5673	.5991	.4009	.9682	20′
50′	.5704	.6028	.3972	.9677	10′
22° 0′	9.5736	9.6064	10.3936	9.9672	**68° 0′**
10′	.5767	.6100	.3900	.9667	50′
20′	.5798	.6136	.3864	.9661	40′
30′	.5828	.6172	.3828	.9656	30′
40′	.5859	.6208	.3792	.9651	20′
50′	.5889	.6243	.3757	.9646	10′
23° 0′	9.5919	9.6279	10.3721	9.9640	**67° 0′**
10′	.5948	.6314	.3686	.9635	50′
20′	.5978	.6348	.3652	.9629	40′
30′	.6007	.6383	.3617	.9624	30′
40′	.6036	.6417	.3583	.9618	20′
50′	.6065	.6452	.3548	.9613	10′
24° 0′	9.6093	9.6486	10.3514	9.9607	**66° 0′**
10′	.6121	.6520	.3480	.9602	50′
20′	.6149	.6553	.3447	.9596	40′
30′	.6177	.6587	.3413	.9590	30′
40′	.6205	.6620	.3380	.9584	20′
50′	.6232	.6654	.3346	.9579	10′
25° 0′	9.6259	9.6687	10.3313	9.9573	**65° 0′**
10′	.6286	.6720	.3280	.9567	50′
20′	.6313	.6752	.3248	.9561	40′
30′	.6340	.6785	.3215	.9555	30′
40′	.6366	.6817	.3183	.9549	20′
50′	.6392	.6850	.3150	.9543	10′
26° 0′	9.6418	9.6882	10.3118	9.9537	**64° 0′**
10′	.6444	.6914	.3086	.9530	50′
20′	.6470	.6946	.3054	.9524	40′
30′	.6495	.6977	.3023	.9518	30′
40′	.6521	.7009	.2991	.9512	20′
50′	.6546	.7040	.2960	.9505	10′
27° 0′	9.6570	9.7072	10.2928	9.9499	**63° 0′**
	L Cos	L Cot	L Tan	L Sin	Angle

Logarithms of Trigonometric Functions

Angle	L Sin	L Tan	L Cot	L Cos	
27° 0′	9.6570	9.7072	10.2928	9.9499	63° 0′
10′	.6595	.7103	.2897	.9492	50′
20′	.6620	.7134	.2866	.9486	40′
30′	.6644	.7165	.2835	.9479	30′
40′	.6668	.7196	.2804	.9473	20′
50′	.6692	.7226	.2774	.9466	10′
28° 0′	9.6716	9.7257	10.2743	9.9459	62° 0′
10′	.6740	.7287	.2713	.9453	50′
20′	.6763	.7317	.2683	.9446	40′
30′	.6787	.7348	.2652	.9439	30′
40′	.6810	.7378	.2622	.9432	20′
50′	.6833	.7408	.2592	.9425	10′
29° 0′	9.6856	9.7438	10.2562	9.9418	61° 0′
10′	.6878	.7467	.2533	.9411	50′
20′	.6901	.7497	.2503	.9404	40′
30′	.6923	.7526	.2474	.9397	30′
40′	.6946	.7556	.2444	.9390	20′
50′	.6968	.7585	.2415	.9383	10′
30° 0′	9.6990	9.7614	10.2386	9.9375	60° 0′
10′	.7012	.7644	.2356	.9368	50′
20′	.7033	.7673	.2327	.9361	40′
30′	.7055	.7701	.2299	.9353	30′
40′	.7076	.7730	.2270	.9346	20′
50′	.7097	.7759	.2241	.9338	10′
31° 0′	9.7118	9.7788	10.2212	9.9331	59° 0′
10′	.7139	.7816	.2184	.9323	50′
20′	.7160	.7845	.2155	.9315	40′
30′	.7181	.7873	.2127	.9308	30′
40′	.7201	.7902	.2098	.9300	20′
50′	.7222	.7930	.2070	.9292	10′
32° 0′	9.7242	9.7958	10.2042	9.9284	58° 0′
10′	.7262	.7986	.2014	.9276	50′
20′	.7282	.8014	.1986	.9268	40′
30′	.7302	.8042	.1958	.9260	30′
40′	.7322	.8070	.1930	.9252	20′
50′	.7342	.8097	.1903	.9244	10′
33° 0′	9.7361	9.8125	10.1875	9.9236	57° 0′
10′	.7380	.8153	.1847	.9228	50′
20′	.7400	.8180	.1820	.9219	40′
30′	.7419	.8208	.1792	.9211	30′
40′	.7438	.8235	.1765	.9203	20′
50′	.7457	.8263	.1737	.9194	10′
34° 0′	9.7476	9.8290	10.1710	9.9186	56° 0′
10′	.7494	.8317	.1683	.9177	50′
20′	.7513	.8344	.1656	.9169	40′
30′	.7531	.8371	.1629	.9160	30′
40′	.7550	.8398	.1602	.9151	20′
50′	.7568	.8425	.1575	.9142	10′
35° 0′	9.7586	9.8452	10.1548	9.9134	55° 0′
10′	.7604	.8479	.1521	.9125	50′
20′	.7622	.8506	.1494	.9116	40′
30′	.7640	.8533	.1467	.9107	30′
40′	.7657	.8559	.1441	.9098	20′
50′	.7675	.8586	.1414	.9089	10′
36° 0′	9.7692	9.8613	10.1387	9.9080	54° 0′
	L Cos	L Cot	L Tan	L Sin	Angle

Logarithms of Trigonometric Functions

Angle	L Sin	L Tan	L Cot	L Cos	
36° 0′	9.7692	9.8613	10.1387	9.9080	**54° 0′**
10′	.7710	.8639	.1361	.9070	50′
20′	.7727	.8666	.1334	.9061	40′
30′	.7744	.8692	.1308	.9052	30′
40′	.7761	.8718	.1282	.9042	20′
50′	.7778	.8745	.1255	.9033	10′
37° 0′	9.7795	9.8771	10.1229	9.9023	**53° 0′**
10′	.7811	.8797	.1203	.9014	50′
20′	.7828	.8824	.1176	.9004	40′
30′	.7844	.8850	.1150	.8995	30′
40′	.7861	.8876	.1124	.8985	20′
50′	.7877	.8902	.1098	.8975	10′
38° 0′	9.7893	9.8928	10.1072	9.8965	**52° 0′**
10′	.7910	.8954	.1046	.8955	50′
20′	.7926	.8980	.1020	.8945	40′
30′	.7941	.9006	.0994	.8935	30′
40′	.7957	.9032	.0968	.8925	20′
50′	.7973	.9058	.0942	.8915	10′
39° 0′	9.7989	9.9084	10.0916	9.8905	**51° 0′**
10′	.8004	.9110	.0890	.8895	50′
20′	.8020	.9135	.0865	.8884	40′
30′	.8035	.9161	.0839	.8874	30′
40′	.8050	.9187	.0813	.8864	20′
50′	.8066	.9212	.0788	.8853	10′
40° 0′	9.8081	9.9238	10.0762	9.8843	**50° 0′**
10′	.8096	.9264	.0736	.8832	50′
20′	.8111	.9289	.0711	.8821	40′
30′	.8125	.9315	.0685	.8810	30′
40′	.8140	.9341	.0659	.8800	20′
50′	.8155	.9366	.0634	.8789	10′
41° 0′	9.8169	9.9392	10.0608	9.8778	**49° 0′**
10′	.8184	.9417	.0583	.8767	50′
20′	.8198	.9443	.0557	.8756	40′
30′	.8213	.9468	.0532	.8745	30′
40′	.8227	.9494	.0506	.8733	20′
50′	.8241	.9519	.0481	.8722	10′
42° 0′	9.8255	9.9544	10.0456	9.8711	**48° 0′**
10′	.8269	.9570	.0430	.8699	50′
20′	.8283	.9595	.0405	.8688	40′
30′	.8297	.9621	.0379	.8676	30′
40′	.8311	.9646	.0354	.8665	20′
50′	.8324	.9671	.0329	.8653	10′
43° 0′	9.8338	9.9697	10.0303	9.8641	**47° 0′**
10′	.8351	.9722	.0278	.8629	50′
20′	.8365	.9747	.0253	.8618	40′
30′	.8378	.9772	.0228	.8606	30′
40′	.8391	.9798	.0202	.8594	20′
50′	.8405	.9823	.0177	.8582	10′
44° 0′	9.8418	9.9848	10.0152	9.8569	**46° 0′**
10′	.8431	.9874	.0126	.8557	50′
20′	.8444	.9899	.0101	.8545	40′
30′	.8457	.9924	.0076	.8532	30′
40′	.8469	.9949	.0051	.8520	20′
50′	.8482	9.9975	.0025	.8507	10′
45° 0′	9.8495	10.0000	10.0000	9.8495	**45° 0′**
	L Cos	L Cot	L Tan	L Sin	Angle

INDEX

D

E

F

G

H

I

J

L

M

N

O

P

Q

R

S

T

V

W

X

Y

Z

ANSWERS

Page x
1. 1:05 $\frac{5}{11}$ p.m. **3.** 120 km

Page 3
1. $\{-4\}$ **3.** $\{-6\}$ **5.** $\{2\}$ **7.** $\{-3\}$ **9.** $\{3\}$
11. $\{-11\}$ **13.** $-3\frac{1}{3}$ **15.** $1\frac{3}{5}$ **17.** $1\frac{1}{3}$ **19.** 10
21. 12 **23.** $9\frac{1}{3}$ **25.** $\{5\}$ **27.** $\left\{\frac{3}{4}\right\}$ **29.** $\{3\}$
31. $\left\{-1\frac{5}{6}\right\}$ **33.** $\frac{a+b}{5}$ **35.** $\frac{c-b}{5a}$ **37.** $\frac{5b-5a}{4}$
39. ϕ

Page 5
1. 3 **3.** -3 **5.** -6 **7.** -2 **9.** -5 **11.** -12
13. $\left\{2\frac{7}{10}\right\}$ **15.** $\left\{-\frac{5}{6}\right\}$ **17.** $\left\{2\frac{2}{5}\right\}$ **19.** 6 **21.** 16
23. ϕ

Page 8
Exercises **1., 3., 5.** require *graphs* and solution sets. **1.** $\{y \mid y \geq -3\}$ **3.** $\{x \mid x \leq 4\}$
5. $\{y \mid y \geq -2\}$ **7.** $\{x \mid x > -2\}$ **9.** $\{x \mid x \leq 1\}$
11. $\{x \mid x \geq -2\}$ **13.** $\{x \mid x \leq -2\}$ **15.** $\{x \mid x < -3\}$
Exercises **17., 19., 21.** require *graphs* and solution sets. **17.** $\{x \mid x > -2\}$ **19.** $\{x \mid x < 1\}$
21. $\{x \mid x \leq 3\}$ **23.** $\{x \mid x > -7\}$

Page 10
1. $6(n + 4) + 7$ **3.** $5 + (2n - 3)$ **5.** -2 **7.** 2
9. -7 **11.** -8 **13.** 7 **15.** 2 **17.** 2 **19.** -7
21. $\{x \mid x > -2\}$ **23.** $\{x \mid x \geq 5\}$

Page 12
1. 639,000 m **3.** 6.39 m **5.** .007839 km
7. 780 m **9.** 83.6 m **11.** 13 L **13.** .086 L
15. .003 L **17.** 300 L **19.** 6,800,000 mg

Page 15
1. \$1,680 **3.** \$8,000 **5.** 12 yr **7.** 18.75 ohms
9. 40 ohms **11.** \$9,780 **13.** 3.84 ohms
15. \$4,000 **17.** 6.5 yr **19.** 157.5 volts

Page 16
1. $\{-4\}$ **3.** $\{-2\}$ **5.** $\{9\}$ **7.** $\{2\}$ **9.** 2 **11.** 1
13. 2 Exercise **15.** requires a *graph* and solution set. **15.** $\{n \mid n \geq -4\}$ **17.** $\{x \mid x < 4\}$
19. $\{a \mid a > -3\}$ **21.** $6 + 2n$ **23.** $5(7 + n) + 3$
25. -3 **27.** 40 ohms **29.** \$4,000

Page 18
1. $ac + cb = ca + cb$ Comm. Prop. Mult.
$= c(a + b)$ Dist. Prop.
$= (a + b)c$ Comm. Prop. Mult.
$ac + cb = (a + b)c$ Trans. Prop. Equality
3. $ac + b = b + ac$ Comm. Prop. Add.
$= b + ca$ Comm. Prop. Mult.
$ac + b = b + ca$ Trans. Prop. Equality
5. $ab + ac = a(b + c)$ Dist. Prop.
$= (b + c)a$ Comm. Prop. Mult.
$= (c + b)a$ Comm. Prop. Add.
$ab + ac = (c + b)a$ Trans. Prop. Equality

Page 21
(Graphs for Exercises **1., 3., 5., 7., 9.** are described.) **1.** between -4 and 2 **3.** left of -2, right of 3 **5.** 6, between 2 and 6 **7.** left of 3, 5, right of 5 **9.** all points except zero
11. $\{y \mid -5 \leq y < 5\}$ **13.** $\{a \mid -3 \leq a \leq 5\}$
15. $\{y \mid -2 < y \leq 3\}$ **17.** $\{a \mid 1 \leq a \leq 2\}$
Exercises **19., 21., 23., 25.** require *graphs* and solution sets. **19.** $\{x \mid x < 2 \text{ or } x > 5\}$
21. $\{a \mid a < -2 \text{ or } a \geq 3\}$
23. $\{x \mid x < -1 \text{ or } x > 3\}$
25. $\{a \mid a \leq -2 \text{ or } a \geq 3\}$ **27.** $\{x \mid -3 < x < 2\}$
29. $\{x \mid -3 < x < 5\}$ **31.** $\{x \mid -6 < x < 2\}$
33. $\{x \mid x \leq -3 \text{ or } x \geq 2\}$
35. $\{x \mid -1 < x < 2 \text{ or } x > 4\}$

Page 24

1. $\{10, -4\}$ **3.** $\{5, -5\}$ **5.** $\{4, -4\}$ **7.** $\{5, -5\}$ **9.** $\left\{5, -1\frac{2}{3}\right\}$ **11.** $\left\{2, -2\frac{2}{5}\right\}$ **13.** $\{x \mid -3 < x < 3\}$ **15.** $\{y \mid -8 \le y \le 8\}$ Exercises **17., 19., 21., 23., 25., 27., 29., 31., 33., 35.** require *graphs* and solution sets. **17.** $\{x \mid 2 < x < 4\}$ **19.** $\{a \mid -2 \le a \le 6\}$ **21.** $\{a \mid -6 < a < 6\}$ **23.** $\{n \mid -3 \le n \le 3\}$ **25.** $\{x \mid -1 < x < 6\}$ **27.** $\{a \mid -1 \le a \le 5\}$ **29.** $\{y \mid -2 < y < 1\}$ **31.** $\{x \mid -3 \le x \le -1\}$ **33.** $\{x \mid -1 < x < 5\}$ **35.** $\{x \mid x < -4 \text{ or } -2 < x < 4 \text{ or } x > 6\}$

Page 26

1. T **3.** F **5.** 3

7.

$\oplus$	0	1	2	3	4
0	0	1	2	3	4
1	1	2	3	4	0
2	2	3	4	0	1
3	3	4	0	1	2
4	4	0	1	2	3

9. $3 \oplus 4 = 2$; $4 \oplus 3 = 2$ **11.** $4 \otimes 2 = 3$; $2 \otimes 4 = 3$ **13.** $(2 \otimes 3) \otimes 1 = (1) \otimes 1 = 1$; $2 \otimes (3 \otimes 1) = 2 \otimes (3) = 1$ **15.** 3 **17.** 3

Page 30

1. 6; 11 **3.** 7; 11 **5.** 2; 7 **7.** 5 m; 11 m **9.** 6 cm; 10 cm; 12 cm **11.** 2; 14 **13.** 3; 18 **15.** 4 cm; 17 cm **17.** 3; 11; 21 **19.** 3 cm; 9 cm

Page 34

1. $15n + 40$ **3.** $130n$ **5.** $20d + 15$ **7.** 7 nickels; 15 dimes **9.** 4 dimes; 12 quarters **11.** 3 quarters; 68 dimes **13.** 5 at 15¢; 10 at 8¢; 14 at 20¢ **15.** 22 nickels; 18 dimes

Page 35

1. no **3.** no **5.** no; yes; no; no

Page 39

1. 12 at 50¢; 8 at 30¢ **3.** 6 kg **5.** 4 kg soft; 8 kg hard **7.** 20 kg milk; 4 kg cocoa **9.** 16 at 60¢; 8 at 80¢ **11.** 3 g chemical A; 6 g B; 9 g C

Page 41

1. Walter 8; Brenda 12 **3.** Conrad 6; Denise 18 **5.** Ted 2; Wanda 22 **7.** Paula 8; Mel 24 **9.** Keith 14; Phyllis 20

Page 42

1. -3 and all points between -3 and 5 **3.** all points to the left of -4; 2 and all points to the right of 2 **5.** $\{x \mid -1 < x < 4\}$ **7.** $\{a \mid a < -4 \text{ or } a > 1\}$ **9.** $\{5, 2\}$ **11.** $\{a \mid -5 < a < 5\}$ Exercises **13., 15.** require *graphs* and solution sets. **13.** $\{x \mid -8 \le x \le 2\}$ **15.** $\{a \mid a < -1 \text{ or } a > 4\}$ **17.** 4; 19 **19.** 2; 11 **21.** 12 nickels; 5 dimes **23.** 3 kg caramels; 5 kg peppermints **25.** Conrad 9; Edna 14

Page 44

1. .36 m² **3.** 420,000 m² **5., 7., 9.** (Answers may vary.) **5.** 30° **7.** 39° **9.** 180°

Page 47

1. $12x^5$ **3.** $20c^6$ **5.** $16a^4$ **7.** $5x^4$ **9.** $40n^5$ **11.** $-5a^{12}$ **13.** $81x^8$ **15.** $100x^7$ **17.** $-5x^3 + 7x$ **19.** -27 **21.** -16 **23.** 1,600 **25.** x^5y^6 **27.** $-10x^5y^7$ **29.** $25x^2y^2$ **31.** a^6b^4 **33.** x^8y^{12} **35.** $-8a^3b^{12}$ **37.** $8a^2 + 5b^2$ **39.** $6x^2 - 4y^2$ **41.** 32 **43.** 900 **45.** $-64{,}000$ **47.** 52 **49.** x^{5a} **51.** x^4 **53.** $x^{2a+1}y^{2a+2}$ **55.** $x^{3a+4}y^{4b}$ **57.** x^{3a} **59.** x^{a^2} **61.** 5^ax^a **63.** $3^{a+1}x^{a+1}$ **65.** $2^{a+2}x^{3a+6}$ **67.** $x^{4ac}y^{4bc}$ **69.** $x^{3a+9}y^{4a+12}$

Page 52

1. -192 **3.** 11 **5.** $2a^2 + 9$ **7.** $5y^2 - y + 2$ **9.** $-9b^2 - 5b - 6$ **11.** $6y^2 + y - 3$ **13.** $-12a^2 + a - 1$ **15.** $5y^2 - 2y + 6$ **17.** $12a - 9$ **19.** $-3n^2 + 3n - 6$ **21.** $-3ab + 4a^3b - 3b^2$ **23.** $2a^2 + 7ab + b^2 - 3a^2b$ **25.** $4y^6 + 4y^4 - 2y^2 - 5$ **27.** $8a^2b^2 - 2ab + 14$ **29.** $7xy + 8x^2$

Page 54

1. $8x^2 - 14x - 15$ **3.** $8x^2 - 2x - 15$ **5.** $2x^2 + 3x - 35$ **7.** $4x^2 - 9$ **9.** $5x^3 - 11x^2 - 2x + 8$ **11.** $9x^3 - 9x^2 + 14x - 8$ **13.** $x^2 + 10x + 25$ **15.** $4x^2 - 12x + 9$ **17.** $6x^2 + 8xy - 8y^2$ **19.** $12x^2 - xy - y^2$ **21.** $4x^2 - 9y^2$ **23.** $12x^2 + 7xy + 9x - 10y^2 - 6y$ **25.** $16x^2 + 8xy + y^2$ **27.** $9x^2 - 24xy + 16y^2$ **29.** $8x^2 + 4x - 60$ **31.** $24x^3 + 76x^2 + 40x$ **33.** $16x^2 - 4$ **35.** $18x^2 - 12x + 2$ **37.** $20x^2 - 2xy - 6y^2$ **39.** $45x^2 - 80y^2$ **41.** $x^3 + 15x^2 + 75x + 125$ **43.** $x^3 - 3x^2y + 3xy^2 - y^3$ **45.** $25a^2 - 20ab + 40a + 4b^2 - 16b + 16$ **47.** $x^{2n} - 2x^n - 15$ **49.** $x^{2c} - y^{4c}$

Page 57

1. $5(t - 2)$ **3.** $5(x^2 - 3x + 2)$ **5.** $2(3a^2 + 4a - 6)$ **7.** $x(4x^2 + x - 7)$ **9.** $a(2a + 1)$ **11.** $c(c + 1)$ **13.** $x(7x^2 + 2x + 1)$ **15.** $3n(n + 2)$ **17.** $5x^2(2x - 3)$ **19.** $4a^3(a - 3)$ **21.** $-1(5a - 3b)$ **23.** $-1(3x^2 - 4x + 2)$ **25.** $-1(x + 4)$ **27.** $3xy(x + 2)$ **29.** $x^2y^2(y^2 - xy + x^2)$ **31.** $2ay(2y + 3a + 4)$ **33.** $4xy(11x^2 - 25y^2)$ **35.** $x^4(x^{a+2} + 1)$ **37.** $x^{n+2}(x^3 + 1)$

Page 61

1. $(3x + 2)(x + 1)$ **3.** $(5x + 7)(x + 1)$ **5.** $(7x - 2)(x - 1)$ **7.** $(2x - 1)(x + 2)$ **9.** $(3x - 2)(x + 1)$ **11.** $(7x + 2)(x - 1)$ **13.** $(x + 1)(x - 3)$ **15.** $(x - 7)(x + 1)$ **17.** $(7x + 1)(x + 3)$ **19.** $(7x - 5)(x - 1)$ **21.** $(x - 17)(x - 1)$ **23.** $(7x + 11)(x - 1)$ **25.** $(5x + 1)(x - 11)$ **27.** $(x - 19)(x + 1)$ **29.** $(5x - 3y)(x - y)$ **31.** $(2x - y)(x + 3y)$ **33.** $(x + 3y)(x - y)$ **35.** $(3x + 2y)(x - y)$ **37.** $(3x - y)(x + 3y)$ **39.** $(2x + 3y)(x - y)$ **41.** $(2x^m - 1)(x^m + 5)$ **43.** $(x^a + y^b)(x^a + 2y^b)$ **45.** $(3x^a + y^{3b})(x^a - 3y^{3b})$

Page 63

1. $(2a + 5)(a - 2)$ **3.** $(2x + 5)(x + 2)$ **5.** $(4n + 3)(2n + 1)$ **7.** $(3a - 8)(a + 1)$ **9.** $(5x - 6)(x - 1)$ **11.** $(2x + 5)(2x - 1)$ **13.** $(7x - 10)(x + 1)$ **15.** $(3x - 10)(x - 1)$ **17.** $(5x + 3)(3x - 1)$ **19.** $(3x - 5)(x - 3)$ **21.** $(7n + 6)(n + 1)$ **23.** $(5a - 3)(3a - 1)$ **25.** $(n - 8)(n + 3)$ **27.** $(n - 4)(n + 3)$ **29.** $(x + 5)(x + 3)$ **31.** $(x - 9)(x + 2)$ **33.** $(n + 6)(n - 4)$ **35.** $(x + 10)(x + 1)$ **37.** $(5x + 4)(2x + 3)$ **39.** $(4c - 3)(c - 3)$ **41.** $(3x + 2)(2x - 5)$ **43.** $(2a - 3b)(a + 2b)$ **45.** $(3m - n)(2m + 3n)$ **47.** $(7a + 4b)(a + 2b)$ **49.** $(5r - 4s)(3r + s)$ **51.** $(2x + 5y)(x - 2y)$ **53.** $(2x + 3y)(x - 2y)$ **55.** $(5x - 3y)(2x - y)$ **57.** $(4a + 3b)(2a - 3b)$ **59.** $(3x^{2m} + 10)(3x^{2m} - 2)$ **61.** $(5x^a + 3y^b)(x^a + 4y^b)$ **63.** $(6x^{3a} + 5y^{2b})(2x^{3a} - 3y^{2b})$ **65.** $(x^{n-1} + 5)(x^{n-1} - 2)$ **67.** $(x^{a+3} + y^{a-2})(x^{a+3} + y^{a-2})$

Page 66

1. $16x^2 - 25$ **3.** $n^2 - 36$ **5.** $a^2 - 9b^2$ **7.** $(2x + 3)(2x - 3)$ **9.** $(c + 1)(c - 1)$ **11.** $(7 + b)(7 - b)$ **13.** $4x^2 + 20x + 25$ **15.** $9 + 24x + 16x^2$ **17.** $a^2 + 14ab + 49b^2$ **19.** no **21.** yes **23.** $(3n - 4)(3n - 4)$ **25.** $(x + 4)(x + 4)$ **27.** $(1 + 10a)(1 + 10a)$ **29.** $(9x + y)(9x - y)$ **31.** $(10x + 7y^2)(10x - 7y^2)$ **33.** $(7cd + 1)(7cd - 1)$ **35.** $(5c - 2d)(5c - 2d)$ **37.** $(6x - 5y)(6x - 5y)$ **39.** $(xy - 1)(xy - 1)$ **41.** $(x + 5)(x^2 - 5x + 25)$ **43.** $(4y + 1)(16y^2 - 4y + 1)$ **45.** $(2x^2 + y)(4x^4 - 2x^2y + y^2)$

Page 68

1. yes **3.** no **5.** yes

Page 71

1. $\{2, 5\}$ **3.** $\{-4, -1\}$ **5.** $\left\{1\frac{2}{3}, -2\right\}$ **7.** $\{0, 4\}$
9. $\{6\}$ **11.** $-5; 3$ **13.** $\frac{1}{2}; 4$ **15.** $\frac{2}{3}; -1$
17. $6; -6$ **19.** $0; 5$ **21.** $0; 3$ **23.** 4 **25.** $5; 8$
27. $-5; -10$ **29.** $-\frac{1}{3}; -\frac{1}{4}$ **31.** $\{-4, 3\}$
33. $\{3, -3\}$ **35.** $\{-3, 5\}$ **37.** $\frac{1}{2}; 3$ **39.** $-\frac{1}{2}; -1$
41. $\left\{1\frac{2}{3}, -3\right\}$ **43.** $\left\{-1\frac{2}{3}, 1\frac{2}{3}\right\}$ **45.** $\left\{-1\frac{1}{5}, 1\frac{1}{5}\right\}$
47. $\left\{-1\frac{3}{4}, 0\right\}$ **49.** $\{9\}$ **51.** $\{7\}$ **53.** $\left\{1\frac{1}{4}, -2\frac{1}{2}\right\}$
55. $2; 5$ **57.** $-6; 1$ **59.** $-\frac{3}{4}; 2$
61. $\{-3, -1, 2, 4\}$ **63.** $\{-4, 4, -10, 10\}$
65. $\{0, 4, -5\}$ **67.** $\{-3, 3, -2, 2\}$

Page 75

1. $\{x|3 < x < 6\}$ **3.** $\{x|-5 \le x \le 2\}$
5. $\{x|x \le 1 \text{ or } x \ge 3\}$ **7.** $\{x|x < -3 \text{ or } x > 6\}$
9. $\{x|x < -5 \text{ or } x > -3\}$
11. $\{x|x < -5 \text{ or } x > 5\}$ **13.** $\{x|0 < x < 5\}$
15. $\{x|x < 0 \text{ or } x > 6\}$ **17.** $\{x|-6 \le x \le 6\}$
19. $\{x|-4 < x < 3\}$ **21.** $\{x|x \le -3 \text{ or } x \ge -1\}$
23. $\{x|-5 < x < 0\}$ **25.** $\{x|x < 1 \text{ or } 4 < x < 7\}$
27. $\{x|x \le -3 \text{ or } 2 \le x \le 5\}$
29. $\{x|x < 2 \text{ or } x > 2\}$ **31.** $\{x|x < -2 \text{ or } x > 4\}$
33. $\{x|x \le -5 \text{ or } x \ge 5\}$

Page 78

1. 57, 58, 59 **3.** 23, 24, 25 **5.** 41, 42, 43
7. 32, 33, 34 **9.** 64, 65, 66 **11.** 70, 71, 72
13. 5, 6, 7; $-8, -7, -6$ **15.** 6, 7, 8; $-8,$
$-7, -6$ **17.** 4, 5, 6 **19.** 7, 8, 9; $-7, -6, -5$
21. 18, 19, 20 **23.** 8, 9, 10; $-10, -9, -8$
25. 3, 4, 5; 9, 10, 11 **27.** any 3 consec. integers

Page 81

1. $3(x + 3)(x + 4)$ **3.** $2(x + 4)^2$
5. $2(3x + 5)(3x - 5)$ **7.** $x(x - 4)(x - 2)$
9. $a(2a + b)^2$ **11.** $x^2(x + y)(x - y)$
13. $2(3x - 1)(x - 2)$ **15.** $25(2x + y)^2$
17. $3a(a + 5b)(a - 5b)$
19. $-1(2x - 3)(2x + 1)$ **21.** $-1(n - 1)^2$
23. $-3(x + 3)(x - 3)$ **25.** $-1(3n + 5)^2$
27. $-a(2c + 1)^2$ **29.** $a(x - 5a)(x + 2a)$

Page 82

1. $15x^5$ **3.** a^4b^5 **5.** x^{12} **7.** $25c^2$ **9.** -144
11. 28 **13.** no **15.** yes **17.** trinom.
19. binom. **21.** $4x^3 + 2x^2 - 5x + 4$
23. $6x^2 - 9x$ **25.** $2x^3 + 2xy^2 + 6x^2y$
27. $4x^2 - 9y^2$
29. $4x^2 - 7xy - 6x + 3y^2 + 6y$
31. $n(3n^2 - 2n + 1)$ **33.** $-1(x^2 - 7x + 3)$
35. $(x - 7)(x + 2)$ **37.** $(3y - 2)(3y - 4)$
39. $(3n + 2)^2$ **41.** $4(x + 1)^2$
43. $x(x - 5)(x + 3)$ **45.** $-1(x - 5)^2$
47. $\left\{\frac{2}{3}, -4\right\}$ **49.** $\{0, -7\}$
51. $\{x|x \le -3 \text{ or } x \ge 5\}$
53. 65; 67; 69

Page 86

1. yes **3.** no **5.** -4 **7.** 0 **9.** does not exist
11. 3 **13.** $\frac{7}{8}$ **15.** does not exist **17.** 0
19. does not exist **21.** 0 **23.** 3 **25.** 3; 4

Page 89

1. $\frac{x^2}{y^4}$ **3.** $\frac{c^3}{a^2}$ **5.** $\frac{3}{2b^6c^3}$ **7.** $\frac{3x + 4}{2}$ **9.** $\frac{2x - 3}{a}$
11. $\frac{a + 2}{a - 3}$ **13.** -2 **15.** $\frac{1}{-2}$ **17.** $\frac{x + 2}{5}$
19. $\frac{3}{3n + 1}$ **21.** $\frac{5}{4a - 5}$ **23.** $\frac{a + 3}{a - 6}$
25. $\frac{5}{-1(x - 4)}$ **27.** $\frac{c + 6}{-1(c - 5)}$ **29.** $\frac{3(a - 2)}{a + 3}$
31. $\frac{a + 8}{4(a - 1)}$ **33.** $\frac{3(c - 2)}{c + 6}$ **35.** $x^{2c+3}y^{3d-4}$
37. $\frac{x^2 + 2x + 4}{3}$ **39.** $\frac{x^2 + 3x + 9}{4}$

Page 92

1. $\frac{3y^2}{4x^2}$ **3.** $\frac{3b}{2c}$ **5.** $-\frac{6}{c}$ **7.** $\frac{3(x - 3)}{2}$ **9.** $\frac{3c}{x - 4}$
11. $\frac{-4ac}{3b^4}$ **13.** $\frac{5x^3}{-6}$ **15.** $\frac{3}{5x(x - 4)}$
17. $\frac{n + 6}{2(n + 4)}$ **19.** $\frac{2(x + 1)}{-1(2x + 3)}$ **21.** $\frac{4x}{-1(x - 5)}$
23. $\frac{2x(x + 2y)}{3}$ **25.** $\frac{2(3x - 1)}{a(2x + 1)}$ **27.** $\frac{-1(c + d)}{2(a + b)}$
29. $\frac{1}{2(5x - 4)}$ **31.** $\frac{(x^2 + 2x + 4)(x + 4)}{3}$
33. $\frac{4(x^a - y^b)}{3(2x^c - 5)}$

Page 95

1. $\frac{a}{2}$ 3. $\frac{5a-3}{4}$ 5. $\frac{2}{b}$ 7. $\frac{3x}{4}$ 9. $\frac{3}{y}$ 11. $\frac{1}{2}$
13. $\frac{3}{5}$ 15. $\frac{3(n+1)}{n+4}$ 17. $\frac{3}{x-5}$ 19. $\frac{n+1}{n+3}$
21. $\frac{2}{x+2}$ 23. $\frac{1}{3(a+6)}$

Page 100

1. $\frac{11c}{6}$ 3. $\frac{12n-1}{35}$ 5. $\frac{7y+4x}{x^2y}$ 7. $\frac{19x-6}{30}$
9. $\frac{3y^3+24+x^2y}{6xy^2}$ 11. $\frac{2a^2-9a+8}{2(a-3)(a-4)}$
13. $\frac{4x^2-12x+1}{4x}$ 15. $\frac{2n^2-3}{n-3}$
17. $\frac{11a+13}{2(a+3)(a-3)}$ 19. $\frac{4}{3}$ 21. $\frac{3x}{2}$ 23. $\frac{a-2}{a}$
25. $\frac{5x+1}{3(x-2)(x-4)}$ 27. $\frac{3x^2-6x+1}{(x+3)(x-1)(x-3)}$
29. $\frac{x^2-7x}{(2x-5)(x+2)(x-1)}$ 31. $\frac{8n+1}{(n+1)(n-1)}$
33. $\frac{12a+3b}{5(a+b)(a-b)}$ 35. $\frac{7x+2y}{3(x-y)(x-y)}$

Page 103

1. $\frac{2x}{3}$ 3. $\frac{2}{x+3}$ 5. $\frac{2(a-3)}{a-4}$ 7. $\frac{5n}{24}$
9. $\frac{2x^2-3x}{2(x-4)(x-3)}$ 11. $\frac{4}{x-5}$ 13. $\frac{13}{x-4}$
15. $\frac{3}{a-4}$ 17. $\frac{4n^2-2n-3}{3(n+2)(n-2)}$ 19. $\frac{3}{x-3}$
21. $\frac{a+9}{2(a+3)(a-1)}$ 23. $\frac{4}{x-5}$ 25. 2
27. $\frac{7yz-3xz-5xy}{xyz}$ 29. $\frac{9x-51}{4(x-3)(x+3)}$
31. $\frac{a+7}{(a+2)(a+2)(a+3)}$

Page 105

1. $\frac{1}{x+5}+\frac{1}{x+3}$ 3. $\frac{5}{4(x+1)}+\frac{11}{4(x+5)}$,
or $\frac{1.25}{x+1}+\frac{2.75}{x+5}$

Page 106

1. yes 3. yes 5. $\frac{3}{10}$ 7. 0 9. 2, 3
11. $\frac{4x-5}{3}$ 13. $\frac{x-5}{2x-1}$ 15. $\frac{1}{-5}$ 17. $\frac{1}{x^2-9}$
19. $\frac{2b^3c^2}{-9a}$ 21. $\frac{3(3c-2)}{4}$ 23. $\frac{5(x+y)}{6(c+2)}$
25. $\frac{2c}{xy}$ 27. $\frac{5}{x-3}$ 29. $\frac{6a^2-16a}{3(a-2)(a-4)}$
31. $\frac{4}{x-4}$ 33. $\frac{11x+7}{3(x+5)(x-3)}$ 35. $\frac{5}{3}$
37. $\frac{8}{2x-3}$ 39. $\frac{4x+18}{(x-5)(x+5)}$

Page 112

1. 1 3. 2 5. $\frac{1}{2}$ 7. {1, 6} 9. {1, 4}
11. {−8} 13. 11 15. 2 17. $\frac{3}{10}$ 19. 2; 5
21. 2 23. 12 25. 40 27. 7 29. $\{\frac{7}{3}\}$
31. {4} 33. {−3, 4} 35. $\frac{3}{2}$ 37. all numbers except −4 and 3

Page 115

1. $\frac{1}{15}$; $\frac{4}{15}$; $\frac{x}{15}$ 3. 6 5. 15 7. $3\frac{3}{7}$ 9. 42
11. $6\frac{6}{11}$

Page 118

1. $\frac{13}{44}$ 3. $\frac{51}{32}$ 5. $\frac{b-a}{b+a}$ 7. $\frac{4xy-6x^2}{7y^2+2xy}$
9. $\frac{4y+20xy}{10xy-3x}$ 11. $\frac{2b^2-20a^2b}{7ab^2+15a^2}$ 13. $\frac{7x-6}{8x-10}$
15. $x+3$ 17. $\frac{6a-2}{5a+6}$ 19. 1 21. $\frac{3c-55}{7c}$
23. $\frac{a+2}{a-3}$ 25. $\frac{x+4y}{x+y}$ 27. $\frac{3}{4}$

Page 121

1. $x = \frac{b + c}{a}$ **3.** $x = \frac{c + ab}{a}$ **5.** $x = a$

7. $x = \frac{b}{a + c}$ **9.** $x = \frac{2b}{3a}$ **11.** $x = b$

13. $x = \frac{ac + d}{ab}$ **15.** $x = \frac{-ab - c}{2a - cn}$

17. $x = a - bc$ **19.** $x = \frac{ab}{3}$ **21.** $x = \frac{3ab}{4a + 2b}$

23. $x = \frac{3a}{2 - a}$ **25.** $p = \frac{i}{rt}$; 600

27. $h = \frac{2A}{b}$; 4.1 **29.** $x = \frac{ab}{c - a}$

31. $x = \frac{a^2 - 2c}{a - 2b}$ **33.** $h = \frac{3V}{B}$; 11

35. $m = \frac{T}{g - f}$; 7 **37.** $x = 8a - 14$ **39.** c; $-d$

41. $5a$; $-5a$ **43.** $-a$; $-b$ **45.** $a - b$; $b - a$

Page 123

1. 80 cm^3 **3.** 450 L **5.** 40 cm^3

Page 126

1. .25; term. **3.** .625; term. **5.** 1.6; term.

7. $\frac{7}{9}$ **9.** $\frac{8}{9}$ **11.** $\frac{8}{99}$ **13.** $\frac{65}{90}$ **15.** $\frac{48}{90}$ **17.** $\frac{74}{9}$

19. $\frac{-4}{9}$ **21.** $\frac{393}{90}$ **23.** 3 **25.** $\frac{1}{2}$ **27.** $\frac{361}{999}$

29. $\frac{346}{990}$

Page 129

1. $5x + 6$ **3.** $n + 7$ **5.** $3n^3 + n^2 - 2$

7. $4n - 5$ **9.** $a^2 - 3a - 4$ **11.** yes; $3x^2 - x - 2$, rem. 0 **13.** no; $2n^2 + n - 1$, rem. 2

15. $2n^2 - 4n + 1$, rem. 6 **17.** $3x + 2$, rem. -3

19. 6; 6; yes

Page 130

1. 1 **3.** 27 **5.** -5 **7.** 300 **9.** 80 **11.** -2, 5

13. $\{-1, 4\}$ **15.** $\{6\}$ **17.** $1\frac{7}{8}$

19. $\frac{2y^2 + 9x^2y}{5xy^2 - 12x^2}$ **21.** $\frac{2n + 4}{8n - 19}$

23. $x = \frac{ab + 3c}{a - 2c}$ **25.** $l = \frac{p - 2w}{2}$; 7 **27.** $\frac{72}{99}$

29. $\frac{-10}{3}$ **31.** $3n^2 + 6n + 1$, rem. 5 **33.** yes; $3x^2 + 2x - 2$, rem. 0

Page 132

1. yes **3.** yes

Page 136

1. $\{-6, 6\}$ **3.** $\{-10, 10\}$ **5.** yes; $4 = 2^2$

7. yes; $81 = 9^2$ **9.** 5 **11.** -2 **13.** $2 < \sqrt{8} < 3$

15. $4 < \sqrt{20} < 5$ **17.** $\{-20, 20\}$

19. $\{-12, 12\}$ **21.** -11 **23.** -30

25. $5 < \sqrt{35} < 6$ **27.** $6 < \sqrt{39} < 7$

Page 140

1. 4.1 **3.** 6.7 **5.** irrat. **7.** rat. **9.** rat.

11. 6.3 **13.** 8.7 **15.** rat. **17.** rat., $.242\overline{242}$

19. rat. **21.** 6.32 **23.** 8.72 **25.** 6.633

Page 144

1. $3\sqrt{3}$ **3.** $-2\sqrt{6}$ **5.** $4\sqrt{2}$ **7.** $-6\sqrt{10}$

9. $8\sqrt{5}$ **11.** $-15\sqrt{7}$ **13.** $3\sqrt{2}$; 4.2

15. $4\sqrt{3}$; 6.9 **17.** $\sqrt{77}$ **19.** $\sqrt{10ab}$ **21.** 8

23. $2\sqrt{3}$ **25.** $3y^3$ **27.** $2b\sqrt{3}$ **29.** $x^4\sqrt{x}$

31. $a^2\sqrt{5a}$ **33.** x^2y^3 **35.** $2c^5d^2\sqrt{6}$

37. $ab\sqrt{a}$ **39.** $xy^3\sqrt{xy}$ **41.** $6x^2$ **43.** $10x^2\sqrt{6}$

45. 28 **47.** $9c^2d$ **49.** $x + 5$ **51.** $4x + 4$

53. x^m **55.** x^{2m} **57.** x^{2m} **59.** $x^{2m}y^n\sqrt{y}$

Page 147

1. $-4\sqrt{10} + 5\sqrt{5}$ **3.** $5\sqrt{5}$ **5.** $3\sqrt{n}$

7. $6\sqrt{3} - 3\sqrt{5}$ **9.** $4\sqrt{3} + 3$ **11.** $6 - 6\sqrt{10}$

13. $4 - 2\sqrt{3}$ **15.** $2 - 3\sqrt{3}$ **17.** $34 - 10\sqrt{14}$

19. 2 **21.** $8 + 2\sqrt{15}$ **23.** $22 - 4\sqrt{30}$

25. $5\sqrt{2} + 5\sqrt{3}$ **27.** $3\sqrt{3} + 5\sqrt{5}$

29. $6b\sqrt{ab}$ **31.** $4c + \sqrt{cd} - 5d$

33. $2c - 5\sqrt{cd} - 3d$ **35.** $9c - 4d$

37. $a + 2\sqrt{ab} + b$ **39.** $4c + 12\sqrt{cd} + 9d$

41. $2x - 14$ **43.** $13 + x - 8\sqrt{x - 3}$

45. $13x - 81 + 12\sqrt{x^2 - 9x}$

Page 151

1. $\sqrt{3}$ **3.** $2\sqrt{2}$ **5.** $2\sqrt{5}$ **7.** $2b$ **9.** $3n^2\sqrt{n}$

11. $\frac{\sqrt{6}}{6}$ **13.** $\frac{\sqrt{6}}{2}$ **15.** $\frac{-\sqrt{15}}{4}$ **17.** $\frac{5\sqrt{3}}{6}$

19. $\frac{-3\sqrt{2}}{8}$ **21.** $\frac{5\sqrt{a}}{a}$ **23.** $\frac{3a\sqrt{c}}{5c}$ **25.** $\frac{\sqrt{xy}}{xy^2}$

27. $\frac{-6+2\sqrt{2}}{7}$ **29.** $\frac{-\sqrt{14}+2\sqrt{2}}{3}$

31. $\frac{3\sqrt{2}}{2}$; 2.1 **33.** $\frac{11\sqrt{2}}{10}$; 1.6 **35.** $\frac{5\sqrt{2}}{4}$; 1.8

37. $\frac{\sqrt{35}}{7}$ **39.** $\frac{\sqrt{10ab}}{5b}$ **41.** $\frac{15\sqrt{2}+5\sqrt{7}}{11}$

43. $\frac{6\sqrt{15}+9\sqrt{6}}{2}$ **45.** $\frac{2\sqrt{a}+b}{4a-b^2}$ **47.** $\frac{2\sqrt{3xy}}{3y^2}$

49. $\frac{a\sqrt{7ab}}{7b^3}$ **51.** $\frac{\sqrt{3ab}}{ab^2}$ **53.** $\frac{2\sqrt{cy}}{x^2y^3}$

55. $\frac{\sqrt{10c+8}}{2}$

Page 152

1. 113 **3.** 98 **5.** .51 **7.** 14.23

Page 155

1. 2 **3.** 4 **5.** 5 **7.** 1 **9.** -4 **11.** -1

13. x^4 **15.** x^3 **17.** $\sqrt[3]{20}$ **19.** $\sqrt[3]{14c^2}$

21. $\sqrt[4]{15}$ **23.** $\sqrt[4]{30c^3}$ **25.** $7\sqrt[3]{4}$ **27.** $6\sqrt[3]{c}$

29. $2\sqrt[3]{2}$ **31.** $-2\sqrt[3]{3}$ **33.** $a\sqrt[3]{a^2}$ **35.** $x^3\sqrt[3]{x}$

37. $6\sqrt[3]{2}$ **39.** $4\sqrt[3]{2}$ **41.** $y^2\sqrt[4]{y}$ **43.** $3a\sqrt[3]{a}$

45. $5x^2\sqrt[3]{x^2}$ **47.** $5c\sqrt[4]{c^3}$

Page 160

1. 1 **3.** 1 **5.** $\frac{1}{16}$ **7.** $\frac{1}{144}$ **9.** 160 **11.** 144

13. $\frac{1}{72}$ **15.** $\frac{64}{25}$ **17.** 49 **19.** $\frac{8}{x^3}$ **21.** $\frac{-10}{c^3}$

23. $\frac{x^3}{4}$ **25.** $\frac{3a^2}{5}$ **27.** $\frac{a^4d^5}{c^2b^3}$ **29.** $\frac{2b^5d^4}{3c^3a^2}$ **31.** x^3

33. $\frac{1}{n^6}$ **35.** x^6 **37.** $\frac{3n^3}{4}$ **39.** $\frac{1}{x^6}$ **41.** n^{12}

43. $\frac{16}{x^{12}}$ **45.** $\frac{n^8}{16}$ **47.** $\frac{x^{12}}{125}$ **49.** $\frac{1}{c^8d^{12}}$

51. $\frac{120y^4}{x^2}$ **53.** $\frac{40y^7}{x^5}$ **55.** $\frac{x^5}{y^6}$ **57.** $\frac{2d^6}{3c^4}$

59. $\frac{1}{a^6b^8}$ **61.** $\frac{y^6}{x^8}$ **63.** $\frac{c^9d^6}{-27}$ **65.** $\frac{16z^4}{x^6y^6}$

67. $\frac{a^6c^4}{b^8d^{10}}$ **69.** $\frac{a^4c^8f^{12}}{b^8d^4e^{12}}$

Page 163

1. 3 **3.** 2 **5.** 4 **7.** 4 **9.** 4 **11.** 16

13. $6^{\frac{1}{2}}$ **15.** $10^{\frac{1}{4}}$ **17.** $5b^{\frac{1}{2}}$ **19.** $\sqrt{10}$ **21.** $5\sqrt{a}$

23. $3\sqrt{b}$ **25.** $\sqrt[3]{7}$ **27.** $6\sqrt[3]{c}$ **29.** $2\sqrt[4]{n}$

31. $c^{\frac{2}{3}}$ **33.** $x^{\frac{3}{4}}$ **35.** $a^{\frac{5}{2}}$ **37.** $7^{\frac{2}{3}}$ **39.** $2^{\frac{3}{2}}$ **41.** $3^{\frac{3}{4}}$

43. $\sqrt{27}$ **45.** $4\sqrt[3]{x^2}$ **47.** $5\sqrt[4]{x^3}$

49. $\frac{1}{5}$ **51.** $\frac{1}{2}$ **53.** $\frac{1}{3}$ **55.** $\frac{1}{125}$ **57.** $\frac{1}{125}$

59. $\frac{1}{16}$

Page 165

1. $7^{\frac{4}{5}}$ **3.** $3^{\frac{3}{7}}$ **5.** $x^{\frac{7}{3}}$ **7.** $n^{\frac{3}{5}}$ **9.** x^3 **11.** n^4

13. $x^{\frac{3}{5}}$ **15.** $n^{\frac{2}{3}}$ **17.** xy^2 **19.** $9mn^5$

21. $\frac{x^2}{y^3}$ **23.** $\frac{5m^2}{3n^3}$ **25.** 3 **27.** 10 **29.** $\frac{1}{x^{\frac{2}{5}}}$ **31.** $\frac{1}{n^{\frac{5}{4}}}$

33. $\frac{1}{x^4}$ **35.** $\frac{1}{n^3}$ **37.** $\frac{1}{x^{\frac{1}{3}}}$ **39.** $\frac{1}{n^{\frac{1}{3}}}$ **41.** $\frac{y^{\frac{1}{3}}}{x^{\frac{1}{2}}}$

43. $m^{\frac{1}{3}}m^{\frac{1}{4}}$ **45.** $\frac{y}{x^2}$ **47.** $\frac{3n}{2m^2}$ **49.** $\frac{3}{4}$ **51.** $\frac{1}{1250}$

53. $\frac{27}{8}$ **55.** $\{2\}$ **57.** $\{-2, 2\}$ **59.** $\{\sqrt[3]{7}\}$

Page 168

1. $\{-10, 10\}$ **3.** $\{-8, 8\}$ **5.** $4 < \sqrt{21} < 5$

7. rat. **9.** irrat. **11.** $-2\sqrt{7}$ **13.** $2a^3b^2\sqrt{b}$

15. $5\sqrt{7}-2\sqrt{5}$ **17.** $6\sqrt{a}$ **19.** $30\sqrt{2}$ **21.** 18

23. 4 **25.** $x+10\sqrt{xy}+25y$ **27.** $\frac{-\sqrt{5a}}{2a}$

29. $\frac{6\sqrt{3}+3\sqrt{10}}{2}$ **31.** $c^2\sqrt{2}$ **33.** $2x^3\sqrt{2}$

35. $\frac{\sqrt{6cd}}{3d}$ **37.** $\frac{\sqrt{6n-3}}{3}$ **39.** $3x^3$ **41.** $3\sqrt[3]{2}$

43. $\frac{8}{9}$ **45.** 25 **47.** $\frac{2}{3}$ **49.** $\frac{6}{x^2}$ **51.** $\frac{1}{a^{\frac{3}{7}}}$ **53.** $\frac{a^3c^4}{b^2d^5}$

55. $\frac{9y^6}{x^4}$ **57.** $4x^2y^4$ **59.** $a^{\frac{2}{3}}$ **61.** $\sqrt{6}$

63. $\sqrt[4]{8a^3}$

Page 173

1. $w = 5$ mm, $l = 15$ mm **3.** $b = 4$ m, $h = 8$ m **5.** 6 dm **7.** $w = 4$ cm, $l = 5$ cm **9.** $b = 6$ dm, $h = 2$ dm **11.** 5 mm; 10 mm **13.** square, 9 mm^2; rect., 36 mm^2 **15.** tri., 36 m^2; rect., 72 m^2

Page 175

1. $\left\{\pm \frac{\sqrt{5}}{3}\right\}$ **3.** $\{7, -1\}$ **5.** $\{4, 1\}$
7. $\left\{\frac{2 \pm \sqrt{10}}{3}\right\}$ **9.** $\left\{\frac{-3 \pm 2\sqrt{2}}{5}\right\}$
11. $\{3\sqrt{5}, \sqrt{5}\}$ **13.** $\{1 \pm \sqrt{5}\}$ **15.** $\{2, -4\}$

Page 178

1. $\{2, -8\}$ **3.** $\{5, 3\}$ **5.** $\{-4 \pm \sqrt{7}\}$
7. $\left\{\frac{-3 \pm \sqrt{21}}{2}\right\}$ **9.** $\left\{\frac{-1 \pm \sqrt{21}}{2}\right\}$
11. $\{5 \pm \sqrt{10}\}$ **13.** $\left\{\frac{5 \pm 3\sqrt{5}}{2}\right\}$
15. $\left\{\frac{1 \pm 3\sqrt{5}}{2}\right\}$ **17.** $4, -\frac{1}{2}$ **19.** $\frac{-5 \pm \sqrt{65}}{4}$
21. $\frac{-2 \pm \sqrt{13}}{3}$ **23.** $-5 \pm 3\sqrt{2}$ **25.** $3 \pm \sqrt{15}$
27. $\frac{5 \pm 2\sqrt{5}}{6}$ **29.** $\frac{3 \pm \sqrt{17}}{4}$

Page 181

1. $\frac{-3 \pm 2\sqrt{2}}{2}$ **3.** $-3 \pm 2\sqrt{2}$ **5.** $\frac{3}{2}$
7. $\frac{-1 \pm \sqrt{33}}{8}$ **9.** $\frac{-7 \pm \sqrt{5}}{2}$ **11.** $\frac{1 \pm \sqrt{17}}{4}$
13. $-2 \pm \sqrt{5}$ **15.** $\frac{3 \pm \sqrt{2}}{2}$ **17.** $\frac{1 \pm 2\sqrt{2}}{3}$
19. $\frac{-1 \pm \sqrt{41}}{10}$ **21.** $\frac{-3 \pm \sqrt{15}}{3}$ **23.** $\frac{-1 \pm \sqrt{6}}{5}$
25. $-\frac{2}{3}$ **27.** 5, 0 **29.** $\frac{7}{2}$, 0 **31.** $\pm 2\sqrt{3}$
33. $\left\{\frac{-1 \pm \sqrt{17}}{2}\right\}$ **35.** $\left\{\frac{2 \pm \sqrt{10}}{2}\right\}$
37. $\left\{\frac{7 \pm \sqrt{37}}{2}\right\}$ **39.** $\left\{\frac{-1 \pm \sqrt{7}}{2}\right\}$
41. $\left\{\frac{7 \pm \sqrt{29}}{2}\right\}$ **43.** $\{3\sqrt{2}, -\sqrt{2}\}$
45. $\left\{\frac{-\sqrt{10} \pm \sqrt{15}}{5}\right\}$

Page 183

1. 36; 2 **3.** −8; none **5.** 0; 1 **7.** −7; none **9.** 13; 2 **11.** 108; 2 **13.** 1; 2 **15.** 33; 2 **17.** −7; none **19.** 60; 2 **21.** 69; 2 **23.** −19; none **25.** 81; 2 **27.** −4; none **29.** 0; 1 **31.** 121; 2 **33.** −8; none

Page 184

1. $5x^3 + 7x^2 + 16x + 30$, rem. 20
3. $3x^2 - x - 2$, rem. 2

Page 187

1. $x^2 - 5x + 6 = 0$ **3.** $x^2 - 3x - 10 = 0$
5. $x^2 - 6x + 9 = 0$ **7.** $x^2 - 10x + 25 = 0$
9. $8x^2 - 6x + 1 = 0$ **11.** $16x^2 - 8x - 3 = 0$
13. $4x^2 - 4x + 1 = 0$ **15.** $25x^2 - 20x + 4 = 0$
17. $2x^2 - 5x + 2 = 0$ **19.** $3x^2 - 5x - 2 = 0$
21. $x^2 - 6x + 4 = 0$ **23.** $x^2 - 4x - 3 = 0$
25. $4x^2 - 8x + 1 = 0$ **27.** $4x^2 - 2x - 1 = 0$
29. $x^2 - 10x + 13 = 0$ **31.** $x^2 - 6x - 11 = 0$
33. $4x^2 - 4x - 7 = 0$ **35.** $4x^2 - 12x + 1 = 0$
37. $x^2 - 15x + 50 = 0$ **39.** $x^2 - 5x - 500 = 0$
41. $10x^2 - 29x + 10 = 0$
43. $10x^2 + 7x - 6 = 0$
45. $\frac{(-b + \sqrt{b^2 - 4ac}) + (-b - \sqrt{b^2 - 4ac})}{2a} =$
$\frac{-2b + 0}{2a} = \frac{-2b}{2a} = \frac{-b}{a} = -\frac{b}{a}$

Page 191

1. $x = 4$; $x - 1 = 3$ **3.** $x = 12$; $x - 7 = 5$ **5.** $x = 8$; $2x - 6 = 10$ **7.** $x = 12$; $x - 3 = 9$; $x + 3 = 15$ **9.** $x = 5$; $2x + 2 = 12$; $2x + 3 = 13$ **11.** 11 **13.** 8 **15.** 41 cm **17.** 3, 11; $-3\frac{2}{3}$, −9 **19.** 4, 5; $\frac{1}{2}$, −2 **21.** $x = 3\sqrt{5}$; $2x = 6\sqrt{5}$ **23.** $x = 2 + \sqrt{5}$; $2x = 4 + 2\sqrt{5}$; $2x + 1 = 5 + 2\sqrt{5}$ **25.** $x = 2 + \sqrt{7}$; $2x + 1 = 5 + 2\sqrt{7}$; $2x + 2 = 6 + 2\sqrt{7}$
27. $4\sqrt{2}$ mm **29.** $3 + 3\sqrt{2}$ dm; $6 + 3\sqrt{2}$ dm
31. $l = 2 + \sqrt{10}$ m; $w = -2 + \sqrt{10}$ m
33. $-2 + \sqrt{5}$, $2 + \sqrt{5}$; $-2 - \sqrt{5}$, $2 - \sqrt{5}$
35. $\frac{-2 + \sqrt{10}}{2}$, $2 + \sqrt{10}$; $\frac{-2 - \sqrt{10}}{2}$, $2 - \sqrt{10}$
37. $w = \frac{-2 + 3\sqrt{2}}{2}$, $l = 2 + 3\sqrt{2}$
39. $w = 6 + 2\sqrt{10}$, $l = 18 + 6\sqrt{10}$, $d = 20 + 6\sqrt{10}$ **41.** $w = 3 + 2\sqrt{3}$, $l = 10 + 6\sqrt{3}$, $d = 11 + 6\sqrt{3}$

Page 195

1. 25 **3.** no sol. **5.** 8 **7.** $-1, -2$ **9.** 3
11. $\frac{1}{2}$ **13.** $\{9\}$ **15.** $\left\{\frac{1}{4}\right\}$ **17.** $\{12\}$ **19.** $\{-2, -3\}$
21. $\{5\}$ **23.** $\{3\}$ **25.** $\{10\}$ **27.** $\left\{-1, \frac{1}{2}\right\}$ **29.** 2
31. 4 **33.** 2, 3 **35.** 2, 6 **37.** 9 **39.** 1
41. 4 **43.** 3 **45.** 5 **47.** no sol.

Page 198

1. $5i$ **3.** $8i$ **5.** $-i\sqrt{2}$ **7.** $2i\sqrt{2}$ **9.** $3i\sqrt{5}$
11. $4 + i\sqrt{2}$ **13.** $3 + 3i\sqrt{2}$ **15.** $6 - 5i\sqrt{2}$
17. $-i\sqrt{2}, i\sqrt{2}$ **19.** $-2i\sqrt{5}, 2i\sqrt{5}$ **21.** $-6i, 6i$
23. $2 + 4i$ **25.** $-1 - 2i$ **27.** $-2 + 4i$
29. $-7 + 4i$ **31.** $6i$ **33.** $8i\sqrt{3}$ **35.** $-2 - 15i$
37. $7 - 6i\sqrt{3}$ **39.** $-4 - 6i\sqrt{5}$ **41.** $-3i\sqrt{5}$, $3i\sqrt{5}$ **43.** $8 + 7i\sqrt{2}$ **45.** $5 + 7i\sqrt{3}$
47. $2 + 4i\sqrt{2}$ **49.** $0 + 0i$; $0 + 0i$

Page 200

1. $-i$ **3.** -1 **5.** -1

Page 203

1. -6 **3.** $-\sqrt{6}$ **5.** $\sqrt{21}$ **7.** $-6\sqrt{10}$
9. $-20\sqrt{30}$ **11.** $3i\sqrt{10}$ **13.** -9 **15.** -5
17. $14 - 2i$ **19.** $14 - 2i$ **21.** $-5 + 5i$
23. $21 - 20i$ **25.** 13 **27.** 25 **29.** 10
31. $\frac{7i}{-2}$ **33.** $\frac{5i}{3}$ **35.** $\frac{i}{6}$ **37.** $\frac{5 + i}{13}$ **39.** i
41. $\frac{-6 - 3i}{10}$ **43.** $4 + i$ **45.** $\frac{7 - i}{10}$ **47.** $6\sqrt{10}$
49. $-8i$ **51.** $-i$ **53.** 21 **55.** $4 - 6i\sqrt{5}$ **57.** 11
59. 1 or $1 + 0i$; $\frac{a - bi}{a^2 + b^2}$
61. $(a + bi) + (a - bi) = 2a + 0 = 2a$; $2a$ is a real number since a is a real number.

Page 207

1. $\frac{1 \pm 2i}{2}$ **3.** $3 \pm i\sqrt{2}$ **5.** $\left\{\frac{3 \pm i\sqrt{7}}{.2}\right\}$
7. $\{-2 \pm i\}$ **9.** $\{4 \pm 2i\}$ **11.** $\left\{\frac{1 \pm i\sqrt{7}}{4}\right\}$
13. $\left\{\frac{-1 \pm i}{2}\right\}$ **15.** $\left\{\frac{1 \pm 5i}{2}\right\}$ **17.** $\left\{\frac{1 \pm i\sqrt{11}}{6}\right\}$
19. $\{-3 \pm i\sqrt{2}\}$ **21.** $\left\{\frac{3 \pm i\sqrt{5}}{2}\right\}$
23. $\left\{\frac{1 \pm i\sqrt{2}}{3}\right\}$ **25.** $\left\{\frac{\pm i\sqrt{10}}{2}\right\}$ **27.** $\left\{\frac{\pm i\sqrt{15}}{5}\right\}$
29. $\left\{\frac{\pm 2i\sqrt{3}}{3}\right\}$ **31.** -7; yes **33.** -23; yes
35. 32; no **37.** $\frac{7 \pm i\sqrt{3}}{2}$ **39.** $\frac{1 \pm i\sqrt{15}}{8}$
41. $3 \pm i$ **43.** $\{-3i, i\}$ **45.** $\left\{\frac{i}{2}, 2i\right\}$ **47.** $\{-3\}$
49. $\{2, -1 \pm i\sqrt{3}\}$ **51.** $\{-4, 2 \pm 2i\sqrt{3}\}$
53.
$$\begin{aligned} &(-1 + i\sqrt{3})(-1 + i\sqrt{3})(-1 + i\sqrt{3}) \\ &= (1 - 2i\sqrt{3} + 3i^2)(-1 \pm i\sqrt{3}) \\ &= (-2 - 2i\sqrt{3})(-1 + i\sqrt{3}) \\ &= 2 - 6i^2 = 2 + 6 = 8 \end{aligned}$$

Page 208

1. $\{-2\sqrt{3}, 2\sqrt{3}\}$ **3.** $\left\{\frac{-1 \pm \sqrt{7}}{3}\right\}$ **5.** $\{3\}$
7. $\frac{3 \pm \sqrt{5}}{2}$ **9.** 0; one **11.** $x^2 + 6x + 2 = 0$
13. $x = 3 + 2\sqrt{6}$, $x + 1 = 4 + 2\sqrt{6}$, $x + 4 = 7 + 2\sqrt{6}$ **15.** 6 **17.** 11 m; 3 m
19. $6 + 10i\sqrt{3}$ **21.** $4 + 2i$ **23.** $512i$ **25.** $8\sqrt{15}$
27. $21 - i$ **29.** $-16 - 30i$ **31.** $\frac{-8 - i}{5}$
33. $-6i, 6i$; yes **35.** $-2 \pm i\sqrt{2}$; yes

Page 210

1. $5 - i$ **3.** $1 + 6i$

Page 213

1. quad. 1 **3.** quad. 3 **5.** quad. 1 **7.** *A* 3 units from vertical axis, 2 units from horizontal axis; *B* 3 units from vertical axis, 1 unit from horizontal axis; *C* 5 units from vertical axis, 1 unit from horizontal axis; *D* 5 units from vertical axis, 4 units from horizontal axis; *E* 5 units from vertical axis, 1 unit from horizontal axis; *F* 5 units from vertical axis, 2 units from horizontal axis; *G* 3 units from vertical axis, 4 units from horizontal axis; *H* 3 units from vertical axis, 4 units from horizontal axis; *I* 1 unit from vertical axis, 4 units from horizontal axis; *J* 1 unit from vertical axis, 3 units from horizontal axis; *K* 2 units from vertical axis, 3 units from horizontal axis.
9. $FE = 3$, $GH = 8$, $JI = 7$, $AB = 3$, $DC = 5$
11. 3 **13.** 4 **15.** 1 **17.** quad. 2 **19.** quad. 3
21. $|b_2 - b_1|$

Page 216
1. -5 **3.** -4 **5.** -9 **7.** 7 **9.** -4 **11.** -2 **13.** $7\frac{3}{4}$ **15.** $7\frac{3}{4}$ **17.** $3a$ **19.** $-12y$ **21.** $-7b$ **23.** $14y$

Page 220
1. negative **3.** undefined **5.** -1 **7.** $\frac{6}{7}$ **9.** 0 **11.** $-\frac{1}{2}$ **13.** undefined **15.** 1 **17.** slopes equal, $\frac{3}{2}$ **19.** slopes equal, $\frac{2}{3}$ **21.** 3 **23.** $-\dfrac{b}{a}$ **25.** $\dfrac{y_2 - y_1}{x_2 - x_1}$

Page 223
1. $y = 4x - 2$ **3.** $y = -\frac{1}{2}x - 2$ **5.** $y = -x - 1$ **7.** $y = 2$ **9.** $y = 4$ **11.** $y = -\frac{1}{2}x - 2$ **13.** $y = \frac{1}{2}x - \frac{1}{3}$ **15.** $y = -4x + 4$ **17.** $y = -2x + 3$ **19.** $m = \frac{1}{2}, b = 2$ **21.** $m = -4, b = 7$ **23.** $m = -\frac{2}{3}, b = 3$ **25.** $m = -\frac{2}{3}, b = \frac{3}{2}$ **27.** $m = 2, b = \frac{4}{3}$ **29.** $m = -\frac{4}{5}, b = 2$ **31.** $m = -2, b = -\frac{4}{3}$ **33.** Solve for $y \cdot y = \dfrac{-A}{B}x - \dfrac{C}{B}$; $m = \dfrac{-A}{B}$

Page 224
1. 30 **3.** 6 **5.** 6 **7.** 3,024

Page 227
1. $y - 3 = 5(x - 2)$; $y = 5x - 7$
3. $y - 1 = -\frac{1}{3}(x - 4)$; $y = -\frac{1}{3}x + \frac{7}{3}$
5. $y - 4 = \frac{1}{2}(x - 8)$; $y = \frac{1}{2}x$
7. $y + 6 = \frac{9}{7}(x + 5)$; $y = \frac{9}{7}x + \frac{3}{7}$
9. $y - 5 = 3(x - 1)$; $y = 3x + 2$
11. $y + 3 = -\frac{1}{2}(x + 2)$; $y = -\frac{1}{2}x - 4$
13. $y - 4 = -4(x + 2)$; $y = -4x - 4$
15. $y - 2 = -3(x + 4)$; $y = -3x - 10$
17. $y + \frac{1}{2} = -\frac{1}{3}(x + 3)$; $y = -\frac{1}{3}x - \frac{3}{2}$
19. $y - 6 = 3\left(x - \frac{5}{3}\right)$; $y = 3x + 1$

Page 230
Three points are given for each line. **1.** (0, 3), (2, 4), (4, 5) **3.** (0, 1), (3, 2), (6, 3) **5.** (0, -1), (2, -2), (4, -3) **7.** (0, -2), (1, -5), (2, -8) **9.** (0, 4), (1, 2), (2, 0) **11.** (0, 3), (1, 7), (2, 11) **13.** (0, 7), (3, 8), (6, 9) **15.** (-1, 0), (-1, 1), (-1, 2) **17.** (0, 2), (1, 2), (2, 2) **19.** yes **21.** no **23.** yes **25.** (0, -2), (3, -1), (6, 0) **27.** (0, 5), (1, 8), (2, 11) **29.** (-7, 0), (-7, 1), (-7, 2) **31.** (0, 2), (7, 0), (14, -2)

Page 231
1. 42 **3.** 35 **5.** 3,060

Page 232
1. quad. 2 **3.** quad. 3 **5.** 11 **7.** 6 **9.** 1 **11.** 4 **13.** $+1$ **15.** $\frac{2}{9}$ **17.** undefined **19.** $y = 8x - 18$ **21.** $m = \frac{2}{3}, b = -1$ **23.** $m = -12, b = 14$ **25.** (0, 4), (2, 7), (4, 10) **27.** (0, -3), (1, -3), (2, -3) **29.** yes

Page 237
1. $4\sqrt{6}$ **3.** $\sqrt{65}$ **5.** $\sqrt{61}$ **7.** $\sqrt{89}$ **9.** $\sqrt{5}$ **11.** no **13.** yes **15.** no **17.** yes **19.** yes **21.** 12 m

Page 239
1. $2\sqrt{5}$ **3.** $\sqrt{97}$ **5.** $3\sqrt{13}$ **7.** $\sqrt{149}$ **9.** $\sqrt{41}$ **11.** $2\sqrt{5}$ **13.** $\sqrt{10}$ **15.** $2\sqrt{2}$ **17.** $6\sqrt{2}$ **19.** $\sqrt{178}$ **21.** $\dfrac{\sqrt{229}}{6}$ **23.** $\dfrac{\sqrt{221}}{6}$ **25.** $\frac{17}{20}$ **27.** $\sqrt{4a^2 + b^2}$ **29.** $\sqrt{16s^2 + 9p^2}$

Page 243
1. (4, 2) **3.** (2, 0) **5.** (4, 5) **7.** (2, -3) **9.** (2, 1) **11.** (6, 4) **13.** (3, 2) **15.** (-4, -3) **17.** (-5, -5) **19.** (-5, -3) **21.** $\left(-\frac{13}{2}, -\frac{13}{2}\right)$ **23.** (9, 8) **25.** (-1, -2) **27.** (4, -2) **29.** (8, 1) **31.** (10, 3) **33.** (9, 9) **35.** (8, -12) **37.** (-9, -14) **39.** (4, 7) **41.** (9, -3) **43.** $M = \left(0, \frac{1}{2}\right)$, $N = (3, 0)$, $T = (3, 2)$, $V = \left(0, 2\frac{1}{2}\right)$; $MN = \dfrac{\sqrt{37}}{2}$ and $VT = \dfrac{\sqrt{37}}{2}$ therefore $MN = VT$; $MV = 2$ and $NT = 2$ therefore $MV = NT$

Page 246

1. $\frac{2}{3}$ **3.** $\frac{1}{2}$ **5.** $-\frac{5}{3}$ **7.** $-\frac{6}{5}$ **9.** $y = \frac{3}{2}x + 4$ **11.** $y = \frac{1}{2}x + 2$ **13.** $y = 2$ **15.** $y = 2x - 5$ **17.** $y = -\frac{4}{3}x + \frac{8}{3}$ **19.** $y = -\frac{3}{2}x + \frac{5}{2}$ **21.** neither **23.** $y = -\frac{3}{2}x + \frac{13}{2}$ **25.** $y = \frac{1}{5}x + 1$

Page 249

1. $m(\overline{BC}) = -\frac{1}{3}$, $m(\overline{AB}) = 3$ **3.** $AB = \sqrt{68}$, $BC = \sqrt{68}$, $AC = \sqrt{136}$; $(\sqrt{136})^2 = (\sqrt{68})^2 + (\sqrt{68})^2$ **5.** yes **7.** yes **9.** no **11.** $m(\overline{AB}) = m(\overline{DC}) = 2$; $m(\overline{AD}) = m(\overline{BC}) = \frac{3}{5}$ **13.** midpoint $\overline{AB}\,(5, 2)$, midpoint $\overline{BC}\,(6, 4)$ **15.** $m(\overline{RV}) = m(\overline{ST}) = \frac{3}{2}$; $m(\overline{RS}) = m(\overline{VT}) = 0$ **17.** $RV = ST = \sqrt{13}$, $VT = RS = 6$ **19.** parallelogram

Page 250

1. $b = 12$ **3.** $a = 4\sqrt{10}$ **5.** $a = 4$ **7.** yes **9.** yes **11.** $15\sqrt{2}$ m **13.** $\sqrt{53}$ **15.** $\sqrt{113}$ **17.** (5.5, .5) **19.** (5.5, 6) **21.** (5, −2) **23.** (12, 7) **25.** $\frac{3}{4}$ **27.** $\frac{1}{3}$ **29.** $-\frac{6}{5}$ **31.** $\frac{3}{2}$ **33.** $y = \frac{3}{5}x + \frac{7}{5}$ **35.** $y = \frac{7}{3}x + \frac{11}{3}$ **37.** $y = \frac{1}{2}x - 8$ **39.** $\overleftrightarrow{PQ} \parallel \overleftrightarrow{RS}$ **41.** $m(\overline{BC}) = \frac{1}{2}$; $m(\overline{AC}) = -2$ **43.** $m(\overline{AB}) = m(\overline{BC}) = m(\overline{AC}) = -2$ **45.** $M\left(\frac{1}{2}, \frac{3}{2}\right)$, $N\left(\frac{9}{2}, 4\right)$, $m(\overline{MN}) = m(\overline{AB}) = \frac{5}{8}$, $MN = \frac{1}{2}\sqrt{89}$, $AB = \sqrt{89}$

Page 255

1. $D = \{0, 2, -1, 3, 4\}$, $R = \{0, 3, 2\}$; yes
3. $D = \{-1, -2, -3\}$, $R = \{-2, -1, -3\}$; no
5. $D = \{4, 6, -2, 5, 0\}$, $R = \{-1, 3, 0\}$; yes
7. $D = \{-1, -2, -3, 4\}$, $R = \{4\}$; yes
9. $D = \{2, 3, -2, -3\}$, $R = \{-3\}$; yes
11. $\{(-2, 1), (-1, 2), (1, 1), (2, 2), (3, -1)\}$, $D = \{-2, -1, 1, 2, 3\}$, $R = \{1, 2, -1\}$; yes
13. $\{(-2, -1), (-1, 1), (1, 1), (2, -1), (4, 3)\}$, $D = \{-2, -1, 1, 2, 4\}$, $R = \{-1, 1, 3\}$; yes
15. $D = \{x \mid 0 \leq x < 3\}$, $R = \{y \mid -2 < y < 2\}$; no
17. $D = \{x \mid -2 < x \leq 2 \text{ and } x \neq -2, -1, 0, 1\}$, $R = \{y \mid -2 \leq y \leq 2\}$; yes **19.** $D = \{x \mid x \neq -2, 2\}$, $R = \{y \mid -3 < y \text{ and } y \neq -2\}$; yes

Page 257

1. 6 **3.** −7 **5.** 1 **7.** 8 **9.** 5 **11.** 5 **13.** {5, 11, 20} **15.** {26, 8, 98} **17.** {9, 4, 1} **19.** {4, 1, 0, 2} **21.** 10 **23.** −9 **25.** $2a + b^2 - 3$ **27.** $2ah + h^2$

Page 260

1. linear function **3.** not a function **5.** linear function **7.** linear function **9.** linear function **11.** not a function **13.** constant function **15.** linear function **17.** function **19.** function **21.** function **23.** function **25.** function **27.** function

Page 263

1. yes; $k = -\frac{1}{2}$ **3.** no **5.** yes; $k = -1$ **7.** −32 **9.** −6 **11.** 6 **13.** 40 **15.** 120 **17.** 24 **19.** decreases, increases **21.** increases, decreases **23.** 360 **25.** 360 km

Page 266

1. yes; $k = 45$ **3.** no **5.** yes; $k = 1$ **7.** 3 **9.** 3 **11.** 1 **13.** $1\frac{7}{9}$ **15.** $\frac{35}{6}$ **17.** decreases, increases **19.** increases **21.** 4 **23.** 12

Page 269

1. joint **3.** no **5.** no **7.** $\frac{20}{3}$ **9.** 128 **11.** 144 **13.** 18 **15.** $\frac{400}{9}$ **17.** 128 **19.** 2.22×10^{-6} dynes

Page 270

1. $D = \{4, -1, 3, -5\}$, $R = \{3, 2\}$; function
3. $D = \{-1, -2, -3\}$, $R = \{2, 3\}$; function
5. $D = \{-1, 1, 2, 3\}$, $R = \{1, -1, 2\}$; not a function **7.** $D = \{x \mid 0 \leq x < 2\}$, $R = \{y \mid -2 < y < 2\}$; not a function **9.** 1 **11.** 7 **13.** 1 **15.** 10 **17.** $R = \{-1, 0, 3\}$ **19.** not a function **21.** function **23.** function **25.** constant function **27.** inverse; $k = 12$ **29.** direct; $k = -3$ or $-\frac{1}{3}$ **31.** 30 **33.** 2 **35.** 180 **37.** 120

Page 276

1. $x = -\frac{3}{2}$; $\left(-\frac{3}{2}, \frac{9}{2}\right)$ **3.** $x = \frac{3}{4}$; $\left(\frac{3}{4}, -\frac{49}{4}\right)$ **5.** $x = -2$; $(-2, -7)$ **7.** $x = 2$; $(2, -1)$ **9.** $x = 0$; $(0, 0)$ **11.** $x = 1$; $(1, 3)$ **13.** .4; -3.9 **15.** 4.2; $-.2$ **17.** 4.4; $-.4$ **19.** $(0, 0)$; $x = 0$ **21.** Turning point is on axis of symmetry $x = -\frac{b}{2a}$. Substitute $x = -\frac{b}{2a}$ in $y = ax^2 + bx + c$. Solve for y.

Page 278

1. $x^2 + y^2 = 64$ **3.** $x^2 + y^2 = 121$ **5.** $x^2 + y^2 = 169$ **7.** $x^2 + y^2 = 16$ **9.** $x^2 + y^2 = 4$ **11.** $x^2 + y^2 = 81$ **13.** $x^2 + y^2 = 100$ **15.** $x^2 + y^2 = 225$ **17.** $x^2 + y^2 = 7$ **19.** $x^2 + y^2 = 169$ **21.** $x^2 + y^2 = 5$ **23.** $x^2 + y^2 = 25$ **25.** $x^2 + y^2 = 13$ **27.** $x^2 + y^2 = 9$ **29.** 4 **31.** 2 **33.** 6 **35.** $\sqrt{17}$ **37.** $\sqrt{6}$ **39.** circle, $r = \sqrt{6}$ **41.** circle, $r = \sqrt{7}$ **43.** circle, $r = \frac{1}{2}\sqrt{35}$ **45.** circle, $r = \frac{1}{2}\sqrt{17}$ **47.** circle, $r = 2$ **49.** circle, $r = \sqrt{2}$ **51.** circle, $r = \frac{1}{4}$ **53.** not a circle **55.** $(x + 1)^2 + (y - 2)^2 = 16$ **57.** $(x + 1)^2 + (y + 2)^2 = 50$

Page 282

1. $\frac{x^2}{25} + \frac{y^2}{9} = 1$ **3.** $\frac{x^2}{4} + \frac{y^2}{3} = 1$ **5.** $\frac{x^2}{8} + \frac{y^2}{36} = 1$ **7.** $x = \pm 5$, $y = \pm 2$ **9.** $x = \pm 5$, $y = \pm 3$ **11.** $x = \pm 4$, $y = \pm 2$ **13.** $x = \pm 5$, $y = \pm 10$ **15.** $x = \pm 5$, $y = \pm 5\sqrt{3}$ **17.** $x = \pm 4$, $y = \pm 4\sqrt{2}$ **19.** $a = \pm\sqrt{10}$, $b = \pm\sqrt{15}$ **21.** $a = \pm\sqrt{\frac{7}{8}}$, $b = \pm\sqrt{\frac{7}{2}}$ **23.** $\frac{(x + 1)^2}{16} + \frac{y^2}{12} = 1$

Page 285

1. $a = \pm 2$, $b = \pm 4$ **3.** $a = \pm 3$, $b = \pm 4$ **5.** $a = \pm 10$, $b = \pm 5$ **7.** $a = \pm 5$, $b = \pm 2$ **9.** $a = \pm 3$, $b = \pm 6$ **11.** $a = \pm 3$, $b = \pm 4$ **13.** 1st and 3rd quadrants **15.** 1st and 3rd quadrants **17.** $a = \pm 2\sqrt{2}$, $b = \pm\sqrt{2}$ **19.** 2nd and 4th quadrants **21.** $y = \pm\frac{b}{a}x$ **23.** $y = \pm\frac{1}{2}x$ **25.** $y = \pm\frac{6}{5}x$ **27.** $y = \pm\frac{4}{9}x$ **29.** $y = \pm\frac{2}{7}x$ **31.** $y = \pm\frac{2}{3}x$ **33.** $y = \pm 2x$

Page 287

1. ellipse **3.** hyperbola **5.** ellipse **7.** ellipse **9.** hyperbola **11.** parabola **13.** hyperbola **15.** ellipse **17.** hyperbola **19.** parabola **21.** hyperbola **23.** hyperbola **25.** ellipse **27.** circle **29.** ellipse **31.** circle **33.** ellipse

Page 290

1. 2 units to left **3.** 3 units to left, 1 unit up **5.** 2 units to right **7.** 3 units to left, 3 units down **9.** 2 units up **11.** 4 units to right, 3 units down **13.** $y = (x + 4)^2 - 12$ **15.** $y = (x + 3)^2 - 13$ **17.** $y = (x + 5)^2 - 3$ **19.** circle, center $(1, -1)$, $r = 4$ **21.** circle, center $(-1, 2)$, $r = 1$ **23.** hyperbola, center $(-1, -2)$, $a = \pm 5$, $b = \pm 2$

Page 291

1. hyperbola

Page 292

1. $(1, 5)$, $x = 1$ **3.** $(-2, -7)$, $x = -2$ **5.** $x = 2$, $(2, 10)$ **7.** -2.9, .2 **9.** -4.7, .7 **11.** $x^2 + y^2 = 169$ **13.** $x^2 + y^2 = 289$ **15.** $x^2 + y^2 = 10$ **17.** $x^2 + y^2 = 5$ **19.** $x^2 + y^2 = 65$ **21.** $x^2 + y^2 = 34$ **23.** circle, $r = 2$ **25.** not a circle **27.** $\frac{x^2}{49} + \frac{y^2}{9} = 1$ **29.** $\frac{x^2}{4} + \frac{y^2}{64} = 1$ **31.** $x = \pm 7$, $y = \pm 4$ **33.** $x = \pm 5$, $y = \pm\sqrt{5}$ **35.** 2nd and 4th quadrants **37.** $a = \pm 2$, $b = \pm 4$. **39.** ellipse **41.** parabola **43.** 2 units to right, 3 units up **45.** 3 units down **47.** $y = (x - 5)^2 - 5$

Page 294

1. 6.58% **3.** .29% **5.** 1.04%

Page 297

1. $(3, 1)$ **3.** $(1, -5)$ **5.** $(0, 1)$ **7.** $(.5, 0)$ **9.** $(2, 1)$ **11.** $(3, -2)$ **13.** inconsistent **15.** inconsistent **17.** same slopes **19.** same line

Page 301

1. $(1, 2)$ **3.** $(2, -1)$ **5.** $(4, 1)$ **7.** $(7, 1)$ **9.** $(3, 0)$ **11.** $(-1, -3)$ **13.** $(2, -3)$ **15.** $(2, 3)$ **17.** $(1, 2)$ **19.** $(3, -2)$ **21.** $(2, 0)$ **23.** $\left(\frac{b - c}{2}, \frac{b - 3c}{2}\right)$ **25.** $(4, 6)$ **27.** $(-21, 21)$ **29.** $\left(-\frac{8}{5}, \frac{18}{5}\right)$

Page 302
1. 3 **3.** 2 **5.** $x = 5, y = -1$ **7.** $x = 3, y = \frac{1}{2}$

Page 305
1. (3, −2, 1) **3.** (2, 1, −1) **5.** (7, −3, 6) **7.** (−1, −1, −2) **9.** (4, −5, 8)

Page 306
1. −63 **3.** −28

Page 310
1. (4, 1), (3, 0) **3.** (−7.8, 4.9), (6.4, −2.2) **5.** (.7, −1.8), (2.5, .9) **7.** no points in common **9.** (1, 2), (−1, −2) **11.** (5, 4), (4, 5) **13.** (5, 3), (2, 0) **15.** (0, −3), (3, 0) **17.** (0, −3), (2, 1) **19.** (1, −1), $\left(-\frac{3}{11}, \frac{17}{11}\right)$

Page 313
1. (4, 0), (−4, 0) **3.** (0, 2), (0, −2) **5.** (3, 2), (−3, 2), (3, −2), (−3, −2) **7.** (5, 0), (−5, 0) **9.** (4, 3), (3, 4), (−4, −3), (−3, −4) **11.** (5, 1), (1, 5), (−5, −1), (−1, −5)

Page 316
1. 52 **3.** 65 **5.** 72 **7.** 35 **9.** 72 **11.** 39 **13.** 27 **15.** 64

Page 322
1. 4 **3.** 6 **5.** 1 **7.** 180 **9.** 4 **11.** 46

Page 326
For each of the following, the two points given lie on the line. They are not necessarily part of the graph. **1.** below and including; (0, −1), (1, 2) **3.** below; (0, 1), (1, 3) **5.** below; (0, −4), (−4, 0) **7.** below and including; (0, 4), (−1, 1) **9.** below and including; (0, 6), (2, 0) **11.** above; (0, 3), (3, 3) **13.** to the right and including; (2, 0), (2, 3) **15.** above and including; (0, −2), (1, 0) **17.** below; (0, 3), (1, 1) **19.** above the line containing (0, 4), (1, 1) *and* below the line containing (0, −3), (1, −1) **21.** above and including the line containing (0, 6), (−3, 0) *and* below and including the line containing (0, −1), (−1, 2) **23.** above and including the line containing (0, 4), (−1, 1) *and* above and including the line containing (0, −1), (1, 2) **25.** below and including; (0, 4), (−4, 2) **27.** below; (0, 2), (−2, 1) **29.** above; (0, 5), (2, 0) **31.** below and including the line containing (0, −6), (−2, 0) *and* above the line containing (0, 1), (2, 5) **33.** below and including the line containing (0, −2), (−3, 0) *and* above the line containing (0, −2), (2, −1) **35.** below the line containing (1, −4), (3, −3) *and* below and including the line containing (0, 2), (4, 0) **37.** below and including the line containing (4, 0), (0, −8) *and* below the line containing $\left(0, \frac{7}{3}\right)$, $\left(3, \frac{4}{3}\right)$ *and* above and including the line containing (0, 2), (4, 0)

Page 328
For each of the following, the two points given lie on the curve. They are not necessarily part of the graph. **1.** Inside the circle; (0, 2), (−2, 0) **3.** outside and including the circle; (0, 4), (−4, 0) **5.** outside the branches of the hyperbola; (4, 2.8), (−3.5, −2) **7.** inside the branches of the hyperbola; (−1, −5), (5, 1) **9.** inside and including the branches of the hyperbola; (4, 5.2), (−4.5, 6) **11.** inside the ellipse; (3, 0), (0, −1.7) **13.** inside and including the circle on which lie the points (3, 0), (0, −3) *and* below and including the line on which lie the points (0, 3), (2, 1) **15.** inside and including the branches of the hyperbola on which lie the points (2, 2), (−1, −4), *and* below the line on which lie the points (0, 4), (2, 3) **17.** inside and including the circle on which lie the points (2, 0), (0, −2) *and* above the line on which lie the points (0, −2), (2, −1) **19.** inside and including the circle on which lie the points (−4, 0), (0, 4) *and* inside and including the branches of the hyperbola on which lie the points (−1, −4), (2, 2) **21.** inside the ellipse on which lie the points (−3, 0), (0, −2) *and* inside and including the circle on which lie the points (−6, 0), and (0, 6)

Page 329
Across **1.** 41; 43; 47 **7.** 4 **9.** 100 **11.** 14,007 **13.** 101 **19.** 3 **21.** 79 **25.** 666 **27.** 1,024 **29.** 102 **31.** 1 Down **1.** 411 **3.** 490 **5.** 4 **7.** 40 **13.** 13 **15.** 1,224 **17.** 196 **23.** 512

Page 330

1. (5, 2) **3.** (4, −1) **5.** (1, 1) **7.** (2, 1) **9.** (−2, −1, 2) **11.** (−3, 0), (5, 4) **13.** (3, 0), (−1, 4) **15.** (3, 2), (2, 3), (−3, −2), (−2, −3) **17.** 93 **19.** 4 **21.** below and including; (0, 1), (1, −1) **23.** below; (0, 2), (2, 1) **25.** below the line containing (0, 1), (1, 3) *and* above the line containing (0, −5), (−3, −2) **27.** above the line containing (0, −1), (3, 1) *and* above and including the line containing (0, 0), (3, 3) **29.** inside and including the circle on which lie the points (0, 4), (−4, 0) **31.** outside and including the ellipse on which lie the points (−3, 0), (0, 2) **33.** outside and including the branches of the hyperbola on which lie the points (2, 3), (−6, −1) *and* above the line containing the points (0, −4), (3, −2)

Page 335

1. 5, 5.5, 6 **3.** 5, 3.5, 2 **5.** −2, $-1\frac{1}{2}$, −1 **7.** $2 + 3\sqrt{5}$, $2 + 4\sqrt{5}$, $2 + 5\sqrt{5}$ **9.** $9 - 3\sqrt{2}$, $11 - 4\sqrt{2}$, $13 - 5\sqrt{2}$ **11.** 9, 3, −3, −9 **13.** 5, 7.4, 9.8, 12.2 **15.** −6, −6.3, −6.6, −6.9 **17.** 3, $4 - \sqrt{3}$, $5 - 2\sqrt{3}$, $6 - 3\sqrt{3}$ **19.** $d = 4$ **21.** not arith. **23.** not arith. **25.** $d = \sqrt{2}$ **27.** $d = 2 - i$ **29.** not arith. **31.** 87 **33.** 52 **35.** −27 **37.** −70 **39.** −85 **41.** −13 **43.** 169 **45.** 98 **47.** 32 **49.** 21 **51.** 33 **53.** 26 **55.** 11 **57.** 46 **59.** $22 + 20\sqrt{5}$ **61.** $23 - 18i$ **63.** $48 + 25\sqrt{3}$ **65.** $22 + 45i$ **67.** $9,000 **69.** 3; $-\frac{1}{2}$ **71.** −4

Page 338

1. 8 **3.** 5.5 **5.** 18, 25 **7.** −6, 2 **9.** 9, 15, 21, 27 **11.** 2, 6, 10, 14 **13.** 10.4, 10.8, 11.2, 11.6 **15.** 6.2, 8.4, 10.6, 12.8 **17.** 2.8 **19.** 7.9 **21.** $3\sqrt{3}$ **23.** $4 + 2\sqrt{2}$ **25.** yes **27.** 2, 3.9 **29.** $3\sqrt{5}$, $4\sqrt{5}$ **31.** $3 - \sqrt{3}$, $5 - 3\sqrt{3}$

Page 341

1. $5 + 8 + 11 + 14 + \cdots$ **3.** $-1 - 2 - 3 - 4 - \cdots$ **5.** −2 **7.** .4 **9.** $2\sqrt{2}$ **11.** −2,260 **13.** −4,815 **15.** −1,475 **17.** −1,809 **19.** 2,000 **21.** 546 **23.** 774 **25.** 170.3 **27.** 385 **29.** −60 **31.** $2{,}501\sqrt{3}$ **33.** 210

Page 344

1. 24, 48, 96 **3.** .008, .0016, .00032 **5.** −4, 2, −1 **7.** 10, 20, 40, 80 **9.** .2, 1, 5, 25 **11.** −9, −3, −1, $-\frac{1}{3}$ **13.** $\frac{1}{2}$, −1, 2, −4 **15.** −2, 6, −18, 54 **17.** −5,120 **19.** $\frac{1}{128}$ **21.** $-\frac{1}{256}$ **23.** −5,120 **25.** $-\frac{1}{128}$ **27.** $\frac{1}{32}$ **29.** −375 **31.** .00009 **33.** −.007 **35.** $-\frac{1}{8}$ **37.** 54 **39.** 80 **41.** $324\sqrt{2}$ **43.** $-3i$ **45.** $160i$ **47.** $\frac{1}{4}$ dm **49.** $10\frac{2}{3}$ m **51.** −2; 2 **53.** $\frac{1}{3}$; 3 **55.** 11th **57.** 11th

Page 347

1. 6 **3.** 2 **5.** 4 **7.** 1 **9.** .4 **11.** .3 **13.** 6, 18 **15.** −8, −32 **17.** 3, 1 **19.** 10, 2 **21.** .6, .06 **23.** .3, .09 **25.** $3\sqrt{2}$ **27.** $4\sqrt{3}$ **29.** $2\sqrt{2}$ **31.** $6\sqrt{2}$ **33.** $2\sqrt{3}$, 6 **35.** $3\sqrt{7}$, 21 **37.** 2, $2\sqrt{2}$ **39.** 7, $7\sqrt{7}$ **41.** $2i$; $-2i$ **43.** $2i$; $-2i$ **45.** $i\sqrt{2}$; $-i\sqrt{2}$ **47.** $2i\sqrt{5}$; $-2i\sqrt{5}$ **49.** $2 + 2i$; $-2 - 2i$ **51.** $\sqrt{xy}$; $-\sqrt{xy}$

Page 348

1. 110 **3.** 870 **5.** 100 **7.** 81 **9.** 10,000 **11.** 5 **13.** 55

Page 353

1. 5,115 **3.** 171 **5.** 5,465 **7.** $31\frac{7}{8}$ **9.** 333,333.3 **11.** 6,666,666 **13.** 77.777 **15.** 24.8 **17.** −3.3333 **19.** 16 **21.** 27 **23.** 1 **25.** $2\frac{2}{9}$ **27.** $\frac{3}{4}$ **29.** $.5 + .05 + .005 + \cdots$; $a = .5$; $r = .1$ **31.** $6 + .6 + .06 + \cdots$; $a = 6$; $r = .1$ **33.** $.02 + .002 + .0002 + \cdots$; $a = .02$; $r = .1$ **35.** $\frac{2}{9}$ **37.** $\frac{5}{90}$ **39.** $\frac{70}{9}$ **41.** $-21\frac{3}{8}$ **43.** $682\frac{1}{2}$ **45.** $728\frac{2}{3}$ **47.** 13,021 **49.** −364 **51.** $393\frac{3}{4}$ **53.** 51 **55.** 6,825 **57.** −547 **59.** $r = 5$; no sum **61.** $-85\frac{1}{3}$ **63.** $-20\frac{1}{4}$ **65.** $-9\frac{1}{7}$ **67.** $16\sqrt{5}$ **69.** $81\sqrt{2}$ **71.** $.38 + .0038 + .000038 + \cdots$; $a = .38$; $r = .01$ **73.** $53 + .53 + .0053 + \cdots$; $a = 53$; $r = .01$ **75.** $.026 + .00026 + .0000026 + \cdots$; $a = .026$; $r = .01$ **77.** $\frac{35}{99}$ **79.** $\frac{1700}{99}$ **81.** $\frac{47}{990}$ **83.** 382 m **85.** $362\frac{1}{3}$ m

Page 356
1. $-3, 2, 7, 12$ **3.** 4, 5.6, 7.2, 8.8 **5.** -99
7. 23 **9.** 5.4, 5.8, 6.2, 6.6 **11.** 105 **13.** -546
15. $5, -10, 20, -40$ **17.** $-16, -8, -4, -2$
19. 3,072 **21.** 512 **23.** -54 **25.** $-4, 8$
27. $10\frac{5}{8}$ **29.** -84 **31.** $25\frac{3}{5}$ **33.** $.35 + .0035 + .000035 + \cdots$; $a = .35$; $r = .01$

Page 358
1. 30; $4

Page 362
1. 10^2 **3.** 10^1 **5.** $10^{1.7372}$ **7.** $10^{0.7372}$
9. $10^{2.3284}$ **11.** $10^{1.6637}$ **13.** $10^{1.4871}$
15. $10^{0.7042}$ **17.** $10^{2.3096}$ **19.** $10^{1.9036}$
21. $10^{0.8142}$ **23.** $10^{4.9159}$ **25.** 8.32 **27.** 832
29. 27.7 **31.** 372 **33.** 439 **35.** 1.68
37. 8,150 **39.** 773 **41.** 9.76 **43.** 12.2
45. $10^{1.6335}$ **47.** $10^{2.7634}$ **49.** $10^{0.7924}$
51. $10^{3.9243}$ **53.** $10^{0.6990}$ **55.** $10^{3.3010}$
57. 240 **59.** 560 **61.** 1,700 **63.** 83,000

Page 364
1. 74.9 **3.** 959 **5.** 23.6 **7.** 384,000
9. 1,780,000 **11.** 300,000 **13.** 17.3 **15.** 14.6
17. 5.93 **19.** 68.6 **21.** 58,100 **23.** 10,400
25. 13,900 **27.** 181 **29.** 30.7 **31.** 9.60
33. 824 **35.** 895,000,000

Page 367
1. 0.5465 **3.** 3.5465 **5.** 0.4456 **7.** 3.9047
9. 0.7810 **11.** 2.35 **13.** 5.78 **15.** 9,770
17. 83.3 **19.** 77.4 **21.** 4,620 **23.** 0; .3345
25. 0; .9671 **27.** 4; .5132 **29.** 0.2788
31. 0.9731 **33.** 3.8062 **35.** 4.9069 **37.** 34
39. 730 **41.** 86 **43.** 430 **45.** 7,000
47. 22,300

Page 369
1. 864 **3.** 26,000 **5.** 210,000 **7.** 141,000
9. 35,500 **11.** 5,700 **13.** $\log l + \log w$
15. $\log p + \log r + \log t$ **17.** $\log 2 + \log \pi + \log r$ **19.** 80,100 **21.** 4,520,000
23. $\log (x^2 - y^2) = \log (x + y)(x - y) = \log (x + y) + \log (x - y)$

Page 373
1. 3.45×10^2 **3.** 2.79×10^4 **5.** 2.16×10^{-2}
7. 3.78×10^{-2} **9.** 2.3×10^1 **11.** 7.6×10^3
13. 817 **15.** 5,630 **17.** .624 **19.** .0352
21. 29,000 **23.** 630 **25.** 2.9258 **27.** 3.3747
29. 0.7993 **31.** 1.7218 **33.** 582 **35.** 64.7
37. 8.49 **39.** 5,200 **41.** 6.7193
43. 6.9345 **45.** 487,000 **47.** 625,000
49. 1.5×10^8 km **51.** 2.998×10^{10} cm per sec

Page 375
1. 14.3 **3.** 15.1 **5.** 6.91 **7.** 49.0 **9.** 173
11. 227 **13.** 4.54 **15.** 1.54 **17.** 26.5

Page 377
1. 3,170 **3.** 329 **5.** 350 **7.** 183,000
9. 29,600 **11.** 575 **13.** 228 **15.** 2,760
17. 694 **19.** 434,000 **21.** 605,000
23. 10.6 **25.** 16.2 **27.** $\log \left(\frac{ab}{c}\right)^r = r \cdot \log \frac{ab}{c} = r(\log ab - \log c) = r(\log a + \log b - \log c)$

Page 379
1. 19.2 **3.** 71.6 **5.** 19.1 **7.** 45.5 or 45.6
9. 12.0 **11.** 3.68 **13.** 13.8 **15.** 5.18
17. 238 **19.** 5.72 **21.** 2.13 **23.** 2.28
25. 1.18

1. 10^3 **3.** $10^{0.7007}$ **5.** 7,730 **7.** 56.7
9. 0.7505 **11.** 1.4314 **13.** 1.67 **15.** 3; .6920
17. 0; .4533 **19.** 4.68×10^1 **21.** 6.2×10^{-3}
23. 503 **25.** .914 **27.** 19,600 **29.** 60.6
31. 18,200 **33.** 135,000 **35.** 1,480 **37.** 1.73
39. 12.2 **41.** 10.4 **43.** $\log V = \log 2 + \log w + \log h$

Page 382
1. 21, 34, 55, 89 **3.** $y_n = y_{n-1} + y_{n-2}$

Page 385
1. 9.5490 − 10 **3.** 6.5490 − 10
5. 8.2227 − 10 **7.** 9.3160 − 10
9. 9.7042 − 10 **11.** .326 **13.** .235
15. .000167 **17.** .123 **19.** .629
21. .000774 **23.** 7.7782 − 10
25. 6.7324 − 10 **27.** 5.3802 − 10 **29.** .034
31. .00018 **33.** .00086 **35.** .0007

Page 388

1. .262 **3.** .0738 **5.** 2.46 **7.** .0233 **9.** .00677 **11.** .00757 **13.** 237 **15.** 45.4 **17.** .0768 **19.** .798 **21.** .189 **23.** .0310 **25.** .00616 **27.** .299 **29.** .0709 **31.** .442 **33.** .566 **35.** .0225 **37.** .000296 **39.** $\log \frac{t^2}{vw}$ **41.** $\log \left(\frac{mn}{p}\right)^3$ **43.** $\log \sqrt[3]{\frac{xy}{z}}$

Page 390

1. .966 **3.** .0677 **5.** 143 **7.** .0645 **9.** .0663 **11.** .00395 **13.** 3.83 **15.** .102 **17.** .00121 **19.** 1.60 **21.** .708 **23.** .664 **25.** $\log 3 + \log f + \log g$ **27.** $\log 2 + \log t + 3 \cdot \log v$ **29.** $\log 5 + \log f - \log g$ **31.** $\log 3 + \log d - 2 \cdot \log t$ **33.** $2(\log 7 + \log f + \log w)$ **35.** $\frac{1}{2}(\log 2 + \log g + \log h)$ **37.** $2(\log 3 + \log d - \log g)$ **39.** $\frac{1}{2}(\log 2 + \log d - \log g)$ **41.** 105 **43.** 30.8 **45.** 378 **47.** .0415 **49.** 2.59 **51.** .370 **53.** no

Page 394

1. 1.2838 **3.** 8.2849 − 10 **5.** 2.5496 **7.** 9.8661 − 10 **9.** 7.3166 − 10 **11.** 2.624 **13.** 4,787 **15.** 896.6 **17.** 4.843 **19.** .1854 **21.** .007268 **23.** 33,880 **25.** 24.67 **27.** .03693 **29.** .1062 **31.** 96.68 **33.** .08382 **35.** 174.4 **37.** .04506 **39.** .3039 **41.** .06813 **43.** 1.526 **45.** .6805 **47.** 18.81 **49.** $2,565

Page 397

1. 2 **3.** 1 **5.** 3 **7.** 4 **9.** 2 **11.** $\frac{1}{3}$ **13.** −1 **15.** −2 **17.** −2 **19.** 16 **21.** 125 **23.** $\frac{1}{25}$ **25.** 2 **27.** 5 **29.** 2 **31.** $\frac{2}{3}$ **33.** $\frac{5}{3}$ **35.** $\frac{2}{5}$ **37.** $\frac{2}{5}$ **39.** 2 **41.** 1

Page 398

1. 2.20 **3.** 1.20 **5.** 1.23 **7.** .816 **9.** .908 **11.** .414

Page 401

1. $a^7 + 7a^6b + 21a^5b^2 + 35a^4b^3 + 35a^3b^4 + 21a^2b^5 + 7ab^6 + b^7$ **3.** $a^9 + 9a^8b + 36a^7b^2 + 84a^6b^3 + 126a^5b^4 + 126a^4b^5 + 84a^3b^6 + 36a^2b^7 + 9ab^8 + b^9$ **5.** $x^5 - 15x^4 + 90x^3 - 270x^2 + 405x - 243$ **7.** $x^5 - 5x^4y + 10x^3y^2 - 10x^2y^3 + 5xy^4 - y^5$ **9.** $x^4 + 12x^3y + 54x^2y^2 + 108xy^3 + 81y^4$ **11.** $243x^5 - 405x^4y + 270x^3y^2 - 90x^2y^3 + 15xy^4 - y^5$ **13.** $x^8 + 4x^6y + 6x^4y^2 + 4x^2y^3 + y^4$ **15.** $x^5 + 5x^4y^2 + 10x^3y^4 + 10x^2y^6 + 5xy^8 + y^{10}$ **17.** $x^4 + .8x^3 + .24x^2 + .032x + .0016$ **19.** $35a^3b^4$ **21.** $-540x^3$ **23.** $15x^4y^4$ **25.** $32x^5 + 240x^4 + 720x^3 + 1{,}080x^2 + 810x + 243$ **27.** $32x^5 - 400x^4 + 2{,}000x^3 - 5{,}000x^2 + 6{,}250x - 3{,}125$ **29.** $x^{12} - 8x^9y + 24x^6y^2 - 32x^3y^3 + 16y^4$ **31.** $243x^5 + 810x^4y + 1{,}080x^3y^2 + 720x^2y^3 + 240xy^4 + 32y^5$ **33.** $32x^5 - 160x^4y + 320x^3y^2 - 320x^2y^3 + 160xy^4 - 32y^5$ **35.** $x^5 - 5x^4\sqrt{3} + 30x^3 - 30x^2\sqrt{3} + 45x - 9\sqrt{3}$ **37.** $x^4 + 2x^3y + \frac{3x^2y^2}{2} + \frac{xy^3}{2} + \frac{y^4}{16}$ **39.** $\frac{1}{m^6} - \frac{6}{m^5n} + \frac{5}{m^4n^2} - \frac{20}{m^3n^3} + \frac{15}{m^2n^4} - \frac{6}{mn^5} + \frac{1}{n^6}$ **41.** $-20{,}000x^3y^3$ **43.** $135x^4y^2$ **45.** $-10x^3$ **47.** $x^{10} - 20x^9y + 180x^8y^2 - 960x^7y^3 + \cdots$ **49.** $x^{12} + 6x^{11}y + \frac{33x^{10}y^2}{2} + \frac{55x^9y^3}{2} + \cdots$

Page 402

1. −2 **3.** 9.9058 − 10 **5.** .00123 **7.** 217 **9.** .0220 **11.** .0768 **13.** 6.96 **15.** .0883 **17.** .891 **19.** $\log 5 + 2 \cdot \log t + \log w$ **21.** $3(\log 7 + 4 \cdot \log t)$ **23.** 1.7732 **25.** 484.3 **27.** 149.3 **29.** 11.23 **31.** −2 **33.** $\frac{4}{3}$ **35.** .315 **37.** $x^6 + 12x^5y + 60x^4y^2 + 160x^3y^3 + 240x^2y^4 + 192xy^5 + 64y^6$ **39.** $32x^5 - 40x^4y + 20x^3y^2 - 5x^2y^3 + \frac{5xy^4}{8} - \frac{y^5}{32}$ **41.** $60x^4$

Page 404

1. $\frac{8}{32}, \frac{13}{64}, \frac{21}{128}, \frac{34}{256}$ **3.** $\frac{y_{n-1} + y_{n-2}}{2^{n-2}}$ where y_{n-1} and y_{n-2} are terms of the Fibonacci sequence.

Page 406

1. 1 **3.** 4 **5.** 3 **7.** 2 **9.** 3 **11.** 2
13. positive y-axis **15.** 4 **17.** 1 **19.** 3
21. 1 **23.** 1 **25.** 405°, −315° **27.** 600°, −120°

Page 408

1. 20° **3.** 45° **5.** 30° **7.** 60° **9.** 65° **11.** 45°
13. 75° − 15° rt. △ **15.** 30° − 60° rt. △
17. 30° − 60° rt. △ **19.** 30° − 60° rt. △
21. 70° − 20° rt. △ **23.** 30° − 60° rt. △
25. 30° **27.** 45° **29.** 70° **31.** 45° **33.** 75°
35. 15°

Page 412

1. $2\sqrt{2}$ **3.** $5\sqrt{3}$ **5.** $b = 5$, $c = 5\sqrt{2}$
7. $b = 7$, $c = 14$ **9.** $x = -3$, $y = 3\sqrt{3}$
11. $x = 2\sqrt{3}$, $y = 2$ **13.** 4, $2\sqrt{3}$ **15.** $3\sqrt{2}$, 3
17. $\frac{5\sqrt{2}}{2}$, $\frac{5\sqrt{2}}{2}$ **19.** In a 45° − 45° rt. △ABC, the hypotenuse $AB = x$. $BC^2 + AC^2 = AB^2 = x^2$. $2BC^2 = x^2$. $BC^2 = \frac{x^2}{2}$. $BC = \sqrt{\frac{x^2}{2}} = \frac{x}{\sqrt{2}}$, or $\frac{x\sqrt{2}}{2}$.

Page 415

1. x-axis **3.** origin **5.** y-axis **7.** origin
9. x-axis and origin **11.** (−3, −5), (3, 5), (3, −5) **13.** (4, 3), (−4, −3), (−4, 3)
15. (−2, −1), (2, 1), (2, −1) **17.** (4, −3), (−4, 3), (−4, −3) **19.** (4, 0), (−4, 0), (−4, 0)
21. (1, −1), (1, 1), (−1, −1), (−1, 1)
23. A graph is symmetric with respect to the x-axis if the portion of the graph above the x-axis is a reflection of the portion of the graph below the x-axis. A graph is symmetric with respect to the y-axis if the portion of the graph to the right of the y-axis is a reflection of the portion of the graph to the left of the y-axis. A graph is symmetric with respect to the origin if the portion of the graph in the 1st quadrant is a reflection of the portion of the graph in the 3rd quadrant, and the portion of the graph in the 2nd quadrant is a reflection of the portion of the graph in the 4th quadrant. **25.** even **27.** even
29. odd

Page 419

1. $\frac{1}{2}$ **3.** $\frac{\sqrt{2}}{2}$ **5.** 0 **7.** 0 **9.** 0 **11.** $\frac{\sqrt{3}}{2}$
13. $-\frac{\sqrt{3}}{2}$ **15.** $-\frac{\sqrt{3}}{2}$ **17.** $-\frac{\sqrt{2}}{2}$ **19.** $\frac{\sqrt{3}}{2}$
21. $\frac{\sqrt{3}}{2}$ **23.** $\frac{1}{2}$ **25.** sin 30° **27.** −cos 45°
29. −sin 30° **31.** −cos 45° **33.** −sin 30°
35. cos 45° **37.** 1, 2 **39.** 1, 4 **41.** 30°, 150°
43. 45°, 135° **45.** 150°, 210° **47.** 135°, 225°
49. 330°, 690°

Page 423

1. $\frac{\sqrt{3}}{3}$ **3.** $\frac{\sqrt{3}}{3}$ **5.** 0 **7.** 0 **9.** $-\sqrt{3}$ **11.** −1
13. $\frac{-\sqrt{3}}{3}$ **15.** undefined **17.** $\frac{\sqrt{3}}{3}$ **19.** 1
21. $\sqrt{3}$ **23.** 0 **25.** $-\sqrt{3}$ **27.** $-\sqrt{3}$ **29.** −1
31. −tan 30° **33.** −cot 45° **35.** tan 30°
37. cot 45° **39.** −tan 30° **41.** −cot 60°
43. −3 **45.** $\tan x = \frac{\sqrt{3}}{3}$, $\cot x = \sqrt{3}$
47. $\tan x = -\frac{5}{12}$, $\cot x = -\frac{12}{5}$ **49.** 2, 4
51. 2, 4 **53.** 135°, 315° **55.** 120°, 300°
57. 150°, 330° **59.** 135°, 315° **61.** 240°, 600° **63.** 45°, 405°

Page 425

1. $\frac{2\sqrt{3}}{3}$ **3.** $\sqrt{2}$ **5.** 2 **7.** $\sqrt{2}$ **9.** $\frac{-2\sqrt{3}}{3}$
11. $-\sqrt{2}$ **13.** −2 **15.** $\frac{-2\sqrt{3}}{3}$ **17.** $\sqrt{2}$
19. −sec 30° **21.** −sec 40° **23.** −sec 50°
25. 210°, 330°

Page 426

1. 1 **3.** 4 **5.** 1 **7.** 2 **9.** 60° **11.** 50°
13. 45° −45° rt. △ **15.** 20° −70° rt. △
17. 60° **19.** 50° **21.** 75° **23.** $4\sqrt{2}$
25. $(6, 6\sqrt{3})$, $r = 12$; $(6, -6\sqrt{3})$, $(-6, 6\sqrt{3})$, $(-6, -6\sqrt{3})$ **27.** y-axis **29.** (6, 1), (−6, −1), (−6, 1) **31.** (−2, 5), (2, −5), (2, 5) **33.** $\sqrt{3}$
35. $\frac{-\sqrt{3}}{2}$ **37.** $\frac{\sqrt{3}}{3}$ **39.** −tan 40°
41. −cos 30° **43.** csc 60° **45.** 1 and 4
47. 1 and 3 **49.** 60°, 300° **51.** 225°, 315°
53. 135°, 225° **55.** 60°, 240° **57.** undefined
59. 0 **61.** undefined

Page 428
1. .77 **3.** .87 **5.** −.77 **7.** horizontally above the circle

Page 430
1. 3, 4 **3.** 1, 4 **5.** 1, 3 **7.** 2, 3 **9.** 2, 4 **11.** 4 **13.** $\cot u = -\frac{5}{12}$, $\sin u = \frac{12}{13}$, $\csc u = \frac{13}{12}$, $\cos u = -\frac{5}{13}$, $\sec u = -\frac{13}{5}$ **15.** $\tan u = -1$, $\sin u = -\frac{\sqrt{2}}{2}$, $\csc u = -\sqrt{2}$, $\cos u = \frac{\sqrt{2}}{2}$, $\sec u = \sqrt{2}$ **17.** $\cot u = -\frac{\sqrt{3}}{3}$, $\sin u = \frac{\sqrt{3}}{2}$, $\cos u = -\frac{1}{2}$, $\csc u = \frac{2\sqrt{3}}{3}$, $\sec u = -2$ **19.** $\cot u = \sqrt{3}$, $\sin u = -\frac{1}{2}$, $\cos u = -\frac{\sqrt{3}}{2}$, $\tan u = \frac{\sqrt{3}}{3}$, $\csc u = -2$ **21.** $\cos u = -\frac{\sqrt{3}}{2}$, $\cot u = \sqrt{3}$, $\sec u = \frac{-2\sqrt{3}}{3}$, $\csc u = -2$ **23.** $\sin u = \frac{1}{2}$, $\tan u = \frac{\sqrt{3}}{3}$, $\cot u = \sqrt{3}$, $\sec u = \frac{2\sqrt{3}}{3}$

Page 433
1. 27° **3.** 37° **5.** 72°10′ **7.** 50°10′ **9.** 2°10′ **11.** 2.7475 **13.** 5.5764 **15.** .7254 **17.** −.8646 **19.** sin 44° **21.** cot 17° **23.** tan 10°50′ **25.** sin 1°20′ **27.** 32°, 212° **29.** 130°10′, 310°10′ **31.** 19°20′, 160°40′ **33.** −cos 40° **35.** cot 10° **37.** cos 10°30′ **39.** −cot 30°30′ **41.** $\frac{1}{\tan 40°}$ **43.** $-\frac{1}{\tan 10°10'}$ **45.** 3.7420 **47.** 53°10′

Page 436
1. 30°, 150° **3.** 45°, 225° **5.** 135°, 315° **7.** 90°, 270° **9.** 210°, 330° **11.** 135°, 225° **13.** 270° **15.** 172°50′, 352°50′ **17.** 56°30′, 123°30′ **19.** not possible **21.** no answer between 0° and 360° **23.** 53°10′, 413°10′, 126°50′, 486°50′ **25.** 120°, 300°, 480°, 660° **27.** 150°, 330°, 510°, 690° **29.** 135°, 315°

Page 439
1. .9791 **3.** .5993 **5.** .9068 **7.** .1239 **9.** .1771 **11.** 1.9883 **13.** 27°58′ **15.** 59°20′ **17.** 57°28′ **19.** 18°38′ **21.** 9°03′ **23.** 203°35′, 336°25′ **25.** 116°34′, 296°34′ **27.** 14°29′, 165°31′ **29.** −.6961 **31.** 3.5340 **33.** 235°18′, 304°42′ **35.** 45°, 225° **37.** all except 90° and 270°

Page 442
1. $\sin G = \frac{g}{i}$, $\cos G = \frac{h}{i}$, $\tan G = \frac{g}{h}$ **3.** $\sin O = \frac{o}{m}$, $\cos O = \frac{n}{m}$, $\tan O = \frac{o}{n}$ **5.** 71° **7.** 70° **9.** 1 **11.** $\frac{3\sqrt{2}}{2}$

Page 446
1. $b = 12$, $c = 14$ **3.** $a = 20$, $c = 21$ **5.** $b = 4$, $c = 7$ **7.** 37°, 53° **9.** 53°, 37° **11.** 42°, 48° **13.** 96 m **15.** 66° **17.** 10.9 **19.** 17.3 **21.** 1,208 m

Page 447
1. $2(\cos 30° + i \sin 30°)$ **3.** $1(\cos 45° + i \sin 45°)$ **5.** $-\frac{5}{2} + \frac{5\sqrt{3}}{2}i$

Page 448
1. 2, 3 **3.** 2, 4 **5.** 3, 4 **7.** $\sin u = \frac{3}{5}$, $\tan u = -\frac{3}{4}$, $\cot u = -\frac{4}{3}$, $\sec u = -\frac{5}{4}$, $\csc u = \frac{5}{3}$ **9.** .4695 **11.** −.7986 **13.** 10°40′ **15.** 106°, 286° **17.** cot 17° **19.** sin 30° **21.** 120°, 240° **23.** 111°50′, 291°50′ **25.** .6060 **27.** 37°43′, 322°17′ **29.** 90° **31.** $1 + \sqrt{3}$ **33.** $\sqrt{6} + 1$ **35.** $b = 9$, $c = 10$ **37.** 36°, 54° **39.** 69°, 21° **41.** 18

Page 450
1. .18 **3.** .70 **5.** 1.43

Page 452
1. −tan 10° **3.** cos 20° **5.** −sin 50° **7.** −cot 18° **9.** −csc 38° **11.** sin 35° **13.** tan 80° **15.** cos 44°27′ **17.** −sin 10° **19.** cos 40° **21.** sec 35° **23.** sec 20° **25.** −cot 35°37′

Page 455

1. 1, 1 **3.** 1, 1 **5.** 1 and 4, 3 and 4 **7.** .9 **9.** .3 **11.** .9 **13.** 45° and 225°

Page 458

9. 360°, 2 and 360°, $\frac{1}{2}$ **11.** 360°, $\frac{1}{2}$ **13.** 360°, $\frac{1}{2}$ **15.** 2

Page 461

1. 360° **3.** 180° **5.** 720° **7.** 120° **9.** 90° **11.** 1,080° **25.** 5 **27.** 180° **29.** 180°

Page 463

1. 720°, 3 **3.** 720°, $\frac{1}{2}$ **5.** 180°, $\frac{1}{2}$ **7.** 720°, 2 **9.** 60°, $\frac{1}{4}$ **11.** 40°, 6 **13.** 720°, 2 **15.** 180°, $\frac{1}{2}$ **17.** 120°, 3 **19.** 120°, $\frac{1}{3}$ **21.** 1,080°, 3 **23.** 60°, 2 **25.** 2 **27.** 2 **29.** 720°, 2 **31.** 720°, 2 **33.** 720°, 2 **35.** 2 **37.** 4

Page 467

1. −180° to 360°, none **3.** 180° **7.** none, none **9.** none **11.** 360° **13.** 30° **15.** 15° **17.** 30° **31.** 4 **33.** 360° **35.** 90° **37.** 2

Page 470

1. .5719 **3.** 9.8699 − 10 **5.** 9.8699 − 10 **7.** 9.9225 − 10 **9.** 9.9519 − 10 **11.** 9.8495 − 10 **13.** 3°0′ **15.** 87°0′ **17.** 82°10′ **19.** 9°49′ **21.** 56°53′ **23.** 2.4 **25.** 1.4 **27.** 3233.1 **29.** 1,790,000.0

Page 473

1. 45°, 135°, 225°, 315° **3.** 45°, 135°, 225°, 315° **5.** 90°, 270°, 135°, 315° **7.** 45°, 225° **9.** no values between 0° and 360° **11.** 190°34′, 349°26′ **13.** 18°35′, 161°25′ **15.** 210°, 330° **17.** 0, b

Page 474

1. 17°0′ **3.** 42°50′

Page 477

1. $\frac{1}{\sin 30°}$ **3.** $\frac{1}{\sin 320°}$ or $\frac{-1}{\sin 40°}$ **5.** $\frac{-\sin 310°}{\cos 310°}$ or $\frac{\sin 50°}{\cos 50°}$ **23.** $\frac{\pm\sqrt{1 - \sin^2 x}}{\sin x}$ **25.** $1 + \frac{\sin^2 x}{1 - \sin^2 x}$ **27.**, **29.** Answers may vary.

Page 479

1., **3.**, **5.**, **7.**, **9.**, **11.**, **13.**, **15.** Answers may vary.

Page 482

1. $\frac{\pi}{12}$ **3.** $\frac{4\pi}{9}$ **5.** $\frac{3\pi}{4}$ **7.** $\frac{7\pi}{4}$ **9.** $\frac{-2\pi}{3}$ **11.** $\frac{-7\pi}{6}$ **13.** $\frac{-5\pi}{3}$ **15.** 60° **17.** 45° **19.** 240° **21.** 162° **23.** −120° **25.** −270° **27.** −360° **29.** 1 **31.** −1 **33.** −1 **35.** $\frac{2 + \sqrt{3}}{\sqrt{3}}$

Page 485

1. 1, −1 **3.** 1 **5.** 2π **7.** 2π, 3 **9.** 2π, $\frac{1}{3}$ **11.** 4π, 2 **13.** π, 2 **15.** 2 **17.** $\frac{\pi}{2}, \frac{-\pi}{2}$ **19.** $\frac{\pi}{6}, \frac{5\pi}{6}$ **21.** $\frac{3\pi}{4}, \frac{7\pi}{4}$ **23.** $\frac{\pi}{6}, \frac{5\pi}{6}, \frac{11\pi}{6}, \frac{7\pi}{6}$ **25.** 0, π, 2π, $\frac{\pi}{6}, \frac{5\pi}{6}$ **27.** $\frac{\pi}{2}, \frac{7\pi}{6}, \frac{11\pi}{6}$ **35.** 2π, $\frac{1}{2}$ **37.** 4π, 1 **39.** π, none **41.** π, 2 **43.** 2

Page 489

1. {(6, −1), (3, −2), (6, 5)}, no **3.** $2y - 3x = 8$, yes **5.** $x = \sin 2y$, no **7.** $\frac{-\pi}{4}$ or −45° **9.** $\frac{\pi}{6}$ **11.** $\frac{-\pi}{2}$ **13.** $\frac{5}{13}$ **15.** −1

Page 490

1. − cos 10° **3.** + sin 40° **5.** cos 30° **7.** 360°, none **9.** 720°, 3 **11.** 4 **13.** 72°08′ **15.** 14,450 **17.** 28°, 332° **19.** 0, π, $\frac{3\pi}{4}, \frac{7\pi}{4}$, 2π **21.** $\frac{1}{\cos 45°}$ **23.** $\frac{\cos 80°}{\sin 80°}$ **27.** $\pm\frac{\sin x}{\sqrt{1 - \sin^2 x}}$ **29.**, **31.** Answers may vary. **33.** $\frac{-\pi}{6}$ **35.** $\frac{-7\pi}{6}$ **37.** −150° **39.** −720° **41.** $\frac{-\sqrt{3}}{2}$ **43.** $\frac{2(3 - \sqrt{3})}{3}$ **45.** {(2, −1), (3, −2), (2, 4)}, no **47.** $x = \sin y$, no **49.** −30° or $-\frac{\pi}{6}$ **51.** $\frac{4}{5}$

Page 492
1. .2955

Page 495
1. $\sqrt{19}$ or 4 **3.** $\sqrt{85 + 42\sqrt{2}}$ **5.** $\sqrt{13 - 6\sqrt{2}}$
7. $\sqrt{29}$ or 5 **9.** $\sqrt{132}$ or 11 **11.** 495 km
13. See Example 2, page 493.

Page 497
1. .759 **3.** .806 **5.** 22° **7.** 67° **9.** 29°
11. See Example 2, page 496.

Page 499
1. 12 **3.** 22 **5.** 28° **7.** 25° **9.** 4 **11.** See Example 2, page 498.

Page 502
1. 18 **3.** 26 **5.** .8999 **7.** 22 **9.** 19
11. See Example 1, page 500.

Page 505
1. 2 **3.** none **5.** 2 **7.** none **9.** 1 **11.** none

Page 509
1. $\frac{\sqrt{2}}{4}(\sqrt{3} + 1)$ **3.** $\frac{\sqrt{2}}{4}(\sqrt{3} + 1)$ **5.** $\frac{\sqrt{3}}{2}$ **7.** $\frac{1}{2}$
9. $\frac{1}{2}$ **11.** 1 **13.** 0 **15.** $\frac{33}{65}, \frac{63}{65}$ **17.** $\frac{84}{85}, -\frac{36}{85}$
19. $-\frac{21}{221}, \frac{171}{221}$ **21.** $\frac{24}{25}$ **23.** $\frac{20\sqrt{6}}{49}$ **25.** sec A
27. 2 tan A **29.** $\sin (90° - u) = \sin 90° \cos u - \cos 90° \sin u = 1(\cos u) - 0(\sin u) = \cos u$
31. $\sin (180° - u) = \sin 180° \cos u - \cos 180° \sin u = 0(\cos u) - (-1)(\sin u) = \sin u$
33. $\sin (360° - u) = \sin 360° \cos u - \cos 360° \sin u = 0(\cos u) - 1(\sin u) = -\sin u$

Page 513
1. 0 **3.** $-\frac{1}{2}$ **5.** $\frac{\sqrt{2}}{4}(\sqrt{3} - 1)$ **7.** 0 **9.** $\frac{1}{2}$
11. $-\frac{1}{2}$ **13.** $\frac{\sqrt{2}(1 - \sqrt{3})}{4}$ **15.** $\frac{\sqrt{2}}{4}(\sqrt{3} + 1)$
17. $\frac{1}{2}$ **19.** $-\frac{1}{2}$ **21.** $-\frac{63}{65}, -\frac{33}{65}$ **23.** 0, $\frac{120}{169}$
25. $-\frac{120}{169}$, 0 **27.** $\frac{119}{169}$ **29.** $\frac{7}{25}$ **31.** 1 **33.** -1
35. $\cos (180° - u) = \cos 180° \cos u + \sin 180° \sin u = -1(\cos u) + 0(\sin u) = -\cos u$
37. $\cos (270° + u) = \cos 270° \cos u - \sin 270° \sin u = 0(\cos u) - (-1)(\sin u) = \sin u$ **39.** $\cos (360° - u) = \cos 360° \cos u + \sin 360° \sin u = 1(\cos u) + 0(\sin u) = \cos u$

Page 516
1. $\frac{3 + \sqrt{3}}{3 - \sqrt{3}}$ **3.** $\frac{3 - \sqrt{3}}{3 + \sqrt{3}}$ **5.** $\frac{3\sqrt{3} + 4}{3 - 4\sqrt{3}}$
7. $\frac{-3 + \sqrt{3}}{3 + \sqrt{3}}$ **9.** $-\frac{95}{59}$ **11.** $\frac{5\sqrt{3} - 36}{15 + 12\sqrt{3}}$ **13.** $\frac{120}{119}$
15. doesn't exist **17.** $-\frac{24}{7}$ **19.** $\frac{3\sqrt{3} - 4}{3 + 4\sqrt{3}}$, $\frac{3\sqrt{3} + 4}{3 - 4\sqrt{3}}$, $-\sqrt{3}$, $\frac{24}{7}$

Page 518
1. $\frac{\sqrt{2 + \sqrt{3}}}{2}$ **3.** $-\frac{\sqrt{2}}{2}$ **5.** $\frac{\sqrt{2 + \sqrt{2}}}{2}$
7. $\frac{\sqrt{2 + \sqrt{2}}}{2}$ **9.** $\frac{\sqrt{2}}{4}, \frac{\sqrt{14}}{4}$ **11.** $\frac{\sqrt{3}}{3}, \frac{\sqrt{6}}{3}$
13. $\frac{\sqrt{10}}{4}, -\frac{\sqrt{6}}{4}$ **15.** $-\sqrt{7 - 4\sqrt{3}}$
17. $\sqrt{7 - 4\sqrt{3}}$

Page 521
1., 3., 5., 7., 9., 11. Answers may vary.

Page 522
1. $\sqrt{22}$ **3.** $\frac{53}{80}$ **5.** 16 **7.** 34° **9.** 37 **11.** 2
13. $\frac{\sqrt{2}}{4}(\sqrt{3} + 1)$ **15.** $\frac{\sqrt{2}}{4}(\sqrt{3} - 1)$
17. $\frac{\sqrt{2}}{4}(\sqrt{3} - 1)$ **19.** $-\frac{33}{65}, \frac{63}{65}$ **21.** $\frac{120}{169}$
23. $-\frac{16}{63}$ **25.** $-\frac{120}{119}$ **27.** csc A **29.** 0, $\frac{24}{7}, \frac{24}{7}$